Pattern Recognition

Fourth Edition

Pattern Recognition

Fourth Edition

Sergios Theodoridis

Konstantinos Koutroumbas

ELSEVIER

AMSTERDAM • BOSTON • HEIDELBERG • LONDON
NEW YORK • OXFORD • PARIS • SAN DIEGO
SAN FRANCISCO • SINGAPORE • SYDNEY • TOKYO

Academic Press is an imprint of Elsevier

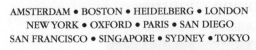

Academic Press is an imprint of Elsevier
30 Corporate Drive, Suite 400, Burlington, MA 01803, USA
525 B Street, Suite 1900, San Diego, California 92101-4495, USA
84 Theobald's Road, London WC1X 8RR, UK

This book is printed on acid-free paper. ♾

Library of Congress Cataloging-in-Publication Data
Application submitted

British Library Cataloguing-in-Publication Data
A catalogue record for this book is available from the British Library.

ISBN: 978-1-59749-272-0

For information on all Academic Press publications visit our Web site at *www.books.elsevier.com*

Contents

Preface

This book is the outgrowth of our teaching advanced undergraduate and graduate courses over the past 20 years. These courses have been taught to different audiences, including students in electrical and electronics engineering, computer engineering, computer science, and informatics, as well as to an interdisciplinary audience of a graduate course on automation. This experience led us to make the book as self-contained as possible and to address students with different backgrounds. As prerequisitive knowledge, the reader requires only basic calculus, elementary linear algebra, and some probability theory basics. A number of mathematical tools, such as probability and statistics as well as constrained optimization, needed by various chapters, are treated in four Appendices. The book is designed to serve as a text for advanced undergraduate and graduate students, and it can be used for either a one- or a two-semester course. Furthermore, it is intended to be used as a self-study and reference book for research and for the practicing scientist/engineer. This latter audience was also our second incentive for writing this book, due to the involvement of our group in a number of projects related to pattern recognition.

SCOPE AND APPROACH

The goal of the book is to present in a unified way the most widely used techniques and methodologies for pattern recognition tasks. Pattern recognition is in the center of a number of application areas, including image analysis, speech and audio recognition, biometrics, bioinformatics, data mining, and information retrieval. Despite their differences, these areas share, to a large extent, a corpus of techniques that can be used in extracting, from the available data, information related to data categories, important "hidden" patterns, and trends. The emphasis in this book is on the most generic of the methods that are currently available. Having acquired the basic knowledge and understanding, the reader can subsequently move on to more specialized application-dependent techniques, which have been developed and reported in a vast number of research papers.

Each chapter of the book starts with the basics and moves, progressively, to more advanced topics'and reviews up-to-date techniques. We have made an effort to keep a balance between mathematical and descriptive presentation. This is not always an easy task. However, we strongly believe that in a topic such as pattern recognition, trying to bypass mathematics deprives the reader of understanding the essentials behind the methods and also the potential of developing new techniques, which fit the needs of the problem at hand that he or she has to tackle. In pattern recognition, the final adoption of an appropriate technique and algorithm is very much a problem-dependent task. Moreover, according to our experience, teaching pattern recognition is also a good "excuse" for the students to refresh and solidify some of the mathematical basics they have been taught in earlier years. "Repetitio est mater studiosum."

NEW TO THIS EDITION

The new features of the fourth edition include the following.

- MATLAB codes and computer experiments are given at the end of most chapters.
- More examples and a number of new figures have been included to enhance the readability and pedagogic aspects of the book.
- New sections on some important topics of high current interest have been added, including:
  - Nonlinear dimensionality reduction
  - Nonnegative matrix factorization
  - Relevance feedback
  - Robust regression
  - Semi-supervised learning
  - Spectral clustering
  - Clustering combination techniques

Also, a number of sections have been rewritten in the context of more recent applications in mind.

SUPPLEMENTS TO THE TEXT

Demonstrations based on MATLAB are available for download from the book Web site, www.elsevierdirect.com/9781597492720. Also available are electronic figures from the text and (for instructors only) a solutions manual for the end-of-chapter problems and exercises. The interested reader can download detailed proofs, which in the book necessarily, are sometimes, slightly condensed. PowerPoint presentations are also available covering all chapters of the book.

Our intention is to update the site regularly with more and/or improved versions of the MATLAB demonstrations. Suggestions are always welcome. Also at this Web site a page will be available for typos, which are unavoidable, despite frequent careful reading. The authors would appreciate readers notifying them about any typos found.

ACKNOWLEDGMENTS

This book would have not been written without the constant support and help from a number of colleagues and students throughout the years. We are especially indebted to Kostas Berberidis, Velissaris Gezerlis, Xaris Georgion, Kristina Georgoulakis, Leyteris Kofidis, Thanassis Liavas, Michalis Mavroforakis, Aggelos Pikrakis, Thanassis Rontogiannis, Margaritis Sdralis, Kostas Slavakis, and Theodoros Yiannakoponlos. The constant support provided by Yannis Kopsinis and Kostas Thernelis from the early stages up to the final stage, with those long nights, has been invaluable. The book improved a great deal after the careful reading and the serious comments and suggestions of Alexandros Bölnn. Dionissis Cavouras, Vassilis Digalakis, Vassilis Drakopoulos, Nikos Galatsanos, George Glentis, Spiros Hatzispyros, Evagelos Karkaletsis, Elias Koutsoupias, Aristides Likas, Gerassimos Mileounis, George Monstakides, George Paliouras, Stavros Perantonis, Takis Stamatoponlos, Nikos Vassilas, Manolis Zervakis, and Vassilis Zissimopoulos.

The book has greatly gained and improved thanks to the comments of a number of people who provided feedback on the revision plan and/or comments on revised chapters:

Tulay Adali, University of Maryland; Mehniet Celenk, Ohio University; Rama Chellappa, University of Maryland; Mark Clements, Georgia Institute of Technology; Robert Duin, Delft University of Technology; Miguel Figneroa, Villanueva University of Puerto Rico; Dimitris Gunopoulos, University of Athens; Mathias Kolsch, Naval Postgraduate School; Adam Krzyzak, Concordia University; Baoxiu Li, Arizona State University; David Miller, Pennsylvania State University; Bernhard Schölkopf, Max Planck Institute; Hari Sundaram, Arizona State University; Harry Wechsler, George Mason University; and Alexander Zien, Max Planck Institute.

We are greatly indebted to these colleagues for their time and their constructive criticisms. Our collaboration and friendship with Nikos Kalouptsidis have been a source of constant inspiration for all these years. We are both deeply indebted to him.

Last but not least, K. Koutroumbas would like to thank Sophia, Dimitris-Marios, and Valentini-Theodora for their tolerance and support and S. Theodoridis would like to thank Despina, Eva, and Eleni, his joyful and supportive "harem."

CHAPTER 1

Introduction

1.1 IS PATTERN RECOGNITION IMPORTANT?

Pattern recognition is the scientific discipline whose goal is the classification of *objects* into a number of categories or *classes*. Depending on the application, these objects can be images or signal waveforms or any type of measurements that need to be classified. We will refer to these objects using the generic term *patterns*. Pattern recognition has a long history, but before the 1960s it was mostly the output of theoretical research in the area of statistics. As with everything else, the advent of computers increased the demand for practical applications of pattern recognition, which in turn set new demands for further theoretical developments. As our society evolves from the industrial to its postindustrial phase, automation in industrial production and the need for information handling and retrieval are becoming increasingly important. This trend has pushed pattern recognition to the high edge of today's engineering applications and research. Pattern recognition is an integral part of most *machine intelligence* systems built for decision making.

Machine vision is an area in which pattern recognition is of importance. A machine vision system captures images via a camera and analyzes them to produce descriptions of what is imaged. A typical application of a machine vision system is in the manufacturing industry, either for automated visual inspection or for automation in the assembly line. For example, in inspection, manufactured objects on a moving conveyor may pass the inspection station, where the camera stands, and it has to be ascertained whether there is a defect. Thus, images have to be analyzed online, and a pattern recognition system has to classify the objects into the "defect" or "nondefect" class. After that, an action has to be taken, such as to reject the offending parts. In an assembly line, different objects must be located and "recognized," that is, classified in one of a number of classes known *a priori*. Examples are the "screwdriver class," the "German key class," and so forth in a tools' manufacturing unit. Then a robot arm can move the objects in the right place.

Character (letter or number) recognition is another important area of pattern recognition, with major implications in automation and information handling. Optical character recognition (OCR) systems are already commercially available and more or less familiar to all of us. An OCR system has a "front-end" device consisting of a *light source*, a *scan lens*, a *document transport*, and a *detector*. At the output of the light-sensitive detector, light-intensity variation is translated into "numbers" and an image array is formed. In the sequel, a series of image processing techniques are applied leading to *line* and *character segmentation*. The pattern recognition software then takes over to recognize the characters—that is, to classify each character in the correct "letter, number, punctuation" class. Storing the recognized document has a twofold advantage over storing its scanned image. First, further electronic processing, if needed, is easy via a word processor, and second, it is much more efficient to store ASCII characters than a document image. Besides the printed character recognition systems, there is a great deal of interest invested in systems that recognize handwriting. A typical commercial application of such a system is in the machine reading of bank checks. The machine must be able to recognize the amounts in figures and digits and match them. Furthermore, it could check whether the payee corresponds to the account to be credited. Even if only half of the checks are manipulated correctly by such a machine, much labor can be saved from a tedious job. Another application is in automatic mail-sorting machines for postal code identification in post offices. Online handwriting recognition systems are another area of great commercial interest. Such systems will accompany *pen computers*, with which the entry of data will be done not via the keyboard but by writing. This complies with today's tendency to develop machines and computers with interfaces acquiring human-like skills.

Computer-aided diagnosis is another important application of pattern recognition, aiming at assisting doctors in making diagnostic decisions. The final diagnosis is, of course, made by the doctor. Computer-assisted diagnosis has been applied to and is of interest for a variety of medical data, such as X-rays, computed tomographic images, ultrasound images, electrocardiograms (ECGs), and electroencephalograms (EEGs). The need for a computer-aided diagnosis stems from the fact that medical data are often not easily interpretable, and the interpretation can depend very much on the skill of the doctor. Let us take for example *X-ray mammography* for the detection of breast cancer. Although mammography is currently the best method for detecting breast cancer, 10 to 30% of women who have the disease and undergo mammography have negative mammograms. In approximately two thirds of these cases with false results the radiologist failed to detect the cancer, which was evident retrospectively. This may be due to poor image quality, eye fatigue of the radiologist, or the subtle nature of the findings. The percentage of correct classifications improves at a second reading by another radiologist. Thus, one can aim to develop a pattern recognition system in order to assist radiologists with a "second" opinion. Increasing confidence in the diagnosis based on mammograms would, in turn, decrease the number of patients with suspected breast cancer who have to undergo surgical breast biopsy, with its associated complications.

Speech recognition is another area in which a great deal of research and development effort has been invested. Speech is the most natural means by which humans communicate and exchange information. Thus, the goal of building intelligent machines that recognize *spoken information* has been a long-standing one for scientists and engineers as well as science fiction writers. Potential applications of such machines are numerous. They can be used, for example, to improve efficiency in a manufacturing environment, to control machines in hazardous environments remotely, and to help handicapped people to control machines by talking to them. A major effort, which has already had considerable success, is to enter data into a computer via a microphone. Software, built around a pattern (spoken sounds in this case) recognition system, recognizes the spoken text and translates it into ASCII characters, which are shown on the screen and can be stored in the memory. Entering information by "talking" to a computer is twice as fast as entry by a skilled typist. Furthermore, this can enhance our ability to communicate with deaf and dumb people.

Data mining and knowledge discovery in databases is another key application area of pattern recognition. Data mining is of intense interest in a wide range of applications such as medicine and biology, market and financial analysis, business management, science exploration, image and music retrieval. Its popularity stems from the fact that in the age of information and knowledge society there is an ever increasing demand for retrieving information and turning it into knowledge. Moreover, this information exists in huge amounts of data in various forms including, text, images, audio and video, stored in different places distributed all over the world. The traditional way of searching information in databases was the description-based model where object retrieval was based on keyword description and subsequent word matching. However, this type of searching presupposes that a manual annotation of the stored information has previously been performed by a human. This is a very time-consuming job and, although feasible when the size of the stored information is limited, it is not possible when the amount of the available information becomes large. Moreover, the task of manual annotation becomes problematic when the stored information is widely distributed and shared by a heterogeneous "mixture" of sites and users. Content-based retrieval systems are becoming more and more popular where information is sought based on "similarity" between an object, which is presented into the system, and objects stored in sites all over the world. In a content-based image retrieval CBIR (system) an image is presented to an input device (e.g., scanner). The system returns "similar" images based on a measured "signature," which can encode, for example, information related to color, texture and shape. In a music content-based retrieval system, an example (i.e., an extract from a music piece), is presented to a microphone input device and the system returns "similar" music pieces. In this case, similarity is based on certain (automatically) measured cues that characterize a music piece, such as the music meter, the music tempo, and the location of certain repeated patterns.

Mining for biomedical and DNA data analysis has enjoyed an explosive growth since the mid-1990s. All DNA sequences comprise four basic building elements; the nucleotides: adenine (A), cytosine (C), guanine (G) and thymine (T). Like the letters in our alphabets and the seven notes in music, these four nucleotides are combined to form long sequences in a twisted ladder form. Genes consist of, usually, hundreds of nucleotides arranged in a particular order. Specific gene-sequence patterns are related to particular diseases and play an important role in medicine. To this end, pattern recognition is a key area that offers a wealth of developed tools for similarity search and comparison between DNA sequences. Such comparisons between healthy and diseased tissues are very important in medicine to identify critical differences between these two classes.

The foregoing are only five examples from a much larger number of possible applications. Typically, we refer to fingerprint identification, signature authentication, text retrieval, and face and gesture recognition. The last applications have recently attracted much research interest and investment in an attempt to facilitate human–machine interaction and further enhance the role of computers in office automation, automatic personalization of environments, and so forth. Just to provoke imagination, it is worth pointing out that the MPEG-7 standard includes a provision for content-based video information retrieval from digital libraries of the type: search and find all video scenes in a digital library showing person "X" laughing. Of course, to achieve the final goals in all of these applications, pattern recognition is closely linked with other scientific disciplines, such as linguistics, computer graphics, machine vision, and database design.

Having aroused the reader's curiosity about pattern recognition, we will next sketch the basic philosophy and methodological directions in which the various pattern recognition approaches have evolved and developed.

1.2 FEATURES, FEATURE VECTORS, AND CLASSIFIERS

Let us first simulate a simplified case "mimicking" a medical image classification task. Figure 1.1 shows two images, each having a distinct region inside it. The two regions are also themselves visually different. We could say that the region of Figure 1.1a results from a benign lesion, class A, and that of Figure 1.1b from a malignant one (cancer), class B. We will further assume that these are not the only patterns (images) that are available to us, but we have access to an image database

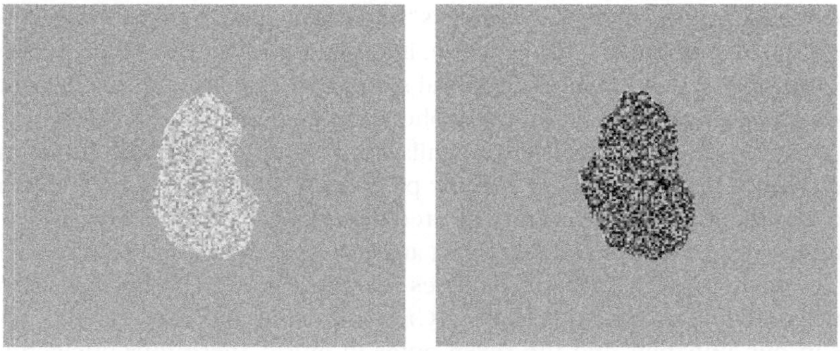

(a) (b)

FIGURE 1.1

Examples of image regions corresponding to (a) class A and (b) class B.

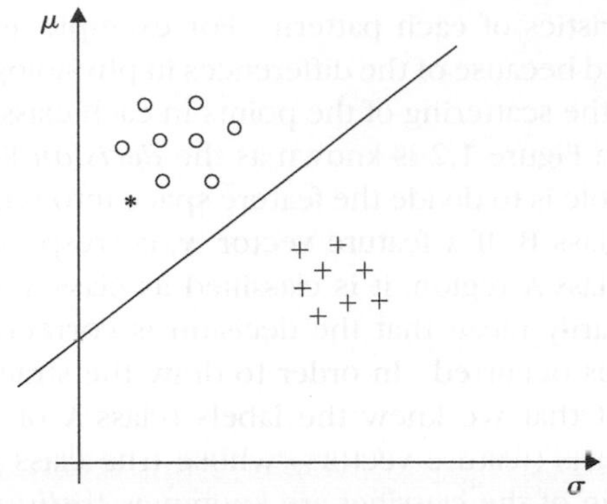

FIGURE 1.2

Plot of the mean value versus the standard deviation for a number of different images originating from class A (o) and class B (+). In this case, a straight line separates the two classes.

with a number of patterns, some of which are known to originate from class A and some from class B.

The first step is to identify the measurable quantities that make these two regions *distinct* from each other. Figure 1.2 shows a plot of the mean value of the intensity in each region of interest versus the corresponding standard deviation around this mean. Each point corresponds to a different image from the available database. It turns out that class A patterns tend to spread in a different area from class B patterns. The straight line seems to be a good candidate for separating the two classes. Let us now assume that we are given a new image with a region in it and that we do not know to which class it belongs. It is reasonable to say that we measure the mean intensity and standard deviation in the region of interest and we plot the corresponding point. This is shown by the asterisk (∗) in Figure 1.2. Then it is sensible to assume that the unknown pattern is *more likely* to belong to class A than class B.

The preceding artificial *classification* task has outlined the rationale behind a large class of pattern recognition problems. The measurements used for the classification, the mean value and the standard deviation in this case, are known as *features*. In the more general case l features x_i, $i = 1, 2, \ldots, l$, are used, and they form the *feature vector*

$$\boldsymbol{x} = [x_1, x_2, \ldots, x_l]^T$$

where T denotes transposition. Each of the feature vectors identifies *uniquely* a single pattern (object). Throughout this book features and feature vectors will be treated as *random variables* and *vectors*, respectively. This is natural, as the measurements resulting from different patterns exhibit a random variation. This is due partly to the measurement noise of the measuring devices and partly to the distinct characteristics of each pattern. For example, in X-ray imaging large variations are expected because of the differences in physiology among individuals. This is the reason for the scattering of the points in each class shown in Figure 1.1.

The straight line in Figure 1.2 is known as the *decision* line, and it constitutes the *classifier* whose role is to divide the feature space into regions that correspond to either class A or class B. If a feature vector $\boldsymbol{x}$, corresponding to an unknown pattern, falls in the class A region, it is classified as class A, otherwise as class B. This does not necessarily mean that the decision is correct. If it is not correct, a *misclassification* has occurred. In order to draw the straight line in Figure 1.2 we exploited the fact that we knew the labels (class A or B) for each point of the figure. The patterns (feature vectors) whose true class is known and which are used for the design of the classifier are known as *training patterns (training feature vectors)*.

Having outlined the definitions and the rationale, let us point out the basic questions arising in a classification task.

- How are the features generated? In the preceding example, we used the mean and the standard deviation, because we knew how the images had been generated. In practice, this is far from obvious. It is problem dependent, and it concerns the *feature generation stage* of the design of a classification system that performs a given pattern recognition task.
- What is the best number l of features to use? This is also a very important task and it concerns the *feature selection stage* of the classification system. In practice, a larger than necessary number of feature candidates is generated, and then the "best" of them is adopted.
- Having adopted the appropriate, for the specific task, features, how does one design the classifier? In the preceding example the straight line was drawn empirically, just to please the eye. In practice, this cannot be the case, and the line should be drawn optimally, with respect to an *optimality criterion*. Furthermore, problems for which a linear classifier (straight line or hyperplane in the l-dimensional space) can result in acceptable performance are not the rule. In general, the surfaces dividing the space in the various class regions are nonlinear. What type of nonlinearity must one adopt, and what type of optimizing criterion must be used in order to locate a surface in the right place in the l-dimensional *feature space*? These questions concern the *classifier design stage*.
- Finally, once the classifier has been designed, how can one assess the performance of the designed classifier? That is, what is the *classification error rate*? This is the task of the *system evaluation stage*.

Figure 1.3 shows the various stages followed for the design of a classification system. As is apparent from the feedback arrows, these stages are not independent. On the contrary, they are interrelated and, depending on the results, one may go back

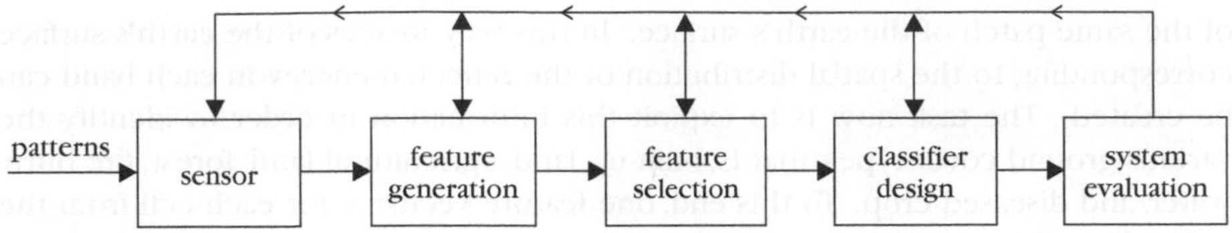

FIGURE 1.3

The basic stages involved in the design of a classification system.

to redesign earlier stages in order to improve the overall performance. Furthermore, there are some methods that combine stages, for example, the feature selection and the classifier design stage, in a common optimization task.

Although the reader has already been exposed to a number of basic problems at the heart of the design of a classification system, there are still a few things to be said.

1.3 SUPERVISED, UNSUPERVISED, AND SEMI-SUPERVISED LEARNING

In the example of Figure 1.1, we assumed that a set of training data were available, and the classifier was designed by exploiting this *a priori* known information. This is known as *supervised pattern recognition* or in the more general context of machine learning as *supervised learning*. However, this is not always the case, and there is another type of pattern recognition tasks for which training data, of known class labels, are not available. In this type of problem, we are given a set of feature vectors $\boldsymbol{x}$ and the goal is to unravel the underlying *similarities* and *cluster* (group) "similar" vectors together. This is known as *unsupervised pattern recognition* or *unsupervised learning* or *clustering*. Such tasks arise in many applications in social sciences and engineering, such as remote sensing, image segmentation, and image and speech coding. Let us pick two such problems.

In *multispectral remote sensing*, the electromagnetic energy emanating from the earth's surface is measured by sensitive scanners located aboard a satellite, an aircraft, or a space station. This energy may be reflected solar energy (passive) or the reflected part of the energy transmitted from the vehicle (active) in order to "interrogate" the earth's surface. The scanners are sensitive to a number of wavelength bands of the electromagnetic radiation. Different properties of the earth's surface contribute to the reflection of the energy in the different bands. For example, in the visible–infrared range properties such as the mineral and moisture contents of soils, the sedimentation of water, and the moisture content of vegetation are the main contributors to the reflected energy. In contrast, at the thermal end of the infrared, it is the thermal capacity and thermal properties of the surface and near subsurface that contribute to the reflection. Thus, each band measures different properties of the same patch of the earth's surface. In this way, images of the earth's surface corresponding to the spatial distribution of the reflected energy in each band can be created. The task now is to exploit this information in order to identify the various ground cover types, that is, built-up land, agricultural land, forest, fire burn, water, and diseased crop. To this end, one feature vector $\boldsymbol{x}$ for each cell from the "sensed" earth's surface is formed. The elements $x_i, i = 1, 2, \ldots, l$, of the vector are the corresponding image pixel intensities in the various spectral bands. In practice, the number of spectral bands varies.

A *clustering* algorithm can be employed to reveal the groups in which feature vectors are clustered in the l-dimensional feature space. Points that correspond to the same ground cover type, such as water, are expected to cluster together and form groups. Once this is done, the analyst can identify the type of each cluster by associating a sample of points in each group with available reference ground data, that is, maps or visits. Figure 1.4 demonstrates the procedure.

Clustering is also widely used in the social sciences in order to study and correlate survey and statistical data and draw useful conclusions, ***which will then lead to the right actions***. Let us again resort to a simplified example and assume that we are interested in studying whether there is any relation between a country's gross national product (GNP) and the level of people's illiteracy, on the one hand, and children's mortality rate on the other. In this case, each country is represented by a three-dimensional feature vector whose coordinates are indices measuring the quantities of interest. A clustering algorithm will then reveal a rather compact cluster corresponding to countries that exhibit low GNPs, high illiteracy levels, and high children's mortality expressed as a population percentage.

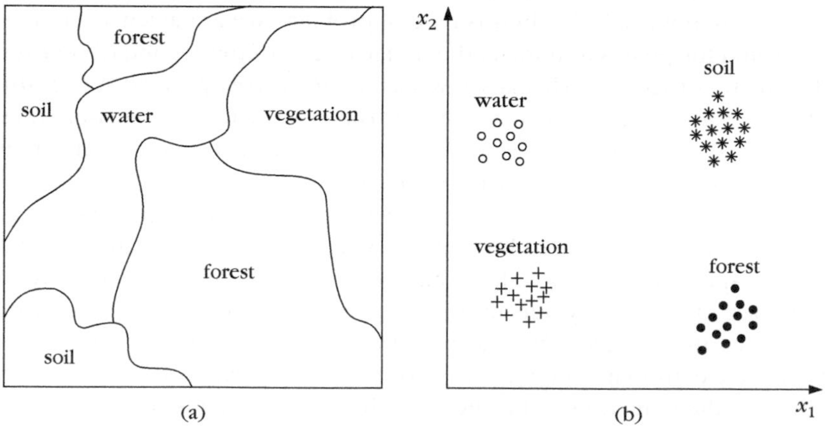

FIGURE 1.4

(a) An illustration of various types of ground cover and (b) clustering of the respective features for multispectral imaging using two bands.

A major issue in unsupervised pattern recognition is that of defining the "similarity" between two feature vectors and choosing an appropriate measure for it. Another issue of importance is choosing an algorithmic scheme that will cluster (group) the vectors on the basis of the adopted similarity measure. In general, different algorithmic schemes may lead to different results, which the expert has to interpret.

Semi-supervised learning/pattern recognition for designing a classification system shares the same goals as the supervised case, however now, the designer has at his or her disposal a set of patterns of unknown class origin, in addition to the training patterns, whose true class is known. We usually refer to the former ones as *unlabeled* and the latter as *labeled* data. Semi-supervised pattern recognition can be of importance when the system designer has access to a rather limited number of labeled data. In such cases, recovering additional information from the unlabeled samples, related to the general structure of the data at hand, can be useful in improving the system design. Semi-supervised learning finds its way also to clustering tasks. In this case, labeled data are used as constraints in the form of *must-links* and *cannot-links*. In other words, the clustering task is constrained to assign certain points in the same cluster or to exclude certain points of being assigned in the same cluster. From this perspective, semi-supervised learning provides an *a priori* knowledge that the clustering algorithm has to respect.

1.4 MATLAB PROGRAMS

At the end of most of the chapters there is a number of MATLAB programs and computer experiments. The MATLAB codes provided are not intended to form part of a software package, but they are to serve a purely pedagogical goal. Most of these codes are given to our students who are asked to play with and discover the "secrets" associated with the corresponding methods. This is also the reason that for most of the cases the data used are simulated data around the Gaussian distribution. They have been produced carefully in order to guide the students in understanding the basic concepts. This is also the reason that the provided codes correspond to those of the techniques and algorithms that, to our opinion, comprise the backbone of each chapter and the student has to understand in a first reading. Whenever the required MATLAB code was available (at the time this book was prepared) in a MATLAB toolbox, we chose to use the associated MATLAB function and explain how to use its arguments. No doubt, each instructor has his or her own preferences, experiences, and unique way of viewing teaching. The provided routines are written in a way that can run on other data sets as well. In a separate accompanying book we provide a more complete list of MATLAB codes embedded in a user-friendly Graphical User Interface (GUI) and also involving more realistic examples using real images and audio signals.

1.5 OUTLINE OF THE BOOK

Chapters 2–10 deal with supervised pattern recognition and Chapters 11–16 deal with the unsupervised case. Semi-supervised learning is introduced in Chapter 10. The goal of each chapter is to start with the basics, definitions, and approaches, and move progressively to more advanced issues and recent techniques. To what extent the various topics covered in the book will be presented in a first course on pattern recognition depends very much on the course's focus, on the students' background, and, of course, on the lecturer. In the following outline of the chapters, we give our view and the topics that we cover in a first course on pattern recognition. No doubt, other views do exist and may be better suited to different audiences. At the end of each chapter, a number of problems and computer exercises are provided.

Chapter 2 is focused on Bayesian classification and techniques for estimating unknown probability density functions. In a first course on pattern recognition, the sections related to Bayesian inference, the maximum entropy, and the expectation maximization (EM) algorithm are omitted. Special focus is put on the Bayesian classification, the minimum distance (Euclidean and Mahalanobis), the nearest neighbor classifiers, and the naive Bayes classifier. Bayesian networks are briefly introduced.

Chapter 3 deals with the design of linear classifiers. The sections dealing with the probability estimation property of the mean square solution as well as the bias variance dilemma are only briefly mentioned in our first course. The basic philosophy underlying the support vector machines can also be explained, although a deeper treatment requires mathematical tools (summarized in Appendix C) that most of the students are not familiar with during a first course class. On the contrary, emphasis is put on the linear separability issue, the perceptron algorithm, and the mean square and least squares solutions. After all, these topics have a much broader horizon and applicability. Support vector machines are briefly introduced. The geometric interpretation offers students a better understanding of the SVM theory.

Chapter 4 deals with the design of nonlinear classifiers. The section dealing with exact classification is bypassed in a first course. The proof of the backpropagation algorithm is usually very boring for most of the students and we bypass its details. A description of its rationale is given, and the students experiment with it using MATLAB. The issues related to cost functions are bypassed. Pruning is discussed with an emphasis on generalization issues. Emphasis is also given to Cover's theorem and radial basis function (RBF) networks. The nonlinear support vector machines, decision trees, and combining classifiers are only briefly touched via a discussion on the basic philosophy behind their rationale.

Chapter 5 deals with the feature selection stage, and we have made an effort to present most of the well-known techniques. In a first course we put emphasis on the t-test. This is because hypothesis testing also has a broad horizon, and at the same time it is easy for the students to apply it in computer exercises. Then, depending on time constraints, divergence, Bhattacharrya distance, and scattered matrices are presented and commented on, although their more detailed treatment is for a more advanced course. Emphasis is given to Fisher's linear discriminant method (LDA) for the two-class case.

Chapter 6 deals with the feature generation stage using transformations. The Karhunen–Loève transform and the singular value decomposition are first introduced as dimensionality reduction techniques. Both methods are briefly covered in the second semester. In the sequel the independent component analysis (ICA), non-negative matrix factorization and nonlinear dimensionality reduction techniques are presented. Then the discrete Fourier transform (DFT), discrete cosine transform (DCT), discrete sine transform (DST), Hadamard, and Haar transforms are defined. The rest of the chapter focuses on the discrete time wavelet transform. The incentive is to give all the necessary information so that a newcomer in the wavelet field can grasp the basics and be able to develop software, based on filter banks, in order to generate features. All these techniques are bypassed in a first course.

Chapter 7 deals with feature generation focused on image and audio classification. The sections concerning local linear transforms, moments, parametric models, and fractals are not covered in a first course. Emphasis is placed on first- and second-order statistics features as well as the run-length method. The chain code for shape description is also taught. Computer exercises are then offered to generate these features and use them for classification for some case studies. In a one-semester course there is no time to cover more topics.

Chapter 8 deals with template matching. Dynamic programming (DP) and the Viterbi algorithm are presented and then applied to speech recognition. In a two-semester course, emphasis is given to the DP and the Viterbi algorithm. The edit distance seems to be a good case for the students to grasp the basics. Correlation matching is taught and the basic philosophy behind deformable template matching can also be presented.

Chapter 9 deals with context-dependent classification. Hidden Markov models are introduced and applied to communications and speech recognition. This chapter is bypassed in a first course.

Chapter 10 deals with system evaluation and semi-supervised learning. The various error rate estimation techniques are discussed, and a case study with real data is treated. The leave-one-out method and the resubstitution methods are emphasized in the second semester, and students practice with computer exercises. Semi-supervised learning is bypassed in a first course.

Chapter 11 deals with the basic concepts of clustering. It focuses on definitions as well as on the major stages involved in a clustering task. The various types of data encountered in clustering applications are reviewed, and the most commonly used proximity measures are provided. In a first course, only the most widely used proximity measures are covered (e.g., l_p norms, inner product, Hamming distance).

Chapter 12 deals with sequential clustering algorithms. These include some of the simplest clustering schemes, and they are well suited for a first course to introduce students to the basics of clustering and allow them to experiment with the computer. The sections related to estimation of the number of clusters and neural network implementations are bypassed.

Chapter 13 deals with hierarchical clustering algorithms. In a first course, only the general agglomerative scheme is considered with an emphasis on single link and complete link algorithms, based on matrix theory. Agglomerative algorithms based on graph theory concepts as well as the divisive schemes are bypassed.

Chapter 14 deals with clustering algorithms based on cost function optimization, using tools from differential calculus. Hard clustering and fuzzy and possibilistic schemes are considered, based on various types of cluster representatives, including point representatives, hyperplane representatives, and shell-shaped representatives. In a first course, most of these algorithms are bypassed, and emphasis is given to the isodata algorithm.

Chapter 15 features a high degree of modularity. It deals with clustering algorithms based on different ideas, which cannot be grouped under a single philosophy. Spectral clustering, competitive learning, branch and bound, simulated annealing, and genetic algorithms are some of the schemes treated in this chapter. These are bypassed in a first course.

Chapter 16 deals with the clustering validity stage of a clustering procedure. It contains rather advanced concepts and is omitted in a first course. Emphasis is given to the definitions of internal, external, and relative criteria and the random hypotheses used in each case. Indices, adopted in the framework of external and internal criteria, are presented, and examples are provided showing the use of these indices.

Syntactic pattern recognition methods are not treated in this book. Syntactic pattern recognition methods differ in philosophy from the methods discussed in this book and, in general, are applicable to different types of problems. In syntactic pattern recognition, the structure of the patterns is of paramount importance, and pattern recognition is performed on the basis of a set of pattern *primitives*, a set of rules in the form of a *grammar*, and a recognizer called *automaton*. Thus, we were faced with a dilemma: either to increase the size of the book substantially, or to provide a short overview (which, however, exists in a number of other books), or to omit it. The last option seemed to be the most sensible choice.

CHAPTER

Classifiers Based on Bayes Decision Theory

2

2.1 INTRODUCTION

This is the first chapter, out of three, dealing with the design of the classifier in a pattern recognition system. The approach to be followed builds upon probabilistic arguments stemming from the statistical nature of the generated features. As has already been pointed out in the introductory chapter, this is due to the statistical variation of the patterns as well as to the noise in the measuring sensors. Adopting this reasoning as our kickoff point, we will design classifiers that classify an unknown pattern in the most probable of the classes. Thus, our task now becomes that of defining what "most probable" means.

Given a classification task of M classes, $\omega_1, \omega_2, \ldots, \omega_M$, and an unknown pattern, which is represented by a feature vector $\boldsymbol{x}$, we form the M conditional probabilities $P(\omega_i|\boldsymbol{x}), i = 1, 2, \ldots, M$. Sometimes, these are also referred to as *a posteriori probabilities*. In words, each of them represents the probability that the unknown pattern belongs to the respective class ω_i, given that the corresponding feature vector takes the value $\boldsymbol{x}$. Who could then argue that these conditional probabilities are not sensible choices to quantify the term *most probable*? Indeed, the classifiers to be considered in this chapter compute either the maximum of these M values or, equivalently, the maximum of an appropriately defined function of them. The unknown pattern is then assigned to the class corresponding to this maximum.

The first task we are faced with is the computation of the conditional probabilities. The Bayes rule will once more prove its usefulness! A major effort in this chapter will be devoted to techniques for estimating probability density functions (pdf), based on the available experimental evidence, that is, the feature vectors corresponding to the patterns of the training set.

2.2 BAYES DECISION THEORY

We will initially focus on the two-class case. Let ω_1, ω_2 be the two classes in which our patterns belong. In the sequel, we assume that the a priori probabilities

$P(\omega_1), P(\omega_2)$ are known. This is a very reasonable assumption, because even if they are not known, they can easily be estimated from the available training feature vectors. Indeed, if N is the total number of available training patterns, and N_1, N_2 of them belong to ω_1 and ω_2, respectively, then $P(\omega_1) \approx N_1/N$ and $P(\omega_2) \approx N_2/N$.

The other statistical quantities assumed to be known are the class-conditional probability density functions $p(\boldsymbol{x}|\omega_i), i = 1, 2$, describing the distribution of the feature vectors in each of the classes. If these are not known, they can also be estimated from the available training data, as we will discuss later on in this chapter. The pdf $p(\boldsymbol{x}|\omega_i)$ is sometimes referred to as the *likelihood function of ω_i with respect to $\boldsymbol{x}$*. Here we should stress the fact that an implicit assumption has been made. That is, the feature vectors can take any value in the l-dimensional feature space. In the case that feature vectors can take only discrete values, density functions $p(\boldsymbol{x}|\omega_i)$ become probabilities and will be denoted by $P(\boldsymbol{x}|\omega_i)$.

We now have all the ingredients to compute our conditional probabilities, as stated in the introduction. To this end, let us recall from our probability course basics the *Bayes rule* (Appendix A)

$$P(\omega_i|\boldsymbol{x}) = \frac{p(\boldsymbol{x}|\omega_i)P(\omega_i)}{p(\boldsymbol{x})} \tag{2.1}$$

where $p(\boldsymbol{x})$ is the pdf of $\boldsymbol{x}$ and for which we have (Appendix A)

$$p(\boldsymbol{x}) = \sum_{i=1}^{2} p(\boldsymbol{x}|\omega_i)P(\omega_i) \tag{2.2}$$

The *Bayes classification rule* can now be stated as

$$\begin{aligned} &\text{If} \quad P(\omega_1|\boldsymbol{x}) > P(\omega_2|\boldsymbol{x}), \quad \boldsymbol{x} \text{ is classified to } \omega_1 \\ &\text{If} \quad P(\omega_1|\boldsymbol{x}) < P(\omega_2|\boldsymbol{x}), \quad \boldsymbol{x} \text{ is classified to } \omega_2 \end{aligned} \tag{2.3}$$

The case of equality is detrimental and the pattern can be assigned to either of the two classes. Using (2.1), the decision can equivalently be based on the inequalities

$$p(\boldsymbol{x}|\omega_1)P(\omega_1) \gtrless p(\boldsymbol{x}|\omega_2)P(\omega_2) \tag{2.4}$$

$p(\boldsymbol{x})$ is not taken into account, because it is the same for all classes and it does not affect the decision. Furthermore, if the *a priori* probabilities are equal, that is, $P(\omega_1) = P(\omega_2) = 1/2$, Eq. (2.4) becomes

$$p(\boldsymbol{x}|\omega_1) \gtrless p(\boldsymbol{x}|\omega_2) \tag{2.5}$$

Thus, the search for the maximum now rests on the values of the conditional pdfs evaluated at $\boldsymbol{x}$. Figure 2.1 presents an example of two equiprobable classes and shows the variations of $p(x|\omega_i), i = 1, 2$, as functions of x for the simple case of a single feature ($l = 1$). The dotted line at x_0 is a threshold partitioning the feature space into two regions, R_1 and R_2. According to the Bayes decision rule, for all values of x in R_1 the classifier decides ω_1 and for all values in R_2 it decides ω_2. However, it is obvious from the figure that decision errors are unavoidable. Indeed, there is

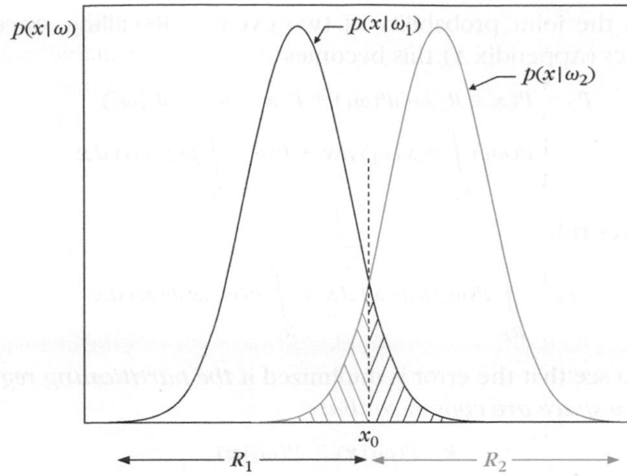

FIGURE 2.1

Example of the two regions R_1 and R_2 formed by the Bayesian classifier for the case of two equiprobable classes.

a finite probability for an x to lie in the R_2 region and at the same time to belong in class ω_1. Then our decision is in error. The same is true for points originating from class ω_2. It does not take much thought to see that the total probability, P_e, of committing a decision error for the case of two equiprobable classes, is given by

$$P_e = \frac{1}{2}\int_{-\infty}^{x_0} p(x|\omega_2)\,dx + \frac{1}{2}\int_{x_0}^{+\infty} p(x|\omega_1)\,dx \tag{2.6}$$

which is equal to the total shaded area under the curves in Figure 2.1. We have now touched on a very important issue. Our starting point to arrive at the Bayes classification rule was rather empirical, via our interpretation of the term *most probable*. We will now see that this classification test, though simple in its formulation, has a sounder mathematical interpretation.

Minimizing the Classification Error Probability

We will show that *the Bayesian classifier is optimal with respect to minimizing the classification error probability*. Indeed, the reader can easily verify, as an exercise, that moving the threshold away from x_0, in Figure 2.1, always increases the corresponding shaded area under the curves. Let us now proceed with a more formal proof.

Proof. Let R_1 be the region of the feature space in which we decide in favor of ω_1 and R_2 be the corresponding region for ω_2. Then an error is made if $\boldsymbol{x} \in R_1$, although it belongs to ω_2 or if $\boldsymbol{x} \in R_2$, although it belongs to ω_1. That is,

$$P_e = P(\boldsymbol{x} \in R_2, \omega_1) + P(\boldsymbol{x} \in R_1, \omega_2) \tag{2.7}$$

where $P(\cdot, \cdot)$ is the joint probability of two events. Recalling, once more, our probability basics (Appendix A), this becomes

$$\begin{aligned} P_e &= P(\boldsymbol{x} \in R_2|\omega_1)P(\omega_1) + P(\boldsymbol{x} \in R_1|\omega_2)P(\omega_2) \\ &= P(\omega_1) \int_{R_2} p(\boldsymbol{x}|\omega_1)\, d\boldsymbol{x} + P(\omega_2) \int_{R_1} p(\boldsymbol{x}|\omega_2)\, d\boldsymbol{x} \end{aligned} \tag{2.8}$$

or using the Bayes rule

$$P_e = \int_{R_2} P(\omega_1|\boldsymbol{x})p(\boldsymbol{x})\, d\boldsymbol{x} + \int_{R_1} P(\omega_2|\boldsymbol{x})p(\boldsymbol{x})\, d\boldsymbol{x} \tag{2.9}$$

It is now easy to see that the error is minimized if *the partitioning regions* R_1 and R_2 *of the feature space are chosen so that*

$$\begin{aligned} R_1 &: P(\omega_1|\boldsymbol{x}) > P(\omega_2|\boldsymbol{x}) \\ R_2 &: P(\omega_2|\boldsymbol{x}) > P(\omega_1|\boldsymbol{x}) \end{aligned} \tag{2.10}$$

Indeed, since the union of the regions R_1, R_2 covers all the space, from the definition of a probability density function we have that

$$\int_{R_1} P(\omega_1|\boldsymbol{x})p(\boldsymbol{x})\, d\boldsymbol{x} + \int_{R_2} P(\omega_1|\boldsymbol{x})p(\boldsymbol{x})\, d\boldsymbol{x} = P(\omega_1) \tag{2.11}$$

Combining Eqs. (2.9) and (2.11), we get

$$P_e = P(\omega_1) - \int_{R_1} (P(\omega_1|\boldsymbol{x}) - P(\omega_2|\boldsymbol{x}))\, p(\boldsymbol{x})\, d\boldsymbol{x} \tag{2.12}$$

This suggests that the probability of error is minimized if R_1 is the region of space in which $P(\omega_1|\boldsymbol{x}) > P(\omega_2|\boldsymbol{x})$. Then, R_2 becomes the region where the reverse is true. □

So far, we have dealt with the simple case of two classes. Generalizations to the multiclass case are straightforward. In a classification task with M classes, $\omega_1, \omega_2, \ldots, \omega_M$, an unknown pattern, represented by the feature vector $\boldsymbol{x}$, is assigned to class ω_i if

$$P(\omega_i|\boldsymbol{x}) > P(\omega_j|\boldsymbol{x}) \quad \forall j \neq i \tag{2.13}$$

It turns out that such a choice also minimizes the classification error probability (Problem 2.1).

Minimizing the Average Risk

The classification error probability is not always the best criterion to be adopted for minimization. This is because it assigns the same importance to all errors. However, there are cases in which some wrong decisions may have more serious implications than others. For example, it is much more serious for a doctor to make a wrong decision and a malignant tumor to be diagnosed as a benign one, than the other way round. If a benign tumor is diagnosed as a malignant one, the wrong decision will be cleared out during subsequent clinical examinations. However, the results from the wrong decision concerning a malignant tumor may be fatal. Thus, in such cases it is more appropriate to assign a penalty term to weigh each error. For our example, let us denote by ω_1 the class of malignant tumors and as ω_2 the class of the benign ones. Let, also, R_1, R_2 be the regions in the feature space where we decide in favor of ω_1 and ω_2, respectively. The error probability P_e is given by Eq. (2.8). Instead of selecting R_1 and R_2 so that P_e is minimized, we will now try to minimize a modified version of it, that is,

$$r = \lambda_{12}P(\omega_1)\int_{R_2} p(\boldsymbol{x}|\omega_1)d\boldsymbol{x} + \lambda_{21}P(\omega_2)\int_{R_1} p(\boldsymbol{x}|\omega_2)d\boldsymbol{x} \tag{2.14}$$

where each of the two terms that contributes to the overall error probability is weighted according to its significance. For our case, the reasonable choice would be to have $\lambda_{12} > \lambda_{21}$. Thus errors due to the assignment of patterns originating from class ω_1 to class ω_2 will have a larger effect on the cost function than the errors associated with the second term in the summation.

Let us now consider an M-class problem and let $R_j, j = 1, 2, \ldots, M$, be the regions of the feature space assigned to classes ω_j, respectively. Assume now that a feature vector $\boldsymbol{x}$ that belongs to class ω_k lies in $R_i, i \neq k$. Then this vector is misclassified in ω_i and an error is committed. A penalty term λ_{ki}, known as *loss*, is associated with this wrong decision. The matrix L, which has at its (k, i) location the corresponding penalty term, is known as the *loss matrix*.[1] Observe that in contrast to the philosophy behind Eq. (2.14), we have now allowed weights across the diagonal of the loss matrix (λ_{kk}), which correspond to correct decisions. In practice, these are usually set equal to zero, although we have considered them here for the sake of generality. The *risk or loss* associated with ω_k is defined as

$$r_k = \sum_{i=1}^{M} \lambda_{ki} \int_{R_i} p(\boldsymbol{x}|\omega_k)\, d\boldsymbol{x} \tag{2.15}$$

Observe that the integral is the overall probability of a feature vector from class ω_k being classified in ω_i. This probability is weighted by λ_{ki}. Our goal now is to choose the partitioning regions R_j so that the *average risk*

$$r = \sum_{k=1}^{M} r_k P(\omega_k)$$

$$= \sum_{i=1}^{M} \int_{R_i} \left(\sum_{k=1}^{M} \lambda_{ki}\, p(\boldsymbol{x}|\omega_k)P(\omega_k) \right) d\boldsymbol{x} \tag{2.16}$$

is minimized. This is achieved if each of the integrals is minimized, which is equivalent to selecting partitioning regions so that

$$\boldsymbol{x} \in R_i \quad \text{if} \quad l_i \equiv \sum_{k=1}^{M} \lambda_{ki}\, p(\boldsymbol{x}|\omega_k)P(\omega_k) < l_j \equiv \sum_{k=1}^{M} \lambda_{kj}\, p(\boldsymbol{x}|\omega_k)P(\omega_k) \quad \forall j \neq i \tag{2.17}$$

[1] The terminology comes from the general decision theory.

It is obvious that if $\lambda_{ki} = 1 - \delta_{ki}$, where δ_{ki} is *Kronecker's delta* (0 if $k \neq i$ and 1 if $k = i$), then minimizing the average risk becomes equivalent to minimizing the classification error probability.

The two-class case. For this specific case we obtain

$$l_1 = \lambda_{11} p(\boldsymbol{x}|\omega_1)P(\omega_1) + \lambda_{21} p(\boldsymbol{x}|\omega_2)P(\omega_2)$$
$$l_2 = \lambda_{12} p(\boldsymbol{x}|\omega_1)P(\omega_1) + \lambda_{22} p(\boldsymbol{x}|\omega_2)P(\omega_2) \tag{2.18}$$

We assign $\boldsymbol{x}$ to ω_1 if $l_1 < l_2$, that is,

$$(\lambda_{21} - \lambda_{22})p(\boldsymbol{x}|\omega_2)P(\omega_2) < (\lambda_{12} - \lambda_{11})p(\boldsymbol{x}|\omega_1)P(\omega_1) \tag{2.19}$$

It is natural to assume that $\lambda_{ij} > \lambda_{ii}$ (correct decisions are penalized much less than wrong ones). Adopting this assumption, the decision rule (2.17) for the two-class case now becomes

$$\boldsymbol{x} \in \omega_1(\omega_2) \quad \text{if} \quad l_{12} \equiv \frac{p(\boldsymbol{x}|\omega_1)}{p(\boldsymbol{x}|\omega_2)} > (<) \frac{P(\omega_2)}{P(\omega_1)} \frac{\lambda_{21} - \lambda_{22}}{\lambda_{12} - \lambda_{11}} \tag{2.20}$$

The ratio l_{12} is known as the *likelihood ratio* and the preceding test as the *likelihood ratio test*. Let us now investigate Eq. (2.20) a little further and consider the case of Figure 2.1. Assume that the loss matrix is of the form

$$L = \begin{bmatrix} 0 & \lambda_{12} \\ \lambda_{21} & 0 \end{bmatrix}$$

If misclassification of patterns that come from ω_2 is considered to have serious consequences, then we must choose $\lambda_{21} > \lambda_{12}$. Thus, patterns are assigned to class ω_2 if

$$p(\boldsymbol{x}|\omega_2) > p(\boldsymbol{x}|\omega_1)\frac{\lambda_{12}}{\lambda_{21}}$$

where $P(\omega_1) = P(\omega_2) = 1/2$ has been assumed. That is, $p(\boldsymbol{x}|\omega_1)$ is multiplied by a factor less than 1 and the effect of this is to move the threshold in Figure 2.1 to the left of x_0. In other words, region R_2 is increased while R_1 is decreased. The opposite would be true if $\lambda_{21} < \lambda_{12}$.

An alternative cost that sometimes is used for two class problems is the Neyman-Pearson criterion. The error for one of the classes is now constrained to be fixed and equal to a chosen value (Problem 2.6). Such a decision rule has been used, for example, in radar detection problems. The task there is to detect a target in the presence of noise. One type of error is the so-called *false alarm*—that is, to mistake the noise for a signal (target) present. Of course, the other type of error is to miss the signal and to decide in favor of the noise (*missed detection*). In many cases the error probability of false alarm is set equal to a predetermined threshold.

Example 2.1
In a two-class problem with a single feature x the pdfs are Gaussians with variance $\sigma^2 = 1/2$ for both classes and mean values 0 and 1, respectively, that is,

$$p(x|\omega_1) = \frac{1}{\sqrt{\pi}} \exp(-x^2)$$

$$p(x|\omega_2) = \frac{1}{\sqrt{\pi}} \exp(-(x-1)^2)$$

If $P(\omega_1) = P(\omega_2) = 1/2$, compute the threshold value x_0 (a) for minimum error probability and (b) for minimum risk if the loss matrix is

$$L = \begin{bmatrix} 0 & 0.5 \\ 1.0 & 0 \end{bmatrix}$$

Taking into account the shape of the Gaussian function graph (Appendix A), the threshold for the minimum probability case will be

$$x_0 : \exp(-x^2) = \exp(-(x-1)^2)$$

Taking the logarithm of both sides, we end up with $x_0 = 1/2$. In the minimum risk case we get

$$x_0 : \exp(-x^2) = 2\exp(-(x-1)^2)$$

or $x_0 = (1 - \ln 2)/2 < 1/2$; that is, the threshold moves to the left of $1/2$. If the two classes are not equiprobable, then it is easily verified that if $P(\omega_1) > (<) P(\omega_2)$ the threshold moves to the right (left). That is, we expand the region in which we decide in favor of the most probable class, since it is better to make fewer errors for the most probable class.

2.3 DISCRIMINANT FUNCTIONS AND DECISION SURFACES

It is by now clear that minimizing either the risk or the error probability or the Neyman-Pearson criterion is equivalent to partitioning the feature space into M regions, for a task with M classes. If regions R_i, R_j happen to be contiguous, then they are separated by a *decision surface* in the multidimensional feature space. For the minimum error probability case, this is described by the equation

$$P(\omega_i|\boldsymbol{x}) - P(\omega_j|\boldsymbol{x}) = 0 \tag{2.21}$$

From the one side of the surface this difference is positive, and from the other it is negative. Sometimes, instead of working directly with probabilities (or risk functions), it may be more convenient, from a mathematical point of view, to work with an equivalent function of them, for example, $g_i(\boldsymbol{x}) \equiv f(P(\omega_i|\boldsymbol{x}))$, where $f(\cdot)$ is a monotonically increasing function. $g_i(\boldsymbol{x})$ is known as a *discriminant function*. The decision test (2.13) is now stated as

$$\text{classify } \boldsymbol{x} \text{ in } \omega_i \quad \text{if} \quad g_i(\boldsymbol{x}) > g_j(\boldsymbol{x}) \quad \forall j \neq i \tag{2.22}$$

The decision surfaces, separating contiguous regions, are described by

$$g_{ij}(\boldsymbol{x}) \equiv g_i(\boldsymbol{x}) - g_j(\boldsymbol{x}) = 0, \quad i,j = 1,2,\ldots,M, \quad i \neq j \tag{2.23}$$

So far, we have approached the classification problem via Bayesian probabilistic arguments and the goal was to minimize the classification error probability or the risk. However, as we will soon see, not all problems are well suited to such approaches. For example, in many cases the involved pdfs are complicated and their estimation is not an easy task. In such cases, it may be preferable to compute decision surfaces *directly by means of alternative costs*, and this will be our focus in Chapters 3 and 4. Such approaches give rise to discriminant functions and decision surfaces, which are entities with no (necessary) relation to Bayesian classification, and they are, in general, suboptimal with respect to Bayesian classifiers.

In the following we will focus on a particular family of decision surfaces associated with the Bayesian classification for the specific case of Gaussian density functions.

2.4 BAYESIAN CLASSIFICATION FOR NORMAL DISTRIBUTIONS

2.4.1 The Gaussian Probability Density Function

One of the most commonly encountered probability density functions in practice is the Gaussian or normal probability density function. The major reasons for its popularity are its computational tractability and the fact that it models adequately a large number of cases. One of the most celebrated theorems in statistics is the *central limit theorem*. The theorem states that if a random variable is the outcome of a summation of a number of *independent* random variables, its pdf approaches the Gaussian function as the number of summands tends to infinity (see Appendix A). In practice, it is most common to assume that the sum of random variables is distributed according to a Gaussian pdf, for a sufficiently large number of summing terms.

The one-dimensional or the univariate Gaussian, as it is sometimes called, is defined by

$$p(x) = \frac{1}{\sqrt{2\pi}\sigma} \exp\left(-\frac{(x-\mu)^2}{2\sigma^2}\right) \tag{2.24}$$

The parameters μ and σ^2 turn out to have a specific meaning. The mean value of the random variable x is equal to μ, that is,

$$\mu = E[x] \equiv \int_{-\infty}^{+\infty} xp(x)dx \tag{2.25}$$

where $E[\cdot]$ denotes the mean (or expected) value of a random variable. The parameter σ^2 is equal to the variance of x, that is,

$$\sigma^2 = E[(x-\mu)^2] \equiv \int_{-\infty}^{+\infty} (x-\mu)^2 p(x)dx \tag{2.26}$$

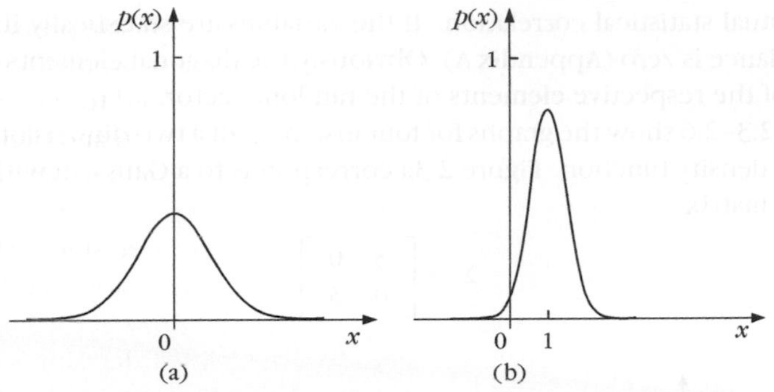

FIGURE 2.2

Graphs for the one-dimensional Gaussian pdf. (a) Mean value $\mu = 0, \sigma^2 = 1$, (b) $\mu = 1$ and $\sigma^2 = 0.2$. The larger the variance the broader the graph is. The graphs are symmetric, and they are centered at the respective mean value.

Figure 2.2a shows the graph of the Gaussian function for $\mu = 0$ and $\sigma^2 = 1$, and Figure 2.2b the case for $\mu = 1$ and $\sigma^2 = 0.2$. The larger the variance the broader the graph, which is symmetric, and it is always centered at μ (see Appendix A, for some more properties).

The multivariate generalization of a Gaussian pdf in the l-dimensional space is given by

$$p(\boldsymbol{x}) = \frac{1}{(2\pi)^{l/2}|\Sigma|^{1/2}} \exp\left(-\frac{1}{2}(\boldsymbol{x}-\boldsymbol{\mu})^T \Sigma^{-1}(\boldsymbol{x}-\boldsymbol{\mu})\right) \tag{2.27}$$

where $\boldsymbol{\mu} = E[\boldsymbol{x}]$ is the mean value and Σ is the $l \times l$ *covariance matrix* (Appendix A) defined as

$$\Sigma = E[(\boldsymbol{x}-\boldsymbol{\mu})(\boldsymbol{x}-\boldsymbol{\mu})^T] \tag{2.28}$$

where $|\Sigma|$ denotes the determinant of Σ. It is readily seen that for $l = 1$ the multivariate Gaussian coincides with the univariate one. Sometimes, the symbol $\mathcal{N}(\boldsymbol{\mu}, \Sigma)$ is used to denote a Gaussian pdf with mean value $\boldsymbol{\mu}$ and covariance Σ.

To get a better feeling on what the multivariate Gaussian looks like, let us focus on some cases in the two-dimensional space, where nature allows us the luxury of visualization. For this case we have

$$\Sigma = E\left[\begin{bmatrix} x_1 - \mu_1 \\ x_2 - \mu_2 \end{bmatrix} \begin{bmatrix} x_1 - \mu_1, & x_2 - \mu_2 \end{bmatrix}\right] \tag{2.29}$$

$$= \begin{bmatrix} \sigma_1^2 & \sigma_{12} \\ \sigma_{12} & \sigma_2^2 \end{bmatrix} \tag{2.30}$$

where $E[x_i] = \mu_i$, $i = 1, 2$, and by definition $\sigma_{12} = E[(x_1 - \mu_1)(x_2 - \mu_2)]$, which is known as the covariance between the random variables x_1 and x_2 and it is a measure of their mutual statistical correlation. If the variables are statistically independent, their covariance is zero (Appendix A). Obviously, the diagonal elements of Σ are the variances of the respective elements of the random vector.

Figures 2.3–2.6 show the graphs for four instances of a two-dimensional Gaussian probability density function. Figure 2.3a corresponds to a Gaussian with a diagonal covariance matrix

$$\Sigma = \begin{bmatrix} 3 & 0 \\ 0 & 3 \end{bmatrix}$$

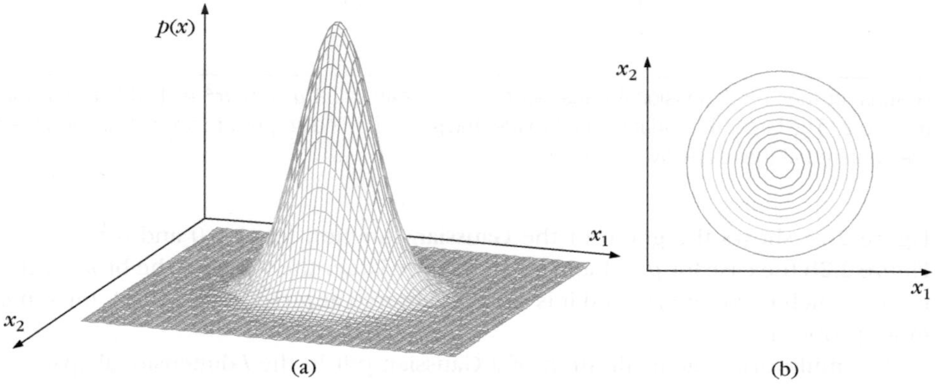

FIGURE 2.3

(a) The graph of a two-dimensional Gaussian pdf and (b) the corresponding isovalue curves for a diagonal Σ with $\sigma_1^2 = \sigma_2^2$. The graph has a spherical symmetry showing no preference in any direction.

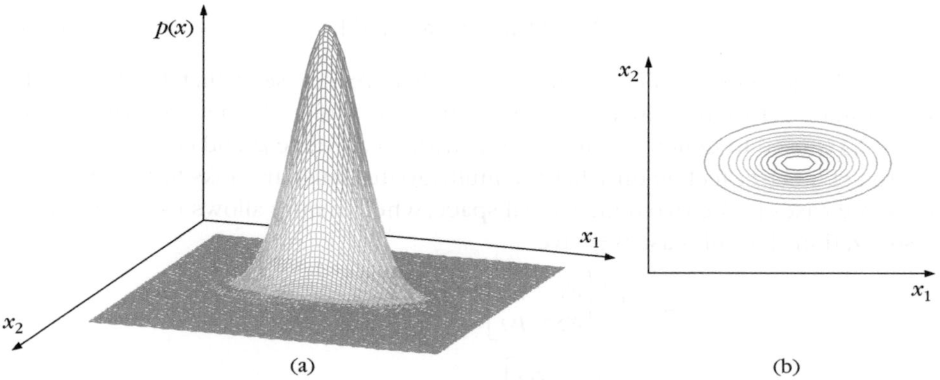

FIGURE 2.4

(a) The graph of a two-dimensional Gaussian pdf and (b) the corresponding isovalue curves for a diagonal Σ with $\sigma_1^2 >> \sigma_2^2$. The graph is elongated along the x_1 direction.

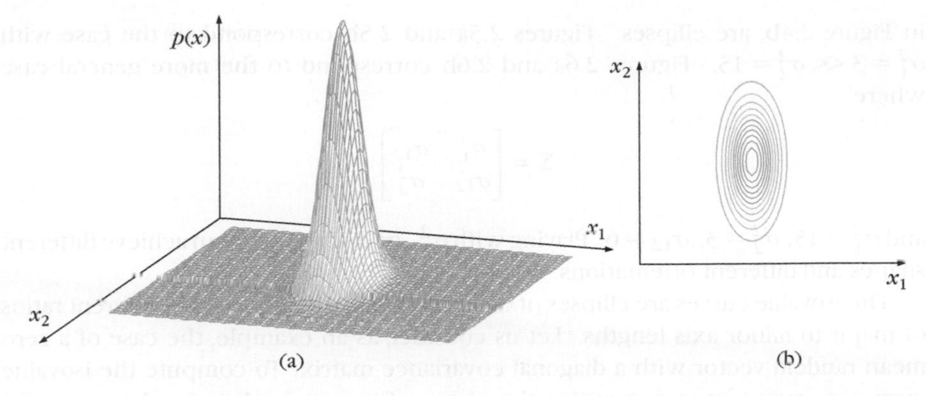

FIGURE 2.5

(a) The graph of a two-dimensional Gaussian pdf and (b) the corresponding isovalue curves for a diagonal Σ with $\sigma_1^2 << \sigma_2^2$. The graph is elongated along the x_2 direction.

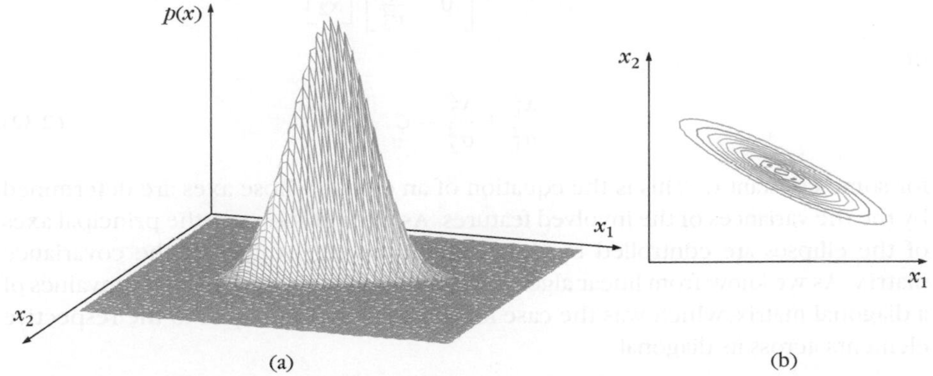

FIGURE 2.6

(a) The graph of a two-dimensional Gaussian pdf and (b) the corresponding isovalue curves for a case of a nondiagonal Σ. Playing with the values of the elements of Σ one can achieve different shapes and orientations.

that is, both features, x_1, x_2 have variance equal to 3 and their covariance is zero. The graph of the Gaussian is symmetric. For this case the isovalue curves (i.e., curves of equal probability density values) are circles (hyperspheres in the general l-dimensional space) and are shown in Figure 2.3b. The case shown in Figure 2.4a corresponds to the covariance matrix

$$\Sigma = \begin{bmatrix} \sigma_1^2 & 0 \\ 0 & \sigma_2^2 \end{bmatrix}$$

with $\sigma_1^2 = 15 >> \sigma_2^2 = 3$. The graph of the Gaussian is now elongated along the x_1-axis, which is the direction of the larger variance. The isovalue curves, shown in Figure 2.4b, are ellipses. Figures 2.5a and 2.5b correspond to the case with $\sigma_1^2 = 3 << \sigma_2^2 = 15$. Figures 2.6a and 2.6b correspond to the more general case where

$$\Sigma = \begin{bmatrix} \sigma_1^2 & \sigma_{12} \\ \sigma_{12} & \sigma_2^2 \end{bmatrix}$$

and $\sigma_1^2 = 15$, $\sigma_2^2 = 3$, $\sigma_{12} = 6$. Playing with σ_1^2, σ_2^2 and σ_{12} one can achieve different shapes and different orientations.

The isovalue curves are ellipses of different orientations and with different ratios of major to minor axis lengths. Let us consider, as an example, the case of a zero mean random vector with a diagonal covariance matrix. To compute the isovalue curves is equivalent to computing the curves of constant values for the exponent, that is,

$$\boldsymbol{x}^T \Sigma^{-1} \boldsymbol{x} = [x_1, x_2] \begin{bmatrix} \frac{1}{\sigma_1^2} & 0 \\ 0 & \frac{1}{\sigma_2^2} \end{bmatrix} \begin{bmatrix} x_1 \\ x_2 \end{bmatrix} = C \tag{2.31}$$

or

$$\frac{x_1^2}{\sigma_1^2} + \frac{x_2^2}{\sigma_2^2} = C \tag{2.32}$$

for some constant C. This is the equation of an ellipse whose axes are determined by the the variances of the involved features. As we will soon see, the principal axes of the ellipses are controlled by the eigenvectors/eigenvalues of the covariance matrix. As we know from linear algebra (and it is easily checked), the eigenvalues of a diagonal matrix, which was the case for our example, are equal to the respective elements across its diagonal.

2.4.2 The Bayesian Classifier for Normally Distributed Classes

Our goal in this section is to study the optimal Bayesian classifier when the involved pdfs, $p(\boldsymbol{x}|\omega_i)$, $i = 1, 2, \ldots, M$ (likelihood functions of ω_i with respect to $\boldsymbol{x}$), describing the data distribution in each one of the classes, are multivariate normal distributions, that is, $\mathcal{N}(\boldsymbol{\mu}_i, \Sigma_i)$, $i = 1, 2, \ldots, M$. Because of the exponential form of the involved densities, it is preferable to work with the following discriminant functions, which involve the (monotonic) logarithmic function $\ln(\cdot)$:

$$g_i(\boldsymbol{x}) = \ln(p(\boldsymbol{x}|\omega_i)P(\omega_i)) = \ln p(\boldsymbol{x}|\omega_i) + \ln P(\omega_i) \tag{2.33}$$

or

$$g_i(\boldsymbol{x}) = -\frac{1}{2}(\boldsymbol{x} - \boldsymbol{\mu}_i)^T \Sigma_i^{-1} (\boldsymbol{x} - \boldsymbol{\mu}_i) + \ln P(\omega_i) + c_i \tag{2.34}$$

where c_i is a constant equal to $-(l/2)\ln 2\pi - (1/2)\ln|\Sigma_i|$. Expanding, we obtain

$$g_i(\boldsymbol{x}) = -\frac{1}{2}\boldsymbol{x}^T \Sigma_i^{-1} \boldsymbol{x} + \frac{1}{2}\boldsymbol{x}^T \Sigma_i^{-1} \boldsymbol{\mu}_i - \frac{1}{2}\boldsymbol{\mu}_i^T \Sigma_i^{-1} \boldsymbol{\mu}_i + \frac{1}{2}\boldsymbol{\mu}_i^T \Sigma_i^{-1} \boldsymbol{x} \quad + \ln P(\omega_i) + c_i \tag{2.35}$$

In general, this is a nonlinear quadratic form. Take, for example, the case of $l = 2$ and assume that

$$\Sigma_i = \begin{bmatrix} \sigma_i^2 & 0 \\ 0 & \sigma_i^2 \end{bmatrix}$$

Then (2.35) becomes

$$g_i(\boldsymbol{x}) = -\frac{1}{2\sigma_i^2}(x_1^2 + x_2^2) + \frac{1}{\sigma_i^2}(\mu_{i1}x_1 + \mu_{i2}x_2) - \frac{1}{2\sigma_i^2}(\mu_{i1}^2 + \mu_{i2}^2) + \ln P(\omega_i) + c_i \quad (2.36)$$

and obviously the associated decision curves $g_i(\boldsymbol{x}) - g_j(\boldsymbol{x}) = 0$ are *quadrics* (i.e., ellipsoids, parabolas, hyperbolas, pairs of lines). That is, in such cases, the Bayesian classifier is a *quadratic classifier*, in the sense that the partition of the feature space is performed via quadric decision surfaces. For $l > 2$ the decision surfaces are *hyperquadrics*. Figure 2.7a shows the decision curve corresponding to $P(\omega_1) = P(\omega_2)$, $\boldsymbol{\mu}_1 = [0, 0]^T$ and $\boldsymbol{\mu}_2 = [4, 0]^T$. The covariance matrices for the two classes are

$$\Sigma_1 = \begin{bmatrix} 0.3 & 0.0 \\ 0.0 & 0.35 \end{bmatrix}, \quad \Sigma_2 = \begin{bmatrix} 1.2 & 0.0 \\ 0.0 & 1.85 \end{bmatrix}$$

For the case of Figure 2.7b the classes are also equiprobable with $\boldsymbol{\mu}_1 = [0, 0]^T$, $\boldsymbol{\mu}_2 = [3.2, 0]^T$ and covariance matrices

$$\Sigma_1 = \begin{bmatrix} 0.1 & 0.0 \\ 0.0 & 0.75 \end{bmatrix}, \quad \Sigma_2 = \begin{bmatrix} 0.75 & 0.0 \\ 0.0 & 0.1 \end{bmatrix}$$

Figure 2.8 shows the two pdfs for the case of Figure 2.7a. The red color is used for class ω_1 and indicates the points where $p(\boldsymbol{x}|\omega_1) > p(\boldsymbol{x}|\omega_2)$. The gray color is similarly used for class ω_2. It is readily observed that the decision curve is an ellipse, as shown in Figure 2.7a. The setup corresponding to Figure 2.7b is shown in Figure 2.9. In this case, the decision curve is a hyperbola.

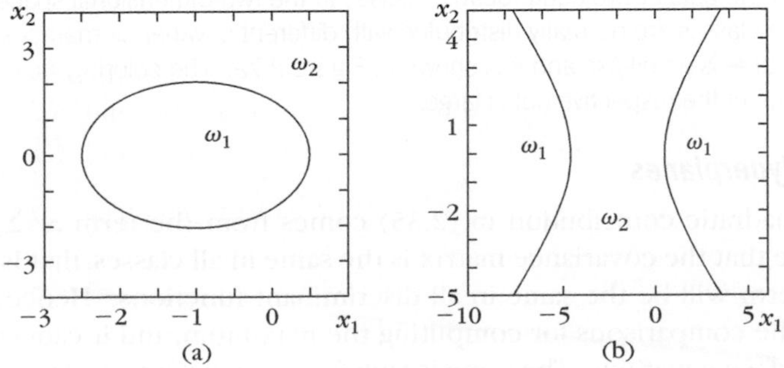

FIGURE 2.7

Examples of quadric decision curves. Playing with the covariance matrices of the Gaussian functions, different decision curves result, that is, ellipsoids, parabolas, hyperbolas, pairs of lines.

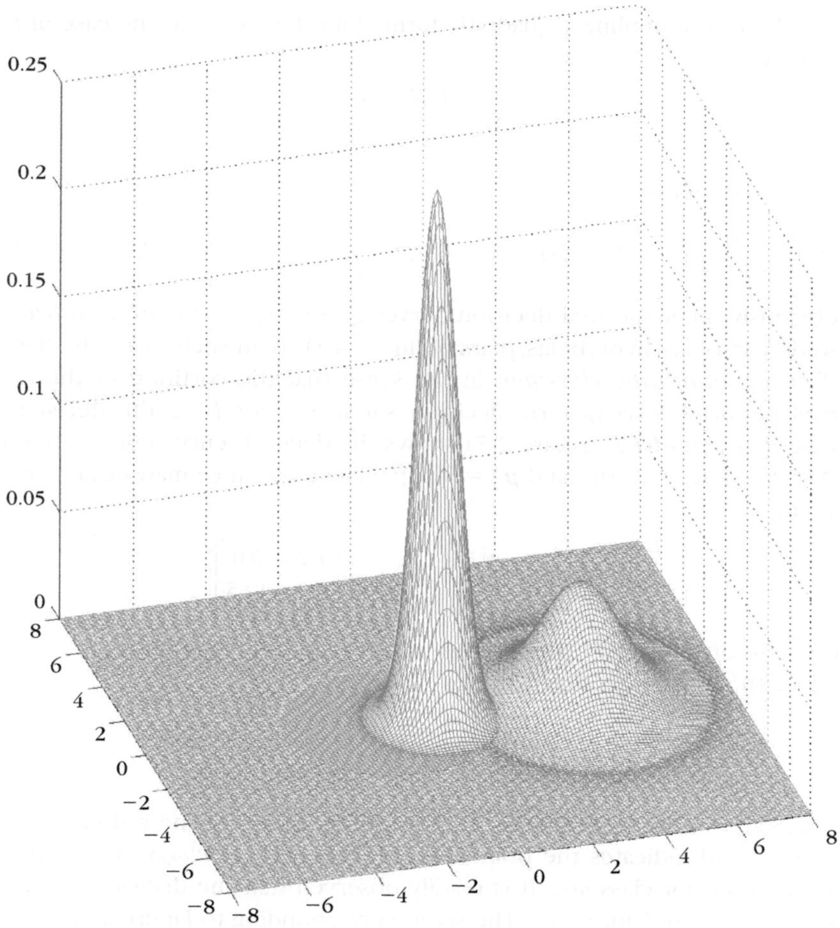

FIGURE 2.8

An example of the pdfs of two equiprobable classes in the two-dimensional space. The feature vectors in both classes are normally distributed with different covariance matrices. In this case, the decision curve is an ellipse and it is shown in Figure 2.7a. The coloring indicates the areas where the value of the respective pdf is larger.

Decision Hyperplanes

The only quadratic contribution in (2.35) comes from the term $\boldsymbol{x}^T \Sigma_i^{-1} \boldsymbol{x}$. If we now assume that the covariance matrix is the same in all classes, that is, $\Sigma_i = \Sigma$, the quadratic term will be the same in all discriminant functions. Hence, it does not enter into the comparisons for computing the maximum, and it cancels out in the decision surface equations. The same is true for the constants c_i. Thus, they can be omitted and we may redefine $g_i(\boldsymbol{x})$ as

$$g_i(\boldsymbol{x}) = \boldsymbol{w}_i^T \boldsymbol{x} + w_{i0} \tag{2.37}$$

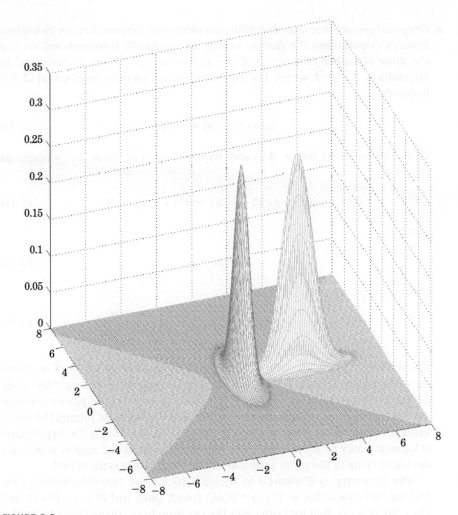

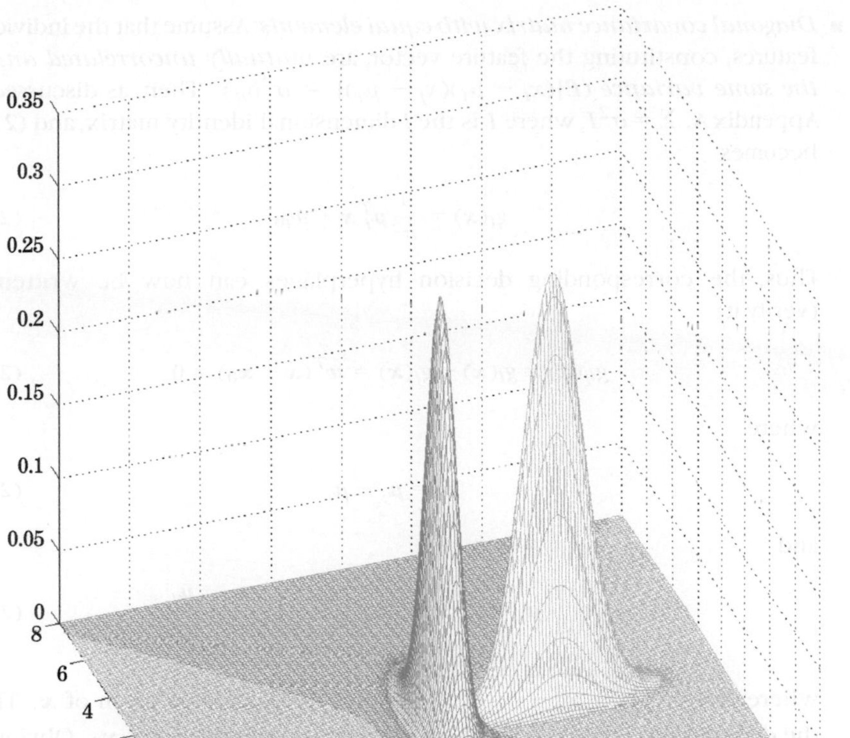

FIGURE 2.9

An example of the pdfs of two equiprobable classes in the two-dimensional space. The feature vectors in both classes are normally distributed with different covariance matrices. In this case, the decision curve is a hyperbola and it is shown in Figure 2.7b.

where

$$\boldsymbol{w}_i = \Sigma^{-1}\boldsymbol{\mu}_i \tag{2.38}$$

and

$$w_{i0} = \ln P(\omega_i) - \frac{1}{2}\boldsymbol{\mu}_i^T\Sigma^{-1}\boldsymbol{\mu}_i \tag{2.39}$$

Hence $g_i(\boldsymbol{x})$ is a *linear function* of $\boldsymbol{x}$ and the respective decision surfaces are *hyperplanes*. Let us investigate this a bit more.

- *Diagonal covariance matrix with equal elements*: Assume that the individual features, constituting the feature vector, are *mutually uncorrelated and of the same variance* ($E[(x_i - \mu_i)(x_j - \mu_j)] = \sigma^2\delta_{ij}$). Then, as discussed in Appendix A, $\Sigma = \sigma^2 I$, where I is the l-dimensional identity matrix, and (2.37) becomes

$$g_i(\boldsymbol{x}) = \frac{1}{\sigma^2}\boldsymbol{\mu}_i^T \boldsymbol{x} + w_{i0} \tag{2.40}$$

Thus, the corresponding decision hyperplanes can now be written as (verify it)

$$g_{ij}(\boldsymbol{x}) \equiv g_i(\boldsymbol{x}) - g_j(\boldsymbol{x}) = \boldsymbol{w}^T(\boldsymbol{x} - \boldsymbol{x}_0) = 0 \tag{2.41}$$

where

$$\boldsymbol{w} = \boldsymbol{\mu}_i - \boldsymbol{\mu}_j \tag{2.42}$$

and

$$\boldsymbol{x}_0 = \frac{1}{2}(\boldsymbol{\mu}_i + \boldsymbol{\mu}_j) - \sigma^2 \ln\left(\frac{P(\omega_i)}{P(\omega_j)}\right)\frac{\boldsymbol{\mu}_i - \boldsymbol{\mu}_j}{\|\boldsymbol{\mu}_i - \boldsymbol{\mu}_j\|^2} \tag{2.43}$$

where $\|\boldsymbol{x}\| = \sqrt{x_1^2 + x_2^2 + \cdots + x_l^2}$ denotes the Euclidean norm of $\boldsymbol{x}$. Thus, the decision surface is a *hyperplane* passing through the point $\boldsymbol{x}_0$. Obviously, if $P(\omega_i) = P(\omega_j)$, then $\boldsymbol{x}_0 = \frac{1}{2}(\boldsymbol{\mu}_i + \boldsymbol{\mu}_j)$, and the hyperplane passes through the average of $\boldsymbol{\mu}_i, \boldsymbol{\mu}_j$, that is, the middle point of the segment joining the mean values. On the other hand, if $P(\omega_j) > P(\omega_i)$ ($P(\omega_i) > P(\omega_j)$) the hyperplane is located closer to $\boldsymbol{\mu}_i(\boldsymbol{\mu}_j)$. In other words, the area of the region where we decide in favor of the more probable of the two classes is increased.

The geometry is illustrated in Figure 2.10 for the two-dimensional case and for two cases, that is, $P(\omega_j) = P(\omega_i)$ (black line) and $P(\omega_j) > P(\omega_i)$ (red line). We observe that for both cases the decision hyperplane (straight line) is orthogonal to $\boldsymbol{\mu}_i - \boldsymbol{\mu}_j$. Indeed, for any point $\boldsymbol{x}$ lying on the decision hyperplane, the vector $\boldsymbol{x} - \boldsymbol{x}_0$ also lies on the hyperplane and

$$g_{ij}(\boldsymbol{x}) = 0 \Rightarrow \boldsymbol{w}^T(\boldsymbol{x} - \boldsymbol{x}_0) = (\boldsymbol{\mu}_i - \boldsymbol{\mu}_j)^T(\boldsymbol{x} - \boldsymbol{x}_0) = 0$$

That is, $\boldsymbol{\mu}_i - \boldsymbol{\mu}_j$ is orthogonal to the decision hyperplane. Furthermore, if σ^2 is small with respect to $\|\boldsymbol{\mu}_i - \boldsymbol{\mu}_j\|$, the location of the hyperplane is rather insensitive to the values of $P(\omega_i), P(\omega_j)$. This is expected, because small variance indicates that the random vectors are clustered within a small radius around their mean values. Thus a small shift of the decision hyperplane has a small effect on the result.

Figure 2.11 illustrates this. For each class, the circles around the means indicate regions where samples have a high probability, say 98%,

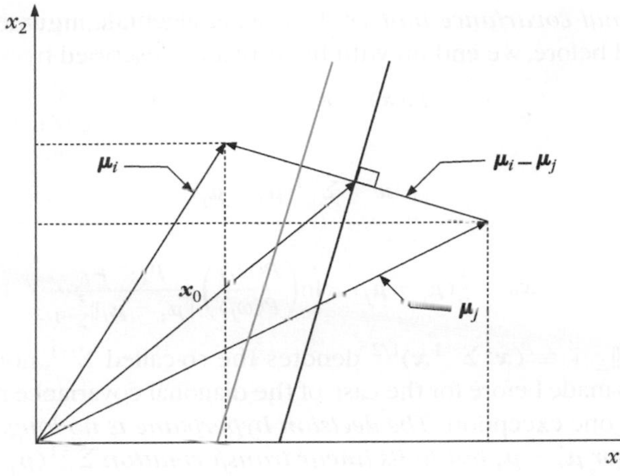

FIGURE 2.10

Decision lines for normally distributed vectors with $\Sigma = \sigma^2 I$. The black line corresponds to the case of $P(\omega_j) = P(\omega_i)$ and it passes through the middle point of the line segment joining the mean values of the two classes. The red line corresponds to the case of $P(\omega_j) > P(\omega_i)$ and it is closer to μ_i, leaving more "room" to the more probable of the two classes. If we had assumed $P(\omega_j) < P(\omega_i)$, the decision line would have moved closer to μ_j.

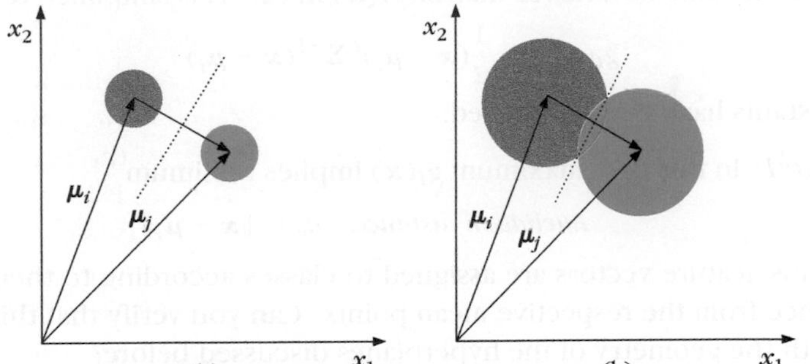

FIGURE 2.11

Decision line (a) for compact and (b) for noncompact classes. When classes are compact around their mean values, the location of the hyperplane is rather insensitive to the values of $P(\omega_1)$ and $P(\omega_2)$. This is not the case for noncompact classes, where a small movement of the hyperplane to the right or to the left may be more critical.

of being found. The case of Figure 2.11a corresponds to small variance, and that of Figure 2.11b to large variance. No doubt the location of the decision hyperplane in Figure 2.11b is much more critical than that in Figure 2.11a.

- *Nondiagonal covariance matrix*: Following algebraic arguments similar to those used before, we end up with hyperplanes described by

$$g_{ij}(\boldsymbol{x}) = \boldsymbol{w}^T(\boldsymbol{x} - \boldsymbol{x}_0) = 0 \tag{2.44}$$

where

$$\boldsymbol{w} = \Sigma^{-1}(\boldsymbol{\mu}_i - \boldsymbol{\mu}_j) \tag{2.45}$$

and

$$\boldsymbol{x}_0 = \frac{1}{2}(\boldsymbol{\mu}_i + \boldsymbol{\mu}_j) - \ln\left(\frac{P(\omega_i)}{P(\omega_j)}\right)\frac{\boldsymbol{\mu}_i - \boldsymbol{\mu}_j}{\|\boldsymbol{\mu}_i - \boldsymbol{\mu}_j\|^2_{\Sigma^{-1}}} \tag{2.46}$$

where $\|\boldsymbol{x}\|_{\Sigma^{-1}} \equiv (\boldsymbol{x}^T\Sigma^{-1}\boldsymbol{x})^{1/2}$ denotes the so-called Σ^{-1} norm of $\boldsymbol{x}$. The comments made before for the case of the diagonal covariance matrix are still valid, with one exception. *The decision hyperplane is no longer orthogonal to the vector $\boldsymbol{\mu}_i - \boldsymbol{\mu}_j$ but to its linear transformation $\Sigma^{-1}(\boldsymbol{\mu}_i - \boldsymbol{\mu}_j)$.*

Figure 2.12 shows two Gaussian pdfs with equal covariance matrices, describing the data distribution of two equiprobable classes. In both classes, the data are distributed around their mean values in *exactly* the same way and the optimal decision curve is a straight line.

Minimum Distance Classifiers

We will now view the task from a slightly different angle. Assuming equiprobable classes with the same covariance matrix, $g_i(\boldsymbol{x})$ in (2.34) is simplified to

$$g_i(\boldsymbol{x}) = -\frac{1}{2}(\boldsymbol{x} - \boldsymbol{\mu}_i)^T\Sigma^{-1}(\boldsymbol{x} - \boldsymbol{\mu}_i) \tag{2.47}$$

where constants have been neglected.

- $\Sigma = \sigma^2 I$: In this case maximum $g_i(\boldsymbol{x})$ implies minimum

$$\textit{Euclidean distance:}\quad d_\epsilon = \|\boldsymbol{x} - \boldsymbol{\mu}_i\| \tag{2.48}$$

Thus, feature vectors are assigned to classes according to their Euclidean distance from the respective mean points. Can you verify that this result ties in with the geometry of the hyperplanes discussed before?

Figure 2.13a shows curves of equal distance $d_\epsilon = c$ from the mean points of each class. They are obviously circles of radius c (hyperspheres in the general case).

- Nondiagonal Σ: For this case maximizing $g_i(\boldsymbol{x})$ is equivalent to minimizing the Σ^{-1} norm, known as the

$$\textit{Mahalanobis distance:}\quad d_m = \left((\boldsymbol{x} - \boldsymbol{\mu}_i)^T\Sigma^{-1}(\boldsymbol{x} - \boldsymbol{\mu}_i)\right)^{1/2} \tag{2.49}$$

In this case, the constant distance $d_m = c$ curves are ellipses (hyperellipses). Indeed, the covariance matrix is symmetric and, as discussed in Appendix B, it can always be diagonalized by a unitary transform

$$\Sigma = \Phi\Lambda\Phi^T \tag{2.50}$$

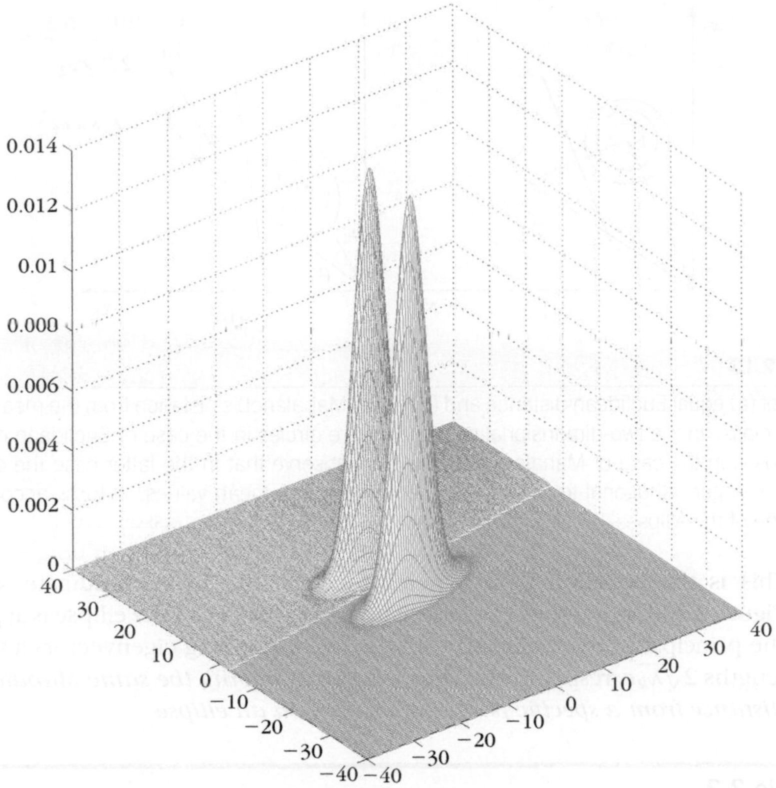

FIGURE 2.12

An example of two Gaussian pdfs with the same covariance matrix in the two-dimensional space. Each one of them is associated with one of two equiprobable classes. In this case, the decision curve is a straight line.

where $\Phi^T = \Phi^{-1}$ and Λ is the diagonal matrix whose elements are the eigenvalues of Σ. Φ has as its columns the corresponding (orthonormal) eigenvectors of Σ

$$\Phi = [\boldsymbol{v}_1, \boldsymbol{v}_2, \ldots, \boldsymbol{v}_l] \tag{2.51}$$

Combining (2.49) and (2.50), we obtain

$$(\boldsymbol{x} - \boldsymbol{\mu}_i)^T \Phi \Lambda^{-1} \Phi^T (\boldsymbol{x} - \boldsymbol{\mu}_i) = c^2 \tag{2.52}$$

Define $\boldsymbol{x}' = \Phi^T \boldsymbol{x}$. The coordinates of $\boldsymbol{x}'$ are equal to $\boldsymbol{v}_k^T \boldsymbol{x}, k = 1, 2, \ldots, l$, that is, the projections of $\boldsymbol{x}$ onto the eigenvectors. In other words, they are the coordinates of $\boldsymbol{x}$ with respect to a new coordinate system whose axes are determined by $\boldsymbol{v}_k, k = 1, 2, \ldots, l$. Equation (2.52) can now be written as

$$\frac{(x_1' - \mu_{i1}')^2}{\lambda_1} + \cdots + \frac{(x_l' - \mu_{il}')^2}{\lambda_l} = c^2 \tag{2.53}$$

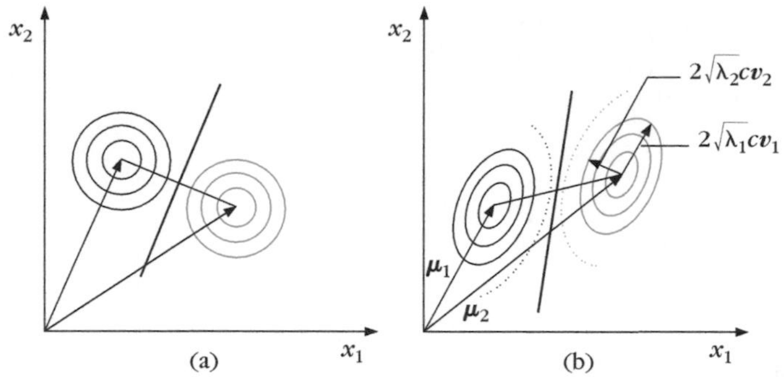

FIGURE 2.13

Curves of (a) equal Euclidean distance and (b) equal Mahalanobis distance from the mean points of each class. In the two-dimensional space, they are circles in the case of Euclidean distance and ellipses in the case of Mahalanobis distance. Observe that in the latter case the decision line is no longer orthogonal to the line segment joining the mean values. It turns according to the shape of the ellipses.

This is the equation of a hyperellipsoid in the new coordinate system. Figure 2.13b shows the $l = 2$ case. The center of mass of the ellipse is at $\boldsymbol{\mu}_i$, and the principal axes are aligned with the corresponding eigenvectors and have lengths $2\sqrt{\lambda_k}c$, respectively. Thus, *all points having the same Mahalanobis distance from a specific point are located on an ellipse.*

Example 2.2

In a two-class, two-dimensional classification task, the feature vectors are generated by two normal distributions sharing the same covariance matrix

$$\Sigma = \begin{bmatrix} 1.1 & 0.3 \\ 0.3 & 1.9 \end{bmatrix}$$

and the mean vectors are $\boldsymbol{\mu}_1 = [0, 0]^T$, $\boldsymbol{\mu}_2 = [3, 3]^T$, respectively.

(a) Classify the vector $[1.0, 2.2]^T$ according to the Bayesian classifier.

It suffices to compute the Mahalanobis distance of $[1.0, 2.2]^T$ from the two mean vectors. Thus,

$$d_m^2(\boldsymbol{\mu}_1, \boldsymbol{x}) = (\boldsymbol{x} - \boldsymbol{\mu}_1)^T \Sigma^{-1} (\boldsymbol{x} - \boldsymbol{\mu}_1)$$

$$= [1.0, 2.2] \begin{bmatrix} 0.95 & -0.15 \\ -0.15 & 0.55 \end{bmatrix} \begin{bmatrix} 1.0 \\ 2.2 \end{bmatrix} = 2.952$$

Similarly,

$$d_m^2(\boldsymbol{\mu}_2, \boldsymbol{x}) = [-2.0, -0.8] \begin{bmatrix} 0.95 & -0.15 \\ -0.15 & 0.55 \end{bmatrix} \begin{bmatrix} -2.0 \\ -0.8 \end{bmatrix} = 3.672 \tag{2.54}$$

Thus, the vector is assigned to the class with mean vector $[0, 0]^T$. *Notice that the given vector* $[1.0, 2.2]^T$ *is closer to* $[3, 3]^T$ *with respect to the Euclidean distance.*

(b) Compute the principal axes of the ellipse centered at $[0, 0]^T$ that corresponds to a constant Mahalanobis distance $d_m = \sqrt{2.952}$ from the center.
To this end, we first calculate the eigenvalues of Σ.

$$\det\left(\begin{bmatrix} 1.1-\lambda & 0.3 \\ 0.3 & 1.9-\lambda \end{bmatrix}\right) = \lambda^2 - 3\lambda + 2 = 0$$

or $\lambda_1 = 1$ and $\lambda_2 = 2$. To compute the eigenvectors we substitute these values into the equation

$$(\Sigma - \lambda I)v = 0$$

and we obtain the unit norm eigenvectors

$$v_1 = \begin{bmatrix} \frac{3}{\sqrt{10}} \\ -\frac{1}{\sqrt{10}} \end{bmatrix}, \quad v_2 = \begin{bmatrix} \frac{1}{\sqrt{10}} \\ \frac{3}{\sqrt{10}} \end{bmatrix}$$

It can easily be seen that they are mutually orthogonal. The principal axes of the ellipse are parallel to v_1 and v_2 and have lengths 3.436 and 4.859, respectively.

Remarks

- In practice, it is quite common to assume that the data in each class are adequately described by a Gaussian distribution. As a consequence, the associated Bayesian classifier is either linear or quadratic in nature, depending on the adopted assumptions concerning the covariance matrices. That is, if they are all equal or different. In statistics, this approach to the classification task is known as *linear discriminant analysis* (LDA) or *quadratic discriminant analysis* (QDA), respectively. *Maximum likelihood* is usually the method mobilized for the estimation of the unknown parameters that define the mean values and the covariance matrices (see Section 2.5 and Problem 2.19).

- A major problem associated with LDA and even more with QDA is the large number of the unknown parameters that have to be estimated in the case of high-dimensional spaces. For example, there are l parameters in each of the mean vectors and approximately $l^2/2$ in each (symmetric) covariance matrix. Besides the high demand for computational resources, obtaining good estimates of a large number of parameters dictates a large number of training points, N. This is a major issue that also embraces the design of other types of classifiers, for most of the cases, and we will come to it in greater detail in Chapter 5. In an effort to reduce the number of parameters to be estimated, a number of approximate techniques have been suggested over the years, including [Kimu 87, Hoff 96, Frie 89, Liu 04]. Linear discrimination will be approached from a different perspective in Section 5.8.

- LDA and QDA exhibit good performance in a large set of diverse applications and are considered to be among the most popular classifiers. No doubt, it is hard to accept that in all these cases the Gaussian assumption provides a reasonable modeling for the data statistics. The secret of the success seems to lie in the fact that linear or quadratic decision surfaces offer a reasonably good partition of the space, from the classification point of view. Moreover, as pointed out in [Hast 01], the estimates associated with Gaussian models have some good statistical properties (i.e., bias variance trade-off, Section 3.5.3) compared to other techniques.

2.5 ESTIMATION OF UNKNOWN PROBABILITY DENSITY FUNCTIONS

So far, we have assumed that the probability density functions are known. However, this is not the most common case. In many problems, the underlying pdf has to be estimated from the available data. There are various ways to approach the problem. Sometimes we may know the type of the pdf (e.g., Gaussian, Rayleigh), but we do not know certain parameters, such as the mean values or the variances. In contrast, in other cases we may not have information about the type of the pdf but we may know certain statistical parameters, such as the mean value and the variance. Depending on the available information, different approaches can be adopted. This will be our focus in the next subsections.

2.5.1 Maximum Likelihood Parameter Estimation

Let us consider an M-class problem with feature vectors distributed according to $p(\boldsymbol{x}|\omega_i)$, $i = 1, 2, \ldots, M$. We assume that these likelihood functions are given in a *parametric* form and that the corresponding parameters form the vectors $\boldsymbol{\theta}_i$ which are unknown. To show the dependence on $\boldsymbol{\theta}_i$ we write $p(\boldsymbol{x}|\omega_i; \boldsymbol{\theta}_i)$. Our goal is to estimate the unknown parameters using a set of known feature vectors in each class. If we further assume that data from one class do not affect the parameter estimation of the others, we can formulate the problem independent of classes and simplify our notation. At the end, one has to solve one such problem for each class independently.

Let $\boldsymbol{x}_1, \boldsymbol{x}_2, \ldots, \boldsymbol{x}_N$ be random samples drawn from pdf $p(\boldsymbol{x}; \boldsymbol{\theta})$. We form the joint pdf $p(X; \boldsymbol{\theta})$, where $X = \{\boldsymbol{x}_1, \ldots, \boldsymbol{x}_N\}$ is the set of the samples. Assuming *statistical independence* between the different samples, we have

$$p(X; \boldsymbol{\theta}) \equiv p(\boldsymbol{x}_1, \boldsymbol{x}_2, \ldots, \boldsymbol{x}_N; \boldsymbol{\theta}) = \prod_{k=1}^{N} p(\boldsymbol{x}_k; \boldsymbol{\theta}) \tag{2.55}$$

This is a function of $\boldsymbol{\theta}$, and it is also known as the likelihood function of $\boldsymbol{\theta}$ with respect to X. *The maximum likelihood (ML) method estimates* $\boldsymbol{\theta}$ *so that the likelihood function takes its maximum value*, that is,

$$\hat{\boldsymbol{\theta}}_{ML} = \arg\max_{\boldsymbol{\theta}} \prod_{k=1}^{N} p(\boldsymbol{x}_k; \boldsymbol{\theta}) \tag{2.56}$$

A necessary condition that $\hat{\boldsymbol{\theta}}_{ML}$ must satisfy in order to be a maximum is the gradient of the likelihood function with respect to $\boldsymbol{\theta}$ to be zero, that is

$$\frac{\partial \prod_{k=1}^{N} p(\boldsymbol{x}_k;\boldsymbol{\theta})}{\partial \boldsymbol{\theta}} = \mathbf{0} \tag{2.57}$$

Because of the monotonicity of the logarithmic function, we define the *log-likelihood function* as

$$L(\boldsymbol{\theta}) \equiv \ln \prod_{k=1}^{N} p(\boldsymbol{x}_k;\boldsymbol{\theta}) \tag{2.58}$$

and (2.57) is equivalent to

$$\frac{\partial L(\boldsymbol{\theta})}{\partial \boldsymbol{\theta}} = \sum_{k=1}^{N} \frac{\partial \ln p(\boldsymbol{x}_k;\boldsymbol{\theta})}{\partial \boldsymbol{\theta}} = \sum_{k=1}^{N} \frac{1}{p(\boldsymbol{x}_k;\boldsymbol{\theta})} \frac{\partial p(\boldsymbol{x}_k;\boldsymbol{\theta})}{\partial \boldsymbol{\theta}} = \mathbf{0} \tag{2.59}$$

Figure 2.14 illustrates the method for the single unknown parameter case. The ML estimate corresponds to the peak of the log-likelihood function.

Maximum likelihood estimation has some very desirable properties. If $\boldsymbol{\theta}_0$ is the true value of the unknown parameter in $p(\boldsymbol{x};\boldsymbol{\theta})$, it can be shown that under generally valid conditions the following are true [Papo 91].

- The ML estimate is *asymptotically unbiased*, which by definition means that

$$\lim_{N\to\infty} E[\hat{\boldsymbol{\theta}}_{ML}] = \boldsymbol{\theta}_0 \tag{2.60}$$

Alternatively, we say that the estimate *converges in the mean* to the true value. The meaning of this is as follows. The estimate $\hat{\boldsymbol{\theta}}_{ML}$ is itself a random vector, because for different sample sets X different estimates will result. An estimate is called *unbiased* if its mean is the true value of the unknown parameter. In the ML case this is true only asymptotically ($N \to \infty$).

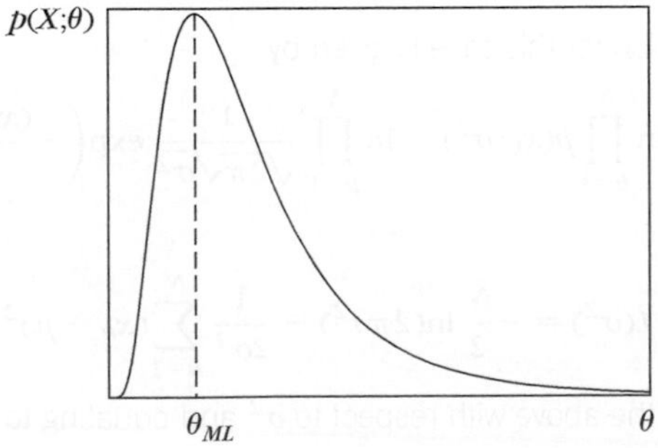

FIGURE 2.14

The maximum likelihood estimator θ_{ML} corresponds to the peak of $p(X;\theta)$.

- The ML estimate is ***asymptotically consistent***, that is, it satisfies

$$\lim_{N\to\infty} \text{prob}\{\|\hat{\boldsymbol{\theta}}_{ML} - \boldsymbol{\theta}_0\| \leq \epsilon\} = 1 \tag{2.61}$$

where ϵ is arbitrarily small. Alternatively, we say that the estimate converges ***in probability***. In other words, for large N it is highly probable that the resulting estimate will be arbitrarily close to the true value. A stronger condition for consistency is also true:

$$\lim_{N\to\infty} E[\|\hat{\boldsymbol{\theta}}_{ML} - \boldsymbol{\theta}_0\|^2] = 0 \tag{2.62}$$

In such cases we say that the estimate converges in the ***mean square***. In words, for large N, the variance of the ML estimates tends to zero.

Consistency is very important for an estimator, because it may be unbiased, but the resulting estimates exhibit large variations around the mean. In such cases we have little confidence in the result obtained from a single set X.

- The ML estimate is asymptotically ***efficient***; that is, it achieves the Cramer–Rao lower bound (Appendix A). This is the lowest value of variance, which ***any*** estimate can achieve.

- The pdf of the ML estimate as $N \to \infty$ approaches the Gaussian distribution with mean $\boldsymbol{\theta}_0$ [Cram 46]. This property is an offspring of (a) the central limit theorem (Appendix A) and (b) the fact that the ML estimate is related to the ***sum*** of random variables, that is, $\partial \ln(p(\boldsymbol{x}_k; \boldsymbol{\theta}))/\partial \boldsymbol{\theta}$ (Problem 2.16).

In summary, the ML estimator is unbiased, is normally distributed, and has the minimum possible variance. However, all these nice properties are valid ***only*** for large values of N.

Example 2.3

Assume that N data points, $x_1, x_2, \ldots, x_N$, have been generated by a one-dimensional Gaussian pdf of known mean, μ, but of unknown variance. Derive the ML estimate of the variance.

The log-likelihood function for this case is given by

$$L(\sigma^2) = \ln \prod_{k=1}^{N} p(x_k; \sigma^2) = \ln \prod_{k=1}^{N} \frac{1}{\sqrt{2\pi}\sqrt{\sigma^2}} \exp\left(-\frac{(x_k - \mu)^2}{2\sigma^2}\right)$$

or

$$L(\sigma^2) = -\frac{N}{2} \ln(2\pi\sigma^2) - \frac{1}{2\sigma^2} \sum_{k=1}^{N} (x_k - \mu)^2$$

Taking the derivative of the above with respect to σ^2 and equating to zero, we obtain

$$-\frac{N}{2\sigma^2} + \frac{1}{2\sigma^4} \sum_{k=1}^{N} (x_k - \mu)^2 = 0$$

and finally the ML estimate of σ^2 results as the solution of the above,

$$\hat{\sigma}^2_{ML} = \frac{1}{N}\sum_{k=1}^{N}(x_k - \mu)^2 \tag{2.63}$$

Observe that, for finite N, $\hat{\sigma}^2_{ML}$ in Eq. (2.63) is a biased estimate of the variance. Indeed,

$$E[\hat{\sigma}^2_{ML}] = \frac{1}{N}\sum_{k=1}^{N}E[(x_k - \mu)^2] = \frac{N-1}{N}\sigma^2$$

where σ^2 is the true variance of the Gaussian pdf. However, for large values of N, we have

$$E[\hat{\sigma}^2_{ML}] = (1 - \frac{1}{N})\sigma^2 \approx \sigma^2$$

which is in line with the theoretical result of asymptotic consistency of the ML estimator.

Example 2.4

Let $\boldsymbol{x}_1, \boldsymbol{x}_2, \ldots, \boldsymbol{x}_N$ be vectors stemmed from a normal distribution with known covariance matrix and unknown mean, that is,

$$p(\boldsymbol{x}_k; \boldsymbol{\mu}) = \frac{1}{(2\pi)^{l/2}|\Sigma|^{1/2}}\exp\left(-\frac{1}{2}(\boldsymbol{x}_k - \boldsymbol{\mu})^T\Sigma^{-1}(\boldsymbol{x}_k - \boldsymbol{\mu})\right)$$

Obtain the ML estimate of the unknown mean vector.

For N available samples we have

$$L(\boldsymbol{\mu}) \equiv \ln\prod_{k=1}^{N}p(\boldsymbol{x}_k; \boldsymbol{\mu}) = -\frac{N}{2}\ln((2\pi)^l|\Sigma|) - \frac{1}{2}\sum_{k=1}^{N}(\boldsymbol{x}_k - \boldsymbol{\mu})^T\Sigma^{-1}(\boldsymbol{x}_k - \boldsymbol{\mu}) \tag{2.64}$$

Taking the gradient with respect to $\boldsymbol{\mu}$, we obtain

$$\frac{\partial L(\boldsymbol{\mu})}{\partial \boldsymbol{\mu}} \equiv \begin{bmatrix} \frac{\partial L}{\partial \mu_1} \\ \frac{\partial L}{\partial \mu_2} \\ \vdots \\ \frac{\partial L}{\partial \mu_l} \end{bmatrix} = \sum_{k=1}^{N}\Sigma^{-1}(\boldsymbol{x}_k - \boldsymbol{\mu}) = 0 \tag{2.65}$$

or

$$\hat{\boldsymbol{\mu}}_{ML} = \frac{1}{N}\sum_{k=1}^{N}\boldsymbol{x}_k \tag{2.66}$$

That is, the ML estimate of the mean, for Gaussian densities, is the sample mean. However, this very "natural approximation" is not necessarily ML optimal for non-Gaussian density functions.

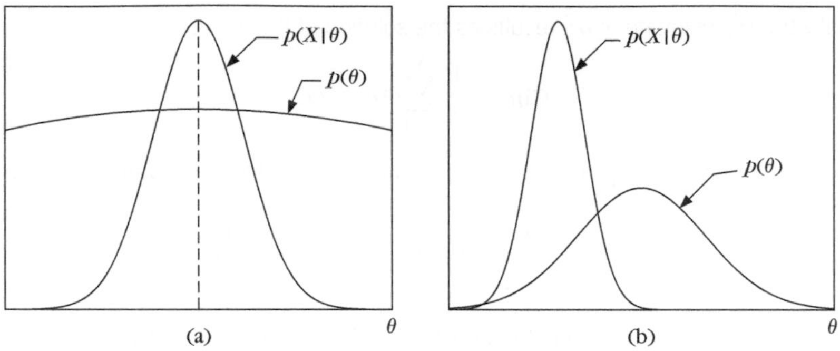

FIGURE 2.15

ML and MAP estimates of θ will be approximately the same in (a) and different in (b).

2.5.2 Maximum *a Posteriori* Probability Estimation

For the derivation of the maximum likelihood estimate, we considered $\boldsymbol{\theta}$ as an unknown parameter. In this subsection we will consider it as a random vector, and we will estimate its value on the condition that samples $\boldsymbol{x}_1, \ldots, \boldsymbol{x}_N$ have occurred. Let $X = \{\boldsymbol{x}_1, \ldots, \boldsymbol{x}_N\}$. Our starting point is $p(\boldsymbol{\theta}|X)$. From our familiar Bayes theorem we have

$$p(\boldsymbol{\theta})p(X|\boldsymbol{\theta}) = p(X)p(\boldsymbol{\theta}|X) \tag{2.67}$$

or

$$p(\boldsymbol{\theta}|X) = \frac{p(\boldsymbol{\theta})p(X|\boldsymbol{\theta})}{p(X)} \tag{2.68}$$

The *maximum a posteriori probability* (MAP) estimate $\hat{\boldsymbol{\theta}}_{MAP}$ is defined at the point where $p(\boldsymbol{\theta}|X)$ becomes maximum,

$$\hat{\boldsymbol{\theta}}_{MAP} : \frac{\partial}{\partial \boldsymbol{\theta}} p(\boldsymbol{\theta}|X) = 0 \quad \text{or} \quad \frac{\partial}{\partial \boldsymbol{\theta}}(p(\boldsymbol{\theta})p(X|\boldsymbol{\theta})) = 0 \tag{2.69}$$

Note that $p(X)$ is not involved since it is independent of $\boldsymbol{\theta}$. The difference between the ML and the MAP estimates lies in the involvement of $p(\boldsymbol{\theta})$ in the latter case. If we assume that this obeys the uniform distribution, that is, is constant for all $\boldsymbol{\theta}$, both estimates yield identical results. This is also approximately true if $p(\boldsymbol{\theta})$ exhibits small variation. However, in the general case, the two methods yield different results. Figures 2.15a and 2.15b illustrate the two cases.

Example 2.5

Let us assume that in Example 2.4 the unknown mean vector $\boldsymbol{\mu}$ is known to be normally distributed as

$$p(\boldsymbol{\mu}) = \frac{1}{(2\pi)^{l/2}\sigma_\mu^l} \exp\left(-\frac{1}{2}\frac{\|\boldsymbol{\mu} - \boldsymbol{\mu}_0\|^2}{\sigma_\mu^2}\right)$$

The MAP estimate is given by the solution of

$$\frac{\partial}{\partial \mu} \ln\left(\prod_{k=1}^{N} p(\boldsymbol{x}_k|\boldsymbol{\mu})p(\boldsymbol{\mu})\right) = \mathbf{0}$$

or, for $\Sigma = \sigma^2 I$,

$$\sum_{k=1}^{N} \frac{1}{\sigma^2}(\boldsymbol{x}_k - \hat{\boldsymbol{\mu}}) - \frac{1}{\sigma_\mu^2}(\hat{\boldsymbol{\mu}} - \boldsymbol{\mu}_0) = \mathbf{0} \Rightarrow$$

$$\hat{\boldsymbol{\mu}}_{MAP} = \frac{\boldsymbol{\mu}_0 + \frac{\sigma_\mu^2}{\sigma^2}\sum_{k=1}^{N} \boldsymbol{x}_k}{1 + \frac{\sigma_\mu^2}{\sigma^2}N}$$

We observe that if $\frac{\sigma_\mu^2}{\sigma^2} >> 1$, that is, the variance σ_μ^2 is very large and the corresponding Gaussian is very wide with little variation over the range of interest, then

$$\hat{\boldsymbol{\mu}}_{MAP} \approx \hat{\boldsymbol{\mu}}_{ML} = \frac{1}{N}\sum_{k=1}^{N} \boldsymbol{x}_k$$

Furthermore, observe that this is also the case for $N \to \infty$, regardless of the values of the variances. Thus, the MAP estimate tends asymptotically to the ML one. This is a more general result. For large values of N, the likelihood term $\prod_{k=1}^{N} p(\boldsymbol{x}_k|\boldsymbol{\mu})$ becomes sharply peaked around the true value (of the unknown parameter) and is the term that basically determines where the maximum occurs. This can be better understood by mobilizing the properties of the ML estimate given before.

2.5.3 Bayesian Inference

Both methods considered in the preceding subsections compute a specific estimate of the unknown parameter vector $\boldsymbol{\theta}$. In the current method, a different path is adopted. Given the set X of the N training vectors and the *a priori* information about the pdf $p(\boldsymbol{\theta})$, the goal is to compute the conditional pdf $p(\boldsymbol{x}|X)$. After all, this is what we actually need to know. To this end, and making use of known identities from our statistics basics, we have the following set of relations at our disposal:

$$p(\boldsymbol{x}|X) = \int p(\boldsymbol{x}|\boldsymbol{\theta})p(\boldsymbol{\theta}|X)\, d\boldsymbol{\theta} \tag{2.70}$$

with

$$p(\boldsymbol{\theta}|X) = \frac{p(X|\boldsymbol{\theta})p(\boldsymbol{\theta})}{p(X)} = \frac{p(X|\boldsymbol{\theta})p(\boldsymbol{\theta})}{\int p(X|\boldsymbol{\theta})p(\boldsymbol{\theta})\, d\boldsymbol{\theta}} \tag{2.71}$$

$$p(X|\boldsymbol{\theta}) = \prod_{k=1}^{N} p(\boldsymbol{x}_k|\boldsymbol{\theta}) \tag{2.72}$$

The conditional density $p(\boldsymbol{\theta}|X)$ is also known as the a posteriori pdf estimate, since it is updated "knowledge" about the statistical properties of $\boldsymbol{\theta}$, after having observed the data set X. Once more, Eq. (2.72) presupposes statistical independence among the training samples.

In general, the computation of $p(\boldsymbol{x}|X)$ requires the integration in the right-hand side of (2.70). However, analytical solutions are feasible only for very special forms of the involved functions. For most of the cases, analytical solutions for (2.70), as well as for the denominator in (2.71), are not possible, and one has to resort to numerical approximations. To this end, a large research effort has been invested in developing efficient techniques for the numerical computation of such statistical quantities. Although a detailed presentation of such approximation schemes is beyond the scope of this book, we will attempt to highlight the main philosophy behind these techniques in relation to our own problem.

Looking more carefully at (2.70) and assuming that $p(\boldsymbol{\theta}|X)$ is known, then $p(\boldsymbol{x}|X)$ is nothing but the average of $p(\boldsymbol{x}|\boldsymbol{\theta})$ with respect to $\boldsymbol{\theta}$, that is,

$$p(\boldsymbol{x}|X) = E_{\boldsymbol{\theta}}\left[p(\boldsymbol{x}|\boldsymbol{\theta})\right]$$

If we assume that a large enough number of samples $\boldsymbol{\theta}_i,\ i = 1, 2 \ldots, L$, of the random vector $\boldsymbol{\theta}$ are available, one can compute the corresponding values $p(\boldsymbol{x}|\boldsymbol{\theta}_i)$ and then approximate the expectation as the mean value

$$p(\boldsymbol{x}|X) \approx \frac{1}{L}\sum_{i=1}^{L} p(\boldsymbol{x}|\boldsymbol{\theta}_i)$$

The problem now becomes that of generating a set of samples, $\boldsymbol{\theta}_i,\ i = 1, 2 \ldots, L$. For example, if $p(\boldsymbol{\theta}|X)$ were a Gaussian pdf, one could use a Gaussian pseudorandom generator to generate the L samples. The difficulty in our case is that, in general, the exact form of $p(\boldsymbol{\theta}|X)$ is not known, and its computation presupposes the numerical integration of the normalizing constant in the denominator of (2.71). This difficulty is bypassed by a set of methods known as *Markov chain Monte Carlo* (MCMC) techniques. The main rationale behind these techniques is that one can generate samples from (2.71) in a sequential manner that *asymptotically* follow the distribution $p(\boldsymbol{\theta}|X)$, even without knowing the normalizing factor. The Gibbs sampler and the Metropolis-Hastings algorithms are two of the most popular schemes of this type. For more details on such techniques, the interested reader may consult, for example, [Bish 06].

Further insight into the Bayesian methods can be gained by focusing on the Gaussian one-dimensional case.

Example 2.6

Let $p(x|\mu)$ be a univariate Gaussian $\mathcal{N}(\mu, \sigma^2)$ with unknown parameter the mean, which is also assumed to follow a Gaussian $\mathcal{N}(\mu_0, \sigma_0^2)$. From the theory exposed before we have

$$p(\mu|X) = \frac{p(X|\mu)p(\mu)}{p(X)} = \frac{1}{\alpha}\prod_{k=1}^{N} p(x_k|\mu)p(\mu)$$

where for a given training data set, X, $p(X)$ is a constant denoted as α, or

$$p(\mu|X) = \frac{1}{\alpha}\prod_{k=1}^{N}\frac{1}{\sqrt{2\pi}\sigma}\exp\left(-\frac{(x_k-\mu)^2}{\sigma^2}\right)\frac{1}{\sqrt{2\pi}\sigma_0}\exp\left(-\frac{(\mu-\mu_0)^2}{2\sigma_0^2}\right)$$

It is a matter of some algebra (Problem 2.25) to show that, given a number of samples, N, $p(\mu|X)$ turns out to be also Gaussian, that is,

$$p(\mu|X) = \frac{1}{\sqrt{2\pi}\sigma_N}\exp\left(-\frac{(\mu-\mu_N)^2}{2\sigma_N^2}\right) \tag{2.73}$$

with mean value

$$\mu_N = \frac{N\sigma_0^2\bar{x}_N + \sigma^2\mu_0}{N\sigma_0^2+\sigma^2} \tag{2.74}$$

and variance

$$\sigma_N^2 = \frac{\sigma^2\sigma_0^2}{N\sigma_0^2+\sigma^2} \tag{2.75}$$

where $\bar{x}_N = \frac{1}{N}\sum_{k=1}^{N} x_k$. Letting N vary from 1 to ∞, we generate a sequence of Gaussians $\mathcal{N}(\mu_N, \sigma_N^2)$, whose mean values move away from μ_0 and tend, in the limit, to the sample mean, which, asymptotically, becomes equal to the true mean value. Furthermore, their variance keeps decreasing at the rate σ^2/N for large N. Hence, for large values of N, $p(\mu|X)$ becomes sharply peaked around the sample mean. Recall that the latter is the ML estimate of the mean value.

Once $p(\mu|X)$ has been computed, it can be shown, by substituting (2.73) into (2.70) (problem 2.25), that

$$p(x|X) = \frac{1}{\sqrt{2\pi(\sigma^2+\sigma_N^2)}}\exp\left(-\frac{1}{2}\frac{(x-\mu_N)^2}{\sigma^2+\sigma_N^2}\right)$$

which is a Gaussian pdf with mean value μ_N and variance $\sigma^2 + \sigma_N^2$.

Observe that as N tends to infinity, the unknown mean value of the Gaussian tends to the ML estimate $\bar{x}_N$ (and asymptotically to the true mean) and the variance to the true value σ^2. For finite values of N, the variance is larger than σ^2 to account for our extra uncertainty about x due to the unknown value of the mean μ. Figure 2.16 shows the posterior pdf estimate $p(\mu|X)$ obtained for different sizes of the training data set. Data were generated using a pseudorandom number generator following a Gaussian pdf with mean value equal to $\mu = 2$ and variance $\sigma^2 = 4$. The mean value was assumed to be unknown, and the prior pdf was adopted to be Gaussian with $\mu_0 = 0$ and $\sigma_0^2 = 8$. We observe that as N increases $p(\mu|X)$ gets narrower (in accordance to (2.75)). The respective mean value estimate (Eq. (2.74)) depends on N and $\bar{x}_N$. For small values of N, the ML estimate of the mean, $\bar{x}_N$, can vary a lot, which has a direct effect in moving around the centers of the Gaussians. However, as N increases, $\bar{x}_N$ tends to the true value of the mean ($\mu = 2$) with a decreasing variance.

It can be shown (Problem 2.27) that the results of this example can be generalized for the case of multivariate Gaussians. More specifically, one can show that Eqs. (2.74) and (2.75) are generalized to the following

$$p(\mu|X) \sim \mathcal{N}(\mu_N, \Sigma_N) \tag{2.76}$$

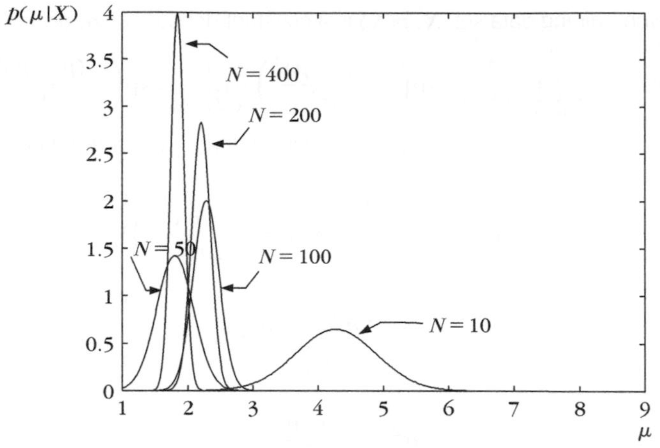

FIGURE 2.16

A sequence of the posterior pdf estimates (Eq. (2.73)), for the case of Example 2.6. As the number of training points increases, the posterior pdf becomes more spiky (the ambiguity decreases) and its center moves toward the true mean value of the data.

where

$$\boldsymbol{\mu}_N = N\Sigma_0[N\Sigma_0 + \Sigma]^{-1}\bar{\boldsymbol{x}}_N + \Sigma[N\Sigma_0 + \Sigma]^{-1}\boldsymbol{\mu}_0 \tag{2.77}$$

and

$$\Sigma_N = \Sigma_0[N\Sigma_0 + \Sigma]^{-1}\Sigma \tag{2.78}$$

and also

$$p(\boldsymbol{x}|X) \sim \mathcal{N}(\boldsymbol{\mu}_N, \Sigma + \Sigma_N) \tag{2.79}$$

Remarks

- If $p(\boldsymbol{\theta}|X)$ in Eq. (2.71) is sharply peaked at a $\hat{\boldsymbol{\theta}}$ and we treat it as a delta function, Eq. (2.70) becomes $p(\boldsymbol{x}|X) \approx p(\boldsymbol{x}|\hat{\boldsymbol{\theta}})$; that is, the parameter estimate is approximately equal to the MAP estimate. This happens, for example, if $p(X|\boldsymbol{\theta})$ is concentrated around a sharp peak and $p(\boldsymbol{\theta})$ is broad enough around this peak. Then the resulting estimate approximates the ML one. The latter was also verified by our previous example. This is a more general property valid for most of the pdfs used in practice, for which the posterior probability of the unknown parameter vector $p(\boldsymbol{\theta}|X)$ tends to a delta function as N tends to $+\infty$. *Thus, all three methods considered so far result, asymptotically, in the same estimate. However, the results are different for small numbers N of training samples.*

- An obvious question concerns the choice of the prior $p(\boldsymbol{\theta})$. In practice, the choice depends on the form of the likelihood function $p(\boldsymbol{x}|\boldsymbol{\theta})$, so that the posterior pdf $p(\boldsymbol{\theta}|X)$ can be of a tractable form. The set of prior distributions for which the adopted model $p(\boldsymbol{x}|\boldsymbol{\theta})$ is of the same functional form as the posterior distribution $p(\boldsymbol{\theta}|X)$ is known as *conjugate* with respect to the model. Some commonly used forms of the conjugate priors are discussed, for example, in [Bern 94].

- For data sets of limited length, ML and MAP estimators are simpler to use, and they result in a single estimate of the unknown parameters vector, which is the outcome of a maximization procedure. On the other hand, Bayesian methods make use of more information and, provided that this information is reliable, these techniques are expected to give better results, albeit at the expense of higher complexity. Due to the advances in computer technology, Bayesian methods have gained a lot of popularity over the recent years.

2.5.4 Maximum Entropy Estimation

The concept of *entropy* is known from Shannon's information theory. It is a measure of the uncertainty concerning an event and, from another viewpoint, a measure of randomness of the messages (feature vectors in our case) occurring at the output of a system. If $p(\boldsymbol{x})$ is the density function, the associated entropy H is given by

$$H = -\int_{\boldsymbol{x}} p(\boldsymbol{x}) \ln p(\boldsymbol{x})\, d\boldsymbol{x} \tag{2.80}$$

Assume now that $p(\boldsymbol{x})$ is unknown but we know a number of related constraints (mean value, variance, etc.). The *maximum entropy* estimate of the unknown pdf is the one that maximizes the entropy, subject to the given constraints. According to the principle of maximum entropy, stated by Jaynes [Jayn 82], such an estimate corresponds to the distribution that exhibits the highest possible randomness, subject to the available constraints.

Example 2.7

The random variable $\boldsymbol{x}$ is nonzero for $x_1 \le x \le x_2$ and zero otherwise. Compute the maximum entropy estimate of its pdf.

We have to maximize (2.80) subject to the constraint

$$\int_{x_1}^{x_2} p(x)\, dx = 1 \tag{2.81}$$

Using Lagrange multipliers (Appendix C), this is equivalent to maximizing

$$H_L = -\int_{x_1}^{x_2} p(x)(\ln p(x) - \lambda)\, dx \tag{2.82}$$

Taking the derivative with respect to $p(x)$, we obtain

$$\frac{\partial H_L}{\partial p(x)} = -\int_{x_1}^{x_2} \left\{(\ln p(x) - \lambda) + 1\right\} dx \tag{2.83}$$

Equating to zero, we obtain

$$\hat{p}(x) = \exp(\lambda - 1) \tag{2.84}$$

To compute λ, we substitute this into the constraint equation (2.81), and we get $\exp(\lambda - 1) = \frac{1}{x_2 - x_1}$. Thus

$$\hat{p}(x) = \begin{cases} \frac{1}{x_2 - x_1} & \text{if } x_1 \leq x \leq x_2 \\ 0 & \text{otherwise} \end{cases} \tag{2.85}$$

That is, the maximum entropy estimate of the unknown pdf is the uniform distribution. This is within the maximum entropy spirit. Since we have imposed no other constraint but the obvious one, the resulting estimate is the one that maximizes randomness and all points are equally probable. It turns out that if the mean value and the variance are given as the second and third constraints, the resulting maximum entropy estimate of the pdf, for $-\infty < x < +\infty$, is the Gaussian (Problem 2.30).

2.5.5 Mixture Models

An alternative way to model an unknown $p(\boldsymbol{x})$ is via a linear combination of density functions in the form of

$$p(\boldsymbol{x}) = \sum_{j=1}^{J} p(\boldsymbol{x}|j)P_j \tag{2.86}$$

where

$$\sum_{j=1}^{J} P_j = 1, \quad \int_{\boldsymbol{x}} p(\boldsymbol{x}|j)\, d\boldsymbol{x} = 1 \tag{2.87}$$

In other words, it is assumed that J distributions contribute to the formation of $p(\boldsymbol{x})$. Thus, this modeling implicitly assumes that each point $\boldsymbol{x}$ may be "drawn" from any of the J model distributions with probability $P_j, j = 1, 2, \ldots, J$. It can be shown that this modeling can approximate arbitrarily closely any continuous density function for a sufficient number of *mixtures* J and appropriate model parameters. The first step of the procedure involves the choice of the set of density components $p(\boldsymbol{x}|j)$ in parametric form, that is, $p(\boldsymbol{x}|j;\boldsymbol{\theta})$, and then the computation of the unknown parameters, $\boldsymbol{\theta}$ and $P_j, j = 1, 2, \ldots, J$, based on the set of the available training samples $\boldsymbol{x}_k$. There are various ways to achieve this. A typical maximum likelihood formulation, maximizing the likelihood function $\prod_k p(\boldsymbol{x}_k; \boldsymbol{\theta}, P_1, P_2, \ldots, P_J)$ with respect to $\boldsymbol{\theta}$ and the P_j's, is a first thought. The difficulty here arises from the fact that the unknown parameters enter the maximization task in a *nonlinear fashion*; thus, nonlinear optimization iterative techniques have to be adopted (Appendix C). A review of related techniques is given in [Redn 84]. The source of this complication is the lack of information concerning the labels of the available training samples, that is, the specific mixture from which each sample is contributed. This is the issue that makes the current problem different from the ML case treated in Section 2.5.1. There, the class labels were known, and this led to a *separate* ML problem for each of the classes. In the same way, if the mixture labels were known, we could collect all data from the same mixture and carry out J separate ML tasks. The missing label information makes our current problem a typical task with an *incomplete data set*.

In the sequel, we will focus on the so-called EM algorithm, which has attracted a great deal of interest over the past few years in a wide range of applications involving tasks with incomplete data sets.

The Expectation Maximization (EM) Algorithm

This algorithm is ideally suited for cases in which the available data set is incomplete. Let us first state the problem in more general terms and then apply it to our specific task. Let us denote by $\boldsymbol{y}$ the *complete data* samples, with $\boldsymbol{y} \in Y \subseteq \mathcal{R}^m$, and let the corresponding pdf be $p_y(\boldsymbol{y}; \boldsymbol{\theta})$, where $\boldsymbol{\theta}$ is an unknown parameter vector. The samples $\boldsymbol{y}$, however, *cannot be directly observed*. What we observe instead are samples $\boldsymbol{x} = \boldsymbol{g}(\boldsymbol{y}) \in X_{ob} \subseteq \mathcal{R}^l, l < m$. We denote the corresponding pdf $p_x(\boldsymbol{x}; \boldsymbol{\theta})$. This is a *many-to-one mapping*. Let $Y(\boldsymbol{x}) \subseteq Y$ be the subset of all the $\boldsymbol{y}$'s corresponding to a specific $\boldsymbol{x}$. Then the pdf of the incomplete data is given by

$$p_x(\boldsymbol{x}; \boldsymbol{\theta}) = \int_{Y(\boldsymbol{x})} p_y(\boldsymbol{y}; \boldsymbol{\theta}) \, d\boldsymbol{y} \tag{2.88}$$

As we already know, the maximum likelihood estimate of $\boldsymbol{\theta}$ is given by

$$\hat{\boldsymbol{\theta}}_{ML}: \sum_k \frac{\partial \ln(p_y(\boldsymbol{y}_k; \boldsymbol{\theta}))}{\partial \boldsymbol{\theta}} = \boldsymbol{0} \tag{2.89}$$

However, the $\boldsymbol{y}$'s are not available. So, the EM algorithm maximizes the *expectation* of the log-likelihood function, *conditioned on the observed samples and the current iteration estimate of* $\boldsymbol{\theta}$. The two steps of the algorithm are:

- *E-step*: At the $(t+1)$th step of the iteration, where $\boldsymbol{\theta}(t)$ is available, compute the expected value of

$$Q(\boldsymbol{\theta}; \boldsymbol{\theta}(t)) \equiv E\left[\sum_k \ln(p_y(\boldsymbol{y}_k; \boldsymbol{\theta}|X; \boldsymbol{\theta}(t))\right] \tag{2.90}$$

  This is the so-called *expectation step* of the algorithm.

- *M-step*: Compute the next $(t+1)$th estimate of $\boldsymbol{\theta}$ by maximizing $Q(\boldsymbol{\theta}; \boldsymbol{\theta}(t))$, that is,

$$\boldsymbol{\theta}(t+1): \frac{\partial Q(\boldsymbol{\theta}; \boldsymbol{\theta}(t))}{\partial \boldsymbol{\theta}} = \boldsymbol{0} \tag{2.91}$$

  This is the *maximization step*, where, obviously, differentiability has been assumed.

To apply the EM algorithm, we start from an initial estimate $\boldsymbol{\theta}(0)$, and iterations are terminated if $\|\boldsymbol{\theta}(t+1) - \boldsymbol{\theta}(t)\| \le \epsilon$ for an appropriately chosen vector norm and ϵ.

Remark

- It can be shown that the successive estimates $\boldsymbol{\theta}(t)$ never decrease the likelihood function. The likelihood function keeps increasing until a maximum (local or global) is reached and the EM algorithm converges. The convergence proof can be found in the seminal paper [Demp 77] and further discussions in [Wu 83, Boyl 83]. Theoretical results as well as practical experimentation confirm that the convergence is slower than the quadratic convergence of Newton-type searching algorithms (Appendix C), although near the optimum a speedup may be possible. However, the great advantage of the algorithm is that its convergence is smooth and is not vulnerable to instabilities. Furthermore, it is computationally more attractive than Newton-like methods, which require the computation of the Hessian matrix. The keen reader may obtain more information on the EM algorithm and some of its applications from [McLa 88, Titt 85, Moon 96].

Application to the Mixture Modeling Problem

In this case, the complete data set consists of the joint events $(\boldsymbol{x}_k, j_k)$, $k = 1, 2, \ldots, N$, and j_k takes integer values in the interval $[1, J]$, and it denotes the mixture from which $\boldsymbol{x}_k$ is generated. Employing our familiar rule, we obtain

$$p(\boldsymbol{x}_k, j_k; \boldsymbol{\theta}) = p(\boldsymbol{x}_k | j_k; \boldsymbol{\theta}) P_{j_k} \tag{2.92}$$

Assuming mutual independence among samples of the data set, the log-likelihood function becomes

$$L(\boldsymbol{\theta}) = \sum_{k=1}^{N} \ln\big(p(\boldsymbol{x}_k | j_k; \boldsymbol{\theta}) P_{j_k}\big) \tag{2.93}$$

Let $\boldsymbol{P} = [P_1, P_2, \ldots, P_J]^T$. In the current framework, the unknown parameter vector is $\boldsymbol{\Theta}^T = [\boldsymbol{\theta}^T, \boldsymbol{P}^T]^T$. Taking the expectation over the *unobserved data*, conditioned on the training samples and the current estimates, $\boldsymbol{\Theta}(t)$, of the unknown parameters, we have

$$\text{E-step: } Q(\boldsymbol{\Theta}; \boldsymbol{\Theta}(t)) = E\left[\sum_{k=1}^{N} \ln(p(\boldsymbol{x}_k | j_k; \boldsymbol{\theta}) P_{j_k})\right]$$

$$= \sum_{k=1}^{N} E[\ln(p(\boldsymbol{x}_k | j_k; \boldsymbol{\theta}) P_{j_k})] \tag{2.94}$$

$$= \sum_{k=1}^{N} \sum_{j_k=1}^{J} P(j_k | \boldsymbol{x}_k; \boldsymbol{\Theta}(t)) \ln(p(\boldsymbol{x}_k | j_k; \boldsymbol{\theta}) P_{j_k}) \tag{2.95}$$

The notation can now be simplified by dropping the index k from j_k. This is because, for each k, we sum up over all possible J values of j_k and these are the same for all k. We will demonstrate the algorithm for the case of Gaussian mixtures with diagonal covariance matrices of the form $\Sigma_j = \sigma_j^2 I$, that is,

$$p(\boldsymbol{x}_k|j;\boldsymbol{\theta}) = \frac{1}{(2\pi\sigma_j^2)^{l/2}} \exp\left(-\frac{\|\boldsymbol{x}_k - \boldsymbol{\mu}_j\|^2}{2\sigma_j^2}\right) \tag{2.96}$$

Assume that besides the prior probabilities, P_j, the respective mean values $\boldsymbol{\mu}_j$ as well as the variances $\sigma_j^2, j = 1, 2, \ldots, J$, of the Gaussians are also unknown. Thus, $\boldsymbol{\theta}$ is a $J(l+1)$-dimensional vector. Combining Eqs. (2.95) and (2.96) and omitting constants, we get

E-step:

$$Q(\boldsymbol{\Theta};\boldsymbol{\Theta}(t)) = \sum_{k=1}^{N}\sum_{j=1}^{J} P(j|\boldsymbol{x}_k;\boldsymbol{\Theta}(t))\left(-\frac{l}{2}\ln\sigma_j^2 - \frac{1}{2\sigma_j^2}\|\boldsymbol{x}_k - \boldsymbol{\mu}_j\|^2 + \ln P_j\right) \tag{2.97}$$

M-step: Maximizing the above with respect to $\boldsymbol{\mu}_j$, σ_j^2, and P_j results in (Problem 2.31)

$$\boldsymbol{\mu}_j(t+1) = \frac{\sum_{k=1}^{N} P(j|\boldsymbol{x}_k;\boldsymbol{\Theta}(t))\boldsymbol{x}_k}{\sum_{k=1}^{N} P(j|\boldsymbol{x}_k;\boldsymbol{\Theta}(t))} \tag{2.98}$$

$$\sigma_j^2(t+1) = \frac{\sum_{k=1}^{N} P(j|\boldsymbol{x}_k;\boldsymbol{\Theta}(t))\|\boldsymbol{x}_k - \boldsymbol{\mu}_j(t+1)\|^2}{l\sum_{k=1}^{N} P(j|\boldsymbol{x}_k;\boldsymbol{\Theta}(t))} \tag{2.99}$$

$$P_j(t+1) = \frac{1}{N}\sum_{k=1}^{N} P(j|\boldsymbol{x}_k;\boldsymbol{\Theta}(t)) \tag{2.100}$$

For the iterations to be complete we need only to compute $P(j|\boldsymbol{x}_k;\boldsymbol{\Theta}(t))$. This is easily obtained from

$$P(j|\boldsymbol{x}_k;\boldsymbol{\Theta}(t)) = \frac{p(\boldsymbol{x}_k|j;\boldsymbol{\theta}(t))P_j(t)}{p(\boldsymbol{x}_k;\boldsymbol{\Theta}(t))} \tag{2.101}$$

$$p(\boldsymbol{x}_k;\boldsymbol{\Theta}(t)) = \sum_{j=1}^{J} p(\boldsymbol{x}_k|j;\boldsymbol{\theta}(t))P_j(t) \tag{2.102}$$

Equations (2.98)–(2.102) constitute the EM algorithm for the estimation of the unknown parameters of the Gaussian mixtures in (2.86). The algorithm starts with valid initial guesses for the unknown parameters. Valid means that probabilities must add to one.

Remark

- Modeling unknown probability density functions via a mixture of Gaussian components and the EM algorithm has been very popular in a number of applications. Besides some convergence issues associated with the EM algorithm,

as previously discussed, another difficulty may arise in deciding about the exact number of components, J. In the context of supervised learning, one may use different values and choose the model that results in the best error probability. The latter can be computed by employing an error estimation technique (Chapter 10).

Example 2.8

Figure 2.17a shows $N = 100$ points in the two-dimensional space, which have been drawn from a multimodal distribution. The samples were generated using two Gaussian random generators $\mathcal{N}(\boldsymbol{\mu}_1, \Sigma_1)$, $\mathcal{N}(\boldsymbol{\mu}_2, \Sigma_2)$, with

$$\boldsymbol{\mu}_1 = \begin{bmatrix} 1.0 \\ 1.0 \end{bmatrix}, \boldsymbol{\mu}_2 = \begin{bmatrix} 2.0 \\ 2.0 \end{bmatrix}$$

and covariance matrices

$$\Sigma_1 = \Sigma_2 = \begin{bmatrix} 0.1 & 0.0 \\ 0.0 & 0.1 \end{bmatrix}$$

respectively. Each time a sample $\boldsymbol{x}_k$, $k = 1, 2, \ldots, N$, is to be generated a coin is tossed. The corresponding probabilities for heads or tails are $P(H) \equiv P = 0.8$, $P(T) = 1 - P = 0.2$, respectively. If the outcome of the coin flip is heads, the sample $\boldsymbol{x}_k$ is generated from $\mathcal{N}(\boldsymbol{\mu}_1, \Sigma_1)$. Otherwise, it is drawn from $\mathcal{N}(\boldsymbol{\mu}_2, \Sigma_2)$. This is the reason that in Figure 2.17a the space around the point $[1.0, 1.0]^T$ is more densely populated. The pdf of the data set can obviously be written as

$$p(\boldsymbol{x}) = g(\boldsymbol{x}; \boldsymbol{\mu}_1, \sigma_1^2)P + g(\boldsymbol{x}; \boldsymbol{\mu}_2, \sigma_2^2)(1 - P) \tag{2.103}$$

where $g(\cdot; \boldsymbol{\mu}, \sigma^2)$ denotes the Gaussian pdf with parameters the mean value $\boldsymbol{\mu}$ and a diagonal covariance matrix, $\Sigma = \text{diag}\{\sigma^2\}$, having σ^2 across the diagonal and zeros

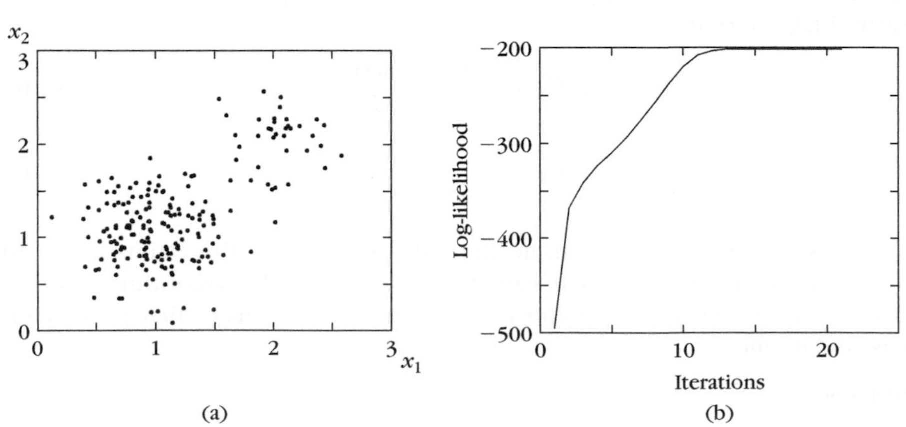

FIGURE 2.17

(a) The data set of Example 2.8 and (b) the log-likelihood as a function of the number of iteration steps.

elsewhere. Equation (2.103) is a special case of the more general formulation given in (2.86). The goal is to compute the maximum likelihood estimate of the unknown parameters vector

$$\boldsymbol{\Theta}^T = [P, \boldsymbol{\mu}_1^T, \sigma_1^2, \boldsymbol{\mu}_2^T, \sigma_2^2]$$

based on the available $N = 100$ points. The full training data set consists of the sample pairs $(\boldsymbol{x}_k, j_k)$, $k = 1, 2, \ldots, N$, where $j_k \in \{1, 2\}$, and it indicates the origin of each observed sample. However, only the points $\boldsymbol{x}_k$ are at our disposal, with the "label" information being hidden from us. To understand this issue better and gain more insight into the rationale behind the EM methodology, it may be useful to arrive at Eq. (2.95) from a slightly different route. Each of the random vectors, $\boldsymbol{x}_k$, can be thought of as the result of a linear combination of two other random vectors; namely,

$$\boldsymbol{x}_k = \alpha_k \boldsymbol{x}_k^1 + (1 - \alpha_k)\boldsymbol{x}_k^2$$

where $\boldsymbol{x}_k^1$ is drawn from $\mathcal{N}(\boldsymbol{\mu}_1, \Sigma_1)$ and $\boldsymbol{x}_k^2$ from $\mathcal{N}(\boldsymbol{\mu}_2, \Sigma_2)$. The binary coefficients $\alpha_k \in \{0, 1\}$ are randomly chosen with probabilities $P(1) = P = 0.8$, $P(0) = 0.2$. If the values of the α_ks, $k = 1, 2, \ldots, N$, were known to us, the log-likelihood function in (2.93) would be written as

$$L(\boldsymbol{\Theta}; \alpha) = \sum_{k=1}^{N} \alpha_k \ln\left\{g(\boldsymbol{x}_k; \boldsymbol{\mu}_1, \sigma_1^2)P\right\} + \sum_{k=1}^{N} (1 - \alpha_k) \ln\left\{g(\boldsymbol{x}_k; \boldsymbol{\mu}_2, \sigma_2^2)(1 - P)\right\} \quad (2.104)$$

since we can split the summation in two parts, depending on the origin of each sample $\boldsymbol{x}_k$. However, this is just an "illusion" since the α_ks are unknown to us. Motivated by the spirit behind the EM algorithm, we substitute in (2.104) the respective mean values $E[\alpha_k|\boldsymbol{x}_k; \hat{\boldsymbol{\Theta}}]$, given an estimate, $\hat{\boldsymbol{\Theta}}$, of the unknown parameter vector. For the needs of our example we have

$$E[\alpha_k|\boldsymbol{x}_k; \hat{\boldsymbol{\Theta}}] = 1 \times P(1|\boldsymbol{x}_k; \hat{\boldsymbol{\Theta}}) + 0 \times (1 - P(1|\boldsymbol{x}_k; \hat{\boldsymbol{\Theta}})) = P(1|\boldsymbol{x}_k; \hat{\boldsymbol{\Theta}}) \quad (2.105)$$

Substitution of (2.105) into (2.104) results in (2.95) for the case of $J = 2$.

We are now ready to apply the EM algorithm [Eqs. (2.98)–(2.102)] to the needs of our example. The initial values were chosen to be

$$\boldsymbol{\mu}_1(0) = [1.37, 1.20]^T, \quad \boldsymbol{\mu}_2(0) = [1.81, 1.62]^T, \quad \sigma_1^2 = \sigma_2^2 = 0.44, \quad P = 0.5$$

Figure 2.17b shows the log-likelihood as a function of the number of iterations. After convergence, the obtained estimates for the unknown parameters are

$$\boldsymbol{\mu}_1 = [1.05, 1.03]^T, \ \boldsymbol{\mu}_2 = [1.90, 2.08]^T, \ \sigma_1^2 = 0.10, \ \sigma_2^2 = 0.06, \ P = 0.844 \quad (2.106)$$

2.5.6 Nonparametric Estimation

So far in our discussion a pdf parametric modeling has been incorporated, in one way or another, and the associated unknown parameters have been estimated. In

the current subsection we will deal with nonparametric techniques. These are basically variations of the ***histogram*** approximation of an unknown pdf, which is familiar to us from our statistics basics. Let us take, for example, the simple one-dimensional case. Figure 2.18 shows two examples of a pdf and its approximation by the histogram method. That is, the x-axis (one-dimensional space) is first divided into successive bins of length h. Then the probability of a sample x being located in a bin is estimated for each of the bins. If N is the total number of samples and k_N of these are located inside a bin, the corresponding probability is approximated by the *frequency ratio*

$$P \approx k_N/N \tag{2.107}$$

This approximation converges to the true P as $N \to \infty$ (Problem 2.32). The corresponding pdf value is assumed constant throughout the bin and is approximated by

$$\hat{p}(x) \equiv \hat{p}(\hat{x}) \approx \frac{1}{h}\frac{k_N}{N}, \quad |x - \hat{x}| \leq \frac{h}{2} \tag{2.108}$$

where $\hat{x}$ is the midpoint of the bin. This determines the amplitude of the histogram curve over the bin. This is a reasonable approximation for continuous $p(x)$ and small enough h so that the assumption of constant $p(x)$ in the bin is sensible. It can be shown that $\hat{p}(x)$ converges to the true value $p(x)$ as $N \to \infty$ provided:

- $h_N \to 0$
- $k_N \to \infty$
- $\frac{k_N}{N} \to 0$

where h_N is used to show the dependence on N. These conditions can be understood from simple reasoning, without having to resort to mathematical details. The first has already been discussed. The other two show the way that k_N must grow

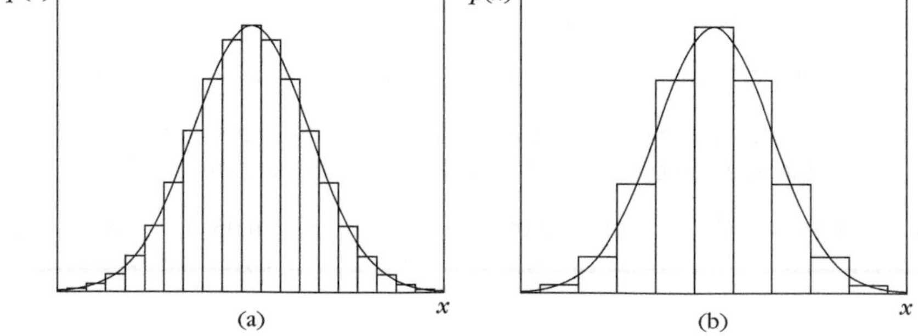

FIGURE 2.18

Probability density function approximation by the histogram method with (a) small and (b) large-size intervals (bins).

to guarantee convergence. Indeed, at all points where $p(x) \neq 0$ fixing the size h_N, however small, the probability P of points occurring in this bin is finite. Hence, $k_N \approx PN$ and k_N tends to infinity as N grows to infinity. On the other hand, as the size h_N of the bin tends to zero, the corresponding probability also goes to zero, justifying the last condition. In practice, the number N of data points is finite. The preceding conditions indicate the way that the various parameters must be chosen. N must be "large enough," h_N "small enough," and the number of points falling in each bin "large enough" too. How small and how large depend on the type of the pdf function and the degree of approximation one is satisfied with. Two popular approaches used in practice are described next.

Parzen Windows

In the multidimensional case, instead of bins of size h, the l-dimensional space is divided into hypercubes with length of side h and volume h^l. Let $\boldsymbol{x}_i$, $i = 1, 2, \ldots, N$, be the available feature vectors. Define the function $\phi(\boldsymbol{x})$ so that

$$\phi(\boldsymbol{x}_i) = \begin{cases} 1 & \text{for } |x_{ij}| \leq 1/2 \\ 0 & \text{otherwise} \end{cases} \tag{2.109}$$

where $x_{ij}, j = 1, \ldots, l$, are the components of $\boldsymbol{x}_i$. In words, the function is equal to 1 for all points inside the unit side hypercube centered at the origin and 0 outside it. This is shown in Figure 2.19(a). Then (2.108) can be "rephrased" as

$$\hat{p}(\boldsymbol{x}) = \frac{1}{h^l}\left(\frac{1}{N}\sum_{i=1}^{N}\phi\left(\frac{\boldsymbol{x}_i - \boldsymbol{x}}{h}\right)\right) \tag{2.110}$$

The interpretation of this is straightforward. We consider a hypercube with length of side h centered at $\boldsymbol{x}$, the point where the pdf is to be estimated. This is illustrated in Figure 2.19(b) for the two-dimensional space. The summation equals k_N, that is, the number of points falling inside this hypercube. Then the pdf estimate results from dividing k_N by N and the respective hypercube volume h^l. However, viewing Eq. (2.110) from a slightly different perspective, we see that we try to approximate a continuous function $p(\boldsymbol{x})$ via an expansion in terms of discontinuous step functions $\phi(\cdot)$. Thus, the resulting estimate will suffer from this "ancestor's sin." This led Parzen [Parz 62] to generalize (2.110) by using smooth functions in the place of $\phi(\cdot)$. It can be shown that, provided

$$\phi(\boldsymbol{x}) \geq 0 \quad \text{and} \tag{2.111}$$

$$\int_{\boldsymbol{x}} \phi(\boldsymbol{x})\, d\boldsymbol{x} = 1 \tag{2.112}$$

the resulting estimate is a legitimate pdf. Such smooth functions are known as ***kernels*** or ***potential functions*** or ***Parzen windows***. A typical example is the Gaussian $\mathcal{N}(\boldsymbol{0}, I)$, kernel. For such a choice, the approximate expansion of the unknown

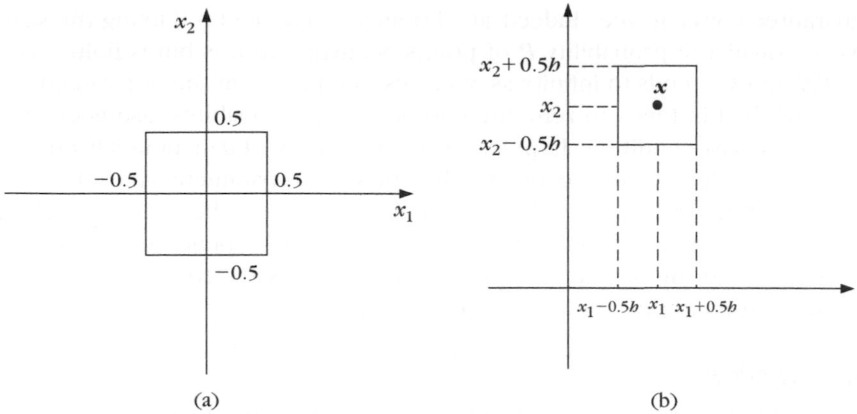

FIGURE 2.19

In the two-dimensional space (a) the function $\phi(\boldsymbol{x}_i)$ is equal to one for every point, $\boldsymbol{x}_i$, inside the square of unit side length, centered at the origin and equal to zero for every point outside it. (b) The function $\phi\left(\frac{\boldsymbol{x}_i-\boldsymbol{x}}{h}\right)$ is equal to unity for every point $\boldsymbol{x}_i$ inside the square with side length equal to h, centered at $\boldsymbol{x}$ and zero for all the other points.

$p(\boldsymbol{x})$ will be

$$\hat{p}(\boldsymbol{x}) = \frac{1}{N}\sum_{i=1}^{N}\frac{1}{(2\pi)^{\frac{l}{2}}h^l}\exp\left(-\frac{(\boldsymbol{x}-\boldsymbol{x}_i)^T(\boldsymbol{x}-\boldsymbol{x}_i)}{2h^2}\right)$$

In other words, the unknown pdf is approximated as an average of N Gaussians, each one centered at a different point of the training set. Recall that as the parameter h becomes smaller, the shape of the Gaussians becomes narrower and more "spiky" (Appendix A) and the influence of each individual Gaussian is more localized in the feature space around the area of its mean value. On the other hand, the larger the value of h, the broader their shape becomes and more global in space their influence is. The expansion of a pdf in a sum of Gaussians was also used in 2.5.5. However, here, the number of Gaussians coincides with the number of points, and the unknown parameter, h, is chosen by the user. In the EM algorithm concept, the number of Gaussians is chosen independently of the number of training points, and the involved parameters are computed via an optimization procedure.

In the sequel, we will examine the limiting behavior of the approximation. To this end, let us take the mean value of (2.110)

$$E[\hat{p}(\boldsymbol{x})] = \frac{1}{h^l}\left(\frac{1}{N}\sum_{i=1}^{N}E\left[\phi\left(\frac{\boldsymbol{x}_i-\boldsymbol{x}}{h}\right)\right]\right)$$

$$\equiv \int_{\boldsymbol{x}'}\frac{1}{h^l}\phi\left(\frac{\boldsymbol{x}'-\boldsymbol{x}}{h}\right)p(\boldsymbol{x}')\,d\boldsymbol{x}' \tag{2.113}$$

Thus, the mean value is a *smoothed version* of the true pdf $p(\boldsymbol{x})$. However as $h \rightarrow 0$ the function $\frac{1}{h^l}\phi\left(\frac{\boldsymbol{x}'-\boldsymbol{x}}{h}\right)$ tends to the delta function $\delta(\boldsymbol{x}' - \boldsymbol{x})$. Indeed, its amplitude goes to infinity, its width tends to zero, and its integral from (2.112) remains equal to one. Thus, in this limiting case and for well-behaved continuous pdfs, $\hat{p}(\boldsymbol{x})$ is an unbiased estimate of $p(\boldsymbol{x})$. *Note that this is independent of the size N of the data set.* Concerning the variance of the estimate (Problem 2.38), the following remarks are valid:

- For fixed N, the smaller the h the higher the variance, and this is indicated by the noisy appearance of the resulting pdf estimate, for example, Figures 2.20a and 2.21a as well as Figures 2.22c and 2.22d. This is because $p(\boldsymbol{x})$ is approximated by a finite sum of δ-like spiky functions, centered at the training sample points. Thus, as one moves $\boldsymbol{x}$ in space the response of $\hat{p}(\boldsymbol{x})$ will be very high near the training points, and it will decrease very rapidly as one moves away, leading to this noiselike appearance. Large values of h smooth out local variations in density.

- For a fixed h, the variance decreases as the number of sample points N increases. This is illustrated in Figures 2.20a and 2.20b as well as in Figures 2.22b and 2.22c. This is because the space becomes dense in points, and the spiky functions are closely located. Furthermore, for a large enough number of samples, the smaller the h the better the accuracy of the resulting estimate, for example, Figures 2.20b and 2.21b.

- It can be shown, for example, [Parz 62, Fuku 90] that, under some mild conditions imposed on $\phi(\cdot)$, which are valid for most density functions, if h tends to zero but in such a way that $hN \rightarrow \infty$, the resulting estimate is both unbiased and asymptotically consistent.

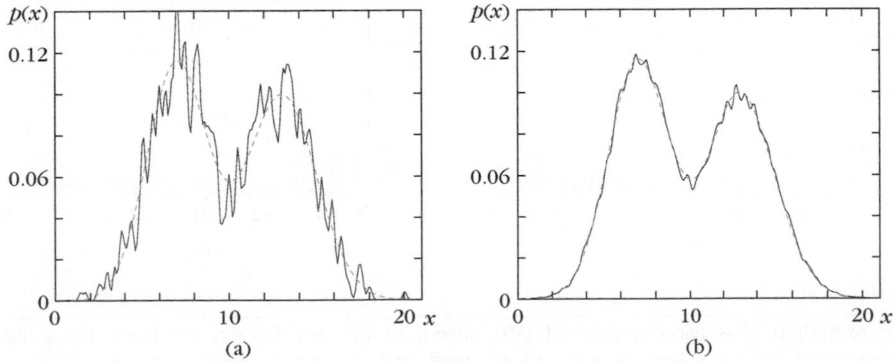

FIGURE 2.20

Approximation (full-black line) of a pdf (dotted-red line) via Parzen windows, using Gaussian kernels with (a) $h = 0.1$ and 1,000 training samples and (b) $h = 0.1$ and 20,000 samples. Observe the influence of the number of samples on the smoothness of the resulting estimate.

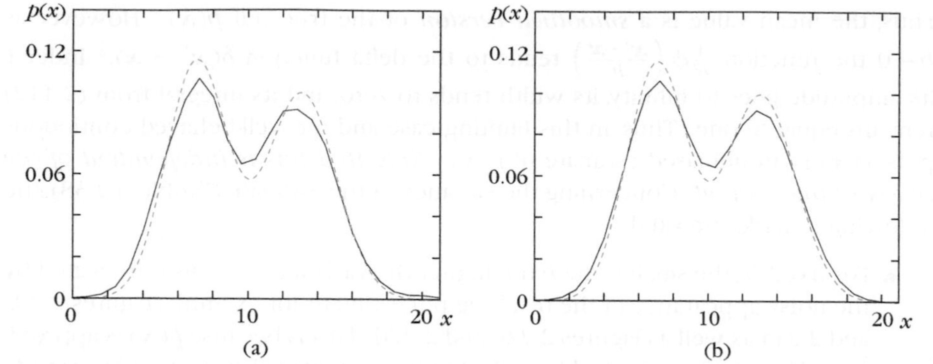

FIGURE 2.21

Approximation (full-black line) of a pdf (dotted-red line) via Parzen windows, using Gaussian kernels with (a) $h = 0.8$ and 1,000 training samples and (b) $h = 0.8$ and 20,000 samples. Observe that, in this case, increasing the number of samples has little influence on the smoothness as well as the approximation accuracy of the resulting estimate.

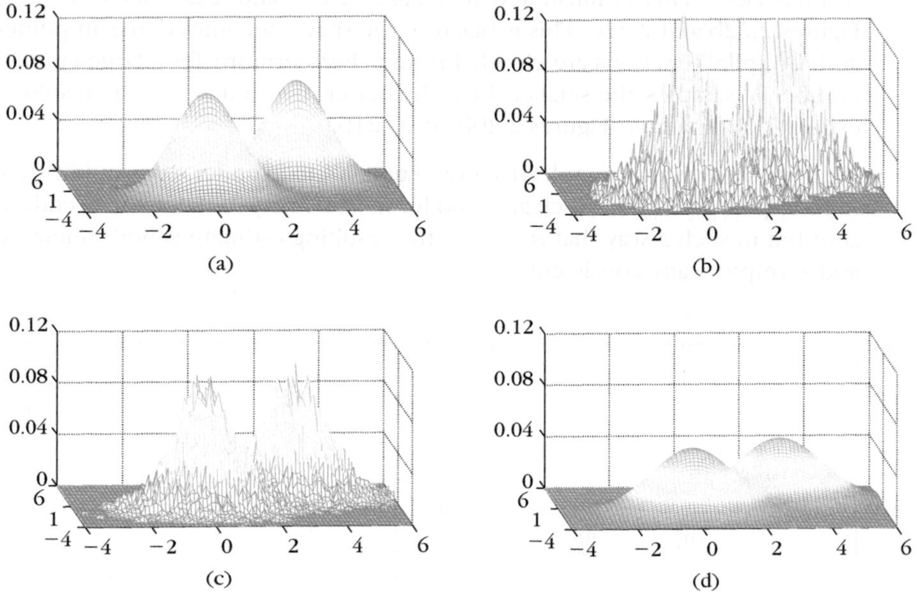

FIGURE 2.22

Approximation of a two-dimensional pdf, shown in (a), via Parzen windows, using two-dimensional Gaussian kernels with (b) $h = 0.05$ and $N = 1000$ samples, (c) $h = 0.05$ and $N = 20000$ samples and (d) $h = 0.8$ and $N = 20000$ samples. Large values of h lead to smooth estimates, but the approximation accuracy is low (the estimate is highly biased), as one can observe by comparing (a) with (d). For small values of h, the estimate is more noisy in appearance, but it becomes smoother as the number of samples increases, (b) and (c). The smaller the h and the larger the N, the better the approximation accuracy.

Remarks

- In practice, where only a finite number of samples is possible, a compromise between h and N must be made. The choice of suitable values for h is crucial, and several approaches have been proposed in the literature, for example, [Wand 95]. A straightforward way is to start with an initial estimate of h and then modify it iteratively to minimize the resulting misclassification error. The latter can be estimated by appropriate manipulation of the training set. For example, the set can be split into two subsets, one for training and one for testing. We will say more on this in Chapter 10.

- Usually, a large N is necessary for acceptable performance. This number grows exponentially with the dimensionality l. If a one-dimensional interval needs, say, N equidistant points to be considered as a densely populated one, the corresponding two-dimensional square will need N^2, the three-dimensional cube N^3, and so on. We usually refer to this as the *curse of dimensionality*. To our knowledge, this term was first used by Bellman in the context of Control theory [Bell 61]. To get a better feeling about the curse of dimensionality problem, let us consider the l-dimensional unit hypercube and let us fill it randomly with N points drawn from a uniform distribution. It can be shown ([Frie 89]) that the average Euclidean distance between a point and its nearest neighbor is given by

$$d(l,N) = 2\left(\frac{l\Gamma(l/2)}{2\pi^{\frac{l}{2}}N}\right)^{\frac{1}{l}}$$

where $\Gamma(\cdot)$ is the gamma function (Appendix A). In words, the average distance to locate the nearest neighbor to a point, for fixed l, shrinks as $N^{-\frac{1}{l}}$. To get a more quantitative feeling, let us fix N to the value $N = 10^{10}$. Then for $l = 2$, 10, 20 and 40, $d(l,N)$ becomes 10^{-5}, 0.18, 0.76, and 1.83, respectively. Figure 2.23a shows 50 points lying within the unit-length segment in the one-dimensional space. The points were randomly generated by the uniform distribution. Figure 2.23b shows the same number of points lying in the unit-length square. These points were also generated by a uniform distribution in the two-dimensional space. It is readily seen that the points in the one-dimensional segment are, on average, more closely located compared to the same number of points in the two-dimensional square.

The large number of data points required for a relatively high-dimensional feature space to be sufficiently covered puts a significant burden on complexity requirements, since one has to consider one Gaussian centered at each point. To this end, some techniques have been suggested that attempt to approximate the unknown pdf by using a reduced number of kernels, see, for example, [Babi 96].

Another difficulty associated with high-dimensional spaces is that, in practice, due to the lack of enough training data points, some regions in the feature space may be sparsely represented in the data set. To cope with

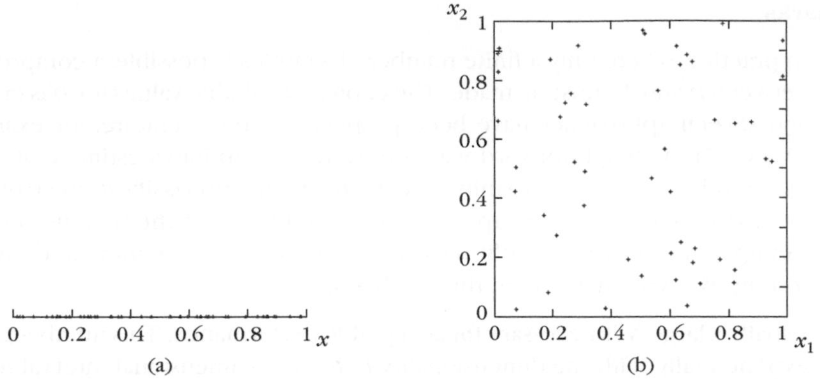

FIGURE 2.23

Fifty points generated by a uniform distribution lying in the (a) one-dimensional unit-length segment and (b) the unit-length square. In the two-dimensional space the points are more spread compared to the same number of points in the one-dimensional space.

such scenarios, some authors have adopted a variable value for h. In regions where data are sparse, a large value of h is used, while in more densely populated areas a smaller value is employed. To this end, a number of mechanisms for adjusting the value of h have been adopted, see, for example, [Brei 77, Krzy 83, Terr 92, Jone 96].

Application to classification: On the reception of a feature vector $\boldsymbol{x}$ the likelihood test in (2.20) becomes

$$\text{assign } \boldsymbol{x} \text{ to } \omega_1(\omega_2) \quad \text{if} \quad l_{12} \approx \left(\frac{\frac{1}{N_1 h^l} \sum_{i=1}^{N_1} \phi\left(\frac{\boldsymbol{x}_i - \boldsymbol{x}}{h}\right)}{\frac{1}{N_2 h^l} \sum_{i=1}^{N_2} \phi\left(\frac{\boldsymbol{x}_i - \boldsymbol{x}}{h}\right)} \right) > (<) \frac{P(\omega_2)}{P(\omega_1)} \frac{\lambda_{21} - \lambda_{22}}{\lambda_{12} - \lambda_{11}} \tag{2.114}$$

where N_1, N_2 are the training vectors in class ω_1, ω_2, respectively. The risk-related terms are ignored when the Bayesian minimum error probability classifier is used. For large N_1, N_2 this computation is a very demanding job, in both processing time and memory requirements.

k Nearest Neighbor Density Estimation

In the Parzen estimation of the pdf in (2.110), the volume around the points $\boldsymbol{x}$ was considered fixed (h^l) and the number of points k_N, falling inside the volume, was left to vary randomly from point to point. Here we will reverse the roles. The number of points $k_N = k$ will be fixed, and the size of the volume around $\boldsymbol{x}$ will be adjusted each time, to include k points. Thus, *in low-density areas the volume will be large and in high-density areas it will be small.* We can also consider more general types of regions, besides the hypercube. The estimator can now be written as

$$\hat{p}(\boldsymbol{x}) = \frac{k}{NV(\boldsymbol{x})} \tag{2.115}$$

where the dependence of the volume $V(\boldsymbol{x})$ on $\boldsymbol{x}$ is explicitly shown. Again it can be shown [Fuku 90] that asymptotically ($\lim k = +\infty$, $\lim N = +\infty$, $\lim(k/N) = 0$) this is an unbiased and consistent estimate of the true pdf, and it is known as the *k Nearest Neighbor (kNN) density estimate*. Results concerning the finite k, N case have also been derived; see [Fuku 90, Butu 93]. A selection of seminal papers concerning NN classification techniques can be found in [Dasa 91].

From a practical point of view, at the reception of an unknown feature vector $\boldsymbol{x}$, we compute its distance d, for example, Euclidean, from *all* the training vectors of the various classes, for example, ω_1, ω_2. Let r_1 be the radius of the hypersphere, centered at $\boldsymbol{x}$, that contains k points from ω_1 and r_2 the corresponding radius of the hypersphere containing k points from class ω_2 (k may not necessarily be the same for all classes). If we denote by V_1, V_2 the respective hypersphere volumes, the likelihood ratio test becomes

$$\text{assign } \boldsymbol{x} \text{ to } \omega_1(\omega_2) \quad \text{if } l_{12} \approx \frac{kN_2V_2}{kN_1V_1} > (<) \frac{P(\omega_2)}{P(\omega_1)} \frac{\lambda_{21} - \lambda_{22}}{\lambda_{12} - \lambda_{11}}$$

$$\frac{V_2}{V_1} > (<) \frac{N_1}{N_2} \frac{P(\omega_2)}{P(\omega_1)} \frac{\lambda_{21} - \lambda_{22}}{\lambda_{12} - \lambda_{11}} \tag{2.116}$$

If the Mahalanobis distance is alternatively adopted, we will have hyperellipsoids in the place of the hyperspheres.

The volume of a hyperellipsoid, corresponding to Mahalanobis distance equal to r, is given by ([Fuku 90])

$$V = V_0 |\Sigma|^{\frac{1}{2}} r^l \tag{2.117}$$

where V_0 is the volume of the hypersphere of unit radius given by

$$V_0 = \begin{cases} \pi^{\frac{l}{2}}/(l/2)!, & l \text{ even} \\ 2^l \pi^{\frac{l-1}{2}} (\frac{l-1}{2})!/l!, & l \text{ odd} \end{cases} \tag{2.118}$$

Verify that Eq. (2.117) results to $4\pi r^3/3$ for the volume of a sphere of radius r in the three-dimensional space.

Remark

- The nonparametric probability density function estimation techniques, discussed in this section, are among the techniques that are still in use in practical applications. It is interesting to note that, although the performance of the methods, as density estimators, degrades in high-dimensional spaces due to the lack of sufficient data, their performance as classifiers may be sufficiently good. After all, lack of enough training data points affects, in one way or another, all the methods.

More recently, the so-called ***probabilistic neural networks*** have been suggested as efficient implementations for the computation of the classifier given in (2.114), by exploiting the intrinsic parallelism of the neural network architectures and will be discussed in Chapter 4.

Example 2.9

The points shown in Figure 2.24 belong to either of two equiprobable classes. Black points belong to class ω_1 and red points belong to class ω_2. For the needs of the example we assume that all points are located at the nodes of a grid. We are given the point denoted by a "star", with coordinates $(0.7, 0.6)$, which is to be classified in one of the two classes. The Bayesian (minimum error probability) classifier and the k-nearest neighbor density estimation technique, for $k = 5$, will be employed.

Adopting the Euclidean distance, we find the five nearest neighbors to the unknown point $(0.7, 0.6)$ from all the points in class ω_2. These are the points $(0.8, 0.6)$, $(0.7, 0.7)$, $(0.6, 0.5)$, $(0.6, 0.6)$, $(0.6, 0.7)$. The full line circle encircles the five nearest neighbors, and its radius is equal to the distance of the point that is furthest from $(0.7, 0.6)$, that is, $\rho = \sqrt{0.1^2 + 0.1^2} = 0.1\sqrt{2}$. In the sequel, we repeat the procedure for the points in class ω_1. The nearest points are the ones with coordinates $(0.7, 0.5)$, $(0.8, 0.4)$, $(0.8, 0.7)$,

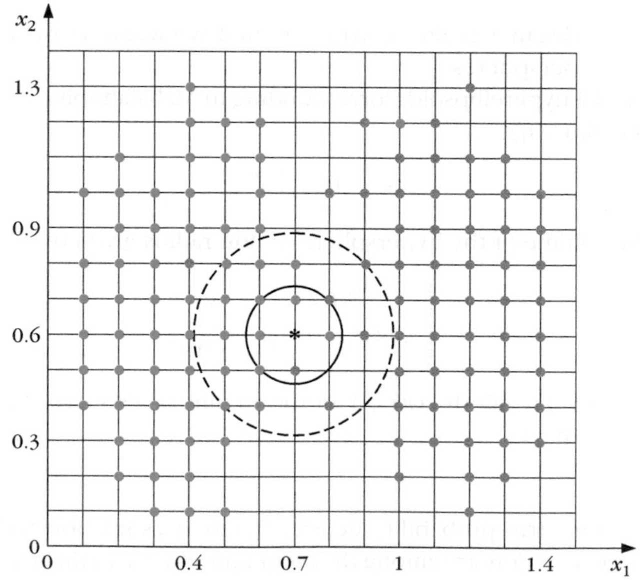

FIGURE 2.24

The setup for the example 2.9. The point denoted by a "star" is classified to the class ω_2 of the red points. The $k = 5$ nearest neighbors from this class lie within a smaller area compared to the five nearest neighbors coming from the other class.

$(0.9, 0.6)$, $(0.9, 0.8)$. The dotted circle is the one that encircles all five points, and its radius is equal to $\sqrt{0.2^2 + 0.2^2} = 0.2\sqrt{2} = 2\rho$.

There are $N_1 = 59$ points in class ω_1 and $N_2 = 61$ in class ω_2. The areas (volumes) of the two circles are $V_1 = 4\pi\rho^2$ and $V_2 = \pi\rho^2$, respectively, for the two classes. Hence, according to Eq. (2.116) and ignoring the risk related terms, we have

$$\frac{V_2}{V_1} = \frac{\pi\rho^2}{4\pi\rho^2} = 0.25$$

and since 0.25 is less than $59/61$ and the classes are equiprobable, the point $(0.7, 0.6)$ is classified to class ω_2.

2.5.7 The Naive-Bayes Classifier

The goal in this section, so far, was to present various techniques for the estimation of the probability density functions $p(\boldsymbol{x}|\omega_i)$, $i = 1, 2, \ldots, M$, required by the Bayes classification rule, based on the available training set, X. As we have already stated, in order to safeguard good estimates of the pdfs the number of training samples, N, must be large enough. To this end, the demand for data increases exponentially fast with the dimension, l, of the feature space. Crudely speaking, if N could be regarded as a good number of training data points for obtaining sufficiently accurate estimates of a pdf in an one-dimensional space, then N^l points would be required for an l-dimensional space. Thus, large values of l make the accurate estimation of a multidimensional pdf a bit of an "illusion" since in practice data is hard to obtain. Loosely speaking, data can be considered to be something like money. It is never enough! Accepting this reality, one has to make concessions about the degree of accuracy that is expected from the pdf estimates. One widely used approach is to assume that individual features x_j, $j = 1, 2, \ldots, l$, are statistically independent. Under this assumption, we can write

$$p(\boldsymbol{x}|\omega_i) = \prod_{j=1}^{l} p(x_j|\omega_i), \quad i = 1, 2, \ldots, M$$

The scenario is now different. To estimate l one-dimensional pdfs, for each of the classes, lN data points would be enough in order to obtain good estimates, instead of N^l. This leads to the so-called *naive-Bayes* classifier, which assigns an unknown sample $\boldsymbol{x} = [x_1, x_2, \ldots, x_l]^T$ to the class

$$\omega_m = \arg\max_{\omega_i} \prod_{j=1}^{l} p(x_j|\omega_i), \quad i = 1, 2, \ldots, M$$

It turns out that the naive-Bayes classifier can be very robust to violations of its independence assumption, and it has been reported to perform well for many real-world data sets. See, for example, [Domi 97].

Example 2.10

The discrete features case: In Section 2.2, it was stated that in the case of discrete-valued features the only required change in the Bayesian classification rule is to replace probability density functions with probabilities. In this example, we will see how the associated with the naive Bayes classifier assumption of statistical independence among the features simplifies the Bayesian classification rule.

Consider the feature vector $\boldsymbol{x} = [x_1, x_2, \ldots, x_l]^T$ with binary features, that is, $x_i \in \{0, 1\}$, $i = 1, 2, \ldots, l$. Also let the respective class-conditional probabilities be $P(x_i = 1|\omega_1) = p_i$ and $P(x_i = 1|\omega_2) = q_i$. According to the Bayesian rule, given the value of $\boldsymbol{x}$, its class is decided according to the value of the likelihood ratio

$$\frac{P(\omega_1)P(\boldsymbol{x}|\omega_1)}{P(\omega_2)P(\boldsymbol{x}|\omega_2)} > (<) 1 \tag{2.119}$$

for the minimum probability error rule (the minimum risk rule could also be used).

The number of values that $\boldsymbol{x}$ can take, for all possible combinations of x_i, amounts to 2^l. If we do not adopt the independence assumption, then one must have enough training data in order to obtain probability estimates for each one of these values (probabilities add to one, thus $2^l - 1$ estimates are required). However, adopting statistical independence among the features, we can write

$$P(\boldsymbol{x}|\omega_1) = \prod_{i=1}^{l} p_i^{x_i}(1 - p_i)^{1-x_i}$$

and

$$P(\boldsymbol{x}|\omega_2) = \prod_{i=1}^{l} q_i^{x_i}(1 - q_i)^{1-x_i}$$

Hence, the number of required probability estimates is now $2l$, that is, the p_i's and q_i's. It is interesting to note that, taking the logarithm of both sides in (2.119), one ends up with a *linear* discriminant function similar to the hyperplane classifier of Section 2.4, that is,

$$g(\boldsymbol{x}) = \sum_{i=1}^{l} \left(x_i \ln \frac{p_i}{q_i} + (1 - x_i) \ln \frac{1 - p_i}{1 - q_i} \right) + \ln \frac{P(\omega_1)}{P(\omega_2)} \tag{2.120}$$

which can easily be brought into the form of

$$g(\boldsymbol{x}) = \boldsymbol{w}^T \boldsymbol{x} + w_0 \tag{2.121}$$

where

$$\boldsymbol{w} = \left[\ln \frac{p_1(1 - q_1)}{q_1(1 - p_1)}, \ldots, \ln \frac{p_l(1 - q_l)}{q_l(1 - p_l)} \right]^T$$

and

$$w_0 = \sum_{i=1}^{l} \ln \frac{1 - p_i}{1 - q_i} + \ln \frac{P(\omega_1)}{P(\omega_2)}$$

Binary features are used in a number of applications where one has to decide based on the presence or not of certain attributes. For example, in medical diagnosis, 1 can represent a normal value in a medical test and a 0 an abnormal one.

2.6 THE NEAREST NEIGHBOR RULE

A variation of the *k*NN density estimation technique results in a *suboptimal*, yet popular in practice, nonlinear classifier. Although this does not fall in the Bayesian framework, it fits nicely at this point. In a way, this section could be considered as a bridge with Chapter 4. The algorithm for the so-called *nearest neighbor rule* is summarized as follows. Given an unknown feature vector $\boldsymbol{x}$ and a distance measure, then:

- Out of the N training vectors, identify the k nearest neighbors, *regardless* of class label. k is chosen to be odd for a two class problem, and in general not to be a multiple of the number of classes M.
- Out of these k samples, identify the number of vectors, k_i, that belong to class $\omega_i, i = 1, 2, \ldots, M$. Obviously, $\sum_i k_i = k$.
- Assign $\boldsymbol{x}$ to the class ω_i with the maximum number k_i of samples.

Figure 2.25 illustrates the k-NN rule for the case of $k = 11$. Various distance measures can be used, including the Euclidean and Mahalanobis distance.

The simplest version of the algorithm is for $k = 1$, known as the *nearest neighbor (NN)* rule. In other words, a feature vector $\boldsymbol{x}$ is assigned to the class of its nearest neighbor! Provided that the number of training samples is large enough, this simple

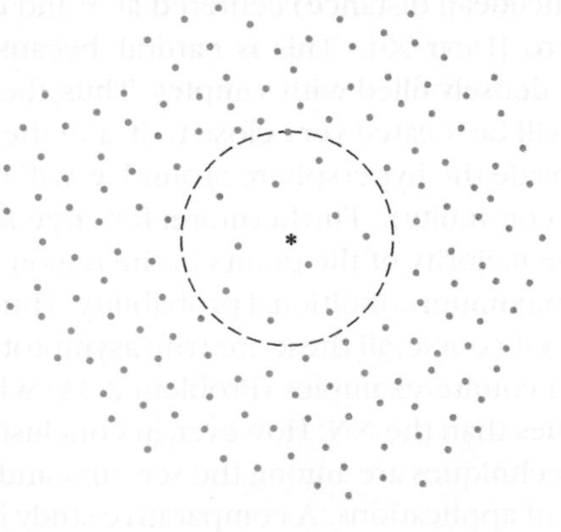

FIGURE 2.25

Using the 11-NN rule, the point denoted by a "star" is classified to the class of the red points. Out of the eleven nearest neighbors seven are red and four are black. The circle indicates the area within which the eleven nearest neighbors lie.

rule exhibits good performance. This is also substantiated by theoretical findings. It can be shown [Duda 73, Devr 96] that, as $N \rightarrow \infty$, the classification error probability, for the NN rule, P_{NN}, is bounded by

$$P_B \leq P_{NN} \leq P_B\left(2 - \frac{M}{M-1}P_B\right) \leq 2P_B \tag{2.122}$$

where P_B is the optimal Bayesian error. Thus, the error committed by the NN classifier is (*asymptotically*) at most twice that of the optimal classifier. The asymptotic performance of the kNN is better than that of the NN, and a number of interesting bounds have been derived. For example, for the two-class case it can be shown, for example, [Devr 96] that

$$P_B \leq P_{kNN} \leq P_B + \frac{1}{\sqrt{ke}} \quad \text{or} \quad P_B \leq P_{kNN} \leq P_B + \sqrt{\frac{2P_{NN}}{k}} \tag{2.123}$$

Both of these suggest that as $k \rightarrow \infty$ the performance of the kNN tends to the optimal one. Furthermore, for small values of Bayesian errors, the following approximations are valid [Devr 96]:

$$P_{NN} \approx 2P_B \tag{2.124}$$

$$P_{3NN} \approx P_B + 3(P_B)^2 \tag{2.125}$$

Thus, for large N and small Bayesian errors, we expect the 3NN classifier to give performance almost identical to that of the Bayesian classifier. As an example, let us say that the error probability of the Bayesian classifier is of the order of 1%; then the error resulting from a 3NN classifier will be of the order of 1.03%! The approximation improves for higher values of k. A little thought can provide justification for this without too much mathematics. Under the assumption of large N, the radius of the hypersphere (Euclidean distance) centered at $\boldsymbol{x}$ and containing its k nearest neighbors tends to zero [Devr 96]. This is natural, because for very large N we expect the space to be densely filled with samples. Thus, the k (a very small portion of N) neighbors of $\boldsymbol{x}$ will be located very close to it, and the conditional class probabilities, at all points inside the hypersphere around $\boldsymbol{x}$, will be approximately equal to $P(\omega_i|\boldsymbol{x})$ (assuming continuity). Furthermore, for large k (yet an infinitesimally small fraction of N), the majority of the points in the region will belong to the class corresponding to the maximum conditional probability. Thus, the kNN rule tends to the Bayesian classifier. Of course, all these are true asymptotically. In the finite sample case there are even counterexamples (Problem 2.34) where the kNN results in higher error probabilities than the NN. However, in conclusion, it can be stated that the nearest neighbor techniques are among the serious candidates to be adopted as classifiers in a number of applications. A comparative study of the various statistical classifiers, considered in this chapter as well as others, can be found in [Aebe 94].

Remarks

- A serious drawback associated with (k)NN techniques is the complexity in search of the nearest neighbor(s) among the N available training samples.

Brute-force searching amounts to operations proportional to kN ($O(kN)$).[2] The problem becomes particularly severe in high-dimensional feature spaces. To reduce the computational burden, a number of efficient searching schemes have been suggested; see, for example, [Fuku 75, Dasa 91, Brod 90, Djou 97, Nene 97, Hatt 00, Kris 00, Same 08]. In [Vida 94, Mico 94] a preprocessing stage is suggested that computes a number of *base prototypes* that are in some sense maximally separated from among the set of training feature vectors. A summary of efficient searching techniques and a comparative study is given in [McNa 01].

- Although, due to its asymptotic error performance, the kNN rule achieves good results when the data set is large (compared to the dimension of the feature space), the performance of the classifier may degrade dramatically when the value of N is relatively small [Devr 96]. Also, in practice, one may have to reduce the number of training patterns due to the constraints imposed by limited computer resources. To this end, a number of techniques, also known as *prototype editing* or *condensing*, have been proposed. The idea is to reduce the number of training points in a way that a cost related to the error performance is optimized; see, for example, [Yan 93, Huan 02, Pare 06a] and the references therein. Besides computational savings, reducing the size of a finite set appropriately may offer performance improvement advantages, by making the classifier less sensitive to outliers. A simple method, which also makes transparent the reason for such a potential improvement, has been suggested in [Wils 72]. This editing procedure tests a sample using a kNN rule against the rest of the data. The sample is discarded if it is misclassified. The edited data set is then used for a NN classification of unknown samples.

  A direction to cope with the performance degradation associated with small values of N is to employ distance measures that are optimized on the available training set. The goal is to find a data-adaptive distance metric that leads to an optimal performance, according to an adopted cost. Such *trained* metrics can be *global* ones (i.e., the same at every point), *class-dependent* (i.e., shared by all points of the same class), and/or *locally dependent* (i.e., the metric varies according to the position in the feature space); see, for example, [Hast 96, Dome 05, Pare 06] and the references therein. An in depth treatment of the topic is given in [Frie 94].

- When the $k = 1$ nearest neighbor rule is used, the training feature vectors $\boldsymbol{x}_i, i=1, 2, \ldots, N$, define a partition of the l-dimensional space into N regions, R_i. Each of these regions is defined by

$$R_i = \{\boldsymbol{x}: d(\boldsymbol{x}, \boldsymbol{x}_i) < d(\boldsymbol{x}, \boldsymbol{x}_j), i \neq j\} \tag{2.126}$$

  that is, R_i contains all points in space that are closer to $\boldsymbol{x}_i$ than any other point of the training set, with respect to the distance d. This partition of the

[2] $O(n)$ denotes order of n calculations.

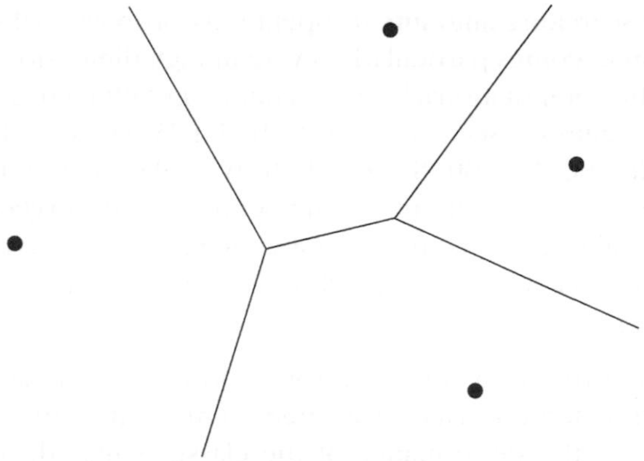

FIGURE 2.26

An example of Voronoi tessellation in the two-dimensional space and for Euclidean distance.

feature space is known as *Voronoi tessellation*. Figure 2.26 is an example of the resulting *Voronoi tessellation* for the case of $l = 2$ and the Euclidean distance.

2.7 BAYESIAN NETWORKS

In Section 2.5.7 the naive-Bayes classifier was introduced as a means of coping with the curse of dimensionality and to exploiting more efficiently the available training data set. However, by adopting the naive-Bayes classifier, one goes from one extreme (fully dependent features) to another (features mutually independent). Common sense drives us to search for approximations that lie between these two extremes.

The essence of the current section is to introduce a methodology that allows one to develop models that can accommodate built-in independence assumptions with respect to the features x_i, $i = 1, 2, \ldots, l$. Recall the well-known probability chain rule [Papo 91, p. 192]

$$p(x_1, x_2, \ldots, x_l) = p(x_l|x_{l-1}, \ldots, x_1)p(x_{l-1}|x_{l-2} \ldots, x_1) \ldots, p(x_2|x_1)p(x_1) \qquad (2.127)$$

This rule applies always and does not depend on the order in which features are presented. The rule states that the joint probability density function can be expressed in terms of the product of conditional pdfs and a marginal one ($p(x_1)$).[3] This important and elegant rule opens the gate through which assumptions will infiltrate the problem. The conditional dependence for each feature, x_i, will be limited into a subset of the features appearing in each term in the product. Under this assumption,

[3] In the study of several random variables, the statistics of each are called marginal.

Eq. (2.127) can now be written as

$$p(x) = p(x_1)\prod_{i=2}^{l} p(x_i|A_i) \tag{2.128}$$

where

$$A_i \subseteq \{x_{i-1}, x_{i-2}, \ldots, x_1\} \tag{2.129}$$

For example, let $l = 6$ and

$$p(x_6|x_5,\ldots,x_1) = p(x_6|x_5,x_4) \tag{2.130}$$

$$p(x_5|x_4,\ldots,x_1) = p(x_5|x_4) \tag{2.131}$$

$$p(x_4|x_3,x_2,x_1) = p(x_4|x_2,x_1) \tag{2.132}$$

$$p(x_3|x_2,x_1) = p(x_3|x_2) \tag{2.133}$$

$$p(x_2|x_1) = p(x_2) \tag{2.134}$$

Then,

$$A_6 = \{x_5, x_4\},\ A_5 = \{x_4\},\ A_4 = \{x_2, x_1\},\ A_3 = \{x_2\}, A_2 = \emptyset$$

where $\emptyset$ denotes the empty set. These assumptions are represented graphically in Figure 2.27. Nodes correspond to features. The *parents* of a feature, x_i, are those features with directed links toward x_i and are the members of the set A_i. In other words, x_i *is conditionally independent* of any combination of its *nondescendants*, *given its parents*. There is a subtle point concerning conditional independence. Take, for example, that $p(x_3|x_2,x_1) = p(x_3|x_2)$. This does not necessarily mean that x_3 and x_1 are independent. They may be dependent while x_2 is unknown, but they become independent once the value of x_2 is disclosed to us. This is not surprising since by measuring the value of a random variable part of the randomness is removed.

Under the previous assumptions, the problem of estimating the joint pdf has broken into the product of simpler terms. Each of them involves, in general, a much smaller number of features compared to the original number. For example,

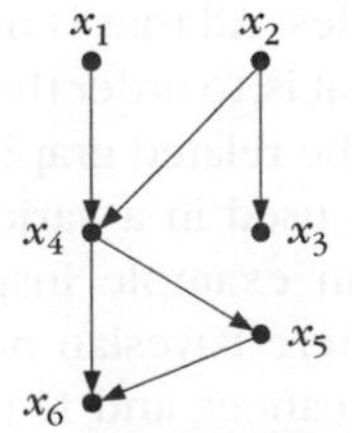

FIGURE 2.27

Graphical model illustrating conditional dependencies.

for the case of Eqs. (2.130)–(2.134) none of the products involves more than three features. Hence, the estimation of each pdf term in the product takes place in a low-dimensional space and the problems arising from the curse of dimensionality can be handled easier. To get a feeling for the computational size reduction implied by the independence assumptions, encoded in the graphical model of Figure 2.27, let us assume that variables x_i, $i = 1, 2, \ldots, 6$, are binary. Then the pdfs in (2.127)–(2.134) become probabilities. Complete knowledge of $P(x_1, \ldots, x_6)$ requires the estimation of 63 ($2^l - 1$) probability values. It is 63 and not 64 due to the constraint that probabilities must add to one. This is also suggested by the right-hand side of Eq. (2.127). The number of the required probability values is $2^{l-1} + 2^{l-2} + \cdots + 1 = 2^l - 1$. In contrast to that, the assumptions in (2.130)–(2.134) reduce the number of the required probability values to be estimated to 13 (Why?). For large values of l, such a saving can be very significant.

The naive-Bayes classifier is a special case for which $A_i = \emptyset$, $i = 2, \ldots, l$, and the product in (2.128) becomes a product of marginal pdfs. Examples of classifiers that exploit the idea of conditional independence with respect to a subset of features are given in, for example, [Frie 97, Webb 05, Roos 05].

Although our original goal was to seek for ways for the approximate estimation of a joint pdf, it turns out that the adopted assumptions (nicely condensed in a graphical representation such as in Figure 2.27), have much more interesting consequences. For the rest of the section and for the sake of simplicity, we will assume that the features can only take values from a discrete set. Thus, pdfs give their place to probabilities.

Definition: A ***Bayesian network*** is a directed acyclic graph (DAG) where the nodes correspond to random variables (features). Each node is associated with a set of conditional probabilities, $P(x_i|A_i)$, where x_i is the variable associated with the specific node and A_i is the set of its parents in the graph.

Acyclic means that there are no cycles in the graph. For example, the graph in Figure 2.27 is an acyclic one, and it will cease to be so if one draws an arc directed from x_6 to, say, x_1. The complete specification of a Bayesian network requires knowledge of (a) the marginal probabilities of the root nodes (those without a parent) and (b) the conditional probabilities of the nonroot nodes, given their parents for *all* possible combinations of their values. The joint probability of the variables can now be obtained by multiplying all conditional probabilities with the prior probabilities of the root nodes. All that is needed is to perform a *topological sorting* of the random variables; that is, to order the variables such that every variable comes before its descendants in the related graph.

Bayesian networks have been used in a variety of applications. The network in Figure 2.28 corresponds to an example inspired by the discipline of medical diagnosis, a scientific area where Bayesian networks have been very popular. S stands for smokers, C for lung cancer, and H for heart disease. H1 and H2 are heart disease medical tests, and C1 and C2 are cancer medical tests. The table of the root node shows the population percentage (probability) of smokers (True) and

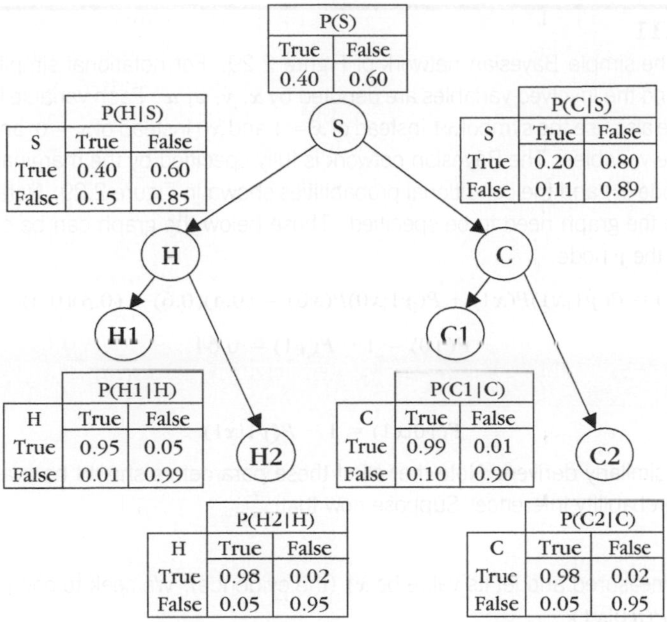

FIGURE 2.28

Bayesian network modeling conditional dependencies for an example concerning smokers (S), tendencies to develop cancer (C), and heart disease (H), together with variables corresponding to heart (H1, H2) and cancer (C1, C2) medical tests.

nonsmokers (False). The tables along the nodes of the tree are the respective conditional probabilities. For example, P(C : True|S : True) = 0.20 is the probability of a smoker (True) to develop cancer (True). (The probabilities used in Figure 2.28 may not correspond to true values having resulted from statistical studies.)

Once a DAG has been constructed, the Bayesian network allows one to calculate *efficiently* the conditional probability of *any* node in the graph, given that the values of some other nodes have been *observed*. Such questions arise in the field of artificial intelligence closely related to pattern recognition. The computational efficiency stems from the existing probability relations encoded in the graph. A detailed treatment of the topic is beyond the scope of this book; the interested reader may consult more specialized texts, such as [Neap 04]. The remainder of this section aims at providing the reader with a flavor of the related theory.

Probability Inference: This is the most common task that Bayesian networks help us to solve efficiently. Given the values of some of the variables, known as *evidence*, the goal is to compute the conditional probabilities for some (or all) of the other variables in the graph, *given* the evidence.

Example 2.11

Let us take the simple Bayesian network of Figure 2.29. For notational simplicity we avoid subscripts, and the involved variables are denoted by x, y, z, w. Each variable is assumed to be binary. We also use the symbol $x1$ instead of $x = 1$ and $x0$ instead of $x = 0$, and similarly for the rest of the variables. The Bayesian network is fully specified by the marginal probabilities of the root node (x) and the conditional probabilities shown in Figure 2.29. Note that only the values above the graph need to be specified. Those below the graph can be derived. Take, for example, the y node.

$$P(y1) = P(y1|x1)P(x1) + P(y1|x0)P(x0) = (0.4)(0.6) + (0.3)(0.4) = 0.36$$

$$P(y0) = 1 - P(y1) = 0.64$$

Also,

$$P(y0|x1) = 1 - P(y1|x1)$$

The rest are similarly derived. Note that all of these parameters should be available prior to performing probability inference. Suppose now that:

(a) x is measured and let its value be $x1$ (the evidence). We seek to compute $P(z1|x1)$ and $P(w0|x1)$.

(b) w is measured and let its value be $w1$. We seek to compute $P(x0|w1)$ and $P(z1|w1)$.

To answer (a), the following calculations are in order.

$$\begin{aligned} P(z1|x1) &= P(z1|y1, x1)P(y1|x1) + P(z1|y0, x1)P(y0|x1) \\ &= P(z1|y1)P(y1|x1) + P(z1|y0)P(y0|x1) \\ &= (0.25)(0.4) + (0.6)(0.6) = 0.46 \end{aligned} \quad (2.135)$$

Though not explicitly required, $P(z0|x1)$ must also be evaluated, as we will soon realize.

$$P(z0|x1) = 1 - P(z1|x1) = 0.54 \quad (2.136)$$

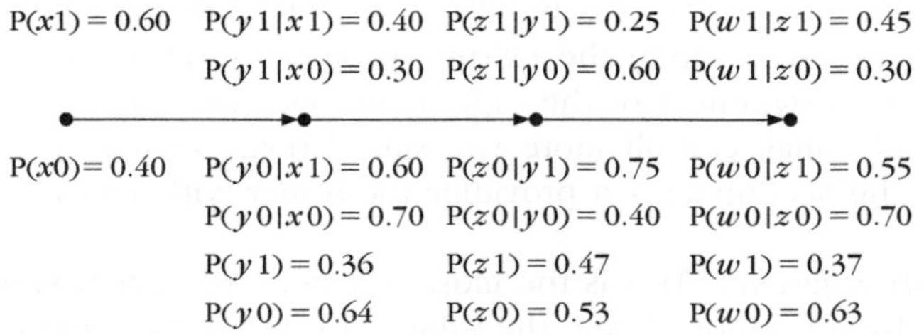

FIGURE 2.29

A simple Bayesian network where conditional dependencies are restricted to a single variable.

In a similar way, we obtain

$$\begin{aligned} P(w0|x1) &= P(w0|z1, x1)P(z1|x1) + P(w0|z0, x1)P(z0|x1) \\ &= P(w0|z1)P(z1|x1) + P(w0|z0)P(z0|x1) \\ &= (0.55)(0.46) + (0.7)(0.54) = 0.63 \end{aligned} \tag{2.137}$$

We can think of the algorithm as a process that *passes messages* (i.e., probabilities) downward from one node to the next. The first two computations, (2.135) and (2.136), "are performed in node z" and then "passed" to the last node, where (2.137) is performed.

To answer (b), the direction of "message propagation" is reversed since, now, the evidence is provided from node w and the required information, $P(x0|w1)$, $P(z1|w1)$ concerns nodes x and z, respectively.

$$P(z1|w1) = \frac{P(w1|z1)P(z1)}{P(w1)} = \frac{(0.45)(0.47)}{0.37} = 0.57$$

The activity is then passed to node y, where the following needs to be performed.

$$P(y1|w1) = \frac{P(w1|y1)P(y1)}{P(w1)}$$

$P(w1|y1)$ is unknown and can be computed as discussed in the "downward" message propagation. That is,

$$\begin{aligned} P(w1|y1) &= P(w1|z1, y1)P(z1|y1) + P(w1|z0, y1)P(z0|y1) \\ &= P(w1|z1)P(z1|y1) + P(w1|z0)P(z0|y1) \\ &= (0.45)(0.25) + (0.3)(0.75) = 0.34 \end{aligned}$$

In a similar way, $P(w1|y0) = 0.39$ is obtained. These values are then "passed" over to node x, and it is left as an exercise to show that $P(x0|w1) = 0.4$.

This idea can be carried out to any net of any size of the form given in Figure 2.29.

For Bayesian networks that have a tree structure, probability inference is achieved via a combination of downward and upward computations propagated through the tree. A number of algorithms have been proposed for the general case of Bayesian networks based on this "message-passing" philosophy. See, for example, [Pear 88, Laur 96]. For the case of singly connected graphs, these algorithms have complexity that is linear in the number of nodes. A singly connected graph is one that has no more than one path between any two nodes. For example, the graph in Figure 2.27 is not singly connected since there are two paths connecting x_1 and x_6. An alternative approach to derive efficient algorithms for probability inference, which exploits the structure of the DAG, has been taken in [Li 94]. Although it is beyond our scope to focus on algorithmic details, it is quite instructive to highlight the basic idea around which this type of algorithm evolves.

Let us take as an example the DAG shown in Figure 2.30, with nodes corresponding to the variables s, u, v, x, y, w, z with the joint probability $P(s, u, v, x, y, w, z)$. This can be obtained, as we have already stated, as the product of all conditional

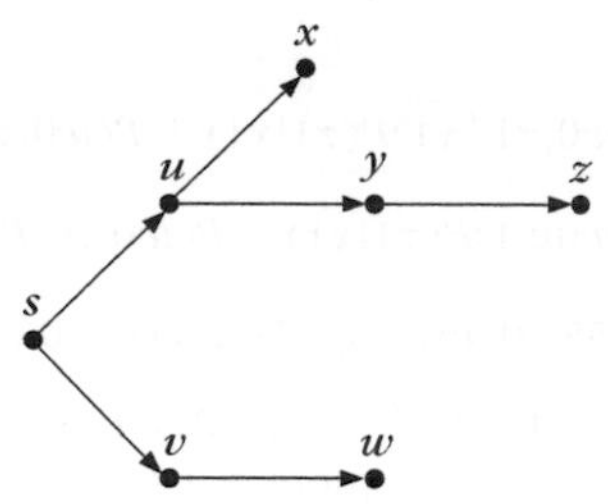

FIGURE 2.30

A Bayesian network with a tree structure.

probabilities defining the network. Suppose one wishes to compute the conditional probability $P(s|z = z_0)$, where $z = z_0$ is the evidence. From the Bayes rule we have

$$P(s|z = z_0) = \frac{P(s, z = z_0)}{P(z = z_0)} = \frac{P(s, z = z_0)}{\sum_s P(s, z = z_0)} \tag{2.138}$$

To obtain $P(s, z = z_0)$, one has to marginalize (Appendix A) the joint probability over all possible values of u, v, x, y, w; that is,

$$P(s, z = z_0) = \sum_{u,v,x,y,w} P(s, u, v, x, y, w, z = z_0) \tag{2.139}$$

Assuming, for simplicity, that each of the discrete variables can take, say, L values, the complexity of the previous computations amounts to L^5 operations. For more variables and a large number of values, L, this can be a prohibitively large number. Let us now exploit the structure of the Bayesian network in order to reduce this computational burden. Taking into account the relations implied by the topology of the graph shown in Figure 2.30 and the Bayes chain rule in (2.128) (for probabilities), we obtain

$$\sum_{u,v,x,y,w} P(s, u, v, x, y, w, z = z_0) =$$

$$\sum_{u,v,x,y,w} P(s)P(u|s)P(v|s)P(w|v)P(x|u)P(y|u)P(z = z_0|y) =$$

$$P(s) \underbrace{\sum_{u,v} P(u|s)P(v|s) \underbrace{\sum_{w} P(w|v)}_{v} \underbrace{\sum_{x} P(x|u)}_{u} \underbrace{\sum_{y} P(y|u)P(z = z_0|y)}_{u}}_{s} \tag{2.140}$$

or

$$\sum_{u,v,x,y,w} P(s, u, v, x, y, w, z = z_0) = P(s) \sum_{u,v} P(u|s)P(v|s)\phi_1(v)\phi_2(u)\phi_3(u) \tag{2.141}$$

where the definitions of $\phi_i(\cdot)$, $i = 1, 2, 3$, are readily understood by inspection. Underbraces indicate what variable the result of each summation depends on. To obtain $\phi_3(u)$ for each value of u, one needs to perform L operations (products and summations). Hence, a total number of L^2 operations is needed to compute $\phi_3(u)$ for all possible values of u. This is also true for the $\phi_2(u)$, $\phi_1(v)$. Thus, the total number of operations required to compute (2.141) is, after the factorization, of the order of L^2, instead of the order of L^5 demanded for the brute-force computation in (2.139). This procedure could be viewed as an effort to decompose a "global" sum into products of "local" sums to make computations tractable. Each summation can be viewed as a processing stage that removes a variable and provides as output a function. The essence of the algorithm given in [Li 94] is to search for the factorization that requires the minimal number of operations. This algorithm also has linear complexity in the number of nodes for singly connected networks. In general, for multiply connected networks the probability inference problem is NP-hard [Coop 90]. In light of this result, one tries to seek approximate solutions, as in [Dagu 93].

Training: Training of a Bayesian network consists of two parts. The first is to learn the network topology. The topology can either be fixed by an expert who can provide knowledge about dependencies or by use of optimization techniques based on the training set. Once the topology has been fixed, the unknown parameters (i.e., conditional probabilities and marginal probabilities) are estimated from the available training data points. For example, the fraction (frequency) of the number of instances that an event occurs over the total number of trials performed is a way to approximate probabilities. In Bayesian networks, other refined techniques are usually encountered. A review of learning procedures can be found in [Heck 95]. For the reader who wishes to delve further into the exciting world of Bayesian networks, the books of [Pear 88, Neap 04, Jens 01] will prove indispensable tools.

2.8 PROBLEMS

2.1 Show that in a multiclass classification task, the Bayes decision rule minimizes the error probability.

Hint: It is easier to work with the probability of correct decision.

2.2 In a two-class one-dimensional problem, the pdfs are the Gaussians $\mathcal{N}(0, \sigma^2)$ and $\mathcal{N}(1, \sigma^2)$ for the two classes, respectively. Show that the threshold x_0 minimizing the average risk is equal to

$$x_0 = 1/2 - \sigma^2 \ln \frac{\lambda_{21}P(\omega_2)}{\lambda_{12}P(\omega_1)}$$

where $\lambda_{11} = \lambda_{22} = 0$ has been assumed.

2.3 Consider a two equiprobable class problem with a loss matrix L. Show that if ϵ_1 is the probability of error corresponding to feature vectors from class ω_1

and ϵ_2 for those from class ω_2, then the average risk r is given by

$$r = P(\omega_1)\lambda_{11} + P(\omega_2)\lambda_{22} + P(\omega_1)(\lambda_{12} - \lambda_{11})\epsilon_1 + P(\omega_2)(\lambda_{21} - \lambda_{22})\epsilon_2$$

2.4 Show that in a multiclass problem with M classes the probability of classification error for the optimum classifier is bounded by

$$P_e \leq \frac{M-1}{M}$$

Hint: Show first that for each $\boldsymbol{x}$ the maximum of $P(\omega_i|\boldsymbol{x}), i = 1, 2, \ldots, M$, is greater than or equal to $1/M$. Equality holds if all $P(\omega_i|\boldsymbol{x})$ are equal.

2.5 Consider a two (equiprobable) class, one-dimensional problem with samples distributed according to the Rayleigh pdf in each class, that is,

$$p(x|\omega_i) = \begin{cases} \frac{x}{\sigma_i^2} \exp\left(\frac{-x^2}{2\sigma_i^2}\right) & x \geq 0 \\ 0 & x < 0 \end{cases}$$

Compute the decision boundary point $g(x) = 0$.

2.6 In a two-class classification task, we constrain the error probability for one of the classes to be fixed, that is, $\epsilon_1 = \epsilon$. Then show that minimizing the error probability of the other class results in the likelihood test

$$\text{decide } \boldsymbol{x} \text{ in } \omega_1 \text{ if } \frac{P(\omega_1|\boldsymbol{x})}{P(\omega_2|\boldsymbol{x})} > \theta$$

where θ is chosen so that the constraint is fulfilled. This is known as the *Neyman-Pearson test*, and it is similar to the Bayesian minimum risk rule.

Hint: Use a Lagrange multiplier to show that this problem is equivalent to minimizing the quantity

$$q = \theta(\epsilon_1 - \epsilon) + \epsilon_2$$

2.7 In a three-class, two-dimensional problem the feature vectors in each class are normally distributed with covariance matrix

$$\Sigma = \begin{bmatrix} 1.2 & 0.4 \\ 0.4 & 1.8 \end{bmatrix}$$

The mean vectors for each class are $[0.1, 0.1]^T, [2.1, 1.9]^T, [-1.5, 2.0]^T$. Assuming that the classes are equiprobable, (a) classify the feature vector $[1.6, 1.5]^T$ according to the Bayes minimum error probability classifier; (b) draw the curves of equal Mahalanobis distance from $[2.1, 1.9]^T$.

2.8 In a two-class, three-dimensional classification problem, the feature vectors in each class are normally distributed with covariance matrix

$$\Sigma = \begin{bmatrix} 0.3 & 0.1 & 0.1 \\ 0.1 & 0.3 & -0.1 \\ 0.1 & -0.1 & 0.3 \end{bmatrix}$$

The respective mean vectors are $[0, 0, 0]^T$ and $[0.5, 0.5, 0.5]^T$. Derive the corresponding linear discriminant functions and the equation describing the decision surface.

2.9 In a two equiprobable class classification problem, the feature vectors in each class are normally distributed with covariance matrix Σ, and the corresponding mean vectors are $\boldsymbol{\mu}_1, \boldsymbol{\mu}_2$. Show that for the Bayesian minimum error classifier, the error probability is given by

$$P_B = \int_{(1/2)d_m}^{+\infty} \frac{1}{\sqrt{2\pi}} \exp(-z^2/2)\, dz$$

where d_m is the Mahalanobis distance between the mean vectors. Observe that this is a decreasing function of d_m.
Hint: Compute the log-likelihood ratio $u = \ln p(\boldsymbol{x}|\omega_1) - \ln p(\boldsymbol{x}|\omega_2)$. Observe that u is also a random variable normally distributed as $\mathcal{N}((1/2)d_m^2, d_m^2)$ if $\boldsymbol{x} \in \omega_1$ and as $\mathcal{N}(-(1/2)d_m^2, d_m^2)$ if $\boldsymbol{x} \in \omega_2$. Use this information to compute the error probability.

2.10 Show that in the case in which the feature vectors follow Gaussian pdfs, the likelihood ratio test in (2.20)

$$\boldsymbol{x} \in \omega_1(\omega_2) \quad \text{if} \quad l_{12} \equiv \frac{p(\boldsymbol{x}|\omega_1)}{p(\boldsymbol{x}|\omega_2)} > (<)\theta$$

is equivalent to

$$d_m^2(\boldsymbol{\mu}_1, \boldsymbol{x}|\Sigma_1) - d_m^2(\boldsymbol{\mu}_2, \boldsymbol{x}|\Sigma_2) + \ln \frac{|\Sigma_1|}{|\Sigma_2|} < (>) - 2\ln\theta$$

where $d_m(\boldsymbol{\mu}_i, \boldsymbol{x}|\Sigma_i)$ is the Mahalanobis distance between $\boldsymbol{\mu}_i$ and $\boldsymbol{x}$ with respect to the Σ_i^{-1} norm.

2.11 If $\Sigma_1 = \Sigma_2 = \Sigma$, show that the criterion of the previous problem becomes

$$(\boldsymbol{\mu}_1 - \boldsymbol{\mu}_2)^T \Sigma^{-1} \boldsymbol{x} > (<)\Theta$$

where

$$\Theta = \ln\theta + 1/2(\|\boldsymbol{\mu}_1\|_{\Sigma^{-1}} - \|\boldsymbol{\mu}_2\|_{\Sigma^{-1}})$$

2.12 Consider a two-class, two-dimensional classification task, where the feature vectors in each of the classes ω_1, ω_2 are distributed according to

$$p(\boldsymbol{x}|\omega_1) = \frac{1}{\left(\sqrt{2\pi\sigma_1^2}\right)^2} \exp\left(-\frac{1}{2\sigma_1^2}(\boldsymbol{x} - \boldsymbol{\mu}_1)^T(\boldsymbol{x} - \boldsymbol{\mu}_1)\right)$$

$$p(\boldsymbol{x}|\omega_2) = \frac{1}{\left(\sqrt{2\pi\sigma_2^2}\right)^2} \exp\left(-\frac{1}{2\sigma_2^2}(\boldsymbol{x} - \boldsymbol{\mu}_2)^T(\boldsymbol{x} - \boldsymbol{\mu}_2)\right)$$

with

$$\boldsymbol{\mu}_1 = [1, 1]^T, \quad \boldsymbol{\mu}_2 = [1.5, 1.5]^T, \quad \sigma_1^2 = \sigma_2^2 = 0.2$$

Assume that $P(\omega_1) = P(\omega_2)$ and design a Bayesian classifier
(a) that minimizes the error probability
(b) that minimizes the average risk with loss matrix

$$\Lambda = \begin{bmatrix} 0 & 1 \\ 0.5 & 0 \end{bmatrix}$$

Using a pseudorandom number generator, produce 100 feature vectors from each class, according to the preceding pdfs. Use the classifiers designed to classify the generated vectors. What is the percentage error for each case? Repeat the experiments for $\boldsymbol{\mu}_2 = [3.0, 3.0]^T$.

2.13 Repeat the preceding experiment if the feature vectors are distributed according to

$$p(\boldsymbol{x}|\omega_i) = \frac{1}{2\pi|\Sigma|^{1/2}} \exp\left(-\frac{1}{2}(\boldsymbol{x} - \boldsymbol{\mu}_i)^T \Sigma^{-1}(\boldsymbol{x} - \boldsymbol{\mu}_i)\right)$$

with

$$\Sigma = \begin{bmatrix} 1.01 & 0.2 \\ 0.2 & 1.01 \end{bmatrix}$$

and $\boldsymbol{\mu}_1 = [1, 1]^T, \boldsymbol{\mu}_2 = [1.5, 1.5]^T$.
Hint: To generate the vectors, recall from [Papo 91, p. 144] that a linear transformation of Gaussian random vectors also results in Gaussian vectors. Note also that

$$\begin{bmatrix} 1.01 & 0.2 \\ 0.2 & 1.01 \end{bmatrix} = \begin{bmatrix} 1 & 0.1 \\ 0.1 & 1 \end{bmatrix} \begin{bmatrix} 1 & 0.1 \\ 0.1 & 1 \end{bmatrix}$$

2.14 Consider a two-class problem with normally distributed vectors with the same Σ in both classes. Show that the decision hyperplane at the point $\boldsymbol{x}_0$, Eq. (2.46), is tangent to the constant Mahalanobis distance hyperellipsoids.
Hint: (a) Compute the gradient of Mahalanobis distance with respect to $\boldsymbol{x}$. (b) Recall from vector analysis that $\frac{\partial f(\boldsymbol{x})}{\partial \boldsymbol{x}}$ is normal to the tangent of the surface $f(\boldsymbol{x}) = $ constant.

2.15 Consider a two-class, one-dimensional problem with $p(x|\omega_1)$ being $\mathcal{N}(\mu, \sigma^2)$ and $p(x|\omega_2)$ a uniform distribution between a and b. Show that the Bayesian error probability is bounded by $G\left(\frac{b-\mu}{\sigma}\right) - G\left(\frac{a-\mu}{\sigma}\right)$, where $G(x) \equiv P(y \le x)$ and y is $\mathcal{N}(0, 1)$.

2.16 Show that the mean value of the random vector $\frac{\partial \ln(p(\boldsymbol{x};\boldsymbol{\theta}))}{\partial \boldsymbol{\theta}}$ is zero.

2.17 In a heads or tails coin-tossing experiment the probability of occurrence of a head (1) is q and that of a tail (0) is $1 - q$. Let $x_i, i = 1, 2, \ldots, N$, be the resulting experimental outcomes, $x_i \in \{0, 1\}$. Show that the ML estimate of q is

$$q_{ML} = \frac{1}{N}\sum_{i=1}^{N} x_i$$

Hint: The likelihood function is

$$P(X : q) = \prod_{i=1}^{N} q^{x_i}(1 - q)^{(1-x_i)}$$

Then show that the ML results from the solution of the equation

$$q^{\sum_i x_i}(1 - q)^{(N-\sum_i x_i)}\left(\frac{\sum_i x_i}{q} - \frac{N - \sum_i x_i}{1 - q}\right) = 0$$

2.18 The random variable x is normally distributed $\mathcal{N}(\mu, \sigma^2)$, where μ is considered unknown. Given N measurements of the variable, compute the Cramer–Rao bound $-E[\frac{\partial^2 L(\mu)}{\partial^2 \mu}]$ (Appendix A). Compare the bound with the variance of the resulting ML estimate of μ. Repeat this if the unknown parameter is the variance σ^2. Comment on the results.

2.19 Show that if the likelihood function is Gaussian with unknowns the mean μ as well as the covariance matrix Σ, then the ML estimates are given by

$$\hat{\mu} = \frac{1}{N}\sum_{k=1}^{N} x_k$$

$$\hat{\Sigma} = \frac{1}{N}\sum_{k=1}^{N}(x_k - \hat{\mu})(x_k - \hat{\mu})^T$$

2.20 Prove that the covariance estimate

$$\hat{\Sigma} = \frac{1}{N-1}\sum_{k=1}^{N}(x_k - \hat{\mu})(x_k - \hat{\mu})^T$$

is an unbiased one, where

$$\hat{\mu} = \frac{1}{N}\sum_{k=1}^{N} x_k$$

2.21 Prove that the ML estimates of the mean value and the covariance matrix (Problem 2.19) can be computed recursively, that is,

$$\hat{\mu}_{N+1} = \hat{\mu}_N + \frac{1}{N+1}(x_{N+1} - \hat{\mu}_N)$$

and

$$\hat{\Sigma}_{N+1} = \frac{N}{N+1}\hat{\Sigma}_N + \frac{N}{(N+1)^2}(\boldsymbol{x}_{N+1} - \hat{\boldsymbol{\mu}}_N)(\boldsymbol{x}_{N+1} - \hat{\boldsymbol{\mu}}_N)^T$$

where the subscript in the notation of the estimates, $\hat{\boldsymbol{\mu}}_N$, $\hat{\Sigma}_N$ indicates the number of samples used for their computation.

2.22 The random variable x follows the Erlang pdf

$$p(x;\theta) = \theta^2 x \exp(-\theta x)u(x)$$

where $u(x)$ is the unit-step function,

$$u(x) = \begin{cases} 1 & \text{if } x > 0 \\ 0 & \text{if } x < 0 \end{cases}$$

Show that the maximum likelihood estimate of θ, given N measurements, $x_1, \ldots, x_N$, of x, is

$$\hat{\theta}_{ML} = \frac{2N}{\sum_{k=1}^{N} x_k}$$

2.23 In the ML estimation, the zero of the derivative of the log pdf derivative was computed. Using a multivariate Gaussian pdf, show that this corresponds to a maximum and not to a minimum.

2.24 Prove that the sum $z = x + y$ of two independent random variables, x and y, where $x \sim \mathcal{N}(\mu_x, \sigma_x^2)$ and $y \sim \mathcal{N}(\mu_y, \sigma_y^2)$, is also a Gaussian one with mean value and variance equal to $\mu_x + \mu_y$ and $\sigma_x^2 + \sigma_y^2$, respectively.

2.25 Show relations (2.74) and (2.75). Then show that $p(x|X)$ is also normal with mean μ_N and variance $\sigma^2 + \sigma_N^2$. Comment on the result.

2.26 Show that the posterior pdf estimate in the Bayesian inference task, for independent variables, can be computed recursively, that is,

$$p(\boldsymbol{\theta}|\boldsymbol{x}_1, \ldots, \boldsymbol{x}_N) = \frac{p(\boldsymbol{x}_N|\boldsymbol{\theta})p(\boldsymbol{\theta}|\boldsymbol{x}_1, \ldots, \boldsymbol{x}_{N-1})}{p(\boldsymbol{x}_N|\boldsymbol{x}_1, \ldots, \boldsymbol{x}_{N-1})}$$

2.27 Show Eqs. (2.76)–(2.79).

2.28 The random variable x is normally distributed as $\mathcal{N}(\mu, \sigma^2)$, with μ being the unknown parameter described by the Rayleigh pdf

$$p(\mu) = \frac{\mu \exp(-\mu^2/2\sigma_\mu^2)}{\sigma_\mu^2}$$

Show that the maximum *a posteriori* probability estimate of μ is given by

$$\hat{\mu}_{MAP} = \frac{Z}{2R}\left(1 + \sqrt{1 + \frac{4R}{Z^2}}\right)$$

where

$$Z = \frac{1}{\sigma^2}\sum_{k=1}^{N} x_k, \quad R = \frac{N}{\sigma^2} + \frac{1}{\sigma_\mu^2}$$

2.29 Show that for the lognormal distribution

$$p(x) = \frac{1}{\sigma x\sqrt{2\pi}} \exp\left(-\frac{(\ln x - \theta)^2}{2\sigma^2}\right), \quad x > 0$$

the ML estimate is given by

$$\hat{\theta}_{ML} = \frac{1}{N}\sum_{k=1}^{N} \ln x_k$$

2.30 Show that if the mean value and the variance of a random variable are known, that is,

$$\mu = \int_{-\infty}^{+\infty} x p(x)\, dx, \quad \sigma^2 = \int_{-\infty}^{+\infty} (x-\mu)^2 p(x)\, dx$$

the maximum entropy estimate of the pdf is the Gaussian $\mathcal{N}(\mu, \sigma^2)$.

2.31 Show Eqs. (2.98), (2.99), and (2.100).
Hint: For the latter, note that the probabilities add to one; thus a Lagrangian multiplier must be used.

2.32 Let P be the probability of a random point x being located in a certain interval h. Given N of these points, the probability of having k of them inside h is given by the binomial distribution

$$\text{prob}\{k\} = \frac{N!}{k!(N-k)!} P^k (1-P)^{N-k}$$

Show that $E[k/N] = P$ and that the variance around the mean is $\sigma^2 = E[(k/N - P)^2] = P(1-P)/N$. That is, the probability estimator $P = k/N$ is unbiased and asymptotically consistent.

2.33 Consider three Gaussian pdfs: $\mathcal{N}(1.0, 0.1)$, $\mathcal{N}(3.0, 0.1)$, and $\mathcal{N}(2.0, 0.2)$. Generate 500 samples according to the following rule. The first two samples are generated from the second Gaussian, the third sample from the first one, and the fourth sample from the last Gaussian. This rule repeats until all 500 samples have been generated. The pdf underlying the random samples is modeled as a mixture

$$\sum_{i=1}^{3} \mathcal{N}(\mu_i, \sigma_i^2) P_i$$

Use the EM algorithm and the generated samples to estimate the unknown parameters μ_i, σ_i^2, P_i.

2.34 Consider two classes ω_1, ω_2 in the two-dimensional space. The data from class ω_1 are uniformly distributed inside a circle of radius r. The data of class ω_2 are also uniformly distributed inside another circle of radius r. The distance between the centers of the circles is greater than $4r$. Let N be the number of the available training samples. Show that the probability of error of the NN classifier is always smaller than that of the kNN, for any $k \geq 3$.

2.35 Generate 50 feature vectors for each of the two classes of Problem 2.12, and use them as training points. In the sequel, generate 100 vectors from each class and classify them according to the NN and 3NN rules. Compute the classification error percentages.

2.36 The pdf of a random variable is given by

$$p(x) = \begin{cases} \frac{1}{2} & \text{for } 0 < x < 2 \\ 0 & \text{otherwise} \end{cases}$$

Use the Parzen window method to approximate it using as the kernel function the Gaussian $\mathcal{N}(0, 1)$. Choose the smoothing parameter to be (a) $h = 0.05$ and (b) $h = 0.2$. For each case, plot the approximation based on $N = 32, N = 256$, and $N = 5000$ points, which are generated from a pseudorandom generator according to $p(x)$.

2.37 Repeat the preceding problem by generating $N = 5000$ points and using k nearest neighbor estimation with $k = 32, 64$, and 256, respectively.

2.38 Show that the variance $\sigma_N^2(\boldsymbol{x})$ of the pdf estimate, given by Eq. (2.110), is upper bounded by:

$$\sigma_N^2(\boldsymbol{x}) \leq \frac{\sup(\phi) E[\hat{p}(\boldsymbol{x})]}{Nh^l}$$

where $\sup(\cdot)$ is the supremum of the associated function. Observe that for large values of h the variance is small. On the other hand, we can make the variance small for small values of h, provided N tends to infinity and if, also, the product Nh^l tends to infinity.

2.39 Recall Equation (2.128)

$$p(\boldsymbol{x}) = p(x_1) \prod_{i=2}^{l} p(x_i|A_i)$$

Assume $l = 6$ and

$$p(x_6|x_5, \ldots, x_1) = p(x_6|x_5, x_1) \tag{2.142}$$

$$p(x_5|x_4, \ldots, x_1) = p(x_5|x_4, x_3) \tag{2.143}$$

$$p(x_4|x_3, x_2, x_1) = p(x_4|x_3, x_2, x_1) \tag{2.144}$$

$$p(x_3|x_2, x_1) = p(x_3) \tag{2.145}$$

$$p(x_2|x_1) = p(x_2) \tag{2.146}$$

Write the respective sets A_i, $i = 1, 2, \ldots, 6$, and construct the corresponding DAG.

2.40 In the DAG defined in Figure 2.29, assume that the variable z is measured to be $z0$. Compute $P(x1|z0)$ and $P(w0|z0)$.

2.41 In the example associated with the tree-structured DAG of Figure 2.28, assume that the patient undergoes the medical test H_1 and that this turns out to be positive (True). Based on this test, compute the probability that the patient has developed cancer. In other words, compute the conditional probability $P(C = \text{True}|H_1 = \text{True})$.

MATLAB PROGRAMS AND EXERCISES

Computer Exercises

A number of MATLAB functions are provided, which will help the interested reader to experiment on some of the most important issues discussed in the present chapter. Needless to say that there may be other implementations of these functions. Short comments are also given along with the code. In addition, we have used the symbols m and S to denote the mean vector (given as a column vector) and the covariance matrix, respectively, instead of the symbols $\boldsymbol{\mu}$ and Σ, which are used in the text. In the following, unless otherwise stated, each class is represented by an integer in $\{1, \ldots, c\}$ where c is the number of classes.

2.1 *Gaussian generator.* Generate N l-dimensional vectors from a Gaussian distribution with mean m and covariance matrix S, using the *mvnrnd* MATLAB function.

Solution

Just type

```
mvnrnd(m,S,N)
```

2.2 *Gaussian function evaluation.* Write a MATLAB function that computes the value of the Gaussian distribution $\mathcal{N}(m, S)$, at a given vector x.

Solution

```
function z=comp_gauss_dens_val(m,S,x)
  [l,q]=size(m);  % l=dimensionality
  z=(1/((2*pi)^ (l/2)*det(S)^ 0.5) )...
      *exp(-0.5*(x-m)'*inv(S)*(x-m));
```

2.3 *Data set generation from Gaussian classes.* Write a MATLAB function that generates a data set of N l-dimensional vectors that stem from c different Gaussian distributions $\mathcal{N}(m_i, S_i)$, with corresponding *a priori* probabilities $P_i, i = 1, \ldots, c$.

Solution

In the sequel:

- m is an $l \times c$ matrix, the i-th column of which is the mean vector of the i-th class distribution.
- S is an $l \times l \times c$ (three-dimensional) matrix, whose ith two-dimensional $l \times l$ component is the covariance of the distribution of the ith class. In MATLAB $S(:, :, i)$ denotes the i-th two-dimensional $l \times l$ matrix of S.
- P is the c dimensional vector that contains the *a priori* probabilities of the classes. m_i, S_i, P_i, and c are provided as inputs.

The following function returns:

- A matrix X with (approximately) N columns, each column of which is an l-dimensional data vector.
- A row vector y whose ith entry denotes the class from which the ith data vector stems.

```
function [X,y]=generate_gauss_classes(m,S,P,N)
  [l,c]=size(m);
  X=[];
  y=[];
  for j=1:c
  % Generating the [p(j)*N)] vectors from each distribution
    t=mvnrnd(m(:,j),S(:,:,j),fix(P(j)*N));
    % The total number of points may be slightly less than N
    % due to the fix operator
    X=[X t];
    y=[y ones(1,fix(P(j)*N))*j];
  end
```

2.4 *Plot of data.* Write a MATLAB function that takes as inputs: (a) a matrix X and a vector y defined as in the previous function, (b) the mean vectors of c class distributions. It plots: (a) the data vectors of X using a different color for each class, (b) the mean vectors of the class distributions. It is assumed that the data live in the two-dimensional space.

Solution

```
% CAUTION: This function can handle up to
% six different classes
```

```
function plot_data(X,y,m)
  [l,N]=size(X); % N=no. of data vectors, l=dimensionality
  [l,c]=size(m); % c=no. of classes
  if(l ~=2)
    fprintf('NO PLOT CAN BE GENERATED\n')
    return
  else
    pale=['r.'; 'g.'; 'b.'; 'y.'; 'm.'; 'c.'];
    figure(1)
    % Plot of the data vectors
    hold on
    for i=1:N
      plot(X(1,i),X(2,i),pale(y(i),:))
    end
    % Plot of the class means
    for j=1:c
    plot(m(1,j),m(2,j),'k+')
    end
  end
```

2.5 *Bayesian classifier (for Gaussian Processes).* Write a MATLAB function that will take as inputs: (a) the mean vectors, (b) the covariance matrices of the class distributions of a c-class problem, (c) the *a priori* probabilities of the c classes, and (d) a matrix X containing column vectors that stem from the above classes. It will give as output an N-dimensional vector whose ith component contains the class where the corresponding vector is assigned, according to the Bayesian classification rule.

Solution

Caution: While inserting the following function, **do not** type the labels (A), (B) and (C). They are used to serve as references, as we will see later on.

```
(A) function z=bayes_classifier(m,S,P,X)
  [l,c]=size(m); % l=dimensionality, c=no. of classes
  [l,N]=size(X); % N=no. of vectors
  for i=1:N
    for j=1:c
      (B) t(j)=P(j)*comp_gauss_dens_val(m(:,j),...
         S(:,:,j),X(:,i));
    end
    % Determining the maximum quantity Pi*p(x|wi)
    (C) [num,z(i)]=max(t);
end
```

2.6 *Euclidean distance classifier.* Write a MATLAB function that will take as inputs: (a) the mean vectors, and (b) a matrix X containing column vectors that stem from the above classes. It will give as output an N-dimensional vector whose ith component contains the class where the corresponding vector is assigned, according to the minimum Euclidean distance classifier.

Solution

The requested function may be obtained by the *bayes_classifier* function by replacing (A), (B), and (C) with

- `function z=euclidean_classifier(m,X)`
- `t(j)=sqrt((X(:,i)-m(:,j))'*(X(:,i)-m(:,j)));`

(computation of the Euclidean distances from all class representatives)

- `[num,z(i)]=min(t);`

(determination of the closest class mean), respectively.

2.7 *Mahalanobis distance classifier.* Write a MATLAB function that will take as inputs: (a) the mean vectors, (b) the covariance matrix of the class distributions of a c-class problem, and (c) a matrix X containing column vectors that stem from the above classes. It will give as output an N-dimensional vector whose ith component contains the class where the corresponding vector is assigned, according to the minimum Mahalanobis distance classifier.

Solution

The requested function may be obtained by the *bayes_classifier* function by replacing (A), (B) and (C) with

- `function z=mahalanobis_classifier(m,S,X)`
- ```
  t(j)=sqrt((X(:,i)-m(:,j))'*inv(S(:,:,j))*...
  (X(:,i)-m(:,j)));
  ```

(computation of the Mahalanobis distances from all class representatives)

- `[num,z(i)]=min(t);`

(determination of the closest class mean), respectively.

2.8 *k-nearest neighbor classifier.* Write a MATLAB function that takes as inputs: (a) a set of N_1 vectors packed as columns of a matrix Z, (b) an N_1-dimensional vector containing the classes where each vector in Z belongs, (c) the value for the parameter k of the classifier, (d) a set of N vectors packed as columns in the
  ```

matrix X. It returns an N-dimensional vector whose ith component contains the class where the corresponding vector of X is assigned, according to the k-nearest neighbor classifier.

Solution

```
function z=k_nn_classifier(Z,v,k,X)
  [l,N1]=size(Z);
  [l,N]=size(X);
  c=max(v); % The number of classes
  % Computation of the (squared) Euclidean distance
  % of a point from each reference vector
  for i=1:N
    dist=sum((X(:,i)*ones(1,N1)-Z).^ 2);
    %Sorting the above distances in ascending order
    [sorted,nearest]=sort(dist);
    % Counting the class occurrences among the k-closest
    % reference vectors Z(:,i)
    refe=zeros(1,c); %Counting the reference vectors per class
    for q=1:k
      class=v(nearest(q));
      refe(class)=refe(class)+1;
    end
    [val,z(i)]=max(refe);
  end
```

2.9 *Classification error evaluation.* Write a MATLAB function that will take as inputs: (a) an N-dimensional vector, each component of which contains the class where the corresponding data vector belongs and (b) a similar N-dimensional vector each component of which contains the class where the corresponding data vector is assigned from a certain classifier. Its output will be the percentage of the places where the two vectors differ (i.e., the classification error of the classifier).

Solution

```
function clas_error=compute_error(y,y_est)
  [q,N]=size(y);    % N= no. of vectors
  c=max(y);    % Determining the number of classes
  clas_error=0; % Counting the misclassified vectors
  for i=1:N
    if(y(i)~ =y_est(i))
      clas_error=clas_error+1;
    end
  end
```

```
% Computing the classification error
clas_error=clas_error/N;
```

Computer Experiments

Notes: In the sequel, it is advisable to use the command

```
randn('seed',0)
```

before generating the data sets, in order to initialize the Gaussian random number generator to 0 (or any other fixed number). This is important for the reproducibility of the results.

2.1 a. Generate and plot a data set of $N = 1{,}000$ two-dimensional vectors that stem from three equiprobable classes modeled by normal distributions with mean vectors $m_1 = [1,\ 1]^T$, $m_2 = [7,\ 7]^T$, $m_3 = [15,\ 1]^T$ and covariance matrices $S_1 = \begin{bmatrix} 12 & 0 \\ 0 & 1 \end{bmatrix}$, $S_2 = \begin{bmatrix} 8 & 3 \\ 3 & 2 \end{bmatrix}$, $S_3 = \begin{bmatrix} 2 & 0 \\ 0 & 2 \end{bmatrix}$.

b. Repeat (a) when the *a priori* probabilities of the classes are given by the vector $P = [0.6,\ 0.3,\ 0.1]^T$.

Solution

Figure (2.31)a–b display the vectors from each class. Note the "shape" of the clusters formed by the vectors of each class. This is directly affected by the corresponding covariance matrix. Also note that, in the first case, each class has roughly the same number of the vectors, while in the latter case,

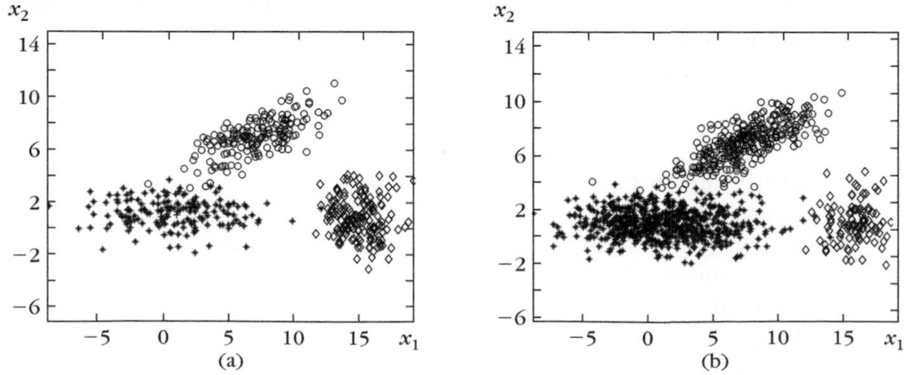

FIGURE 2.31

(a) The equiprobable classes case. (b) The case where the a-priori probabilities differ.

the leftmost and the rightmost classes are more "dense" and more "sparse" compared to the previous case, respectively.

2.2 **a.** Generate a data set X_1 of $N = 1{,}000$ two-dimensional vectors that stem from three equiprobable classes modeled by normal distributions with mean vectors $m_1 = [1, 1]^T$, $m_2 = [12, 8]^T$, $m_3 = [16, 1]^T$ and covariance matrices $S_1 = S_2 = S_3 = 4I$, where I is the 2×2 identity matrix.

b. Apply the Bayesian, the Euclidean, and the Mahalanobis classifiers on X_1.

c. Compute the classification error for each classifier.

2.3 **a.** Generate a data set X_2 of $N = 1{,}000$ two-dimensional vectors that stem from three equiprobable classes modeled by normal distributions with mean vectors $m_1 = [1, 1]^T$, $m_2 = [14, 7]^T$, $m_3 = [16, 1]^T$ and covariance matrices $S_1 = S_2 = S_3 = \begin{bmatrix} 5 & 3 \\ 3 & 4 \end{bmatrix}$.

(b)–(c) Repeat steps b) and (c) of experiment 2.2, for X_2.

2.4 **a.** Generate a data set X_3 of $N = 1{,}000$ two-dimensional vectors that stem from three equiprobable classes modeled by normal distributions with mean vectors $m_1 = [1, 1]^T$, $m_2 = [8, 6]^T$, $m_3 = [13, 1]^T$ and covariance matrices $S_1 = S_2 = S_3 = 6I$, where I is the 2×2 identity matrix.

(b)–(c) Repeat (b) and (c) from experiment 2.2, for X_3.

2.5 **a.** Generate a data set X_4 of $N = 1{,}000$ two-dimensional vectors that stem from three equiprobable classes modeled by normal distributions with mean vectors $m_1 = [1, 1]^T$, $m_2 = [10, 5]^T$, $m_3 = [11, 1]^T$ and covariance matrices $S_1 = S_2 = S_3 = \begin{bmatrix} 7 & 4 \\ 4 & 5 \end{bmatrix}$.

(b)–(c) Repeat steps (b) and (c) of experiment 2.2, for X_4.

2.6 Study carefully the results obtained by experiments (2.2)–(2.5) and draw your conclusions.

2.7 **a.** Generate two data sets X_5 and X_5' of $N = 1{,}000$ two-dimensional vectors each that stem from three classes modeled by normal distributions with mean vectors $m_1 = [1, 1]^T$, $m_2 = [4, 4]^T$, $m_3 = [8, 1]^T$ and covariance matrices $S_1 = S_2 = S_3 = 2I$. In the generation of X_5, the classes are assumed to be equiprobable, while in the generation of X_5', the *a priori* probabilities of the classes are given by the vector $P = [0.8, 0.1, 0.1]^T$.

b. Apply the Bayesian and the Euclidean classifiers on both X_5 and X_5'.

c. Compute the classification error for each classifier for both data sets and draw your conclusions.

2.8 Consider the data set X_3 (from experiment (2.4)). Using the same settings, generate a data set Z, where the class from which a data vector stems is known. Apply the k nearest neighbor classifier on X_3 for $k = 1$ and $k = 11$ using Z as the training set and draw your conclusions.

REFERENCES

[Aebe 94] Aeberhard S., Coomans D., Devel O. "Comparative analysis of statistical pattern recognition methods in high dimensional setting," *Pattern Recognition*, Vol. 27(8), pp. 1065–1077, 1994.

[Babi 96] Babich G.A., Camps O.I. "Weighted Parzen windows for pattern classification," *IEEE Transactions on Pattern Analysis and Machine Intelligence*, Vol. 18(5), pp. 567–570, 1996.

[Bell 61] Bellman R. *Adaptive Control Processes: A Guided Tour*, Princeton University Press, 1961.

[Bern 94] Bernardo J.M., Smith A.F.M *Bayesian Theory*, John Wiley, 1994.

[Bish 06] Bishop C.M. *Pattern Recognition and Machine Learning*, Springer, 2006.

[Boyl 83] Boyles R.A. "On the convergence of the EM algorithm," *J. Royal Statistical Society B*, Vol. 45(1), pp. 47–55, 1983.

[Brei 77] Breiman L., Meisel W., Purcell E. "Variable kernel estimates of multivariate densities," *Technometrics*, Vol. 19(2), pp. 135–144, 1977.

[Brod 90] Broder A. "Strategies for efficient incremental nearest neighbor search," *Pattern Recognition*, Vol. 23, pp. 171–178, 1990.

[Butu 93] Buturovic L.J. "Improving *k*-nearest neighbor density and error estimates," *Pattern Recognition*, Vol. 26(4), pp. 611–616, 1993.

[Coop 90] Cooper G.F. "The computational complexity of probabilistic inference using Bayesian belief networks," *Artifical Intelligence*, Vol. 42, pp. 393–405, 1990.

[Cram 46] Cramer H. *Mathematical Methods of Statistics*, Princeton University Press, 1941.

[Dagu 93] Dagum P., Chavez R.M. "Approximating probabilistic inference in Bayesian belief networks," *IEEE Transactions on Pattern Analysis and Machine Intelligence*, Vol. 15(3), pp. 246–255, 1993.

[Dasa 91] Dasarasthy B. *Nearest Neighbor Pattern Classification Techniques*, IEEE Computer Society Press, 1991.

[Demp 77] Dempster A.P., Laird N.M., Rubin D.B. "Maximum likelihood from incomplete data via the EM algorithm," *J. Royal Statistical Society*, Vol. 39(1), pp. 1–38, 1977.

[Devr 96] Devroye L., Gyorfi L., Lugosi G. *A Probabilistic Theory of Pattern Recognition*, Springer-Verlag, 1996.

[Djou 97] Djouadi A., Bouktache E. "A fast algorithm for the nearest neighbor classifier," *IEEE Transactions on Pattern Analysis and Machine Intelligence*, Vol. 19(3), pp. 277–282, 1997.

[Dome 05] Domeniconi C., Gunopoulos D., Peng J. "Large margin nearest neighbor classifiers," *IEEE Transactions on Neural Networks*, Vol. 16(4), pp. 899–909, 2005.

[Domi 97] Domingos P., Pazzani M. "Beyond independence: Conditions for the optimality of the simple Bayesian classifier," *Machine Learning*, Vol. 29, pp. 103–130, 1997.

[Duda 73] Duda R., Hart P.E. *Pattern Classification and Scene Analysis*, John Wiley & Sons, 1973.

[Frie 94] Friedman J.H. "Flexible metric nearest neighbor classification," *Technical Report, Department of Statistics, Stanford University*, 1994.

[Frie 89] Friedman J.H. "Regularized discriminant analysis," *Journal of American Statistical Association*, Vol. 84(405), pp. 165–175, 1989.

[Frie 97] Friedman N., Geiger D., Goldszmidt M. "Bayesian network classifiers," *Machine Learning*, Vol. 29, pp. 131–163, 1997.

[Fuku 75] Fukunaga F., Narendra P.M. "A branch and bound algorithm for computing *k*-nearest neighbors," *IEEE Transactions on Computers*, Vol. 24, pp. 750–753, 1975.

[Fuku 90] Fukunaga F. *Introduction to Statistical Pattern Recognition*, 2nd ed., Academic Press, 1990.

[Hast 96] Hastie T., Tibshirani R. "Discriminant adaptive nearest neighbor classification," *IEEE Transactions on Pattern Analysis and Machine Intelligence*, Vol. 18(6), pp. 607–616, 1996.

[Hast 01] Hastie T., Tibshirani R., Friedman J. *The Elements of Statistical Learning: Data Mining, Inference and Prediction*, Springer, 2001.

[Hatt 00] Hattori K., Takahashi M. "A new edited *k*-nearest neighbor rule in the pattern classification problem," *Pattern Recognition*, Vol. 33, pp. 521–528, 2000.

[Heck 95] Heckerman D. "A tutorial on learning Bayesian networks," *Technical Report #MSR-TR-95-06*, Microsoft Research, Redmond, Washington, 1995.

[Hoff 96] Hoffbeck J.P., Landgrebe D.A. "Covariance matrix estimation and classification with limited training data," *IEEE Transactions on Pattern Analysis and Machine Intelligence*, Vol. 18(7), pp. 763–767, 1996.

[Huan 02] Huang Y.S., Chiang C.C., Shieh J.W., Grimson E. "Prototype optimization for nearest-neighbor classification," *Pattern Recognition*, Vol. 35, pp. 1237–1245, 2002.

[Jayn 82] Jaynes E.T. "On the rationale of the maximum entropy methods," *Proceedings of the IEEE*, Vol. 70(9), pp. 939–952, 1982.

[Jens 01] Jensen F.V. *Bayesian Networks and Decision Graphs*, Springer, 2001.

[Jone 96] Jones M.C., Marron J.S., Seather S.J. "A brief survey of bandwidth selection for density estimation," *Journal of the American Statistical Association*, Vol. 91, pp. 401–407, 1996.

[Kimu 87] Kimura F., Takashina K., Tsuruoka S., Miyake Y. "Modified quadratic discriminant functions and the application to Chinese character recognition," *IEEE Transactions on Pattern Analysis and Machine Intelligence*, Vol. 9(1), pp. 149–153, 1987.

[Kris 00] Krishna K., Thathachar M.A.L., Ramakrishnan K.R. "Voronoi networks and their probability of misclassification," *IEEE Transactions on Neural Networks*, Vol. 11(6), pp. 1361–1372, 2000.

[Krzy 83] Krzyzak A. "Classification procedures using multivariate variable kernel density estimate," *Pattern Recognition Letters*, Vol. 1, pp. 293–298, 1983.

[Laur 96] Lauritzen S.L. *Graphical Models*, Oxford University Press, 1996.

[Li 94] Li Z., D'Abrosio B. "Efficient inference in Bayes' networks as a combinatorial optimization problem," *International Journal of Approximate Inference*, Vol. 11, 1994.

[Liu 04] Liu C.-L., Sako H., Fusisawa H. "Discriminative learning quadratic discriminant function for handwriting recognition," *IEEE Transactions on Neural Networks*, Vol. 15(2), pp. 430–444, 2004.

[McLa 88] McLachlan G.J., Basford K.A. *Mixture Models: Inference and Applications to Clustering*, Marcel Dekker, 1988.

[McNa 01] McNames J. "A Fast nearest neighbor algorithm based on principal axis search tree," *IEEE Transactions on Pattern Analysis and Machine Intelligence*, Vol. 23(9), pp. 964–976, 2001.

[Mico 94] Mico M.L., Oncina J., Vidal E. "A new version of the nearest neighbor approximating and eliminating search algorithm (AESA) with linear preprocessing time and memory requirements," *Pattern Recognition Letters*, Vol. 15, pp. 9–17, 1994.

[Moon 96] Moon T. "The expectation maximization algorithm," *Signal Processing Magazine*, Vol. 13(6), pp. 47–60, 1996.

[Neap 04] Neapolitan R.D. *Learning Bayesian Networks*, Prentice Hall, 2004.

[Nene 97] Nene S.A., Nayar S.K. "A simple algorithm for nearest neighbor search in high dimensions," *IEEE Transactions on Pattern Analysis and Machine Intelligence*, Vol. 19(9), pp. 989–1003, 1997.

[Papo 91] Papoulis A. *Probability Random Variables and Stochastic Processes*, 3rd ed., McGraw-Hill 1991.

[Pare 06] Paredes R., Vidal E. "Learning weighted metrics to minimize nearest neighbor classification error," *IEEE Transactions on Pattern Analysis and Machine Intelligence*, Vol. 28(7), pp. 1100–1111, 2006.

[Pare 06a] Paredes R., Vidal E. "Learning prototypes and distances: A prototype reduction technique based on nearest neighbor error minimization," *Pattern Recognition*, Vol. 39, pp. 180–188, 2006.

[Parz 62] Parzen E. "On the estimation of a probability density function and mode," *Ann. Math. Stat.* Vol. 33, pp. 1065–1076, 1962.

[Pear 88] Pearl J. *Probabilistic Reasoning in Intelligent Systems*, Morgan Kaufmann, 1988.

[Redn 84] Redner R.A., Walker H.F. "Mixture densities, maximum likelihood and the EM algorithm," *SIAM Review*, Vol. 26(2), pp. 195–239, 1984.

[Roos 05] Roos T., Wettig H., Grunwald P., Myllymaki P., Tirri H. "On discriminative Bayesian network classifiers and logistic regression," *Machine Learning*, Vol. 59, pp. 267–296, 2005.

[Same 08] Samet H. "*k*-Nearest neighbor finding using MaxNearestDist," *IEEE Transactions on Pattern Analysis and Machine Intelligence*, Vol. 30(2), pp. 243–252, 2008.

[Terr 92] Terrell G.R., Scott D.W. "Variable kernel density estimation," *Annals of Statistics*, Vol. 20(3), pp. 1236–1265, 1992.

[Titt 85] Titterington D.M., Smith A.F.M., Makov U.A. *Statistical Analysis of Finite Mixture Distributions*, John Wiley & Sons, 1985.

[Vida 94] Vidal E. "New formulation and improvements of the nearest neighbor approximating and eliminating search algorithm (AESA)," *Pattern Recognition Letters*, Vol. 15, pp. 1–7, 1994.

[Wand 95] Wand M., Jones M. *Kernel Smoothing*, Chapman & Hall, London, 1995.

[Webb 05] Webb G.I., Boughton J.R., Wang Z. "Not so naive Bayes: Aggregating one dependence estimators," *Machine Learning*, Vol. 58, pp. 5–24, 2005.

[Wils 72] Wilson D.L. "Asymptotic properties of NN rules using edited data," *IEEE Transactions on Systems, Man, and Cybernetics*, Vol. 2, pp. 408–421, 1972.

[Wu 83] Wu C. "On the convergence properties of the EM algorithm," *Annals of Statistics*, Vol. 11(1), pp. 95–103, 1983.

[Yan 93] Yan H. "Prototype optimization for nearest neighbor classifiers using a two layer perceptron," *Pattern Recognition*, Vol. 26(2), pp. 317–324, 1993.

CHAPTER

3 Linear Classifiers

3.1 INTRODUCTION

Our major concern in Chapter 2 was to design classifiers based on probability density or probability functions. In some cases, we saw that the resulting classifiers were equivalent to a set of linear discriminant functions. In this chapter, we will focus on the design of linear classifiers, ***regardless of the underlying distributions describing the training data***. The major advantage of linear classifiers is their simplicity and computational attractiveness. The chapter starts with the assumption that *all* feature vectors from the available classes can be classified correctly using a linear classifier, and we will develop techniques for the computation of the corresponding linear functions. In the sequel we will focus on a more general problem, in which a linear classifier cannot correctly classify all vectors, yet we will seek ways to design an ***optimal linear*** classifier by adopting an appropriate optimality criterion.

3.2 LINEAR DISCRIMINANT FUNCTIONS AND DECISION HYPERPLANES

Let us once more focus on the two-class case and consider linear discriminant functions. Then the respective decision hypersurface in the l-dimensional feature space is a hyperplane, that is

$$g(\boldsymbol{x}) = \boldsymbol{w}^T\boldsymbol{x} + w_0 = 0 \tag{3.1}$$

where $\boldsymbol{w} = [w_1, w_2, \ldots, w_l]^T$ is known as the ***weight vector*** and w_0 as the ***threshold***. If $\boldsymbol{x}_1, \boldsymbol{x}_2$ are two points on the decision hyperplane, then the following is valid

$$\begin{aligned} 0 = \boldsymbol{w}^T\boldsymbol{x}_1 + w_0 = \boldsymbol{w}^T\boldsymbol{x}_2 + w_0 \quad &\Rightarrow \\ \boldsymbol{w}^T(\boldsymbol{x}_1 - \boldsymbol{x}_2) = 0 & \end{aligned} \tag{3.2}$$

Since the difference vector $\boldsymbol{x}_1 - \boldsymbol{x}_2$ obviously lies on the decision hyperplane (for any $\boldsymbol{x}_1, \boldsymbol{x}_2$), it is apparent from Eq. (3.2) that the vector $\boldsymbol{w}$ is *orthogonal* to the decision hyperplane.

Figure 3.1 shows the corresponding geometry (for $w_1 > 0, w_2 > 0, w_0 < 0$). Recalling our high school math, it is easy to see that the quantities entering in the figure are given by

$$d = \frac{|w_0|}{\sqrt{w_1^2 + w_2^2}} \tag{3.3}$$

and

$$z = \frac{|g(\boldsymbol{x})|}{\sqrt{w_1^2 + w_2^2}} \tag{3.4}$$

In other words, $|g(\boldsymbol{x})|$ is a measure of the Euclidean distance of the point $\boldsymbol{x}$ from the decision hyperplane. On one side of the plane $g(\boldsymbol{x})$ takes positive values and on the other negative. In the special case that $w_0 = 0$, the hyperplane passes through the origin.

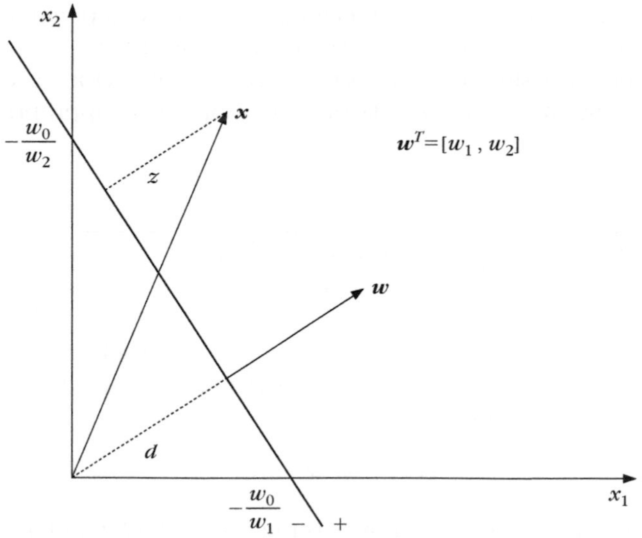

FIGURE 3.1

Geometry for the decision line. On one side of the line it is $g(\boldsymbol{x}) > 0(+)$ and on the other $g(\boldsymbol{x}) < 0(-)$.

3.3 THE PERCEPTRON ALGORITHM

Our major concern now is to compute the unknown parameters $w_i, i = 0,\ldots,l$, defining the decision hyperplane. In this section, we assume that the two classes ω_1, ω_2 are *linearly separable*. In other words, we assume that *there exists* a hyperplane, defined by $\boldsymbol{w}^{*T}\boldsymbol{x} = 0$, such that

$$\begin{aligned} \boldsymbol{w}^{*T}\boldsymbol{x} > 0 \quad \forall \boldsymbol{x} \in \omega_1 \\ \boldsymbol{w}^{*T}\boldsymbol{x} < 0 \quad \forall \boldsymbol{x} \in \omega_2 \end{aligned} \tag{3.5}$$

The formulation above also covers the case of a hyperplane not crossing the origin, that is, $\boldsymbol{w}^{*T}\boldsymbol{x} + w_0^* = 0$, since this can be brought into the previous formulation by defining the extended $(l+1)$-dimensional vectors $\boldsymbol{x}' \equiv [\boldsymbol{x}^T, 1]^T$, $\boldsymbol{w}' \equiv [\boldsymbol{w}^{*T}, w_0^*]^T$. Then $\boldsymbol{w}^{*T}\boldsymbol{x} + w_0^* = \boldsymbol{w}'^T\boldsymbol{x}'$.

We will approach the problem as a typical optimization task (Appendix C). Thus we need to adopt (a) an appropriate cost function and (b) an algorithmic scheme to optimize it. To this end, we choose the *perceptron cost* defined as

$$J(\boldsymbol{w}) = \sum_{\boldsymbol{x} \in Y} (\delta_x \boldsymbol{w}^T \boldsymbol{x}) \tag{3.6}$$

where Y is the subset of the training vectors, which are misclassified by the hyperplane defined by the weight vector $\boldsymbol{w}$. The variable δ_x is chosen so that $\delta_x = -1$ if $\boldsymbol{x} \in \omega_1$ and $\delta_x = +1$ if $\boldsymbol{x} \in \omega_2$. Obviously, the sum in (3.6) is always positive, and it becomes zero when Y becomes the empty set, that is, if there are not misclassified vectors $\boldsymbol{x}$. Indeed, if $\boldsymbol{x} \in \omega_1$ and it is misclassified, then $\boldsymbol{w}^T\boldsymbol{x} < 0$ and $\delta_x < 0$, and the product is positive. The result is the same for vectors originating from class ω_2. When the cost function takes its minimum value, 0, a solution has been obtained, since all training feature vectors are correctly classified.

The perceptron cost function in (3.6) is *continuous and piecewise linear*. Indeed, if we change the weight vector smoothly, the cost $J(\boldsymbol{w})$ changes linearly until the point at which there is a change in the number of misclassified vectors (Problem 3.1). At these points the gradient is not defined, and the gradient function is discontinuous.

To derive the algorithm for the iterative minimization of the cost function, we will adopt an iterative scheme in the spirit of the *gradient descent* method (Appendix C), that is,

$$\boldsymbol{w}(t+1) = \boldsymbol{w}(t) - \rho_t \left.\frac{\partial J(\boldsymbol{w})}{\partial \boldsymbol{w}}\right|_{\boldsymbol{w}=\boldsymbol{w}(t)} \tag{3.7}$$

where $\boldsymbol{w}(t)$ is the weight vector estimate at the tth iteration step and ρ_t is a sequence of positive real numbers. However, we must be careful here. This is not defined at the points of discontinuity. From the definition in (3.6), and at the points where this is valid, we get

$$\frac{\partial J(\boldsymbol{w})}{\partial \boldsymbol{w}} = \sum_{\boldsymbol{x}\in Y} \delta_x \boldsymbol{x} \tag{3.8}$$

Substituting (3.8) into (3.7), we obtain

$$\boldsymbol{w}(t+1) = \boldsymbol{w}(t) - \rho_t \sum_{\boldsymbol{x}\in Y} \delta_x \boldsymbol{x} \tag{3.9}$$

The algorithm is known as the ***perceptron algorithm*** and is quite simple in its structure. *Note that Eq. (3.9) is defined at all points.* The algorithm is initialized from an arbitrary weight vector $\boldsymbol{w}(0)$, and the correction vector $\sum_{\boldsymbol{x}\in Y} \delta_x \boldsymbol{x}$ is formed using the misclassified features. The weight vector is then corrected according to the preceding rule. This is repeated until the algorithm converges to a solution, that is, all features are correctly classified. A pseudocode for the perceptron algorithm is given below.

The Perceptron Algorithm

- Choose $\boldsymbol{w}(0)$ randomly
- Choose ρ_0
- $t = 0$
- Repeat
  - $Y = \emptyset$
  - For $i = 1$ to N
    - If $\delta_{x_i} \boldsymbol{w}(t)^T \boldsymbol{x}_i \geq 0$ then $Y = Y \cup \{\boldsymbol{x}_i\}$
  - End {For}
  - $\boldsymbol{w}(t+1) = \boldsymbol{w}(t) - \rho_t \sum_{\boldsymbol{x}\in Y} \delta_x \boldsymbol{x}$
  - Adjust ρ_t
  - $t = t + 1$
- Until $Y = \emptyset$

Figure 3.2 provides a geometric interpretation of the algorithm. It has been assumed that at step t there is only one misclassified sample, $\boldsymbol{x}$, and $\rho_t = 1$. The perceptron algorithm corrects the weight vector in the direction of $\boldsymbol{x}$. Its effect is to turn the corresponding hyperplane so that $\boldsymbol{x}$ is classified in the correct class ω_1. Note that in order to achieve this, it may take more than one iteration step, depending on the value(s) of ρ_t. No doubt, this sequence is critical for the convergence. We will now show that the perceptron algorithm converges

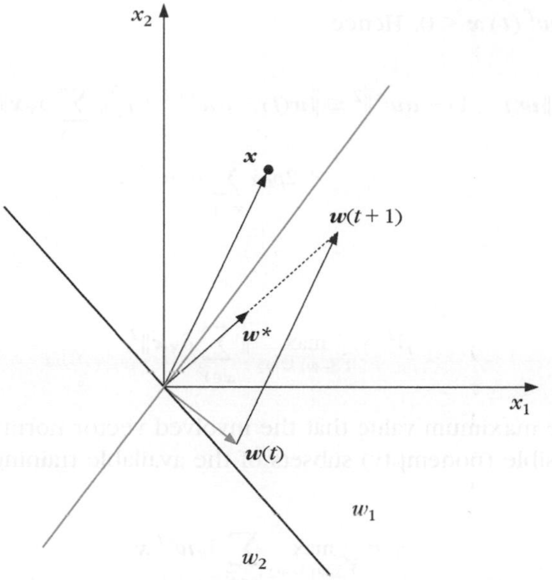

FIGURE 3.2
Geometric interpretation of the perceptron algorithm. The update of the weight vector is in the direction of $\boldsymbol{x}$ in order to turn the decision hyperplane to include $\boldsymbol{x}$ in the correct class.

to a solution in a ***finite number of iteration steps***, provided that the sequence ρ_t is properly chosen. The solution is not unique, because there are more than one hyperplanes separating two linearly separable classes. The convergence proof is necessary because the algorithm is not a true gradient descent algorithm and the general tools for the convergence of gradient descent schemes cannot be applied.

Proof of the Perceptron Algorithm Convergence

Let α be a positive real number and $\boldsymbol{w}^*$ a solution. Then from (3.9) we have

$$\boldsymbol{w}(t+1) - \alpha\boldsymbol{w}^* = \boldsymbol{w}(t) - \alpha\boldsymbol{w}^* - \rho_t \sum_{\boldsymbol{x}\in Y} \delta_x \boldsymbol{x} \tag{3.10}$$

Squaring the Euclidean norm of both sides results in

$$\|\boldsymbol{w}(t+1) - \alpha\boldsymbol{w}^*\|^2 = \|\boldsymbol{w}(t) - \alpha\boldsymbol{w}^*\|^2 + \rho_t^2 \|\sum_{\boldsymbol{x}\in Y} \delta_x \boldsymbol{x}\|^2$$
$$- 2\rho_t \sum_{\boldsymbol{x}\in Y} \delta_x (\boldsymbol{w}(t) - \alpha\boldsymbol{w}^*)^T \boldsymbol{x} \tag{3.11}$$

But $-\sum_{\boldsymbol{x}\in Y}\delta_x \boldsymbol{w}^T(t)\,\boldsymbol{x} < 0$. Hence

$$\|\boldsymbol{w}(t+1)-\alpha\boldsymbol{w}^*\|^2 \leq \|\boldsymbol{w}(t)-\alpha\boldsymbol{w}^*\|^2 + \rho_t^2\|\sum_{\boldsymbol{x}\in Y}\delta_x\boldsymbol{x}\|^2 + 2\rho_t\alpha\sum_{\boldsymbol{x}\in Y}\delta_x\boldsymbol{w}^{*T}\boldsymbol{x} \tag{3.12}$$

Define

$$\beta^2 = \max_{\tilde{Y}\subseteq\omega_1\cup\omega_2}\|\sum_{\boldsymbol{x}\in\tilde{Y}}\delta_x\boldsymbol{x}\|^2 \tag{3.13}$$

That is, β^2 is the maximum value that the involved vector norm can take by considering *all* possible (nonempty) subsets of the available training feature vectors. Similarly, let

$$\gamma = \max_{\tilde{Y}\subseteq\omega_1\cup\omega_2}\sum_{\boldsymbol{x}\in\tilde{Y}}\delta_x\boldsymbol{w}^{*T}\boldsymbol{x} \tag{3.14}$$

Recall that the summation in this equation is negative; thus, its maximum value over all possible subsets of $\boldsymbol{x}$'s will also be a negative number. Hence, (3.12) can now be written as

$$\|\boldsymbol{w}(t+1)-\alpha\boldsymbol{w}^*\|^2 \leq \|\boldsymbol{w}(t)-\alpha\boldsymbol{w}^*\|^2 + \rho_t^2\beta^2 - 2\rho_t\alpha|\gamma| \tag{3.15}$$

Choose $\alpha = \frac{\beta^2}{2|\gamma|}$ and apply (3.15) successively for steps $t, t-1, \ldots, 0$. Then

$$\|\boldsymbol{w}(t+1)-\alpha\boldsymbol{w}^*\|^2 \leq \|\boldsymbol{w}(0)-\alpha\boldsymbol{w}^*\|^2 + \beta^2\left(\sum_{k=0}^{t}\rho_k^2 - \sum_{k=0}^{t}\rho_k\right) \tag{3.16}$$

If the sequence ρ_t is chosen to satisfy the following two conditions:

$$\lim_{t\to\infty}\sum_{k=0}^{t}\rho_k = \infty \tag{3.17}$$

$$\lim_{t\to\infty}\sum_{k=0}^{t}\rho_k^2 < \infty \tag{3.18}$$

then there will be a constant t_0 such that the right-hand side of (3.16) becomes nonpositive. Thus

$$0 \leq \|\boldsymbol{w}(t_0+1)-\alpha\boldsymbol{w}^*\| \leq 0 \tag{3.19}$$

or

$$w(t_0 + 1) = \alpha w^* \tag{3.20}$$

That is, the algorithm converges to a solution in a finite number of steps. An example of a sequence satisfying conditions (3.17), (3.18) is $\rho_t = c/t$, where c is a constant. In other words, the corrections become increasingly small. What these conditions basically state is that ρ_t should vanish as $t \to \infty$ [Eq. (3.18)] but on the other hand should not go to zero very fast [Eq. (3.17)]. Following arguments similar to those used before, it is easy to show that the algorithm also converges for constant $\rho_t = \rho$ (Problem 3.2). In practice, the proper choice of the sequence ρ_t is vital for the convergence speed of the algorithm.

Example 3.1

Figure 3.3 shows the dashed line

$$x_1 + x_2 - 0.5 = 0$$

corresponding to the weight vector $[1, 1, -0.5]^T$, which has been computed from the latest iteration step of the perceptron algorithm (3.9), with $\rho_t = \rho = 0.7$. The line classifies correctly

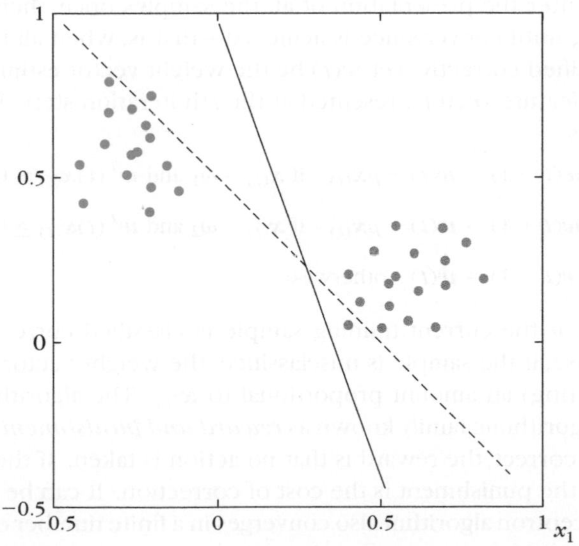

FIGURE 3.3

An example of the perceptron algorithm. After the update of the weight vector, the hyperplane is turned from its initial location (dotted line) to the new one (full line), and all points are correctly classified.

all the vectors except $[0.4, 0.05]^T$ and $[-0.20, 0.75]^T$. According to the algorithm, the next weight vector will be

$$w(t+1) = \begin{bmatrix} 1 \\ 1 \\ -0.5 \end{bmatrix} - 0.7(-1)\begin{bmatrix} 0.4 \\ 0.05 \\ 1 \end{bmatrix} - 0.7(+1)\begin{bmatrix} -0.2 \\ 0.75 \\ 1 \end{bmatrix}$$

or

$$w(t+1) = \begin{bmatrix} 1.42 \\ 0.51 \\ -0.5 \end{bmatrix}$$

The resulting new (solid) line $1.42x_1 + 0.51x_2 - 0.5 = 0$ classifies all vectors correctly, and the algorithm is terminated.

Variants of the Perceptron Algorithm

The algorithm we have presented is just one form of a number of variants that have been proposed for the training of a linear classifier in the case of linearly separable classes. We will now state another simpler and also popular form. The N training vectors enter the algorithm cyclically, one after the other. If the algorithm has not converged after the presentation of all the samples once, then the procedure keeps repeating until convergence is achieved—that is, when all training samples have been classified correctly. Let $w(t)$ be the weight vector estimate and $x_{(t)}$ the corresponding feature vector, presented at the tth iteration step. The algorithm is stated as follows:

$$w(t+1) = w(t) + \rho x_{(t)} \quad \text{if } x_{(t)} \in \omega_1 \text{ and } w^T(t)x_{(t)} \leq 0 \tag{3.21}$$

$$w(t+1) = w(t) - \rho x_{(t)} \quad \text{if } x_{(t)} \in \omega_2 \text{ and } w^T(t)x_{(t)} \geq 0 \tag{3.22}$$

$$w(t+1) = w(t) \quad \text{otherwise} \tag{3.23}$$

In other words, if the current training sample is classified correctly, no action is taken. Otherwise, if the sample is misclassified, the weight vector is corrected by adding (subtracting) an amount proportional to $x_{(t)}$. The algorithm belongs to a more general algorithmic family known as *reward and punishment* schemes. If the classification is correct, the reward is that no action is taken. If the current vector is misclassified, the punishment is the cost of correction. It can be shown that this form of the perceptron algorithm also converges in a finite number of iteration steps (Problem 3.3).

The perceptron algorithm was originally proposed by Rosenblatt in the late 1950s. The algorithm was developed for training the *perceptron*, the basic unit used for modeling neurons of the brain. This was considered central in developing powerful models for machine learning [Rose 58, Min 88].

Example 3.2

Figure 3.4 shows four points in the two-dimensional space. Points $(-1, 0), (0, 1)$ belong to class ω_1, and points $(0, -1), (1, 0)$ belong to class ω_2. The goal of this example is to design a linear classifier using the perceptron algorithm in its reward and punishment form. The parameter ρ is set equal to one, and the initial weight vector is chosen as $\boldsymbol{w}(0) = [0, 0, 0]^T$ in the extended three-dimensional space. According to (3.21)–(3.23), the following computations are in order:

Step 1.

$$\boldsymbol{w}^T(0) \begin{bmatrix} -1 \\ 0 \\ 1 \end{bmatrix} = 0, \quad \boldsymbol{w}(1) = \boldsymbol{w}(0) + \begin{bmatrix} -1 \\ 0 \\ 1 \end{bmatrix} = \begin{bmatrix} -1 \\ 0 \\ 1 \end{bmatrix}$$

Step 2.

$$\boldsymbol{w}^T(1) \begin{bmatrix} 0 \\ 1 \\ 1 \end{bmatrix} = 1 > 0, \quad \boldsymbol{w}(2) = \boldsymbol{w}(1)$$

Step 3.

$$\boldsymbol{w}^T(2) \begin{bmatrix} 0 \\ -1 \\ 1 \end{bmatrix} = 1 > 0, \quad \boldsymbol{w}(3) = \boldsymbol{w}(2) - \begin{bmatrix} 0 \\ -1 \\ 1 \end{bmatrix} = \begin{bmatrix} -1 \\ 1 \\ 0 \end{bmatrix}$$

Step 4.

$$\boldsymbol{w}^T(3) \begin{bmatrix} 1 \\ 0 \\ 1 \end{bmatrix} = -1 < 0, \quad \boldsymbol{w}(4) = \boldsymbol{w}(3)$$

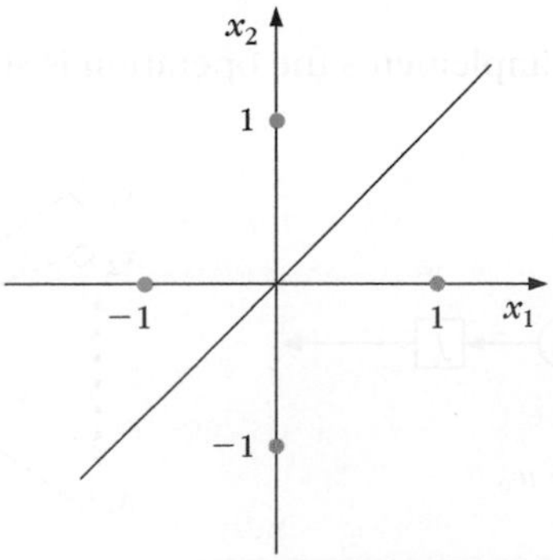

FIGURE 3.4

The setup for Example 3.2. The line $x_1 = x_2$ is the resulting solution.

Step 5.

$$w^T(4)\begin{bmatrix}-1\\0\\1\end{bmatrix} = 1 > 0, \quad w(5) = w(4)$$

Step 6.

$$w^T(5)\begin{bmatrix}0\\1\\1\end{bmatrix} = 1 > 0, \quad w(6) = w(5)$$

Step 7.

$$w^T(6)\begin{bmatrix}0\\-1\\1\end{bmatrix} = -1 < 0, \quad w(7) = w(6)$$

Since for four consecutive steps no correction is needed, all points are correctly classified and the algorithm terminates. The solution is $\boldsymbol{w} = [-1, 1, 0]^T$. That is, the resulting linear classifier is $-x_1 + x_2 = 0$, and it is the line passing through the origin shown in Figure 3.4.

The Perceptron

Once the perceptron algorithm has converged to a weight vector $\boldsymbol{w}$ and a threshold w_0, our next goal is the classification of an unknown feature vector to either of the two classes. Classification is achieved via the simple rule

$$\begin{aligned}&\text{If} \quad \boldsymbol{w}^T\boldsymbol{x} + w_0 > 0 \quad \text{assign } \boldsymbol{x} \text{ to } \omega_1\\ &\text{If} \quad \boldsymbol{w}^T\boldsymbol{x} + w_0 < 0 \quad \text{assign } \boldsymbol{x} \text{ to } \omega_2\end{aligned} \tag{3.24}$$

A basic network unit that implements the operation is shown in Figure 3.5a.

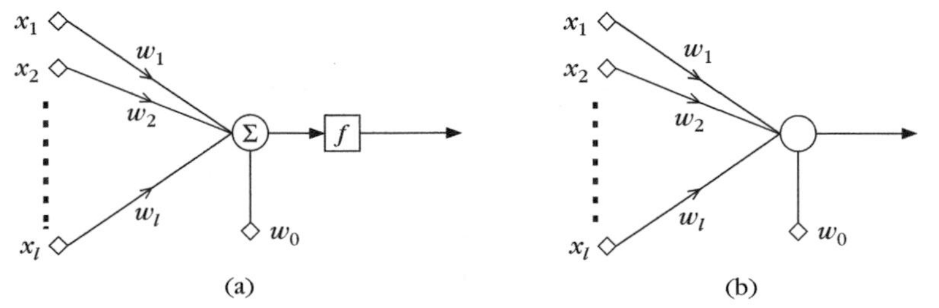

FIGURE 3.5

The basic perceptron model. (a) A linear combiner is followed by the activation function. (b) The combiner and the activation function are merged together.

The elements of the feature vector $x_1, x_2, \ldots, x_l$ are applied to the *input nodes* of the network. Then each one is multiplied by the corresponding weights $w_i, i = 1, 2, \ldots, l$. These are known as *synaptic weights* or simply *synapses*. The products are summed up together with the *threshold* value w_0. The result then goes through a nonlinear device, which implements the so-called *activation function*. A common choice is a hard limiter; that is, $f(\cdot)$ is the step function [$f(x) = -1$ if $x < 0$ and $f(x) = 1$ if $x > 0$]. The corresponding feature vector is classified in one of the classes depending on the sign of the output. Besides $+1$ and -1, other values (class labels) for the hard limiter are also possible. Another popular choice is 1 and 0, and it is achieved by choosing the two levels of the step function appropriately. This basic network is known as a *perceptron* or *neuron*. Perceptrons are simple examples of the so-called *learning machines*—that is, structures whose free parameters are updated by a *learning algorithm*, such as the perceptron algorithm, in order to "learn" a specific task, based on a set of training data. Later on we will use the perceptron as the basic building element for more complex learning networks. Figure 3.5b is a simplified graph of the neuron where the summer and nonlinear device have been merged for notational simplification. Sometimes a neuron with a hard limiter device is referred to as a McCulloch–Pitts neuron. Other types of neurons will be considered in Chapter 4.

The Pocket Algorithm

A basic requirement for the convergence of the perceptron algorithm is the linear separability of the classes. If this is not true, as is usually the case in practice, the perceptron algorithm does not converge. A variant of the perceptron algorithm was suggested in [Gal 90] that converges to an optimal solution even if the linear separability condition is not fulfilled. The algorithm is known as the *pocket* algorithm and consists of the following two steps

- Initialize the weight vector $\boldsymbol{w}(0)$ randomly. Define a stored (in the pocket!) vector $\boldsymbol{w}_s$. Set a history counter h_s of the $\boldsymbol{w}_s$ to zero.
- At the tth iteration step compute the update $\boldsymbol{w}(t+1)$, according to the perceptron rule. Use the updated weight vector to test the number h of training vectors that are classified correctly. If $h > h_s$ replace $\boldsymbol{w}_s$ with $\boldsymbol{w}(t+1)$ and h_s with h. Continue the iterations.

It can be shown that this algorithm converges with probability one to the optimal solution, that is, the one that produces the minimum number of misclassifications [Gal 90, Muse 97]. Other related algorithms that find reasonably good solutions when the classes are not linearly separable are the *thermal perceptron algorithm* [Frea 92], the *loss minimization algorithm* [Hryc 92], and the barycentric correction procedure [Poul 95].

Kesler's Construction

So far we have dealt with the two-class case. The generalization to an M-class task is straightforward. A linear discriminant function $\boldsymbol{w}_i, i = 1, 2, \ldots, M$, is defined for each of the classes. A feature vector x (in the $(l+1)$-dimensional space to account for the threshold) is classified in class ω_i if

$$w_i^T x > w_j^T x, \quad \forall j \neq i \tag{3.25}$$

This condition leads to the so-called Kesler's construction. For each of the training vectors from class $\omega_i, i = 1, 2, \ldots, M$, we construct $M-1$ vectors $x_{ij} = [0^T, 0^T, \ldots, x^T, \ldots, -x^T, \ldots, 0^T]^T$ of dimension $(l+1)M \times 1$. That is, they are block vectors having zeros everywhere except at the ith and jth block positions, where they have x and $-x$, respectively, for $j \neq i$. We also construct the block vector $w = [w_1^T, \ldots, w_M^T]^T$. If $x \in \omega_i$, this imposes the requirement that $w^T x_{ij} > 0, \forall j = 1, 2, \ldots, M, j \neq i$. The task now is to design a linear classifier, in the extended $(l+1)M$-dimensional space, so that each of the $(M-1)N$ training vectors lies in its positive side. The perceptron algorithm will have no difficulty in solving this problem for us, provided that such a solution is possible—that is, if all the training vectors can be correctly classified using a set of linear discriminant functions.

Example 3.3

Let us consider a three-class problem in the two-dimensional space. The training vectors for each of the classes are the following:

$$\omega_1\colon [1, 1]^T, [2, 2]^T, [2, 1]^T$$

$$\omega_2\colon [1, -1]^T, [1, -2]^T, [2, -2]^T$$

$$\omega_3\colon [-1, 1]^T, [-1, 2]^T, [-2, 1]^T$$

This is obviously a linearly separable problem, since the vectors of different classes lie in different quadrants.

To compute the linear discriminant functions, we first extend the vectors to the three-dimensional space, and then we use Kesler's construction. For example,

For $[1, 1]^T$ we get $[1, 1, 1, -1, -1, -1, 0, 0, 0]^T$ and

$[1, 1, 1, 0, 0, 0, -1, -1, -1]^T$

For $[1, -2]^T$ we get $[-1, 2, -1, 1, -2, 1, 0, 0, 0]^T$ and

$[0, 0, 0, 1, -2, 1, -1, 2, -1]^T$

For $[-2, 1]^T$ we get $[2, -1, -1, 0, 0, 0, -2, 1, 1]^T$ and

$[0, 0, 0, 2, -1, -1, -2, 1, 1]^T$

Similarly, we obtain the other twelve vectors. To obtain the corresponding weight vectors

$$w_1 = [w_{11}, w_{12}, w_{10}]^T$$

$$w_2 = [w_{21}, w_{22}, w_{20}]^T$$

$$w_3 = [w_{31}, w_{32}, w_{30}]^T$$

we can run the perceptron algorithm by requiring $w^T x > 0$, $w = [w_1^T, w_2^T, w_3^T]^T$, for each of the eighteen 9-dimensional vectors. That is, we require all the vectors to lie on the same side of the decision hyperplane. The initial vector of the algorithm $w(0)$ is computed using the uniform pseudorandom sequence generator in $[0, 1]$. The learning sequence ρ_t was chosen to be constant and equal to 0.5. The algorithm converges after four iterations and gives

$$w_1 = [5.13, 3.60, 1.00]^T$$

$$w_2 = [-0.05, -3.16, -0.41]^T$$

$$w_3 = [-3.84, 1.28, 0.69]^T$$

3.4 LEAST SQUARES METHODS

As we have already pointed out, the attractiveness of linear classifiers lies in their simplicity. Thus, in many cases, although we know that the classes are not linearly separable, we still wish to adopt a linear classifier, despite the fact that this will lead to *suboptimal* performance from the classification error probability point of view. The goal now is to compute the corresponding weight vector under a suitable optimality criterion. The least squares methods are familiar to us, in one way or another, from our early college courses. Let us then build upon them.

3.4.1 Mean Square Error Estimation

Let us once more focus on the two-class problem. In the previous section, we saw that the perceptron output was ± 1, depending on the class ownership of x. Since the classes were linearly separable, these outputs were correct for all the training feature vectors, after, of course, the perceptron algorithm's convergence. In this section we will attempt to design a linear classifier so that its *desired* output is again ± 1, depending on the class ownership of the input vector. However, we will have to live with errors; that is, the true output will not always be equal to the desired one. Given a vector x, the output of the classifier will be $w^T x$ (thresholds can be accommodated by vector extensions). The desired output will be denoted as $y(x) \equiv y = \pm 1$. The weight vector will be computed so as to minimize the mean square error (MSE) between the desired and true outputs, that is,

$$J(w) = E[|y - x^T w|^2] \tag{3.26}$$

$$\hat{w} = \arg\min_{w} J(w) \tag{3.27}$$

The reader can easily check that $J(w)$ is equal to

$$J(w) = P(\omega_1) \int (1 - x^T w)^2 p(x|\omega_1)\, dx + P(\omega_2) \int (1 + x^T w)^2 p(x|\omega_2)\, dx \tag{3.28}$$

Minimizing (3.27) easily results in

$$\frac{\partial J(\boldsymbol{w})}{\partial \boldsymbol{w}} = 2E[\boldsymbol{x}(y - \boldsymbol{x}^T \boldsymbol{w})] = 0 \tag{3.29}$$

Then

$$\hat{\boldsymbol{w}} = R_x^{-1} E[\boldsymbol{x}y] \tag{3.30}$$

where

$$R_x \equiv E[\boldsymbol{x}\boldsymbol{x}^T] = \begin{bmatrix} E[x_1x_1] & \cdots & E[x_1x_l] \\ E[x_2x_1] & \cdots & E[x_2x_l] \\ \vdots & \vdots & \vdots \\ E[x_lx_1] & \cdots & E[x_lx_l] \end{bmatrix} \tag{3.31}$$

is known as the *correlation* or autocorrelation matrix and is equal to the covariance matrix, introduced in the previous chapter, if the respective mean values are zero. The vector

$$E[\boldsymbol{x}y] = E\left[\begin{bmatrix} x_1y \\ \vdots \\ x_ly \end{bmatrix}\right] \tag{3.32}$$

is known as the *cross-correlation* between the desired output and the (input) feature vectors. Thus, the mean square optimal weight vector results as the solution of a *linear set of equations*, provided, of course, that the correlation matrix is invertible.

It is interesting to point out that there is a geometrical interpretation of this solution. Random variables can be considered as points in a vector space. It is straightforward to see that the expectation operation $E[xy]$ between two random variables satisfies the properties of the inner product. Indeed, $E[x^2] \geq 0, E[xy] = E[yx], E[x(c_1y + c_2z)] = c_1E[xy] + c_2E[xz]$. In such a vector space $\boldsymbol{w}^T\boldsymbol{x} = w_1x_1 + \cdots + w_lx_l$ is a linear combination of vectors, and thus it lies in the subspace defined by the x_i's.

This is illustrated by an example in Figure 3.6. Then, if we want to approximate y by this linear combination, the resulting error is $y - \boldsymbol{w}^T\boldsymbol{x}$. Equation (3.29) states that the minimum mean square error solution results if the error is orthogonal to each x_i; thus it is *orthogonal* to the vector subspace spanned by $x_i, i = 1, 2, \ldots, l$—in other words, if y is approximated by its *orthogonal projection* on the subspace (Figure 3.6). Equation (3.29) is also known as the *orthogonality condition*.

Multiclass Generalization

In the multiclass case, the task is to design the M linear discriminant functions $g_i(\boldsymbol{x}) = \boldsymbol{w}_i^T\boldsymbol{x}$ according to the MSE criterion. The corresponding desired output responses (i.e., class labels) are chosen so that $y_i = 1$ if $\boldsymbol{x} \in \omega_i$ and $y_i = 0$ otherwise. This is in agreement with the two-class case. Indeed, for such

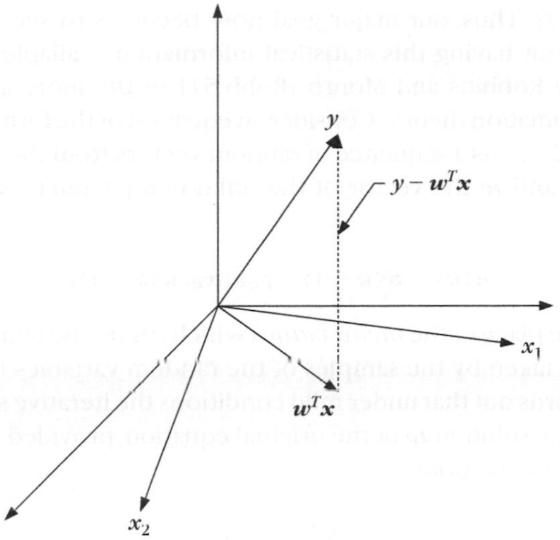

FIGURE 3.6

Interpretation of the MSE estimate as an orthogonal projection on the input vector elements' subspace.

a choice and if $M = 2$, the design of the decision hyperplane $\boldsymbol{w}^T\boldsymbol{x} \equiv (\boldsymbol{w}_1 - \boldsymbol{w}_2)^T\boldsymbol{x}$ corresponds to ± 1 desired responses, depending on the respective class ownership.

Let us now define $\boldsymbol{y}^T = [y_1, \ldots, y_M]$, for a given vector $\boldsymbol{x}$, and $W = [\boldsymbol{w}_1, \ldots, \boldsymbol{w}_M]$. That is, matrix W has as columns the weight vectors $\boldsymbol{w}_i$. The MSE criterion in (3.27) can now be generalized to minimize the norm of the error vector $\boldsymbol{y} - W^T\boldsymbol{x}$, that is,

$$\hat{W} = \arg\min_W E[\|\boldsymbol{y} - W^T\boldsymbol{x}\|^2] = \arg\min_W E\left[\sum_{i=1}^{M}\left(y_i - \boldsymbol{w}_i^T\boldsymbol{x}\right)^2\right] \tag{3.33}$$

This is equivalent to M MSE independent minimization problems of the (3.27) type, with scalar desired responses. *In other words, in order to design the MSE optimal linear discriminant functions, it suffices to design each one of them so that its desired output is* 1 *for vectors belonging to the corresponding class and* 0 *for all the others.*

3.4.2 Stochastic Approximation and the LMS Algorithm

The solution of (3.30) requires the computation of the correlation matrix and cross-correlation vector. This presupposes knowledge of the underlying distributions, which in general are not known. After all, if they were known, why not then use

Bayesian classifiers? Thus, our major goal now becomes to see if it is possible to solve (3.29) without having this statistical information available. The answer has been provided by Robbins and Monro [Robb 51] in the more general context of stochastic approximation theory. Consider an equation of the form $E[F(\boldsymbol{x}_k, \boldsymbol{w})] = \mathbf{0}$, where $\boldsymbol{x}_k, k = 1, 2, \ldots$, is a sequence of random vectors from the same distribution, $F(\cdot, \cdot)$ a function, and $\boldsymbol{w}$ the vector of the unknown parameters. Then adopt the iterative scheme

$$\hat{\boldsymbol{w}}(k) = \hat{\boldsymbol{w}}(k-1) + \rho_k F(\boldsymbol{x}_k, \hat{\boldsymbol{w}}(k-1)) \tag{3.34}$$

In other words, the place of the *mean value* (which cannot be computed due to lack of information) is taken by the samples of the random variables resulting from the experiments. It turns out that under mild conditions the iterative scheme converges in probability to the solution $\boldsymbol{w}$ of the original equation, provided that the sequence ρ_k satisfies the two conditions

$$\sum_{k=1}^{\infty} \rho_k \rightarrow \infty \tag{3.35}$$

$$\sum_{k=1}^{\infty} \rho_k^2 < \infty \tag{3.36}$$

and which implies that

$$\rho_k \rightarrow 0 \tag{3.37}$$

That is,

$$\lim_{k\to\infty} \text{prob}\{\hat{\boldsymbol{w}}(k) = \boldsymbol{w}\} = 1 \tag{3.38}$$

The stronger, in the mean square sense, convergence is also true

$$\lim_{k\to\infty} E[\|\hat{\boldsymbol{w}}(k) - \boldsymbol{w}\|^2] = 0 \tag{3.39}$$

Conditions (3.35), (3.36) have already been met before and guarantee that the corrections of the estimates in the iterations tend to zero. Thus, for large values of k (in theory at infinity) iterations freeze. However, this must not happen too early (first condition) to make sure that the iterations do not stop away from the solution. The second condition guarantees that the accumulated noise, due to the stochastic nature of the variables, remains finite and the algorithm can cope with it [Fuku 90]. The proof is beyond the scope of the present text. However, we will demonstrate its validity via an example.

Let us consider the simple equation $E[x_k - w] = 0$. For $\rho_k = 1/k$ the iteration becomes

$$\hat{w}(k) = \hat{w}(k-1) + \frac{1}{k}[x_k - \hat{w}(k-1)] = \frac{(k-1)}{k}\hat{w}(k-1) + \frac{1}{k}x_k$$

For large values of k it is easy to see that

$$\hat{w}(k) = \frac{1}{k}\sum_{r=1}^{k} x_r$$

That is, the solution is the sample mean of the measurements. Most natural!

Let us now return to our original problem and apply the iteration to solve (3.29). Then (3.34) becomes

$$\hat{\boldsymbol{w}}(k) = \hat{\boldsymbol{w}}(k-1) + \rho_k \boldsymbol{x}_k \left(y_k - \boldsymbol{x}_k^T \hat{\boldsymbol{w}}(k-1)\right) \tag{3.40}$$

where $(y_k, \boldsymbol{x}_k)$ are the desired output (± 1)-input training sample pairs, successively presented to the algorithm. The algorithm is known as the least mean squares (LMS) or Widrow–Hoff algorithm, after those who suggested it in the early 1960s [Widr 60, Widr 90]. The algorithm converges asymptotically to the MSE solution.

A number of variants of the LMS algorithm have been suggested and used. The interested reader may consult, for example, [Hayk 96, Kalou 93]. A common variant is to use a constant ρ in the place of ρ_k. However, in this case the algorithm does not converge to the MSE solution. It can be shown, for example, [Hayk 96], that if $0 < \rho < 2/\text{trace}\{R_x\}$ then

$$E[\hat{\boldsymbol{w}}(k)] \rightarrow \boldsymbol{w}_{\text{MSE}} \quad \text{and} \quad E[\|\hat{\boldsymbol{w}}(k) - \boldsymbol{w}_{\text{MSE}}\|^2] \rightarrow \text{constant} \tag{3.41}$$

where $\boldsymbol{w}_{\text{MSE}}$ denotes the MSE optimal estimate and trace$\{\cdot\}$ the trace of the matrix. That is, the mean value of the LMS estimate is equal to the MSE solution, and also the corresponding variance remains finite. It turns out that the smaller the ρ, the smaller the variance around the desired MSE solution. However, the smaller the ρ, the slower the convergence of the LMS algorithm. The reason for using Constant ρ in place of a vanishing sequence is to keep the algorithm "alert" to track variations when the statistics are not stationary but are *slowly* varying, that is, when the underlying distributions are time dependent.

Remarks

- Observe that in the case of the LMS, the parameters' update iteration step, k, coincides with the index of the current input sample $\boldsymbol{x}_k$. In case k is a time index, LMS is a time-*adaptive* scheme, which adapts to the solution as successive samples become available to the system.

- Observe that Eq. (3.40) can be seen as the training algorithm of a *linear neuron*, that is, a neuron without the nonlinear activation function. This type of training, which neglects the nonlinearity *during training* and applies the desired response just after the adder of the linear combiner part of the neuron (Figure 3.5a), was used by Widrow and Hoff. The resulting neuron architecture is known as *adaline* (adaptive linear element). After training and once the weights have been fixed, the model is the same as in Figure 3.5, with the hard limiter following the linear combiner. In other words, the adaline is a neuron that is trained according to the LMS instead of the perceptron algorithm.

3.4.3 Sum of Error Squares Estimation

A criterion closely related to the MSE is the *sum of error squares* or simply the *least squares* (LS) criterion defined as

$$J(\boldsymbol{w}) = \sum_{i=1}^{N} (y_i - \boldsymbol{x}_i^T \boldsymbol{w})^2 \equiv \sum_{i=1}^{N} e_i^2 \tag{3.42}$$

In other words, the errors between the desired output of the classifier (± 1 in the two-class case) and the true output are summed up over all the available training feature vectors, instead of averaging them out. In this way, we overcome the need for explicit knowledge of the underlying pdfs. Minimizing (3.42) with respect to $\boldsymbol{w}$ results in

$$\sum_{i=1}^{N} \boldsymbol{x}_i (y_i - \boldsymbol{x}_i^T \hat{\boldsymbol{w}}) = 0 \Rightarrow \left(\sum_{i=1}^{N} \boldsymbol{x}_i \boldsymbol{x}_i^T \right) \hat{\boldsymbol{w}} = \sum_{i=1}^{N} (\boldsymbol{x}_i y_i) \tag{3.43}$$

For the sake of mathematical formulation let us define

$$X = \begin{bmatrix} \boldsymbol{x}_1^T \\ \boldsymbol{x}_2^T \\ \vdots \\ \boldsymbol{x}_N^T \end{bmatrix} = \begin{bmatrix} x_{11} & x_{12} & \dots & x_{1l} \\ x_{21} & x_{22} & \dots & x_{2l} \\ \vdots & \vdots & \ddots & \vdots \\ x_{N1} & x_{N2} & \dots & x_{Nl} \end{bmatrix}, \quad \boldsymbol{y} = \begin{bmatrix} y_1 \\ y_2 \\ \vdots \\ y_N \end{bmatrix} \tag{3.44}$$

That is, X is an $N \times l$ matrix whose rows are the available training feature vectors, and $\boldsymbol{y}$ is a vector consisting of the corresponding desired responses. Then $\sum_{i=1}^{N} \boldsymbol{x}_i \boldsymbol{x}_i^T = X^T X$ and also $\sum_{i=1}^{N} \boldsymbol{x}_i y_i = X^T \boldsymbol{y}$. Hence, (3.43) can now be written as

$$(X^T X)\hat{\boldsymbol{w}} = X^T \boldsymbol{y} \Rightarrow \hat{\boldsymbol{w}} = (X^T X)^{-1} X^T \boldsymbol{y} \tag{3.45}$$

Thus, the optimal weight vector is again provided as the solution of a linear set of equations. Matrix $X^T X$ is known as the *sample correlation matrix*. Matrix $X^+ \equiv (X^T X)^{-1} X^T$ is known as the *pseudoinverse* of X, and it is meaningful only if $X^T X$ is invertible, that is, X is of rank l. X^+ is a generalization of the inverse of an invertible square matrix. Indeed, if X is an $l \times l$ square and invertible matrix, then it is straightforward to see that $X^+ = X^{-1}$. In such a case the estimated weight vector is the solution of the linear system $X\hat{w} = y$. If, however, there are more equations than unknowns, $N > l$, as is the usual case in pattern recognition, there is not, in general, a solution. The solution obtained by the pseudoinverse is the vector that minimizes the sum of error squares. It is easy to show that (under mild assumptions) the sum of error squares tends to the MSE solution for large values of N (Problem 3.8).

Remarks

- So far we have restricted the desired output values to be ± 1. Of course, this is not necessary. All we actually need is a desired response positive for ω_1 and negative for ω_2. Thus, in place of ± 1 in the y vector we could have any positive (negative) values. Obviously, all we have said so far is still applicable. However, the interesting aspect of this generalization would be to compute these desired values in an optimal way, in order to obtain a better solution. The Ho–Kashyap algorithm is such a scheme solving for both the optimal w and optimal desired values y_i. The interested reader may consult [Ho 65, Tou 74].
- Generalization to the multi-class case follows the same concept as that introduced for the MSE cost, and it is easily shown that it reduces to M equivalent problems of scalar desired responses, one for each discriminant function (Problem 3.10).

Example 3.4

Class ω_1 consists of the two-dimensional vectors $[0.2, 0.7]^T$, $[0.3, 0.3]^T$, $[0.4, 0.5]^T$, $[0.6, 0.5]^T$, $[0.1, 0.4]^T$ and class ω_2 of $[0.4, 0.6]^T$, $[0.6, 0.2]^T$, $[0.7, 0.4]^T$, $[0.8, 0.6]^T$, $[0.7, 0.5]^T$. Design the sum of error squares optimal linear classifier $w_1x_1 + w_2x_2 + w_0 = 0$.

We first extend the given vectors by using 1 as their third dimension and form the 10×3 matrix X, which has as rows the transposes of these vectors. The resulting sample correlation 3×3 matrix X^TX is equal to

$$X^TX = \begin{bmatrix} 2.8 & 2.24 & 4.8 \\ 2.24 & 2.41 & 4.7 \\ 4.8 & 4.7 & 10 \end{bmatrix}$$

The corresponding y consists of five 1's and then five -1's and

$$X^Ty = \begin{bmatrix} -1.6 \\ 0.1 \\ 0.0 \end{bmatrix}$$

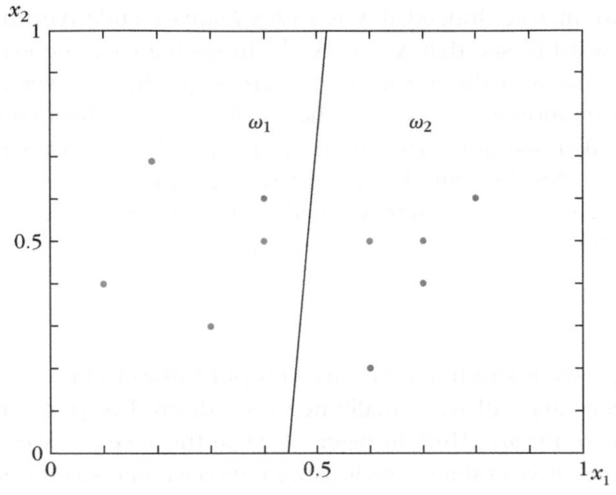

FIGURE 3.7

Least sum of error squares linear classifier. The task is not linearly separable. The linear LS classifier classifies some of the points in the wrong class. However, the resulting sum of error squares is minimum.

Solving the corresponding set of equations results in $[w_1, w_2, w_0] = [-3.218, 0.241, 1.431]$. Figure 3.7 shows the resulting geometry.

3.5 MEAN SQUARE ESTIMATION REVISITED

3.5.1 Mean Square Error Regression

In this subsection, we will approach the MSE task from a slightly different perspective and in a more general framework.

Let $\boldsymbol{y}, \boldsymbol{x}$ be two random vector variables of dimensions $M \times 1$ and $l \times 1$, respectively, and assume that they are described by the joint pdf $p(\boldsymbol{y}, \boldsymbol{x})$. The task of interest is to estimate the value of $\boldsymbol{y}$, given the value of $\boldsymbol{x}$ that is obtained from an experiment. No doubt the classification task falls under this more general formulation. For example, when we are given a feature vector $\boldsymbol{x}$, our goal is to estimate the value of the class label y, which is ± 1 in the two-class case. In a more general setting, the values of $\boldsymbol{y}$ may not be discrete. Take, as an example, the case where $y \in \mathcal{R}$ is generated by an unknown rule, i.e.,

$$y = f(\boldsymbol{x}) + \epsilon$$

where $f(\cdot)$ is some unknown function and ϵ is a noise source. The task now is to estimate (predict) the value of y, given the value of $\boldsymbol{x}$. Once more, this is a problem of designing a function $g(\boldsymbol{x})$, based on a set of training data points $(\boldsymbol{y}_i, \boldsymbol{x}_i)$, $i = 1, 2, \ldots, N$, so that the predicted value

$$\hat{\boldsymbol{y}} = \boldsymbol{g}(\boldsymbol{x})$$

to be as close as possible to the true value $\boldsymbol{y}$ in some optimal sense. This type of problem is known as a *regression* task. One of the most popular optimality criteria for regression is the mean square error (MSE). In this section, we will focus on the MSE regression and highlight some of its properties.

The mean square estimate $\hat{\boldsymbol{y}}$ of the random vector $\boldsymbol{y}$, given the value $\boldsymbol{x}$, is defined as

$$\hat{\boldsymbol{y}} = \arg\min_{\tilde{\boldsymbol{y}}} E[\|\boldsymbol{y} - \tilde{\boldsymbol{y}}\|^2] \tag{3.46}$$

Note that the mean value here is with respect to the conditional pdf $p(\boldsymbol{y}|\boldsymbol{x})$. We will show that the optimal estimate is the mean value of $\boldsymbol{y}$, that is,

$$\hat{\boldsymbol{y}} = E[\boldsymbol{y}|\boldsymbol{x}] \equiv \int_{-\infty}^{\infty} \boldsymbol{y} p(\boldsymbol{y}|\boldsymbol{x})\, d\boldsymbol{y} \tag{3.47}$$

Proof. Let $\tilde{\boldsymbol{y}}$ be another estimate. It will be shown that it results in higher mean square error. Indeed,

$$E[\|\boldsymbol{y} - \tilde{\boldsymbol{y}}\|^2] = E[\|\boldsymbol{y} - \hat{\boldsymbol{y}} + \hat{\boldsymbol{y}} - \tilde{\boldsymbol{y}}\|^2] = E[\|\boldsymbol{y} - \hat{\boldsymbol{y}}\|^2] + E[\|\hat{\boldsymbol{y}} - \tilde{\boldsymbol{y}}\|^2] + 2E[(\boldsymbol{y} - \hat{\boldsymbol{y}})^T(\hat{\boldsymbol{y}} - \tilde{\boldsymbol{y}})] \tag{3.48}$$

where the dependence on $\boldsymbol{x}$ has been omitted for notational convenience. Note now that $\hat{\boldsymbol{y}} - \tilde{\boldsymbol{y}}$ is a constant. Thus,

$$E[\|\boldsymbol{y} - \tilde{\boldsymbol{y}}\|^2] \geq E[\|\boldsymbol{y} - \hat{\boldsymbol{y}}\|^2] + 2E[(\boldsymbol{y} - \hat{\boldsymbol{y}})^T](\hat{\boldsymbol{y}} - \tilde{\boldsymbol{y}}) \tag{3.49}$$

and from the definition of $\hat{\boldsymbol{y}} = E[\boldsymbol{y}]$ it follows that

$$E[\|\boldsymbol{y} - \tilde{\boldsymbol{y}}\|^2] \geq E[\|\boldsymbol{y} - \hat{\boldsymbol{y}}\|^2] \tag{3.50}$$

□

Remark

- This is a very elegant result. Given a measured value of $\boldsymbol{x}$, the best (in the MSE sense) estimate of $\boldsymbol{y}$ is given by the function $\boldsymbol{y}(\boldsymbol{x}) \equiv E[\boldsymbol{y}|\boldsymbol{x}]$. In general, this is a *nonlinear* vector-valued function of $\boldsymbol{x}$ (i.e., $\boldsymbol{g}(\cdot) \equiv [g_1(\cdot), \ldots, g_M(\cdot)]^T$), and it is known as the *regression* of $\boldsymbol{y}$ *conditioned* on $\boldsymbol{x}$. It can be shown (Problem 3.11) that if $(\boldsymbol{y}, \boldsymbol{x})$ are jointly Gaussian, then the MSE optimal regressor is a linear function.

3.5.2 MSE Estimates Posterior Class Probabilities

In the beginning of the chapter we promised to "emancipate" ourselves from the Bayesian classification. However, the nice surprise is that a Bayesian flavor still remains, although in a disguised form. Let us reveal it—it can only be beneficial.

We will consider the multiclass case. Given $\boldsymbol{x}$, we want to estimate its class label. Let $g_i(\boldsymbol{x})$ be the discriminant functions to be designed. The cost function in Eq. (3.33) now becomes

$$J = E\left[\sum_{i=1}^{M}(g_i(\boldsymbol{x}) - y_i)^2\right] \equiv E[\|\boldsymbol{g}(\boldsymbol{x}) - \boldsymbol{y}\|^2] \tag{3.51}$$

where the vector $\boldsymbol{y}$ consists of zeros and a single 1 at the appropriate place. Note that each $g_i(\boldsymbol{x})$ depends *only* on $\boldsymbol{x}$, whereas the y_i's depend on the specific class to which $\boldsymbol{x}$ belongs. Let $p(\boldsymbol{x}, \omega_i)$ be the joint probability density of the feature vector belonging to the ith class. Then (3.51) is written as

$$J = \int_{-\infty}^{+\infty} \sum_{j=1}^{M}\left\{\sum_{i=1}^{M}(g_i(\boldsymbol{x}) - y_i)^2\right\} p(\boldsymbol{x}, \omega_j)\, d\boldsymbol{x} \tag{3.52}$$

Taking into account that $p(\boldsymbol{x}, \omega_j) = P(\omega_j|\boldsymbol{x})p(\boldsymbol{x})$, (3.52) becomes

$$\begin{aligned} J &= \int_{-\infty}^{\infty}\left\{\sum_{j=1}^{M}\sum_{i=1}^{M}\left(g_i(\boldsymbol{x}) - y_i\right)^2 P(\omega_j|\boldsymbol{x})\right\} p(\boldsymbol{x})\, d\boldsymbol{x} \\ &= E\left[\sum_{j=1}^{M}\sum_{i=1}^{M}\left(g_i(\boldsymbol{x}) - y_i\right)^2 P(\omega_j|\boldsymbol{x})\right] \end{aligned} \tag{3.53}$$

where the mean is taken with respect to $\boldsymbol{x}$. Expanding this, we get

$$J = E\left[\sum_{j=1}^{M}\sum_{i=1}^{M}\left(g_i^2(\boldsymbol{x})P(\omega_j|\boldsymbol{x}) - 2g_i(\boldsymbol{x})y_iP(\omega_j|\boldsymbol{x}) + y_i^2P(\omega_j|\boldsymbol{x})\right)\right] \tag{3.54}$$

Exploiting the fact that $g_i(\boldsymbol{x})$ is a function of $\boldsymbol{x}$ only and $\sum_{j=1}^{M} P(\omega_j|\boldsymbol{x}) = 1$, (3.54) becomes

$$\begin{aligned} J &= E\left[\sum_{i=1}^{M}\left(g_i^2(\boldsymbol{x}) - 2g_i(\boldsymbol{x})\sum_{j=1}^{M} y_iP(\omega_j|\boldsymbol{x}) + \sum_{j=1}^{M} y_i^2P(\omega_j|\boldsymbol{x})\right)\right] \\ &= E\left[\sum_{i=1}^{M}\left(g_i^2(\boldsymbol{x}) - 2g_i(\boldsymbol{x})E[y_i|\boldsymbol{x}] + E\left[y_i^2|\boldsymbol{x}\right]\right)\right] \end{aligned} \tag{3.55}$$

where $E[y_i|\boldsymbol{x}]$ and $E[y_i^2|\boldsymbol{x}]$ are the respective mean values conditioned on $\boldsymbol{x}$. Adding and subtracting $(E[y_i|\boldsymbol{x}])^2$, Eq. (3.55) becomes

$$J = E\left[\sum_{i=1}^{M}(g_i(\boldsymbol{x}) - E[y_i|\boldsymbol{x}])^2\right] + E\left[\sum_{i=1}^{M}(E[y_i^2|\boldsymbol{x}] - (E[y_i|\boldsymbol{x}])^2)\right] \tag{3.56}$$

The second term in (3.56) does not depend on the functions $g_i(\boldsymbol{x})$, $i = 1, 2, \ldots, M$. Thus, minimization of J with respect to (the parameters of) $g_i(\cdot)$ affects only the first of the two terms. Let us concentrate and look at it more carefully. Each of the M summands involves two terms: the unknown discriminant function $g_i(\cdot)$ and the conditional mean of the corresponding desired response. Let us now write $g_i(\cdot) = g_i(\cdot; \boldsymbol{w}_i)$, to state explicitly that the functions are defined in terms of a set of parameters, to be determined optimally during training. Minimizing J with respect to $\boldsymbol{w}_i$, $i = 1, 2, \ldots, M$, results in *the mean square estimates of the unknown parameters, $\hat{\boldsymbol{w}}_i$, so that the discriminant functions approximate optimally the corresponding conditional means—that is, the regressions of y_i conditioned on $\boldsymbol{x}$.* Moreover, for the M-class problem and the preceding definitions we have

$$E[y_i|\boldsymbol{x}] \equiv \sum_{j=1}^{M} y_i P(\omega_j|\boldsymbol{x}) \tag{3.57}$$

However $y_i = 1(0)$ if $\boldsymbol{x} \in \omega_i (\boldsymbol{x} \in \omega_j, j \neq i)$. Hence

$$g_i(\boldsymbol{x}, \hat{\boldsymbol{w}}_i) \text{ is the MSE estimate of } P(\omega_i|\boldsymbol{x}) \tag{3.58}$$

This is an important result. Training the discriminant functions g_i with desired outputs 1 or 0 in the MSE sense, Eq. (3.51) is equivalent to obtaining the MSE estimates of the class posterior probabilities, without using any statistical information or pdf modeling! It suffices to say that these estimates may in turn be used for Bayesian classification. An important issue here is to assess *how good the resulting estimates are. It all depends on how well the adopted functions $g_i(\cdot; \boldsymbol{w}_i)$ can model the desired (in general) nonlinear functions $P(\omega_i|\boldsymbol{x})$.* If, for example, we adopt linear models, as was the case in Eq. (3.33), and $P(\omega_i|\boldsymbol{x})$ is highly nonlinear, the resulting MSE optimal approximation will be a bad one. Our focus in the next chapter will be on developing modeling techniques for nonlinear functions.

Finally, it must be emphasized that the *conclusion above is an implication of the cost function itself and not of the specific model function used.* The latter plays its part when the approximation accuracy issue comes into the scene. MSE cost is just one of the costs that have this important property. Other cost functions share this property too, see, for example, [Rich 91, Bish 95, Pear 90, Cid 99]. In [Guer 04] a procedure is developed to design cost functions that provide more accurate estimates of the probability values, taking into account the characteristics of each classification problem.

3.5.3 The Bias–Variance Dilemma

So far we have touched on some very important issues concerning the interpretation of the output of an optimally designed classifier. Also, we saw that a regressor or a classifier can be viewed as *learning machines* realizing a function or a set of functions $g(\boldsymbol{x})$, which attempt to estimate the corresponding value or class label y and make a decision based on these estimates. In practice, the functions $g(\cdot)$ are estimated using a *finite* training data set $\mathcal{D} = \{(y_i, \boldsymbol{x}_i), i = 1, 2, \dots, N\}$ and a suitable methodology (e.g., mean square error, sum of error squares, LMS). To emphasize the explicit dependence on $\mathcal{D}$, we write $g(\boldsymbol{x}; \mathcal{D})$. This subsection focuses on the capabilities of $g(\boldsymbol{x}; \mathcal{D})$ to approximate the MSE optimal regressor $E[y|\boldsymbol{x}]$ and on how this is affected by the *finite* size, N, of the training data set.

The key factor here is the dependence of the approximation on $\mathcal{D}$. The approximation may be very good for a specific training data set but very bad for another. The effectiveness of an estimator can be evaluated by computing its mean square deviation from the desired optimal value. This is achieved by averaging over all possible sets $\mathcal{D}$ of size N, that is,

$$E_{\mathcal{D}}\left[\left(g(\boldsymbol{x}; \mathcal{D}) - E[y|\boldsymbol{x}]\right)^2\right] \tag{3.59}$$

If we add and subtract $E_{\mathcal{D}}[g(\boldsymbol{x}; \mathcal{D})]$ and follow a procedure similar to that in the proof of (3.47), we easily obtain

$$\begin{aligned} E_{\mathcal{D}}\left[\left(g(\boldsymbol{x}; \mathcal{D}) - E[y|\boldsymbol{x}]\right)^2\right] &= \left(E_{\mathcal{D}}[g(\boldsymbol{x}; \mathcal{D})] - E[y|\boldsymbol{x}]\right)^2 \\ &\quad + E_{\mathcal{D}}\left[\left(g(\boldsymbol{x}; \mathcal{D}) - E_{\mathcal{D}}[g(\boldsymbol{x}; \mathcal{D})]\right)^2\right] \end{aligned} \tag{3.60}$$

The first term is the contribution of the *bias* and the second that of the *variance*. In other words, even if the estimator is unbiased, it can still result in a large mean square error due to a large variance term. For a finite data set, it turns out that there is a trade-off between these two terms. Increasing the bias decreases the variance and vice versa. This is known as the *bias–variance dilemma*. This behavior is reasonable. The problem at hand is similar to that of a curve fitting through a given data set. If, for example, the adopted model is complex (many parameters involved) with respect to the number N, the model will fit the idiosyncrasies of the specific data set. *Thus, it will result in low bias but will yield high variance, as we change from one data set to another*. The major issue now is to seek ways to make both bias and variance low at the same time. It turns out that this may be possible only asymptotically, as the number N grows to infinity. Moreover, N has to grow in such a way as *to allow more complex models, g, to be fitted (which reduces bias) and at the same time to ensure low variance*. However, in practice N is finite, and one should aim at the best compromise. If, on the other hand, some *a priori* knowledge is available, this must be exploited in the form of constraints that the classifier/regressor has to satisfy. This can lead to lower values of both the variance and the bias, compared with a more general type of classifier/regressor. This is natural, because one takes advantage of the available information and helps the optimization process.

Let us now use two simplified "extreme" example cases, which will help us grasp the meaning of the bias–variance dilemma using common-sense reasoning. Let us assume that our data are generated by the following mechanism

$$y = f(x) + \epsilon$$

where $f(\cdot)$ is an unknown function and ϵ a noise source of zero mean and known variance equal to, say, σ_ϵ^2. Obviously, for any x, the optimum MSE regressor is $E[y|x] = f(x)$. To make our point easier to understand, let us further assume that the randomness in the different training sets, $\mathcal{D}$, is due to the y_i's (whose values are affected by the noise), while the respective points, x_i, are fixed. Such an assumption is not an unreasonable one. Since our goal is to obtain an estimate of $f(\cdot)$, it is sensible for one to divide the interval $[x_1, x_2]$, in which x lies, in equally spaced points. For example, one can choose $x_i = x_1 + \frac{x_2 - x_1}{N-1}(i-1)$, $i = 1, 2, \ldots, N$.

- *Case 1.* Choose the estimate of $f(x)$, $g(x; \mathcal{D})$, to be independent of $\mathcal{D}$, for example,

$$g(x) = w_1 x + w_0$$

  for some fixed values of w_1 and w_0. Figure 3.8 illustrates this setup showing a line $g(x)$ and $N = 11$ training pairs (y_i, x_i), which spread around $f(x), x \in [0, 1]$. Since $g(x)$ is fixed and does not change, we have $E_\mathcal{D}[g(x; \mathcal{D})] = g(x; \mathcal{D}) \equiv g(x)$, and the variance term in (3.60) is zero. On the other hand, since $g(x)$ has been chosen arbitrarily, in general, one expects the bias term to be large.

- *Case 2.* In contrast to $g(x)$, the function $g_1(x)$, shown in Figure 3.8, corresponds to a polynomial of high degree and with a large enough number of free parameters so that, for each one of the different training sets $\mathcal{D}$, the respective graphs of $g_1(x)$ pass through the training points (y_i, x_i), $i = 1, 2, \ldots, 11$. For this case, due to the zero mean of the noise source, we have that $E_\mathcal{D}[g_1(x; \mathcal{D})] = f(x) = E[y|x]$, for any $x = x_i$. That is, at the training points, the bias is zero. Due to the continuity of $f(x)$ and $g_1(x)$, one expects similar behavior and at the points that lie in the vicinity of the training points x_i. Thus, if N is large enough we can expect the bias to be small for all the points in the interval $[0, 1]$. However, now the variance increases. Indeed, for this case we have that

$$E_\mathcal{D}\left[\left(g_1(x; \mathcal{D}) - E_\mathcal{D}[g_1(x; \mathcal{D})]\right)^2\right] = E_\mathcal{D}\left[\left(f(x) + \epsilon - f(x)\right)^2\right]$$

$$= \sigma_\epsilon^2, \text{ for } x = x_i,\ i = 1, 2, \ldots, N$$

  In other words, the bias becomes zero (or approximately zero) but the variance is now equal to the variance of the noise source.

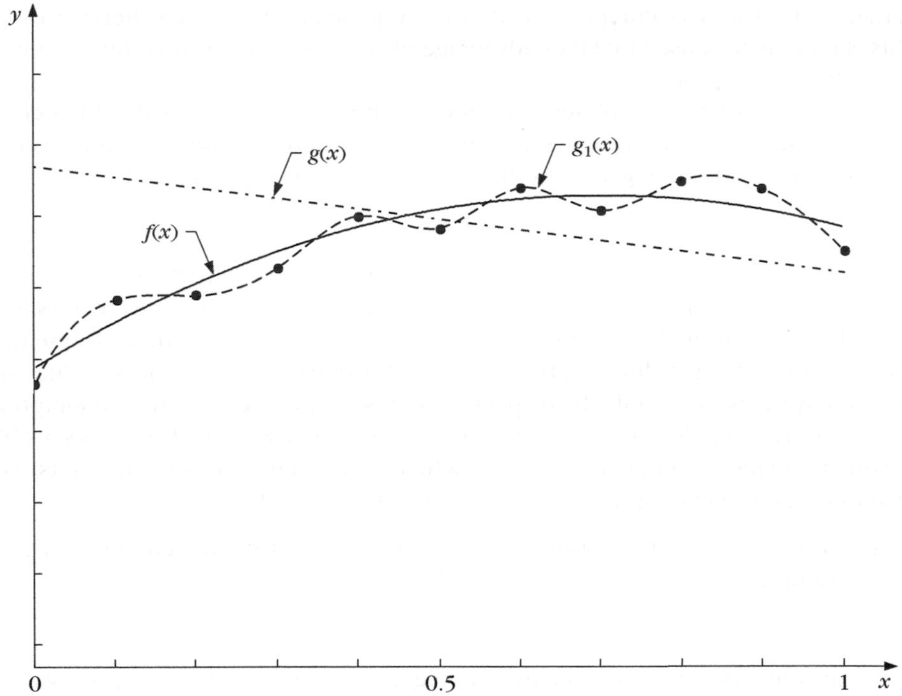

FIGURE 3.8

The data points are spread around the $f(x)$ curve. The line $g(x) = 0$ exhibits zero variance but high bias. The high degree polynomial curve, $g_1(x) = 0$, always passes through the training points and leads to low bias (zero bias at the training points) but to high variance.

The reader will notice that everything that has been said so far applies to both the regression and the classification tasks. The reason that we talked only for regression is that the mean square error is not the best criterion to validate the performance of a classifier. After all, we may have a classifier that results in high mean square error, yet its error performance can be very good. Take, as an example, the case of a classifier $g(x)$ resulting in relatively high mean square error, but predicts the correct class label y for most of the values of x. That is, for all points originating from class ω_1 (ω_2) the predicted values lie, for most of the cases, on the correct side of the classifier, albeit with a lot of variation (for the different training sets) and away from the desired values of ± 1. From the classification point of view, such a designed classifier would be perfectly acceptable. Concerning the preceding theory, in order to get more meaningful results, for the classification task, one has to rework the previous theory in terms of the probability of error. However, now the algebra gets a bit more involved, and some further assumptions need to be adopted (e.g., Gaussian data), in order to make the algebra more tractable. We will not delve into that, since more recent and elegant theories are now available, which study the trade-off between model complexity and the accuracy of the resulting classifier for finite data sets in a generalized framework (see Chapter 5).

A simple and excellent treatment of the bias–variance dilemma task can be found in [Gema 92]. As for ourselves, this was only the beginning. We will come to the finite data set issue and its implications many times throughout this book and from different points of view.

3.6 LOGISTIC DISCRIMINATION

In *logistic discrimination* the logarithm of the likelihood ratios [Eq. (2.20)] is modeled via linear functions. That is,

$$\ln \frac{P(\omega_i|\boldsymbol{x})}{P(\omega_M|\boldsymbol{x})} = w_{i,0} + \boldsymbol{w}_i^T \boldsymbol{x},\ i = 1, 2, \ldots, M-1 \tag{3.61}$$

In the denominator, any class other than ω_M can also be used. The unknown parameters, $w_{i,0}, \boldsymbol{w}_i,\ i = 1, 2, \ldots, M-1$, must be chosen to ensure that probabilities add to one. That is,

$$\sum_{i=1}^{M} P(\omega_i|\boldsymbol{x}) = 1 \tag{3.62}$$

Combining (3.61) and (3.62), it is straightforward to see that this type of linear modeling is equivalent to an exponential modeling of the *a posteriori* probabilities

$$P(\omega_M|\boldsymbol{x}) = \frac{1}{1 + \sum_{i=1}^{M-1} \exp\left(w_{i,0} + \boldsymbol{w}_i^T \boldsymbol{x}\right)} \tag{3.63}$$

$$P(\omega_i|\boldsymbol{x}) = \frac{\exp\left(w_{i,0} + \boldsymbol{w}_i^T \boldsymbol{x}\right)}{1 + \sum_{i=1}^{M-1} \exp\left(w_{i,0} + \boldsymbol{w}_i^T \boldsymbol{x}\right)},\ i = 1, 2, \ldots M-1 \tag{3.64}$$

For the two-class case, the previous equations are simplified to

$$P(\omega_2|\boldsymbol{x}) = \frac{1}{1 + \exp(w_0 + \boldsymbol{w}^T \boldsymbol{x})} \tag{3.65}$$

$$P(\omega_1|\boldsymbol{x}) = \frac{\exp(w_0 + \boldsymbol{w}^T \boldsymbol{x})}{1 + \exp(w_0 + \boldsymbol{w}^T \boldsymbol{x})} \tag{3.66}$$

To estimate the set of the unknown parameters, a maximum likelihood approach is usually employed. Optimization is performed with respect to all parameters, which we can think of as the components of a parameter vector $\boldsymbol{\theta}$. Let $\boldsymbol{x}_k,\ k = 1, 2, \ldots, N$, be the training feature vectors with known class labels. Let us denote by $\boldsymbol{x}_k^{(m)}$, $k = 1, 2, \ldots, N_m$, the vectors originating from class $m = 1, 2, \ldots, M$. Obviously, $\sum_m N_m = N$. The log-likelihood function to be optimized is given by

$$L(\boldsymbol{\theta}) = \ln \left\{ \prod_{k=1}^{N_1} p(\boldsymbol{x}_k^{(1)}|\omega_1; \boldsymbol{\theta}) \prod_{k=1}^{N_2} p(\boldsymbol{x}_k^{(2)}|\omega_2; \boldsymbol{\theta}) \ldots \prod_{k=1}^{N_M} p(\boldsymbol{x}_k^{(M)}|\omega_M; \boldsymbol{\theta}) \right\} \tag{3.67}$$

Taking into account that

$$p(\boldsymbol{x}_k^{(m)}|\omega_m; \boldsymbol{\theta}) = \frac{p(\boldsymbol{x}_k^{(m)}) P(\omega_m|\boldsymbol{x}_k^{(m)}; \boldsymbol{\theta})}{P(\omega_m)} \tag{3.68}$$

(3.67) becomes

$$L(\boldsymbol{\theta}) = \sum_{k=1}^{N_1} \ln P(\omega_1|\boldsymbol{x}_k^{(1)}) + \sum_{k=1}^{N_2} \ln P(\omega_2|\boldsymbol{x}_k^{(2)}) + \ldots + \sum_{k=1}^{N_M} \ln P(\omega_M|\boldsymbol{x}_k^{(M)}) + C \tag{3.69}$$

where the explicit dependence on $\boldsymbol{\theta}$ has been suppressed for notational simplicity and C is a parameter independent on $\boldsymbol{\theta}$ equal to

$$C = \ln \frac{\prod_{k=1}^{N} p(\boldsymbol{x}_k)}{\prod_{m=1}^{M} P(\omega_m)^{N_m}} \tag{3.70}$$

Inserting Eqs. (3.63) and (3.64) in (3.69), any optimization algorithm can then be used to perform the required maximization (Appendix C). More on the optimization task and the properties of the obtained solution can be found in, for example, [Ande 82, McLa 92].

There is a close relationship between the method of logistic discrimination and the LDA method, discussed in Chapter 2. It does not take much thought to realize that under the Gaussian assumption and for equal covariance matrices across all classes the following holds true.

$$\ln \frac{P(\omega_1|\boldsymbol{x})}{P(\omega_2|\boldsymbol{x})} = \frac{1}{2}(\boldsymbol{\mu}_2^T \Sigma^{-1} \boldsymbol{\mu}_2 - \boldsymbol{\mu}_1 \Sigma^{-1} \boldsymbol{\mu}_1) + (\boldsymbol{\mu}_1 - \boldsymbol{\mu}_2)^T \Sigma^{-1} \boldsymbol{x}$$
$$\equiv w_0 + \boldsymbol{w}^T \boldsymbol{x}$$

Here, the equiprobable two-class case was considered for simplicity. However, LDA and logistic discrimination are not identical methods. Their (subtle) difference lies in the way the unknown parameters are estimated. In LDA, the class probability densities are assumed to be Gaussian and the unknown parameters are, basically, estimated by maximizing (3.67) directly. In this maximization, the marginal probability densities ($p(\boldsymbol{x}_k)$) play their own part, since they enter implicitly into the game. However, in the case of logistic discrimination, marginal densities contribute to C and do not affect the solution. Thus, if the Gaussian assumption is a reasonable one for the problem at hand, LDA is the natural approach since it exploits all available information. On the other hand, if this is not a good assumption, then logistic discrimination seems to be a better candidate, since it relies on fewer assumptions. However, in practice it has been reported [Hast 01] that there is little difference between the results obtained by the two methods. Generalizations of the logistic discrimination method to include nonlinear models have also been suggested. See, for example, [Yee 96, Hast 01].

3.7 SUPPORT VECTOR MACHINES

3.7.1 Separable Classes

In this section, an alternative rationale for designing linear classifiers will be adopted. We will start with the two-class linearly separable task, and then we will extend the method to more general cases where data are not separable.

Let $\boldsymbol{x}_i, i = 1, 2, \ldots, N$, be the feature vectors of the training set, X. These belong to either of two classes, ω_1, ω_2, which are assumed to be linearly separable. The goal, once more, is to design a hyperplane

$$g(\boldsymbol{x}) = \boldsymbol{w}^T \boldsymbol{x} + w_0 = 0 \tag{3.71}$$

that classifies correctly all the training vectors. As we have already discussed in Section 3.3, such a hyperplane is not unique. The perceptron algorithm may converge to any one of the possible solutions. Having gained in experience, this time we will be more demanding. Figure 3.9 illustrates the classification task with

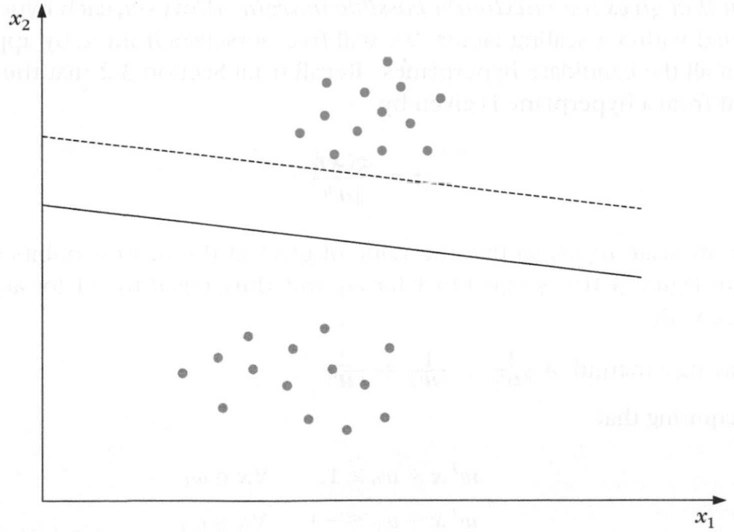

FIGURE 3.9

An example of a linearly separable two-class problem with two possible linear classifiers.

two possible hyperplane[1] solutions. Both hyperplanes do the job for the training set. However, which one of the two would any sensible engineer choose as the classifier for operation in practice, where data outside the training set will be fed to it? No doubt the answer is: the full-line one. The reason is that this hyperplane leaves more "room" on either side, so that data in both classes can move a bit more freely, with less risk of causing an error. Thus such a hyperplane can be trusted more, when it is faced with the challenge of operating with unknown data. Here we have touched a very important issue in the classifier design stage. It is known as the *generalization performance of the classifier*. This refers to the capability of the classifier, designed using the training data set, to operate satisfactorily with data outside this set. We will come to this issue over and over again.

After the above brief discussion, we are ready to accept that a very sensible choice for the hyperplane classifier would be the one that leaves the maximum margin from both classes. Later on, at the end of Chapter 5, we will see that this sensible choice has a deeper justification, springing from the elegant mathematical formulation that Vapnik and Chervonenkis have offered to us.

Let us now quantify the term *margin* that a hyperplane leaves from both classes. Every hyperplane is characterized by its direction (determined by $\boldsymbol{w}$) and its exact position in space (determined by w_0). Since we want to give no preference to either of the classes, then it is reasonable for each direction to select that hyperplane which has the same distance from the respective nearest points in ω_1 and ω_2. This is illustrated in Figure 3.10. The hyperplanes shown with dark lines are the selected ones from the infinite set in the respective direction. The margin for direction "1" is $2z_1$, and the margin for direction "2" is $2z_2$. *Our goal is to search for the direction that gives the maximum possible margin*. However, each hyperplane is determined within a scaling factor. We will free ourselves from it, by appropriate scaling of all the candidate hyperplanes. Recall from Section 3.2 that the distance of a point from a hyperplane is given by

$$z = \frac{|g(\boldsymbol{x})|}{\|\boldsymbol{w}\|}$$

We can now scale $\boldsymbol{w}, w_0$ so that the value of $g(\boldsymbol{x})$, at the nearest points in ω_1, ω_2 (circled in Figure 3.10), is equal to 1 for ω_1 and, thus, equal to -1 for ω_2. This is equivalent with

1. Having a margin of $\frac{1}{\|\boldsymbol{w}\|} + \frac{1}{\|\boldsymbol{w}\|} = \frac{2}{\|\boldsymbol{w}\|}$
2. Requiring that

$$\begin{aligned} \boldsymbol{w}^T\boldsymbol{x} + w_0 &\geq 1, \quad \forall \boldsymbol{x} \in \omega_1 \\ \boldsymbol{w}^T\boldsymbol{x} + w_0 &\leq -1, \quad \forall \boldsymbol{x} \in \omega_2 \end{aligned}$$

[1] We will refer to lines as hyperplanes to cover the general case.

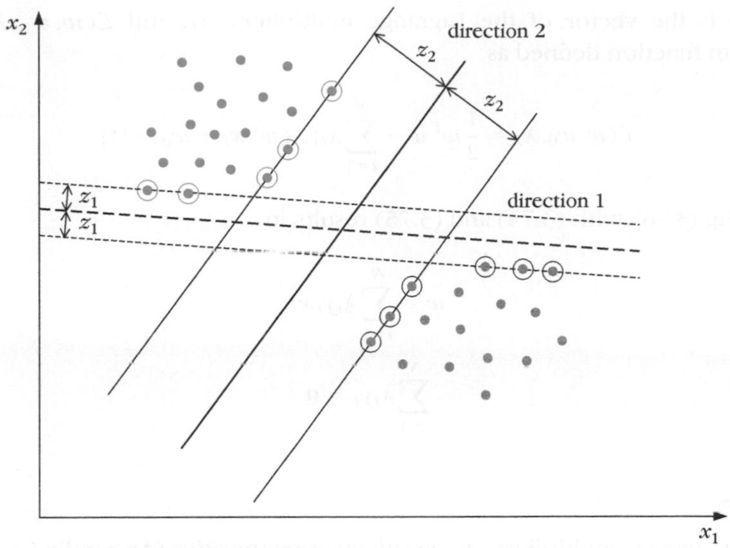

FIGURE 3.10

An example of a linearly separable two-class problem with two possible linear classifiers.

We have now reached the point where mathematics will take over. For each $\boldsymbol{x}_i$, we denote the corresponding class indicator by y_i (+1 for ω_1, −1 for ω_2.) Our task can now be summarized as: Compute the parameters $\boldsymbol{w}, w_0$ of the hyperplane so that to:

$$\text{minimize} \quad J(\boldsymbol{w}, w_0) \equiv \frac{1}{2}\|\boldsymbol{w}\|^2 \tag{3.72}$$

$$\text{subject to} \quad y_i(\boldsymbol{w}^T\boldsymbol{x}_i + w_0) \geq 1, \quad i = 1, 2, \ldots, N \tag{3.73}$$

Obviously, minimizing the norm makes the margin maximum. This is a nonlinear (quadratic) optimization task subject to a set of linear inequality constraints. The Karush–Kuhn–Tucker (KKT) conditions (Appendix C) that the minimizer of (3.72), (3.73) has to satisfy are

$$\frac{\partial}{\partial \boldsymbol{w}}\mathcal{L}(\boldsymbol{w}, w_0, \boldsymbol{\lambda}) = \boldsymbol{0} \tag{3.74}$$

$$\frac{\partial}{\partial w_0}\mathcal{L}(\boldsymbol{w}, w_0, \boldsymbol{\lambda}) = 0 \tag{3.75}$$

$$\lambda_i \geq 0, \quad i = 1, 2, \ldots, N \tag{3.76}$$

$$\lambda_i[y_i(\boldsymbol{w}^T\boldsymbol{x}_i + w_0) - 1] = 0, \quad i = 1, 2, \ldots, N \tag{3.77}$$

where $\boldsymbol{\lambda}$ is the vector of the Lagrange multipliers, λ_i, and $\mathcal{L}(\boldsymbol{w}, w_0, \boldsymbol{\lambda})$ is the Lagrangian function defined as

$$\mathcal{L}(\boldsymbol{w}, w_0, \boldsymbol{\lambda}) = \frac{1}{2}\boldsymbol{w}^T\boldsymbol{w} - \sum_{i=1}^{N} \lambda_i [y_i(\boldsymbol{w}^T\boldsymbol{x}_i + w_0) - 1] \tag{3.78}$$

Combining (3.78) with (3.74) and (3.75) results in

$$\boldsymbol{w} = \sum_{i=1}^{N} \lambda_i y_i \boldsymbol{x}_i \tag{3.79}$$

$$\sum_{i=1}^{N} \lambda_i y_i = 0 \tag{3.80}$$

Remarks

- The Lagrange multipliers can be either zero or positive (Appendix C). Thus, the vector parameter $\boldsymbol{w}$ of the optimal solution is a linear combination of $N_s \leq N$ feature vectors that are associated with $\lambda_i \neq 0$. That is,

$$\boldsymbol{w} = \sum_{i=1}^{N_s} \lambda_i y_i \boldsymbol{x}_i \tag{3.81}$$

  These are known as *support vectors* and the optimum hyperplane classifier as a *support vector machine* (SVM). As it is pointed out in Appendix C, a nonzero Lagrange multiplier corresponds to a so called active constraint. Hence, as the set of constraints in (3.77) suggests for $\lambda_i \neq 0$, *the support vectors lie on either of the two hyperplanes*, that is,

$$\boldsymbol{w}^T\boldsymbol{x} + w_0 = \pm 1 \tag{3.82}$$

  In other words, they are the training vectors that are closest to the linear classifier, and they constitute the *critical elements of the training set*. Feature vectors corresponding to $\lambda_i = 0$ can either lie outside the "class separation band," defined as the region between the two hyperplanes given in (3.82), or they can also lie on one of these hyperplanes (degenerate case, Appendix C). The resulting hyperplane classifier is insensitive to the number and position of such feature vectors, provided they do not cross the class separation band.

- Although $\boldsymbol{w}$ is explicitly given, w_0 can be implicitly obtained by any of the *(complementary slackness)* conditions (3.77), satisfying strict complementarity (i.e., $\lambda_i \neq 0$, Appendix C). In practice, w_0 is computed as an average value obtained using all conditions of this type.

- The cost function in (3.72) is a strict convex one (Appendix C), a property that is guaranteed by the fact that the corresponding Hessian matrix is positive definite [Flet 87]. Furthermore, the inequality constraints consist of linear functions. As discussed in Appendix C, these two conditions guarantee that any local minimum is also global and unique. This is most welcome. *The optimal hyperplane classifier of a support vector machine is unique.*

Having stated all these very interesting properties of the optimal hyperplane of a support vector machine, we next need to compute the involved parameters. From a computational point of view this is not always an easy task, and a number of algorithms exist, for example, [Baza 79]. We will move to a path, which is suggested to us by the special nature of our optimization task, given in (3.72) and (3.73). It belongs to the *convex programming* family of problems, since the cost function is convex and the constraints are linear and define a convex set of feasible solutions. As we discuss in Appendix C, such problems can be solved by considering the so-called *Lagrangian duality*. The problem can be stated equivalently by its Wolfe dual representation form, that is,

$$\text{maximize} \quad \mathcal{L}(\boldsymbol{w}, w_0, \boldsymbol{\lambda}) \tag{3.83}$$

$$\text{subject to} \quad \boldsymbol{w} = \sum_{i=1}^{N} \lambda_i y_i \boldsymbol{x}_i \tag{3.84}$$

$$\sum_{i=1}^{N} \lambda_i y_i = 0 \tag{3.85}$$

$$\boldsymbol{\lambda} \geq \boldsymbol{0} \tag{3.86}$$

The two equality constraints are the result of equating to zero the gradient of the Lagrangian, with respect to $\boldsymbol{w}, w_0$. We have already gained something. The training feature vectors enter into the problem via equality constraints and not inequality ones, which can be easier to handle. Substituting (3.84) and (3.85) into (3.83) and after a bit of algebra we end up with the equivalent optimization task

$$\max_{\boldsymbol{\lambda}} \left(\sum_{i=1}^{N} \lambda_i - \frac{1}{2} \sum_{i,j} \lambda_i \lambda_j y_i y_j \boldsymbol{x}_i^T \boldsymbol{x}_j \right) \tag{3.87}$$

$$\text{subject to} \quad \sum_{i=1}^{N} \lambda_i y_i = 0 \tag{3.88}$$

$$\boldsymbol{\lambda} \geq \boldsymbol{0} \tag{3.89}$$

Once the optimal Lagrange multipliers have been computed, by maximizing (3.87), the optimal hyperplane is obtained via (3.84), and w_0 via the complementary slackness conditions, as before.

Remarks

- Besides the more attractive setting of the involved constraints in (3.87), (3.88), there is another important reason that makes this formulation popular. The training vectors enter into the game in pairs, in the form of inner products. This is most interesting. *The cost function does not depend explicitly on the dimensionality of the input space!* This property allows for efficient generalizations in the case of nonlinearly separable classes. We will return to this at the end of Chapter 4.
- Although the resulting optimal hyperplane is unique, there is no guarantee about the uniqueness of the associated Lagrange multipliers λ_i. In words, the expansion of $\boldsymbol{w}$ in terms of support vectors in (3.84) may not be unique, although the final result is unique (Example 3.5).

3.7.2 Nonseparable Classes

When the classes are not separable, the above setup is no longer valid. Figure 3.11 illustrates the case in which the two classes are not separable. Any attempt to draw a hyperplane will never end up with a class separation band with no

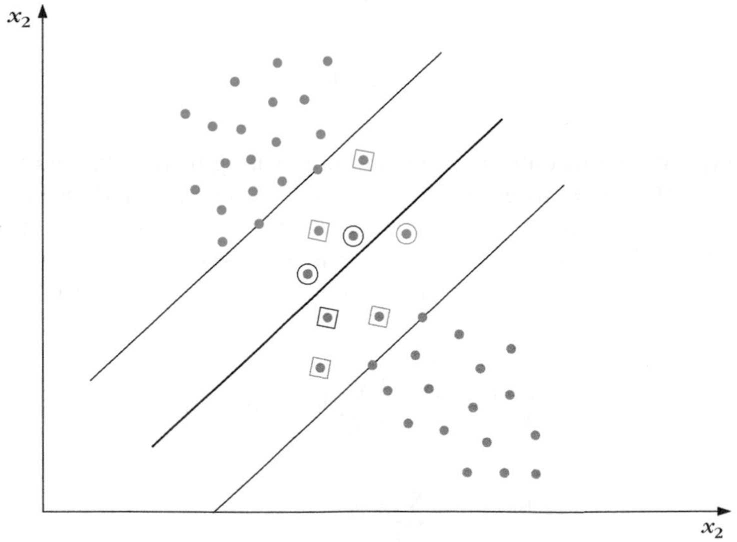

FIGURE 3.11

In the nonseparable class case, points fall inside the class separation band.

data points inside it, as was the case in the linearly separable task. Recall that the margin is defined as the distance between the pair of parallel hyperplanes described by

$$w^T x + w_0 = \pm 1$$

The training feature vectors now belong to one of the following three categories:

- Vectors that fall outside the band and are correctly classified. These vectors comply with the constraints in (3.73).
- Vectors falling inside the band and are correctly classified. These are the points placed in squares in Figure 3.11, and they satisfy the inequality

$$0 \le y_i(w^T x + w_0) < 1$$

- Vectors that are misclassified. They are enclosed by circles and obey the inequality

$$y_i(w^T x + w_0) < 0$$

All three cases can be treated under a single type of constraints by introducing a new set of variables, namely,

$$y_i[w^T x + w_0] \ge 1 - \xi_i \tag{3.90}$$

The first category of data corresponds to $\xi_i = 0$, the second to $0 < \xi_i \le 1$, and the third to $\xi_i > 1$. The variables ξ_i are known as *slack variables*. The optimizing task becomes more involved, yet it falls under the same rationale as before. The goal now is to make the margin as large as possible but at the same time to keep the number of points with $\xi > 0$ as small as possible. In mathematical terms, this is equivalent to adopting to minimize the cost function

$$J(w, w_0, \boldsymbol{\xi}) = \frac{1}{2}\|w\|^2 + C\sum_{i=1}^{N} I(\xi_i) \tag{3.91}$$

where $\boldsymbol{\xi}$ is the vector of the parameters ξ_i and

$$I(\xi_i) = \begin{cases} 1 & \xi_i > 0 \\ 0 & \xi_i = 0 \end{cases} \tag{3.92}$$

The parameter C is a positive constant that controls the relative influence of the two competing terms. However, optimization of the above is difficult since it involves a discontinuous function $I(\cdot)$. As it is common in such cases, we choose to optimize a closely related cost function, and the goal becomes

$$\text{minimize} \quad J(\boldsymbol{w}, w_0, \boldsymbol{\xi}) = \frac{1}{2}\|\boldsymbol{w}\|^2 + C\sum_{i=1}^{N}\xi_i \tag{3.93}$$

$$\text{subject to} \quad y_i[\boldsymbol{w}^T\boldsymbol{x}_i + w_0] \geq 1 - \xi_i, \quad i = 1, 2, \ldots, N \tag{3.94}$$

$$\xi_i \geq 0, \quad i = 1, 2, \ldots, N \tag{3.95}$$

The problem is again a convex programming one, and the corresponding Lagrangian is given by

$$\mathcal{L}(\boldsymbol{w}, w_0, \boldsymbol{\xi}, \boldsymbol{\lambda}, \boldsymbol{\mu}) = \frac{1}{2}\|\boldsymbol{w}\|^2 + C\sum_{i=1}^{N}\xi_i - \sum_{i=1}^{N}\mu_i\xi_i - \sum_{i=1}^{N}\lambda_i[y_i(\boldsymbol{w}^T\boldsymbol{x}_i + w_0) - 1 + \xi_i] \tag{3.96}$$

The corresponding Karush–Kuhn–Tucker conditions are

$$\frac{\partial\mathcal{L}}{\partial\boldsymbol{w}} = \boldsymbol{0} \quad \text{or} \quad \boldsymbol{w} = \sum_{i=1}^{N}\lambda_i y_i \boldsymbol{x}_i \tag{3.97}$$

$$\frac{\partial\mathcal{L}}{\partial w_0} = 0 \quad \text{or} \quad \sum_{i=1}^{N}\lambda_i y_i = 0 \tag{3.98}$$

$$\frac{\partial\mathcal{L}}{\partial\xi_i} = 0 \quad \text{or} \quad C - \mu_i - \lambda_i = 0, \quad i = 1, 2, \ldots, N \tag{3.99}$$

$$\lambda_i[y_i(\boldsymbol{w}^T\boldsymbol{x}_i + w_0) - 1 + \xi_i] = 0, \quad i = 1, 2, \ldots, N \tag{3.100}$$

$$\mu_i\xi_i = 0, \quad i = 1, 2, \ldots, N \tag{3.101}$$

$$\mu_i \geq 0, \quad \lambda_i \geq 0, \quad i = 1, 2, \ldots, N \tag{3.102}$$

The associated Wolfe dual representation now becomes

$$\text{maximize} \quad \mathcal{L}(\boldsymbol{w}, w_0, \boldsymbol{\lambda}, \boldsymbol{\xi}, \boldsymbol{\mu})$$

$$\text{subject to} \quad \boldsymbol{w} = \sum_{i=1}^{N}\lambda_i y_i \boldsymbol{x}_i$$

$$\sum_{i=1}^{N}\lambda_i y_i = 0$$

$$C - \mu_i - \lambda_i = 0, \quad i = 1, 2, \ldots, N$$

$$\lambda_i \geq 0, \ \mu_i \geq 0, \quad i = 1, 2, \ldots, N$$

Substituting the above equality constraints into the Lagrangian, we end up with

$$\max_{\boldsymbol{\lambda}} \left(\sum_{i=1}^{N} \lambda_i - \frac{1}{2} \sum_{i,j} \lambda_i \lambda_j y_i y_j \boldsymbol{x}_i^T \boldsymbol{x}_j \right) \tag{3.103}$$

$$\text{subject to} \quad 0 \leq \lambda_i \leq C, \quad i = 1, 2, \ldots, N \tag{3.104}$$

$$\sum_{i=1}^{N} \lambda_i y_i = 0 \tag{3.105}$$

Note that the Lagrange multipliers corresponding to the points residing either within the margin or on the wrong side of the classifier, that is, $\xi_i > 0$, are all equal to the maximum allowable value C. Indeed, at the solution, for $\xi_i \neq 0$ the KKT conditions give $\mu_i = 0$ leading to $\lambda_i = C$. In other words, these points have the largest possible "share" in the final solution $\boldsymbol{w}$.

3.7.3 The Multiclass Case

In all our discussions, so far, we have been involved with the two-class classification task. In an M-class problem, a straightforward extension is to consider it as a set of M two-class problems (*one-against-all*). For each one of the classes, we seek to design an optimal discriminant function, $g_i(\boldsymbol{x})$, $i = 1, 2, \ldots, M$, so that $g_i(\boldsymbol{x}) > g_j(\boldsymbol{x})$, $\forall j \neq i$, if $\boldsymbol{x} \in \omega_i$. Adopting the SVM methodology, we can design the discriminant functions so that $g_i(\boldsymbol{x}) = 0$ to be the optimal hyperplane separating class ω_i from all the others. Thus, each classifier is designed to give $g_i(\boldsymbol{x}) > 0$ for $\boldsymbol{x} \in \omega_i$ and $g_i(\boldsymbol{x}) < 0$ otherwise. Classification is then achieved according to the following rule:

$$\text{assign } \boldsymbol{x} \text{ in } \omega_i \text{ if } i = \arg\max_k \{g_k(\boldsymbol{x})\}$$

This technique, however, may lead to indeterminate regions, where more than one $g_i(\boldsymbol{x})$ is positive (Problem 3.15). Another drawback of the technique is that each binary classifier deals with a rather asymmetric problem in the sense that training is carried out with many more negative than positive examples. This becomes more serious when the number of classes is relatively large.

An alternative technique is the *one-against-one*. In this case, $M(M-1)/2$ binary classifiers are trained and each classifier separates a pair of classes. The decision is made on the basis of a majority vote. The obvious disadvantage of the technique is that a relatively large number of binary classifiers has to be trained. In [Plat 00] a methodology is suggested that may speed up the procedure.

A different and very interesting rationale has been adopted in [Diet 95]. The multiclass task is treated in the context of error correcting coding, inspired by the coding schemes used in communications. For a M-class problem a number of, say, L binary classifiers are used, where L is appropriately chosen by the designer. Each class is now represented by a binary code word of length L. During training, for the ith classifier, $i = 1, 2, \ldots, L$, the desired labels, y, for each class are chosen to be either $+1$ or -1. For each class, the desired labels may be different for the various classifiers. This is equivalent to constructing a matrix $M \times L$ of desired labels. For example, if $M = 4$ and $L = 6$, such a matrix can be

$$\begin{bmatrix} -1 & -1 & -1 & +1 & -1 & +1 \\ +1 & -1 & +1 & +1 & -1 & -1 \\ +1 & +1 & -1 & -1 & -1 & +1 \\ -1 & -1 & +1 & -1 & +1 & +1 \end{bmatrix} \tag{3.106}$$

In words, during training, the first classifier (corresponding to the first column of the previous matrix) is designed in order to respond $(-1,+1,+1,-1)$ for patterns originating from classes ω_1, ω_2, ω_3, ω_4, respectively. The second classifier will be trained to respond $(-1,-1,+1,-1)$, and so on. The procedure is equivalent to grouping the classes into L different pairs, and, for each pair, we train a binary classifier accordingly. For the case of our example and for the first binary classifier, class ω_1 has been grouped together with ω_4 and class ω_2 with class ω_3. Each row must be distinct and corresponds to a class. For our example, and in the absence of errors, the outputs of the L classifiers for a pattern from class ω_1 will result in the code word $(-1,-1,-1,+1,-1,+1)$, and so on. When an unknown pattern is presented, the output of each one of the binary classifiers is recorded, resulting in a code word. Then, the Hamming distance (number of places where two code words differ) of this code word is measured against the M code words, and the pattern is classified to the class corresponding to the smallest distance.

Here in lies the power of the technique. If the code words are designed so that the minimum Hamming distance between any pair of them is, say, d, then a correct decision will still be reached even if the decisions of at most $\lfloor \frac{d-1}{2} \rfloor$, out of the L, classifiers are wrong, where $\lfloor \cdot \rfloor$ is the floor operation. For the matrix in (3.106) the minimum Hamming distance, between any pair, is equal to three. In [Diet 95], the method has been applied for numerical digit classification, and the grouping of the ten classes is done in such a way as to be meaningful. For example, one grouping is based on the existence in the numeric digits of a horizontal line (e.g., "4" and "2"), or the existence of a vertical line (e.g., "1" and "4"), and so on. An extension of this method, which is proposed in [Allw 00], takes into consideration the resulting values of the margin (when an SVM or another type of margin classifier, e.g., boosting classifiers discussed in Chapter 4, is used). In [Zhou 08], the composition of the individual binary problems and their number (code word length, L) is the result of a data-adaptive procedure that designs the code words by taking into account the inherent structure of the training data.

All previous techniques are appropriate for any classifier. Another alternative, specific for SVMs, is to extentd the two class SVM mathematical formulation to the M-class problem, see, for example, [Vapn 98, Liu 06]. Comparative studies of the various methods for multiclass SVM classification can be found in [Rifk 04, Hsu 02, Fei 06].

Remarks

- The only difference between the linearly separable and nonseparable cases lies in the fact that for the latter one the Lagrange multipliers need to be bounded above by C. The linearly separable case corresponds to $C \rightarrow \infty$, see Eqs. (3.104) and (3.89). The slack variables, ξ_i, and their associated Lagrange multipliers, μ_i, do not enter into the problem explicitly. Their presence is indirectly reflected through C.
- A major limitation of support vector machines is the high computational burden required, both during training and in the test phase. A naive implementation of a quadratic programming (QP) solver takes $O(N^3)$ operations, and its memory requirements are of the order of $O(N^2)$. For problems with a relatively small number of training data, any general purpose optimization algorithm can be used. However, for a large number of training points (e.g., of the order of a few thousands), a naive QP implementation does not scale up well, and a special treatment is required. Training of SVM is usually performed in batch mode. For large problems this sets high demands on computer memory requirements. To attack such problems, a number of procedures have been devised. Their philosophy relies on the decomposition, in one way or another, of the optimization problem into a sequence of smaller ones, for example, [Bose 92, Osun 97, Chan 00]. The main rationale behind such algorithms is to start with an arbitrary data subset (chunk of data, working set) that can fit in the computer memory. Optimization is, then, performed on this subset via a general optimizer. Support vectors remain in the working set while others are replaced by new ones, outside the current working set, that violate severely the KKT conditions. It can be shown that this iterative procedure guarantees that the cost function is decreasing at each iteration step.

  In [Plat 99, Matt 99], the so called *Sequential Minimal Optimization* (SMO) algorithm is proposed where the idea of decomposition is pushed to its extreme and each working set consists of only two points. Its great advantage is that the optimization can now be performed analytically. In [Keer 01], a set of heuristics is used for the choice of the pair of points that constitute the working set. To this end, it is suggested that the use of two thresholded parameters can lead to considerable speedups. As suggested in [Plat 99, Platt 98], efficient implementations of such a scheme have an empirical training time complexity that scales between $O(N)$ and $O(N^{2.3})$.

  Theoretical issues related to the algorithm, such as convergence, are addressed in [Chen 06] and the references therein. The parallel implementation of the algorithm is considered in [Cao 06]. In [Joac 98] the working set is the result of a search for the steepest feasible direction. More recently, [Dong 05] suggested a technique to quickly remove most of the nonsupport vectors, using a parallel optimization step, and the original problem can be split into many subproblems that can be solved more efficiently. In [Mavr 06], the geometric interpretation of SVMs (Section 3.7.5) is exploited, and the optimization task is treated as a minimum distance points search between convex sets. It is reported that substantial computational savings can be obtained compared to the SMO algorithm. A sequential algorithm, which operates on the primal problem formulation, has been proposed in [Navi 01], where an iterative reweighted least squares procedure is employed and alternates weight optimization with constraint forcing. An advantage of the latter technique is that it naturally leads to online implementations. Another trend is to employ an algorithm that aims at an approximate solution to the problem. In [Fine 01] a low-rank approximation is used in place of the the so-called kernel matrix, which is involved in the computations. In [Tsan 06, Hush 06] the issues of complexity and accuracy of the approximation are considered together. For example, in [Hush 06] polynomial-time algorithms are derived that produce approximate solutions with a guaranteed accuracy for a class of QP problems that include the SVM classifiers.

For large problems, the test phase can also be quite demanding, if the number of support vectors is excessively high. Methods that speed up computations have also been suggested, for example, [Burg 97, Nguy 06].

Example 3.5

Consider the two-class classification task that consists of the following points:

$$\omega_1: [1, 1]^T, [1, -1]^T$$

$$\omega_2: [-1, 1]^T, [-1, -1]$$

Using the SVM approach, we will demonstrate that the optimal separating hyperplane (line) is $x_1 = 0$ and that this is obtained via different sets of Lagrange multipliers.

The points lie on the corners of a square, as shown in Figure 3.12. The simple geometry of the problem allows for a straightforward computation of the SVM linear classifier. Indeed, a careful observation of Figure 3.12 suggests that the optimal line

$$g(\boldsymbol{x}) = w_1 x_1 + w_2 x_2 + w_0 = 0$$

is obtained for $w_2 = w_0 = 0$ and $w_1 = 1$, that is,

$$g(\boldsymbol{x}) = x_1 = 0$$

Hence for this case, all four points become support vectors, and the margin of the separating line from both classes is equal to 1. For any other direction, e.g., $g_1(\boldsymbol{x}) = 0$, the margin is smaller. It must be pointed out that the same solution is obtained if one solves the associated KKT conditions (Problem 3.16.)

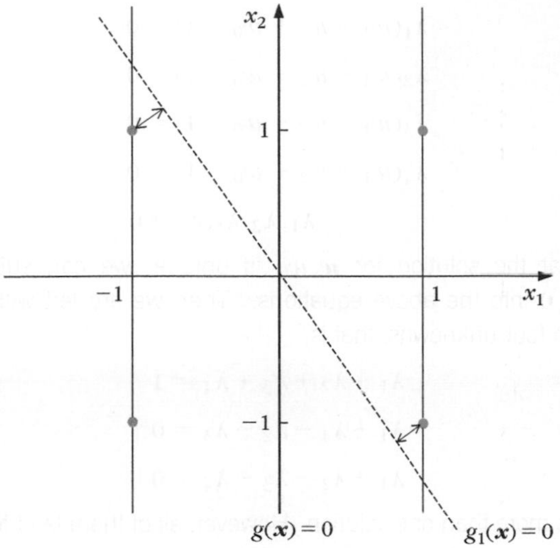

FIGURE 3.12

In this example all four points are support vectors. The margin associated with $g_1(\boldsymbol{x}) = 0$ is smaller compared to the margin defined by the optimal $g(\boldsymbol{x}) = 0$.

Let us now consider the mathematical formulation of our problem. The linear inequality constraints are

$$w_1 + w_2 + w_0 - 1 \geq 0$$
$$w_1 - w_2 + w_0 - 1 \geq 0$$
$$w_1 - w_2 - w_0 - 1 \geq 0$$
$$w_1 + w_2 - w_0 - 1 \geq 0$$

and the associated Lagrangian function becomes

$$\begin{aligned}\mathcal{L}(w_2, w_1, w_0, \boldsymbol{\lambda}) = \frac{w_1^2 + w_2^2}{2} &- \lambda_1(w_1 + w_2 + w_0 - 1) \\ &- \lambda_2(w_1 - w_2 + w_0 - 1) \\ &- \lambda_3(w_1 - w_2 - w_0 - 1) \\ &- \lambda_4(w_1 + w_2 - w_0 - 1)\end{aligned}$$

The KKT conditions are given by

$$\frac{\partial \mathcal{L}}{\partial w_1} = 0 \Rightarrow w_1 = \lambda_1 + \lambda_2 + \lambda_3 + \lambda_4 \tag{3.107}$$

$$\frac{\partial \mathcal{L}}{\partial w_2} = 0 \Rightarrow w_2 = \lambda_1 + \lambda_4 - \lambda_2 - \lambda_3 \tag{3.108}$$

$$\frac{\partial \mathcal{L}}{\partial w_0} = 0 \Rightarrow \lambda_1 + \lambda_2 - \lambda_3 - \lambda_4 = 0 \tag{3.109}$$

$$\lambda_1(w_1 + w_2 + w_0 - 1) = 0 \tag{3.110}$$

$$\lambda_2(w_1 - w_2 + w_0 - 1) = 0 \tag{3.111}$$

$$\lambda_3(w_1 - w_2 - w_0 - 1) = 0 \tag{3.112}$$

$$\lambda_4(w_1 + w_2 - w_0 - 1) = 0 \tag{3.113}$$

$$\lambda_1, \lambda_2, \lambda_3, \lambda_4 \geq 0 \tag{3.114}$$

Since we know that the solution for $\boldsymbol{w}, w_0$ is unique, we can substitute the solution $w_1 = 1, w_2 = w_0 = 0$ into the above equations. Then we are left with a linear system of three equations with four unknowns, that is,

$$\lambda_1 + \lambda_2 + \lambda_3 + \lambda_4 = 1 \tag{3.115}$$

$$\lambda_1 + \lambda_4 - \lambda_2 - \lambda_3 = 0 \tag{3.116}$$

$$\lambda_1 + \lambda_2 - \lambda_3 - \lambda_4 = 0 \tag{3.117}$$

which obviously has more than one solution. However, all of them lead to the unique optimal separating line.

Example 3.6

Figure 3.13 shows a set of training data points residing in the two-dimensional space and divided into two nonseparable classes. The full line in Figure 3.13a is the resulting hyperplane using Platt's algorithm and corresponds to the value $C = 0.2$. Dotted lines meet the conditions given in (3.82) and define the margin that separates the two classes, for those points with

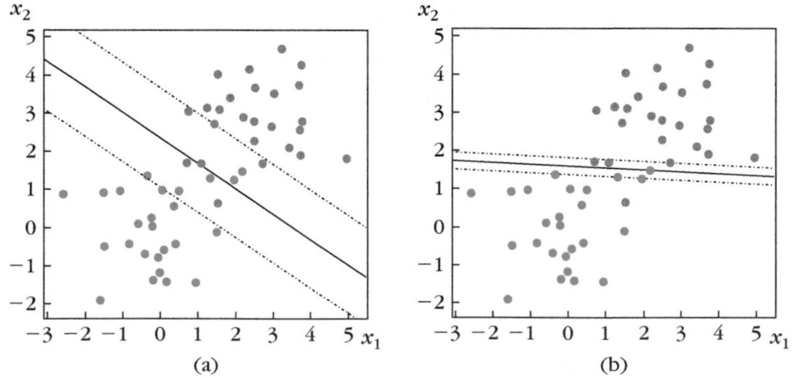

FIGURE 3.13

An example of two nonseparable classes and the resulting SVM linear classifier (full line) with the associated margin (dotted lines) for the values (a) $C = 0.2$ and (b) $C = 1000$. In the latter case, the location and direction of the classifier as well as the width of the margin have changed in order to include a smaller number of points inside the margin.

$\xi_i = 0$. The setting in Figure 3.13b corresponds to $C = 1000$ and has been obtained with the same algorithm and the same set of trimming parameters (e.g., stopping criteria).

It is readily observed that the margin associated with the classifier corresponding to the larger value of C is smaller. This is because the second term in (3.91) has now more influence in the cost, and the optimization process tries to satisfy this demand by reducing the margin and consequently the number of points with $\xi_i > 0$. In other words, the width of the margin does not depend entirely on the data distribution, as was the case with the separable class case, but is heavily affected by the choice of C. This is the reason SVM classifiers, defined by (3.91), are also known as *soft margin classifiers*.

3.7.4 ν-SVM

Example 3.6 demonstrated the close relation that exists between the parameter C and the width of the margin obtained as a result of the optimization process. However, since the margin is such an important entity in the design of SVM (after all, the essence of the SVM methodology is to maximize it), a natural question that arises is why not involve it in a more direct way in the cost function, instead of leaving its control to a parameter (i.e., C) whose relation with the margin, although strong, is not transparent to us. To this end, in [Scho 00] a variant of the soft margin SVM was introduced. The margin is defined by the pair of hyperplanes

$$\boldsymbol{w}^T \boldsymbol{x} + w_0 = \pm\, \rho \tag{3.118}$$

and $\rho \geq 0$ is left as a free variable to be optimized. Under this new setting, the primal problem given in (3.93)–(3.95) can now be cast as

$$\text{minimize} \quad J(\boldsymbol{w}, w_0, \boldsymbol{\xi}, \rho) = \frac{1}{2}\|\boldsymbol{w}\|^2 - \nu\rho + \frac{1}{N}\sum_{i=1}^{N} \xi_i \tag{3.119}$$

$$\text{subject to} \quad y_i[\boldsymbol{w}^T \boldsymbol{x}_i + w_0] \geq \rho - \xi_i, \quad i = 1, 2, \ldots, N \tag{3.120}$$

$$\xi_i \geq 0, \quad i = 1, 2, \ldots, N \tag{3.121}$$

$$\rho \geq 0 \tag{3.122}$$

To understand the role of ρ, note that for $\xi_i = 0$ the constraints in (3.120) state that the margin separating the two classes is equal to $\frac{2\rho}{\|\boldsymbol{w}\|}$. In the previous formulation, also known as ν-SVM, we simply count and average the number of points with $\xi_i > 0$, whose number is now controlled by the margin variable ρ. The larger the ρ the wider the margin and the higher the number of points within the margin, for a specific direction $\boldsymbol{w}$. The parameter ν controls the influence of the second term in the cost function, and its value lies in the range [0, 1]. (We will revisit this issue later on.)

The Lagrangian function associated with the task (3.119)-(3.122) is given by

$$\mathcal{L}(\boldsymbol{w}, w_0, \boldsymbol{\lambda}, \boldsymbol{\xi}, \boldsymbol{\mu}, \rho, \delta) = \frac{1}{2}\|\boldsymbol{w}\|^2 - \nu\rho + \frac{1}{N}\sum_{i=1}^{N}\xi_i - \sum_{i=1}^{N}\mu_i\xi_i$$
$$- \sum_{i=1}^{N}\lambda_i\left[y_i(\boldsymbol{w}^T\boldsymbol{x}_i + w_0) - \rho + \xi_i\right] - \delta\rho \tag{3.123}$$

Adopting similar steps as in Section 3.7.2, the following KKT conditions result:

$$\boldsymbol{w} = \sum_{i=1}^{N}\lambda_i y_i \boldsymbol{x}_i \tag{3.124}$$

$$\sum_{i=1}^{N}\lambda_i y_i = 0 \tag{3.125}$$

$$\mu_i + \lambda_i = \frac{1}{N}, \quad i = 1, 2, \ldots, N \tag{3.126}$$

$$\sum_{i=1}^{N}\lambda_i - \delta = \nu \tag{3.127}$$

$$\lambda_i\left[y_i(\boldsymbol{w}^T\boldsymbol{x}_i + w_0) - \rho + \xi_i\right] = 0, \quad i = 1, 2, \ldots, N \tag{3.128}$$

$$\mu_i\xi_i = 0, \quad i = 1, 2, \ldots, N \tag{3.129}$$

$$\delta\rho = 0 \tag{3.130}$$

$$\mu_i \geq 0, \quad \lambda_i \geq 0, \quad \delta \geq 0, \quad i = 1, 2, \ldots, N \tag{3.131}$$

The associated Wolfe dual representation is easily shown to be

$$\text{maximize} \quad \mathcal{L}(\boldsymbol{w}, w_0, \boldsymbol{\lambda}, \boldsymbol{\xi}, \boldsymbol{\mu}, \delta) \tag{3.132}$$

$$\text{subject to} \quad \boldsymbol{w} = \sum_{i=1}^{N}\lambda_i y_i \boldsymbol{x}_i \tag{3.133}$$

$$\sum_{i=1}^{N}\lambda_i y_i = 0 \tag{3.134}$$

$$\mu_i + \lambda_i = \frac{1}{N}, \quad i = 1, 2, \ldots, N \tag{3.135}$$

$$\sum_{i=1}^{N}\lambda_i - \delta = \nu \tag{3.136}$$

$$\lambda_i \geq 0, \ \mu_i \geq 0, \ \delta \geq 0, \quad i = 1, 2, \ldots, N \tag{3.137}$$

If we substitute the equality constraints (3.133)-(3.136) in the Lagrangian, the dual problem becomes equivalent to (Problem 3.17)

$$\max_{\boldsymbol{\lambda}} \left(-\frac{1}{2} \sum_{i,j} \lambda_i \lambda_j y_i y_j \boldsymbol{x}_i^T \boldsymbol{x}_j \right) \tag{3.138}$$

$$\text{subject to} \quad 0 \leq \lambda_i \leq \frac{1}{N}, \quad i = 1, 2, \ldots, N \tag{3.139}$$

$$\sum_{i=1}^{N} \lambda_i y_i = 0 \tag{3.140}$$

$$\sum_{i=1}^{N} \lambda_i \geq \nu \tag{3.141}$$

Once more, only the Lagrange multipliers $\boldsymbol{\lambda}$ enter into the problem explicitly, and ρ and the slack variables, ξ_i, make their presence felt through the bounds appearing in the constraints. Observe that in contrast to (3.103) the cost function is now quadratically homogeneous and the linear term $\sum_{i=1}^{N} \lambda_i$ is not present. Also, the new formulation has an extra constraint.

Remarks

- [Chan 01] shows that the ν-SVM and the more standard SVM formulation [(3.103)-(3.105)], sometimes referred to as C-SVM, lead to the same solution for appropriate values of C and ν. Also, it is shown that in order for the optimization problem to be feasible, the constant ν must lie in a range $0 \leq \nu_{\min} \leq \nu \leq \nu_{\max} \leq 1$.

- Although both SVM formulations result in the same solution, for appropriate choices of ν and C the ν-SVM offers certain advantages to the designer. As we will see in the next section, it leads to a geometric interpretation of the SVM task for nonseparable classes. Furthermore, the constant ν, controlled by the designer, offers itself to serve two important bounds concerning (a) the error rate and (b) the number of the resulting support vectors.

  At the solution, the points lying either within the margin or outside it but on the wrong side of the separating hyperplane correspond to $\xi_i > 0$ and hence to $\mu_i = 0$ [Eq. (3.129)], forcing the respective Lagrange multipliers to be $\lambda_i = \frac{1}{N}$ [Eq. (3.126)]. Also, since at the solution, for $\rho > 0$, $\delta = 0$ [Eq. (3.130)], it turns out that $\sum_{i=1}^{N} \lambda_i = \nu$ [Eq. (3.127)]. Combining these and taking into account that all points that lie in the wrong side of the classifier correspond to $\xi_i > 0$, the total number of errors can, at most, be equal to $N\nu$. Thus, the error rate, P_e, on the training set is upper-bounded as

$$P_e \leq \nu. \tag{3.142}$$

Also, at the solution, from the constraints (3.127) and (3.126) we have that

$$\nu = \sum_{i=1}^{N} \lambda_i = \sum_{i=1}^{N_s} \lambda_i \leq \sum_{i=1}^{N_s} \frac{1}{N} \tag{3.143}$$

or

$$N\nu \leq N_s \tag{3.144}$$

Thus, the designer, by controlling the value of ν, may have a feeling for both the error rate on the training set and the number of the support vectors to result from the optimization process. The number of the support vectors, N_s, is very important for the performance of the classifier in practice. First, as we have already commented, it directly affects the computational load, since large N_s means that a large number of inner products are to be computed for classifying an unknown pattern. Second, as we will see at the end of Section 5.10, a large number of support vectors can limit the error performance of the SVM classifier when it is fed with data outside the training set (this is also known as the generalization performance of the classifier). For more on the ν-SVM, the interested reader can consult [Scho 00, Chan 01, Chen 03], where implementation issues are also discussed.

3.7.5 Support Vector Machines: A Geometric Viewpoint

In this section, we will close the circle around the SVM design task via a path that is very close to what we call common sense. Figure 3.14a illustrates the case of two separable data classes together with their respective convex hulls. The convex hull of a data set X is denoted as conv$\{X\}$ and is defined as the intersection of all convex sets (see Appendix C.4) containing X. It can be shown (e.g., [Luen 69]) that conv$\{X\}$ consists of all the convex combinations of the N elements of X. That is,

$$\text{conv}\{X\} = \left\{ \boldsymbol{y} : \boldsymbol{y} = \sum_{i=1}^{N} \lambda_i \boldsymbol{x}_i : \boldsymbol{x}_i \in X, \right.$$

$$\left. \sum_{i=1}^{N} \lambda_i = 1,\ 0 \leq \lambda_i \leq 1,\ i = 1, 2, \ldots, N \right\} \tag{3.145}$$

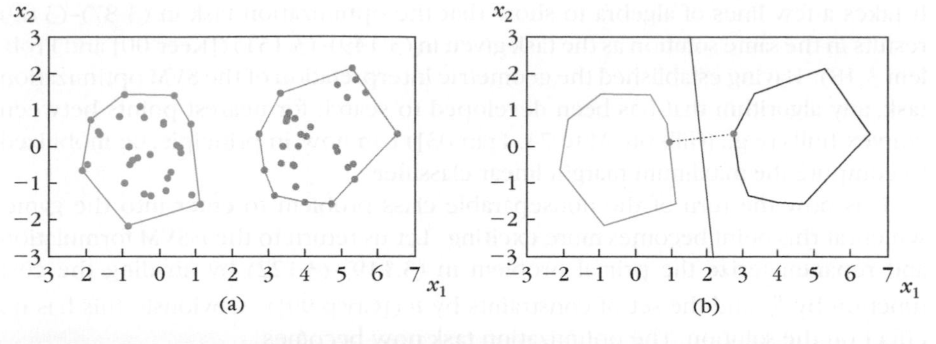

FIGURE 3.14

(a) A data set for two separable classes with the respective convex hulls. (b) The SVM optimal hyperplane bisects the segment joining the two nearest points between the convex hulls.

It turns out that solving the dual optimization problem in (3.87)–(3.89) for the linearly separable task results in the hyperplane that bisects the linear segment joining two ***nearest points between the convex hulls*** of the data classes [Figure 3.14b]. In other words, ***searching for the maximum margin hyperplane is equivalent to searching for two nearest points between the corresponding convex hulls!*** Let us investigate this a bit further.

Denote the convex hull of the vectors in class ω_1 as conv$\{X^+\}$ and the convex hull corresponding to class ω_2 as conv$\{X^-\}$. Following our familiar notation, any point in conv$\{X^+\}$, being a convex combination of all the points in ω_1, can be written as $\sum_{i:y_i=1} \lambda_i \boldsymbol{x}_i$, and any point in conv$\{X^-\}$ as $\sum_{i:y_i=-1} \lambda_i \boldsymbol{x}_i$, provided that λ_i fulfill the convexity constraints in (3.145). Searching for the closest points, it suffices to find the specific values of λ_i, $i = 1, 2, \ldots N$, such that

$$\min_{\boldsymbol{\lambda}} \ \| \sum_{i:y_i=1} \lambda_i \boldsymbol{x}_i - \sum_{i:y_i=-1} \lambda_i \boldsymbol{x}_i \|^2 \tag{3.146}$$

$$\text{subject to} \quad \sum_{i:y_i=1} \lambda_i = 1, \quad \sum_{i:y_i=-1} \lambda_i = 1 \tag{3.147}$$

$$\lambda_i \geq 0, \quad i = 1, 2, \ldots N \tag{3.148}$$

Elaborating the norm in (3.146) and reshaping the constraints in (3.147), we end up with the following equivalent formulation.

$$\text{minimize} \quad \sum_{i,j} y_i y_j \lambda_i \lambda_j \boldsymbol{x}_i^T \boldsymbol{x}_j \tag{3.149}$$

$$\text{subject to} \quad \sum_{i=1}^{N} y_i \lambda_i = 0, \ \sum_{i=1}^{N} \lambda_i = 2 \tag{3.150}$$

$$\lambda_i \geq 0, \quad i = 1, 2, \ldots N \tag{3.151}$$

It takes a few lines of algebra to show that the optimization task in (3.87)-(3.89) results in the same solution as the task given in (3.149)-(3.151) ([Keer 00] and Problem 3.18). Having established the geometric interpretation of the SVM optimization task, any algorithm that has been developed to search for nearest points between convex hulls (e.g., [Gilb 66, Mitc 74, Fran 03]) can now, in principle, be mobilized to compute the maximum margin linear classifier.

It is now the turn of the nonseparable class problem to enter into the game, which, at this point becomes more exciting. Let us return to the ν-SVM formulation and reparameterize the primal problem in (3.119)-(3.122) by dividing the cost function by $\frac{\nu^2}{2}$ and the set of constraints by ν ([Crisp 99]). Obviously, this has no effect on the solution. The optimization task now becomes

$$\text{minimize} \quad J(\boldsymbol{w}, w_0, \boldsymbol{\xi}, \rho) = \|\boldsymbol{w}\|^2 - 2\rho + \mu \sum_{i=1}^{N} \xi_i \tag{3.152}$$

$$\text{subject to} \quad y_i[\boldsymbol{w}^T \boldsymbol{x}_i + w_0] \geq \rho - \xi_i, \quad i = 1, 2, \ldots, N \tag{3.153}$$

$$\xi_i \geq 0, \quad i = 1, 2, \ldots, N \tag{3.154}$$

$$\rho \geq 0 \tag{3.155}$$

where $\mu = \frac{2}{\nu N}$ and we have kept, for economy, the same notation, although the parameters in (3.152)-(3.155) are scaled versions of those in (3.119)-(3.122). That is, $\boldsymbol{w} \to \frac{\boldsymbol{w}}{\nu}$, $w_0 \to \frac{w_0}{\nu}$, $\rho \to \frac{\rho}{\nu}$, $\xi_i \to \frac{\xi_i}{\nu}$. Hence, the solution obtained via (3.152)-(3.155) is a scaled version of the solution resulting via (3.119)-(3.122). The Wolfe dual representation of the primal problem in (3.152)-(3.155) is easily shown to be equivalent to

$$\text{minimize} \quad \sum_{i,j} y_i y_j \lambda_i \lambda_j \boldsymbol{x}_i^T \boldsymbol{x}_j \tag{3.156}$$

$$\text{subject to} \quad \sum_i y_i \lambda_i = 0, \ \sum_i \lambda_i = 2 \tag{3.157}$$

$$0 \leq \lambda_i \leq \mu, \ i = 1, 2, \ldots N \tag{3.158}$$

This set of relations is almost the same as those defining the nearest points between the convex hulls in the separable class case, (3.149)-(3.151), with a small, yet significant, difference. The Lagrange multipliers are bounded by μ, and for $\mu < 1$ they are not permitted to span their entire allowable range (i.e., [0, 1]).

3.7.6 Reduced Convex Hulls

The *reduced convex hull* (RCH) of a (finite) vector set, X, is denoted as $R(X, \mu)$ and is defined as the convex set

$$R(X,\mu) = \left\{ y : y = \sum_{i=1}^{N} \lambda_i x_i : x_i \in X, \right.$$

$$\left. \sum_{i=1}^{N} \lambda_i = 1,\ 0 \leq \lambda_i \leq \mu,\ i = 1,2,\ldots,N \right\} \tag{3.159}$$

It is apparent from the previous definition that $R(X,1) \equiv \text{conv}\{X\}$ and that

$$R(X,\mu) \subseteq \text{conv}\{X\} \tag{3.160}$$

Figure 3.15a shows the respective convex hulls for the case of two intersecting data classes. In Figure 3.15b, full lines indicate the convex hulls, $\text{conv}\{X^+\}$ and $\text{conv}\{X^-\}$, and the dotted lines the reduced convex hulls $R(X^+,\mu)$, $R(X^-,\mu)$, for two different values of $\mu = 0.4$ and $\mu = 0.1$, respectively. It is readily apparent that the smaller the value of μ, the smaller the size of the reduced convex hull. For small enough values of μ, one can make $R(X^+,\mu)$ and $R(X^-,\mu)$ nonintersecting. Adopting a procedure similar to the one that led to (3.149)–(3.151), it is not difficult to see that finding two nearest points between $R(X^+,\mu)$ and $R(X^-,\mu)$ results in the ν-SVM dual optimization task given in (3.156)–(3.158). Observe that the only difference between the latter and the task for the separable case, defined in (3.149)–(3.151), lies in the range in which the Lagrange multipliers are allowed to be. In the separable class case, the constraints (3.150) and (3.151) imply that $0 \leq \lambda_i \leq 1$, which in its geometric interpretation means that the full convex hulls are searched for the nearest points. In contrast, in the nonseparable class case a lower upper bound (i.e.,

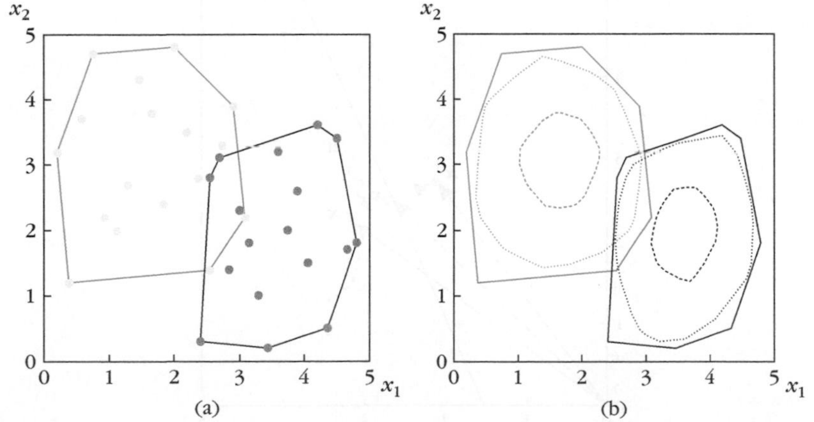

FIGURE 3.15

(a) Example of a data set with two intersecting classes and their respective convex hulls. (b) The convex hulls (indicated by full lines) and the resulting reduced convex hulls (indicated by dotted lines) corresponding to $\mu = 0.4$ and $\mu = 0.1$, respectively, for each class. The smaller the value of μ the smaller the RCH size.

$\mu \leq 1$) is imposed for the Lagrange multipliers. From the geometry point of view, this means that the search for the nearest points is limited within the respective reduced convex hulls.

Having established the geometric interpretation of the ν-SVM dual representation form, let us follow pure geometric arguments to draw the separating hyperplane. It is natural to choose it as the one bisecting the line segment joining two nearest points between the reduced convex hulls. Let $\boldsymbol{x}^+$ and $\boldsymbol{x}^-$ be two nearest points, with $\boldsymbol{x}^+ \in R(X^+, \mu)$ and $\boldsymbol{x}^- \in R(X^-, \mu)$. Also, let λ_i, $i = 1, 2, \ldots, N$, be the optimal set of multipliers resulting from the optimization task. Then, as can be deduced from Figure 3.16,

$$\boldsymbol{w} = \boldsymbol{x}^+ - \boldsymbol{x}^- = \sum_{i:y_i=1} \lambda_i \boldsymbol{x}_i - \sum_{i:y_i=-1} \lambda_i \boldsymbol{x}_i \tag{3.161}$$

$$= \sum_{i=1}^{N} \lambda_i y_i \boldsymbol{x}_i \tag{3.162}$$

This is the same (within a scaling factor) as the $\boldsymbol{w}$ obtained from the KKT conditions associated with the ν-SVM task [Eq. (3.124)]. Thus, both approaches result in a separating hyperplane pointing in the same direction (recall from Section 3.2 that $\boldsymbol{w}$ defines the direction of the hyperplane). However, it is early to say that the two solutions are exactly the same. The hyperplane bisecting the line segment

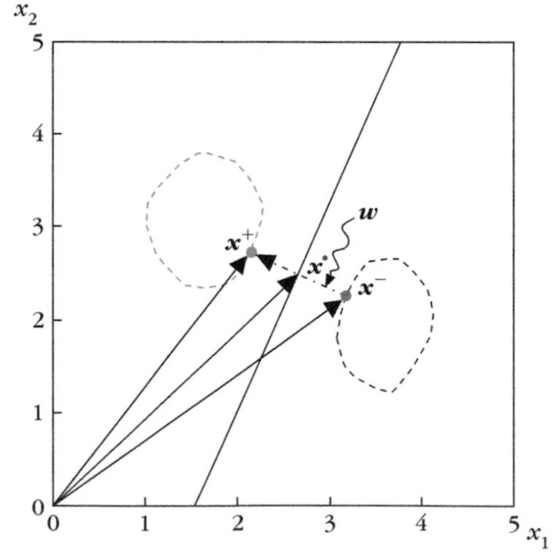

FIGURE 3.16

The optimal linear classifier resulting as the bisector of the segment joining the two closest points between the reduced convex hulls of the classes, for the case of the data set shown in Figure 3.15 and for $\mu = 0.1$.

joining the nearest points crosses the middle of this segment; that is, the point $x^* = \frac{1}{2}(x^+ + x^-)$. Thus,

$$w^T x^* + w_0 = 0 \tag{3.163}$$

from which we get

$$w_0 = -\frac{1}{2} w^T \left(\sum_{i:y_i=1} \lambda_i x_i + \sum_{i:y_i=-1} \lambda_i x_i \right) \tag{3.164}$$

This value for w_0 is, in general, different from the value resulting from the KKT conditions in (3.128). In conclusion, the geometric approach in the case of the nonseparable problem is equivalent to the ν-SVM formulation only to the extent that both approaches result in hyperplanes pointing in the same direction. However, note that the value in Eq. (3.128) can be obtained from that given in Eq. (3.164) in a trivial way [Crisp 99].

Remarks

- The choice of μ and consequently of $\nu = \frac{2}{\mu N}$ must guarantee that the feasible region is nonempty (i.e., a solution exists, Appendix C) and also that the solution is a nontrivial one (i.e., $w \neq 0$). Let N^+ be the number of points in X^+ and N^- the number of points in X^-, where $N^+ + N^- = N$. Let $N_{\min} = \min\{N^+, N^-\}$. Then it is readily seen from the crucial constraint $0 \leq \lambda_i \leq \mu$ and the fact that $\sum_i \lambda_i = 1$, in the definition of the reduced convex hull, that $\mu \geq \mu_{\min} = \frac{1}{N_{\min}}$. This readily suggests that ν cannot take any value but must be upper-bounded as

$$\nu \leq \nu_{\max} = 2\frac{N_{\min}}{N} \leq 1$$

Also, if the respective reduced convex hulls intersect, then the distance between the closest points is zero, leading to the trivial solution (Problem 3.19). Thus, nonintersection is guaranteed for some value $\mu_{\max}$ such that $\mu \leq \mu_{\max} \leq 1$, which leads to

$$\nu \geq \nu_{\min} = \frac{2}{\mu_{\max} N}$$

From the previous discussion it is easily deduced that for the feasible region to be nonempty it is required that

$$R(X^+, \mu_{\min}) \cap R(X^-, \mu_{\min}) = \emptyset$$

If $N^+ = N^- = \frac{N}{2}$, it is easily checked out that in this case each of the reduced convex hulls is shrunk to a point, which is the centroid of the respective class (e.g., $\frac{2}{N}\sum_{i:y_i=1}\boldsymbol{x}_i$). In other words, a solution is feasible if the centroids of the two classes do not coincide. Most natural!

- Computing the nearest points between reduced convex hulls turns out not to be a straightforward extension of the algorithms that have been developed for computing nearest points between convex hulls. This is because, for the latter case, such algorithms rely on the extreme points of the involved convex hulls. However, in this case, extreme points coincide with points in the original data sets, that is, X^+, X^-. This is not the case for the reduced convex hulls, where extreme points are *combinations* of points of the original data sets. The lower the value of μ, the higher the number of data samples that contribute to an extreme point in the respective reduced convex hull. A neat solution to this problem is given in [Mavr 05, Mavr 06, Mavr 07, Tao 04, Theo 07]. The developed nearest point algorithms are reported to offer computational savings, which in some cases can be significant, compared to the more classical algorithms in [Plat 99, Keer 01].

3.8 PROBLEMS

3.1 Explain why the perceptron cost function is a *continuous* piecewise linear function.

3.2 Show that if $\rho_k = \rho$ in the perceptron algorithm, the algorithm converges after $k_0 = \frac{\|\boldsymbol{w}(0)-\alpha\boldsymbol{w}^*\|}{\beta^2\rho(2-\rho)}$ steps, where $\alpha = \frac{\beta^2}{|\gamma|}$ and $\rho < 2$.

3.3 Show that the reward and punishment form of the perceptron algorithm converges in a finite number of iteration steps.

3.4 Consider a case in which class ω_1 consists of the two feature vectors $[0, 0]^T$ and $[0, 1]^T$ and class ω_2 of $[1, 0]^T$ and $[1, 1]^T$. Use the perceptron algorithm in its reward and punishment form, with $\rho = 1$ and $\boldsymbol{w}(0) = [0, 0]^T$, to design the line separating the two classes.

3.5 Consider the two-class task of Problem 2.12 of the previous chapter with

$$\boldsymbol{\mu}_1^T = [1, 1], \quad \boldsymbol{\mu}_2^T = [0,0], \quad \sigma_1^2 = \sigma_2^2 = 0.2$$

Produce 50 vectors from each class. To guarantee linear separability of the classes, disregard vectors with $x_1 + x_2 < 1$ for the [1, 1] class and vectors with $x_1 + x_2 > 1$ for the [0, 0] class. In the sequel, use these vectors to design a linear classifier using the perceptron algorithm of (3.21)–(3.23). After convergence, draw the corresponding decision line.

3.6 Consider once more the classification task of Problem 2.12. Produce 100 samples for each of the classes. Use these data to design a linear classifier via the LMS algorithm. Once all samples have been presented to the algorithm, draw the corresponding hyperplane to which the algorithm has converged. Use $\rho_k = \rho = 0.01$.

3.7 Show, using Kesler's construction, that the tth iteration step of the reward and punishment form of the perceptron algorithm (3.21)–(3.23), for an $\boldsymbol{x}_{(t)} \in \omega_i$, becomes

$$\begin{aligned} \boldsymbol{w}_i(t+1) &= \boldsymbol{w}_i(t) + \rho \boldsymbol{x}_{(t)} \quad \text{if } \boldsymbol{w}_i^T(t)\boldsymbol{x}_{(t)} \le \boldsymbol{w}_j^T(t)\boldsymbol{x}_{(t)}, j \ne i \\ \boldsymbol{w}_j(t+1) &= \boldsymbol{w}_j(t) - \rho \boldsymbol{x}_{(t)} \quad \text{if } \boldsymbol{w}_i^T(t)\boldsymbol{x}_{(t)} \le \boldsymbol{w}_j^T(t)\boldsymbol{x}_{(t)}, j \ne i \\ \boldsymbol{w}_k(t+1) &= \boldsymbol{w}_k(t), \quad \forall k \ne j \text{ and } k \ne i \end{aligned}$$

3.8 Show that the sum of error squares optimal weight vector tends asymptotically to the MSE solution.

3.9 Repeat Problem 3.6 and design the classifier using the sum of error squares criterion.

3.10 Show that the design of an M class linear, sum of error squares optimal, classifier reduces to M equivalent ones, with scalar desired responses.

3.11 Show that, if x, y are jointly Gaussian, the regression of y on x is given by

$$E[y|x] = \frac{\alpha\sigma_y x}{\sigma_x} + \mu_y - \frac{\alpha\sigma_y\mu_x}{\sigma_x}, \quad \text{where } \Sigma = \begin{bmatrix} \sigma_x^2 & \alpha\sigma_x\sigma_y \\ \alpha\sigma_x\sigma_y & \sigma_y^2 \end{bmatrix} \tag{3.165}$$

3.12 Let an M class classifier be given in the form of parameterized functions $g(\boldsymbol{x}; \boldsymbol{w}_k)$. The goal is to estimate the parameters $\boldsymbol{w}_k$ so that the outputs of the classifier give desired response values, depending on the class of $\boldsymbol{x}$. Assume that as $\boldsymbol{x}$ varies randomly in each class, the classifier outputs vary around the corresponding desired response values, according to a Gaussian distribution of known variance, assumed to be the same for all outputs. Show that in this case the sum of error squares criterion and the ML estimation result in identical estimates.

Hint: Take N training data samples of known class labels. For each of them form $y_i = g(\boldsymbol{x}_i; \boldsymbol{w}_k) - d_k^i$, where d_k^i is the desired response for the kth class of the ith sample. The y_i's are normally distributed with zero mean and variance σ^2. Form the likelihood function using the y_i's.

3.13 In a two-class problem, the Bayes optimal decision surface is given by $g(\boldsymbol{x}) = P(\omega_1|\boldsymbol{x}) - P(\omega_2|\boldsymbol{x}) = 0$. Show that if we train a decision surface $f(\boldsymbol{x}; \boldsymbol{w})$ in the MSE so as to give $+1(-1)$ for the two classes, respectively, this is equivalent to approximating $g(\cdot)$ in terms of $f(\cdot; \boldsymbol{w})$, in the MSE optimal sense.

3.14 Consider a two-class classification task with jointly Gaussian distributed feature vectors and with the same variance Σ in both classes. Design the linear MSE classifier and show that in this case the Bayesian classifier (Problem 2.11) and the resulting MSE one differ only in the threshold value. For simplicity, consider equiprobable classes.

Hint: To compute the MSE hyperplane $\boldsymbol{w}^T\boldsymbol{x}+w_0=0$, increase the dimension of $\boldsymbol{x}$ by one and show that the solution is provided by

$$\begin{bmatrix} R & E[\boldsymbol{x}] \\ E[\boldsymbol{x}]^T & 1 \end{bmatrix} \begin{bmatrix} \boldsymbol{w} \\ w_0 \end{bmatrix} = \begin{bmatrix} \frac{1}{2}(\boldsymbol{\mu}_1 - \boldsymbol{\mu}_2) \\ 0 \end{bmatrix}$$

Then relate R with Σ and show that the MSE classifier takes the form

$$(\boldsymbol{\mu}_1 - \boldsymbol{\mu}_2)^T\Sigma^{-1}\left(\boldsymbol{x} - \frac{1}{2}(\boldsymbol{\mu}_1 + \boldsymbol{\mu}_2)\right) \geq 0$$

3.15 In an M class classification task, the classes can be linearly separated. Design M hyperplanes, so that hyperplane $g_i(\boldsymbol{x}) = 0$ leaves class ω_i on its positive side and the rest of the classes on its negative side. Demonstrate via an example, for example, $M = 3$, that the partition of the space using this rule creates indeterminate regions (where no training data exist) for which more than one $g_i(\boldsymbol{x})$ is positive or all of them are negative.

3.16 Obtain the optimal line for the task of Example 3.5, via the KKT conditions. Restrict the search for the optimum among the lines crossing the origin.

3.17 Show that if the equality constraints (3.133)-(3.136) are substituted in the Lagrangian (3.123), the dual problem is described by the set of relations in (3.138)-(3.141).

3.18 Show that for the case of two linearly separable classes the hyperplane obtained as the SVM solution is the same as that bisecting the segment joining two closest points between the convex hulls of the classes.

3.19 Show that if ν in the ν-SVM is chosen smaller than $\nu_{\min}$, it leads to the trivial zero solution.

3.20 Show that if the soft margin SVM cost function is chosen to be

$$\frac{1}{2}||\boldsymbol{w}||^2 + \frac{C}{2}\sum_{i=1}^{N}\xi_i^2$$

the task can be transformed into an instance of the class-separable case problem [Frie 98].

MATLAB PROGRAMS AND EXERCISES

Computer Programs

3.1 *Perceptron algorithm.* Write a MATLAB function for the perceptron algorithm. This will take as inputs: (a) a matrix X containing N l-dimensional column vectors, (b) an N-dimensional row vector y, whose ith component contains the class (−1 or +1) where the corresponding vector belongs, and (c) an initial value vector *w_ini* for the parameter vector. It returns the estimated parameter vector.

Solution

```
function w=perce(X,y,w_ini)
  [l,N]=size(X);
  max_iter=10000;   % Maximum allowable number of iterations
  rho=0.05;         % Learning rate
  w=w_ini;          % Initialization of the parameter vector
  iter=0;           % Iteration counter
  mis_clas=N;       % Number of misclassified vectors
  while (mis_clas>0) && (iter<max_iter)
    iter=iter+1;
    mis_clas=0;
    gradi=zeros(l,1);% Computation of the "gradient"
    % term
    for i=1:N
      if((X(:,i)'*w)*y(i)<0)
        mis_clas=mis_clas+1;
        gradi=gradi+rho*(-y(i)*X(:,i));
      end
    end
    w=w-rho*gradi; % Updating the parameter vector
  end
```

3.2 *Sum of error squares classifier:* Write a MATLAB function that implements the sum of error squares classifier for two classes. This will take as inputs: (a) a matrix X containing N l-dimensional column vectors, and (b) an N-dimensional row vector y whose ith component contains the class (−1 or +1) where the corresponding vector belongs. It returns the estimated parameter vector.

Solution

```
function w=SSErr(X,y)
  w=inv(X*X')*(X*y');
```

3.3 *LMS algorithm.* Write a MATLAB function for the LMS algorithm. This will take as inputs: (a) a matrix X containing N l-dimensional column vectors, (b) an N-dimensional row vector y whose ith component contains the class (−1 or +1) where the corresponding vector is assigned, and (c) an initial value vector *w_ini* for the parameter vector. It returns the estimated parameter vector.

Solution

```
function w=LMSalg(X,y,w_ini)
  [l,N]=size(X);
  rho=0.1;       % Learning rate initialization
  w=w_ini;       % Initialization of the parameter vector
  for i=1:N
    w=w+(rho/i)*(y(i)-X(:,i)'*w)*X(:,i);
  end
```

Computer Experiments

Note: In the sequel, it is advisable to use the command

```
randn('seed',0)
```

before generating the data sets, in order to initialize the Gaussian random number generator to 0 (or any other fixed number). This is important for the reproducibility of the results.

3.1 **a.** Generate two data sets $X1$ and X_1' of $N = 200$ two-dimensional vectors each. The first half of the vectors stem from the normal distribution with $m1 = [-5, 0]^T$ and $S1 = I$, while the second half of the vectors stem from the normal distribution with $m1 = [5, 0]^T$ and $S1 = I$, where I is the identity 2×2 matrix. Append each vector of both $X1$ and X_1' by inserting an additional coordinate, which is set equal to 1.

b. Apply the perceptron algorithm, the sum of error squares classifier, and the LMS algorithm on the previous data set, using various initial values for the parameter vector (where necessary).

c. Measure the performance of each one of the above methods on both $X1$ and $X1'$.

d. Plot the data sets $X1$ and $X1'$ as well as the line corresponding to the parameter vector w.

3.2 Repeat experiment 1 using now the sets $X2$ and $X2'$ whose first half of their vectors stem from the normal distribution with $m1 = [-2, 0]^T$ and $S1 = I$, while the second half of their vectors stem from the normal distribution with $m1 = [2, 0]^T$ and $S1 = I$.

3.3 Repeat experiment 3.1 using now the sets $X3$ and $X3'$ whose first half of the vectors stem from the normal distribution with $m1 = [-1, 0]^T$ and $S1 = I$, while the second half of the vectors stem from the normal distribution with $m1 = [1, 0]^T$ and $S1 = I$.

3.4 Discuss the results obtained by the previous experiments.

REFERENCES

[Allw 00] Allwein E.L., Schapire R.E., Singer Y. "Reducing multiclass to binary: aunifying approach for margin classifiers," *Journal of Machine Learning Research*, Vol. 1, pp. 113–141, 2000.

[Ande 82] Anderson J.A. "Logistic discrimination," in *Handbook of Statistics* (Krishnaiah R.P., Kanal L.N., eds.), North Holland, 1982.

[Baza 79] Bazaraa M.S., Shetty C.M. *Nonlinear Programming*, John Wiley & Sons, 1979.

[Bish 95] Bishop C. *Neural Networks for Pattern Recognition*, Oxford University Press, 1995.

[Bose 92] Bose B.E., Guyon I.M., Vapnik, V.N. "A training algorithm for optimal margin classifiers," *Proceedings of the 5th Annual Workshop on Computational Learning Theory*, pp. 144–152, Morgan Kaufman, 1992.

[Burg 97] Burges C.J.C., Schölkoff B. "Improving the accuracy and speed of support vectors learning machines," in *Advances in Neural Information Processing Systems 9* (Mozer M. Jordan M., Petsche T., eds.), pp. 375–381, MIT Press, 1997.

[Cao 06] Cao L.J., Keerthi S.S., Ong C.-J., Zhang J.Q., Periyathamby V., Fu X.J., Lee H.P. "Parallel sequential minimal optimization for the training of support vector machines," *IEEE Transactions on Neural Networks*, Vol. 17(4), pp. 1039–1049, 2006.

[Chan 00] Chang C.C., Hsu C.W., Lin C.J. "The analysis of decomposition methods for SVM," *IEEE Transactions on Neural Networks*, Vol. 11(4), pp. 1003–1008, 2000.

[Chan 01] Chang C.C., Lin C.J. "Training ν-support vector classifiers: Theory and algorithms," *Neural Computation* Vol. 13(9), pp. 2119–2147, 2001.

[Chen 06] Chen P.-H., Fan R.-E., Lin C.-J. "A study on SMO-type decomposition for support vector machines," *IEEE Transactions on Neural Networks*, Vol. 17(4), pp. 893–908, 2006.

[Chen 03] Chen P.-H., Lin C.J., Schölkopf B. "A tutorial on ν-support vector machines," *Applied Stochastic Models in Business and Industry*, Vol. 21, pp. 111–136, 2005.

[Cid 99] Cid-Sueiro J., Arribas J.I., Urban-Munoz S., Figuieras-Vidal A.R. "Cost functions to estimate a-posteriori probabilities in multi-class problems," *IEEE Transactions on Neural Networks*, Vol. 10(3), pp. 645–656, 1999.

[Crisp 99] Crisp D.J., Burges C.J.C. "A geometric interpretation of ν-SVM classifiers," *Proceedings of Neural Information Processing*, Vol. 12, MIT Press, 1999.

[Diet 95] Dietterich T.G., Bakiri G. "Solving multi-class learning problems via error-correcting output codes," *Journal of Artificial Intelligence Research*, Vol. 2, pp. 263–286, 1995.

[Dong 05] Dong J.X., Krzyzak A., Suen C.Y. "Fast SVM training algorithm with decomposition on very large data sets," *IEEE Transactions on Pattern Analysis and Machine Intelligence*, Vol. 27(4), pp. 603–618, 2005.

[Fei 06] Fei B., Liu J. "Binary tree of SVM: A new fast multi-class training and classification algorithm," *IEEE Transactions on Neural Networks*, Vol. 17(3), pp. 696–704, 2006.

[Fine 01] Fine S., Scheinberg K. "Efficient SVM training using low rank kernel representations," *Journal of Machine Learning Research*, Vol. 2, pp. 243–264, 2001.

[Flet 87] Fletcher R. *Practical Methods of Optimization*, 2nd ed., John Wiley & Sons, 1987.

[Fran 03] Franc V., Hlaváč V. "An iterative algorithm learning the maximal margin classifier," *Pattern Recognition*, Vol. 36, pp. 1985–1996, 2003.

[Frea 92] Frean M. "A thermal perceptron learning rule," *Neural Computation*, Vol. 4, pp. 946–957, 1992.

[Frie 98] Friess T.T. "The kernel adatron with bias and soft margin," Technical Report, The University of Sheffield, Dept. of Automatic Control, England, 1998.

[Fuku 90] Fukunaga K. *Introduction to Statistical Pattern Recognition*, 2nd ed., Academic Press, 1990.

[Gal 90] Gallant S.I. "Perceptron based learning algorithms," *IEEE Transactions on Neural Networks*, Vol. 1(2), pp. 179–191, 1990.

[Gema 92] Geman S., Bienenstock E., Doursat R. "Neural networks and the bias/variance dilemma," *Neural Computation*, Vol. 4, pp. 1–58, 1992.

[Gilb 66] Gilbert E.G. "An iterative procedure for computing the minimum of a quadratic form on a convex set," *SIAM Journal on Control*, Vol. 4(1), pp. 61–79, 1966.

[Guer 04] Guerrero-Curieses A., Cid-Sueiro J., Alaiz-Rodriguez R., Figueiras-Vidal A.R. "Local estimation of posterior class probabilities to minimize classification errors," *IEEE Transactions on Neural Networks*, Vol. 15(2), pp. 309–317, 2004.

[Hast 01] Hastie T., Tibsharini R., Friedman J. *The Elements of Statistical Learning: Data Mining, Inference and Prediction*, Springer, 2001.

[Hayk 96] Haykin S. *Adaptive Filter Theory*, 3rd ed., Prentice Hall, 1996.

[Ho 65] Ho Y.H., Kashyap R.L. "An algorithm for linear inequalities and its applications," *IEEE Transactions on Electronic Computers*, Vol. 14(5), pp. 683–688, 1965.

[Hsu 02] Hsu C.W., Lin C.J. "A comparison of methods for multi-class SVM," *IEEE Transactions on Neural Networks*, Vol. 13, pp. 415–425, 2002.

[Hryc 92] Hrycej T., *Modular Learning in Neural Networks*, John Wiley & Sons, 1992.

[Hush 06] Hush D., Kelly P. Scovel C., Steinwart I. "QP algorithms with guaranteed accuracy and run time for support vector machines," *Journal of Machine Learning Research*, Vol. 7, pp. 733–769, 2006.

[Joac 98] Joachims T. "Making large scale support vector machines practical," in *Advances in Kernel Methods*, (Schölkoph B., Burges C.J.C., Smola A. eds.), MIT Press, 1998.

[Kalou 93] Kalouptsidis N., Theodoridis S. *Adaptive System Identification and Signal Processing Algorithms*, Prentice Hall, 1993.

[Keer 00] Keerthi S.S., Shevade S.K., Bhattacharyya C., Murthy K.R.K. "A fast iterative nearest point algorithm for support vector machine classifier design," *IEEE Transactions on Neural Networks*, Vol. 11(1), pp. 124–136, 2000.

[Keer 01] Keerthi S.S., Shevade S.K., Bhattacharyya C., Murth K.R.K. "Improvements to Platt's SMO algorithm for SVM classifier design," *Neural Computation*, Vol. 13, pp. 637–649, 2001.

[Liu 06] Liu Y., You Z., Cao L. "A novel and quick SVM-based multi-class classifier," *Pattern Recognition*, Vol. 39(11), pp. 2258–2264, 2006.

[Luen 69] Luenberger D.G. *Optimization by Vector Space Methods*, John Wiley & Sons, New York, 1969.

[McLa 92] McLachlan G. J. *Discriminant Analysis and Statistical Pattern Recognition*, John Wiley & Sons, 1992.

[Matt 99] Mattera D., Palmieri F., Haykin S. "An explicit algorithm for training support vector machines," *IEEE Signal Processing Letters*, Vol. 6(9), pp. 243–246, 1999.

[Mavr 07] Mavroforakis M., Sdralis M., Theodoridis S. "A geometric nearest point algorithm for the efficient solution of the SVM classification task," *IEEE Transactions on Neural Networks*, Vol. 18(5), pp. 1545–1550, 2007.

[Mavr 06] Mavroforakis M., Theodoridis S. "A geometric approach to Support Vector Machine (SVM) classification," *IEEE Transactions on Neural Networks*, Vol. 17(3), pp. 671–682, 2006.

[Mavr 05] Mavroforakis M., Theodoridis S. "Support Vector Machine classification through geometry," *Proceedings of the XII European Signal Processing Conference (EUSIPCO)*, Antalya, Turkey, 2005.

[Mitc 74] Mitchell B.F., Demyanov V.F., Malozemov V.N. "Finding the point of a polyhedron closest to the origin," *SIAM Journal on Control*, Vol. 12, pp. 19–26, 1974.

[Min 88] Minsky M. L., Papert S.A. *Perceptrons*, expanded edition, MIT Press, MA, 1988.

[Muse 97] Muselli M. "On convergence properties of pocket algorithm," *IEEE Transactions on Neural Networks*, Vol. 8(3), pp. 623–629, 1997.

[Navi 01] Navia-Vasquez A., Perez-Cuz F., Artes-Rodriguez A., Figueiras-Vidal A. "Weighted least squares training of support vector classifiers leading to compact and adaptive schemes," *IEEE Transactions on Neural Networks*, Vol. 12(5), pp. 1047–1059, 2001.

[Nguy 06] Nguyen D., Ho T. "A bottom-up method for simplifying support vector solutions," *IEEE Transactions on Neural Networks*, Vol. 17(39), pp. 792–796, 2006.

[Osun 97] Osuna E., Freund R., Girosi F., "An improved training algorithm for support vector machines," *Proceedings of IEEE Workshop on Neural Networks for Signal Processing*, pp. 276–285, Amelia Island, FL, 1997.

[Papo 91] Papoulis A. *Probability, Random Variables and Stochastic Processes*, 3rd ed., McGraw-Hill, 1991.

[Pear 90] Pearlmutter B., Hampshire J. "Equivalence proofs for multilayer perceptron classifiers and the Bayesian discriminant function," *Proceedings Connectionists Models Summer School*, Morgan Kauffman, 1990.

[Plat 00] Platt J.C., Cristianini N., Shawe-Taylor J. "Large margin DAGs for the multiclass classification," in *Advances in Neural Information Processing*, (Smola S.A., Leen T.K., Müller K.R., eds.), Vol. 12, pp. 547–553, MIT Press, 2000.

[Plat 99] Platt J. "Fast training of support vector machines using sequential minimal optimization," in *Advances in Kernel Methods: Support Vector Learning* (Scholkopf B., Burges C.J.C., Smola A. J., eds), pp. 185–208, MIT Press, 1999.

[Platt 98] Platt J. "Sequential minimal optimization: A fast algorithm for training support vector machines," *Technical Report, Microsoft Research*, MSR-TR-98-14, April 21, 1998.

[Poul 95] Poulard H., "Barycentric correction procedure: A fast method of learning threshold units," *Proc. WCNN '95*, Vol. 1, Washington, DC, pp. 710–713, July, 1995.

[Rich 91] Richard M.D., Lippmann R.P. "Neural network classifiers estimate Bayesian a posteriori probabilities," *Neural Computation*, Vol. 3, pp. 461–483, 1991.

[Rifk 04] Rifkiy R., Klautau A. "In defense of one-vs-all classification," *Journal of Machine Learning Research*, Vol. 5, pp. 101–141, 2004.

[Robb 51] Robbins H., Monro S. "A stochastic approximation method," *Annals of Mathematical Statistics*, Vol. 22, pp. 400–407, 1951.

[Rose 58] Rosenblatt F. "The perceptron: only A probabilistic model for information storage and organization in the brain," *Psychological Review*, Vol. 65, pp. 386–408, 1958.

[Scho 00] Schölkoph B., Smola A.J., Williamson R.C., Bartlett P.L. "New support vector algorithms," *Neural Computation*, Vol. 12, pp. 1207–1245, 2000.

[Tao 04] Tao Q., Wu G.-W., Wang J. "A generalized S-K algorithm for learning ν-SVM classifiers," *Pattern Recognition Letters*, Vol. 25(10), pp. 1165–1171, 2004.

[Theo 07] Theodoridis S., Mavroforakis M. "Reduced convex hulls: A geometric approach to support vector machines," *IEEE Signal Processing Magazine*, Vol. 24(3), pp. 119–122, 2007.

[Tou 74] Tou J., Gonzalez R.C. *Pattern Recognition Principles*, Addison-Wesley, 1974.

[Tsan 06] Tsang I.W.-H., Kwok J.T. -Y., Zurada J.M. "Generalized core vector machines," *IEEE Transactions on Neural Networks*, Vol. 17(5), pp. 1126–1140, 2006.

[Vapn 98] Vapnik V.N. *Statistical Learning Theory*, John Wiley & Sons, 1998.

[Widr 60] Widrow B., Hoff M.E., Jr. "Adaptive switching circuits," *IRE WESCON Convention Record*, pp. 96–104, 1960.

[Widr 90] Widrow B., Lehr M.A. "30 years of adaptive neural networks: Perceptron, madaline, and backpropagation," *Proceedings of the IEEE*, Vol. 78(9), pp. 1415–1442, 1990.

[Yee 96] Yee T., Wild C. "Vector generalized additive models," *Journal of the Royal Statistical Society, Series B*, Vol. 58, pp. 481–493, 1996.

[Zhou 08] Zhou J., Peng H., Suen C.Y. "Data-driven decomposition for multi-class classification," *Pattern Recognition*, Vol. 41, pp. 67–76, 2008.

CHAPTER

Nonlinear Classifiers

4

4.1 INTRODUCTION

In the previous chapter we dealt with the design of linear classifiers described by linear discriminant functions (hyperplanes) $g(\boldsymbol{x})$. In the simple two-class case, we saw that the perceptron algorithm computes the weights of the linear function $g(\boldsymbol{x})$, provided that the classes are linearly separable. For nonlinearly separable classes, linear classifiers were optimally designed, for example, by minimizing the squared error. In this chapter we will deal with problems that are not linearly separable and for which the design of a linear classifier, even in an optimal way, does not lead to satisfactory performance. The design of nonlinear classifiers emerges now as an inescapable necessity.

4.2 THE XOR PROBLEM

To seek nonlinearly separable problems one does not need to go into complicated situations. The well-known *Exclusive OR (XOR)* Boolean function is a typical example of such a problem. Boolean functions can be interpreted as classification tasks. Indeed, depending on the values of the input binary data $\boldsymbol{x} = [x_1, x_2, \ldots, x_l]^T$, the output is either 0 or 1, and $\boldsymbol{x}$ is classified into one of the two classes $A(1)$ or $B(0)$. The corresponding truth table for the XOR operation is shown in Table 4.1.

Figure 4.1 shows the position of the classes in space. It is apparent from this figure that no single straight line exists that separates the two classes. In contrast, the other two Boolean functions, AND and OR, are linearly separable. The corresponding truth tables for the AND and OR operations are given in Table 4.2 and the respective class positions in the two-dimensional space are shown in Figure 4.2a and 4.2b. Figure 4.3 shows a perceptron, introduced in the previous chapter, with synaptic weights computed so as to realize an OR gate (verify).

Our major concern now is first to tackle the XOR problem and then to extend the procedure to more general cases of nonlinearly separable classes. Our kickoff point will be geometry.

Table 4.1 Truth Table for the XOR Problem

x_1	x_2	XOR	Class
0	0	0	B
0	1	1	A
1	0	1	A
1	1	0	B

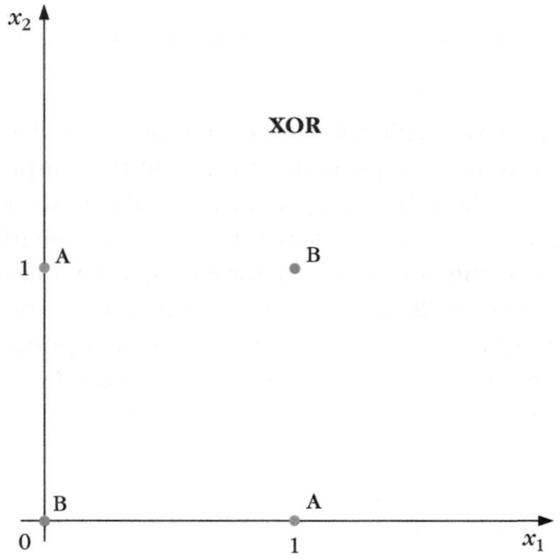

FIGURE 4.1

Classes A and B for the XOR problem.

Table 4.2 Truth Table for AND and OR Problems

x_1	x_2	AND	Class	OR	Class
0	0	0	B	0	B
0	1	0	B	1	A
1	0	0	B	1	A
1	1	1	A	1	A

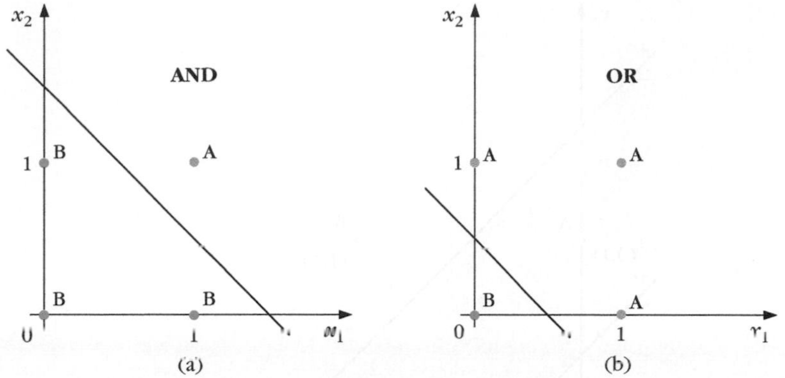

FIGURE 4.2

Classes A and B for (a) the AND and (b) OR problems.

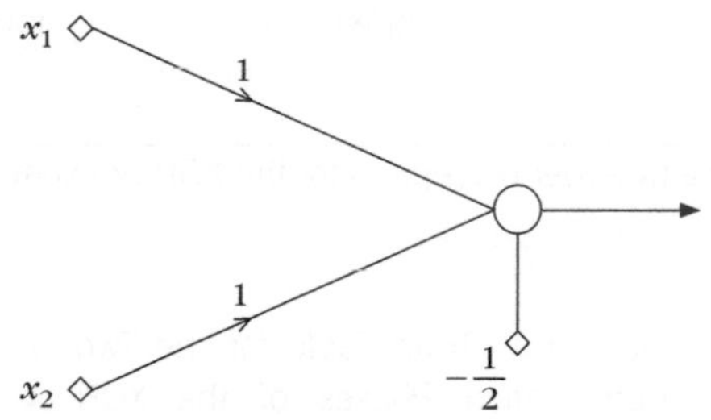

FIGURE 4.3

A perceptron realizing an OR gate.

4.3 THE TWO-LAYER PERCEPTRON

To separate the two classes A and B in Figure 4.1, a first thought that comes to mind is to draw two, instead of one, straight lines.

Figure 4.4 shows two such possible lines, $g_1(\boldsymbol{x}) = g_2(\boldsymbol{x}) = 0$, as well as the regions in space for which $g_1(\boldsymbol{x}) \gtrless 0, g_2(\boldsymbol{x}) \gtrless 0$. The classes can now be separated. Class A is to the right $(+)$ of $g_1(\boldsymbol{x})$ and to the left $(-)$ of $g_2(\boldsymbol{x})$. The region corresponding to class B lies either to the left or to the right of both lines. What we have really done is to attack the problem in *two* successive phases. During the first phase, we calculate the position of a feature vector $\boldsymbol{x}$ with respect to *each* of the two decision lines. In the second phase, we combine the results of the previous phase and we find the position of $\boldsymbol{x}$ with respect to *both* lines, that is, outside or inside the shaded area. We will now view this from a slightly different perspective, which will subsequently lead us easily to generalizations.

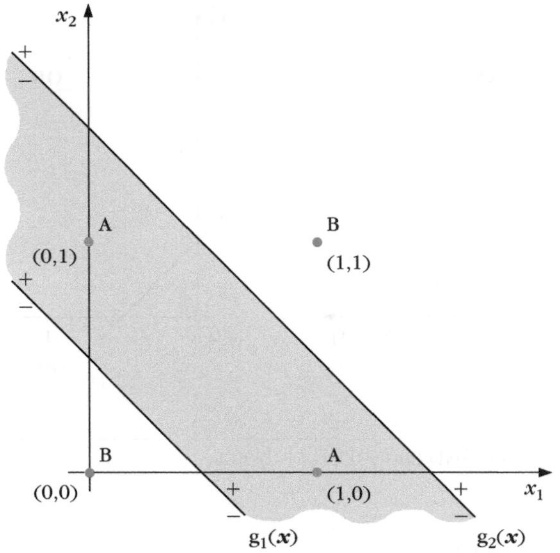

FIGURE 4.4

Decision lines realized by a two-layer perceptron for the XOR problem.

Table 4.3 Truth Table for the Two Computation Phases of the XOR Problem

1st Phase				
x_1	x_2	y_1	y_2	2nd Phase
0	0	0 (−)	0 (−)	B (0)
0	1	1 (+)	0 (−)	A (1)
1	0	1 (+)	0 (−)	A (1)
1	1	1 (+)	1 (+)	B (0)

Realization of the two decision lines (hyperplanes), $g_1(\cdot)$ and $g_2(\cdot)$, during the first phase of computations is achieved with the adoption of two perceptrons with inputs x_1, x_2 and appropriate synaptic weights. The corresponding outputs are $y_i = f(g_i(\boldsymbol{x})), i = 1, 2$, where the activation function $f(\cdot)$ is the step function with levels 0 and 1. Table 4.3 summarizes the y_i values for all possible combinations of the inputs. These are nothing else than the relative positions of the input vector $\boldsymbol{x}$ with respect to each of the two lines. From another point of view, the computations during the first phase *perform a mapping* of the input vector $\boldsymbol{x}$ to a new one $y = [y_1, y_2]^T$. The decision during the second phase is now based on the transformed data; that is, our goal is now to separate $[y_1, y_2] = [0, 0]$ and $[y_1, y_2] = [1, 1]$, which correspond to class B vectors, from the $[y_1, y_2] = [1, 0]$, which corresponds to class A vectors. As is apparent from Figure 4.5, this is easily achieved by drawing a third line $g(y)$, which can be realized via a third neuron. *In other words, the mapping of the first phase transforms the nonlinearly separable problem to a linearly separable one.* We will return to this important issue later on. Figure 4.6 gives a possible realization of these steps. Each of the three lines is realized via a neuron with appropriate synaptic weights. The resulting *multilayer* architecture can be considered as a generalization of the perceptron, and it is known as a *two-layer perceptron*

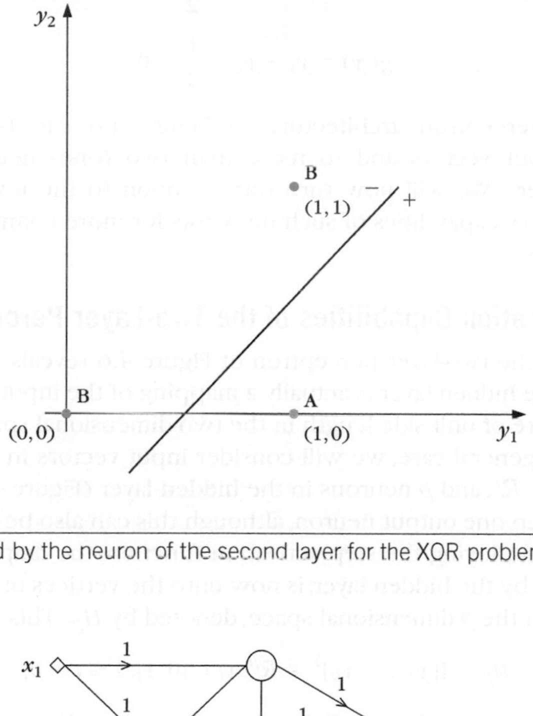

FIGURE 4.5

Decision line formed by the neuron of the second layer for the XOR problem.

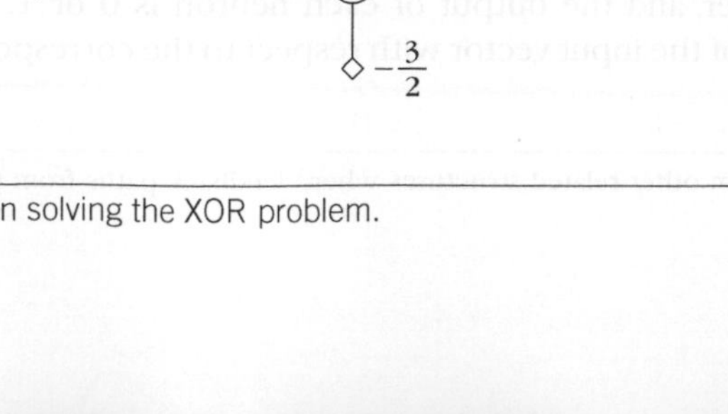

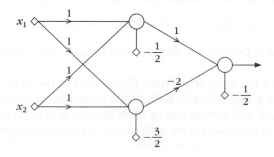

FIGURE 4.6

A two-layer perceptron solving the XOR problem.

or a *two-layer feedforward*[1] *neural network*. The two neurons (nodes) of the first layer perform computations of the first phase and they constitute the so-called *hidden layer*. The single neuron of the second layer performs the computations of the final phase and constitutes the *output layer*. In Figure 4.6 the *input layer* corresponds to the (nonprocessing) nodes where input data are applied. Thus, the number of input layer nodes equals the dimension of the input space. Note that at the input layer nodes no processing takes place. The lines that are realized by the two-layer perceptron of the figure are

$$g_1(\boldsymbol{x}) = x_1 + x_2 - \frac{1}{2} = 0$$

$$g_2(\boldsymbol{x}) = x_1 + x_2 - \frac{3}{2} = 0$$

$$g(\boldsymbol{y}) = y_1 - y_2 - \frac{1}{2} = 0$$

The multilayer perceptron architecture of Figure 4.6 can be generalized to l-dimensional input vectors and to more than two (one) neurons in the hidden (output) layer. We will now turn our attention to the investigation of the class discriminatory capabilities of such networks for more complicated nonlinear classification tasks.

4.3.1 Classification Capabilities of the Two-Layer Perceptron

A careful look at the two-layer perceptron of Figure 4.6 reveals that the action of the neurons of the hidden layer is actually a mapping of the input space $\boldsymbol{x}$ onto the vertices of a square of unit side length in the two-dimensional space (Figure 4.5).

For the more general case, we will consider input vectors in the l-dimensional space, that is, $\boldsymbol{x} \in \mathcal{R}^l$, and p neurons in the hidden layer (Figure 4.7). For the time being, we will keep one output neuron, although this can also be easily generalized to many. Again employing the step activation function, the mapping of the input space, performed by the hidden layer, is now onto the vertices of the hypercube of unit side length in the p-dimensional space, denoted by H_p. This is defined as

$$H_p = \{[y_1, \ldots, y_p]^T \in \mathcal{R}^p, y_i \in [0, 1], 1 \leq i \leq p\}$$

The vertices of the hypercube are all the points $[y_1, \ldots, y_p]^T$ of H_p with $y_i \in \{0, 1\}$, $1 \leq i \leq p$.

The mapping of the input space onto the vertices of the hypercube is achieved via the creation of p hyperplanes. Each of the hyperplanes is created by a neuron in the hidden layer, and the output of each neuron is 0 or 1, depending on the relevant position of the input vector with respect to the corresponding hyperplane.

[1] To distinguish it from other related structures where feedback paths from the output back to the input exist.

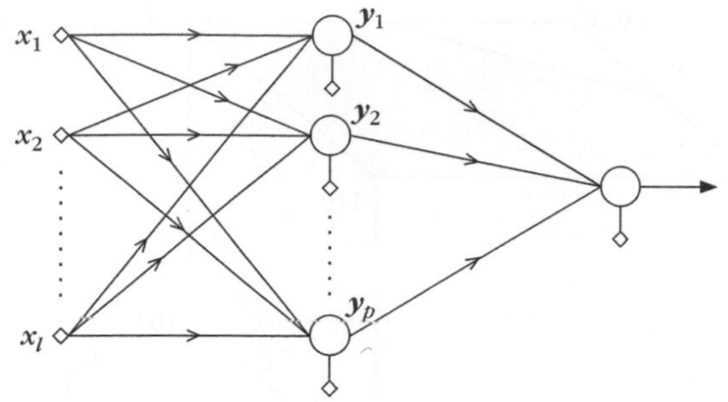

FIGURE 4.7

A two-layer perceptron.

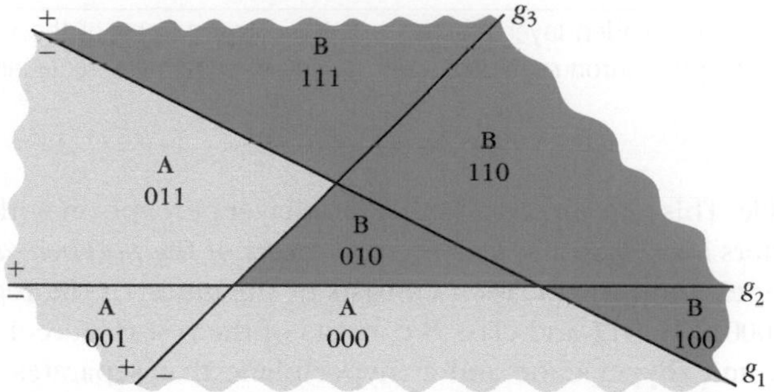

FIGURE 4.8

Polyhedra formed by the neurons of the first hidden layer of a multilayer perceptron.

Figure 4.8 is an example of three intersecting hyperplanes (three neurons) in the two-dimensional space. Each region defined by the intersections of these hyperplanes corresponds to a vertex of the unit three-dimensional hypercube, depending on its position with respect to each of these hyperplanes. The ith dimension of the vertex shows the position of the region with respect to the g_i hyperplane. For example, the 001 vertex corresponds to the region that is in the (−) side of g_1, in the (−) side of g_2, and in the (+) side of g_3. Thus, the conclusion we reach is that *the first layer of neurons divides the input l-dimensional space into polyhedra,*[2] *which are formed by hyperplane intersections. All vectors located within one of these polyhedral regions are mapped onto a specific vertex of the unit H_p hypercube*. The output neuron subsequently realizes another hyperplane, which separates the hypercube into two parts, having some of its vertices on one and some

[2] A polyhedron or polyhedral set is the finite intersection of closed half-spaces of $\mathcal{R}^l$, which are defined by a number of hyperplanes.

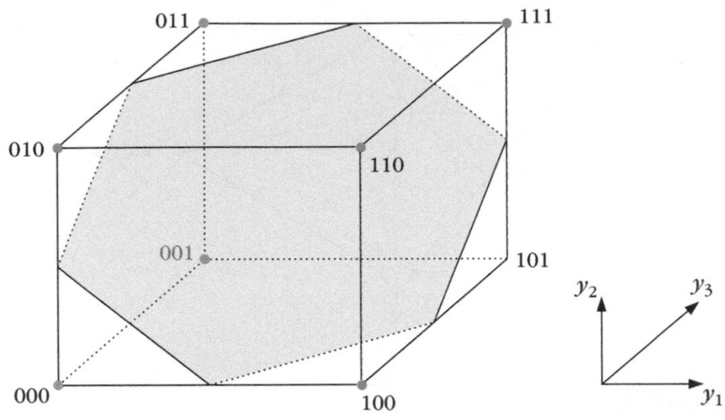

FIGURE 4.9

The neurons of the first hidden layer map an input vector onto one of the vertices of a unit (hyper)cube. The output neuron realizes a (hyper)plane to separate vertices according to their class label.

on the other side. This neuron provides the multilayer perceptron with the potential to classify vectors into *classes consisting of unions of the polyhedral regions*. Let us consider, for example, that class A consists of the union of the regions mapped onto vertices 000, 001, 011 and class B consists of the rest (Figure 4.8). Figure 4.9 shows the H_3 unit (hyper)cube and a (hyper)plane that separates the space $\mathcal{R}^3$ into two regions with the (class A) vertices 000, 001, 011 on one side and (class B) vertices 010, 100, 110, 111 on the other. This is the $-y_1 - y_2 + y_3 + 0.5 = 0$ plane, which is realized by the output neuron. With such a configuration all vectors from class A result in an output of 1(+) and all vectors from class B in 0(−). On the other hand, if class A consists of the union $000 \cup 111 \cup 110$ and class B of the rest, it is not possible to construct a single plane that separates class A from class B vertices. Thus, we can conclude that *a two-layer perceptron can separate classes each consisting of unions of polyhedral regions but not any union of such regions*. It all depends on the relative positions of the vertices of H_p, where the classes are mapped, and on whether or not these are linearly separable. Before we proceed further to see ways to overcome this shortcoming, it should be pointed out that vertex 101 of the cube does not correspond to any of the polyhedral regions. Such vertices are said to correspond to *virtual polyhedra*, and they do not influence the classification task.

4.4 THREE-LAYER PERCEPTRONS

The inability of the two-layer perceptrons to separate classes resulting from *any* union of polyhedral regions springs from the fact that the output neuron can realize

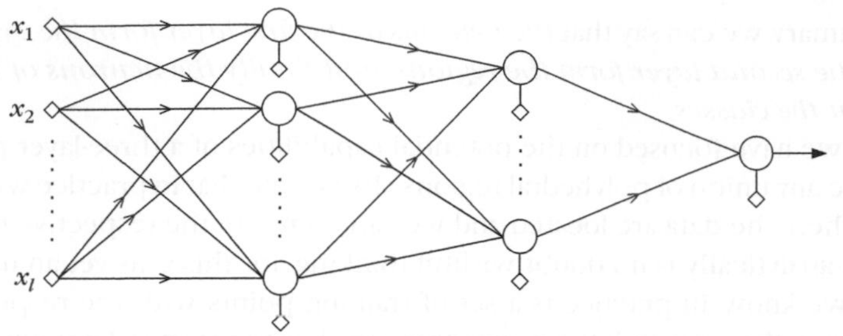

FIGURE 4.10

Architecture of a multilayer perceptron with two hidden layers of neurons and a single output neuron.

only a single hyperplane. This is the same situation confronting the basic perceptron when dealing with the XOR problem. The difficulty was overcome by constructing two lines instead of one. A similar escape path will be adopted here.

Figure 4.10 shows a three-layer perceptron architecture with two layers of hidden neurons and one output layer. We will show, constructively, that such an architecture can *separate classes resulting from any union of polyhedral regions*. Indeed, let us assume that all regions of interest are formed by intersections of p l-dimensional half-spaces defined by the p hyperplanes. These are realized by the p neurons of the first hidden layer, which also perform the mapping of the input space onto the vertices of the H_p hypercube of unit side length. In the sequel let us assume that class A consists of the union of K of the resulting polyhedra and class B of the rest. We then use K neurons in the second hidden layer. Each of these neurons realizes a hyperplane in the p-dimensional space. The synaptic weights for each of the second-layer neurons are chosen so that the realized hyperplane leaves only one of the H_p vertices on one side and *all* the rest on the other. For each neuron a different vertex is isolated, that is, one of the K A class vertices. In other words, each time an input vector from class A enters the network, one of the K neurons of the second layer results in a 1 and the remaining $K - 1$ give 0. In contrast, for class B vectors all neurons in the second layer output a 0. Classification is now a straightforward task. Choose the output layer neuron to realize an OR gate. Its output will be 1 for class A and 0 for class B vectors. The proof is now complete.

The number of neurons in the second hidden layer can be reduced by exploiting the geometry that results from each specific problem—for example, whenever two of the K vertices are located in a way that makes them separable from the rest, using a single hyperplane. Finally, the multilayer structure can be generalized to more than two classes. To this end, the output layer neurons are increased in number, realizing one OR gate for each class. Thus, one of them results in 1 every time a vector from the respective class enters the network, and all the others give 0. The number of second-layer neurons is also affected (why?).

In summary, we can say that *the neurons of the first layer form the hyperplanes, those of the second layer form the regions, and finally the neurons of the output layer form the classes.*

So far, we have focused on the potential capabilities of a three-layer perceptron to separate any union of polyhedral regions. To assume that in practice we know the regions where the data are located and we can compute the respective hyperplane equations analytically is no doubt wishful thinking, for this is as yet an unrealizable goal. All we know in practice is a set of training points with the respective class labels. As was the case with the perceptron, one has to resort to learning algorithms that learn the synaptic weights from the available training data vectors. We will focus our attention on two major directions. In one of them the network is constructed in a way that classifies correctly *all* the available training data, by building it as a succession of linear classifiers. The other direction relieves itself of the correct classification constraint and computes the synaptic weights so as to minimize a preselected cost function.

4.5 ALGORITHMS BASED ON EXACT CLASSIFICATION OF THE TRAINING SET

The starting point of these techniques is a small architecture (usually unable to solve the problem at hand), which is successively augmented until the correct classification of all N feature vectors of the training set X is achieved. Different algorithms follow different ways to augment their architectures. Thus, some algorithms expand their architectures in terms of the number of layers [Meza 89, Frea 90], whereas others use one or two hidden layers and expand them in terms of the number of their nodes (neurons) [Kout 94, Bose 96]. Moreover, some of these algorithms [Frea 90] allow connections between nodes of nonsuccessive layers. Others allow connections between nodes of the same layer [Refe 91]. A general principle adopted by most of these techniques is the decomposition of the problem into smaller problems that are easier to handle. For each smaller problem, a single node is employed. Its parameters are determined either iteratively using appropriate learning algorithms, such as the pocket algorithm or the LMS algorithm (Chapter 3), or directly via analytical computations. From the way these algorithms build the network, they are sometimes referred to as *constructive techniques*.

The *tiling algorithm* [Meza 89] constructs architectures with many (usually more than three) layers. We describe the algorithm for the two-class (A and B) case. The algorithm starts with a single node, $n(X)$, in the first layer, which is called the *master unit* of this layer.

This node is trained using the pocket algorithm (Chapter 3), and, after the completion of the training, it divides the training data set X into two subsets X^+ and X^- (line 1 in Figure 4.11). If X^+ (X^-) contains feature vectors from both classes, we introduce an additional node, $n(X^+)$ ($n(X^-)$), which is called the *ancillary*

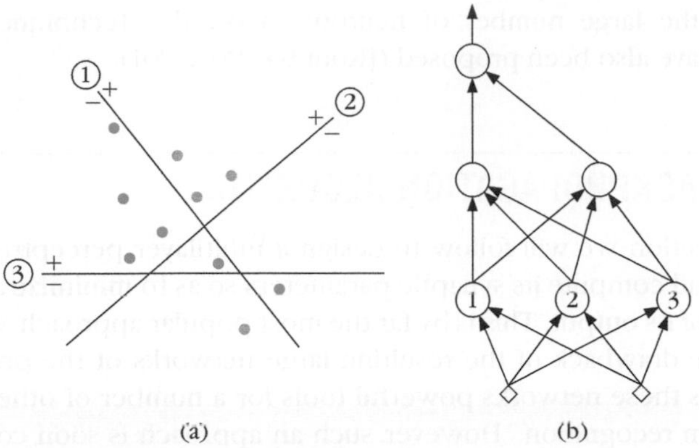

FIGURE 4.11

Decision lines and the corresponding architecture resulting from the tiling algorithm. The black (red) dots correspond to class A (B).

unit. This node is trained using only the feature vectors in X^+ (X^-) (line 2). If one of the X^{++}, X^{+-} (X^{-+}, X^{--}) produced by neuron $n(X^+)$ ($n(X^-)$) contains vectors from both classes, more ancillary nodes are added. This procedure stops after a finite number of steps, since the number of vectors a newly added (ancillary) unit has to discriminate decreases at each step. Thus, the first layer consists of a single master unit and, in general, more than one ancillary units. It is easy to show that in this way we succeed so that no two vectors from *different* classes give the same first-layer outputs.

Let $X_1 = \{\boldsymbol{y} : \boldsymbol{y} = f_1(\boldsymbol{x}), \boldsymbol{x} \in X\}$, where f_1 is the mapping implemented by the first layer. Applying the procedure just described to the set X_1 of the transformed $\boldsymbol{y}$ samples, we construct the second layer of the architecture and so on. In [Meza 89] it is shown that proper choice of the weights between two adjacent layers ensures that each newly added master unit classifies correctly all the vectors that are correctly classified by the master unit of the previous layer, plus at least one more vector. Thus, the tiling algorithm produces an architecture that classifies correctly all patterns of X in a finite number of steps.

An interesting observation is that all but the first layer treat binary vectors. This reminds us of the unit hypercube of the previous section. Mobilizing the same arguments as before, we can show that this algorithm may lead to correct classification architectures having three layers of nodes at the most.

Another family of constructive algorithms builds on the idea of the nearest neighbor classification rule, discussed in Chapter 2. The neurons of the first layer implement the hyperplanes bisecting the line segments that join the training feature vectors [Murp 90]. The second layer forms the regions, using an appropriate number of neurons that implement AND gates, and the classes are formed via the neurons of the last layer, which implement OR gates. The major drawback of this technique is the large number of neurons involved. Techniques that reduce this number have also been proposed ([Kout 94, Bose 96]).

4.6 THE BACKPROPAGATION ALGORITHM

The other direction we will follow to design a multilayer perceptron is to fix the architecture and compute its synaptic parameters so as to minimize an appropriate cost function of its output. This is by far the most popular approach, which not only overcomes the drawback of the resulting large networks of the previous section but also makes these networks powerful tools for a number of other applications, beyond pattern recognition. However, such an approach is soon confronted with a serious difficulty. This is the discontinuity of the step (activation) function, prohibiting differentiation with respect to the unknown parameters (synaptic weights). Differentiation enters into the scene as a result of the cost function minimization procedure. In the sequel we will see how this difficulty can be overcome.

The multilayer perceptron architectures we have considered so far have been developed around the McCulloch–Pitts neuron, employing as the activation function the step function

$$f(x) = \begin{cases} 1 & x > 0 \\ 0 & x < 0 \end{cases}$$

A popular family of continuous differentiable functions, which approximate the step function, is the family of *sigmoid functions*. A typical representative is the *logistic function*

$$f(x) = \frac{1}{1 + \exp(-ax)} \tag{4.1}$$

where a is a slope parameter.

Figure 4.12 shows the sigmoid function for different values of a, along with the step function. Sometimes a variation of the logistic function is employed that is antisymmetric with respect to the origin, that is, $f(-x) = -f(x)$. It is defined as

$$f(x) = \frac{2}{1 + \exp(-ax)} - 1 \tag{4.2}$$

It varies between 1 and -1, and it belongs to the family of hyperbolic tangent functions,

$$f(x) = c\frac{1 - \exp(-ax)}{1 + \exp(-ax)} = c \tanh\left(\frac{ax}{2}\right) \tag{4.3}$$

All these functions are also known as *squashing functions* since their output is limited in a finite range of values. In the sequel, we will adopt multilayer neural architectures like the one in Figure 4.10, and we will assume that the activation

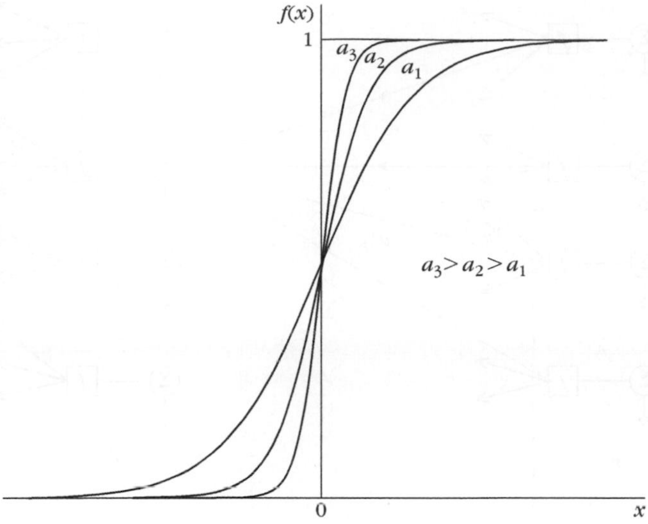

FIGURE 4.12

The logistic function. The larger the value of the slope parameter, a, the better the approximation of the unit-step function achieved.

functions are of the form given in (4.1)–(4.3). Our goal is to derive an iterative training algorithm that computes the synaptic weights of the network so that an appropriately chosen cost function is minimized. Before going into the derivation of such a scheme, an important point must be clarified. From the moment we move away from the step function, all we have said before about mapping the input vectors onto the vertices of a unit hypercube is no longer valid. It is now the cost function that takes on the burden for correct classification.

For the sake of generalization, let us assume that the network consists of a fixed number of L layers of neurons, with k_0 nodes in the input layer and k_r neurons in the rth layer, for $r = 1, 2, \ldots, L$. Obviously, k_0 equals l. All the neurons employ the same sigmoid activation function. As was the case in Section 3.3, we assume that N training pairs are available $(\boldsymbol{y}(i), \boldsymbol{x}(i)), i = 1, 2, \ldots, N$.[3] Because we have now assumed k_L output neurons, the output is no longer a scalar but a k_L-dimensional vector, $\boldsymbol{y}(i) = [y_1(i), \ldots, y_{k_L}(i)]^T$. The input (feature) vectors are k_0-dimensional vectors, $\boldsymbol{x}(i) = [x_1(i), \ldots, x_{k_0}(i)]^T$. During training, when vector $\boldsymbol{x}(i)$ is applied to the input, the output of the network will be $\hat{\boldsymbol{y}}(i)$, which is different from the desired value, $\boldsymbol{y}(i)$. *The synaptic weights are computed such that an appropriate (for each problem) cost function J, which is dependent on the values $\boldsymbol{y}(i)$ and*

[3] In contrast to other chapters, we use i in parentheses and not as an index. This is because, for the needs of the chapter, the latter notation can become very cumbersome.

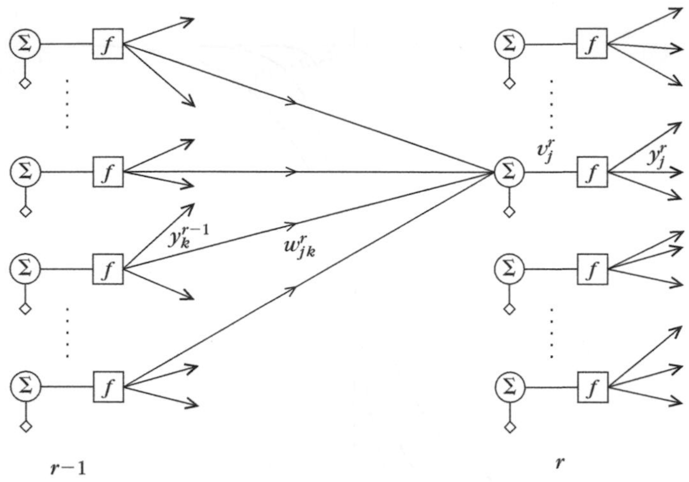

FIGURE 4.13

Definition of variables involved in the backpropagation algorithm.

$\hat{y}(i), i = 1, 2, \ldots, N$, *is minimized*. It is obvious that J depends, through $\hat{y}(i)$, on the weights and that this is a nonlinear dependence, due to the nature of the network itself. Thus, minimization of the cost function can be achieved via iterative techniques. In this section we will adopt the gradient descent scheme (Appendix C), which is the most widely used approach. Let $\boldsymbol{w}_j^r$ be the weight vector (including the threshold) of the jth neuron in the rth layer, which is a vector of dimension $k_{r-1} + 1$ and is defined as (Figure 4.13) $\boldsymbol{w}_j^r = [w_{j0}^r, w_{j1}^r, \ldots, w_{jk_{r-1}}^r]^T$. The basic iteration step will be of the form

$$\boldsymbol{w}_j^r(\text{new}) = \boldsymbol{w}_j^r(\text{old}) + \Delta\boldsymbol{w}_j^r$$

with

$$\Delta\boldsymbol{w}_j^r = -\mu\frac{\partial J}{\partial \boldsymbol{w}_j^r} \tag{4.4}$$

where $\boldsymbol{w}_j^r(\text{old})$ is the current estimate of the unknown weights and $\Delta\boldsymbol{w}_j^r$ the corresponding correction to obtain the next estimate $\boldsymbol{w}_j^r(\text{new})$.

In Figure 4.13 v_j^r is the weighted summation of the inputs to the jth neuron of the rth layer and y_j^r the corresponding output after the activation function. In the sequel we will focus our attention on cost functions of the form

$$J = \sum_{i=1}^{N} \mathcal{E}(i) \tag{4.5}$$

where $\mathcal{E}$ is an appropriately defined function depending on $\hat{y}(i)$ and $y(i), i = 1, 2, \ldots, N$. In other words, J is expressed as a sum of the N values that function $\mathcal{E}$

takes for each of the training pairs $(y(i), x(i))$. For example, we can choose $\mathcal{E}(i)$ as the sum of squared errors in the output neurons

$$\mathcal{E}(i) = \frac{1}{2}\sum_{m=1}^{k_L} e_m^2(i) \equiv \frac{1}{2}\sum_{m=1}^{k_L}(y_m(i) - \hat{y}_m(i))^2, \quad i = 1, 2, \ldots, N \tag{4.6}$$

For the computation of the correction term in (4.4) the gradient of the cost function J with respect to the weights is required and, consequently, the evaluation of $\partial\mathcal{E}(i)/\partial \boldsymbol{w}_j^r$.

Computation of the Gradients

Let $y_k^{r-1}(i)$ be the output of the kth neuron, $k = 1, 2, \ldots, k_{r-1}$, in the $(r-1)$th layer for the ith training pair and w_{jk}^r the current estimate of the corresponding weight leading to the jth neuron in the rth layer, with $j = 1, 2, \ldots, k_r$ (Figure 4.13). Thus, the argument of the activation function $f(\cdot)$ of the latter neuron will be

$$v_j^r(i) = \sum_{k=1}^{k_{r-1}} w_{jk}^r y_k^{r-1}(i) + w_{jo}^r \equiv \sum_{k=0}^{k_{r-1}} w_{jk}^r y_k^{r-1}(i) \tag{4.7}$$

where by definition $y_0^r(i) \equiv +1, \forall r, i$; so as to include the thresholds in the weights. For the output layer, we have $r = L$, $y_k^r(i) = \hat{y}_k(i), k = 1, 2, \ldots, k_L$, that is, the outputs of the neural network, and for $r = 1, y_k^{r-1}(i) = x_k(i), k = 1, 2, \ldots, k_0$, that is, the network inputs.

As is apparent from (4.7), the dependence of $\mathcal{E}(i)$ on $\boldsymbol{w}_j^r$ passes through $v_j^r(i)$. By the chain rule in differentiation, we have

$$\frac{\partial\mathcal{E}(i)}{\partial \boldsymbol{w}_j^r} = \frac{\partial\mathcal{E}(i)}{\partial v_j^r(i)}\frac{\partial v_j^r(i)}{\partial \boldsymbol{w}_j^r} \tag{4.8}$$

From (4.7) we obtain

$$\frac{\partial}{\partial \boldsymbol{w}_j^r} v_j^r(i) \equiv \begin{bmatrix} \frac{\partial}{\partial w_{j0}^r} v_j^r(i) \\ \vdots \\ \frac{\partial}{\partial w_{jk_{r-1}}^r} v_j^r(i) \end{bmatrix} = \boldsymbol{y}^{r-1}(i) \tag{4.9}$$

where

$$\boldsymbol{y}^{r-1}(i) = \begin{bmatrix} +1 \\ y_1^{r-1}(i) \\ \vdots \\ y_{k_{r-1}}^{r-1}(i) \end{bmatrix} \tag{4.10}$$

Let us define

$$\frac{\partial \mathcal{E}(i)}{\partial v_j^r(i)} \equiv \delta_j^r(i) \tag{4.11}$$

Then (4.4) becomes

$$\Delta \boldsymbol{w}_j^r = -\mu \sum_{i=1}^{N} \delta_j^r(i) \boldsymbol{y}^{r-1}(i) \tag{4.12}$$

Relation (4.12) is general for *any* differentiable cost function of the form (4.5). In the sequel we will compute $\delta_j^r(i)$ for the special case of least squares (4.6). *The procedure is similar for alternative cost function choices.*

Computation of $\delta_j^r(i)$ for the Cost Function in (4.6)

The computations start from $r = L$ and *propagate backward* for $r = L-1$, $L-2, \ldots, 1$. This is why the algorithm that will be derived is known as *the backpropagation algorithm*.

1. $r = L$

$$\delta_j^L(i) = \frac{\partial \mathcal{E}(i)}{\partial v_j^L(i)} \tag{4.13}$$

$$\mathcal{E}(i) \equiv \frac{1}{2} \sum_{m=1}^{k_L} e_m^2(i) \equiv \frac{1}{2} \sum_{m=1}^{k_L} (f(v_m^L(i)) - y_m(i))^2 \tag{4.14}$$

Hence

$$\delta_j^L(i) = e_j(i) f'(v_j^L(i)) \tag{4.15}$$

where f' is the derivative of $f(\cdot)$. In the last layer, the dependence of $\mathcal{E}(i)$ on $v_j^L(i)$ is explicit, and the computation of the derivative is straightforward. This is not true, however, for the hidden layers, where the computations of the derivatives need more elaboration.

2. $r < L$. Due to the successive dependence among the layers, the value of $v_j^{r-1}(i)$ influences all $v_k^r(i), k = 1, 2, \ldots, k_r$, of the next layer. Employing the chain rule in differentiation once more, we obtain

$$\frac{\partial \mathcal{E}(i)}{\partial v_j^{r-1}(i)} = \sum_{k=1}^{k_r} \frac{\partial \mathcal{E}(i)}{\partial v_k^r(i)} \frac{\partial v_k^r(i)}{\partial v_j^{r-1}(i)} \tag{4.16}$$

and from the respective definition (4.11)

$$\delta_j^{r-1}(i) = \sum_{k=1}^{k_r} \delta_k^r(i) \frac{\partial v_k^r(i)}{\partial v_j^{r-1}(i)} \tag{4.17}$$

But

$$\frac{\partial v_k^r(i)}{\partial v_j^{r-1}(i)} = \frac{\partial\left[\sum_{m=0}^{k_{r-1}} w_{km}^r y_m^{r-1}(i)\right]}{\partial v_j^{r-1}(i)} \tag{4.18}$$

with

$$y_m^{r-1}(i) = f(v_m^{r-1}(i)) \tag{4.19}$$

Hence,

$$\frac{\partial v_k^r(i)}{\partial v_j^{r-1}(i)} = w_{kj}^r f'(v_j^{r-1}(i)) \tag{4.20}$$

From (4.20) and (4.17) the following results:

$$\delta_j^{r-1}(i) = \left[\sum_{k=1}^{k_r} \delta_k^r(i) w_{kj}^r\right] f'(v_j^{r-1}(i)) \tag{4.21}$$

and for uniformity with (4.15)

$$\delta_j^{r-1}(i) = e_j^{r-1}(i) f'(v_j^{r-1}(i)) \tag{4.22}$$

where

$$e_j^{r-1}(i) = \sum_{k=1}^{k_r} \delta_k^r(i) w_{kj}^r \tag{4.23}$$

Relations (4.15), (4.22), and (4.23) constitute the iterations leading to the computation of $\delta_j^r(i), r = 1, 2, \ldots, L, j = 1, 2, \ldots, k_r$. The only quantity that is not yet computed is $f'(\cdot)$. For the function in (4.1) we have

$$f'(x) = af(x)(1 - f(x))$$

The algorithm has now been derived. The algorithmic scheme was first presented in [Werb 74] in a more general formulation.

The Backpropagation Algorithm

- *Initialization:* Initialize all the weights with small random values from a pseudorandom sequence generator.
- *Forward computations:* For each of the training feature vectors $\boldsymbol{x}(i), i = 1, 2, \ldots, N$, compute all the $v_j^r(i)$, $y_j^r(i) = f(v_j^r(i)), j = 1, 2, \ldots, k_r$, $r = 1, 2, \ldots, L$, from (4.7). Compute the cost function for the current estimate of weights from (4.5) and (4.14).

- *Backward computations:* For each $i = 1, 2, \ldots, N$ and $j = 1, 2, \ldots, k_L$ compute $\delta_j^L(i)$ from (4.15) and in the sequel compute $\delta_j^{r-1}(i)$ from (4.22) and (4.23) for $r = L, L-1, \ldots, 2$, and $j = 1, 2, \ldots, k_r$
- *Update the weights:* For $r = 1, 2, \ldots, L$ and $j = 1, 2, \ldots, k_r$

$$\boldsymbol{w}_j^r(\text{new}) = \boldsymbol{w}_j^r(\text{old}) + \Delta\boldsymbol{w}_j^r$$

$$\Delta\boldsymbol{w}_j^r = -\mu \sum_{i=1}^{N} \delta_j^r(i)\boldsymbol{y}^{r-1}(i)$$

Remarks

- A number of criteria have been suggested for terminating the iterations. In [Kram 89] it is suggested that we terminate the iterations either when the cost function J becomes smaller than a certain threshold or when its gradient with respect to the weights becomes small. Of course, the latter has a direct effect on the rate of change of the weights between successive iteration steps.
- As with all the algorithms that spring from the gradient descent method, the convergence speed of the backpropagation scheme depends on the value of the learning constant μ. Its value must be sufficiently small to guarantee convergence but not too small, because the convergence speed becomes very slow. The best choice of μ depends very much on the problem and the cost function shape in the weight space. Broad minima yield small gradients; thus large values of μ lead to faster convergence. On the other hand, for steep and narrow minima small values of μ are required to avoid overshooting the minimum. As we will soon see, scenarios with adaptive μ are also possible.
- The cost function minimization for a multilayer perceptron is a nonlinear minimization task. Thus, the existence of local minima in the corresponding cost function surface is an expected reality. Hence, the backpropagation algorithm runs the risk of being trapped in a local minimum. If the local minimum is deep enough, this may still be a good solution. However, in cases in which this is not true, getting stuck in such a minimum is an undesirable situation, and the algorithm should be reinitialized from a different set of initial conditions.
- The algorithm described in this section updates the weights once *all* the training (input–desired output) pairs have appeared in the network. This mode of operation is known as the *batch mode*. A variation of this approach is to update the weights for each of the training pairs. This is known as the *pattern* or *online mode*. This is analogous to the LMS, where, according to

the Robbins–Monro approach, the instantaneous value of the gradient is computed instead of its mean. In the backpropagation case, the sum of $\delta_j^r(i)$ over all i is substituted with *each* of them. The algorithm in its pattern mode of operation then becomes

$$w_j^r(i+1) = w_j^r(i) - \mu\delta_j^r(i)y^{r-1}(i)$$

Compared with the pattern mode, the batch mode is an inherent averaging process. This leads to a better estimate of the gradient, and thus to more well-behaved convergence. On the other hand, the pattern mode presents a higher degree of randomness during training. This may help the algorithm to avoid being trapped in a local minimum. In [Siet 91] it is suggested that the beneficial effects that randomness may have on training can be further emphasized by adding a (small) white noise sequence in the training data. Another commonly used practice focuses on the way the training data are presented in the network. *During training, the available training vectors are used in the update equation more than once* until the algorithm converges. One complete presentation of all N training pairs constitutes an *epoch*. As successive epochs are applied, it is good practice from the convergence point of view to randomize the order of presentation of the training pairs. Randomization can again help the pattern mode algorithm to jump out of regions around local minima, when this occurs. However, the final choice between the batch and pattern modes of operation depends on the specific problem [Hert 91, p. 119].

- Once training of the network has been achieved, the values to which the synapses and thresholds have converged are frozen, and the network is ready for classification. This is a much easier task than training. An unknown feature vector is presented in the input and is classified in the class that is indicated by the output of the network. The computations performed by the neurons are of the multiply–add type followed by a nonlinearity. This has led to various hardware implementations ranging from optical to VLSI chip design. Furthermore, neural networks have a natural built-in parallelism, and computations in each layer can be performed in parallel. These distinct characteristics of neural networks have led to the development of special *neurocomputers*. A number of those are already commercially available; see, for example, [Koli 97].

4.7 VARIATIONS ON THE BACKPROPAGATION THEME

Both versions of the backpropagation scheme—the batch and the pattern modes—inherit the disadvantage of all methods built on the gradient descent approach: *their convergence to the cost function minimum is slow*. Appendix C discusses the fact that this trait becomes more prominent if the eigenvalues of the corresponding

Hessian matrix exhibit large spread. In such cases, the change of the cost function gradient between successive iteration steps is not smooth but oscillatory, leading to slow convergence. One way to overcome this problem is to use a *momentum term* that smoothes out the oscillatory behavior and speeds up the convergence. The backpropagation algorithm with momentum term takes the form

$$\Delta \boldsymbol{w}_j^r(\text{new}) = \alpha \Delta \boldsymbol{w}_j^r(\text{old}) - \mu \sum_i^N \delta_j^r(i) \boldsymbol{y}^{r-1}(i) \tag{4.24}$$

$$\boldsymbol{w}_j^r(\text{new}) = \boldsymbol{w}_j^r(\text{old}) + \Delta \boldsymbol{w}_j^r(\text{new}) \tag{4.25}$$

Compared with (4.4), we see that the correction vector $\Delta \boldsymbol{w}_j^r$ depends not only on the gradient term but also on its value in the previous iteration step. The constant α is called the *momentum factor* and in practice is chosen between 0.1 and 0.8. To see the effect of the momentum factor, let us look at the correction term for a number of successive iteration steps. At the tth iteration step we have

$$\Delta \boldsymbol{w}_j^r(t) = \alpha \Delta \boldsymbol{w}_j^r(t-1) - \mu \boldsymbol{g}(t) \tag{4.26}$$

where the last term denotes the gradient. For a total of T successive iteration steps we obtain

$$\Delta \boldsymbol{w}_j^r(T) = -\mu \sum_{t=0}^{T-1} \alpha^t \boldsymbol{g}(T-t) + \alpha^T \Delta \boldsymbol{w}_j^r(0) \tag{4.27}$$

Since $\alpha < 1$, the last term gets close to zero after a few iteration steps and the smoothing (averaging) effect of the momentum term becomes apparent. Let us now assume that the algorithm is at a low-curvature point of the cost function surface in the weight space. We can then assume that the gradient is approximately constant over a number of iteration steps. Applying this, we can write that

$$\Delta \boldsymbol{w}_j^r(T) \simeq -\mu(1 + \alpha + \alpha^2 + \alpha^3 + \cdots)\boldsymbol{g} = -\frac{\mu}{1-\alpha}\boldsymbol{g}$$

In other words, in such cases the effect of the momentum term is to effectively increase the learning constant. In practice, improvements in converging speed by a factor of 2 or even more have been reported [Silv 90].

A heuristic variation of the previous technique is to use an adaptive value for the learning factor μ, depending on the cost function values at successive iteration steps. A possible procedure is the following: Let $J(t)$ be the value of the cost at the tth iteration step. If $J(t) < J(t-1)$, then increase the learning rate by a factor of r_i. If, on the other hand, the new value of the cost is larger than the old one by a factor c, then decrease the learning rate by a factor of r_d. Otherwise use the same value. In summary

$$\frac{J(t)}{J(t-1)} < 1, \quad \mu(t) = r_i \mu(t-1)$$

$$\frac{J(t)}{J(t-1)} > c, \quad \mu(t) = r_d\mu(t-1)$$

$$1 \leq \frac{J(t)}{J(t-1)} \leq c, \quad \mu(t) = \mu(t-1)$$

Typical values of the parameters which are adopted in practice are $r_i = 1.05, r_d = 0.7, c = 1.04$. For iteration steps where the cost increases, it may be advantageous not only to decrease the learning rate but also to set the momentum term equal to 0. Others suggest that the update of the weighting not to be done at this step.

Another strategy for updating the learning factor μ is followed in the so-called *delta-delta* rule and in its modification *delta-bar-delta* rule [Jaco 88]. The idea here is to use a different learning factor for each weight and to increase the particular learning factor if the gradient of the cost function with respect to the corresponding weight has the same sign on two successive iteration steps. Conversely, if the sign changes, this is an indication of a possible oscillation and the learning factor should be reduced. A number of alternative techniques for speeding up convergence have also been suggested. In [Cich 93] a more extensive review of such techniques is provided.

The other option for faster convergence is to free ourselves from the gradient descent rationale and to adopt alternative searching schemes, usually at the expense of increased complexity. A number of such algorithmic techniques have appeared in the related literature. For example, [Kram 89, Barn 92, Joha 92] present algorithmic schemes based on the conjugate gradient algorithm; [Batt 92, Rico 88, Barn 92, Watr 88] provide schemes of the Newton family; [Palm 91, Sing 89] propose algorithms based on the Kalman filtering approach; and [Bish 95] a scheme based on the Levenberg–Marquardt algorithm. In many of these algorithms, elements of the Hessian matrix need to be computed, that is, the second derivatives of the cost function with respect to the weights

$$\frac{\partial^2 J}{\partial w_{jk}^q \partial w_{nm}^r}$$

The computations of the Hessian matrix are performed by adopting, once more, the backpropagation concept (see also Problems 4.12, 4.13). More on these issues can be found in [Hayk 99, Zura 92].

A popular scheme that is loosely based on Newton's method is the *quickprop* scheme [Fahl 90]. It is a heuristic method and treats the weights as if they were quasi-independent. It then approximates the error surface, as a function of each weight, by a quadratic polynomial. If this has its minimum at a sensible value, the latter is used as the new weight for the iterations; otherwise a number of heuristics are used. A usual form of the algorithm, for the weights in the various layers, is

$$\Delta w_{ij}(t) = \begin{cases} \alpha_{ij}(t)\Delta w_{ij}(t-1), & \text{if } \Delta w_{ij}(t-1) \neq 0 \\ \mu \frac{\partial J}{\partial w_{ij}}, & \text{if } \Delta w_{ij}(t-1) = 0 \end{cases} \tag{4.28}$$

where

$$\alpha_{ij}(t) = \min\left\{ \frac{\frac{\partial J}{\partial w_{ij}}(t)}{\frac{\partial J}{\partial w_{ij}}(t-1) - \frac{\partial J}{\partial w_{ij}}(t)}, \alpha_{\max} \right\} \tag{4.29}$$

with typical values of the involved variables being $0.01 \leq \mu \leq 0.6, \alpha_{\max} \approx 1.75$ [Cich 93]. An algorithm similar in spirit to quickprop has been proposed in [Ried 93]. It is reported that it is as fast as quickprop, and it requires less adjustment of the parameters to be stable.

4.8 THE COST FUNCTION CHOICE

It will not come as a surprise that the least squares cost function in (4.6) is not the unique choice available to the user. Depending on the specific problem, other cost functions can lead to better results. Let us look, for example, at the least squares cost function more carefully. Since all errors in the output nodes are first squared and summed up, large error values influence the learning process much more than the small errors. Thus, if the dynamic ranges of the desired outputs are not *all* of the same order, the least squares criterion will result in weights that have "learned" via a process of unfair provision of information. Furthermore, in [Witt 00] it is shown that for a class of problems, the gradient descent algorithm with the squared error criterion can be trapped in a local minimum and fail to find a solution, although (at least) one exists. In the current context, a solution is assumed to be a classifier that classifies correctly all training samples. In contrast, it is shown that there is an alternative class of functions, satisfying certain criteria, which guarantee that the gradient descent algorithm converges to such a solution, provided that one exists. This class of cost functions is known as *well-formed functions*. We will now present a cost function of this type, which is well suited for pattern recognition tasks.

The multilayer network performs a nonlinear mapping of the input vectors $\boldsymbol{x}$ to the output values $\hat{y}_k = \phi_k(\boldsymbol{x}; \boldsymbol{w})$ for each of the output nodes $k = 1, 2, \ldots k_L$, where the dependence of the mapping on the values of the weights is explicitly shown. In Chapter 3 we have seen that, if we adopt the least squares cost function and the desired outputs y_k are binary (belong to or not in class ω_k), then for the optimal values of the weights $\boldsymbol{w}^*$ the corresponding output of the network, $\hat{y}_k$, is *the least squares optimal estimate of the posterior probability* $P(\omega_k|\boldsymbol{x})$. (The question of how good or bad this estimate is will be of interest to us soon.) At this point we will adopt this probabilistic interpretation of the real outputs $\hat{y}_k$ as the basis on which our cost function will be built. Let us assume that the desired output values, y_k, are independent binary random variables and that $\hat{y}_k$ are the respective posterior probabilities that these random variables are 1 [Hint 90, Baum 88].

The *cross-entropy* cost function is then defined by

$$J = -\sum_{i=1}^{N}\sum_{k=1}^{k_L}(y_k(i)\ln\hat{y}_k(i) + (1 - y_k(i))\ln(1 - \hat{y}_k(i))) \tag{4.30}$$

J takes its minimum value when $y_k(i) = \hat{y}_k(i)$, and for binary desired response values the minimum is zero. There are various interpretations of this cost function [Hint 90, Baum 88, Gish 90, Rich 91]. Let us consider, for example, the output vector $\boldsymbol{y}(i)$ when $\boldsymbol{x}(i)$ appears at the input. This consists of a 1 at the true class node and zero elsewhere. If we take into account that the probability of the kth node to be 1(0) is $\hat{y}_k(i)(1 - \hat{y}_k(i))$ and by considering nodes *independently*, then

$$p(y) = \prod_{k=1}^{k_L}(\hat{y}_k)^{y_k}(1 - \hat{y}_k)^{1-y_k} \tag{4.31}$$

where the dependence on i has been suppressed for notational convenience. Then it is straightforward to check that J results from the negative log-likelihood of the training sample pairs. If $y_k(i)$ were true probabilities in (0, 1) then subtracting the minimum value from J (4.30) becomes

$$J = -\sum_{i=1}^{N}\sum_{k=1}^{k_L}\left(y_k(i)\ln\frac{\hat{y}_k(i)}{y_k(i)} + (1 - y_k(i))\ln\frac{1 - \hat{y}_k(i)}{1 - y_k(i)}\right) \tag{4.32}$$

For binary valued y_ks the above is still valid if we use the limiting value $0 \ln 0 = 0$.

It is not difficult to show (Problem 4.5) that the *cross-entropy cost function depends on the relative errors and not on the absolute errors, as its least squares counterpart; thus it gives the same weight to small and large values*. Furthermore, it has been shown that it satisfies the conditions of the well-formed functions [Adal 97]. Finally, it can be shown that adopting the cross-entropy cost function and binary values for the desired responses, *the outputs $\hat{y}_k$ corresponding to the optimal weights $\boldsymbol{w}^*$ are indeed estimates of $P(\omega_k|\boldsymbol{x})$*, as in the least squares case [Hamp 90].

A major advantage of the cross-entropy cost function is that it diverges if one of the outputs converges to the wrong extreme, hence the gradient descent reacts fast. On the other hand, the squared error cost function approaches a constant in this case, and the gradient descent on the LS will wander on a plateau, even though the error may not be small. This advantage of the cross-entropy cost function is demonstrated in the channel equalization context in [Adal 97].

A different cost function results if we treat $\hat{y}_k(i)$ and $y_k(i)$ as the true and desired probabilities, respectively. Then a measure of their similarity is given by the cross-entropy function (Appendix A)

$$J = -\sum_{i=1}^{N}\sum_{k=1}^{k_L} y_k(i)\ln\frac{\hat{y}_k(i)}{y_k(i)} \tag{4.33}$$

This is also valid for binary target values (using the limiting form). However, although we have interpreted the outputs as probabilities, there is no guarantee that they sum up to unity. This can be imposed onto the network by adopting an alternative activation function for the output nodes. In [Brid 90] the so-called *softmax* activation function was suggested, given by

$$\hat{y}_k = \frac{\exp(v_k^L)}{\sum_{k'} \exp(v_{k'}^L)} \tag{4.34}$$

This guarantees that the outputs lie in the interval [0, 1] and that they sum up to unity (note that in contrast to (4.32) the output probabilities are not considered independently). It is easy to show, (Problem 4.7) that in this case the quantity δ_j^L required by the backpropagation is equal to $\hat{y}_k - y_k$.

Besides the cross-entropy cost function in (4.33) a number of alternative cost functions has been proposed. For example, in [Kara 92] a generalization of the quadratic error cost function is utilized with the aim of speeding up convergence. Another direction is to minimize the classification error, which after all is the major goal in pattern recognition. A number of techniques have been suggested with this philosophy [Nede 93, Juan 92, Pado 95], which is known as *discriminative learning*. The basic potential advantage of discriminative learning is that essentially it tries to move the decision surfaces so as to reduce classification error. To achieve this goal, it puts more emphasis on the largest of the class *a posteriori* probability estimates. In contrast, the squared error cost function, for example, assigns the same importance to all posterior probability estimates. In other words, it tries to learn more than what is necessary for classification, which may limit its performance for a fixed size network. Most of the discriminative learning techniques use a smoothed version of the classification error, so as to be able to apply differentiation in association with gradient descent approaches. This, of course, presents the danger that the minimization procedure will be trapped in a local minimum. In [Mill 96] a *deterministic annealing* procedure is employed to train the networks, with an enhanced potential to avoid local minima (see Chapter 15).

The final choice of the cost function depends on the specific problem under consideration. However, as is pointed out in [Rich 91], in a number of practical situations the use of alternative, to least squares, cost functions did not necessarily lead to substantial performance improvements.

So far, the task of training multilayer perceptrons has been approached via the *unconstrained* optimization route (Appendix C). However, more recent research has shown that it is often beneficial to incorporate *additional knowledge* in the learning rule using *constrained* optimization. This has been shown to lead to the formulation of efficient learning algorithms with accelerated learning properties. The additional knowledge can be encoded in the form of objectives leading to single- or multi-objective optimization criteria that have to be satisfied simultaneously, with the demand for a long-term decrease of the cost function [Pera 00].

In this approach, an optimization problem is formulated at each epoch of the learning process. For example, the requirement of partial alignment of current and previous epoch weight vector updates (respectively denoted by $\Delta\boldsymbol{w}(t)$ and $\Delta\boldsymbol{w}(t-1)$), which basically underlies the use of the momentum term in the backpropagation algorithm, is enhanced by requiring maximization of the quantity $\Phi = \Delta\boldsymbol{w}(t)^T\Delta\boldsymbol{w}(t-1)$ and simultaneously allowing for a controlled decrease of the cost function. This leads to a constrained first-order algorithm, and it has been reported that it outperforms backpropagation and several of its variants in different benchmark learning tasks [Pera 95].

Generalizations of the method involving the Hessian matrix have also been proposed and used successfully in several benchmarks and applications [Ampa 02, Huan 04]. Under certain conditions, the solution of the problem for each epoch is provided analytically, leading to a closed formula for the weight update rule [Pera 03].

A Bayesian Framework for Network Training

All the cost functions considered so far aim at computing a single set of optimal values for the unknown parameters of the network. An alternative rationale is to look at the probability distribution function of the unknown weights, $\boldsymbol{w}$, in the weight space. The idea behind this approach stems from the Bayesian inference technique used for the estimating an unknown parametric pdf, as we discussed in Chapter 2. The basic steps followed for this type of network training, known as *Bayesian learning*, are (e.g., [Mack 92a]):

- Assume a model for the prior distribution $p(\boldsymbol{w})$ of the weights. This must be rather broad in shape in order to provide equal chance to a rather large range of values.
- Let $Y = \{\boldsymbol{y}(i), i = 1, 2, \dots, N\}$ be the set of the desired output training vectors for a given input data set $X = \{\boldsymbol{x}(i), i = 1, 2, \dots, N\}$. Assume a model for the likelihood function $p(Y|\boldsymbol{w})$, for example, Gaussian.[4] This basically models the error distribution between the true and desired output values, and it is the stage at which the input training data come into the scene.
- Using Bayes's theorem, we obtain

$$p(\boldsymbol{w}|Y) = \frac{p(Y|\boldsymbol{w})p(\boldsymbol{w})}{p(Y)} \tag{4.35}$$

  where $p(Y) = \int p(Y|\boldsymbol{w})p(\boldsymbol{w})\,d\boldsymbol{w}$. The resulting posterior pdf will be more sharply shaped around a value $\boldsymbol{w}_0$, since it has learned from the available training data.

[4] Strictly speaking we should write $P(Y|\boldsymbol{w}, X)$. However, all probabilities and pdf's are conditioned on X, and we omit it for notational convenience.

- Interpreting the true outputs of a network, $\hat{y}_k = \phi_k(\boldsymbol{x}; \boldsymbol{w})$, as the respective class probabilities, conditioned on the input $\boldsymbol{x}$ and the weight vector $\boldsymbol{w}$, the conditional class probability is computed by averaging over all $\boldsymbol{w}$ [Mack 92b]:

$$P(\omega_k|\boldsymbol{x}; Y) = \int \phi_k(\boldsymbol{x}; \boldsymbol{w}) p(\boldsymbol{w}|Y)\, d\boldsymbol{w} \tag{4.36}$$

The major computational cost associated with this type of technique is due to the required integration in the multidimensional space. This is not an easy task, and various practical implementations have been suggested in the literature. Further discussion of these issues is beyond the scope of this book. A good introduction to Bayesian learning, including a discussion of related practical implementations, is provided in [Bish 95].

4.9 CHOICE OF THE NETWORK SIZE

In the previous sections, we assumed the number of layers and neurons for each layer to be known and fixed. How one determines the appropriate number of layers and neurons was not of interest to us. This task will become our major focus now.

One answer to the problem could be to choose the size of the network large enough and leave the training to decide about the weights. A little thought reveals that such an approach is rather naive. Besides the associated computational complexity problems, there is a major reason why the size of the network should be kept as small as possible. This is imposed by the generalization capabilities that the network must possess. As has already been pointed out in Section 3.7, the term *generalization* refers to the capability of the multilayer neural network (and of any classifier) to classify correctly feature vectors that were not presented to it during the training phase—that is, the capability of a network to decide upon data unknown to it, based on what it has learned from the training set. Taking for granted the finite (and in many cases small) number N of training pairs, the number of free parameters (synaptic weights) to be estimated should be *(a) large enough to learn what makes "similar" the feature vectors within each class and at the same time what makes one class different from the other and (b) small enough, with respect to N, so as not to be able to learn the underlying differences among the data of the same class*. When the number of free parameters is large, the network tends to adapt to the particular details of the specific training data set. This is known as *overfitting* and leads to poor generalization performance, when the network is called to operate on feature vectors unknown to it. In conclusion, the network should have the smallest possible size to adjust its weights to the largest regularities in the data and ignore the smaller ones, which might also be the result of noisy measurements. Some theoretical touches concerning the generalization aspects of a classifier will

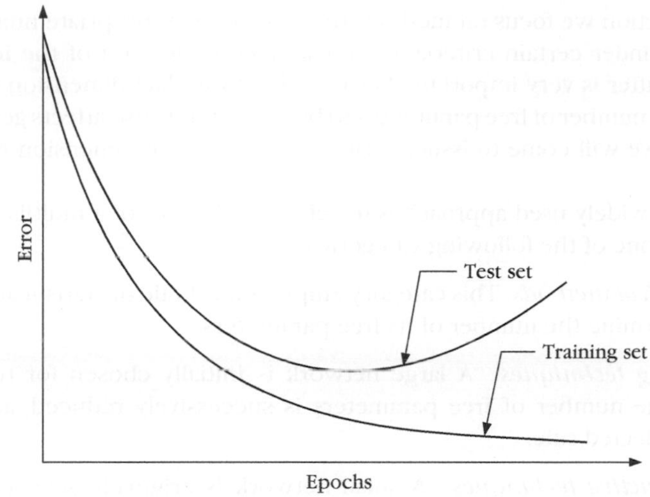

FIGURE 4.14

Trend of the output error versus the number of epochs illustrating overtraining of the training set.

be presented in Chapter 5, when we discuss the Vapnik–Chervonenkis dimension. The bias–variance dilemma, discussed in Chapter 3, is another side of the same problem.

Adaptation of the free parameters to the peculiarities of the specific training set may also occur as the result of *overtraining* (e.g., see [Chau 90]). Let us assume that we can afford the luxury of having a large set of training data. We divide this set into two subsets, one for training and one for test. The latter is known as the *validation* or *test set*. Figure 4.14 shows the trend of two curves of the output error as a function of iteration steps. One corresponds to the training set, and we observe that the error keeps decreasing as the weights converge. The other corresponds to the error of the validation set. Initially, the error decreases, but at some later stage it starts increasing. This is because the weights, computed from the training set, adapt to the idiosyncrasies of the specific training set, thus affecting the generalization performance of the network. This behavior could be used in practice to determine the point where the learning process iterations must terminate. This is the point where the two curves start departing. However, this methodology assumes the existence of a large number of data sets, which is not usually the case in practice.

Besides generalization, other performance factors also demand to keep the size of a network as small as possible. Small networks are computationally faster and cheaper to build. Furthermore, their performance is easier to understand, which is important in some critical applications.

In this section we focus on methods that select the appropriate number of free parameters, under certain criteria and for a given dimension of the input vector space. The latter is very important, because the input data dimension is no doubt related to the number of free parameters to be used; thus it also affects generalization properties. We will come to issues related to input space dimension reduction in Chapter 5.

The most widely used approaches to selecting the size of a multilayer network come under one of the following categories:

- *Analytical methods*. This category employs algebraic or statistical techniques to determine the number of its free parameters.
- *Pruning techniques*. A large network is initially chosen for training, and then the number of free parameters is successively reduced, according to a preselected rule.
- *Constructive techniques*. A small network is originally selected, and neurons are successively added, based on an appropriately adopted learning rule.

Algebraic Estimation of the Number of Free Parameters

We have already discussed in Section 4.3.1 the capabilities of a multilayer perceptron, with one hidden layer and units of the McCulloch–Pitts type, to divide the input l-dimensional space into a number of polyhedral regions. These are the result of intersections of hyperplanes formed by the neurons. In [Mirc 89] it is shown that in the l-dimensional space a multilayer perceptron of a single hidden layer with K neurons can form *a maximum* of M polyhedral regions with M given by

$$M = \sum_{m=0}^{l} \binom{K}{m}, \quad \text{where} \binom{K}{m} = 0, \quad \text{for } K < m \tag{4.37}$$

and

$$\binom{K}{m} \equiv \frac{K!}{m!(K-m)!}$$

For example, if $l = 2$, and $K = 2$, this results in $M = 4$; thus, the XOR problem, with $M = 3(< 4)$, can be solved with two neurons. The disadvantage of this method is that it is static and does not take into consideration the cost function used as well as the training procedure.

Pruning Techniques

These techniques start training a sufficiently large network, and then they remove, in a stepwise procedure, the free parameters that have little influence on the cost function. There are two major methodological directions:

Methods Based on Parameter Sensitivity Calculations

Let us take for example the technique suggested in [Lecu 90]. Using a Taylor series expansion, the variation imposed on the cost function by parameter perturbations is

$$\delta J = \sum_i g_i \delta w_i + \frac{1}{2}\sum_i h_{ii}\delta w_i^2 + \frac{1}{2}\sum_{\substack{i,j \\ i \neq j}} h_{ij}\delta w_i \delta w_j$$
$$+ \text{ higher order terms}$$

where

$$g_i = \frac{\partial J}{\partial w_i}, \quad h_{ij} = \frac{\partial^2 J}{\partial w_i \partial w_j}$$

and i, j runs over all the weights. The derivatives can be computed via the backpropagation methodology (Problem 4.13). In practicc, the derivatives are computed after some initial period of training. This allows us to adopt the assumption that a point near a minimum has been reached and the first derivatives can be set equal to zero. A further computational simplification is to assume that the Hessian matrix is diagonal. Under these assumptions, the cost function sensitivity is approximately given by

$$\delta J = \frac{1}{2}\sum_i h_{ii}\delta w_i^2 \tag{4.38}$$

and the contribution of each parameter is determined by the *saliency* value s_i, given approximately by

$$s_i = \frac{h_{ii} w_i^2}{2} \tag{4.39}$$

where we assume that a weight of value w_i is changed to zero. Pruning is now achieved in an iterative fashion according to the following steps:

- The network is trained using the backpropagation algorithm for a number of iteration steps so that its cost function is reduced to a sufficient percentage.
- For the current weight estimates, the respective saliency values are computed and weights with small salencies are removed.
- The training process is continued with the remaining weights, and the process is repeated after some iteration steps. The process is stopped when a chosen stopping criterion is met.

In [Hass 93] the full Hessian matrix has been employed for the pruning procedure. It should be stressed that, although the backpropagation concept is present in this technique, the learning procedure is distinctly different from the backpropagation training algorithm of Section 4.6. *There, the number of free parameters was fixed throughout the training. In contrast, the philosophy here is exactly the opposite.*

Methods Based on Cost Function Regularization

These methods achieve the reduction of the originally large size of the network by including *penalty terms* in the cost function. The cost function now has the form

$$J = \sum_{i=1}^{N} \mathcal{E}(i) + \alpha \mathcal{E}_p(\boldsymbol{w}) \tag{4.40}$$

The first term is the performance cost function, and it is chosen according to what we have already discussed (e.g., least squares, cross entropy). The second depends on the weight vector, and it is chosen to *favor small values for the weights*. The constant α is the so-called *regularization parameter*, and it controls the relative significance of these two terms. A popular form for the penalty term is

$$\mathcal{E}_p(\boldsymbol{w}) = \sum_{k=1}^{K} h(w_k^2) \tag{4.41}$$

with K being the total number of weights in the network and $h(\cdot)$ an appropriately chosen differentiable function. According to such a choice, weights that do not contribute significantly in the formation of the network output do not materially affect much the first term of the cost function. Hence, the existence of the penalty term drives them to small values. Thus, pruning is achieved. In practice, a threshold is preselected and weights are compared against it after a number of iteration steps. Weights that become smaller than it are removed, and the process is continued. This type of pruning is known as *weight elimination*. Function $h(\cdot)$ can take various forms. For example, in [Wein 90] the following is suggested:

$$h(w_k^2) = \frac{w_k^2}{w_0^2 + w_k^2} \tag{4.42}$$

where w_0 is a preselected parameter close to unity. Closer observation of this penalty term reveals that it goes to zero very fast for values $w_k < w_0$; thus such weights become insignificant. In contrast, the penalty term tends to unity for $w_k > w_0$.

A variation of (4.41) is to include in the regularized cost function another penalty term that favors small values of y_k^r, that is, small neuron outputs. Such techniques lead to removal of insignificant neurons as well as weights. A summary and discussion of various pruning techniques can be found in [Refe 91, Russ 93].

Keeping the size of the weights small is in line with the theoretical results obtained in [Bart 98]. Assume that a large multilayer perceptron is used as a classifier and that the learning algorithm finds a network with (a) small weights and (b) small squared error on the training patterns. Then, it can be shown that the generalization performance depends on the size of the weights rather than the number of weights. More specifically, for a two-layer perceptron with sigmoid activation functions if A is an upper bound of the sum of the magnitudes of the weights associated with each neuron, then the associated classification error probability is no more than a certain error estimate (related to the output squared error) plus $A^3\sqrt{(\log l)/N}$.

This is a very interesting result indeed. It confronts the generally accepted fact that the generalization performance is directly related to the number of training points and the number of free parameters. It also explains why sometimes the generalization error performance of multilayer perceptrons is good, although the training has been performed with a relatively small (compared to the size of the network) number of training points. Further discussion concerning the generalization performance of a classifier and some interesting theoretical results is found at the end of Chapter 5.

Constructive Techniques

In Section 4.5 we have already discussed such techniques for training neural networks. However, the activation function was the unit-step function, and also the emphasis was put on classifying *correctly all* input training data and not on the generalization properties of the resulting network. In [Fahl 90] an alternative constructive technique for training neural networks, with a single hidden layer and sigmoid activation functions, was proposed, known as *cascade correlation*. The network starts with input and output units only. Hidden neurons are added one by one and are connected to the network with two types of weights. The first type connects the new unit with the input nodes as well as the outputs of previously added hidden neurons. Each time a new hidden neuron is added in the network, these weights are trained so as to maximize the correlation between the new unit's output and the residual error signal in the network outputs prior to the addition of the new unit. Once a neuron is added, these weights are computed once and then they remain fixed. The second type of synaptic weights connects the newly added neuron with the output nodes. These weights are not fixed and are trained adaptively, each time a new neuron is installed, in order to minimize a sum of squares error cost function. The procedure stops when the performance of the network meets the prespecified goals. Discussion of constructive techniques with an emphasis on pattern recognition can be found in [Pare 00].

4.10 A SIMULATION EXAMPLE

This section demonstrates the capability of a multilayer perceptron to classify nonlinearly separable classes. The classification task consists of two distinct classes, each being the union of four regions in the two-dimensional space. Each region consists of normally distributed random vectors with statistically independent components and each with variance $\sigma^2 = 0.08$. The mean values are different for each of the regions. Specifically, the regions of the class denoted by red "o" (see Figure 4.15) are formed around the mean vectors

$$[0.4, 0.9]^T, [2.0, 1.8]^T, [2.3, 2.3]^T, [2.6, 1.8]^T$$

and those of the class denoted by black "+" around the values

$$[1.5, 1.0]^T, [1.9, 1.0]^T, [1.5, 3.0]^T, [3.3, 2.6]^T$$

A total of 400 training vectors were generated, 50 from each distribution. A multilayer perceptron, with three neurons in the first and two neurons in the second hidden layer, were used, with a single output neuron. The activation function was the logistic one with $a = 1$ and the desired outputs 1 and 0, respectively, for the two classes. Two different algorithms were used for the training, namely, the momentum and the adaptive momentum. After some experimentation the algorithmic parameters employed were (a) for the momentum $\mu = 0.05, \alpha = 0.85$ and (b) for the adaptive momentum $\mu = 0.01, \alpha = 0.85, r_i = 1.05, c = 1.05, r_d = 0.7$. The weights were initialized by a uniform pseudorandom distribution between 0 and 1. Figure 4.15a shows the respective output error convergence curves for the two algorithms as a function of the number of epochs (each epoch consisting of the 400 training feature vectors). The respective curves can be considered typical and the adaptive momentum algorithm leads to faster convergence. Both curves correspond to the batch mode of operation. Figure 4.15b shows the resulting decision surface using

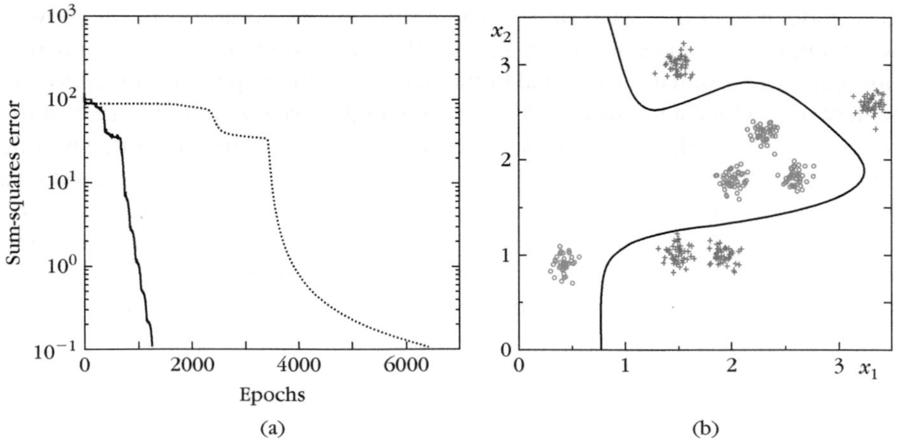

FIGURE 4.15

(a) Error convergence curves for the adaptive momentum (dark line) and the momentum algorithms. Note that the adaptive momentum leads to faster convergence. (b) The decision curve formed by the multilayer perceptron.

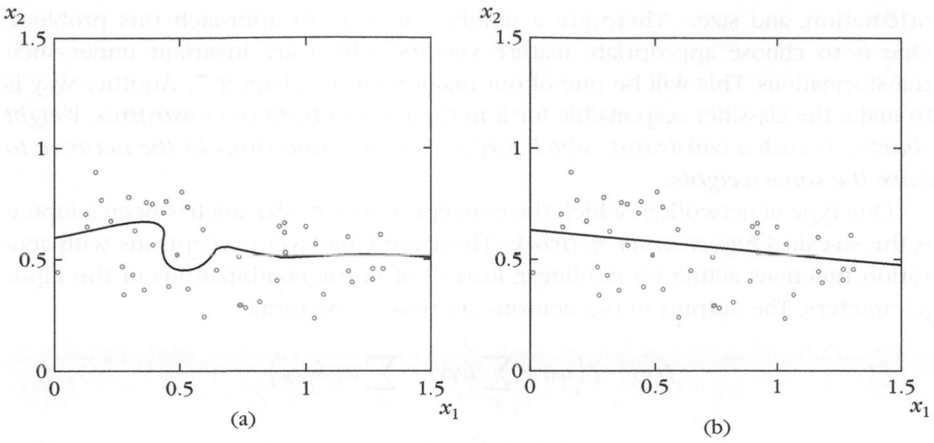

FIGURE 4.16

Decision curve (a) before pruning and (b) after pruning.

the weights estimated from the adaptive momentum training. Once the weights of the network have been estimated, the decision surface can easily be drawn. To this end, a two-dimensional grid is constructed over the area of interest, and the points of the grid are given as inputs to the network, row by row. The decision surface is formed by the points where the output of the network changes from 0 to 1 or vice versa.

A second experiment was conducted in order to demonstrate the effect of the pruning. Figure 4.16 shows the resulting decision surfaces separating the samples of the two classes, denoted by black and red "o," respectively. Figure 4.16a corresponds to a multilayer perceptron (MLP), with two hidden layers and 20 neurons in each of them, amounting to a total of 480 weights. Training was performed via the backpropagation algorithm. The overfitting nature of the resulting curve is readily observed. Figure 4.16b corresponds to the same MLP trained with a pruning algorithm. Specifically, the method based on parameter sensitivity was used, testing the saliency values of the weights every 100 epochs and removing weights with saliency value below a chosen threshold. Finally, only 25 of the 480 weights were left, and the curve is simplified to a straight line.

4.11 NETWORKS WITH WEIGHT SHARING

One major issue encountered in many pattern recognition applications is that of transformation invariance. This means that the pattern recognition system should classify correctly, independent of transformations performed on the input space, such as translation, rotation, and scaling. For example, the character "5" should "look the same" to an optical character recognition system, regardless of its position, orientation, and size. There are a number of ways to approach this problem. One is to choose appropriate feature vectors, which are invariant under such transformations. This will be one of our major goals in Chapter 7. Another way is to make the classifier responsible for it in the form of *built-in constraints. Weight sharing is such a constraint, which forces certain connections in the network to have the same weights.*

One type of network in which the concept of weight sharing has been adopted is the so-called *higher order network*. These are multilayer perceptrons with activation functions acting on nonlinear, instead of linear, combinations of the input parameters. The outputs of the neurons are now of the form

$$f(v) = f\Big(w_0 + \sum_i w_i x_i + \sum_{jk} w_{jk} x_j x_k\Big)$$

This can be generalized to include higher order products. Let us now assume that the inputs to the network originate from a two-dimensional grid (image). Each point of the grid corresponds to a specific x_i and each pair (x_i, x_j) to a line segment. Invariance to translation is built in by constraining the weights $w_{jk} = w_{rs}$, whenever the respective line segments, defined by the points (x_j, x_k) and (x_r, x_s), are of the same gradient. Invariance to rotation can be built in by sharing weights corresponding to segments of equal length. Of course, all these are subject to inaccuracies caused by the resolution coarseness of the grid. Higher order networks can accommodate more complex transformations [Kana 92, Pera 92, Delo 94]. Because of the weight sharing, the number of free parameters for optimization is substantially reduced. However, it must be pointed out that, so far, such networks have not been widely used in practice. A special type of network called the *circular backpropagation model* results if

$$f(v) = f\Big(w_0 + \sum_i w_i x_i + w_s \sum_i x_i^2\Big)$$

The increase in the number of parameters is now small, and in [Ride 97] it is claimed that the involvement of the nonlinear term offers the network increased representation power without affecting its generalization capabilities.

Besides the higher order networks, weight sharing has been used to impose invariance on first-order networks used for specific applications [Fuku 82, Rume 86, Fuku 92, Lecu 89]. The last, for example, is a system for handwritten zip code recognition. It is a hierarchical structure with three hidden layers and inputs the gray levels of the image pixels. Nodes in the first two layers form groups of two-dimensional arrays known as *feature maps*. Each node in a given map receives inputs from a specific window area of the previous layer, known as the *receptive field*. Translation invariance is imposed by forcing corresponding nodes in the same map, looking at different receptive fields, to share weights. Thus, if an object moves from one input receptive field to the other, the network responds in the same way.

4.12 GENERALIZED LINEAR CLASSIFIERS

In Section 4.3, dealing with the nonlinearly separable XOR problem, we saw that the neurons of the hidden layer performed a mapping that transformed the problem to a linearly separable one. The actual mapping was

$$\boldsymbol{x} \longrightarrow \boldsymbol{y}$$

with

$$\boldsymbol{y} = \begin{bmatrix} y_1 \\ y_2 \end{bmatrix} = \begin{bmatrix} f(g_1(\boldsymbol{x})) \\ f(g_2(\boldsymbol{x})) \end{bmatrix} \tag{4.43}$$

where $f(\cdot)$ is the activation function and $g_i(\boldsymbol{x}), i = 1, 2$, the linear combination of the inputs performed by each neuron. This will be our kickoff point for this section.

Let us consider our feature vectors to be in the l-dimensional space $\mathcal{R}^l$ and assume that they belong to either of the two classes A, B, which are nonlinearly separable. Let $f_1(\cdot), f_2(\cdot), \ldots, f_k(\cdot)$ be nonlinear (in the general case) functions

$$f_i\colon \mathcal{R}^l \to \mathcal{R}, \quad i = 1, 2, \ldots, k$$

which define the mapping $\boldsymbol{x} \in \mathcal{R}^l \to \boldsymbol{y} \in \mathcal{R}^k$

$$\boldsymbol{y} \equiv \begin{bmatrix} f_1(\boldsymbol{x}) \\ f_2(\boldsymbol{x}) \\ \vdots \\ f_k(\boldsymbol{x}) \end{bmatrix} \tag{4.44}$$

Our goal now is to investigate whether there is an appropriate value for k and functions f_i so that classes A, B are linearly separable in the k-dimensional space of the vectors $\boldsymbol{y}$. In other words, we investigate whether there exists a k-dimensional space where we can construct a hyperplane $\boldsymbol{w} \in \mathcal{R}^k$ so that

$$w_0 + \boldsymbol{w}^T\boldsymbol{y} > 0, \quad \boldsymbol{x} \in A \tag{4.45}$$

$$w_0 + \boldsymbol{w}^T\boldsymbol{y} < 0, \quad \boldsymbol{x} \in B \tag{4.46}$$

Assuming that in the original space the two classes were separable by a (nonlinear) hypersurface $g(\boldsymbol{x}) = 0$, relations (4.45), (4.46) are basically equivalent to approximating the nonlinear $g(\boldsymbol{x})$ as a linear combination of $f_i(\boldsymbol{x})$, that is,

$$g(\boldsymbol{x}) = w_0 + \sum_{i=1}^{k} w_i f_i(\boldsymbol{x}) \tag{4.47}$$

This is a typical problem of function approximation in terms of a preselected class of *interpolation functions* $f_i(\cdot)$. This is a well-studied task in numerical analysis,

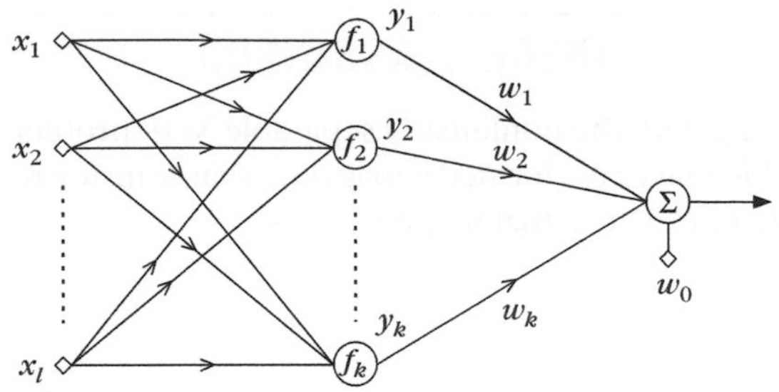

FIGURE 4.17

Generalized linear classifier.

and a number of different interpolating functions have been proposed (exponential, polynomial, Tchebyshev, etc.). In the next sections, we will focus on two such classes of functions, which have been widely used in pattern recognition.

Once the functions f_i have been selected, the problem becomes a typical design of a linear classifier, that is, to estimate the weights w_i in the k-dimensional space. This justifies the term *generalized linear classification*. Figure 4.17 shows the corresponding block diagram. The first layer of computations performs the mapping to the $\boldsymbol{y}$ space; the second layer performs the computation of the decision hyperplane. In other words, (4.47) corresponds to a *two-layer network* where the nodes of the hidden layer have different activation functions, $f_i(\cdot)$, $i = 1, 2, \ldots, k$. For an M-class problem we need to design M such weight vectors $\boldsymbol{w}_r$, $r = 1, 2, \ldots, M$, one for each class, and select the rth class according to the maximum output $\boldsymbol{w}_r^T \boldsymbol{y} + w_{r0}$.

A similar expression to (4.47) expansion of $g(\boldsymbol{x})$ is known as *projection pursuit*, introduced in [Fried 81], and it is defined as

$$g(\boldsymbol{x}) = \sum_{i=1}^{k} f_i(\boldsymbol{w}_i^T \boldsymbol{x})$$

Observe that the argument in each of the functions $f_i(\cdot)$ is not the feature vector $\boldsymbol{x}$ but its *projection* on the direction determined by the respective $\boldsymbol{w}_i$. Optimization with respect to f_i and $\boldsymbol{w}_i$, $i = 1, 2, \ldots, k$, results in the best choice for directions to project as well as the interpolation functions. If $f_i(\cdot)$ are all chosen *a priori* to be sigmoid functions, the projection pursuit method becomes identical to a neural network with a single hidden layer. Projection pursuit models *are not* members of the generalized linear models family, and their optimization is carried out iteratively in a two-stage fashion. Given the functions $f_i(\cdot)$, $\boldsymbol{w}_i$s are estimated and in the next stage optimization is performed with respect to the f_is. See, for example, [Fried 81]. Although interesting from a theoretical point of view, it seems that in practice projection pursuit methods have been superseded by the multilayer perceptrons.

In the sequel, we will first try to justify our expectations, that by going to a higher dimensional space the classification task may be transformed into a linear one, and then study popular alternatives for the choice of functions $f_i(\cdot)$ in (4.44).

4.13 CAPACITY OF THE l-DIMENSIONAL SPACE IN LINEAR DICHOTOMIES

Let us consider N points in the l-dimensional space. We will say that these points are *in general position* or *well distributed* if there is no subset of $l+1$ of them that lie on an $(l-1)$-dimensional hyperplane. Such a definition excludes detrimental cases, such as in the two-dimensional space having three points on a straight line (a one-dimensional hyperplane). The number $O(N,l)$ of groupings that can be formed by $(l-1)$-dimensional hyperplanes to separate the N points in two classes, taking all possible combinations, is given by ([Cove 65] and Problem 4.18):

$$O(N,l) = 2\sum_{i=0}^{l}\binom{N-1}{i} \tag{4.48}$$

where

$$\binom{N-1}{i} = \frac{(N-1)!}{(N-1-i)!\,i!} \tag{4.49}$$

Each of these two class groupings is also known as a (linear) *dichotomy*. From the properties of the binomial coefficients, it turns out that for $N \leq l+1, O(N,l) = 2^N$. Figure 4.18 shows two examples of such hyperplanes resulting in $O(4,2) = 14$ and $O(3,2) = 8$ two-class groupings, respectively. The seven lines of Figure 4.18a form the following groupings. [(ABCD)], [A,(BCD)], [B,(ACD)], [C,(ABD)], [D,(ABC)], [(AB),(CD)], and [(AC),(BD)]. Each grouping corresponds to two possibilities. For example, (ABCD) can belong to either class ω_1 or ω_2. Thus, the total number of

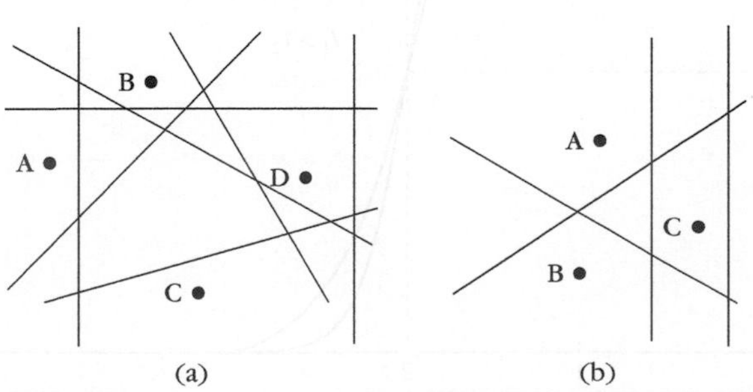

FIGURE 4.18

Number of linear dichotomies (a) for four and (b) for three points.

combinations of assigning four points in the two-dimensional space in two *linearly separable classes is* 14. This number is obviously smaller than the total number of combinations of assigning N points in two classes, which is known to be 2^N. This is because the latter also involves nonlinearly separable combinations. In the case of our example, this is 16, which arises from the two extra possibilities of the grouping [(AD), (BC)]. We are now ready to write the probability (percentage) of grouping N points in the l-dimensional space in two *linearly* separable classes [Cove 65]. This is given by:

$$P_N^l = \frac{O(N,l)}{2^N} = \begin{cases} \frac{1}{2^{N-1}} \sum_{i=0}^{l} \binom{N-1}{i} & N > l+1 \\ 1 & N \leq l+1 \end{cases} \tag{4.50}$$

A practical way to study the dependence of P_N^l on N and l is to assume that $N = r(l+1)$ and investigate the probability for various values of r. The curve in Figure 4.19 shows the probability of having linearly separable classes for various values of l. It is readily observed that there are two regions, one to the left of $r = 2$, that is, $N = 2(l+1)$, and one to the right. Furthermore, all curves go through the point $(P_N^l, r) = (1/2, 2)$, since $O(2l+2, l) = 2^{2l+1}$ (Problem 4.19). The transition from one region to the other becomes sharper as $l \to \infty$. Thus, for large values of l and if $N < 2(l+1)$ the probability of any two groups of the N points being linearly separable approaches unity. The opposite is true if $N > 2(l+1)$. In practice, where we cannot afford the luxury of very large values of N and l, our findings

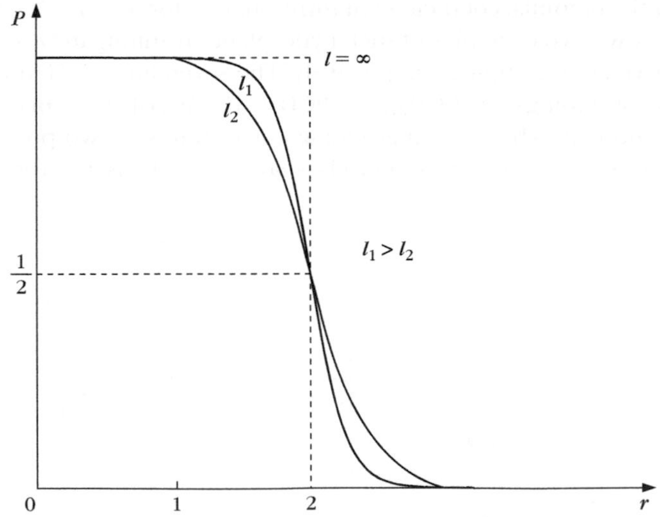

FIGURE 4.19

Probability of linearly separable groupings of $N = r(l+1)$ points in the l-dimensional space.

guarantee that *if we are given N points, then mapping into a higher dimensional space increases the probability of locating them in linearly separable two-class groupings.*

4.14 POLYNOMIAL CLASSIFIERS

In this section we will focus on one of the most popular classes of interpolation functions $f_i(\boldsymbol{x})$ in (4.47). Function $g(\boldsymbol{x})$ is approximated in terms of up to order r polynomials of the $\boldsymbol{x}$ components, for large enough r. For the special case of $r = 2$ we have

$$g(\boldsymbol{x}) = w_0 + \sum_{i=1}^{l} w_i x_i + \sum_{i=1}^{l-1} \sum_{m=i+1}^{l} w_{im} x_i x_m + \sum_{i=1}^{l} w_{ii} x_i^2 \tag{4.51}$$

If $\boldsymbol{x} = [x_1, x_2]^T$, then the general form of $\boldsymbol{y}$ will be

$$\boldsymbol{y} = [x_1, x_2, x_1 x_2, x_1^2, x_2^2]^T$$

and

$$g(\boldsymbol{x}) = \boldsymbol{w}^T \boldsymbol{y} + w_0$$

$$\boldsymbol{w}^T = [w_1, w_2, w_{12}, w_{11}, w_{22}]$$

The number of free parameters determines the new dimension k. The generalization of (4.51) for rth-order polynomials is straightforward, and it will contain products of the form $x_1^{p_1} x_2^{p_2} \ldots x_l^{p_l}$ where $p_1 + p_2 + \cdots + p_l \leq r$. For an rth-order polynomial and l-dimensional $\boldsymbol{x}$ it can be shown that

$$k = \frac{(l+r)!}{r!\,l!}$$

For $l = 10$ and $r = 10$ we obtain $k = 184{,}756$ (!!). That is, even for medium-size values of the network order and the input space dimensionality the number of free parameters gets very high.

Let us consider, for example, our familiar nonlinearly separable XOR problem. Define

$$\boldsymbol{y} = \begin{bmatrix} x_1 \\ x_2 \\ x_1 x_2 \end{bmatrix} \tag{4.52}$$

The input vectors are mapped onto the vertices of a three-dimensional unit (hyper) cube, as shown in Figure 4.20a ((00) → (000), (11) → (111), (10) → (100), (01) → (010)). These vertices are separable by the plane

$$y_1 + y_2 - 2y_3 - \frac{1}{4} = 0$$

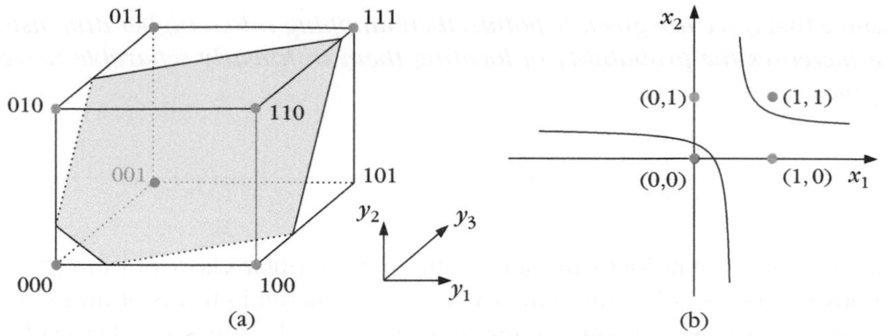

FIGURE 4.20

The XOR classification task, via the polynomial generalized linear classifier. (a) Decision plane in the three-dimensional space and (b) decision curves in the original two-dimensional space.

The plane in the three-dimensional space is equivalent to the decision function

$$g(\boldsymbol{x}) = -\frac{1}{4} + x_1 + x_2 - 2x_1x_2 \begin{array}{ll} >0 & \boldsymbol{x} \in A \\ <0 & \boldsymbol{x} \in B \end{array}$$

in the original two-dimensional space, which is shown in Figure 4.20b.

4.15 RADIAL BASIS FUNCTION NETWORKS

The interpolation functions (kernels) that will be considered in this section are of the general form

$$f(\|\boldsymbol{x} - \boldsymbol{c}_i\|)$$

That is, the argument of the function is the Euclidean distance of the input vector $\boldsymbol{x}$ from a center $\boldsymbol{c}_i$, which justifies the name *radial basis function (RBF)*. Function f can take various forms, for example,

$$f(\boldsymbol{x}) = \exp\left(-\frac{1}{2\sigma_i^2}\|\boldsymbol{x} - \boldsymbol{c}_i\|^2\right) \tag{4.53}$$

$$f(\boldsymbol{x}) = \frac{\sigma^2}{\sigma^2 + \|\boldsymbol{x} - \boldsymbol{c}_i\|^2} \tag{4.54}$$

The Gaussian form is more widely used. For a large enough value of k, it can be shown that the function $g(\boldsymbol{x})$ is sufficiently approximated by [Broo 88, Mood 89]

$$g(\boldsymbol{x}) = w_0 + \sum_{i=1}^{k} w_i \exp\left(-\frac{(\boldsymbol{x} - \boldsymbol{c}_i)^T(\boldsymbol{x} - \boldsymbol{c}_i)}{2\sigma_i^2}\right) \tag{4.55}$$

That is, the approximation is achieved via a summation of RBFs, where each is located on a different point in the space. One can easily observe the close relation that exists between this and the Parzen approximation method for the probability density functions of Chapter 2. Note, however, that there the number of the kernels was chosen to be equal to the number of training points $k = N$. In contrast, in (4.55) $k \ll N$. Besides the gains in computational complexity, this reduction in the number of kernels is beneficial for the generalization capabilities of the resulting approximation model.

Coming back to Figure 4.17, we can interpret (4.55) as the output of a network with *one* hidden layer of RBF activation functions (e.g., (4.53), (4.54)) and a *linear* output node. As has already been said in Section 4.12, for an M-class problem there will be M linear output nodes. At this point, it is important to stress one basic difference between RBF networks and multilayer perceptrons. In the latter, the inputs to the activation functions, of the first hidden layer, are linear combinations of the input feature parameters $\left(\sum_j w_j x_j\right)$. That is, the output of each neuron is the same for all $\{\boldsymbol{x}: \sum_j w_j x_j = c\}$, where c is a constant. *Hence, the output is the same for all points on a hyperplane.* In contrast, in the RBF networks the output of each RBF node, $f_i(\cdot)$, is the same for all points having *the same Euclidean distance from the respective center* $\boldsymbol{c}_i$ and decreases exponentially (for Gaussians) with the distance. *In other words, the activation responses of the nodes are of a local nature in the RBF and of a global nature in the multilayer perceptron networks.* This intrinsic difference has important repercussions for both the convergence speed and the generalization performance. In general, multilayer perceptrons learn slower than their RBF counterparts. In contrast, multilayer perceptrons exhibit improved generalization properties, especially for regions that are not represented sufficiently in the training set [Lane 91]. Simulation results in [Hart 90] show that, in order to achieve performance similar to that of multilayer perceptrons, an RBF network should be of much higher order. This is due to the locality of the RBF activation functions, which makes it necessary to use a large number of centers to fill in the space in which $g(\boldsymbol{x})$ is defined, and this number exhibits an exponential dependence on the dimension of the input space (curse of dimensionality) [Hart 90].

Let us now come back to our XOR problem and adopt an RBF network to perform the mapping to a linearly separable class problem. Choose $k = 2$, the centers $\boldsymbol{c}_1 = [1, 1]^T$, $\boldsymbol{c}_2 = [0, 0]^T$, and $f(\boldsymbol{x}) = \exp(-\|\boldsymbol{x} - \boldsymbol{c}_i\|^2)$. The corresponding $\boldsymbol{y}$ resulting from the mapping is

$$\boldsymbol{y} = \boldsymbol{y}(\boldsymbol{x}) = \begin{bmatrix} \exp(-\|\boldsymbol{x} - \boldsymbol{c}_1\|^2) \\ \exp(-\|\boldsymbol{x} - \boldsymbol{c}_2\|^2) \end{bmatrix}$$

Hence $(0, 0) \rightarrow (0.135, 1)$, $(1, 1) \rightarrow (1, 0.135)$, $(1, 0) \rightarrow (0.368, 0.368)$, $(0, 1) \rightarrow (0.368, 0.368)$. Figure 4.21a shows the resulting class position after the mapping

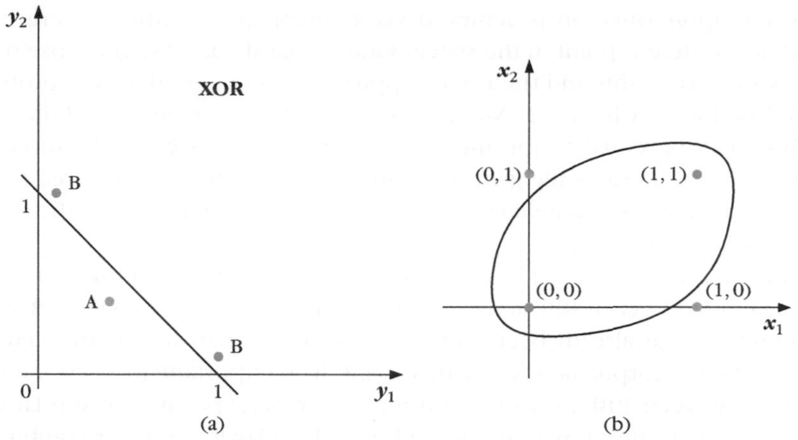

FIGURE 4.21

Decision curves formed by an RBF generalized linear classifier for the XOR task. The decision curve is linear in the transformed space (a) and nonlinear in the original space (b).

in the $\boldsymbol{y}$ space. Obviously, the two classes are now linearly separable and the straight line

$$g(\boldsymbol{y}) = y_1 + y_2 - 1 = 0$$

is a possible solution. Figure 4.21b shows the equivalent decision curve,

$$g(\boldsymbol{x}) = \exp(-\|\boldsymbol{x} - \boldsymbol{c}_1\|^2) + \exp(-\|\boldsymbol{x} - \boldsymbol{c}_2\|^2) - 1 = 0$$

in the input vector space. In our example we selected the centers $\boldsymbol{c}_1, \boldsymbol{c}_2$ as $[0, 0]^T$ and $[1, 1]^T$. The question now is, why these specific ones? *This is an important issue for RBF networks.* Some basic directions on how to tackle this problem are given in the following.

Fixed Centers

Although in some cases the nature of the problem suggests a specific choice for the centers [Theo 95], in the general case these centers can be selected randomly from the training set. Provided that the training set is distributed in a representative manner over all the feature vector space, this seems to be a reasonable way to choose the centers. Having now selected k centers for the RBF functions, the problem has become a typical linear one in the k-dimensional space of the vectors $\boldsymbol{y}$,

$$\boldsymbol{y} = \begin{bmatrix} \exp\left(\frac{-\|\boldsymbol{x}-\boldsymbol{c}_1\|^2}{2\sigma_1^2}\right) \\ \vdots \\ \exp\left(\frac{-\|\boldsymbol{x}-\boldsymbol{c}_k\|^2}{2\sigma_k^2}\right) \end{bmatrix}$$

where the variances are also considered to be known, and

$$g(\boldsymbol{x}) = w_0 + \boldsymbol{w}^T \boldsymbol{y}$$

All methods described in Chapter 3 can now be recalled to estimate w_0 and $\boldsymbol{w}$.

Training of the Centers

If the centers are not preselected, they have to be estimated during the training phase along with the weights w_i and the variances σ_i^2, if the latter are also considered unknown. Let N be the number of input-desired output training pairs $(\boldsymbol{x}(j), y(j), j = 1, \ldots, N)$. We select an appropriate cost function of the output error

$$J = \sum_{j=1}^{N} \phi(e(j))$$

where $\phi(\cdot)$ is a differentiable function (e.g., the square of its argument) of the error

$$e(j) = y(j) - g(\boldsymbol{x}(j))$$

Estimation of the weights w_i, the centers $\boldsymbol{c}_i$, and the variances σ_i^2 becomes a typical task of a nonlinear optimization process. For example, if we adopt the gradient descent approach, the following algorithm results:

$$w_i(t+1) = w_i(t) - \mu_1 \left.\frac{\partial J}{\partial w_i}\right|_t, \quad i = 0, 1, \ldots, k \tag{4.56}$$

$$\boldsymbol{c}_i(t+1) = \boldsymbol{c}_i(t) - \mu_2 \left.\frac{\partial J}{\partial \boldsymbol{c}_i}\right|_t, \quad i = 1, 2, \ldots, k \tag{4.57}$$

$$\sigma_i(t+1) = \sigma_i(t) - \mu_3 \left.\frac{\partial J}{\partial \sigma_i}\right|_t, \quad i = 1, 2, \ldots, k \tag{4.58}$$

where t is the current iteration step. The computational complexity of such a scheme is prohibitive for a number of practical situations. To overcome this drawback, alternative techniques have been suggested.

One way is to choose the centers in a manner that is representative of the way data are distributed in space. This can be achieved by unraveling the clustering properties of the data and choosing a representative for each cluster as the corresponding center [Mood 89]. This is a typical problem of *unsupervised learning*, and algorithms discussed in the relevant chapters later in the book can be employed. The unknown weights, w_i, are then learned via a supervised scheme (i.e., gradient descent algorithm) to minimize the output error. Thus, such schemes use a combination of supervised and unsupervised learning procedures.

An alternative strategy is described in [Chen 91]. A large number of candidate centers is initially chosen from the training vector set. Then, a forward linear regression technique is employed, such as orthogonal least squares, which leads to a parsimonious set of centers. This technique also provides a way to estimate the order of the model k. A recursive form of the method, which can lead to computational savings, is given in [Gomm 00].

Another method has been proposed based on support vector machines. The idea behind this methodology is to look at the RBF network as a mapping machine, through the kernels, into a high-dimensional space. Then we design a hyperplane classifier using the vectors that are closest to the decision boundary. These are the support vectors and correspond to the centers of the input space. The training consists of a quadratic programming problem and guarantees a global optimum [Scho 97]. The nice feature of this algorithm is that it automatically computes all the unknown parameters including the number of centers. We will return to it later in this chapter.

In [Plat 91] an approach similar in spirit to the constructive techniques, discussed for the multilayer perceptrons, has been suggested. The idea is to start training the RBF network with a few nodes (initially one) and keep growing the network by allocating new ones, based on the "novelty" in the feature vectors that arrive sequentially. The novelty of each training input–desired output pair is determined by two conditions: (a) the input vector to be very far (according to a threshold) from all already existing centers *and* (b) the corresponding output error (using the RBF network trained up to this point) greater than another predetermined threshold. If both conditions are satisfied, then the new input vector is assigned as the new center. If not, the input–desired output pair is used to update the parameters of the network according to the adopted training algorithm, for example, the gradient descent scheme. A variant of this scheme that allows removal of previously assigned centers has also been suggested in [Ying 98]. This is basically a combination of the constructive and pruning philosophies. The procedure suggested in [Kara 97] also moves along the same direction. However, the assignment of the new centers is based on a procedure of progressive splitting (according to a splitting criterion) of the feature space using clustering or learning vector quantization techniques (Chapter 14). The representatives of the regions are then assigned as the centers of the RBF's. As was the case with the aforementioned techniques, network growing and training is performed concurrently. In [Yang 06] a weight structure is imposed that binds the weights to a specified probability density function, and estimation is achieved in the Bayesian framework rationale.

A number of other techniques have also been suggested. For a review see, for example, [Hush 93]. A comparison of RBF networks with different center selection strategies versus multilayer perceptrons in the context of speech recognition is given in [Wett 92]. Reviews involving RBF networks and related applications are given in [Hayk 96, Mulg 96].

4.16 UNIVERSAL APPROXIMATORS

In this section we provide the basic guidelines concerning the approximation properties of the nonlinear functions used throughout this chapter—that is, sigmoid, polynomial, and radial basis functions. The theorems that are stated justify the use of the corresponding networks as decision surface approximators as well as probability function approximators, depending on how we look at the classifier.

In (4.51) the polynomial expansion was used to approximate the nonlinear function $g(\boldsymbol{x})$. This choice for the approximation functions is justified by the Weierstrass theorem.

Theorem *Let $g(\boldsymbol{x})$ be a continuous function defined in a compact (closed) subset $S \subset \mathcal{R}^l$, and $\epsilon > 0$. Then there are an integer $r = r(\epsilon)$ and a polynomial function $\phi(\boldsymbol{x})$ of order r so that*

$$|g(\boldsymbol{x}) - \phi(\boldsymbol{x})| < \epsilon, \quad \forall \boldsymbol{x} \in S$$

In other words, function $g(\boldsymbol{x})$ can be approximated arbitrarily closely for sufficiently large r. A major problem associated with polynomial expansions is that good approximations are usually achieved for large values of r. That is, the convergence to $g(\boldsymbol{x})$ is slow. In [Barr 93] it is shown that the approximation error is reduced according to an $O(\frac{1}{r^{2/l}})$ rule, where $O(\cdot)$ denotes order of magnitude. *Thus, the error decreases more slowly with increasing dimension l of the input space*, and large values of r are necessary for a given approximation error. However, large values of r, besides the computational complexity and generalization issues (due to the large number of free parameters required), also lead to poor numerical accuracy behavior in the computations, because of the large number of products involved. On the other hand, the polynomial expansion can be used effectively for piecewise approximation, where smaller r's can be adopted.

The slow decrease of the approximation error with respect to the system order and the input space dimension is common to all expansions of the form (4.47) with fixed basis functions $f_i(\cdot)$. The scenario becomes different if data-adaptive functions are chosen, as is the case with the multilayer perceptrons. In the latter, the argument in the activation functions is $f(\boldsymbol{w}^T\boldsymbol{x})$, with $\boldsymbol{w}$ computed in an optimal fashion from the available data.

Let us now consider a two-layer perceptron with one hidden layer, having k nodes with activation functions $f(\cdot)$ and an output node with *linear* activation. The output of the network is then given by

$$\phi(\boldsymbol{x}) = \sum_{j=1}^{k} w_j^o f(\boldsymbol{w}_j^{hT}\boldsymbol{x}) + w_o^o \tag{4.59}$$

where h refers to the weights, including the thresholds, of the hidden layer and o to the weights of the output layer. Provided that $f(\cdot)$ is a squashing function, the following theorem establishes the universal approximation properties of such a network [Cybe 89, Funa 89, Horn 89, Ito 91, Kalo 97].

Theorem *Let $g(\boldsymbol{x})$ be a continuous function defined in a compact subset $S \subset \mathcal{R}^l$ and $\epsilon > 0$. Then there exist $k = k(\varepsilon)$ and a two-layer perceptron (4.59) so that*

$$|g(\boldsymbol{x}) - \phi(\boldsymbol{x})| < \epsilon, \quad \forall \boldsymbol{x} \in S$$

In [Barr 93] is shows that, in contrast to the polynomial expansion, the approximation error decreases according to an $O(\frac{1}{k})$ rule. *In other words, the input space dimension does not enter explicitly into the scene and the error is inversely proportional to the system order, that is, the number of neurons.* Obviously, the price we pay for it is that the optimization process is now nonlinear, with the associated disadvantage of the potential for convergence to local minima. The question that now arises is whether we gain anything by using more than one hidden layer, since a single one is sufficient for the function approximation. An answer is that using more than one layer may lead to a more efficient approximation; that is, the same accuracy is achieved with fewer neurons in the network.

The universal approximation property is also true for the class of RBF functions. For sufficiently large values of k in (4.55) the resulting expansion can approximate arbitrarily closely any continuous function in a compact subset S [Park 91, Park 93].

4.17 PROBABILISTIC NEURAL NETWORKS

In Section 2.5.6 we have seen that the Parzen estimate of an unknown pdf, using a Gaussian kernel, is given by

$$\hat{p}(\boldsymbol{x}|\omega_i) = \frac{1}{N_i}\sum_{i=1}^{N_i}\frac{1}{(2\pi)^{\frac{l}{2}}h^l}\exp\left(-\frac{(\boldsymbol{x}-\boldsymbol{x}_i)^T(\boldsymbol{x}-\boldsymbol{x}_i)}{2h^2}\right) \tag{4.60}$$

where now we have explicitly included in the notation the class dependence, since decisions according to the Bayesian rule rely on the maximum value, with respect to ω_i, of $P(\omega_i)\hat{p}(\boldsymbol{x}|\omega_i)$. Obviously, in Eq. (4.60) only the training samples, $\boldsymbol{x}_i,\ i = 1, 2, \ldots, N_i$, of class ω_i are involved.

The objective of this section is to develop an efficient architecture for implementing Eq. (4.60), which is inspired by the multilayer NN rationale. The critical computation involving the unknown feature vector, $\boldsymbol{x}$, in Eq. (4.60) is the inner product norm

$$(\boldsymbol{x}-\boldsymbol{x}_i)^T(\boldsymbol{x}-\boldsymbol{x}_i) = ||\boldsymbol{x}||^2 + ||\boldsymbol{x}_i||^2 - 2\boldsymbol{x}_i^T\boldsymbol{x} \tag{4.61}$$

Let us now normalize *all* the feature vectors, which are involved in the game, to unit norm. This is achieved by dividing each vector $\boldsymbol{x}$ by its norm $||\boldsymbol{x}|| = \sqrt{\sum_{i=1}^{l}x_i^2}$. After normalization, and combining Eqs. (4.61) and (4.60), Bayesian classification now relies on searching for the maximum of the following discriminant functions

$$g(\omega_i) = \frac{P(\omega_i)}{N_i}\sum_{i=1}^{N_i}\exp\left(\frac{\boldsymbol{x}_i^T\boldsymbol{x}-1}{h^2}\right) \tag{4.62}$$

where the constant multiplicative weights have been omitted. The above can be efficiently implemented by the network of Figure 4.22, when parallel processing resources are available. The input consists of nodes where an unknown feature

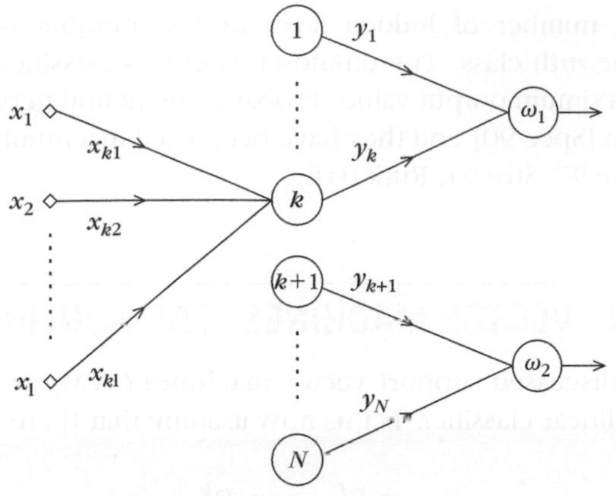

FIGURE 4.22

A probabilistic neural network architecture with N training data points. Each node corresponds to a training point, and it is numbered accordingly. Only the synaptic weights for the kth node are drawn. We have assumed that there are two classes and that the first k points originate from class ω_1 and the rest from class ω_2.

vector $\boldsymbol{x} = [x_1, x_2, \ldots, x_l]^T$ is applied. The number of hidden layer nodes is equal of the number of training data, $N = \sum_{i=1}^{M} N_i$, where M is the number of classes. In the figure, for the sake of simplicity, we have assumed two classes, although generalizing to more classes is obvious. The synaptic weights, leading to the kth hidden node, consist of the components of the respective *normalized* training feature vector $\boldsymbol{x}_k$, i.e., $x_{k,j}$, $j = 1, 2, \ldots, l$ and $k = 1, 2, \ldots, N$. In other words, the training of this type of network is very simple and is directly dictated by the values of the training points. Hence, the input presented to the activation function of the kth hidden layer node is given by

$$input_k = \sum_{j=1}^{l} x_{k,j} x_j = \boldsymbol{x}_k^T \boldsymbol{x}$$

Using as activation function for each node the Gaussian kernel, the output of the kth node is equal to

$$y_k = \exp\left(\frac{input_k - 1}{b^2}\right)$$

There are M output nodes, one for each class. Output nodes are linear combiners. Each output node is connected to *all* hidden layer nodes *associated with the respective class*. The output for the mth output node, $m = 1, 2, \ldots, M$, will be

$$output_m = \frac{P(\omega_m)}{N_m} \sum_{i=1}^{N_m} y_i$$

where N_m is the number of hidden layer nodes (number of training points) associated with the mth class. The unknown vector is classified according to the class giving the maximum output value. Probabilistic neural network architectures were introduced in [Spec 90], and they have been used in a number of applications, for example, [Rome 97, Stre 94, Rutk 04].

4.18 SUPPORT VECTOR MACHINES: THE NONLINEAR CASE

In Chapter 3, we discussed support vector machines (SVM) as an optimal design methodology of a linear classifier. Let us now assume that there exists a mapping

$$\boldsymbol{x} \in \mathcal{R}^l \longrightarrow \boldsymbol{y} \in \mathcal{R}^k$$

from the input feature space into a k-dimensional space, where the classes can satisfactorily be separated by a hyperplane. Then, in the framework discussed in Section 4.12, the SVM method can be mobilized for the design of the hyperplane classifier in the new k-dimensional space. However, there is an elegant property in the SVM methodology that can be exploited for the development of a more general approach. This will also allow us for (implicit) mappings in infinite dimensional spaces, if required.

Recall from Chapter 3 that, in the computations involved in the Wolfe dual representation the feature vectors participate in pairs, via the inner product operation. Also, once the optimal hyperplane $(\boldsymbol{w}, w_0)$ has been computed, classification is performed according to whether the sign of

$$g(\boldsymbol{x}) = \boldsymbol{w}^T \boldsymbol{x} + w_0$$

$$= \sum_{i=1}^{N_s} \lambda_i y_i \boldsymbol{x}_i^T \boldsymbol{x} + w_0$$

is + or −, where N_s is the number of support vectors. Thus, once more, only inner products enter into the scene. If the design is to take place in the new k-dimensional space, the only difference is that the involved vectors will be the k-dimensional mappings of the original input feature vectors. A naive look at it would lead to the conclusion that now the complexity is much higher, since, usually, k is much higher than the input space dimensionality l, in order to make the classes linearly separable. However, there is a nice surprise just waiting for us. Let us start with a simple example. Assume that

$$\boldsymbol{x} \in \mathcal{R}^2 \longrightarrow \boldsymbol{y} = \begin{bmatrix} x_1^2 \\ \sqrt{2}x_1x_2 \\ x_2^2 \end{bmatrix}$$

Then, it is a matter of simple algebra to show that

$$y_i^T y_j = (x_i^T x_j)^2$$

In words, the inner product of the vectors in the new (higher dimensional) space has been expressed as a function of the inner product of the corresponding vectors in the original feature space. Most interesting!

Theorem *Mercer's Theorem. Let $x \in \mathcal{R}^l$ and a mapping ϕ*

$$x \mapsto \phi(x) \in H$$

where H is a Hilbert space.[5] *Let the inner product operation have an equivalent representation*

$$\langle \phi(x), \phi(z) \rangle = K(x, z) \tag{4.63}$$

where $\langle \cdot, \cdot \rangle$ denotes the inner product operation in H. Then $K(x, z)$ is a symmetric continuous function satisfying the following condition:

$$\int_C \int_C K(x, z) g(x) g(z) \, dx \, dz \geq 0 \tag{4.64}$$

for any $g(x)$, $x \in C \subset \mathcal{R}^l$ such that

$$\int_C g(x)^2 \, dx < +\infty \tag{4.65}$$

where C is a compact (finite) subset of $\mathcal{R}^l$. The opposite is always true; that is, for any *symmetric, continuous function $K(x, z)$ satisfying (4.64) and (4.65) there exists a space in which $K(x, z)$ defines an inner product! Such functions are also known as kernels and the space H as* Reproducing kernel Hilbert space (RKHS) *(e.g., [Shaw 04, Scho 02]). What Mercer's theorem does not disclose to us, however, is how to find this space. That is, we do not have a general tool to construct the mapping $\phi(\cdot)$ once we know the inner product of the corresponding space. Furthermore, we lack the means to know the dimensionality of the space, which can even be infinite. This is the case, for example, for the radial basis (Gaussian) kernel ([Burg 99]). For more on these issues, the mathematically inclined reader is referred to [Cour 53].*

Typical examples of kernels used in pattern recognition applications are as follows:

Polynomials

$$K(x, z) = (x^T z + 1)^q, \quad q > 0 \tag{4.66}$$

[5] A Hilbert space is a complete linear space equipped with an inner product operation. A finite dimensional Hilbert space is a Euclidean space.

Radial Basis Functions

$$K(\boldsymbol{x}, \boldsymbol{z}) = \exp\left(-\frac{\|\boldsymbol{x}-\boldsymbol{z}\|^2}{\sigma^2}\right) \tag{4.67}$$

Hyperbolic Tangent

$$K(\boldsymbol{x}, \boldsymbol{z}) = \tanh\left(\beta \boldsymbol{x}^T \boldsymbol{z} + \gamma\right) \tag{4.68}$$

for appropriate values of β and γ so that Mercer's conditions are satisfied. One possibility is $\beta = 2,\ \gamma = 1$. In [Shaw 04] a unified treatment of kernels is presented, focusing on their mathematical properties as well as methods for pattern recognition and regression that have been developed around them.

Once an appropriate kernel has been adopted that implicitly defines a mapping into a higher dimensional space (RKHS), the Wolfe dual optimization task (Eqs. (3.103)–(3.105)) becomes

$$\max_{\boldsymbol{\lambda}} \left(\sum_i \lambda_i - \frac{1}{2}\sum_{i,j} \lambda_i \lambda_j y_i y_j K(\boldsymbol{x}_i, \boldsymbol{x}_j)\right) \tag{4.69}$$

$$\text{subject to} \quad 0 \le \lambda_i \le C, \quad i = 1, 2, \ldots, N \tag{4.70}$$

$$\sum_i \lambda_i y_i = 0 \tag{4.71}$$

and the resulting linear (in the RKHS) classifier is

$$\text{assign } \boldsymbol{x} \text{ in } \omega_1(\omega_2) \text{ if } g(\boldsymbol{x}) = \sum_{i=1}^{N_s} \lambda_i y_i K(\boldsymbol{x}_i, \boldsymbol{x}) + w_0 > (<)\, 0 \tag{4.72}$$

Due to the nonlinearity of the kernel function, the resulting classifier is a nonlinear one in the original $\mathcal{R}^l$ space. Similar arguments hold true for the ν-SVM formulation.

Figure 4.23 shows the corresponding architecture. This is nothing else than a special case of the generalized linear classifier of Figure 4.17. The number of nodes is determined by the number of support vectors N_s. The nodes perform the inner products between the mapping of $\boldsymbol{x}$ and the corresponding mappings of the support vectors in the high-dimensional space, via the kernel operation.

Figure 4.24 shows the resulting SVM classifier for two nonlinearly separable classes, where the Gaussian radial basis function kernel, with $\sigma = 1.75$, has been used. Dotted lines mark the margin and circled points the support vectors.

Remarks

- Notice that if the kernel function is the RBF, then the architecture is the same as the RBF network architecture of Figure 4.17. However, the approach followed here is different. In Section 4.15, a mapping in a k-dimensional space

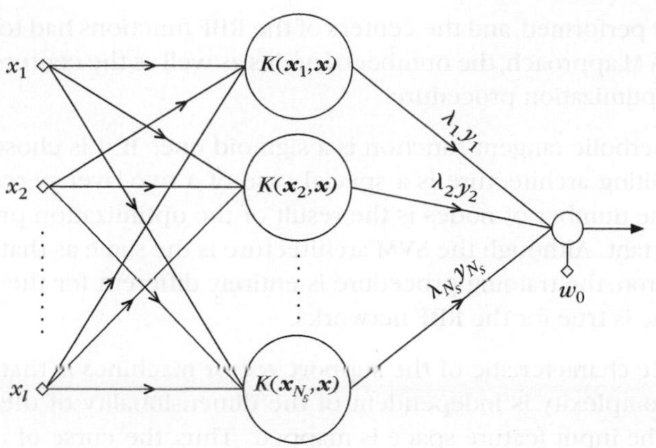

FIGURE 4.23

The SVM architecture employing kernel functions.

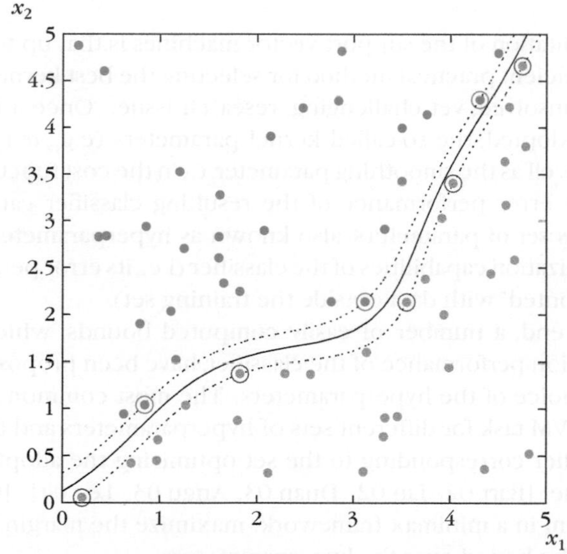

FIGURE 4.24

Example of a nonlinear SVM classifier for the case of two nonlinearly separable classes. The Gaussian RBF kernel was used. Dotted lines mark the margin and circled points the support vectors.

was first performed, and the centers of the RBF functions had to be estimated. In the SVM approach, the number of nodes as well as the centers are the result of the optimization procedure.

- The hyperbolic tangent function is a sigmoid one. If it is chosen as a kernel, the resulting architecture is a special case of a two-layer perceptron. Once more, the number of nodes is the result of the optimization procedure. This is important. Although the SVM architecture is the same as that of a two-layer perceptron, the training procedure is entirely different for the two methods. The same is true for the RBF networks.

- A notable characteristic of the support vector machines is that the computational complexity is independent of the dimensionality of the kernel space, where the input feature space is mapped. Thus, the curse of dimensionality is bypassed. In other words, one designs in a high-dimensional space without having to adopt explicit models using a large number of parameters, as this would be dictated by the high dimensionality of the space. This also has an influence on the generalization properties, and indeed, SVMs tend to exhibit *good generalization performance*. We will return to this issue at the end of Chapter 5.

- A major limitation of the support vector machines is that up to now there has been no efficient practical method for selecting the best kernel function. This is still an unsolved, yet challenging, research issue. Once a kernel function has been adopted, the so-called kernel parameters (e.g., σ for the Gaussian kernel) as well as the smoothing parameter, C, in the cost function are selected so that the error performance of the resulting classifier can be optimized. Indeed, this set of parameters, also known as hyperparameters, is crucial for the generalization capabilities of the classifier (i.e., its error performance when it is "confronted" with data outside the training set).

  To this end, a number of easily computed bounds, which relate to the generalization performance of the classifier, have been proposed and used for the best choice of the hyperparameters. The most common procedure is to solve the SVM task for different sets of hyperparameters and finally select the SVM classifier corresponding to the set optimizing the adopted bound. See, for example, [Bart 02, Lin 02, Duan 03, Angu 03, Lee 04]. [Chap 02] treats this problem in a minimax framework: maximize the margin over the $\boldsymbol{w}$ and minimize the bound over the hyperparameters.

  A different approach to the task of data-adaptive kernel tuning, with the same goal of improving the error performance, is to use information geometry arguments [Amar 99]. The basic idea behind this approach is to introduce a conformal mapping into the Riemannian geometry induced by the chosen kernel function, aiming at enhancing the margin. [Burg 99] points out that the feature vectors, which originally lie in the l-dimensional space, after the mapping induced by the kernel function lie in an l-dimensional surface, S, in the high-dimensional space. It turns out that (under some very general assumptions) S is a Riemannian manifold with a metric that can be expressed solely in terms of the kernel.

- Support vector machines have been applied to a number of diverse applications, ranging from handwritten digit recognition ([Cort 95]), to object recognition ([Blan 96]), person identification ([Ben 99]), spam categorization ([Druc 99]), channel equalization ([Seba 00]), and medical imaging [ElNa 02, Flao 06]. The results from these applications indicate that SVM classifiers exhibit enhanced generalization performance, which seems to be the power of support vector machines. An extensive comparative study concerning the performance of SVM against sixteen other popular classifiers, using twenty-one different data sets, is given in [Meye 03]. The results verify that SVM classifiers rank at the very top among these classifiers, although there are cases for which other classifiers gave lower error rates.

4.19 BEYOND THE SVM PARADIGM

One of the most attractive properties of the support vector machines, which has contributed to their popularity, is that their computational structure allows for the use of a kernel function, as discussed in the previous section. Sometimes this is also known as the ***kernel trick***. This powerful tool makes the design of a linear classifier in the high-dimensional space independent of the dimensionality of this space. Moreover, due to the implicit nonlinear mapping, dictated by the adopted kernel function, the designed classifier is a nonlinear one in the original space. The success of the SVMs in practice has inspired a research effort to extend a number of linear classifiers to nonlinear ones, by embedding the kernel trick in their structure. This is possible if all the computations can be expressed in terms of inner product operations. Let us illustrate the procedure for the case of the classical Euclidean distance classifier.

Assume two classes ω_1 and ω_2, with N_1 and N_2 training pairs, $(y_i, \boldsymbol{x}_i)$, respectively, with $y_i = \pm 1$ being the class label of the ith sample. Let $K(\cdot,\cdot)$ be a kernel function associated with an implicit mapping $\boldsymbol{x} \mapsto \boldsymbol{\phi}(\boldsymbol{x})$ from the original $\mathcal{R}^l$ space to a high-dimensional RKHS. Given an unknown $\boldsymbol{x}$, the Euclidean classifier, in the RKHS, classifies it to the ω_2 class if

$$||\boldsymbol{\phi}(\boldsymbol{x}) - \boldsymbol{\mu}_1||^2 > ||\boldsymbol{\phi}(\boldsymbol{x}) - \boldsymbol{\mu}_2||^2 \tag{4.73}$$

or, after some basic algebra, if

$$\langle \boldsymbol{\phi}(\boldsymbol{x}), (\boldsymbol{\mu}_2 - \boldsymbol{\mu}_1) \rangle > \frac{1}{2}\left(||\boldsymbol{\mu}_2||^2 - ||\boldsymbol{\mu}_1||^2\right) \equiv \theta \tag{4.74}$$

where $\langle \cdot,\cdot \rangle$ denotes the inner product operation in the RKHS and

$$\boldsymbol{\mu}_1 = \frac{1}{N_1} \sum_{i:y_i=+1} \boldsymbol{\phi}(\boldsymbol{x}_i) \quad \text{and} \quad \boldsymbol{\mu}_2 = \frac{1}{N_2} \sum_{i:y_i=-1} \boldsymbol{\phi}(\boldsymbol{x}_i) \tag{4.75}$$

Combining Eqs. (4.74) and (4.75), we conclude that we assign $\boldsymbol{x}$ in ω_2 if

$$\frac{1}{N_2}\sum_{i:y_i=-1} K(\boldsymbol{x},\boldsymbol{x}_i) - \frac{1}{N_1}\sum_{i:y_i=+1} K(\boldsymbol{x},\boldsymbol{x}_i) > \theta \tag{4.76}$$

where

$$2\theta = \frac{1}{N_2^2}\sum_{i:y_i=-1}\sum_{j:y_j=-1} K(\boldsymbol{x}_i,\boldsymbol{x}_j) - \frac{1}{N_1^2}\sum_{i:y_i=+1}\sum_{j:y_j=+1} K(\boldsymbol{x}_i,\boldsymbol{x}_j)$$

The left-hand side in formula (4.76) reminds us of the Parzen pdf estimate. Adopting the Gaussian kernel, the first term can be taken as the pdf estimator associated with the class ω_2 and the second one with the ω_1 one. Besides the Euclidean classifier, other classical cases, including Fisher's linear discriminant (Chapter 5), have been extended to nonlinear ones by employing the kernel trick, see, for example, [Mull 01, Shaw 04]. Another notable and pedagogically attractive example of a "kernelized" version of a linear classifier is the kernel perceptron algorithm.

The perceptron rule was introduced in Section 3.3. There it was stated that the perceptron algorithm converges in a finite number of steps, provided that the two classes are linearly separable. This drawback has prevented the perceptron algorithm to be used in realistic practical applications. However, after mapping the original feature space to a high-dimensional (even of infinite dimensionality) space and utilizing Cover's theorem (Section 4.13), one expects the classification task to be linearly separable, with high probability, in the RKHS space. In this perspective, the kernelized version of the perceptron rule transcends its historical, theoretical, and educational role and asserts a more practical value as a candidate for solving linearly separable tasks in the RKHS. We will choose to work on the perceptron algorithm in its reward and punishment form, given in Eqs. (3.21)–(3.23).

The heart of the method is the update given by the Eqs. (3.21), and (3.22). These recursions take place in the extended RKHS (its dimension is increased by one to account for the bias term), and they are compactly written as

$$\begin{bmatrix} \boldsymbol{w}(t+1) \\ w_0(t+1) \end{bmatrix} = \begin{bmatrix} \boldsymbol{w}(t) \\ w_0(t) \end{bmatrix} + y_{(t)} \begin{bmatrix} \boldsymbol{\phi}(\boldsymbol{x}_{(t)}) \\ 1 \end{bmatrix}$$

each time a misclassification occurs—that is, if $y_{(t)}(\langle \boldsymbol{w}(t), \boldsymbol{\phi}(\boldsymbol{x}_{(t)})\rangle + w_0) \leq 0$, where the coefficient ρ has been taken to be equal to one. Let α_i, $i = 1, 2, \ldots, N$, be a counter corresponding to each one of the training points. The counter α_i is increased by one every time $\boldsymbol{x}_{(t)} = \boldsymbol{x}_i$, *and* a misclassification occurs leading to a respective update of the classifier. If one starts from a zero initial vector, then the solution, after all points have been correctly classified, can be written as

$$\boldsymbol{w} = \sum_{i=1}^{N} \alpha_i y_i \boldsymbol{\phi}(\boldsymbol{x}_i), \quad w_0 = \sum_{i=1}^{N} \alpha_i y_i$$

The final resulting nonlinear classifier, in the original feature space, then becomes

$$g(\boldsymbol{x}) \equiv \langle \boldsymbol{w}, \boldsymbol{\phi}(\boldsymbol{x}) \rangle + w_0 = \sum_{i=1}^{N} \alpha_i y_i K(\boldsymbol{x}, \boldsymbol{x}_i) + \sum_{i=1}^{N} \alpha_i y_i$$

A pseudocode for the kernel perceptron algorithm follows.

The Kernel Perceptron Algorithm

- Set $\alpha_i = 0,\ i = 1, 2, \ldots, N$
- Repeat
  - count_misclas = 0
  - For $i = 1$ to N
    - If $y_i \left(\sum_{j=1}^{N} \alpha_j y_j K(\boldsymbol{x}_i, \boldsymbol{x}_j) + \sum_{j=1}^{N} \alpha_j y_j \right) \leq 0$ then
      - $\alpha_i = \alpha_i + 1$
      - count_misclas = count_misclas + 1
  - End {For}
- Until count_misclas = 0

4.19.1 Expansion in Kernel Functions and Model Sparsification

In this subsection, we will briefly discuss classifiers that resemble to (or, even are inspired by) the SVMs, in an effort to establish bridges among different methodologies. We have already done so in Section 4.18 for the SVM, RBF, and multilayer neural networks. After all, it is a small world! Using the Gaussian kernel in Eq. (4.72), we obtain

$$g(\boldsymbol{x}) = \sum_{i=1}^{N_s} a_i \exp\left(-\frac{\|\boldsymbol{x} - \boldsymbol{x}_i\|^2}{2\sigma^2} \right) + w_0 \tag{4.77}$$

where we have used $a_i = \lambda_i y_i$. Equation (4.77) is very similar to the Parzen expansion of a pdf, discussed in Chapter 2. There are a few differences, however. In contrast to the Parzen expansion, $g(\boldsymbol{x})$ in (4.77) is not a pdf function; that is, in general, it does not integrate to unity. Moreover, from the practical point of view, the most important difference lies in the different number of terms involved in the summation. In the Parzen expansion *all* the training samples offer their contribution to the final solution. In contrast, in the solution provided by the SVM formulation only the support vectors, that is, the points lying either in the margin or in the wrong side of the resulting classifier, are assigned as the "significant" ones and are selected to contribute to the solution. In practice, a small fraction of the training points enter in the summation in Eq. (4.77), that is $N_s \ll N$. In fact, as we will discuss at the end of Section 5.10, if the number of support vectors gets large, the generalization performance of the classifier is expected to degrade. If $N_s \ll N$, we say that the solution is *sparse*. A sparse solution spends computational resources only on the most relevant of the training patterns. Besides the computational complexity aspects, having a sparse solution is in line with our desire to avoid overfitting (see also Section 4.9). In real-world data, the presence of noise in regression and the overlap of classes in classification, as well as the presence of outliers, imply that the modeling must be such that to avoid overfitting to the specific training data set.

A closer look behind the SVM philosophy reveals that the source of sparsity in the obtained solution is the presence of the margin term in the cost function. Another way to view the term $\|\boldsymbol{w}\|^2$ in the cost function in (3.93), that is,

$$J(\boldsymbol{w}, w_0, \boldsymbol{\xi}) = \frac{1}{2}\|\boldsymbol{w}\|^2 + C\sum_{i=1}^{N} I(\xi_i)$$

is as a *regularization* term, whose presence satisfies our will to keep the norm of the solution as "small" and "simple" as possible, while, at the same time, trying to minimize the number of margin errors $\left(\sum_{i=1}^{N} I(\xi_i)\right)$. This implicitly forces most of the λ_is in the solution to be zero, keeping only the most significant of the samples, that is, the support vectors. In Section 4.9, regularization was also used in order to keep the size of the neural networks small. For a deeper and an elegant discussion of the use of regularization in the context of regression/classification the interested reader can refer to [Vapn 00].

With the sparsification goal in mind, a major effort has been invested to develop techniques, both for classification and for regression tasks, which lead to classifiers/regressors of the form

$$g(\boldsymbol{x}) = \sum_{j=1}^{N} a_j K(\boldsymbol{x}, \boldsymbol{x}_j) \tag{4.78}$$

for an appropriately chosen kernel function. A bias constant term can also be added, but it has been omitted for simplicity. The task is to estimate the unknown weights a_j, $j = 1, 2 \ldots, N$, of the expansion. Functions of the form in (4.78) are justified by the following theorem ([Kime 71, Scho 02]):

Representer Theorem

Let $\mathcal{L}(\cdot,\cdot) : \mathcal{R}^2 \mapsto [0, \infty)$ be an arbitrary nonnegative loss function, measuring the deviation between a desired response, y, and the value of $g(\boldsymbol{x})$. Then the minimizer $g(\cdot) \in H$, where H is a RKHS defined by a kernel function $K(\cdot,\cdot)$, of the regularized cost

$$\sum_{i=1}^{N} \mathcal{L}(g(\boldsymbol{x}_i), y_i) + \Omega(\|g\|) \tag{4.79}$$

admits a representation of the form in (4.78). In (4.79), $(y_i, \boldsymbol{x}_i)$, $i = 1, 2, \ldots, N$, are the training data, $\Omega(\cdot) : [0, \infty) \mapsto \mathcal{R}$ is a strictly monotonic increasing function and $||\cdot||$ is the norm operation in H. For a more mathematical treatment of this result, the interested reader may refer to, for example, [Scho 02]. For those who are not familiar with functional analysis and some of the mathematical secrets behind RKHS, recall that the set of functions $\mathcal{R}^l \mapsto \mathcal{R}$ form a linear space, which can be equipped with an inner product operation to become a Hilbert space. Hence, by restricting $g(\cdot) \in H$, we limit our search for solutions in (4.79) among the points in an RKHS (function space) defined by the specific kernel function.

This is an important theorem because it states that, although working in a high (even infinite) dimensional space, the optimal solution, minimizing (4.79), is expressed as a linear combination of only N kernels *placed at the training points*! In order to see how this theorem can simplify the search for the optimal solution in practice, let us consider the following example.

Example 4.1

The kernel least squares solution. Let $(y_i, \boldsymbol{x}_i)$, $i = 1, 2, \ldots, N$, be the training points. The goal is to design the optimal linear least squares classifier (regressor) in a RKHS space, which is defined by the kernel function $K(\cdot, \cdot)$.

According to the definition of the least squares cost in Section 3.4.3, we have to minimize, with respect to $g \in H$, the cost

$$\sum_{i=1}^{N} (y_i - g(\boldsymbol{x}_i))^2 \tag{4.80}$$

According to the Representer Theorem, we can write

$$g(\boldsymbol{x}) = \sum_{j=1}^{N} a_j K(\boldsymbol{x}, \boldsymbol{x}_j) \tag{4.81}$$

Substituting (4.81) into (4.80), we get the equivalent task of minimizing with respect to a *finite* number of parameters, a_i, $i = 1, 2, \ldots, N$, the cost

$$J(\boldsymbol{a}) = \sum_{i=1}^{N} \left(y_i - \sum_{j=1}^{N} a_j K(\boldsymbol{x}_i, \boldsymbol{x}_j) \right)^2 \tag{4.82}$$

The cost in (4.82) can be written in terms of the Euclidean norm in the $\mathcal{R}^N$ space, that is,

$$J(\boldsymbol{a}) = (\boldsymbol{y} - \mathcal{K}\boldsymbol{a})^T(\boldsymbol{y} - \mathcal{K}\boldsymbol{a}) \tag{4.83}$$

where $\boldsymbol{y} = [y_1, y_2, \ldots, y_N]^T$ and $\mathcal{K}$ is the $N \times N$ matrix known as the ***Gram*** matrix, which is defined as

$$\mathcal{K}(i, j) \equiv K(\boldsymbol{x}_i, \boldsymbol{x}_j) \tag{4.84}$$

Expanding (4.83) and taking the gradient with respect to $\boldsymbol{a}$ to be equal to zero, we obtain

$$\boldsymbol{a} = \mathcal{K}^{-1}\boldsymbol{y} \tag{4.85}$$

provided that the Gram matrix is invertible. Hence, recalling (4.81), the kernel least squares estimate can be written compactly as

$$g(\boldsymbol{x}) = \boldsymbol{a}^T \boldsymbol{p} = \boldsymbol{y}^T \mathcal{K}^{-1} \boldsymbol{p} \tag{4.86}$$

where

$$\boldsymbol{p} \equiv [K(\boldsymbol{x}, \boldsymbol{x}_1), \ldots, K(\boldsymbol{x}, \boldsymbol{x}_N)]^T \tag{4.87}$$

The Representer Theorem has been exploited in [Tipp 01] in the context of the so-called *relevance vector machine* (RVM) methodology. Based on (4.78), a conditional probability model is built for the desired response (label) given the values of $\boldsymbol{a}$. The computation of the unknown weights is carried out in the Bayesian framework rationale (Chapter 2). Sparsification is achieved by constraining the unknown weight parameters and imposing an explicit prior probability distribution over each one of them. It is reported that RVMs lead to sparser solutions compared to the SVMs, albeit at a higher complexity. Memory requirements scale with the square, and the required computational resources scale with the cube of the number of basis functions, which makes the algorithm less practical for large data sets. In contrast, the amount of memory requirements for the SVMs is linear, and the number of computations is somewhere between linear and (approximately) quadratic in the size of the training set ([Plat 99]).

A more recent trend is to obtain solutions of the form in Eq. (4.78) in an online time-adaptive fashion. That is, the solution is updated each time a new training pair $(y_i, \boldsymbol{x}_i)$ is received. This is most important when the statistics of the involved data are slowly time varying. To this end, in [Kivi 04] a kernelized online LMS-type algorithm (see Section 3.4.2) is derived that minimizes the cost function

$$J(g_t) = \sum_{i=1}^{t} \mathcal{L}(g_t(\boldsymbol{x}_i), y_i) + \lambda ||g_t||^2 \tag{4.88}$$

where the index t has been used in g_t to denote the time dependence explicitly. $\mathcal{L}(\cdot, \cdot)$ is a loss function that quantifies the deviation between the desired output value, y_i, and the true one that is provided by the current estimate, $g_t(\cdot)$, of the unknown function. The summation accounts for the total number of errors committed on all the samples that have been received up to the time instant t. Sparsification is achieved by regularizing the cost function by the square norm $||g_t||^2$ of the required solution.

Another way to look at the regularization term and better understand how its presence beneficially affects the sparsification process is the following. Instead of minimizing, for example, (4.88) one can choose to work with an alternative formulation of the optimization task, that is,

$$\text{minimize} \quad \sum_{i=1}^{t} \mathcal{L}(g_t(\boldsymbol{x}_i), y_i) \tag{4.89}$$

$$\text{subject to} \quad ||g_t||^2 \leq s \tag{4.90}$$

The use of Lagrange multipliers leads to minimizing $J(g_t)$ in (4.88). For a wide class of functions, it can be shown (see, e.g., [Vapn 00]) that the two problems are equivalent for appropriate choices of the parameters s and λ. However, formulating the optimization task as in (4.89)–(4.90) makes our desire for constraining the size of the solution explicitly stated.

In [Slav 08] an adaptive solution of the cost in Eq. (4.78) is given based on projections and convex set arguments. Sparsification is achieved by constraining the solution to lie within a (hyper)sphere in the RKHS. It is shown that such a constraint becomes equivalent to imposing a forgetting factor that forces the algorithm to forget data in the remote past and adaptation focuses on the most recent samples. The algorithm scales linearly with the number of data corresponding to its effective memory (due to the forgetting factor). An interesting feature of this algorithmic scheme is that it provides as special cases a number of well-known algorithms, such as the kernelized normalized LMS (NLMS) [Saye 04] and the kernelized affine projection [Slav 08a] algorithms. Another welcome feature of this methodology is that it can accommodate differentiable as well as nondifferentiable cost functions, in a unified way, due to the possibility of employing subdifferentials of the cost function, in place of the gradient in the correction term, in each time-update recursion.

A different root to the online sparsification is followed in [Enge 04, Slav 08a]. A dictionary of basis functions is adaptively formed. For each received sample, its dependence on the samples that are already contained in the dictionary is tested, according to a predefined criterion. If the dependence measure is below a threshold value, the new sample is included in the dictionary whose cardinality is now increasing by one; otherwise the dictionary remains unaltered. It is shown that the size of the dictionary does not increase indefinitely and that it remains finite. The expansion of the solution is carried out by using only the basis functions associated with the samples in the dictionary. A pitfall of this technique is that the complexity scales with the square of the size of the dictionary, as opposed to the linear complexity of the two previous adaptive techniques.

In our discussion so far we have assumed the use of a loss function. The choice of the loss function is user-dependent. Some typical choices that can be used and have frequently been adopted in classification tasks are as follows:

- *Soft margin loss*

$$\mathcal{L}(g(\boldsymbol{x}), y) = \max(0, \rho - yg(\boldsymbol{x}))$$

  where ρ defines the margin parameter. In words, a (margin) error is committed if $yg(\boldsymbol{x})$ cannot achieve a value of at least ρ. For smaller values, the loss function becomes positive, and it is also linearly increasing as the value of $yg(\boldsymbol{x})$ becomes smaller moving toward negative values. That is, it provides a measure of how far from the margin the estimate lies. Figure 4.25 shows the respective graph for $\rho = 0$.

- *Exponential loss*

$$\mathcal{L}(g(\boldsymbol{x}), y) = \exp(-yg(\boldsymbol{x}))$$

As shown in Figure 4.25, this loss function penalizes heavily nonpositive values of $yg(\boldsymbol{x})$, which lead to wrong decisions. We will use this loss function very soon in Section 4.22.

- *Logistic loss*

$$\mathcal{L}(g(\boldsymbol{x}), y) = \ln\big(1 + \exp(-yg(\boldsymbol{x}))\big)$$

The logistic loss function is basically the negative log-likelihood of a logistic-like probabilistic model, discussed in Section 3.6, operating in the RKHS. Indeed, interpreting $g(\boldsymbol{x})$ as a linear function in the RKHS, that is, $g(\boldsymbol{x}) = \langle \boldsymbol{w}, \boldsymbol{\phi}(\boldsymbol{x})\rangle + w_0$, and modeling the probability of the class label as

$$P(y|\boldsymbol{x}) = \frac{1}{1 + \exp\big(-y(w_0 + \langle \boldsymbol{w}, \boldsymbol{\phi}(\boldsymbol{x})\rangle)\big)}$$

then the logistic loss is the respective negative log-likelihood function. This loss function has also been used in the context of support vector machines, see [Keer 05].

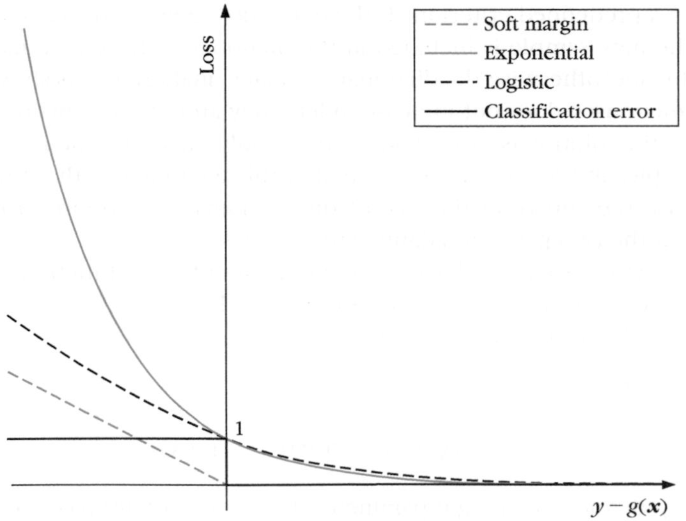

FIGURE 4.25

Typical loss functions used in classification tasks. The margin parameter ρ for the soft margin loss has been set equal to 0. The logistic loss has been normalized in order to pass through the [1,0] point, to facilitate the comparison among the different loss functions. The classification error loss function scores a one if an error is committed and zero otherwise.

4.19.2 Robust Statistics Regression

The regression task was introduced in Section 3.5.1. Let $y \in \mathcal{R}, \boldsymbol{x} \in \mathcal{R}^l$ be two statistically dependent random entities. Given a set of training samples $(y_i, \boldsymbol{x}_i)$, the goal is to compute a function $g(\boldsymbol{x})$ that optimally estimates the value of y when $\boldsymbol{x}$ is measured. In a number of cases, mean square or least squares type of costs are not the most appropriate ones. For example, in cases where the statistical distribution of the data has long tails, then using the least squares criterion will lead to a solution dominated by a small number of points that have very large values (outliers). A similar situation can occur from incorrectly labeled data. Take, for example, a single training data point whose target value has been incorrectly labeled by a large amount. This point will have an unjustifiably (by the true statistics of the data) strong influence on the solution. Such situations can be handled more efficiently by using alternative cost functions, which are known as *robust statistics* loss functions. Typical examples of such loss functions are:

- *Linear ϵ-insensitive loss*

$$\mathcal{L}(g(\boldsymbol{x}), y) = |y - g(\boldsymbol{x})|_\epsilon \equiv \max(0, |y - g(\boldsymbol{x})| - \epsilon)$$

- *Quadratic ϵ-insensitive loss*

$$\mathcal{L}(g(\boldsymbol{x}), y) = |y - g(\boldsymbol{x})|_\epsilon^2 \equiv \max(0, |y - g(\boldsymbol{x})|^2 - \epsilon)$$

- *Huber loss*

$$\mathcal{L}(g(\boldsymbol{x}), y) = \begin{cases} c\,|y - g(\boldsymbol{x})| - \frac{c^2}{2} & \text{if } |y - g(\boldsymbol{x})| > c \\ \frac{1}{2}\left(y - g(\boldsymbol{x})\right)^2 & \text{if } |y - g(\boldsymbol{x})| \leq c \end{cases}$$

where ϵ and c are user-defined parameters. Huber's loss function reduces from quadratic to linear the contributions of samples with absolute error values greater than c. Such a choice makes the optimization task less sensitive to outliers. Figure 4.26 shows the curves associated with the previous loss functions. In the sequel, we will focus on the linear ϵ-insensitive loss.

We are by now experienced enough to solve for the nonlinear $g(\boldsymbol{x})$ case by expressing the problem as a linear one in an RKHS. For the linear ϵ-insensitive case, nonzero contributions to the cost have samples with error values $|y - g(\boldsymbol{x})|$ larger than ϵ. This setup can be compactly expressed by adopting two slack variables, ξ, ξ^*, and the optimization task is now cast as

$$\text{minimize} \quad J(\boldsymbol{w}, w_0, \boldsymbol{\xi}, \boldsymbol{\xi}^*) = \frac{1}{2}\|\boldsymbol{w}\|^2 + C\left(\sum_{i=1}^{N} \xi_i + \sum_{i=1}^{N} \xi_i^*\right) \tag{4.91}$$

$$\text{subject to} \quad y_i - \langle \boldsymbol{w}, \boldsymbol{\phi}(\boldsymbol{x}_i) \rangle - w_0 \leq \epsilon + \xi_i^*, \quad i = 1, 2, \ldots, N \tag{4.92}$$

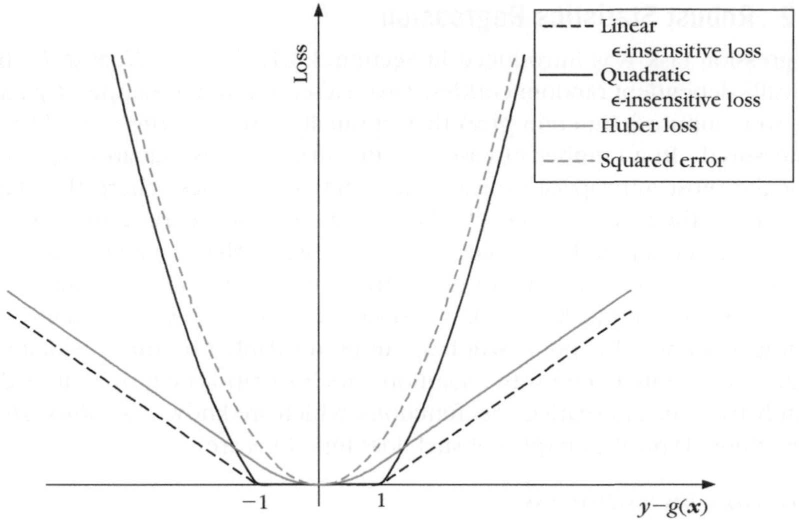

FIGURE 4.26

Loss functions used for regression tasks. The parameters ϵ and c have been set equal to one. In Huber's loss, observe the change from quadratic to linear beyond $\pm c$.

$$\langle \boldsymbol{w}, \boldsymbol{\phi}(\boldsymbol{x}_i)\rangle + w_0 - y_i \le \epsilon + \xi_i, \quad i = 1, 2, \ldots, N \tag{4.93}$$

$$\xi_i \ge 0, \quad \xi_i^* \ge 0, \quad i = 1, 2, \ldots, N \tag{4.94}$$

The above setup guarantees that ξ_i, ξ_i^* are zero if $|y_i - \langle \boldsymbol{w}, \boldsymbol{\phi}(\boldsymbol{x}_i)\rangle - w_0| \le \epsilon$ and contribution to the cost function occurs if either $y_i - \langle \boldsymbol{w}, \boldsymbol{\phi}(\boldsymbol{x}_i)\rangle - w_0 > \epsilon$ or if $y_i - \langle \boldsymbol{w}, \boldsymbol{\phi}(\boldsymbol{x}_i)\rangle - w_0 < -\epsilon$. The presence of the norm $||\boldsymbol{w}||$ guards against overfitting, as has already been discussed. Following similar arguments made in Section 3.7.2, it turns out that the solution is given by

$$\boldsymbol{w} = \sum_{i=1}^{N} (\lambda_i^* - \lambda_i)\boldsymbol{\phi}(\boldsymbol{x}_i) \tag{4.95}$$

where λ_i^*, λ_i are the Lagrange multipliers associated with the set of constraints in (4.92)-(4.93), respectively. The corresponding KKT conditions are (in analogy of (3.98)-(3.102))

$$\lambda_i^*(y_i - \langle \boldsymbol{w}, \boldsymbol{\phi}(\boldsymbol{x}_i)\rangle - w_0 - \epsilon - \xi_i^*) = 0, \quad i = 1, 2, \ldots, N \tag{4.96}$$

$$\lambda_i(\langle \boldsymbol{w}, \boldsymbol{\phi}(\boldsymbol{x}_i)\rangle + w_0 - y_i - \epsilon - \xi_i) = 0, \quad i = 1, 2, \ldots, N \tag{4.97}$$

$$C - \lambda_i - \mu_i = 0, \; C - \lambda_i^* - \mu_i^* = 0, \quad i = 1, 2, \ldots, N \tag{4.98}$$

$$\mu_i \xi_i = 0, \ \mu_i^* \xi_i^* = 0, \quad i = 1, 2, \ldots, N \tag{4.99}$$

$$\lambda_i \geq 0, \ \lambda_i^* \geq 0, \ \mu_i \geq 0, \ \mu_i^* \geq 0, \quad i = 1, 2, \ldots, N \tag{4.100}$$

$$\sum_{i=1}^{N} \lambda_i = \sum_{i=1}^{N} \lambda_i^* \tag{4.101}$$

$$\xi_i \xi_i^* = 0, \ \lambda_i \lambda_i^* = 0, \quad i = 1, 2, \ldots, N \tag{4.102}$$

where μ_i^*, μ_i are the Lagrange multipliers associated with the set of constraints in (4.94). Note that ξ_i, ξ_i^* cannot be nonzero simultaneously, and the same applies for the Lagrange multipliers λ_i^*, λ_i. Furthermore, a careful look at the KKT conditions reveals that:

- The points with absolute error values strictly less than ϵ, i.e., $|y_i - \langle \boldsymbol{w}, \boldsymbol{\phi}(\boldsymbol{x}_i) \rangle - w_0| < \epsilon$ result in *zero* Lagrange multipliers, λ_i, λ_i^*. This is a direct consequence of (4.96) and (4.97). These points are the counterparts of the points that lie strictly outside the margin in the SVM classification task.
- Support vectors are those points satisfying the inequality $|y_i - \langle \boldsymbol{w}, \boldsymbol{\phi}(\boldsymbol{x}_i) \rangle - w_0| \geq \epsilon$.
- The points associated with errors satisfying the strict inequality $|y_i - \langle \boldsymbol{w}, \boldsymbol{\phi}(\boldsymbol{x}_i) \rangle - w_0| > \epsilon$ result in either $\lambda_i = C$ or $\lambda_i^* = C$. This is a consequence of (4.99) and (4.98) and of the fact that, in this case, ξ_i (or ξ_i^*) is nonzero. For those of the points that equality holds, that is, $|y_i - \langle \boldsymbol{w}, \boldsymbol{\phi}(\boldsymbol{x}_i) \rangle - w_0| = \epsilon$, the respective ξ_i (ξ_i^*) $= 0$ and from (4.97) (or (4.96)) the respective λ_i (λ_i^*) can be nonzero. Then from (4.99), (4.100), and (4.98) it turns out that $0 \leq \lambda_i$ (λ_i^*) $\leq C$.

The Lagrange multipliers can be obtained by writing the problem in its equivalent dual representation form, that is,

$$\text{maximize} \quad \sum_{i=1}^{N} y_i(\lambda_i^* - \lambda_i) - \epsilon \sum_{i=1}^{N} (\lambda_i^* + \lambda_i) - \frac{1}{2} \sum_{i,j} (\lambda_i^* - \lambda_i)(\lambda_j^* - \lambda_j) \langle \boldsymbol{\phi}(\boldsymbol{x}_i), \boldsymbol{\phi}(\boldsymbol{x}_j) \rangle \tag{4.103}$$

$$\text{subject to} \quad 0 \leq \lambda_i \leq C, \quad 0 \leq \lambda_i^* \leq C, \quad i = 1, 2, \ldots, N \tag{4.104}$$

$$\sum_{i=1}^{N} \lambda_i^* = \sum_{i=1}^{N} \lambda_i \tag{4.105}$$

where maximization is with respect to the Lagrange multipliers λ_i, λ_i^*, $i = 1, 2, \ldots, N$. This optimization task is similar to the problem defined by (3.103) and (3.105).

Once the Lagrange multipliers have been computed, the nonlinear regressor is obtained as

$$g(x) \equiv \left\langle \phi(x), \sum_{i=1}^{N}(\lambda_i^* - \lambda_i)\phi(x_i) \right\rangle + w_0 = \sum_{i=1}^{N}(\lambda_i^* - \lambda_i)K(x, x_i) + w_0$$

where w_0 is computed from the KKT conditions in (4.97) and (4.96) for $0 < \lambda_i < C$, $0 < \lambda_i^* < C$. Sparsification is achieved via the points associated with zero Lagrange multipliers, that is, points resulting in absolute error values *strictly* less than ϵ.

If instead of the linear ϵ-insensitive loss one adopts the quadratic ϵ-insensitive loss or the Huber loss functions, the resulting sets of formulas are similar to the ones derived here, see, for example, [Vapn 00].

Throughout the derivations in this subsection, we kept referring to the optimization of the SVM classification task considered in Section 3.7.2. The similarity is not accidental. Indeed, it is a matter of a few simple arithmetic manipulations to see that if we set $\epsilon = 0$ and $y_i = \pm 1$, depending on the class origin, our regression task becomes the same as the problem considered in Section 3.7.2.

Ridge Regression

We will close this section by establishing the connection of the regression task, which was considered before, with the classical regression problem known as *ridge regression*. This concept has been used extensively in statistical learning and has been rediscovered under different names. If in the quadratic ϵ-insensitive loss we set $\epsilon = 0$ and, for simplicity, $w_0 = 0$, the result is the standard sum of squared errors cost function. Substituting in the associated constraints the inequalities with equalities and slightly rephrasing the cost (to bring it in its classical formulation), we end up with the following

$$\text{minimize} \quad J(w, \xi) = C\|w\|^2 + \sum_{i=1}^{N} \xi_i^2 \tag{4.106}$$

$$\text{subject to} \quad y_i - \langle w, \phi(x_i) \rangle = \xi_i, \quad i = 1, 2, \ldots, N \tag{4.107}$$

where $C = \frac{1}{2C}$. The task defined in (4.106)–(4.107) is a regularized version of the least squares cost function expressed in an RKHS. If we work on the dual Wolfe representation, it turns out that the solution of the kernel ridge regression is expressed in closed form (see Problem 4.25), that is,

$$w = \frac{1}{2C} \sum_{i=1}^{N} \lambda_i \phi(x_i) \tag{4.108}$$

$$[\lambda_1, \ldots, \lambda_N]^T = 2C\,(\mathcal{K} + CI)^{-1} y \tag{4.109}$$

and

$$g(\boldsymbol{x}) \equiv \langle \boldsymbol{w}, \boldsymbol{\phi}(\boldsymbol{x}) \rangle = \boldsymbol{y}^T(\mathcal{K} + CI)^{-1}\boldsymbol{p} \tag{4.110}$$

where I is the $N \times N$ identity matrix and $\mathcal{K}$ is the $N \times N$ Gram matrix, defined in (4.84) and $\boldsymbol{p}$ the N-dimensional vector defined in (4.87). Observe that the only difference from the kernel least squares solution is the presence of the CI factor.

An advantage of the (kernel) ridge regression, compared to the robust statistics regression, is that a neat closed form solution results. However, by having adopted $\epsilon = 0$ we have lost in model sparseness. As we have already pointed out for the case of the linear ϵ-insensitive loss (the same is true for the quadratic version), training points that result in error with absolute value strictly less than ϵ ***do not contribute*** in the solution. There is no free lunch in real life!

To establish another bridge with Chapter 3, let us employ the linear kernel, that is, $K(\boldsymbol{x}_i, \boldsymbol{x}_j) = \boldsymbol{x}_i^T \boldsymbol{x}_j$ (which implies that one works in the input low-dimensional space and no mapping in a high-dimensional RKHS is performed), and solve the primal instead of the dual task ridge regression task. It is easy to show (Problem 4.26) that the solution becomes

$$\boldsymbol{w} = (X^T X + CI)^{-1} X^T \boldsymbol{y} \tag{4.111}$$

In other words, the solution is the same as the least squares error solution, given in (3.45). The only difference lies in the presence of the CI factor. The latter is the result of the regularization term in the minimized cost function (4.106). In practice, the term CI is used in the LS solution in cases where $X^T X$ has a small determinant and matrix inversion problems arise. Adding a small positive value across the diagonal acts beneficially from the numerical stability point of view.

As a last touch on this section, let us comment on (4.111) and (4.108)-(4.109). For the linear kernel case, the Gram matrix becomes XX^T, and the solution resulting from the dual formulation is given by

$$\boldsymbol{w} = X^T(XX^T + CI)^{-1}\boldsymbol{y} \tag{4.112}$$

Since this is a convex programming task, both solutions, in (4.111) and (4.112), must be the same. This can be verified by simple algebra (Problem 4.27).

4.20 DECISION TREES

In this section we briefly review a large class of nonlinear classifiers known as ***decision trees***. These are ***multistage*** decision systems in which classes are sequentially rejected until we reach a finally accepted class. To this end, the feature space is split into unique regions, corresponding to the classes, ***in a sequential manner***. Upon the arrival of a feature vector, the searching of the region to which the feature vector will be assigned is achieved via a sequence of decisions along a path of

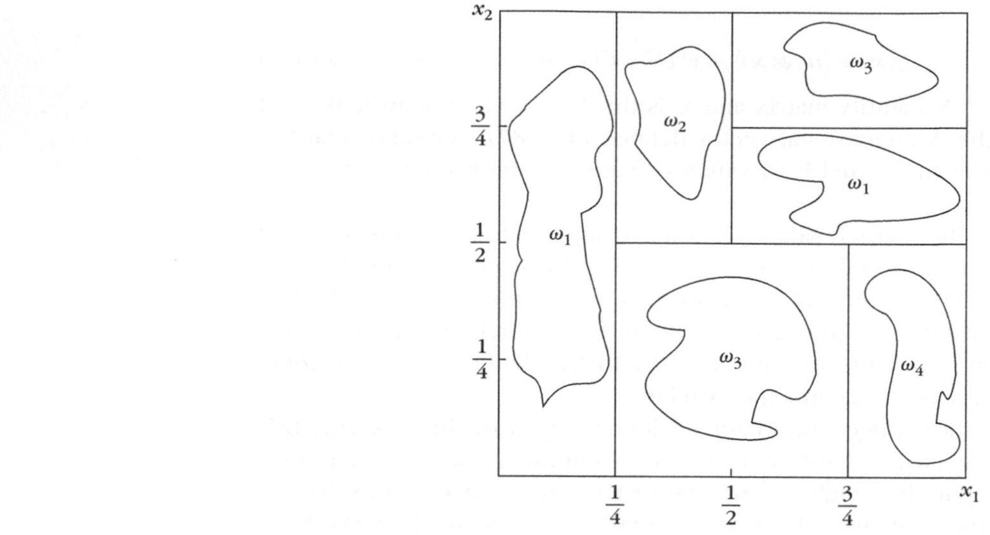

FIGURE 4.27

Decision tree partition of the space.

nodes of an appropriately constructed *tree*. Such schemes offer advantages when a large number of classes are involved. The most popular decision trees are those that split the space into hyperrectangles with sides parallel to the axes. The sequence of decisions is applied to individual features, and the questions to be answered are of the form "*is feature* $x_i \leq \alpha$?" where α is a threshold value. Such trees are known as *ordinary binary classification trees (OBCTs)*. Other types of trees are also possible that split the space into convex polyhedral cells or into pieces of spheres.

The basic idea behind an OBCT is demonstrated via the simplified example of Figure 4.27. By a successive sequential splitting of the space, we have created regions corresponding to the various classes.

Figure 4.28 shows the respective binary tree with its decision nodes and leaves. Note that it is possible to reach a decision without having tested *all* the available features.

The task illustrated in Figure 4.27 is a simple one in the two-dimensional space. The thresholds used for the *binary splits* at each node of the tree in Figure 4.28 were dictated by a simple observation of the geometry of the problem. However, this is not possible in higher dimensional spaces. Furthermore, we started the queries by testing x_1 against $\frac{1}{4}$. An obvious question is why to consider x_1 first and not another feature. In the general case, in order to develop a binary decision tree, the designer has to consider the following design elements in the training phase:

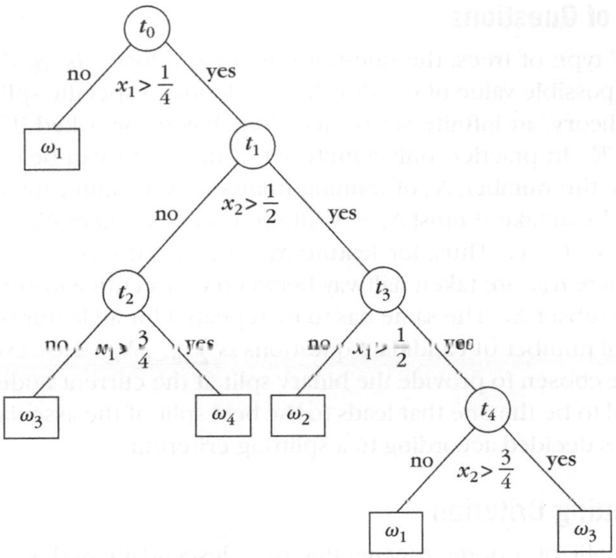

FIGURE 4.28

Decision tree classification for the case of Figure 4.27.

- At each node, the set of candidate questions to be asked has to be decided. Each question corresponds to a specific binary split into two *descendant* nodes. Each node, t, is associated with a specific subset X_t of the training set X. Splitting of a node is equivalent to the split of the subset X_t into two *disjoint* descendant subsets, X_{tY}, X_{tN}. The first of the two consists of the vectors in X_t that correspond to the answer "Yes" of the question and those of the second to the "No." The first (*root*) node of the tree is associated with the training set X. For every split, the following is true:

$$X_{tY} \bigcap X_{tN} = \emptyset$$

$$X_{tY} \bigcup X_{tN} = X_t$$

- A *splitting criterion* must be adopted according to which the best split from the set of candidate ones is chosen.
- A stop-splitting rule is required that controls the growth of the tree, and a node is declared as a terminal one (*leaf*).
- A rule is required that assigns each leaf to a specific class.

We are now experienced enough to understand that more than one method can be used to approach each of the above design elements.

4.20.1 Set of Questions

For the OBCT type of trees, the questions are of the form "*Is* $x_k \leq \alpha$?" For *each* feature, every possible value of the threshold α defines a specific split of the subset X_t. Thus in theory, an infinite set of questions has to be asked if α varies in an interval $Y_\alpha \subseteq \mathcal{R}$. In practice, only a finite set of questions can be considered. For example, since the number, N, of training points in X is finite, any of the features $x_k, k = 1, \ldots, l$, can take at most $N_t \leq N$ different values, where N_t is the cardinality of the subset $X_t \subseteq X$. Thus, for feature x_k, one can use $\alpha_{kn}, n = 1, 2, \ldots, N_{tk}$ ($N_{tk} \leq N_t$), where α_{kn} are taken halfway between consecutive distinct values of x_k in the training subset X_t. The same has to be repeated for all features. Thus in such a case, the total number of candidate questions is $\sum_{k=1}^{l} N_{tk}$. However, only one of them has to be chosen to provide the binary split at the current node, t, of the tree. This is selected to be the one that leads to the best split of the associated subset X_t. The best split is decided according to a splitting criterion.

4.20.2 Splitting Criterion

Every binary split of a node, t, generates two descendant nodes. Let us denote them by t_Y and t_N according to the "Yes" or "No" answer to the single question adopted for the node t, also referred as the *ancestor* node. As we have already mentioned, the descendant nodes are associated with two new subsets, that is, X_{tY}, X_{tN}, respectively. In order for the tree growing methodology, from the root node down to the leaves, to make sense, every split must generate subsets that are more "class homogeneous" compared to the ancestor's subset X_t. This means that the training feature vectors in each one of the new subsets show a higher preference for specific class(es), whereas data in X_t are more equally distributed among the classes. As an example, let us consider a four-class task and assume that the vectors in subset X_t are distributed among the classes with equal probability (percentage). If one splits the node so that the points that belong to ω_1, ω_2 classes form the X_{tY} subset, and the points from ω_3, ω_4 classes form the X_{tN} subset, then the new subsets are more homogeneous compared to X_t or "purer" in the decision tree terminology. The goal, therefore, is to define a measure that quantifies node impurity and split the node so that the overall impurity of the descendant nodes is optimally decreased with respect to the ancestor node's impurity.

Let $P(\omega_i|t)$ denote the probability that a vector in the subset X_t, associated with a node t, belongs to class ω_i, $i = 1, 2, \ldots, M$. A commonly used definition of *node impurity*, denoted as $I(t)$, is given by

$$I(t) = -\sum_{i=1}^{M} P(\omega_i|t) \log_2 P(\omega_i|t) \tag{4.113}$$

where $\log_2$ is the logarithm with base 2. This is nothing else than the entropy associated with the subset X_t, known from Shannon's Information Theory. It is not difficult to show that $I(t)$ takes its maximum value if all probabilities are equal to

$\frac{1}{M}$ (highest impurity) and it becomes zero (recall that $0 \log 0 = 0$) if all data belong to a single class, that is, if only one of the $P(\omega_i|t) = 1$ and all the others are zero (least impurity). In practice, probabilities are estimated by the respective percentages, N_t^i/N_t, where N_t^i is the number of points in X_t that belong to class ω_i. Assume now that performing a split, N_{tY} points are sent into the "Yes" node (X_{tY}) and N_{tN} into the "No" node (X_{tN}). The *decrease in node impurity* is defined as

$$\Delta I(t) = I(t) - \frac{N_{tY}}{N_t} I(t_Y) - \frac{N_{tN}}{N_t} I(t_N) \tag{4.114}$$

where $I(t_Y)$, $I(t_N)$ are the impurities of the t_Y and t_N nodes, respectively. *The goal now becomes to adopt, from the set of candidate questions, the one that performs the split leading to the highest decrease of impurity.*

4.20.3 Stop-Splitting Rule

The natural question that now arises is when one decides to stop splitting a node and declares it as a leaf of the tree. A possibility is to adopt a threshold T and stop splitting if the maximum value of $\Delta I(t)$, over all possible splits, is less than T. Other alternatives are to stop splitting either if the cardinality of the subset X_t is small enough or if X_t is pure, in the sense that all points in it belong to a single class.

4.20.4 Class Assignment Rule

Once a node is declared to be a leaf, then it has to be given a class label. A commonly used rule is the majority rule, that is, the leaf is labeled as ω_j where

$$j = arg \max_i P(\omega_i|t)$$

In words, we assign a leaf, t, to that class to which the majority of the vectors in X_t belong.

Having discussed the major elements needed for the growth of a decision tree, we are now ready to summarize the basic algorithmic steps for constructing a binary decision tree

- Begin with the root node, that is, $X_t = X$
- For each new node t
  - For every feature $x_k, k = 1, 2, \ldots, l$
    - For every value $\alpha_{kn}, n = 1, 2, \ldots, N_{tk}$
      - Generate X_{tY} and X_{tN} according to the answer in the question: is $x_k(i) \leq \alpha_{kn}$, $i = 1, 2, \ldots, N_t$
      - Compute the impurity decrease
    - End
    - Choose α_{kn_0} leading to the maximum decrease w.r. to x_k

- End
- Choose x_{k_0} and associated $\alpha_{k_0 n_0}$ leading to the overall maximum decrease of impurity
- If the stop-splitting rule is met, declare node t as a leaf and designate it with a class label
- If not, generate two descendant nodes t_Y and t_N with associated subsets X_{tY} and X_{tN}, depending on the answer to the question: is $x_{k_0} \le \alpha_{k_0 n_0}$

- End

Remarks

- A variety of node impurity measures can be defined. However, as pointed out in [Brei 84], the properties of the resulting final tree seem to be rather insensitive to the choice of the splitting criterion. Nevertheless, this is very much a problem-dependent task.
- A critical factor in designing a decision tree is its size. As was the case with the multilayer perceptrons, the size of a tree must be large enough but not too large; otherwise it tends to learn the particular details of the training set and exhibits poor generalization performance. Experience has shown that use of a threshold value for the impurity decreases as the stop-splitting rule does not lead to trees of the right size. Many times it stops tree growing either too early or too late. The most commonly used approach is to grow a tree up to a large size first and then prune nodes according to a pruning criterion. This philosophy is similar to that for pruning multilayer perceptrons. A number of pruning criteria have been suggested in the literature. A commonly used criterion is to combine an estimate of the error probability with a complexity measuring term (e.g., number of terminal nodes). For more on this issue the interested reader may refer to [Brei 84, Ripl 94].
- A drawback associated with tree classifiers is their high variance. In practice it is not uncommon for a small change in the training data set to result in a very different tree. The reason for this lies in the hierarchical nature of the tree classifiers. An error that occurs in a node high in the tree propagates all the way down to the leaves below it. *Bagging* (bootstrap aggregating) [Brei 96, Gran 04] is a technique that can reduce variance and improve the generalization error performance. The basic idea is to create a number of, say, B variants, $X_1, X_2, \ldots, X_B$, of the training set, X, using *bootstrap* techniques, by uniformly sampling from X with replacement (see also Section 10.3). For each of the training set variants, X_i, a tree, T_i, is constructed. The final decision is in favor of the class predicted by the majority of the subclassifiers, T_i, $i = 1, 2, \ldots, B$.

Random forests use the idea of bagging in tandem with random feature selection [Brei 01]. The difference with bagging lies in the way the decision trees are constructed. The feature to split in each node is selected as the best among a set of F *randomly* chosen features, where F is a user-defined parameter. This extra introduced randomness is reported to have a substantial effect in performance improvement.

- Our discussion so far was focused on the OBCT type of tree. More general partition of the feature space, via hyperplanes not parallel to the axis, is possible via questions of the type: *Is* $\sum_{k=1}^{l} c_k x_k \leq \alpha$*?* This can lead to a better partition of the space. However, the training now becomes more involved; see, for example, [Quin 93].
- Constructions of fuzzy decision trees have also been suggested, by allowing the possibility of partial membership of a feature vector in the nodes that make up the tree structure. Fuzzification is achieved by imposing a fuzzy structure over the basic skeleton of a standard decision tree; see, for example, [Suar 99] and the references therein.
- Decision trees have emerged as one of the most popular methods for classification. An OBCT performs binary splits on single variables, and classifying a pattern may only require a few tests. Moreover, they can naturally treat mixtures of numeric and categorical variables. Also, due to their structural simplicity, they are easily interpretable.

Example 4.2

In a tree classification task, the set X_t, associated with node t, contains $N_t = 10$ vectors. Four of these belong to class ω_1, four to class ω_2, and two to class ω_3, in a three-class classification task. The node splitting results into two new subsets X_{tY}, with three vectors from ω_1, and one from ω_2, and X_{tN} with one vector from ω_1, three from ω_2, and two from ω_3. The goal is to compute the decrease in node impurity after splitting.
We have that

$$I(t) = -\frac{4}{10}\log_2\frac{4}{10} - \frac{4}{10}\log_2\frac{4}{10} - \frac{2}{10}\log_2\frac{2}{10} = 1.521$$

$$I(t_Y) = -\frac{3}{4}\log_2\frac{3}{4} - \frac{1}{4}\log_2\frac{1}{4} = 0.815$$

$$I(t_N) = -\frac{1}{6}\log_2\frac{1}{6} - \frac{3}{6}\log_2\frac{3}{6} - \frac{2}{6}\log_2\frac{2}{6} = 1.472$$

Hence, the impurity decrease after splitting is

$$\Delta I(t) = 1.521 - \frac{4}{10}(0.815) - \frac{6}{10}(1.472) = 0.315$$

For further information and a deeper study of decision tree classifiers, the interested reader may consult the seminal book [Brei 84]. A nonexhaustive sample of later contributions in the area is [Datt 85, Chou 91, Seth 90, Graj 86, Quin 93]. A comparative guide for a number of well-known techniques is provided in [Espo 97].

Finally, it must be stated that there are close similarities between the decision trees and the neural network classifiers. Both aim at forming complex decision boundaries in the feature space. A major difference lies in the way decisions are made. Decision trees employ a hierarchically structured decision function in a sequential fashion. In contrast, neural networks utilize a set of soft (not final) decisions in a parallel fashion.

Furthermore, their training is performed via different philosophies. However, despite their differences, it has been shown that linear tree classifiers (with a linear splitting criterion) can be adequately mapped to a multilayer perceptron structure [Seth 90, Seth 91, Park 94].

So far, from the performance point of view, comparative studies seem to give an advantage to the multilayer perceptrons with respect to the classification error, and an advantage to the decision trees with respect to the required training time [Brow 93].

4.21 COMBINING CLASSIFIERS

The present chapter is the third one concerning the classifier design phase. Although we have not exhausted the list (a few more cases will be discussed in the chapters to follow), we feel that we have presented to the reader the most popular directions currently used for the design of a classifier.

Another trend that offers more possibilities to the designer is to *combine different classifiers*. Thus, one can exploit their individual advantages in order to reach an overall better performance than could be achieved by using each of them separately. An important observation that justifies such an approach is the following. From the different (candidate) classifiers we design in order to choose the one that fits our needs, one results in the best performance; that is, minimum classification error rate. However, different classifiers may fail (to classify correctly) on different patterns. That is, even the "best" classifier can fail on patterns that other classifiers succeed on.

Combining classifiers aims at exploiting this complementary information that seems to reside in the various classifiers. This is illustrated in Figure 4.29. Many interesting design issues have now come onto the scene. What is the strategy that one has to adopt for combining the individual outputs in order to reach the final conclusion? Should one combine the results following the product rule, the sum rule, the min rule, the max rule, or the median rule? Should all classifiers be fed with the same feature vectors, or must different feature vectors be selected for the different classifiers? Let us now highlight some of these issues a bit further.

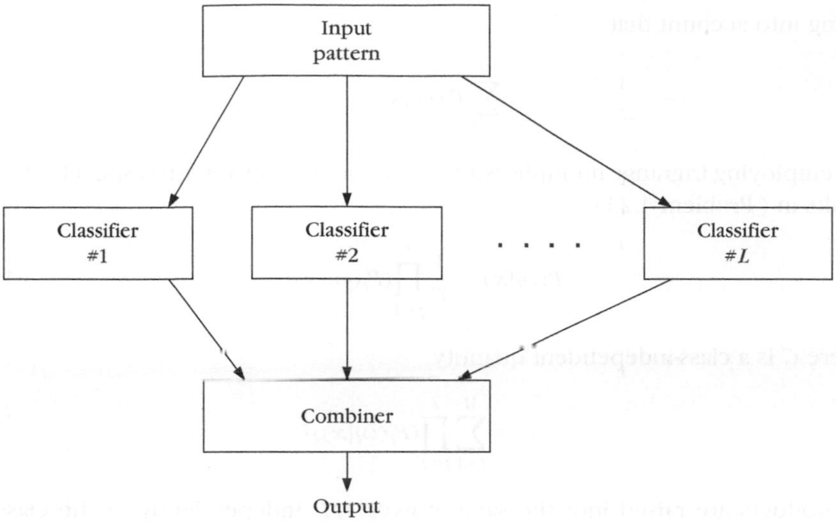

FIGURE 4.29

L classifiers are combined in order to provide the final decision for an input pattern. The individual classifiers may operate in the same or in different feature spaces.

Assume that we are given a set of L classifiers, which have already been trained (in one way or another) to provide as outputs the class *a posteriori* probabilities. For a classification task of M classes, given an unknown feature vector $\boldsymbol{x}$ each classifier produces estimates of the *a posteriori* class probabilities; that is, $P_j(\omega_i|\boldsymbol{x})$, $i = 1, 2, \ldots, M$, $j = 1, 2, \ldots, L$. Our goal is to devise a way to come up with an improved estimate of a "final" *a posteriori* probability $P(\omega_i|\boldsymbol{x})$ based on all the resulting estimates from the individual classifiers, $P_j(\omega_i|\boldsymbol{x})$, $j = 1, 2, \ldots, L$. An elegant way is to resort to information theoretic criteria [Mill 99] by exploiting the Kullback-Leibler (KL) (Appendix A) probability distance measure.

4.21.1 Geometric Average Rule

According to this rule, one chooses $P(\omega_i|\boldsymbol{x})$ in order to minimize the *average* KL distance between probabilities. That is,

$$D_{av} = \frac{1}{L}\sum_{j=1}^{L} D_j \tag{4.115}$$

where

$$D_j = \sum_{i=1}^{M} P(\omega_i|\boldsymbol{x}) \ln \frac{P(\omega_i|\boldsymbol{x})}{P_j(\omega_i|\boldsymbol{x})} \tag{4.116}$$

Taking into account that

$$\sum_{i=1}^{M} P_j(\omega_i|\boldsymbol{x}) = 1$$

and employing Lagrange multipliers, optimization of (4.115) with respect to $P(\omega_i|\boldsymbol{x})$ results in (Problem 4.23)

$$P(\omega_i|\boldsymbol{x}) = \frac{1}{C}\prod_{j=1}^{L}(P_j(\omega_i|\boldsymbol{x}))^{\frac{1}{L}} \tag{4.117}$$

where C is a class-independent quantity

$$C = \sum_{i=1}^{M}\prod_{j=1}^{L}(P_j(\omega_i|\boldsymbol{x}))^{\frac{1}{L}}$$

All products are raised into the same power $1/L$, independently of the class ω_i. Thus, neglecting all the terms common to all classes, the classification rule becomes equivalent to assigning the unknown pattern to the class maximizing the product. That is,

$$\max_{\omega_i}\prod_{j=1}^{L} P_j(\omega_i|\boldsymbol{x}) \tag{4.118}$$

4.21.2 Arithmetic Average Rule

As pointed out in Appendix A, the KL probability dissimilarity cost is not a true distance measure (according to the strict mathematical definition), in the sense that it is not symmetric. A different (from the product) combination rule results if we choose to measure the probability distance via the alternative KL distance formulation. That is,

$$D_j = \sum_{i=1}^{M} P_j(\omega_i|\boldsymbol{x}) \ln \frac{P_j(\omega_i|\boldsymbol{x})}{P(\omega_i|\boldsymbol{x})} \tag{4.119}$$

Using (4.119) in (4.115), optimization leads to (Problem 4.24)

$$P(\omega_i|\boldsymbol{x}) = \frac{1}{L}\sum_{j=1}^{L} P_j(\omega_i|\boldsymbol{x}) \tag{4.120}$$

There is no theoretical basis for preferring to maximize (4.120) instead of (4.118) with respect to ω_i, and it has been reported (e.g., [Mill 99]), that although the product rule often produces better results than the sum rule, it may lead to less reliable results when the outputs of some of the classifiers result in values close to zero.

4.21.3 Majority Voting Rule

The product and the summation schemes of combining classifiers belong to the so-called soft type rules. Hard type combination rules are also very popular, owing to their simplicity and their robust performance. According to the majority vote scheme, one decides in favor of the class for which there is a consensus, or when at least l_c of the classifiers agree on the class label of the unknown pattern, where

$$l_c = \begin{cases} \frac{L}{2} + 1, & L \text{ even} \\ \frac{L+1}{2}, & L \text{ odd} \end{cases} \tag{4.121}$$

Otherwise, the decision is *rejection* (i.e., no decision is taken). In other words, the combined decision is correct when the decisions of the majority of classifiers are correct, and it is wrong when the decisions of the majority of classifiers are wrong and they *agree* on the wrong label. A rejection is considered neither correct nor wrong. Assume now that we are given L individually trained classifiers, as previously, and in addition

1. The number L is odd.
2. Each classifier has the same probability p of correct classification.
3. The decision of each classifier is taken independently of the others.

Of these three assumptions the third is the strongest. In reality, the decisions cannot be independent. The other two assumptions can (fairly) easily be relaxed (e.g., [Mill 99, Lam 97]).

Let $P_c(L)$ be the probability of correct decision, after the majority vote. Then, this is given by the binomial distribution (see [Lam 97])

$$P_c(L) = \sum_{m=l_c}^{L} \binom{L}{m} p^m (1-p)^{L-m}$$

where l_c is defined in (4.121). Assuming $L \geq 3$, then the following are true.

- If $p > 0.5$, $P_c(L)$ is monotonically increasing in L and $P_c(L) \to 1$ as $L \to \infty$.
- If $p < 0.5$, $P_c(L)$ is monotonically decreasing in L and $P_c(L) \to 0$ as $L \to \infty$.
- If $p = 0.5$, $P_c(L) = 0.5$ for all L.

In other words, using L classifiers combined with the majority vote scheme and *under the assumptions made* the probability of correct classification increases with L and tends to 1, provided $p > 0.5$. This is slightly surprising! Does it mean that by combining classifiers with the majority vote scheme one can do better than the optimal Bayesian classifier? The answer is obviously no, and the secret lies in the last of the three assumptions, which is not valid in practice. On the contrary, one can deduce that even if this is approximately valid for small values of L as L increases the independence assumption becomes more and more unrealistic. However, the previously cited analysis provides a theoretical framework that justifies the general trend observed from experimental studies; that is, increasing the number of classifiers increases the probability of a correct decision.

In [Kunc 03] it is pointed out that, in the case of combining dependent classifiers, there is no guarantee that a majority vote combination improves performance. Based on an artificially generated data set, an upper and a lower limit for the accuracy of the majority vote combiner are derived, in terms of the accuracy p of the individual classifiers, the number L of the classifiers, and the degree of dependence among the individual classifiers. Furthermore, it is shown that dependency is not necessarily detrimental and that training-dependent classifiers with a certain pattern of dependency may be beneficial. Similar results have been obtained in [Nara 05], where lower and upper bounds for the performance of combining classifiers through the majority voting rule have been theoretically derived for the binary classification problem. The analysis involves no assumptions about independence, and the majority voting problem is treated as a constraint optimization task. Other attempts to extend the theory to deal with dependent voters have also appeared in the literature, for example, [Berg 93, Bola 89], and some results are available that give performance predictions closer to the experimental evidence [Mill 99].

In practice, a number of scenarios have been proposed aiming to make the decisions of the individual classifiers more independent. One approach is to train individual classifiers using different data points that reside in the same feature space. This can be done using various resampling techniques from the original training set, such as bootstrapping. Bagging (Section 4.20) belongs to this family of methods. These types of combination approaches are most appropriate for *unstable* classifiers—that is, classifiers whose output(s) exhibit large variations for small changes in their input data. Tree classifiers and large (with respect to the number of training points) neural networks are typical examples of unstable classifiers. *Stacking* [Wolpe 92] is an alternative attempt toward independence that constructs the combiner using for its training the outputs of the individual classifiers. However, these outputs correspond to data points that have been *excluded* from the training set, which was used to train the classifiers. This is done in a rotated fashion. Each time, different points are excluded from the training set and are kept to be used for testing. The outputs of the classifiers, obtained from these tests, are then employed to train the combiner. The rationale is basically the same as that behind the leave-one-out method, discussed in Chapter 10.

An alternative route that takes us closer to independence is to let each classifier operate in a different feature subspace. That is, each classifier is trained by employing a different subset from an original set of selected features (e.g., [Ho 98]). The majority vote scheme needs no modification in operating under such a scenario, in that all that is required is a counting of hard decisions. In contrast, the situation is different for the soft-type combination rules considered previously. Now, each pattern is represented in each classifier by a different input vector, and the resulting class posterior probabilities at the outputs of the classifiers can no longer be considered estimates of the same functional value, as is the case for Eqs. (4.118) and (4.120). The classifiers operate on different feature spaces. In [Kitt 98], a Bayesian framework is adopted to justify soft-type combination rules for such scenarios.

4.21.4 A Bayesian Viewpoint

Let $\boldsymbol{x}_i$, $i = 1, 2, \ldots, L$, be the feature vector representing the *same* pattern at the input of the ith classifier, with $\boldsymbol{x}_i \in \mathcal{R}^{l_i}$, where l_i is the dimensionality of the respective feature space, which may not be the same for all feature vectors. The task is now cast in the Bayesian rationale. Given L measurements, $\boldsymbol{x}_i, i = 1, 2, \ldots, L$, compute the maximum *a posteriori* joint probability

$$P(\omega_i|\boldsymbol{x}_1, \ldots, \boldsymbol{x}_L) = \max_{k=1}^{M} P(\omega_k|\boldsymbol{x}_1, \ldots, \boldsymbol{x}_L) \tag{4.122}$$

However,

$$P(\omega_k|\boldsymbol{x}_1, \ldots, \boldsymbol{x}_L) = \frac{P(\omega_k)p(\boldsymbol{x}_1, \ldots, \boldsymbol{x}_L|\omega_k)}{p(\boldsymbol{x}_1, \ldots, \boldsymbol{x}_L)} \tag{4.123}$$

For the problem to become tractable, we will once more adopt the statistical independence assumption, thus

$$p(\boldsymbol{x}_1, \ldots, \boldsymbol{x}_L|\omega_k) = \prod_{j=1}^{L} p(\boldsymbol{x}_j|\omega_k) \tag{4.124}$$

Combining Eqs. (4.122) through (4.124) and dropping out the class-independent quantities, the classification rule becomes equivalent to

$$P(\omega_i|\boldsymbol{x}_1, \ldots, \boldsymbol{x}_L) = \max_{k=1}^{M} P(\omega_k) \prod_{j=1}^{L} p(\boldsymbol{x}_j|\omega_k) \tag{4.125}$$

Substituting in the previous the Bayes rule

$$p(\boldsymbol{x}_j|\omega_k) = \frac{P(\omega_k|\boldsymbol{x}_j)p(\boldsymbol{x}_j)}{P(\omega_k)}$$

and getting rid of the class-independent terms, the classification rule finally becomes equivalent to searching for

$$\max_{k=1}^{M} (P(\omega_k))^{1-L} \prod_{j=1}^{L} P(\omega_k|\boldsymbol{x}_j) \tag{4.126}$$

If we adopt the assumption that each class posterior probability, $P(\omega_k|\boldsymbol{x}_j)$, $k = 1, 2, \ldots, M$, is provided (as an estimate) at the output of the respective classifier, then Eq. (4.126) is, once more, the product rule, this time with a Bayesian "blessing." Although such an approach seems to provide the optimal classifier, it is built on two assumptions. The first is that of statistical independence. The second is that the true $P(\omega_k|\boldsymbol{x}_j)$ is approximated sufficiently well by the output of the jth classifier. The accuracy of the final result depends on how good this approximation is. A sensitivity analysis in [Kitt 98] shows that in many cases the product rule is very sensitive in such approximation errors. In contrast, the sum rule seems to be more resilient to errors and in many cases in practice outperforms the product rule. It can easily be shown that the sum rule can be obtained from Eq. (4.126) by assuming that $P(\omega_k|\boldsymbol{x}) \approx P(\omega_k)(1 + \delta)$, where δ is a small value. This is a very strong assumption, since it implies that the *a posteriori* probability is approximately equal to the *a priori* one. This implicitly states that the classification task is very hard and no extra information about the class label is gained after the value of $\boldsymbol{x}$ becomes available. No doubt, from a theoretical point of view this is not a very pleasing assumption!

An alternative viewpoint of the summation rule is given in [Tume 95], through the bias–variance dilemma looking glass. Assuming that the individual classifiers result in low bias estimates of the *a posteriori* class probabilities and under the mutual independence assumption, averaging of the outputs reduces variance, leading to a reduction in error rate. This point of view tempts one to choose large classifiers (large number of free parameters) with respect to the number of training data, N, since such a choice favors low bias at the expense of high variance (Section 3.5.3), which is then reduced by the action of averaging. These results have been extended to include the weighted average case in [Fume 05].

In addition to the product and sum, other combination rules have been suggested in the literature, such as the max, min, and median rules. These rules are justified by the valid inequalities

$$\prod_{j=1}^{L} P(\omega_k|\boldsymbol{x}_j) \leq \min_{j=1}^{L} P(\omega_k|\boldsymbol{x}_j) \leq \frac{1}{L}\sum_{j=1}^{L} P(\omega_k|\boldsymbol{x}_j) \leq \max_{j=1}^{L} P(\omega_k|\boldsymbol{x}_j)$$

and classification is achieved by maximizing the respective bounds instead of the product or the summation [Kitt 98]. In some cases, the existence of outliers may lead the sum average value to be very wrong, since the value of the outlier is the dominant one in the summation. In such cases, the median value is well known to provide a more robust estimate, and the combination rule decides in favor of the class that gives the maximum median value. That is,

$$\max_{k=1}^{M} \ \text{median}\{P(\omega_k|\boldsymbol{x}_j)\}$$

In the published literature, a number of variants of the previous methods have also been suggested, such as in [Kang 03, Lin 03, Ho 94, Levi 02]. The choice of the specific combination rule is, in general, problem dependent. In [Jain 00], a set of experiments is reported concerning the results of combining twelve different classifiers using five different combination rules. For the same data set (handwritten numerals Ø–9), six different feature sets were generated. Each classifier was trained separately for each of the six feature sets, resulting in six different variants. Two types of combinations were performed: (a) all classifiers, trained on the same feature set, were combined using the five different combiners and (b) the outputs of the six variants of the same classifier were also combined via the five different combiners. The results show the following.

1. There is not a single type of combination (e.g., product rule, majority voting) that scores best for all cases. Each case seems to "prefer" its own combining rule.
2. For every case, some, out of the five, combining rules result in a higher error rate compared to that obtained by the best individual classifier. This means that *combining does not necessarily lead to improved performance*.
3. There are cases where none of the combining rules does better than the best individual classifier.
4. Improvements obtained by combining the variants of the same classifier, each trained on a different feature set, are substantially better than those obtained by combining different classifiers but trained on the same set. This seems to be a more general trend. That is, training each individual classifier on a different feature set offers the combiner better chances for improvements.

In practice, one tries to combine classifiers that are as "diverse" as possible, expecting to improve performance by exploiting the complementary information residing in the outputs of the individual classifiers. Take, for example, the extreme case where all classifiers agree on their predictions. Any attempt to combine the classifiers for improving the overall performance would obviously be meaningless. As there is no formal definition of classifier diversity, a number of different measures have been suggested to quantify diversity for the purpose of classifier combining. For example, in [Germ 92] the variance is adopted as a diversity measure. For the case of hard decisions, let $\omega_i(\boldsymbol{x}_j)$ be the class label predicted by the ith classifier for pattern $\boldsymbol{x}_j$. Let $\bar{\omega}(\boldsymbol{x}_j)$ also be the respective "mean" class label, computed over all classifiers. The mean must be defined in a meaningful way. For hard decisions, one possibility is to adopt as the mean value the most frequent one among all (L) classifiers. Define

$$d\left(\omega_i(\boldsymbol{x}_j), \bar{\omega}(\boldsymbol{x}_j)\right) = \begin{cases} 1 & \text{if } \omega_i(\boldsymbol{x}_j) \neq \bar{\omega}(\boldsymbol{x}_j) \\ 0 & \text{otherwise} \end{cases}$$

The variance of the combined classifiers can be computed as

$$V = \frac{1}{NL} \sum_{j=1}^{N} \sum_{i=1}^{L} d\left(\omega_i(\boldsymbol{x}_j), \bar{\omega}(\boldsymbol{x}_j)\right)$$

A large variance is taken to be indicative for large diversity. Besides the variance, other measures have also been suggested and used. For example, in [Kang 00] the mutual information among the outputs of the classifiers is used, and in [Kunc 03] the Q statistics test is employed. For a review of diversity measures and comparative studies see, for example, [Kunc 03a, Akse 06]. In [Rodr 06] the issue of designing diverse classifiers is considered together with the issue of accuracy. A methodology, called *Rotation Forest*, is proposed, which aims at designing classifiers that are both accurate and diverse. The classifiers are then combined to boost the overall performance.

Experimental comparative studies that demonstrate the performance improvement that may be gained by combining classifiers can be found in [Mill 99, Kitt 98, Tax 00, Dzer 04]. It seems that the sum average and the majority vote rules are more popular and used the most frequently. Which of the two is to be adopted depends on the application. In [Kitt 03], it is shown that for normally distributed error probabilities the sum rule outperforms the voting rule. In contrast, for heavy tail error distributions the voting scheme may be better. More theoretical results, concerning combinations of classifiers, can be found in [Kunc 02] and [Klei 90]. However, it is true that most of the available theoretical results have been developed under rather restrictive assumptions. More recently, a "fresh" look at the theoretical study of the performance of combiners has been presented in [Evge 04] and theoretical nonasymptotic bounds on the combiner' s generalization error are derived, for the case of combining SVM classifiers, via weighted averaging. The so called *no panacea theorem* is stated in [Hu 08]. It is shown that if the combination function is continuous and diverse one can always construct probability density distributions that describe the data and lead the combination scheme to poor performance. In other words, this theorem points out that combining classifiers has to be considered carefully.

The common characteristic of all combination techniques presented so far is that the individual classifiers are separately trained and the combiner relies on a simple rule. Besides these techniques, a number of other schemes have been developed, which rely on optimizing the combiner and in some cases jointly with the individual classifiers, for example, [Ueda 97, Rose 96, Kunc 01]. The price one pays for such procedures obviously is complexity, which in some cases can become impractical, see [Rose 96]. Moreover, there is no guarantee that optimization leads to improved performance compared to the simpler nonoptimal methods considered previously. More recently, Bayesian approaches ([Tres 01]) and Bayesian networks ([Garg 02, Pikr 08]) have been mobilized to construct combiners. A game-theoretic approach has been adopted in [Geor 06].

The so-called *mixture of experts* [Jaco 91, Hayk 96, Avni 99] are structures that share some of the ideas exposed in this section. The rationale behind such models is to assign different classifiers for different regions in space and then use an extra "gating" network, which also sees the input feature vector, to decide which classifier (expert) should be used each time. All classifiers as well as the gating network are jointly trained.

4.22 THE BOOSTING APPROACH TO COMBINE CLASSIFIERS

Boosting is a general approach to improve the performance of a given classifier and is one of the most powerful techniques, together with the support vector machines, that blossomed in the 1990s. Although boosting can be considered an approach to combine classifiers, it is conceptually different from the techniques presented in the previous section, and it deserves a separate treatment. The roots of boosting go back to the original work of Viliant and Kearns [Vali 84, Kear 94], who posed the question whether a "weak" learning algorithm (i.e., one that performs just slightly better than a random guessing) can be *boosted* into a "strong" algorithm with good error performance. At the heart of a boosting method lies the so-called *base* classifier, which is a weak classifier. A series of classifiers is then designed iteratively, employing each time the base classifier but using a different subset of the training set, according to an iteratively computed distribution, or a different weighting over the samples of the training set. At each iteration, the computed weighting distribution gives emphasis to the "hardest" (incorrectly classified) samples.

The final classifier is obtained as a weighted average of the previously *hierarchically* designed classifiers. It turns out that given a sufficient number of iterations the classification error of the final combination *measured on the training set* can become arbitrarily low [Scha 98]. This is very impressive indeed. Using a weak classifier as the base, one can achieve an arbitrarily low training error rate by appropriate manipulation of the training data set in harmony with the performance of the sequence of the designed classifiers (Problem 4.28). In this section we will focus on one such algorithm, the so-called *AdaBoost* (adaptive boosting), which is sometimes known as the discrete AdaBoost, to emphasize the fact that it returns a binary discrete label. This is the most popular algorithm of the family and one that has been extensively studied. The treatment follows the approach introduced in [Frie 00].

We concentrate on the two-class classification task and let the set of the training data be $\{(\boldsymbol{x}_1, y_1), (\boldsymbol{x}_2, y_2) \ldots, (\boldsymbol{x}_N, y_N)\}$ with $y_i \in \{-1, 1\}, i = 1, 2, \ldots, N$. The goal is to construct an optimally designed classifier of the form

$$f(\boldsymbol{x}) = \text{sign}\,\{F(\boldsymbol{x})\} \tag{4.127}$$

where

$$F(\boldsymbol{x}) = \sum_{k=1}^{K} \alpha_k \phi(\boldsymbol{x}; \boldsymbol{\theta}_k) \tag{4.128}$$

where $\phi(\boldsymbol{x}; \boldsymbol{\theta})$ denotes the base classifier that returns a binary class label; that is, $\phi(\boldsymbol{x}; \boldsymbol{\theta}) \in \{-1, 1\}$. The base classifier is described by the corresponding parameter vector $\boldsymbol{\theta}$, whose value is allowed to be different in each of the summand terms, as will become apparent soon. The values of the unknown parameters result from the following optimization.

$$\arg \min_{\alpha_k; \boldsymbol{\theta}_k, k:1,K} \sum_{i=1}^{N} \exp\left(-y_i F(\boldsymbol{x}_i)\right) \tag{4.129}$$

This cost function is common in learning theory. It penalizes the samples that are wrongly classified ($y_i F(\boldsymbol{x}_i) < 0$) much more heavily than those correctly classified

($y_i F(\boldsymbol{x}_i) > 0$). However, direct optimization of (4.129) is a highly complex task. A suboptimal method commonly employed in optimization theory for complex problems is to carry out the optimization in a stage-wise fashion. At each step, a new parameter is considered and optimization is carried out with respect to this parameter, leaving unchanged the previously optimized ones. To this end, let us define $F_m(\boldsymbol{x})$ to denote the result of the partial sum up to m terms. That is,

$$F_m(\boldsymbol{x}) = \sum_{k=1}^{m} \alpha_k \phi(\boldsymbol{x}; \boldsymbol{\theta}_k), \; m = 1, 2, \ldots, K \tag{4.130}$$

Based on the this definition, the following recursion becomes obvious.

$$F_m(\boldsymbol{x}) = F_{m-1}(\boldsymbol{x}) + \alpha_m \phi(\boldsymbol{x}; \boldsymbol{\theta}_m) \tag{4.131}$$

Let us now employ a stage-wise optimization in our problem. At step m, $F_{m-1}(\boldsymbol{x})$ is the part that has been optimized in the previous step, and the current task is to compute the optimal values for α_m, $\boldsymbol{\theta}_m$. In other words, the task at step m is to compute

$$(\alpha_m, \boldsymbol{\theta}_m) = \arg\min_{\alpha, \boldsymbol{\theta}} J(\alpha, \boldsymbol{\theta})$$

where the cost function is defined as

$$J(\alpha, \boldsymbol{\theta}) = \sum_{i=1}^{N} \exp\left(-y_i (F_{m-1}(\boldsymbol{x}_i) + \alpha \phi(\boldsymbol{x}_i; \boldsymbol{\theta}))\right) \tag{4.132}$$

Once more, optimization will be carried out in two steps. First, α will be considered constant, and the cost will be optimized with respect to the base classifier $\phi(\boldsymbol{x}; \boldsymbol{\theta})$. That is, the cost to be minimized is now simplified to

$$\boldsymbol{\theta}_m = \arg\min_{\boldsymbol{\theta}} \sum_{i=1}^{N} w_i^{(m)} \exp\left(-y_i \alpha \phi(\boldsymbol{x}_i; \boldsymbol{\theta})\right) \tag{4.133}$$

where

$$w_i^{(m)} \equiv \exp\left(-y_i F_{m-1}(\boldsymbol{x}_i)\right) \tag{4.134}$$

Since each $w_i^{(m)}$ depends neither on α nor on $\phi(\boldsymbol{x}_i; \boldsymbol{\theta})$, it can be regarded as a weight associated with the sample point $\boldsymbol{x}_i$. Due to the binary nature of the base classifier ($\phi(\boldsymbol{x}; \boldsymbol{\theta}) \in \{-1, 1\}$), it is easy to see that minimizing (4.133) is equivalent to designing the optimal classifier $\phi(\boldsymbol{x}; \boldsymbol{\theta}_m)$ so that the *weighted* empirical error (the fraction of the training samples that are wrongly classified) is minimum. That is,

$$\boldsymbol{\theta}_m = \arg\min_{\boldsymbol{\theta}} \left\{ P_m = \sum_{i=1}^{N} w_i^{(m)} I(1 - y_i \phi(\boldsymbol{x}_i; \boldsymbol{\theta})) \right\} \tag{4.135}$$

Function $I(\cdot)$ is either 0 or 1, depending on its argument, whether it is zero or positive, respectively. To guarantee that the value of the weighted empirical error rate remains in the interval [0, 1], the weights must sum to one. This is easily achieved by appropriate normalization; that is, dividing each weight by the respective sum, $\sum_{i=1}^{N} w_i^{(m)}$, which does not affect the optimization and can easily be incorporated in the final iterative algorithm. Having computed the optimal classifier at step m, $\phi(\boldsymbol{x}; \boldsymbol{\theta}_m)$, the following are easily established from the respective definitions.

$$\sum_{y_i\phi(\boldsymbol{x}_i;\boldsymbol{\theta}_m)<0} w_i^{(m)} = P_m \tag{4.136}$$

$$\sum_{y_i\phi(\boldsymbol{x}_i;\boldsymbol{\theta}_m)>0} w_i^{(m)} = 1 - P_m \tag{4.137}$$

Combining Eqs. (4.137) and (4.136) with (4.134) and (4.132), the optimum value, α_m, results from

$$\alpha_m = arg \min_{\alpha} \{\exp(-\alpha)(1 - P_m) + \exp(\alpha)P_m\} \tag{4.138}$$

Taking the derivative with respect to α and equating to zero, we obtain

$$\alpha_m = \frac{1}{2} \ln \frac{1 - P_m}{P_m} \tag{4.139}$$

Once α_m and $\phi(\boldsymbol{x}; \boldsymbol{\theta}_m)$ have been computed, the weights for the next step are readily available via the iteration

$$w_i^{(m+1)} = \frac{\exp(-y_i F_m(\boldsymbol{x}_i))}{Z_m} = \frac{w_i^{(m)} \exp\left(-y_i \alpha_m \phi(\boldsymbol{x}_i; \boldsymbol{\theta}_m)\right)}{Z_m} \tag{4.140}$$

where Z_m is the normalizing factor

$$Z_m \equiv \sum_{i=1}^{N} w_i^{(m)} \exp\left(-y_i \alpha_m \phi(\boldsymbol{x}_i; \boldsymbol{\theta}_m)\right) \tag{4.141}$$

Observe that the value of the weight corresponding to sample $\boldsymbol{x}_i$ is increased (decreased) with respect to its value at the previous iteration step if the classifier $\phi(\boldsymbol{x}_i; \boldsymbol{\theta}_m)$ fails (wins) at the respective point. Moreover, the percentage of the increase or decrease depends on the value of α_m, which also controls the relative importance of the term $\phi(\boldsymbol{x}; \boldsymbol{\theta}_m)$ in building up the final classifier $F(\boldsymbol{x})$ in (4.128). Hard examples (i.e., samples that fail to be classified correctly by a number of successive classifiers) gain an increased importance in the weighted empirical error rate as they insist on failing! A pseudocode for the AdaBoost algorithm follows.

The AdaBoost Algorithm

- Initialize: $w_i^{(1)} = \frac{1}{N}$, $i = 1, 2 \ldots, N$
- Initialize: $m = 1$
- Repeat
  - Compute optimum $\boldsymbol{\theta}_m$ in $\phi(\cdot; \boldsymbol{\theta}_m)$ by minimizing P_m; (4.135)
  - Compute the optimum P_m; (4.135)
  - $\alpha_m = \frac{1}{2} \ln \frac{1-P_m}{P_m}$
  - $Z_m = 0.0$
  - For $i = 1$ to N
    - $w_i^{(m+1)} = w_i^{(m)} \exp\left(-y_i \alpha_m \phi(\boldsymbol{x}_i; \boldsymbol{\theta}_m)\right)$
    - $Z_m = Z_m + w_i^{(m+1)}$
  - End{For}
  - For $i = 1$ to N
    - $w_i^{(m+1)} = w_i^{(m+1)} / Z_m$
  - End {For}
  - $K = m$
  - $m = m + 1$
- Until a termination criterion is met.
- $f(\cdot) = \text{sign}(\sum_{k=1}^{K} \alpha_k \phi(\cdot, \boldsymbol{\theta}_k))$

One of the main and very interesting properties of boosting is its relative immunity to overfitting, which was defined in Section 4.9. In practice, it has been verified that, although the number of terms, K, and consequently the associated number of parameters can be quite high, the error rate on a test set does not increase but keeps decreasing and finally *levels off at a certain value*. It has been observed that the test error continues to decrease long after the error on the training set has become zero. A mathematically pleasing explanation is offered in [Scha 98], where an upper bound for the error probability (also known as generalization error) is derived in terms of the margins of the training points with respect to the designed classifier. Note that the test error rate is an estimate of the error probability (more formal definitions of these quantities are provided in Section 5.9). The bound is *independent of the number of iterations*, K. More specifically, it is shown that with high probability the generalization error is upper-bounded by the quantity

$$\text{prob}\{\text{margin}_f(\boldsymbol{x}, y) < \gamma\} + O\left(\sqrt{\frac{V_c}{N\gamma^2}}\right) \tag{4.142}$$

for $\gamma > 0$, where V_c is a parameter measuring the complexity of the base classifier and is known as the Vapnic–Chervonenkis dimension (we will discuss it later in the book). The margin of a training example with respect to a classifier f [Eq. (4.127)] is defined as

$$\text{margin}_f(\boldsymbol{x}, y) = \frac{yF(\boldsymbol{x})}{\sum_{k=1}^{K} \alpha_k} = \frac{y \sum_{k=1}^{K} \alpha_k \phi_k(\boldsymbol{x}; \theta_k)}{\sum_{k=1}^{K} \alpha_k}$$

The margin lies in the interval $[-1, 1]$ and is positive if and only if the respective pattern is classified correctly.

The bound implies that if (a) the margin probability is small for large values of margin γ *and* (b) N is large enough with respect to V_c, one expects the generalization error to be small, and this does not depend on the number of iterations that were used to design $f(\boldsymbol{x})$. The bound suggests that if for most of the training points the margin is large, the generalization error is expected to be small. This is natural, since the magnitude of the margin can be interpreted as a measure of confidence about the decision of the classifier with respect to a sample. Hence, if for a large training data set the resulting margin for most of the training points is large, it is not beyond common sense to expect that a low training error rate may also suggest a low generalization error.

Furthermore, as pointed out in [Maso 00, Scha 98], boosting is particularly aggressive at improving the margin distribution, since it concentrates on examples with the smallest margins, as one can readily dig out by looking carefully at the cost (4.129). From this point of view, there is an affinity with the support vectors machines, which also try to maximize the margin of the training samples from the decision surface. See, for example, [Scha 98, Rats 02]. The major criticism about the bound in (4.142) lies in the fact that it is very loose (unless the number, N, of the training points is very large; i.e., of the order of tens of thousands!), so it can only be used as a qualitative rather than as a quantitative explanation of the commonly encountered experimental evidence.

Another explanation for this overfitting immunity associated with the boosting algorithms could be that parameter optimization is carried out in a stage-wise fashion, each time with respect to a single parameter. Some very interesting and enlightening discussions, among leading experts in the field, regarding the overfitting as well as other issues concerning boosting and related algorithmic families can be found in the papers [Frie 00, Brei 98]. A comparative study of the performance of boosting and other related algorithms can be found in [Baue 99].

Remarks

- Obviously, Adaboost is not the only boosting algorithm available. For example, one can come up with other algorithms by adopting alternatives to (4.129) cost functions or growing mechanisms to build up the final classifier. In fact, it has been observed that in difficult tasks corresponding to relatively high Bayesian error probabilities (i.e., attained by using the optimal Bayesian classifier), the performance of the AdaBoost can degrade dramatically. An explanation for it is that the exponential cost function over-penalizes "bad" samples that correspond to large negative margins, and this affects the overall performance. More on these issues can be obtained from [Hast 01, Frie 00] and the references therein.

A variant of the AdaBoost has been proposed in [Viol 01] and later generalized in [Yin 05]. Instead of training a single base classifier, a number of base classifiers are trained simultaneously, each on a different set of features. At each iteration step, the classifier $\phi(\cdot)$ results by combining these base classifiers. In principle, any of the combination rules can be used. [Scha 05] presents a modification of the AdaBoost that allows for incorporation of prior knowledge into boosting as a means of compensating for insufficient data. The so called AdaBoost$^*_\nu$ version was introduced in [Rats 05], where the margin is explicitly brought into the game and the algorithm maximizes the minimum margin of the training setup. The algorithm incorporates a current estimate of the achievable margin which is used for computation of the optimal combining coefficients of the base classifiers.

Multiple additive regression trees (MART) is a possible alternative that overcomes some of the drawbacks related to AdaBoost. In this case, the additive model in (4.128) consists of an expansion in a series of classification trees (CART), and the place of the exponential cost in (4.129) can be taken by any differentiable function. MART classifiers have been reported to perform well in a number of real cases, such as in [Hast 01, Meye 03].

- For the multiclass case problem there are several extensions of AdaBoost. A straightforward extension is given in [Freu 97, Eibl 06]. However, this extension fails if the base classifier results in error rates higher than 50%. This means that the base classifier will not be a weak one, since in the multiclass case random guessing means a success rate equal to $\frac{1}{M}$, where M is the number of classes. Thus, for large M 50% rate of correct classification can be a strong requirement. To overcome this difficulty, other (more sophisticated) extensions have been proposed. See [Scha 99, Diet 95].

Example 4.3

Let us consider a two-class classification task. The data reside in the 20-dimensional space and obey a Gaussian distribution of unit covariance matrix and mean values $[-a, -a, \ldots, -a]^T$, $[a, a, \ldots, a]^T$, respectively, for each class, where $a = 2/\sqrt{20}$. The training set consists of 200 points (100 from each class) and the test set of 400 points (200 from each class) independently generated from the points of the training set.

To design a classifier using the AdaBoost algorithm, we chose as a seed the weak classifier known as *stump*. This is a very "naive" type of tree, consisting of a single node, and classification of a feature vector $\boldsymbol{x}$ is achieved on the basis of the value of only one of its features, say, x_i. Thus, if $x_i < 0$, $\boldsymbol{x}$ is assigned to class A. If $x_i > 0$, it is assigned to class B. The decision

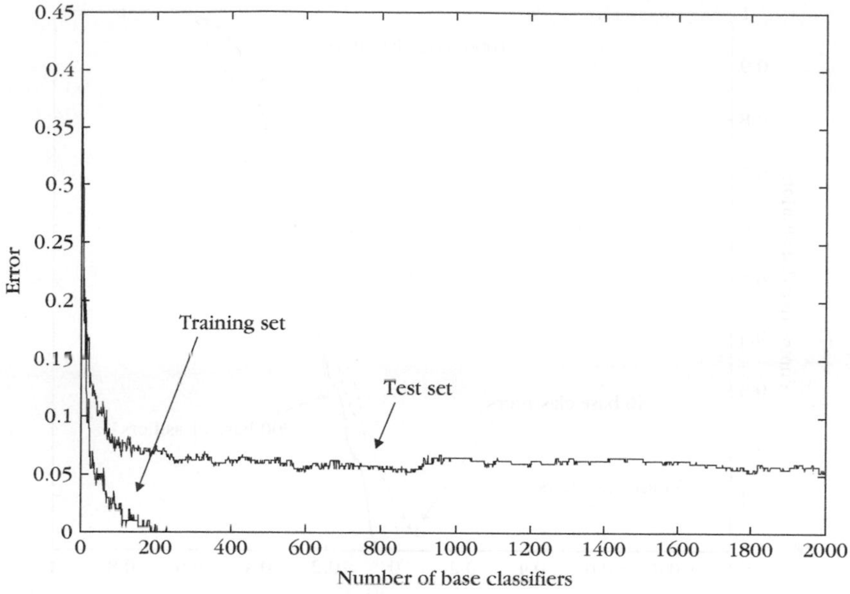

FIGURE 4.30

Training and test error rate curves as functions of the number of iteration steps for the AdaBoost algorithm, using a stump as the weak base classifier. The test error keeps decreasing even after the training error becomes zero.

about the choice of the specific feature, x_i, to be used in the classifier was randomly made. Such a classifier results in a training error rate slightly better than 0.5.

The AdaBoost algorithm was run on the training data for 2000 iteration steps. Figure 4.30 verifies the fact that the training error rate converges to zero very fast. The test error rate keeps decreasing even after the training error rate becomes zero and then levels off at around 0.05.

Figure 4.31 shows the margin distributions, over the training data points, for four different training iteration steps. It is readily observed that the algorithm is indeed greedy in increasing the margin. Even when only 40 iteration steps are used for the AdaBoost training, the resulting classifier classifies the majority of the training samples with large margins. Using 200 iteration steps, all points are correctly classified (positive margin values), and the majority of them with large margin values. From then on, more iteration steps further improve the margin distribution by pushing it to higher values.

4.23 THE CLASS IMBALANCE PROBLEM

In practice there are cases in which one class is represented by a large number of training points while another by only of few. This is usually referred to as the *class imbalance problem*. Such situations occur in a number of applications

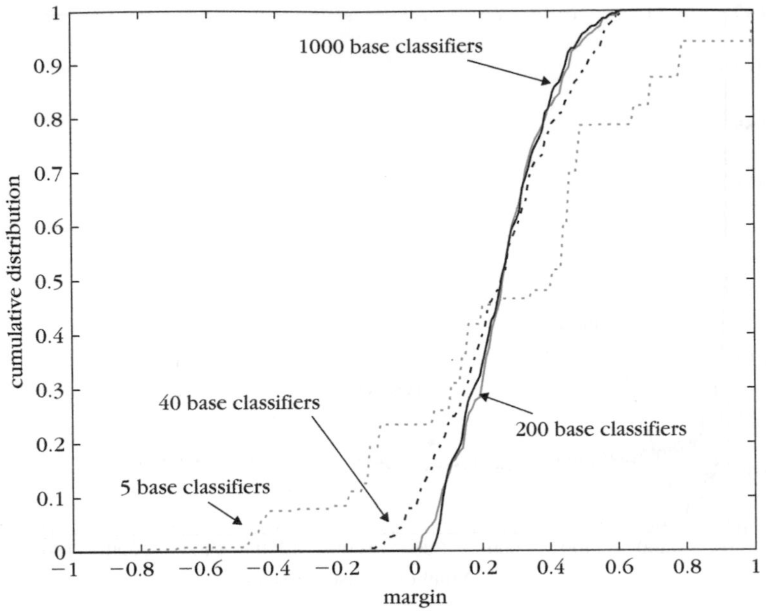

FIGURE 4.31

Margin distribution for the AdaBoost classifier corresponding to different numbers of training iteration steps. Even when only 40 iteration steps are used, the resulting classifier classifies the majority of the training samples with large margins.

such as text classification, diagnosis of rare medical conditions, and detection of oil spills in satellite imaging. It is by now well established that class imbalances may severely hinder the performance of a number of standard classifiers, for example, decision trees, multilayer neural networks, SVMs, and boosting classifiers. This does not come as a surprise, since our desire for a good generalization performance dictates the design of classifiers that are as "simple" as possible. A simple hypothesis, however, will not pay much attention to the rare cases in imbalanced data sets. A study of the imbalance class problem is given in [Japk 02]. There it is stated that the class imbalance may not necessarily be a hindrance to the classification, and it has to be considered in relation to the number of training points as well as the complexity and the nature of the specific classification task. For example, a large class imbalance may not be a problem in the case of an easy to learn task, for example, well separable classes, or in cases where a large training data set is available. On the other hand, there are cases where a small imbalance may be very harmful in difficult-to-learn tasks with overlapping classes and/or in the absence of a sufficient number of training points. To cope with this problem, a number of approaches have been proposed that evolve along two major directions.

Data-level Approaches

The aim here is to "rebalance" the classes by either oversampling the small class and/or undersampling the large class. Resampling can be either *random* or *focused*. The focus can be on points that lie close to the boundaries of the decision surfaces (oversampling) or far away (undersampling); see, for example, [Chaw 02, Zhou 06]. A major problem with this method is how to decide the class distribution given the data set; see, for example, [Weis 03].

Cost-sensitive Approaches

According to this line of "thought", standard classifiers are modified appropriately to account for the unfair data representation in the training set. For example, in SVMs, one way is to use different parameters C in the cost function for the two classes, for example, [Lin 02a]. According to the geometric interpretation, given in Section 3.7.5, this is equivalent to reducing the convex hulls at a different rate paying more respect to the smaller class, for example, [Mavr 07]. In [Sun 07], cost-sensitive modifications of the AdaBoost algorithm are proposed, where, during the iterations, samples from the small class are more heavily weighted than those coming from the more prevalent class.

Class imbalance is a very important issue in practice. The designer of any classification system must be aware of the problems that may arise and alert of the ways to cope with it.

4.24 DISCUSSION

The number of available techniques is large, and the user has to choose what is more appropriate for the problem at hand. There are no magic recipes. A large research effort has been focused on comparative studies of various classifiers in the context of different applications; see also the review in [Jain 00]. One of the most extensive efforts was the Statlog project [Mich 94], in which a wide range of classifiers was tested using a large number of different data sets. Furthermore, research effort has been devoted to unraveling relations and affinities between the different techniques. Many of these techniques have their origin in different scientific disciplines. Therefore, until a few years ago, they were considered independently. Recently, researchers have started to recognize underlying similarities among various approaches. For readers who want to dig a bit deeper into these questions, the discussions and results presented in [Chen 94, Ripl 94, Ripl 96, Spec 90, Holm 97, Josh 97, Reyn 99] will be quite enlightening. In [Zhan 00] a survey on applications of neural networks in pattern recognition is presented, and links between neural and more conventional classifiers are discussed.

In summary, the only tip that can be given to the designer is that all of the techniques presented in this book are still serious players in the classifier design game. The final choice depends on the specific task. *The proof of the pudding is in the eating!*

4.25 PROBLEMS

4.1 We are given 10 feature vectors that originate from two classes ω_1 and ω_2 as follows

$$\omega_1\text{: } [0.1, -0.2]^T, [0.2, 0.1]^T, [-0.15, 0.2]^T, [1.1, 0.8]^T, [1.2, 1.1]^T$$
$$\omega_2\text{: } [1.1, -0.1]^T, [1.25, 0.15]^T, [0.9, 0.1]^T, [0.1, 1.2]^T, [0.2, 0.9]^T$$

Check whether these are linearly separable and, if not, design an appropriate multilayer perceptron with nodes having step function activation to classify the vectors in the two classes.

4.2 Using the computer, generate four two-dimensional Gaussian random sequences with covariance matrices

$$\Sigma = \begin{bmatrix} 0.01 & 0.0 \\ 0.0 & 0.01 \end{bmatrix}$$

and mean values $\boldsymbol{\mu}_1 = [0, 0]^T, \boldsymbol{\mu}_2 = [1, 1]^T, \boldsymbol{\mu}_3 = [0, 1]^T, \boldsymbol{\mu}_4 = [1, 0]^T$. The first two form class ω_1, and the other two class ω_2. Produce 100 vectors from each distribution. Use the batch mode backpropagation algorithm of Section 4.6 to train a two-layer perceptron with two hidden neurons and one in the output. Let the activation function be the logistic one with $a = 1$. Plot the error curve as a function of iteration steps. Experiment yourselves with various values of the learning parameter μ. Once the algorithm has converged, produce 50 more vectors from each distribution and try to classify them using the weights you have obtained. What is the percentage classification error?

4.3 Draw the three lines in the two-dimensional space

$$x_1 + x_2 = 0$$
$$x_2 = \frac{1}{4}$$
$$x_1 - x_2 = 0$$

For each of the polyhedra that are formed by their intersections, determine the vertices of the cube into which they will be mapped by the first layer of a multilayer perceptron, realizing the preceding lines. Combine the regions into two classes so that (a) a two-layer network is sufficient to classify them and (b) a three-layer network is necessary. For both cases compute analytically the corresponding synaptic weights.

4.4 Show that if $\boldsymbol{x}_1$ and $\boldsymbol{x}_2$ are two points in the l-dimensional space, the hyperplane bisecting the segment with end points $\boldsymbol{x}_1, \boldsymbol{x}_2$, leaving $\boldsymbol{x}_1$ at its positive side, is given by

$$(\boldsymbol{x}_1 - \boldsymbol{x}_2)^T \boldsymbol{x} - \frac{1}{2}\|\boldsymbol{x}_1\|^2 + \frac{1}{2}\|\boldsymbol{x}_2\|^2 = 0$$

4.5 For the cross-entropy cost function of (4.33)

- Show that its minimum value for binary desired response values is zero and it occurs when the true outputs are equal to the desired ones.
- Show that the cross-entropy cost function depends on the relative output errors.

4.6 Show that if the cost function, optimized by a multilayer perceptron, is the cross entropy (4.33) and the activation function is the sigmoid (4.1), then the gradient $\delta_j^L(i)$ of (4.13) becomes

$$\delta_j^L(i) = a(1 - \hat{y}_j(i))y_j(i)$$

4.7 Repeat Problem 4.6 for the softmax activation function and show that $\delta_j^L(i) = \hat{y}_j(i) - y_j(i)$.

4.8 Show that for the cross-entropy cost function (4.30) the outputs of the network, corresponding to the optimal weights, approximate the conditional probabilities $P(\omega_i|\boldsymbol{x})$.

4.9 Using formula (4.37), show that if $l \geq K$ then $M = 2^K$.

4.10 Develop a program to repeat the simulation example of Section 4.10.

4.11 For the same example start with a network consisting of six neurons in the first hidden layer and nine neurons in the second. Use a pruning algorithm to reduce the size of the network.

4.12 Let the sum of error squares

$$J = \frac{1}{2}\sum_{i=1}^{N}\sum_{m=1}^{k_L}(\hat{y}_m(i) - y_m(i))^2$$

be the minimized function for a multilayer perceptron. Compute the elements of the Hessian matrix

$$\frac{\partial^2 J}{\partial w_{kj}^r \partial w_{k'j'}^{r'}}$$

Show that near a minimum, this can be approximated by

$$\frac{\partial^2 J}{\partial w_{kj}^r \partial w_{k'j'}^{r'}} = \sum_{i=1}^{N}\sum_{m=1}^{k_L}\frac{\partial \hat{y}_m(i)}{\partial w_{kj}^r}\frac{\partial \hat{y}_m(i)}{\partial w_{k'j'}^{r'}}$$

Thus, the second derivatives can be approximated by products of the first-order derivatives. Following arguments similar to those used for the derivation of the backpropagation algorithm, show that

$$\frac{\partial \hat{y}_m(i)}{\partial w_{kj}^r} = \hat{\delta}_{jm}^r y_k^{r-1}$$

where

$$\hat{\delta}_{jm}^r = \frac{\partial \hat{y}_m(i)}{\partial v_j^r(i)}$$

Its computation takes place recursively in the backpropagation philosophy. This has been used in [Hass 93].

4.13 In Section 4.4 it was pointed out that an approximation to the Hessian matrix, which is often employed in practice, is to assume that it is diagonal. Prove that under this assumption

$$\frac{\partial^2 \mathcal{E}}{(\partial w_{kj}^r)^2}$$

is propagated via a backpropagation concept according to the formulas:

(1) $$\frac{\partial^2 \mathcal{E}}{(\partial w_{kj}^r)^2} = \frac{\partial^2 \mathcal{E}}{(\partial v_j^r)^2}(y_k^{r-1})^2$$

(2) $$\frac{\partial^2 \mathcal{E}}{(\partial v^{L_j})^2} = f''(v^{L_j})e_j + (f'(v^{L_j}))^2$$

(3) $$\frac{\partial^2 \mathcal{E}}{(\partial v_j^{r-1})^2} = (f'(v_j^{r-1}))^2 \sum_{k=1}^{k_l} (w_{kj}^r)^2 \frac{\partial^2 \mathcal{E}}{(\partial v_j^r)^2} + f''(v_j^{r-1}) \sum_{k=1}^{k_r} w_{kj}^r \delta_k^r$$

where all off-diagonal terms of the Hessian matrix have been neglected and the dependence on i has been suppressed for notational convenience.

4.14 Derive the full Hessian matrix for a simple two-layer network with two hidden neurons and an output one. Generalize to the case of more than two hidden neurons.

4.15 Rederive the backpropagation algorithm of Section 4.6 with activation function

$$f(x) = c \tanh(bx)$$

4.16 In [Dark 91] the following scheme for adaptation of the learning parameter μ has been proposed:

$$\mu = \mu_0 \frac{1}{1 + \frac{t}{t_0}}$$

Verify that, for large enough values of t_0 (e.g., $300 \leq t_0 \leq 500$), the learning parameter is approximately constant for the early stages of training (small values of iteration step t) and decreases in inverse proportion to t for large values. The first phase is called *search phase* and the latter *converge phase*. Comment on the rationale of such a procedure.

4.17 Use a two-layer perceptron with a linear output unit to approximate the function $y(x) = 0.3 + 0.2\cos(2\pi x), x \in [0, 1]$. To this end, generate a sufficient number of data points from this function for the training. Use the backpropagation algorithm in one of its forms to train the network. In the sequel produce 50 more samples, feed them into the trained network, and plot the resulting outputs. How does it compare with the original curve? Repeat the procedure for a different number of hidden units.

4.18 Show Eq. (4.48).

Hint: Show first that the following recursion is true:

$$O(N+1, l) = O(N, l) + O(N, l-1)$$

To this end, start with the N points and add an extra one. Then show that the difference in the dichotomies, as we go from N to $N+1$ points, is due to the dichotomizing hyperplanes (for the N points case) that could have been drawn via this new point.

4.19 Show that if $N = 2(l+1)$ the number of dichotomies is given by 2^{N-1}.

Hint: Use the identity that

$$\sum_{i=0}^{J} \binom{J}{i} = 2^J$$

and recall that

$$\binom{2n+1}{n-i+1} = \binom{2n+1}{n+i}$$

4.20 Repeat the experiments of Problem 4.17 using an RBF network. Select the k centers regularly spaced within [0, 1]. Repeat the experiments with different numbers of Gaussian functions and σ. Estimate the unknown weights using the least squares method.

4.21 Using your experience from the previous problem, repeat the procedure for the two-dimensional function

$$y(x_1, x_2) = 0.3 + 0.2\cos(2\pi x_1)\cos(2\pi x_2)$$

4.22 Let the mapping from the input space to a higher dimensional space be

$$x \in \mathcal{R} \longrightarrow y \equiv \phi(x) \in \mathcal{R}^{2k+1}$$

where

$$\phi(x) = \left[\frac{1}{\sqrt{2}}, \cos x, \cos 2x, \ldots, \cos kx, \sin x, \sin 2x, \ldots, \sin kx\right]^T$$

Then show that the corresponding inner product kernel is

$$\begin{aligned} y_i^T y_j &= K(x_i, x_j) \\ &= \frac{\sin\left(\left(k+\frac{1}{2}\right)(x_i - x_j)\right)}{2\sin\left(\frac{x_i - x_j}{2}\right)} \end{aligned}$$

4.23 Show (4.117).

4.24 Show (4.120).

4.25 Prove Eqs. (4.108) and (4.109) for the ridge regression task in its dual representation form.

4.26 Show that if the linear kernel is used, the primal ridge regression task results in Eq. (4.111).

4.27 Prove that Eqs. (4.111) and (4.112) are equivalent.

4.28 Prove that the the error rate on the training set corresponding to the final boosting classifier tends to zero exponentially fast.

MATLAB PROGRAMS AND EXERCISES

Computer Programs

4.1 *Data generator.* Write a MATLAB function named *data_generator* that generates a two-class, two-dimensional data set using four normal distributions, with covariance matrices $S_i = s * I, i = 1, \ldots, 4$, where I is the 2×2 identity matrix. The vectors that stem from the first two distributions belong to class $+1$, while the vectors originating from the other two distributions belong to class -1. The inputs for this function are: (a) a 2×4 matrix, m, whose ith column is the mean vector of the ith distribution, (b) the variance parameter s, mentioned before, and (c) the number of the points, N, which will be generated from each distribution. The output of the function consists of (a) an array, X, of dimensionality $2 \times 4 * N$, whose first group of N vectors stem from the first distribution, the second group from the second distribution and so on, (b) a $4 * N$ dimensional row vector y with values $+1$ or -1, indicating the classes to which the corresponding data vectors in X belong.

Solution

```
function [x,y]=data_generator(m,s,n)
  S = s*eye(2);
```

```
[l,c]=size(m);
x=[]; % Creating the training set
for i=1:c
  x=[x mvnrnd(m(:,i)',S,N)'];
end
y=[ones(1,N) ones(1,N) -ones(1,N) -ones(1,N)];
```

4.2 *Neural network training.* Write a MATLAB function, named *NN_training*, which uses the least squares criterion to train a two-layer feed-forward neural network with a single node in the output layer. The activation function for all the nodes is the hyperbolic tangent one. For training, one may select one of the following algorithms: a) the standard gradient descent backpropagation algorithm (code 1), (b) the backpropagation algorithm with momentum (code 2), and (c) the backpropagation algorithm with adaptive learning rate (code 3). The inputs of this function are:

(a) The data set (X, y), where the ith column of the array matrix X is the data vector and the ith element of the row vector y contains the class label (-1 or $+1$), indicating the corresponding class to which the ith data vector belongs.

(b) The number of first layer nodes.

(c) The code number of the training method to be adopted.

(d) The number of iterations for which the algorithm will run.

(e) A parameter vector that contains the values of the parameters required for the adopted training method. This has the form

$$[lr,\ mc,\ lr\_inc,\ lr\_dec,\ max\_perf\_inc]$$

where lr is the learning rate, mc is the momentum parameter and the remaining three parameters, which are used in the backpropagation algorithm with variable learning rate, correspond to r_i, r_d, and c, as defined in Section 4.7, respectively. For the standard backpropagation algorithm, the last four components of the parameter vector are 0, for the momentum variant the last three parameters are 0, while for the adaptive learning rate case only the second component is 0.

The output of the network is the object *net* that corresponds to the trained neural network. To make the results reproducible for comparison purposes, ensure that every time this function is called it begins from the same initial condition.

Solution

```
function net=NN_training(x,y,k,code,iter,par_vec)
  rand('seed',0) % Initialization of the random number
  % generators
```

```
randn('seed',0) % for reproducibility of net initial
% conditions
% List of training methods
methods_list = {'traingd'; 'traingdm'; 'traingda'};
% Limits of the region where data lie
limit = [min(x(:,1)) max(x(:,1)); min(x(:,2)) max(x(:,2))];
% Neural network definition
net = newff(limit,[k 1],{'tansig','tansig'},...
methods_list{code,1});
% Neural network initialization
net = init(net);
% Setting parameters
net.trainParam.epochs = iter;
net.trainParam.lr=par_vec(1);
if(code == 2)
  net.trainParam.mc=par_vec(2);
elseif(code == 3)
  net.trainParam.lr_inc = par_vec(3);
  net.trainParam.lr_dec = par_vec(4);
  net.trainParam.max_perf_inc = par_vec(5);
end
% Neural network training
net = train(net,x,y);
%NOTE: During training, the MATLAB shows a plot of the
% MSE vs the number of iterations.
```

4.3 Write a MATLAB function, named *NN_evaluation*, which takes as inputs: (a) a neural network object and (b) a data set (X, y) and returns the probability of error that this neural network gives when it runs over this data set. The two-class case (-1 and $+1$) is considered.

Solution

```
function pe = NN_evaluation(net,x,y)
  y1 = sim(net,x); %Computation of the network outputs
  pe=sum(y.*y1<0)/length(y);
```

4.4 Write a MATLAB function, named *plot_dec_regions*, that plots the decision regions produced by a neural network trained on data sets produced by the *data_generator* function. The inputs to this function are: (a) a neural network object, (b) the lower and the upper bounds in the horizontal and vertical directions (*lh*, *uh*, *lv*, *uv*, respectively) of the region of space where the decision regions will be plotted as well as the resolution parameters (*rh*, *rv*) in both directions (the lower their value the finer the plot) and (c) the matrix *m* with the mean vectors of the normal distributions. Its output is a decision region plot, where each region is marked with a red star or with a blue circle according to whether it belongs to class +1 or −1, respectively.

Solution

For an implementation of this function see in the book website www.elsevierdirect.com/9781597492720.

4.5 Write a MATLAB function, named *SVM_clas*, which (i) generates an SVM classifier that uses RBF kernels, utilizing a given training set and (ii) measures its performance on both the training set and a given test set. The function takes as inputs (a) a training set $(X1, y1)$, where each column of $X1$ is a data vector and each element of the row vector $y1$ indicates the category of the corresponding data vector in $X1$ (−1 or +1 in this case), (b) a similarly defined test set $(X2, y2)$, (c) a tolerance parameter that is used in the termination condition of the SVM training procedure, (d) the parameter C in the SVM cost function (Section 3.7.2), (e) the parameter *sigma* for the RBF kernels. The function returns as outputs (a) the structure of the SVM classifier in the *SVMstruct* object, (b) the indices of the support vectors in the *svIndex* vector, (c) the error probability for the training set *pe_tr*, and (d) the error probability of the test set *pe_te*.

Solution

```
function [SVMstruct,svIndex,pe_tr,pe_te]=...
  SVM_clas(X1,y1,X2,y2,tol,C,sigma)
  options = svmsmoset('TolKKT',tol,'Display','iter',...
  'MaxIter',20000,'KernelCacheLimit',10000);
  %Training and Ploting parameters
  [SVMstruct,svIndex]=svmtrain(X1', y1','...
  KERNEL_FUNCTION','rbf',...
  'RBF_SIGMA',sigma,'BOXCONSTRAINT',C,'showplot',true,...
  'Method','SMO','SMO_Opts',options);
  %Computation of the error probability
  train_res=svmclassify(SVMstruct,X1');
  pe_tr=sum(y1'~=train_res)/length(y1);
  test_res=svmclassify(SVMstruct,X2');
  pe_te=sum(y2'~=test_res)/length(y2);
```

Computer Experiments

4.1 (a) After initializing the seed of the *randn* function of MATLAB to, say, 0 (*randn*('*seed*', 0)), use the *data_generator* function to create the data set $(X1, y1)$, with $m = \begin{bmatrix} -5 & +5 & +5 & -5 \\ +5 & -5 & +5 & -5 \end{bmatrix}$, $s = 2$ and $N = 100$.

(b) Initialize the seed of *randn* to, say, 10 and repeat (a) to produce the data set $(X2, y2)$.

(c) Repeat the above two steps using the corresponding seeds for *randn* (although this is not necessary), for $s = 5$, and produce the $(X3, y3)$ and $(X4, y4)$ data sets, respectively (m and N are as before).

(d) Plot the data sets.

4.2 (a) Run the standard backpropagation algorithm with $lr = 0.01$ and 2, 4, and 15 first layer nodes, for 1000 iterations, using the data set $(X1, y1)$ as training set.

(b) Evaluate the performance of the designed neural networks for both $(X1, y1)$ (training set) and $(X2, y2)$ (test set) and plot the decision regions (use $lh = lv = -10, uh = uv = 10, rh = rv = 0.2$).

(c) Comment on the results.

4.3 (a) Run the backpropagation algorithm with 4 first layer nodes with (i) $lr = 0.01$ for 300 iterations, (ii) $lr = 0.001$, for 300 iterations, (iii) $lr = 0.01$, for 1000 iterations and (iv) $lr = 0.001$, for 1000 iterations, using the $(X1, y1)$ as training set.

(b) Evaluate the performance of the designed neural networks for both $(X1, y1)$ (training set) and $(X2, y2)$ (test set) and plot the decision regions.

(c) Comment on the results.

4.4 (a) Run the adaptive learning rate variation of the backpropagation algorithm with $lr = 0.001, lr\_inc = 1.05, lr\_dec = 0.7, max\_perf\_inc = 1.04$, for 300 iterations.

(b) Evaluate the performance of the designed neural networks for both $(X1, y1)$ (training set) and $(X2, y2)$ (test set) and plot the decision regions.

(c) Compare the above results with whose obtained for the standard backpropagation algorithm with $lr = 0.001$, for 300 iterations.

4.5 Repeat 4.2–4.4 using the $(X3, y3)$ and $(X4, y4)$ as training and test sets, respectively.

4.6 (a) Generate SVM classifiers using $(X1, y1)$ and $(X2, y2)$ defined in 4.1 as training and test sets, respectively, for $C = 1, 100, 1000$, $sigma = 0.5, 1, 2, 4$, and tolerance parameter equal to 0.001.

(b) For each case, evaluate the respective classifier' s performance on the training and the test sets.

(c) Comment on the number of support vectors produced in each case.

4.7 Repeat 4.6 where now $(X1, y1)$, $(X2, y2)$ are replaced by $(X3, y3)$ $(X4, y4)$, respectively.

REFERENCES

[Adal 97] Adali T., Liu X., Sonmez K. "Conditional distribution learning with neural networks and its application to channel equalization," *IEEE Transactions on Signal Processing*, Vol. 45(4), pp. 1051-1064, 1997.

[Akse 06] Aksela M., Laaksonen J. "Using diversity of errors for selecting members of a committee classifier," *Pattern Recognition*, Vol. 39, pp. 608-623, 2006.

[Amar 99] Amari S., Wu S. "Improving support vector machines by modifying kernel functions," *Neural Networks*, Vol. 12(6), pp. 783-789, 1999.

[Ampa 02] Ampazis N., Perantonis S.J. "Two highly efficient second order algorithms for feed-forward networks," *IEEE Transactions on Neural Networks*, Vol. 13(5), pp. 1064-1074, 2002.

[Angu 03] Anguita D., Ridella S., Rivieccio F., Zunino R. "Hyperparameter design criteria for support vector classifiers," *Neurocomputing*, Vol. 55, pp. 109-134, 2003.

[Avni 99] Avnimelech R., Intrator N. "Boosted mixture of experts: An ensemble learning scheme," *Neural Computation*, Vol. 11, pp. 475-490, 1999.

[Barn 92] Barnard E. "Optimization for training neural networks," *IEEE Transactions on Neural Networks*, Vol. 3(2), pp. 232-240, 1992.

[Barr 93] Barron A.R. "Universal approximation bounds for superposition of a sigmoid function," *IEEE Transactions on Information Theory*, Vol. 39(3), pp. 930-945, 1993.

[Bart 02] Bartlett P., Boucherou S., Lugosi G. "Model selection and error estimation," *Machine Learning*, Vol. 48, pp. 85-113, 2002.

[Bart 98] Bartlett P.L. "The sample complexity of pattern classification with neural networks: The size of the weights is more important than the size of the network," *IEEE Transactions on Information Theory*, Vol. 44(2), pp. 525-536, 1998.

[Batt 92] Battiti R. "First and second order methods for learning: Between steepest descent and Newton's method," *Neural Computation*, Vol. 4, pp. 141-166, 1992.

[Baue 99] Bauer E., Kohavi R. "An empirical comparison of voting classification algorithms: Bagging, boosting, and variants," *Machine Learning*, Vol. 36, pp. 105-139, 1999.

[Baum 88] Baum E.B., Wilczek F. "Supervised learning of probability distributions by neural networks," in *Neural Information Processing Systems* (Anderson D., ed.), pp. 52-61, American Institute of Physics, New York, 1988.

[Ben 99] Ben-Yakoub S., Abdeljaoued, Mayoraj E. "Fusion of face and speech data for person identity verification," *IEEE Transactions on Neural Networks*, Vol. 10(5), pp. 1065-1075, 1999.

[Berg 93] Berg S. "Condorcet's jury theorem, dependency among jurors," *Social Choice Welfare*, Vol. 10, pp. 87-95, 1993.

[Bish 95] Bishop C.M. *Neural Networks for Pattern Recognition*, Oxford University Press, 1995.

[Blan 96] Blanz V., Schölkopf B., Bülthoff H., Burges C., Vapnik V., Vetter T. "Comparison of view based object recognition using realistic 3D models," *Proceedings of International Conference on Artificial Neural Networks*, pp. 251-256, Berlin, 1996.

[Bola 89] Boland P.J. "Majority systems and the Condorcet jury theorem," *Statistician*, Vol. 38, pp. 181–189, 1989.

[Bose 96] Bose N.K., Liang P. *Neural Network Fundamentals with Graphs, Algorithms and Applications*, McGraw-Hill, 1996.

[Brei 84] Breiman L., Friedman J., Olshen R., Stone C. *Classification and Regression Trees*, Wadsworth International, pp. 226–239, Belmont, CA, 1984.

[Brei 96] Breiman L. "Bagging predictors," *Machine Learning*, Vol. 24, pp. 123–140, 1996.

[Brei 98] Breiman L. "Arcing classifiers," *The Annals of Statistics*, Vol. 26(3), pp. 801–849, 1998.

[Brei 01] Breiman L. "Random Forests," *Machine Learning*, Vol. 45, pp. 5–32, 2001.

[Brid 90] Bridle J.S. "Training stochastic model recognition algorithms as networks can lead to maximum mutual information estimation parameters," in *Neural Information Processing Systems 2* (Touretzky D.S., ed.), pp. 211–217, Morgan Kaufmann, 1990.

[Broo 88] Broomhead D.S., Lowe D. "Multivariable functional interpolation and adaptive networks," *Complex Systems*, Vol. 2, pp. 321–355, 1988.

[Brow 93] Brown D., Corrnble V., Pittard C.L. "A comparison of decision tree classifiers with backpropagation neural networks for multimodal classification problems," *Pattern Recognition*, Vol. 26(6), pp. 953–961, 1993.

[Burg 99] Burges C.J.C. "Geometry and invariance in kernel based methods," in *Advances in Kernel Methods: Support Vector Learning* (Schólkopf B., Burges C.J.C., Smola A.J., eds.), MIT Press, 1999.

[Chap 02] Chapelle O., Vapnik V., Bousquet O., Mukherjee S. "Choosing multiple parameters for support vector machines," *Machine Learning*, Vol. 46, pp. 131–159, 2002.

[Chau 90] Chauvin Y. "Generalization performance of overtrained backpropagation networks," in *Neural Networks, Proc. EURASIP Workshop* (Almeida L.B., Wellekens C.J., eds.), pp. 46–55, 1990.

[Chaw 02] Chawla N.V., Bowyer K., Hall L., Kelgemeyer W.P. "SMOTE: Synthetic minority oversampling technique," *Journal of Artificial Intelligence Research*, Vol. 16, pp. 321–357, 2002.

[Chen 91] Chen S., Cowan C.F.N., Grant P.M. "Orthogonal least squares learning algorithm for radial basis function networks," *IEEE Transactions on Neural Networks*, Vol. 2, pp. 302–309, 1991.

[Chen 94] Cheng B., Titterington D.M. "Neural networks: A review from a statistical perspective," *Statistical Science*, Vol. 9(1), pp. 2–30, 1994.

[Chou 91] Chou P. "Optimal partitioning for classification and regression trees," *IEEE Transactions on Pattern Analysis and Machine Intelligence*, Vol. 13, pp. 340–354, 1991.

[Cich 93] Cichocki A., Unbehauen R. *Neural Networks for Optimization and Signal Processing*, John Wiley & Sons, 1993.

[Cort 95] Cortes C., Vapnik V.N. "Support vector networks," *Machine Learning*, Vol. 20, pp. 273–297, 1995.

[Cour 53] Courant R., Hilbert D. *Methods of Mathematical Physics*, Interscience, 1953.

[Cove 65] Cover T.M. "Geometrical and statistical properties of systems of linear inequalities with applications in pattern recognition," *IEEE Transactions on Electronic Computers*, Vol. 14, pp. 326–334, 1965.

[Cybe 89] Cybenko G. "Approximation by superpositions of a sigmoidal function," *Mathematics of Control, Signals and Systems*, Vol. 2, pp. 304–314, 1989.

[Czyz 04] Czyz J., Kittler J., Vandendorpe L. "Multiple classifier combination for face-based identity verification," *Pattern Recognition*, Vol. 37, pp. 1459–1469, 2004.

[Dark 91] Darken C., Moody J. "Towards faster stochastic gradient search," *Advances in Neural Information Processing Systems 4*, pp. 1009–1016, Morgan Kaufmann, San Mateo, CA, 1991.

[Datt 85] Dattatreya G.R., Kanal L.N. "Decision trees in pattern recognition," in *Progress in Pattern Recognition 2* (Kanal L.N., Rosenfeld A., eds.), North Holland, 1985.

[Delo 94] Delopoulos A., Tirakis A., Kollias S. "Invariant image classification using triple correlation based neural networks," *IEEE Transactions on Neural Networks*, Vol. 5, pp. 392–408, 1994.

[Diet 95] Dietterich T.G., Bakiri G. "Solving multiclass learning problems via error-correcting output codes," *Journal of Artificial Intelligence Research*, Vol. 2, pp. 263–286, 1995.

[Druc 99] Drucker H., Wu D., Vapnik, V.N. "Support vector machines for spam categorization," *IEEE Transactions on Neural Networks*, Vol. 10(5), pp. 1048–1055, 1999.

[Duan 03] Duan K., Keerthi S.S., Poo A.N. "Evaluation of simple performance measures for tuning SVM hyperparameters," *Neurocomputing*, Vol. 51, pp. 41–59, 2003.

[Dzer 04] Džeroski S., Ženko B. "Is combining classifiers with stacking better than selecting the best one?" *Machine Learning*, Vol. 54, pp. 255–273, 2004.

[Eibl 06] Eibl G., Pfeifer K.P. "Multiclass boosting for weak classifiers," *Journal of Machine Learning Research*, Vol. 6, pp. 189–210, 2006.

[ElNa 02] El-Naga I., Yang Y., Wernick M.N., Galatsanos N. "A support vector machine approach for detection of microcalcifications," *IEEE Transactions on Medical Imaging*, Vol. 21, pp. 1552–1563, 2002.

[Enge 04] Engel Y., Mannor S. "The kernel least-squares algorithm," *IEEE Transactions on Signal Processing*, Vol. 52(8), pp. 2275–2285, 2004.

[Espo 97] Esposito F., Malebra D., Semeraro G. "A comparative analysis of methods for pruning decision trees," *IEEE Transactions on Pattern Analysis and Machine Intelligence*, Vol. 19(5), pp. 476–491, 1997.

[Evge 04] Evgeniou T., Pontil M., Elisseeff A. "Leave one out error, stability, and generalization of voting combinations of classifiers," *Machine Learning*, Vol. 55, pp. 71–97, 2004.

[Fahl 90] Fahlman S.E., Lebiere C. "The cascade-correlation learning architecture," in *Advances in Neural Information Processing Systems 2* (Touretzky D.S., ed.), pp. 524–532, Morgan Kaufmann, 1990.

[Flao 06] Flaounas I.N., Iakovidis D.K., Marvoulis D.E. "Cascading SVMs as a tool for medical diagnosis using multi-class gene expression data," International Journal of Artificial Intelligence Tools.

[Frea 90] Frean M. "The Upstart algorithm: A method for constructing and training feedforward networks," *Neural Computation*, Vol. 2(2), pp. 198–209, 1990.

[Freu 97] Freund Y., Schapire R.E. "A decision theoretic generalization of on-line learning and an application to boosting," *Journal of Computer and System Sciences*, Vol. 55(1), pp. 119–139, 1997.

[Frie 00] Friedman J., Hastie T., Tibshirani R. "Additive logistic regression: A statistical view of boosting," *The Annals of Statistics*, Vol. 28(2), pp. 337–407, 2000.

[Fried 81] Friedman J., Stuetzle W. "Projection pursuit regression," *Journal of American Statistical Association*, Vol. 76, pp. 817–823, 1981.

[Fuku 92] Fukumi M., Omatu S., Takeda F., Kosaka T. "Rotational invariant neural pattern recognition system with application to coin recognition," *IEEE Transactions on Neural Networks*, Vol. 3, pp. 272–279, 1992.

[Fuku 82] Fukushima K., Miyake S., "Neogognitron: A new algorithm for pattern recognition tolerant of deformations and shifts in position," *Pattern Recognition*, Vol. 15(6), pp. 445–469, 1982.

[Fume 05] Fumera G., Roli F. "A theoretical and experimental analysis of linear combiners for multiple classifier systems," *IEEE Transactions on Pattern Analysis and Machine Intelligence*, Vol. 27(6), pp. 942–956, 2005.

[Funa 89] Funahashi K. "On the approximate realization of continuous mappings by neural networks," *Neural Networks*, Vol. 2(3), pp. 183–192, 1989.

[Gall 90] Gallant S.I. "Perceptron-based learning algorithms," *IEEE Transactions on Neural Networks*, Vol. 1, pp. 179–191, 1990.

[Garg 02] Garg A., Pavlovic V., Huang T.S. "Bayesian networks as enseble of classifiers," *Proceedings 16th International Conference on Pattern Recognition*, Vol. 2, pp. 779–784, 2002.

[Geor 06] Georgion H., Mavroforakis M., Theodoridis S. "A game theoretic approach to weighted majority voting for combining SVM classifiers," in Proceedings of the International Conference on Artificial Neural Networks, pp. 284–292, LNCS 4131, Springer, 2006.

[Germ 92] German S., Vienenstock E., Doursat R. "Neural networks and bias/variance dilemma," *Neural Computation*, Vol. 4(1), pp. 1–58, 1992.

[Gibs 90] Gibson G.J., Cowan C.F.N. "On the decision regions of multilayer perceptrons," *Proceedings of the IEEE*, Vol. 78(10), pp. 1590–1594, 1990.

[Gish 90] Gish H. "A probabilistic approach to the understanding and training neural classifiers," in *Proceedings of the IEEE Conference on Acoustics Speech and Signal Processing*, pp. 1361–1364, April 1990.

[Gomm 00] Gomm J.B., Yu D.L. "Selecting radial basis function network centers with recursive orthogonal least squares training," *IEEE Transactions on Neural Networks*, Vol. 11(2), pp. 306–314, 2000.

[Graj 86] Grajki K.A. et al. "Classification of EEG spatial patterns with a tree structured methodology," *IEEE Transactions on Biomedical Engineering*, Vol. 33(12), pp. 1076–1086, 1986.

[Gran 04] Grandvalet Y. "Bagging equalizes influence," *Machine Learning*, Vol. 55, pp. 251–270, 2004.

[Hamp 90] Hampshire J.B. II, Perlmutter B.A. "Equivalence proofs for multilayer perceptron classifiers and the Bayesian discriminant function," in *Proceedings of the 1990 Connectionist Models Summer School* (Touretzky D, et al., eds.), Morgan Kaufmann, 1990.

[Hart 90] Hartman E.J., Keeler J.D., Kowalski J.M. "Layered neural networks with Gaussian hidden units as universal approximations," *Neural Computations*, Vol. 2(2), pp. 210–215, 1990.

[Hass 93] Hassibi B., Stork D.G., Wolff G.J. "Optimal brain surgeon and general network pruning," *Proceedings IEEE Conference on Neural Networks*, Vol. 1, pp. 293–299, San Francisco, 1993.

[Hast 01] Hastie T., Tibshirani R., Friedman J. *The Elements of Statistical Learning*, Springer, 2001.

[Hayk 96] Haykin S. "Neural networks expand SP's horizons," *IEEE Signal Processing Magazine*, Vol. 13(2), pp. 24–49, 1996.

[Hayk 99] Haykin S. *Neural Networks: A Comprehensive Foundation*, 2nd ed., Prentice Hall, 1999.

[Hert 91] Hertz J., Krogh A., Palmer R.G. *Introduction to the Theory of Neural Computation*, Addison-Wesley, 1991.

[Hint 90] Hinton G.E. "Connectionist learning procedures," in *Machine Learning: Paradigms and Methods* (Carbonell J.G., ed.), pp. 185–234, MIT Press, 1990.

[Ho 98] Ho T.K. "The random subspace method for constructing decision forests," *IEEE Transactions on Pattern Analysis and Machine Intelligence*, Vol. 20(8), pp. 832–844, 1998.

[Ho 94] Ho T.K., Hull J.J., Srihari S.N. "Decision combination in multiple classifier design," *IEEE Transactions on Pattern Analysis and Machine Intelligence*, Vol. 16(1), pp. 66–75, 1994.

[Holm 97] Holmstrom L., Koistinen P., Laaksonen J., Oja E. "Neural and statistical classifiers—taxonomy and two case studies," *IEEE Transactions on Neural Networks*, Vol. 8(1), pp. 5–17, 1997.

[Horn 89] Hornik K., Stinchcombe M., White H. "Multilayer feedforward networks are universal approximators," *Neural Networks*, Vol. 2(5), pp. 359–366, 1989.

[Hu 08] Hu R., Damper R.I. "A no panacea theorem for classifier combination," Pattern Recognition, Vol. 41, pp. 2665–2673, 2008.

[Huan 04] Huang D-S., Ip H., Chi Z. "A neural root finder of polynomials based on root moments," *Neural Computation*, Vol. 16, pp. 1721–1762, 2004.

[Hush 93] Hush D.R., Horne B.G. "Progress in supervised neural networks," *IEEE Signal Processing Magazine*, Vol. 10(1), pp. 8–39, 1993.

[Ito 91] Ito Y. "Representation of functions by superpositions of a step or sigmoid function and their applications to neural network theory," *Neural Networks*, Vol. 4(3), pp. 385–394, 1991.

[Jaco 88] Jacobs R.A. "Increased rates of convergence through learning rate of adaptation," *Neural Networks*, Vol. 2, pp. 359–366, 1988.

[Jaco 91] Jacobs R.A., Jordan M.I., Nowlan S.J., Hinton G.E. "Adaptive mixtures of local experts," *Neural Computation*, Vol. 3, pp. 79–87, 1991.

[Jain 00] Jain A.K., Duin P.W., Mao J. "Statistical pattern recognition: A review," *IEEE Transactions on Pattern Analysis and Machine Intelligence*, Vol. 22(1), pp. 4–37, 2000.

[Japk 02] Japkowicz N., Stephen S. "The class imbalance problem: a systematic study," *Intelligent Data Analysis Journal*, Vol. 6(5), pp. 429–450, 2002.

[Joha 92] Johansson E.M., Dowla F.U., Goodman D.M. "Backpropagation learning for multilayer feedforward neural networks using conjugate gradient method," *International Journal of Neural Systems*, Vol. 2(4), pp. 291–301, 1992.

[Josh 97] Joshi A., Ramakrishman N., Houstis E.N., Rice J.R. "On neurobiological, neuro-fuzzy, machine learning, and statistical pattern recognition techniques," *IEEE Transactions on Neural Networks*, Vol. 8(1), pp. 18–31, 1997.

[Juan 92] Juang B.H., Katagiri S. "Discriminative learning for minimum error classification," *IEEE Transactions on Signal Processing*, Vol. 40(12), pp. 3043–3054, 1992.

[Kalo 97] Kalouptsidis N. *Signal Processing Systems, Theory and Design*, John Wiley & Sons, 1997.

[Kana 92] Kanaoka T., Chellapa R., Yoshitaka M., Tomita S. "A higher order neural network for distortion invariant pattern recognition," *Pattern Recognition Letters*, Vol. 13, pp. 837–841, 1992.

[Kang 00] Kang H., Lee S. "An information-theoretic strategy for constructing multiple classifier systems," *Proceedings of the 15th International Conference on Pattern Recognition (ICPR)*, Vol. 2, pp. 483–486, 2000.

[Kang 03] Kang H.J. "Combining multiple classifiers based on third-order dependency for handwritten numeral recognition," *Pattern Recognition Letters*, Vol. 24, pp. 3027-3036, 2003.

[Kara 92] Karayiannis N.B., Venetsanopoulos A.N. "Fast learning algorithm for neural networks," *IEEE Transactions on Circuits and Systems*, Vol. 39, pp. 453–474, 1992.

[Kara 97] Karayiannis N.B., Mi G.W. "Growing radial basis neural networks. Merging supervised and unsupervised learning with network growth techniques," *IEEE Transactions on Neural Networks*, Vol. 8(6), pp. 1492–1506, 1997.

[Kear 94] Kearns M., Valiant L.G. "Cryptographic limitations on learning Boolean formulae and finite automata," *Journal of the ACM*, Vol. 41(1), pp. 67–95, 1994.

[Keer 05] Keerthi S.S., Duan K.B., Shevade S.K., Poo A.N. "A fast dual algorithm for kernel logistic regression," *Machine Learning*, Vol. 61, pp. 151–165, 2005.

[Kime 71] Kimeldorf G.S., Wahba G. "Some results on Tchebycheffian spline functions," *Journal of Mathematical Analysis and Applications*, Vol. 33, pp. 88–95, 1971.

[Kitt 03] Kittler J., Alkoot F.M. "Sum versus vote fusion in multiple classifiers," *IEEE Transactions on Pattern Analysis and Machine Intelligence*, Vol. 25(1), pp. 110–115, 2003.

[Kitt 98] Kittler J., Hatef M., Duin R. Matas J. "On combining classifiers," *IEEE Transactions on Pattern Analysis and Machine Intelligence*, Vol. 20(3), pp. 226–234, 1998.

[Kivi 04] Kivinen J.K., Smola A.L., Williamson R.C. "Online learning with kernels," *IEEE Transactions on Signal Processing*, Vol. 52(8), pp. 2165–2176, 2004.

[Klei 90] Kleinberg R.M. "Stochastic discrimination," *Annals of Mathematics and Artificial Intelligence*, Vol. 1, pp. 207–239, 1990.

[Koli 97] Kolinummi P., Hamalainen T., Kaski K. "Designing a digital neurocomputer," *IEEE Circuits and Systems Magazine*, Vol. 13(2), pp. 19–27, 1997.

[Kout 94] Koutroumbas K., Kalouptsidis N. "Nearest neighbor pattern classification neural networks," *Proc. of IEEE World Congress of Computational Intelligence*, pp. 2911–2915, Orlando, FL, July 1994.

[Kram 89] Kramer A.H., Sangiovanni-Vincentelli A. "Efficient parallel learning algorithms for neural networks," in *Advances in Neural Information Processing Systems 1* (Touretzky D.S., ed.), pp. 40–48, Morgan Kaufmann, 1989.

[Kunc 01] Kuncheva L.I., Bezdek J.C., Duin R.P.W. "Decision templates for multiple classifier fusion: an experimental comparison," *Pattern Recognition*, Vol. 34, pp. 299–314, 2001.

[Kunc 02] Kunchera L.I. "A theoretical study on six classifier fusion strategies," *IEEE Transactions on Pattern Analysis and Machine Intelligence*, Vol. 24(2), pp. 281–286, 2002.

[Kunc 03] Kuncheva L.I., Whitaker C.J., Shipp C.A., Duin R.P.W. "Limits on the majority vote accuracy in classifier fusion," *Pattern Analysis and Applications*, Vol. 6, pp. 22–31, 2003.

[Kunc 03a] Kuncheva L.I., Whitaker C.J. "Measures of diversity in classifier ensembles," *Machine Learning*, Vol. 51, pp. 181–207, 2003.

[Lam 97] Lam L., Suen Y. "Application of majority voting to pattern recognition: An analysis of its behaviour and performance," *IEEE Transactions on Systems, Man, and Cybernetics* Vol. 27(5), pp. 553–568, 1997.

[Lane 91] Lane S.H., Flax M.G., Handelman D.A., Gelfand J.J. "Multilayer perceptrons with B-spline receptive field functions," in *Advances in Neural Information Processing Systems 3* (Lippmann R.P., Moody J., Touretzky D.S., eds.), pp. 684–692, Morgan Kaufmann, 1991.

[Lecu 89] Le Cun Y., Boser B., Denker J.S., Henderson D., Howard R.E., Hubbard W., Jackel L.D. "Backpropagation applied to handwritten zip code recognition," *Neural Computation*, Vol. 1(4), pp. 541–551, 1989.

[Lecu 90] Le Cun Y., Denker J.S., Solla S.A. "Optimal brain damage," in *Advances in Neural Information Systems 2* (Touretzky D.S., ed.), pp. 598–605, Morgan Kaufmann, 1990.

[Lee 04] Lee M.S., Keerthi S.S., Ong C.J. "An efficient method for computing Leave-One-Out error in support vector machines with Gaussian kernels," *IEEE Transactions on Neural Networks*, Vol. 15(3), pp. 750–757, 2004.

[Levi 02] Levitin G. "Evaluating correct classification probability for weighted voting classifiers with plurality voting," *European Journal of Operational Research*, Vol. 141, pp. 596–607, 2002.

[Lin 03] Lin X., Yacoub S., Burns J., Simske S. "Performance analysis of pattern classifier combination by plurality voting," *Pattern Recognition Letters*, Vol. 24, pp. 1959–1969, 2003.

[Lin 02] Lin Y., Wahba G., Zhang H., Lee Y. "Statistical properties and adaptive tuning of support vector machines," *Machine Learning*, Vol. 48, pp. 115–136, 2002.

[Lin 02a] Lin Y., Lee Y., Wahba G. "Support vector machines for classification in nonstandard situations," *Machine Learning*, Vol. 46, pp. 191–202, 2002.

[Lipp 87] Lippmann R.P. "An introduction to computing with neural networks," *IEEE ASSP Magazine*, Vol. 4(2), pp. 4–22, 1987.

[Mack 92a] MacKay D.J.C. "A practical Bayesian framework for backpropagation networks," *Neural Computation*, Vol. 4(3), pp. 448–472, 1992.

[Mack 92b] MacKay D.J.C. "The evidence framework applied to classification networks," *Neural Computation*, Vol. 4(5), pp. 720–736, 1992.

[Maso 00] Mason L., Baxter J., Bartleet P., Frean M. "Boosting algorithms as gradient descent," in *Neural Information Processing Systems*, Vol. 12, 2000.

[Mavr 07] Mavroforakis M., Sdralis M., Theodoridis S. "A geometric nearest point algorithm for the efficient solution of the SVM classification task," *IEEE Transactions on Neural Networks*, Vol. 18(5), pp. 1545–1550, 2007.

[Meye 03] Meyer D., Leisch F., Hornik K. "The support vector machine under test," *Neurocomputing*, Vol. 55, pp. 169–186, 2003.

[Meza 89] Mezard M., Nadal J.P. "Learning in feedforward layered networks: The tilling algorithm," *Journal of Physics*, Vol. A 22, pp. 2191–2203, 1989.

[Mich 94] Michie D., Spiegelhalter D.J., Taylor C.C., eds. *Machine Learning, Neural, and Statistical Classification*, Ellis Horwood Ltd., London, 1994.

[Mill 96] Miller D., Rao A., Rose K., Gersho A. "A global optimization technique for statistical classifier design," *IEEE Transactions on Signal Processing*, Vol. 44(12), pp. 3108–3122, 1996.

[Mill 99] Miller D.J., Yan L. "Critic-driven ensemble classification," *IEEE Transactions on Pattern Analysis and Machine Intelligence*, Vol. 47(10), pp. 2833–2844, 1999.

[Mirc 89] Mirchandini G., Cao W. "On hidden nodes in neural nets," *IEEE Transactions on Circuits and Systems*, Vol. 36(5), pp. 661–664, 1989.

[Mood 89] Moody J., Darken C.J. "Fast learning in networks of locally tuned processing units," *Neural Computation*, Vol. 6(4), pp. 281–294, 1989.

[Mulg 96] Mulgrew B. "Applying radial basis functions," *IEEE Signal Processing Magazine*, Vol. 13(2), pp. 50–65, 1996.

[Mull 01] Müller R.M., Mika S., Rätsch G., Tsuda K., Schölkopf B. "An introduction to kernel-based learning algorithms," *IEEE Transactions on Neural Networks*, Vol. 12(2), pp. 181–201, 2001.

[Murp 90] Murphy O.J. "Nearest neighbor pattern classification perceptrons," *Proceedings of the IEEE*, Vol. 78(10), October 1990.

[Nara 05] Narasimhamurthy A. "Theoretical bounds of majority voting performance for a binary classification problem," *IEEE Transactions on Pattern Analysis and Machine Intelligence*, Vol. 27(12), pp. 1988–1995, 2005.

[Nede 93] Nedeljkovic V. "A novel multilayer neural networks training algorithm that minimizes the probability of classification error," *IEEE Transactions on Neural Networks*, Vol. 4(4), pp. 650–659, 1993.

[Pado 95] Pados D.A., Papantoni-Kazakos P. "New non least squares neural network learning algorithms for hypothesis testing," *IEEE Transactions on Neural Networks*, Vol. 6, pp. 596–609, 1995.

[Palm 91] Palmieri F., Datum M., Shah A., Moiseff A. "Sound localization with a neural network trained with the multiple extended Kalman algorithm," *International Joint Conference on Neural Networks*, Vol. 1, pp. 125–131, Seattle, 1991.

[Pare 00] Parekh R., Yang J., Honavar V. "Constructive neural network learning algorithms for pattern classification," *IEEE Transactions on Neural Networks*, Vol. 11(2), pp. 436–451, 2000.

[Park 91] Park J., Sandberg I.W. "Universal approximation using radial basis function networks," *Neural Computation*, Vol. 3(2), pp. 246–257, 1991.

[Park 93] Park J., Sandberg I.W. "Approximation and radial basis function networks," *Neural Computation*, Vol. 5(2), pp. 305–316, 1993.

[Park 94] Park Y. "A comparison of neural net classifiers and linear tree classifiers: Their similarities and differences," *Pattern Recognition*, Vol. 27(11), pp. 1493–1503, 1994.

[Pera 03] Perantonis S.J. "Neural networks: nonlinear optimization for constrained learning and its applications," *Proceedings of NOLASC 2003*, pp. 589–594, Athens, Greece, December 27–29, 2003.

[Pera 00] Perantonis S.J., Ampazis N., Virvilis V. "A learning framework for neural networks using constrained optimization methods," *Annals of Operations Research*, Vol. 99, pp. 385–401, 2000.

[Pera 95] Perantonis S.J., Karras D.A. "An efficient learning algorithm with momentum acceleration," *Neural Networks*, Vol. 8, pp. 237–249, 1995.

[Pera 92] Perantonis S.J., Lisboa P.J.G. "Translation, rotation, and scale invariant pattern recognition by high-order neural networks and moment classifiers," *IEEE Transactions on Neural Networks*, Vol. 3(2), pp. 241–251, 1992.

[Pikr 08] Pikrakis A., Ganakopoulos T., Theodoridis S. "A speech/music discriminator of radio recordings based on dynamic programming and Bayesian networks. IEEE Transactions mulitmedia, Vol. 10(5), pp. 846–856, 2008.

[Plat 91] Platt J. "A resource allocating network for function interpolation," *Neural Computation*, Vol. 3, pp. 213–225, 1991.

[Plat 99] Platt J. "Fast training of support vector machines using sequential minimal optimization," in *Advances in Kernel Methods: Support Vector Learning* (Scholkopf B., Burges C.J.C., Smola A.J. eds.), pp. 185–208, MIT Press, 1999.

[Quin 93] Quinlan J.R. *C4.5: Programs for Machine Learning*, Morgan Kaufmann, 1993.

[Rats 02] Rätsch G., Mika S., Schölkopf B., Müller K.R. "Constructing boosting algorithms from SVMS: An application to one class classification," *IEEE Transactions on Pattern Analysis and Machine Intelligence*, Vol. 24(9), pp. 1184–1199, 2002.

[Rats 05] Rätsch G., Warmuth M.K. "Efficient margin maximizing with boosting," *Journal of Machine Learning Research*, Vol. 6, pp. 2131–2152, 2005.

[Refe 91] Refenes A., Chen L. "Analysis of methods for optimal network construction," *University College London Report*, CC30/080:DCN, 1991.

[Reyn 99] Reyneri L. "Unification of neural and wavelet networks and fuzzy systems," *IEEE Transactions on Neural Networks*, Vol. 10(4), pp. 801–814, 1999.

[Rich 91] Richard M., Lippmann R.P. "Neural network classifiers estimate Bayesian a posteriori probabilities," *Neural Computation*, Vol. 3, pp. 461–483, 1991.

[Rico 88] Ricotti L.P., Ragazinni S., Martinelli G. "Learning of word stress in a suboptimal second order backpropagation neural network," in *Proceedings of the IEEE International Conference on Neural Networks*, Vol. 1, pp. 355–361, San Diego, 1988.

[Ride 97] Ridella S., Rovetta S., Zunino R. "Circular backpropagation networks for classification," *IEEE Transactions on Neural Networks*, Vol. 8(1), pp. 84–97, 1997.

[Ried 93] Riedmiller M., Brau H. "A direct adaptive method for faster backpropagation learning: The rprop algorithm," *Proceedings of the IEEE Conference on Neural Networks*, San Francisco, 1993.

[Ripl 94] Ripley B.D. "Neural networks and related methods for classification," *Journal of Royal Statistical Society*, Vol. B, 56(3), pp. 409–456, 1994.

[Ripl 96] Ripley B.D. *Pattern Recognition and Neural Networks*, Cambridge University Press, 1996.

[Rodr 06] Rodriguez J.J., Kuncheva L.I., Alonso C.J. "Rotation forests: A new classifier ensemble method," *IEEE Transactions on Pattern Analysis and Machine Intelligence*, Vol. 28(10), pp. 1619–1631, 2006.

[Rome 97] Romero R.D., Touretzky D.S., Thibadeau G.H. "Optical character recognition using probabilistic neural networks," *Pattern Recognition*, Vol. 3, pp. 1279–1292, 1997.

[Rose 96] Rosen B.E. "Ensemble learning using decorrelated neural networks," *Connections Science* Vol. 8(3), pp. 373–384, 1996.

[Rume 86] Rumelhart D.E., Hinton G.E., Williams R.J. "Learning internal representations by error propagation," *Parallel Distributed Processing: Explorations in the Microstructures of Cognition* (Rumelhart D.E., McClelland J.L., eds.), Vol. 1, pp. 318–362, MIT Press, 1986.

[Russ 93] Russell R. "Pruning algorithms. A survey," *IEEE Transactions on Neural Networks*, Vol. 4(5), pp. 740–747, 1993.

[Rutk 04] Rutkowski L. "Adaptive probabilistic neural networks in time-varying environments," *IEEE Transactions on Neural Networks*, Vol. 15, pp. 811–827, 2004.

[Saye 04] Sayed A. *Fundamentals of Adaptive Filtering*, John. Wiley & Sons, 2003.

[Scha 98] Schapire R.E., Freund V., Bartlett P., Lee W.S. "Boosting the margin: A new explanation for the effectiveness of voting methods," *The Annals of Statistics*, Vol. 26(5), pp. 1651–1686, 1998.

[Scha 05] Schapire R.E., Rochery M., Rahim M., Gupta N. "Boosting with prior knowledge for call classification," *IEEE Transactions on Speech and Audio Processing*, Vol. 13(2), pp. 174-181, 2005.

[Scha 99] Schapire R.E., Singer Y. "Improved boosting algorithms using confidence-rated predictions," *Machine Learning*, Vol. 37(3), pp. 297-336, 1999.

[Scho 97] Schölkopf B., Sung K.-K., Burges C.J.C., Girosi F., Niyogi P., Poggio T., Vapnic V. "Comparing support vector machines with Gaussian kernels to RBF classifiers," *IEEE Transactions on Signal Processing*, Vol. 45(11), pp. 2758-2766, 1997.

[Scho 02] Schölkoph B., Smola A.J. *Learning with Kernels*, MIT Press , 2002.

[Seba 00] Sebald D.J., Bucklew J.A. "Support vector machine techniques for nonlinear equalization," *IEEE Transactions on Signal Processing*, Vol. 48(11), pp. 3217-3227, 2000.

[Seth 90] Sethi I.K. "Entropy nets: From decision trees to neural networks," *Proceedings of the IEEE*, Vol. 78, pp. 1605-1613, 1990.

[Seth 91] Sethi I.K. "Decision tree performance enhancement using an artificial neural network interpretation," in *Artificial Neural Networks and Statistical Pattern Recognition* (Sethi I., Jain A., eds.), Elsevier Science Publishers, 1991.

[Shaw 04] Shawe-Taylor J., Cristianini N. *Kernel Methods for Pattern Analysis*, Cambridge University Press, 2004.

[Siet 91] Sietsma J., Dow R.J.F. "Creating artificial neural networks that generalize," *Neural Networks*, Vol. 4, pp. 67-79, 1991.

[Silv 90] Silva F.M., Almeida L.B. "Accelaration technique for the backpropagation algorithm," *Proceedings of the EURASIP Workshop on Neural Networks* (Almeida L.B. et al., eds.), pp. 110-119, Portugal, 1990.

[Sing 89] Singhal S., Wu L. "Training feedforward networks with the extended Kalman filter," *Proceedings of the IEEE International Conference on Acoustics Speech and Signal Processing*, pp. 1187-1190, Glasgow, 1989.

[Slav 08] Slavakis K., Theodoridis S., I. Yamada "Online Kernel-Based Classification and Adaptive Projection Algorithms," *IEEE Transactions on Signal Processing*, Vol. 56(7), pp. 2781-2797, 2008.

[Slav 08a] Slavakis K., Theodoridis S. "Sliding Window Generalized Kernel Affine Projection Algorithm using Projection Mappings," *EURASIP Journal on Advances on Signal Processing, JASP*, To appear 2008.

[Spec 90] Specht D. "Probabilistic neural networks," *Neural Networks*, Vol. 3, pp. 109-118, 1990.

[Stre 94] Streit R.L., Luginbuhl T.E. "Maximum likelihood training of probabilistic neural networks," *IEEE Transactions on Neural Networks*, Vol. 5, pp. 764-783, 1994.

[Spec 90] Specht D.F. "Probabilistic neural networks," *Neural Networks*, Vol. 3, pp. 109-118, 1990.

[Suar 99] Suarez A., Lutsko J.F. "Globally optimal fuzzy decision trees for classification and regression," *IEEE Transactions on Pattern Analysis and Machine Intelligence*, Vol. 21(12), pp. 1297-1311, 1999.

[Sun 07] Sun Y., Kamel M.S., Wong A.K.C., Wang T. "Cost-effective boosting for classification of imbalanced data," *Pattern Recognition*, Vol. 40, pp. 3358-3378, 2007.

[Tax 00] Tax D.M.J., Breukelen M., Duin R.P.W., Kittler J. "Combining multiple classifiers by averaging or by multiplying?" *Pattern Recognition*, Vol. 33, pp. 1475-1485, 2000.

[Theo 95] Theodoridis S., Cowan C.F.N., Callender C., Lee C.M.S. "Schemes for equalization in communication channels with nonlinear impairments," *IEE Proceedings on Communications*, Vol. 61(3), pp. 268–278, 1995.

[Tipp 01] Tipping M.E. "Sparse Bayesiay learning and the relevance vector machine," Journal of Machine Learning Research, Vol. 1, pp. 211–244, 2001.

[Tres 01] Tresp V. "Committee Machines," in *Handbook for Neural Network Signal Processing* (Hu Y.H., Hwang J.N., eds), CRC Press, 2001.

[Tume 95] Tumer K., Ghosh J. "Analysis of decision boundaries in linearly combined classifiers," *Pattern Recognition*, Vol. 29(2), pp. 341–348, 1995.

[Ueda 97] Ueda N. "Optimal linear combination of neural networks for improving classification performance," *IEEE Transactions on Pattern Analysis and Machine Intelligence*, Vol. 22(2), pp. 207–215, 2000.

[Vali 84] Valiant L.G. "A theory of the learnable," *Communications of the ACM*, Vol. 27(11), pp. 1134–1142, 1984.

[Vapn 00] Vapnik V.N. *The Nature of Statistical Learning Theory*, Springer Verlag, 2000.

[Viol 01] Viola P., Jones M. "Robust real-time object detection," *Proceedings IEEE Workshop on Statistical and Computational Theories of Vision*, Vancouver, Canada, 2001.

[Watr 88] Watrous R.L. "Learning algorithms for connectionist networks: Applied gradient methods of nonlinear optimization," in *Proceedings of the IEEE International Conference on Neural Networks*, Vol. 2, pp. 619–627, San Diego, 1988.

[Wein 90] Weigend A.S., Rumelhart D.E., Huberman B.A. "Backpropagation, weight elimination and time series prediction," in *Proceedings of the Connectionist Models Summer School* (Touretzky D., Elman J., Sejnowski T., Hinton G., eds.), pp. 105–116, 1990.

[Weis 03] Weiss G, Provost F. "Learning when training data are costly: the effect of class distribution on tree induction," *Journal of Artificial Intelligence Research*, Vol. 19, pp. 315–354, 2003.

[Werb 74] Werbos P.J. "Beyond regression: New tools for prediction and analysis in the behavioral sciences," Ph.D. Thesis, Harvard University, Cambridge, MA, 1974.

[Wett 92] Wettschereck D., Dietterich T. "Improving the performance of radial basis function networks by learning center locations," in *Advances in Neural Information Processing Systems*, 4th ed. (Moody J.E., Hanson S.J., Lippmann R.P., eds.), pp. 1133–1140, Morgan Kaufmann, 1992.

[Witt 00] Witten I., Frank E. *Data Mining: Practical Machine Learning Tools and Techniques with JAVA Implementations*, Morgan Kaufmann, 2000.

[Wolpe 92] Wolpet D.H. "Stacked generalization," *Neural Networks*, Vol. 5(2), pp. 241–260, 1992.

[Yang 06] Yang Z.R. "A novel radial basis function neural network for discriminant analysis," *IEEE Transactions on Neural Networks*, Vol. 17(3), pp. 604–612, 2006.

[Yin 05] Yin X.C., Liu C.P., Han Z. "Feature combination using boosting," *Pattern Recognition Letters*, Vol. 25(14), pp. 2195–2205, 2005.

[Ying 98] Yingwei L., Sundararajan N., Saratihandran P. "Performance evaluation of a sequential minimal RBF neural network learning algorithm," *IEEE Transactions on Neural Networks*, Vol. 9(2), pp. 308–318, 1998.

[Zhan 00] Zhang G.P. "Neural networks for classification: A survey," *IEEE Transactions on Systems Man and Cybernetics - Part C*, Vol. 30(4), pp. 451–462, 2000.

[Zhou 06] Zhou Z.H., Liu X.Y. "Training cost sensitive neural networks with methods addressing the class imbalance problem," *IEEE Transactions on Knowledge Data Engineering*, Vol. 18(1), pp. 63–77, 2006.

[Zura 92] Zurada J. *Introduction to Artificial Neural Networks*, West Publishing Company, St. Paul, MN., 1992.

CHAPTER

Feature Selection

5

5.1 INTRODUCTION

In all previous chapters, we considered the features that should be available prior to the design of the classifier. The goal of this chapter is to study methodologies related to the selection of these variables. As we pointed out very early in the book, a major problem associated with pattern recognition is the so-called curse of dimensionality (Section 2.5.6). The number of features at the disposal of the designer of a classification system is usually very large. As we will see in Chapter 7, this number can easily reach the order of a few dozens or even hundreds.

There is more than one reason to reduce the number of features to a sufficient minimum. Computational complexity is the obvious one. A related reason is that, although two features may carry good classification information when treated separately, there is little gain if they are combined into a feature vector because of a high mutual correlation. Thus, complexity increases without much gain. Another major reason is that imposed by the required generalization properties of the classifier, as discussed in Section 4.9 of Chapter 4. As we will state more formally at the end of this chapter, the higher the ratio of the number of training patterns N to the number of free classifier parameters, the better the generalization properties of the resulting classifier.

A large number of features are directly translated into a large number of classifier parameters (e.g., synaptic weights in a neural network, weights in a linear classifier). Thus, for a finite and usually limited number N of training patterns, keeping the number of features as small as possible is in line with our desire to design classifiers with good generalization capabilities. Furthermore, the ratio N/l enters the scene from another nearby corner. One important step in the design of a classification system is the performance evaluation stage, in which the classification error probability of the designed classifier is estimated. We not only need to design a classification system, but we must also assess its performance. As is pointed out in Chapter 10, the classification error estimate improves as this ratio becomes higher. In [Fine 83] it is pointed out that in some cases ratios as high as 10 to 20 were considered necessary.

The major task of this chapter can now be summarized as follows. *Given a number of features, how can one select the most important of them so as to reduce their number and at the same time retain as much as possible of their class discriminatory information?* The procedure is known as *feature selection* or *reduction*. This step is very crucial. If we selected features with little discrimination power, the subsequent design of a classifier would lead to poor performance. On the other hand, if information-rich features are selected, the design of the classifier can be greatly simplified. In a more quantitative description, we should aim to select features leading to *large between-class distance and small within-class variance* in the feature vector space. This means that features should take distant values in the different classes and closely located values in the same class. To this end, different scenarios will be adopted. One is to examine the features individually and discard those with little discriminatory capability. A better alternative is to examine them in combinations. Sometimes the application of a linear or nonlinear transformation to a feature vector may lead to a new one with better discriminatory properties. All these paths will be our touring directions in this chapter.

Finally, it must be pointed out that there is some confusion in the literature concerning the terminology of this stage. In some texts the term *feature extraction* is used, but we feel that this may be confused with the feature generation stage treated in Chapter 7. Others prefer to call it a *preprocessing stage*. We have kept the latter term to describe the processing performed on the features prior to their utilization. Such processing involves removing outliers, scaling of the features to safeguard comparable dynamic range of their respective values, treating missing data, and so forth.

5.2 PREPROCESSING

5.2.1 Outlier Removal

An *outlier* is defined as a point that lies very far from the mean of the corresponding random variable. This distance is measured with respect to a given threshold, usually a number of times the standard deviation. For a normally distributed random variable, a distance of two times the standard deviation covers 95% of the points, and a distance of three times the standard deviation covers 99% of the points. Points with values very different from the mean value produce large errors during training and may have disastrous effects. These effects are even worse when the outliers are the result of noisy measurements. If the number of outliers is very small, they are usually discarded. However, if this is not the case and they are the result of a distribution with long tails, then the designer may have to adopt cost functions that are not very sensitive in the presence of outliers. For example, the least squares criterion is very sensitive to outliers, because large errors dominate the cost function due to the squaring of the terms. A review of related techniques that attempt to address such problems is given in [Hube 81].

5.2.2 Data Normalization

In many practical situations a designer is confronted with features whose values lie within different dynamic ranges. Thus, features with large values may have a larger influence in the cost function than features with small values, although *this does not necessarily reflect their respective significance in the design of the classifier*. The problem is overcome by normalizing the features so that their values lie within similar ranges. A straightforward technique is normalization via the respective estimates of the mean and variance. For N available data of the kth feature we have

$$\bar{x}_k = \frac{1}{N}\sum_{i=1}^{N} x_{ik}, \quad k = 1, 2, \ldots, l$$

$$\sigma_k^2 = \frac{1}{N-1}\sum_{i=1}^{N}(x_{ik} - \bar{x}_k)^2$$

$$\hat{x}_{ik} = \frac{x_{ik} - \bar{x}_k}{\sigma_k}$$

In words, all the resulting normalized features will now have zero mean and unit variance. This is obviously a linear method. Other linear techniques limit the feature values in the range of $[0, 1]$ or $[-1, 1]$ by proper scaling. Besides the linear methods, nonlinear methods can also be employed in cases in which the data are not evenly distributed around the mean. In such cases transformations based on nonlinear (i.e., logarithmic or sigmoid) functions can be used to map data within specified intervals. The so-called softmax scaling is a popular candidate. It consists of two steps

$$y = \frac{x_{ik} - \bar{x}_k}{r\sigma_k}, \quad \hat{x}_{ik} = \frac{1}{1 + \exp(-y)} \tag{5.1}$$

This is basically a squashing function limiting data in the range of $[0, 1]$. Using a series expansion approximation, it is not difficult to see that for small values of y this is an approximately linear function with respect to x_{ik}. The range of values of x_{ik} that correspond to the linear section depends on the standard deviation and the factor r, which is user defined. Values away from the mean are squashed exponentially.

5.2.3 Missing Data

In practice, certain features may be missing from some feature vectors. Such incomplete-data cases are common in social sciences due, for example, to partial response in surveys. Remote sensing is another area where incomplete-data may occur when certain regions are covered by a subset of sensors. Sensor networks, which rely on a set of distributed information sources and on the data fusion from a number of sensors, is also a discipline where incomplete-data may arise.

The most traditional techniques in dealing with missing data include schemes that "complete" the missing values by (a) zeros or (b) the unconditional mean, computed from the available values of the respective feature or (c) the conditional mean, if one has an estimate of the pdf of the missing values given the observed data. Completing the missing values in a set of data is also known as *imputation*. Another approach is to discard feature vectors with missing values. Although such a technique can be useful in cases of large data sets, in most cases it is considered a "luxury" to afford to drop available information.

Since the mid-1970s ([Rubi 76]), a large research effort has been invested to cope efficiently with the missing data task, and a number of sophisticated methods have been developed and used successfully. A popular alternative to the previously exposed, rather naive, approaches is known as *imputing from a conditional distribution*. The idea here is to impute by respecting the statistical nature of the missing values. Under this rationale, missing values are not replaced by statistical means or zeros but by a random draw from a distribution. Let us denote the complete feature vector as $\boldsymbol{x}_{com}$ and assume that some of its components are missing ($\boldsymbol{x}_{mis}$) and the rest are observed ($\boldsymbol{x}_{obs}$). Then the complete feature vector is written as

$$\boldsymbol{x}_{com} = \begin{bmatrix} \boldsymbol{x}_{obs} \\ \boldsymbol{x}_{mis} \end{bmatrix}$$

Under the assumption that the probability of missing a value does not depend on the value itself (this is known as the *missing at random* (MAR) assumption), imputing from a conditional distribution means to simulate a draw from the following conditional pdf

$$p(\boldsymbol{x}_{mis}|\boldsymbol{x}_{obs}; \boldsymbol{\theta}) = \frac{p(\boldsymbol{x}_{obs}, \boldsymbol{x}_{mis}; \boldsymbol{\theta})}{p(\boldsymbol{x}_{obs}; \boldsymbol{\theta})} \tag{5.2}$$

where

$$p(\boldsymbol{x}_{obs}; \boldsymbol{\theta}) = \int p(\boldsymbol{x}_{com}; \boldsymbol{\theta}) d\boldsymbol{x}_{mis} \tag{5.3}$$

where $\boldsymbol{\theta}$ is an unknown set of parameters. In practice, an estimate $\hat{\boldsymbol{\theta}}$ of $\boldsymbol{\theta}$ must first be obtained from $\boldsymbol{x}_{obs}$. The celebrated EM algorithms (Chapter 2) is a popular choice for parameter estimation under the missing data setting, for example, [Ghah 94, Tsud 03].

The *multiple imputation* (MI) procedure ([Rubi 87]) is one step beyond the previous methodology, usually referred to as single imputation (SI). In MI, for each missing value, $m > 1$ samples are generated. The results are then combined appropriately so that certain statistical properties are fulfilled. MI can be justified as an attempt to overcome uncertainties associated with the estimation of the parameter $\boldsymbol{\theta}$. Hence, instead of drawing a single point from $p(\boldsymbol{x}_{mis}|\boldsymbol{x}_{obs}; \hat{\boldsymbol{\theta}})$ one can use different parameters, $\hat{\boldsymbol{\theta}}_i$, $i = 1, 2, \ldots, m$, and draw the m samples from

$$p(\boldsymbol{x}_{mis}|\boldsymbol{x}_{obs}; \hat{\boldsymbol{\theta}}_i), \; i = 1, 2, \ldots, m$$

A way to approach this problem is via Bayesian inference arguments (Chapter 2), where the unknown parameter vector is treated as a random one described by a posterior probability, see, for example, [Gelm 95].

In a more recent paper ([Will 07]), the missing data problem is treated in the context of logistic regression classification (Section 3.6) and explicit imputation is bypassed. This is achieved by integrating out the missing values and predicting the binary class label, y_i, of the ith pattern based on the value of

$$P(y_i|\boldsymbol{x}_{i,obs}) = \int P(y_i|\boldsymbol{x}_{i,obs}, \boldsymbol{x}_{i,mis})p(\boldsymbol{x}_{i,mis}|\boldsymbol{x}_{i,obs})d\boldsymbol{x}_{i,mis}$$

Under the assumption that $p(\boldsymbol{x}_{i,mis}|\boldsymbol{x}_{i,obs})$ is sufficiently modeled by a Gaussian mixture model (Section 2.5.5) the previous integration can be performed analytically. For more on the problem of missing data the interested reader may refer to the excellent review article [Scha 02]. We will return to the missing data problem in Chapter 11.

5.3 THE PEAKING PHENOMENON

As stated in the introduction of this chapter, in order to design a classifier with good generalization performance, the number of training points, N, must be large enough with respect to the number of features, l, that is, the dimensionality of the feature space. Take as an example the case of designing a linear classifier, $\boldsymbol{w}^T\boldsymbol{x} + w_0$. The number of the unknown parameters is $l + 1$. In order to get a good estimate of these parameters, the number of data points must be larger than $l + 1$. The larger the N the better the estimate, since we can filter out the effects of the noise and also minimize the effects of the outliers.

In [Trun 79], an elegant simple example has been given that reveals the interplay between the number of features and the size of the training data set and elucidates the way these two parameters influence the performance of a classifier. Consider a two-class classification task with equal prior probabilities, $P(\omega_1) = P(\omega_2) = \frac{1}{2}$, in the l-dimensional space. Both classes, ω_1, ω_2, are described by Gaussian distributions of the same covariance matrix $\Sigma = I$, where I is the identity matrix and mean values $\boldsymbol{\mu}$ and $-\boldsymbol{\mu}$, respectively, where

$$\boldsymbol{\mu} = \left[\ 1, \frac{1}{\sqrt{2}}, \frac{1}{\sqrt{3}}, \ldots, \frac{1}{\sqrt{l}}\ \right]^T \tag{5.4}$$

Since the features are jointly Gaussian and $\Sigma = I$, the involved features are statistically independent (see Appendix A.9). Moreover, the optimal Bayesian rule is equivalent to the minimum Euclidean distance classifier. Given an unknown feature vector $\boldsymbol{x}$, we classify it to, say, ω_1 if

$$||\boldsymbol{x} - \boldsymbol{\mu}||^2 < ||\boldsymbol{x} + \boldsymbol{\mu}||^2$$

or after performing the algebra, if

$$z \equiv \boldsymbol{x}^T\boldsymbol{\mu} > 0$$

If $z < 0$, we decide in favor of the class ω_2. Thus, the decision relies on the value of the inner product z. In the sequel we will consider two cases.

Known Mean Value $\boldsymbol{\mu}$

The inner product z, being a linear combination of independent Gaussian variables, is also a Gaussian variable (see, e.g., [Papo 91]) with mean value $E[z] = ||\boldsymbol{\mu}||^2 = \sum_{i=1}^{l} \frac{1}{i}$ and variance $\sigma_z^2 = ||\boldsymbol{\mu}||^2$ (Problem 5.1). The probability of commiting an error turns out to be equal to (Problem 5.1)

$$P_e = \int_{b_l}^{\infty} \frac{1}{\sqrt{2\pi}} \exp\left(-\frac{z^2}{2}\right) dz \tag{5.5}$$

where

$$b_l = \sqrt{\sum_{i=1}^{l} \frac{1}{i}} \tag{5.6}$$

Note that the series in (5.6) tends to infinity as $l \longrightarrow \infty$; hence the probability of error tends to zero as the number of features increases.

Unknown mean value $\boldsymbol{\mu}$

In this case, the mean value has to be estimated from the training data set. Adopting the maximum likelihood estimate we obtain

$$\hat{\boldsymbol{\mu}} = \frac{1}{N} \sum_{k=1}^{N} s_k \boldsymbol{x}_k$$

where $s_k = 1$ if $\boldsymbol{x}_k \in \omega_1$ and $s_k = -1$ if $\boldsymbol{x}_k \in \omega_2$. Decisions are taken depending on the inner product $z = \boldsymbol{x}^T \hat{\boldsymbol{\mu}}$. However, z is no more a Gaussian variable, since $\hat{\boldsymbol{\mu}}$ is not a constant but a random vector. By the definition of the inner product, $z = \sum_{i=1}^{l} x_i \hat{\mu}_i$ and for large enough l and the central limit theorem (Appendix A) z can be considered approximately Gaussian. Its mean value and variance are given by (Problem 5.2)

$$E[z] = \sum_{i=1}^{l} \frac{1}{i} \tag{5.7}$$

and

$$\sigma_z^2 = \left(1 + \frac{1}{N}\right) \sum_{i=1}^{l} \frac{1}{i} + \frac{l}{N} \tag{5.8}$$

The probability of error is given by (5.5) with

$$b_l = \frac{E[z]}{\sigma_z} \tag{5.9}$$

It can now be shown that $b_l \longrightarrow 0$ as $l \longrightarrow \infty$ and the probability of error tends to $\frac{1}{2}$ for any finite number N (Problem 5.2).

The above example demonstrates that:

- If for any l the corresponding pdf is known, then we can perfectly discriminate the two classes by arbitrarily increasing the number of features.
- If the pdfs are not known and the associated parameters must be estimated using a finite training set, then the arbitrary increase of the number of features leads to the maximum possible value of the error rate, that is, $P_e = 0.5$. This implies that under a limited number of training data we must try to keep the number of features to a relatively low number.

In practice, for a finite N, by increasing the number of features one obtains an initial improvement in performance, but after a critical value further increase of the number of features results in an increase of the probability of error. This phenomenon is also known as the *peaking phenomenon*. Figure 5.1 illustrates the general trend that one expects to experience in practice by playing with the number of features, l, and the size of the training data set, N. For $N_2 >> N_1$, the error values corresponding to N_2 are lower than those resulting for N_1, and the peaking phenomenon occurs for a value $l_2 > l_1$. For each value of N, the probability of error starts decreasing with increasing l till a critical value where the error starts increasing. The minimum in the curves occurs at some number $l = \frac{N}{\alpha}$, where α, usually, takes values in the range of 2 to 10. Consequently, in practice, for a small number of training data, a small number of features must be used. If a large number of training data is available, a larger number of features can be selected that yield better performance.

Although the above scenario covers a large set of "traditional" classifiers, it is not always valid. We have already seen that adopting an appropriate kernel function to design a nonlinear SVM classifier implies a mapping to a high-dimensional space,

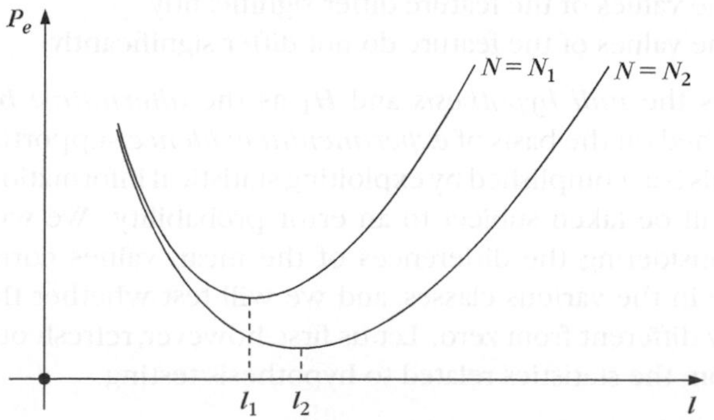

FIGURE 5.1

For a given value of N, the probability of error decreases as the number of features increases till a critical value. Further increase in the number of features forces the error probability to increase. If the number of points increases, $N_2 >> N_1$, the peaking phenomenon occurs for larger values, $l_2 > l_1$.

which can even be of infinite dimension. In spite of the fact that one now works in almost empty spaces (N is much less than the dimensionality of the space), the generalization performance of the SVM classifiers can be very good. The secret to that was disclosed to us fairly recently. It is not the number of parameters that really controls the generalization performance, under finite N, but another quantity. For some types of classifiers, this quantity is directly related to the number of parameters to be estimated and the dimensionality of the feature space. However, for some classifiers, such as the SVM, this quantity can be controlled independent of the dimensionality of the feature space. These issues are discussed at the end of this chapter. More on the peaking phenomenon and the small sample size problem can be found in, for example, [Raud 80, Raud 91, Duin 00].

5.4 FEATURE SELECTION BASED ON STATISTICAL HYPOTHESIS TESTING

A first step in feature selection is to look at each of the generated features *independently* and test their discriminatory capability for the problem at hand. Although looking at the features independently is far from optimal, this procedure helps us to discard easily recognizable "bad" choices and keeps the more elaborate techniques, which we will consider next, from unnecessary computational burden.

Let x be the random variable representing a specific feature. We will try to investigate whether the values it takes for the different classes, say ω_1, ω_2, *differ significantly*. To give an answer to this question, we will formulate the problem in the context of statistical *hypothesis testing*. That is, we will try to answer which of the following hypotheses is correct:

H_1: The values of the feature differ significantly
H_0: The values of the feature do not differ significantly

H_0 is known as the *null hypothesis* and H_1 as the *alternative hypothesis*. The decision is reached on the basis of *experimental evidence* supporting the rejection or not of H_0. This is accomplished by exploiting statistical information, and obviously any decision will be taken subject to an error probability. We will approach the problem by considering the differences of the mean values corresponding to a specific feature in the various classes, and we will test whether these differences are significantly different from zero. Let us first, however, refresh our memory with some basics from the statistics related to hypothesis testing.

5.4.1 Hypothesis Testing Basics

Let x be a random variable with a probability density function, which is assumed to be known within an unknown parameter θ. As we have already seen in Chapter 2, in the case of a Gaussian, this parameter may be the mean value or its variance. Our interest here lies in the following hypothesis test:

$$H_1 : \theta \neq \theta_0$$
$$H_0 : \theta = \theta_0$$

The decision on this test is reached in the following context. Let $x_i, i = 1, 2, \ldots, N$, be the experimental samples of the random variable x. A function $f(\cdot, \ldots, \cdot)$ is selected, depending on the specific problem, and let $q = f(x_1, x_2, \ldots, x_N)$. The function is selected so that the probability density function of q is easily parameterized in terms of the unknown θ, that is, $p_q(q; \theta)$. Let D be the interval of q in which it has a high probability of lying *under hypothesis* H_0. Let $\bar{D}$ be its complement, that is, the interval of low probability, also under hypothesis H_0. Then, if the value of q that results from the available samples, $x_i, i = 1, 2, \ldots, N$, lies in D we will accept H_0, and if it lies in $\bar{D}$ we will reject it. D is known as the *acceptance interval* and $\bar{D}$ as the *critical interval*. The variable q is known as *test statistic*. The obvious question now refers to the probability of reaching a wrong decision. Let H_0 be true. Then the probability of an error in our decision is $P(q \in \bar{D}|H_0) \equiv \rho$. This probability is obviously the integral of $p_q(q|H_0)$ $(p_q(q; \theta_o))$ over $\bar{D}$ (Figure 5.2). In practice, we preselect this value of ρ, which is known as the *significance level*, and we sometimes denote the corresponding critical (acceptance) interval as $\bar{D}_\rho$ (D_ρ). Let us now apply this procedure in the case in which the unknown parameter is the mean of x.

The Known Variance Case

Let x be a random variable and $x_i, i = 1, 2, \ldots, N$, the resulting experimental samples, which we will assume to be *mutually independent*. Let

$$E[x] = \mu$$
$$E[(x - \mu)^2] = \sigma^2$$

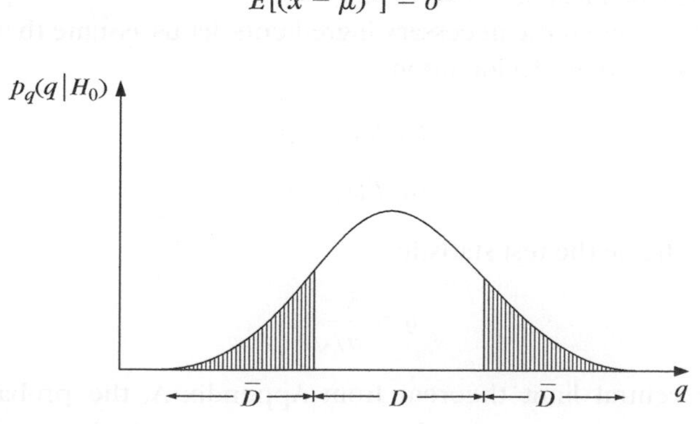

FIGURE 5.2

Acceptance and critical regions for hypothesis testing. The area of the shaded region is the probability of an erroneous decision.

A popular estimate of μ based on the known samples is the *sample mean*

$$\bar{x} = \frac{1}{N}\sum_{i=1}^{N} x_i$$

Using a different set of N samples, a different estimate will result. Thus, $\bar{x}$ is also a random variable, and it is described in terms of a probability density function $p_{\bar{x}}(\bar{x})$. The corresponding mean is

$$E[\bar{x}] = \frac{1}{N}E\left[\sum_{i=1}^{N} x_i\right] = \frac{1}{N}\sum_{i=1}^{N} E[x_i] = \mu \tag{5.10}$$

Thus, $\bar{x}$ is an *unbiased* estimate of the mean μ of x. The variance $\sigma_{\bar{x}}^2$ of $\bar{x}$ is

$$E[(\bar{x}-\mu)^2] = E\left[\left(\frac{1}{N}\sum_{i=1}^{N} x_i - \mu\right)^2\right]$$

$$= \frac{1}{N^2}\sum_{i=1}^{N} E[(x_i-\mu)^2] + \frac{1}{N^2}\sum_{i}\sum_{j\neq i} E[(x_i-\mu)(x_j-\mu)]$$

The statistical independence of the samples dictates

$$E[(x_i-\mu)(x_j-\mu)] = E[x_i-\mu]E[x_j-\mu] = 0$$

Hence

$$\sigma_{\bar{x}}^2 = \frac{1}{N}\sigma^2 \tag{5.11}$$

In words, the larger the number of measurement samples, the smaller the variance of $\bar{x}$ around the true mean μ.

Having now acquired the necessary ingredients, let us assume that we are given a value $\hat{\mu}$ and we have to decide upon

$$H_1 : E[x] \neq \hat{\mu}$$

$$H_0 : E[x] = \hat{\mu}$$

To this end we define the test statistic

$$q = \frac{\bar{x}-\hat{\mu}}{\sigma/\sqrt{N}} \tag{5.12}$$

Recalling the central limit theorem from Appendix A, the probability density function of $\bar{x}$ *under* H_0 (i.e., given $\hat{\mu}$) is (approximately) the Gaussian $\mathcal{N}(\hat{\mu}, \frac{\sigma^2}{N})$

$$p_{\bar{x}}(\bar{x}) = \frac{\sqrt{N}}{\sqrt{2\pi}\sigma}\exp\left(-\frac{N(\bar{x}-\hat{\mu})^2}{\sigma^2}\right)$$

Table 5.1 Acceptance Intervals $[-x_\rho, x_\rho]$ Corresponding to Various Probabilities for an $\mathcal{N}(0, 1)$ Normal Distribution

$1-\rho$	0.8	0.85	0.9	0.95	0.98	0.99	0.998	0.999
x_ρ	1.282	1.440	1.645	1.967	2.326	2.576	3.090	3.291

Hence, the probability density function of q under H_0 is approximately $\mathcal{N}(0, 1)$. For a significance level ρ the acceptance interval $D \equiv [-x_\rho, x_\rho]$, is chosen as the interval in which the random variable q lies with probability $1-\rho$ (ρ the probability of being in $\bar{D}$). This is readily provided from available tables.

An example for normally distributed $\mathcal{N}(0, 1)$ variables is given in Table 5.1. The decision on the test hypothesis can now be reached by the following steps.

- Given the N experimental samples of x, compute $\bar{x}$ and then q.
- Choose the significance level ρ.
- Compute from the corresponding tables for $\mathcal{N}(0, 1)$ the acceptance interval $D = [-x_\rho, x_\rho]$, corresponding to probability $1 - \rho$.
- If $q \in D$ decide H_0, if not decide H_1.

Basically, all we say is that we *expect* the resulting value q to lie in the high-percentage $1-\rho$ interval. If it does not, then we decide that this is because the assumed mean value is not "correct." Of course, the assumed mean value may be correct, but it so happens that the resulting q lies in the least probable area because of the specific set of experimental samples available. In such a case our decision is erroneous, and the probability of committing such an error is ρ.

Example 5.1

Let us consider an experiment with a random variable x of $\sigma = 0.23$, and assume N to be equal to 16 and the resulting $\bar{x}$ equal to 1.35. Adopt the significance level $\rho = 0.05$. We will test if the hypothesis $\hat{\mu} = 1.4$ is true.

From Table 5.1 we have

$$\text{prob}\left\{-1.97 < \frac{\bar{x} - \hat{\mu}}{0.23/4} < 1.97\right\} = 0.95$$

$$\text{prob}\{-0.113 < \bar{x} - \hat{\mu} < 0.113\} = 0.95$$

Thus, since the value of $\hat{\mu}$, which we have assumed, is in the interval

$$1.237 = 1.35 - 0.113 < \hat{\mu} < 1.35 + 0.113 = 1.463$$

we accept it, as *there is no evidence at the 5% level that the mean value is not equal to* $\hat{\mu}$. The interval [1.237, 1463] is also known as the *confidence interval at the* $1-\rho = 0.95$ *level.*

The Unknown Variance Case

If the variance of x is not known, it must be estimated. The estimate

$$\hat{\sigma}^2 \equiv \frac{1}{N-1}\sum_{i=1}^{N}(x_i - \bar{x})^2 \tag{5.13}$$

is an unbiased estimate of the variance. Indeed,

$$\begin{aligned} E[\hat{\sigma}^2] &= \frac{1}{N-1}\sum_{i=1}^{N} E[(x_i - \bar{x})^2] \\ &= \frac{1}{N-1}\sum_{i=1}^{N} E\left[((x_i - \mu) - (\bar{x} - \mu))^2\right] \\ &= \frac{1}{N-1}\sum_{i=1}^{N}\left(\sigma^2 + \frac{\sigma^2}{N} - 2E[(x_i - \mu)(\bar{x} - \mu)]\right) \end{aligned}$$

Due to the independence of the experimental samples

$$E[(x_i - \mu)(\bar{x} - \mu)] = \frac{1}{N}E\big[(x_i - \mu)\big((x_1 - \mu) + \cdots + (x_N - \mu)\big)\big] = \frac{\sigma^2}{N}$$

Thus,

$$E[\hat{\sigma}^2] = \frac{N}{N-1}\frac{N-1}{N}\sigma^2 = \sigma^2$$

The test statistic is now defined as

$$q = \frac{\bar{x} - \mu}{\hat{\sigma}/\sqrt{N}} \tag{5.14}$$

However, this is no longer a Gaussian variable. Following Appendix A, and if we assume that x *is a Gaussian random variable*, then q is described by the so-called t-distribution with $N - 1$ degrees of freedom. Table 5.2 shows the confidence interval $D = [-x_\rho, x_\rho]$ for various significance levels and degrees of freedom of the t-distribution.

Example 5.2

For the case of Example 5.1 let us assume that the estimate of the standard deviation $\hat{\sigma}$ is 0.23. Then, according to Table 5.2 for 15 degrees of freedom ($N = 16$) and significance level $\rho = 0.025$

$$\text{prob}\left\{-2.49 < \frac{\bar{x} - \hat{\mu}}{0.23/4} < 2.49\right\} = 0.975$$

and the confidence interval for $\hat{\mu}$ at the 0.975 level is

$$1.207 < \hat{\mu} < 1.493$$

Table 5.2 Interval Values at Various Significance Levels and Degrees of Freedom for a t-Distribution

Degrees of Freedom	$1-\rho$	0.9	0.95	0.975	0.99	0.995
10		1.81	2.23	2.63	3.17	3.58
11		1.79	2.20	2.59	3.10	3.50
12		1.78	2.18	2.56	3.05	3.43
13		1.77	2.16	2.53	3.01	3.37
14		1.76	2.15	2.51	2.98	3.33
15		1.75	2.13	2.49	2.95	3.29
16		1.75	2.12	2.47	2.92	3.25
17		1.74	2.11	2.46	2.90	3.22
18		1.73	2.10	2.44	2.88	3.20
19		1.73	2.09	2.43	2.86	3.17
20		1.72	2.09	2.42	2.84	3.15

5.4.2 Application of the t-Test in Feature Selection

We will now see how all of this is specialized for the case of feature selection in a classification problem. Our major concern now will be to test, against zero, the difference $\mu_1 - \mu_2$ between the means of the values taken by a feature in two classes. Let x_i, $i = 1, 2, \ldots, N$, be the sample values of the feature in class ω_1 with mean μ_1. Correspondingly, for the other class ω_2 we have y_i, $i = 1, 2, \ldots, N$, with mean μ_2. *Let us now assume that the variance of the feature values is the same in both classes,* $\sigma_1^2 = \sigma_2^2 = \sigma^2$. To decide about the closeness of the two mean values, we will test for the hypothesis

$$\begin{aligned} H_1 &: \Delta\mu = \mu_1 - \mu_2 \neq 0 \\ H_0 &: \Delta\mu = \mu_1 - \mu_2 = 0 \end{aligned} \tag{5.15}$$

To this end, let

$$z = x - y \tag{5.16}$$

where x, y denote the random variables corresponding to the values of the feature in the two classes ω_1, ω_2, respectively, for which *statistical independence* has been assumed. Obviously, $E[z] = \mu_1 - \mu_2$, and due to the independence assumption $\sigma_z^2 = 2\sigma^2$. Following arguments similar to those used before, we now have

$$\bar{z} = \frac{1}{N}\sum_{i=1}^{N}(x_i - y_i) = \bar{x} - \bar{y} \tag{5.17}$$

and for the known variance case $\bar{z}$ follows the normal $\mathcal{N}(\mu_1 - \mu_2, \frac{2\sigma^2}{N})$ distribution for large N. Thus, Table 5.1 can be used to decide about (5.15). If the variance is not known, then we choose the test statistic

$$q = \frac{(\bar{x} - \bar{y}) - (\mu_1 - \mu_2)}{s_z\sqrt{\frac{2}{N}}} \tag{5.18}$$

where

$$s_z^2 = \frac{1}{2N-2}\left(\sum_{i=1}^{N}(x_i - \bar{x})^2 + \sum_{i=1}^{N}(y_i - \bar{y})^2\right)$$

It can be shown that $\frac{s_z^2(2N-2)}{\sigma^2}$ follows a chi-square distribution with $2N - 2$ degrees of freedom (Appendix A and Problem 5.3). As is pointed out in Appendix A, if x, y are ***normally distributed*** variables of the same variance σ^2, then the random variable q turns out to follow the t-distribution with $2N - 2$ degrees of freedom. Thus, Table 5.2 has to be adopted for the test. When the available number of samples is not the same in all classes, a slight modification is required (Problem 5.4). Furthermore, in practice the variances may not be the same in the two classes. Sometimes this becomes the object of another hypothesis test, concerning the ratio $F = \frac{\hat{\sigma}_1^2}{\hat{\sigma}_2^2}$, to check whether it is close to unity. It can be shown that F, being the ratio of two chi-square distributed variables, follows the so-called F-distribution and the related tables should be used [Fras 58]. Finally, if the Gaussian assumption about x is not a valid one, other criteria can be used to check the equality hypothesis of the means, such as the Kruskal–Wallis statistic [Walp 78, Fine 83].

Example 5.3

The sample measurements of a feature in two classes are

class ω_1:	3.5	3.7	3.9	4.1	3.4	3.5	4.1	3.8	3.6	3.7
class ω_2:	3.2	3.6	3.1	3.4	3.0	3.4	2.8	3.1	3.3	3.6

The question is to check whether this feature is informative enough. If not, it will be discarded during the selection phase. To this end, we will test whether the values of the feature in the two classes differ significantly. We choose the significance level $\rho = 0.05$.
From the foregoing we have

$$\omega_1: \bar{x} = 3.73 \quad \hat{\sigma}_1^2 = 0.0601$$

$$\omega_2: \bar{y} = 3.25 \quad \hat{\sigma}_2^2 = 0.0672$$

For $N = 10$ we have

$$s_z^2 = \frac{1}{2}(\hat{\sigma}_1^2 + \hat{\sigma}_2^2)$$

$$q = \frac{(\bar{x} - \bar{y} - 0)}{s_z\sqrt{\frac{2}{N}}}$$

$$q = 4.25$$

From Table 5.2 and for $20 - 2 = 18$ degrees of freedom and significance level 0.05, we obtain $D = [-2.10, 2.10]$. Since 4.25 lies outside the interval D, we decide in favor of H_1; that is, the mean values differ significantly at the 0.05 level. *Hence, the feature is selected.*

5.5 THE RECEIVER OPERATING CHARACTERISTIC (ROC) CURVE

The hypothesis tests we have presented offer statistical evidence about the difference of the mean values of a single feature in the various classes. However, although this is useful information for *discarding* features, if the corresponding mean values are closely located, this information may not be sufficient to guarantee good discrimination properties of a feature passing the test. The mean values may differ significantly yet the spread around the means may be large enough to blur the class distinction. We will now focus on techniques providing information about the *overlap* between the classes.

Figure 5.3a illustrates an example of two overlapping probability density functions describing the distribution of a feature in two classes, together with a threshold (one pdf has been inverted for illustration purposes). We decide class ω_1 for values on the left of the threshold and class ω_2 for the values on the right. This decision is associated with an error probability, a, of reaching a wrong decision concerning class ω_1 (the probability of a correct decision is $1 - a$). This is equal to the

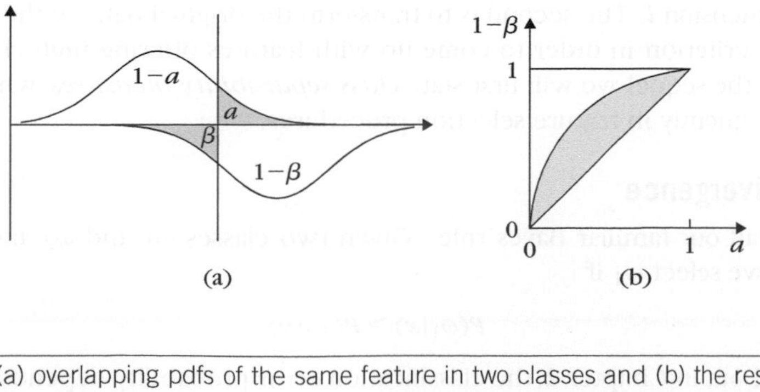

FIGURE 5.3

Example of (a) overlapping pdfs of the same feature in two classes and (b) the resulting ROC curve. The larger the shaded area in (b) the less the overlap of the respective pdfs.

shaded area under the corresponding curve. Similarly, let β $(1-\beta)$ be the probability of a wrong (correct) decision concerning class ω_2. By moving the threshold over "all" possible positions, different values of a and β result. It takes little thought to realize that if the two distributions have complete overlap, then for *any* position of the threshold we get $a = 1 - \beta$. Such a case corresponds to the straight line in Figure 5.3b, where the two axes are a and $1-\beta$. As the two distributions move apart, the corresponding curve departs from the straight line, as Figure 5.3b demonstrates. Once more, a little thought reveals that the less the overlap of the classes, the larger the area between the curve and the straight line. At the other extreme of two completely separated class distributions, moving the threshold to sweep the whole range of values for a in [0, 1], $1 - \beta$ remains equal to unity. Thus, the aforementioned area varies between zero, for complete overlap, and 1/2 (the area of the upper triangle), for complete separation, and ***it is a measure of the class discrimination capability of the specific feature***. In practice, the ROC curve can easily be constructed by sweeping the threshold and computing percentages of wrong and correct classifications over the available training feature vectors. Other related criteria that test the overlap of the classes have also been suggested (see Problem 5.7).

More recently, the area under the receiver operating characteristic curve (AUC) has been used as an effective criterion to design classifiers. This is because larger AUC values indicate on average better classifier performance, see, for example, [Brad 97, Marr 08, Land 08].

5.6 CLASS SEPARABILITY MEASURES

The emphasis in the previous section was on techniques referring to the discrimination properties of ***individual*** features. However, such methods neglect to take into account the correlation that unavoidably exists among the various features and influences the classification capabilities of the feature vectors that are formed. Measuring the discrimination effectiveness of feature ***vectors*** will now become our major concern. This information will then be used in two ways. The first is to allow us to combine features appropriately and end up with the "best" feature vector for a given dimension l. The second is to transform the original data on the basis of an optimality criterion in order to come up with features offering high classification power. In the sequel we will first state ***class separability measures***, which will be used subsequently in feature selection procedures.

5.6.1 Divergence

Let us recall our familiar Bayes rule. Given two classes ω_1 and ω_2 and a feature vector $\boldsymbol{x}$, we select ω_1 if

$$P(\omega_1|\boldsymbol{x}) > P(\omega_2|\boldsymbol{x})$$

As pointed out in Chapter 2, the classification error probability depends on the difference between $P(\omega_1|\boldsymbol{x})$ and $P(\omega_2|\boldsymbol{x})$, e.g., Eq. (2.12). Hence, the ratio $\frac{P(\omega_1|\boldsymbol{x})}{P(\omega_2|\boldsymbol{x})}$ can convey useful information concerning the discriminatory capabilities associated with an adopted feature vector $\boldsymbol{x}$, with respect to the two classes ω_1, ω_2. Alternatively (for given values of $P(\omega_1), P(\omega_2)$), the same information resides in the ratio $\ln \frac{p(\boldsymbol{x}|\omega_1)}{p(\boldsymbol{x}|\omega_2)} \equiv D_{12}(\boldsymbol{x})$, and this can be used as a measure of the underlying discriminating information of class ω_1 with respect to ω_2. Clearly, for completely overlapped classes, we get $D_{12}(\boldsymbol{x}) = 0$. Since $\boldsymbol{x}$ takes different values, it is natural to consider the mean value over class ω_1, that is,

$$D_{12} = \int_{-\infty}^{+\infty} p(\boldsymbol{x}|\omega_1) \ln \frac{p(\boldsymbol{x}|\omega_1)}{p(\boldsymbol{x}|\omega_2)}\, d\boldsymbol{x} \tag{5.19}$$

Similar arguments hold for class ω_2, and we define

$$D_{21} = \int_{-\infty}^{+\infty} p(\boldsymbol{x}|\omega_2) \ln \frac{p(\boldsymbol{x}|\omega_2)}{p(\boldsymbol{x}|\omega_1)}\, d\boldsymbol{x} \tag{5.20}$$

The sum

$$d_{12} = D_{12} + D_{21}$$

is known as the *divergence* and can be used as a separability measure for the classes ω_1, ω_2, with respect to the adopted feature vector $\boldsymbol{x}$. For a multiclass problem, the divergence is computed for every class pair ω_i, ω_j

$$d_{ij} = D_{ij} + D_{ji}$$

$$= \int_{-\infty}^{+\infty} (p(\boldsymbol{x}|\omega_i) - p(\boldsymbol{x}|\omega_j)) \ln \frac{p(\boldsymbol{x}|\omega_i)}{p(\boldsymbol{x}|\omega_j)}\, d\boldsymbol{x} \tag{5.21}$$

and the average class separability can be computed using the average divergence

$$d = \sum_{i=1}^{M} \sum_{j=1}^{M} P(\omega_i)P(\omega_j)d_{ij}$$

Divergence is basically a form of the Kullback–Leibler distance measure between density functions [Kulb 51] (Appendix A). The divergence has the following easily shown properties:

$$d_{ij} \geq 0$$

$$d_{ij} = 0 \quad \text{if } i = j$$

$$d_{ij} = d_{ji}$$

If the components of the feature vector are statistically independent, then it can be shown (Problem 5.10) that

$$d_{ij}(x_1, x_2, \ldots, x_l) = \sum_{r=1}^{l} d_{ij}(x_r)$$

Assuming now that the density functions are Gaussians $\mathcal{N}(\boldsymbol{\mu}_i, \Sigma_i)$ and $\mathcal{N}(\boldsymbol{\mu}_j, \Sigma_j)$, respectively, the computation of the divergence is simplified, and it is not difficult to show that

$$d_{ij} = \frac{1}{2}\text{trace}\{\Sigma_i^{-1}\Sigma_j + \Sigma_j^{-1}\Sigma_i - 2I\} + \frac{1}{2}(\boldsymbol{\mu}_i - \boldsymbol{\mu}_j)^T(\Sigma_i^{-1} + \Sigma_j^{-1})(\boldsymbol{\mu}_i - \boldsymbol{\mu}_j) \tag{5.22}$$

For the one-dimensional case this becomes

$$d_{ij} = \frac{1}{2}\left(\frac{\sigma_j^2}{\sigma_i^2} + \frac{\sigma_i^2}{\sigma_j^2} - 2\right) + \frac{1}{2}(\mu_i - \mu_j)^2\left(\frac{1}{\sigma_i^2} + \frac{1}{\sigma_j^2}\right)$$

As already pointed out, a class separability measure cannot depend only on the difference of the mean values; it must also be variance dependent. Indeed, divergence does depend explicitly on both the difference of the means and the respective variances. Furthermore, d_{ij} can be large even for equal mean values, *provided the variances differ significantly*. Thus, class separation is still possible even if the class means coincide. We will come to this later on.

Let us now investigate (5.22). If the covariance matrices of the two Gaussian distributions are equal, $\Sigma_i = \Sigma_j = \Sigma$, then the divergence is further simplified to

$$d_{ij} = (\boldsymbol{\mu}_i - \boldsymbol{\mu}_j)^T\Sigma^{-1}(\boldsymbol{\mu}_i - \boldsymbol{\mu}_j)$$

which is nothing other than the Mahalanobis distance between the corresponding mean vectors. This has another interesting implication. Recalling Problem 2.9 of Chapter 2, it turns out that in this case we have a direct relation between the divergence d_{ij} and the Bayes error—that is, the minimum error we can achieve by adopting the specific feature vector. This is a most desirable property for any class separability measure. Unfortunately, such a direct relation of the divergence with the Bayes error is not possible for more general distributions. Furthermore, in [Swai 73, Rich 95] it is pointed out that the specific dependence of the divergence on the difference of the mean vectors may lead to misleading results, in the sense that small variations in the difference of the mean values can produce large changes in the divergence, which, however, are not reflected in the classification error. To overcome this, a variation of the divergence is suggested, called the *transformed divergence:*

$$\hat{d}_{ij} = 2\left(1 - \exp(-d_{ij}/8)\right)$$

In the sequel, we will try to define class separability measures with a closer relationship to the Bayes error.

5.6.2 Chernoff Bound and Bhattacharyya Distance

The minimum attainable classification error of the Bayes classifier for two classes ω_1, ω_2 can be written as:

$$P_e = \int_{-\infty}^{\infty} \min\left[P(\omega_i)p(\boldsymbol{x}|\omega_i), P(\omega_j)p(\boldsymbol{x}|\omega_j)\right] d\boldsymbol{x} \tag{5.23}$$

Analytic computation of this integral in the general case is not possible. However, an upper bound can be derived. The derivation is based on the inequality

$$\min[a,b] \leq a^s b^{1-s} \quad \text{for} \quad a,b \geq 0, \quad \text{and} \quad 0 \leq s \leq 1 \tag{5.24}$$

Combining (5.23) and (5.24), we get

$$P_e \leq P(\omega_i)^s P(\omega_j)^{1-s} \int_{-\infty}^{\infty} p(\boldsymbol{x}|\omega_i)^s p(\boldsymbol{x}|\omega_j)^{1-s}\, d\boldsymbol{x} \equiv \epsilon_{CB} \tag{5.25}$$

ϵ_{CB} is known as the *Chernoff bound*. The minimum bound can be computed by minimizing ϵ_{CB} with respect to s. A special form of the bound results for $s = 1/2$:

$$P_e \leq \epsilon_{CB} = \sqrt{P(\omega_i)P(\omega_j)} \int_{-\infty}^{\infty} \sqrt{p(\boldsymbol{x}|\omega_i)p(\boldsymbol{x}|\omega_j)}\, d\boldsymbol{x} \tag{5.26}$$

For Gaussian distributions $\mathcal{N}(\boldsymbol{\mu}_i, \Sigma_i), \mathcal{N}(\boldsymbol{\mu}_j, \Sigma_j)$ and after a bit of algebra, we obtain

$$\epsilon_{CB} = \sqrt{P(\omega_i)P(\omega_j)}\exp(-B)$$

where

$$B = \frac{1}{8}(\boldsymbol{\mu}_i - \boldsymbol{\mu}_j)^T \left(\frac{\Sigma_i + \Sigma_j}{2}\right)^{-1} (\boldsymbol{\mu}_i - \boldsymbol{\mu}_j) + \frac{1}{2}\ln \frac{\left|\frac{\Sigma_i+\Sigma_j}{2}\right|}{\sqrt{|\Sigma_i||\Sigma_j|}} \tag{5.27}$$

and $|\cdot|$ denotes the determinant of the respective matrix. The term B is known as the *Bhattacharyya distance*, and it is used as a class separability measure. It can be shown (Problem 5.11) that it corresponds to the optimum Chernoff bound when $\Sigma_i = \Sigma_j$. It is readily seen that in this case the Bhattacharyya distance becomes proportional to the Mahalanobis distance between the means. In [Lee 00] an equation that relates the optimal Bayesian error and the Bhattacharyya distance is proposed, based on an empirical study involving normal distributions. This was subsequently used for feature selection in [Choi 03].

A comparative study of various distance measures for feature selection in the context of multispectral data classification in remote sensing can be found in [Maus 90]. A more detailed treatment of the topic is given in [Fuku 90].

Example 5.4

Assume that $P(\omega_1) = P(\omega_2)$ and that the corresponding distributions are Gaussians $\mathcal{N}(\boldsymbol{\mu}, \sigma_1^2 I)$ and $\mathcal{N}(\boldsymbol{\mu}, \sigma_2^2 I)$. The Bhattacharyya distance becomes

$$B = \frac{1}{2}\ln \frac{\left(\frac{\sigma_1^2+\sigma_2^2}{2}\right)^l}{\sqrt{\sigma_1^{2l}\sigma_2^{2l}}} = \frac{1}{2}\ln\left(\frac{\sigma_1^2+\sigma_2^2}{2\sigma_1\sigma_2}\right)^l \tag{5.28}$$

For the one-dimensional case $l = 1$ and for $\sigma_1 = 10\sigma_2$, $B = 0.8097$ and

$$P_e \leq 0.2225$$

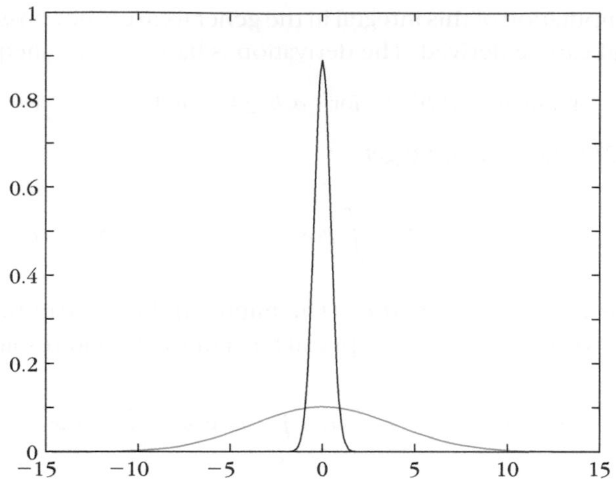

FIGURE 5.4

Gaussian pdfs with the same mean and different variances.

If $\sigma_1 = 100\sigma_2$, $B = 1.9561$ and

$$P_e \leq 0.0707$$

Thus, the greater the difference of the variances, the smaller the error bound. The decrease is bigger for higher dimensions due to the dependence on l. Figure 5.4 shows the pdfs for the same mean and $\sigma_1 = 1$, $\sigma_2 = 0.01$. The figure is self-explanatory as to how the Bayesian classifier discriminates between two classes of the same mean and significantly different variances. Furthermore, as $\sigma_2/\sigma_1 \rightarrow 0$, the probability of error tends to zero (why?)

5.6.3 Scatter Matrices

A major disadvantage of the class separability criteria considered so far is that they are not easily computed, unless the Gaussian assumption is employed. We will now turn our attention to a set of simpler criteria, built upon information related to the way feature vector samples are scattered in the l-dimensional space. To this end, the following matrices are defined:

Within-class scatter matrix

$$S_w = \sum_{i=1}^{M} P_i \Sigma_i$$

where Σ_i is the covariance matrix for class ω_i

$$\Sigma_i = E[(\boldsymbol{x} - \boldsymbol{\mu}_i)(\boldsymbol{x} - \boldsymbol{\mu}_i)^T]$$

and P_i the *a priori* probability of class ω_i. That is, $P_i \simeq n_i/N$, where n_i is the number of samples in class ω_i, out of a total of N samples. Obviously, trace $\{S_w\}$ is a measure of the average, over all classes, variance of the features.

Between-class scatter matrix

$$S_b = \sum_{i=1}^{M} P_i(\boldsymbol{\mu}_i - \boldsymbol{\mu}_0)(\boldsymbol{\mu}_i - \boldsymbol{\mu}_0)^T$$

where $\boldsymbol{\mu}_0$ is the global mean vector

$$\boldsymbol{\mu}_0 = \sum_{i}^{M} P_i \boldsymbol{\mu}_i$$

Trace$\{S_b\}$ is a measure of the average (over all classes) distance of the mean of each individual class from the respective global value.

Mixture scatter matrix

$$S_m = E[(\boldsymbol{x} - \boldsymbol{\mu}_0)(\boldsymbol{x} - \boldsymbol{\mu}_0)^T]$$

That is, S_m is the covariance matrix of the feature vector with respect to the global mean. It is not difficult to show (Problem 5.12) that

$$S_m = S_w + S_b$$

Its trace is the sum of variances of the features around their respective global mean. From these definitions it is straightforward to see that the criterion

$$J_1 = \frac{\text{trace}\{S_m\}}{\text{trace}\{S_w\}}$$

takes large values when samples in the l-dimensional space are well clustered around their mean, within each class, and the clusters of the different classes are well separated. Sometimes S_b is used in place of S_m. An alternative criterion results if determinants are used in the place of traces. This is justified for scatter matrices that are symmetric positive definite, and thus their eigenvalues are positive (Appendix B). The trace is equal to the sum of the eigenvalues, while the determinant is equal to their product. Hence, large values of J_1 also correspond to large values of the criterion

$$J_2 = \frac{|S_m|}{|S_w|} = |S_w^{-1} S_m|$$

A variant of J_2 commonly encountered in practice is

$$J_3 = \text{trace}\{S_w^{-1} S_b\}$$

As we will see later on, criteria J_2 and J_3 have the advantage of being invariant under linear transformations, and we will adopt them to derive features in an optimal way.

In [Fuku 90] a number of different criteria are also defined by using various combinations of S_w, S_b, S_m in a "trace" or "determinant" formulation. However, whenever a determinant is used, one should be careful with S_b, since $|S_b| = 0$ for $M < l$. This is because S_b is the sum of M $l \times l$ matrices, of rank one each. In practice, all three matrices are approximated by appropriate averaging using the available data samples.

These criteria take a special form in the one-dimensional, two-class problem. In this case, it is easy to see that for equiprobable classes $|S_w|$ is proportional to $\sigma_1^2 + \sigma_2^2$ and $|S_b|$ proportional to $(\mu_1 - \mu_2)^2$. Combining S_b and S_w, the so-called ***Fisher's discriminant ratio (FDR)*** results

$$FDR = \frac{(\mu_1 - \mu_2)^2}{\sigma_1^2 + \sigma_2^2}$$

FDR is sometimes used to quantify the separability capabilities of individual features. It reminds us of the test statistic q appearing in the hypothesis statistical tests dealt with before. However, here the use of *FDR* is suggested in a more "primitive" fashion, independent of the underlying statistical distributions. For the multiclass case, averaging forms of *FDR* can be used. One possibility is

$$FDR_1 = \sum_{i}^{M} \sum_{j \neq i}^{M} \frac{(\mu_i - \mu_j)^2}{\sigma_i^2 + \sigma_j^2}$$

where the subscripts i, j refer to the mean and variance corresponding to the feature under investigation for the classes ω_i, ω_j, respectively.

Example 5.5

Figure 5.5 shows three cases of classes at different locations and within-class variances. The resulting values for the J_3 criterion involving the S_w and S_m matrices are 164.7, 12.5, and

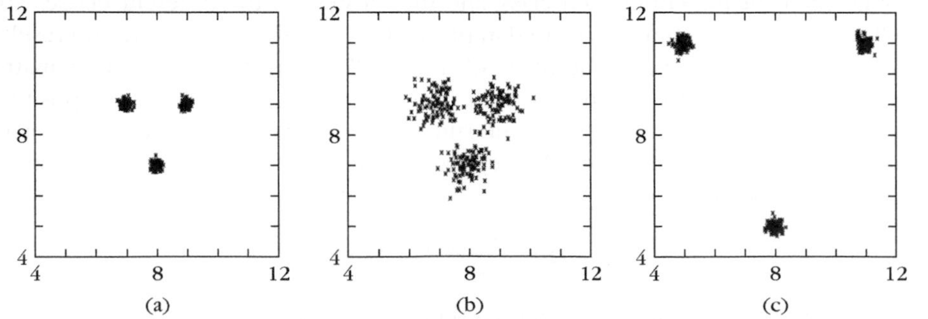

FIGURE 5.5

Classes with (a) small within-class variance and small between-class distances, (b) large within-class variance and small between-class distances, and (c) small within-class variance and large between-class distances.

620.9 for the cases in Figures 5.5a, b, and c, respectively. That is, the best is for distant well-clustered classes and the worst for the case of closely located classes with large within-class variance.

5.7 FEATURE SUBSET SELECTION

Having defined a number of criteria, measuring the classification effectiveness of individual features and/or feature vectors, we come to the heart of our problem, that is, to select a subset of l features out of m originally available. There are two major directions to follow.

5.7.1 Scalar Feature Selection

Features are treated individually. Any of the class separability measuring criteria can be adopted, for example, *ROC*, *FDR*, one-dimensional divergence, and so on. The value of the criterion $C(k)$ is computed for each of the features, $k = 1, 2, \ldots, m$. Features are then ranked in order of descending values of $C(k)$. The l features corresponding to the l best values of $C(k)$ are then selected to form the feature vector.

All the criteria we have dealt with in the previous sections measure the classification capability with respect to a two-class problem. As we have already pointed out in a couple of places, in a multiclass situation a form of average or "total" value, over all classes, is used to compute $C(k)$. However, this is not the only possibility. In [Su 94] the one-dimensional divergence d_{ij} was used and computed for every pair of classes. Then, for each of the features, the corresponding $C(k)$ was set equal to

$$C(k) = \min_{i,j} d_{ij}$$

that is, the minimum divergence value over all class pairs, instead of an average value. Thus, selecting the features with the largest $C(k)$ values is equivalent to choosing features with the best "worst-case" class separability capability, giving a "*maxmin*" flavor to the feature selection task. Such an approach may lead to more robust performance in certain cases.

The major advantage of dealing with features individually is computational simplicity. However, such approaches do not take into account existing correlations between features. Before we proceed to techniques dealing with vectors, we will comment on some *ad hoc* techniques that incorporate correlation information combined with criteria tailored for scalar features.

Let $x_{nk}, n = 1, 2, \ldots, N$ and $k = 1, 2, \ldots, m$, be the kth feature of the nth pattern. The cross-correlation coefficient between any two of them is given by

$$\rho_{ij} = \frac{\sum_{n=1}^{N} x_{ni} x_{nj}}{\sqrt{\sum_{n=1}^{N} x_{ni}^2 \sum_{n=1}^{N} x_{nj}^2}} \tag{5.29}$$

It can be shown that $|\rho_{ij}| \leq 1$ (Problem 5.13). The selection procedure evolves along the following steps:

- Select a class separability criterion C and compute its values for all the available features $x_k, k = 1, 2, \ldots, m$. Rank them in descending order and choose the one with the best C value. Let us say that this is x_{i_1}.
- To select the second feature, compute the cross-correlation coefficient defined in Eq. (5.29) between the chosen x_{i_1} and each of the remaining $m-1$ features, that is, $\rho_{i_1 j}, j \neq i_1$.
- Choose the feature x_{i_2} for which

$$i_2 = arg \max_j \left\{\alpha_1 C(j) - \alpha_2 |\rho_{i_1 j}|\right\}, \quad \text{for all } j \neq i_1$$

  where α_1, α_2 are weighting factors that determine the relative importance we give to the two terms. In words, for the selection of the next feature, we take into account not only the class separability measure C but also the correlation with the already chosen feature. This is then generalized for the kth step

- Select $x_{i_k}, k = 3, \ldots, l$, so that

$$i_k = arg \max_j \left\{\alpha_1 C(j) - \frac{\alpha_2}{k-1} \sum_{r=1}^{k-1} |\rho_{i_r j}|\right\} \quad \text{for } j \neq i_r,$$

$$r = 1, 2, \ldots, k-1$$

  That is, the average correlation with all previously selected features is taken into account.

There are variations of this procedure. For example, in [Fine 83] more than one criterion is adopted and averaged out. Hence, the best index is found by optimizing

$$\left\{\alpha_1 C_1(j) + \alpha_2 C_2(j) - \frac{\alpha_3}{k-1} \sum_{r=1}^{k-1} |\rho_{i_r j}|\right\}$$

5.7.2 Feature Vector Selection

Treating features individually, that is, as scalars, has the advantage of computational simplicity but may not be effective for complex problems and for features with high mutual correlation. We will now focus on techniques measuring classification capabilities of feature vectors. It does not require much thought to see that computational burden is the major limiting factor of such an approach. Indeed, if we want to act according to what "optimality" suggests, we should form *all* possible vector combinations of l features out of the m originally available. According to the type of optimality rule that one chooses to work with, the feature selection task is classified into two categories:

Filter approach. In this approach, the optimality rule for feature selection is independent of the classifier, which will be used in the classifier design stage. For each combination we should use one of the separability criteria introduced previously (e.g., Bhattacharrya distance, J_2) and select the best feature vector combination. Recalling our combinatorics basics, we obtain the total number of vectors as

$$\binom{m}{l} = \frac{m!}{l!(m-l)!} \tag{5.30}$$

This is a large number even for small values of l, m. Indeed, for $m = 20, l = 5$, the number equals 15,504. Furthermore, in many practical cases the number l is not even known *a priori*. Thus, one has to try feature combinations for different values of l and select the "best" value for it (beyond which no gain in performance is obtained) and the corresponding "best" l-dimensional feature vector.

Wrapper approach. As we will see in Chapter 10, sometimes it is desirable to base our feature selection decision not on the values of an adopted class separability criterion but on the performance of the classifier itself. That is, for each feature vector combination the classification error probability of the classifier has to be estimated and the combination resulting in the minimum error probability is selected. This approach may increase the complexity requirements even more, depending, of course, on the classifier type.

For both approaches, in order to reduce complexity, a number of efficient searching techniques have been suggested. Some of them are suboptimal and some optimal (under certain assumptions or constraints).

Suboptimal Searching Techniques

Sequential Backward Selection

We will demonstrate the method via an example. Let $m = 4$, and the originally available features are x_1, x_2, x_3, x_4. We wish to select two of them. The selection procedure consists of the following steps:

- Adopt a class separability criterion, C, and compute its value for the feature vector $[x_1, x_2, x_3, x_4]^T$.
- Eliminate one feature and for each of the possible resulting combinations, that is, $[x_1, x_2, x_3]^T$, $[x_1, x_2, x_4]^T$, $[x_1, x_3, x_4]^T$, $[x_2, x_3, x_4]^T$, compute the corresponding criterion value. Select the combination with the best value, say $[x_1, x_2, x_3]^T$.
- From the selected three-dimensional feature vector eliminate one feature and for each of the resulting combinations, $[x_1, x_2]^T$, $[x_1, x_3]^T$, $[x_2, x_3]^T$, compute the criterion value and select the one with the best value.

Thus, starting from m, at each step we drop out one feature from the "best" combination until we obtain a vector of l features. Obviously, this is a *suboptimal* searching procedure, since nobody can guarantee that the optimal two-dimensional vector has to originate from the optimal three-dimensional one. The number of combinations searched via this method is $1 + 1/2((m + 1)m - l(l + 1))$ (Problem 5.15), which is substantially less than that of the full search procedure.

Sequential Forward Selection

Here, the reverse to the preceding procedure is followed:

- Compute the criterion value for each of the features. Select the feature with the best value, say x_1.
- Form all possible two-dimensional vectors that contain the winner from the previous step, that is, $[x_1, x_2]^T, [x_1, x_3]^T, [x_1, x_4]^T$. Compute the criterion value for each of them and select the best one, say $[x_1, x_3]^T$.

If $l = 3$, then the procedure must continue. That is, we form all three-dimensional vectors springing from the two-dimensional winner, that is, $[x_1, x_3, x_2]^T, [x_1, x_3, x_4]^T$, and select the best one. For the general l, m case, it is simple algebra to show that the number of combinations searched with this procedure is $lm - l(l - 1)/2$. Thus, from a computational point of view, the backward search technique is more efficient than the forward one for l closer to m than to 1.

Floating Search Methods

The preceding two methods suffer from the so-called *nesting effect*. That is, once a feature is discarded in the backward method, there is no possibility for it to be reconsidered again. The opposite is true for the forward procedure; once a feature is chosen, there is no way for it to be discarded later on. In [Pudi 94] a technique is suggested that offers the flexibility to reconsider features previously discarded and, vice versa, to discard features previously selected. The technique is called the *floating search method*. Two schemes implement this technique. One springs from the forward selection, and the other from the backward selection rationale. We will focus on the former. We consider a set of m features, and the idea is to search for the best subset of k of them for $k = 1, 2, \ldots, l \leq m$ so that a cost criterion C is optimized. Let $X_k = \{x_1, x_2, \ldots, x_k\}$ be the set of the best combination of k of the features and Y_{m-k} the set of the remaining $m - k$ features. We also keep all the lower dimension best subsets $X_2, X_3, \ldots, X_{k-1}$ of $2, 3, \ldots, k - 1$ features, respectively. The rationale at the heart of the method is summarized as follows: At the next step the $k + 1$ best subset X_{k+1} is formed by "borrowing" an element from Y_{m-k}. Then, return to the previously selected lower dimension subsets to check whether the inclusion of this new element improves the criterion C. If it does, the new element replaces one of the previously selected features. The steps of the algorithm, when maximization of C is required are:

- *Step I: Inclusion* $x_{k+1} = arg\ \max_{y \in Y_{m-k}} C(\{X_k, y\})$; that is, choose that element from Y_{m-k} which, combined with X_k, results in the best value of C. $X_{k+1} = \{X_k, x_{k+1}\}$
- *Step II: Test*

  1. $x_r = arg\ \max_{y \in X_{k+1}} C(X_{k+1} - \{y\})$; that is, find the feature that has the least effect on the cost when it is removed from X_{k+1}.
  2. If $r = k+1$, change $k = k+1$ and go to step I.
  3. If $r \neq k+1$ AND $C(X_{k+1} - \{x_r\}) < C(X_k)$ go to step I; that is, if removal of x_r does not improve upon the cost of the previously selected best group of k, no further backward search is performed.
  4. If $k = 2$ put $X_k = X_{k+1} - \{x_r\}$ and $C(X_k) = C(X_{k+1} - \{x_r\})$; go to step I.

- *Step III: Exclusion*

  1. $X_k' = X_{k+1} - \{x_r\}$; that is, remove x_r.
  2. $x_s = arg\ \max_{y \in X_k'} C(X_k' - \{y\})$; that is, find the least significant feature in the new set.
  3. If $C(X_k' - \{x_s\}) < C(X_{k-1})$ then $X_k = X_k'$ and go to step I; no further backward search is performed.
  4. Put $X_{k-1}' = X_k' - \{x_s\}$ and $k = k - 1$.
  5. If $k = 2$ put $X_k = X_k'$ and $C(X_k) = C(X_k')$ and go to step I.
  6. Go to step III.1.

The algorithm is initialized by running the sequential forward algorithm to form X_2. The algorithm terminates when l features have been selected. Although the algorithm does not guarantee finding all the best feature subsets, it results in substantially improved performance compared with its sequential counterpart, at the expense of increased complexity. The backward floating search scheme operates in the reverse direction but with the same philosophy.

Optimal Searching Techniques

These techniques are applicable when the *separability criterion is monotonic*, that is,

$$C(x_1, \ldots, x_i) \leq C(x_1, \ldots, x_i, x_{i+1})$$

This property allows identifying the optimal combination but at a considerably reduced computational cost with respect to (5.30). Algorithms based on the *dynamic programming* concept (Chapter 8) offer one possibility to approaching the problem. A computationally more efficient way is to formulate the problem as a combinatorial optimization task and employ the so-called *branch and bound* methods to obtain the optimal solution [Lawe 66, Yu 93]. These methods compute the optimal value without involving exhaustive enumeration of all possible combinations. A more detailed description of the branch and bound methods is given in Chapter 15 and can also be found in [Fuku 90]. However, the complexity of these techniques is still higher than that of the previously mentioned suboptimal techniques.

Remark

- The separability measures and feature selection techniques presented above, although they indicate the major directions followed in practice, do not cover the whole range of methods that have been suggested. For example, in [Bati 94, Kwak 02, Leiv 07] the mutual information between the input features and the classifier's outputs is used as a criterion. The features that are selected maximize the input-output mutual information. In [Sind 04] the mutual information between the class labels of the respective features and those predicted by the classifier is used as a criterion. This has the advantage that only discrete random variables are involved. The existence of bounds that relate the probability of error to the mutual information function, for example, [Erdo 03, Butz 05], could offer a theoretically pleasing flavor to the adoption of information theoretic criteria for feature selection. In [Seti 97] a feature selection technique is proposed based on a decision tree by excluding features one by one and retraining the classifier. In [Zhan 02] the tabu combinatorial optimization technique is employed for feature selection.

Comparative studies of various feature selection searching schemes can be found in [Kitt 78, Devi 82, Pudi 94, Jain 97, Brun 00, Wang 00, Guyo 03]. The task of *selection bias*, when using the wrapper approach and how to overcome it is treated in [Ambr 02]. This is an important issue, and it has to be carefully considered in practice in order to avoid biased estimates of the error probability.

5.8 OPTIMAL FEATURE GENERATION

So far, the class separability measuring criteria have been used in a rather "passive" way, that is, to measure the classification effectiveness of features generated in *some* way. In this section we will employ these measuring criteria in an "active" manner, as an integral part of the feature generation process itself. From this point of view, this section can be considered as a bridge between this chapter and the following one. The method goes back to the pioneering work of Fisher ([Fish 36]) on *linear discrimination*, and it is also known as *linear discriminant analysis (LDA)*. We will first focus on the simplest form of the method in order to get a better feeling and physical understanding of its basic rationale.

The Two-class Case

Let our data points, $\boldsymbol{x}$, be in the m-dimensional space and assume that they originate from two classes. Our goal is to generate a feature y as a linear combination of the components of $\boldsymbol{x}$. In such a way, we expect to "squeeze" the classification-related information residing in $\boldsymbol{x}$ in a smaller number (in this case only one) of features. In this section, this goal is achieved by seeking the direction $\boldsymbol{w}$ in the m-dimensional space, *along which the two classes are best separated in some way*. This is not the only possible path for generating features via linear combination of measurements, and a number of alternative techniques will be studied in the next chapter.

Given an $\boldsymbol{x} \in \mathcal{R}^m$ the scalar

$$y = \frac{\boldsymbol{w}^T \boldsymbol{x}}{||\boldsymbol{w}||} \tag{5.31}$$

is the projection of $\boldsymbol{x}$ along $\boldsymbol{w}$. Since scaling all our feature vectors by the same factor does not add any classification-related information, we will ignore the scaling factor $||\boldsymbol{w}||$. We adopt the Fisher's discriminant ratio (FDR) (Section 5.6.3)

$$FDR = \frac{(\mu_1 - \mu_2)^2}{\sigma_1^2 + \sigma_2^2} \tag{5.32}$$

where μ_1, μ_2 are the mean values and σ_1^2, σ_2^2 the variances of y in the two classes ω_1 and ω_2, respectively, after the projection along $\boldsymbol{w}$. Using the definition in (5.31) and omitting $||\boldsymbol{w}||$, it is readily seen that

$$\mu_i = \boldsymbol{w}^T \boldsymbol{\mu}_i, \; i = 1, 2 \tag{5.33}$$

where $\boldsymbol{\mu}_i$, $i = 1, 2$, is the mean value of the data in ω_i in the m-dimensional space. Assuming the classes to be equiprobable and recalling the definition of S_b in Section 5.6.3, it is easily shown that

$$(\mu_1 - \mu_2)^2 = \boldsymbol{w}^T (\boldsymbol{\mu}_1 - \boldsymbol{\mu}_2)(\boldsymbol{\mu}_1 - \boldsymbol{\mu}_2)^T \boldsymbol{w} \propto \boldsymbol{w}^T S_b \boldsymbol{w} \tag{5.34}$$

where $\propto$ denotes proportionality. We now turn our attention to the denominator of (5.32). We have

$$\sigma_i^2 = E[(y - \mu_i)^2] = E[\boldsymbol{w}^T (\boldsymbol{x} - \boldsymbol{\mu}_i)(\boldsymbol{x} - \boldsymbol{\mu}_i)^T \boldsymbol{w}] = \boldsymbol{w}^T \Sigma_i \boldsymbol{w} \tag{5.35}$$

where for each $i = 1, 2$, samples $y(\boldsymbol{x})$ from the respective class ω_i have been used. Σ_i is the covariance matrix corresponding to the data of class ω_i in the m-dimensional space. Recalling the definition of S_w from Section 5.6.3, we get

$$\sigma_1^2 + \sigma_2^2 \propto \boldsymbol{w}^T S_w \boldsymbol{w} \tag{5.36}$$

Combining (5.36), (5.34), and (5.32), we end up that the optimal direction is obtained by maximizing Fisher's criterion

$$FDR(\boldsymbol{w}) = \frac{\boldsymbol{w}^T S_b \boldsymbol{w}}{\boldsymbol{w}^T S_w \boldsymbol{w}} \tag{5.37}$$

with respect to w. This is the celebrated generalized Rayleigh quotient, which, as it is known from linear algebra (Problem 5.16), is maximized if w is chosen such that

$$S_b w = \lambda S_w w \tag{5.38}$$

where λ is the largest eigenvalue of $S_w^{-1} S_b$. However, for our simple case we do not have to worry about any eigen decomposition. By the definition of S_b we have that

$$\lambda S_w w \propto (\mu_1 - \mu_2)(\mu_1 - \mu_2)^T w = \alpha(\mu_1 - \mu_2)$$

where α is a scalar. Solving the previous equation with respect to w, and since we are only interested in the direction of w, we can write

$$w = S_w^{-1}(\mu_1 - \mu_2) \tag{5.39}$$

assuming, of course, that S_w is invertible. As has already been discussed, in practice, S_w and S_b are approximated by averaging using the available data samples.

Figures 5.6a and 5.6b correspond to two examples for the special case of the two-dimensional space ($m = 2$). In both cases, the classes are assumed equiprobable and have the same covariance matrix Σ. Thus $S_w = \Sigma$. In Figure 5.6a, Σ is diagonal with equal diagonal elements, and w turns out to be parallel to $\mu_1 - \mu_2$. In Figure 5.6b, Σ is no more diagonal, and the data distribution does not have a spherical symmetry. In this case, the optimal direction for projection (the line on the left) is no more parallel to $\mu_1 - \mu_2$, and its direction changes in order to account for the shape of the data distribution. This simple example once again demonstrates that the right choice of the features is of paramount importance. Take as an example the case of generating a feature by projecting along the direction of the line on the right in Figure 5.6b. Then, the values that this feature takes for the two classes exhibit a heavy overlap.

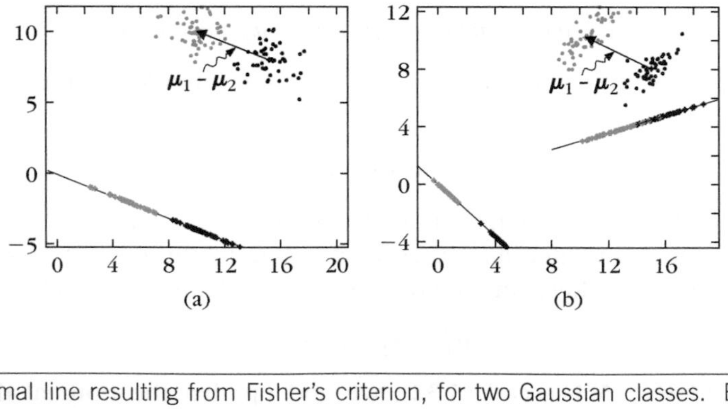

FIGURE 5.6

(a) The optimal line resulting from Fisher's criterion, for two Gaussian classes. Both classes share the same diagonal covariance matrix, with equal elements on the diagonal. The line is parallel to $\mu_1 - \mu_2$. (b) The covariance matrix for both classes is nondiagonal. The optimal line is on the left. Observe that it is no more parallel to $\mu_1 - \mu_2$. The line on the right is not optimal and the classes, after the projection, overlap.

Thus, we have reduced the number of features from m to 1 in an optimal way. Classification can now be performed based on y. Optimality guarantees that the class separability, with respect to y, is as high as possible, as this is measured by the FDR criterion.

In the case where both classes are described by Gaussian pdfs with equal covariance matrices, Eq. (5.39) corresponds to nothing else but the optimal Bayesian classifier with the exception of a threshold value (Problem 2.11 and Eqs. (2.44)–(2.46)). Moreover, recall from Problem 3.14 that this is also directly related to the linear MSE classifier. In other words, although our original goal was to generate a single feature (y) by linearly combining the m components of $\boldsymbol{x}$, we obtained something extra for free. Fisher's method performed feature generation and at the same time the design of a (linear) classifier; it combined the stages of feature generation and classifier design into a single one. The resulting classifier is

$$g(\boldsymbol{x}) = (\boldsymbol{\mu}_1 - \boldsymbol{\mu}_2)^T S_w^{-1} \boldsymbol{x} + w_0 \tag{5.40}$$

However, Fisher's criterion does not provide a value for w_0, which has to be determined. For example, for the case of two Gaussian classes with the same covariance matrix the optimal classifier is shown to take the form (see also Problem 3.14)

$$g(\boldsymbol{x}) = (\boldsymbol{\mu}_1 - \boldsymbol{\mu}_2)^T S_w^{-1} \left(\boldsymbol{x} - \frac{1}{2}(\boldsymbol{\mu}_1 + \boldsymbol{\mu}_2) \right) - \ln \frac{P(\omega_2)}{P(\omega_1)} \tag{5.41}$$

It has to be emphasized, however, that in the context of Fisher's theory the Gaussian assumption was not necessary to derive the direction of the optimal hyperplane. In practice, sometimes the rule in (5.41) is used even if we know that the data are non-Gaussian. Of course, other values of w_0 may be devised, according to the problem at hand.

Multiclass Case

The previous results, obtained for the two-class case, are readily generalized for the case of $M > 2$ classes. The multiclass LDA has been adopted as a tool for optimal feature generation in a number of applications, including biometrics and bioinformatics, where an original large number of features has to be compactly reduced. Our major task can be summarized as follows: If $\boldsymbol{x}$ is an m-dimensional vector of measurement samples, transform it into another l-dimensional vector $\boldsymbol{y}$ so that an adopted class separability criterion is optimized. We will confine ourselves to linear transformations,

$$\boldsymbol{y} = A^T \boldsymbol{x}$$

where A^T is an $l \times m$ matrix. Any of the criteria exposed so far can be used. Obviously, the degree of complexity of the optimization procedure depends heavily on the chosen criterion. We will demonstrate the method via the J_3 scattering matrix criterion, involving S_w and S_b matrices. Its optimization is straightforward, and at the same time it has some interesting implications. Let S_{xw}, S_{xb} be the within-class and between-class scatter matrices of $\boldsymbol{x}$. From the respective definitions, the corresponding matrices of $\boldsymbol{y}$ become

$$S_{yw} = A^T S_{xw} A, \quad S_{yb} = A^T S_{xb} A$$

Thus, the J_3 criterion in the $\boldsymbol{y}$ subspace is given by

$$J_3(A) = \text{trace}\{(A^T S_{xw} A)^{-1}(A^T S_{xb} A)\}$$

Our task is to compute the elements of A so that this is maximized. Then A must necessarily satisfy

$$\frac{\partial J_3(A)}{\partial A} = 0$$

It can be shown that (Problem 5.17)

$$\begin{aligned}\frac{\partial J_3(A)}{\partial A} &= -2S_{xw} A(A^T S_{xw} A)^{-1}(A^T S_{xb} A)(A^T S_{xw} A)^{-1} + 2S_{xb} A(A^T S_{xw} A)^{-1} \\ &= 0\end{aligned}$$

or

$$(S_{xw}^{-1} S_{xb})A = A(S_{yw}^{-1} S_{yb}) \tag{5.42}$$

An experienced eye will easily identify the affinity of this with an eigenvalue problem. It suffices to simplify its formulation slightly. Recall from Appendix B that the matrices S_{yw}, S_{yb} can be diagonalized simultaneously by a linear transformation

$$B^T S_{yw} B = I, \quad B^T S_{yb} B = D \tag{5.43}$$

which are the within- and between-class scatter matrices of the transformed vector

$$\hat{\boldsymbol{y}} = B^T \boldsymbol{y} = B^T A^T \boldsymbol{x}$$

B is an $l \times l$ matrix and D an $l \times l$ diagonal matrix. Note that in going from $\boldsymbol{y}$ to $\hat{\boldsymbol{y}}$ there is no loss in the value of the cost J_3. This is because J_3 is invariant under linear transformations, within the l-dimensional subspace. Indeed,

$$\begin{aligned}J_3(\hat{\boldsymbol{y}}) = \text{trace}\{S_{\hat{y}w}^{-1} S_{\hat{y}b}\} &= \text{trace}\{(B^T S_{yw} B)^{-1}(B^T S_{yb} B)\} \\ &= \text{trace}\{B^{-1} S_{yw}^{-1} S_{yb} B\} \\ &= \text{trace}\{S_{yw}^{-1} S_{yb} BB^{-1}\} = J_3(\boldsymbol{y})\end{aligned}$$

Combining (5.42) and (5.43), we finally obtain

$$(S_{xw}^{-1}S_{xb})C = CD \tag{5.44}$$

where $C = AB$ is an $m \times l$ dimensional matrix. Equation (5.44) is a typical eigenvalue-eigenvector problem, with the diagonal matrix D having the eigenvalues of $S_{xw}^{-1}S_{xb}$ on its diagonal and C having the corresponding eigenvectors as its columns. However, $S_{xw}^{-1}S_{xb}$ is an $m \times m$ matrix, and the question is which l out of a total of m eigenvalues we must choose for the solution of (5.44). From its definition, matrix S_{xb} is of rank $M-1$, where M is the number of classes (Problem 5.18). Thus, $S_{xw}^{-1}S_{xb}$ is also of rank $M-1$ and there are $M-1$ nonzero eigenvalues. Let us focus on the two possible alternatives separately.

- $l = M - 1$: We first form matrix C so that its columns are the unit norm $M - 1$ eigenvectors of $S_{xw}^{-1}S_{xb}$. Then we form the transformed vector

$$\hat{\boldsymbol{y}} = C^T \boldsymbol{x} \tag{5.45}$$

This guarantees the maximum J_3 value. *In reducing the number of data from m to $M - 1$, there is no loss in class separability power, as this is measured by J_3.* Indeed, recalling from linear algebra that the trace of a matrix is equal to the sum of its eigenvalues, we have

$$J_{3,x} = \text{trace}\{S_{xw}^{-1}S_{xb}\} = \lambda_1 + \cdots + \lambda_{M-1} + 0 \tag{5.46}$$

Also

$$J_{3,\hat{y}} = \text{trace}\{(C^T S_{xw} C)^{-1}(C^T S_{xb} C)\} \tag{5.47}$$

Rearranging (5.44), we get

$$C^T S_{xb} C = C^T S_{xw} C D \tag{5.48}$$

Combining (5.47) and (5.48), we obtain

$$J_{3,\hat{y}} = \text{trace}\{D\} = \lambda_1 + \cdots + \lambda_{M-1} = J_{3,x} \tag{5.49}$$

It is most interesting to view this from a slightly different perspective. Let us recall the Bayesian classifier for an M class problem. Of the M conditional class probabilities, $P(\omega_i|\boldsymbol{x})$, $i = 1, 2, \ldots, M$, only $M - 1$ are independent, since they all add up to one. In general, $M - 1$ is the *minimum* number of discriminant functions needed for an M-class classification task (Problem 5.19). *The linear operation $C^T \boldsymbol{x}$, which computes the $M - 1$ components of $\hat{\boldsymbol{y}}$, can be seen as an optimal linear rule that provides $M - 1$ discriminant functions, where optimality is with respect to J_3.* This was clearly demonstrated in the two-class case, where Fisher's method was also used as a classifier (subject to an unknown threshold).

Investigating the specific form that Eq. (5.45) takes for the two-class problem, one can show that for $M = 2$ there is only one nonzero eigenvalue, and it turns out that (Problem 5.20)

$$\hat{y} = (\boldsymbol{\mu}_1 - \boldsymbol{\mu}_2)^T S_{xw}^{-1} \boldsymbol{x}$$

which is our familiar Fisher's linear discriminant.

- $l < M-1$: In this case C is formed from the eigenvectors corresponding to the l largest eigenvalues of $S_{xw}^{-1} S_{xb}$. The fact that J_3 is given as the sum of the corresponding eigenvalues guarantees its maximization. Of course, in this case there is loss of the available information because now $J_{3,\hat{y}} < J_{3,x}$.

A geometric interpretation of (5.45) reveals that $\hat{\boldsymbol{y}}$ is the projection of the original vector $\boldsymbol{x}$ onto the subspace spanned by the eigenvectors $\boldsymbol{v}_i$ of $S_w^{-1} S_b$. It must be pointed out that these *are not* necessarily mutually orthogonal. Indeed, although matrices S_w, S_b (S_m) are symmetric, products of the form $S_w^{-1} S_b$ are not; thus, the eigenvectors are not mutually orthogonal (Problem 5.21). Furthermore, as we saw during the proof, once we decide on which subspace to project (by selecting the appropriate combination of eigenvectors) *the value of J_3 remains invariant under any linear transformation within this subspace*. That is, it is independent of the coordinate system, and its value depends only on the particular subspace. In general, projection of the original feature vectors onto a lower dimensional subspace is associated with some information loss. An extreme example is shown in Figure 5.7, where the two classes coincide after projection on the v_2 axis. On the other hand, from all possible projection directions, Fisher's linear discrimination rule leads to

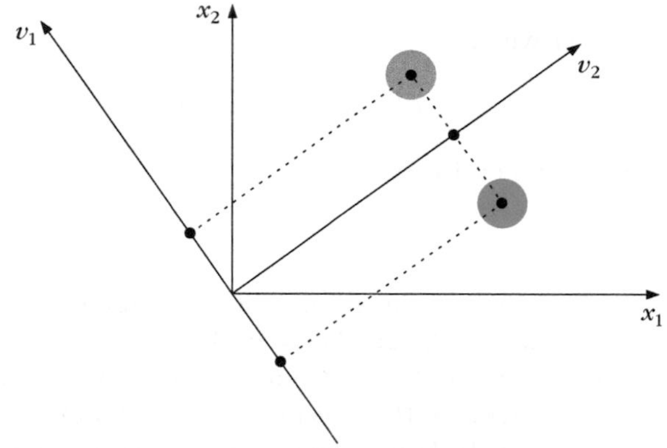

FIGURE 5.7

Geometry illustrating the loss of information associated with projections in lower dimensional subspaces. Projecting onto the direction of the principle eigenvector, $\boldsymbol{v}_1$, there is no loss of information. Projection on the orthogonal direction results in a complete class overlap.

the choice of the one-dimensional subspace v_1, which corresponds to the optimal J_3 value, that guarantees no loss of information for $l = M - 1 = 1$ (as this is measured by the J_3 criterion). Thus, this is a good choice, provided that J_3 is a good criterion for the problem of interest. Of course, this is not always the case; it depends on the specific classification task. For example, in [Hams 08] the criterion used is the probability of error for a multiclass task involving normally distributed data. A more extensive treatment of the topic, also involving other optimizing criteria, can be found in [Fuku 90].

Remarks

- If J_3 is used with another combination of matrices, such as S_w and S_m, then, in general, the rank of the corresponding matrix product involved in the trace is m and there are m nonzero eigenvalues. In such cases, the transformation matrix C is formed so that its columns are the eigenvectors corresponding to the *l largest eigenvalues*. According to (5.49), this guarantees the maximum value of J_3.

- In practice, one may encounter cases in which S_w is not invertible. This occurs in applications where the available size of the training set, N, is smaller than the dimensionality, m, of the original feature space. In such cases the resulting estimate of S_w, which is obtained as the mean of N outer vector products, has rank lower than m; hence it is singular. This is known as the small sample size (SSS) problem. Web document classification, face recognition, and disease classification based on gene-expression profiling are some examples where the small sample size problem occurs frequently in practice.
  One way to overcome this difficulty is to use the pseudoinverse S_w^+ in place of S_w^{-1} [Tian 86]. However, now, there is no guarantee that the J_3 criterion is maximized by selecting the eigenvectors of $S_w^+ S_b$ corresponding to the largest eigenvalues. An alternative route is to employ regularization techniques, in one way or another, for example, [Frie 89, Hast 95]. For example, S_w may be replaced by $S_w + \sigma\Omega$, where Ω can be any positive definite and symmetric matrix. The specific choice depends on the problem. The choice of σ is also a critical factor here. Another drawback of these techniques is that they do not scale well for problems with large dimensionality. For example, in certain tasks of face recognition, the resulting covariance matrices can be as high as a few thousand making matrix inversion a computationally thirsty task.
  Another way to deal with the small sample size problem is to adopt a two-stage approach. One such technique is the so-called PCA+LDA technique. In the first stage, principle component analysis (PCA, see Chapter 6) is performed to reduce, appropriately, the dimensionality of the feature space and linear discriminant analysis (LDA) is then performed in the low-dimensional space, for example, [Belh 97]. A drawback of this technique is that during the dimension reduction phase part of the discriminatory information may be lost.

In [Yang 02] the mixture scatter matrix, S_m, is used in the J criterion in the place of S_w. It is shown that in this case, applying first a PCA on S_m, to reduce the dimensionality to the value of the rank of S_m, followed by an LDA in the reduced space, does not lead to any loss of information. In [Chen 00] the null space of the within-class scatter matrix is brought into the game. It has been observed that the null space of S_w contains useful discriminant information. The method first projects onto the null space and, then, in the projected space the transformation that maximizes the between-class scatter is computed. A disadvantage of this approach is that it may lose information by considering the null space instead of S_w. A second problem is that the complexity of determining the null space of S_w is very high. Computational difficulties of the method are addressed in [Cevi 05]. In [Ye 05], in the first stage, dimensionality reduction is achieved by maximizing the between-class cluster (S_b), via a QR decomposition technique. In the second stage, a refinement is achieved by focusing on the within-class scatter issue, following arguments similar to the classical LDA. A unifying treatment of a number from the previous techniques is considered in [Zhan 07].

A different approach is proposed in [Li 06]. Instead of the J_3 criterion, another criterion is introduced that involves the trace of the difference of the involved matrices, thus bypassing the need for inversions.

Besides the small sample size problem, another issue associated with the LDA is that the number of features that can be generated is at most one less than the number of classes. As we have seen, this is due to the rank of the matrix product $S_w^{-1}S_b$. For an M-class problem, there are only $M-1$ nonzero eigenvalues. All the J_3 related discriminatory information can be recovered by projecting onto the subspace generated by the eigenvectors associated with these nonzero eigenvalues. Projecting on any other direction adds no information.

Good insight into it can be gained through geometry by considering a simple example. Let us assume, for simplicity, a two-class task with classes normally distributed with covariance matrices equal to the identity matrix. Then by its definition, S_w is also an identity matrix. It is easy to show (Problem 5.20) that in this case the eigenvector corresponding to the only nonzero eigenvalue is equal to $\boldsymbol{\mu}_1 - \boldsymbol{\mu}_2$. The (Euclidean) distance between the mean values of the projection points in the (nonzero) eigenvector direction is the same as the distance between the mean values of the classes in the original space, i.e., $||(\boldsymbol{\mu}_1 - \boldsymbol{\mu}_2)||$. This can easily be deduced by visual inspection of Figure 5.7, which corresponds to a case such as is discussed our example. Projecting on the orthogonal direction adds no information since the classes coincide. All the scatter information, with respect to both classes, is obtained from a single direction.

Due to the previous drawback, there are cases where the number of classes M is small, and the resulting number of, at most, $M-1$ features is insufficient. An attempt to overcome this difficulty is given in [Loog 04]. The main idea is to employ a different to S_b measure to quantify the between-class scatter. The Chernoff distance (closely related to the Bhattacharyya distance of Section 5.6.2) is employed. This change offers the possibility of reducing the dimensionality to any dimension l smaller than the original m. A different path is followed in [Kim 07]. From the original m features, the authors build a number of so-called composite vectors. Each vector consists of a subset of the m features. Different composite vectors are allowed to share some of the original features. LDA is then performed on this new set of feature vectors. This procedure enhances the range of the rank of the involved matrix product beyond $M-1$. In [Nena 07], the shortcomings of LDA are overcome by defining a new class-separability measure based on an information-theoretic cost inspired by the concept of mutual information.

- No doubt, scattering matrix criteria are not the only ones that can be used to compute the optimal transformation matrix. For example, [Wata 97] suggested using ta different transformation matrix for each class and optimizing with respect to the classification error. This is within the spirit of the recent trend, to optimize directly with respect to the quantity of interest, which is the classification error probability. For the optimization, smooth versions of the error rate are used to guarantee differentiability. Other ways to compute the transformation matrix will be discussed in the next chapter.

- Besides the linear nonlinear transformations can also be employed for optimal feature selection. For example, in [Samm 69] a nonlinear technique is proposed that attempts to preserve maximally all the distances between vectors. Let $\boldsymbol{x}_i, \boldsymbol{y}_i, i = 1, 2, \ldots, N$, be the feature vectors in the original m-dimensional and the transformed l-dimensional space, respectively. The transformation into the lower dimensional space is performed so as to maximize

$$J = \frac{1}{\sum_{i=1}^{N-1}\sum_{j=i+1}^{N} d^o(i,j)} \sum_{i=1}^{N-1}\sum_{j=i+1}^{N} \frac{(d^o(i,j) - d(i,j))^2}{d^o(i,j)} \tag{5.50}$$

where $d^o(i,j)$, $d(i,j)$ are the (Euclidean) distances between vectors $\boldsymbol{x}_i$, and $\boldsymbol{x}_j$ in the original space and $\boldsymbol{y}_i$, $\boldsymbol{y}_j$ in the transformed space, respectively.

- Another nonlinear generalization of the method consists of two (implicit) steps. First, one employs a nonlinear vector function to transform the input feature space into a higher-dimensional one, which can even be of infinite dimension. Then, the linear discriminant method is applied in this high-dimensionality space. However, the problem formulation is done so that vectors appear only via inner products. This allows the use of kernel functions to facilitate computations, as was the case with the nonlinear support vector machines presented in Chapter 4 [Baud 00, Ma 03].

5.9 NEURAL NETWORKS AND FEATURE GENERATION/SELECTION

Recently, efforts have been made to use neural networks for feature generation and selection. A possible solution is via the so-called *auto-associative networks*. A network is employed having m input and m output nodes and a single hidden layer with l nodes with linear activations. During training, the desired outputs are the same as the inputs. That is,

$$\mathcal{E}(i) = \sum_{k=1}^{m} (\hat{y}_k(i) - x_k(i))^2$$

where the notation of the previous chapter has been adopted. Such a network has a unique minimum, and the outputs of the hidden layer constitute the projection of the input m-dimensional space onto an l-dimensional subspace. In [Bour 88] it is shown that this is basically a projection onto the subspace spanned by the l principal eigenvectors of the input correlation matrix, a topic that will be our focus in the next chapter. An extension of this idea is to use three hidden layers [Kram 91]. Such a network performs a nonlinear principal component analysis. The major drawback of such an architecture is that nonlinear optimization techniques have to be used for the training. Besides the computational load, the risk of being trapped in local minima is always present.

An alternative is to use neural networks, or any other (non)linear structure, to exploit properties of the LS cost function. In Chapter 3, we saw that the outputs of a network approximate posterior probabilities, provided that the weights have been trained so that the outputs match, in the LS sense, the class labels. In [Lowe 91] it is pointed out that, besides this property, another very interesting one is also valid. A multilayer perceptron was considered with linear output nodes. The network was trained to minimize the squared error between the actual and desired responses (i.e., class labels 1 and 0). It was shown that minimizing the squared error is equivalent to maximizing the criterion

$$J = \text{trace}\{S_m^{-1} S_b\} \tag{5.51}$$

where S_m is the mixture scatter matrix of the vectors formed by the outputs of the last hidden layer nodes and S_b the corresponding between-class scatter matrix in a weighted form (Problem 5.22). If the inverse of S_m does not exist, it is replaced by its pseudoinverse. In other words, such a network can be used as a *J-optimal nonlinear* transformer of the input m-dimensional vectors into l-dimensional vectors, where l is the number of nodes in the last hidden layer.

Another approach is to employ neural networks to perform the computations associated with the optimization of various class separability criteria discussed in this chapter. Although these techniques do not necessarily provide new approaches, the incorporation of neural networks offers the capability of adaptation in case the statistics of the input data are slowly varying. In [Chat 97, Mao 95] a number of such techniques are developed. The idea behind most of these techniques is to use a network that iteratively computes eigenvectors of correlation matrices, a step that, as we have seen, is at the heart of a number of optimality criteria.

An alternative technique has been suggested in [Lee 93, Lee 97]. They have shown that the discriminantly informative feature vectors have a component that is normal to the decision surface at least at one point on the surface. Furthermore, the less informative vectors are orthogonal to a vector normal to the decision surface at every point of the surface. This is natural, because vectors that do not have a component normal to the decision surface cannot cross it (and hence change classes) whatever their value is. Based on this observation, they estimate normal vectors to the decision boundary, using gradient approximation techniques, which are then used to formulate an appropriate eigenvalue-eigenvector problem leading to the computation of the transformation matrix.

Finally, pruning a neural network is a form of feature selection integrated into the classifier design stage. Indeed, the weights of the input nodes corresponding to less important features are expected to be small. As discussed in Chapter 4, the incorporation of appropriate regularization terms in the cost function encourages such weights to converge to zero and ultimately to be eliminated. This approach was followed, for example, in [Seti 97].

5.10 A HINT ON GENERALIZATION THEORY

So far in this book, two major issues have occupied us: the design of the classifier and its generalization capabilities. The design of the classifier involved two stages: the choice of the classifier type and the choice of the optimality criterion. The generalization capabilities led us to seek ways to reduce the feature space dimensionality. In this section we will point out some important theoretical results that relate the size N of the training data set and the generalization performance of the designed classifier.

To this end, let us summarize a few necessary basic steps and definitions.

- Let $\mathcal{F}$ be the set of all the functions f that can be realized by the adopted classifier scheme. For example, if the classifier is a multilayer perceptron with a given number of neurons, then $\mathcal{F}$ is the set of all the functions that can be realized by the specific network structure. Functions f are mappings from $\mathcal{R}^l \rightarrow \{0, 1\}$. Thus, the response is either 1 or 0; that is, the two-class problem is considered, and the mapping is either $f(\boldsymbol{x}) = 1$ or $f(\boldsymbol{x}) = 0$.

- Let $P_e^N(f)$ be the *empirical* classification error probability, based on the available input—desired output training pairs $(\boldsymbol{x}_i, y_i)$, $i = 1, 2, \ldots, N$, which are considered to be independent and identically distributed (i.i.d.). Thus, $P_e^N(f)$ is the fraction of training samples for which an error occurs, that is, $f(\boldsymbol{x}_i) \neq y_i$. Obviously, this depends on the specific function f and the size N. The optimal function that results from minimizing this empirical cost is denoted by f^* and belongs to the set $\mathcal{F}$.

- $P_e(f)$ is the true classification error probability when a function f is realized. The corresponding empirical $P_e^N(f)$ can be very small, even zero, since a classifier can be designed to classify all training feature vectors correctly. However, $P_e(f)$ is the important performance measure, because it measures error probability based on the statistical nature of the data and not on the specific training set only. *For a classifier with good generalization capabilities, we expect the empirical and the true error probabilities to be close*. $P_e(f)$ is sometimes known as the *generalization error* probability.
- P_e denotes the minimum error probability over all the functions of the set, that is, $P_e = \min_{f \in \mathcal{F}} P_e(f)$.[1] Again, in practice we would like the optimal empirical error $P_e^N(f^*)$ to be close to P_e.

The Vapnik–Chervonenkis theorem is as follows.

Theorem *Let $\mathcal{F}$ be the class of functions of the form $\mathcal{R}^l \rightarrow \{0,1\}$. Then the empirical and true error probabilities corresponding to a function f in the class, satisfy*

$$prob\{\max_{f \in \mathcal{F}} |P_e^N(f) - P_e(f)| > \epsilon\} \leq 8\mathcal{S}(\mathcal{F}, N)\exp(-N\epsilon^2/32) \tag{5.52}$$

The term $\mathcal{S}(\mathcal{F}, N)$ is called the *shatter* coefficient of the class $\mathcal{F}$. This is defined as the *maximum* number of dichotomies of N points that can be formed by the functions in $\mathcal{F}$. From our combinatorics basics, we know that the maximum number of dichotomies on a set of N points (separating them into two distinct subsets) is 2^N. However, not all these combinations can be implemented by a function $f: \mathcal{R}^l \rightarrow \{0,1\}$. For example, we know that, in the two-dimensional space, the set of functions realized by a perceptron (hyperplane) can form only fourteen distinct dichotomies on four points out of the $16 = 2^4$ possibilities. The two XOR combinations cannot be realized. However, the class of functions realized by the perceptron can form all possible $8 = 2^3$ dichotomies for $N = 3$ points. This leads us to the following definition

Definition 1. *The largest integer $k \geq 1$ for which $\mathcal{S}(\mathcal{F}, k) = 2^k$ is called the Vapnik–Chervonenkis, or VC dimension of the class $\mathcal{F}$, and is denoted by V_c. If $\mathcal{S}(\mathcal{F}, N) = 2^N$ for every N, then the VC dimension is infinite.*

Thus, in the two dimensional space, the VC dimension of a single perceptron is 3. In the general l-dimensional space case, the VC dimension of a perceptron is $l + 1$, as is easily verified from Section 4.13. It will not come as a surprise to say that the VC dimension and the shatter coefficient are related, because they have common origins. Indeed, this is true. It turns out that if the VC dimension is finite, then the following bound is valid

$$\mathcal{S}(\mathcal{F}, N) \leq N^{V_c} + 1 \tag{5.53}$$

[1] Strictly speaking, in this section *inf* must be used instead of *min* and *sup* instead of *max*.

That is, the shatter coefficient is either 2^N or is bounded as given in (5.53). This bound has a very important implication for the usefulness of (5.52). Indeed, for finite VC dimensions (5.53) guarantees that for large enough N the shatter coefficient is bounded by *polynomial growth*. Then the bound in (5.52) is dominated by its exponential decrease, and it tends to zero as $N \to \infty$. In words, *for large N the probability of having large differences between the empirical and the true probability errors is very small! Thus, the network guarantees good generalization performance for large N*. Furthermore, the theory guarantees another strong result [Devr 96]

$$\text{prob}\{P_e(f^*) - \min_{f \in \mathcal{F}} P_e(f) > \epsilon\} \leq 8\mathcal{S}(\mathcal{F}, N) \exp(-N\epsilon^2/128) \tag{5.54}$$

That is, for large N we expect with high probability the performance of the empirical error optimal classifier to be close to the performance of the optimal one, over the specific class of functions.

Let us look at these theoretical results from a more intuitive perspective. Consider two different networks with VC dimensions $V_{c1} << V_{c2}$. Then if we fix N and ϵ, we expect the first network to have better generalization performance, because the bound in (5.52) will be much tighter. Thus, the probability that the respective empirical and true errors will differ more than the predetermined quantity will be much smaller. We can think of the VC dimension as an *intrinsic capacity of a network, and only if the number of training vectors exceeds this number sufficiently can we expect good generalization performance*.

Learning theory is rich in bounds that have been derived and that relate quantities such as the empirical error, the true error probability, the number of training vectors, and the VC dimension or a VC related quantity. In his elegant theory of learning, Valiant [Vali 84] proposed to express such bounds in the flavor of statistical tests. That is, the bounds involve an error ϵ, such as in Eqs. (5.52) and (5.54), and a confidence probability level that the bound holds true. Such bounds are known as PAC bounds, which stands for Probably (the probability of the bound to fail is small) Approximately Correct (when the bound holds, the error is small). A very interesting (for our purposes) bound that can be derived refers to the minimum number of training points that guarantee, with high probability, the design of a classifier with good error performance. Let us denote this minimum number of points as $N(\epsilon, \rho)$. It can be shown that if

$$N(\epsilon, \rho) \leq \max\left(\frac{k_1 V_c}{\epsilon^2} \ln \frac{k_2 V_c}{\epsilon^2}, \frac{k_3}{\epsilon^2} \ln \frac{8}{\rho}\right) \tag{5.55}$$

then for any number of training points $N \geq N(\epsilon, \rho)$ the optimal classifier, f^*, resulting by minimizing the empirical error probability $P_e^N(f)$ satisfies the bound

$$P\{P_e(f^*) - P_e > \epsilon\} \leq \rho \tag{5.56}$$

where k_1, k_2, k_3 are constants [Devr 96]. In other words, for small values of ϵ and ρ, if $N \geq N(\epsilon, \rho)$, the performance of the optimum empirical error classifier is guaranteed, with high probability, to be close to the optimal classifier in the class of functions $\mathcal{F}$, realized by the specific classification scheme. The number $N(\epsilon, \rho)$ is also known as *sample complexity*. Observe that the first of the two terms in the bound has a linear dependence on the VC dimension and an inverse quadratic dependence on the error ϵ. Doubling, for example, the VC dimension roughly requires that we need to double the number of training points in order to keep the same ϵ and confidence level. On the other hand, doubling the accuracy (i.e., $\epsilon/2$) requires us to quadruple the size of the training set. The confidence level ρ has a little influence on the bound, due to its logarithmic dependence. Thus, high VC dimension sets high demands on the number of training points required to guarantee, with high probability, a classifier with good performance.

Another related bound of particular interest to us that holds with a probability at least $1 - \rho$ is the following:

$$P_e(f) \leq P_e^N(f) + \phi\left(\frac{V_c}{N}\right) \tag{5.57}$$

where V_c is the VC dimension of the corresponding class and

$$\phi\left(\frac{V_c}{N}\right) \equiv \sqrt{\frac{V_c\left(\ln\left(\frac{2N}{V_c}+1\right)\right) - \ln(\rho/4)}{N}} \tag{5.58}$$

The interested reader may obtain more bounds and results concerning the Vapnik-Chervonenkis theory from [Devr 96, Vapn 95]. It will take some effort, but it is worth it! In some of the published literature, the constants in the bounds are different. This depends on the way the bounds are derived. However, this is not of major practical importance, since the essence of the theory remains the same.

Due to the importance of the VC dimension, efforts have been made to compute it for certain classes of networks. In [Baum 89] it has been shown that the VC dimension of a multilayer perceptron with hard limiting activation functions in the nodes is bounded by

$$2\left\lfloor\frac{K_n^h}{2}\right\rfloor l \leq V_c \leq 2K_w \log_2(eK_n) \tag{5.59}$$

where K_n^h is the total number of hidden layer nodes, K_n the total number of nodes, K_w the total number of weights, l the input space dimension, e the base of the natural logarithm, and $\lfloor\cdot\rfloor$ the floor operator that gives the largest integer less than its argument. The lower bound holds only for networks with a single hidden layer and full connectivity between the layers. A similar upper bound is true for RBF networks too. Looking at this more carefully, one can say that for such networks the VC dimension is roughly given by the number of weights of the network, that is, the number of its free parameters to be determined! In practice, good generalization performance is expected if the number of training samples is a few times the VC dimension. A good rule of thumb is to choose N to be of the order of 10 times the VC dimension or more [Hush 93].

Besides the Vapnik–Chervonenkis theory, the published literature is rich in results concerning aspects of designing various classifiers using a finite data set N. Although they lack the elegance of the generality of the Vapnik–Chervonenkis theory, they provide further insight into this important task. For example, in [Raud 91] asymptotic analytic results are derived for a number of classifiers (linear, quadratic, etc.) under the Gaussian assumption of the involved densities. The classification error probability of a classifier designed using a finite set of N training samples is larger, by an amount Δ_N, than the error of the same classifier designed using an infinite ($N \to \infty$) set of data. It is shown that the mean of Δ_N (over different design sets) decreases as N tends to infinity. The rate of decrease depends on the specific type of classifier, on the dimensionality of the problem, and also on the value of the asymptotic ($N \to \infty$) error. It turns out that in order to keep the mismatch Δ_N within certain limits, *the number N of design samples must be a number of times larger than the dimension l*. Depending on the type of classifier, this proportionality constant can range from low values (e.g., 1.5) up to a few hundred! Furthermore, in [Fuku 90] it is shown that keeping N constant and increasing l, beyond a point, results in an increase of the classification error. This is known as the *Hughes phenomenon*, and it was also discussed in Section 5.3.

All these theoretical results provide useful guidelines in selecting appropriate values for N and l for certain types of classifiers. Moreover, they *make crystal clear the urge to keep the number of features as small as possible with respect to N and the importance of the feature selection stage in the design of a classification system*. In the fringes of this theoretical "happening," a substantial research effort has also been devoted to experimental comparisons, involving different classifiers, with respect to their generalization capabilities; see, for example, [Mama 96] and the references therein. In practice, however, experience and empirical results retain an important part in the final decision. Engineering still has a flavor of art!

Structural Risk Minimization

In our discussion so far, we have focused on the effects of the finite size of the training data set, N, for a given class of functions, that is, a given classifier structure. Let us now touch on another important issue. If we allow N to grow indefinitely, does this luxury provide us with the means not only to have good generalization properties but also to improve our classification error so as to approach the optimal Bayesian performance? Recall that as N grows, we can expect to obtain the optimal performance with respect to all *allowable* sets of classifiers that can be implemented by the chosen network structure. However, the error rate of the corresponding optimal classifier may still be very far from that of the Bayesian classifier. Let us denote by P_B the Bayesian error probability. Then we can write

$$P_e(f^*) - P_B = \left(P_e(f^*) - P_e\right) + (P_e - P_B) \tag{5.60}$$

A diagrammatic interpretation of Eq. (5.60) is given in Figure 5.8. The right-hand side in Eq. (5.60) consists of two conflicting terms. If the class $\mathcal{F}$ is too small, then

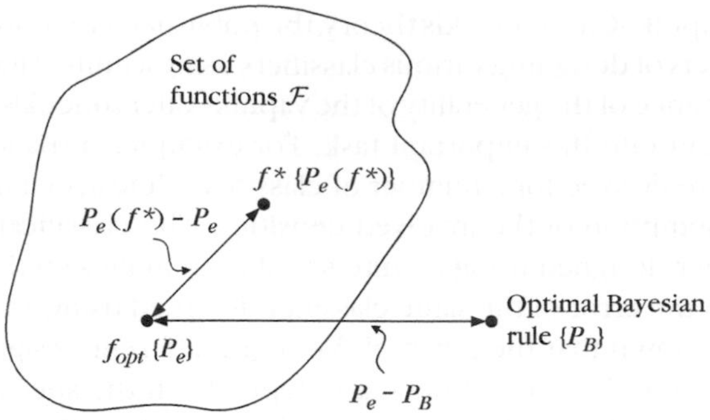

FIGURE 5.8

Diagrammatic interpretation of Eq. (5.60). The optimal function over the set $\mathcal{F}$, associated with the minimum error P_e, is denoted as f_{opt}, and f^* is the optimal function resulting from the empirical cost for a given N.

the first term is expected to be small, but the second term is likely to be large. If, on the other hand, the class of functions $\mathcal{F}$ is large, then the second term is expected to be small but the first term is probably large. This is natural, because the larger the set of functions, the higher the probability of including in the set a good approximation of the Bayesian classifier. Moreover, the smaller the class, the less the variation between its members. This reminds us of the bias–variance dilemma we discussed in Chapter 3. A little thought suffices to reveal that the two problems are basically the same, seen from a different viewpoint. Then the natural question arises once more, can we make both terms small and how? The answer is that this is possible only asymptotically, provided that *at the same time* the size of the class $\mathcal{F}$ grows appropriately. An elegant strategy to achieve this has been suggested by Vapnik and Chervonenkis [Vapn 82].

Let $\mathcal{F}^{(1)}, \mathcal{F}^{(2)}, \ldots$ be a sequence of nested classes of functions, that is,

$$\mathcal{F}^{(1)} \subset \mathcal{F}^{(2)} \subset \mathcal{F}^{(3)} \subset \cdots \tag{5.61}$$

with an increasing, yet finite, VC dimension,

$$V_{c,\mathcal{F}^{(1)}} \leq V_{c,\mathcal{F}^{(2)}} \leq V_{c,\mathcal{F}^{(3)}} \leq \cdots \tag{5.62}$$

Also let

$$\lim_{i\to\infty} \inf_{f \in \mathcal{F}^{(i)}} P_e(f) = P_B \tag{5.63}$$

For each N and class of functions $\mathcal{F}^{(i)}$, $i = 1, 2, \ldots$, compute the optimum, $f^*_{N,i}$, with respect to the *empirical* error using the N training pairs of input–output samples. Vapnik and Chervonenkis suggest choosing for each N the function f^*_N according to the *structural risk minimization principle* (SRM). This consists of the following two steps. First we select the classifier $f^*_{N,i}$ from every class $\mathcal{F}^{(i)}$ that minimizes the corresponding empirical error over the class of functions. Then, from all these classifiers, we choose the one that minimizes the upper bound in (5.57), over all i. More precisely, form the so-called *guaranteed error bound*,

$$\tilde{P}_e(f^*_{N,i}) \equiv P^N_e(f^*_{N,i}) + \phi\left(\frac{V_{c,\mathcal{F}^{(i)}}}{N}\right) \tag{5.64}$$

and choose

$$f^*_N = arg \min_i \tilde{P}_e(f^*_{N,i}) \tag{5.65}$$

Then, as $N \to \infty$, $P_e(f^*_N)$ tends to P_B with probability one. Note that the second term in the minimized bound, $\phi\left(\frac{V_{c,\mathcal{F}^{(i)}}}{N}\right)$, is a *complexity penalty term* that increases as the network complexity increases (i.e., with the size of the class of functions and $V_{c,\mathcal{F}^{(i)}}$). If on one hand the classifier model is too simple, the penalty term is small but the empirical error term will be large in (5.64). On the other hand, if the model is complex, the empirical error is small but the penalty term large. The structural risk minimization criterion aims at achieving the best trade-off between these two terms. This is illustrated in Figure 5.9.

From this point of view, the structural risk minimization principle belongs to a more general class of approaches that try to estimate the order of a system, by considering simultaneously the model complexity and a performance index. Depending

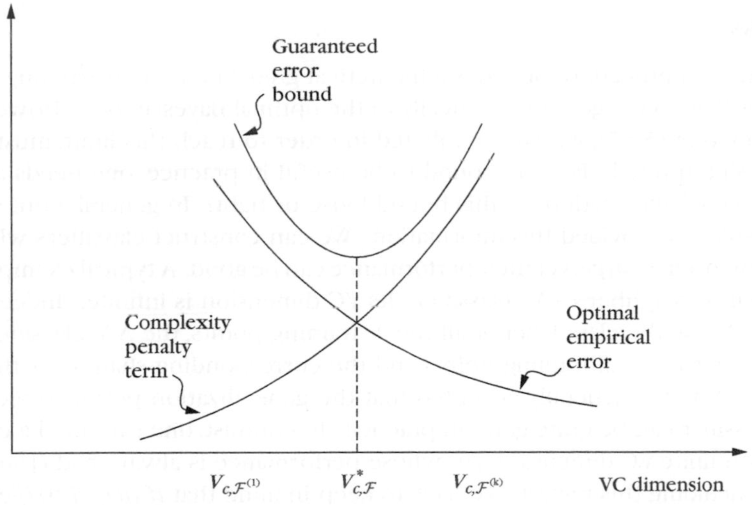

FIGURE 5.9

For a fixed N, the complexity penalty term increases and the optimal empirical error decreases as the VC dimension of the model increases. Choosing the model according to the SRM principle aims at achieving the best trade-off between these two terms that corresponds to the minimum of the guaranteed error bound. Note that, $V_{c,\mathcal{F}^{(1)}} < V^*_{c,\mathcal{F}} < V_{c,\mathcal{F}^{(k)}}$, which implies $\mathcal{F}^{(1)} \subset \mathcal{F}^* \subset \mathcal{F}^{(k)}$.

on the function used to measure the model complexity and the corresponding performance index, different criteria result. For example, in the Akaike Information Criterion [Akai 74], the place of the empirical error is taken by the value of the log-likelihood function, corresponding to the maximum likelihood estimates of the unknown parameters, and the complexity term is proportional to the number of free parameters to be estimated. See also Sections 5.11 and 16.4.1.

An alternative interpretation of the SVM cost function in Eq. (3.93) is given in [Lin 02]. It is treated as a typical regularization method with two components a data fit functional term $\left(\sum_i I(\xi_i)\right)$ and a regularization penalty term ($||\boldsymbol{w}||^2$). The latter is the complexity-related component and is used to guard against overfitting. In general, the data-fit term approaches a limiting functional as $N \to \infty$. Under some general conditions, the estimate resulting from the regularization method is the minimizer of this data-fit-limiting functional, as $N \to \infty$. It turns out that in the SVM case the minimizer of the limiting data-fit functional is the Bayes optimal rule and the SVM solution approaches it, as $N \to \infty$, provided that the kernel choice guarantees a rich enough space (RKHS) and the smoothing parameter, C, is chosen appropriately. This interesting result nicely ties the SVM and the Bayesian optimal rule. Would it be an exaggeration to say that a good theory is like a good piece of art, in the sense that both support more than one interpretation?

Remarks

- The SRM procedure provides a theoretical guideline for constructing a classifier that converges asymptotically to the optimal Bayesian one. However, the bound in (5.57), which is exploited in order to reach this limit, must not be misinterpreted. For any bound to be useful in practice, one needs an extra piece of information. Is this bound loose or tight? In general, until now, no result has provided this information. We can construct classifiers whose VC dimension is large, yet their performance can be good. A typical example is the nearest neighbor (*NN*) classifier. Its VC dimension is infinite. Indeed, since we know the class label of all the N training points, the *NN* classifier classifies correctly all training points and the corresponding shatter coefficient is 2^N. Yet, it is generally accepted that the generalization performance of this classifier can be quite good in practice. In contrast, one can build a classifier with finite VC dimension, yet whose performance is always bad ([Burg 98]). Concluding this remark, we have to keep in mind that *if two classifiers have the same empirical error, it does not, necessarily, mean that the one with the smaller VC dimension leads to better performance.*

- Observe that in all bounds given previously no assumptions have been made about the statistical distribution underlying the data set. That is, they are *distribution-free* bounds.

Support Vector Machines: A Last Touch

We have already discussed that the VC dimension of a linear classifier in the l-dimensional space is $l+1$. However, hyperplanes that are *constrained* to leave the maximum margin between the classes *may have a smaller VC dimension*.

Let us assume that r is the radius of the *smallest* (hyper)sphere that encloses all the data (Problem 5.23), that is,

$$\|\boldsymbol{x}_i\| \leq r, \ i = 1, 2, \ldots, N$$

Then if a hyperplane satisfies the conditions in Eq. (3.73) and

$$\|\boldsymbol{w}\|^2 \leq c$$

where c is a constant, then its VC dimension, V_c, is bounded by ([Vapn 98])

$$V_c \leq \min(r^2 c, l) + 1 \tag{5.66}$$

That is, the capacity of the classifier can be *controlled independently of the dimensionality* of the feature space. This is very interesting indeed. It basically states that the capacity of a classifier may not, necessarily, be related to the number of unknown parameters! This is a more general result. To emphasize it further, note that it is possible to construct a classifier with only one free parameter, yet with infinite VC dimension; see, for example, [Burg 98]. Let us now consider a sequence of bounds

$$c_1 < c_2 < c_3 < \ldots$$

This defines the following sequence of classifiers:

$$\mathcal{F}^i : \left\{ \boldsymbol{w}^T \boldsymbol{x} + w_0 : \|\boldsymbol{w}\|^2 \leq c_i \right\} \tag{5.67}$$

where

$$\mathcal{F}^i \subset \mathcal{F}^{i+1}$$

If the classes are separable, then the empirical error is zero. Minimizing the norm $\|\boldsymbol{w}\|$ is equivalent to minimizing the VC dimension (to be fair, the upper bound of the VC dimension). Thus, we can conclude that, the design of an SVM classifier senses the spirit of the SRM principle. Hence, keeping the VC dimension minimum suggests that we can expect support vector machines to exhibit good generalization performance. More on these issues can be found in [Vapn 98, Burg 98].

The essence of all formulas and discussion in this section is that the generalization performance and accuracy of a classifier depend heavily on two parameters: the VC dimension and the number of the available feature vectors used for the training. The VC dimension may or may not be related to the number of free parameters describing the classifier. For example, in the case of the perceptron linear classifier the VC dimension coincides with the number of free parameters. However, one can construct nonlinear classifiers whose VC dimension can be either lower or higher than the number of free parameters [Vapn 98, p. 159]. The design methodology of the SVM allows one to "play" with the VC dimension (by minimizing $\|\boldsymbol{w}\|$, Eq. (5.66)), leading to good generalization performance, although the design may be carried out in a high- (even infinite) dimensional space.

Digging this fertile ground in a slightly different direction, using tools from the PAC theory of learning one can derive a number of distribution-free and *dimension-free* bounds. These bounds bring into the surface a key property underlying the SVM design; that is, that of the *maximum margin* (SVMs are just one example of this larger family of classifiers, which are designed with an effort to maximize the margin the training points leave from the corresponding decision surface). (See also the discussion at the end of Chapter 4.) Although a more detailed treatment of this topic is beyond the scope of this book, we will provide two related bounds that reinforce this, at first surprising, property of the "emancipation" of the generalization performance from the feature space dimensionality.

Assume that all available feature vectors lie within a sphere of radius R (i.e., $\|\boldsymbol{x}\| \leq R$). Let, also, the classifier be a *linear* one, normalized so that $\|\boldsymbol{w}\| = 1$, designed using N randomly chosen training vectors. If the resulting classifier has a margin of 2γ (according to the margin definition in Section 3.7.1) and *all training vectors* lie outside the margin, the corresponding true error probability (generalization error) is no more than

$$\frac{c}{N}\left(\frac{R^2}{\gamma^2}\ln^2 N + \ln\left(\frac{1}{\rho}\right)\right) \tag{5.68}$$

where c is a constant, and this bound holds true with a probability at least $1 - \rho$. Thus, adjusting the margin, as the SVM does, to be maximum we improve the bound, and this can be carried out even in an infinite dimensional space if the adopted kernel so dictates [Bart 99, Cris 00]. This result is logical. If the margin is large on a set of training points randomly chosen, this implies a classifier with large confidence, thus leading with high probability to good performance.

The bound given previously was derived under the assumption that all training points are correctly classified. Furthermore, the margin constraint implies that for all training points $y_i f(\boldsymbol{x}_i) \geq \gamma$, where $f(\boldsymbol{x})$ denotes the linear classifier (the decision is taken according to sign($f(\boldsymbol{x})$)). A very interesting related bound refers to the more realistic case, where some of the training points are misclassified. Let k be the number of points with $y_i f(\boldsymbol{x}_i) < \gamma$. (The product $yf(\boldsymbol{x})$ is also known as the *functional* margin of the pair $(y, \boldsymbol{x})$ with respect to classifier $f(\boldsymbol{x})$) Obviously, this also allows for negative values of the product. It can be shown that with probability at least $1 - \rho$ the true error probability is upper bounded by ([Bart 99, Cris 00])

$$\frac{k}{N} + \sqrt{\frac{c}{N}\left(\frac{R^2}{\gamma^2}\ln^2 N + \ln\left(\frac{1}{\rho}\right)\right)} \tag{5.69}$$

Another bound relates the error performance of the SVM classifier with the number of support vectors. It can be shown [Bart 99] that if N is the number of training vectors and N_s the number of support vectors, the corresponding true error probability is bounded by

$$\frac{1}{N - N_s}\left(N_s \log_2 \frac{eN}{N_s} + \log_2 \frac{N}{\rho}\right) \tag{5.70}$$

where e is the base of the natural logarithm and the bound holds true with a probability at least $1 - \rho$. Note that this bound is also independent of the dimension of the feature space, where the design takes place. The bound increases with N_s and this must make the user, who has designed an SVM that results in a relatively large number (with respect to N) of support vectors, cautious and "suspicious" about the performance of the resulting SVM classifier.

The previous three bounds indicate that the error performance is controlled by both N_s and γ. In practice, one may end up, for example, with a large number of support vectors and at the same time with a large margin. In such a case, the error performance could be assessed, with high confidence, depending on which of the two bounds has lower value.

5.11 THE BAYESIAN INFORMATION CRITERION

The structural risk minimization principle, discussed in the previous section, belongs to a more general class of methods that estimate the order of a system by considering, *simultaneously*, the model complexity and a performance index. Depending on the function used to measure the model complexity and the corresponding performance index, different criteria result. In this section we will focus on one of such criteria, which provides a Bayesian theory flavor to the model selection problem. Moreover, it has a structural form that resembles a number of other popular criteria that have been proposed over the years. Although the criteria of this "family" lack the elegance and generality of the SRM principle, they can be useful in a number of cases. Furthermore, they shed light on the "performance versus complexity" trade-off task from another perspective.

Let $\mathcal{D} = \{(\boldsymbol{x}_i, y_i),\ i = 1, 2, \ldots, N\}$ be the training data set. We will focus on the Bayesian classification problem, and the goal is to adopt a *parametric* model for the class posterior probabilities; that is, $P(y_i|\boldsymbol{x}; \boldsymbol{\theta}),\ y_i \in \{1, 2, \ldots, M\}$ for an M class task. The cost used for the optimal choice of the unknown parameter $\boldsymbol{\theta}$ is the log-likelihood function computed over the training set; that is, $L(\boldsymbol{\theta}) = \ln p(\mathcal{D}|\boldsymbol{\theta})$.

Let $\mathcal{M}_m$ denote one of the possible models described by the set of parameters $\boldsymbol{\theta}_m$, where m runs over all candidate models. Let us also assume that for each model we know the prior information with respect to the distribution of $\boldsymbol{\theta}_m$, that is, the pdf $p(\boldsymbol{\theta}_m|\mathcal{M}_m)$. Our goal is to choose the model for which the posterior probability

$P(\mathcal{M}_m|\mathcal{D})$ becomes maximum, over all candidate models. Using the Bayes theorem, we have

$$P(\mathcal{M}_m|\mathcal{D}) = \frac{P(\mathcal{M}_m)p(\mathcal{D}|\mathcal{M}_m)}{p(\mathcal{D})} \tag{5.71}$$

If we further assume that all models are equiprobable, $P(\mathcal{M}_m)$ can be dropped out. The joint data pdf $p(\mathcal{D})$, which is the same for all models, can also be neglected and our goal becomes to maximize

$$p(\mathcal{D}|\mathcal{M}_m) = \int p(\mathcal{D}|\boldsymbol{\theta}_m, \mathcal{M}_m)p(\boldsymbol{\theta}_m|\mathcal{M}_m)d\boldsymbol{\theta}_m \tag{5.72}$$

Employing a series of assumptions (e.g., Gaussian distribution for $\boldsymbol{\theta}_m$) and the so-called Laplacian approximation to the integral ([Schw 79, Ripl 96]), and taking the logarithm of both sides in Eq. (5.72) results in

$$\ln p(\mathcal{D}|\mathcal{M}_m) = L(\hat{\boldsymbol{\theta}}_m) - \frac{K_m}{2}\ln N \tag{5.73}$$

where $L(\hat{\boldsymbol{\theta}}_m)$ is the log-likelihood function computed at the ML estimate, $\hat{\boldsymbol{\theta}}_m$, and K_m is the number of free parameters (i.e., the dimensionality of $\boldsymbol{\theta}_m$). Equivalently, one can search for the minimum of the quantity

$$\text{BIC} = -2L(\hat{\boldsymbol{\theta}}_m) + K_m \ln N \tag{5.74}$$

The criterion is known as the Bayesian information criterion (BIC) or the Schwartz criterion. In other words, the best model indicated by this criterion depends (a) on the value of the log-likelihood function at its maximum (i.e., the adopted performance index) and (b) on a term that depends on the complexity of the model and the number of data points. If the model is too simple to describe the distribution underlying the given data set, the first term in the criterion will have a large value, since the probability of having obtained the set $\mathcal{D}$ from such a model will be small. On the other hand, if the model is complex, with a large number of free parameters that can adequately describe the data, the first term will have a small value, which however, is penalized by a large value of the second term. BIC provides a trade-off between these two terms. It can be shown that BIC is asymptotically consistent. This means that if the family of the candidate models contains the true one, then as $N \longrightarrow \infty$ the probability that BIC will select the correct model tends to one.

The Akaike information criterion (AIC) [Akai 74], though derived in a different way, has similar structure, and the only difference lies in the second term, which is $2K_m$ instead of $K_m \ln N$ (see also Section 16.4.1). In practice, it is not clear which model is to be used. It has been reported that for large values of N AIC tends to choose models that are too complex. On the other hand, for smaller values of N BIC tends to choose models that are too simple [Hast 01]. Besides the previous two criteria, a number of alternatives have also been suggested, such as [Riss 83, Mood 92, Leth 96, Wang 98]. For a review of such techniques, see, for example, [Ripl 96, Hast 01, Stoi 04a, Stoi 04b].

5.12 PROBLEMS

5.1 In Trunk's example, discussed in Section 5.3, prove that the mean value and the variance of the variable z are given by $E[z] = ||\boldsymbol{\mu}||^2 = \sum_{i=1}^{l} \frac{1}{i}$ and $\sigma_z^2 = ||\boldsymbol{\mu}||^2$ respectively. Also show that the probability of error is

$$P_e = \int_{b_l}^{\infty} \frac{1}{\sqrt{2\pi}} \exp\left(-\frac{z^2}{2}\right) dz \tag{5.75}$$

where

$$b_l = \sqrt{\sum_{i=1}^{l} \frac{1}{i}} \tag{5.76}$$

5.2 In Trunk's example, as in Problem 5.1, show that the mean value and variance of z, in the case of unknown mean value, are given the Eqs. (5.7) and (5.8), respectively. Derive the formula for the probability of error and show that it tends to 0.5 as the number of features tends to infinity.

5.3 If x_i, y_i, $i = 1, 2, \ldots, N$ are independent samples of two Gaussian distributions of the same variance σ^2, show that the random variable $\frac{(2N-2)s_z^2}{\sigma^2}$, where

$$s_z^2 = \frac{1}{2N-2}\left(\sum_{i=1}^{N}(x_i - \bar{x})^2 + \sum_{i=1}^{N}(y_i - \bar{y})^2\right)$$

where $\bar{x}, \bar{y}$ are the respective sample mean values, is chi-square distributed with $2N - 2$ degrees of freedom.

5.4 Let N_1, N_2 be the available values of a feature in two classes, respectively. The feature is assumed to follow a Gaussian distribution with the same variance in each class. Define the test statistic

$$q = \frac{(\bar{x} - \bar{y}) - (\mu_1 - \mu_2)}{s_z\sqrt{\frac{1}{N_1} + \frac{1}{N_2}}} \tag{5.77}$$

where

$$s_z^2 = \frac{1}{N_1 + N_2 - 2}\left(\sum_{i=1}^{N_1}(x_i - \bar{x})^2 + \sum_{i=1}^{N_2}(y_i - \bar{y})^2\right)$$

and μ_1, μ_2 are the respective true mean values. Show that q follows the t-distribution with $N_1 + N_2 - 2$ degrees of freedom.

5.5 Show that the matrix

$$A = \begin{bmatrix} \frac{1}{\sqrt{n}} & \frac{1}{\sqrt{n}} & \frac{1}{\sqrt{n}} & \cdots & \frac{1}{\sqrt{n}} \\ \frac{-1}{\sqrt{2}} & \frac{1}{\sqrt{2}} & 0 & \cdots & 0 \\ \frac{-1}{\sqrt{6}} & \frac{-1}{\sqrt{6}} & \frac{2}{\sqrt{6}} & \cdots & 0 \\ \vdots & \vdots & \vdots & \vdots & \vdots \\ \frac{-1}{\sqrt{n(n-1)}} & \frac{-1}{\sqrt{n(n-1)}} & \frac{-1}{\sqrt{n(n-1)}} & \cdots & \frac{n-1}{\sqrt{n(n-1)}} \end{bmatrix}$$

is orthogonal, that is, $AA^T = I$.

5.6 Show that if $x_i, i = 1, 2, \ldots, l$, are jointly Gaussian, then the l variables $y_i, i = 1, 2, \ldots, l$, resulting from a linear transformation of them are also jointly Gaussian. Furthermore, if x_i are mutually independent and the transformation is orthogonal, then y_i are also mutually independent and Gaussian.

5.7 Let $\omega_i, i = 1, 2, \ldots, M$, be the classes for a classification task. Divide the interval of the possible values of a feature into subintervals $\Delta_j, j = 1, 2, \ldots, K$. If $P(\Delta_j)$ is the probability of having values in the respective subinterval and $P(\omega_i|\Delta_j)$, the probability of occurrence of ω_i in this interval, show that the so-called *ambiguity function*

$$A = -\sum_i \sum_j P(\Delta_j) P(\omega_i|\Delta_j) \log_M(P(\omega_i|\Delta_j))$$

is equal to 1 for completely overlapped distributions and is equal to 0 for perfectly separated ones. For all other cases it takes intermediate values. Thus, it can be used as a distribution overlap criterion [Fine 83].

5.8 Show that if $d_{ij}(x_1, x_2, \ldots, x_m)$ is the class divergence based on m features, adding a new one x_{m+1} cannot decrease the divergence, that is,

$$d_{ij}(x_1, x_2, \ldots, x_m) \le d_{ij}(x_1, x_2, \ldots, x_m, x_{m+1})$$

5.9 Show that if the density functions are Gaussian in both classes with the same covariance matrix Σ, then on adding a new feature x_{m+1} to the feature vector the new divergence is recursively computed by

$$d_{ij}(x_1, \ldots, x_{m+1}) = d_{ij}(x_1, \ldots, x_m) + \frac{[(\mu_i - \mu_j) - (\boldsymbol{\mu}_i - \boldsymbol{\mu}_j)^T \Sigma^{-1} \boldsymbol{r}]^2}{\sigma^2 - \boldsymbol{r}^T \Sigma^{-1} \boldsymbol{r}}$$

where μ_i, μ_j are the mean values of x_{m+1} for the two classes, σ^2 is its variance, $\boldsymbol{r}$ is its cross-covariance vector with the other elements of $\boldsymbol{x}$, and $\boldsymbol{\mu}_i, \boldsymbol{\mu}_j$ are the mean vectors of $\boldsymbol{x}$ prior to x_{m+1}. If x_{m+1} is now uncorrelated with the previously selected features $x_1, \ldots, x_m$, then this becomes

$$d_{ij}(x_1, \ldots, x_{m+1}) = d_{ij}(x_1, \ldots, x_m) + \frac{(\mu_i - \mu_j)^2}{\sigma^2}$$

5.10 Show that if the features are statistically independent, then the divergence is given by

$$d_{ij}(x_1, x_2, \ldots, x_l) = \sum_{i=1}^{l} d_{ij}(x_i)$$

5.11 Show that in the case of Gaussian distributions the Chernoff bound becomes

$$\epsilon_{CB} = \exp(-b(s))$$

where

$$b(s) = \frac{s(1-s)}{2}(\boldsymbol{\mu}_i - \boldsymbol{\mu}_j)^T [s\Sigma_j + (1-s)\Sigma_i]^{-1}(\boldsymbol{\mu}_i - \boldsymbol{\mu}_j) + \frac{1}{2}\ln\frac{|s\Sigma_j + (1-s)\Sigma_i|}{|\Sigma_j|^s |\Sigma_i|^{1-s}}$$

Then take the derivative with respect to s and show that for equal covariance matrices the optimum is achieved for $s = 1/2$. Thus, in this case $b(s)$ equals the Bhattacharyya distance.

5.12 Show that the mixture scatter matrix is the sum of the within-class and between-class scatter matrices.

5.13 Show that the cross-correlation coefficient in (5.29) lies in the interval $[-1, 1]$. *Hint:* Use Schwartz's inequality $|\boldsymbol{x}^T \boldsymbol{y}| \leq \|\boldsymbol{x}\| \|\boldsymbol{y}\|$.

5.14 Show that for a two-class problem and Gaussian distributed feature vectors, with the same covariance matrix in the two classes, which are assumed equiprobable, the divergence is equal to

$$\text{trace}\{S_w^{-1} S_b\}$$

5.15 Show that the number of combinations to be searched using the backward search technique is given by

$$1 + 1/2((m+1)m - l(l+1))$$

5.16 Show that the optimal solution of the generalized Rayleigh quotient in (5.37) satisfies (5.38).

5.17 Show that

$$\frac{\partial}{\partial A}\text{trace}\{(A^T S_1 A)^{-1}(A^T S_2 A)\} = -2S_1 A(A^T S_1 A)^{-1}(A^T S_2 A)(A^T S_1 A)^{-1} + 2S_2 A(A^T S_1 A)^{-1}$$

5.18 Show that for an M-class problem the matrix S_b is of rank $M - 1$. *Hint:* Recall that $\boldsymbol{\mu}_0 = \sum_i P_i \boldsymbol{\mu}_i$.

5.19 Show that if $f_i(\boldsymbol{x}), i = 1, \ldots, M$, are the discriminant functions of an M-class problem, we can construct from them $M - 1$ new functions that are, in principle, sufficient for the classification.
Hint: Consider the differences $f_i(\boldsymbol{x}) - f_j(\boldsymbol{x})$.

5.20 Show that for a two-class problem the nonzero eigenvalue of matrix $S_w^{-1} S_b$ is equal to

$$\lambda_1 = P_1 P_2 (\boldsymbol{\mu}_1 - \boldsymbol{\mu}_2)^T S_{xw}^{-1} (\boldsymbol{\mu}_1 - \boldsymbol{\mu}_2)$$

and the corresponding eigenvector

$$\boldsymbol{v}_1 = S_{xw}^{-1} (\boldsymbol{\mu}_1 - \boldsymbol{\mu}_2)$$

where P_1, P_2 are the respective class probabilities.

5.21 Show that if matrices Σ_1, Σ_2 are two covariance matrices, then the eigenvectors of $\Sigma_1^{-1}\Sigma_2$ are orthogonal with respect to Σ_1, that is,

$$\boldsymbol{v}_i^T \Sigma_1 \boldsymbol{v}_j = \delta_{ij}$$

Hint: Use the fact that Σ_1, Σ_2 can be simultaneously diagonalized (Appendix B).

5.22 Show that in a multilayer perceptron with a linear output node, minimizing the squared error is equivalent to maximizing (5.51).
Hint: Assume the weights of the nonlinear nodes fixed and compute first the LS optimal weights driving the linear output nodes. Then substitute these values into the sum of error squares cost function.

5.23 Compute the minimal enclosure (hyper)sphere, that is, the radius as well as its center, of a set of points $\boldsymbol{x}_i$, $i = 1, 2, \ldots, N$.

MATLAB PROGRAMS AND EXERCISES

Computer Programs

5.1 *Scatter matrices.* Write a MATLAB function named *scatter_mat* that computes (a) the within-class (*Sw*), (b) the between-class (*Sb*) and the mixture (*Sm*) scatter matrices for a *c*-class classification problem, taking as inputs (a) an $l \times N$ dimensional matrix X, whose ith row is the ith data vector and (b) an N dimensional row vector y whose ith element contains the class label for the ith vector in X (the jth class is denoted by the integer $j, j = 1, \ldots, c$).

Solution

```
function [Sw,Sb,Sm]=scatter_mat(X,y)
  [l,N]=size(X);
```

```
c=max(y);
%Computation of class mean vectors, a priori prob. and
%Sw
m=[];
Sw=zeros(l);
for i=1:c
  y_temp=(y==i);
  X_temp=X(:,y_temp);
  P(i)=sum(y_temp)/N;
  m(:,i)=(mean(X_temp'))';
  Sw=Sw+P(i)*cov(X_temp');
end
%Computation of Sb
m0=(sum(((ones(l,1)*P).*m)'))';
Sb=zeros(l);
for i=1:c
  Sb=Sb+P(i)*((m(:,i)-m0)*(m(:,i)-m0)');
end
%Computation of Sm
Sm=Sw+Sb;
```

5.2 ***J3 criterion.*** **Write a MATLAB function named *J3_comp* that takes as inputs the within-class (*Sw*) and the mixture (*Sm*) scatter matrices and returns the value of the *J*3 criterion**

Solution

```
function J3=J3_comp(Sw,Sm)
J3=trace(inv(Sw)*Sm);
```

5.3 ***Best features combination.*** **Write a MATLAB function named *features_best_combin* that takes as inputs (a) an $l \times N$ dimensional matrix X, whose ith row is the ith data vector, (b) an N dimensional row vector y, whose ith element contains the class label for the ith vector in X (the jth class is denoted by the integer j, $j = 1, \ldots, c$), and (c) an integer q, the number of required features. It returns the best combination of q, out of the l, available features, according to the $J3$ criterion.**

Solution

```
function id=features_best_combin(X,y,q)
  [l,N]=size(X);
  J3_max=0;
  id=[];
  combin=nchoosek(1:l,q);
  for j=1:size(combin,1)
```

```
    X1=X(combin(j,:),:);
    [Sw,Sb,Sm]=scatter_mat(X1,y);
    J3=J3_comp(Sw,Sm)
    if(J3>J3_max)
      J3_max=J3;
      id=combin(j,:);
    end
  end
```

5.4 *FDR criterion.* Write a MATLAB function named *FDR_comp* that returns the FDR index for a c class problem taking as inputs (a) an $l \times N$ dimensional matrix X, whose ith row is the ith data vector, (b) an N dimensional row vector y, whose ith element contains the class label for the ith vector in X (the jth class is denoted by the integer $j, j = 1, \ldots, c$), and (c) the index *ind* of the feature over which the FDR will be computed.

Solution

```
function FDR=FDR_comp(X,y,ind)
  [l,N]=size(X);
  c=max(y);
  for i=1:c
    y_temp=(y==i);
    X_temp=X(ind,y_temp);
    m(i)=mean(X_temp);
    vari(i)=var(X_temp);
  end
  a=nchoosek(1:c,2);
  q=(m(a(:,1))-m(a(:,2))).^ 2 ./ (vari(a(:,1))+vari(a(:,2)))';
  FDR=sum(q);
```

Computer Experiments

5.1 a. Generate $N1 = 100$ random numbers from the zero mean unit variance normal distribution and another $N2 = 100$ random numbers from the unit variance normal distribution with mean value equal to 2. Assume that these numbers correspond to the values a specific feature takes in the framework of a two-class problem. Use the t-test to check whether or not the hypothesis that the mean values for this feature, for the two classes, differ significantly, at a 5% significance level.

b. Repeat (a) when the mean value for the second distribution is 0.2.

c. Repeat (a) and (b) when $N1 = 150$ and $N2 = 200$.
Comment on the results.

Hint: Use the *normrnd* MATLAB function to generate the random numbers and the *ttest*2, to perform the *t*-test.

5.2 a. (i) Generate four sets, each one consisting of 100 two-dimensional vectors, from the normal distributions with mean values $[-10,\ -10]^T$, $[-10,\ 10]^T$, $[10,\ -10]^T$, $[10,\ 10]^T$ and covariance matrices equal to $0.2 * I$. These sets constitute the data set for a four-class two-dimensional classification problem (each set corresponds to a class).

a. (ii) Compute the *Sw*, *Sb*, and *Sm* scatter matrices.

a. (iii) Compute the value for the criterion $J3$.

b. Repeat (a) when the mean vectors of the normal distributions that generate the data are $[-1,\ -1]^T$, $[-1,\ 1]^T$, $[1,\ -1]^T$, $[1,\ 1]^T$.

c. Repeat (a) when the covariance matrices of the normal distributions that generate the data are equal to $3 * I$.

5.3 Generate two sets, each one consisting of 100 five-dimensional vectors, from the normal distributions with mean values $[0,\ 0,\ 0,\ 0,\ 0]^T$ and $[0,\ 2,\ 2,\ 3,\ 3]^T$ and covariance matrices equal to

$$\begin{bmatrix} 0.5 & 0 & 0 & 0 & 0 \\ 0 & 0.5 & 0 & 0 & 0 \\ 0 & 0 & 1 & 0 & 0 \\ 0 & 0 & 0 & 1 & 0 \\ 0 & 0 & 0 & 0 & 1.5 \end{bmatrix}.$$

Their composition forms the data set for a two-class two-dimensional classification problem (each set corresponds to a class). Using the J_3 criterion find the best combination of features if:

a. they are considered individually.

b. they are considered in pairs.

c. they are considered in triples.

d. Justify the results.

5.4 a. (i) Generate two sets, each one consisting of 100 two-dimensional vectors, from the normal distributions with mean values $[2,\ 4]^T$ and $[2.5,\ 10]^T$ and covariance matrices equal to the 2×2 identity matrix I. Their composition forms the data set for a two class two dimensional classification problem (each set corresponds to a class).

a. (ii) Compute the value of the FDR index for both features.

b. Repeat (a) when the covariance matrices of the normal distributions that generate the data are both equal to $0.25 * I$.

c. Discuss the results.

REFERENCES

[Akai 74] Akaike H. "A new look at the statistical model identification," *IEEE Transactions on Automatic Control*, Vol. 19(6), pp. 716–723, 1974.

[Ambr 02] Ambroise C., McLachlan G.J. "Selection bias in gene extraction on the basis of microarray gene-expression data," *Proceedings of the National Academy of Sciences*, Vol. 99(10), pp. 6562–6566, 2002.

[Bart 99] Bartlett P., Shawe-Taylor J. "Generalization performance of support vector machines and other pattern classifiers," in *Advances in Kernel Methods: Support Vector Learning* (Schölkopf S., Burges J.C., Smola A., eds.), MIT Press, 1999.

[Bati 94] Batiti R. "Using mutual information for selecting features in supervised neural network learning," *IEEE Transactions on Neural Networks*, Vol. 5(8), pp. 537–550, 1994.

[Baud 00] Baudat G., Anouar F. "Generalized discriminant analysis using a kernel approach," *Neural Computation*, Vol. 12(10), pp. 2385–2404, 2000.

[Baum 89] Baum E.B., Haussler D. "What size net gives valid generalization," *Neural Computation*, Vol. 1(1), pp. 151–160, 1989.

[Belh 97] Belhumeour P.N., Hespanha J.P., Kriegman D.J. "Eigenfaces vs Fisherfaces: Recognition using class specific linear projection," *IEEE Transactions on Pattern Analysis and Machine Intelligence*, Vol. 19(7), pp. 711–720, 1997.

[Bish 95] Bishop C. *Neural Networks for Pattern Recognition*, Oxford University Press, 1995.

[Bour 88] Bourland H., Kamp Y. "Auto-association by multilayer perceptrons and singular value decomposition," *Biological Cybernetics*, Vol. 59, pp. 291–294, 1988.

[Brad 97] Bradley A. "The use of the area under the ROC curve in the evaluation of machine learning algorithms," *Pattern Recognition*, Vol. 30(7), pp. 1145–1159, 1997.

[Brun 00] Brunzell H., Erikcson J. "Feature reduction for classification of multidimensional data," *Pattern Recognition*, Vol. 33, pp. 1741–1748, 2000.

[Burg 98] Burges C.J.C. "A tutorial on support vector machines for pattern recognition," *Data Mining and Knowledge Discovery*, Vol. 2(2), pp. 1–47, 1998.

[Butz 05] Butz T., Thiran J.P. "From error probability to information theoretic (multi-modal) signal processing," *Signal Processing*), Vol. 85(5), pp. 875–902, 2005.

[Cevi 05] Cevikalp H., Neamtu M., Wilkes M., Barkana A. "Discriminative common vectors for face recognition," *IEEE Trans. on Pattern Analysis and Machine Intelligence*, Vol. 27(1), pp. 4–13, 2005.

[Chat 97] Chatterjee C., Roychowdhury V. "On self-organizing algorithms and networks for class-separability features," *IEEE Transactions on Neural Networks*, Vol. 8(3), pp. 663–678, 1997.

[Chen 00] Chen L.-F., Liao H.-Y.M., Ko M.-T., Lin J.-C., Yu G.-J. "A new LDA-based face recognition system which can solve the small sample size problem," *Pattern Recognition*, Vol. 33(10), pp. 1713–1726, 2000.

[Choi 03] Choi E., Lee C. "Feature extraction based on the Bhattacharyya distance," *Pattern Recognition Letters*, Vol. 36, pp. 1703–1709, 2003.

[Cris 00] Cristianini N., Shawe-Taylor J. *An Introduction to Support Vector Machines and Other Kernel-Based Learning Methods*, Cambridge University Press, Cambridge, MA, 2000.

[Devi 82] Devijver P.A., Kittler J. *Pattern Recognition; A Statistical Approach*, Prentice Hall, 1982.

[Devr 96] Devroye L., Gyorfi L., Lugosi G. *A Probabilistic Theory of Pattern Recognition*, Springer-Verlag, 1996.

[Duin 00] Duin R.P.W. "Classifiers in almost empty spaces," *Proceedings of the 15th Int. Conference on Pattern Recognition (ICPR)*, vol. 2, Pattern Recognition and Neural Networks, IEEE Computer Society Press, 2000.

[Erdo 03] Erdogmus D., Principe J. "Lower and upper bounds for misclassification probability based on Renyi' s information," *Journal of VLSI Signal Processing*, 2003.

[Fine 83] Finette S., Bleier A., Swindel W. "Breast tissue classification using diagnostic ultrasound and pattern recognition techniques: I. Methods of pattern recognition," *Ultrasonic Imaging*, Vol. 5, pp. 55–70, 1983.

[Fish 36] Fisher R.A. "The use of multiple measurements in taxonomic problems," *Annals of Eugenics*, Vol. 7, pp. 179–188, 1936.

[Fras 58] Fraser D.A.S. *Statistics: An Introduction*, John Wiley & Sons, 1958.

[Frie 89] Friedman J.H. "Regularized discriminant analysis," *Journal of American Statistical Association*, Vol. 84, pp. 165–175, 1989.

[Fuku 90] Fukunaga K. *Introduction to Statistical Pattern Recognition*, 2nd ed., Academic Press, 1990.

[Gelm 95] Gelman A., Rubin D.B., Carlin J., Stern H. *Bayesian Data Analysis*, Chapman & Hall, London, 1995.

[Ghah 94] Ghaharamani Z., Jordan M.I. "Supervised learning from incomplete data via the EM approach," in *Advances in Neural Information Processing Systems* (Cowan J.D., Tesauro G.T., Alspector J., eds), Vol. 6, pp. 120–127, Morgan Kaufmann, San Mateo, CA, 1994.

[Guyo 03] Guyon I, Elisseeff A. "An introduction to variable and feature selection," *Journal of Machine Learning Research*, Vol. 3, pp. 1157–1182, 2003.

[Hams 08] Hamsici O. C., Martinez A. M. " Bayes optimality in LDA," *IEEE Transactions on Pattern Analysis and Machine Intelligence*, Vol. 30(4), pp. 647–657, 2008.

[Hast 95] Hastie T., Tibshirani R. "Penalized discriminant analysis," *Annals of Statistics*, Vol. 23, pp. 73–102, 1995.

[Hast 01] Hastie T., Tibshirani R., Friedman J. *The Elements of Statistical Learning*, Springer, 2001.

[Hube 81] Huber P.J. *Robust Statistics*, John Wiley & Sons, 1981.

[Hush 93] Hush D.R., Horne B.G. "Progress in supervised neural networks," *Signal Processing Magazine*, Vol. 10(1), pp. 8–39, 1993.

[Jain 97] Jain A., Zongker D. "Feature selection: Evaluation, application, and small sample performance," *IEEE Transactions on Pattern Analysis and Machine Intelligence*, Vol. 19(2), pp. 153–158, 1997.

[Kim 07] Kim C., Choi C.-H. " A discriminant analysis using composite features for classification problems," *Pattern Recognition*, Vol. 40(11), pp. 2958–2967, 2007.

[Kitt 78] Kittler J. "Feature set search algorithms," in *Pattern Recognition and Signal Processing* (Chen C.H., ed.), pp. 41–60, Sijthoff and Noordhoff, Alphen aan den Rijn, The Netherlands, 1978.

[Kram 91] Kramer M.A. "Nonlinear principal component analysis using auto-associative neural networks," *AIC Journal*, Vol. 37(2), pp. 233–243, 1991.

[Kulb 51] Kullback S., Liebler R.A. "On information and sufficiency," *Annals of Mathematical Statistics*, Vol. 22, pp. 79–86, 1951.

[Kwak 02] Kwak N., Choi C.-H. "Input feature selection for classification problems," *IEEE Transactions on Neural Networks*, Vol. 13(1), pp. 143-159, 2002.

[Land 08] Landgrebe T. C. W., Duiy R. P. W. "Efficient multiclasss ROC approximation by decomposing via confusion matrix pertubation analysis," *IEEE Transactions on Pattern Analysis and Machine Intelligence*, Vol. 30(5), pp. 810-822, 2008.

[Lawe 66] Lawer E.L., Wood D.E. "Branch and bound methods: A survey," *Operational Research*, Vol. 149(4), 1966.

[Lee 00] Lee C., Choi E. "Bayes error evaluation of the Gaussian ML classifier," *IEEE Transactions on Geoscience Remote Sensing*, Vol. 38(3), pp. 1471-1475, 2000.

[Lee 93] Lee C., Landgrebe D.A. "Decision boundary feature extraction for nonparametric classifiers," *IEEE Transactions on Systems Man and Cybernetics*, Vol. 23, pp. 433-444, 1993.

[Lee 97] Lee C., Landgrebe D. "Decision boundary feature extraction for neural networks," *IEEE Transactions on Neural Networks*, Vol. 8(1), pp. 75-83, 1997.

[Leiv 07] Leiva-Murillo J.M., Artes-Rodriguez A. "Maximization of mutual information for supervised linear feature extraction," *IEEE Transactions on Neural Networks*, Vol. 18(5), pp. 1433-1442, 2007.

[Leth 96] Lethtokanga S.M., Saarinen J., Huuhtanen P., Kaski K. "Predictive minimum description length criterion for time series modeling with neural networks," *Neural Computation*, Vol. 8, pp. 583-593, 1996.

[Li 06] Li H., Jiang T., Zhang K. "Efficient and robust extraction by maximum margin criterion," *IEEE Transactions on Neural Networks*, Vol. 17(1), pp. 157-165, 2006.

[Lin 02] Lin Y., Wahba G., Zhang H., Lee Y. "Statistical properties and adaptive tuning of support vector machines," *Machine Learning*, Vol. 48, pp. 115-136, 2002.

[Loog 04] Loog M., Duin P.W. "Linear Dimensionality reduction via a heteroscedastic extension of LDA: The Chernoff criterion," *IEEE Transactions on Pattern Analysis and Machine Intelligence*, Vol. 26(6), pp. 732-739, 2004.

[Lowe 90] Lowe D., Webb A.R. "Exploiting prior knowledge in network optimization: An illustration from medical prognosis," *Network: Computation in Neural Systems*, Vol. 1(3), pp. 299-323, 1990.

[Lowe 91] Lowe D., Webb A.R. "Optimized feature extraction and the Bayes decision in feedforward classifier networks," *IEEE Transactions in Pattern Analysis and Machine Intelligence*, Vol. 13(4), pp. 355-364, 1991.

[Ma 03] Ma J., Jose L. S., Ahalt S. "Nonlinear multiclass discriminant analysis," *IEEE Signal Processing Letters*, Vol. 10(33), pp. 196-199, 2003.

[Mama 96] Mamamoto Y., Uchimura S., Tomita S. "On the behaviour of artificial neural network classifiers in high dimensional spaces," *IEEE Transactions on Pattern Analysis and Machine Intelligence*, Vol. 18(5), pp. 571-574, 1996.

[Mao 95] Mao J., Jain A.K. "Artificial neural networks for feature extraction and multivariate data projection," *IEEE Transactions on Neural Networks*, Vol. 6(2), pp. 296-317, 1997.

[Marr 08] Marroco C., Duin R. P. W., Tortorella F. "Maximizing the area under the ROC curve by pairwise feature combination," *Pattern Recognition*, Vol. 41, pp. 1961-1974, 2008.

[Maus 90] Mausel P.W., Kramber W.J., Lee J.K. "Optimum band selection for supervised classification of multispectra data," *Photogrammetric Engineering and Remote Sensing* Vol. 56, pp. 55-60, 1990.

[Mood 92] Moody J.E. "The effective number of parameters: An analysis of generalization and regularization in nonlinear learning systems" in *Advances in Neural Computation* (Moody J.E., Hanson S.J., Lippman R.R., eds.), pp. 847–854, Morgan Kaufman, San Mateo, CA, 1992.

[Nena 07] Nenadic Z. "Information discriminant analysis: feature extraction with an information-theoretic objective," *IEEE Transactions on Pattern Analysis and Machine Intelligence*, Vol. 29(8), pp. 1394–1408, 2007.

[Papo 91] Papoulis A. *Probability Random Variables and Stochastic Processes*, 3rd ed., McGraw-Hill, 1991.

[Pudi 94] Pudil P., Novovicova J., Kittler J. "Floating search methods in feature selection," *Pattern Recognition Letters*, Vol. 15, pp. 1119–1125, 1994.

[Raud 91] Raudys S.J., Jain A.K. "Small size effects in statistical pattern recognition: Recommendations for practitioners," *IEEE Transactions on Pattern Analysis and Machine Intelligence*, Vol. 13(3), pp. 252–264, 1991.

[Raud 80] Raudys S.J., Pikelis V. "On dimensionality, sample size, classification error, and complexity of classification algorithms in pattern recognition," *IEEE Transactions on Pattern Analysis and Machine Intelligence*, Vol. 2(3), pp. 243–251, 1980.

[Rich 95] Richards J. *Remote Sensing Digital Image Analysis*, 2nd ed., Springer-Verlag, 1995.

[Ripl 96] Ripley B.D. *Pattern Recognition And Neural Networks*, Cambridge University Press, Cambridge, MA, 1996.

[Riss 83] Rissanen J. "A universal prior for integers and estimation by minimum description length," *The Annals of Statistics*, Vol. 11(2), pp. 416–431, 1983.

[Rubi 76] Rubin D.B. "Inference and missing data," *Biometrika*, Vol. 63, pp. 581–592, 1976.

[Rubi 87] Rubin D.B. *Multiple Imputation for Nonresponse in Surveys*, John Wiley & Sons, 1987.

[Samm 69] Sammon J.W. "A nonlinear mapping for data structure analysis," *IEEE Transactions on Computers*, Vol. 18, pp. 401–409, 1969.

[Scha 02] Schafer J., Graham J. "Missing data: Our view of the state of the art," *Psychological Methods*, vol. 7(2), pp. 67–81, 2002.

[Schw 79] Schwartz G. "Estimating the dimension of the model," *Annals of Statistics*, Vol. 6, pp. 461–464, 1978.

[Seti 97] Setiono R., Liu H. "Neural network feature selector," *IEEE Transactions on Neural Networks*, Vol. 8(3), pp. 654–662, 1997.

[Sind 04] Sindhwami V., Rakshit S., Deodhare D., Erdogmus D., Principe J.C., Niyogi P. "Feature selection in MLPs and SVMs based on maximum output information," *IEEE Transactions on Neural Networks*, Vol. 15(4), pp. 937–948, 2004.

[Stoi 04b] Stoica P., Moses R. *Spectral Analysis of Signals*, Prentice Hall, 2004.

[Stoi 04a] Stoica P., Selén Y. "A review of information criterion rules," *Signal Processing Magazine*, Vol. 21(4), pp. 36–47, 2004.

[Su 94] Su K.Y, Lee C.H. "Speech recognition using weighted HMM and subspace projection approaches," *IEEE Transactions on Speech and Audio Processing*, Vol. 2(1), pp. 69–79, 1994.

[Swai 73] Swain P.H., King R.C. "Two effective feature selection criteria for multispectral remote sensing," *Proceedings of the 1st International Conference on Pattern Recognition*, pp. 536-540, 1973.

[Tian 86] Tian Q., Marbero M., Gu Z.H., Lee S.H. "Image classification by the Folley-Sammon transform," *Optical Engineering*, Vol. 25(7), pp. 834-840, 1986.

[Tou 74] Tou J., Gonzalez R.C. *Pattern Recognition Principles*, Addison-Wesley, 1974.

[Trun 79] Trunk G.V. "A problem of dimensionality: A simple example," *IEEE Transactions on Pattern Analysis and Machine Intelligence*, Vol. 1(3), pp. 306-307, 1979.

[Tsud 03] Tsuda K., Akaho S., Asai K. "The EM algorithm for kernel matrix completion with auxiliary data," *Journal of Machine Learning Research*, Vol. 4, pp. 67-81, 2003.

[Vali 84] Valiant L. "A theory of the learnable," *Communications of the ACM*, Vol. 27(11), pp. 1134-1142, 1984.

[Vapn 82] Vapnik V.N. *Estimation of Dependencies Based on Empirical Data*, Springer-Verlag, 1982.

[Vapn 95] Vapnik V.N. *The Nature of Statistical Learning Theory*, Springer-Verlag, 1995.

[Vapn 98] Vapnik, V.N. *Statistical Learning Theory*, John Wiley & Sons, 1998.

[Walp 78] Walpole R.E., Myers R.H. *Probability and Statistics for Engineers and Scientists*, Macmillan, 1978.

[Wang 00] Wang W., Jones P., Partridge D. "A comparative study of feature salience ranking techniques," *Neural Computation*, Vol. 13(7), pp. 1603-1623, 2000.

[Wang 98] Wang Y., Adali T., Kung S.Y., Szabo Z. "Quantization and segmentation of brain tissues from MR images: A probabilistic neural network approach," *IEEE Transactions on Image Processing*, Vol. 7(8), 1998.

[Wata 97] Watanabe H., Yamaguchi T., Katagiri S. "Discriminative metric for robust pattern recognition," *IEEE Transactions on Signal Processing*, Vol. 45(11), pp. 2655-2663, 1997.

[Will 07] Williams D., Liao X., Xue Y., Carin L., Krishnapuram B. "On classification with incomplete data," *IEEE Transactions on Pattern Analysis and Machine Intelligence* Vol. 29(3), pp. 427-436, 2007.

[Yang 02] Yang J., Yang J.-Y. "Why can LDA be performed in PCA transformed space?" *Pattern Recognition*, Vol. 36, pp. 563-566, 2002.

[Ye 05] Ye J., Li Q. "A two stage linear discriminant analysis via QR decomposition," *IEEE Transactions on Pattern Analysis and Machine Intelligence*, Vol. 27(6), pp. 929-941, 2005.

[Yu 93] Yu B., Yuan B. "A more efficient branch and bound algorithm for feature selection," *Pattern Recognition*, Vol. 26(6), pp. 883-889, 1993.

[Zhan 02] Zhang H., Sun G. "Feature selection using tabu search method," *Pattern Recognition*, Vol. 35, pp. 701-711, 2002.

[Zhan 07] Zhang S., Sim T. "Discriminant subspace analysis: A Fukunaga-Koontz approach," *IEEE Transactions on Pattern Analysis and Machine Intelligence*, Vol. 29(10), pp. 1732-1745, 2007.

CHAPTER

Feature Generation I: Data Transformation and Dimensionality Reduction

6

6.1 INTRODUCTION

Feature generation is of paramount importance in any pattern recognition task. Given a set of measurements, the goal is to discover compact and informative representations of the obtained data. A similar process is also taking place in the human perception apparatus. Our mental representation of the world is based on a relatively small number of perceptually relevant features. These are generated after processing a large amount of sensory data, such as the intensity and the color of the pixels of the images sensed by our eyes, and the power spectra of the sound signals sensed by our ears.

The basic approach followed in this chapter is to transform a given set of measurements to a new set of features. If the transform is suitably chosen, transform domain features can exhibit high *information packing* properties compared with the original input samples. This means that most of the classification-related information is "squeezed" in a relatively small number of features, leading to a reduction of the necessary feature space dimension. Sometimes we refer to such processing tasks as *dimensionality reduction* techniques.

The basic reasoning behind transform-based features is that an appropriately chosen transform can exploit and remove information redundancies, which usually exist in the set of samples obtained by the measuring devices. Let us take for example an image resulting from a measuring device, for example, X-rays or a camera. The pixels (i.e., the input samples) at the various positions in the image have a large degree of correlation, due to the internal morphological consistencies of real-world images that distinguish them from noise. Thus, if one uses the pixels as features, there will be a large degree of redundant information. Alternatively, if one obtains the Fourier transform, for example, of a typical real-world image, it turns out that most of the energy lies in the low-frequency components, due to the high correlation between the pixels' gray levels. Hence, using the Fourier coefficients as features seems a reasonable choice, because the low-energy, high-frequency coefficients can be neglected, with little loss of information. In this chapter we will see that the Fourier transform is just one of the tools from a palette of possible transforms.

6.2 BASIS VECTORS AND IMAGES

Let $x(0), x(1), \ldots, x(N-1)$ be a set of input samples and $\boldsymbol{x}$ be the $N \times 1$ corresponding vector,

$$\boldsymbol{x}^T = [x(0), \ldots, x(N-1)]$$

Given a unitary $N \times N$ matrix A,[1] we define the transformed vector $\boldsymbol{y}$ of $\boldsymbol{x}$ as

$$\boldsymbol{y} = A^H \boldsymbol{x} \equiv \begin{bmatrix} \boldsymbol{a}_0^H \\ \vdots \\ \boldsymbol{a}_{N-1}^H \end{bmatrix} \boldsymbol{x} \tag{6.1}$$

where H denotes the Hermitian operation, that is, complex conjugation and transposition. From (6.1) and the definition of unitary matrices we have

$$\boldsymbol{x} = A\boldsymbol{y} = \sum_{i=0}^{N-1} y(i)\boldsymbol{a}_i \tag{6.2}$$

The columns of A, $\boldsymbol{a}_i, i = 0, 1, \ldots, N-1$, are called the *basis vectors* of the transform. The elements $y(i)$ of $\boldsymbol{y}$ are nothing but the projections of $\boldsymbol{x}$ onto these basis vectors. Indeed, taking the inner product of $\boldsymbol{x}$ with $\boldsymbol{a}_j$ we have

$$\langle \boldsymbol{a}_j, \boldsymbol{x} \rangle \equiv \boldsymbol{a}_j^H \boldsymbol{x} = \sum_{i=0}^{N-1} y(i)\langle \boldsymbol{a}_j, \boldsymbol{a}_i \rangle = \sum_{i=0}^{N-1} y(i)\delta_{ij} = y(j) \tag{6.3}$$

This is due to the unitary property of A, that is, $A^H A = I$ or $\langle \boldsymbol{a}_i, \boldsymbol{a}_j \rangle = \boldsymbol{a}_i^H \ \boldsymbol{a}_j = \delta_{ij}$.

In many problems, such as in image analysis, the input set of samples is a two-dimensional sequence $X(i,j), i,j = 0, 1, \ldots, N-1$, defining an $N \times N$ matrix X instead of a vector. In such cases, one can define an equivalent N^2 vector $\boldsymbol{x}$, for example, by ordering the rows of the matrix one after the other (*lexicographic ordering*)

$$\boldsymbol{x}^T = [X(0,0), \ldots, X(0,N-1), \ldots, X(N-1,0), \ldots, X(N-1,N-1)]$$

and then transform this equivalent vector. However, this is not the most efficient way to work. The number of operations required to multiply an $N^2 \times N^2$ square matrix (A) with an $N^2 \times 1$ vector $\boldsymbol{x}$ is of the order of $O(N^4)$, which is prohibitive

[1] A complex matrix is called unitary if $A^{-1} = A^H$. Real matrices are equivalently called *orthogonal* if $A^{-1} = A^T$.

for many applications. An alternative possibility is to transform matrix X via a set of *basis matrices* or *basis images*. Let U and V be unitary $N \times N$ matrices. Define the transformed matrix Y of X as

$$Y = U^H X V \tag{6.4}$$

or

$$X = U Y V^H \tag{6.5}$$

The number of operations is now reduced to $O(N^3)$. Equation (6.5) can alternatively be written (Problem 6.1) as

$$X = \sum_{i=0}^{N-1} \sum_{j=0}^{N-1} Y(i,j) \boldsymbol{u}_i \boldsymbol{v}_j^H \tag{6.6}$$

where $\boldsymbol{u}_i$ are the column vectors of U and $\boldsymbol{v}_j$ the column vectors of V. Each of the outer products $\boldsymbol{u}_i \boldsymbol{v}_j^H$ is an $N \times N$ matrix

$$\boldsymbol{u}_i \boldsymbol{v}_j^H = \begin{bmatrix} u_{i0} v_{j0}^* & \cdots & u_{i0} v_{jN-1}^* \\ \vdots & \vdots & \vdots \\ u_{iN-1} v_{j0}^* & \cdots & u_{iN-1} v_{jN-1}^* \end{bmatrix} \equiv \mathcal{A}_{ij}$$

and (6.6) is an expansion of matrix X in terms of these N^2 basis images (matrices). The $*$ denotes complex conjugation. Furthermore, if Y turns out to be diagonal, then (6.6) becomes

$$X = \sum_{i=0}^{N-1} Y(i,i) \boldsymbol{u}_i \boldsymbol{v}_i^H$$

and the number of basis images is reduced to N. An interpretation similar to (6.3) is also possible. To this end, let us define the inner product between two matrices as

$$\langle A, B \rangle \equiv \sum_{m=0}^{N-1} \sum_{n=0}^{N-1} A^*(m,n) B(m,n) \tag{6.7}$$

Then it is not difficult to show that (Problem 6.1)

$$Y(i,j) = \langle \mathcal{A}_{ij}, X \rangle \tag{6.8}$$

In words, the (i,j) element of the transformed matrix results from multiplying each element of X by the conjugate of the corresponding element of $\mathcal{A}_{ij}$ and summing up all products.

Transformations of the type (6.4) are also known as *separable* (Problem 6.2). The reason is that one can look at them as a succession of one-dimensional transforms, first applied on column vectors and then on row vectors. For example, the intermediate result in (6.4), $Z = U^H X$, is equivalent to N transforms applied to the column vectors of X, and $(U^H X)V = (V^H Z^H)^H$ is equivalent to a second sequence of N transforms acting upon the rows of Z. All the two-dimensional transforms that we will deal with in this chapter are separable ones.

Example 6.1

Given the image X and the orthogonal transform matrix U

$$X = \begin{bmatrix} 1 & 2 \\ 2 & 3 \end{bmatrix}, \quad U = \frac{1}{\sqrt{2}} \begin{bmatrix} 1 & 1 \\ 1 & -1 \end{bmatrix}$$

the transformed image $Y = U^T X U$ is

$$Y = \frac{1}{2} \begin{bmatrix} 1 & 1 \\ 1 & -1 \end{bmatrix} \begin{bmatrix} 1 & 2 \\ 2 & 3 \end{bmatrix} \begin{bmatrix} 1 & 1 \\ 1 & -1 \end{bmatrix} = \begin{bmatrix} 4 & -1 \\ -1 & 0 \end{bmatrix}$$

The corresponding basis images are

$$\mathcal{A}_{00} = \frac{1}{2} \begin{bmatrix} 1 \\ 1 \end{bmatrix} [1, 1] = \frac{1}{2} \begin{bmatrix} 1 & 1 \\ 1 & 1 \end{bmatrix},$$

$$\mathcal{A}_{11} = \frac{1}{2} \begin{bmatrix} 1 \\ -1 \end{bmatrix} [1, -1] = \frac{1}{2} \begin{bmatrix} 1 & -1 \\ -1 & 1 \end{bmatrix}$$

and similarly

$$\mathcal{A}_{01} = \mathcal{A}_{10}^T = \frac{1}{2} \begin{bmatrix} 1 & -1 \\ 1 & -1 \end{bmatrix}$$

Now verify that the elements of Y are obtained via the matrix inner products $\langle \mathcal{A}_{ij}, X \rangle$.

6.3 THE KARHUNEN–LOÈVE TRANSFORM

In Section 5.8 the problem of the linear transformation of a feature vector was considered in the spirit of linear discriminant analysis (LDA). The class labels of the feature vectors were assumed known, and this information was optimally exploited to compute the transformation matrix. The linear transform task will also be considered in this section but from a different perspective. Here, the computation of the transformation matrix will exploit the statistical information describing the data, and it will take place in an unsupervised mode. The Karhunen–Loève transform or principal component analysis (PCA), as it is also known, is one of the most popular methods for feature generation and dimensionality reduction in pattern recognition. Though an old technique, it is still in use, and it forms the basis for a number of more advanced approaches.

Let $\boldsymbol{x}$ be the vector of input samples. In the case of an image array, $\boldsymbol{x}$ may be formed by lexicographic ordering of the array elements. In order to simplify the presentation, we will assume that the data samples have zero mean. If this is not the case, we can always subtract the mean value. We have already mentioned that a desirable property of the generated features is to be mutually uncorrelated in an effort to avoid information redundancies. We begin this section by first developing a method that generates mutually uncorrelated features, that is, $E[y(i)y(j)] = 0$, $i \neq j$. Let[2]

$$y = A^T x \tag{6.9}$$

Since we have assumed that $E[x] = \mathbf{0}$, it is readily seen that $E[y] = \mathbf{0}$. From the definition of the correlation matrix we have

$$R_y \equiv E[yy^T] = E[A^T xx^T A] = A^T R_x A \tag{6.10}$$

In practice, R_x is estimated as an average over the given set of training vectors. For example, if we are given n data vectors x_k, $k = 1, 2, \ldots, n$, then

$$R_x \approx \frac{1}{n} \sum_{k=1}^{n} x_k x_k^T \tag{6.11}$$

Note that R_x is a symmetric matrix, and hence its eigenvectors are mutually orthogonal (Appendix B). Thus, if matrix A is chosen so that its columns are the orthonormal eigenvectors a_i, $i = 0, 1, \ldots, N-1$, of R_x, then R_y is diagonal (Appendix B)

$$R_y = A^T R_x A = \Lambda \tag{6.12}$$

where Λ is the diagonal matrix having as elements on its diagonal the respective eigenvalues λ_i, $i = 0, 1, \ldots, N-1$, of R_x. (Recall that in Section 5.8 a linear transform of the form in (6.9) was also considered, but there the elements of matrix A were computed so that a class separability criterion could be optimized.) Furthermore, assuming R_x to be positive definite (Appendix B) the eigenvalues are positive. The resulting transform is known as the *Karhunen–Loève (KL)* transform, and it achieves our original goal of generating mutually uncorrelated features. The KL transform was introduced in [Karh 46] in the context of representing a random process in terms of orthogonal functions and in the discrete form used in this section in [Hote 33]. Other classical references of the topic are [Diam 96, Joll 86].

It has to be emphasized that the solution provided by the KL transform is not a unique one, and it was obtained by imposing an orthogonal structure on matrix A ($A^T A = I$). Also, note that for zero mean variables the correlation matrix R coincides with the covariance matrix Σ. As a matter of fact, a direct consequence of the respective definitions is that

$$\Sigma_x = R_x - E[x]E[x]^T$$

[2] We deal with real data. The complex case is a straightforward extension.

In case the zero mean assumption is not valid, the condition for uncorrelated variables becomes $E[(y(i)-E[y(i)])(y(j)-E[y(j)])]=0,\ i\neq j$, and the problem results in the eigendecomposition of the covariance matrix, that is,

$$\Sigma_y = A^T \Sigma_x A = \Lambda \tag{6.13}$$

Although our starting point was to generate mutually uncorrelated features, the KL transform turns out to have a number of other important properties, which provide different ways for its interpretation and also the secret for its popularity.

Mean Square Error Approximation

From Eqs. (6.2) and (6.3) we have

$$\boldsymbol{x} = \sum_{i=0}^{N-1} y(i)\boldsymbol{a}_i \quad \text{and} \quad y(i) = \boldsymbol{a}_i^T \boldsymbol{x} \tag{6.14}$$

Let us now define a new vector in the m-dimensional subspace

$$\hat{\boldsymbol{x}} = \sum_{i=0}^{m-1} y(i)\boldsymbol{a}_i \tag{6.15}$$

where only m of the basis vectors are involved. Obviously, this is nothing but the projection of $\boldsymbol{x}$ onto the subspace spanned by the m (orthonormal) eigenvectors involved in the summation. If we try to approximate $\boldsymbol{x}$ by its projection $\hat{\boldsymbol{x}}$, the resulting mean square error is given by

$$E\left[\|\boldsymbol{x}-\hat{\boldsymbol{x}}\|^2\right] = E\left[\left\|\sum_{i=m}^{N-1} y(i)\boldsymbol{a}_i\right\|^2\right] \tag{6.16}$$

Our goal now is to choose the eigenvectors that result in the minimum MSE. From (6.16) and taking into account the orthonormality property of the eigenvectors, we have

$$E\left[\left\|\sum_{i=m}^{N-1} y(i)\boldsymbol{a}_i\right\|^2\right] = E\left[\sum_i \sum_j (y(i)\boldsymbol{a}_i^T)(y(j)\boldsymbol{a}_j)\right] \tag{6.17}$$

$$= \sum_{i=m}^{N-1} E[y^2(i)] = \sum_{i=m}^{N-1} \boldsymbol{a}_i^T E[\boldsymbol{x}\boldsymbol{x}^T]\boldsymbol{a}_i \tag{6.18}$$

Combining this with (6.16) and the eigenvector definition, we finally get

$$E\left[\|\boldsymbol{x}-\hat{\boldsymbol{x}}\|^2\right] = \sum_{i=m}^{N-1} \boldsymbol{a}_i^T \lambda_i \boldsymbol{a}_i = \sum_{i=m}^{N-1} \lambda_i \tag{6.19}$$

Thus, if we choose in (6.15) the eigenvectors corresponding to the m *largest* eigenvalues of the *correlation matrix*, then the error in (6.19) is *minimized, being the sum of the* $N-m$ *smallest eigenvalues. Furthermore, it can be shown*

(Problem 6.3) that this is also the minimum MSE, compared with any other approximation of $\boldsymbol{x}$ by an m-dimensional vector. This is the reason that the KL transform is also known as *principal component analysis* (PCA).

A difficulty in practice is how to choose the m principal components. One way is to rank the eigenvalues in descending order, $\lambda_0 \geq \lambda_1 \geq \ldots \geq \lambda_{m-1} \geq \lambda_m \geq \ldots \geq \lambda_{N-1}$, and determine m so that the gap between the values λ_{m-1} and λ_m is "large." For more on this issue, see [Jack 91].

Note that the previous analysis concerning the MSE property of the KL transform, after projecting onto the m principal components of R_x, is still valid even if the mean of the data is not zero. However, in this case, although a minimum MSE solution is obtained, the approximation is not, in general, a good one (why?) In such cases, one tries to find the optimum m-dimensional subspace, so that the MSE between $\boldsymbol{x}$ and its following approximation

$$\hat{\boldsymbol{x}} = \sum_{i=0}^{m-1} y(i)\hat{\boldsymbol{a}}_i + \sum_{i=m}^{N-1} b_i\hat{\boldsymbol{a}}_i, \quad y(i) \equiv \hat{\boldsymbol{a}}_i^T \boldsymbol{x} \tag{6.20}$$

to be minimum, where b_i, $i = m, \ldots, N-1$, are constants independent of $\boldsymbol{x}$. It turns out (Problem 6.4) that the resulting orthonormal basis consists of the eigenvectors of the covariance matrix, Σ_x, where $\hat{\boldsymbol{a}}_i$, $i = 0, 2, \ldots, m-1$, correspond to the principal eigenvalues of Σ_x, and the constants are equal to

$$b_i = E[y(i)] = \hat{\boldsymbol{a}}_i^T E[\boldsymbol{x}], \quad i = m, \ldots, N-1$$

In other words, $\boldsymbol{x}$ is projected onto the subspace spanned by the m principal components of Σ_x and the rest $N - m$ components are frozen to the respective mean values, in order to bring the estimate closer to its mean. Note, however, that the number of free parameters remains equal to m. The optimality of the KL transform, with respect to the MSE approximation, leads to excellent information packing properties and offers us a tool to select the m dominant features out of N measurement samples. However, although this may be a good criterion, in many cases it does not necessarily lead to maximum class separability in the lower dimensional subspace. This is reasonable, since the dimensionality reduction is not optimized with respect to class separability, as was, for example, the case with the scattering matrix criteria of the previous chapter. This is demonstrated via the example of Figure 6.1. The feature vectors in the two classes follow the Gaussian distribution with the same covariance matrix. The ellipses show the curves of constant pdf values. We have computed the eigenvectors of the overall correlation matrix, and the resulting eigenvectors are shown in the figure. Eigenvector $\boldsymbol{a}_0$ is the one that corresponds to the largest eigenvalue. It does not take time for someone to realize that projection on $\boldsymbol{a}_0$ makes the two classes almost coincide. However, projecting on $\boldsymbol{a}_1$ keeps the two class separable.

Total Variance

Let $E[\boldsymbol{x}]$ be zero. Let $\boldsymbol{y}$ be the KL transformed vector of $\boldsymbol{x}$. From the respective definitions we have that $\sigma_{y(i)}^2 \equiv E[y^2(i)] = \lambda_i$. That is, *the eigenvalues of the input*

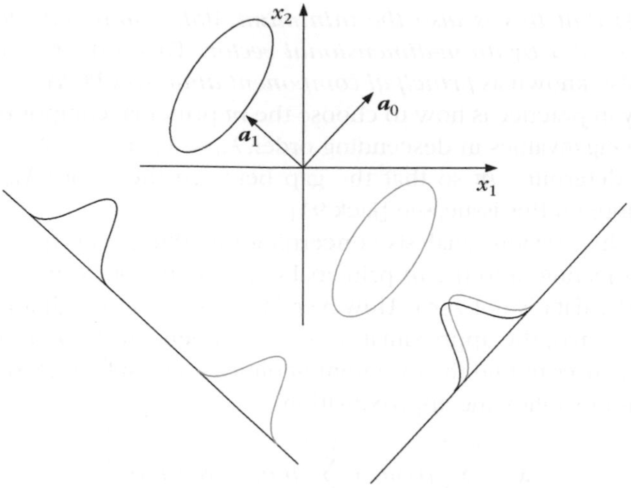

FIGURE 6.1

The KL transform is not always best for pattern recognition. In this example, projection on the eigenvector with the larger eigenvalue makes the two classes coincide. On the other hand, projection on the other eigenvector keeps the classes separated.

correlation matrix are equal to the variances of the transformed features. Thus, selecting those features, $y(i) \equiv \boldsymbol{a}_i^T \boldsymbol{x}$, corresponding to the m largest eigenvalues makes their sum variance $\sum_i \lambda_i$ maximum. In other words, the selected m features retain most of the total variance associated with the original random variables $x(i)$. Indeed, the latter is equal to the trace of R_x, which we know from linear algebra to be equal to the sum of the eigenvalues $\sum_{i=0}^{N-1} \lambda_i$ [Stra 80]. It can be shown that this is a more general property. That is, from all possible sets of m features, obtained via any orthogonal linear transformation on $\boldsymbol{x}$, the ones resulting from the KL transform have the largest sum variance (Problem 6.3). If the mean value is not zero, to maximize the sum variance, we use Σ_x in place of R_x.

Entropy

We know from Chapter 2 that the entropy of a process is defined as

$$H_y = -E[\ln p_y(\boldsymbol{y})]$$

and it is a measure of the randomness of the process. For a zero mean Gaussian multivariable m-dimensional process, the entropy becomes

$$H_y = \frac{1}{2}E[\boldsymbol{y}^T R_y^{-1} \boldsymbol{y}] + \frac{1}{2}\ln |R_y| + \frac{m}{2}\ln(2\pi) \tag{6.21}$$

However,

$$E[\boldsymbol{y}^T R_y^{-1} \boldsymbol{y}] = E[\text{trace}\{\boldsymbol{y}^T R_y^{-1} \boldsymbol{y}\}] = E[\text{trace}\{R_y^{-1} \boldsymbol{y}\boldsymbol{y}^T\}] = \text{trace}(I) = m$$

and using the known property from linear algebra the determinant is

$$\ln|R_y| = \ln(\lambda_0\lambda_1 \ldots \lambda_{m-1})$$

In words, selection of the m features that correspond to the m largest eigenvalues maximizes the entropy of the process. This is expected because variance and randomness are directly related.

Dimensionality Reduction

PCA achieves a *linear transformation* of a high-dimensional input vector into a low-dimensional one whose components are uncorrelated. As already stated, we can assume that $E[\boldsymbol{x}]$ is zero, without loss of generality. Assuming that the $N - m$ smallest eigenvalues of the correlation matrix are zero, then Eq. (6.19) suggests that $\boldsymbol{x} = \hat{\boldsymbol{x}}$. In other words, vector $\boldsymbol{x}$ lies in an m-dimensional subspace ([Eq. (6.15)]) of the original N-dimensional space. This brings us to the notion of *intrinsic dimensionality*.

A data set $X \subset \mathcal{R}^N$ is said to have intrinsic dimensionality (ID) $m < N$, if X can be described in terms of m free parameters. For example, if X consists of vectors whose components are functions of m random variables, $x_i = g_i(u_1, u_2, \ldots, u_m)$, $i = 1, 2, \ldots, N$, $u_i \in \mathcal{R}$, then the intrinsic dimensionality of X is m. The geometric interpretation of this is that the entire data set lies on a m-dimensional hypersurface (manifold) in $\mathcal{R}^N$. Take as an example the case of a random variable θ and the functions

$$x_1 = r\cos\theta,\ x_2 = r\sin\theta$$

It does not take time to see that $\boldsymbol{x} = [x_1, x_2]^T$ lies on the perimeter of the circle with radius equal to r. This is a one-dimensional surface (manifold) since one parameter suffices to describe the data (the length across the circumference from a point, origin, on the perimeter of the circle). From a statistical point of view, the fact that the intrinsic or "effective" dimension is smaller than the "apparent" one means that the features in the data set are correlated.

The PCA method has been used extensively for dimensionality reduction and for estimation of the ID of a data set. If ID $= m < N$, then in theory there will be $N - m$ zero eigenvalues. In practice, one has to ignore the eigenvalues with small values, and thus an approximation of the ID is obtained. PCA, being a linear projection method, works well if the data points are distributed, more or less, throughout a hyperplane. The eigenvalue–eigenvector decomposition of the correlation (covariance) matrix reveals the dimensionality of this hyperplane across which data are spread; in other words, dimensionality is a measure of the number of underlying modes of variability. Recall that PCA projects across the directions of maximum variance.

For more general cases, however where the generation mechanism of the data is highly nonlinear and they lie on more complicated manifolds, PCA fails, and it tends to overestimate the true value of the ID. For example, for the case considered before, where all the data points lie on the perimeter of a circle in the two-dimensional space, PCA would result in ID = 2, although the true value is ID = 1. For such cases, nonlinear dimensionality reduction techniques have been developed and used. For example, in [Karh 94], a special type of neural network with three hidden layers is proposed to perform nonlinear PCA. We will return to this issue in Section 6.7.

As we have seen in Chapter 5, LDA is another linear method that has been used for dimensionality reduction. However, in the case of LDA, this is achieved in a supervised manner; that is, the lower dimensional space is chosen in order to preserve a class separability measure. Another linear technique used to project in a lower dimensional space, while respecting certain constraints, is the *metric multidimensional scaling* (MDS). Given the set $X \subset \mathcal{R}^N$, the goal is to project into a lower dimensional space, $Y \subset \mathcal{R}^m$, so that inner products are optimally preserved, that is,

$$E = \sum_i \sum_j \left(\boldsymbol{x}_i^T \boldsymbol{x}_j - \boldsymbol{y}_i^T \boldsymbol{y}_j\right)^2$$

is minimized, where $\boldsymbol{y}_i$ is the image of $\boldsymbol{x}_i$ and the sum runs over all the training points in X. The problem is similar to the PCA, and it can be shown that the solution is given by the eigendecomposition of the Gram matrix, whose elements are defined as

$$\mathcal{K}(i,j) = \boldsymbol{x}_i^T \boldsymbol{x}_j$$

Another side of the same coin is to require the Euclidean distances, instead of the inner products, to be optimally preserved. A Gram matrix, consistent with the squared Euclidean distances can then be formed, leading to the same solution as before. It turns out that the solutions obtained by PCA and MDS are equivalent. We can see this by the following simple reasoning. PCA performs the eigen decomposition of the correlation matrix R_x, which is approximated by

$$R_x = E[\boldsymbol{x}\boldsymbol{x}^T] \approx \frac{1}{n}\sum_{k=1}^{n} \boldsymbol{x}_k \boldsymbol{x}_k^T = \frac{1}{n} X^T X \tag{6.22}$$

where, as we have defined in (3.44),

$$X^T = [\boldsymbol{x}_1, \boldsymbol{x}_2, \ldots, \boldsymbol{x}_n]$$

On the other hand, the Gram matrix can also be written as

$$\mathcal{K} = XX^T$$

However, as we will see in more detail in Section 6.4, the two matrices $X^T X$ and XX^T are of the same rank and have the same eigenvalues. Their eigenvectors, although different, are related.

More on these issues can be found in, for example, [Cox 94, Burg 04]. As we will see in Section 6.6, the main idea behind MDS of preserving the distances is used, in one way or another, in a number of more recently developed nonlinear dimensionality reduction techniques.

Remarks

- The concept of principal eigenvector subspace has also been exploited as a classifier. First, the sample mean of the whole training set is subtracted from the feature vectors. For each class, ω_i, the correlation matrix R_i is estimated and the principal m eigenvectors (corresponding to the m largest eigenvalues) are computed. A matrix A_i is then formed using the respective eigenvectors as columns. An unknown feature vector $\boldsymbol{x}$ is then classified in the class ω_j for which

$$\|A_j^T \boldsymbol{x}\| > \|A_i^T \boldsymbol{x}\|, \quad \forall i \neq j \tag{6.23}$$

that is, the class corresponding to the maximum norm subspace projection of $\boldsymbol{x}$ [Wata 73]. From the Pythagoras theorem this is equivalent to classifying a vector in its *nearest class subspace*. The decision surfaces are hyperplanes if all the subspaces have the same dimension or quadric surfaces in the more general case. *Subspace classification integrates the stages of feature generation/selection and classifier design*.

If this approach results in a relatively high classification error, the performance may be improved by suitable modifications known as *learning subspace methods*. For example, one can iteratively rotate the subspaces to adjust the lengths of the projections of the training vectors. The basic idea is to increase the length of a projection in the subspace of the correct class and decrease it for the rest. Such techniques have been applied successfully in a number of applications, such as speech recognition, texture classification, and character recognition. The interested reader may consult, for example, [Oja 83, Koho 89, Prak 97].

- For the computation of the correlation matrix eigenvectors, a number of iterative schemes have been developed. The computation is performed working directly with the vectors, without having to estimate the corresponding correlation matrix, using neural network concepts [Oja 83, Diam 96].

Example 6.2

The correlation matrix of a vector $\boldsymbol{x}$ is given by

$$R_x = \begin{bmatrix} 0.3 & 0.1 & 0.1 \\ 0.1 & 0.3 & -0.1 \\ 0.1 & -0.1 & 0.3 \end{bmatrix}$$

Compute the KL transform of the input vector.

The eigenvalues of R_x are $\lambda_0 = \lambda_1 = 0.4$, $\lambda_2 = 0.1$. Since the matrix R_x is symmetric, we can always construct orthonormal eigenvectors. For this case we have

$$\boldsymbol{a}_0 = \frac{1}{\sqrt{6}} \begin{bmatrix} 2 \\ 1 \\ 1 \end{bmatrix}, \quad \boldsymbol{a}_1 = \frac{1}{\sqrt{2}} \begin{bmatrix} 0 \\ 1 \\ -1 \end{bmatrix}, \quad \boldsymbol{a}_2 = \frac{1}{\sqrt{3}} \begin{bmatrix} 1 \\ -1 \\ -1 \end{bmatrix}$$

The KL transform is then given by

$$\begin{bmatrix} y(0) \\ y(1) \\ y(2) \end{bmatrix} = \begin{bmatrix} 2/\sqrt{6} & 1/\sqrt{6} & 1/\sqrt{6} \\ 0 & 1/\sqrt{2} & -1/\sqrt{2} \\ 1/\sqrt{3} & -1/\sqrt{3} & -1/\sqrt{3} \end{bmatrix} \begin{bmatrix} x(0) \\ x(1) \\ x(2) \end{bmatrix}$$

where $y(0)$, $y(1)$ correspond to the two largest eigenvalues.

Example 6.3

Figure 6.2 shows 100 points in the two-dimensional space. The points spread around the $x_2 = x_1$ line, and they have been generated by the model $x_2 = x_1 + \epsilon$, where ϵ is a noise source following the uniform distribution in $[-0.5,\ 0.5]$.

We first compute the covariance matrix and perform an eigendecomposition. The resulting eigenvectors are

$$\boldsymbol{a}_0 = [0.7045,\ 0.7097]^T, \quad \boldsymbol{a}_1 = [-0.7097,\ 0.7045]^T$$

corresponding to the eigenvalues

$$\lambda_0 = 17.26, \quad \lambda_1 = 0.04$$

respectively. Observe that $\lambda_0 >> \lambda_1$. Figure 6.2 shows the two eigenvectors. $\boldsymbol{a}_0$, which correspond to the largest eigenvalue, points in the direction where data show maximum variability. Projecting along this direction retains most of the variance. Moreover, according to PCA, the dimensionality of the set is approximately one, due to the large gap between λ_0 and λ_1, which is the correct answer. Also, note, that $\boldsymbol{a}_0$, is (approximately) parallel to the line $x_2 = x_1$.

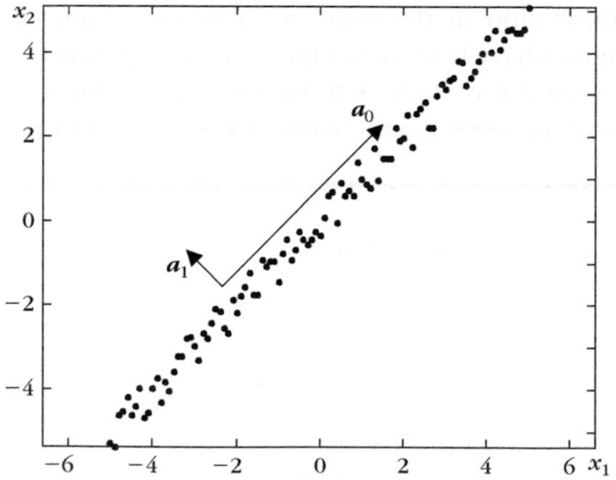

FIGURE 6.2

Points around the $x_2 = x_1$ line. The eigenvectors of the associated covariance matrix are $\boldsymbol{a}_0$ and $\boldsymbol{a}_1$. The principal eigenvector $\boldsymbol{a}_0$ points in the direction of maximum variance.

6.4 THE SINGULAR VALUE DECOMPOSITION

The singular value decomposition of a matrix is one of the most elegant and powerful algorithms in linear algebra, and it has been extensively used for rank and dimension reduction in pattern recognition and information retrieval applications. Given a $l \times n$ matrix X of rank r (obviously $r \leq \min\{l, n\}$), we will show that there exist unitary matrices U and V of dimensions $l \times l$ and $n \times n$, respectively, so that

$$X = U\begin{bmatrix} \Lambda^{\frac{1}{2}} & O \\ O & 0 \end{bmatrix}V^H \quad \text{or} \quad Y \equiv \begin{bmatrix} \Lambda^{\frac{1}{2}} & O \\ O & 0 \end{bmatrix} = U^H X V \tag{6.24}$$

where $\Lambda^{\frac{1}{2}}$ is the $r \times r$ diagonal matrix with elements $\sqrt{\lambda_i}$, and λ_i are the r nonzero eigenvalues of the associated matrix $X^H X$. O denotes a zero element matrix. *In other words, there exist unitary matrices U and V that transform X into the special diagonal structure of Y.* If $\boldsymbol{u}_i$, $\boldsymbol{v}_i$ denote the column vectors of matrices U and V, respectively, then Eq. (6.24) is written as

$$X = [\boldsymbol{u}_0, \boldsymbol{u}_1, \ldots, \boldsymbol{u}_{r-1}]\begin{bmatrix} \sqrt{\lambda_0} & & & \\ & \sqrt{\lambda_1} & & \\ & & \ddots & \\ & & & \sqrt{\lambda_{r-1}} \end{bmatrix}\begin{bmatrix} \boldsymbol{v}_0^H \\ \boldsymbol{v}_1^H \\ \vdots \\ \boldsymbol{v}_{r-1}^H \end{bmatrix} \tag{6.25}$$

or

$$X = \sum_{i=0}^{r-1} \sqrt{\lambda_i}\,\boldsymbol{u}_i \boldsymbol{v}_i^H \tag{6.26}$$

Sometimes, the above is also written as

$$X = U_r \Lambda^{\frac{1}{2}} V_r^H \tag{6.27}$$

where U_r denotes the $l \times r$ matrix that consists of the first r columns of U and V_r the $r \times n$ matrix formed by using the first r columns of V. More precisely, $\boldsymbol{u}_i$, $\boldsymbol{v}_i$ are the eigenvectors corresponding to the nonzero eigenvalues of the matrices XX^H and $X^H X$, respectively. The eigenvalues λ_i are known as *singular values* of X and the expansion in (6.26) as the *singular value decomposition* (SVD) of X or the *spectral representation* of X.

Proof Given a matrix X of rank r, it is known from linear algebra [Stra 80] that the $n \times n$ matrix $X^H X$ as well as the $l \times l$ matrix XX^H are of the same rank r. Furthermore, both matrices have the same nonzero eigenvalues but different (yet related) eigenvectors (Problem 6.5),

$$XX^H \boldsymbol{u}_i = \lambda_i \boldsymbol{u}_i \tag{6.28}$$

$$X^H X \boldsymbol{v}_i = \lambda_i \boldsymbol{v}_i \tag{6.29}$$

Since both matrices are Hermitian and nonnegative (i.e., $(XX^H)^H = XX^H$), they have nonnegative real eigenvalues and orthogonal eigenvectors (Appendix B). The eigenvectors, given in (6.28) and (6.29), can also be normalized to become orthonormal, that is, $\boldsymbol{u}_i^H \boldsymbol{u}_i = 1$ and $\boldsymbol{v}_i^H \boldsymbol{v}_i = 1$. It is straightforward to see from (6.28) and (6.29) that

$$\boldsymbol{u}_i = \frac{1}{\sqrt{\lambda_i}} X \boldsymbol{v}_i, \quad \text{for } \lambda_i \neq 0 \tag{6.30}$$

Indeed, premultiplying (6.29) by X results in

$$(XX^H) X \boldsymbol{v}_i = \lambda_i X \boldsymbol{v}_i$$

That is, $\boldsymbol{u}_i = \alpha X \boldsymbol{v}_i$, where (without loss of generality) the scaling factor α can be taken as positive and it is found from

$$\|\boldsymbol{u}_i\|^2 = 1 = \alpha^2 \boldsymbol{v}_i^H X^H X \boldsymbol{v}_i = \alpha^2 \lambda_i \|\boldsymbol{v}_i\|^2 \Rightarrow \alpha = \frac{1}{\sqrt{\lambda_i}}$$

Let us now assume that $\boldsymbol{u}_i, \boldsymbol{v}_i,\ i = 0, 1, \ldots, r-1$, are the eigenvectors corresponding to the nonzero eigenvalues and $\boldsymbol{u}_i,\ i = r, \ldots, l-1,\ \boldsymbol{v}_i,\ i = r, \ldots, n-1$, to the zero ones. Then, for the latter case we have

$$X^H X \boldsymbol{v}_i = 0 \Rightarrow \boldsymbol{v}_i^H X^H X \boldsymbol{v}_i = 0 \Rightarrow \|X \boldsymbol{v}_i\|^2 = 0$$

Hence

$$X \boldsymbol{v}_i = \boldsymbol{0}, \quad i = r, \ldots, n-1 \tag{6.31}$$

In a similar way one can show that

$$X^H \boldsymbol{u}_i = \boldsymbol{0}, \quad i = r, \ldots, l-1 \tag{6.32}$$

Combining (6.30) and (6.31), we show that the right-hand side of (6.26) is

$$\sum_{i=0}^{r-1} \sqrt{\lambda_i} \boldsymbol{u}_i \boldsymbol{v}_i^H = X \sum_{i=0}^{r-1} \sqrt{\lambda_i} \frac{1}{\sqrt{\lambda_i}} \boldsymbol{v}_i \boldsymbol{v}_i^H = X \sum_{i=0}^{n-1} \boldsymbol{v}_i \boldsymbol{v}_i^H \tag{6.33}$$

Let us now define a matrix V that has as columns the orthonormal eigenvectors $\boldsymbol{v}_i$,

$$V = [\boldsymbol{v}_0, \ldots, \boldsymbol{v}_{n-1}]$$

Orthonormality of the columns results in $V^H V = I$; that is, V is unitary and hence $VV^H = I$. Thus, it turns out that

$$I = VV^H = [\boldsymbol{v}_0, \ldots, \boldsymbol{v}_{n-1}] \begin{bmatrix} \boldsymbol{v}_0^H \\ \vdots \\ \boldsymbol{v}_{n-1}^H \end{bmatrix} = \sum_{i=0}^{n-1} \boldsymbol{v}_i \boldsymbol{v}_i^H \tag{6.34}$$

From (6.33) and (6.34) we obtain

$$X = \sum_{i=0}^{r-1} \sqrt{\lambda_i} \boldsymbol{u}_i \boldsymbol{v}_i^H \tag{6.35}$$

and X can be written as

$$X = U \begin{bmatrix} \Lambda^{\frac{1}{2}} & O \\ O & 0 \end{bmatrix} V^H \tag{6.36}$$

where U is the unitary matrix with columns the orthonormal eigenvectors $\boldsymbol{u}_i$. □

Low Rank Approximation

The expansion in (6.26) is an exact representation of matrix X. A very interesting implication occurs if one uses less than r (the rank of X) terms in the summation. Let X be approximated by

$$X \simeq \hat{X} = \sum_{i=0}^{k-1} \sqrt{\lambda_i} \boldsymbol{u}_i \boldsymbol{v}_i^H, \quad k \leq r \tag{6.37}$$

Matrix $\hat{X}$, being the sum of $k \leq r$ rank-one independent $l \times n$ matrices, is of rank k. If the k largest eigenvalues are involved, it can be shown that the squared error

$$\epsilon^2 = \sum_{i=0}^{l-1} \sum_{j=0}^{n-1} |X(i,j) - \hat{X}(i,j)|^2 \tag{6.38}$$

is the minimum one with respect to all rank-k $l \times n$ matrices. The square root of the right-hand side in (6.38) is also known as the Frobenius norm $\|X - \hat{X}\|_F$ of the difference matrix $X - \hat{X}$. The error in the approximation turns out to be (Problem 6.6)

$$\epsilon^2 = \sum_{i=k}^{r-1} \lambda_i \tag{6.39}$$

Hence, if we order the eigenvalues in descending order, $\lambda_0 \geq \lambda_1 \geq \cdots \geq \lambda_{r-1}$, then for a given number of k terms in the expansion, the SVD leads to the minimum square error. *Thus, $\hat{X}$ is the best rank-k approximation of X in the Frobenius norm sense*. This reminds us of the Karhunen–Loève expansion. However, in the latter case the optimality was with respect to the mean square error. This is a major difference in philosophy between SVD and KL. *The former is related to a single set of samples and the latter to an ensemble of them.*

Dimensionality Reduction

SVD has been used extensively for dimension reduction in pattern recognition and information retrieval, and it forms the basis of what is known as *latent semantics indexing*, see, for example, [Berr 95]. Adopting the notation used in (6.25) and

(6.27), Eq. (6.37) can be written as

$$X \simeq \hat{X} = [\boldsymbol{u}_0, \boldsymbol{u}_1, \ldots, \boldsymbol{u}_{k-1}] \begin{bmatrix} \sqrt{\lambda_0}\boldsymbol{v}_0^H \\ \sqrt{\lambda_1}\boldsymbol{v}_1^H \\ \vdots \\ \sqrt{\lambda_{k-1}}\boldsymbol{v}_{k-1}^H \end{bmatrix}$$

$$= U_k[\boldsymbol{a}_0, \boldsymbol{a}_1, \ldots, \boldsymbol{a}_{n-1}] \tag{6.40}$$

where U_k consists of the first k columns of U and the k-dimensional vectors $\boldsymbol{a}_i$, $i = 0, 1, \ldots, n-1$, are the column vectors of the $k \times n$ product matrix $\Lambda_k^{\frac{1}{2}} V_k^H$, where V_k^H consists of the first k rows of V^H and $\Lambda_k^{\frac{1}{2}}$ is the diagonal matrix having elements the square roots of the respective k singular values. Figure 6.3 gives a diagrammatic interpretation of the matrix products involved in SVD. The formulation given in (6.40) suggests that each column vector, $\boldsymbol{x}_i$ of X, is approximated as

$$\boldsymbol{x}_i \simeq U_k \boldsymbol{a}_i = \sum_{m=0}^{k-1} \boldsymbol{u}_m a_i(m), \quad i = 0, 2, \ldots, n-1 \tag{6.41}$$

where $a_i(m)$, $m = 0, 1, \ldots, k-1$, denote the elements of the respective vector $\boldsymbol{a}_i$. In words, the l-dimensional vector $\boldsymbol{x}_i$ is approximated by the k-dimensional vector $\boldsymbol{a}_i$, lying in the subspace spanned by $\boldsymbol{u}_i$, $i = 0, 1, \ldots, k-1$ ($\boldsymbol{a}_i$ is the projection of $\boldsymbol{x}_i$ on this subspace; Problem 6.6.) Furthermore, due to the orthonormality of the columns $\boldsymbol{u}_i$, $i = 0, 1, \ldots, k-1$, of U_k it is straightforward to see that

$$||\boldsymbol{x}_i - \boldsymbol{x}_j|| \simeq ||U_k(\boldsymbol{a}_i - \boldsymbol{a}_j)|| = ||\sum_{m=0}^{k-1} \boldsymbol{u}_m (a_i(m) - a_j(m))||$$

$$= ||\boldsymbol{a}_i - \boldsymbol{a}_j||, \quad i, j = 0, 1, \ldots, n-1 \tag{6.42}$$

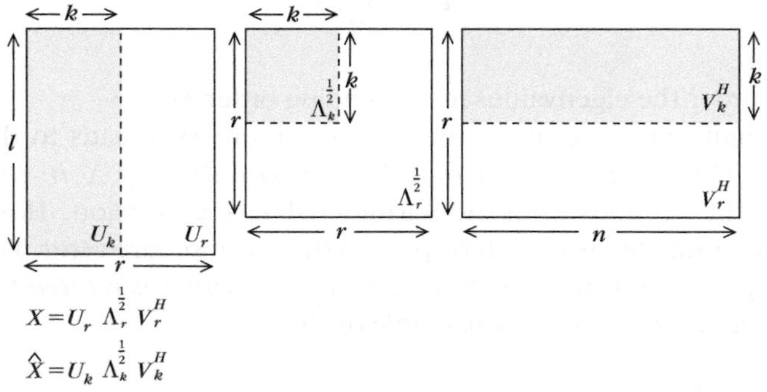

FIGURE 6.3

Diagrammatic interpretation of the matrix products involved in SVD. In the approximation of X by $\hat{X}$, the first k columns of U_r and the first k rows of V_r^H are involved.

where $||\cdot||$ represents the Euclidean norm of a vector. That is, using the previous projection and assuming the approximation to be reasonably good, the Euclidean distance between $\boldsymbol{x}_i$ and $\boldsymbol{x}_j$ in the high l-dimensional space is (approximately) preserved under the projection in the lower k-dimensional subspace.

The previous observation has important implications in applications such as information retrieval. Let us take as an example the simple case where we are given a set of n patterns each represented by a l-dimensional feature vector. These patterns constitute the available database. Given an unknown pattern, the goal is to search for and recover from the database the pattern that is most similar to the unknown one, by computing its Euclidean distance from each vector in the database. When l and n are large numbers this can be a very time-consuming task. A procedure to simplify computations is the following. We form the $l \times n$ data matrix, X,[3] having as columns the n feature vectors. Perform a SVD on X and represent each feature vector, $\boldsymbol{x}_i$, by its lower dimensional projection, $\boldsymbol{a}_i$, as described before. Given the unknown vector, one projects it on the subspace spanned by the columns of U_k and performs Euclidean distance computations in the k-dimensional space. Since Euclidean distances are approximately preserved, one can decide about the proximity of vectors by working in a lower dimensional space. If $k << l$ substantial computational savings are obtained (see, e.g., [Berr 95, Deer 90, Sebr 03]).

SVD builds upon *global* information spread over all the data vectors in X. Indeed, a crucial part of the algorithm is the computation of the eigenvalues of $X^H X$ or XX^H, which, for zero mean data, is directly related to the respective covariance matrix (Eq. 6.22). Hence, the performance of the SVD, as a dimensionality reduction technique, is most effective for cases where data can sufficiently be described in terms of the covariance matrix, for example, to be Gaussian-like distributed. In [Cast 03] a modification of the simple SVD is suggested to account for data with a clustered structure. In Section 6.7, nonlinear dimensionality techniques will be reviewed, where more than a simple linear projection on a subspace is required to reduce dimensionality.

Remarks

- Due to its optimal approximation properties, the SVD transform also has excellent "information packing" properties, and an image array can be represented efficiently by a few of its singular values. Thus, SVD is a natural candidate as a tool for feature generation/selection in classification.
- Performing SVD of large matrices is a computationally expensive task. In order to overcome this drawback, a number of computationally efficient schemes have been developed, see, for example, [Ye 04, Achl 01].

[3] Note that X is defined here as the transpose of the data matrix in (3.44), to comply with the notation used in latent semantics indexing.

Example 6.4

Consider the matrix

$$X = \begin{bmatrix} 6 & 6 \\ 0 & 1 \\ 4 & 0 \\ 0 & 6 \end{bmatrix}$$

The goal is to compute its singular value decomposition.

- Step 1: Find the eigenvalues and eigenvectors of

$$X^T X = \begin{bmatrix} 52 & 36 \\ 36 & 73 \end{bmatrix}$$

These are $\lambda_0 = 100$, $\lambda_1 = 25$, and the corresponding eigenvectors are $\boldsymbol{v}_0 = [0.6, 0.8]^T$, $\boldsymbol{v}_1 = [0.8, -0.6]^T$.

- Step 2: Compute the eigenvectors of XX^T. This is a 4×4 matrix of rank 2. The eigenvectors corresponding to the nonzero eigenvalues λ_0, λ_1 are computed via (6.30), that is, $\boldsymbol{u}_0 = 0.1X\boldsymbol{v}_0$, $\boldsymbol{u}_1 = 0.2X\boldsymbol{v}_1$ or $[0.84, 0.08, 0.24, 0.48]^T$ and $[0.24, -0.12, 0.64, -0.72]^T$ respectively.

- Step 3: Compute the SVD of X

$$X = 10[0.84, 0.08, 0.24, 0.48]^T [0.6, 0.8] + 5[0.24, -0.12, 0.64, -0.72]^T [0.8, -0.6]$$

If we keep the first of the two terms, then the resulting approximation is the best, in the Frobenius sense, rank-1 approximation of X.

Example 6.5

The goal of this example is to demonstrate the power of the SVD as a dimensionality reduction tool, in the context used in latent semantics indexing in information retrieval.

(a) Let our data set consist of 1000 three-dimensional vectors

$$\boldsymbol{x}_i = [x_1(i), x_2(i), x_3(i)]^T, \quad i = 1, 2, \ldots, 1000$$

This set of points comprises our database. We form the 3×1000 matrix X having these data vectors as columns. For the needs of this example, the components $x_1(i)$, $x_2(i)$ are randomly generated using the uniform distribution in $[-10, 10]$. The value of the third dimension of each point is given by $x_3(i) = -x_1(i) - x_2(i) + \varepsilon$, where ε is a noise source following the uniform distribution in $[-1, 1]$. In other words, our data are crowded around the plane

$$H : x_1 + x_2 + x_3 = 0 \tag{6.43}$$

Performing SVD analysis, it turns out that the singular values are

$$\lambda_0 = 158.43,\ \lambda_1 = 89.01,\ \lambda_2 = 10.50.$$

The relatively small value of λ_2 is the consequence of the fact that our data are approximately two-dimensional. Recall that the singular values are eigenvalues of XX^T, which is the same (within a scaling factor) with the estimate of the correlation matrix used in PCA (Eq. (6.11)). The corresponding (orthonormal) eigenvectors, which are also the column vectors of the U matrix, are (after rounding to the second decimal point)

$$\boldsymbol{u}_0 = \begin{bmatrix} -0.39 \\ -0.42 \\ 0.82 \end{bmatrix},\ \boldsymbol{u}_1 = \begin{bmatrix} 0.71 \\ -0.70 \\ -0.01 \end{bmatrix},\ \boldsymbol{u}_2 = \begin{bmatrix} 0.58 \\ 0.58 \\ 0.57 \end{bmatrix}$$

It is not difficult to verify that the plane (subspace) formed by the two principal eigenvectors is

$$H_1:\ 14.26x_1 + 14.10x_2 + 13.95x_3 = 0$$

which is very close to the hyperplane H in (6.43), around which our data cluster.

(b) We now select randomly six of the points in the data set X and project them along the H_1 plane. The 6×6 distance matrix D whose (i,j) element is the squared Euclidean distance between the ith and jth points for $i, j = 1, 2, \ldots, 6$, is

$$D = \begin{bmatrix} 0 & 26.17 & 24.70 & 112.25 & 11.92 & 4.81 \\ 26.17 & 0 & 61.46 & 43.96 & 38.33 & 49.25 \\ 24.70 & 61.46 & 0 & 107.97 & 4.34 & 14.51 \\ 112.25 & 43.96 & 107.97 & 0 & 88.72 & 140.18 \\ 11.92 & 38.33 & 4.34 & 88.72 & 0 & 9.95 \\ 4.81 & 49.25 & 14.51 & 140.18 & 9.95 & 0 \end{bmatrix}$$

The corresponding distance matrix, D', for the respective projections on H_1 is

$$D' = \begin{bmatrix} 0 & 25.85 & 24.32 & 112.21 & 11.72 & 4.57 \\ 25.85 & 0 & 61.46 & 43.83 & 37.29 & 49.24 \\ 24.32 & 61.46 & 0 & 107.80 & 3.20 & 14.49 \\ 112.21 & 43.83 & 107.80 & 0 & 88.29 & 140.10 \\ 11.72 & 37.29 & 3.20 & 88.29 & 0 & 9.06 \\ 4.57 & 49.24 & 14.49 & 140.10 & 9.06 & 0 \end{bmatrix}$$

which is in close agreement with D. Note that this good agreement is a consequence of the fact that the true dimensionality of the data is very close to 2. Increasing the variance of the noise source ε, the spread of the data around the plane H would increase and the data would become more and more "three-dimensional." In other words, the higher the variance of ε, the less the agreement between D and D' that one expects to get.

6.5 INDEPENDENT COMPONENT ANALYSIS

As we have already seen, the principal component analysis (PCA) performed by the Karhunen–Loève transform produces features $y(i)$, $i = 0, 1, \ldots, N-1$, that are mutually uncorrelated. The solution obtained by the KL transform solution is optimal when dimensionality reduction is the goal and one wishes to minimize the approximation mean square error. However, for certain applications, such as the one illustrated in Figure 6.1, the obtained solution falls short of the expectations. In contrast, the more recently developed ***independent component analysis*** (ICA) theory, for example, [Hyva 01, Como 94, Jutt 91, Hayk 00, Lee 98], tries to achieve much more than simple decorrelation of the data. The ICA task is casted as follows: Given the set of input samples $\boldsymbol{x}$, determine an $N \times N$ invertible matrix W such that the entries $y(i)$, $i = 0, 1, \ldots, N-1$, of the transformed vector

$$\boldsymbol{y} = W\boldsymbol{x} \tag{6.44}$$

are mutually independent. The goal of statistical independence is a stronger condition than the uncorrelatedness required by the PCA. The two conditions are equivalent *only* for Gaussian random variables.

Searching for independent rather than uncorrelated features gives us the means of exploiting a lot more information, hidden in the higher order statistics of the data. As the example of Figure 6.1 suggests, constraining the search by digging information in the second-order statistics only results in the least interesting, for our problem, projection direction, that is, that of $\boldsymbol{a}_0$. However, $\boldsymbol{a}_1$ is, no doubt, the most interesting direction from the class separation point of view. In contrast, employing ICA can unveil from the higher order statistics of the data the piece of information that points $\boldsymbol{a}_1$ as the most interesting one. Furthermore, searching for statistically independent features is in line with the way nature builds up the "cognitive" maps of the outside world in the brain, by processing the (input) sensory data. Barlow [Barl 89], in the so-called Barlow's hypothesis, suggests that the outcome of the early processing performed in our visual cortical feature detectors might be the result of a ***redundancy reduction*** process. In other words, the neural outputs are mutually as statistically independent as possible, conditioned, of course, on the received sensory messages. The interested reader can find more on issues related to redundancy reduction and also to a number of methodologies inspired by it in [Atti 92, Fiel 94, Deco 95, Bell 00]. The potential of the ICA as an optimal feature generator technique, in the context of pattern recognition, has been demonstrated in [Cao 03, Bell 97, Hoy 00, Jang 99, Bart 02, Kwon 04].

Before we proceed to develop techniques for performing ICA, we need to be sure that such a problem is well defined and has a solution and under what conditions. To this end, let us assume that our input random data vector $\boldsymbol{x}$ is indeed generated by a linear combination of statistically independent and *stationary in the strict sense* components (sources), that is,

$$\boldsymbol{x} = A\boldsymbol{y} \tag{6.45}$$

The task now is under what conditions a matrix, W, can be computed so as to recover the components of $\boldsymbol{y}$ from Eq. (6.44), by exploiting information hidden in $\boldsymbol{x}$. Usually $\mathcal{A}$ is known as the mixing and W as the demixing matrix, respectively. The following condition is proved in [Como 94].

Identifiability Condition of the ICA Model

All independent components $y(i)$, $i = 1, 2, \ldots, N$, with the possible exception of one, must be non-Gaussian. A second condition is that matrix $\mathcal{A}$ must be invertible. In the more general case where $\mathcal{A}$ is a nonsquare $l \times N$ matrix, then l must be greater than N and $\mathcal{A}$ must be of full column rank.

In other words, in contrast to PCA which can always be performed, ICA is meaningful only if the involved random variables are non-Gaussian. Indeed, as has already been stated, for Gaussian random variables independence is equivalent to uncorrelatedness and PCA suffices. From a mathematical point of view, the ICA problem is ill-posed for Gaussian processes. Indeed, if we assume that the obtained independent components $y(i)$, $i = 0, 1, \ldots, N-1$, are all Gaussian, then a linear transformation of them by *any* unitary matrix will also be a solution (see Problem 5.4). PCA achieves a unique solution by imposing a *specific orthogonal structure* onto the transformation matrix.

Under the above stated conditions, it can be shown that each one of the resulting independent components is ***uniquely*** estimated up to a multiplicative constant, which is a rather insignificant indeterminacy associated with the method. This is the reason that many times the components are considered to be of unit variance. Finally, it is interesting that the independent components result in no specific ordering, in contrast to the PCA, where a specific ordering is associated with the values of the corresponding eigenvalues. However, in practice, some form of ordering can be adopted. For example, the components can be ordered according to the degree of "non-Gaussianity," measured by an appropriate index, for example, the fourth-order cumulant (see Appendix A). Although such an index may seem a bit strange to a newcomer, its physical interpretation will become clearer as we go on. After all, from a common-sense point of view, a Gaussian pdf must be the least interesting one. Recall from Chapter 2 that maximizing the entropy, constraining the solution to be within the family of random variables with given mean and variance, the result is a Gaussian pdf. That is, the Gaussian is the most "random" of all the pdfs describing this family of random variables and from this point of view the least informative one with respect to the underlying structure of the data. In contrast, distributions that have the "least resemblance" to the Gaussian are more interesting since they display some structure associated with the data. This observation is at the heart of a closely related, to ICA, family of techniques known as *projection pursuit*; see also Section 4.12. The essence behind such techniques is to search for directions (subspaces) in the feature space so that the corresponding data vector projections are described by "interesting" non-Gaussian distributions. For a more rigorous discussion on such issues the reader may refer to, for example, [Hube 85, Jone 87].

6.5.1 ICA Based on Second- and Fourth-Order Cumulants

This approach in performing ICA is a direct generalization of the PCA technique. The Karhunen–Loève transform focuses on the second-order statistics and demands the cross-correlations $E[y(i)y(j)]$ to be zero. Since in ICA we demand that the components of $\boldsymbol{y}$ be statistically independent, this is equivalent to demanding that all the higher order cross-cumulants to be zero (see Appendix A). In [Como 94] it is suggested that restricting ourselves up to the fourth-order cumulants is sufficient for many applications. Appendix A shows the first three cumulants are equal to the first three moments, that is,

$$\kappa_1(y(i)) = E[y(i)] = 0$$

$$\kappa_2(y(i)y(j)) = E[y(i)y(j)]$$

$$\kappa_3(y(i)y(j)y(k)) = E[y(i)y(j)y(k)]$$

and the fourth-order cumulants are given by

$$\begin{aligned}\kappa_4(y(i)y(j)y(k)y(r)) = {} & E[y(i)y(j)y(k)y(r)] - E[y(i)y(j)]E[y(k)y(r)] \\ & - E[y(i)y(k)]E[y(j)y(r)] \\ & - E[y(i)y(r)]E[y(j)y(k)]\end{aligned}$$

where zero mean processes have been assumed. Another assumption that is usually encountered in practice, and will be adopted here, is that the associated pdfs are symmetric. This makes all odd order cumulants zero. Thus the problem has now been reduced to finding a matrix, W, so that the second-order (cross-correlations) and fourth-order cross-cumulants of the transformed variables are zero. In [Como 94] this is achieved by the following steps:

Step 1: Perform a PCA on the input data, that is,

$$\hat{\boldsymbol{y}} = A^T \boldsymbol{x} \tag{6.46}$$

A is our familiar unitary transformation matrix of the Karhunen–Loève transform. The components of the transformed random vector $\hat{\boldsymbol{y}}$ are thus uncorrelated.

Step 2: Compute another unitary matrix, $\hat{A}$, so that the fourth-order cross-cumulants of the components of the transformed random vector

$$\boldsymbol{y} = \hat{A}^T \hat{\boldsymbol{y}} \tag{6.47}$$

are zero. This is equivalent to searching for a matrix $\hat{A}$ that makes the sum of the squares of the fourth-order auto-cumulants maximum, that is,

$$\max_{\hat{A}\hat{A}^T = I} \Psi(\hat{A}) \equiv \sum_{i=0}^{N-1} \kappa_4(y(i))^2 \tag{6.48}$$

Step 2 is justified as follows. It can be shown [Como 94] that the sum of the squares of the fourth-order cumulants is invariant under a linear transformation by a unitary matrix. Therefore, since the sum of squares of the fourth-order cumulants is fixed for $\hat{\boldsymbol{y}}$, maximizing the sum of squares of the auto-cumulants of $\boldsymbol{y}$ will force the corresponding cross-cumulants to zero. Observe that this is basically a diagonalization problem of the fourth-order cumulant multidimensional array. In practice, this is achieved by generalizing the method of Givens rotations, used for matrix diagonalization [Como 94]. Note that the right hand side in Eq. (6.48) is a function of (a) the elements of the unknown matrix $\hat{A}$, (b) the elements of the known (for this step) matrix A, and (c) the cumulants of the random components of the input data vector $\boldsymbol{x}$, which have to be estimated prior to the application of the method. In practice, it may turn out that the nulling of cross-cumulants is only approximately achieved. This is because (a) the input data may not obey the linear model of Eq. (6.45); (b) the input data are corrupted by noise, which has not been taken into account; and (c) the cumulants of the input are only approximately known, since they are estimated by the available input data set.

Once the two steps have been completed, the final feature vector with (approximately) independent components is given by the combined transform

$$\boldsymbol{y} = (A\hat{A})^T \boldsymbol{x} \equiv W\boldsymbol{x} \tag{6.49}$$

Notice that since $\hat{A}$ is unitary, the uncorrelatedness achieved in the first step is inherited by the elements of $\boldsymbol{y}$, which now has its second- and fourth-order cross-cumulants (at least approximately) zero.

6.5.2 ICA Based on Mutual Information

The approach based on nulling the second- and fourth-order cross-cumulants, though one of the most widely used in practice, somehow lacks in generality and also imposes, externally, a structure in the transformation matrix. An alternative, theoretically more pleasing approach is estimating W by minimizing the *mutual information* between the transformed random variables. The mutual information, $I(\boldsymbol{y})$, between the components of $\boldsymbol{y}$ is defined as

$$I(\boldsymbol{y}) = -H(\boldsymbol{y}) + \sum_{i=0}^{N-1} H(y(i)) \tag{6.50}$$

where $H(y(i))$ is the associated entropy of $y(i)$, defined as ([Papo 91])

$$H(y(i)) = -\int p_i(y(i)) \ln p_i(y(i))\, dy(i) \tag{6.51}$$

where $p_i(y(i))$ is the marginal pdf of $y(i)$. In Appendix A, it is shown that $I(\boldsymbol{y})$ is equal to the Kullback-Leibler probability distance between the joint pdf $p(\boldsymbol{y})$ and the product of the respective marginal probability densities $\prod_{i=0}^{N-1} p_i(y(i))$. This distance (and hence the associated mutual information $I(\boldsymbol{y})$) is zero if the components $y(i)$ are statistically independent. This is because only in this case is the joint pdf equal to the product of the corresponding marginal pdfs and the Kullback–Leibler distance becomes zero. Hence, what is more natural than trying to compute W so as to force $I(\boldsymbol{y})$ to be minimum, since this will make the components of $\boldsymbol{y}$ *as independent as possible*? Combining Eqs. (6.44), (6.50), and (6.51) and taking into account the formula that relates the two pdfs associated with $\boldsymbol{x}$ and $\boldsymbol{y}$ ($\boldsymbol{y}$ is a function of $\boldsymbol{x}$), e.g., [Papo 91], we end up with

$$I(\boldsymbol{y}) = -H(\boldsymbol{x}) - \ln|det(W)| - \sum_{i=0}^{N-1}\int p_i(y(i))\ln p_i(y(i))\,dy(i) \tag{6.52}$$

where $det(W)$ denotes the determinant of W. The elements of the unknown matrix W are hidden in the marginal pdfs of the transformed variables, $y(i)$. However, it is not easy to express this dependence explicitly. An approach currently used is to expand each of the marginal probabilities around the Gaussian pdf, $g(y)$, following Edgeworth's expansion (Appendix A), and truncate the series to a reasonable approximation. For example, keeping the first two terms in the Edgeworth expansion, we have

$$p(y) = g(y)\left(1 + \frac{1}{3!}\kappa_3(y)H_3(y) + \frac{1}{4!}\kappa_4(y)H_4(y)\right) \tag{6.53}$$

where $H_k(y)$ is the Hermite polynomial of order k (Appendix A). To obtain an approximate expression for $I(\boldsymbol{y})$ in terms of cumulants of $y(i)$ and W, we can (a) insert in Eq. (6.52) the pdf approximation in Eq. (6.53), (b) adopt the approximation $\ln(1+y) \simeq y - y^2$, and (c) perform the integrations. This is no doubt a rather painful task! For the case of Eq. (6.53) and constraining W to be unitary, the following is obtained ([Hyva 01]):

$$I(\boldsymbol{y}) \simeq C - \sum_{i=0}^{N-1}\left(\frac{1}{12}\kappa_3^2(y(i)) + \frac{1}{48}\kappa_4^2(y(i)) + \frac{7}{48}\kappa_4^4(y(i)) - \frac{1}{8}\kappa_3^2(y(i))\kappa_4(y(i))\right) \tag{6.54}$$

where C is a variable independent of W. Under the assumption that the pdfs are symmetric (thus, third-order cumulants are zero,) it can be shown that minimizing the approximate expression of the mutual information in Eq. (6.54) is equivalent to maximizing the sum of the squares of the fourth-order cumulants. Of course, the unitary W constraint is not necessary, and in this case other approximate expressions for $I(\boldsymbol{y})$ result, e.g., [Hayk 99].

Minimization of $I(\boldsymbol{y})$ in Eq. (6.54) can be carried out by a gradient descent technique (Appendix C), where the involved expectations (associated with the cumulants) are replaced by the respective instantaneous values. Although a detailed treatment of the optimization procedure is beyond the scope of this book, it is worth pointing out some of its aspects.

Before we apply the approximations, let us go back to Eq. (6.52). Since $H(\boldsymbol{x})$ does not depend on W, minimizing $I(\boldsymbol{y})$ is equivalent to maximization of

$$J(W) = \ln|det(W)| + E\left[\sum_{i=0}^{N-1}\ln p_i(y(i))\right] \tag{6.55}$$

Taking the gradient of the cost function with respect to W results in

$$\frac{\partial J(W)}{\partial W} = W^{-T} - E[\phi(y)x^T] \tag{6.56}$$

where

$$\phi(y) \equiv \left[-\frac{p'_0(y(0))}{p_0(y(0))}, \ldots, -\frac{p'_{N-1}(y(N-1))}{p_{N-1}(y(N-1))} \right]^T \tag{6.57}$$

and

$$p'_i(y(i)) \equiv \frac{dp_i(y(i))}{dy(i)} \tag{6.58}$$

Obviously, the derivatives of the marginal probability densities depend on the type of approximation adopted in each case. The general gradient ascent scheme at the tth iteration step can now be written as

$$W(t) = W(t-1) + \mu(t)\left(W^{-T}(t-1) - E[\phi(y)x^T]\right)$$

$$W(t) = W(t-1) + \mu(t)\left(I - E[\phi(y)y^T]\right)W^{-T}(t-1) \tag{6.59}$$

In practice, the expectation operator is neglected, in the spirit of the stochastic approximation rationale (Section 3.4.2).

Remarks

- From the gradient in Eq. (6.56) it is easy to see that at a stationary point the following is true:

$$\frac{\partial J(W)}{\partial W} W^T = E[I - \phi(y)y^T] = 0 \tag{6.60}$$

  In other words, what we achieve with ICA is a nonlinear generalization of PCA. Recall that for the latter, the uncorrelatedness condition can be written as

$$E[I - yy^T] = 0 \tag{6.61}$$

  The presence of the nonlinear function $\phi(\cdot)$ takes us beyond simple uncorrelatedness, and brings the cumulants into the scene. In fact, Eq. (6.60) was the one that inspired the early pioneering work on ICA, as a direct nonlinear generalization of PCA [Jutt 91].

- The updated equation in Eq. (6.59) involves the inversion of the transpose of the current estimate of W. Besides the computational complexity issues, there is no guarantee of invertibility in the process of adaptation. Use of the so-called natural gradient [Doug 00], instead of the gradient in Eq. (6.56), results in

$$W(t) = W(t-1) + \mu(t)\left(I - E[\phi(y)y^T]\right)W(t-1) \tag{6.62}$$

which does not involve matrix inversion and at the same time improves convergence. A more detailed treatment of this issue is beyond the scope of the present book. Just to give an incentive to the mathematically inclined reader for indulging more deeply this field, it suffices to say that our familiar gradient, that is, Eq. (6.56), points to the steepest ascent direction if the space is Euclidean. However, in our case the parameter space consists of all the nonsingular $N \times N$ matrices, which is a multiplicative group. The space is Riemannian, and it turns out that the natural gradient, pointing to the steepest ascent direction, results if we multiply the gradient in Eq. (6.56) by $W^T W$, which is the corresponding Riemannian metric tensor [Doug 00].

6.5.3 An ICA Simulation Example

The example is a realization of the case shown in Figure 6.4. A total of 1024 samples of a two-dimensional normal distribution were generated.

The mean and covariance matrix of the normal pdf were

$$\boldsymbol{\mu} = [-2.6042, 2.5]^T, \quad \Sigma = \begin{bmatrix} 10.5246 & 9.6313 \\ 9.6313 & 11.3203 \end{bmatrix}$$

Similarly, 1024 samples from a second normal pdf were generated with the same covariance and mean $-\boldsymbol{\mu}$. For the ICA, the method based on the second- and

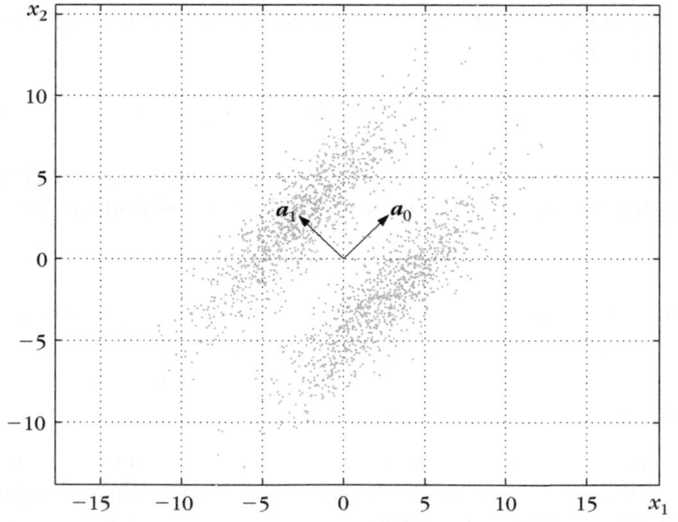

FIGURE 6.4

The setup for the ICA simulation example. The two vectors point to the projection directions resulting from the analysis. The optimal direction for projection, resulting from the ICA analysis, is that of $\boldsymbol{a}_1$.

fourth-order cumulants, presented in this section, was used. The resulting transformation matrix W is

$$W = \begin{bmatrix} -0.7088 & 0.7054 \\ 0.7054 & 0.7088 \end{bmatrix} \equiv \begin{bmatrix} \boldsymbol{a}_1^T \\ \boldsymbol{a}_0^T \end{bmatrix}.$$

The vectors $\boldsymbol{a}_0$ and $\boldsymbol{a}_1$ point in the principal and minor axis directions, respectively, obtained from the PCA analysis. However, the most interesting direction for projection, according to the ICA analysis, is that of $\boldsymbol{a}_1$ and not of $\boldsymbol{a}_0$. Indeed, the kurtosis of the obtained transformed variables $[y_1, y_2]^T = W\boldsymbol{x}$ is

$$\kappa_4(y_1) = -1.7$$

$$\kappa_4(y_2) = 0.1$$

Thus, projection in the principal axis direction results in a variable with a pdf close to a Gaussian. The projection to the minor axis direction results in a variable with a pdf that deviates from the Gaussian (Figures 6.1, 6.4) and is more appropriate from the classification point of view.

6.6 NONNEGATIVE MATRIX FACTORIZATION

In the PCA as well as the SVD analysis, the underlying constraints were the orthogonality of the involved basis vectors, in the PCA, and of the column vectors in the U and V matrices in the SVD. This becomes crystal clear from the problem 6.3. PCA is formulated as a task minimizing the mean square error subject to the orthogonality constraint of the basis vectors. Although, in some cases, the resulting expansion is useful, for some other cases such a constraint turns out to be very "weak" in representing the data. More recently, a new matrix factorization was suggested in [Paat 91, Paat 94], which guarantees the nonnegativity of the elements of the resulting matrix factors. Such a constraint is enforced in certain applications since negative elements contradict physical reality. For example, in image analysis the intensity values of the pixels cannot be negative. Also, probability values cannot be negative. The resulting factorization is known as *nonnegative matrix factorization* (NMF) and it has been used successfully in a number of applications including document clustering ([Xu 03]), molecular pattern discovery ([Brun 04]), image analysis ([Lee 01]), clustering ([Szym 06]), music transcription and music instrument classification ([Smar 03, Benn 06]) and face verification ([Zafe 06]).

Given a $l \times n$ matrix X, the task of NMF consists of finding an approximate factorization of X, that is,

$$X \approx WH \tag{6.63}$$

where W and H are $l \times r$ and $r \times n$ matrices, respectively, $r < \min(n, l)$ and all the matrix elements are nonnegative, that is, $W(i,k) \geq 0$, $H(k,j) \geq 0$,

$i = 1, 2, \ldots, l$, $k = 1, 2, \ldots, r$, $j = 1, 2, \ldots, n$. Clearly, matrices W and H are of rank at most r and their product is a low rank, at most r, approximation of X. The significance of the above is that every column vector in X is represented by the expansion

$$\boldsymbol{x}_i = \sum_{k=1}^{r} H(k,i)\boldsymbol{w}_k, \quad i = 1, 2, \ldots, n$$

where $\boldsymbol{w}_k$, $k = 1, 2, \ldots, r$, are the column vectors of W and constitute the basis of the expansion. The number of vectors in the basis is less than the dimensionality of the vector itself. Hence, NMF can also be seen as a method for *dimensionality reduction*.

To get a good approximation in (6.63) one can adopt different costs. The most conventional cost is the Frobenius norm of the error matrix. In such a setting, the NMF task is casted as follows:

$$\text{minimize} \quad ||X - WH||_F \equiv \sum_{i=1}^{l}\sum_{j=1}^{n} \left(X(i,j) - [WH](i,j)\right)^2 \tag{6.64}$$

$$\text{subject to} \quad W(i,k) \geq 0,\ H(k,j) \geq 0. \quad H(k,j) \geq 0\ \forall i,k,j \tag{6.65}$$

where $[WH](i,j)$ is the (i,j) element of matrix WH. Minimization is performed with respect to W and H.

Another cost function has also been suggested, which is a close relative of the Kullback–Leibler distance (see Appendix A) and the task now becomes

$$\text{minimize} \quad \sum_{i=1}^{l}\sum_{j=1}^{n} \left(X(i,j) \ln \frac{X(i,j)}{[WH](i,j)} - X(i,j) + [WH](i,j) \right) \tag{6.66}$$

$$\text{subject to} \quad W(i,k) \geq 0,\ H(k,j) \geq 0\ \forall i,k,j \tag{6.67}$$

It is readily seen that if $X = WH$ the previous cost becomes zero. Also, observe that if $\sum_{i,j} X(i,j) = \sum_{i,j} [WH](i,j) = 1$ then the cost becomes identical to the Kullback–Leibler (KL) distance formulation. Note, however, that the previous KL-like cost is not well defined if either $X(i,j)$ or $[WH](i,j)$ are zero. For more information on this topic the interested reader may consult, for example, [Sra 06].

Once the problem has been formulated, the major issue rests on the solution of the optimization task. To this end, a number of algorithms have been proposed, for example, Newton-type or gradient descent type (Appendix C). Such algorithmic issues, as well as a number of related theoretic ones, are beyond the scope of the current book, and the interested reader may consult [Chu 04, Dono 04, Trop 03].

6.7 NONLINEAR DIMENSIONALITY REDUCTION

All the techniques that have been discussed so far in this chapter, as well as the LDA in the previous chapter, share a common goal: dimensionality reduction. In other words, given a high-dimensional data set $X = \{\boldsymbol{x}_1, \boldsymbol{x}_2, \ldots, \boldsymbol{x}_n\} \subset \mathcal{R}^N$ of input patterns,[4] the goal is to compute n corresponding patterns, $Y = \{\boldsymbol{y}_1, \boldsymbol{y}_2, \ldots, \boldsymbol{y}_n\} \subset \mathcal{R}^m$, $m < N$, that provide an "informative" representation of the input patterns. The word "informative" is interpreted in a different way for different methods; for example, PCA and ICA adopt different views on the issue. Another common characteristic of all the previous methods is that they respect linearity. Once a transformation matrix is computed, for each method, points in Y are obtained by projecting the points in X along the rows of this matrix.

The emphasis of dimensionality reduction has, so far, focussed to the domain of feature generation, in order to bypass the curse of dimensionality and cope efficiently with the generalization aspects of a classifier. However, the significance of such techniques goes much beyond and embraces a number of other applications. Data visualization is an area where dimensionality reduction techniques are employed in order to transform the original data from a high-dimensional into two or three dimensions, thereby offering additional insight into the problem at hand. As stated in [Tene 00], the human brain confronts the same problem extracting from the high-dimensional sensory system (i.e., 10^6 optic nerve fibers) a reduced number of perceptually relevant features. Data mining and information retrieval is another area where searching can be substantially facilitated if it is performed in a lower dimensional space. This is a typical area where data usually lie in a very high-dimensional space, although its intrinsic dimensionality is low. Dimensionality reduction has been used extensively in clustering and in semi-supervised learning, an area that is gaining in importance over the last years. So, in a way, this section is a bridge to the chapters dedicated to these topics.

The aim of this section is to discuss nonlinear dimensionality reduction techniques. We will discuss the main directions that are currently popular, and we will not delve into many details.

6.7.1 Kernel PCA

As its name suggests, this is a kernelized version of the classical PCA, introduced in [Scho 98]. Given the data set, X, we make an implicit mapping into a RKHS H,

$$\boldsymbol{x} \in X \mapsto \boldsymbol{\phi}(\boldsymbol{x}) \in H$$

Let $\boldsymbol{x}_i$, $i = 1, 2, \ldots, n$, be the available training points. We will work with an estimate of the correlation matrix in H obtained as an average over the known sample points[5]

$$R = \frac{1}{n}\sum_{i=1}^{n} \boldsymbol{\phi}(\boldsymbol{x}_i)\boldsymbol{\phi}(\boldsymbol{x}_i)^T \tag{6.68}$$

[4] To serve the specific needs of this chapter, we have reserved the symbol N to denote dimensionality of the input space. For the rest of the book, N denotes the number of data points.

[5] If the dimensionality of H is infinite, the definition of the correlation matrix needs a special interpretation, but we will not bother about it.

The goal is to perform the eigendecomposition of R, that is,

$$Rv = \lambda v \tag{6.69}$$

Let us make the assumption that the data are centered $\left(\sum_{i=1}^{n} \phi(x_i) = \mathbf{0}\right.$. This assumption is only to simplify the discussion, and it can be relaxed.$\left.\right)$ By the definition of R, it can be shown that v lies in the span of $\{\phi(x_1), \phi(x_2), \ldots, \phi(x_n)\}$. Indeed,

$$\lambda v = \left(\frac{1}{n}\sum_{i=1}^{n} \phi(x_i)\phi(x_i)^T\right) v = \frac{1}{n}\sum_{i=1}^{n} \left(\phi(x_i)^T v\right) \phi(x_i)$$

and for $\lambda \neq 0$ we can write

$$v = \sum_{i=1}^{n} a(i)\phi(x_i) \tag{6.70}$$

Combining (6.69) and (6.70), it turns out ([Scho 98]) that the problem is equivalent to performing an eigendecomposition of the Gram matrix

$$\mathcal{K}a = n\lambda a \tag{6.71}$$

where

$$a \equiv [a(1), a(2), \ldots, a(n)]^T \tag{6.72}$$

As we already know (Section 4.19.1), the elements of the Gram matrix are $\mathcal{K}(i,j) = K(x_i, x_j)$, with $K(\cdot,\cdot)$ being the adopted kernel function. Thus, the kth eigenvector of R, corresponding to the kth (nonzero) eigenvector of $\mathcal{K}$ in (6.71), is expressed as

$$v_k = \sum_{i=1}^{n} a_k(i)\phi(x_i), \; k = 1, 2, \ldots, p \tag{6.73}$$

where $\lambda_1 \geq \lambda_2 \geq \ldots \geq \lambda_p$ denote the respective eigenvalues in descending order and λ_p is the smallest nonzero one and $a_k^T \equiv [a_k(1), \ldots, a_k(n)]$ is the kth eigenvector of the Gram matrix. The latter is assumed to be normalized so that $\langle v_k, v_k\rangle = 1, \; k = 1, 2, \ldots, p$, where $\langle\cdot,\cdot\rangle$ is the dot product in the Hilbert space H. This imposes an equivalent normalization on the respective a_k's, resulting from

$$\begin{aligned} 1 = \langle v_k, v_k\rangle &= \left\langle \sum_{i=1}^{n} a_k(i)\phi(x_i), \sum_{j=1}^{n} a_k(j)\phi(x_j)\right\rangle \\ &= \sum_{i=1}^{n}\sum_{j=1}^{n} a_k(i)a_k(j)\mathcal{K}(i,j) \\ &= a_k^T \mathcal{K} a_k = n\lambda_k a_k^T a_k, \quad k = 1, 2, \ldots, p \end{aligned} \tag{6.74}$$

We are now ready to summarize the basic steps for performing a kernel PCA. Given a vector $\boldsymbol{x} \in \mathcal{R}^N$ and a kernel function $K(\cdot,\cdot)$:

- Compute the Gram matrix $\mathcal{K}(i,j) = K(\boldsymbol{x}_i, \boldsymbol{x}_j), i,j = 1,2,\ldots,n$.
- Compute the m dominant eigenvalues/eigenvectors λ_k, $\boldsymbol{a}_k$, $k = 1,2,\ldots,m$, of $\mathcal{K}$ (Eq. (6.71)).
- Perform the required normalization (Eq. (6.74)).
- Compute the m projections onto each one of the dominant eigenvectors,

$$y(k) \equiv \langle \boldsymbol{v}_k, \boldsymbol{\phi}(\boldsymbol{x})\rangle = \sum_{i=1}^{n} a_k(i) K(\boldsymbol{x}_i, \boldsymbol{x}), \; k = 1,2,\ldots,m \tag{6.75}$$

The operations given in (6.75) correspond to a *nonlinear mapping* in the input space. Note that, in contrast to the linear PCA, the dominant eigenvectors $\boldsymbol{v}_k$, $k = 1,2,\ldots,m$, are not computed explicitly. All we know are the respective (nonlinear) projections, $y(k)$ along them. After all, this is what we are finally interested in.

Remarks

- Kernel PCA is equivalent to performing a standard PCA in the RKHS H. It can be shown that all the properties associated with the dominant eigenvectors, as discussed for the PCA, are still valid for the kernel PCA. That is, (a) the dominant eigenvector directions optimally retain most of the variance, (b) the MSE in approximating a point in H in terms of the m dominant eigenvectors is minimal, with respect to any other m directions, (c) projections onto the eigenvectors are uncorrelated, and (d) the entropy (under Gaussian assumption) is maximized ([Scho 98]).
- Recall from Section 6.3 that the eigendecomposition of the Gram matrix was required for the metric multidimensional scaling (MDS) method. Hence, kernel PCA can be considered to be a kernelized version of MDS, where inner products in the input space have been replaced by kernel operations in the Gram matrix.
- Note that the kernel PCA method does not explicitly consider the underlying structure of the manifold on which the data reside.

6.7.2 Graph-Based Methods

Laplacian eigenmaps

The starting point of this method is the assumption that the data points lie on a smooth manifold (hypersurface) $\mathcal{M} \supset X$, whose intrinsic dimension is equal to $m < N$ and it is embedded in $\mathcal{R}^N$, that is, $\mathcal{M} \subset \mathcal{R}^N$. The dimension m is given as a parameter by the user. In contrast, this is not required in the kernel PCA, where m is the number of dominant components, which, in practice, is determined so that the gap between λ_m and λ_{m+1} has a "large" value.

The main philosophy behind the method is to compute the low-dimensional representation of the data so that *local neighborhood information* in $X \subset \mathcal{M}$ is optimally preserved. In this way, one attempts to get a solution that reflects the geometric structure of the manifold. To achieve this, the following steps are in order:

Step 1: Construct a graph $G = (V, E)$, where $V = \{v_i,\ i = 1, 2, \ldots, n\}$ is a set of vertices and $E = \{e_{ij}\}$ is a set of edges connecting vertices (v_i, v_j) (see also Section 13.2.5). Each node v_i of the graph corresponds to a point $\boldsymbol{x}_i$ in our data set X. We connect v_i, v_j, that is, insert the edge e_{ij} between the respective nodes, if points $\boldsymbol{x}_i, \boldsymbol{x}_j$ are "close" to each other. According to the method, there are two ways of quantifying "closeness." Vertices v_i, v_j are connected with an edge if:

1. $||\boldsymbol{x}_i - \boldsymbol{x}_j||^2 < \epsilon$, for some user-defined parameter ϵ, where $||\cdot||$ is the Euclidean norm operation in $\mathcal{R}^N$, or

2. $\boldsymbol{x}_j$ is among the k-nearest neighbors of $\boldsymbol{x}_i$ or $\boldsymbol{x}_i$ is among the k-nearest neighbors of $\boldsymbol{x}_j$, where k is a user-defined parameter and neighbors are chosen according to the Euclidean distance in $\mathcal{R}^N$. The use of Euclidean distance is justified by the smoothness of the manifold that allows one to approximate, locally, manifold geodesics by Euclidean distances in the space where the manifold is embedded. The latter is a known result from differential geometry.

   For those who are unfamiliar with such concepts, think of a sphere embedded in three-dimensional space. If somebody is constrained to live on the surface of the sphere, the shortest path to go from one point to another is the geodesic between these two points. Obviously, this is not a straight line but rather an arc across the surface of the sphere. However, if these points are close enough, their geodesic distance can be approximated by their Euclidean distance, computed in the three-dimensional space.

Step 2: Each edge, e_{ij}, is associated with a weight, $W(i,j)$. For nodes that are not connected, the respective weights are zero. Each weight, $W(i,j)$, is a measure of the "closeness" of the respective neighbors, $\boldsymbol{x}_i, \boldsymbol{x}_j$. A typical choice is

$$W(i,j) = \begin{cases} \exp\left(-\frac{||\boldsymbol{x}_i - \boldsymbol{x}_j||^2}{\sigma^2}\right), & \text{if } v_i, v_j \text{ correspond to neighbors} \\ 0 & \text{otherwise} \end{cases}$$

where σ^2 is a user-defined parameter. We form the $n \times n$ weight matrix W having as elements the weights $W(i,j)$. Note that W is symmetric, and it is *sparse* since, in practice, many of its elements turn out to be zero.

Step 3: Define the diagonal matrix D with elements $D_{ii} = \sum_j W(i,j), i = 1, 2, \ldots, n$, and also the matrix $L = D - W$. The latter is known as the *Laplacian matrix of the graph* $G(V, E)$. Perform the generalized eigen-decomposition

$$Lv = \lambda Dv$$

Let $0 = \lambda_0 \leq \lambda_1 \leq \lambda_2 \leq \ldots \leq \lambda_m$ be the smallest $m+1$ eigenvalues.[6] Ignore the $\boldsymbol{v}_o$ eigenvector corresponding to $\lambda_0 = 0$ and choose the next m eigenvectors $\boldsymbol{v}_1, \boldsymbol{v}_2, \ldots, \boldsymbol{v}_m$. Then map

$$\boldsymbol{x}_i \in \mathcal{R}^N \mapsto \boldsymbol{y}_i \in \mathcal{R}^m, \quad i = 1, 2, \ldots, n$$

where

$$\boldsymbol{y}_i^T = [v_1(i), v_2(i), \ldots, v_m(i)], \quad i = 1, 2, \ldots, n \tag{6.76}$$

The computational complexity of a general eigendecomposition solver amounts to $O(n^3)$ operations. However, for sparse matrices, such as L, efficient schemes, e.g., the Lanczos algorithm ([Golu 89]), can be employed to reduce complexity to subquadratic in n.

We will prove the statement of step 3 for the case of $m = 1$. That is, the low dimensional space is the real axis. Our path evolves along the lines adopted in [Belk 03]. The goal is to compute $y_i \in \mathcal{R}$, $i = 1, 2, \ldots, n$, so that connected points (in the graph, i.e., neighbors) stay as close as possible after the mapping onto the one-dimensional subspace. The criterion used to satisfy the closeness after the mapping is

$$E_L = \sum_{i=1}^{n} \sum_{j=1}^{n} (y_i - y_j)^2 W(i,j) \tag{6.77}$$

to become minimum. Observe that if $W(i,j)$ has a large value (i.e., $\boldsymbol{x}_i, \boldsymbol{x}_j$ are close in $\mathcal{R}^N$), then if the respective y_i, y_j are far apart in $\mathcal{R}$ it incurs a heavy penalty in the cost function. Also, points that are not neighbors do not affect the minimization since the respective weights are zero. For the more general case where $1 < m < N$ the cost function becomes

$$E_L = \sum_{i=1}^{n} \sum_{j=1}^{n} ||\boldsymbol{y}_i - \boldsymbol{y}_j||^2 W(i,j)$$

[6] In contrast to the notation used for PCA, the eigenvalues here are marked in ascending order. This is because, in this subsection, we are interested in determining the smallest values, and such a choice is notationally more convenient.

Let us now reformulate (6.77). After some obvious algebra, we obtain

$$
\begin{aligned}
E_L &= \sum_i y_i^2 \sum_j W(i,j) + \sum_j y_j^2 \sum_i W(i,j) - 2\sum_i \sum_j y_i y_j W(i,j) \\
&= \sum_i y_i^2 D_{ii} + \sum_j y_j^2 D_{jj} - 2\sum_i \sum_j y_i y_j W(I,j) \\
&= 2\boldsymbol{y}^T L \boldsymbol{y}
\end{aligned}
\tag{6.78}
$$

where

$$
L \equiv D - W \tag{6.79}
$$

and $\boldsymbol{y}^T = [y_1, y_2, \ldots, y_n]$. The Laplacian matrix L is symmetric and nonnegative definite. The latter is readily seen from the definition in (6.78), where E_L is always a nonnegative scalar. Note that the larger the value of D_{ii}, the more "important" is the sample $\boldsymbol{x}_i$. This is because it implies large values for $W(i,j)$, $j = 1, 2, \ldots, n$, and plays a dominant role in the minimization process. Obviously, the minimum of E_L is achieved by the trivial solution $y_i = 0$, $i = 1, 2, \ldots, n$. To avoid this, as it is common in such cases, we constrain the solution to a prespecified norm. Hence our problem now becomes

$$
\begin{aligned}
&\text{minimize} \quad \boldsymbol{y}^T L \boldsymbol{y} \\
&\text{subject to} \quad \boldsymbol{y}^T D \boldsymbol{y} = 1
\end{aligned}
$$

Although we can work directly on the previous task, we will reshape slightly it in order to use tools that are more familiar to us. Define

$$
\boldsymbol{z} = D^{1/2} \boldsymbol{y} \tag{6.80}
$$

and

$$
\tilde{L} = D^{-1/2} L D^{-1/2} \tag{6.81}
$$

which is known as the *normalized graph Laplacian* matrix. It is now readily seen that our optimization problem becomes

$$
\text{minimize} \quad \boldsymbol{z}^T \tilde{L} \boldsymbol{z} \tag{6.82}
$$

$$
\text{subject to} \quad \boldsymbol{z}^T \boldsymbol{z} = 1 \tag{6.83}
$$

Using Lagrange multipliers and equating the gradient of the Lagrangian to zero (Appendix C) it turns out that the solution is given by

$$
\tilde{L} \boldsymbol{z} = \lambda \boldsymbol{z} \tag{6.84}
$$

In other words, computing the solution becomes equivalent to solving an eigenvalue problem. By substituting (6.84) into the cost function (6.82) and taking into account the constraint (6.83), it turns out that the value of the cost associated with the optimal $\boldsymbol{z}$ is equal to λ. Hence, the solution is the eigenvector corresponding to the minimum eigenvalue. However, the minimum eigenvalue of $\tilde{L}$ is zero and the corresponding eigenvector corresponds to a trivial solution. Indeed, observe that

$$\tilde{L}D^{1/2}\mathbf{1} = D^{-1/2}LD^{-1/2}D^{1/2}\mathbf{1} = D^{-1/2}(D - W)\mathbf{1} = \mathbf{0}$$

where $\mathbf{1}$ is the vector having all its elements equal 1. In words, $\boldsymbol{z} = D^{1/2}\mathbf{1}$ is an eigenvector corresponding to the zero eigenvalue, and it results to the trivial solution, $y_i = 1$, $i = 1, 2, \ldots, n$. That is, all the points are mapped onto the same point in the real line. To exclude this undesired solution, recall that $\tilde{L}$ is a nonnegative matrix, and, hence, 0 is its smallest eigenvalue (if the graph is connected, that is, at least one path (see Section 13.2.5) connects any pair of vertices, $D^{1/2}\mathbf{1}$ is the only eigenvector associated with the zero eigenvalue, λ_0, [Belk 03]. This is an assumption we adopt here.) Also, since $\tilde{L}$ is a symmetric matrix, we know (Appendix B) that its eigenvectors are orthogonal to each other. In the sequel, we impose an extra constraint, and we now require the solution to be *orthogonal* to $D^{1/2}\mathbf{1}$. Constraining the solution to be orthogonal to the eigenvector corresponding to the smallest (zero) eigenvalue drives the solution to the next eigenvector corresponding to the next smallest (nonzero) eigenvalue λ_1. Note that the eigendecomposition of $\tilde{L}$ is equivalent to what we called generalized eigendecomposition of L in step 3 earlier.

For the more general case of $m > 1$, we have to compute the m eigenvectors associated with $\lambda_1 \leq \ldots \leq \lambda_m$. For this case, the constraints prevent us from mapping into a subspace of dimension less than the desired m. For example, we do not want to project in a three-dimensional space and the points to lie on a two-dimensional plane or on an one-dimensional line. For more details, the interested reader is referred to the insightful paper [Belk 03].

Local Linear Embedding (LLE)

As was the case with the Laplacian eigenmap method, LLE assumes that our data rest on a smooth enough manifold of dimension m, which is embedded in the $\mathcal{R}^N$ subspace, with $m < N$ ([Rowe 00]). The smoothness assumption allows us to further assume that, provided there is sufficient data and the manifold is "well" sampled, nearby points lie on (or close to) a "locally" *linear* patch of the manifold. The algorithm in its simplest form is summarized in the following three steps:

Step 1: For each point $\boldsymbol{x}_i$, $i = 1, 2, \ldots, n$, search for its nearest neighbors.

Step 2: Compute the weights $W(i,j)$, $i,j = 1, 2, \ldots, n$, that best reconstruct each point, $\boldsymbol{x}_i$, from its nearest neighbors, so that to minimize the cost

$$\arg\min_{W} E_W = \sum_{i=1}^{n} ||\boldsymbol{x}_i - \sum_{j=1}^{n} W(i,j)\boldsymbol{x}_{i_j}||^2 \tag{6.85}$$

where, $\boldsymbol{x}_{i_j}$ denotes the jth neighbor of the ith point. The weights are constrained: (a) to be zero for points that are not neighbors and (b) the rows of the weight matrix add to one, that is,

$$\sum_{j=1}^{n} W(i,j) = 1 \tag{6.86}$$

That is, the sum of the weights, over all neighbors, must be equal to one.

Step 3: Use the weights obtained from the previous step to compute the corresponding points $\boldsymbol{y}_i \in \mathcal{R}^m$, $i = 1, 2, \ldots, n$, so that to minimize the cost with respect to the unknown points $Y = \{\boldsymbol{y}_i,\ i = 1, 2, \ldots, n\}$

$$arg \min_{Y} E_Y = \sum_{i=1}^{n} ||\boldsymbol{y}_i - \sum_{j} W(i,j)\boldsymbol{y}_j||^2 \tag{6.87}$$

The above minimization takes place subject to two constraints so as to avoid degenerate results: (a) the outputs are centered, $\sum_i \boldsymbol{y}_i = \mathbf{0}$ and (b) the outputs have unit covariance matrix ([Saul 01]). The nearest points, in step 1, are searched in the same way as carried out for the Laplacian eigenmap method. Once again, use of the Euclidean distance is justified by the smoothness of the manifold, as long as the search is limited "locally" among neighboring points. For the second step, the method exploits the local linearity of a smooth manifold and tries to predict *linearly* each point by its neighbors using the least squares error criterion. Minimizing the cost subject to the constraint given in (6.86) results in a solution that satisfies the following three properties:

1. Rotation invariance.
2. Scale invariance.
3. Translation invariance.

The first two properties can easily be verified by the form of the cost function and the third one is the consequence of the constraint equation. The implication of this is that the computed weights encode information about the intrinsic characteristics of each neighborhood and they do not depend on the particular point.

The resulting weights, $W(i,j)$, reflect the intrinsic properties of the local geometry underlying the data. Since our goal is to retain the local information after the mapping, these weights are used to reconstruct each point in the $\mathcal{R}^m$ subspace by its neighbors. As is nicely stated in [Saul 01], it is as if we take a pair of scissors to cut small linear patches of the manifold and place them in the low-dimensional subspace.

It turns out that solving for (6.87) for the unknown points $\boldsymbol{y}_i$, $i = 1, 2, \ldots, n$, is equivalent to:

- Performing an eigendecomposition of the matrix $(I - W)^T(I - W)$.
- Discarding the eigenvector that corresponds to the smallest eigenvalue.
- Taking the eigenvectors that correspond to the next (lower) eigenvalues. These yield the low-dimensional outputs $\boldsymbol{y}_i$, $i = 1, 2, \ldots, n$.

Once again, the involved matrix W is sparse, and if this is taken into account, the eigenvalue problem scales relatively well to large data sets with complexity subquadratic in n. The complexity for step 2 scales as $O(nk^3)$, and it is contributed by the solver of the linear set of equations with k unknowns for each point. The method requires that the user provides two parameters, the number of nearest neighbors, k (or ϵ) and the dimensionality m. The interested reader can find more on the LLE method in [Saul 01].

Isometric Mapping (ISOMAP)

In contrast to the two previous methods that unravel the geometry of the manifold on a local basis, the ISOMAP algorithm adopts the view that only the geodesic distances between all pairs of the data points can reflect the true structure of the manifold. Euclidean distances between points in a manifold cannot represent it properly, since points that lie far apart, as measured by their geodesic distance, may be close when measured in terms of their Euclidean distance, (see Figure 6.5). ISOMAP is basically a variant of the MDS algorithm in which the Euclidean distances are substituted by the respective geodesic distances along the manifold. The essence of the method is to estimate geodesic distances between points that lie faraway. To this end, a two-step procedure is adopted:

Step 1: For each point, x_i, $i = 1, 2, \ldots, n$, compute the nearest neighbors and construct a graph $G(V, E)$ whose vertices represent input patterns and the edges connect nearest neighbors (nearest neighbors are computed with

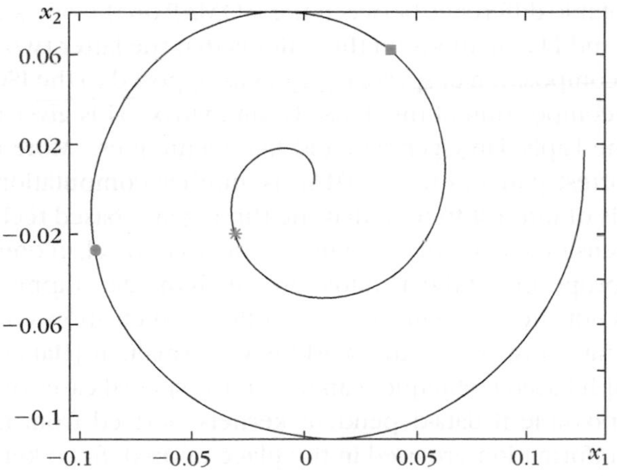

FIGURE 6.5

The point denoted by a "star" is deceptively closer to the point denoted by a "dot" than to the point denoted by a "box," if distance is measured in terms of the Euclidean distance. However, if one is constrained to travel along the spiral, the geodesic distance is the one that determines closeness and it is the "box" point that is closer to the "star."

either of the two alternatives used for the Laplacian eigenmap method. The parameters k or ϵ are user-defined parameters.) The edges are assigned weights based on the respective Euclidean distance. (For nearest neighbors this is a good approximation of the respective geodesic distance.)

Step 2: Compute the pairwise geodesic distances among all pairs (i, j), $i, j = 1, 2, \ldots, n$, along shortest paths through the graph. The key assumption is that the geodesic between any two points on the manifold can be approximated by the *shortest path* connecting the two points along the graph $G(V, E)$. To this end, efficient algorithms can be used to achieve it with complexity $O(n^2 \ln n + n^2 k)$ (e.g., Djikstar's algorithm, [Corm 01]). This cost can be prohibitive for large values of n.

Having estimated the geodesic distances between all pairs of point, the MDS method is mobilized. Thus, the problem becomes equivalent to performing the eigendecomposition of the respective Gram matrix and selecting the m most significant eigenvectors to represent the low-dimensional space. After the mapping, Euclidean distances between points in the low-dimensional subspace match the respective geodesic distances on the manifold in the original high-dimensional space. As is the case in PCA and MDS, m is estimated by the number of significant eigenvalues. It can be shown that ISOMAP is guaranteed asymptotically ($n \longrightarrow \infty$) to recover the true dimensionality of a class of nonlinear manifolds [Tene 00, Dono 04].

All three graph-based methods share a common step for computing nearest neighbors in a graph. This is a problem of complexity $O(n^2)$ but more efficient search techniques can be used by employing a special type of data structures, for example, [Beyg 06]. A notable difference between the ISOMAP on the one side and the Laplacian eigenmap and LLE methods on the other is that the latter two approaches rely on the eigendecomposition of sparse matrices as opposed to the ISOMAP that relies on the eigendecomposition of the dense Gram matrix. This gives a computational advantage to the Laplacian eigenmap and LLE techniques. Moreover, the calculation of the shortest paths in the ISOMAP is another computationally demanding task. Finally, it is of interest to note that the three graph-based techniques perform the goal of dimensionality reduction while trying to unravel, in one way or another, the geometric properties of the manifold on which the data (approximately) lie. In contrast, this is not the case with the kernel PCA, which shows no interest in any manifold learning. However, as the world is very small, in [Ham 04], it is pointed out that the graph-based techniques can be seen as special cases of the kernel PCA! This becomes possible if data-dependent kernels, derived from graphs encoding neighborhood information, are used in the place of predefined kernel functions.

The goal of this section was to present some of the most basic directions that have been suggested for nonlinear dimensionality reduction. Besides the previous basic schemes, a number of improved updates have been proposed in the literature, for example, [Desi 03, Sha 05, Beng 04]. In [Lafo 06, Qui 07], the low-dimensional embedding is achieved to preserve certain measures that reflect the connectivity of the graph $G(V,E)$. In [He 03, Cai 05, Koki 07] the idea of preserving the local information in the manifold has been carried out to define linear transforms of the form $\boldsymbol{y} = A^T\boldsymbol{x}$, and the optimization is now carried out with respect to the elements of A. The task of incremental manifold learning for dimensionality reduction was more recently considered in [Law 06]. In [Wein 05, Sun 06], the *maximum variance unfolding* method is introduced. The variance of the outputs is maximized under the constraint that (local) distances and angles are preserved among neighbors in the graph. Like the ISOMAP, it turns out that the top eigenvectors of a Gram matrix have to be computed, albeit avoiding the computationally demanding step of estimating geodesic distances, as required by the ISOMAP. In [Shui 07] a general framework, called *graph embedding*, is presented that offers a unified view for understanding and explaining a number of known (including PCA and nonlinear PCA) dimensionality reduction techniques, and it also offers a platform for developing new ones. In [Lin 08] a manifold learning technique is adopted that constructs coordinate charts for a given Riemannian manifold. For a more detailed and insightful treatment of the topic the interested reader is referred to [Burg 04]. A review of nonlinear dimensionality reduction techniques can be found in [Cama 03].

Example 6.6

A data set consists of 30 points in the two-dimensional space. The points result from sampling the spiral of Archimedes (see Figure 6.6a), described by

$$x_1 = a\theta\cos\theta, \quad x_2 = a\theta\sin\theta$$

The points of the data set correspond to the values $\theta = 0.5\pi$, and $0.7\pi,\ 0.9\pi,\ldots,2.05\pi$ (θ is expressed in radians) and $a = 0.1$. For illustration purposes and in order to keep track of the "neighboring" information, we have used a sequence of six symbols, that is, "×", "+", "?", "□", "◇", "∘"—with black color—followed by the same sequence of symbols in red color, repeatedly.

To study the performance of PCA for this case, where data lie on a nonlinear manifold, we first performed the eigendecomposition of the covariance matrix, estimated from the data set. The resulting eigenvalues are

$$\lambda_0 = 0.089 \text{ and } \lambda_1 = 0.049$$

Observe that, in contrast to the linear case of Example 6.3, here the eigenvalues are comparable in size. Thus, if one would trust the "verdict" coming from PCA, the answer concerning the dimensionality of the data would be that it is equal to 2. Moreover, after projecting along the direction of the principal component (the straight line in Figure 6.6b), corresponding to λ_0, neighboring information is lost since points from different locations are mixed together.

Next, the Laplacian eigenmap technique for dimensionality reduction is employed, with $\varepsilon = 0.2$ and $\sigma = \sqrt{0.5}$. The results obtained are shown in Figure 6.6c. Looking from right to left, we see that the Laplacian method nicely "unfolds" the spiral in an one-dimensional straight line. Furthermore, neighboring information is retained in this one-dimensional

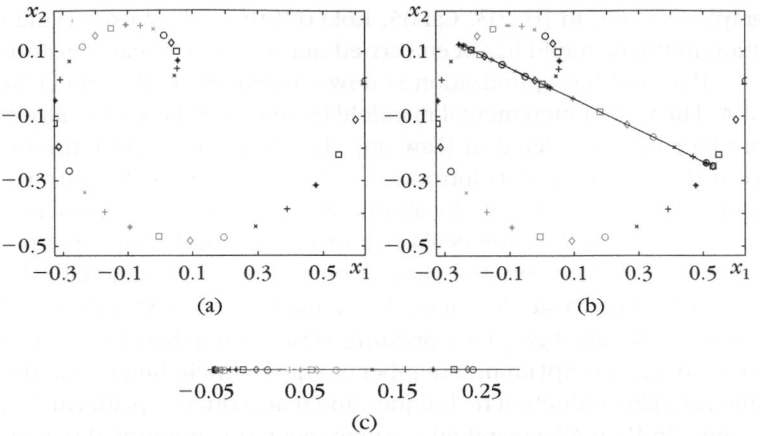

FIGURE 6.6

(a) A spiral of Archimedes in the two-dimensional space. (b) The previous spiral together with the projections of the sampled points on the direction of the first principal component, resulting from PCA. It is readily seen that neighboring information is lost after the projection and points corresponding to different parts of the spiral overlap. (c) The one-dimensional map of the spiral using the Laplacian method. In this case, the neighboring information is retained after the nonlinear projection and the spiral nicely unfold to a one-dimensional line.

representation of the data. Black and red areas are succeeding each other in the right order, and also, observing the symbols, one can see that neighbors are mapped to neighbors.

Example 6.7

Figure 6.7 shows samples from a three-dimensional spiral, parameterized as $x_1 = a\theta\cos\theta$, $x_2 = a\theta\sin\theta$, and sampled at $\theta = 0.5\pi,\ 0.7\pi,\ 0.9\pi, \ldots, 2.05\pi$ (θ is expressed in radians), $a = 0.1$ and $x_3 = -1,\ -0.8,\ -0.6, \ldots, 1$.

For illustration purposes and in order to keep track of the "identity" of each point, we have used red crosses and dots interchangeably, as we move upward in the x_3 dimension. Also, the first, the middle and the last points for each level of x_3 are denoted by black "◇," black "?" and black "□," respectively. Basically, all points at the same level lie on a two-dimensional spiral.

Figure 6.8 shows the two-dimensional mapping of the three-dimensional spiral using the Laplacian method for dimensionality reduction, with parameter values $\epsilon = 0.35$ and $\sigma = \sqrt{0.5}$. Comparing Figures 6.7 and 6.8, we see that all points corresponding to the same level x_3 are mapped across the same line, with the first point being mapped to the first one and so on. That is, as was the case in Example 6.6, the Laplacian method unfolds the three-dimensional spiral into a two-dimensional surface, while retaining neighboring information.

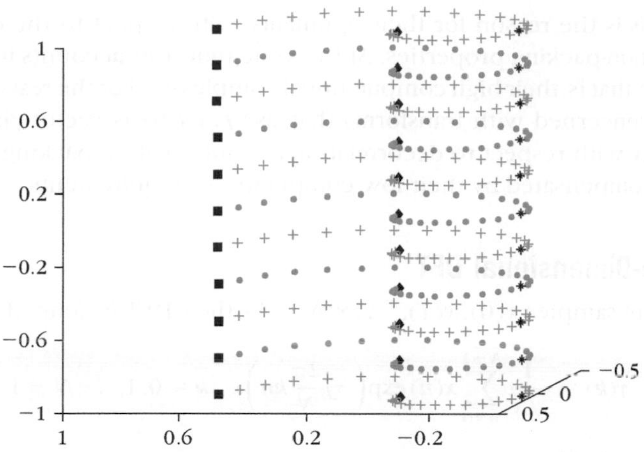

FIGURE 6.7

Samples from a three-dimensional spiral. One can think of it as a number of two-dimensional spirals one above the other. Different symbols have been used in order to track neighboring information.

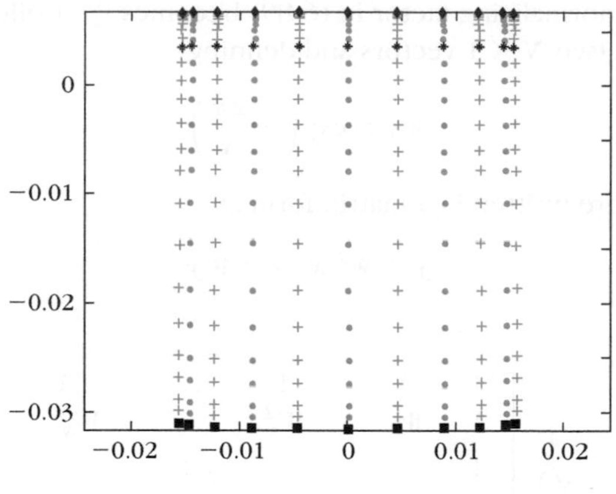

FIGURE 6.8

Two-dimensional mapping of the spiral of Figure 6.7 using the Laplacian eigenmap method. The three-dimensional structure is unfolded to the two-dimensional space by retaining the neighboring information.

6.8 THE DISCRETE FOURIER TRANSFORM (DFT)

We have already seen that the basis vectors/images for the KL and SVD expansions are not fixed but are "problem dependent" and are the result of an optimization process. This is the reason for their optimality with respect to the decorrelation and information-packing properties. At the same time, this accounts for their major disadvantage, that is, their high computational complexity. For the rest of the chapter we will be concerned with transforms that use *fixed* basis vectors/images. Their suboptimality with respect to decorrelation and information packing properties is most often compensated by their low computational requirements.

6.8.1 One-Dimensional DFT

Given N input samples $x(0), x(1), \ldots, x(N-1)$, their DFT is defined as

$$y(k) = \frac{1}{\sqrt{N}} \sum_{n=0}^{N-1} x(n) \exp\left(-j\frac{2\pi}{N}kn\right), \quad k = 0, 1, \ldots, N-1 \tag{6.88}$$

and the inverse DFT as

$$x(n) = \frac{1}{\sqrt{N}} \sum_{k=0}^{N-1} y(k) \exp\left(j\frac{2\pi}{N}kn\right), \quad n = 0, 1, \ldots, N-1 \tag{6.89}$$

where $j \equiv \sqrt{-1}$. Sometimes, (6.88) is given without the normalizing factor $\frac{1}{\sqrt{N}}$. In such cases the normalizing factor in (6.89) becomes $\frac{1}{N}$. Collecting all $x(n)$ and $y(k)$ together into two $N \times 1$ vectors and defining

$$W_N \equiv \exp\left(-j\frac{2\pi}{N}\right) \tag{6.90}$$

(6.88) and (6.89) are written in a matrix form as

$$\boldsymbol{y} = W^H \boldsymbol{x}, \ \boldsymbol{x} = W \boldsymbol{y} \tag{6.91}$$

where

$$W^H = \frac{1}{\sqrt{N}} \begin{bmatrix} 1 & 1 & 1 & \cdots & 1 \\ 1 & W_N & W_N^2 & \cdots & W_N^{N-1} \\ \vdots & \vdots & \vdots & \vdots & \vdots \\ 1 & W_N^{N-1} & W_N^{2(N-1)} & \cdots & W_N^{(N-1)(N-1)} \end{bmatrix} \tag{6.92}$$

It is not difficult to see that W is a unitary and symmetric matrix

$$W^{-1} = W^H = W^*$$

The basis vectors are the columns of W. For example, for $N = 4$

$$W = \frac{1}{2} \begin{bmatrix} 1 & 1 & 1 & 1 \\ 1 & j & -1 & -j \\ 1 & -1 & 1 & -1 \\ 1 & -j & -1 & j \end{bmatrix}$$

and the basis vectors are

$$w_0 = \frac{1}{2}[1, 1, 1, 1]^T$$

$$w_1 = \frac{1}{2}[1, j, -1, -j]^T$$

$$w_2 = \frac{1}{2}[1, -1, 1, -1]^T$$

$$w_3 = \frac{1}{2}[1, -j, -1, j]^T$$

and

$$\boldsymbol{x} = \sum_{i=0}^{3} y(i)\boldsymbol{w}_i$$

The direct computation of (6.91) requires $O(N^2)$ computations. However, taking advantage of the specific structure of the matrix W, a substantial saving in computations is possible via the celebrated Fast Fourien Transform (FFT) algorithm, which computes each equation of (6.91) in $O(N \log_2 N)$ operations [Proa 92].

So far, the DFT has been introduced as a special type of a linear unitary transform of one vector to another. Another point of view, which will be useful later in this chapter, is to see the DFT as a means of expanding a sequence $x(n)$ into a set of N *basis sequences* $h_k(n)$

$$x(n) = \sum_{k=0}^{N-1} y(k)h_k(n)$$

where

$$h_k(n) = \begin{cases} \frac{1}{\sqrt{N}} \exp(j\frac{2\pi}{N}kn), & n = 0, 1, \ldots, N-1 \\ 0, & \text{otherwise} \end{cases}$$

and $y(k)$ are the coefficients of the expansion. The DFT basis sequences belong to a more general class of sequences known as orthonormal, that is,

$$\langle h_l(n), h_k(n)\rangle \equiv \sum_n h_k(n)h_l^*(n) = \delta_{kl} \tag{6.93}$$

where $\langle .\,,.\rangle$ is known as the inner product of the sequences $h_k(n), h_l(n)$. For the DFT expansion, we have

$$\langle h_k(n), h_l(n)\rangle = \frac{1}{N}\sum_{n=0}^{N-1} \exp\left(j\frac{2\pi}{N}kn\right)\exp\left(-j\frac{2\pi}{N}ln\right)$$

$$= \frac{1}{N}\sum_{n=0}^{N-1} \exp\left(j\frac{2\pi}{N}(k-l)n\right), \quad k, l = 0, 1, \ldots, N-1$$

However, it can easily be shown (Problem 6.8) that

$$\frac{1}{N}\sum_{n=0}^{N-1}\exp\left(j\frac{2\pi}{N}(k-l)n\right)=\begin{cases}1, & l=k+rN, \quad r=0,\pm1,\pm2,\ldots\\ 0, & \text{otherwise}\end{cases} \tag{6.94}$$

Hence

$$\langle h_k(n), h_l(n)\rangle = \delta_{kl}$$

6.8.2 Two-Dimensional DFT

Given an $N \times N$ matrix/image, its two-dimensional DFT is defined as

$$Y(k,l)=\frac{1}{N}\sum_{m=0}^{N-1}\sum_{n=0}^{N-1}X(m,n)W_N^{km}W_N^{ln} \tag{6.95}$$

and the inverse DFT as

$$X(m,n)=\frac{1}{N}\sum_{k=0}^{N-1}\sum_{l=0}^{N-1}Y(k,l)W_N^{-km}W_N^{-ln} \tag{6.96}$$

It is readily seen that this can be written in a compact form as

$$Y = W^H X W^H, \quad X = WYW \tag{6.97}$$

Thus, the two-dimensional DFT is a separable transform with basis images $\boldsymbol{w}_i\boldsymbol{w}_j^T$, $i,j = 0,1,\ldots,N-1$. It is apparent from (6.97) that the number of operations required for the respective computations is $O(N^2 \log_2 N)$, that is, the number of additions and multiplications needed for $2N$ one-dimensional DFTs, via the FFT algorithm.

6.9 THE DISCRETE COSINE AND SINE TRANSFORMS

Given N input samples $x(0), x(1), \ldots, x(N-1)$ their *discrete cosine transform* (DCT) is defined as

$$y(k)=\alpha(k)\sum_{n=0}^{N-1}x(n)\cos\left(\frac{\pi(2n+1)k}{2N}\right), \quad k=0,1,\ldots,N-1 \tag{6.98}$$

and the inverse DCT is given by

$$x(n)=\sum_{k=0}^{N-1}\alpha(k)y(k)\cos\left(\frac{\pi(2n+1)k}{2N}\right), \quad n=0,1,\ldots,N-1 \tag{6.99}$$

where

$$\alpha(k) = \begin{cases} \sqrt{\frac{1}{N}}, & k = 0 \\ \sqrt{\frac{2}{N}}, & k \neq 0 \end{cases}$$

In vector form the transform is written as

$$y = C^T x$$

where the elements of the matrix C are given by

$$C(n,k) = \frac{1}{\sqrt{N}}, \quad k = 0,\ 0 \leq n \leq N-1$$

$$C(n,k) = \sqrt{\frac{2}{N}} \cos\left(\frac{\pi(2n+1)k}{2N}\right), \quad 1 \leq k \leq N-1, 0 \leq n \leq N-1$$

Matrix C has real elements, and it is easy to see that it is orthogonal,

$$C^{-1} = C^T$$

The two-dimensional DCT is the separable transform defined as

$$Y = C^T X C, \quad X = C Y C^T \tag{6.100}$$

The *discrete sine transform* (DST) is defined via the transform matrix

$$S(k,n) = \sqrt{\frac{2}{N+1}} \sin\left(\frac{\pi(k+1)(n+1)}{N+1}\right), \quad k, n = 0, 1, \ldots, N-1$$

and it is also an orthogonal transform. The DCT and DST belong to the family of transforms that can be computed via a fast method in $O(N \log_2 N)$ operations [Jain 89, Lim 90].

Remark

- The DCT and DST have very good information-packing properties for most of the images, in the sense that they concentrate most of the energy in a few coefficients. An explanation for this property is that both offer a close approximation to the (KL) transform for a class of random signals, known as first-order Markov processes, which can approximately model a number of real-world images [Jain 89]. Figure 6.9 shows an image and the resulting DFT (magnitude), DST, and DCT transforms. It is apparent from the figure that the high-intensity coefficients of the transforms (dark) are concentrated in a small area, whose size depends on the energy compaction properties of the respective transform. This area is smaller for DCT and DST than for DFT.

FIGURE 6.9

Example of an image and its magnitude DFT, DST, and DCT transforms, shown from top left to bottom left in the clockwise sense.

6.10 THE HADAMARD TRANSFORM

The Hadamard transform and the Haar transform, to be considered in the next section, share a serious computational advantage over the previously considered DFT, DCT, and DST transforms. Their unitary matrices consist of ± 1, and the transforms are computed via additions and subtractions only, with no multiplications being involved. Hence, for processors for which multiplication is a time-consuming operation, a substantial saving is obtained.

The Hadamard unitary matrix of order n is the $N \times N$ matrix, $N = 2^n$, generated by the following iteration rule:

$$H_n = H_1 \otimes H_{n-1} \tag{6.101}$$

where

$$H_1 = \frac{1}{\sqrt{2}} \begin{bmatrix} 1 & 1 \\ 1 & -1 \end{bmatrix} \tag{6.102}$$

and $\otimes$ denotes the Kronecker product of two matrices

$$A \otimes B = \begin{bmatrix} A(1,1)B & A(1,2)B & \dots & A(1,N)B \\ \vdots & \vdots & \vdots & \vdots \\ A(N,1)B & A(N,2)B & \dots & A(N,N)B \end{bmatrix}$$

where $A(i,j)$ is the (i,j) element of A, $i,j = 1,2,\dots,N$. Thus, according to (6.101), (6.102), it is

$$H_2 = H_1 \otimes H_1 = \frac{1}{2}\begin{bmatrix} 1 & 1 & 1 & 1 \\ 1 & -1 & 1 & -1 \\ 1 & 1 & -1 & -1 \\ 1 & -1 & -1 & 1 \end{bmatrix}$$

and for $n = 3$

$$H_3 = H_1 \otimes H_2 = \frac{1}{\sqrt{2}}\begin{bmatrix} H_2 & H_2 \\ H_2 & -H_2 \end{bmatrix}$$

It is not difficult to show the orthogonality of H_n, $n = 1,2,\dots$, that is,

$$H_n^{-1} = H_n^T = H_n$$

For a vector $\boldsymbol{x}$ of N samples and $N = 2^n$ the transform pair is

$$\boldsymbol{y} = H_n \boldsymbol{x}, \quad \boldsymbol{x} = H_n \boldsymbol{y} \tag{6.103}$$

The two-dimensional Hadamard transform is given by

$$Y = H_n X H_n, \quad X = H_n Y H_n \tag{6.104}$$

The Hadamard transform has good to very good energy packing properties. Fast algorithms for its computation in $O(N \log_2 N)$ subtractions and/or additions are also available [Jain 89].

Remark

- Experimental results using the DCT, DST, and Hadamard transforms for texture discrimination have shown that the performance obtained was close to that of the optimal KL transform [Unse 86, Unse 89]. At the same time, this near-optimal performance is obtained at substantially reduced complexity, due to the availability of fast computational schemes as reported before.

6.11 THE HAAR TRANSFORM

The starting point for the definition of the Haar transform is the Haar functions $h_k(z)$, which are defined in the closed interval [0, 1]. The order k of the function is

Table 6.1 Parameters for the Haar functions

k	0	1	2	3	4	5	6	7
p	0	0	1	1	2	2	2	2
q	0	1	1	2	1	2	3	4

uniquely decomposed into two integers p, q

$$k = 2^p + q - 1, \quad k = 0, 1, \ldots, L - 1, \text{ and } L = 2^n$$

where

$$0 \leq p \leq n - 1, 0 \leq q \leq 2^p \quad \text{for} \quad p \neq 0 \text{ and } q = 0 \text{ or } 1 \text{ for } p = 0$$

Table 6.1 summarizes the respective values for $L = 8$. The Haar functions are

$$h_0(z) \equiv h_{00}(z) = \frac{1}{\sqrt{L}}, \quad z \in [0, 1] \tag{6.105}$$

$$h_k(z) \equiv h_{pq}(z) = \frac{1}{\sqrt{L}} \begin{cases} 2^{\frac{p}{2}} & \frac{q-1}{2^p} \leq z < \frac{q-\frac{1}{2}}{2^p} \\ -2^{\frac{p}{2}} & \frac{q-\frac{1}{2}}{2^p} \leq z < \frac{q}{2^p} \\ 0 & \text{otherwise in } [0, 1] \end{cases} \tag{6.106}$$

The Haar transform matrix of order L consists of rows resulting from the preceding functions computed at the points $z = \frac{m}{L}, m = 0, 1, 2, \ldots, L - 1$. For example, the 8×8 transform matrix is

$$H = \frac{1}{\sqrt{8}} \begin{bmatrix} 1 & 1 & 1 & 1 & 1 & 1 & 1 & 1 \\ 1 & 1 & 1 & 1 & -1 & -1 & -1 & -1 \\ \sqrt{2} & \sqrt{2} & -\sqrt{2} & -\sqrt{2} & 0 & 0 & 0 & 0 \\ 0 & 0 & 0 & 0 & \sqrt{2} & \sqrt{2} & -\sqrt{2} & -\sqrt{2} \\ 2 & -2 & 0 & 0 & 0 & 0 & 0 & 0 \\ 0 & 0 & 2 & -2 & 0 & 0 & 0 & 0 \\ 0 & 0 & 0 & 0 & 2 & -2 & 0 & 0 \\ 0 & 0 & 0 & 0 & 0 & 0 & 2 & -2 \end{bmatrix} \tag{6.107}$$

It is not difficult to see that

$$H^{-1} = H^T$$

that is, H is orthogonal. The energy packing properties of the Haar transform are not very good. However, its importance for us lies beyond that. We will use it as the vehicle to take us from the world of unitary transforms to that of multiresolution analysis. To this end, let us look carefully at the Haar transform matrix. We readily observe its sparse nature with a number of zeros, whose location reveals an underlying cyclic shift mechanism. To satisfy our curiosity as to why this happens, let us look at the Haar transform from a different perspective.

6.12 THE HAAR EXPANSION REVISITED

Let us split our original set of N input samples (N even) $x(0), x(1), \ldots, x(N-1)$ into successive blocks of two, that is, $(x(2k), x(2k+1)), k = 0, 1, \ldots, \frac{N}{2} - 1$, and apply the Haar transform of order $L = 2$. For each pair of input samples, a pair of transformed samples is obtained,

$$\begin{bmatrix} y_1(k) \\ y_0(k) \end{bmatrix} = \frac{1}{\sqrt{2}} \begin{bmatrix} 1 & 1 \\ 1 & -1 \end{bmatrix} \begin{bmatrix} x(2k) \\ x(2k+1) \end{bmatrix}, \quad k = 0, 1, \ldots, \frac{N}{2} - 1 \tag{6.108}$$

That is,

$$y_1(k) = \frac{1}{\sqrt{2}}(x(2k) + x(2k+1)) \tag{6.109}$$

$$y_0(k) = \frac{1}{\sqrt{2}}(x(2k) - x(2k+1)), \quad k = 0, 1, \ldots, \frac{N}{2} - 1 \tag{6.110}$$

This can be interpreted as the action—on the sequence of N input samples—of two (noncausal) filters with impulse responses $(h_1(0) = \frac{1}{\sqrt{2}}, h_1(-1) = \frac{1}{\sqrt{2}})$ and $(h_0(0) = \frac{1}{\sqrt{2}}, h_0(-1) = -\frac{1}{\sqrt{2}})$, respectively. The corresponding transfer functions (Appendix D) are

$$H_1(z) = \frac{1}{\sqrt{2}}(1 + z) \tag{6.111}$$

$$H_0(z) = \frac{1}{\sqrt{2}}(1 - z) \tag{6.112}$$

In other words, the order $L = 2$ Haar transform computes the output samples of the two filters when they are fed with the input sequence $x(n), n = 0, 1, 2, \ldots, N-1$. *Furthermore, the output sequence samples are computed for every other sample of the input sequence, at even time instants* $0, 2, 4, \ldots,$ *as* (6.109) and (6.110) suggest. This operation is portrayed in Figure 6.10b. The operation at the output of the two filters is known as *subsampling by* M, in this case $M = 2$, and it is defined in Figure 6.10a. In other words, from the samples generated at the filter output we keep one every $M(= 2)$. In the time domain and for an input sequence consisting of eight samples, the output, $y_0(k)$, of the H_0 branch of Figure 6.10b will consist of

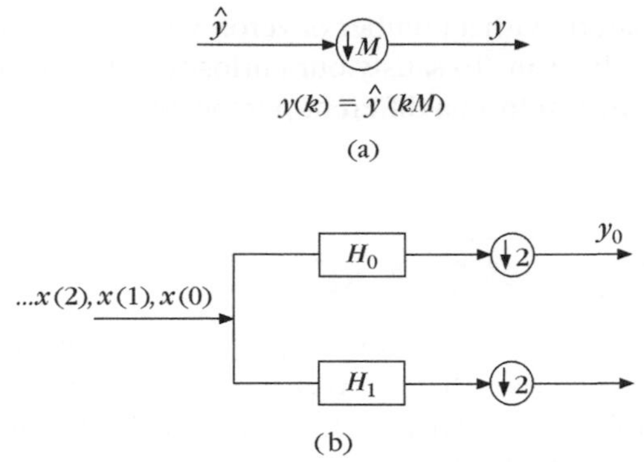

FIGURE 6.10

(a) Subsampling operation and (b) filtering interpretation of the Haar transform.

four samples given by

$$\begin{bmatrix} y_0(0) \\ y_0(1) \\ y_0(2) \\ y_0(3) \end{bmatrix} = \left[\begin{array}{cc|cc|cc|cc} \frac{1}{\sqrt{2}} & -\frac{1}{\sqrt{2}} & 0 & 0 & 0 & 0 & 0 & 0 \\ 0 & 0 & \frac{1}{\sqrt{2}} & -\frac{1}{\sqrt{2}} & 0 & 0 & 0 & 0 \\ 0 & 0 & 0 & 0 & \frac{1}{\sqrt{2}} & -\frac{1}{\sqrt{2}} & 0 & 0 \\ 0 & 0 & 0 & 0 & 0 & 0 & \frac{1}{\sqrt{2}} & -\frac{1}{\sqrt{2}} \end{array}\right] \begin{bmatrix} x(0) \\ x(1) \\ -- \\ x(2) \\ x(3) \\ -- \\ x(4) \\ x(5) \\ -- \\ x(6) \\ x(7) \end{bmatrix} \tag{6.113}$$

This is nothing other than the action of the last four rows of the 8×8 Haar transform in (6.107)! What about the rest? Let us carry on the splitting of Figure 6.10b one step further, as shown in Figure 6.11. Using the easily shown Noble identity illustrated in Figure 6.12a (Problem 6.17), the structure of Figure 6.11 turns out to be equivalent to that of Figure 6.12b. Taking into account the subsampling operation of the lower branch after the filters H_0 and H_1, the Noble identity leads to

$$\hat{F}_1(z) = \frac{1}{2}(1+z)(1-z^2) = \frac{1}{2}(1+z-z^2-z^3) \tag{6.114}$$

$$\hat{F}_2(z) = \frac{1}{2}(1+z)(1+z^2) = \frac{1}{2}(1+z+z^2+z^3) \tag{6.115}$$

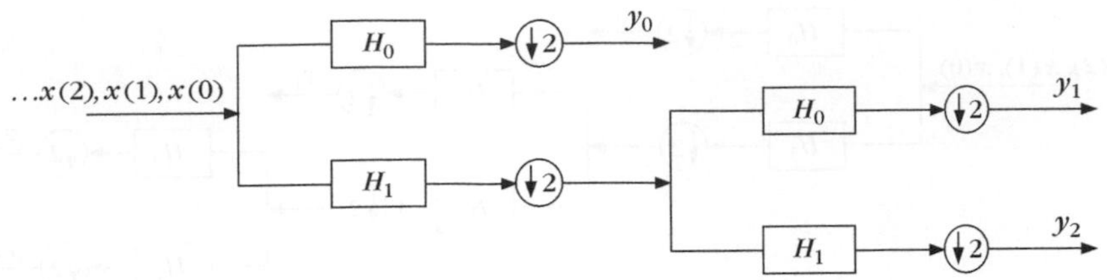

FIGURE 6.11

Two-stage filtering followed by subsampling operation.

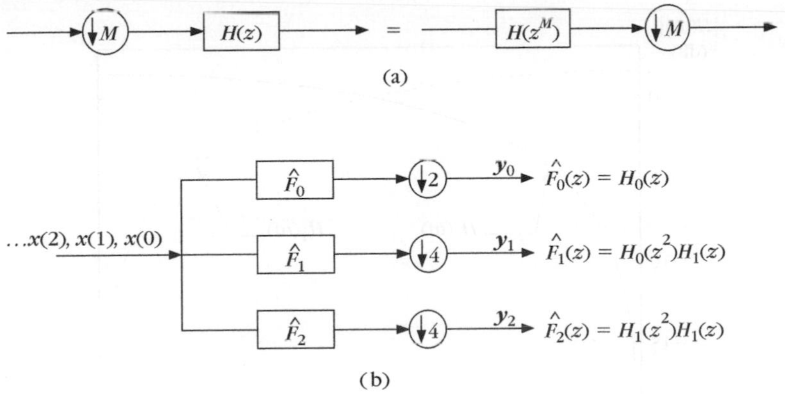

FIGURE 6.12

(a) Noble identity I and (b) equivalent filter bank of Figure 6.11.

From the transfer function $\hat{F}_1(z)$ and taking into account the subsampling by 4 (2×2) operation, the first two samples of the $y_1(k)$ sequence are given by

$$\begin{bmatrix} y_1(0) \\ y_1(1) \end{bmatrix} = \begin{bmatrix} \frac{1}{2} & \frac{1}{2} & -\frac{1}{2} & -\frac{1}{2} & 0 & 0 & 0 & 0 \\ 0 & 0 & 0 & 0 & \frac{1}{2} & \frac{1}{2} & -\frac{1}{2} & -\frac{1}{2} \end{bmatrix} \begin{bmatrix} x(0) \\ x(1) \\ \vdots \\ x(7) \end{bmatrix} \tag{6.116}$$

This is nothing but the action of the third and fourth rows of the 8×8 Haar transform on the input vector. If we now carry on the splitting one step further, as in Figure 6.13, it is straightforward to show by repeating the preceding arguments that

$$y_2(0) = \frac{1}{\sqrt{8}} \begin{bmatrix} 1 & 1 & 1 & 1 & -1 & -1 & -1 & -1 \end{bmatrix} \begin{bmatrix} x(0) \\ x(1) \\ \vdots \\ x(7) \end{bmatrix} \tag{6.117}$$

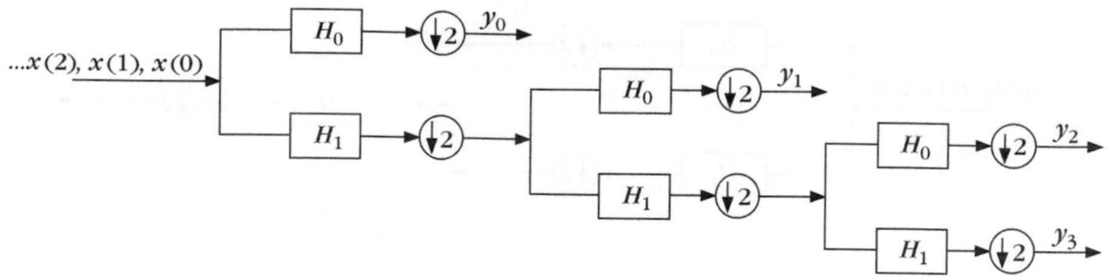

FIGURE 6.13

Tree-structured filter bank.

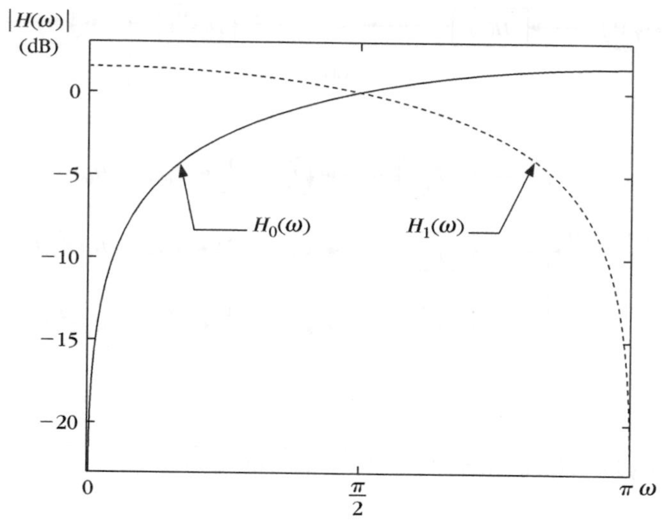

FIGURE 6.14

Magnitude of the frequency response for the two Haar filters. H_1 is a low-pass and H_0 high-pass filter.

and

$$y_3(0) = \frac{1}{\sqrt{8}} \begin{bmatrix} 1 & 1 & 1 & 1 & 1 & 1 & 1 & 1 \end{bmatrix} \begin{bmatrix} x(0) \\ x(1) \\ \vdots \\ x(7) \end{bmatrix} \tag{6.118}$$

These equations are the actions of the second and first rows of the Haar transform on the input vector. The structure of Figure 6.13 is known as a (three-level) ***tree-structured filter bank*** generated by the filters $H_0(z)$ and $H_1(z)$. Figure 6.14 shows the frequency responses of these two filters. One ($H_0(z)$) is a high-pass and the other a low-pass filter. Herein lies the importance of the filter bank interpretation of the Haar transform. The input sequence $x(n)$ is first split into two versions *of lower resolution* with respect to the original one: a low-pass (average) *coarser* resolution version and a high-pass (difference) *detailed* resolution one. In the sequel the coarser resolution version is further split into two versions, and so on. This leads to a number of versions with a hierarchy of resolutions. This decomposition is known as *multiresolution decomposition*.

The idea of multiresolution decomposition has been around for some time [Burt 83, Akan 93] and has been exploited in various applications and for a number of reasons. Its popularity as a tool in pattern recognition is mainly due to the information compaction capabilities associated with such a decomposition, *provided filters* H_0, H_1 *are properly designed*. For many types of signals, such as speech and images, most of the information is localized in certain resolution levels. Thus, most of the energy is concentrated in a (relatively) small number of samples, which carry most of the necessary information [Este 77, Mall 89]. Some fundamental issues related to the multiresolution decomposition and the design of filter banks will be highlighted next.

6.13 DISCRETE TIME WAVELET TRANSFORM (DTWT)

The goal of this section is twofold. We first free ourselves from the Haar functions, and we seek the possibility of using other filters in place of H_0, H_1. There is more than one reason for this generalization. An obvious reason is that the frequency responses of the Haar filters are far from ideal. If our aim is to split the original sequence into a hierarchy of "coarse" and "detailed" versions, we should require the filters that perform the splitting to be as close as possible to the ideal low/high-pass ones (Figure 6.15). Our next concern in this section is the inversion problem.

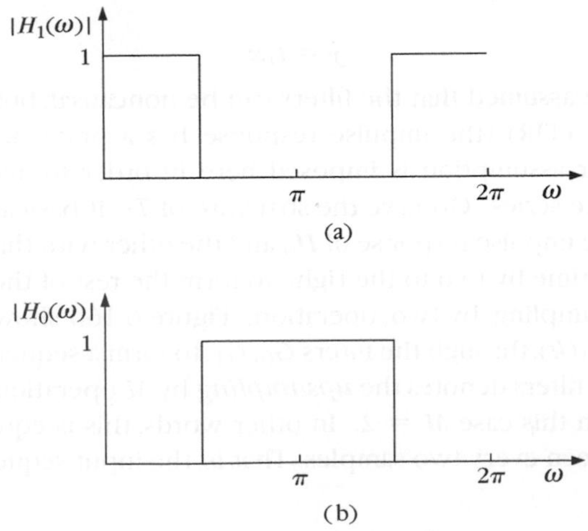

FIGURE 6.15

Ideal frequency responses for (a) low-pass and (b) high-pass filters.

That is, if we know the lower resolution versions, can we obtain the original sequence, $x(n)$, as was the case with the unitary transforms? We will show that under certain constraints in the design of H_0 and H_1 this is indeed possible.

The Two-Band Case

Let us start with the simple two-band case of Figure 6.10b, where we now assume that the filters are not the Haar ones. If $h_0(k), h_1(k)$ are the respective impulse responses, then we can write

$$y_0(k) = \sum_l x(l)h_0(n-l)|_{n=2k}$$

$$y_1(k) = \sum_l x(l)h_1(n-l)|_{n=2k}$$

where $y_1(k)$ is the output of the lower branch of Figure 6.10b. Collecting all $y_0(k), y_1(k), k = 0, 1, 2, \ldots,$ together in a vector, we have

$$\begin{bmatrix} \vdots \\ y_0(0) \\ y_1(0) \\ y_0(1) \\ y_1(1) \\ y_0(2) \\ y_1(2) \\ \vdots \end{bmatrix} = \begin{bmatrix} \vdots & \vdots & \vdots & \vdots & \vdots & \vdots & \vdots \\ \cdots & h_0(2) & h_0(1) & h_0(0) & h_0(-1) & h_0(-2) & \cdots \\ \cdots & h_1(2) & h_1(1) & h_1(0) & h_1(-1) & h_1(-2) & \cdots \\ \cdots & \cdots & \cdots & h_0(2) & h_0(1) & h_0(0) & \cdots \\ \cdots & \cdots & \cdots & h_1(2) & h_1(1) & h_1(0) & \cdots \\ \cdots & \cdots & \cdots & \cdots & \cdots & h_0(2) & \cdots \\ \cdots & \cdots & \cdots & \cdots & \cdots & h_1(2) & \cdots \\ \vdots & \vdots & \vdots & \vdots & \vdots & \vdots & \vdots \end{bmatrix} \begin{bmatrix} \vdots \\ x(0) \\ x(1) \\ x(2) \\ \vdots \end{bmatrix}$$

or

$$\boldsymbol{y} = T_i \boldsymbol{x} \tag{6.119}$$

Here, we have assumed that the filters can be noncausal, but they are of *finite impulse response* (FIR) (the impulse response has a finite nonzero number of terms). The latter assumption is imposed here in order to avoid issues of convergence of infinite series. Observe the structure of T_i. It basically consists of two rows, one with the impulse response of H_0 and the other with that of H_1, which are then shifted each time by two to the right, to form the rest of the rows. This is the result of the subsampling, by two, operation. Figure 6.16b shows a structure that combines $y_0(k), y_1(k)$, through the filters G_0, G_1, to form a sequence $\hat{x}$. The symbol at the input of the filters denotes the *upsampling* by M operation, which is defined in Figure 6.16a. In this case $M = 2$. In other words, this is equivalent to stuffing $M - 1$ zeros between every two samples. That is, the input sequences of the filters G_0, G_1 will be

$$\ldots 0\ y_0(0)\ 0\ y_0(1)\ 0\ y_0(2)\ 0 \ldots$$

$$\ldots 0\ y_1(0)\ 0\ y_1(1)\ 0\ y_1(2)\ 0 \ldots$$

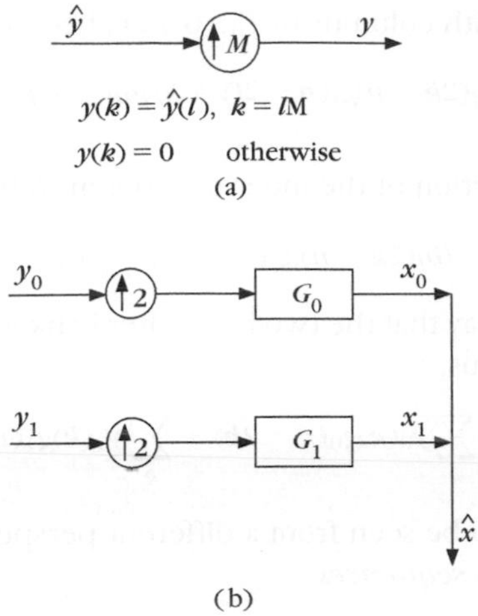

FIGURE 6.16

(a) The upsampling operation. (b) Tree-structured synthesis filter bank.

respectively. Thus, every other sample of the impulse response hits a zero and

$$x_0(n) = \sum_k y_0(k) g_0(n - 2k)$$

$$x_1(n) = \sum_k y_1(k) g_1(n - 2k)$$

$$\hat{x}(n) = x_0(n) + x_1(n)$$

Filters G_i are known as the ***synthesis filters*** and the corresponding H_i, of Figure 6.10b, as the ***analysis filters***. Collecting all $\hat{x}(n)$ together, it is not difficult to see that

$$\begin{bmatrix} \vdots \\ \hat{x}(0) \\ \hat{x}(1) \\ \hat{x}(2) \\ \vdots \end{bmatrix} = \begin{bmatrix} \vdots & \vdots & \vdots & \vdots & \vdots & \vdots \\ \cdots & g_0(0) & g_1(0) & g_0(-2) & g_1(-2) & \cdots \\ \cdots & g_0(1) & g_1(1) & g_0(-1) & g_1(-1) & \cdots \\ \cdots & g_0(2) & g_1(2) & g_0(0) & g_1(0) & \cdots \\ \cdots & g_0(3) & g_1(3) & g_0(1) & g_1(1) & \cdots \\ \vdots & \vdots & \vdots & \vdots & \vdots & \vdots \end{bmatrix} \begin{bmatrix} \vdots \\ y_0(0) \\ y_1(0) \\ y_0(1) \\ y_1(1) \\ \vdots \end{bmatrix}$$

or

$$\hat{\boldsymbol{x}} = T_o \boldsymbol{y} \tag{6.120}$$

In order that $\hat{\boldsymbol{x}} = \boldsymbol{x}$ we require that [Vett 92]

$$T_o T_i = I = T_i T_o \tag{6.121}$$

Multiplying rows of T_i with columns of T_o, (6.121) becomes equivalent to

$$\sum_n h_i(2k-n)g_j(n-2l) = \delta_{kl}\delta_{ij}, \quad i,j=0,1 \tag{6.122}$$

or according to the definition of the inner product in (6.93),

$$\langle h_i(2k-n), g_j(n-2l)\rangle = \delta_{kl}\delta_{ij}$$

If (6.122) is satisfied, we say that the two-band filter bank is a *perfect reconstruction* one and $\hat{x}(n) = x(n)$. Thus,

$$x(n) = \sum_k y_0(k)g_0(n-2k) + \sum_k y_1(k)g_1(n-2k) \tag{6.123}$$

Equation (6.123) can also be seen from a different perspective. It is an expansion of $x(n)$ into a set of *basis sequences*

$$\{g_0(n-2k), g_1(n-2k)\}, \quad k \in Z$$

where Z is the set of the integer numbers. From such a point of view $y_0(k), y_1(k)$ are the respective coefficients of the expansion. This is known as the *discrete time wavelet transform (DTWT)* and the coefficients $y_0(k), y_1(k)$ as the *discrete time wavelet coefficients*. Thus, given a perfect reconstruction two-band filter bank (i.e., condition (6.122) is satisfied), the following transform pair is defined:

$$y_i(k) = \sum_n x(n)h_i(2k-n) \ (a)$$
$$x(n) = \sum_{i=0}^{1}\sum_k y_i(k)g_i(n-2k) \ (b) \tag{6.124}$$

Remarks

- Two sets of basis functions are involved, namely,

  $$h_i(2k-n) \equiv \phi_{ik}(n), \ g_j(n-2l) \equiv \psi_{jl}(n) \quad i,j=0,1 \text{ and } k,l \in Z$$

  Equation (6.122) is an orthogonality condition between $\phi_{ik}(n)$ and $\psi_{jl}(n)$, that is,

  $$\langle \phi_{ik}(n), \psi_{jl}(n)\rangle = \delta_{ij}\delta_{kl}$$

  and it is known as the *biorthogonality condition*. The discrete time wavelet transform pair in (6.124) is a *biorthogonal expansion*.

- The basis sequences $\phi_{ik}(n)$ and $\psi_{jl}(n)$ of the expansion are *shifts by an even number of samples* of four basic mother sequences $g_0(n)$, $g_1(n)$, $h_0(-n)$, $h_1(-n)$, which are the impulse responses of the synthesis and the time-reversed analysis filters. For the recovery of $x(n)$ from its discrete time wavelet coefficients, each coefficient $y_i(k)$ weighs and adds a copy of the mother sequences $g_i(n)$ shifted by $2k$.

- When the sequences $\phi_{ik}(n) = h_i(2k - n)$ are themselves orthogonal, that is,

$$\sum_n h_j^i(2k - n)h_j(2l - n) = \delta_{kl}\delta_{ij}, \quad i, j = 0, 1 \text{ and } k, l \in Z$$

then

$$g_i(n) = h_i(-n)$$

That is, the synthesis filters are the time reverse of the analysis ones. Such a filter bank is known as orthogonal or *paraunitary*, and we have the same set of mother sequences (h_i only) involved in both equations of the discrete time wavelet transform (6.124).

- A number of orthogonal and biorthogonal perfect reconstruction filter pairs have been proposed in the literature [Daub 90, Vett 95]. Table 6.2 gives the coefficients for the first four of Daubechies's maximally flat orthogonal filters. The low-pass version $h_1(n)$ is shown. The high-pass versions are obtained as $h_0(n) = (-1)^n h_1(-n + 2L - 1)$, where L is the length of the filters.
- Besides the case of wavelet basis sequences with predefined values, a large research effort has been devoted to constructing such sequences that are optimized to the specific problem of interest. This has also been used in pattern recognition applications. For example, in [Mall 97] it is proposed to design the filters of the bank to optimize a class discriminant criterion. A different approach is followed in [Szu 92], where an optimal linear combination of predefined bases is sought for classification of speech signals.

- When implementing filter banks in practice, noncausal filters have to be appropriately delayed to make them realizable (Appendix D). This makes it

Table 6.2 Daubechies's Low-pass Filters of Length 4, 6, 8, and 10

$h_1(0)$	0.4829629	0.33267	0.2303778	0.1601024
$h_1(1)$	0.8365163	0.806891	0.7148466	0.6038293
$h_1(2)$	0.2241439	0.459877	0.6308808	0.7243085
$h_1(3)$	−0.1294095	−0.135011	−0.0279838	0.1384281
$h_1(4)$		−0.08544	−0.1870348	−0.2422949
$h_1(5)$		0.03522	0.0308414	−0.0322449
$h_1(6)$			0.0328830	0.0775715
$h_1(7)$			−0.0105974	−0.0062415
$h_1(8)$				−0.0125807
$h_1(9)$				0.0033357

necessary to involve certain delay elements at different points, in order to safeguard the perfect reconstruction property of the analysis–synthesis bank (Problem 6.19).

- In practice, the number of input samples $x(n)$ is finite, that is, $n = 0, 1, \ldots, N - 1$. Thus, for the computation of (6.124) some initial conditions are required. Zero, periodic, or symmetric extensions of the data are popular alternatives. Such implementation issues as well as algorithms for the efficient computation of the DTWT coefficients are discussed in [Vett 95, Chapter 6].

Many Bands Case

Figure 6.17 shows the synthesis part corresponding to the analysis bank of Figure 6.13, and it is a generalization of the two-band synthesis concept. Using the easily shown Noble identity given in Figure 6.18 (Problem 6.17), we end up with the equivalent structure of the synthesis part, shown in Figure 6.19. Let $f_i(n)$ be the impulse responses of the F_i filters. It is easy to see that the respective contribution of each $y_i(k)$ sequence to the output $\hat{x}(n)$ is

$$x_i(n) = \sum_k y_i(k) f_i(n - 2^{i+1}k) \quad i = 0, 1, \ldots, J - 2$$

$$x_{J-1}(n) = \sum_k y_{J-1}(k) f_{J-1}(n - 2^{J-1}k)$$

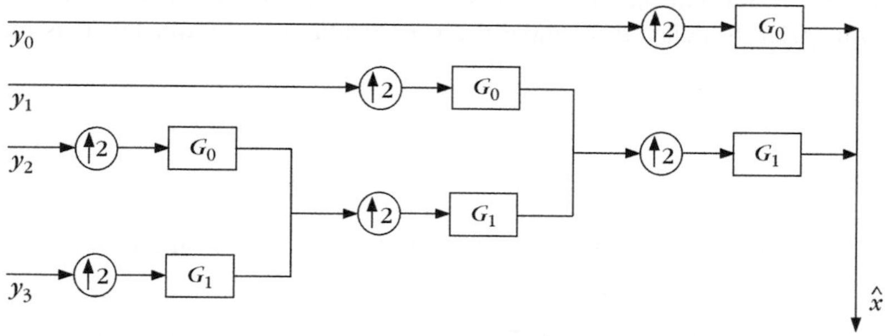

FIGURE 6.17

Tree-structured synthesis filter bank.

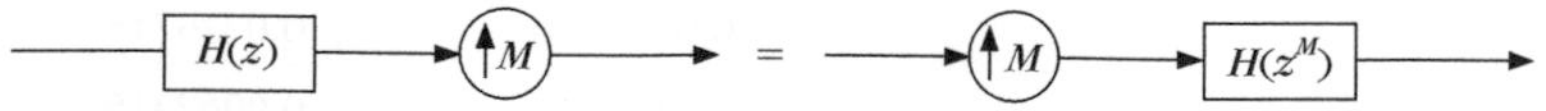

FIGURE 6.18

Noble identity II.

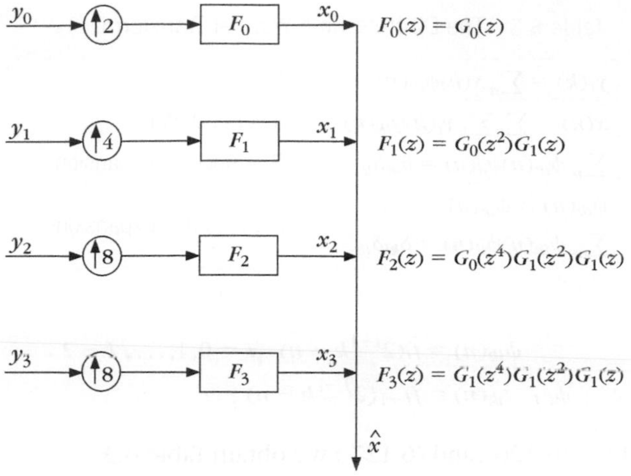

FIGURE 6.19

Equivalent of the tree-structured filter bank of Figure 6.17.

$$\hat{x}(n) = \sum_{i=0}^{J-1} x_i(n)$$

where J is the number of bands, with $J = 4$ in the case of Figures 6.17 and 6.19. It can be shown [Vaid 93] that if the mother filters G_0, G_1, which generate the synthesis part, and the mother filters H_0, H_1 of the analysis part *satisfy the biorthogonality condition* (6.122), *then the J-level analysis-synthesis bank is also a perfect reconstruction filter bank*, that is,

$$\hat{x}(n) = x(n) = \sum_{i=0}^{J-2}\sum_{k} y_i(k) f_i(n - 2^{i+1}k) + \sum_{k} y_{J-1}(k) f_{J-1}(n - 2^{J-1}k) \tag{6.125}$$

where

$$y_i(k) = \sum_{n} x(n)\hat{f}_i(2^{i+1}k - n), \quad i = 0, 1, \ldots, J-2 \tag{6.126}$$

$$y_{J-1}(k) = \sum_{n} x(n)\hat{f}_{J-1}(2^{J-1}k - n) \tag{6.127}$$

with $\hat{f}_i(k)$ being the impulse responses of the corresponding analysis band, in analogy with Figure 6.12b. To summarize our findings, let us define

$$\psi_{ik}(n) = f_i(n - 2^{i+1}k), \quad i = 0, 1, \ldots, J-2$$
$$\psi_{(J-1)k}(n) = f_{J-1}(n - 2^{J-1}k)$$

Table 6.3 The Discrete Time Wavelet Transform

$y_i(k) = \sum_n x(n)\phi_{ik}(n)$	DTWT
$x(n) = \sum_i \sum_k y_i(k)\psi_{ik}(n)$	Inverse DTWT
$\sum_n \phi_{ik}(n)\psi_{jl}(n) = \delta_{kl}\delta_{ij}$	Biorthogonal expansion
$\psi_{ik}(n) = \phi_{ik}(n)$ $\sum_n \phi_{ik}(n)\phi_{jl}(n) = \delta_{kl}\delta_{ij}$	Orthonormal expansion

$$\phi_{ik}(n) = \hat{f}_i(2^{i+1}k - n) \quad i = 0, 1, \ldots, J-2$$
$$\phi_{(J-1)k}(n) = \hat{f}_{J-1}(2^{J-1}k - n)$$

Then from (6.125), (6.126), and (6.127) we obtain Table 6.3.

Remarks

- A notable characteristic of the DTWT is that the basis sequences for each level i are the power of two shifts of a corresponding mother sequence:

$$\begin{aligned} \psi_{ik}(n) &= \psi_{i0}(n - 2^r k), && r = i+1 && \text{for } i \neq J-1 \\ & && \text{or } r = J-1 && \text{for } i = J-1 \\ \phi_{ik}(n) &= \phi_{i0}(n - 2^r k), && r = i+1 && \text{for } i \neq J-1 \\ & && \text{or } r = J-1 && \text{for } i = J-1 \end{aligned}$$

  In the more elegant theory of continuous wavelet transform, all analysis (synthesis) basis functions are produced from a *single* analysis (synthesis) mother function by dilations (time scaling) and shifts [Meye 93, Daub 90, Vett 95].

- The magic number 2, whose powers determine the shifts in the mother basis sequences, results from the successive splitting by two in the tree-structured filter banks, which we have adopted to introduce the DTWT. Filter banks of this type are known as octave-band filter banks. Their characteristic is that the bandwidth of each of the filters in the bank is the same in a logarithmic scale. Sometimes they are also called constant-Q filter banks to stress the fact that the ratio of the filters' bandwidth to the respective central frequency is constant. Generalizations of DTWT with another integer M in place of 2 can also be defined and used [Stef 93].

Example 6.8

The Haar Transform—The Epilogue

We have already seen that the Haar transform is equivalent to a tree-structured analysis filter bank. Let us now look at the synthesis problem. For the 8×8 Haar transform and after a row reshuffling of the corresponding Haar matrix, we have

$$\begin{bmatrix} y_0(0) \\ y_0(1) \\ y_0(2) \\ y_0(3) \\ y_1(0) \\ y_1(1) \\ y_2(0) \\ y_3(0) \end{bmatrix} = \frac{1}{\sqrt{8}} \begin{bmatrix} 2 & -2 & 0 & 0 & 0 & 0 & 0 & 0 \\ 0 & 0 & 2 & -2 & 0 & 0 & 0 & 0 \\ 0 & 0 & 0 & 0 & 2 & -2 & 0 & 0 \\ 0 & 0 & 0 & 0 & 0 & 0 & 2 & -2 \\ \sqrt{2} & \sqrt{2} & -\sqrt{2} & -\sqrt{2} & 0 & 0 & 0 & 0 \\ 0 & 0 & 0 & 0 & \sqrt{2} & \sqrt{2} & -\sqrt{2} & -\sqrt{2} \\ 1 & 1 & 1 & 1 & -1 & -1 & -1 & -1 \\ 1 & 1 & 1 & 1 & 1 & 1 & 1 & 1 \end{bmatrix} \begin{bmatrix} x(0) \\ x(1) \\ \vdots \\ x(7) \end{bmatrix}$$

or

$$\mathbf{y} = \hat{H}\mathbf{x}$$

Thus the 8×8 Haar transform gives four coefficients at the finest resolution level 0, two at level 1 and one for each of the coarsest resolution levels 2 and 3. We will now design the corresponding synthesis bank to obtain $x(n)$ from these coefficients. The impulse responses of the Haar analysis filters are

$$h_1(n) = \begin{cases} \frac{1}{\sqrt{2}} & n = 0 \text{ or } n = -1 \\ 0 & \text{otherwise} \end{cases}$$

$$h_0(n) = \begin{cases} \frac{1}{\sqrt{2}} & n = 0 \\ -\frac{1}{\sqrt{2}} & n = -1 \\ 0 & \text{otherwise} \end{cases}$$

It is readily seen that

$$\sum_n h_i(2k - n)h_j(2l - n) = \delta_{ij}\delta_{kl}, \quad i, j = 0, 1$$

That is, the Haar filter bank is paraunitary. Thus, the synthesis filters can be defined as

$$g_i(n) = h_i(-n), \quad i = 0, 1$$

Hence

$$G_0(z) = \frac{1}{\sqrt{2}}(1 - z^{-1})$$

$$G_1(z) = \frac{1}{\sqrt{2}}(1 + z^{-1})$$

From the equivalent structure of the synthesis bank of Figure 6.19 we have

$$F_1(z) = G_1(z)G_0\left(z^2\right) = \frac{1}{2}\left(1 + z^{-1} - z^{-2} - z^{-3}\right)$$

and the respective impulse response is

$$f_1(n) = \begin{cases} \frac{1}{2} & n = 0, 1 \\ -\frac{1}{2} & n = 2, 3 \\ 0 & \text{otherwise} \end{cases}$$

Following similar arguments, we have

$$f_2(n) = \begin{cases} \frac{1}{\sqrt{8}} & n = 0, 1, 2, 3 \\ -\frac{1}{\sqrt{8}} & n = 4, 5, 6, 7 \\ 0 & \text{otherwise} \end{cases}$$

$$f_3(n) = \begin{cases} \frac{1}{\sqrt{8}} & n = 0, 1, \ldots, 7 \\ 0 & \text{otherwise} \end{cases}$$

If we now insert these values in (6.125) and collect the values of $x(n)$ together, we get

$$\begin{bmatrix} x(0) \\ x(1) \\ \vdots \\ x(7) \end{bmatrix} = \frac{1}{\sqrt{8}} \begin{bmatrix} 2 & 0 & 0 & 0 & \sqrt{2} & 0 & 1 & 1 \\ -2 & 0 & 0 & 0 & \sqrt{2} & 0 & 1 & 1 \\ 0 & 2 & 0 & 0 & -\sqrt{2} & 0 & 1 & 1 \\ 0 & -2 & 0 & 0 & -\sqrt{2} & 0 & 1 & 1 \\ 0 & 0 & 2 & 0 & 0 & \sqrt{2} & -1 & 1 \\ 0 & 0 & -2 & 0 & 0 & \sqrt{2} & -1 & 1 \\ 0 & 0 & 0 & 2 & 0 & -\sqrt{2} & -1 & 1 \\ 0 & 0 & 0 & -2 & 0 & -\sqrt{2} & -1 & 1 \end{bmatrix} \begin{bmatrix} y_0(0) \\ y_0(1) \\ y_0(2) \\ y_0(3) \\ y_1(0) \\ y_1(1) \\ y_2(0) \\ y_3(0) \end{bmatrix}$$

or

$$\boldsymbol{x} = \hat{H}^T \boldsymbol{y}$$

That is, we reobtain the inverse (within a permutation) Haar transform. Hence, we can now state that *the Haar transform and its inverse form a DTWT pair, using the orthogonal Haar sequences as the basis for the wavelet expansion.*

6.14 THE MULTIRESOLUTION INTERPRETATION

The goal of this section is to highlight, without resorting to mathematical details, an important aspect of the wavelet transform that accounts for its success as a tool in pattern recognition as well as in numerous other applications. Let us assume for simplicity that the two filters in the analysis–synthesis bank of a paraunitary filter bank are ideal low/high pass. Figure 6.20 shows the magnitude responses of the respective filters in the equivalent of the tree-structured octave band filter bank of Figure 6.19. The width of the frequency response (bandwidth) is halved for each level of the tree (Figure 6.20d). That is, the "detail" resolution (high-pass) filters have a wide bandwidth, and the "coarse" resolution (low-pass) filters are of narrow bandwidth. Filters F_3 and F_2, the two coarser resolution ones, are of the same bandwidth. These observations are true for any octave band filter bank of any

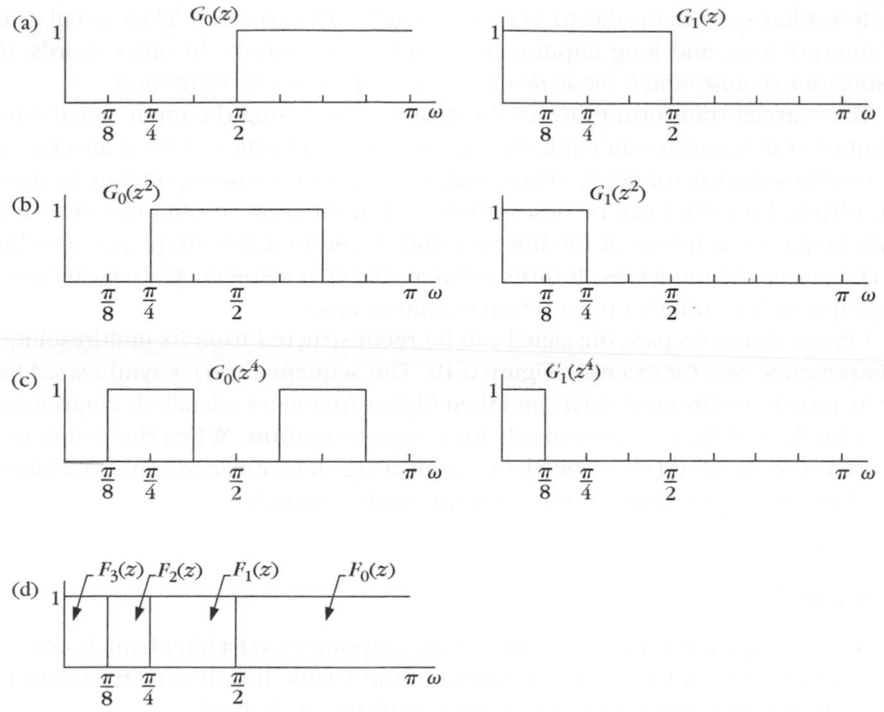

FIGURE 6.20

Filter bandwidths in an octave band filter bank.

number of levels J. That is, the width of $F_i(z)$ is half of the width of $F_{i-1}(z)$, and the widths of F_{J-2} and F_{J-1} are equal. This multiresolution viewpoint of the DTWT is a source of its power as a tool, and it is worth spending some time on it.

It is known (uncertainty principle) that filters with narrow bandwidth have long impulse responses and that filters with wide bandwidth have short impulse responses. Let us take for example the Haar filter bank. The filter impulse response at level zero, $h_0(n) = \hat{f}_0(n)$, is $(\frac{1}{\sqrt{2}}, -\frac{1}{\sqrt{2}})$ and at level three [i.e., corresponding to Eq. (6.118)], $\hat{f}_3(n)$, is $\frac{1}{\sqrt{8}}(1, 1, 1, 1, 1, 1, 1, 1)$. Since the output of each filter of the analysis bank, that is, the DTWT coefficients, is the convolution of the input sequence with the respective impulse response, the filter tends to spread out the input activity. For example, a single impulse in the input of the $\hat{F}_3(z)$ filter produces a sequence of eight samples at the output. Thus, if our goal is to identify sudden (short in time) changes in an input signal, it is apparent that one should use filters of short impulse response to be able to obtain good time locality. Otherwise, the sudden activity will spread in time. Hence, for sudden changes in time (rich in high-frequency components), one needs a "detailed" analysis filter, that is, a short impulse response. For slowly time-varying activities, rich in low frequencies, one needs a "less detailed" analysis filter, so as to be able to "see" the longer scale variations. Thus, detail is not of interest here, and long impulse responses are required. In other words, *the resolution should match the scale of the activity under investigation*.

The wavelet transform provides the means of analyzing the input signal into a number of different resolution levels in a hierarchical fashion. This is also known as *multiresolution analysis*. Thus, signal components corresponding to different physical activities can be best represented at different resolution levels: short high-frequency activities at the finer resolution and long low-frequency activities at coarser resolution levels. It turns out that this coarse-to-fine analysis strategy is appropriate for a number of pattern recognition tasks.

On the synthesis part, the signal can be reconstructed from its multiresolution components. See, for example, Figure 6.19. The sequence $x(n)$ is synthesized first by its coarser component $x_3(n)$, and then higher frequency (detailed) components are added, resulting in a *successively finer approximation*. When the component of the finest detail, $x_0(n)$, is added, the original signal is obtained. This philosophy is at the heart of a number of signal compression schemes.

Remarks

- The analysis of a signal in a number of components via a filter bank is not new and goes back to the work of Gabor in the 1940s. It is directly related to the short-time Fourier transform defined as [Gabo 46, Vett 95]

$$X_s(\omega, m) = \sum_{n=-\infty}^{\infty} x(n)\theta(n-m)\exp(-j\omega n) \tag{6.128}$$

  where $\theta(n)$ is a window sequence, whose center is successively moved to the different points m. Thus, each time, the part of the sequence $x(n)$ around m (depending on the window's effective width) is selected and Fourier transformed. It can be shown that this is equivalent to filtering the signal $x(n)$ by a bank of filters, each centered at a different frequency but all of them having the same bandwidth (Problem 6.20). This is its drawback, because low- and high-frequency signal components are "looked" at through the same window in time, resulting in poor overall localization of the events. What is really needed is a long window to analyze slowly time-varying low-frequency components and a narrow window to detect high-frequency short-time activities. As we saw, this is offered by a tree-structured octave-band filter bank, associated with the DTWT.

- All we have said about wavelet transforms and multiresolution analysis is just a glimpse of the whole story, a story that is really worth further effort; see, for example, [Daub 90].

6.15 WAVELET PACKETS

The DTWT has been introduced via an octave-band filter bank, and the wavelet coefficients result at the outputs of the bank, when its input is fed with the signal of interest. The octave-band filter bank is constructed by successively splitting by two the lowest frequency band (leaf) of the tree-structured bank (Figure 6.13). However, in many cases most of the activity is not in the low-frequency band but in the middle or high-frequency parts of the spectrum. In such cases, it may be useful to be able to allocate finer frequency bandwidths in the bands where the activity occurs. As we will see later on in the chapter, this can boost the discriminatory power of our system from a classification point of view, which is always our main interest. Figure 6.21a shows an example of a tree-structured filter bank but with the finer frequency splitting occurring at a midfrequency band. Figure 6.21b shows the resulting bandwidths for each of the (ideal) filters in the bank (f-axis) and the respective window length of the impulse responses in the time domain (n-axis). In other words, filters 2 and 3 have half the bandwidth and twice the impulse response of 4. Furthermore, they have one fourth of the bandwidth and a four times longer impulse response than that of 1. For comparison, Figure 6.22 shows

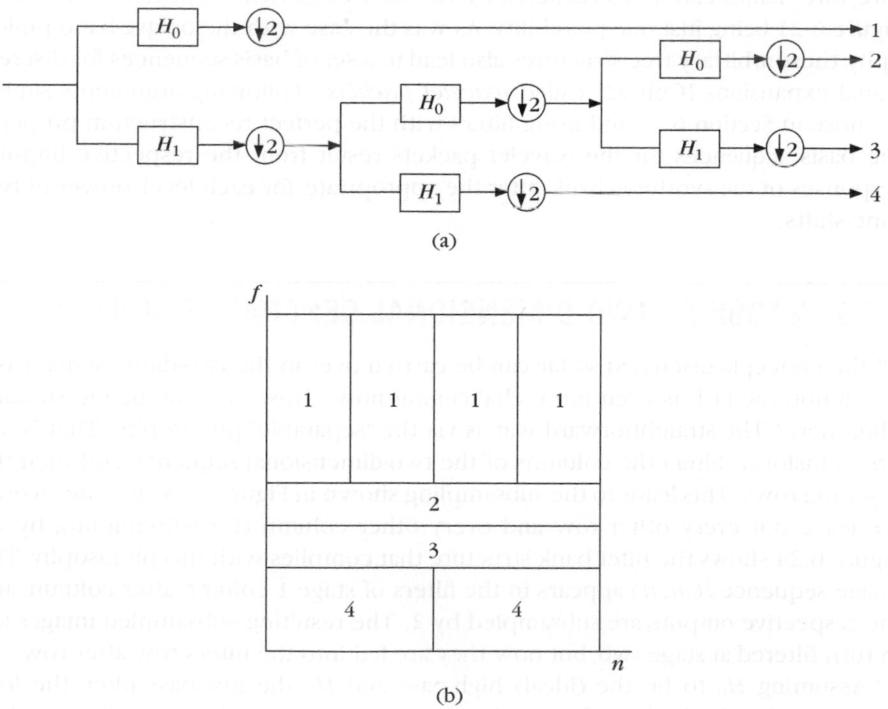

FIGURE 6.21

(a) Wavelet packet tree structure and (b) the corresponding frequency versus time resolution.

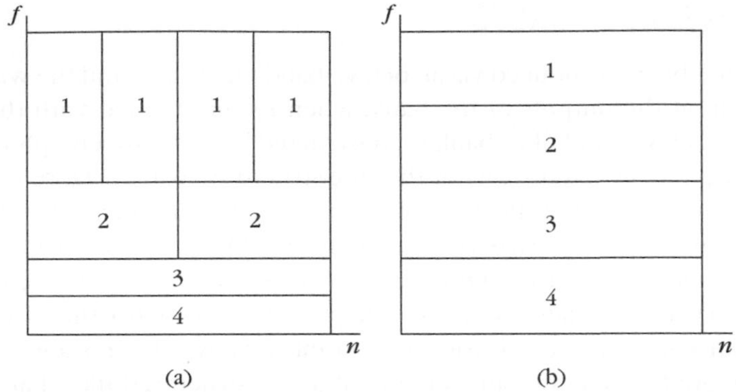

FIGURE 6.22

Frequency versus time resolution for (a) octave band and (b) equal bandwidth filter banks.

the frequency–time resolution plots for an octave-band filter bank (a) and for a bank with equal bandwidths (b), associated with the DTWT and the short-time Fourier transform, respectively. Having freed ourselves from the octave-band tree structure, filter banks can be constructed by various tree growth strategies, with that of Figure 6.21 being just one possibility. As was the case with the octave-band philosophy, these arbitrary tree structures also lead to a set of basis sequences for discrete signal expansions [Coif 92] called *wavelet packets*. Following arguments similar to those in Section 6.13 and using filters with the perfect reconstruction property, the basis sequences for the wavelet packets result from the respective impulse responses of the synthesis bank, after the appropriate, for each level, power of two time shifts.

6.16 A LOOK AT TWO-DIMENSIONAL GENERALIZATIONS

All the concepts discussed so far can be carried over in the two-dimensional case. No doubt, the task is even more challenging now. How can one define subsampling here? The straightforward way is via the "separable" philosophy. That is, we first transform (filter) the columns of the two-dimensional sequence and then the resulting rows. This leads to the subsampling shown in Figure 6.23. In other words, we leave out every other row and every other column (for subsampling by 2). Figure 6.24 shows the filter bank structure that complies with this philosophy. The image sequence $I(m, n)$ appears in the filters of stage 1 column after column, and the respective outputs are subsampled by 2. The resulting subsampled images are in turn filtered at stage two, but now they are fed into the filters row after row.

Assuming H_0 to be the (ideal) high-pass and H_1 the low-pass filter, the four frequency bands that are formed by the previous procedure are illustrated in Figure 6.25a. The area H_1H_1 corresponds to low-pass columns and rows, H_1H_0 to low-pass columns and high-pass rows, and so forth. Figure 6.25b shows the

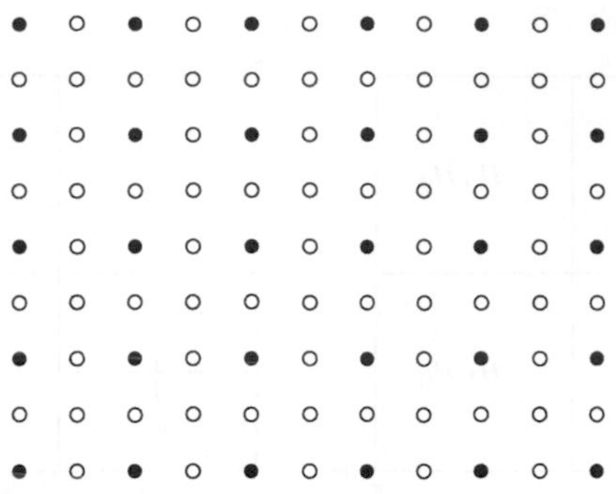

FIGURE 6.23

Separable subsampling by 2 for images.

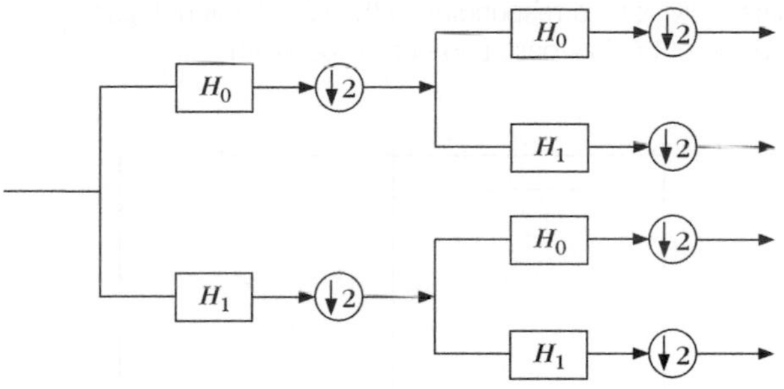

FIGURE 6.24

Basic element for a two-dimensional filter bank, leading to separable subsampling by 2.

resulting segmentation of the frequency domain when the low-pass area H_1H_1 is successively split by repeating the procedure.

Example 6.9

Figure 6.26 shows a 64×64 image of a triangle. The three "line" images are the 32×32 images that result when passing the triangle image through the structure of Figure 6.24. In the columnwise filtering of the first stage, the vertical line goes through the low-pass H_1 filter (no variation across it) and the horizontal and diagonal lines go through the high-pass H_0. This is because in a columnwise filtering, these appear as impulses in each column, and thus are rich in high frequencies. In the row scanning of the second stage, it will be the horizontal line that will go through the low-pass filter. Similar reasoning explains the position of the various parts of the triangle in the different bands.

Although this is obviously a very simplified example, it is quite instructive. It demonstrates how the original image can be obtained from its multiresolution components and also how different characteristics (directional in this case) of the whole may be isolated at different bands.

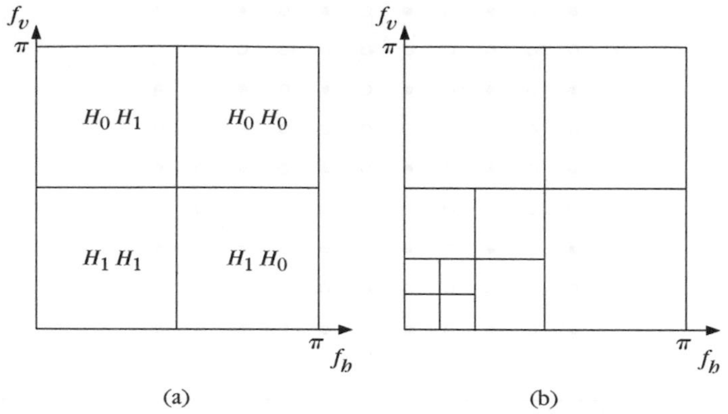

FIGURE 6.25

(a) Frequency domain division corresponding to the filter bank of Figure 6.24 and (b) the result of a successive division of the low-pass part of the spectrum.

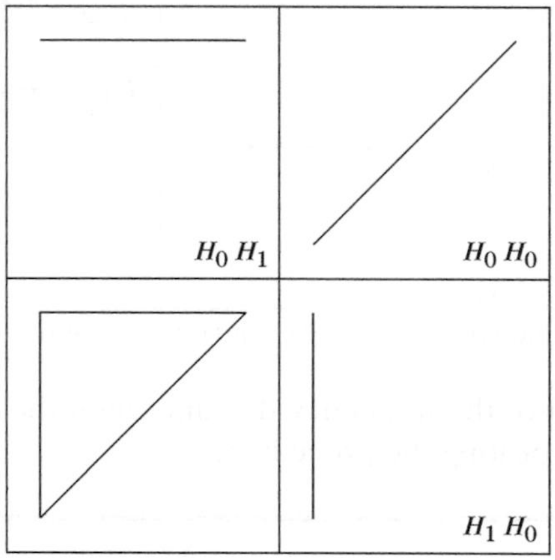

FIGURE 6.26

A triangle image and its filtered versions through the filter bank of Figure 6.24.

6.17 APPLICATIONS

All the transforms we have studied in this chapter are good candidates for feature generation, and they have been used extensively in various pattern recognition tasks. However, the wavelet transform offers an extra advantage, which in some cases can be beneficially exploited. *Its multiresolution properties conform to the way perception is achieved by humans, through their hearing and visual systems.* The human ear exhibits decreasing resolution at higher frequencies, in a way that is uniform on a logarithmic scale (octave bands) [Flan 72]. Experiments in psychophysics and physiology show that the human visual cortex perceives by decomposing the stimuli in a number of frequency bands [Camp 68, Levi 85], which are also dependent on the spatial orientation [Camp 66]. Experimental results in [Nach 75] indicate that these frequency bands have the approximate bandwidth of an octave. The similarities between the mechanism with which human perception systems treat the respective stimuli and the processing techniques that split the signal into various (spatial) frequency bands, in a way similar to the wavelet transform, justify the use of the latter in pattern recognition tasks [Mall 89].

The following examples come from two of the most important areas of interest in today's pattern recognition applications.

Recognition of Handwritten Characters

The development of OCR systems is of particular importance in various application areas. One of the most challenging among them is the recognition of handwritten characters. Figure 6.27 shows the character "3" as well as its boundary contour after the application of a contour tracing algorithm [Pita 92]. The task now becomes one of shape recognition. As discussed in more detail in Chapter 7, the boundary can

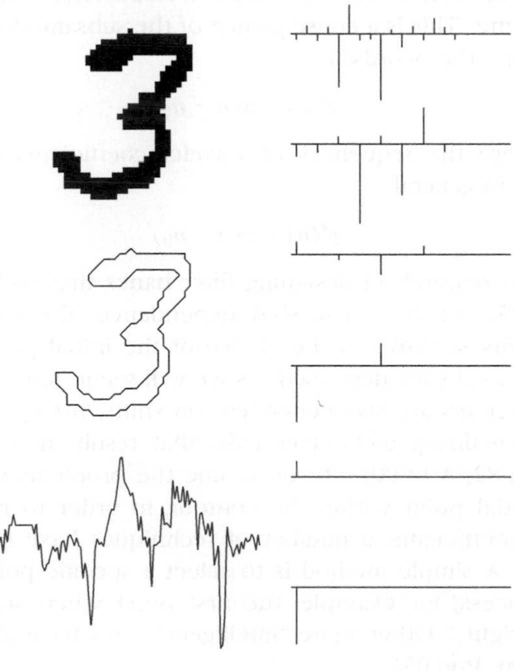

FIGURE 6.27

Wavelet coefficients corresponding to the curvature of the boundary of number "3."

be represented as a closed parametric curve in the complex plane

$$u(n) = x(n) + jy(n), \quad 0 \le n \le N-1 \tag{6.129}$$

with N being the number of samples (pixels) found tracing the contour and $x(n), y(n)$ the corresponding coordinates. The first point $(x(0), y(0))$ of the sequence is considered as the origin. Fourier methods have been used extensively in such classification tasks, by obtaining the DFT of $u(n)$ and keeping a sufficient number, from the total of N, of Fourier components as features. An alternative way is to extract the features from the wavelet domain. In other words, $x(n)$ and $y(n)$ are independently filtered through a tree-structured filter bank of appropriate resolution depth, and the resulting wavelet coefficients are used as features. The low-frequency components account for the basic shape of the character and are less sensitive to varying writing styles. The high-frequency components account for the details and are more sensitive to the specific handwriting style. In [Wuns 95], a comparative study was carried out using the same number of DTWT coefficients and Fourier-based features, with a neural network classifier. The experiments showed that classification based on wavelet coefficients resulted in reduced error rates. Furthermore, it was pointed out that Fourier-based features exhibited larger within-class variance and weaker between-class separation than the wavelet-based ones. A major disadvantage associated with the wavelet coefficients is that *they are not shift invariant*. That is, if we rotate/translate a character, the resulting coefficients will not be the same. This is a consequence of the subsampling process [Mall 89] (Problem 6.21). In other words, if

$$x'(n) = x(n - n_0)$$

and $y'(n)$, $y(n)$ are the sequences of wavelet coefficients of $x'(n)$ and $x(n)$, respectively, then, in general

$$y'(n) \neq y(n - n_0)$$

This has led to research in designing filter banks that seek to overcome this property [Marc 95, Hui 96]. The shift dependence obviously also makes the wavelet coefficients sensitive to the choice of the initial point from which the contour is traced, as already described. As we will see in more detail in Chapter 7, the Fourier coefficients are also dependent on shifts *but* in a deterministic way. Thus, various normalizing techniques exist that result in shift-invariant feature parameters [Crim 82, Arbt 90]. To overcome the problems associated with the choice of the initial point within the contour, in order to minimize its effects on the wavelet coefficients, a number of techniques have been suggested and used in practice. A simple method is to select a specific point resulting during the scanning process, for example, the first pixel when scanning the character from left to right. Other more "intelligent" ways have also been suggested [Wuns 95, Chua 96, Pun 03].

An alternative approach to that of Eq. (6.129) is to describe a contour in terms of the arc length between a given point and the origin, within the contour. As arc length t at a given point, we define the number of consecutive pixels between the given point and the point considered as the origin. The contour description can now be achieved via a one-parameter real-valued function, the arc tangent angle $\theta(t)$ or the corresponding curvature $\kappa(t)$, defined as

$$\theta(t) = \tan^{-1}\left[\frac{dy(t)}{dx(t)}\right]$$

$$\kappa(t) = \frac{d\theta(t)}{dt}$$

where $x(t), y(t)$ are the coordinates of the respective point as a function of length t from the origin and $dt = \sqrt{dy^2 + dx^2}$. A further discussion of this is provided in Chapter 7.

[Kapo 96] suggests wavelet transforming the curvature of the contour and using the corresponding wavelet coefficients as features. Figure 6.27 shows the coefficients resulting from the wavelet analysis of the curvature (bottom left) of the boundary contour of number 3. The wavelet basis used for the analysis was Daubechies's biorthogonal pairs (3,9) [Vett 95]. Six successive resolution levels are shown, the finest being on top and the coarsest at the bottom. Extensive experimentation with a number of different characters shows that the use of wavelet coefficients from more than six resolution levels adds no further discriminatory information to the system. Thus, each of the resulting feature vectors has 32 components. A different philosophy is followed in [Geze 00, Geze 02], in the context of an OCR system for the Greek Orthodox Byzantine music notation. The wavelet transform is applied to the vector combining the four projections (horizontal, vertical, left-diagonal, right-diagonal) of the characters. It turns out that such an approach leads to an efficient coding of the directional properties of the characters.

Texture Classification

Texture characterization in image analysis tasks is another area where the wavelet transform, as well as the other transforms discussed in this chapter, has been heavily utilized. The basic approach is similar to that in the OCR example. However, because texture is a property of the image region and not of its boundary, the two-dimensional variants of the transforms are used. The two-dimensional wavelet transform offers the advantages of spatial frequency and orientation selectivity, provided the appropriate bases are chosen. The information compaction properties result from the fact that most of the energy activity is concentrated in certain resolution levels, as was the case with the OCR example considered earlier. The wavelet coefficients of these levels are then selected as features to form the feature vectors. Sometimes a function of the features is employed, such as the energy, $\sum_i y_i^2$, or the entropy, $\sum_i y_i^2 \log y_i^2$, where y_i are the respective wavelet coefficients at each resolution level [Lain 93].

In many cases, the underlying image texture exhibits a great deal of activity in middle or high-frequency bands.

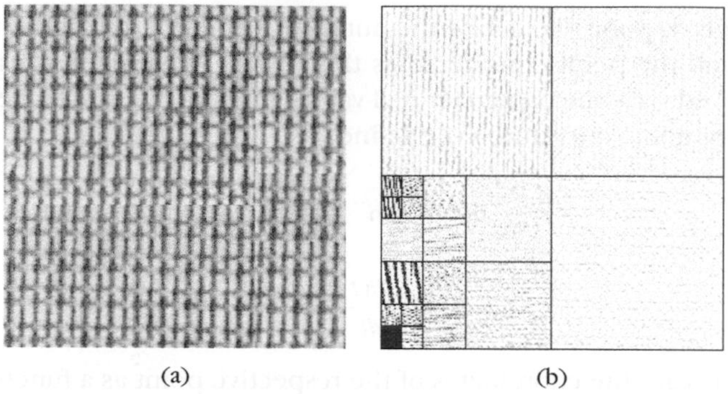

FIGURE 6.28

An example of (a) a textured image and (b) its corresponding wavelet packet transform.

Figure 6.28a is an example of a textured image taken from [Brod 66]. In such cases, one must pay attention to the bands of high energy, instead of looking with fine frequency bandwidths at low-frequency bands of low energy. This leads to the adoption of wavelet packets, discussed in Section 6.15. There, we saw that there are a number of choices for splitting the frequency bands, leading to different wavelet packets. In [Chan 93] a dynamic procedure is suggested, depending on the particular texture image. A threshold C is selected prior to the analysis. If the output energy in a band is less than C, no splitting of the band is carried out. If it is higher than C, the band is split further. Splitting terminates if subimages, after filtering and subsampling, become small, for example, 16×16. It is then suggested to use as features the energy values at the J (a preselected number) most dominant (energywise) bands. In [Mojs 00] the effects of the properties of the analysis band on texture characterization are considered. It is demonstrated that the choice of the analysis filters may have a significant influence on the classification performance. In [Unse 95, Lain 96] another variation of the DTWT is adopted, called the *discrete wavelet frame*. The difference from DTWT is that the filter outputs in the bank are not subsampled. Although this leads to a redundant representation, it results in a texture description tolerant to translations.

A procedure similar in concept to the DTWT one is to employ a bank of two-dimensional Gabor filters to perform the splitting of the image into a number of frequency bands. The impulse response (point spread function, Appendix D) of the complex two-dimensional *Gabor filter* is given as the product of a Gaussian low-pass filter with a complex exponential, that is [Bovi 91],

$$h(x,y) = g'(x,y)\exp(j(\omega_x x + \omega_y y)) \tag{6.130}$$

where

$$g'(x,y) = \frac{1}{\lambda\sigma^2}\, g\left(\frac{x'}{\lambda\sigma}, \frac{y'}{\sigma}\right), g(x,y) = \frac{1}{2\pi}\exp\left(-\frac{x^2+y^2}{2}\right) \tag{6.131}$$

and

$$x' = x\cos\theta + y\sin\theta$$
$$y' = -x\sin\theta + y\cos\theta \tag{6.132}$$

That is, $g'(x,y)$ is a version of the Gaussian $g(x,y)$ that is spatially scaled and rotated by θ. The parameter σ is the spatial scaling, which controls the width of the filter impulse response, and λ defines the aspect ratio of the filter, which determines the directionality of the filter, which is no longer circularly symmetric. The orientation angle θ is usually chosen to be equal to the direction of the filter's center circular frequency

$$\omega = \sqrt{\omega_x^2 + \omega_y^2} \tag{6.133}$$

That is,

$$\theta = \tan^{-1}\frac{\omega_y}{\omega_x} \tag{6.134}$$

By varying the free parameters σ, λ, ω, and θ, filters of arbitrary orientation and bandwidth characteristics are obtained (Problem 6.22). Figure 6.29a shows the magnitude of the complex Gabor filter for $\sigma = 1.0, \lambda = 0.3$. Figure 6.29b shows the magnitude of the corresponding spatial frequency response. The choice of the Gabor filters is justified by the fact that these filters offer the optimal trade-off between spectral bandwidth and spatial localization. In Section 6.14 we saw that the shorter the filter's impulse response (spatial localization), the wider its frequency bandwidth and vice versa, according to the *uncertainty principle* [Papo 91],

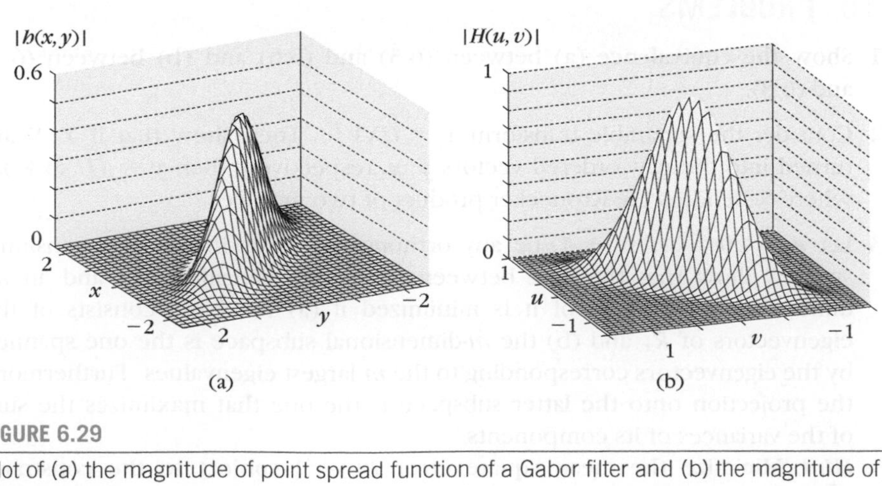

FIGURE 6.29

Plot of (a) the magnitude of point spread function of a Gabor filter and (b) the magnitude of its Fourier transform.

that is,

$$\Delta x \Delta \omega_x \geq \frac{1}{2}$$
$$\Delta y \Delta \omega_y \geq \frac{1}{2} \tag{6.135}$$

In [Daug 85] it has been shown that the two-dimensional Gabor filters attain the minimum uncertainty bound. For digital images, a sampling of the above functions has to be performed, which introduces aliasing errors regardless of the sampling interval. This happens because Gabor filters are not bandlimited, but have a Gaussian-shaped frequency response (Problem 6.23). The topic is treated in [Bovi 90]. Gabor filter banks for analyzing the image in a number of bands, in the context of texture classification, have been successfully employed in a number of cases, [Jain 91, Turn 86, Bovi 90, Hale 95, Hale 99, Weld 96]. A set of Gabor filters centered at different frequencies and having different orientations are used to cover the frequency range of interest, using various frequency and orientation bandwidths. Images are then filtered through this set of filters, and the features are generated from the resulting output samples. For example, the output energies of the Gabor filters may be chosen to be the respective features. Thus, using this strategy, the generated features encode classification information related to the spatial frequency as well as the orientation activity of the various textures. In order to grasp most of the textural information, techniques have been proposed to place the centers of the Gabor filters in the most "important" image frequencies [Pich 96]. In [Chan 93, Grig 02] a comparative study of various transform-based features is provided in the context of texture classification.

6.18 PROBLEMS

6.1 Show the equivalence (a) between (6.5) and (6.6) and (b) between (6.7) and (6.8).

6.2 Consider the separable transform $Y = UXV^T$. Then show that if Y, X are turned into the row-ordered vectors $\boldsymbol{y}, \boldsymbol{x}$, respectively, then $\boldsymbol{y} = (U \otimes V)\boldsymbol{x}$, where $\otimes$ denotes the Kronecker product of two matrices.

6.3 Let $\boldsymbol{e}_i, i = 0, 1, \ldots, N-1$, be any orthonormal basis in the N-dimensional space. Show that the MSE between an N-dimensional vector and an m-dimensional projection of it is minimized if (a) the basis consists of the eigenvectors of R_x and (b) the m-dimensional subspace is the one spanned by the eigenvectors corresponding to the m largest eigenvalues. Furthermore, the projection onto the latter subspace is the one that maximizes the sum of the variances of its components.
Hint: Minimize the mean square error $E[\|\epsilon\|^2]$ subject to the constraint $\boldsymbol{e}_i^T \boldsymbol{e}_i = 1$.

6.4 Consider an N-dimensional random vector $\boldsymbol{x}$, which is approximated by

$$\hat{\boldsymbol{x}} = \sum_{i=0}^{m-1} y_i \boldsymbol{e}_i + \sum_{i=m}^{N-1} c_i \boldsymbol{e}_i$$

where c_i are nonrandom constants and $\boldsymbol{e}_i, i = 0, 1, 2, \ldots, N-1$, constitute an orthonormal basis. Show that the minimum mean square error $E\|\boldsymbol{x} - \hat{\boldsymbol{x}}\|^2$ is achieved if (a) $c_i = E[y_i], i = m, \ldots, N-1$; (b) the orthonormal basis consists of the eigenvectors of Σ_x; and (c) $\boldsymbol{e}_i, i = m, \ldots, N-1$, correspond to the $N - m$ smallest eigenvalues.

6.5 If X is a rank r matrix, show that the two square matrices XX^H and $X^H X$ have the same nonzero eigenvalues.

6.6 (a) Show Eq. (6.39).
(b) Show that the expansion in the right-hand side of (6.41) is the projection of $\boldsymbol{x}_i$ on the subspace spanned by the first k columns of U.

6.7 Given the matrix

$$\begin{bmatrix} 1 & 2 \\ 2 & 1 \\ 1 & 3 \end{bmatrix}$$

compute its SVD representation.

6.8 Show the orthogonality of the DFT matrix W and also identity (6.94).

6.9 Given the image array

$$\begin{bmatrix} 1 & 2 & 1 \\ 0 & 1 & 1 \\ 2 & 1 & 2 \end{bmatrix}$$

compute its two-dimensional DFT transform.

6.10 For one of the images available at the Web site of the book, write a program to compute its DFT transform. Use a routine to implement the fast Fourier transform and plot the magnitude of the resulting Fourier transform.

6.11 Show the orthogonality of the DCT transform.

6.12 Compute the DCT of the image array of Problem 6.9.

6.13 Develop a program to compute the DCT for one of the images available from the Web site of the book.

6.14 Show the orthogonality of the Hadamard transform.

6.15 Compute the Hadamard transform for a 2×2 submatrix of the matrix of Problem 6.9.

6.16 Show the orthogonality of the Haar transform.

6.17 Show the two Noble identities of Figures 6.12a and 6.18.

6.18 Show the equivalence between the tree structure of Figure 6.11 and that of Figure 6.12b.

6.19 Consider the perfect reconstruction two-band Haar bank. Show that

a. If we make the analysis filters causal by delaying each of them by one sample, then the reconstructed sequence $\hat{x}(n)$ is delayed by one sample, that is, $\hat{x}(n) = x(n-1)$. In the more general case, if both the analysis filters have to be delayed by L, then the output of the band is also delayed by L.

b. If this is repeated with the more general N-band case, show that the output is delayed as $\hat{x}(n) = x(n - (2^{N-1} - 1)L)$ and a delay must be inserted in each band to safeguard perfect reconstruction. Figure 6.30 shows the three-band case. In the general case, the delay element in each band is $z^{-(2^{N-i-1}-1)L}$, $i = 0, 1, \ldots, N-1$.

6.20 Show that the short-time Fourier transform defined in (6.128) is equal to

$$X_s(\omega, m) = \exp(-j\omega m) \sum_{n=-\infty}^{\infty} x(n)\theta(n-m)\exp(j\omega(m-n))$$

Verify that this is equivalent to filtering the sequence $x(n)$ with different filters of the same bandwidth but centered at different frequencies.

6.21 Show that the process of filtering and then subsampling is equivalent to the action of a linear but time-varying system. In fact, it is a periodically time-varying linear system. The same is true for the process of upsampling followed by linear filtering.

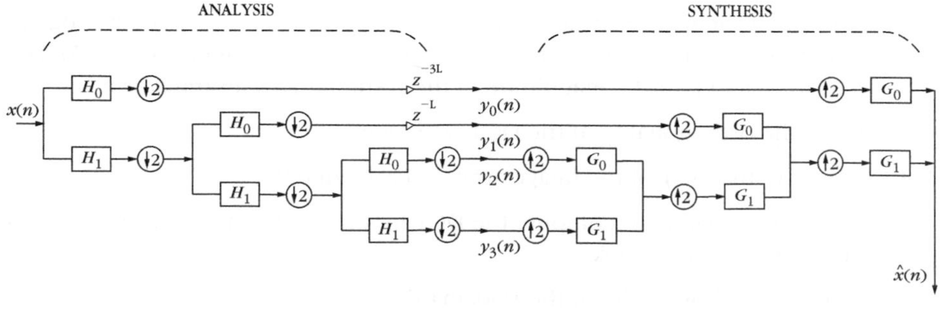

FIGURE 6.30

A three-band perfect reconstruction filter bank with causal analysis and synthesis filters.

6.22 Show that the frequency (octave) and orientation (radians) half-peak bandwidths for Gabor filters are given by B_f, B_θ, respectively, where

$$B_f = \log_2 \frac{\omega\lambda\sigma + \sqrt{2\ln 2}}{\omega\lambda\sigma - \sqrt{2\ln 2}}$$

$$B_\theta = 2\tan^{-1}\frac{\sqrt{2\ln 2}}{\omega\sigma}$$

Compute these for different values of λ, σ.

6.23 Show that the Fourier transform of the two-dimensional Gabor filter response $h(x, y)$ is given by

$$H(u, v) = \exp\left(-\frac{\sigma^2}{2}\{(u' - \omega_x')^2\lambda^2 + (v' - \omega_y')^2\}\right)$$

where

$$u' = u\cos\theta + v\sin\theta$$

$$v' = -u\sin\theta + v\cos\theta$$

and ω_x', ω_y' the corresponding versions of ω_x, ω_y rotated by θ. Draw its magnitude versus frequency (u, v) for different values of λ and θ.

MATLAB PROGRAMS AND EXERCISES

Computer Programs

6.1 *Generation of points around an $(l-1)$-dimensional hyperplane:* Write a MATLAB function named *generate_hyper* that generates randomly l-dimensional points $\boldsymbol{x}_i = [x_1(i), x_2(i), \ldots, x_l(i)]^T$ around an $(l-1)$-dimensional hyperplane H: $\boldsymbol{w}^T\boldsymbol{x} + w_0 = 0$, where $\boldsymbol{w} = [w_1, w_2, \ldots, w_l]^T$. More specifically, the function takes as inputs: (a) the parameter (column) vector $\boldsymbol{w}$ for H ($w_l \neq 0$), (b) the offset w_0 for H, (c) a positive parameter a that defines the range $[-a, a]$, where each one of the first $(l-1)$ coordinates of the points is uniformly distributed, (d) the positive parameter e that defines the range $[-e, e]$ of a uniformly distributed noise source, which is added to the term $(-w_0 - \sum_{i=0}^{l-1} w_i x_i)/w_l$ to produce the lth coordinate, (e) the number N of points to be generated, and (f) the seed *sed* for the *rand* MATLAB function. It returns an $l \times N$ dimensional matrix, X, whose columns contain the generated data points. In addition, the function plots the data points for $l = 2, 3$.

Solution

```
function X=generate_hyper(w,w0,a,e,N,sed)
  l=length(w);
```

```
t=(rand(l-1,N)-.5)*2*a;
t_last=-(w(1:l-1)/w(l))'*t + 2*e*(rand(1,N)-.5)-(w0/w(l));
X=[t; t_last];
%Plots for the 2d and 3d case
if(l==2)
  figure(1), plot(X(1,:),X(2,:),'.b')
elseif(l==3)
  figure(1), plot3(X(1,:),X(2,:),X(3,:),'.b')
end
figure(1), axis equal
```

6.2 *PCA analysis:* Write MATLAB commands to compute the principal components of the covariance matrix of an $l \times N$ dimensional data matrix X as well as the corresponding variances.

Solution

Just write

```
[pc,variances]=pcacov(cov(X'))
```

In the above command (a) $cov(X')$ computes the covariance matrix of X' and (b) *pcacov* returns the principal components (eigenvectors of the covariance matrix) in the columns of *pc* as well as the corresponding variances (eigenvalues) in the column vector *variances* (note that *pcacov* assumes the the data vectors lie in the rows of corresponding data matrix).

6.3 *Distance matrix computation:* Write a MATLAB function named *compute_distances* that takes as input an $l \times N$ matrix X and returns the $N \times N$ dimensional matrix *distX* whose (i,j) entry contains the squared Euclidean distance between the ith and jth column vectors of X.

Solution

```
function distX=compute_distances(X)
  [l,N]=size(X);
  distX=zeros(N);
  for i=1:N
    for j=i+1:N
      distX(i,j)=(X(:,i)-X(:,j))'*(X(:,i)-X(:,j));
      distX(j,i)=distX(i,j);
    end
  end
```

6.4 *Singular Value Decomposition:* Write MATLAB commands to perform SVD on an $l \times N$ dimensional data matrix X whose columns are the data vectors.

Solution

Just write

```
[U,S,V]=svd(X)
```

The above command returns: (a) a diagonal matrix S of the same size with X, containing in its diagonal the singular values of X in decreasing order and (b) the unitary matrices U, V such that $U * S * V' = X$.

6.5 *Dimensionality reduction using SVD*: Write a MATLAB function named *SVD_eval* that evaluates the performance of the SVD method when applied on a data matrix X. More specifically, this function takes as inputs: (a) an $l \times N$ dimensional matrix X, whose columns contain the data vectors, (b) the dimensionality $k(<l)$ of the reduced space (k-dimensional hyperplane), h, generated by the k column vectors of the matrix U_r, which correspond to the k largest singular values of X. It returns: (a) an $l \times 1$ column vector containing the singular values of X, (b) the corresponding U_r matrix, denoted by Ur, (c) the $1 \times l$ dimensional parameter vector w of h, (d) the offset $w0$ of h, (e) the distance matrix *distX* for X, and (f) the distance matrix *distX_proj* of the projections of the vectors of X on h.

Solution

```
function [s,Ur,w,w0,distX,distX_proj]=SVD_eval(X,k)
  [l,N]=size(X);
  [Ur,S,Vr] = svd(X);
  s=diag(S(1:l,1:l));
  a=S(1:k,1:k)*Vr(:,1:k)';
  X_proj=Ur(:,1:k)*a;
  % Deterimnation of the estimated by the SVD hypeprlane
  P=X_proj(:,1:l)';
  w=[];
  for i=1:l
    w=[w (-1)^(i+1)*det([P(:,1:i-1) P(:,i+1:l) ones(l,1)])];
  end
  w0=(-1)^(l+2)*det(P(:,1:l));
  % Computation of distances
  distX=compute_distances(X);
  distX_proj=compute_distances(X_proj);
```

Computer Experiments

6.1 **a.** Generate an $l \times N$ dimensional matrix X ($l = 2$ and $N = 1000$), whose columns are two-dimensional points lying around the line h: $x_1 + x_2 = 0$

(i.e., $w = [1, 1]^T$ and $w_0 = 0$), using the *generate_hyper* function with parameters $a = 10, e = 1$ and $sed = 0$.

b. Compute the principal components of the covariance of X as well as the corresponding variances (eigenvalues). Compare the direction of the first principal component with the direction vector of h (which is perpendicular to w) and draw your conclusions.

6.2 Repeat 6.1 with $e = 5$.

6.3 Repeat 6.1 and 6.2 where now $l = 3$ and the line h is replaced by the plane H: $x_1 - 5x_2 + 2x_3 = 0$.

6.4 **a.** Generate an $l \times N$ dimensional matrix X ($l = 3$ and $N = 1000$), whose columns are two-dimensional points lying around the three dimensional hyperplane with $w = [1, 1, 1]^T, w_0 = 0$, using the *generate_hyper* function, with parameters $a = 10, e = 1$ and $sed = 0$.

b. Use the *SVD_eval* function to compute (a) the singular values of X,(b) $k = 2$ column vectors of the matrix U_r, that correspond to the two largest singular values of X. Also, compare the distances between the points of X and the distances of their corresponding projections on h.

c. Repeat (b) for $k = 1$.

6.5 Repeat 6.4 with $e = 6$. Comment on the results.

REFERENCES

[Achl 01] Achlioptas D., McSherry F. "Fast computation of low rank approximations," *Proceedings of the ACM STOC Conference*, pp. 611–618, 2001.

[Akan 93] Akansu A.N., Hadda R.A. *Multiresolution Signal Decomposition*, Academic Press, 1992.

[Arbt 90] Arbter K., Snyder W.E., Burkhardt H., Hirzinger G. "Application of affine-invariant Fourier descriptors to recognition of 3-D objects," *IEEE Transactions on Pattern Analysis and Machine Intelligence*, Vol. 12(7), pp. 640–647, 1990.

[Atti 92] Attick J.J. "Entropy minimization: A design principle for sensory perception," *International Journal of Neural Systems*, Vol. 3, pp. 81–90, 1992.

[Barl 89] Barlow H.B. "Unsupervised learning," *Neural Computation*, Vol. 1, pp. 295–311, 1989.

[Bart 02] Bartlett M.S., Movellan J.R., Sejnowski T.J. "Face recognition by independent component analysis," *IEEE Transactions on Neural Networks*, Vol. 13(6), pp. 1450–1464, 2002.

[Bell 00] Bell A.J. "Information theory, independent component analysis, and applications," in *Unsupervised Adaptive Filtering, Part I: Blind Source Separation* (Haykin S., ed.), pp. 237–264, John Wiley & Sons, 2000.

[Bell 97] Bell A.J., Sejnowski T.J. "The independent components of natural scenes are edge filters," *Vision Research*, Vol. 37(23), pp. 3327–3338, 1997.

[Belk 03] Belikn M., Niyogi P. "Laplacian eigenmaps for dimensionality reduction and data representation," *Neural Computation*, Vol. 15(6), pp. 1373–1396, 2003.

[Beng 04] Bengio Y., Paiement J.-F., Vincent P., Delalleau O., Le Roux N., Quimet M. "Out of sample extensions for LLE, Isomap, MDS, Eigenmaps and Spectral clustering," *Advances in Neural Information Processing Systems Conference*, (Thrun S., Saul L., Schölkopf B., eds.), MIT Press, 2004.

[Benn 06] Benetos E., Kotti M., Kotropoulos C. "Applying supervised classifiers based on non-negative matrix factorization to musical instrument classification," *Proceedings IEEE Intl. Conference on Multimedia and Expo*, pp. 2105–2108, Toronto, Canada, 2006.

[Berr 95] Berry M., Dumais S., O'Brie G. "Using linear algebra for intelligent information retrieval," *SIAM Review*, Vol. 37, pp. 573–595, 1995.

[Beyg 06] Beygelzimer A., Kakade S., Langford J. "Cover trees for nearest neighbor," *Proceedings of the 23rd International Conference on Machine Learning*, Pittsburgh, PA, 2006.

[Bovi 91] Bovic A.C. "Analysis of multichannel narrow-band filters for image texture segmentation," *IEEE Transactions on Signal Processing*, Vol. 39(9), pp. 2025–2044, 1991.

[Bovi 90] Bovic A.C., Clark M., Geisler W.S. "Multichannel texture analysis using localized spatial filters," *IEEE Transactions on Pattern Analysis and Machine Intelligence*, Vol. 12(1), pp. 55–73, 1990.

[Brod 66] Brodatz P. *Textures: A Photographic Album for Artists and Designers*, Dover, 1966.

[Brun 04] Brunet J.-P. Tamayo P., Golub T.R., Mesirov J.P. "Meta-genes and molecular pattern discovery using matrix factorization," *Proceedings of the National Academy of Science*, Vol. 101(2), pp. 4164–4169, 2004.

[Burg 04] Burges C.J.C. "Geometric methods for feature extraction and dimensional reduction: A guided tour," *Technical Report MSR-TR-2004-55*, Microsoft Research, 2004.

[Burt 83] Burt P.J., Adelson E.H. "The Laplacian pyramid as a compact image code," *IEEE Transactions on Communications*, Vol. 31(4), pp. 532–540, 1983.

[Cai 05] Cai D., He X. "Orthogonal locally preserving indexing," *Proceedings 28th Annual International Conference on Research and Development in Information Retrieval*, 2005.

[Cama 03] Camastra F. "Data dimensionality estimation methods: A survey," *Pattern Recognition*, Vol. 36, pp. 2945–2954, 2003.

[Camp 66] Campell F., Kulikowski J. "Orientation selectivity of the human visual system," *Journal of Physiology*, Vol. 197, pp. 437–441, 1966.

[Camp 68] Campell F., Robson J. "Application of Fourier analysis to the visibility of gratings," *Journal of Physiology*, Vol. 197, pp. 551–566, 1968.

[Cao 03] Cao J.J., Chua K.S., Chong W.K., Lee H.P., Gu Q.M. " A comparison of PCA, KPCA and ICA for dimensionality reduction," *Neurocomputing*, Vol. 55, pp. 321–336, 2003.

[Cast 03] Casteli V., Thomasian A., Li C.-S. "CSVD: Clustering and singular value decomposition for approximate similarity searches in high-dimensional space," *IEEE Transactions on Knowledge and Data Engineering*, Vol. 15(3), pp. 671–685, 2003.

[Chan 93] Chang T., Kuo C.C.J. "Texture analysis and classification with tree structured wavelet transform," *IEEE Transactions on Image Processing*, Vol. 2(4), pp. 429–442, 1993.

[Chu 04] Chu M., Diele F., Plemmons R., Ragni S. "Optimality, computation and interpretation of the nonnegative matrix factorization," available at http://www.wfu.edu/p̃lemmons, 2004.

[Chua 96] Chuang G.C.H., Kuo C.C.J. "Wavelet descriptor of planar curves: Theory and applications," *IEEE Transactions on Image Processing*, Vol. 5(1), pp. 56–71, 1996.

[Coif 92] Coifman R.R., Meyer Y., Wickerhauser M.V. "Wavelet analysis and signal processing," in *Wavelets and Their Applications* (Ruskai M.B. *et al.*, eds.), pp. 153–178, Jones and Barlett, 1992.

[Como 94] Comon P. "Independent component analysis—A new concept?" *Signal Processing*, Vol. 36, pp. 287–314, 1994.

[Corm 01] Cormen T.H., Leiserson C.E., Rivest R.L., Stein C. *Introduction to Algorithms*, Second Edition, MIT Press and McGraw-Hill, 2001.

[Cox 94] Cox T., Cox M. Multidimensional Scaling, Chapmay & Hall, London, 1994.

[Crim 82] Crimmins T.R. "A complete set of Fourier descriptors for two dimensional shapes," *IEEE Transactions on Systems, Man Cybernetics*, Vol. 12(6), pp. 848–855, 1982.

[Daub 90] Daubechies I. *Ten Lectures on Wavelets*, SIAM, Philadelphia, 1991.

[Daug 85] Daugman J.G. "Uncertainty relation for resolution in space, spatial frequency, and orientation optimized by two dimensional visual cortical filters," *Journal of Optical Society of America*, Vol. 2, pp. 1160–1169, 1985.

[Deco 95] Deco G., Obradovic D. "Linear redundancy reduction learning," *Neural Networks*, Vol. 8(5), pp. 751–755, 1995.

[Deer 90] Deerwester S., Dumais S., Furnas G., Landauer T., Harshman R. "Indexing by latent semantic analysis," *Journal of the Society for Information Science*, Vol. 41, pp. 391–407, 1990.

[Desi 03] De Silva V., Tenenbaum J.B. "Global versus local methods in nonlinear dimensionality reduction," in *Advances in Neural Information Processing Systems* Becker S., Thrun S., Obermayer K. (eds.), Vol. 15, pp. 721–728, MIT Press, 2003.

[Diam 96] Diamantaras K.I., Kung S.Y. *Principal Component Neural Networks*, John Wiley Sons, 1996.

[Dono 02] Donoho D.L., Grimes C.E. "When does ISOMAP recover the natural parameterization of families of articulated images?" *Technical Report 2002-27*, Department of Statistics, Stanford University, 2002.

[Dono 04] Donoho D., Stodden V. "When does nonnegative matrix factorization give a correct decomposition into parts?" in *Advances in Neural Information Processing Systems* (Thrun S., Saul L., Schölkopf B., eds.), MIT Press, 2004.

[Doug 00] Douglas S.C., Amari S. "Natural gradient adaptation," in *Unsupervised Adaptive Filtering, Part I: Blind Source Separation* (Haykin S., ed.), pp. 13–61, John Wiley & Sons, 2000.

[Este 77] Esteban D., Galand C. "Application of quadrature mirror filters to split band voice coding schemes," *Proceedings of the IEEE Conference on Acoustics Speech and Signal Procesing*, pp. 191–195, May 1977.

[Fiel 94] Field D.J. "What is the goal of sensory coding?" *Neural Computation*, Vol. 6, pp. 559–601, 1994.

[Flan 72] Flanagan J.L. *Speech Analysis, Synthesis and Perception*, Springer-Verlag, New York, 1972.

[Fuku 90] Fukunaga K. *Introduction to Statistical Pattern Recognition*, 2nd ed., Academic Press, 1990.

[Gabo 46] Gabor D. "Theory of communications," *Journal of the Institute of Elec. Eng.*, Vol. 93, pp. 429–457, 1946.

[Geze 00] Gezerlis V., Theodoridis, S. "An optical music recognition system for the notation of Orthodox Hellenic Byzantine music," *Proceedings of the International Conference on Pattern Recognition (ICPR)*, Barcelona, 2000.

[Geze 02] Gezerlis V., Theodoridis S. "Optical character recognition of the Orthodox Hellenic Byzantine music," *Pattern Recognition*, Vol. 35(4), pp. 895–914, 2002.

[Golu 89] Golub G.H., Van Loan C.F. *Matrix Computations*, Johns Hopkins Press, 1989.

[Grig 02] Grigorescu S.E., Petkov N., Kruizinga P. "Comparison of texture features based on Gabor filters," *IEEE Transactions on Image Processing*, Vol. 11(10), pp. 1160–1167, 2002.

[Hale 95] Haley G., Manjunath B.S. "Rotation-invariant texture classification using modified Gabor filters," *IEEE International Conference on Image Processing*, pp. 262–265, 1995.

[Hale 99] Haley G., Manjunath B.S. "Rotation-invariant texture classification using complete space frequency model," *IEEE Transactions on Image Processing*, Vol. 8(2), pp. 255–269, 1999.

[Ham 04] Ham J., Lee D.D., Mika S., Schölkopf B. "A kernel view of the dimensionality reduction of manifolds," *Proceedings of the 21st International Conference on Machine Learning*, pp. 369–376, Banff, Canada, 2004.

[Hayk 99] Haykin S. *Neural Networks—A Comprehensive Foundation*, 2nd ed., Prentice Hall, 1999.

[Hayk 00] Haykin S. (ed.) *Unsupervised Adaptive Filtering, Part I: Blind Source Separation*, John Wiley & Sons, 2000.

[He 03] He X., Niyogi P. "Locally preserving projections," *Proceedings Advances in Neural Information Processing Systems Conference*, 2003.

[Hote 33] Hotelling H. "Analysis of a complex of statistical variables into principal components," *Journal of Educational Psychology*, Vol. 24, pp. 417–441, 1933.

[Hoy 00] Hoyer P.O., Hyvärien A. "Independent component analysis applied to feature extraction from color and stereo images," *Network: Comput. Neural Systems*, Vol. 11(3), pp. 191–210, 2000.

[Hube 85] Huber P.J. "Projection pursuit," *The Annals of Statistics*, Vol. 13(2), pp. 435–475, 1985.

[Hui 96] Hui Y., Kok C.W., Nguyen T.Q. "Theory and design of shift invariant filter banks," *Proceeding of IEEE TFTS'96*, June 1996.

[Hyva 01] Hyvärien A., Karhunen J., Oja E. *Independent Component Analysis*, Wiley Interscience, 2001.

[Jack 91] Jackson J.E *A User's Guide to Principle Components*, John Wiley & Sons, 1991.

[Jain 89] Jain A.K. *Fundamentals of Digital Image Processing*, Prentice Hall, 1989.

[Jain 91] Jain A.K., Farrokhnia F. "Unsupervised texture segmentation using Gabor filters," *Pattern Recognition*, Vol. 24(12), pp. 1167–1186, 1991.

[Jang 99] Jang G.J., Yun S.J., Hwan Y. "Feature vector transformation using independent component analysis and its application to speaker identification," *Proceedings of Eurospeech*, pp. 767–770, Hungary, 1999.

[Joll 86] Jollife I.T. *Principal Component Analysis*, Springer-Verlag, 1986.

[Jone 87] Jones M.C., Sibson R. "What is projection pursuit?" *Journal of the Royal Statistical Society, Ser. A*, Vol. 150, pp. 1-36, 1987.

[Jutt 91] Jutten C., Herault J. "Blind separation of sources, Part I: An adaptive algorithm based on neuromimetic architecture," *Signal Processing*, Vol. 24, pp. 1-10, 1991.

[Kann 04] Kannan R., Vempala S., Vetta A. "On clustering: good, bad and spectral," *Journal of the ACM*, Vol. 51(3), pp. 497-515, 2004.

[Kapo 96] Kapogiannopoulos G., Papadakis M. "Character recognition using biorthogonal discrete wavelet transform," *Proceedings of the 41st Annual SPIE Meeting*, Vol. 2825, August 1996.

[Karh 46] Karhunen K. "Zur spektraltheorie stochastischer prozesse," *Annales Academiae Scientiarum Fennicae*, Vol. 37, 1946.

[Karh 94] Karhunen J., Joutsensalo J. "Representation and separation of signals using nonlinear PCA type learning," *Neural Networks*, Vol. 7(1), pp. 113-127, 1994.

[Koho 89] Kohonen T. *Self-Organization and Associative Memory*, 3rd ed., Springer-Verlag, 1989.

[Koki 07] Kokiopoulou E., Saad Y. "Orthogonal neighborhood preserving projections: A projection-based dimensionality reduction technique," *IEEE Transactions on Pattern Analysis and Machine Intelligence*, Vol. 29(12), pp. 2143-2156, 2007.

[Kwon 04] Kwon O.W., Lee T.W. "Phoneme recognition using the ICA-based feature extraction and transformation," *Signal Processing*, Vol. 84(6), pp. 1005-1021, 2004.

[Lafo 06] Lafon S., Lee A.B. "Diffusion maps and coarse-graining: A unified framework for dimensionality reduction, graph partitioning and data set parameterization," *IEEE Transactions on Pattern Analysis and Machine Intelligence*, Vol. 28(9), pp. 1393-1403, 2006.

[Lain 93] Laine A., Fan J. "Texture classification by wavelet packet signatures," *IEEE Transactions on Pattern Analysis and Machine Intelligence*, Vol. 15(11), pp. 1186-1191, 1993.

[Lain 96] Laine A., Fan J. "Frame representations for texture segmentation," *IEEE Transcaction on Image Processing*, Vol. 5(5), pp. 771-780, 1996.

[Law 06] Law M.H.C., Jain A.K. "Incremental nonlinear dimensionality reduction by manifold learning," *IEEE Transactions on Pattern Analysis and Machine Intelligence*, Vol. 28(3), pp. 377-391, 2006.

[Lee 98] Lee T.-W. *Independent Component Analysis*, Kluwer Academic Publishers, 1998.

[Lee 01] Lee D.D., Seung S. "Learning the parts of objects by nonnegative matrix factorization," *Nature*, Vol. 401, pp. 788-791, 1999.

[Levi 85] Levine M.D. *Vision in Man and Machine*, McGraw-Hill, 1985.

[Lim 90] Lim J.S. *Two-Dimensional Signal Processing*, Prentice Hall, 1990.

[Lin 08] Lin T., Zha H. "Riemannian manifold learning," *IEEE Transactions on Pattern Analysis and Machine Intelligence*, Vol. 30(5), pp. 796-810, 2008.

[Mall 89] Mallat S. "Multifrequency channel decompositions of images and wavelet models," *IEEE Transactions on Acoustics, Speech, and Signal Processing*, Vol. 37(12), pp. 2091-2110, 1989.

[Mall 97] Mallet Y., Coomans D., Kautsky J., De Vel O. "Classification using adaptive wavelets for feature extraction," *IEEE Transactions for Pattern Analysis and Machine Intelligence*, Vol. 19(10), pp. 1058-1067, 1997.

[Marc 95] Marco S.D., Heller P.N., Weiss J. "An M-band two dimensional translation-invariant wavelet transform and its applivations," *Proceedings of the IEEE Conference on Acoustics Speech and Signal Processing*, pp. 1077–1080, 1995.

[Meye 93] Meyer Y. *Wavelets, Algorithms and Applications*, SIAM, Philadelphia, 1993.

[Mojs 00] Mojsilovic A., Popovic M.V., Rackov D.M. "On the selection of an optimal wavelet basis for texture characterization," *IEEE Transactions Image Processing*, Vol. 9(12), 2000.

[Nach 75] Nachmais J., Weber A. "Discrimination of simple and complex gratings," *Vision Research*, Vol. 15, pp. 217–223, 1975.

[Oja 83] Oja E. *Subspace Methods for Pattern Recognition*, Res. Studies Press, Letchworth, U.K., 1983.

[Paat 91] Paatero P., Tapper U., aalto R., Kulmala M. "Matrix factorization methods for analysis diffusion battery data," *Journal of Aerosol Science*, Vol. 22 (Supplement 1), pp. 273–276, 1991.

[Paat 94] Paatero P., Tapper U. "Positive matrix factor model with optimal utilization of error," *Environmetrics*, Vol. 5, pp. 111–126, 1994.

[Papo 91] Papoulis A. *Probability, Random Variables, and Stochastic Processes*, 3rd ed., McGraw-Hill, 1991.

[Pich 96] Pichler O., Teuner A., Hosticka, B. "A comparison of texture feature extraction using adaptive Gabor filtering, pyramidal and tree structured wavelet transforms," *Pattern Recognition*, Vol. 29(5), pp. 733–742, 1996.

[Pita 92] Pitas I. *Digital Image Processing Algorithms*, Prentice Hall, 1992.

[Prak 97] Prakash M., Murty M.N. "Growing subspace pattern recognition methods and their neural network models," *IEEE Transactions on Neural Networks*, Vol. 8(1), pp. 161–168, 1997.

[Proa 92] Proakis J., Manolakis D. *Digital Signal Processing*, 2nd ed., Macmillan, 1992.

[Pun 03] Pun C.-M., Lee M.-C. "Log-Polar wavelet energy signatures for rotation and scale invariant texture classification," *IEEE Transactions on Pattern Analysis and Machine Intelligence*, Vol. 25(5), pp. 590–603, 2003.

[Qui 07] Qui H., Hancock E.R. "Clustering and embedding using commute times," *IEEE Transactions on Pattern Analysis and Machine Intelligence*, Vol. 29(11), pp. 1873–1890, 2007.

[Rowe 00] S.T. Roweis S.T., Saul L.K. "Nonlinear dimensionality reduction by locally linear embedding," *Science*, Vol. 290, pp. 2323–2326, 2000.

[Saul 01] Saul L.K., Roweis S.T. "An introduction to locally linear embedding," http://www.cs.toronto.edu/~ roweis/lle/papers/lleintro.pdf

[Scho 98] Schölkopf B., Smola A., Muller K.R. "Nonlinear component analysis as a kernel eigenvalue problem," *Neural Computation*, Vol. 10, pp. 1299–1319, 1998.

[Sebr 03] Sebro N., Jaakola T. "Weighted low-rank approximations," *Proceedings of the ICML Conference*, pp. 720–727, 2003.

[Sha 05] Sha F., Saul L.K. "Analysis and extension of spectral methods for nonlinear dimensionality reduction," *Proceedings of the 22nd International Conference on Machine Learning*, Bonn, Germany, 2005.

[Shui 07] Shuicheng Y., Xu D., Zhang B., Zhang H.-J., Yang Q., Lin S. "Graph embedding and extensions: A general framework for dimensionality reduction," *IEEE Transactions on Pattern Analysis and Machine Intelligence*, Vol. 29(1), pp. 40–51, 2007.

[Shaw 04] Shawe-Taylor J., Cristianini N. *Kernel Methods for Pattern Analysis*, Cambridge University Press, Cambridge, MA, 2004.

[Smar 03] Smaragdis P., Brown J.C. "Nonnegative matrix factorization for polyphonic music transcription," *Proceedings IEEE Workshop on Applications of Signal Processing to Audio and Acoustics*, 2003.

[Sra 06] Sra S., Dhillon I.S. "Non-negative matrix approximation: Algorithms and applications," Technical Report TR-06-27, University of Texas at Austin, 2006.

[Stef 93] Steffen P., Heller P.N., Gopinath R.A., Burrus C.S. "Theory of regular M-band wavelet bases," *IEEE Tansactions on Signal Processing*, Vol. 41(12), pp. 3497–3511, 1993.

[Stra 80] Strang G. *Linear Algebra and Its Applications*, 2nd ed., Harcourt Brace Jovanovich, 1980.

[Szu 92] Szu H.H., Telfer B.A., Katambe S. "Neural network adaptive wavelets for signal representation and classification," *Optical Eng.*, Vol. 31, pp. 1907–1916, 1992.

[Szym 06] Szymkowiak-Have A., Girolami M.A., Larsen J. "Clustering via kernel decomposition," *IEEE Transactions on Neural Networks*, Vol. 17(1), pp. 256–264, 2006.

[Sun 06] Sun J., Boyd S., Xiao L., Diaconis P. "The fastest mixing Markov process on a graph and a connection to a maximum variance unfolding problem," *SIAM Review*, Vol. 48(4), pp. 681–699, 2006.

[Tene 00] Tenenbaum J.B., De Silva V., Langford J.C. "A global geometric framework for dimensionality reduction," *Science*, Vol. 290, pp. 2319–2323, 2000.

[Trop 03] Tropp J.A. "Literature survey: Nonnegative matrix factorization," Unpublished note, http://www-personal.umich.edu/j̃tropp/, 2003.

[Turn 86] Turner M.R. "Texture discrimination by Gabor functions," *Biol. Cybern.*, Vol. 55, pp. 71–82, 1986.

[Unse 86] Unser M. "Local linear transforms for texture measurements," *Signal Processing*, Vol. 11(1), pp. 61–79, 1986.

[Unse 95] Unser M. "Texture classification and segmentation using wavelet frames," *IEEE Transactions on Image Processing*, Vol. 4(11), pp. 1549–1560, 1995.

[Unse 89] Unser M., Eden M. "Multiresolution feature extraction and selection for texture segmentation," *IEEE Transations on Pattern Analysis and Machine Intelligence*, Vol. 11(7), pp. 717–728, 1989.

[Vaid 93] Vaidyanathan P.P. *Multirate Systems and Filter Banks*, Prentice Hall, 1993.

[Vett 92] Vetterli M., Herley C. "Wavelets and filter banks: Theory and design," *IEEE Transactions on Signal Processing*, Vol. 40(9), pp. 2207–2232, 1992.

[Vett 95] Vetterli M., Kovacevic J. *Wavelets and Subband Coding*, Prentice Hall, 1995.

[Wata 73] Watanabe S., Pakvasa N. "Subspace method in pattern recognition," *Proceedings of the International Joint Conference on Pattern Recognition*, pp. 25–32, 1973.

[Weld 96] Weldon T., Higgins W., Dunn D. "Efficient Gabor filter design for texture segmentation," *Pattern Recognition*, Vol. 29(2), pp. 2005–2025, 1996.

[Wein 05] Weinberger K.Q., Saul L.K. "Unsupervised learning of image manifolds by semidefinite programming," *Proceedings of the IEEE Conference on Computer Vision and Pattern Recognition*, Vol. 2, pp. 988–995, Washington D.C., USA, 2004.

[Wuns 95] Wuncsh P., Laine A. "Wavelet descriptors for multiresolution recognition of handwritten characters," *Pattern Recognition*, Vol. 28(8), pp. 1237–1249, 1995.

[Xu 03] Xu W., Liu X., Gong Y. "Document clustering based on nonnegative matrix factorization," *Proceedings 26th Annual International ACM SIGIR Conference*, pp. 263-273, ACM Press, 2003.

[Ye 04] Ye J. "Generalized low rank approximation of matrices," *Proceedings of the 21st International Conference on Machine Learning*, pp. 887-894, Banff, Alberta, Canada, 2004.

[Zafe 06] Zafeiriou S., Tefas A., Buciu I., Pitas I. "Exploiting discriminant information in non-negative matrix factorization with application to frontal face verification," *IEEE Transactions on Neural Networks*, Vol. 17(3), pp. 683-695, 2006.

CHAPTER

Feature Generation II

7

7.1 INTRODUCTION

In the previous chapter we dealt with the task of feature generation via linear or nonlinear transformation techniques. This is just one of the possibilities available to the designer. There are a number of alternatives, however, that are very much application dependent. Although similarities among various applications do exist, there are also major differences. We will start by focusing on one major application area; that of *image analysis*. Clearly, we cannot review all techniques that have been suggested and used. Their number is really large. Instead, we will focus on basic directions, with a wide range of applications in mind, such as medical imaging, remote sensing, robot vision, and optical character recognition.

The major goal may be summarized as follows: *given an image, or a region within an image, generate the features that will subsequently be fed to a classifier in order to classify the image in one of the possible classes*. A digital (monochrome) image is usually the result of a discretization process (sampling) of a continuous image function $I(x, y)$ and is stored in the computer as a two-dimensional array $I(m, n)$ with $m = 0, 1, \ldots, N_x - 1$ and $n = 0, 1, \ldots, N_y - 1$. That is, it is stored as an $N_x \times N_y$ array. Every (m, n) element of the array corresponds to a *pixel* (picture element or image element) of the image, whose brightness or intensity is equal to $I(m, n)$. Furthermore, when the intensity $I(m, n)$ is quantized in N_g discrete (gray) levels N_g is known as the depth of the image. Then, the gray-level sequence $I(m, n)$ can take one of the integer values $0, 1, \ldots, N_g - 1$. The depth N_g is usually a power of 2 and can take large values (e.g., 64, 256) when the image is stored in the computer. However, for the human eye it is difficult to discern detailed intensity differences, and in practice $N_g = 32$ or 16 is a sufficient choice for image representation.

The need for feature generation stems from our inability to use the raw data. Even for a small 64×64 image the number of pixels is 4096. For most classification tasks this number is too large, raising computational as well as generalization problems, as discussed in earlier chapters. Feature generation is a procedure that computes new variables that in one way or another originate from the stored values

and μ_4 from left to right are

$$\begin{array}{lrrrrrr} \mu_3: & 587 & 0 & -169 & 169 & 0 & 0 \\ \mu_4: & 16609 & 7365 & 7450 & 7450 & 9774 & 1007 \end{array}$$

Other quantities that result from the first-order histogram are:

Absolute moments:

$$\hat{\mu}_i = E[| I - E[I] |^i] = \sum_{I=0}^{N_g-1} | I - E[I] |^i P(I) \tag{7.4}$$

Entropy:

$$H = -E[\log_2 P(I)] = -\sum_{I=0}^{N_g-1} P(I) \log_2 P(I) \tag{7.5}$$

Entropy is a measure of histogram uniformity. The closer to the uniform distribution ($P(I) = \text{constant}$), the higher the H. For the six images of Figure 7.1 the corresponding values are

$$H: \quad 4.61 \quad 4.89 \quad 4.81 \quad 4.81 \quad 4.96 \quad 4.12$$

Second-Order Statistics Features—Co-occurrence Matrices

The features resulting from the first-order statistics provide information related to the gray-level distribution of the image, but they do not give any information about the relative positions of the various gray levels within the image. Are all low-value gray levels positioned together, or are they interchanged with the high-value ones? This type of information can be extracted from the second-order histograms, where the pixels are considered in pairs. Two more parameters now enter into the scene. These are the relative distance among the pixels and their relative orientation. Let d be the relative distance measured in pixel numbers ($d = 1$ for neighboring pixels, etc.). The orientation ϕ is quantized in four directions: horizontal, diagonal, vertical, and antidiagonal (0°, 45°, 90°, 135°), as shown in Figure 7.2. For each combination of d and ϕ a two-dimensional histogram is defined

$$\begin{aligned} 0°: \; & P\big(I(m,n) = I_1,\ I(m \pm d, n) = I_2\big) \\ &= \frac{\textit{number of pairs of pixels at distance d with values } (I_1, I_2)}{\textit{total number of possible pairs}} \end{aligned} \tag{7.6}$$

In a similar way

$$45° : P\big(I(m,n) = I_1, I(m \pm d, n \mp d) = I_2\big)$$

$$90° : P\big(I(m,n) = I_1, I(m, n \mp d) = I_2\big)$$

$$135° : P\big(I(m,n) = I_1, I(m \pm d, n \pm d) = I_2\big)$$

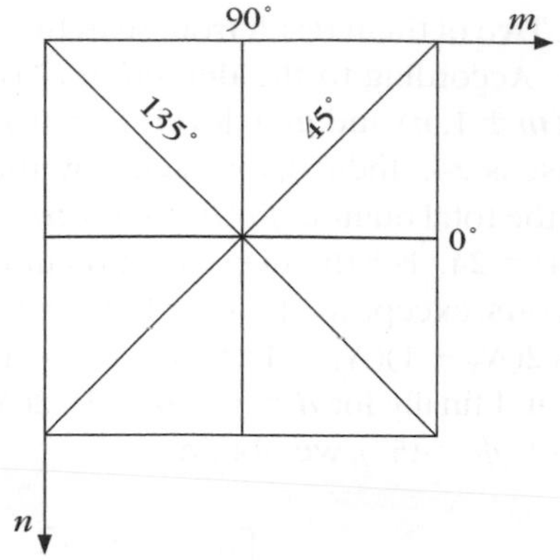

FIGURE 7.2

The four orientations used to construct co-occurrence matrices.

For each of these histograms an array is defined, known as the *co-occurrence or spatial dependence matrix*. Let, for example, an image array $I(m,n)$ be

$$I = \begin{bmatrix} 0 & 0 & 2 & 2 \\ 1 & 1 & 0 & 0 \\ 3 & 2 & 3 & 3 \\ 3 & 2 & 2 & 2 \end{bmatrix} \tag{7.7}$$

which corresponds to a 4×4 image. We have also assumed that $N_g = 4$ ($I(m,n) \in \{0,1,2,3\}$). The co-occurrence matrix for a pair (d,ϕ) is defined as the $N_g \times N_g$ matrix

$$A = \frac{1}{R} \begin{bmatrix} \eta(0,0) & \eta(0,1) & \eta(0,2) & \eta(0,3) \\ \eta(1,0) & \eta(1,1) & \eta(1,2) & \eta(1,3) \\ \eta(2,0) & \eta(2,1) & \eta(2,2) & \eta(2,3) \\ \eta(3,0) & \eta(3,1) & \eta(3,2) & \eta(3,3) \end{bmatrix}$$

where $\eta(I_1,I_2)$ is the number of pixel pairs, at relative position (d,ϕ), which have gray-level values I_1, I_2, respectively. R is the total number of possible pixel pairs. Hence $\frac{1}{R}\eta(I_1,I_2) = P(I_1,I_2)$. For the image of (7.7) and relative pixel position $(1, 0°)$ we have

$$A^0(d=1) = \frac{1}{24} \begin{bmatrix} 4 & 1 & 1 & 0 \\ 1 & 2 & 0 & 0 \\ 1 & 0 & 6 & 3 \\ 0 & 0 & 3 & 2 \end{bmatrix}$$

In words, for each of the intensity pairs, such as $(0,0)$, we count the number of pixel pairs at relative distance $d = 1$ and orientation $\phi = 0°$ that take these values.

For our example this is 4. Two of them result from searching in the positive direction and two in the negative. According to the definition (7.6), these pixel pairs have coordinates (m, n) and $(m \pm 1, n)$ and gray levels $I_1 = 0, I_2 = 0$. The total number of pixel pairs for this case is 24. Indeed, for each row there are $N_x - 1$ pairs and there are N_y rows. Thus, the total number *for both positive and negative directions* is $2(N_x - 1)N_y = 2(3 \times 4) = 24$. For the diagonal direction 45° and $d = 1$ for each row we have $2(N_x - 1)$ pairs, except the first (or last) one, for which no pairs exist. Thus, the total number is $2(N_x - 1)(N_y - 1) = 2(3 \times 3) = 18$. For $d = 1$ and 90° we have $2(N_y - 1)N_x$ pairs, and finally for $d = 1$ and 135° $2(N_x - 1)(N_y - 1)$. For our example image and ($d = 1$, $\phi = 45°$), we obtain

$$A^{45}(d = 1) = \frac{1}{18}\begin{bmatrix} 0 & 1 & 2 & 1 \\ 1 & 0 & 1 & 1 \\ 2 & 1 & 0 & 3 \\ 1 & 1 & 3 & 0 \end{bmatrix}$$

From the definition of the co-occurrence matrix, it is apparent that it is a symmetric one, something that can be used to reduce subsequent computations.

Having defined the probabilities of occurrence of gray levels with respect to relative spatial pixel position, we will go ahead to define the corresponding features. Some of them have a direct physical interpretation with respect to texture, for example, to quantify coarseness, smoothness, and so on. On the other hand, others do not possess such a property, but they still encode texture-related information with high discriminatory power.

- *Angular second moment*

$$ASM = \sum_{i=0}^{N_g-1} \sum_{j=0}^{N_g-1} (P(i,j))^2 \tag{7.8}$$

This feature is a measure of the smoothness of the image. Indeed, if all pixels are of the same gray-level $I = k$, then $P(k, k) = 1$ and $P(i,j) = 0, i \neq k$ or $j \neq k$, and $ASM = 1$. At the other extreme, if we could have all possible pairs of gray levels with equal probability $\frac{1}{R}$, then $ASM = \frac{R}{R^2} = \frac{1}{R}$. The less smooth the region is, the more uniformly distributed $P(i,j)$ and the lower the ASM (Problem 7.5).

- *Contrast*

$$CON = \sum_{n=0}^{N_g-1} n^2 \left\{ \sum_{i=0}^{N_g-1} \sum_{\substack{j=0 \\ |i-j|=n}}^{N_g-1} P(i,j) \right\} \tag{7.9}$$

This is a measure of the image contrast—that is, a measure of local gray-level variations. Indeed, $\sum_i \sum_j P(i,j)$ is the percentage of pixel pairs whose intensity differs by n. The n^2 dependence weighs the big differences more; thus, CON takes high values for images of high contrast.

- *Inverse difference moment*

$$IDF = \sum_{i=0}^{N_g-1} \sum_{j=0}^{N_g-1} \frac{P(i,j)}{1+(i-j)^2} \tag{7.10}$$

This feature takes high values for low-contrast images due to the inverse $(i-j)^2$ dependence.

- *Entropy*

$$H_{xy} = -\sum_{i=0}^{N_g-1} \sum_{j=0}^{N_g-1} P(i,j) \log_2 P(i,j) \tag{7.11}$$

Entropy is a measure of randomness and takes low values for smooth images.

These features are only a few from a larger set that can be derived. In the classical [Hara 73] paper, fourteen of those are summarized. They are repeated in Table 7.1. $P_x(P_y)$ (and related quantities) refer to the statistics with respect to the $x(y)$-axis. All features in the table are functions of the distance d and the orientation ϕ. Thus, if an image is rotated, the values of the features will be different. In practice, for each d the resulting values for the four directions are averaged out. In this way, we make these textural features *rotation tolerant*.

Besides the previous list of features, a number of other statistics-related features have been proposed. For example, in [Tamu 78] textural features are generated with an emphasis on the human visual perception. A set of features is suggested corresponding to texture coarseness, contrast, regularity, and so on. In [Davi 79] features based on a generalized definition of co-occurrence matrices are suggested, which are more appropriate for textures with long scale variations (macrotextures). An extensive treatment of texture is given in [Petr 06].

Example 7.1

Figure 7.3 shows two texture images, one coarse, known as grass [Brod 66], and the other smooth. Table 7.2 summarizes the values of some of the features for both of them.

Features Using Gray-Level Run Lengths

A gray-level *run* is a set of *consecutive* pixels having the *same gray-level value*. The *length of the run* is the number of pixels in the run [Gall 75, Tang 98]. Run length features encode textural information related to the number of times each gray-level,

Table 7.1 Features for Texture Characterization

Angular Second Moment:

$$f_1 = \sum_i \sum_j (P(i,j))^2$$

Contrast:

$$f_2 = \sum_{n=0}^{N_g-1} n^2 \left\{ \sum_i \sum_{\substack{j \\ |i-j|=n}} P(i,j) \right\}$$

Correlation:

$$f_3 = \frac{\{\sum_i \sum_j (ij)P(i,j)\} - \mu_x \mu_y}{\sigma_x \sigma_y}$$

Variance:

$$f_4 = \sum_i \sum_j (i-\mu)^2 P(i,j)$$

Inverse Difference Moment:

$$f_5 = \sum_i \sum_j \frac{P(i,j)}{1+(i-j)^2}$$

Sum (Difference) Average:

$$f_6(\hat{f}_6) = \sum_{i=0}^{2N_x-2\ (N_g-1)} iP_{x+(-)y}(i)$$

Sum Variance:

$$f_7 = \sum_{i=0}^{2N_g-2} (i - f_6)^2 P_{x+y}(i)$$

Sum Entropy:

$$f_8 = -\sum_{i=0}^{2N_g-2} P_{x+y}(i) \log P_{x+y}(i)$$

Entropy:

$$f_9 = -\sum_i \sum_j P(i,j) \log P(i,j) \equiv H_{xy}$$

Difference Variance:

$$f_{10} = \sum_{i=0}^{N_g-1} (i - \hat{f}_6)^2 P_{x-y}(i)$$

Difference Entropy:

$$f_{11} = -\sum_{i=0}^{N_g-1} P_{x-y}(i) \log P_{x-y}(i)$$

Information Measure I:

$$f_{12} = \frac{H_{xy} - H_{xy}^1}{max\{H_x, H_y\}}$$

Information Measure II:

$$f_{13} = \sqrt{1 - \exp(-2(H_{xy}^2 - H_{xy}))}$$

Maximal Correlation Coefficient:

$$f_{14} = (\text{2nd largest eigenvalue of } Q)^{\frac{1}{2}}$$

Definitions: $Q(i,j) = \sum_k \frac{P(i,k)P(j,k)}{P_x(i)P_y(k)}$

$$H_{xy}^1 = -\sum_i \sum_j P(i,j) \log(P_x(i)P_y(j))$$

$$H_{xy}^2 = -\sum_j \sum_i P_x(i)P_y(j) \log(P_x(i)P_y(j))$$

$$P_x(i) = \sum_j P(i,j)$$

$$P_y(j) = \sum_i P(i,j)$$

$$P_{x\pm y}(k) = \sum_i \sum_{j, |i\pm j|=k} P(i,j)$$

$\mu, \mu_x, \mu_y, \sigma_x, \sigma_y; H_x, H_y$

means, st. deviations and entropies.

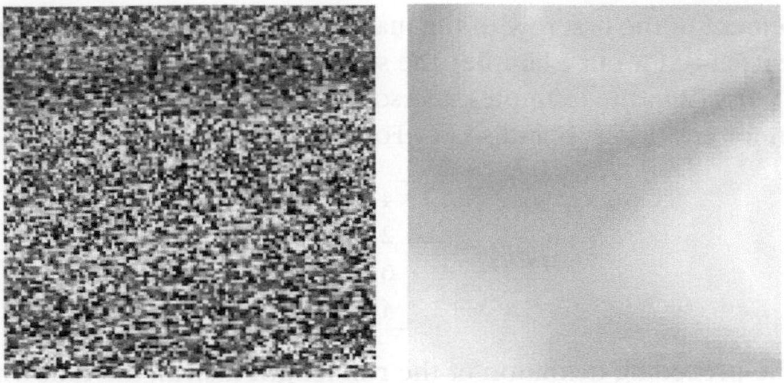

(a) (b)

FIGURE 7.3

Examples of (a) coarse and (b) smooth images.

Table 7.2 Second-Order Histogram Features for the Two Images of Figure 7.3

	Coarse	Smooth
ASM	0.0066	0.0272
CON	989.5	0.613
IDF	0.117	0.783
H_{xy}	8.352	5.884

for example, "1," appears in the image by itself, the number of times it appears in pairs, and so on. Take, for example, the image

$$I = \begin{bmatrix} 0 & 0 & 2 & 2 \\ 1 & 1 & 0 & 0 \\ 3 & 2 & 3 & 3 \\ 3 & 2 & 2 & 2 \end{bmatrix}$$

with four possible levels of gray ($N_g = 4$). For each of the four directions (0°, 45°, 90°, 135°) we define the corresponding run length matrix Q_{RL}. Its (i,j) element gives the number of times a gray-level $i - 1, i = 1, \ldots, N_g$, appears in the image with run length $j, j = 1, 2, \ldots, N_r$. This is an $N_g \times N_r$ array, where N_r is the largest possible run length in the image. For 0° we obtain

$$Q_{RL}(0°) = \begin{bmatrix} 0 & 2 & 0 & 0 \\ 0 & 1 & 0 & 0 \\ 1 & 1 & 1 & 0 \\ 2 & 1 & 0 & 0 \end{bmatrix} \tag{7.12}$$

The first element of the first row of the matrix is the number of times gray-level "0" appears by itself (0 for our example), the second element is the number of times it appears in pairs (2 in the example), and so on. The second row provides the same information for gray-level "1" and so on. For the 45° direction we have

$$Q_{RL}(45°) = \begin{bmatrix} 4 & 0 & 0 & 0 \\ 2 & 0 & 0 & 0 \\ 6 & 0 & 0 & 0 \\ 4 & 0 & 0 & 0 \end{bmatrix} \tag{7.13}$$

Based on the preceding definition of the run length matrix, the following features are defined.

- *Short-run emphasis*

$$SRE = \frac{\sum_{i=1}^{N_g} \sum_{j=1}^{N_r} (Q_{RL}(i,j)/j^2)}{\sum_{i=1}^{N_g} \sum_{j=1}^{N_r} Q_{RL}(i,j)} \tag{7.14}$$

The denominator is the total number of run lengths in the matrix, 9 for (7.12) and 16 for (7.13). This feature emphasizes small run lengths, due to the division by j^2.

- *Long-run emphasis*

$$LRE = \frac{\sum_{i=1}^{N_g} \sum_{j=1}^{N_r} (Q_{RL}(i,j)j^2)}{\sum_{i=1}^{N_g} \sum_{j=1}^{N_r} Q_{RL}(i,j)} \tag{7.15}$$

This gives emphasis to long-run lengths. Thus, we expect *SRE* to be large for coarser and *LRE* to be large for smoother images.

- *Gray-level nonuniformity*

$$GLNU = \frac{\sum_{i=1}^{N_g} \left[\sum_{j=1}^{N_r} Q_{RL}(i,j)\right]^2}{\sum_{i=1}^{N_g} \sum_{j=1}^{N_r} Q_{RL}(i,j)} \tag{7.16}$$

The term in the brackets is the total number of run lengths for each gray-level. Large run length values contribute a great deal because of the square. When runs are uniformly distributed among the gray levels, *GNLU* takes small values.

- *Run length nonuniformity*

$$RLN = \frac{\sum_{j=1}^{N_r} \left[\sum_{i=1}^{N_g} Q_{RL}(i,j)\right]^2}{\sum_{i=1}^{N_g} \sum_{j=1}^{N_r} Q_{RL}(i,j)} \tag{7.17}$$

In a similar way, *RLN* is a measure of run length nonuniformity.

Table 7.3 Run Length Features for the Images of Figure 7.3		
	Coarse	Smooth
SRE	0.932	0.563
LRE	1.349	16.929
GLNU	255.6	71.6
RLN	3108	507
RP	0.906	0.4

- ***Run percentage***

$$RP = \frac{\sum_{i=1}^{N_g}\sum_{j=1}^{N_r} Q_{RL}(i,j)}{L} \tag{7.18}$$

where L is the total possible number of runs in the image, if *all* runs had length equal to one, that is, the total number of pixels. RP takes low values for smooth images.

Example 7.2

For the two images of Figure 7.3 the values of Table 7.3 have resulted.

7.2.2 Local Linear Transforms for Texture Feature Extraction

Second-order statistics features were introduced in order to exploit the spatial dependencies that characterize the texture of an image region. We will now focus on an alternative possibility, which has been used extensively in practice. Let us consider a neighborhood of size $N \times N$ centered at pixel location (m, n). Let $\boldsymbol{x}_{mn}$ be the vector with elements the N^2 points within the area, arranged in a row-by-row mode. A *local linear transform or local feature extractor* is defined as

$$\boldsymbol{y}_{mn} = A^T \boldsymbol{x}_{mn} \equiv \begin{bmatrix} \boldsymbol{a}_1^T \\ \boldsymbol{a}_2^T \\ \vdots \\ \boldsymbol{a}_{N^2}^T \end{bmatrix} \boldsymbol{x}_{mn}, \tag{7.19}$$

The respective correlation matrices are related via the $N^2 \times N^2$ nonsingular transformation matrix A as

$$R_y \equiv E[\boldsymbol{y}_{mn}\boldsymbol{y}_{mn}^T] = A^T R_x A \tag{7.20}$$

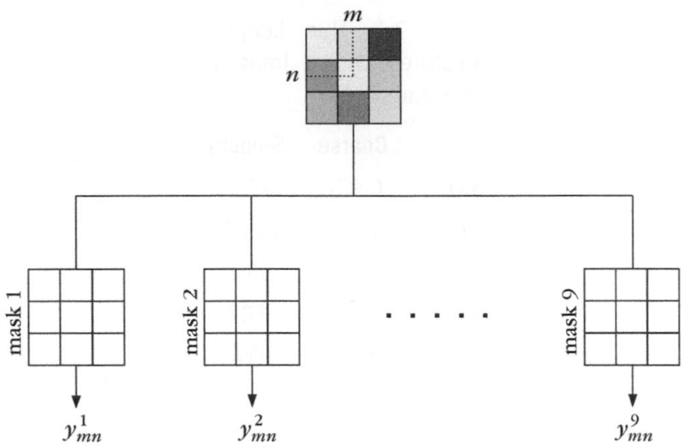

FIGURE 7.4

Filtering the image with each of the masks results in new transformed images/channels.

From these definitions it is readily seen that each element of $\boldsymbol{y}$ contains information about *all* the elements of $\boldsymbol{x}$. This becomes clearer if we look more closely at the way the two correlation matrices are related. Indeed, the diagonal elements of R_y are the respective variances of the elements of $\boldsymbol{y}$. These are first-order statistics quantities, yet their values contain information about the spatial dependencies (second-order statistics) of the original image. *Here lies the essence of the technique. Texture-related spatial dependencies of an image can be accommodated in the first-order statistics of the transformed image*. Using appropriately defined local transform matrices, various aspects of texture properties can be extracted. Of course, the philosophy does not change if instead of transforming vectors we use two-dimensional (separable) transforms of the corresponding subimage region.

One way to look at (7.19) is to interpret it as a series of N^2 filtering operations (convolutions, Appendix D), with a common input vector, $\boldsymbol{x}_{mn}$, that is, the $N \times N$ subimage centered at (m, n). The elements of $\boldsymbol{y}_{mn}$ are the respective filter output samples. This is illustrated in Figure 7.4, where the $N \times N$ subimage ($N = 3$) is filtered through 9 equivalent two-dimensional filters, each characterized by a different coefficient matrix, known as *mask*. In [Laws 80] it is suggested that the corresponding masks be constructed from three basic vectors, namely, $[1, 2, 1]^T, [-1, 0, 1]^T, [-1, 2, -1]^T$, for $N = 3$. The first corresponds to a local averaging operator, the second to an edge detection operator, and the third to a spot detector. These form a complete (nonorthogonal) set of vectors in the $\mathcal{R}^3$ space. The respective nine masks are formed by their cross-products, that is,

$$\begin{bmatrix} 1 & 2 & 1 \\ 2 & 4 & 2 \\ 1 & 2 & 1 \end{bmatrix} \begin{bmatrix} -1 & 0 & 1 \\ -2 & 0 & 2 \\ -1 & 0 & 1 \end{bmatrix} \begin{bmatrix} -1 & 2 & -1 \\ -2 & 4 & -2 \\ -1 & 2 & -1 \end{bmatrix}$$

$$\begin{bmatrix} -1 & -2 & -1 \\ 0 & 0 & 0 \\ 1 & 2 & 1 \end{bmatrix} \begin{bmatrix} 1 & 0 & -1 \\ 0 & 0 & 0 \\ -1 & 0 & 1 \end{bmatrix} \begin{bmatrix} 1 & -2 & 1 \\ 0 & 0 & 0 \\ -1 & 2 & -1 \end{bmatrix}$$

$$\begin{bmatrix} -1 & -2 & -1 \\ 2 & 4 & 2 \\ -1 & -2 & -1 \end{bmatrix} \begin{bmatrix} 1 & 0 & -1 \\ -2 & 0 & 2 \\ 1 & 0 & -1 \end{bmatrix} \begin{bmatrix} 1 & -2 & 1 \\ -2 & 4 & -2 \\ 1 & -2 & 1 \end{bmatrix}$$

Each element of the vector y_{mn} is the result of filtering the local image neighborhood centered at (m, n) with each of the masks. By moving the masks around at the various (m, n) positions, nine different images, *channels*, will be obtained, *each encoding different aspects of the texture of the original image.* First-order-statistics quantities, such as variance and kurtosis, computed from each of these images, can then be used as features for texture classification. Masks larger than 3×3 have also been used. In some cases, an attempt to optimize the masks has been made, so that the resulting variances of the channels for the different classes are as different as possible [Unse 86]. This turns out to be an eigenvalue–eigenvector task, similar to the ones we have already met in Chapter 5. A comparative study of a number of optimal or suboptimal local linear transforms, including orthogonal ones, such as DCT, DST, and Karhunen-Loève, is given in [Unse 86, Unse 89, Rand 99]. Finally, it must be pointed out that all these techniques are closely related to the Gabor filtering approach of the previous chapter.

7.2.3 Moments

Geometric Moments

Let $I(x, y)$ be a continuous image function. Its *geometric moment* of order $p + q$ is defined as

$$m_{pq} = \int_{-\infty}^{\infty}\int_{-\infty}^{\infty} x^p y^q I(x, y)\, dx\, dy \tag{7.21}$$

Geometric moments provide rich information about the image and are popular features for pattern recognition. Their information content stems from the fact that moments provide an equivalent representation of an image, in the sense that an image can be reconstructed from its moments (of all orders) [Papo 91, p. 115]. Thus, each moment coefficient conveys a certain amount of the information residing in an image.

It is by now commonplace to state that a desirable property in pattern recognition is invariance in geometric transformations. Moments, as defined in (7.21), depend on the coordinates of the object of interest within an image; thus, they lack the invariance property. This problem can be circumvented by defining appropriate combinations of normalized versions of the moments. Specifically, our goal will be to define moments that are invariant to:

Translations:

$$x' = x + a, \quad y' = y + b$$

Scaling:

$$x' = \alpha x, \quad y' = \alpha y$$

Rotations:

$$\begin{bmatrix} x' \\ y' \end{bmatrix} = \begin{bmatrix} \cos\theta & \sin\theta \\ -\sin\theta & \cos\theta \end{bmatrix} \begin{bmatrix} x \\ y \end{bmatrix}$$

To this end, let us define

Central moments:

$$\mu_{pq} = \iint I(x,y)(x-\bar{x})^p (y-\bar{y})^q \, dx\, dy \tag{7.22}$$

where

$$\bar{x} = \frac{m_{10}}{m_{00}}, \quad \bar{y} = \frac{m_{01}}{m_{00}}$$

Central moments are invariant to translations.

Normalized central moments:

$$\eta_{pq} = \frac{\mu_{pq}}{\mu_{00}^{\gamma}}, \quad \gamma = \frac{p+q+2}{2} \tag{7.23}$$

These are easily shown to be invariant to both translation and scaling (Problem 7.6).

The Seven Moments of Hu

Hu [Hu 62] has defined a set of seven moments that are invariant under the actions of translation, scaling, and rotation. These are

$p+q=2$

$$\phi_1 = \eta_{20} + \eta_{02}$$

$$\phi_2 = (\eta_{20} - \eta_{02})^2 + 4\eta_{11}^2$$

$p+q=3$

$$\phi_3 = (\eta_{30} - 3\eta_{12})^2 + (\eta_{03} - 3\eta_{21})^2$$

$$\phi_4 = (\eta_{30} + \eta_{12})^2 + (\eta_{03} + \eta_{21})^2$$

$$\begin{aligned}\phi_5 = {} & (\eta_{30} - 3\eta_{12})(\eta_{30} + \eta_{12})[(\eta_{30} + \eta_{12})^2 - 3(\eta_{21} + \eta_{03})^2] \\ & + (\eta_{03} - 3\eta_{21})(\eta_{03} + \eta_{21})[(\eta_{03} + \eta_{21})^2 - 3(\eta_{12} + \eta_{30})^2]\end{aligned}$$

$$\phi_6 = (\eta_{20} - \eta_{02})[(\eta_{30} + \eta_{12})^2 - (\eta_{21} + \eta_{03})^2] + 4\eta_{11}(\eta_{30} + \eta_{12})(\eta_{03} + \eta_{21})$$

$$\phi_7 = (3\eta_{21} - \eta_{03})(\eta_{30} + \eta_{12})[(\eta_{30} + \eta_{12})^2 - 3(\eta_{21} + \eta_{03})^2]$$
$$+ (\eta_{30} - 3\eta_{12})(\eta_{21} + \eta_{03})[(\eta_{03} + \eta_{21})^2 - 3(\eta_{30} + \eta_{12})^2]$$

The first six of these moments are also invariant under the action of reflection, while ϕ_7 changes sign. The values of these quantities can be quite different. In practice, in order to avoid precision problems, the logarithms of their absolute values are usually used as features. A number of other moment-based features that are invariant to more general transformations have also been proposed [Reis 91, Flus 93, Flus 94]. The case of moment invariants in the general l-dimensional space is treated in [Mami 98].

For a digital image $I(i,j)$, with $i = 0, 1, \ldots, N_x - 1, j = 0, 1, \ldots, N_y - 1$, the preceding moments can be *approximated* by replacing integrals by summations,

$$m_{pq} = \sum_i \sum_j I(i,j) i^p j^q \tag{7.24}$$

In order to keep the dynamic range of the moment values consistent for different-sized images, normalization of the $x - y$ axis can be performed, prior to computation of the moments. The moments are then approximated by

$$m_{pq} = \sum_i I(x_i, y_i) x_i^p y_i^q \tag{7.25}$$

where the sum is over all image pixels. Then x_i, y_i are the coordinates of the center point of the ith pixel and are no longer integers but real numbers in the interval $x_i \in [-1, +1]$, $y_i \in [-1, +1]$. *For digital images, the invariance properties of the moments we have defined are only approximately true.* An analysis in [Liao 96] reveals that the approximation error increases with the coarseness of the sampling grid as well as with the order of the moments.

Example 7.3

Figure 7.5 shows the Byzantine music symbol known as "petasti," resulting from a scanner, in scaled and various rotated versions. From left to right in the clockwise sense we have the original version, the scaled, the mirrored, and the rotated by 15°, 90°, and 180° versions, respectively.

Table 7.4 shows the resulting Hu moments for each of the version. The (approximate) invariance of the moments is apparent. Note the minus sign in ϕ_7 for the reflected (mirror) version.

Zernike Moments

The geometric moments defined in (7.21) can also be viewed as projections (Chapter 6) of $I(x, y)$ on the basis functions formed by the monomials $x^p y^q$. These monomials are not orthogonal; thus, the resulting geometric moment features are not optimal from an information redundancy point of view. In this subsection we will derive moments based on alternative complex polynomial functions, known as

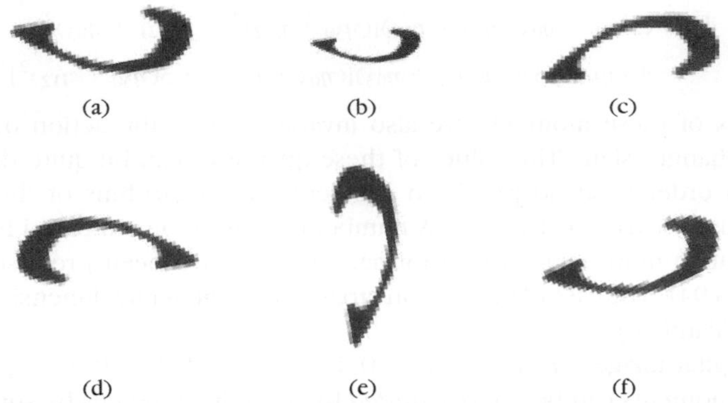

FIGURE 7.5

The Byzantine symbol "petasti" in various scaled and rotated versions, from (a) to (f).

Table 7.4 The Invariant Moments of Hu for the Versions of the "Petasti" Symbol

Moments	0°	Scaled	180°	15°	Mirror	90°
ϕ_1	93.13	91.76	93.13	94.28	93.13	93.13
ϕ_2	58.13	56.60	58.13	58.59	58.13	58.13
ϕ_3	26.70	25.06	26.70	27.00	26.70	26.70
ϕ_4	15.92	14.78	15.92	15.83	15.92	15.92
ϕ_5	3.24	2.80	3.24	3.22	3.24	3.24
ϕ_6	10.70	9.71	10.70	10.57	10.70	10.70
ϕ_7	0.53	0.46	0.53	0.56	−0.53	0.53

Zernike polynomials. These form a *complete orthogonal set over the interior of the unit circle* $x^2 + y^2 \leq 1$ (Problem 7.7) and are defined as

$$V_{pq}(x, y) = V_{pq}(\rho, \theta) = R_{pq}(\rho)\exp(jq\theta)$$

where:

p is a nonnegative integer

q is an integer subject to the constraint $p - |q|$ even, $|q| \leq p$

$\rho = \sqrt{x^2 + y^2}$

$\theta = \tan^{-1}\frac{y}{x}$

$$R_{pq}(\rho) = \sum_{s=0}^{(p-|q|)/2} \frac{(-1)^s [(p-s)!] \rho^{p-2s}}{s! \left(\frac{p+|q|}{2} - s\right)! \left(\frac{p-|q|}{2} - s\right)!}$$

The Zernike moments of a function $I(x, y)$ are given by

$$A_{pq} = \frac{p+1}{\pi} \iint_{x^2+y^2 \leq 1} I(x, y) V^*(\rho, \theta)\, dx\, dy$$

where the $*$ denotes complex conjugation. For a digital image, the respective Zernike moments are computed as

$$A_{pq} = \frac{p+1}{\pi} \sum_i I(x_i, y_i) V^*(\rho_i, \theta_i), x_i^2 + y_i^2 \leq 1$$

where i runs over all the image pixels. The computation of the corresponding moments of an image considers the center of the image as the origin and pixels are mapped into the unit circle, that is, $x_i^2 + y_i^2 \leq 1$. The pixels falling outside the unit circle are not taken into consideration. The magnitude of the Zernike moments is invariant to rotations [Teag 80] (Problem 7.8). Translation and scaling invariance is treated in [Khot 90a, Chon 03]. A drawback of the Zernike moments is the computational complexity associated with the computation of the radial polynomials. A common approach used in reducing complexity includes the application of recurrence relations between successive radial polynomials and coefficients. Computational aspects of the Zernike moments are examined in [Muku 95, Wee 06, Huan 06]. Numerical error issues associated with the computations of the Zernike moments are treated in [Sing 06]. Comparative studies of the performance of the Zernike moments against the moments of Hu, in the context of character recognition, have demonstrated that the former behave better, especially in noisy environments [Khot 90b]. In [Wang 98], Zernike moments are used to cope with both geometry and illumination invariance, in the context of multispectral texture classification. Variants of the Zernike moments, called pseudo-Zernike moments, have also been proposed and used. Comparative studies can be found in [Teh 88, Heyw 95]. Besides Zernike moments, other types of moments have also been suggested and used, such as the Fourier–Mellin moments and moments based on Legendre polynomials, as in [Kan 02, Chon 04, Muku 98].

7.2.4 Parametric Models

So far, in various parts of the book, we have treated the gray levels as random variables and looked at aspects of their first- and second-order statistics. In this subsection, their randomness will be approached from a different perspective. We will assume that $I(m, n)$ is a real *nondiscrete* random variable, and we will try *to model* its underlying generation mechanism by adopting an appropriate *parametric model*. The parameters of the resulting models encode useful information and lend themselves as powerful feature candidates for a number of pattern recognition tasks.

We will move in two directions. One is to treat an image as a successive sequence of rows or columns. That is, our random variables will be considered as successive realization samples from a one-dimensional random process $I(n)$. The alternative looks at the image as a two-dimensional random process $I(m, n)$, also known as *random field*.

One-Dimensional Parametric Models

Let $I(n)$ denote the random sequence. We will assume that it is stationary in the wide sense. This means that its autocorrelation sequence $r(k)$ exists and is of the form

$$r(k) = E[I(n)I(n-k)]$$

and the Fourier transform of $r(k)$ also exists and is a *positive* function (power spectral density)

$$I(\omega) = \sum_{k=-\infty}^{+\infty} r(k)\exp(-j\omega k)$$

Under certain assumptions, which are met in practice most of the time [Papo 91, Theo 93], it can be shown that such a random sequence can be generated at the output of a linear, causal, stable, time-invariant system with impulse response $h(n)$, whose input is excited by a white noise sequence, as shown in Figure 7.6. In simple terms, this means that we can write

$$I(n) = \sum_{k=0}^{\infty} h(k)\eta(n-k)$$

where $h(n)$ satisfies the stability condition $\sum_n |h(n)| < \infty$. The sequence $\eta(n)$ is a white noise sequence, that is, $E[\eta(n)] = 0$ and $E[\eta(n)\eta(n-l)] = \sigma^2\delta(l) : \delta(l) = 1$ for $l = 0$ and zero otherwise. Such processes of a special type are the so-called *autoregressive processes (AR)*, which are generated by systems of the form

$$I(n) = \sum_{k=1}^{p} a(k)I(n-k) + \eta(n) \tag{7.26}$$

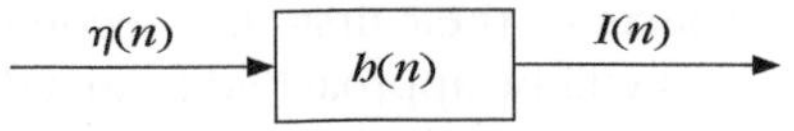

FIGURE 7.6

Generation model of a stationary random process at the output of a stable, linear, time-invariant system excited by a white noise sequence.

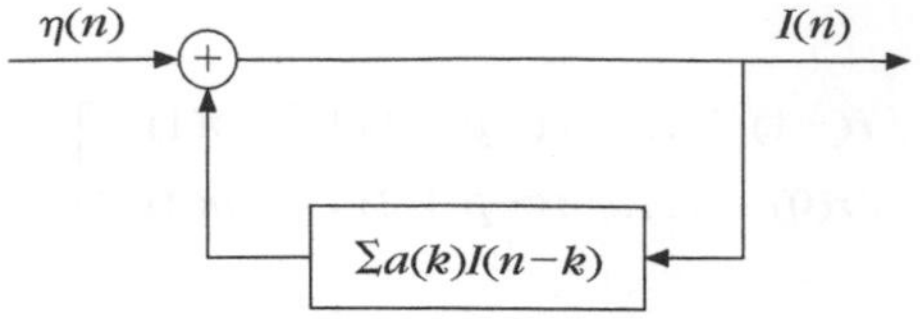

FIGURE 7.7

Generation model of an AR stationary random process.

In words, the random sequence $I(n)$ is given as a linear combination of previous samples $I(n-k)$ and the current input sample $\eta(n)$ (Figure 7.7). Here p is the order of the AR model, and we write AR(p). The coefficients $a(k)$, $k = 1, 2, \ldots, p$ are the AR model parameters. AR models are a special case of a more general class of models, known as *autoregressive-moving average (ARMA(p,m))*, for which

$$I(n) = \sum_{k=1}^{p} a(k)I(n-k) + \sum_{l=0}^{m} b(l)\eta(n-l) \tag{7.27}$$

That is, the model is regressive with respect to both input and output sequences. The major advantage of the AR models, compared with their ARMA relatives, is that the former lead to *linear* systems of equations for the estimation of the model parameters.

Estimation of the AR Parameters

Another way to look at (7.26) is to interpret the coefficients $a(k), k = 1, \ldots, p$, as the predictor parameters of the sequence $I(n)$. That is, the parameters weigh previous samples, $I(n-1), \ldots, I(n-p)$, in order to predict the value of the current sample $I(n)$, and $\eta(n)$ is the prediction error,

$$\hat{I}_n = \sum_{k=1}^{p} a(k)I(n-k) \equiv \boldsymbol{a}_p^T \boldsymbol{I}_p(n-1) \tag{7.28}$$

where $\boldsymbol{I}_p^T(n-1) \equiv [I(n-1), \ldots, I(n-p)]$. The unknown parameter vector $\boldsymbol{a}_p^T = [a(1), a(2), \ldots, a(p)]$ is optimally estimated, for example, by minimizing the mean square prediction error,

$$E[\eta^2(n)] = E[(I(n) - \hat{I}(n))^2] = E[(I(n) - \boldsymbol{a}_p^T \boldsymbol{I}_p(n-1))^2] \tag{7.29}$$

The problem is exactly the same as that of the mean square linear classifier estimation of Chapter 3, and the unknown parameters result from the solution of

$$E[\boldsymbol{I}_p(n-1)\boldsymbol{I}_p^T(n-1)]\boldsymbol{a}_p = E[I(n)\boldsymbol{I}_p(n-1)] \tag{7.30}$$

or

$$\begin{bmatrix} r(0) & r(-1) & \dots & r(-p+1) \\ r(1) & r(0) & \dots & r(-p+2) \\ \vdots & \vdots & \vdots & \vdots \\ r(p-2) & r(p-3) & \dots & r(-1) \\ r(p-1) & r(p-2) & \dots & r(0) \end{bmatrix} \begin{bmatrix} a(1) \\ a(2) \\ \vdots \\ a(p-1) \\ a(p) \end{bmatrix} = \begin{bmatrix} r(1) \\ r(2) \\ \vdots \\ r(p-1) \\ r(p) \end{bmatrix}$$

or

$$R\boldsymbol{a}_p = \boldsymbol{r}_p \tag{7.31}$$

with $\boldsymbol{r}_p \equiv [r(1), \dots, r(p)]^T$. The relation of the optimal parameters $a(k)$ with the mean square error (variance of generating noise) is obtained from (7.29) and (7.31) and is given by

$$\sigma_\eta^2 = E[\eta^2(n)] = r(0) - \sum_{k=1}^{p} a(k)r(k) \tag{7.32}$$

The autocorrelation matrix has a computationally rich structure. It is symmetric ($r(k) = r(-k)$) and Toeplitz—that is, all the elements across any of its diagonals are the same. Exploitation of these properties leads to the development of a computationally efficient scheme for the solution of (7.31). This is *Levinson's algorithm*, which solves the linear system of equations in $O(p^2)$ multiplications and additions, as opposed to $O(p^3)$ required by more classical algorithmic schemes [Theo 93, Hayk 96]. In Chapter 3, we saw that when the autocorrelation sequence is not known, it is often preferable to adopt the least sum of squares instead of the mean square criterion. Then the AR parameters are still provided by a linear system of equations, but the associated matrix is no longer Toeplitz. However, it is still computationally rich, and Levinson-type $O(p^2)$ algorithms for the efficient solution of such systems have also been derived [Theo 93].

Besides images, AR (ARMA) models have been used extensively to model other type of random sequences, such as those resulting from digitizing speech signals and electroencephalographic signals. *For all these cases the resulting AR parameters can be used as features to classify one type of signal from another.*

Example 7.4

Let the AR random sequence of order $p = 2$ be

$$I(n) = \sum_{k=1}^{2} a(k)I(n-k) + \eta(k)$$

with $r(0) = 1$, $r(1) = 0.5$, $r(2) = 0.85$. Computing the mean square estimates of $a(k)$, $k = 1, 2$, we obtain

$$\begin{bmatrix} 1 & 0.5 \\ 0.5 & 1 \end{bmatrix} \begin{bmatrix} a(1) \\ a(2) \end{bmatrix} = \begin{bmatrix} 0.5 \\ 0.85 \end{bmatrix}$$

and its solution gives $a(1) = 0.1$, $a(2) = 0.8$.

Two-Dimensional AR Models

A two-dimensional AR random sequence $I(m,n)$ is defined as

$$\hat{I}(m,n) = \sum_k \sum_l a(k,l) I(m-k, n-l), (k,l) \in W \tag{7.33}$$

$$I(m,n) = \hat{I}(m,n) + \eta(m,n) \tag{7.34}$$

Figure 7.8 shows the region W of the pixels that contribute to the prediction of $\hat{I}(m,n)$, for a number of possible choices. The case in Figure 7.8a corresponds to what is known as a *strongly causal predictor model*. This is because all pixels in the contributing area have coordinates smaller than the coordinates m, n of the predicted pixel, which is represented by the unshaded node in the figure. The corresponding window is $W_1 = \{0 \le k \le p, 0 \le l \le q, (k,l) \ne (0,0)\}$. However, the notions of past and present have no real meaning for an image, and alternative windows can also be used. A *noncausal predictor* is defined as

$$I(m,n) = \sum_{k=-p}^{p} \sum_{l=-q}^{q} a(k,l) I(m-k, n-l) + \eta(m,n)$$

In Figure 7.8d the corresponding window is shown for the case of $p = q = 2$. Figure 7.8c shows a third possibility, which is known as a *semicausal predictor*, and Figure 7.8b shows the case of a *causal* predictor. Next we summarize the last three cases, which are the most common in practice:

$$\begin{aligned} \textit{Causal}: W_2 &= \{(-p \le k \le p, 1 \le l \le q) \cup (1 \le k \le p, l = 0)\} \\ \textit{Semicausal}: W_3 &= \{-p \le k \le p, 0 \le l \le q\,,\ (k,l) \ne (0,0)\} \\ \textit{Noncausal}: W_4 &= \{-p \le k \le p, -q \le l \le q\,,\ (k,l) \ne (0,0)\} \end{aligned}$$

AR Parameter Estimation

We have

$$\hat{I}(m,n) = \sum_k \sum_l a(k,l) I(m-k, n-l)$$

Recalling the orthogonality condition from Chapter 3, in its two-dimensional generalization, we obtain that the minimum mean square error solution satisfies

$$E\left[I(m-i, n-j) \left(I(m,n) - \sum_k \sum_l a(k,l) I(m-k, n-l) \right) \right] = 0, \quad (i,j) \in W \tag{7.35}$$

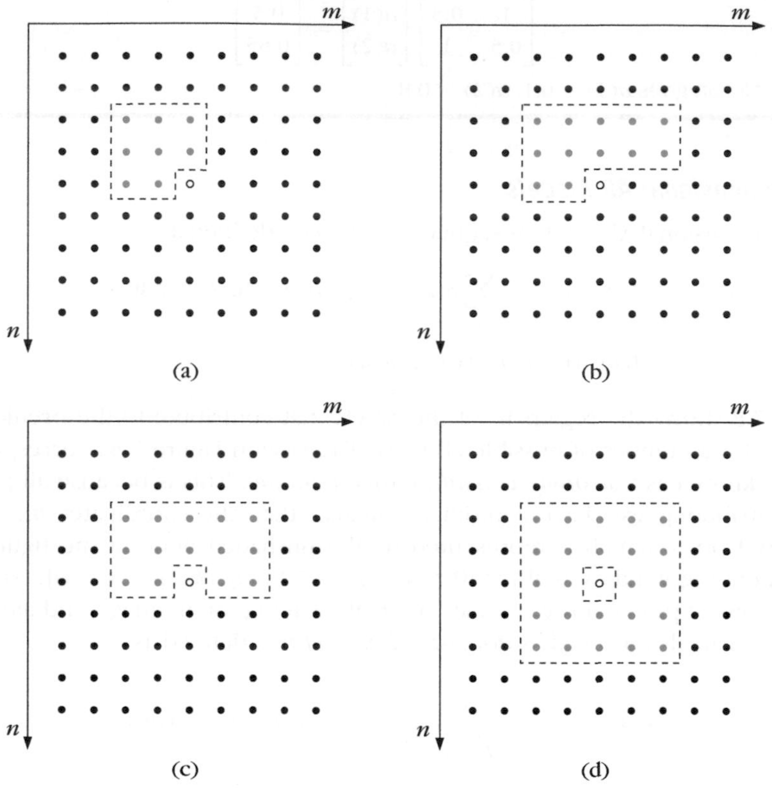

FIGURE 7.8

Different types of two-dimensional prediction models. The predicted pixel is represented by the unshaded node. The red pixels are those that take part in the prediction and the corresponding window W is the area enclosed by the dotted line. (a) Strictly causal, (b) causal, (c) semicausal, and (d) noncausal.

or

$$r(i,j) = \sum_k \sum_l a(k,l) r(i-k, j-l), \quad (i,j) \in W \tag{7.36}$$

where $r(i,j) \equiv E[I(m,n)I(m-i, n-j)]$ is the two-dimensional autocorrelation sequence of the random field $I(m,n)$. The set of equations in (7.36) constitutes a linear system of equations leading to the estimates of $a(k,l)$. The associated matrix also has a computationally rich structure, which can be exploited to develop efficient schemes to compute the solution. Let us take, for example, the noncausal window for $p = q$. This is a symmetric window, in the sense that for each index pair (i,j), the $(-i,-j)$ is also present. Combining (7.36) with the equation of the variance of the minimum error, which is given by (Problem 7.11)

$$\sigma_\eta^2 = r(0,0) - \sum_k \sum_l a(k,l) r(k,l) \tag{7.37}$$

the following system results:

$$Ra = -\begin{bmatrix} \mathbf{0} \\ \sigma_n^2 \\ \mathbf{0} \end{bmatrix} \tag{7.38}$$

where $\mathbf{0}$ is the zero vector of appropriate dimension and

$$\boldsymbol{a}^T = [a(p,p), \ldots, a(p,-p), \ldots, a(0,0), \ldots, a(-p,p), \ldots, a(-p,-p)]$$

where $a(0,0) \equiv 1$ and R is the corresponding autocorrelation matrix. The dimension of $\boldsymbol{a}$ is $(2p+1)^2$. The correlation of a homogeneous (i.e., $E[I(m,n)I(m-k,n-l)] = r(k,l)$) and isotropic (no direction dependence) image depends only on the relative distance between pixels,

$$r(k,l) = r\left(\sqrt{k^2+l^2}\right)$$

and the resulting autocorrelation matrix is easily shown to be *symmetric and block Toeplitz with each block being itself a Toeplitz matrix*,

$$R = \begin{bmatrix} R_0 & R_1 & \cdots & R_{2p} \\ R_1 & R_0 & \cdots & R_{2p-1} \\ \vdots & \vdots & \vdots & \vdots \\ R_{2p} & R_{2p-1} & \cdots & R_0 \end{bmatrix} \tag{7.39}$$

where

$$R_i = \begin{bmatrix} r(i,0) & \cdots & r(i,2p) \\ \vdots & \vdots & \vdots \\ r(i,2p) & \cdots & r(i,0) \end{bmatrix} \tag{7.40}$$

For homogeneous images and symmetric windows it is easy to show that the AR parameters are symmetric $a(k,l) = a(-k,-l)$ and the system can be solved efficiently by a Levinson-type algorithm [Kalo 89]. If the image is homogeneous but anisotropic, the resulting system's associated matrix is block Toeplitz, but the elements are no longer Toeplitz. Furthermore, more general windows than the ones introduced in this section have also been suggested and used. Efficient Levinson-type algorithms for such cases have also been developed (e.g., [Glen 94]). Finally, besides the squared error criteria, maximum likelihood techniques can be employed for the estimation of the unknown parameters, which can lead to more accurate estimates. Of course, in such cases assumptions about the underlying statistics have to be adopted (e.g., [Kash 82]).

Remarks

- The AR modeling of images has been used in the classification context in a number of cases [Chel 85, Cros 83, Kash 82, Sark 97]. In [Kash 86, Mao 92] extensions have been proposed for rotation-invariant models.
- The AR random field models are related to a class of models known as *Markov random fields*. The essence of these fields is that for each pixel (m, n) the image is divided into three areas: Ω^+ ("future"), Ω ("present"), and Ω^- ("past"). It is then assumed that the random variable $I(m, n), (m, n) \in \Omega^+$, is independent of its values in Ω^- and depends only on the values in Ω; thus, the conditional density function satisfies
$$p\left(I(m,n),(m,n) \in \Omega^+ | I(m,n),(m,n) \in \Omega^- \cup \Omega\right)$$
$$= p\left(I(m,n),(m,n) \in \Omega^+ | I(m,n),(m,n) \in \Omega\right)$$
In words, the "future" depends only on the "present" and not on the "past"; that is, the value of the random variable at a pixel depends on the values that the random variable takes in a specific (neighboring) area only, and it does not depend on the values in the remaining regions of the image.
- It can be shown that every Gaussian AR model is a Markov random field. [Wood 72, Chel 85].

Example 7.5

For an image whose autocorrelation sequence obeys

$$r(k, l) = 0.8^{\sqrt{k^2+l^2}}$$

estimate the AR parameters for a noncausal $p = q = 1$ window.

From the definition we have

$$\begin{aligned}\hat{I}(m,n) &= a(1,1)I(m-1,n-1) + a(1,0)I(m-1,n)\\ &+ a(1,-1)I(m-1,n+1) + a(0,1)I(m,n-1)\\ &+ a(0,-1)I(m,n+1) + a(-1,1)I(m+1,n-1)\\ &+ a(-1,0)I(m+1,n) + a(-1,-1)I(m+1,n+1)\end{aligned}$$

The resulting matrix R is a block $(2p+1) \times (2p+1) = 3 \times 3$ matrix with elements the 3×3 matrices

$$R = \begin{bmatrix} R_0 & R_1 & R_2 \\ R_1 & R_0 & R_1 \\ R_2 & R_1 & R_0 \end{bmatrix}$$

where

$$R_0 = \begin{bmatrix} r(0,0) & r(0,1) & r(0,2) \\ r(0,1) & r(0,0) & r(0,1) \\ r(0,2) & r(0,1) & r(0,0) \end{bmatrix}$$

$$R_1 = \begin{bmatrix} r(1,0) & r(1,1) & r(1,2) \\ r(1,1) & r(1,0) & r(1,1) \\ r(1,2) & r(1,1) & r(1,0) \end{bmatrix}$$

$$R_2 = \begin{bmatrix} r(2,0) & r(2,1) & r(2,2) \\ r(2,1) & r(2,0) & r(2,1) \\ r(2,2) & r(2,1) & r(2,0) \end{bmatrix}$$

For this specific model the linear system in (7.38) has nine unknowns and the solution gives

$$a(1,1) = a(-1,-1) = -0.011, \quad a(1,0) = a(-1,0) = -0.25$$
$$a(1,-1) = a(-1,1) = -0.011, \quad a(0,1) = a(0,-1) = -0.25$$
$$\sigma_\eta^2 = 0.17$$

7.3 FEATURES FOR SHAPE AND SIZE CHARACTERIZATION

In a number of image analysis applications, an important piece of information is the shape and size of an object of interest within the image. For example, in medical applications the shape and size of nodules are crucial in classifying them as malignant, or benign. Nodules with an irregular boundary have a high probability of being malignant, and those with a more regular boundary are usually benign. Also, it has been observed that in certain cases nodules with a perimeter of more than 3 cm are usually malignant [Cavo 92].

Another example in which the shape of the object is of major importance is the automatic character recognition in an *optical character recognition (OCR)* system [Mori 92, Plam 00, Vinc 02]. Although OCR systems employing our already familiar regional features, there is a large class of techniques that use the shape information residing in the *boundary curve* of the characters.

Figure 7.9a shows the character "5" as seen from the scanner of an OCR system. An appropriate image segmentation algorithm (e.g., [Pita 94]) has first been applied to separate the character from the rest of the image. The character in Figure 7.9b is in binary form. This is the result of the binarization phase, in which all gray

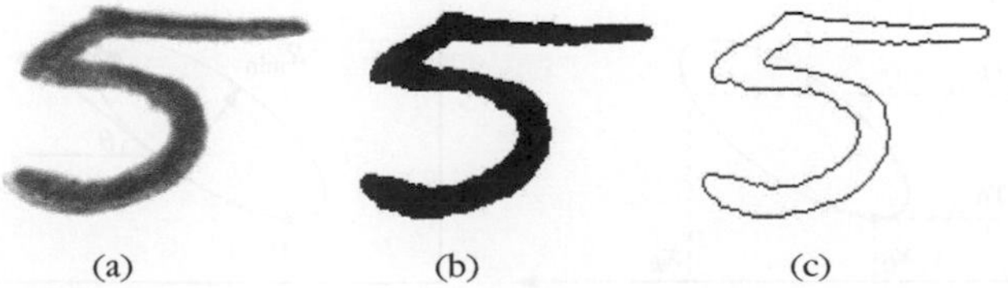

FIGURE 7.9

The character "5" after (a) the segmentation of the scanned image and then (b) the application of a binarization algorithm and (c) its boundary after the application of a boundary extraction algorithm in the binarized version.

levels of the character region below a certain threshold become 0 and all above it become 1 [Trie 95]. Figure 7.9c shows the resulting boundary, after the application of a boundary extraction algorithm (e.g., [Pita 94]) on the binary version. Thus, in the last version there is no texture of interest inside the character. What is of paramount importance in such systems is feature invariance in geometric transformations. The recognition of the character must be insensitive to its position, size, and orientation. A review of various methodologies for invariant pattern recognition techniques can be found in [Wood 96].

The shape characterization of a region or an object can be achieved in various ways. Two are the major directions along which we will proceed. One is to develop techniques that provide a full description of the boundary of the object in a regenerative manner. In words, the boundary can be reobtained from the description coefficients, such as by using a Fourier expansion of the boundary, which in turn can be reconstructed from its Fourier coefficients. The other direction is to use features that are descriptive of the characteristics of the shape of the region but are not regenerative. Examples of such features are the number of corners in the boundary and the perimeter. They provide useful information about the boundary, but they are not sufficient to reproduce it. In the following we will focus on some basic techniques, which have in turn given birth to a large number of variants shaped to fit specific application requirements (for example, see [Trie 96] for a review).

7.3.1 Fourier Features

Let (x_k, y_k), $k = 0, 1, \ldots, N-1$, be the coordinates of N samples on the boundary of an image region, Figure 7.10a. For each pair (x_k, y_k) we define the complex variable

$$u_k = x_k + jy_k$$

For the N u_k points we obtain the DFT f_l

$$f_l = \sum_{k=0}^{N-1} u_k \exp\left(-j\frac{2\pi}{N}lk\right), \quad l = 0, 1, \ldots, N-1$$

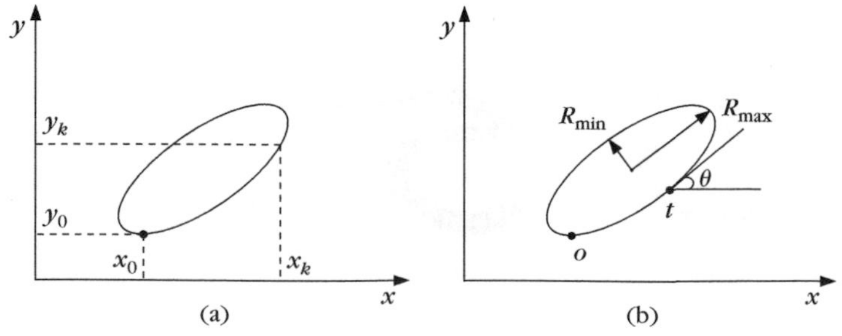

FIGURE 7.10

Boundary of an image region (a) and associated parameters (b).

The coefficients f_l are also known as the *Fourier descriptors* of the boundary. Once the f_l are available, the u_k can be recovered and the boundary can be reconstructed. However, our goal in pattern recognition is not to reconstruct the boundary. Thus, a smaller number of coefficients (or descriptors) is usually used, enough to include sufficient discriminatory information. In the sequel, we will investigate how the Fourier descriptors are affected by the actions of translation, rotation, and scaling. For translation we have

$$x'_k = x_k + \Delta x$$

$$y'_k = y_k + \Delta y$$

Then

$$u'_k = u_k + (\Delta x + j\Delta y) \equiv u_k + \Delta u'$$

For rotation, it is not difficult to verify that in rotating all points of the region by θ, with respect to the origin, the rotated coordinates correspond to (Problem 7.13)

$$u'_k = u_k \exp(j\theta)$$

If f_l, f'_l are the DFTs of u_k, u'_k, respectively, then from the DFT definition we get

$$\textit{Translation}: u'_k = u_k + \Delta u' \Longrightarrow f'_l = f_l + \Delta u' \delta(l)$$

$$\textit{Rotation}: u'_k = u_k \exp(j\theta) \Longrightarrow f'_l = f_l \exp(j\theta)$$

$$\textit{Scaling}: u'_k = a u_k \Longrightarrow f'_l = a f_l$$

$$\textit{Translation of the sampling origin}: u'_k = u_{k-k_0} \Longrightarrow$$

$$f'_l = f_l \exp\left(-j2\pi k_0 \frac{l}{N}\right)$$

In words, translation affects only the f'_0 coefficient. Rotation affects *the phase* of all the coefficients by the *same factor*, and it has *no effect on their magnitude*. Scaling affects all coefficients in the same way, and thus it has no effect on the ratios $\frac{f_i}{f_j}$. The sampling point origin, within the boundary, affects the phase but leaves invariant the magnitude $|f_l|$.

This *deterministic* manner, in which the three geometric transformations affect the Fourier coefficients, allows the development of appropriate normalized versions that are invariant to these actions [Crim 82, Arbt 90, Gran 72]. Let us demonstrate the rationale of such approaches via an example, by considering the boundary of an object. The first decision to be taken, prior to the computation of the Fourier coefficients, is to define the first sampling point (x_0, y_0) on the boundary. In practice, the choice of this point for each character has a degree of randomness. The choice of a different sampling origin corresponds to a relative translation of, say, $k_0 < N$ samples (since the boundary is a closed curve, the relative translation will always be ($k - k_0$) modulo $N < N$). As we have seen earlier, this affects the Fourier descriptors

$$u_k' = u_{k-k_0} \Longrightarrow f_l' = f_l \exp\left(-j2\pi k_0 \frac{l}{N}\right) \tag{7.41}$$

hence

$$f_1' = f_1 \exp\left(-j2\pi \frac{k_0}{N}\right) \Longrightarrow f_1' = |f_1| \exp(-j\phi_1) \exp\left(-j2\pi \frac{k_0}{N}\right)$$

where $|f_1|, \phi_1$ are the magnitude and phase of f_1, respectively. Hence, the phase of f_1' is $\phi_1' = \phi_1 + 2\pi\frac{k_0}{N}$. In the sequel we define the following normalized Fourier coefficients:

$$\hat{f}_l = f_l \exp(jl\phi_1) \tag{7.42}$$

The corresponding normalized coefficient with shifted origin will be

$$\hat{f}'_l = f_l' \exp(jl\phi'_1) = f_l' \exp\left(jl\phi_1 + j2\pi k_0 \frac{l}{N}\right) \tag{7.43}$$

Taking into account (7.41), we obtain

$$\hat{f}'_l = \hat{f}_l$$

Thus, the preceding normalization generates features that are *invariant to the choice of the sampling origin* (x_0, y_0).

This method of exploiting the power of the Fourier transform as a tool for boundary description is not the only possibility. An alternative is to express the coordinates of the boundary contour points as functions of the boundary length t, measured from an origin within the boundary, that is, $(x(t), y(t))$. Since the boundary is a closed curve, these are periodic functions and they can be expanded in their Fourier series. Invariant versions of the Fourier coefficients can then be computed and used as features for pattern recognition [Kuhl 82, Lin 87]. Comparative performance studies of a number of invariant Fourier-based features, in the context of handwritten character recognition, can be found in [Pers 77, Taxt 90].

Another way is to generate Fourier descriptors from the curvature $k(t)$ function of the boundary, defined as

$$k(t) = \frac{d\theta(t)}{dt}$$

where $\theta(t)$ is the tangent angle (Figure 7.10b) at a point a distance t from the origin, which is marked "o" in the figure. Such a description is justified by Gauss's theorem, stating that every curvature function corresponds to one and only one curve in space (with the exception of its position in space). The advantage of such a description stems from its obvious scale invariance property. If we measure the length of the boundary at a point by the number of pixels n

between this point and the origin of the curve, the curvature of the boundary is approximated by

$$\theta_n = \tan^{-1}\frac{y_{n+1} - y_n}{x_{n+1} - x_n}, \quad n = 0, 1, \ldots, N-1$$
$$k_n = \theta_{n+1} - \theta_n, \quad n = 0, 1, \ldots, N-1 \tag{7.44}$$

In the previous chapter we have seen that an alternative to Fourier descriptors is to use wavelet coefficients. However, as we pointed out there, the definition of invariant wavelet descriptors is not a straightforward task, and invariance is attempted via indirect methods.

7.3.2 Chain Codes

Chain coding is among the most widely used techniques for boundary shape description. In [Free 61], the boundary curve is approximated via a sequence of connected straight line segments of preselected direction and length. Every line segment is coded with a specific coding number depending on its direction.

In Figure 7.11 two possible choices, usually encountered in practice, are shown. In this way a *chain code* $[d_i]$ is created, where d_i is the coding number of the direction of the line segment that connects boundary pixel (x_i, y_i) with the next one (x_{i+1}, y_{i+1}), sweeping the boundary in, say, the clockwise sense. A disadvantage of this description is that the resulting chain codes are usually long and at the same time are very sensitive in the presence of noise. This leads to chain codes with variations due to noise and not necessarily to the boundary curve. A way out is to resample the boundary curve by selecting a grid of larger dimensions. For each of the boxes of the grid all points inside a box are assigned the value of the respective box center. In Figure 7.12a the original samples are shown alongside the larger sampling grid. Figure 7.12b is the resulting resampled version. The chain

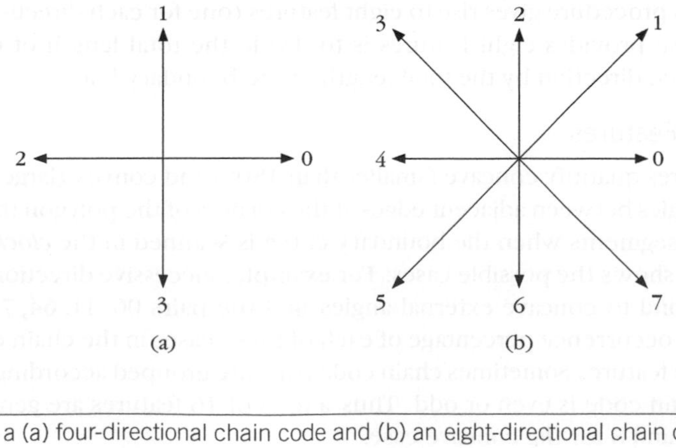

FIGURE 7.11

Directions for a (a) four-directional chain code and (b) an eight-directional chain code.

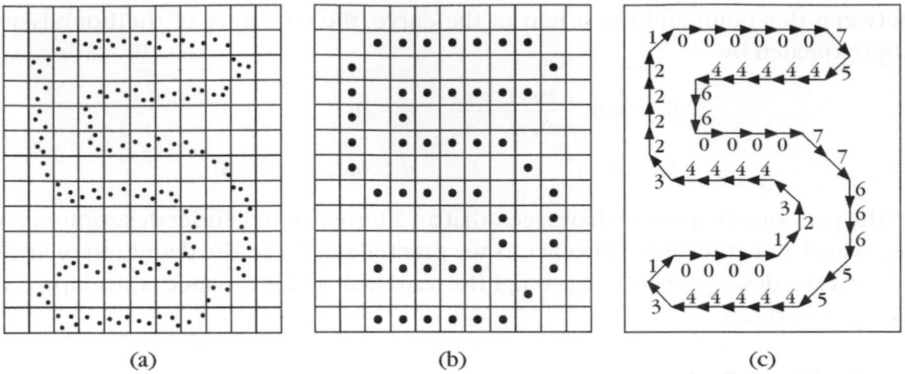

FIGURE 7.12

The character "5" and (a) its original sampled image, (b) its resampled version on a coarser grid, and (c) the resulting chain code.

code is formed from the sequential connection of these pixels, Figure 7.12c. If we consider the length of the grid side as the basic measurement unit, then for even-coded directions, 0, 2, 4, 6, the length of the corresponding straight line segment is 1, and for the odd-coded directions, 1, 3, 5, 7, it is $\sqrt{2}$ (from the Pythagoras theorem). For the case of Figure 7.12b and for a coding with eight possible directions, the resulting chain code is shown in the Figure 7.12c. This sequence of numbers constitutes the spine on which a number of shape-related features are built. Two possibilities, for example, are the following [Lai 81, Mahm 94].

Direction and Direction Length Features

For each direction we count the number of times a specific chain code number appears in the chain. Then this number is divided by the total number of chain codes that appear in the chain description of the boundary. For an eight-code scheme this procedure gives rise to eight features (one for each direction). Another way that also provides eight features is to divide the total length of the line segments in each direction by the total length of the boundary line.

Curvature Features

These features quantify concave (smaller than 180°) and convex (larger than 180°) external angles between adjacent edges at the corners of the polygon that is formed by the line segments when the boundary curve is scanned in the *clockwise sense*. Figure 7.13 shows the possible cases. For example, successive directions 01, 02, 23, 71 correspond to concave external angles, and the pairs 06, 41, 64, 75 to convex angles. The occurrence percentage of each of these cases in the chain code defines a respective feature. Sometimes chain code pairs are grouped according to whether the first chain code is even or odd. Thus, a total of 16 features are generated, 8 for the convex and 8 for the concave case.

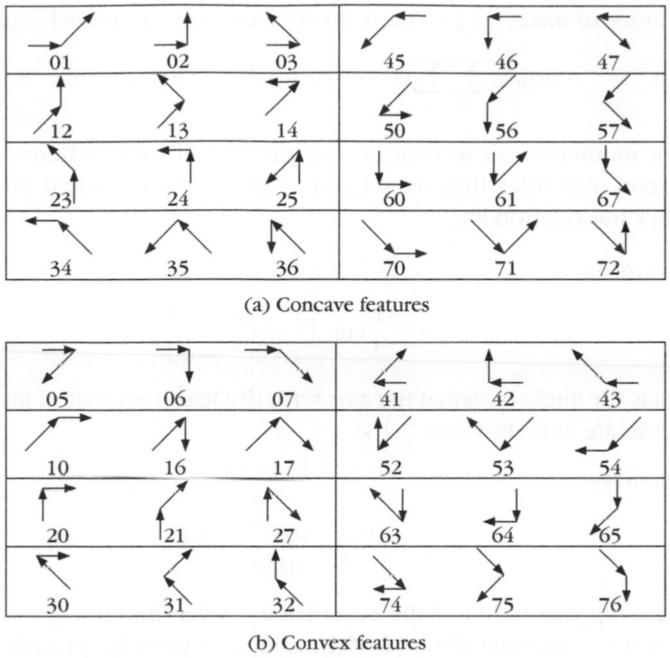

FIGURE 7.13

Curvature features characterizing the boundary polygon that results from an eight-directional chain code description of the boundary.

7.3.3 Moment-Based Features

In (7.21) and (7.25) the geometric moments and central moments were defined. If in the place of $I(i,j)$ we consider the sequence

$$I(i,j) = \begin{cases} 1 & (i,j) \in C \\ 0 & (i,j) \text{ otherwise} \end{cases}$$

where C is the set of points (i,j) *inside* the object of interest, then we obtain a way to describe the shape of the object through the moments. Indeed, in such a case only the limits in the summations (hence the object's shape) are taken into account, whereas the details inside the object (i.e., texture) do not participate. Hence

$$m_{pq} = \sum_i \sum_j i^p j^q, \quad (i,j) \in C$$

with $m_{00} = N$, the total number of pixels inside the region. The features

$$\bar{x} = \frac{m_{10}}{m_{00}} \quad \text{and} \quad \bar{y} = \frac{m_{01}}{m_{00}}$$

define the *center of mass* $(\bar{x}, \bar{y})$. The respective central moments become

$$\mu_{pq} = \sum_i \sum_j (i - \bar{x})^p (j - \bar{y})^q, \quad (i, j) \in C$$

The invariant moments can in turn be computed and used, whenever appropriate. Two useful quantities that are related to these moments and provide useful discriminatory information are:

1. *Orientation*

$$\theta = \frac{1}{2} \tan^{-1} \left[\frac{2\mu_{11}}{\mu_{20} - \mu_{02}} \right]$$

which is the angle between the axis with the least moment of inertia and the x-coordinate axis (Problem 7.18).

2. *Eccentricity*

$$\epsilon = \frac{(\mu_{20} - \mu_{02})^2 + 4\mu_{11}}{area}$$

Another representation of the eccentricity is via the ratio $\frac{R_{max}}{R_{min}}$ of the maximum to the minimum distance of the center of mass $(\bar{x}, \bar{y})$ from the object's boundary (Figure 7.10b).

7.3.4 Geometric Features

The features of this subsection are derived directly from the geometry of the object's shape. The *perimeter P* of the object and its *area A* are two widely used features. If $\boldsymbol{x}_i, i = 1, 2, \ldots, N$, are the samples of the boundary, then the perimeter is given by

$$P = \sum_{i=1}^{N-1} \|\boldsymbol{x}_{i+1} - \boldsymbol{x}_i\| + \|\boldsymbol{x}_N - \boldsymbol{x}_1\|$$

If we consider the area of a pixel as the measuring unit, a straightforward way to compute the area enclosed by a boundary is by counting the number of pixels inside the region of the object. The *roundness ratio* is a third quantity, defined as

$$\gamma = \frac{P^2}{4\pi A}$$

A useful feature that is related to the curvature of the boundary, as defined in (7.44), is the so-called *bending energy* at a point n, given by

$$E(n) = \frac{1}{P} \sum_{i=0}^{n-1} |k_i|^2$$

Another popular feature is the *number of corners* in the boundary contour. These correspond to points where the curvature k_i takes large values (infinity in theory). In [Ghos 97] corners as well as other topological features are detected via the use of Zernike moments and appropriate parametric modeling of the respective topological image intensity profile.

The *number of holes* inside the region of an object is another useful quantity. For example, a large error percentage in handwriting character recognition tasks is related to the difficulty of the classifiers in distinguishing "8" from "0," because their boundaries look alike. The detection of the presence of holes inside the object, using appropriate algorithms, is extra information that can be beneficially used for recognition [Lai 81, Mahm 94].

In our study so far, we have demonstrated how to derive geometric features from the boundary curve. However, this is not the only possibility. For example, in [Wang 98] geometric features are extracted directly from the gray-level variation within the image region. In this way, the binarization phase is avoided, which in some cases can become tricky. Another direction that has been used extensively in OCR is to work on the thinned version of the binarized character.

Figure 7.14 illustrates the procedure via an example. Figure 7.14b is the result of the application of a thinning algorithm (e.g., [Pita 94]) on the binary version of the character "5" of Figure 7.14a. Also in Figure 7.14b the so-called *key points* are denoted. These can be node points where one or more lines (strokes) of the character are crossed or corner points or end points. These can be computed by processing neighboring pixels. For example, in order to identify an end point, we look at its eight neighboring pixels. An end point has only one neighbor at gray-level 1, and the rest are 0. In the sequel, the thinned version of the character is simplified as a set of line segments (edges) connecting the key points, Figure 7.14c. Each edge can then be characterized by its direction, for example, using the chain code; its length, for example, long or short; and its relation to its neighboring edges. In the sequel each character is described by an array providing this information in a coded form. Classification is then based on these coded matrices by defining appropriate costs. The interested reader may consult for example [Lu 91, Alem 90] for more details.

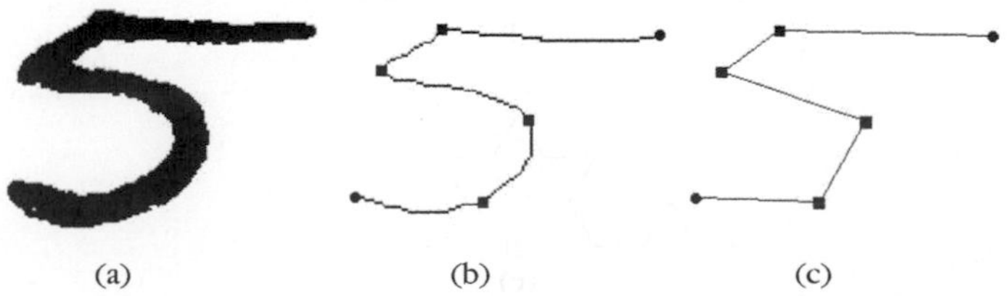

FIGURE 7.14

(a) Binarized version of 5 (b) the thinned version with the key points and (c) version with edges connecting key points.

7.4 A GLIMPSE AT FRACTALS

We have already seen that the 1980s was the decade in which two major tools were introduced into the realm of pattern recognition applications (among others): neural networks and wavelets. The same decade was also the time when another tool was adopted in many application areas to offer its potential power. ***Fractals*** and ***fractal dimension*** have become the focus of considerable research effort. This section aims at giving the basic definitions and outlining the basic concepts behind the use of fractals in pattern recognition. A deeper study of the area is beyond the goals of a short section, and the interested reader may refer to a number of books and articles available [Mand 77, Tson 92, Falc 90].

7.4.1 Self-Similarity and Fractal Dimension

Let us consider the straight-line segment of length L in Figure 7.15a. Divide L into N (two for the example of the figure) equal parts of length l. Each of the resulting parts is still a straight-line segment, and its length has been scaled down by a factor $m = \frac{l}{L} = \frac{1}{N}$. Magnification of any of these parts by the same factor will reproduce the original line segment. We refer to such types of structures as ***self-similar***. If instead of a straight-line segment we had a square of side L (Figure 7.15b), then scaling

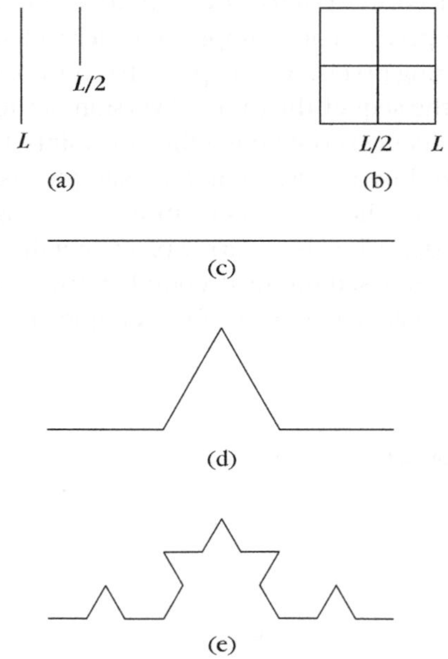

FIGURE 7.15

Self similar structures (a) line segment, (b) square, (c)–(e) three stages in the generation of Koch's curve.

down the side by $m = \frac{1}{N^{1/2}}$ would result into N square parts. Each part looks like the original square, which can be reobtained from the parts after magnification. The same is true for all dimensions, where N similar parts result after scaling the sides of the D-dimensional (hyper)cube by $m = \frac{1}{N^{1/D}}, D = 1, 2, \ldots$. That is, the Euclidean dimension D is directly related to the scaling and the number N of the resulting self-similar parts. We can write

$$N = \left(\frac{L}{l}\right)^D \equiv m^{-D} \tag{7.45}$$

If we now want to measure the length (area, volume) of the original segment (hypercube) using as a measurement unit a scaled element of length l (l^2, l^3, etc.), then the result is independent of the size l of the measuring unit. Indeed, the resulting metric property (length, area, etc.) is given by

$$M = N(l)l^D \tag{7.46}$$

where $N(l)$ is the number of parts that cover the curve (area, etc.) to be measured and l is the size of the measuring unit. Combination of (7.45) and (7.46) leads to a metric M that is always constant (L^D) for the same structure, and it is independent of the size l of the chosen unit, as expected.

Let us now turn our attention to some more interesting structures, such as the one in Figure 7.15d. The curve in Figure 7.15d results from the straight-line segment of Figure 7.15c, known as the *initiator*, by the following strategy: (a) divide the segment into three equal parts and (b) replace the central one by the two sides of an equilateral triangle, with sides of size equal to the size of the scaled parts. The procedure is then repeated for each of the line segments in Figure 7.15d, and this results in the structure of Figure 7.15e. This process can go on indefinitely, and the limit curve is the so-called Koch curve [Mand 77]. Such a curve is everywhere continuous but nowhere differentiable. It is readily observed that at each step of the iteration, the resulting structure is part of the structure that will result in the next iteration, after a scaling by 3. The curve therefore has a self-similar structure. In the sequel we will try to measure the length of the curve. Using as a (measuring) unit a segment of length $l = \frac{L}{3}$ (i.e., Figure 7.15d), the resulting length is 4. For a unit segment $l = \frac{L}{3^2}$ (i.e., Figure 7.15e), the measured length is 4^2. It is not difficult to see that the length keeps increasing with decreasing unit size and tends to infinity as the size of the measuring unit tends to zero! That is, the length of the curve depends not only on the curve itself but also on the adopted measurement unit! This strange result is the outcome of an "unfair" measurement process. Indeed, in the case of the Koch curve, scaling by 3 results in four similar parts. In contrast, in the case of a straight-line segment, scaling by $m = \frac{1}{N}$ results in the same number N of similar parts. In higher dimensional Euclidean space, scaling by $m = N^{-1/D}$ results in N parts. The measurement process involves this number N, the scaled side length l, and the *Euclidean dimension D*, as (7.46) suggests. From this discussion, the Euclidean dimension can also be seen as the ratio $\frac{\ln N}{-\ln m} = D$. Starting from this observation, let us now define the *similarity dimension* of a general self-similar structure as

$$D = \frac{\ln N}{-\ln m} \tag{7.47}$$

where N is the number of the resulting similar parts, when scaling by a factor m. For hypercube structures the similarity dimension is the respective Euclidean dimension, which is an *integer*. In contrast, the corresponding similarity dimension of the Koch curve $D = \frac{\ln 4}{-\ln(\frac{1}{3})}$ is a *fraction* and not an integer. Such structures are called *fractals*, and the corresponding similarity dimension is called a *fractal dimension*. Measuring a fractal structure, we can adopt (7.46) with the corresponding fractal dimension in the place of D. The result of the measurement process now becomes independent of the measuring tool l. Indeed, it is easy to see that using the definition in (7.47), (7.46) results in a constant $M = L^D$ for $m = \frac{l}{L}$. *The use of similarity dimension, therefore, results in a consistent description of the metric properties of such self-similar structures.* For a deeper treatment and other definitions of the dimension the interested reader may consult more specialized texts (e.g., [Falc 90]).

7.4.2 Fractional Brownian Motion

A major part of our effort in this chapter was dedicated to the description of statistical properties of signals and images and to the ways these can be exploited to extract information-rich features for classification (e.g., co-occurrence matrices, AR models). In this section we will focus our attention on whether the notion of self-similarity is extendable to stochastic processes and, if it is, how useful it can be for our interests. In the previous section "similarity" referred to the shape of a curve. For statistics such a view would be of no interest. From a statistical point of view it would be more reasonable and justifiable to interpret similarity from the perspective of "similar statistical properties," that is, mean, standard deviation, and so forth. Indeed, it can be shown that stochastic processes that are self-similar under scaling do exist. Furthermore, such processes can model adequately a number of processes met in practice.

Let $\eta(n)$ be a white (Gaussian) noise sequence with variance $\sigma_\eta^2 = 1$. The process defined as

$$x(n) = \sum_{i=1}^{n} \eta(i)$$

is known as a *random walk* sequence, and it belongs to a more general class of stochastic processes known as *Brownian motion* processes [Papo 91, p. 350]. It is straightforward to see that

$$E[x(n)] = 0$$

and that its variance is given by

$$E[x^2(n)] = n\sigma_\eta^2$$

Thus, the process is a nonstationary one because its variance is time dependent. A direct generalization of the previous result is

$$E[\Delta^2 x(n)] \equiv E[(x(n+n_0) - x(n_0))^2] = n\sigma_\eta^2 \tag{7.48}$$

where by definition $\Delta x(n)$ is the sequence of increments. Scaling the time axis by m results in

$$E[\Delta^2 x(mn)] \equiv E[(x(mn+n_0) - x(n_0))^2] = mn\sigma_\eta^2 \tag{7.49}$$

Hence, if the sequence of increments is to retain the same variance after scaling, it should be scaled by $\sqrt{m}$. Furthermore, it is easy to see that the sequence of increments, as well as the scaled versions, follow a Gaussian distribution (e.g., [Falc 90]). Recalling that Gaussian processes are completely specified by their mean and variance, we conclude that the increments $\Delta x(n)$ of $x(n)$ are *statistically self-similar in the sense that*

$$\Delta x(n) \quad \text{and} \quad \frac{1}{\sqrt{m}}\Delta x(mn) \tag{7.50}$$

are described by the same probability density functions, for any n_0 and m. Figure 7.16 shows three curves of the scaled random walk increments for $m = 1, 3, 6$. It is readily observed that they indeed "look" alike. Such curves, for which different scaling has been used for the coordinates $(\Delta x, n)$, are also known as *statistically self-affine*.

The random walk Brownian motion is a special case of a more general class of processes known as *fractional Brownian motion sequences* (fBm), introduced in [Mand 68]. The increments of this type of processes have variance of the general form

$$E[\Delta^2 x] \propto (\Delta n)^{2H} \tag{7.51}$$

with $0 < H < 1, \Delta n \equiv n - n_0$ and $\propto$ denoting proportionality. The parameter H is also known as the *Hurst parameter*. As in the case of Brownian motion, the increments of such processes are *statistically self-affine* in the sense that the processes

$$\Delta x(n) \quad \text{and} \quad \frac{1}{m^H}\Delta x(mn)$$

are described by the same probability density functions. The parameter H relates to the visual smoothness or coarseness of the respective graph of the increments versus time. This is an implication of (7.51). Let us start from a maximum interval Δn, corresponding to an incremental variance σ^2. In the sequel we halve the interval to $\Delta n/2$. The respective variance will be reduced by the factor $(1/2)^{2H}$.

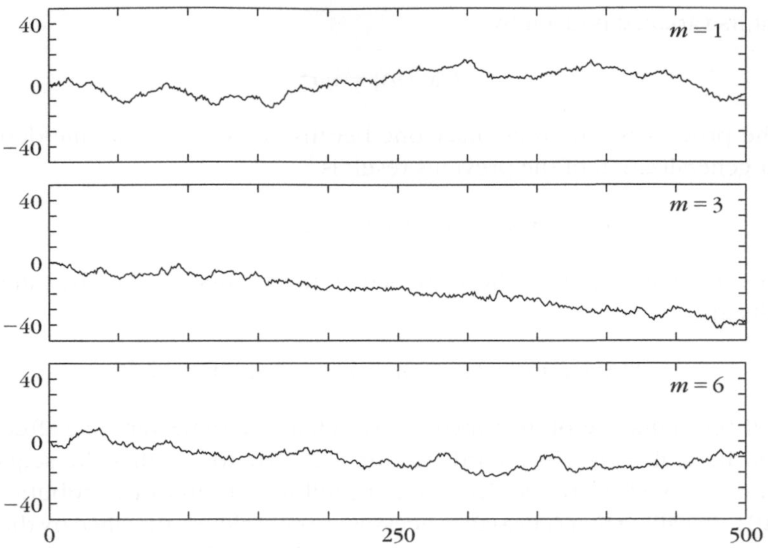

FIGURE 7.16

Time evolution of the random walk ($m = 1$) sequence and two of its self-affine versions. They all do look alike.

This process can go on. Each time we reduce the interval Δn by half, we look at increments between points located closer in time. The higher the value of H, the greater the reduction of the variance of the increments between these points, indicating smoother curves. For $H = 0$, the variance of the increments remains constant and independent of Δn. This process is not an fBm, and it corresponds to a white noise stationary process, with no dependence between adjacent time instants. Hence, it exhibits the most erratic behavior, and its graph has the most coarse appearance. *This observation indicates that the parameter H could be used as a measure of the "smoothness" of such curves.* By varying H one can get curves of varying degree of smoothness [Saup 91]. Figure 7.17 indeed verifies that the curve for $H = 0.8$ is smoother than the one for $H = 0.2$, and both are smoother than the top one, which corresponds to a white noise sequence. As was the case with the fractal curves of the previous subsection, a dimension can also be defined for curves resulting from fBm processes. It can be shown [Falc 90, p. 246] *that an fBm process with parameter H corresponds to a curve with fractal dimension* $2-H$. In general, if l is the number of free parameters of the graph, the corresponding fractal dimension is $l + 1 - H$. For a graph as in Figure 7.17, $l = 1$ and for an image $l = 2$.

The question now is how all these no doubt mind-stimulating points can be of use to us in the context of pattern recognition. The terms *smoothness* and *coarseness* have been used in association with the parameter H and equivalently with the dimension D of an fBm process. On the other hand, the terms *smoothness*

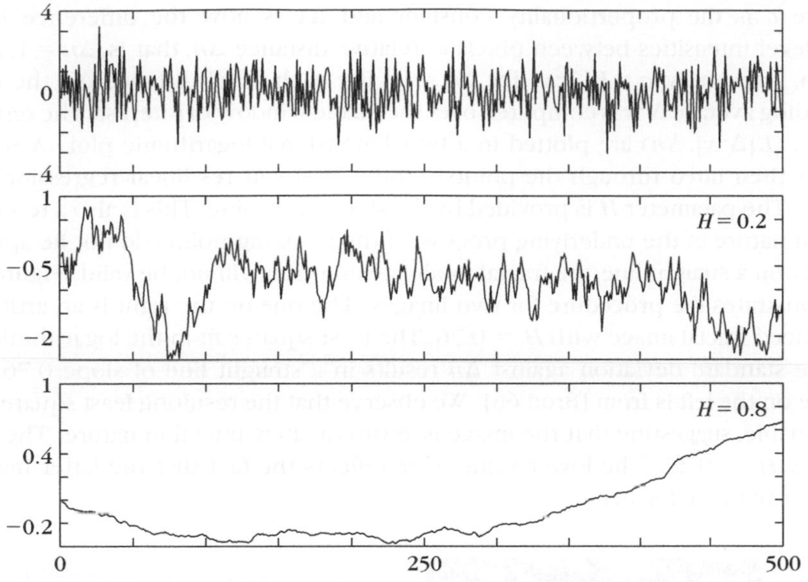

FIGURE 7.17

Time evolution of a white noise sequence (top) and two fBm processes of different Hurst parameters H. The lower the value of H, the coarser the appearance of the graph.

and *roughness*, were in a central position when dealing with feature generation for texture classification. We have now reached our crucial point. Provided that we can describe the sequence of gray levels of an image as an fBm process, the corresponding fractal dimension can be used as a potential feature for texture classification. In [Pent 84] it is reported that this is indeed true for a number of cases. Using a number of textured images from [Brod 66], as well as images from natural scenes, it was found that a large percentage of them exhibited fBm behavior. This was easily verified by constructing the histogram of differences (increments) of the gray-level intensities for various relative pixel distances Δn. It turned out that for each value of Δn the corresponding histogram was close to a Gaussian pdf centered at zero. Furthermore, the widths of the Gaussian-like histograms were different for the different relative pixel distances Δn. The larger the Δn, the wider the resulting histogram. However, we know that the width of a Gaussian is directly related to its variance. The plot of the variance as a function of relative pixel distance revealed an underlying fBm nature of the intensity processes, at least over a 10:1 range of relative distances measured. The parameter H, or equivalently the fractal dimension D, was then used successfully to distinguish a number of different textured regions in an image. The estimation of the H can take place via its definition in (7.51). Taking the logarithm, we get

$$\ln E[\Delta^2 x] = \ln c + 2H \ln \Delta n$$

where c is the proportionality constant and Δx is now the difference in the gray-level intensities between pixels at relative distance Δn, that is, $\Delta n = 1, 2$, and so on. Obviously, $c = E[\Delta^2 x]$ for $\Delta n = 1$. For each pixel distance Δn the corresponding average $\Delta^2 x$ is computed over the image window of interest. The resulting points $(E[\Delta^2 x], \Delta n)$ are plotted in a two-dimensional logarithmic plot. A straight line is then fitted through the points using a least squares linear regression technique. The parameter H is provided by the slope of the line. This is also a test of the fractal nature of the underlying process. If the resulting points do not lie approximately on a straight line, the fractal model assumption will not be valid. Figure 7.18 demonstrates the procedure for two images. The one on the right is an artificially produced fractal image with $H = 0.76$. The least squares fit in the logarithmic plot of the standard deviation against Δn results in a straight line of slope 0.76. The image on the left is from [Brod 66]. We observe that the resulting least squares fit is reasonable, suggesting that the image is approximately fractal in nature. The slope is now $H = 0.27$. The lower value of H reflects the fact that the latter image is coarser than the former.

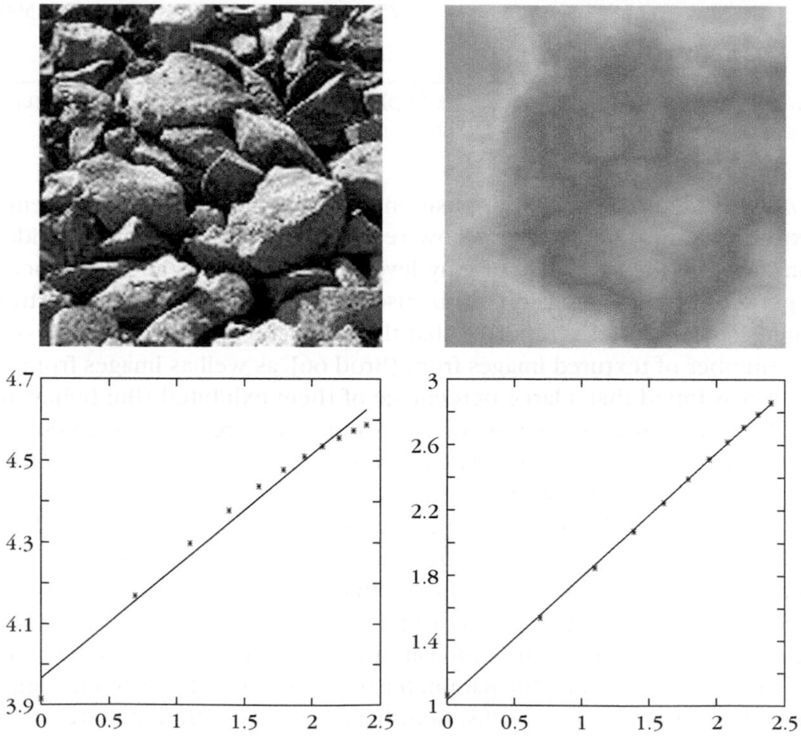

FIGURE 7.18

Examples of images with corresponding logarithmic plots of the standard deviation of increments (vertical axis) versus relative distance (horizontal axis).

The method presented previously for the computation of the Hurst parameter is not the only one, and a number of other techniques have been suggested. A popular alternative is based on the wavelet analysis of the underlying fBm process. The basis sequences (functions) used in the wavelet analysis are scaled and translated versions of a mother sequence (function). This underlying notion of scale invariance is shown to relate to fBm processes, whose statistical properties are scale invariant [Flan 92]. It turns out that the wavelet coefficients of an fBm process, at a given resolution level, i, form a stationary sequence, with a variance proportional to $2^{-i(2H+1)}$, see [Worn 96]. This leads to a simple method for estimating the associated Hurst parameter. Other methods for estimating the fractal dimension include the box-counting and variation method [Huan 94, Kell 87]; maximum likelihood estimates, as in [Lund 86, Deri 93, Fieg 96]; morphological covers [Mara 93]; methods in the spectral domain, as in [Gewe 83]; and fractal interpolation function models [Penn 97].

Fractional modeling and the use of fractal dimension D as a feature for classification have been demonstrated in a number of applications [Chen 89, Lund 86, Rich 95]. However, the method is not without drawbacks. Indeed, it may happen that different textures result in the same fractal dimension, thus limiting the classification potential of the method. Another shortcoming is that in practice physical processes retain their fractal characteristics over a range of distances but not over all ranges. Thus, the fractal dimension may change as we pass from one range of scales to another [Pent 84, Pele 84], and the use of a single Hurst parameter may not lead to sufficient modeling. To overcome these drawbacks, a number of possible solutions have been suggested. The *multifractional Brownian motion* (mBm) is an extension of an fBm process with a parameter H, which is allowed to vary, as in [Ayac 00]. *Extended self similar* (ESS) processes allow in (7.51) for a more general dependence on Δn, via a so-called structure function [Kapl 94]. For more on these issues the interested reader is referred, for example, to [Bass 92, Ardu 92, Kapl 95, Kapl 99, Pesq 02]. A comparative study of various textural features, including fractal modeling techniques, can be found in [Ohan 92, Ojal 96].

7.5 TYPICAL FEATURES FOR SPEECH AND AUDIO CLASSIFICATION

As we have already commented in the Preface, speech recognition is a major application area of pattern recognition, and a number of speech-recognizing systems are already available in the market. Audio classification and recognition have also received a lot of attention in recent years. A great number of commercial applications are envisaged for the future in the field of multimedia databases. Techniques for automatic indexing tools, intelligent browsers, and search engines with content-based retrieval capabilities are currently the focus of a major research effort. In this context, *audiovisual* data segmentation and indexing, based not only on visual information but on the accompanying audio signal, greatly enhance the performance. For instance, classifying video scenes of gun fights using the audio information related to shooting and explosions will, no doubt, enhance the performance of the classifier compared to a system that is based on the corresponding visual information only.

Content-based retrieval from music databases is another application that attracts the interest of current research. It is very likely that not far in the future a large corpus of the recorded music in human history will be available on the Web. Automatic music analysis is envisaged to be one of the main services to facilitate content distribution. Automatic music genre classification, querying music databases by *humming* the tune of a song or querying by *example* (i.e., providing a music extract of short duration in order to locate and retrieve the complete recording) are examples of services that vendors would very much like to offer in such systems. More on these issues can be found in [Wold 96, Wang 00, Zhan 01, Pikr 03, Pikr 06, Pikr 08, Frag 01, Clau 04].

This section focuses on the generation of some typical and commonly used features for speech recognition and audio classification/recognition. However, as has already been stated elsewhere in the book one must keep in mind that feature generation is very much a problem-dependent task. Thus, the combination of the designer's imagination with his or her good knowledge of the peculiarities of the specific task can only benefit the generation of informative features.

7.5.1 Short Time Processing of Signals

Speech and audio signals are statistically nonstationary; that is, their statistical properties vary with time. A way to circumvent this problem and be able to use analysis tools that have been developed and make sense for stationary signals only, such as the Fourier transform, is to divide the time signal into a series of successive *frames*. Each frame consists of a *finite* number, N, of samples. During the time interval of a frame, the signal is assumed to be "reasonably stationary." Such signals are also known as *quasistationary*. Figure 7.19 shows a signal segment and three successive frames, each consisting of $N = 20$ samples, with an overlap among neighboring frames of 5 samples. Choosing the length, N, of the frames is a "bit of an art" and is a problem-dependent task. First, each frame must be long enough for an analysis method to have enough "data resources" to build up the required information. For example, if we are interested in estimating the period of a periodic signal the number of samples in each frame must be large enough to allow the signal periodicity to be revealed. Of course, this will depend on the value of the period. For short periods, a few samples can be sufficient. On the other hand, for long periods more samples will be necessary. Second, N must be short enough to guarantee the approximate stationarity of the signal within the time scale of each frame in order for the results to be meaningful. For speech signals sampled at a frequency of $f_s = 10$ KHz, reasonable frame sizes range from 100 to 200 samples, corresponding to 10–20 msecs time duration. For music signals sampled at 44.1 KHz, reasonable frame sizes range from 2048 to 4096 samples, corresponding to 45–95 msecs, approximately.

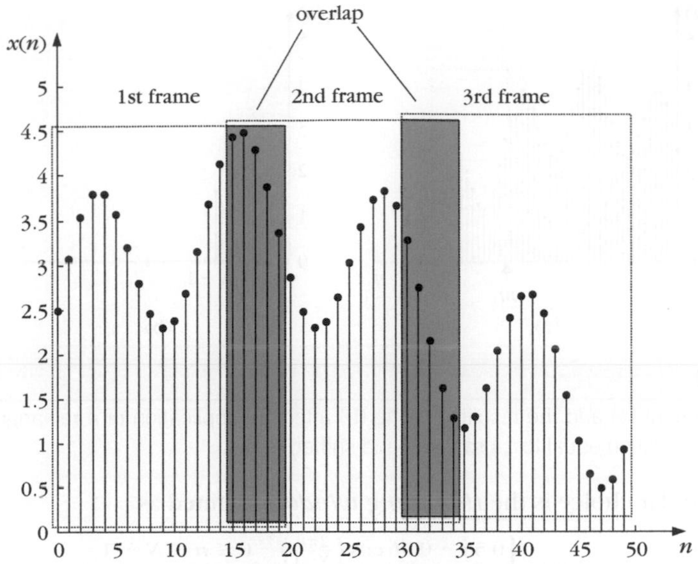

FIGURE 7.19

Three successive frames, each of length $N = 20$ samples. The overlap between successive frames is 5 samples.

From a mathematical formulation point of view, dividing the time signal in a sequence of successive frames is equivalent to multiplying the signal segment by a *window* sequence, $w(n)$, of finite duration N. The simplest window sequence is the *rectangular window*, defined as

$$w(n) = \begin{cases} 1 & 0 \leq n \leq N-1 \\ 0 & \text{elsewhere} \end{cases}$$

For different frames, the window is shifted to different points, m_i, in the time axis. Hence, if $x(n)$ denotes the signal sequence, the samples of the ith frame can be written as

$$x_i(n) = x(n + m_i)w(n)$$

where m_i is the corresponding window shift associated with the ith frame. This implies that all samples in the ith frame are identically zero, except for the time instants $n = 0, \ldots, N-1$, which correspond to the original signal samples, $x(n)$, with $n \in [m_i, m_i + N - 1]$. The procedure is illustrated in Figure 7.20. As it is known from the Fourier transform theory basics, multiplying a sequence by a window in the time domain smooths out its Fourier transform by convolving it with the Fourier transform of the window sequence. Some of the effects of this smoothing action can be minimized by using a different window sequence, with a smoother decay to

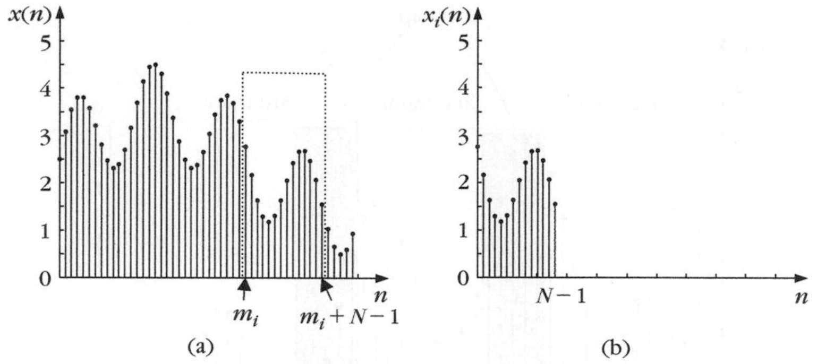

FIGURE 7.20

A signal segment (a) and the resulting frame (b) after the application of a rectangular window sequence of duration equal to 14 samples and shifted at m_i.

zero. A popular choice is the *Hamming window*, defined as

$$w(n) = \begin{cases} 0.54 - 0.46\cos\left(\frac{2\pi n}{N-1}\right) & 0 \leq n \leq N-1 \\ 0 & \text{elsewhere} \end{cases}$$

More on these issues can be found in [Rabi 78, Dell 00].

As an example, let us assume that we have divided a speech segment into a sequence of I frames, each of length N. Then, for each frame, $i = 1, 2, \ldots, I$, we can compute the discrete Fourier transform (DFT) as

$$X_i(m) = \sum_{n=0}^{N-1} x_i(n) \exp\left(-j\frac{2\pi}{N}mn\right), \quad m = 0, 1, \ldots, N-1$$

It is common to refer to this DFT as the *short-time DFT*. It must be pointed out that this definition implies much more theory and interesting implementation issues (which, however, are not of interest to us and we will not delve into this topic any deeper). Selecting $l \leq N$ DFT coefficients from each frame, we can construct a sequence of feature vectors

$$\boldsymbol{x}_i = \begin{bmatrix} X_i(0) \\ X_i(1) \\ \vdots \\ X_i(l) \end{bmatrix}, \quad i = 1, 2, \ldots, I \tag{7.52}$$

Thus, the pattern of interest (i.e., the speech segment) is not represented by a single feature vector but by a *sequence* of feature vectors. We will see how to attack such problems in Chapters 8 and 9.

The autocorrelation sequence is another very important statistical quantity, which is also defined for stationary processes. Recall from Section 7.2.4 that, if $x(n)$ is a stationary process, the autocorrelation sequence is defined as

$$r(k) = E[x(n)x(n-k)] = E[x(n)x(n+k)] = r(-k) \tag{7.53}$$

In other words, it is the expectation of the product of $x(n)$ with its shifted version $x(n \pm k)$. In practice, the expectation can be obtained as

$$r(k) = \frac{1}{2N+1} \sum_{n=-N}^{N} x(n)x(n+k)$$

which under mild assumptions (i.e., ergodicity) tends to the true value of $r(k)$ as N tends to infinity. In the case of a quasistationary process, the *short-time autocorrelation* sequence, $r_i(k)$, is defined for each of the frames as

$$r_i(k) = \frac{1}{N} \sum_{n=0}^{N-1-|k|} x_i(n)x_i(n+|k|) \tag{7.54}$$

where $|\cdot|$ denotes the absolute value operator. The limits in the sum indicate that outside the interval $[0, N-1-|k|]$ the product $x_i(n)x_i(n+|k|)$ is identically zero, due to the finite duration, N, of the frame. This definition complies with the definition in (7.53) in the sense that for stationary processes $r_i(k)$ is an asymptotically unbiased estimate of the autocorrelation as the length of the frame $N \Longrightarrow \infty$. Indeed, viewing $r_i(k)$ as an estimate of $r(k)$, its mean value for different realizations is easily shown to be

$$E[r_i(k)] = \frac{N-|k|}{N} r(k) \tag{7.55}$$

Thus, for finite frame length, N, Eq. (7.54) results in a biased estimate of $r(k)$. However, for small values of the lag k, with respect to N, the bias is small. On the other hand, for values of k close to N we expect $r_i(k)$ to get values close to zero, something that is verified in practice. To remedy this drawback, other definitions of the short-time autocorrelation have been proposed. See, for example, [Rabi 78]. Another important property of the definition in (7.54), from the computational point of view, is that the corresponding (short-time) autocorrelation matrix retains the computationally elegant properties of being symmetric and having a Toeplitz structure (Section 7.2.4).

7.5.2 Cepstrum

Let $x(0), x(1), \ldots, x(N-1)$ be the samples from the current data frame (the index i has been dropped for notational simplicity). The Fourier transform (FT) of this finite-length sequence of data is defined as the periodic complex function

$$X(\omega) = \sum_{n=0}^{N-1} x(n) \exp\left(-j\omega T n\right) \tag{7.56}$$

with period (in the frequency domain) $2\pi/T$, where T is the sampling period. It is well known from the basic theory of signal processing (e.g., [Proa 92] and Chapter 6), that the coefficients of the DFT transform

$$X(m) = \sum_{n=0}^{N-1} x(n) \exp\left(-j\frac{2\pi}{N}mn\right), \quad m = 0, 1, \ldots, N-1 \tag{7.57}$$

are the samples of the FT taken at the frequency points $0, \frac{2\pi}{NT}, \ldots, \frac{2\pi}{NT}(N-1)$. Assuming, without loss of generality, $T = 1$, the inverse FT is defined as

$$x(n) = \frac{1}{2\pi}\int_{-\pi}^{\pi} X(\omega)\exp\left(j\omega n\right)d\omega, \quad n = 0, 1, \ldots, N-1 \tag{7.58}$$

That is, the resulting samples are equal to the samples of the original sequence and identical to what is obtained by the inverse DFT. That is,

$$x(n) = \frac{1}{N}\sum_{m=0}^{N-1} X(m)\exp\left(j\frac{2\pi}{N}mn\right), \quad n = 0, 1, \ldots, N-1 \tag{7.59}$$

The *cepstrum*, $c(n)$, of a sequence, $x(n)$, is the sequence resulting from the inverse FT of the logarithm of the magnitude of its FT. That is,

$$c(n) = \frac{1}{2\pi}\int_{-\pi}^{\pi} \log_{10}|X(\omega)|\exp\left(j\omega n\right)d\omega \tag{7.60}$$

Although the base 10 logarithm has been used, the logarithm of any base can be adopted. Another way of looking at the cepstral coefficients, $c(n)$, is the following. Since the FT function $X(\omega)$ is a periodic function of ω, with period 2π, the function $\log_{10}|X(\omega)|$ is also periodic with the same period. Therefore, it can be written in terms of its Fourier series expansion

$$\log_{10}|X(\omega)| = \sum_{n=-\infty}^{\infty} c(n)\exp\left(-j\omega\frac{2\pi}{2\pi}n\right) = \sum_{n=-\infty}^{\infty} c(n)\exp\left(-j\omega n\right) \tag{7.61}$$

Hence, Eq. (7.60) is the formula that provides the coefficients of the Fourier series expansion in (7.61). However, the function $\log_{10}|X(\omega)|$ is defined in the frequency and not in the time domain. Its Fourier transform domain is known as the *quefrency* domain, and the respective Fourier series coefficients, $c(n)$, are known as *cepstral coefficients*. This is only to remind us that the original transformed function is in the frequency domain. Otherwise, all Fourier transform/series properties are still valid. Thus, since $\log_{10}|X(\omega)|$ is a real and even function, ($|X(\omega)|$ is even for a real sequence $x(n)$), the cepstral coefficients are real and even. That is,

$$c^*(n) = c(n) = c(-n)$$

Cepstral coefficients have very good information-packing properties, from the class discrimination point of view, and are very popular candidates as features, both for speech recognition and audio classification tasks [Rabi 93, Tzan 02].

Computation of the cepstral coefficients is achieved via the DFT (using the FFT) of $X(\omega)$. However, this computation is not as innocent as it is for the case of $X(\omega)$. As we have already seen, the inverse transforms of $X(\omega)$ and $X(m)$ coincide [Eqs. (7.58) and (7.59)]. This is because the input sequence is of finite length, N, and the sampling period ω_s, in the frequency domain, is chosen to obey the Nyquist criterion; that is, $\omega_s = \frac{2\pi}{N}$. This is not the case with the cepstral coefficients. Using $\log_{10}|X(m)|$, $m = 0, 1, \ldots, N-1$, and taking the inverse DFT results in

$$\hat{c}(n) = \frac{1}{N}\sum_{m=0}^{N-1} \log_{10}|X(m)|\exp\left(j\frac{2\pi}{N}mn\right), \quad n = 0, 1, \ldots, N-1 \tag{7.62}$$

where $\hat{c}(n)$ and $c(n)$ are related as

$$\hat{c}(n) = \sum_{r=-\infty}^{\infty} c(n + rN) \tag{7.63}$$

For those readers familiar with basic digital signal processing this is most natural. The sequence $c(n)$ is not of finite duration. Thus, sampling its FT ($C(\omega) \equiv \log_{10} |X(\omega)|$) with a sampling period of $\frac{2\pi}{N}$ does not satisfy Nyquist's criterion. Hence, taking the inverse DFT [Eq. (7.62)] will result in aliasing with a periodic repetition of the aliased sequence ($c(n)$) every N samples, which is expressed via (7.63). See, for example, [Proa 92]. In practice, the effects of this aliasing are minimized if one extends the length of the frame from N to M by appending $M - N$ zeros at the end of it. That is,

$$x(n): \; x(0), \ldots, x(N-1), x(N) = 0, \ldots, x(M-1) = 0 \tag{7.64}$$

These zeros (as it can easily be checked out) *have no effect on the* FT $X(\omega)$. However, the DFT is now of length M (corresponding to sampling the FT every $\frac{2\pi}{M}$ frequency points), which makes the implicit repetition period in (7.63) equal to M. Assuming that the cepstral coefficients decay fast enough, with respect to the repetition period M, we can assume that $\hat{c}(n) \approx c(n), n = 0, 1, \ldots, N-1$, since successive repetitions in (7.63) now have little overlap. In practice, a number of 512 or 1024 zeros may be necessary. For further information related to cepstrum, the interested reader may refer to [Rabi 78, Dell 00].

In summary, the computational steps to obtain the cepstral coefficients of a frame, $x_i(n), n = 0, 1, \ldots, N-1$, are the following:

- Extend the length of the frame by appending $M - N$ zeros at the end of the frame.
- Obtain the DFT of length M of the extended frame.
- Compute the logarithm of the magnitude of the DFT coefficients.
- Compute the inverse DFT of length M.

The obtained coefficients are the (approximate) cepstral coefficients of the sequence $x_i(n)$.

7.5.3 The Mel-Cepstrum

Human perception of audio has often been studied from a psychophysical point of view. Experiments have suggested that perception of the frequency content of pure tones does not follow a linear scale. This led to the idea of mapping acoustic frequency content to a linear "perceptual" frequency scale. A popular approximation to this type of mapping is known as the *mel scale* [Pico 93, Rabi 93]:

$$f_{mel} = 2595 \log_{10} 10(1 + f/700.0) \tag{7.65}$$

Equation (7.65) suggests that an actual frequency of 1 KHz is mapped to 1000 mel units. The plot of (7.65) is shown in Figure 7.21, and as can be seen the mapping

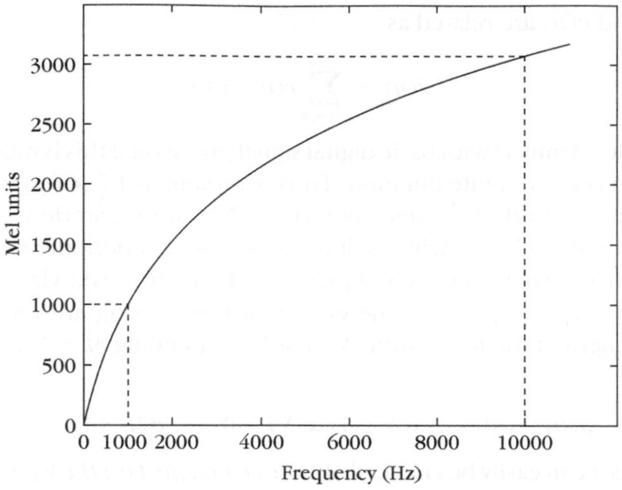

FIGURE 7.21

Subjectively perceived pitch, in mel units, as a function of the measured frequency.

is approximately linear from 0 Hz up to 1000 Hz, and its logarithmic nature prevails for frequencies above 1000 Hz. As an example, Eq. (7.65) suggests that the perceived frequency of a pure tone at the frequency of 10 KHz will be approximately 3000 mel units (mels). That is, a tenfold increase in the frequency will be subjectively perceived as a threefold increase.

Another psychoacoustics phenomenon is related to the way our auditory system perceives the differences in frequency among different tones that contribute to the formation of a more complex sound. It turns out that the tones cannot be individually distinguished if they fall within a certain bandwidth around the center frequency of the sound. We refer to this bandwidth as the *critical bandwidth* [Pico 93, Rabi 93]. Furthermore, if the bandwidth of a complex sound is less than the critical bandwidth around its center frequency the ear would perceive it as a single tone at the center of the critical band and with its loudness being equal to a weighted average of the loudness of each one of the contributing tones. The critical bandwidth around a frequency f can approximately be given by

$$BW_{critical} = 25 + 75[1 + 1.4(f/1000)^2]^{0.69} \tag{7.66}$$

A plot of this equation reveals that the critical bandwidth is approximately linear below 1 KHz and increases logarithmically for frequencies above 1 KHz.

In an effort to generate features that are rich in information, we will try to "manipulate" the frequency content of a sound segment by imitating nature, in the way our auditory system perceives and recognizes sounds. To this end the following steps are adopted:

- A sound segment, of length N, is analyzed via the DFT transform. As we have already observed, it is useful to append a number of zeros at the end and let M denote the number of samples after the extension. If $f_s = 1/T$ is the sampling frequency, each DFT coefficient $X(m)$ corresponds to a real frequency of $\frac{mf_s}{M}$.
- In the sequel, a number, L, of critical bands are "spread" over the frequency range up to $f_s/2$. Figure 7.22 is an example, where only the first 17 of $L = 35$ such bands are shown (for illustration simplicity) to occupy the frequency range from 0 Hz up to approximately 3700 Hz. The shape of each frequency band is a graphical representation of the weighting imposed on the corresponding frequency sample (frequency bin) within the bandwidth of the band. The sampling frequency is $f_s = 44.1$ KHz. These bands are uniformly distributed in a mel scale, and their bandwidth has been chosen to be approximately equal to 110 mels. In the frequency scale, these bands are almost uniformly spaced below 1 KHz and logarithmically above it. The shape of the bands has been chosen to be triangular. In general, the shape, bandwidth, and number of bands are critical design issues, and several approaches have been suggested throughout the years, depending on the application domain (see, for example, [Pico 93, Rabi 93, Davi 80]). In our example, we chose nonoverlapping bands, although in some cases an overlap between successive bands is allowed to exist.
- Since, in general, the center frequencies of these frequency bands do not coincide with the frequency quantization performed by the DFT, each band is moved so that its center frequency coincides with the nearest DFT frequency

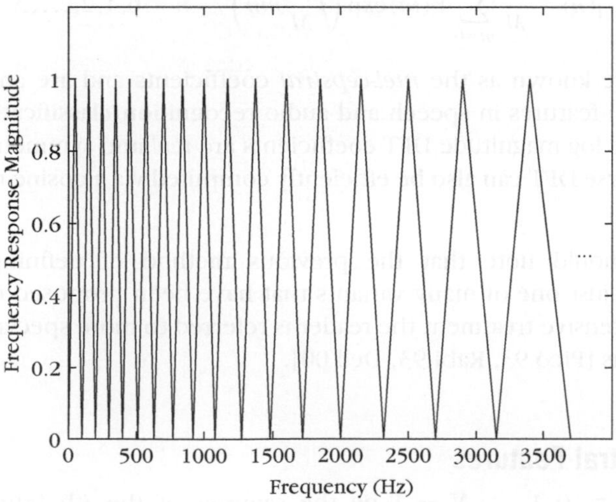

FIGURE 7.22

A critical-band filter bank consisting of nonoverlapping triangular bandpass filters.

bin ($\frac{mf_s}{M}$). Denote by m_i this (center) frequency bin of the ith band in the bank, $i = 1, 2, \ldots, L$.

- For each of the bands in the bank, compute the weighted average of the log-magnitude of the DFT coefficients that fall within the frequency band. That is,

$$Y(m_i) \equiv \sum_m \log_{10} |X(m)| H_i\left(\frac{m}{M} f_s\right), \quad i = 1, 2, \ldots, M \tag{7.67}$$

  where $H_i(\cdot)$ is the corresponding weighting value. Note that since the width of each band is finite, this summation, over m, is restricted to the DFT coefficients located within the bandwidth of the respective critical band.

- Define the sequence

$$Y(m) = \begin{cases} Y(m_i) & \text{if } m = m_i, \quad i = 1, 2, \ldots, L \\ 0 & \text{otherwise} \end{cases} \tag{7.68}$$

  In other words, this sequence is zero everywhere except at the frequency bins corresponding to the centers of the bands in the bank, where the value is equal to the weighted average of the log-magnitude of the DFT coefficients at the frequency bins within the bandwidth of the respective band. We can think of this new sequence as the psychologically perceived log-magnitude spectrum equivalent to the physically measured one.

- Take the inverse DFT

$$c_{\text{mel}}(n) = \frac{1}{M} \sum_{m=0}^{M-1} Y(m) \exp\left(j\frac{2\pi}{M}mn\right), \quad n = 0, 1, 2, \ldots, N-1 \tag{7.69}$$

  These are known as the *mel-cepstral* coefficients and are among the most powerful features in speech and audio recognition/classification. Note that since the log-magnitude DFT coefficients are real and symmetrical, the previous inverse DFT can also be efficiently computed via a cosine transform, as in [Proa 92].

The reader should note that the previous method of defining mel-cepstral coefficients is just one of many variants that have been proposed over the years. For a more extensive treatment the reader is referred to more specialized texts and articles, such as [Pico 93, Rabi 93, Dell 00].

7.5.4 Spectral Features

Let $x_i(n)$, $n = 0, 1, \ldots, N-1$ be the samples of the ith frame and $X_i(m)$, $m = 0, 1, \ldots, N-1$, the corresponding DFT coefficients. The following features are quite common in speech recognition and audio classification/recognition, each providing information for different acoustic qualities.

Spectral Centroid

$$C(i) = \frac{\sum_{m=0}^{N-1} m|X_i(m)|}{\sum_{m=0}^{N-1} |X_i(m)|}$$

The centroid is a measure of the spectral shape. High values of the centroid correspond to "brighter" acoustic structures with more energy in the high frequencies.

Spectral Roll-off

The spectral roll-off is the frequency sample, $m_c^R(i)$, below which the c% (e.g., $c =$ 85 or 90) of the magnitude distribution of the DFT coefficients is concentrated. That is, for this frequency sample the following is true:

$$\sum_{m=0}^{m_c^R(i)} |X_i(m)| = \frac{c}{100} \sum_{m=0}^{N-1} |X_i(m)|$$

This is another measure indicating where most of the spectral energy is concentrated. It is a measure of skewness of the spectral shape, with right-skewed shapes (brighter sounds) resulting in higher values.

Spectral Flux

$$F(i) = \sum_{m=0}^{N-1} (N_i(m) - N_{i-1}(m))^2$$

Here, $N_i(m)$ is the normalized (by its maximum value) magnitude of the respective DFT coefficient of the ith frame and is a measure of the *local* spectral change between successive frames.

Fundamental Frequency

Speech and audio signals can be either noise-like in nature, such as unvoiced speech segments or audio segments corresponding to an applause or footstep recordings, or can exhibit a periodic nature. In the latter case, we talk about *harmonic* signals to distinguish them from their noise-like *inharmonic* counterparts. Audio signals produced from musical instruments and voiced speech segments are two examples of harmonic signals. A distinct characteristic of a harmonic sound signal is its *fundamental frequency*.

In the case of voiced speech signals, this is the frequency of successive vocal fold openings and is also known as the *pitch* of the signal. For men this lies in the range of 80 to 200 Hz and for women in the range of 150 Hz to 350 Hz. For musical instruments, the fundamental frequency may vary a lot, and in some cases the fundamental frequency may not be present in the frequency spectrum, although the ear can have the ability to perceive it, by processing the information provided by the higher harmonics. This is a *psychoacoustics* phenomenon. Psychoacousticians as well as musicologists use the term *pitch* to define the frequency *perceived* by the ear, which in some cases may even be different from the fundamental.

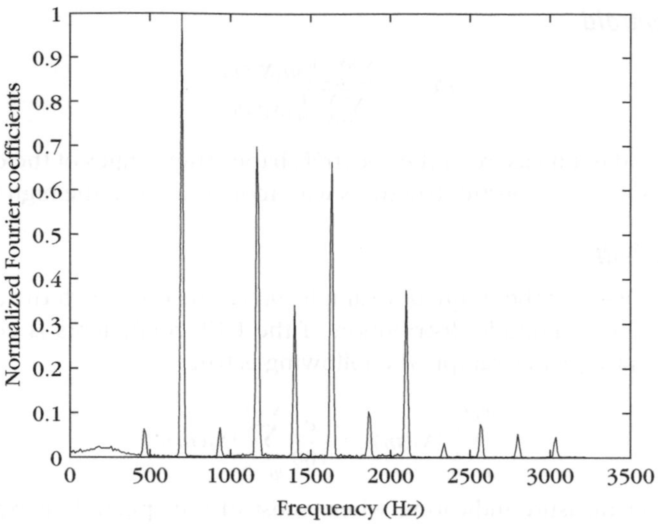

FIGURE 7.23

Normalized DFT coefficients of a clarinet sound, whose fundamental frequency is absent.

The fundamental frequency estimation is not an easy task, and a number of techniques have been proposed in the published literature, including [Schr 68, Brow 91, Wu 03, Tolo 00, Klap 03, Goto 04] and the references therein. Figure 7.23 is an example of the (normalized) amplitude of the DFT of a signal segment corresponding to a harmonic sound produced by a clarinet. The length of the frame is 4096 samples long. The spectrum consists of the regularly spaced harmonics of the fundamental frequency, which is equal to 230 Hz but is missing from the spectrum. Note that, listening to this sound, a trained ear will perceive that the pitch of this signal is indeed 230 Hz. It can be observed that odd and even multiples of the fundamental frequency are present as peaks. (For this frame, it turned out that the amplitudes of the odd multiples are considerably smaller than the amplitudes of the even multiples of the fundamental frequency.) Application of the algorithm given in [Schr 68], for the estimation of the fundamental frequency, returns the true value of 230 Hz. The method builds on the idea of exploiting the greatest common divisor of all peaks present in the spectrum.

7.5.5 Time Domain Features

Zero-Crossing Rate

The zero-crossing rate is defined as

$$Z(i) = \frac{1}{2N}\sum_{n=0}^{N-1} |\text{sgn}[x_i(n)] - \text{sgn}[x_i(n-1)]|$$

where

$$\text{sgn}[x_i(n)] = \begin{cases} 1 & x_i(n) \geq 0 \\ -1 & x_i(n) < 0 \end{cases}$$

This is a measure of the noisiness of the signal. Thus, unvoiced speech signals have higher zero-crossing values compared to the voiced ones. Temporal curves of variation of the zero-crossing rate from frame to frame may also be informative of the type of signal.

Energy

This is a very simple feature, defined as

$$E(i) = \frac{1}{N} \sum_{n=0}^{N-1} |x_i(n)|^2$$

and it can be used to discriminate voiced from unvoiced speech signals, since the latter tend to have much lower energy values than the former. It is also useful to discriminate silent periods in a recording and can be useful during the segmentation process.

All features discussed in this section are also known as *frame-level* features. They provide local information with respect to individual frames, and their goal is to capture short-term characteristics. However, to extract semantic content information, one needs to follow how the previously cited features change from frame to frame over a longer time scale. To this end, one can develop various methods to quantify this variation. In [Tzan 02], the mean and variances of the frame-level features have been used as features for music genre classification.

Besides the above features, other features—such as wavelet coefficients, fractal dimension, AR modeling, and independent components (ICA), presented in the previous and the present chapter—are also popular candidates. For music signals, based on early studies on the human perception of pitch, it has been proposed to use a 12-element representation of the spectral energy of a music signal, known as the chroma vector [Bart 05]. Each element of the vector corresponds to one of the 12 traditional pitch classes (i.e., 12 notes) of the equal-tempered scale of Western music. The chroma vector can encode and represent harmonic relationships within a particular music signal.

7.5.6 An Example

To demonstrate the classification power of two of the previously discussed features, let us take a simple example. Figure 7.24 shows the variation of the zero-crossing rate from frame to frame as time evolves, for a clapping sound. The sampling frequency was 44.1 KHz, and the length of each frame was equal to 1024 samples, with a successive frame overlap of 512 samples. For each frame the Hamming window was used. One can observe the noisy look of the resulting graph, with a large change of the feature values from frame to frame. This noisy nature of the clapping sound can also be revealed if the fundamental frequency is adopted as a

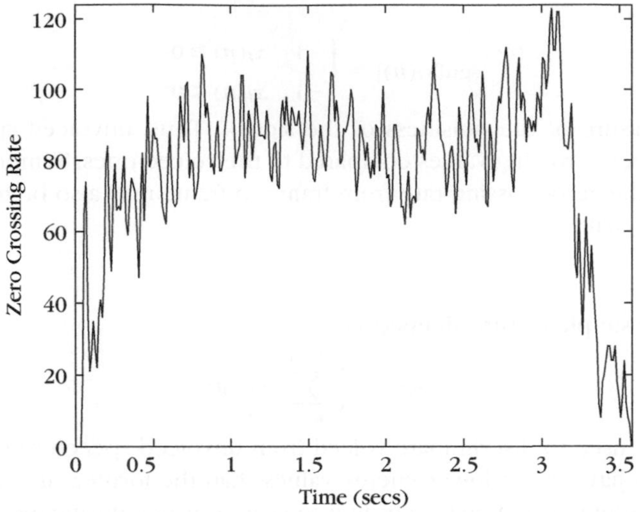

FIGURE 7.24

Zero-crossing rate results for a clapping sound recording, using a Hamming moving window technique.

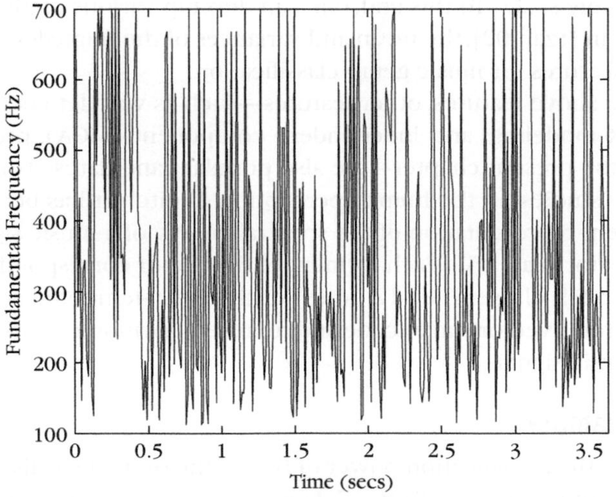

FIGURE 7.25

Fundamental frequency tracking results for the same clapping sound recording as in Figure 7.24.

feature. Figure 7.25 shows the change of the fundamental frequency from frame to frame. The algorithm used for the fundamental frequency tracking was that proposed in [Brow 91]. In contrast to the noisy nature of the previous audio recording, Figures 7.26 and 7.27 show the graphs of the change, from frame to frame, of the

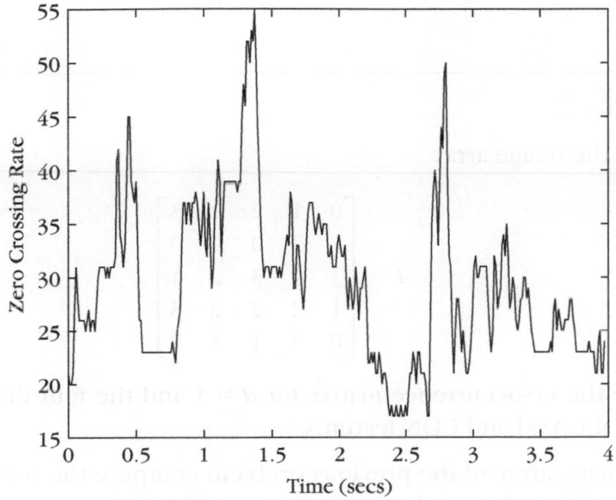

FIGURE 7.26

Zero-crossing rate results for the piano music recording, using a Hamming moving window technique.

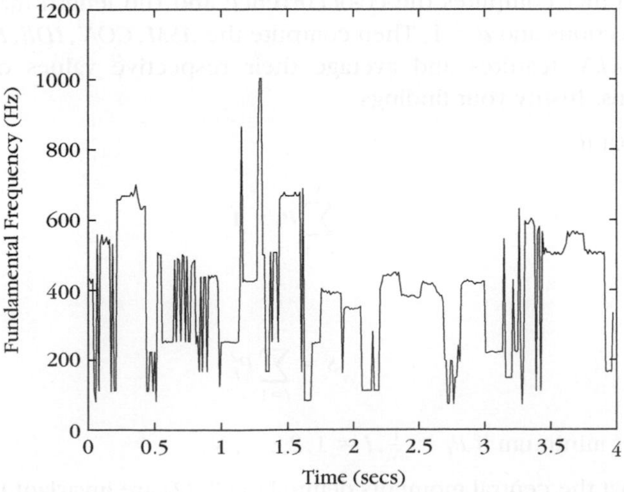

FIGURE 7.27

Fundamental frequency tracking results for the same piano music recording as in Figure 7.26.

zero-crossing rate and fundamental frequency, respectively, for a piano recording (from Bach's English suite in A major). The frame parameters used for the analysis were the same as before. One can observe that both graphs are now much less noisy. The variation from frame to frame is much smaller, and at some points of the graph the values of both features remain almost constant for relatively long periods of time, revealing the "structured" harmonic nature of the analyzed sound.

7.6 PROBLEMS

7.1 Consider the image array

$$I = \begin{bmatrix} 0 & 1 & 2 & 2 & 3 \\ 1 & 2 & 0 & 2 & 0 \\ 3 & 0 & 3 & 2 & 1 \\ 1 & 2 & 2 & 2 & 3 \\ 0 & 0 & 1 & 1 & 2 \end{bmatrix}$$

Compute the co-occurrence matrix for $d = 1$ and the four directions. Then compute the ASM and CON features.

7.2 For the image array of the previous problem compute the run length matrix for the four directions and then the features SRE, LRE.

7.3 Construct a 4×4 array that will have a high contrast (CON) value in the $0°$ direction and low CON value in the $45°$ direction.

7.4 For two of the test images provided in the Web site of the book, develop a program that computes the co-occurrence and run length matrices for the four directions and $d = 1$. Then compute the $ASM, CON, IDF, H_{xy}, SRE, LRE, GLNU, RLN$ features and average their respective values over the four directions. Justify your findings.

7.5 Show that if

$$\sum_{i=1}^{N} P_i = 1$$

then

$$S = \sum_{i=1}^{N} P_i^2$$

becomes minimum if $P_i = \frac{1}{N}, i = 1, 2, \ldots, N$.

7.6 Show that the central moments defined in (7.22) are invariant to translations and the normalized central moments in (7.23) are invariant to both translation and scaling.

7.7 Show that the Zernike polynomials are orthogonal on the unit circle disk, that is,

$$\iint_{x^2+y^2\leq 1} V^*_{nm}(x,y)V_{pq}(x,y)\,dx\,dy = \frac{\pi}{n+1}\delta_{np}\delta_{mq}$$

7.8 Show that if a region in an image is rotated by an angle θ with respect to the origin, the Zernike moments of the rotated region are related to the unrotated ones by

$$A'_{pq} = A_{pq}\exp(-jq\theta)$$

7.9 Write a program to compute the moments of Hu. Then apply it to two of the test images involving objects, available from the web, and compute the respective moments.

7.10 Repeat Problem 7.9 for the Zernike moments of order A_{11}, A_{20}, A_{02}.

7.11 Show Eq. (7.37).

7.12 Write a program that computes the AR parameters for a noncausal prediction model. Apply it to a homogeneous and isotropic image whose autocorrelation sequence is given by

$$r(k,l) = \exp(-\sqrt{k^2+l^2})$$

for a window W of order $p = q = 1$.

7.13 Let $u_k = x_k + jy_k$, where (x_k, y_k) are the coordinates of the points on the boundary of an object in an image. Show that if the object is rotated by an angle θ with respect to the origin of the axis, the new complex sequence is

$$u'_k = u_k\exp(j\theta)$$

7.14 Let t denote the length along a closed boundary curve measured from an origin within the curve, and $x(t), y(t)$ the coordinates as functions of t. If T is the total length of the curve, then the following Fourier series expansion holds.

$$x(t) = a_0 + \sum_{n=1}^{\infty}\left[a_n\cos\frac{2\pi nt}{T} + b_n\sin\frac{2\pi nt}{T}\right]$$

$$y(t) = c_0 + \sum_{n=1}^{\infty}\left[c_n\cos\frac{2\pi nt}{T} + d_n\sin\frac{2\pi nt}{T}\right]$$

Prove that if $x(t), y(t)$ are piecewise linear functions between the sampled points $(x(t), y(t)), t = 0, 1, \ldots, m-1$, the Fourier coefficients a_n, b_n, c_n, d_n are given by

$$a_n = \frac{T}{2\pi^2 n^2} \sum_{i=1}^{m} \frac{\Delta x_i}{\Delta t_i} [\cos\phi_i - \cos\phi_{i-1}]$$

$$b_n = \frac{T}{2\pi^2 n^2} \sum_{i=1}^{m} \frac{\Delta x_i}{\Delta t_i} [\sin\phi_i - \sin\phi_{i-1}]$$

$$c_n = \frac{T}{2\pi^2 n^2} \sum_{i=1}^{m} \frac{\Delta y_i}{\Delta t_i} [\cos\phi_i - \cos\phi_{i-1}]$$

$$d_n = \frac{T}{2\pi^2 n^2} \sum_{i=1}^{m} \frac{\Delta y_i}{\Delta t_i} [\sin\phi_i - \sin\phi_{i-1}]$$

where

$$\Delta x_i = x_i - x_{i-1}, \quad \Delta y_i = y_i - y_{i-1}$$

$$\Delta t_i = \sqrt{\Delta x_i^2 + \Delta y_i^2}, \quad t_i = \sum_{j=1}^{i} \Delta t_j$$

$$T = t_m, \quad \phi_i = \frac{2n\pi t_i}{T}$$

7.15 For the Fourier coefficients in Problem 7.14, prove that the following parameters are rotation invariant:

$$I_n = a_n^2 + b_n^2 + c_n^2 + d_n^2$$

$$J_n = a_n d_n - b_n c_n$$

$$K_{1,n} = (a_1^2 + b_1^2)(a_n^2 + b_n^2) + (c_1^2 + d_1^2)(c_n^2 + d_n^2) + 2(a_1 c_1 + b_1 d_1)(a_n c_n + b_n d_n)$$

7.16 If $(x(t), y(t))$ are defined as in Problem 7.14 and

$$z(t) = x(t) + jy(t)$$

the respective complex exponential Fourier series is given as

$$z(t) = \sum_{n=-\infty}^{\infty} a_n \exp(j2\pi nt/T)$$

$$a_n = \frac{1}{T} \int_0^T z(t) \exp(-j2\pi nt/T)\, dt$$

Prove that the following parameters are scale and rotation invariant [Gran 72]:

$$b_n = \frac{a_{1+n}a_{1-n}}{a_1^2}, \quad d_{mn} = \frac{a_{1+m}^n a_{1-n}^m}{a_1^{(m+n)}}$$

where $n \neq 1$.

7.17 Prove that if $x(t), y(t)$ of the previous problem are piecewise linear functions between the points $(x(t), y(t)), t = 0, 1, \ldots, m-1$, then the Fourier coefficients a_n are given by [Lai 81]

$$a_n = \frac{T}{(2\pi n)^2} \sum_{i=1}^{m} (b_{i-1} - b_i) \exp(-jn2\pi t_i/T)$$

where

$$b_i = \frac{V_{i+1} - V_i}{|V_{i+1} - V_i|}, \quad t_i = \sum_{k=1}^{i} |V_k - V_{k-1}|, \quad i > 0, t_0 = 0$$

and $V_i, i = 1, 2, \ldots, m$, the phasors at the respective points.

7.18 Show that the orientation θ in Section 7.3.3 results from minimizing

$$I(\theta) = \sum_i \sum_j [(i - \bar{x}) \cos\theta - (j - \bar{y}) \sin\theta]^2$$

7.19 Show that the power spectrum of an fdm process with Hurst parameter H is given by

$$S(f) \propto \frac{1}{f^{(2H+1)}}$$

7.20 Show that the definition of M in (7.46) results in a consistent metric for the Koch curve.

7.21 Assuming that $x(0) = 0$, show that

$$E[x(n)(x(n+n_0) - x(n))] = \frac{1}{2}\{(n+n_0)^{2H} - n^{2H} - n_0^{2H}\}$$

For the case of a Brownian motion ($H = \frac{1}{2}$) this suggests that $x(n)$ is uncorrelated to the increment. This is not true for $H \neq \frac{1}{2}$, where a nonzero correlation exists, *positive* for $H > 1/2$ and *negative* for $H < 1/2$. To prove this, use, the definition in (7.51). This can be generalized. That is, if $n_1 \leq n_2 \leq n_3 \leq n_4$ and the process is Brownian then

$$E[(x(n_2) - x(n_1))(x(n_4) - x(n_3))] = 0$$

MATLAB PROGRAMS AND EXERCISES

Computer Exercises

7.1 *First-order image statistics.* Write a MATLAB function named *first_order_stats* that computes the first-order statistics of a set of images. Specifically, the function takes as inputs (a) a $num\_in \times q$ dimensional array *name_images*, whose ith row contains the name of the ith image file, (b) the number *N_gray* specifying the range [0, $N\_gray - 1$] in which the intensity of the pixels will be scaled. It returns a $num\_im \times 4$ dimensional matrix *features*, the ith row of which contains the mean, the standard deviation, the skewness and the kurtosis of the intensity of the pixels of the ith image.

Solution

In practice, it is more convenient to work with images where the intensity of the pixels lie in a rather small range of values (e.g. [0, 31]), in order to produce smoother histograms and smaller (nonsparse and easier to handle) co-occurence matrices (see next program). This is the reason for the *N_gray* input argument in the functions of this section.

```
function features=first_order_stats(name_images,N_gray)
  [num_im,q]=size(name_images);
  features=zeros(num_im,4);
  for i=1:num_im
    A=imread(name_images(i,:));
    A=double(A);
    %Normalization of the pixels intensity in [0,N_gray-1]
    A=round((N_gray-1)*((A-min(A(:)))/(max(A(:))-min(A(:)))));
    features(i,1)=mean2(A);
    features(i,2)=std2(A);
    features(i,3)=skewness(A(:));
    features(i,4)=kurtosis(A(:));
  end
```

7.2 *Second-order image statistics.* Write a MATLAB function named *second_order_stats* that computes second-order statistics of a set of images. Specifically, the function takes as inputs: (a) a $num\_in \times q$ dimensional array *name_images*, whose ith row contains the name of the ith image file, (b) the number *N_gray* specifying the range [0, $N\_gray - 1$] in which the intensity of the pixels will be scaled. For each image, four co-occurence matrices determined by the pixel pairs that are at relative positions $(1, 0)^\circ$, $(1, 45^\circ)$, $(1, 90^\circ)$, $(1, 135^\circ)$ should be computed. The function returns a $num\_im \times 8$ dimensional matrix *features*, the ith row of which contains (a) *in its first four entries*: the means of the contrast, the correlation, the angular second moment (in MATLAB termed "Energy") and the inverse difference moment (in MATLAB termed

"Homogeneity") computed from the four co-occurence matrices of the *i*th image and (b) *in its last four entries*: the ranges of the values of the previous features of the *i*th image.

Solution

In the following implementation, we make use of the *graycomatrix* and *graycoprops* built-in MATLAB functions. The first computes the co-occurence matrices of an image, while the second is applied on co-occurence matrices and computes the features 'Contrast', 'Correlation', 'Energy', 'Homogeneity'.

```
function features=second_order_stats(name_images,N_gray)
  [num_im,q]=size(name_images);
  features=zeros(num_im,8);
  for i=1:num_im
    A=imread(name_images(i,:));
    A=double(A);
    %Normalization of the pixels intensity in [0, N_gray-1]
    A=round((N_gray-1)*((A-min(A(:)))/(max(A(:))-min(A(:)))));
    [glcm,SI]=graycomatrix(A,'GrayLimits',[0,N_gray-1],...
    'NumLevels',...
    N_gray,'Offset',[0 1;-1 0;-1 1; -1 -1],'Symmetric',true);
    stats=graycoprops(glcm,{'Contrast','Correlation',...
    'Energy','Homogeneity'});
    features(i,1)=mean(stats.Contrast);
    features(i,2)=mean(stats.Correlation);
    features(i,3)=mean(stats.Energy);
    features(i,4)=mean(stats.Homogeneity);
    features(i,5)=range(stats.Contrast);
    features(i,6)=range(stats.Correlation);
    features(i,7)=range(stats.Energy);
    features(i,8)=range(stats.Homogeneity);
  end
```

7.3 *Second-order image statistics (masks).* Write a MATLAB function named *mask_order_stats* that takes as input a set of original images. Each one of them is convolved with nine masks, and for each one of the nine resulting images first order statistics are computed. Specifically, the function takes as inputs: (a) a $num\_in \times q$ dimensional array *name_images*, whose *i*th row contains the name of the *i*th original image file, (b) the number *N_gray* specifying the range $[0, N\_gray - 1]$ in which the intensity of the pixels will be scaled. The function should convolve each one of the original images with each one of the nine masks, defined in Section 7.2.2, producing nine (convolved) images for each original image. Then, the first-order statistics (mean, standard deviation, skewness, kurtosis) for each one of the convolved images is computed. The function returns a $num\_in \times 4 \times 9$ three-dimensional matrix *features*, where its ith *num_im* $\times$ 4 two-dimensional component corresponds to the results produced when the ith mask is applied to each one of the original images. Each one of the two-dimensional components is defined as in the *first_order_stats* function.

Solution

```
function features=mask_stats(name_images,N_gray)
  [num_im,q]=size(name_images);
  features=zeros(num_im,4,9);
  %Definition of the masks
  mask(:,:,1)=[1 2 1; 2 4 2; 1 2 1];
  mask(:,:,2)=[-1 0 1; -2 0 2; -1 0 1];
  mask(:,:,3)=[-1 2 -1; -2 4 -2; -1 2 -1];
  mask(:,:,4)=[-1 -2 -1; 0 0 0; 1 2 1];
  mask(:,:,5)=[1 0 -1; 0 0 0; -1 0 1];
  mask(:,:,6)=[1 -2 1; 0 0 0; -1 2 -1];
  mask(:,:,7)=[-1 -2 -1; 2 4 2; -1 -2 -1];
  mask(:,:,8)=[1 0 -1; -2 0 2; 1 0 -1];
  mask(:,:,9)=[1 -2 1; -2 4 -2; 1 -2 1];
  % The following is useful in normalizing the convolution result
  sum_mask=sum(sum(mask))+(sum(sum(mask))==0);
  for i=1:num_im
  A=imread(name_images(i,:));
  A=double(A);
  %Normalization of the pixels intensity in [0, N_gray-1]
  A=round((N_gray-1)*((A-min(A(:)))/(max(A(:))-min(A(:)))));
  for j=1:9
  B=conv2(A,mask(:,:,j),'same')/sum_mask(j);
  features(i,1,j)=mean2(B);
  features(i,2,j)=std2(B);
  features(i,3,j)=skewness(B(:));
  features(i,4,j)=kurtosis(B(:));
  end
  end
```

Computer Experiments

Notes:

- All filenames included in the rows of the *name_images* array should have the same number of characters.
- Test images on which the above programs can be applied can be found in www.elsevierdirect.com/9781597492720 ('*ROI_01_seeds.bmp*', '*ROI_02_seeds.bmp*', ..., '*ROI_10_seeds.bmp*'). In the sequel, this set of images is called "set of seeds".

7.1 Compute the first-order statistics for the set of seeds, using $N\_gray = 32$ and comment on the results.

7.2 Compute the second-order statistics for the set of seeds, using $N\_gray = 32$ and comment on the results.

7.3 Compute the first-order statistics of the nine images produced by each image of the set of seeds, after its convolution with each one of the nine masks, given in Section 7.2.2 and comment on the results. Use $N\_gray = 32$.

REFERENCES

[Alem 90] Al-Emami S., Usher M. "On-line recognition of handwritten Arabic characters," *IEEE Transactions on Pattern Analysis and Machine Intelligence*, Vol. 12(7), pp. 704–710, 1990.

[Arbt 89] Arbter K. "Affine-invariant Fourier descriptors," in *From Pixel to Features* (Simon J.C., ed.), pp. 153–164, Elsevier, 1989.

[Arbt 90] Arbter K., Snyder W.E., Burkhardt H., Hirzinger G. "Application of affine-invariant Fourier descriptors to recognition of 3-D objects," *IEEE Transactions on Pattern Analysis and Machine Intelligence*, Vol. 12, pp. 640–647, 1990.

[Ardu 92] Ardunini F., Fioravanti S., Giusto D.D., Inzirillo F. "Multifractals and texture classification," *IEEE International Conference on Image Processing*, pp. 454–457, 1992.

[Ayac 00] Ayache A., Véhel J.L. "The generalized multifractional Brownian motion," *Statistical Inference for Stochastic Processes*, Vol. 3, pp. 7–18, 2000.

[Bart 05] Bartch M., Wakefield G.H. "Audio thumbnailing of popular music using chroma-based representations," *IEEE Transactions on Multimedia*, Vol. 7(1), pp. 96–104, February 2005.

[Bass 92] Bassevile M., Benveniste A., Chou K., Golden S.A., Nikoukhah R., Willsky A.S. "Modeling and estimation of multiresolution stochastic processes," *IEEE Transactions on Information Theory*, Vol. 38, pp. 766–784, 1992.

[Brod 66] Brodatz P. *Textures—A Photographic Album for Artists and Designers*, Dover, 1966.

[Brow 91] Brown J.C., Zhang B. "Musical frequency tracking using the methods of conventional and narrowed autocorrelation," *Journal of the Acoustical Society of America*, Vol. 89(5), 1991.

[Cavo 92] Cavouras D., Prassopoulos P., Pantelidis N. "Image analysis methods for solitary pulmonary nodule characterization by CT," *European Journal of Radiology*, Vol. 14, pp. 169–172, 1992.

[Chel 85] Chellapa R. "Two dimensional discrete Gaussian Markov random field models for image processing," in *Progress in Pattern Recognition* (Kanal L.N., Rosenfeld A., eds.), Vol. 2, pp. 79–112, North Holland, 1985.

[Chen 89] Chen C.C., Daponee J.S., Fox M.D. "Fractal feature analysis and classification in medical imaging," *IEEE Transactions on Medical Imaging*, Vol. 8, pp. 133–142, 1989.

[Chon 03] Chong C.-W., Raveendran P., Mukundan R. "Translation invariants of Zernike moments," *Pattern Recognition*, Vol. 36, pp. 1765–1773, 2003.

[Chon 04] Chong C.-W., Raveendran P., Mukundan R. "Translation and scale invariants of Legendre moments," *Pattern Recognition*, Vol. 37, pp. 119–129, 2004.

[Clau 04] Clausen M., Kurth F. "A unified approach to content-based and fault-tolerant music recognition," *IEEE Transactions on Multimedia*, Vol. 6(5), pp. 717–731, October 2004.

[Crim 82] Crimmins T.R. "A complete set of Fourier descriptors," *IEEE Transactions on Systems Man and Cybernetics*, Vol. 12, pp. 236–258, 1982.

[Cros 83] Cross G.R., Jain A.K. "Markov random field texture models," *IEEE Transactions on Pattern Analysis and Machine Intelligence*, Vol. 5(1), pp. 25–39, 1983.

[Davi 79] Davis L., Johns S., Aggrawal J.K. "Texture analysis using generalized co-occurrence matrices," *IEEE Transactions on Pattern Analysis and Machine Intelligence*, Vol. 1(3), pp. 251–259, 1979.

[Davi 80] Davis S.B., Mermelstein P. "Comparison of parametric representations of monosyllabic word recognition in continuously spoken sentences," *IEEE Transactions on Acoustics, Speech, and Signal Processing*, Vol. 28(4), pp. 357–366, 1980.

[Dell 00] Deller J.R., Hansen J.H.L., Proakis J.G. *Discrete Processing of Speech Signals*, John Wiley & Sons, New York, 2000.

[Deri 93] Deriche M., Tewfik A.H. "Signal modeling with filtered discrete fractional noise processes," *IEEE Transactions on Signal Processing*, Vol. 41, pp. 2839–2850, 1993.

[Falc 90] Falconer K. *Fractal Geometry: Mathematical Foundations and Applications*, John Wiley & Sons, 1990.

[Fieg 96] Fieguth P.W., Willsky A.S. "Fractal estimation using models on multiscale trees," *IEEE Transactions on Signal Processing*, Vol. 41, pp. 1297–1300, 1996.

[Flan 92] Flandrin P. "Wavelet analysis and synthesis of fractional Brownian motion," *IEEE Transactions on Information Theory*, Vol. 38, pp. 910–917, 1992.

[Flus 93] Flusser J., Suk T. "Pattern recognition by affine moment invariants," *Pattern Recognition*, Vol. 26(1), pp. 167–174, 1993.

[Flus 94] Flusser J., Suk T. "Affine moment invariants: A new tool for character recognition," *Pattern Recognition*, Vol. 15(4), pp. 433–436, 1994.

[Frag 01] Fragoulis D., Rousopoulos G., Panagopoulos T., Alexiou C., Papaodysseus C. "On the automated recognition of seriously distorted musical recordings," *IEEE Transactions on Signal Processing*, Vol. 49(4), pp. 898–908, 2001.

[Free 61] Freeman H. "On the encoding of arbitrary geometric configurations," *IRE Transactions on Electronic Computers*, Vol. 10(2), pp. 260–268, 1961.

[Gall 75] Galloway M. "Texture analysis using gray-level run lengths," *Computer Graphics and Image Processing*, Vol. 4, pp. 172–179, 1975.

[Gewe 83] Geweke J., Porter-Hudak S. "The estimation and application of long memory time series," *Journal of Time Series Analysis*, Vol. 4, pp. 221–237, 1983.

[Ghos 97] Ghosal S., Mehrotra R. "A moment based unified approach to image feature detection," *IEEE Transactions on Image Processing*, Vol. 6(6), pp. 781–794, 1997.

[Glen 94] Glentis G., Slump C., Herrmann O. "An efficient algorithm for two-dimensional FIR filtering and system identification," *SPIE Proceedings*, VCIP, pp. 220–232, Chicago, 1994.

[Goto 04] Goto M. "A real-time music-scene-description system: Predominant - F0 estimation for detecting melody and bass lines in real-world audio signals," *Speech Communication (ISCA) Journal*, Vol. 43(4), pp. 311–329, 2004.

[Gran 72] Granlund G.H. "Fourier preprocessing for hand print character recognition," *IEEE Transactions on Computers*, Vol. 21, pp. 195–201, 1972.

[Hara 73] Haralick R., Shanmugam K., Distein I. "Textural features for image classification," *IEEE Transactions on Systems Man and Cybernetics*, Vol. 3(6), pp. 610–621, 1973.

[Hayk 96] Haykin S. *Adaptive Filter Theory*, 3rd ed., Prentice Hall, 1996.

[Heyw 95] Heywodd M.I., Noakes P.D. "Fractional central moment method for movement-invariant object classification," *IEE Proceedings Vision, Image and Signal Processing*, Vol. 142 (4), pp. 213–219, 1995.

[Hu 62] Hu M.K. "Visual pattern recognition by moment invariants," *IRE Transactions on Information Theory*, Vol. 8(2), pp. 179–187, 1962.

[Huan 94] Huang Q., Lorch J.R., Dubes R.C. "Can the fractal dimension of images be measured?" *Pattern Recognition*, Vol. 27(3), pp. 339–349, 1994.

[Huan 06] Huang S.-K., Kim W.-Y. "A novel approach to the fast computation of Zernike moments," *Pattern Recognition*, Vol. 39(11), pp. 2065–2076, 2006.

[Kalo 89] Kalouptsidis N., Theodoridis S. "Concurrent algorithms for a class of 1-D and 2-D Wiener filters with symmetric impulse response," *IEEE Transactions on Signal Processing*, Vol. ASSP-37, pp. 1780–1782, 1989.

[Kan 02] Kan C., Srinath M.D. "Invariant character recognition with Zernike moments and orthogonal Fourier-Mellin moments," *Pattern Recognition*, Vol. 35, pp. 143–154, 2003.

[Kapl 99] Kaplan L.M. "Extended fractal analysis for the texture classification and segmentation," *IEEE Transactions on Image Processing*, Vol. 8(11), pp. 1572–1585, 1999.

[Kapl 94] Kaplan L.M., Kuo C.-C.J. "Extending self similarity for fractional Brownian motion," *IEEE Transactions on Signal Processing*, Vol. 42(12), pp. 3526–3530, 1994.

[Kapl 95] Kaplan L.M., Kuo C.C.J. "Texture roughness analysis and synthesis via extended self-similar model," *IEEE Transactions on Pattern Analysis and Machine Intelligence*, Vol. 17(11), pp. 1043–1056, 1995.

[Kara 95] Karayannis Y.A., Stouraitis T. "Texture classification using fractal dimension as computed in a wavelet decomposed image," *IEEE Workshop on Nonlinear Signal and Image Processing*, pp. 186–189, Neos Marmaras, Halkioliki, June 95.

[Kash 82] Kashyap R.L., Chellapa R., Khotanzad A. "Texture classification using features derived from random field models," *Pattern Recognition Letters*, Vol. 1, pp. 43–50, 1982.

[Kash 86] Kashyap R.L., Khotanzad A. "A model based method for rotation invariant texture classification," *IEEE Transactions on Pattern Analysis and Machine Intelligence*, Vol. 8(4), pp. 472–481, 1986.

[Kell 87] Keller J.M., Crownover R., Chen R.Y. "Characteristics of natural scenes related to fractal dimension," *IEEE Transactions on Pattern Analysis and Machine Intelligence*, Vol. 9, pp. 621–627, 1987.

[Khot 90a] Khotanzad A., Hong Y.H. "Invariant image recognition by Zernike moments," *IEEE Transactions on Pattern Analysis and Machine Intelligence*, Vol. 12(5), pp. 489–497, 1990.

[Khot 90b] Khotanzad A., Lu J.H. "Classification of invariant image representations using a neural network," *IEEE Transactions on Acoustics Speech and Signal Processing*, Vol. 38(6), pp. 1028–1038, 1990.

[Klap 03] Klapuri A. "Multiple fundamental frequency estimation by harmonicity and spectral smoothness," *IEEE Transactions on Speech and Audio Processing*, Vol. 11(6), pp. 804–816, 2003.

[Kuhl 82] Kuhl F.P., Giardina C.R. "Elliptic Fourier features of a closed contour," *Comput. Vis. Graphics Image Processing*, Vol. 18, pp. 236–258, 1982.

[Lai 81] Lai M., Suen Y.C. "Automatic recognition of characters by Fourier descriptors and boundary line encoding," *Pattern Recognition*, Vol. 14, pp. 383–393, 1981.

[Laws 80] Laws K.I. "Texture image segmentation" Ph.D. Thesis, University of Southern California, 1980.

[Liao 96] Liao S., Pawlak M. "On image analysis by moments," *IEEE Transactions on Pattern Analysis and Machine Intelligence*, Vol. 18(3), pp. 254–266, March 1996.

[Lin 87] Lin C.S., Hwang C.L. "New forms of shape invariants from elliptic Fourier descriptors," *Pattern Recognition*, Vol. 20(5), pp. 535–545, 1987.

[Lu 91] Lu S.Y., Ren Y., Suen C.Y. "Hierarchical attributed graph representation and recognition of handwritten Chinese characters," *Pattern Recognition*, Vol. 24(7), pp. 617–632, 1991.

[Lund 86] Lundahl T., Ohley W.J., Kay S.M., Siffer R. "Fractional Brownian motion: A maximum likelihood estimator and its application to image texture," *IEEE Transactions on Medical Imaging*, Vol. 5, pp. 152–161, 1986.

[Mahm 94] Mahmoud S. "Arabic character recognition using Fourier descriptors and character contour encoding," *Pattern Recognition*, Vol. 27(6), pp. 815–824, 1994.

[Mami 98] Mamistvalov A.G. "n-Dimensional moment invariants and the conceptual mathematical theory of recognition n-dimensional objects," *IEEE Transactions on Pattern Analysis and Machine Intelligence*, Vol. 20(8), pp. 819–831, 1998.

[Mand 68] Mandelbrot B.B, Van Ness J.W. "Fractional Brownian motion, fractional noises and applications," *SIAM Review*, Vol. 10, pp. 422–437, 1968.

[Mand 77] Manderbrot B.B. *The Fractal Geometry of Nature*, W.H. Freeman, New York, 1982.

[Mao 92] Mao J., Jain A.K. "Texture classification and segmentation using multiresolution simultaneous autoregressive models," *Pattern Recognition*, Vol. 25(2), pp. 173–188, 1992.

[Mara 93] Maragos P., Sun F.K. "Measuring the fractal dimension of signals: Morphological covers and iterative optimization," *IEEE Transactions on Signal Processing*, Vol. 41, pp. 108–121, 1993.

[Mori 92] Mori S., Suen C. "Historical review of OCR research and development," *Proceedings of IEEE*, Vol. 80(7), pp. 1029–1057, 1992.

[Muku 95] Mukundan R., Ramakrshnan J. "Fast computation of Legendre and Zernike moments," *Pattern Recognition*, Vol. 28(9), pp. 1433–1442, 1995.

[Muku 98] Mukundan R., Ramakrshnan J. *Moment Functions in Image Analysis-Theory and Applications*, World Scientific, Singapore, 1998.

[Ohan 92] Ohanian P., Dubes R. "Performance evaluation for four classes of textural features," *Pattern Recognition*, Vol. 25(8), pp. 819–833, 1992.

[Ojal 96] Ojala T., Pietikainen M., Harwood D. "A comparative study of texture measures with classification based on feature distributions," *Pattern Recognition*, Vol. 29(1), pp. 51–59, 1996.

[Papo 91] Papoulis A. *Probability, Random Variables, and Stochastic Processes*, 3rd ed., McGraw-Hill, 1991.

[Pele 84] Peleg S., Naor J., Hartley R., Anvir D. "Multiple resolution texture analysis and classification," *IEEE Transactions on Pattern Analysis and Machine Intelligence*, Vol. 6, pp. 818–523, 1984.

[Penn 97] Penn A.I., Loew M.H. "Estimating fractal dimension with fractal interpolation function models," *IEEE Transcations on Medical Imaging*, Vol. 16, pp. 930–937, 1997.

[Pent 84] Pentland A. "Fractal based decomposition of natural scenes," *IEEE Transactions on Pattern Analysis and Machine Intelligence*, Vol. 6(6), pp. 661–674, 1984.

[Pers 77] Persoon E., Fu K.S. "Shape discrimination using Fourier descriptors," *IEEE Transactions on Systems Man and Cybernetics*, Vol. 7, pp. 170–179, 1977.

[Petr 06] Petrou M., Sevilla P.G. *Image Processing: Dealing with Texture*, John Wiley & Sons, 2006.

[Pesq 02] Pesquet-Popescu B., Vehel J.L. "Stochastic fractal models for image processing," *IEEE Signal Processing Magazine*, Vol. 19(5), pp. 48–62, 2002.

[Pico 93] Picone J. "Signal modeling techniques in speech recognition," *Proceedings of the IEEE*, Vol. 81(9), pp. 1215–1247, 1993.

[Pikr 03] Pikrakis A., Theodoridis S., Kamarotos D. "Recognition of isolated musical patterns using context dependent dynamic time warping," *IEEE Transactions on Speech and Audio Processing*, Vol. 11(3), pp. 175–183, 2003.

[Pikr 06] Pikrakis A., Theodoridis S., Kamarotos D. "Classification of musical patterns using variable duration hidden Markov models," *IEEE Transactions on Speech and Audio Processing*, to appear in 2006.

[Pikr 08] Pikrakis A., Gannakopoulos T., Theodoridis S. "A speech-music discriminator of radio recordings based on dynamic programming and Bayesian networks," *IEEE Transactions on Multimedia*, Vol. 10(5), pp. 846–856, 2008.

[Pita 94] Pitas I. *Image Processing Algorithms*, Prentice Hall, 1994.

[Plam 00] Plamondon R., Srihari S.N. "On-line and off-line handwriting recognition: A comprehensive survey," *IEEE Transactions on Pattern Analysis and Machine Intelligence*, Vol. 22(1), pp. 63–84, 2000.

[Proa 92] Proakis J., Manolakis D. *Digital Signal Processing: Principles, Algorithms, and Applications*, 2nd ed., Macmillan, 1992.

[Rabi 93] Rabiner L., Juang B.H. *Fundamentals of Speech Recognition*, Prentice Hall, Englewood Cliffs, NJ, 1993.

[Rabi 78] Rabiner L.R., Schafer R.W. *Digital Processing of Speech Signals*, Prentice Hall, 1978.

[Rand 99] Randen T., Husoy H.H. "Filtering for texture classification: A comparative study," *IEEE Transactions on Pattern Analysis and Machine Intelligence*, Vol. 21(4), pp. 291–310, 1999.

[Reis 91] Reiss T.H. "The revised fundamental theorem of moment invariants," *IEEE Transactions on Pattern Analysis and Machine Intelligence*, Vol. 13, pp. 830–834, 1991.

[Rich 95] Richardson W. "Applying wavelets to mammograms," *IEEE Engineering in Medicine and Biology*, Vol. 14, pp. 551–560, 1995.

[Sark 97] Sarkar A., Sharma K.M.S., Sonak R.V. "A new approach for subset 2-D AR model identification for describing textures," *IEEE Transactions on Image Processing*, Vol. 6(3), pp. 407–414, 1997.

[Saup 91] Saupe D. "Random fractals in image processing," in *Fractals and Chaos* (Crilly A.J., Earnshaw R.A., Jones H., eds.), pp. 89–118, Springer-Verlag, 1991.

[Schr 68] Schroeder M.R. "Period histogram and product spectrum: New methods for fundamental frequency measurement," *Journal of Acoustical Society of America*, Vol. 43(4), pp. 829–834, 1968.

[Sing 06] Singh C. "Improved quality of reconstructed images using floating point arithmetic for moment calculation," *Pattern Recognition*, Vol. 39(11), pp. 2047–2064, 2006.

[Tamu 78] Tamura H., Mori S., Yamawaki T. "Textural features corresponding to visual Perception," *IEEE Transactions on Systems, Man, and Cybernetics*, Vol. 8(6), pp. 460–473, 1978.

[Tang 98] Tang X. "Texture information in run-length matrices," *IEEE Transactions on Image Processing*, Vol. 7(11), pp. 1602–1609, 1998.

[Taxt 90] Taxt T., Olafsdottir J.B., Daechlen M. "Recognition of hand written symbols," *Pattern Recognition*, Vol. 23(11), pp. 1155–1166, 1990.

[Teag 80] Teague M. "Image analysis via the general theory of moments," *Journal of Optical Society of America*, Vol. 70(8), pp. 920–930, 1980.

[Teh 88] Teh C.H., Chin R.T. "On image analysis by the method of moments," *IEEE Transactions on Pattern Analysis and Machine Intelligence*, Vol. 10(4), pp. 496–512, 1988.

[Theo 93] Theodoridis S., Kalouptsidis N. "Spectral analysis," in *Adaptive System Identification and Signal Processing Algorithms* (Kalouptsidis N., Theodoridis S., eds.), Prentice Hall, 1993.

[Tolo 00] Tolonen T., Karjalainen M. "A computationally efficient multipitch analysis model," *IEEE Transactions on Speech and Audio Processing*, Vol. 8(6), pp. 708–716, November 2000.

[Trie 95] Trier O.D., Jain A.K. "Goal-directed evaluation of binarization methods," *IEEE Transactions on Pattern Analysis and Machine Intelligence*, Vol. 17(12), pp. 1191–1201, 1995.

[Trie 96] Trier O.D., Jain A.K., Taxt T. "Feature extraction methods for character recognition—A survey," *Pattern Recognition*, Vol. 29(4), pp. 641–661, 1996.

[Tson 92] Tsonis A. *Chaos: From Theory to Applications*, Plenum Press, 1992.

[Tzan 02] Tzanetakis G., Cook P. "Musical genre classification of audio signals," *IEEE Transactions on Speech and Audio Processing*, Vol. 10(5), pp. 293–302, 2002.

[Unse 86] Unser M. "Local linear transforms for texture measurements," *Signal Processing*, Vol. 11, pp. 61–79, 1986.

[Unse 89] Unser M., Eden M. "Multiresolution feature extraction and selection for texture segmentation," *IEEE Transactions on Pattern Analysis and Machine Intelligence*, Vol. 11(7), pp. 717–728, 1989.

[Vinc 02] Vinciarelli A. "A survey on off-line cursive word recognition," *Pattern Recognition*, Vol. 35, pp. 1433–1446, 2002.

[Wang 98] Wang L., Healey G. "Using Zernike moments for the illumination and geometry invariant classification of multispectral textures," *IEEE Transactions on Pattern Analysis and Machine Intelligence*, Vol. 7(2), pp. 196–203, 1998.

[Wang 93] Wang L., Pavlidis T. "Direct gray-scale extraction of features for character recognition," *IEEE Transactions on Pattern Analysis and Machine Intelligence*, Vol. 15(10), pp. 1053–1067, 1993.

[Wang 00] Wang Y., Huang J.C. "Multimedia content analysis," *IEEE Signal Processing Magazine*, Vol. 17(6), pp. 12–36, 2000.

[Wee 06] Wee C.-Y., Paramesran R. "Efficient computation of radial moment functions using symmetrical property," *Pattern Recognition*, Vol. 39(11), pp. 2036–2046, 2006.

[Wold 96] Wold E., Blum T., Keislar D., Wheaton J. "Content-based classification, search, and retrieval of audio," *IEEE Multimedia Magazine*, Vol. 22, pp. 27–36, 1996.

[Wood 72] Woods J. "Markov image modeling," *IEEE Transactions on Information Theory*, Vol. 18(3), pp. 232–240, 1972.

[Wood 96] Wood J. "Invariant pattern recognition," *Pattern Recognition*, Vol. 29(1), pp. 1–17, 1996.

[Worn 96] Wornell W.G. *Signal Processing with Fractals. A Wavelet Based Approach*, Prentice Hall, 1996.

[Wu 03] Wu M., Wang D., Brown G.J. "A multipitch tracking algorithm for noisy speech," *IEEE Transactions on Speech and Audio Processing*, Vol. 11(3), pp. 229–241, May 2003.

[Zhan 01] Zhang T., Kuo C.C.J. "Audio content analysis for online audiovisual data segmentation and classification," *IEEE Transactions on Speech and Audio Processing*, Vol. 9(4), pp. 441–458, 2001.

CHAPTER

Template Matching

8

8.1 INTRODUCTION

In all previous chapters, the major concern was to assign an unknown pattern to one of the possible classes. The problem that will accompany us throughout this chapter is of a slightly different nature. We will assume that a set of reference patterns (*templates*) are available to us, and we have to decide which one of these reference patterns an unknown one (the *test pattern*) matches best. These templates may be certain objects in a scene or can be strings of patterns, such as letters forming words in a written text or words or phrases in a spoken text. Typically, such problems arise in speech recognition, in automation using robot vision, in motion estimation for video coding, and in image database retrieval systems, to name but a few. A reasonable first step to approaching such a task is to define a *measure* or a *cost* measuring the "distance" or the "similarity" between the (known) reference patterns and the (unknown) test pattern, in order to perform the matching operation known as *template matching*. We know by now that each pattern is expressed in terms of a vector or a matrix with elements the set of the selected features. Then why not use one of the already known distance measures, that is, Euclidean or Frobenius norms, and perform the matching operation based on the minimum distance? A little more thinking reveals that such a straightforward approach is not enough and something more is needed. This is the crucial point that makes template matching different and at the same time interesting.

To understand this issue better, let us consider a written text matching problem, that is, to identify which one from a set of written words is the word, say, "beauty." However, because of errors in the reading sensors, the specific test pattern may appear, for example, as "beety" or "beaut." In a speech recognition task, if a specific word is spoken by the same speaker a number of times, it will be spoken differently every time. Sometimes it will be spoken quickly, and the resulting pattern will be of short duration in time, sometimes slowly, and the pattern will be longer. Yet in all cases it is the "same" word spoken by the same person. In a scene analysis application, the object to be identified may be present in an image, but its location within the image is not known. In content-based image database retrieval systems, queries often include the shape of an object. However, the shape provided by the user, most often, does not match exactly the shape of the object residing in the database images. The major goal of this chapter is to define measures that accommodate the distinct characteristics for each category of these problems. As is always the case with a textbook, only general directions and typical cases will be treated.

We will begin with the problem of string pattern matching and will deal with the scene analysis and shape recognition problems later on. The tasks, although they share the same goal, require different tools because of their different nature.

8.2 MEASURES BASED ON OPTIMAL PATH SEARCHING TECHNIQUES

We will first focus on a category of template matching, where the involved patterns consist of strings of identified symbols or feature vectors (string patterns). That is, each of the reference and test patterns is represented as a sequence (string) of measured parameters, and one has to decide which reference sequence the test-pattern matches best.

Let $\boldsymbol{r}(i), i=1,2,\ldots,I$, and $\boldsymbol{t}(j), j=1,2,\ldots,J$, be the respective feature vector sequences for a specific pair of reference and test patterns, where in general $I \neq J$. The objective is to develop an appropriate distance measure between the two sequences. To this end, we form a two-dimensional grid with the elements of the two sequences as points on the respective axes, that is, the reference string at the abscissa (i-axis) and the test one at the ordinate (j-axis). Figure 8.1 is an example for $I = 6$, $J = 5$. Each point of the grid (node) marks a correspondence between the respective elements of the two sequences. For example, node $(3, 2)$ maps the element $\boldsymbol{r}(3)$ to $\boldsymbol{t}(2)$. Each node (i, j) of the grid is associated with a *cost*, which is an appropriately defined function $d(i, j)$ measuring the "distance" between the respective elements of the strings, $\boldsymbol{t}(j)$ and $\boldsymbol{r}(i)$. A path through the grid from an initial node (i_0, j_0) to a final one (i_f, j_f) is an *ordered* set of nodes of the form

$$(i_0, j_0), (i_1, j_1), (i_2, j_2), \ldots, (i_f, j_f)$$

Each path is associated with an overall cost D defined as

$$D = \sum_{k=0}^{K-1} d(i_k, j_k)$$

where K is the number of nodes along the path. For the example of Figure 8.1, $K = 8$. The overall cost up to node (i_k, j_k) will be denoted by $D(i_k, j_k)$, and by convention we assume $D(0, 0) = 0$ and also $d(0, 0) = 0$. The path is said to be *complete* if

$$(i_0, j_0) = (0, 0), \ (i_f, j_f) = (I, J)$$

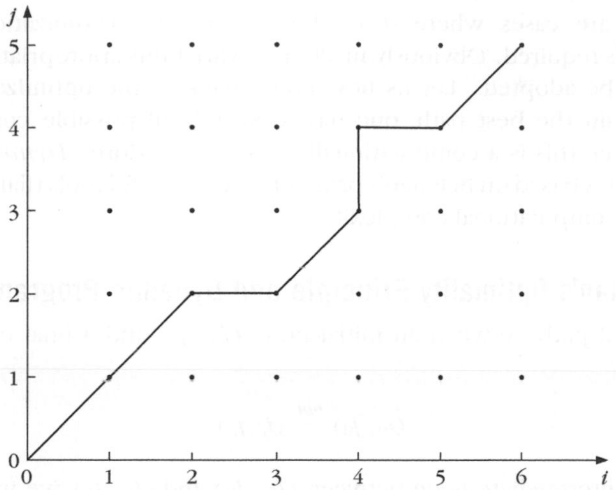

FIGURE 8.1

Each point along the path marks a correspondence between the respective elements of the test and reference patterns.

The distance[1] *between the two sequences is defined as the minimum D over all possible paths.* At the same time, the minimum cost path unravels the optimal pairwise correspondence between the elements of the two sequences, which is the crucial part, since the two sequences are of different lengths. In other words, the optimal path procedure makes the *alignment or warping* of the elements of the test string to the elements of the reference string, corresponding to the best matching score. Before we talk about the optimization procedure, we must point out that there are a number of variations of this scheme. For example, one may not impose the constraint of having necessarily a complete path but may adopt more relaxed constraints known as *end point constraints*. Furthermore, one could associate a cost not only with each node but also with each transition between nodes, making certain transitions more costly than others. In such cases, the cost at a node (i_k, j_k) also depends on the specific transition, that is, from which node (i_{k-1}, j_{k-1}) the (i_k, j_k) node was reached. Thus, the cost d is now of the form $d(i_k, j_k|i_{k-1}, j_{k-1})$ and the overall path cost is

$$D = \sum_k d(i_k, j_k|i_{k-1}, j_{k-1})$$

In some cases the overall path cost is defined as the product

$$D = \prod_k d(i_k, j_k|i_{k-1}, j_{k-1})$$

[1] The term *distance* here must not be interpreted with its strict mathematical definition.

Finally, there are cases where d is chosen so that maximization instead of minimization is required. Obviously, in all these variations appropriate initial conditions have to be adopted. Let us now come back to the optimization problem itself. To obtain the best path, one has to search all possible combinations of paths. However, this is a computationally costly procedure. *Dynamic programming algorithms* based on Bellman's principle are powerful tools that we will adopt to reduce the computational complexity.

8.2.1 Bellman's Optimality Principle and Dynamic Programming

Let the optimal path between an initial node (i_0, j_0) and a final one (i_f, j_f) be denoted as

$$(i_0, j_0) \xrightarrow{opt} (i_f, j_f)$$

If (i, j) is an intermediate node between (i_0, j_0) and (i_f, j_f), we will denote the optimal path constrained to pass through (i, j) as

$$(i_0, j_0) \underset{(i,j)}{\xrightarrow{opt}} (i_f, j_f)$$

Bellman's principle states that [Bell 57]

$$(i_0, j_0) \underset{(i,j)}{\xrightarrow{opt}} (i_f, j_f) = (i_0, j_0) \xrightarrow{opt} (i, j) \oplus (i, j) \xrightarrow{opt} (i_f, j_f)$$

where $\oplus$ denotes concatenation of paths. In other words, Bellman's principle states that the overall optimal path from (i_0, j_0) to (i_f, j_f) through (i, j) is the concatenation of the optimal path from (i_0, j_0) to (i, j) and the optimal path from (i, j) to (i_f, j_f). The consequence of this principle is that once we are at (i, j) through the optimal path, then to reach (i_f, j_f) optimally we need *to search only* for the optimal path from (i, j) to (i_f, j_f).

Let us now express this in a way that will be useful to us later on. Assume that we have departed from (i_0, j_0) and let the kth node of the path be (i_k, j_k). The goal is to compute the minimum cost required to reach the latter node. The transition to (i_k, j_k) has to take place from one of the possible nodes that are allowed to be in the $(k-1)$th position of the path (that is, the (i_{k-1}, j_{k-1}) node). This is important. For *each node* of the grid we assume that there is a set of allowed *predecessors*, defining the so-called *local constraints*. Bellman's principle readily leads to

$$D_{\min}(i_k, j_k) = \min_{i_{k-1}, j_{k-1}} [D_{\min}(i_{k-1}, j_{k-1}) + d(i_k, j_k | i_{k-1}, j_{k-1})] \tag{8.1}$$

Indeed, the overall minimum cost to reach node (i_k, j_k) is the minimum cost up to node (i_{k-1}, j_{k-1}) plus the extra cost of the transition from (i_{k-1}, j_{k-1}) to (i_k, j_k). Furthermore, the search for the minimum is constrained only within the set of allowable predecessors for the (i_k, j_k) node. This procedure is carried out for all the nodes of the grid. However, in many cases not all nodes of the grid are involved, and the optimal path searching takes place among a subset of the nodes, which are defined via the so-called *global constraints*. The resulting algorithm is known as *dynamic programming*. Equation (8.1) has to be modified accordingly if the cost D is given in its multiplicative form and/or if maximization is required.

Let us now focus on our string pattern matching task and see how the ***recursive*** equation (8.1) is used to construct the optimal complete path.

Figure 8.2 illustrates the procedure. The set of nodes involved in the optimization (global constraints) is denoted as dark dots, and the local constraints, defining the allowable transitions among these nodes, are shown in the figure by the black lines. Having decided to search for the complete path and assuming $D(0, 0) = 0$, the respective costs $D(i_1, j_1)$ for all the allowed nodes involved in step $k = 1$ are computed, via (8.1) (in our case there are only two allowable nodes, ((1, 1) and (1, 2)). Subsequently, the costs of the (three) nodes at step $k = 2$ are computed, and the procedure is repeated until we arrive at the final node (I, J). The sequence of transitions leading to the minimum $D(I, J)$ of the final node defines the minimum cost path, denoted by the red line. The optimal node correspondence, between the test and reference patterns, can then be unraveled by ***backtracking the optimal path***. In the example of Figure 8.2, each step k of the recursion involves only nodes with the same abscissa coordinate, which reflects the local constraints

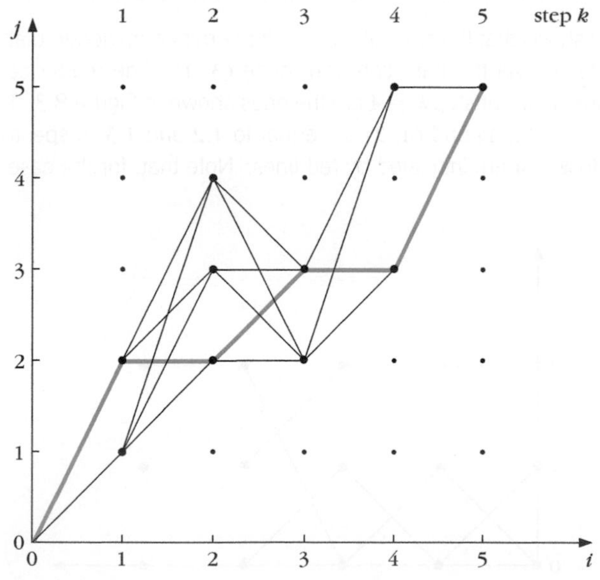

FIGURE 8.2

The optimal path (red line) is constructed by searching among all allowable paths, as defined by the global and local constraints. The optimal node correspondence, between the test and reference patterns, is unraveled by backtracking the optimal path.

adopted. In general, this is not necessary, and more involved topologies may be used. However, the philosophy of the search for the minimum remains the same. In the following subsections, we will apply the procedure in two different popular application areas.

Example 8.1

Figure 8.3 shows the optimal paths (black lines) to reach the nodes at step $k = 3$ starting from the nodes at step $k = 0$. The grid contains three nodes per step. Only the optimal paths, up to step $k = 3$, have been drawn. The goal of this example is to extend the previous paths to the next step and compute the optimal paths terminating at the three nodes at step $k = 4$. Bellman's principle will be employed. Assume that the accumulated costs of the optimal paths $D_{\min}(3, j_3), j_3 = 0, 1, 2$ at the respective nodes are:

$$D_{\min}(3, 0) = 0.8, D_{\min}(3, 1) = 1.2, D_{\min}(3, 2) = 1.0 \tag{8.2}$$

We are also given the transition costs $d(4, j_4|3, j_3), j_3 = 0, 1, 2, j_4 = 0, 1, 2$, in the form of a transition matrix in Table 8.1. In other words, the transition cost, for example, from node (3, 1) to node (4, 2) is equal to 0.2. To obtain the optimal path to node (4, 0) one has to combine the values given in (8.2) and the corresponding transition costs provided in Table 8.1. Thus

Total cost for the transition from (3, 0) to (4, 0) is equal to $0.8 + 0.8 = 1.6$.
Total cost for the transition from (3, 1) to (4, 0) is equal to $1.2 + 0.2 = 1.4$.
Total cost for the transition from (3, 2) to (4, 0) is equal to $1.0 + 0.7 = 1.7$.

Applying Eq. (8.1) shows that the optimal path, with the minimum accumulated cost, to reach node (4, 0) is obtained via the transition from node (3, 1). The reader can verify that the optimal paths to the nodes at step $k = 4$ are the ones shown in Figure 8.3. The optimal costs associated with nodes (4, 1) and (4, 2) are equal to 1.2 and 1.3, respectively. Transitions from step $k = 3$ to $k = 4$ are indicated by red lines. Note that, for the case of our example,

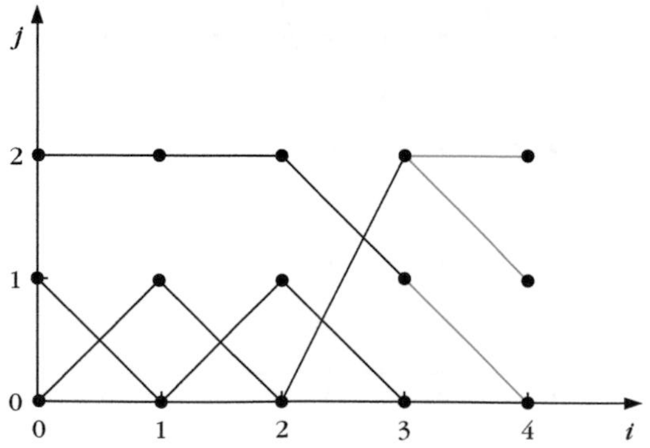

FIGURE 8.3

Optimal paths for the grid of example 8.1. Red lines correspond to the extensions of the optimal paths from step $k = 3$ to step $k = 4$.

Table 8.1 Transition Costs Between Nodes for the Example 8.1

Nodes	(4, 0)	(4, 1)	(4, 2)
(3, 0)	0.8	0.6	0.8
(3, 1)	0.2	0.3	0.2
(3, 2)	0.7	0.2	0.3

the path originating from node (0,1) will not take place in the computations if more steps are added, that is, $k = 5, 6, \ldots$. As we say, this path does not survive beyond step $k = 3$.

8.2.2 The Edit Distance

In this section, we will be concerned with patterns that consist of sets of *ordered symbols*. For example, if these symbols are letters, then the patterns are words from a written text. Such problems arise in automatic editing and text retrieval applications. Other examples of symbol strings occur in structural pattern recognition. Once the symbols of a (test) pattern have been identified, for example, via a reading device, the task is to recognize the pattern, searching for the best match of it against a set of reference patterns. The measure to be adopted for the matching procedure should take into account the following errors, which may occur during the symbol identification phase.

- Wrongly identified symbol (e.g., "befuty" instead of "beauty")
- Insertion error (e.g., "bearuty")
- Deletion error (e.g., "beuty")

Obviously, a combination of these errors may also occur. For the matching procedure we will adopt the philosophy behind the so-called *variational similarity*. In other words, the similarity between two patterns is based on the "cost" associated with converting one pattern to the other. If the patterns are of the same length, then the cost is directly related to the number of symbols that have to be changed in one of them so that the other pattern results. More interest arises when the two patterns are not of equal length. In such cases symbols have to be either deleted or inserted at certain places of the test string. The location where deletions or insertions are to be made presupposes an optimal alignment (warping) among the symbols of the two patterns. The Edit distance [Dame 64, Leven 66] between two string patterns A and B, denoted $D(A, B)$, is defined as *the minimum total number of changes C, insertions I, and deletions R* required to change pattern A into pattern B,

$$D(A, B) = \min_{j}[C(j) + I(j) + R(j)] \tag{8.3}$$

where j runs over all possible combinations of symbol variations in order to obtain B from A. To elaborate a bit, note that there is more than one way to change, say, "beuty" to "beauty." For example, one can either insert "a" after "e" or change "u" to "a" and then insert "u."

We will employ the dynamic programming methodology to compute the required minimum in (8.3). To this end, we form the grid by placing the symbols of the reference pattern in the abscissa axis and the test pattern in the ordinate one. Figure 8.4 demonstrates the procedure via four examples. As already pointed out, the first step in an optimal path searching procedure via dynamic programming

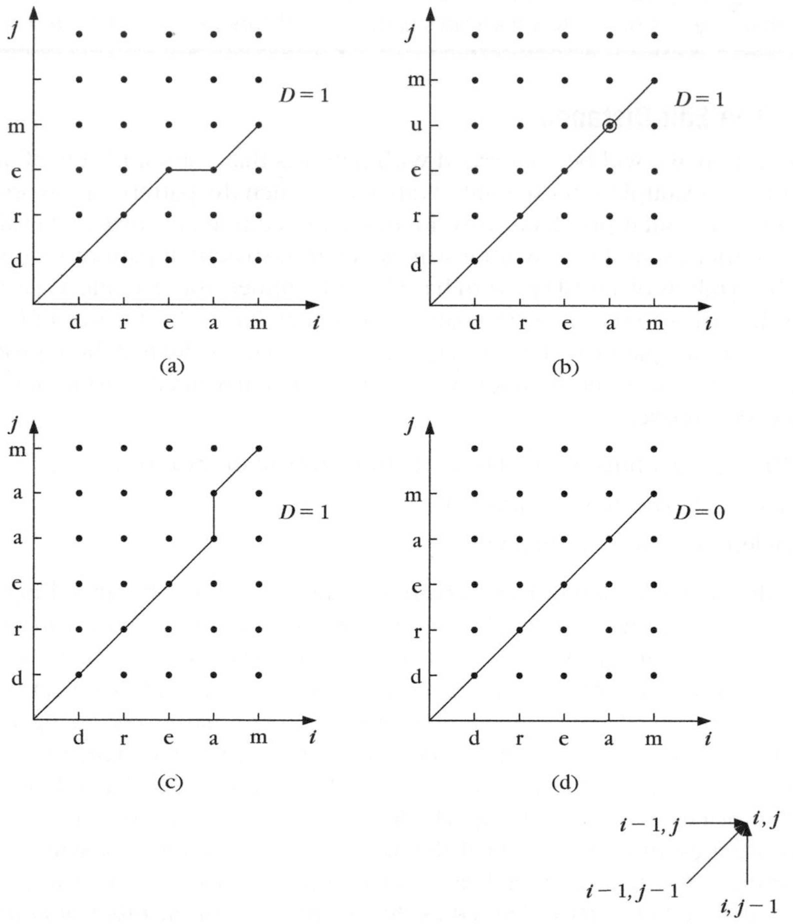

FIGURE 8.4

Computation of the Edit distance with (a) an insertion, (b) a change, (c) a deletion, and (d) an equality. The local constraints are shown at the bottom right corner.

techniques is to state the node transition constraints imposed by the problem. For our case of interest, the following constraints are adopted.

- The cost $D(0, 0)$ of the $(0, 0)$ node is zero.
- A complete path is searched.
- Each node (i, j) can be reached only through three allowable predecessors, that is,

$$(i-1, j), \quad (i-1, j-1), \quad (i, j-1)$$

as indicated at the bottom of Figure 8.4.

The costs associated with the above three transitions are:

1. Diagonal transitions:

$$d(i, j|i-1, j-1) = \begin{cases} 0 & \text{if } r(i) = t(j) \\ 1 & \text{if } r(i) \neq t(j) \end{cases}$$

That is, the cost of a transition is zero if the symbols corresponding to the (i, j) node are the same and unity if they are different; hence a symbol change has to take place.

2. Horizontal and vertical transitions:

$$d(i, j|i-1, j) = d(i, j|i, j-1) = 1$$

The meaning of horizontal transitions is that they attempt alignment of the two strings by insertion of a symbol; see Figure 8.4a. Thus, they add to the cost, because they imply local mismatch. Similarly, vertical transitions add to the cost because they imply deletions, Figure 8.4c.

Incorporating these constraints and the distance (8.3) in a dynamic programming procedure, the following algorithm results.

Algorithm for Computing the Edit Distance

- $D(0, 0) = 0$
- For $i = 1$ to I
  - $D(i, 0) = D(i-1, 0) + 1$
- End { For }
- For $j = 1$ to J
  - $D(0, j) = D(0, j-1) + 1$
- End { For }

- For $i = 1$ to I
  - For $j = 1$ to J
    - $c1 = D(i-1, j-1) + d(i,j|i-1,j-1)$
    - $c2 = D(i-1, j) + 1$
    - $c3 = D(i, j-1) + 1$
    - $D(i,j) = \min(c1, c2, c3)$
  - End { For }
- End { For }
- $D(A, B) = D(I, J)$

In other words, we first compute the minimum cost for reaching *each node* of the grid, starting at (0, 0), and the optimal (complete) path is subsequently constructed. Figure 8.4 shows the respective minimum cost paths and the resulting Edit distances for each of the cases. Verify that any other path for the examples of Figure 8.4 results in a higher cost.

The Edit distance is also known as *Levenstein distance*. Over the years a number of variants of the previous basic Edit distance scheme have been suggested to better address problems rising in various applications. In [Ocud 76] the cost for a change of one symbol to another is allowed to take values different to one, depending on the dependence between different symbols in different applications. For example, for the spelling correction task, it is reasonable to assume that changing an "a" to a "q" results in lower cost than changing an "a" to a "b." This is because in touch typing the letters "a" and "q" are typed using the same finger whereas "a" and "b" are not. Another generalization as suggested in [Seni 96] allows for merges, splits, and two-letter substitutions in the context of handwriting recognition.

A drawback of the basic Edit distance scheme is that it takes no account of the length of the string sequences that are compared. Thus, for example, if two string sequences differ in one symbol, their Edit distance will be equal to one regardless of their length being, say, equal to two or fifty. However, common sense leads us to assume that in the latter case the two strings are more similar than in the former, for which the two sequences share only one out of two symbols. In [Marz 93] the normalization by the length of the corresponding optimal path in the grid is proposed to account for the length of the involved sequences.

In [Mei 04] a variant called *Markov Edit distance* is defined that accounts for the local interactions among adjacent symbols. For example, this modified Edit distance assigns a lower cost to symbol changes in the test pattern if these are reshuffles of the corresponding subpattern in the reference pattern. This is natural since in practice it is not uncommon for one to mess up locally in typing. Taking this into consideration, comparing the reference pattern "beauty" with the test pattern "beuaty" will result in lower Markov Edit distance than comparing the same reference pattern with "besrty," in contrast to the basic Edit distance that is equal to two for both cases.

The Edit distance and its variants have been used in a number of applications, where the problem can be posed as a string matching, such as polygon matching ([Koch 89]), OCR ([Tsay 93, Seni 96]), stereo vision ([Wang 90]), computational biology, and genome sequence matching ([Durb 97]).

8.2.3 Dynamic Time Warping in Speech Recognition

In this section, we highlight the application of dynamic programming techniques in speech recognition. We will focus on the simpler form of the task, known as *discrete or isolated word recognition* (IWR). That is, we will assume that the spoken text consists of discrete words, well isolated with sufficient silent periods between them. In tasks of this type, it is fairly straightforward to decide where, in time, one word finishes and another one starts. This is not, however, the case in the more complex *continuous speech recognition* (CSR) systems, where the speaker speaks in a natural way and temporal boundaries between words are not well defined. In the latter case, more elaborate schemes are required (e.g., [Silv 90, Desh 99, Neg 99]). When words are spoken by a single speaker and the purpose of the recognition system is to recognize words spoken by this person, then the recognition task is known as *speaker-dependent recognition*. A more complex task is *speaker-independent recognition*. In the latter case, the system must be trained using a number of speakers and the system must be able to generalize and recognize words spoken by people outside the training population.

At the heart of any IWR system are a set of known reference patterns and a distance measure (recall the footnote 1 in Section 8.2). Recognition of an unknown test pattern is achieved by searching for the best match between the test and each of the reference patterns, on the basis of the adopted measure.

Figures 8.5a and 8.6a show the plots of two time sequences resulting from the sampling of the word "love," spoken twice by the same speaker. The samples were taken at the output of a microphone at a sampling rate of 22,050 Hz. Although it is difficult to describe the differences, these are readily noticeable. Moreover, the two spoken words are of different duration. The arrows indicate (approximately) the intervals in which the spoken segments lie. The intervals outside the arrows correspond to silent periods. Specifically, the sequence in Figure 8.6a is 0.4 second long, and the sequence in Figure 8.5a is 0.45 second long. Furthermore, it is important to say that this is not the result of a simple linear time scaling. On the contrary, a ***highly nonlinear mapping*** is required to obtain a match between these two "*same*" words spoken by the *same* person. For comparison, Figure 8.6b shows the plot of the time sequence corresponding to another word, "kiss," spoken by the same speaker.

We will resort to dynamic programming techniques to unravel the nonlinear mapping (warping) required to achieve the optimal match between a test and a reference pattern. To this end, we must first express the spoken words as sequences (strings) of appropriate feature vectors, $\boldsymbol{r}(i)$ $i = 1, \ldots, I$, for the reference pattern and $\boldsymbol{t}(j)$, $j = 1, \ldots, J$, for the test one. Obviously, there is more than one way to choose the feature vectors. We will focus on Fourier transform

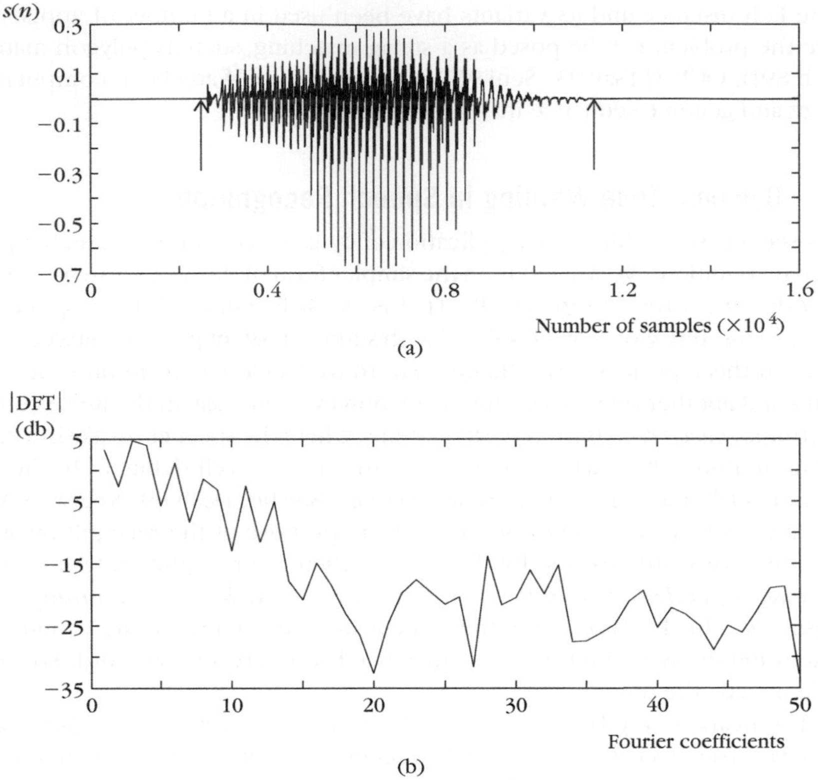

FIGURE 8.5

Plots of (a) the time sequence corresponding to the word "love" and (b) the magnitude of the DFT, in dB, for one of its frames.

features. Each of the time sequences involved is divided into successive overlapped time frames. In our case, each frame is chosen to be $t_f = 512$ samples long and the overlap between successive frames is $t_0 = 100$ samples, as shown in Figure 8.7. The resulting number of frames for the speech segment shown in Figure 8.5a is $I = 24$, and for the other two they are $J = 21$ (Figure 8.6a) and $J = 23$ (Figure 8.6b), respectively. We assume that the former is the reference pattern and the latter two the test patterns. Let $x_i(n), n = 0, \ldots, 511$, be the samples for the ith frame of the reference pattern, with $i = 1, \ldots, I$. The corresponding DFT is given as

$$X_i(m) = \frac{1}{\sqrt{512}} \sum_{n=0}^{n=511} x_i(n) \exp\left(-j\frac{2\pi}{512}mn\right), \quad m = 0, \ldots, 511$$

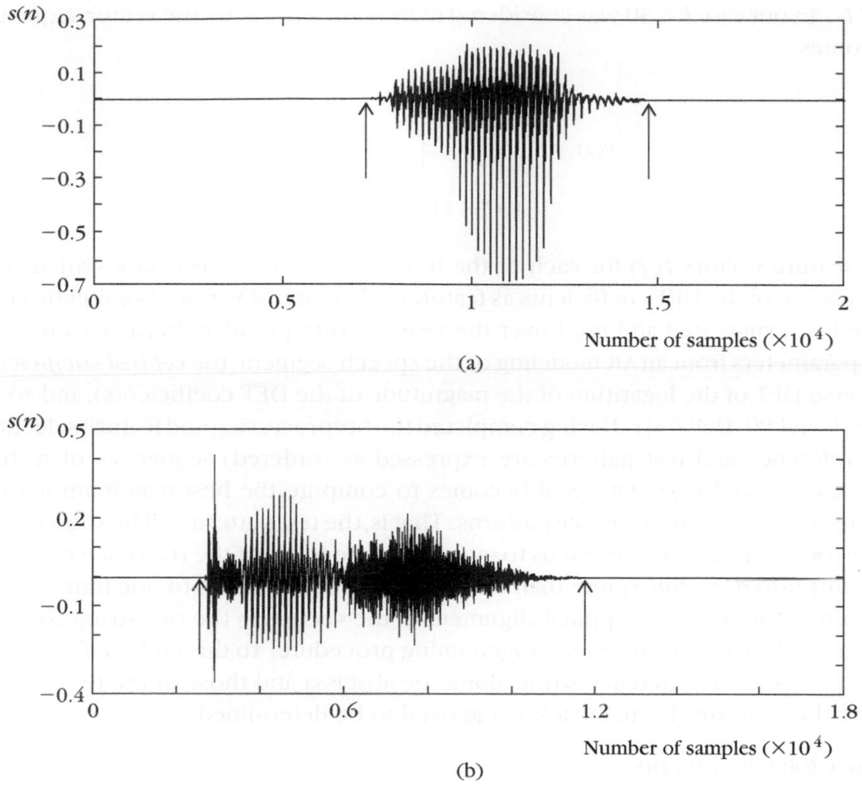

FIGURE 8.6

Plots of the time sequences resulting from the words (a) "love" and (b) "kiss," spoken by the same speaker.

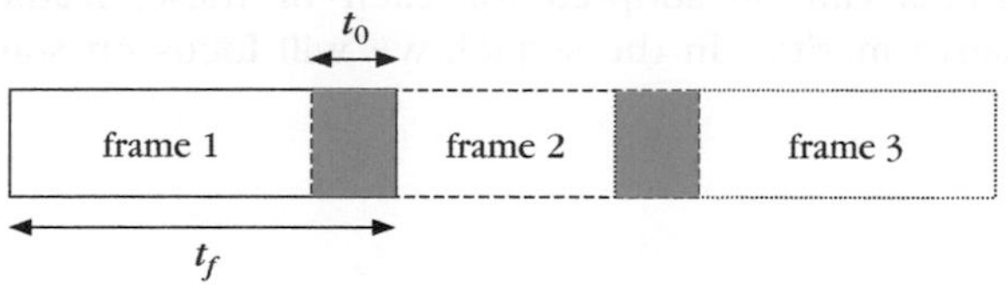

FIGURE 8.7

Successive overlapping frames for computation of the DFT feature vectors.

Figure 8.5b shows the magnitude of the DFT coefficients for one of the I frames of the reference pattern. The plot is a typical one for speech segments. The magnitude of the higher DFT coefficients is very small, with little contribution to the signal. This justifies use of the first l DFT coefficients as features, where usually

$l \ll t_f$. In our case $l = 50$ was considered to be sufficient. Thus, the vector sequence becomes

$$\boldsymbol{r}(i) = \begin{bmatrix} X_i(0) \\ X_i(1) \\ \vdots \\ X_i(l-1) \end{bmatrix}, \quad i = 1, \ldots, I \tag{8.4}$$

The feature vectors $\boldsymbol{t}(j)$ for each of the test patterns are formed in a similar way. The choice of the DFT coefficients as features is just one of various possibilities that have been suggested and used over the years. Other popular alternatives include the parameters from an AR modeling of the speech segment, the *ceptral coefficients* (inverse DFT of the logarithm of the magnitude of the DFT coefficients), and so on (e.g., [Davi 80, Dell 93]). Having completed the preprocessing and feature selection, the reference and test patterns are expressed as (ordered) sequences of feature vectors $\boldsymbol{r}(i)$ and $\boldsymbol{t}(j)$. Our goal becomes to compute the best match among the frames of the test and reference patterns. That is, the test pattern will be *stretched in time* (one test frame corresponds to more than one frame of the reference patterns) or *compressed in time* (more than one test frame corresponds to one frame of the reference pattern). This optimal alignment of the vectors in the two string patterns will take place via the dynamic programming procedure. To this end, we first locate the vectors of the reference string along the abscissa and those of the test pattern along the ordinate. Then, the following need to be determined:

- Global constraints
- Local constraints
- End-point constraints
- The cost d for the transitions

Various assumptions can be adopted for each of these, leading to different results with relative merits. In the sequel, we will focus on some widely used cases.

End-Point Constraints

In our example, we will look for the optimal complete path that starts at (0, 0) and ends at (I, J) and whose first transition is to the node (1, 1). Thus, it is implicitly assumed that the end points of the speech segments (i.e., $\boldsymbol{r}(1), \boldsymbol{t}(1)$ and $\boldsymbol{r}(I), \boldsymbol{t}(J)$) match to a fair degree. These can be the vectors resulting from the silent periods just before and just after the speech segments, respectively. A simple variation of the complete path constraints results if the end points of the path are not specified *a priori* and are assumed to be located within a distance ϵ from points (1, 1) and (I, J). It is left to the optimizing algorithm to locate them.

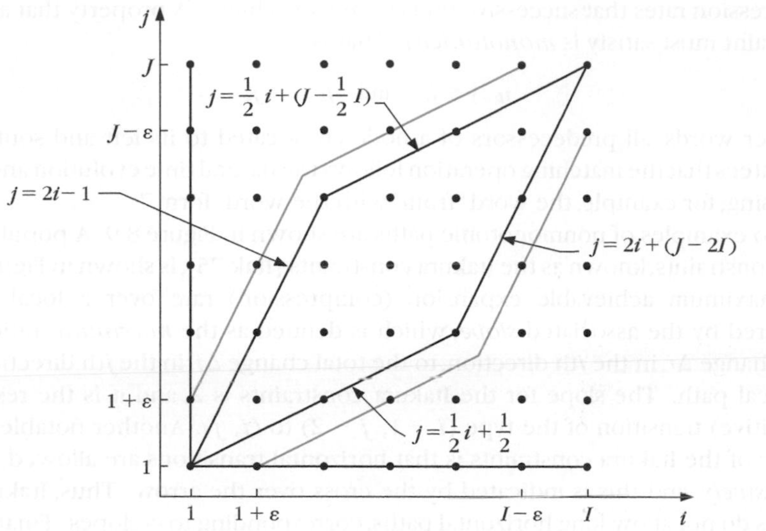

FIGURE 8.8

Itakura global constraints. The maximum compression/expansion factor is 2, and it determines the slope of the boundary line segments. The red lines correspond to the same global constraints when the relaxed end-point constraints are adopted.

Global Constraints

The global constraints define the region of nodes that are searched for the optimal path. Nodes outside this region are not searched. Basically, the global constraints define the overall stretching or compression allowed for the matching procedure. An example is shown in Figure 8.8. They are known as *Itakura constraints* and impose a maximum factor of 2 for any expansion or compression of the test with respect to the reference pattern. The allowable nodes are then located within the parallelogram shown in Figure 8.8 by the black line. The red lines correspond to the same global constraints when the relaxed end-point constraints, mentioned before, are adopted. Observe from the figure that paths across the sides of the parallelogram compress or expand corresponding frame intervals by a factor of 2, and this is the maximum possible factor attained. This constraint is usually reasonable and at the same time it reduces the number of nodes to be searched for the optimal path substantially. If $I \approx J$, then it is not difficult to show that the number of grid points to be searched is reduced by approximately one-third.

Local Constraints

These constraints define the set of predecessors and the allowable transitions to a given node of the grid. Basically, they impose limits for the maximum expansion/

compression rates that successive transitions can achieve. A property that any local constraint must satisfy is *monotonicity*. That is,

$$i_{k-1} \leq i_k \quad \text{and} \quad j_{k-1} \leq j_k$$

In other words, all predecessors of a node are located to its left and south. This guarantees that the matching operation follows the natural time evolution and avoids confusing, for example, the word "from" with the word "form."

Two examples of nonmonotonic paths are shown in Figure 8.9. A popular set of local constraints, known as the Itakura constraints [Itak 75], is shown in Figure 8.10. The maximum achievable expansion (compression) rate over a local path is measured by the associated *slope*, which is defined as the *maximum* ratio of the total change Δi, in the ith direction, to the total change Δj, in the jth direction, over the local path. The slope for the Itakura constraints is 2, and it is the result of a (repetitive) transition of the type $(i-1, j-2)$ to (i, j). Another notable characteristic of the Itakura constraints is that horizontal transitions are allowed, but *not successively*, and this is indicated by the cross over the arrow. Thus, Itakura constraints do not allow long horizontal paths, corresponding to ∞ slopes. Finally, these constraints allow the path to skip at *most one* feature vector in the test pattern string, that is, the one at the $j-1$ position of the ordinate axis, and the path jumps from $(i-1, j-2)$ to (i, j). In contrast, feature vectors in the reference string are not skipped, and all take part in the optimal path. Such constraints are known as *asymmetrical*.

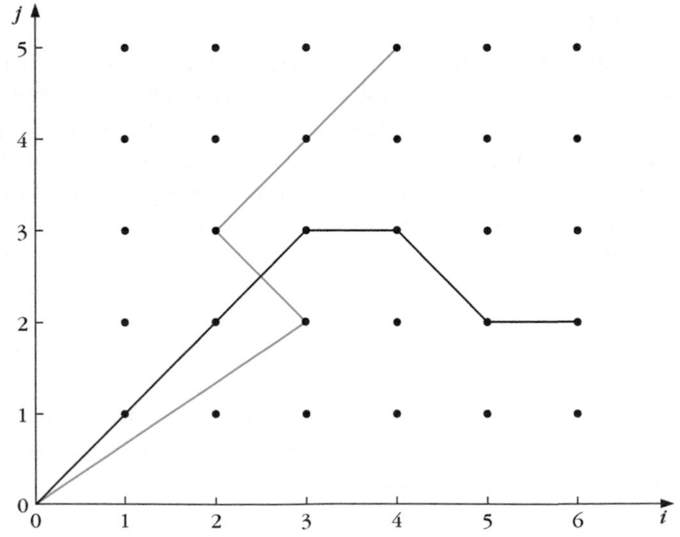

FIGURE 8.9

Examples of nonmonotonic paths. Such paths are not allowed and are not considered in the search for the optimum.

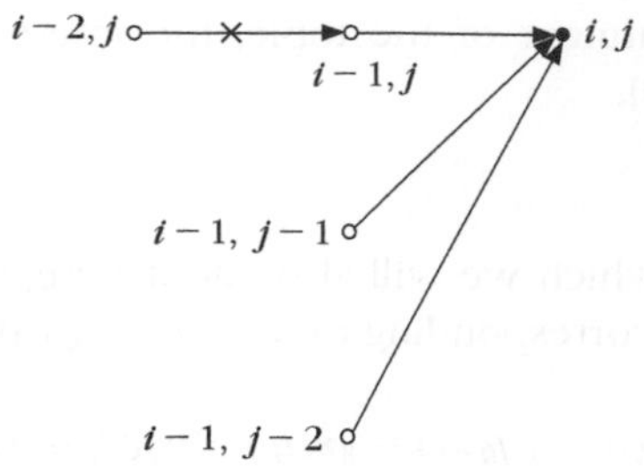

FIGURE 8.10

The Itakura local constraints. Two successive horizontal transitions are not allowed.

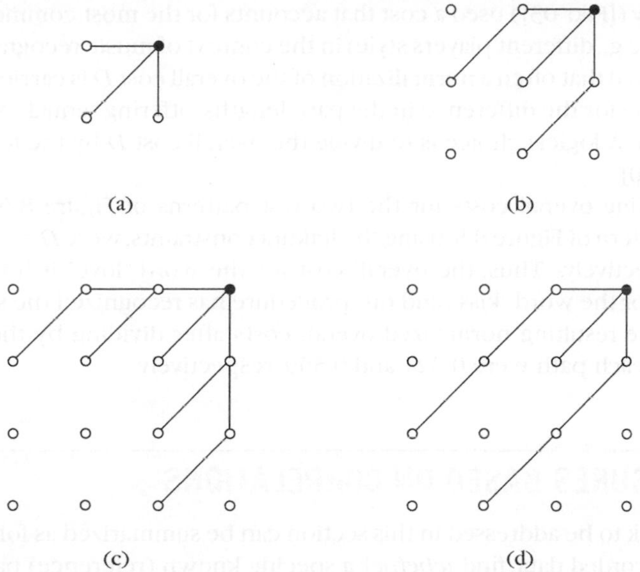

FIGURE 8.11

The Sakoe and Chiba local constraints.

A number of alternative local constraints have also been suggested and used in practice. Figure 8.11 shows four different types of constraints considered by Sakoe and Chiba [Sako 78]. For the type in Figure 8.11a, there is no limit in the rate of expansion/compression, since successive horizontal or vertical transitions can take place, until of course one falls outside the region defined by the global constraints. In contrast, in Figure 8.11b horizontal (vertical) transitions are allowed *only after* a diagonal transition and in Figure 8.11d after two successive diagonal transitions. In Figure 8.11c, at most two successive horizontal (vertical) transitions are allowed only after a diagonal one. The slopes for each of the constraints in Figures 8.11a, b, c and d are ∞, 2, 3, and 3/2, respectively (Problem 8.2).

For a more detailed treatment of the topic, the interested reader may consult [Dell 93, Silv 90, Myer 80].

The Cost

A commonly used cost, which we will also adopt here, is the Euclidean distance between $\boldsymbol{r}(i_k)$ and $\boldsymbol{t}(j_k)$, corresponding to node (i_k, j_k), that is,

$$d(i_k, j_k|i_{k-1}, j_{k-1}) = \|\boldsymbol{r}(i_k) - \boldsymbol{t}(j_k)\| \equiv d(i_k, j_k)$$

In this, we assume that no cost is associated with the transitions to a specific node, and the cost depends entirely on the feature vectors corresponding to the respective node. Other costs have also been suggested and used [Gray 76, Gray 80, Rabi 93]. More recently ([Pikr 03]) used a cost that accounts for the most commonly encountered errors (e.g., different players style) in the context of music recognition. Finally, it must be stated that often a normalization of the overall cost D is carried out. This is to compensate for the difference in the path lengths, offering "equal" opportunities to all of them. A logical choice is to divide the overall cost D by the length of each path [Myer 80].

The resulting overall costs for the two test patterns of Figure 8.6, against the reference pattern of Figure 8.5, using the Itakura constraints, were $D = 11.473$, $D = 25.155$, respectively. Thus, the overall cost for the word "love" is lower than the overall cost for the word "kiss," and our procedure has recognized the spoken word correctly. The resulting normalized overall costs after dividing by the number of nodes along each path were 0.221 and 0.559, respectively.

8.3 MEASURES BASED ON CORRELATIONS

The major task to be addressed in this section can be summarized as follows: "Given a block of recorded data, find *whether* a specific known (reference) pattern is contained within the block and *where* it is located."A typical application of this is found in scene analysis, when we want to search for specific objects within the image. Such problems arise in many applications, such as target detection, robot vision, and video coding. For example, in video coding a major step is that of *motion estimation*—that is, the process of locating *corresponding pixels* (of the same moved object) among successive image frames at different time instants. This step is then followed by the *motion compensation* stage, which compensates for the displacement of moving objects from one frame to another. One then codes the frame difference

$$e(i, j, t) = r(i, j, t) - r(i - m, j - n, t - 1)$$

where $r(i, j, t)$ are the pixel gray levels of the image frame at time t and $r(i - m, j - n, t - 1)$ the *corresponding* pixel values at spatial locations $i - m$,

$j - n$ of the previous frame at time $t - 1$. In this way, we code only the *new information* contained at the latest frame, avoiding redundancies.

Let us assume that we are given a reference pattern expressed as an $M \times N$ image array $r(i, j), i = 0, \ldots, M - 1, j = 0, \ldots, N - 1$, and an $I \times J$ image array $t(i, j), i = 0, \ldots, I - 1, j = 0, \ldots, J - 1$, where $M \leq I, N \leq J$. The goal is to develop a measure for detecting an $M \times N$ subimage within $t(i, j)$ that matches best the reference pattern $r(i, j)$. To this end, the reference image $r(i, j)$ is superimposed on the test image, and *it is translated* to all possible positions (m, n) within it. For each of the points (m, n), the mismatch between $r(i, j)$ and the $M \times N$ subimage of $t(i, j)$ is computed according to

$$D(m,n) = \sum_{i=m}^{m+M-1} \sum_{j=n}^{n+N-1} |t(i,j) - r(i-m, j-n)|^2 \tag{8.5}$$

Template matching is conducted by searching for the location (m, n) for which $D(m, n)$ is minimum. Let us now give this a computationally more attractive form. Equation (8.5) is equivalent to

$$D(m,n) = \sum_i \sum_j |t(i,j)|^2 + \sum_i \sum_j |r(i,j)|^2 - 2\sum_i \sum_j t(i,j)r(i-m, j-n) \tag{8.6}$$

The second summand is constant for a given reference pattern. *Assuming that the first one does not change much across the image*, that is, there is not much variation of the pixel gray levels over the test image, the minimum of $D(m, n)$ is achieved when

$$c(m,n) = \sum_i \sum_j t(i,j)r(i-m, j-n) \tag{8.7}$$

is maximum for all possible locations (m, n). The quantity $c(m, n)$ is nothing other than a cross-correlation sequence between $t(i, j)$ and $r(i, j)$ computed at the point (m, n). In cases for which the assumption of little gray-level variation is not valid, this measure is very sensitive to gray-level variations within $t(i, j)$. In such cases the *cross-correlation coefficient*, defined as

$$c_N(m,n) = \frac{c(m,n)}{\sqrt{\sum_i \sum_j |t(i,j)|^2 \sum_i \sum_j |r(i,j)|^2}} \tag{8.8}$$

is a more appropriate measure. Here, $c_N(m, n)$ is a normalized version of $c(m, n)$, and variations in $t(i, j)$ tend to cancel out. Recall now the Cauchy–Schwarz inequality

$$\left| \sum_i \sum_j t(i, j)r(i-m, j-n) \right| \leq \sqrt{\sum_i \sum_j |t(i, j)|^2 \sum_i \sum_j |r(i, j)|^2}$$

Equality holds *if and only if*

$$t(i,j) = \alpha r(i-m, j-n), \quad i = m, \ldots, m+M-1 \text{ and}$$
$$j = n, \ldots, n+N-1$$

with α being an arbitrary constant. Hence, $c_N(m, n)$ is always less than unity and achieves its maximum value of one only if the (test) subimage is the same (within a scaling factor) as the reference pattern.

In our discussion so far, we have assumed that the reference pattern has only been translated within $t(i, j)$ and no rotation or scaling has been involved. In applications such as video coding, this is a valid assumption and it has been adopted in the video coding standards [Bhas 95]. However, this is not always the case and the technique has to be modified. One way is to describe the reference and test subimages in terms of invariant moments and measure the similarity using correlations involving these moments [Hall 79] (Problem 8.4). Another rotation- and scale-invariant technique, using a combination of the Fourier and Mellin transforms, is described in [Scha 89]. This technique tries to exploit the translation invariance of the magnitude of the Fourier transform (already discussed in Chapter 7) and the scale invariance of the Mellin transform ([Ravi 95], Problem 8.5). Another path, which of course demands high computational resources, is to have a set of distorted (e.g., rotated and scaled) reference templates to cover all possibilities. Correlation matching will then reveal the best match between a test pattern and one of the reference templates. A computationally more attractive technique is to employ the Karhunen–Loève transform [Ueno 97]. The main idea is that rotated templates are highly correlated, and each of them can be approximated by its projection onto a lower dimension eigenspace, using the most significant eigenvectors of their correlation matrix. Matching of an unknown pattern with the template of the right orientation is performed in the lower dimensional space, leading to substantial computational savings.

Example 8.2

The image $t(i, j)$ in Figure 8.12a contains two objects, a screwdriver and a hammer. The latter is the object that we want to search for in the image. The reference image is shown at the top right corner of Figure 8.12a. The dotted area represents the general (m, n) position of the reference image when it is superimposed on the test one. Figure 8.12b shows the cross-correlation $c(m, n)$ between the two images. We readily observe that the maximum (black) occurs at the position (13, 66), that is, where the hammer is located in $t(i, j)$.

Computational Considerations

- In some cases, it is more efficient to compute the cross-correlation via its Fourier transform. Recall that in the frequency domain (8.7) is written as

$$C(k, l) = T(k, l)R^*(k, l) \tag{8.9}$$

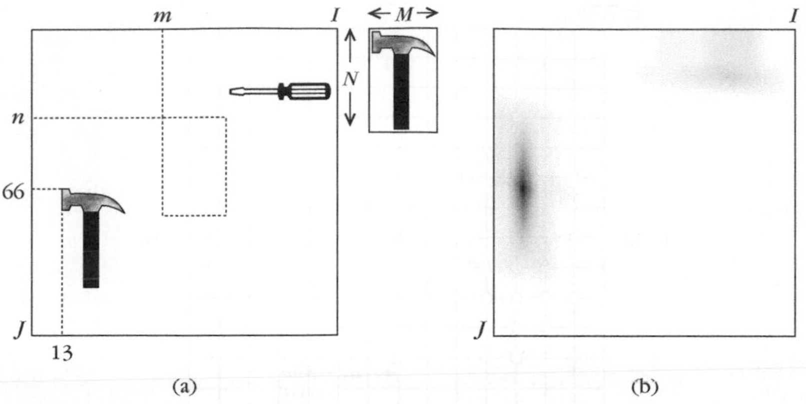

FIGURE 8.12
Example of (a) reference and test images and (b) their respective correlation.

where $T(k, l), R(k, l)$ are the DFT transforms of $t(i, j)$ and $r(i, j)$, respectively, with "$*$" denoting complex conjugation. Of course, in order to write (8.9), both images must be of the same size. If they are not, which is usually the case, a number of zeros must be appended to extend the smaller sized image. $c(m, n)$ is obtained via the inverse DFT of $C(k, l)$. Taking into account the computational efficiency of the FFT, this procedure may lead to savings depending, of course, on the relative size of M, N and I, J.

- A major computational load in correlation-based template matching is searching over the pixels of $t(i, j)$ in order to locate the maximum correlation. Usually, the search is restricted within a rectangle $[-p, p] \times [-p, p]$ centered at a point (x, y) in $t(i, j)$. For example, in video coding, if the $M \times N$ block is centered at a position (x, y) in the frame at time $t-1$, then the current frame is searched within $(x \pm p, y \pm p)$. The value of p depends on the application. For broadcast TV $p = 15$ is sufficient. For sporting events (high motion) $p = 63$ is more appropriate. Thus, an exhaustive search for the maximum of $c(m, n)$, defined in (8.7), will require a number proportional to $(2p + 1)^2 MN$ additions and multiplications. This leads to a huge number of operations indeed (Problem 8.6). Thus, in practice, suboptimal heuristic searching techniques are usually adopted, which, although they do not guarantee locating the maximum, reduce the required number of operations substantially. There are two major directions. One is to reduce the search points and the other is to reduce the size of the involved images.

Two-Dimensional Logarithmic Search

Logarithmic Search Figure 8.13 shows the rectangular $[-p, p] \times [-p, p]$ searching area for the case of $p = 7$. The center of the rectangle is assumed to be the point $(0, 0)$. The cross-correlation computation is first performed at the center as well

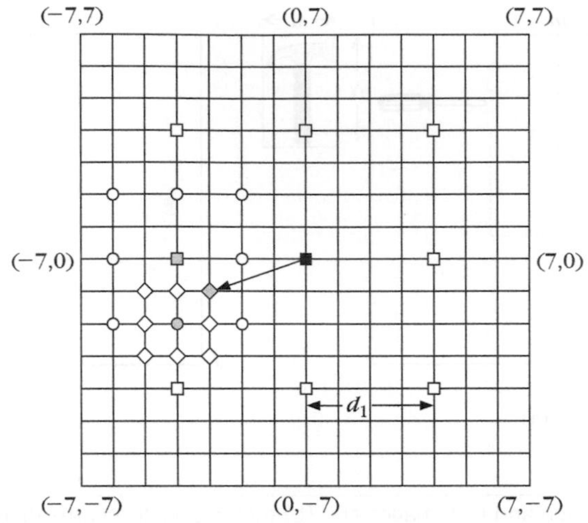

FIGURE 8.13

Logarithmic search to find the point of maximum cross-correlation.

as the eight points located on the perimeter of the inner $[-p/2, p/2] \times [-p/2, p/2]$ area ($p/2$ rounded to an integer). These points are denoted by a square. The spacing between these points is $d_1 = 4$ pixels, that is, $d_1 = 2^{k-1}$ and $k = \lceil \log_2 p \rceil$, where $\lceil \cdot \rceil$ denotes rounding to the first larger integer. For $p = 7$ we get $k = 3$ and $d_1 = 4$. We will demonstrate the procedure via an example. Let us assume that the largest cross-correlation value resulted at the position $(-4, 0)$ (shaded square). Then we consider this point as the center of a rectangle of size $[-p/4, p/4] \times [-p/4, p/4]$ ($[-2, 2] \times [-2, 2]$ in our case) and compute the correlation at the eight points of its perimeter. These points are denoted by a circle, and the spacing between them is now $d_2 = d_1/2$ (2). The process is repeated, and finally the computation is performed on the eight (diamond) points on the perimeter of the rectangle of size $[-1, 1] \times [-1, 1]$, which is centered at the optimum (of the previous step) shaded circle point. The spacing between the diamond points is $d_3 = 1$. The shaded diamond corresponds to the point with the maximum cross-correlation, and the process is terminated. The number of computations has now been reduced to $MN(8k + 1)$ operations, which is a substantial saving compared with the exhaustive search.

A variant of the two-dimensional logarithmic search is to search the i and j directions independently. The point whose coordinates are the resulting best values of i and j becomes the new origin of the coordinate system, and the search is repeated in the new i, j directions, with smaller spacing d. The process is repeated until the spacing becomes unity.

Hierarchical Search

The hierarchical search technique springs from the multiresolution concept considered in Chapter 6. Let us again consider an example.

- Step 1: A reference block of, say, 16×16 is given, and the search area is assumed to be the rectangle $[-p, p] \times [-p, p]$, centered at the point (x, y) in the test image. We refer to level 0 versions of the images. Both the reference block and the test image are low-pass filtered and subsampled by 2, resulting in their level 1 versions. The total number of pixels in the level 1 versions has been reduced by 4. Level 1 versions are in turn low-pass filtered and subsampled, resulting in level 2 versions. In general, this process can continue.

- Step 2: At level 2 the search for the maximum takes place with the 4×4 low-pass version of the reference block. The search area in the level 2 low-pass version of the test image is the rectangle $[-p/4, p/4] \times [-p/4, p/4]$ centered at $(x/4, y/4)$. Either a full or a logarithmic search can be employed. Let (x_1, y_1) be the coordinates of the optimum, with respect to $(x/4, y/4)$.

- Step 3: At level 1, the search for the maximum is performed using the corresponding 8×8 version of the reference block. The search area, within the level 1 version of the test image, is the rectangle of size $[-1, 1] \times [-1, 1]$ centered at $(x/2 + 2x_1, y/2 + 2y_1)$, that is, nine pixels in total. This is because the eight pixels at the perimeter of this area were not involved at level 2, due to the subsampling (see Figure 6.23 of Chapter 6). The center point must also be included in order to have a fair comparison at this level for the search for the maximum. Let the maximum occur at (x_2, y_2) with respect $(x/2, y/2)$.

- Step 4: At level 0 the search is performed using the original reference template, within the area of size $[-1, 1] \times [-1, 1]$ centered at $(x + 2x_2, y + 2y_2)$ in the test image. The location of the maximum is the final one and the process terminates.

The computational saving with this method depends on the number of levels, as well as the type of search adopted at the highest level (Problem 8.5). In general, hierarchical methods are very efficient from a computational point of view. This is gained at the expense of increased memory requirements, due to the various image versions that must be kept. A disadvantage of the method is that if small objects are present in the templates, they may disappear at the lowest resolution images due to the subsampling. Furthermore, the method cannot guarantee to find the global best match. In [Alkh 01] an alternative philosophy is suggested that results in the global best match. Computational savings are achieved by pruning the number of candidates for the best match in a level, using the results in a higher level and an appropriately chosen threshold value.

Sequential Method

A number of other techniques have also been suggested. For example, the *sequential search* method computes a variant of (8.5) directly. Specifically, define

$$D_{pq}(m, n) = \sum_{i=0}^{p-1} \sum_{j=0}^{q-1} |t(i + m, j + n) - r(i, j)| \tag{8.10}$$

Thus, the error is computed in a smaller and sequentially increasing window area, for $p, q = 1, 2, \ldots$ and $p \leq M, q \leq N$. The computations stop when $D_{pq}(m, n)$ becomes larger than a predetermined threshold. Then computations start in a different direction (m, n). Hence, saving is achieved, because for bad positions only a small number of computations need to be performed.

8.4 DEFORMABLE TEMPLATE MODELS

In the previous section, we considered the problem of searching for a known reference pattern (template) within a test image. We assumed that the template and the object, residing in the image, were identical. The only differences that were allowed to enter into our discussion were those imposed by a different orientation and/or scaling. However, there are many problems where we know *a priori* that the available template and the object we search for in the image may not look exactly the same. This may be due to varying imaging conditions, occlusion, and imperfect image segmentation. Furthermore, in a content-based image database retrieval system, the user may provide the system with a sketch of the shape of the object to be retrieved. Obviously, the sketch will not match exactly the corresponding object in the database images. Our goal in this section is to allow the template matching procedure to account for deviations between the reference template and the corresponding test pattern in the image. In our discussion, we will assume that the reference template is available in the form of an image array containing the object's boundary information (contour). That is, we will focus on shape information only. Extensions incorporating more information, for example, texture, are also possible.

Let us denote the reference template image array as $r(i, j)$. This is also known as *prototype*. The basic idea behind the *deformable template matching* procedure is simple: *Deform* the prototype and produce *deformed* variants of it. From a mathematical point of view a deformation consists of the application of a *parametric transform* $T_{\boldsymbol{\xi}}$ on $r(i,j)$ to produce a deformed version $T_{\boldsymbol{\xi}}[r(i,j)]$. Different values of the vector parameter $\boldsymbol{\xi}$ lead to different versions. From the set of the deformed prototype variants that can be generated, there will be one that "best" matches the test pattern. The goodness of fit is measured via a cost, which we will call the *matching energy* $E_m(\boldsymbol{\xi})$. Obviously, the goal is to choose $\boldsymbol{\xi}$ so that $E_m(\boldsymbol{\xi})$ is *minimum*. However, this is not enough. If, for example, the optimal set of parameters is such that the corresponding deformed template has been deformed to such an extent that it bears little resemblance to the original prototype, the method will be meaningless. Thus, one more term has to be taken into account in the optimization process. This is the cost measuring the "deformation," which the prototype needs to undergo in order to fit the test pattern. We will call this term of the cost *deformation energy* $E_d(\boldsymbol{\xi})$. Then the optimal vector parameter is computed so that

$$\boldsymbol{\xi} : \min_{\boldsymbol{\xi}} \{E_m(\boldsymbol{\xi}) + E_d(\boldsymbol{\xi})\} \tag{8.11}$$

In words, one could think of the boundary curve of the prototype as made by rubber. Then with the help of a pencil we deform the shape of the rubber curve to match the test pattern. The more we deform the shape of the prototype, the higher the energy we have to spend for it. This energy, quantified by $E_d(\boldsymbol{\xi})$, depends on the shape of the prototype. That is, it is an intrinsic property of the prototype, and this is the reason that it is also known as ***internal energy***. The other energy term, $E_m(\boldsymbol{\xi})$, depends on the input data (test image), and we usually refer to it as ***external energy***. The optimal vector parameter, $\boldsymbol{\xi}$, is chosen so that the best trade-off between these two energy terms is achieved. Sometimes, a weighting factor C is used to give preference to one of the two terms, and $\boldsymbol{\xi}$ is computed so that

$$\boldsymbol{\xi} : \min_{\boldsymbol{\xi}} \{E_m(\boldsymbol{\xi}) + CE_d(\boldsymbol{\xi})\} \tag{8.12}$$

Hence, in order to apply the above procedure in practice, one must have at one's disposal the following ingredients:

- A prototype
- A transformation procedure to deform the prototype
- The two energy function terms

Choice of the Prototype

This should be carefully chosen so that it is a (typical) representative of the various instances in which this object is expected to appear in practice. In a way, the prototype should encode the "mean shape" characteristics of the corresponding "shape class."

Deformation Transformation

This consists of a set of parametric operations. Let (x, y) be the (continuous) coordinates of a point in a two-dimensional image. Without loss of generality, assume that the image is defined in a square $[0, 1] \times [0, 1]$. Then each point (x, y) is mapped using a continuous mapping function, as

$$(x, y) \longrightarrow (x, y) + (D^x(x, y), D^y(x, y)) \tag{8.13}$$

For discrete image arrays a quantization step is necessary after the transformation. Different functions can be used to perform the above mapping. A set that has successfully been used in practice is ([Amit 91])

$$D^x(x, y) = \sum_{m=1}^{M} \sum_{n=1}^{N} \xi_{mn}^x e_{mn}^x(x, y) \tag{8.14}$$

$$D^y(x, y) = \sum_{m=1}^{M} \sum_{n=1}^{N} \xi_{mn}^y e_{mn}^y(x, y) \tag{8.15}$$

$$e_{mn}^x(x, y) = \alpha_{mn} \sin \pi n x \cos \pi m y \tag{8.16}$$

$$e_{mn}^y(x, y) = \alpha_{mn} \cos \pi m x \sin \pi n y \tag{8.17}$$

for appropriately chosen values of M, N. The normalizing constants α_{mn} can be taken as

$$\alpha_{mn} = \frac{1}{\pi^2(n^2 + m^2)}$$

Other basis functions can also be used, such as splines or wavelets. Figure 8.14 shows a prototype for an object and three deformed versions obtained for the simplest case of the transformation model in (8.14)–(8.17), that is, $M = N = 1$.

Internal Energy

This should be minimum for no deformation, that is, for $\boldsymbol{\xi} = \mathbf{0}$. A reasonable choice, associated with the transformation functions considered above, is

$$E_d(\boldsymbol{\xi}) = \sum_m \sum_n ((\xi^x_{mn})^2 + (\xi^y_{mn})^2) \quad (8.18)$$

External Energy

Here again a number of choices are possible, measuring the goodness of fit between the test pattern and each one of the deformed template variants. For example, for a specific position, orientation, and scale of a deformed template, this energy term can be measured in terms of the distance of each point in the contour of the deformed

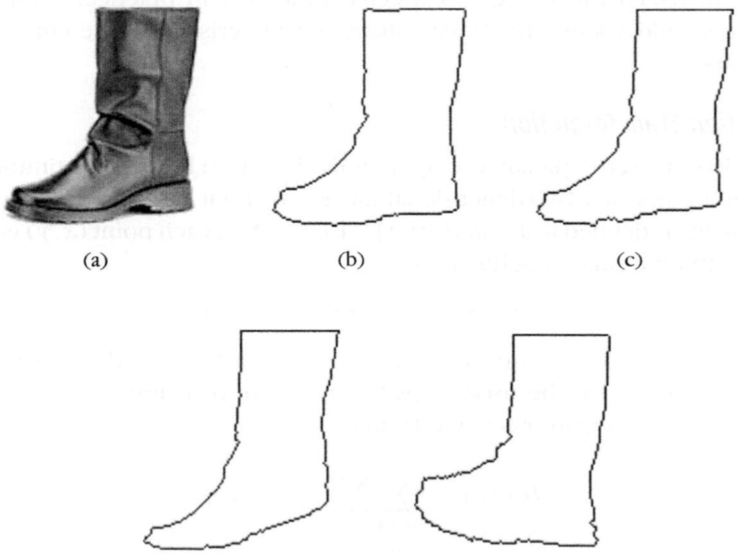

(a) (b) (c) (d) (e)

FIGURE 8.14

(a) A reference pattern, (b) its contour used as prototype, and (c), (d), (e) three of its deformed variants.

template from the nearest point in the contour of the test image, I. One way to achieve this is via the following energy function:

$$E_m(\boldsymbol{\xi}, \boldsymbol{\theta}, I) = \frac{1}{N_d} \sum_{i,j} (1 + \Phi(i, j)) \tag{8.19}$$

where $\boldsymbol{\theta}$ is the vector of the parameters defining the position, orientation, and scaling and N_d the number of contour pixels of the corresponding deformed template and

$$\Phi(i,j) = -\exp\left(-\rho(\delta_i^2 + \delta_j^2)^{1/2}\right) \tag{8.20}$$

where ρ a constant and (δ_i, δ_j) is the displacement of the (i,j) pixel of the deformed template from the nearest point in the test image. In [Jain 96] directional information is also incorporated into the cost.

Remarks

- One can arrive at (8.11) in a more systematic way via probabilistic arguments, that is, by seeking to maximize the a posteriori probability density of $(\boldsymbol{\xi}, \boldsymbol{\theta})$ given the test image array, that is,

$$p(\boldsymbol{\xi}, \boldsymbol{\theta}|I) = \frac{p(\boldsymbol{\xi}, \boldsymbol{\theta}) p(I|\boldsymbol{\xi}, \boldsymbol{\theta})}{p(I)} \tag{8.21}$$

  where the Bayes rule has been employed. In this framework, (8.18) results if one assumes that the various parameters ξ^x_{mn}, ξ^y_{mn} are statistically independent and are normally distributed, for example, $p(\xi) = \mathcal{N}(0, \sigma^2)$. The higher the variance σ^2, the wider the range of the values that occur with high probability. To obtain (8.19)–(8.20) the model

$$p(I|\boldsymbol{\xi}, \boldsymbol{\theta}) = \alpha \exp\left(-E_m(\boldsymbol{\xi}, \boldsymbol{\theta}, I)\right) \tag{8.22}$$

  is adopted, where α is a normalizing constant [Jain 96].

- The cost in (8.11) is a nonlinear function, and its minimization can be achieved via any nonlinear optimization scheme. Besides complexity, the major drawback is the omnipresent danger that the algorithm will be trapped in a local minimum. In [Jain 96] a procedure has been suggested that builds around the gradient descent methodology (Appendix C). The idea is to adopt a multiresolution iterative approach and use larger values of ρ in (8.20) for the coarser levels and smaller values for the finer ones. This procedure seems to help the algorithm to avoid local minima, at an affordable computing cost.

- The methodology we described in this section belongs to a more general class of deformable template matching techniques, which builds around an available prototype. This is not the only available path. Another class of deformable models stems from an analytic description of the prototype shape, using a set of parameterized geometrical shapes, for example, ellipses or parabolic curves.

To delve deeper into these issues the reader may refer to the review articles [Jain 98, McIn 96, Cheu 02] and the references therein. [Widr 73, Fisc 73] seem to be the first attempts to introduce the concept of deformable models in computer vision.

- In the pattern recognition context, given an unknown test pattern, we seek to see to which one from a known set of different prototypes this matches best. For each prototype, the best deformed variant is selected, based on the minimum energy cost. The prototype that wins is the one whose best deformed variant results in the overall minimum energy cost.

8.5 CONTENT-BASED INFORMATION RETRIEVAL: RELEVANCE FEEDBACK

With the rapid development and spread of the Internet, a large corpus of information is stored and distributed over the Web. Search engines have become indispensable tools for searching and retrieving information in all possible forms, including text, images, audio and more recently video. The more traditional way of information retrieval is text-based; stored information is manually annotated by text descriptors, which are in turn used by a distributed database system to perform the information retrieval task. Such a procedure has the obvious drawback of requiring manual annotation, which, besides being time consuming and costly, is vulnerable to annotation inaccuracies and also to the subjectivity of the human perception. Due to the advances in pattern recognition, an alternative search procedure, known as *content-based* retrieval, is gaining in importance, and it is becoming more and more popular. Stored information is now indexed and searched based on its content. For example, in image retrieval, images can be indexed automatically by using features "describing" image content qualities such as color, texture and shapes. A music or speech segment could be represented in terms of a number of features such as those described in Section 7.5.

Content-based information retrieval is similar in concept to template matching, as introduced in this chapter. The goal is to search for and retrieve stored pieces of information, that is, patterns/templates that are most "similar" to the pattern presented as input to the search engine system. Similarity is quantified in terms of a similarity measure defined in the feature space. The similarity measures described at the beginning of this chapter may be possible candidates for some content-based retrieval tasks. A popular metric that has extensively been used is the weighted l_p metric between two feature vectors $\boldsymbol{x}$, $\boldsymbol{y}$, given by:

$$d(\boldsymbol{x}, \boldsymbol{y}) = \left(\sum_{i=1}^{l} w_i |x_i - y_i|^p \right)^{\frac{1}{p}}$$

Obviously, for $p = 2$ and $w_i = 1,\ i = 1, 2, \ldots, l$, this becomes the Euclidean distance and for $p = 1$ the so called weighted l_1 (Manhattan) norm. A more detailed treatment of similarity/dissimilarity measures is given in Chapter 11.

A major disadvantage of such a contend-based approach is that search and subsequent retrieval is based solely on the derived features, which are usually referred to as *low-level* features. Humans, being much more intelligent than the machines they themselves develop, utilize a number of so-called *high-level* concepts when they come to recognize objects (patterns). The notion of *semantic gap* is usually adopted to express this discrepancy between the low-level features, derived from and describing the patterns, and the high-level descriptions that are meaningful to a human. Artificial Intelligence is the discipline that focuses on developing methodologies and techniques for machine high-level reasoning in association with the low-level derived features. However, so far, such techniques are feasible and applicable only to restricted domains and applications.

The goal of the current section is a more humble one, compared to the goals originally set in the field of artificial intelligence, and yet very interesting. Since learning machines cannot compete with the high-level concepts and reasoning of a human being, let the user be involved and become part of the learning "game". Such a methodology offers the system the advantage of exploiting the user' s own way of conceptualizing the patterns, which are experienced through his or her senses. To this end, the search/retrieval session is divided into a number of consecutive loops. At every loop, the user provides *feedback* regarding the results by characterizing the retrieved patterns as either *relevant* or *irrelevant*. Relevance is usually defined by a characteristic that is shared by some of the patterns. It can be a perceptual characteristic or a more semantic one [Cruc 04]. Such a methodology is known as *relevance feedback* (RF). Since the user is directly involved, for the learning process to be useful in practice, convergence should be achieved within a few iteration steps and the search engine must operate in real time. In turn, this imposes the constraint that the selected (low-level) features must be as informative as possible. Thus the feature selection techniques, as exposed in Chapter 5, are of significance here too. One of the most successful paradigms for RF is the case of *content-based image retrieval* (CBIR), for which commercial products are already available; see, for example, [Liu 07].

A typical scenario, met in a number of RF tasks, is given below and it is schematically presented in Figure 8.15.

1. The system provides an initial set of patterns "similar" to the pattern presented by the user to the search engine (e.g., an image, a Web page, or an audio segment from a piece of music).

2. The user marks a number from the returned patterns as "relevant" or "irrelevant."

3. A classification procedure is used to "learn" the user' s feedback.

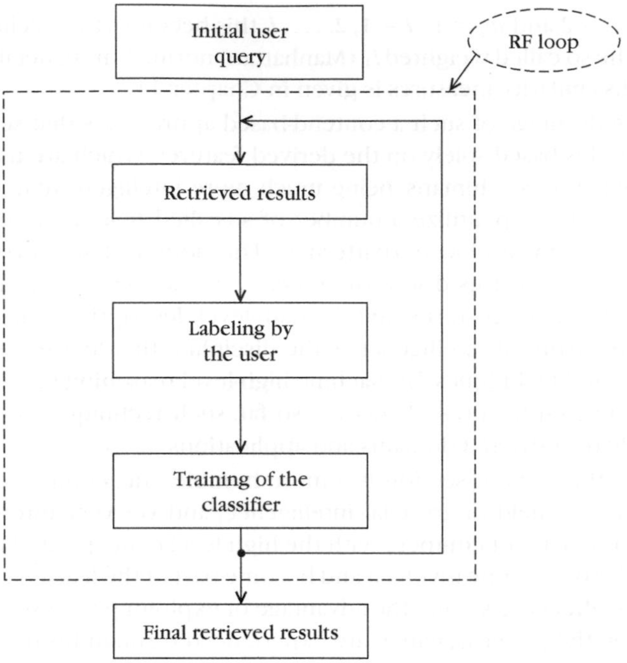

FIGURE 8.15

Block diagram illustrating the basic steps involved in a relevance feedback task.

Steps 2–3 are repeated untill the RF algorithm converges to a level that satisfies the user; that is, enough of truly relevant patterns have been retrieved by the search engine. Obviously, different techniques have been suggested and used for all three steps in the loop.

For Step 1, one can initialize the system either randomly or, preferably, by retrieving an initial set of "similar" patterns based on a similarity metric, such as the l_p metric. For Step 3, a popular choice is to employ a binary classifier that is trained to classify the marked patterns, in Step 2, either to the "relevant" or the "irrelevant" class, according to their label as judged by the user. Support vector machines have enjoyed a high popularity among the researchers, although other classifiers can also be used, see, [Druc 01, Cruc 04, Liu 07].

Interestingly enough, Step 2 is of critical importance. A *selection strategy* is first adopted concerning the *type* of the patterns that the system returns and on which the user must apply the query concept and label these patterns accordingly. Obviously, the patterns that the search engine returns in each round depend on the current knowledge of the "learner" in Step 3. In each iteration step, the system performs a ranking of the retrieved patterns, according to a confidence measure associated with the classifier' s decisions. For example, in the case an SVM has been adopted, this confidence measure can be the distance of the feature vector, representing the pattern, from the decision hyperplane in the RKHS feature space, which is proportional to the value of $|g(\boldsymbol{x})|$ in Eq. (4.72) (Section 4.18). In [Druc 01], the remotest instance to the positive (relevant) side is ranked at the top of the list, and the remotest one to the negative side (irrelevant) is ranked at the bottom. The user selects a number of patterns (say, 10 to 20) among the top ranked in the list. That is, selection is done among patterns that have been classified (by the current classifier) as relevant and with high confidence. Obviously, if the system has not yet converged, some of the returned patterns will not satisfy the user and will be judged as being irrelevant. This strategy seems to be the most popular one. The strong point of such a selection procedure is that the user gets a few good relevant patterns quite early in the iteration process. On the other hand, such a philosophy turns out to lead to a relatively slow convergence of the method.

Another point of view has been adopted in the strategy proposed in [Tong 01]. This is inspired by the notion of *active learning* used in pattern recognition (see, [Lewi 94, Scho 00]). Active learning is an approach that trains the classifier by using a *subset* of the data, that are considered the most *informative* ones. Hence, one can achieve better performance with less training data. The most informative data points are taken to be the most *uncertain* instances. Thus, in such a RF setting, the user is asked to label a number of pool patterns that lie closest to the classifier' s boundary decision. In [Tong 01] a strong theoretical justification is also provided for such a scenario. In other words, such a selection criterion forces the system to elicit from the user a crucial part of information: what makes distinct the "relevant" from the "irrelevant." This is because, once the user labels as relevant or irrelevant such "uncertain" patterns, a significant part of uncertainty is removed from the system. The advantage of this selection criterion is that it speeds up the convergence of the RF scheme.

Figure 8.16 compares the convergence performance of an RF system for the two previous strategies and for two different users; a "lazy" user, who only marks up to two relevant and two irrelevant patterns (if a single relevant pattern is returned by the system, the user marks only one) and a "patient" user who marks all the relevant and all the irrelevant patterns, among the patterns that the engine returns at each iteration step. The horizontal axis corresponds to the number of iteration steps and the vertical one to the *precision*, denoted as Pr. As precision, we define the ratio of the relevant patterns to the total number of returned (at each iteration round) patterns. In all cases, the curves start from the same point. This is the precision value corresponding to the initialization step, where a simple similarity measure was used. Also, all curves tend to $Pr = 1$; that is, as the learning process advances, more and more of the returned patterns are judged by the user as being relevant. For both users, the curve corresponding to the active learning strategy tends to $Pr = 1$ faster. As it is natural, the system tends to $Pr = 1$ much faster for the cases of the patient user, since, at each iteration step, more patterns are available for training the classifier in Step 3. The curves in the figure have been obtained using the Wang image database [Wang 01] and an SVM classifier. More details about the features and

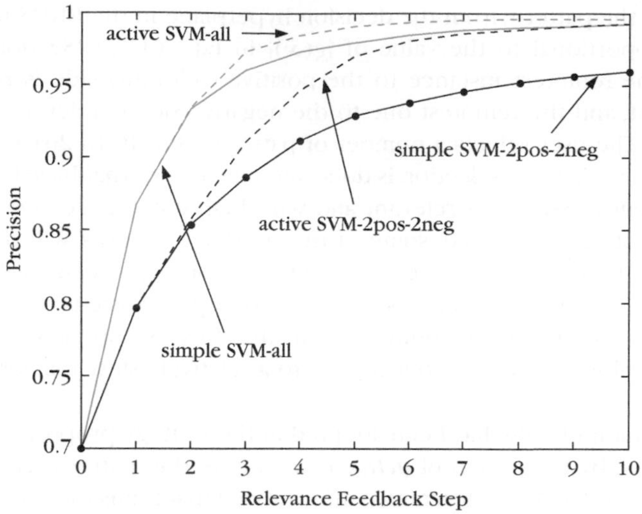

FIGURE 8.16

Dotted lines correspond to the active learning scenario. Learning curves for the "lazy" user, who, in each iteration round, marks up to two positive (relevant) and two negative (irrelevant) images, converge slower compared to those associated with the user who labels all the returned images as either positive or negative. For both users, active learning leads to faster convergence. We have used the terms *active* and *simple* to annotate the curves corresponding to the two strategies.

the parameters used are available in the book's Web site, and the interested reader can also perform a set of related experiments.

For the experienced researcher, it will not come as a surprise to say that a third route to the selection strategy has been proposed by combining the previous two "extremes." In [Xu 03], a hybrid approach has been suggested consisting of the following steps:

1. An SVM is trained by the user' s initial query.
2. The system returns a set of M patterns consisting of the $K \leq M$ remotest ones, from the decision hyperplane in the RKHS and on the positive side, and the M–K closest ones to it.
3. The user labels the patterns as relevant or irrelevant.
4. The SVM classifier is retrained using all the patterns labeled, so far, by the user. The algorithm is either terminated, if the user is satisfied, or it is redirected to Step 2.

In [Xu 03], it is claimed that such a scheme shares the advantages of both previous selection criteria. It speeds up convergence and at the same time presents to the user some truly relevant patterns early enough in the iteration process.

Although in this section we have focused on the use of a classifier in Step 3, other alternatives are also possible. For example, in the so-called *query point method* (QPM) the existence of an ideal query point in the feature space is assumed, which, if found, would provide the appropriate answer to the user's query. Each feature in the feature vector is weighted and the task of the learner is to adjust the weights so that to move the point in the feature space appropriately. Learning can be based either on the positive examples (e.g., [Scla 97]) or on both positive and negative ones (e.g., [Rui 98]). For a more detailed reference survey on the topic the interested reader may consult [Liu 07, Long 03, Cruc 04].

8.6 PROBLEMS

8.1 Find the Edit distance between the word "poem" and its misspelled version "poten." Draw the optimal path.

8.2 Derive the slopes for the Sakoe–Chiba constraints and draw paths that achieve the maximum expansion/compression rates as well as paths corresponding to intermediate rates.

8.3 Develop a computer program for dynamic time warping, for a complete optimum path, and for the Itakura constraints. Verify the algorithm, using speech segments available from the book's web site.

8.4 Let the seven Hu moments of the reference $M \times N$ image block be $\phi_i, i = 1, 2, \ldots, 7$. Also, denote by $\psi_i(m, n), i = 1, 2, \ldots, 7$, the respective moments resulting from the test subimage located at (m, n). Explain why

$$\mathcal{M}(m,n) = \frac{\sum_{i=1}^{7} \phi_i \psi_i(m,n)}{\left(\sum_{i=1}^{7} \phi_i^2 \sum_{i=1}^{7} \psi_i^2(m,n)\right)^{1/2}}$$

is a reasonable measure of similarity.

8.5 Show that the Mellin transform $M(u, v)$ of a function $f(x, y)$, defined as

$$M(u,v) = \iint f(x,y) x^{-ju-1} y^{-jv-1} \, dx \, dy$$

is scale invariant.

8.6 For the motion compensation step in a video coding application, the image frame of size $I \times J$ is divided into subblocks of size $M \times N$. For each subblock the search for its corresponding one, in the current frame, is restricted in an area of size $[-p, p] \times [-p, p]$. Find the required number of operations per second for the computation of the maximum cross-correlation for a full search, a two-dimensional and a one-dimensional (independent i, j) logarithmic search, and a hierarchical search of three levels. Typical values for broadcast TV are $M = N = 16, I = 720, J = 480$, and $p = 15$. The number of frames transmitted per second is $f = 30$.

MATLAB PROGRAMS AND EXERCISES

Computer Programs

8.1 *Edit distance.* Write a MATLAB function named *edit_distance* that computes the Edit distance between two strings. This function takes as inputs (a) the reference string *ref_str* and (b) the test string *test_str*. It returns (a) the matching cost *edit_cost* and (b) the node predecessors matrix *pred*. Each element of the last matrix stores the coordinates of the predecessor of a node as a complex number, the real part of which stands for the row index and the imaginary part for the column index.

Solution

In the following implementation it is assumed that the reference pattern is associated with the horizontal axis.

```
function [edit_cost,pred]=edit_distance(ref_str,test_str)
  I = length(ref_str);
  J = length(test_str);
  D = zeros(J,I);
  %Initialization
  D(1,1) = ~(ref_str(1) == test_str(1));
  pred(1,1) = 0;
  for j = 2:J
    D(j,1) = D(j-1,1)+1;
    pred(j,1) = (j-1) + sqrt(-1)*1;
  end
  for i = 2:I
    D(1,i) = D(1,i-1)+1;
    pred(1,i) = 1+ sqrt(-1)*(i-1);
  end
  %Main Loop
  for i = 2:I
    for j = 2:J
      if(ref_str(i) == test_str(j))
        d(j,i) = 0;
      else
        d(j,i) = 1;
      end
      c1 = D(j-1,i-1)+d(j,i); c2 = D(j,i-1)+1; c3 = D(j-1,i)+1;
      [D(j,i),ind] = min([c1 c2 c3]);
      if(ind == 1)
        pred(j,i) = (j-1)+sqrt(-1)*(i-1);
      elseif(ind == 2)
```

```
            pred(j,i) = j+sqrt(-1)*(i-1);
          else
            pred(j,i) = (j-1)+sqrt(-1)*i;
          end
          end
      end
      edit_cost = D(J,I);
```

8.2 *Backtracking.* Write a MATLAB function named *back_tracking*, which takes as inputs (a) a node predecessors matrix *pred*, (b) the coordinates *k* and *l* of the node from which the backtracking will start. It returns the best path on the cost grid, *best_path*, and also plots the best path. It is assumed that backtracking always terminates when a node whose predecessor is (0, 0) is reached.

Solution

```
function best_path=back_tracking(pred,k,l)
  Node = k+sqrt(-1)*l;
  best_path = [Node];
  while (pred(real(Node),imag(Node)) = 0)
    Node = pred(real(Node),imag(Node));
    best_path = [Node;best_path];
  end
  %Plot the best path
  [I,J] = size(pred);
  clf;
  hold
  for j = 1:J
    for i = 1:I
      plot(j,i,'r.')
    end
  end
  plot(imag(best_path),real(best_path),'g')
  axis off
```

8.3 *Dynamic time warping with Sakoe-Chiba local path constraints.* Write a MATLAB function named *Dtw_Sakoe* that implements a dynamic time-warping scheme where the Sakoe-Chiba local path constraints are adopted. More specifically, the function takes as input (a) a row vector *ref* that corresponds to the reference sequence and (b) a row vector *test* that corresponds to the test sequence. It returns (a) the time-warping matching cost, *matching_cost* and (b) the best path *best_ path*. It is assumed that (i) the cost assigned to each node in the grid upon initialization is equal to the Euclidean distance of the respective pattern elements and (ii) the cost of a transition depends only on the cost that has been assigned to the node at the end of the transition.

Solution

In the following implementation, function *back_tracking* introduced before is utilized to determine the best path.

```
function [matching_cost,best_path]=Dtw_Sakoe(ref,test)
  I = length(ref);
  J = length(test);
  for i = 1:I
    for j = 1:J
      %Euclidean distance
      node_cost(i,j) = sqrt(sum((ref(:,i)-test(:,j)).^2));
    end
  end
  %Initialization
  D(1,1) = node_cost(1,1);
  pred(1,1) = 0;
  for i = 2:I
    D(i,1) = D(i-1,1)+node_cost(i,1);
    pred(i,1) = i-1 + sqrt(-1)*1;
  end
  for j = 2:J
    D(1,j) = D(1,j-1)+node_cost(1,j);
    pred(1,j) = 1 + sqrt(-1)*(j-1);
  end
  %Main Loop
  for i = 2:I
    for j = 2:J
      [D(i,j),ind] = min([D(i-1,j-1) D(i-1,j) ...
         D(i,j-1)]+node_cost(i,j));
      if (ind == 1)
        pred(i,j) = (i-1)+sqrt(-1)*(j-1);
      elseif (ind == 2)
        pred(i,j) = (i-1)+sqrt(-1)*(j);
      else
        pred(i,j) = (i)+sqrt(-1)*(j-1);
      end
    end %for j
  end %for i
  %End of Main Loop
  matching_cost = D(I,J);
  best_path = back_tracking(pred,I,J);
```

Computer Experiments

8.1 **a.** Compute the Edit distance between the following pairs of strings: (i) (*beauty*, *beaty*), (ii) (*beauty*, *biauty*), (iii) (*beauty*, *betty*), using the first element of each pair as the reference string.

b. Plot the respective matching paths using the function *back_tracking*.

8.2 Use the *Dtw_Sakoe* function to compute the time-warping cost between the sequences $\{1, 2, 3\}$ and $\{1, 1, 2, 2, 2, 3, 3, 3\}$, using the former as reference sequence. Comment on the results.

8.3 Let $r1 = [1, 0]^T$, $r2 = [0, 1]^T$ and $ref = [r1, r2]$. Generate a sequence of 10 two-dimensional vectors with the following MATLAB command $test = [1 + rand(1,4)/2 \;\; rand(1,6)/3; \;\; rand(1,4)/2 \;\; 1 + rand(1,6)/3]$. Use the *Dtw_Sakoe* function to compute the time-warping cost and the respective best path between *ref* and *test*, taking the former as the reference sequence. Comment on the results.

REFERENCES

[Alkh 01] Ghavari-Alkhansavi M. "A fast globally optimal algorithm for template matching using low resolution pruning," *IEEE Transactions on Image Processing*, Vol. 10(4), pp. 526–533, 2001.

[Amit 91] Amit Y., Grenander U., Piccioni M. "Structural image restoration through deformable template," *J. Amer. Statist. Association*, Vol. 86(414), pp. 376–387, 1991.

[Bell 57] Bellman R.E. *Dynamic Programming*, Princeton University Press, 1957.

[Bhas 95] Bhaskaran V., Konstantinides K. *Image and Video Compression Standards*, Kluwer Academic Publishers, 1995.

[Cheu 02] Cheung K.-W., Yeung D.-Y., Chin R.T. "On deformable models for visual pattern recognition," *Pattern Recognition*, Vol. 35, pp. 1507–1526, 2002.

[Cruc 04] Crucianu M., Ferecatu M., Boujemaa N. "Relevance feedback for image retrieval: a short survey," in *Audiovisual Content-based Retrieval, Information Universal Access and Interaction, Including Datamodels and Languages*, Report of the DELOS2 European Network of Excellence, FP6, 2004.

[Dame 64] Damerau F.J. "A technique for computer detection and correction of spelling errors," *Commun. ACM*, Vol. 7(3), pp. 171–176, 1964.

[Davi 80] Davis S.B., Mermelstein P. "Comparison of parametric representations for monosyllabic word recognition in continuously spoken sentences," *IEEE Transactions on Acoustics Speech and Signal Processing*, Vol. 28(4), pp. 357–366, 1980.

[Dell 93] Deller J., Proakis J., Hansen J.H.L. *Discrete-Time Processing of Speech Signals*, Macmillan, 1993.

[Desh 99] Deshmukh N., Ganapathirajn A., Picone J. "Hierarchical search for large vocabulary conversational speech recognition," *IEEE Signal Processing Magazine*, Vol. 16(5), pp. 84–107, 1999.

[Druc 01] Drucker H., Shahraray B., Gibbon D. "Relevance feedback using support vector machines," *Proceedings of th 18th International Conference on Machine Learning*, pp. 122–129, 2001.

[Durb 97] Durbin K., Eddy S., Krogh A., Mitchison G. *Biological Sequence Analysis: Probabilistic Models of Proteins and Nucleic Acids*, Cambridge University Press, Cambridge, MA, 1997.

[Fisc 73] Fischler M., Elschlager R. "The representation and matching of pictorial structures," *IEEE Transations on Computers*, Vol. 22(1), pp. 67–92, 1973.

[Gray 76] Gray A.H., Markel J.D. "Distance measures for speech processing," *IEEE Transactions on Acoustics Speech and Signal Processing*, Vol. 24(5), pp. 380–391, 1976.

[Gray 80] Gray R.M., Buzo A., Gray A.H., Matsuyama Y. "Distortion measures for speech processing," *IEEE Transactions on Acoustics Speech and Signal Processing*, Vol. 28(4), pp. 367–376, 1980.

[Hall 79] Hall E. *Computer Image Processing and Recognition*, Academic Press, 1979.

[Itak 75] Itakura F. "Minimum prediction residual principle applied to speech recognition," *IEEE Transactions on Acoustics Speech and Signal Processing*, Vol. 23(2), pp. 67–72, 1975.

[Jain 98] Jain A.K., Zhong Y., Dubuisson-Jolly M.P. "Deformable template models: A review," *Signal Processing*, Vol. 71, pp. 109–129, 1998.

[Jain 96] Jain A.K., Zhong Y., Lakshmanan S. "Object matching using deformable templates," *IEEE Transactions on Pattern Analysis and Machine Itelligence*, Vol. 18(3), pp. 267–277, 1996.

[Koch 89] Koch M.W., Kashyap R.L. "Matching polygon fragments," *Pattern Recognition Letters*, Vol. 10, pp. 297–308, 1989.

[Leven 66] Levenshtein V.I. "Binary codes capable of correcting deletions, insertions and reversals," *Soviet phys. Dokl.*, Vol. 10(8), pp. 707–710, 1966.

[Lewi 94] Lewis D., Gale W. "A sequential algorithm for training text classifiers," *Proceedings of the 11th International Conference on Machine Learning*, pp. 148–156, Morgan Kaufmann, 1994.

[Liu 07] Liu Y., Zhang D., Lu G., Ma W.-Y. "A survey of content-based image retrieval with high-level semantics," *Pattern Recognition*, Vol. 40, pp. 262–282, 2007.

[Long 03] Long F., Zang H.J., Feng D.D. "Fundamentals of content-based image retrieval," in *Multimedia Information Retrieval and Management* (Feng D. ed.), Springer, Berlin, 2003.

[Marz 93] Marzal A., Vidal E. "Computation of normalized edit distance and applications," *IEEE Transactions on Pattern Analysis and Machine Intelligence*, Vol. 15(9), 1993.

[McIn 96] McInerney T., Terzopoulos D. "Deformable models in medical image analysis: A survey," *Med. Image Anal.*, Vol. 1(2), pp. 91–108, 1996.

[Mei 04] Mei J. "Markov edit distance," *IEEE Transactions on Pattern Analysis and Machine Intelligence*, Vol. 6(3), pp. 311–320, 2004.

[Myer 80] Myers C.S., Rabiner L.R., Rosenberg A.E. "Performance tradeoffs in dynamic time warping algorithms for isolated word recognition," *IEEE Transactions on Acoustics Speech and Signal Processing*, Vol. 28(6), pp. 622–635, 1980.

[Neg 99] Neg H., Ortmanns S. "Dynamic programming search for continuous speech recognition," *IEEE Signal Processing Magazine*, Vol. 16(5), pp. 64–83, 1999.

[Ocud 76] Ocuda T., Tanaka E., Kasai T. "A method for correction of garbled words based on the Levenstein metric," *IEEE Transactions on Computers*, pp. 172–177, 1976.

[Pikr 03] Pikrakis A., Theodoridis S., Kamarotos D. "Recognition of isolated musical patterns using context dependent dynamic time warping," *IEEE Transactions on Speech and Audio Processing*, Vol 11(3), pp. 175–183, 2003.

[Rabi 93] Rabiner L., Juang B.H. *Fundamentals of Speech Recognition*, Prentice Hall, 1993.

[Ravi 95] Ravichandran G., Trivedi M.M. "Circular-Mellin features for texture segmentation," *IEEE Transactions on Image Processing*, Vol. 2(12), pp. 1629–1641, 1995.

[Rui 98] Rui Y., Huang T.S., Ortega M., Mehrotra S. "Relevance feedback: power tool in interactive content-based image retrieval," *IEEE Transactions on Circuits and Systems for Video Technology*, Vol. 8(5), pp. 644–655, 1998.

[Sako 78] Sakoe H., Chiba S. "Dynamic programming algorithm optimization for spoken word recognition," *IEEE Transactions on Acoustics Speech and Signal Processing*, Vol. 26(2), pp. 43–49, 1978.

[Scha 89] Schalkoff R. *Digital Image Processing and Computer Vision*, John Wiley & Sons, 1989.

[Scho 00] Schohn G., Cohn D. "Less is more: Active learning with support vector machines," *Proceedings of the 17th International Conference on Machine Learning*, pp. 839–846, Morgan Kaufmann, 2000.

[Scla 97] Sclaroff S., Taycher L. Cascia M "Imagerover: A content-based image browser for the world wide web," *Proceedings of the 1997 Workshop on Content-Based Access of Image and Video Libraries (CBAIVL'97)*, pp. 2–9, IEEE Computer Society, 1997.

[Seni 96] Seni G., Kripasundar V., Srihari R. "Generalizing Edit distance to incorporate domain information: Handwritten text recognition as a case study," *Pattern Recognition*, Vol. 29(3), pp. 405–414, 1996.

[Silv 90] Silverman H., Morgan D.P. "The application of the dynamic programming to connected speech recognition," *IEEE Signal Processing Magazine*, Vol. 7(3), pp. 7–25, 1990.

[Tong 01] Tong S., Chang E. "Support vector machine active learning for image retrieval," *Proceedings of the 9th ACM International Conference on Multimedia*, pp. 107–118, ACM Press, 2001.

[Tsay 93] Tsay Y.T., Tsai W.H. "Attributed string matching split and merge for on-line Chinese character recognition," *IEEE Transactions on Pattern Analysis and Machine Intelligence*, Vol. 15(2), pp. 180–185, 1993.

[Ueno 97] Uenohara M., Kanade T. "Use of the Fourier and Karhunen–Loève decomposition for fast pattern matching with a large set of templates," *IEEE Transactions on Pattern Analysis and Machine Intelligence*, Vol. 19(8), pp. 891–899, 1997.

[Wang 01] Wang J.Z., Li J., Wiederhold G. "SIMPLIcity: Semantics-sensitive Integrated Matching for Picture LIbraries," *IEEE Transactions on Pattern Analysis and Machine Intelligence*, vol 23, no.9, pp. 947–963, 2001.

[Wang 90] Wang Y.P., Pavlidis T. "Optimal correspondence of string subsequences," *IEEE Transactions on Pattern Analysis and Machine Intelligence*, Vol. 12(11), pp. 1080–1086, 1990.

[Widr 73] Widrow B. "The rubber mask technique, Parts I and II," *Pattern Recognition*, Vol. 5, pp. 175–211, 1973.

[Xu 03] Xu Z., Xu X., Yu K., Tresp V. "A hybrid relevance-feedback approach to text retrieval," *Proceedings of the 25th European Conference on Information Retrieval Research*, Lecture Notes in Computer Science, Vol. 2633, Springer Verlag, 2003.

CHAPTER

Context-Dependent Classification

9

9.1 INTRODUCTION

The classification tasks considered so far have assumed that no relation exists among the various classes. In other words, having obtained a feature vector $\boldsymbol{x}$ from a class ω_i, the next feature vector could belong to any other class. In this chapter we will remove this assumption, and we will assume that the various classes are closely related. That is, successive feature vectors are not independent. Under such an assumption, classifying each feature vector separately from the others obviously has no meaning. The class to which a feature vector is assigned depends (a) on its own value, (b) on the values of the other feature vectors, and (c) on the existing relation among the various classes. Such problems appear in various applications such as communications, speech recognition, and image processing.

In the context-free classification, our starting point was the Bayesian classifier. In other words, a feature vector was classified to a class ω_i if

$$P(\omega_i|\boldsymbol{x}) > P(\omega_j|\boldsymbol{x}), \quad \forall j \neq i$$

The Bayesian point of view will also be adopted here. However, the dependence among the various classes sets demands for a more general formulation of the problem. The mutual information that resides within the feature vectors requires the classification to be performed using *all* feature vectors simultaneously and also to be arranged in the same sequence in which they occurred from the experiments. For this reason, in this chapter we will refer to the feature vectors as *observations* occurring in sequence, one after the other, with $\boldsymbol{x}_1$ being the first and $\boldsymbol{x}_N$ the last from a set of N observations.

9.2 THE BAYES CLASSIFIER

Let $X: \boldsymbol{x}_1, \boldsymbol{x}_2, \ldots, \boldsymbol{x}_N$ be a sequence of N (feature vectors) observations and $\omega_i, i = 1, 2, \ldots, M$, the classes in which these vectors must be classified. Let

$\Omega_i : \omega_{i_1}, \omega_{i_2}, \ldots, \omega_{i_N}$ be one of the possible sequences of these classes corresponding to the observation sequence, with $i_k \in \{1, 2, \ldots, M\}$ for $k = 1, 2, \ldots, N$. The total number of these class sequences Ω_i is M^N, that is, the number of possible *ordered* combinations of M distinct objects taken in groups of N. *Our classification task is to decide to which class sequence Ω_i a sequence of observations X corresponds*. This is equivalent to appending $\boldsymbol{x}_1$ to class ω_{i_1}, $\boldsymbol{x}_2$ to ω_{i_2}, and so on. A way to approach the problem is to view each specific sequence X as an (extended) feature vector and $\Omega_i, i = 1, 2, \ldots, M^N$, as the available classes. Having observed a specific X, the Bayes rule assigns it to Ω_i for which

$$P(\Omega_i|X) > P(\Omega_j|X), \quad \forall i \neq j \tag{9.1}$$

and following our already familiar arguments, this is equivalent to

$$P(\Omega_i)p(X|\Omega_i) > P(\Omega_j)p(X|\Omega_j), \quad \forall i \neq j \tag{9.2}$$

In the following we will investigate the specific form that Eq. (9.2) takes for some typical class dependence models.

9.3 MARKOV CHAIN MODELS

One of the most widely used models describing the underlying class dependence is the Markov chain rule. If $\omega_{i_1}, \omega_{i_2}, \ldots$ is a sequence of classes, then the Markov model assumes that

$$P(\omega_{i_k}|\omega_{i_{k-1}}, \omega_{i_{k-2}}, \ldots, \omega_{i_1}) = P(\omega_{i_k}|\omega_{i_{k-1}}) \tag{9.3}$$

The meaning of this is that the class dependence is limited only within two successive classes. This type of model is also called a first-order Markov model, to distinguish it from obvious generalizations (second, third, etc.). In other words, given that the observations $\boldsymbol{x}_{k-1}, \boldsymbol{x}_{k-2}, \ldots, \boldsymbol{x}_1$ belong to classes $\omega_{i_{k-1}}, \omega_{i_{k-2}}, \ldots, \omega_{i_1}$, respectively, the probability of the observation $\boldsymbol{x}_k$, at stage k, belonging to class ω_{i_k} *depends only on the class from which observation $\boldsymbol{x}_{k-1}$, at stage $k-1$, has occurred*. Now combining (9.3) with the probability chain rule [Papo 91, p. 192],

$$\begin{aligned} P(\Omega_i) &\equiv P(\omega_{i_1}, \omega_{i_2}, \ldots, \omega_{i_N}) \\ &= P(\omega_{i_N}|\omega_{i_{N-1}}, \ldots, \omega_{i_1})P(\omega_{i_{N-1}}|\omega_{i_{N-2}}, \ldots, \omega_{i_1}) \ldots P(\omega_{i_1}) \end{aligned}$$

we obtain

$$P(\Omega_i) = P(\omega_{i_1}) \prod_{k=2}^{N} P(\omega_{i_k}|\omega_{i_{k-1}}) \tag{9.4}$$

where $P(\omega_{i_1})$ is the prior probability for class $\omega_{i_1}, i_1 \in \{1, 2, \ldots, M\}$, to occur. Furthermore, two commonly adopted assumptions are that (a) given the sequence of classes, the observations are statistically independent, and (b) the probability density function in one class does not depend on the other classes. *That is,*

dependence exists only on the sequence in which classes occur, but within a class observations "obey" the class' own rules. This assumption implies that

$$p(X|\Omega_i) = \prod_{k=1}^{N} p(\boldsymbol{x}_k|\omega_{i_k}) \tag{9.5}$$

Combining Eqs. (9.4) and (9.5), the Bayes rule for Markovian models becomes equivalent to the statement:

Statement

Having observed the sequence of feature vectors $X : \boldsymbol{x}_1, \ldots, \boldsymbol{x}_N$, classify them in the respective sequence of classes $\Omega_i : \omega_{i_1}, \omega_{i_2}, \ldots, \omega_{i_N}$, so that the quantity

$$p(X|\Omega_i)P(\Omega_i) = P(\omega_{i_1})p(\boldsymbol{x}_1|\omega_{i_1}) \prod_{k=2}^{N} P(\omega_{i_k}|\omega_{i_{k-1}})p(\boldsymbol{x}_k|\omega_{i_k}) \tag{9.6}$$

becomes maximum.

As we have already stated, searching for this maximum requires the computation of Eq. (9.6) for each of the $\Omega_i, i = 1, 2, \ldots, M^N$. This amounts to a total of $O(NM^N)$ multiplications, which is usually a large number indeed. However, this direct computation is a brute-force task. Let us take for example two sequences Ω_i and Ω_j, which we assume differ only in the last class, that is, $\omega_{i_k} = \omega_{j_k}, k = 1, 2, \ldots, N-1$ and $\omega_{i_N} \neq \omega_{j_N}$. It does not take much "scientific thought" to realize that the computation of (9.6) for these two sequences shares all multiplications (which need not be repeated) but one. Furthermore, closer observation of (9.6) reveals that it has a rich computational structure, which can be exploited in order to maximize it in a much more efficient way. This now becomes our next concern.

9.4 THE VITERBI ALGORITHM

Figure 9.1 shows a diagram with N dot columns, where each dot in a column represents one of the M possible classes, $\omega_1, \omega_2, \ldots, \omega_M$. Successive columns correspond to successive observations $\boldsymbol{x}_k, k = 1, 2, \ldots, N$. The arrows determine transitions from one class to another, as observation vectors are obtained in sequence. Thus, *each of the class sequences Ω_i corresponds to a specific path of successive transitions*. Each transition from one class ω_i to another ω_j is characterized by a *fixed* probability $P(\omega_j|\omega_i)$, which is assumed to be known for the adopted model of class dependence. Furthermore, we assume that these probabilities are the same for all successive stages k. That is, the probabilities depend only on the respective class transitions and not on the stage at which they occur. We will further assume that the conditional probability densities $p(\boldsymbol{x}|\omega_i), i = 1, 2, \ldots, M$, are also known to the model. The task of maximizing (9.6) can now be stated as follows. Given a sequence of observations $\boldsymbol{x}_1, \boldsymbol{x}_2, \ldots, \boldsymbol{x}_N$, find the path of successive (class) transitions that

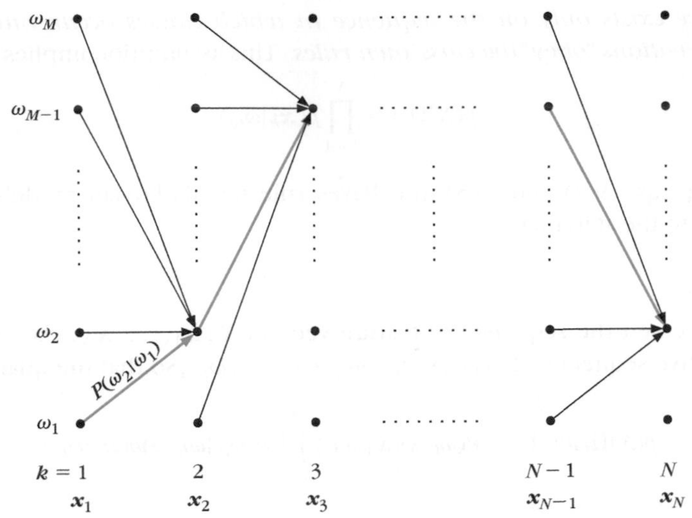

FIGURE 9.1

Trellis diagram for the Viterbi algorithm. The red line denotes the optimal path. The classes along this optimal path will be the ones in which the respective observations are classified.

maximizes (9.6) (e.g., red line in the figure). *The classes along this optimal path will be the ones in which the respective observations are classified.* To search for the optimal path we have to associate a cost with each of the transitions, in agreement with the cost function given in (9.6). A careful look at (9.6) suggests adopting

$$\hat{d}(\omega_{i_k}, \omega_{i_{k-1}}) = P(\omega_{i_k}|\omega_{i_{k-1}})p(\boldsymbol{x}_k|\omega_{i_k}) \tag{9.7}$$

as the cost associated with a transition of a path i from node (class) $\omega_{i_{k-1}}$, at stage $k-1$, to node ω_{i_k}, at stage k *and* at the same time observation $\boldsymbol{x}_k$ occurring. The initial condition for $k = 1$ is given by

$$\hat{d}(\omega_{i_1}, \omega_{i_0}) = P(\omega_{i_1})p(\boldsymbol{x}_1|\omega_{i_1})$$

Using this notation, the overall cost to be optimized becomes

$$\hat{D} \equiv \prod_{k=1}^{N} \hat{d}(\omega_{i_k}, \omega_{i_{k-1}}) \tag{9.8}$$

It will not come as a surprise to us if instead of (9.8) one chooses to maximize

$$\begin{aligned} \ln(\hat{D}) &= \sum_{k=1}^{N} \ln \hat{d}(\omega_{i_k}, \omega_{i_{k-1}}) \\ &\equiv \sum_{k=1}^{N} d(\omega_{i_k}, \omega_{i_{k-1}}) \equiv D \end{aligned} \tag{9.9}$$

where $d(\cdot,\cdot) \equiv \ln \hat{d}(\cdot,\cdot)$. Looking carefully at (9.9) or (9.8), it will not take us long to realize that Bellman's principle can again offer us the means for efficient optimization. Indeed, let us define, in accordance with D, the cost for reaching class ω_{i_k} at stage k via a path i as

$$D(\omega_{i_k}) = \sum_{r=1}^{k} d(\omega_{i_r}, \omega_{i_{r-1}}) \tag{9.10}$$

Then, Bellman's principle states that

$$D_{\max}(\omega_{i_k}) = \max_{i_{k-1}}[D_{\max}(\omega_{i_{k-1}}) + d(\omega_{i_k}, \omega_{i_{k-1}})], \quad i_k, i_{k-1} = 1, 2, \ldots, M \tag{9.11}$$

with

$$D_{\max}(\omega_{i_0}) = 0 \tag{9.12}$$

It is now straightforward to see that the optimal path, which leads to the maximum D in (9.9), is the one that ends at the final stage N in the class $\omega^*_{i_N}$ for which

$$\omega^*_{i_N} = \arg\max_{\omega_{i_N}} D_{\max}(\omega_{i_N}) \tag{9.13}$$

Going back to Figure 9.1, we see that at each stage $k, k = 1, 2, \ldots, N$, there are M possible transitions to each of the nodes ω_{i_k}. The recursive relation in (9.11) suggests that in searching for the maximum we need only keep one of these transitions for every node, *the one that leads to the maximum respective cost* $D_{\max}(\omega_{i_k})$ (red lines in the figure). *Hence, at each stage there are only M surviving paths*. Therefore, the number of operations is M for each node, thus M^2 for all the nodes at each stage, and NM^2 in total. The latter number compares very favorably with the NM^N of the "brute-force task." The resulting algorithm is known as the Viterbi algorithm. This dynamic programming algorithm was first suggested for optimal decoding in convolutional codes in communications [Vite 67].

In previous chapters, we stated that for a number of reasons an alternative to the optimal Bayes classifier can be used. No doubt context-dependent classification can also be emancipated from its Bayesian root. This can easily be achieved by adopting different transition costs $d(\omega_{i_k}, \omega_{i_{k-1}})$, which are not necessarily related to probability densities. In the following sections we will present two typical application areas of the Viterbi algorithm.

Example 9.1

In this example we apply the Viterbi algorithm to compute the optimal paths up to stage $k = 4$, once observation x_4 has been received. Assume that $x_4 = 1.2$ and that the observations reside in the one-dimensional space. Let the task involve three classes, namely, ω_1, ω_2, ω_3. We will further assume that the optimal paths up to stage $k = 3$ have been computed and

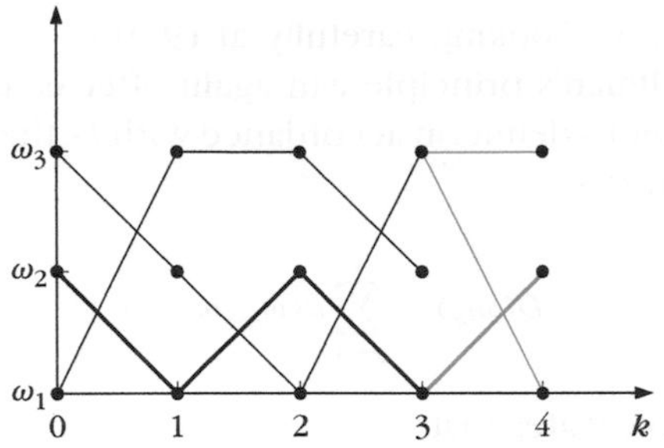

FIGURE 9.2

Optimal paths for the grid corresponding to Example 9.1. The red lines correspond to the extensions of the optimal paths from stage $k = 3$ to stage $k = 4$.

Table 9.1 Transition Costs Between Nodes for Example 9.1

Classes	$\omega_{i_k} = \omega_1$	$\omega_{i_k} = \omega_2$	$\omega_{i_k} = \omega_3$
$\omega_{i_{k-1}} = \omega_1$	0.1	0.7	0.2
$\omega_{i_{k-1}} = \omega_2$	0.4	0.3	0.3
$\omega_{i_{k-1}} = \omega_3$	0.3	0.1	0.6

are shown in black lines in Figure 9.2. Let the optimal costs associated with each class at stage $k = 3$ be equal to

$$D(\omega_1) = -0.5, \quad D(\omega_2) = -0.6, \quad D(\omega_3) = -0.2 \tag{9.14}$$

All the values are negative, since, as Eqs. (9.9) and (9.10) suggest, costs result by summing up logarithms of probability products. The previous costs are assumed known and they have been computed based on the initial class probability values and (a) the transition probability costs among the three classes, which are given in Table 9.1 and (b) the values of the received observations, x_0, x_1, x_2, x_3. We have further assumed that the probability density function describing the emission of an observation from each one of the classes follows a Gaussian distribution

$$p(x|\omega_i) = \frac{1}{\sqrt{2\pi}\sigma_i} \exp\left(-\frac{(x-\mu_i)^2}{2\sigma_i^2}\right)$$

where $\mu_1 = 1.0$ and $\sigma_1^2 = 0.03$, $\mu_2 = 1.5$ and $\sigma_2^2 = 0.02$, $\mu_3 = 0.5$ and $\sigma_3^2 = 0.01$.

We will first compute the optimal path reaching class ω_1 at stage $k = 4$. According to Eqs. (9.11), (9.10), and (9.7), the following calculations are in order:

$$\ln p(x_4 = 1.2|\omega_{i_4} = \omega_1) = -0.1578$$

Total cost for the transition from $\omega_{i_3} = \omega_1$ to $\omega_{i_4} = \omega_1$ is equal to $-0.5 + \ln(0.1) - 0.1578 = -2.9604$.

Total cost for the transition from $\omega_{i_3} = \omega_2$ to $\omega_{i_4} = \omega_1$ is equal to

$$-0.6 + \ln(0.4) - 0.1578 = -1.6741.$$

Total cost for the transition from $\omega_{i_3} = \omega_3$ to $\omega_{i_4} = \omega_1$ is equal to

$$-0.2 + \ln(0.3) - 0.1578 = -1.5617.$$

Hence, the optimal path reaching class ω_1 at stage $k = 4$ is through ω_3 at stage $k = 3$.

For the transitions to ω_2 at $k = 4$, we have

$$\ln p(x_4 = 1.2|\omega_{i_4} = \omega_2) = -0.2591$$

and the respective values for the paths reaching class ω_2 from classes ω_1, ω_2 and ω_3 at $k = 3$ are -1.1158, -2.0631, -2.7617. Thus the optimal path reaching ω_2 at $k = 4$ is through ω_1 at $k = 3$.

Finally, the respective values for the paths reaching ω_3 at $k = 4$ are

$$\ln p(x_4 = 1.2|\omega_{i_4} = \omega_3) = -2.2176$$

and -4.3271, -4.0216, -2.9285. As a result, the best path reaching node ω_3 at stage $k = 4$ goes through class ω_3 at stage $k = 3$ (self-transition).

If $k = 4$ is the final stage, that is, only four observations are available, then the optimal path, denoted by a bold line in Figure 9.2, is the one ending at stage ω_2 with an overall cost equal to -1.1158. Going backwards along the optimal path (backtracking), we assign: x_4 to ω_2, x_3 to ω_1, x_2 to ω_2, x_1 to ω_1 and x_0 to ω_2.

9.5 CHANNEL EQUALIZATION

Channel equalization is the task of recovering a transmitted sequence of information bits I_k (e.g., 1 or 0) after they have been corrupted by the transmission channel and noise sources. The samples received at the receiver end are, thus, given by

$$x_k = f(I_k, I_{k-1}, \ldots, I_{k-n+1}) + \eta_k \tag{9.15}$$

where the function $f(\cdot)$ represents the action of the channel and η_k the noise sequence. The channel contribution to the overall corruption is called the ***intersymbol interference (ISI)*** and it spans n successive information bits. The equalizer is the ***inverse*** system whose task is to provide decisions $\hat{I}_k$ about the transmitted information bits I_k, based on l successively received samples $[x_k, x_{k-1}, \ldots, x_{k-l+1}] \equiv \boldsymbol{x}_k^T$. Usually, a delay r must be used in order to accommodate the (possible) noncausal nature of the inverse system. In such cases the decisions made at time k correspond to the I_{k-r} transmitted information bit (Figure 9.3). A simple example will reveal to us how the equalization problem comes under the umbrella of a Markovian context-dependent classification task. Assume a simple linear channel

$$x_k = 0.5I_k + I_{k-1} + \eta_k \tag{9.16}$$

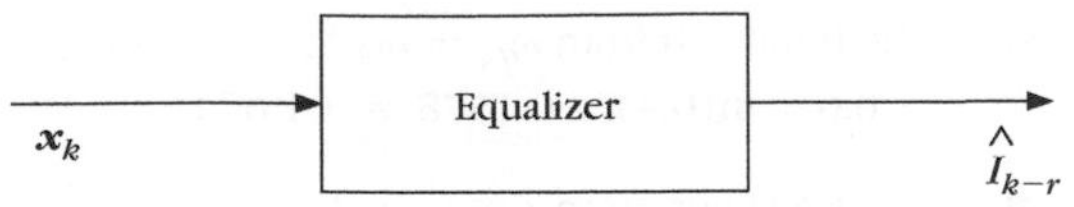

FIGURE 9.3

Block diagram of an equalizer.

For $l = 2$, successively received samples are combined in vectors in the two-dimensional space, that is,

$$\boldsymbol{x}_k^T = [x_k, x_{k-1}]$$

Let us further assume that there are N such observation vectors available. From (9.16) it is readily seen that each $\boldsymbol{x}_k, k = 1, 2, \ldots, N$, depends on the values of three successive information bits, namely I_k, I_{k-1}, I_{k-2}. Neglecting the effects of noise, the possible values of the received samples x_k are given in Table 9.1, together with the respective information bits. Figure 9.4a shows the geometry in the two-dimensional space with (x_k, x_{k-1}) on its axis. When the effect of noise is taken into account, the received vectors are clustered around these points (for example, shaded area). For the specific channel of (9.16) there are eight possible clusters, $\omega_i, i = 1, 2, \ldots, 8$ (Table 9.1). Clusters (red) around "+" correspond to $I_k = 1$ and those (gray) around "o" to $I_k = 0$. In general, the total number of clusters for a binary information sequence is $m = 2^{n+l-1}$. On the reception of each $\boldsymbol{x}_k = [x_k, x_{k-1}]^T$, the equalizer has to decide whether the corresponding transmitted information bit I_k was either a "1" (i.e., class "A") or a "0" (class "B"). In other words, this is nothing other than a two-class classification problem (M class for the M-ary case), and each class consists of a *union of clusters*. Thus, various techniques, from those we have already studied in previous chapters, can be used. A simple way, which was followed in [Theo 95], consists of two steps. During the training period a sequence of known information bits is transmitted, and the representative center $\boldsymbol{\mu}_i$ for each of the clusters is computed by a simple averaging of all the vectors $\boldsymbol{x}_k$ belonging to the respective cluster. This is possible during the *training period*, since the transmitted information bits are known, and thus we know to which of the clusters each received $\boldsymbol{x}_k$ belongs. For example, in the case of Table 9.1 if the sequence of transmitted bits is $(I_k = 1, I_{k-1} = 0, I_{k-2} = 1)$, then the received $\boldsymbol{x}_k$ belongs to ω_6. At the same time, the clusters are labeled as "1" or "0" depending on the value of the I_k bit. At the so-called *decision directed mode* the transmitted information bits are unknown, and the decision about the transmitted I_k is based on which cluster ("1" or "0" label) the received vector $\boldsymbol{x}_k$ is closest to. For this purpose a metric is adopted to define distance. The Euclidean distance of the received vector $\boldsymbol{x}_k$ from the representatives $\boldsymbol{\mu}_i$ of the clusters is an obvious candidate. Although such an equalizer can result in reasonable performance, measured in *bit error rate (BER)* (the percentage of information bits wrongly identified), there is still a great deal of information that has not been exploited. Let us search for it!

Table 9.2 Received Samples and Respective Information Bits for the Channel of Eq. (9.16)

I_k	I_{k-1}	I_{k-2}	x_k	x_{k-1}	Cluster
0	0	0	0	0	ω_1
0	0	1	0	1	ω_2
0	1	0	1	0.5	ω_3
0	1	1	1	1.5	ω_4
1	0	0	0.5	0	ω_5
1	0	1	0.5	1	ω_6
1	1	0	1.5	0.5	ω_7
1	1	1	1.5	1.5	ω_8

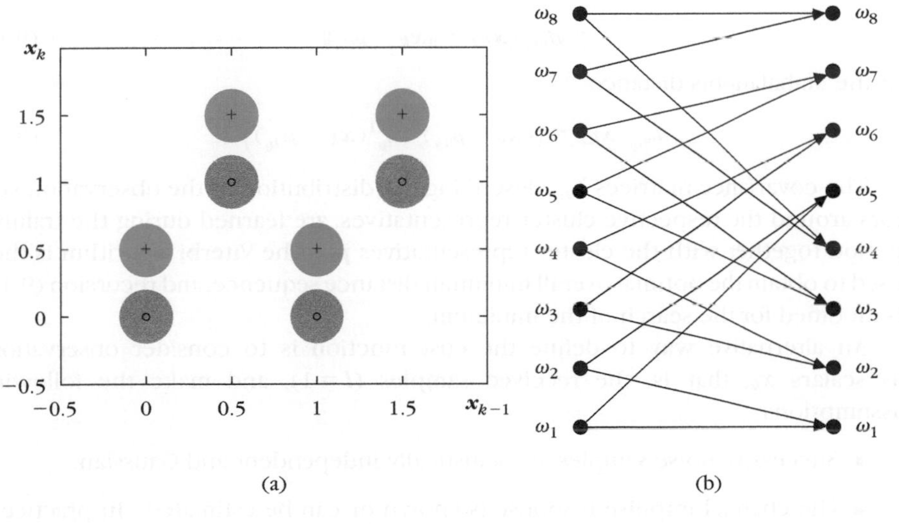

FIGURE 9.4

Plot (a) of the eight possible clusters associated with the channel of Eq. (9.16) and (b) the allowable transitions among them.

From the definition of ISI in (9.15) we know that the channel spans a number of successive information bits. *Thus, only certain transitions among clusters are possible*. Indeed, let us assume, for example, that at time k the transmitted information bits (I_k, I_{k-1}, I_{k-2}) were $(0, 0, 1)$; hence, the corresponding observation vector $\boldsymbol{x}_k$ belongs to cluster ω_2. The next received observation vector $\boldsymbol{x}_{k+1}$ will depend on the triple (I_{k+1}, I_k, I_{k-1}) bits, which can be either $(1, 0, 0)$ or $(0, 0, 0)$. Thus, $\boldsymbol{x}_{k+1}$

will belong either to ω_5 or to ω_1. That is, there are only two possible transitions from ω_2. Figure 9.4b shows the possible transitions among the various clusters. Assuming equiprobable information bits, then from Figure 9.4b we can easily conclude that all *allowable* transitions have probability 0.5, that is,

$$P(\omega_1|\omega_1) = 0.5 = P(\omega_5|\omega_1)$$

and the rest are zero (not allowable). We now have all the ingredients to define the equalization problem as a context-dependent classification task.

Given N observation vectors $\boldsymbol{x}_k, k = 1, 2, \ldots, N$, classify them in a sequence of clusters $\omega_{i_1}, \omega_{i_2}, \ldots, \omega_{i_N}$. This automatically classifies each $\boldsymbol{x}_k$ in one of the two classes "A" and "B," which is equivalent to deciding that I_k is 1 or 0. For this goal a cost function has to be adopted. In [Theo 95, Geor 97] the cost $d(\omega_{i_k}, \omega_{i_{k-1}})$ in (9.9) used in the Viterbi algorithm, for the allowable transitions, was taken to be

$$d(\omega_{i_k}, \omega_{i_{k-1}}) = d_{\omega_{i_k}}(\boldsymbol{x}_k) \tag{9.17}$$

where $d_{\omega_{i_k}}(\boldsymbol{x}_k)$ is the distance of $\boldsymbol{x}_k$ from the representative of the ω_{i_k} cluster. This can be either the Euclidean

$$d_{\omega_{i_k}}(\boldsymbol{x}_k) = \|\boldsymbol{x}_k - \boldsymbol{\mu}_{i_k}\| \tag{9.18}$$

or the Mahalanobis distance

$$d_{\omega_{i_k}}(\boldsymbol{x}_k) = \left((\boldsymbol{x}_k - \boldsymbol{\mu}_{i_k})^T \Sigma_{i_k}^{-1} (\boldsymbol{x}_k - \boldsymbol{\mu}_{i_k})\right)^{1/2} \tag{9.19}$$

The covariance matrices Σ_{i_k}, describing the distribution of the observation vectors around the respective cluster representatives, are learned during the training period together with the cluster representatives $\boldsymbol{\mu}_i$. The Viterbi algorithm is then used to obtain the optimal overall minimum distance sequence, and recursion (9.11) is modified for the search of the minimum.

An alternative way to define the cost function is to consider observations as scalars x_k, that is, the received samples ($l = 1$), and make the following assumptions:

- Successive noise samples are statistically independent and Gaussian.
- The channel impulse response is known or can be estimated. In practice, a specific channel model is assumed, that is, $\hat{f}(\cdot)$, and its parameters are estimated via an optimization method, for example, least squares [Proa 89].

Under the preceding assumptions, the cost for the allowable state transitions in (9.9) becomes

$$\begin{aligned} d(\omega_{i_k}, \omega_{i_{k-1}}) &= \ln p(x_k|\omega_{i_k}) \equiv \ln(p(\eta_k)) \\ &= -(x_k - \hat{f}(I_k, \ldots, I_{k-n+1}))^2 \end{aligned} \tag{9.20}$$

where η_k is the respective Gaussian distributed noise sample. Obviously, in (9.20) the constants in the Gaussian density function have been omitted. If the Gaussian and independence assumptions are valid, this is obviously a Bayesian optimal classification to the clusters (from which the "0" and "1" classes result). However, if this is not true, the cost in (9.20) is no longer the optimal choice. This is, for example, the case when the so-called cochannel (nonwhite) interference is present. In such cases, the cluster-based approach is justifiable, and indeed it leads to equalizers with more robust performance [Geor 97]. Furthermore, the fact that in the clustering approach no channel estimation is required can be very attractive in a number of cases, where nonlinear channels are involved and their estimation is not a straightforward task [Theo 95]. In [Kops 03] equalization is performed in the one-dimensional space, that is, $l = 1$. Although this increases the probability of having clusters with different labels to overlap, this is not crucial to the performance, since the Viterbi algorithm has the power to detect the correct label, by exploiting the history in the path. Furthermore, it is pointed out that one needs not determine directly all the 2^n cluster representatives; it suffices to learn, during the training phase, only n of the clusters, and the rest can be obtained by simple arithmetic operations. This is achieved by exploiting the mechanism underlying the cluster formation and the associated symmetries. Both of these observations have a substantial impact on reducing the computational complexity as well as the required length of the training sequence.

The discussion so far has been based on the development of a trellis diagram associated with the transitions among clusters, and the goal has been to unravel the optimal path, using the Viterbi algorithm. However, although this came as a natural consequence of the context-dependent Bayesian classifier, it turns out this is not the most efficient way from a computational point of view. From Figure 9.4b one can easily observe that pairs of clusters jump to the same clusters after transition. For example, the allowable transitions from ω_1 and ω_2 are the same and lead to either ω_1 or ω_5. The same is true for ω_3 and ω_4, and so on. This is because the allowable transitions are determined by the $n + l - 2$ most recent bits. For the example of Figure 9.4, transitions are determined by the pair (I_k, I_{k-1}), which, however, is shared by two clusters, that is, (I_k, I_{k-1}, I_{k-2}), depending on the value of I_{k-2} if it is 0 or 1. The pair (I_k, I_{k-1}) is known as the *state* at time k. This is because, knowing the state at time k and the transmitted bit I_{k+1} at time $k + 1$, we can determine the next state (I_{k+1}, I_k) at time $k + 1$, as is the case in the finite state machines. Since transitions are determined by the states, one can construct a trellis diagram based on the states instead of on clusters. For the example of Figure 9.5, where eight clusters are present, there is a total of four states, that is, $s_1 : (0, 0)$, $s_2 : (0, 1)$, $s_3 : (1, 0)$, $s_4 : (1, 1)$. Figure 9.5 shows these states and the allowable transitions among them. Each transition is associated with one bit, which is the current transmitted bit. Obviously, there is a close relationship between states and clusters. If we know the state transition, that is, $(I_k, I_{k-1}) \rightarrow (I_{k+1}, I_k)$, then the current cluster at time $k + 1$ will be determined by the corresponding values of (I_{k+1}, I_k, I_{k-1}), and this automatically determines the cost of the respective transition, for example, (9.18) or (9.19). Hence the estimates of the transmitted bits are obtained from the sequence of bits along the optimal path, that is, the one with the minimum total cost, in the state trellis diagram instead of the larger cluster trellis diagram.

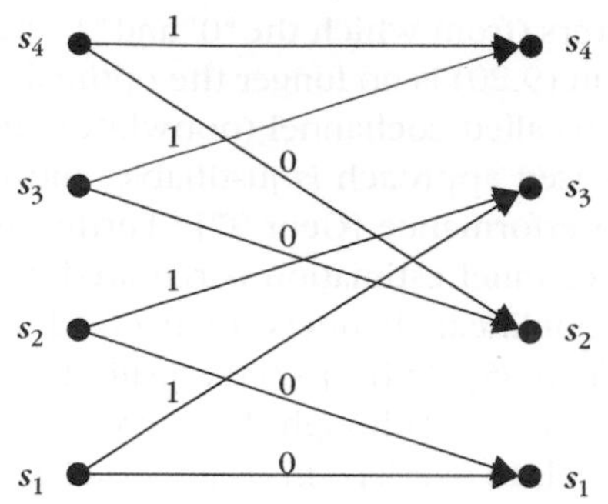

FIGURE 9.5

Plot of the four states associated with the channel of Eq. (9.16) and a two-dimensional equalizer showing the allowable transitions among them.

9.6 HIDDEN MARKOV MODELS

In the channel equalization application of the previous section, the states of the Markov chain were *observable*. That is, given the $l+n-1$ most recent information bits (i.e., I_k, I_{k-1}, I_{k-2}, in the given example), the state to which the corresponding observation vector $\boldsymbol{x}_k$ belongs is readily known. Thus, during the training period these states can be "labeled," and we can estimate their associated parameters (i.e., the related clusters). In this section, we will be concerned with systems where the states cannot be directly observed. The observations will be considered as the result of an action associated with each state and which is described by a set of probabilistic functions. Moreover, the sequence in which the different states are visited by successive observations is itself the result of another stochastic process, *which is hidden to us*, and the associated parameters describing it can only be inferred by the set of the received observations.

These types of Markov models are known as *hidden Markov models* (HMMs). Let us consider some simple examples of such processes inspired by the well-known coin-tossing problem. We will assume that in all experiments the coin tossing takes place behind a curtain, and all that is disclosed to us is the outcome of each experiment. That is, each time an experiment is performed we cannot know the specific coin (in the case of multiple coins) whose tossing resulted in the current observation (heads or tails). Thus, a crucial part of the probabilistic process is hidden to us.

In the first experiment, a single coin is tossed to produce a sequence of heads (H) and tails (T). This experiment is characterized by a single parameter indicating the propensity of the coin for landing heads. This is quantified by the probability $P(H)$ of H ($P(T) = 1 - P(H)$). A straightforward modeling of this statistical process is to associate one state with the outcome H and one with the outcome T. Hence, this is another example of a process with observable states, since states coincide with the

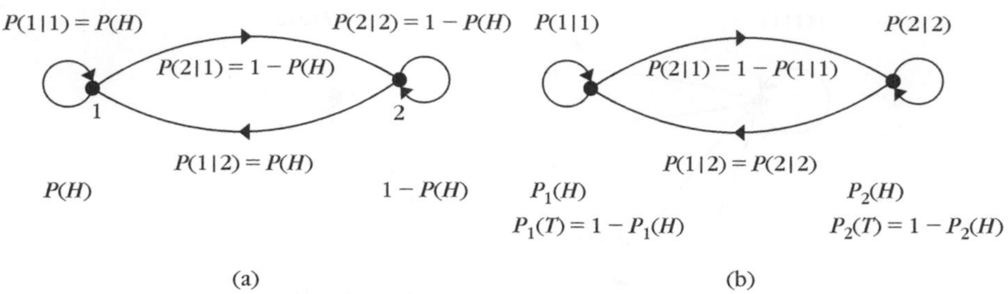

FIGURE 9.6

Markov models for hidden coin-tossing experiments: (a) single coin and (b) two coins.

observations. Figure 9.6a illustrates the underlying generation mechanism of the sequence of observations. $P(i|j)$ denotes the transition probability from state s_j to state s_i, once the coin has been tossed and an observation has been made available to us. For simplicity, states are shown simply as j and i, respectively. For this specific example, state $i = 1$ represents H and state $j = 2$ represents T. Also, for this case, all transition probabilities are expressed in terms of one parameter; that is, $P(H)$. For example, assume that the coin is in state $i = 1$ (H). Tossing the coin again it can either result in H (the coin stays in the same state and $P(1|1) = P(H)$) or it can jump into the other state resulting in T ($P(2|1) = P(T) = 1 - P(H)$).

In the second experiment, we assume that two coins are used behind the curtain. Although, again, the sequence of observations consists of a random succession of heads or tails, it is apparent that this process cannot be sufficiently modeled by a single parameter. To model the process we assume two states, corresponding to the two coins. The model is shown in Figure 9.6b. Two sets of parameters are involved. One of them consists of the probabilities $P_1(H)$ and $P_2(H)$; that is, the probabilities of H for coins 1 and 2, respectively. The other set of parameters are the transition probabilities, $P(i|j)$, $i, j = 1, 2$. For example, $P(1|2)$ is the probability that the current observation (which can be either H or T) is the outcome of an experiment performed using coin 1 (state $i = 1$) and that the previous observation was the result of tossing coin 2 (state $j = 2$). Self-transition probabilities, for example, $P(1|1)$, mean that the same coin (1) is tossed twice and the process remains in the same state ($i = 1$) for two successive times. Taking into account that probabilities of an event add to one, two of the transition parameters are sufficient, and this amounts to a total of four parameters (i.e., $P_1(H)$, $P_2(H)$, $P(1|1)$, $P(2|2)$). It is important to point out that the states of this process are not observable, since we have no access to the information related to which coin is tossed each time.

Figure 9.7 shows the Markov model for the case of tossing three coins behind the curtain. Nine parameters are now required to describe the process; that is, the probabilities $P_i(H)$, $i = 1, 2, 3$, one for each coin and six transition probabilities (the number of transition probabilities is nine but there are three constraints;

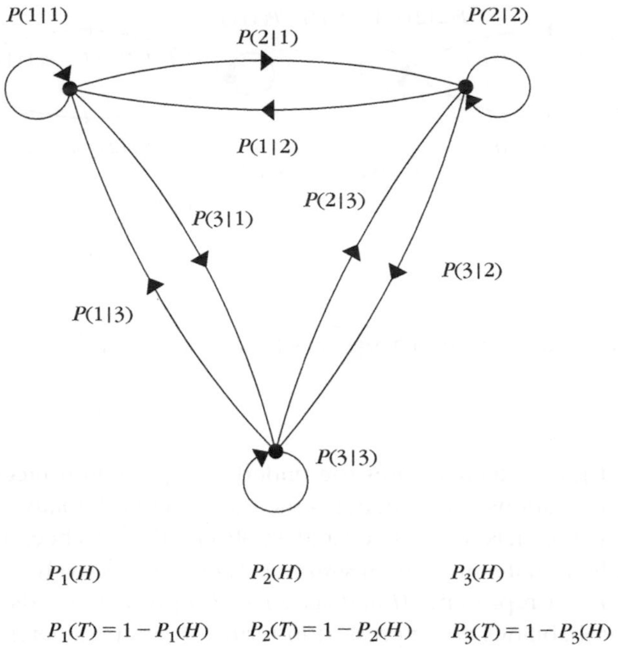

FIGURE 9.7

Markov model for three hidden coins.

i.e., $\sum_{i=1}^{3} P(i|j) = 1, j = 1, 2, 3$). The case of three coins is another example of a probabilistic process with hidden states.

As we will soon see, a major task associated with HMM is first to adopt a model for the underlying process that produces the sequence of observations and then to estimate the unknown set of parameters based on these observations. For the tossing coins examples, these parameters are the transition probabilities as well as the head or tail probabilities for each of the coins. No doubt, adopting the right model is crucial. If, for example, the head or tail observations were produced by tossing two coins and we selected, wrongly, a three-coin model, the resulting estimates of the parameters would lead to an overall poor modeling of the received observations.

In general, an HMM is a type of stochastic modeling appropriate for nonstationary stochastic sequences, with statistical properties that undergo distinct random transitions among a set of different stationary processes. In other words, an HMM models a sequence of observations as a *piecewise stationary process*. Such models have been used extensively in speech recognition to model speech utterances [Bake 75, Jeli 76]. An utterance may be a spoken word, part of a word, or even a complete sentence or a paragraph. The statistical properties of the speech signal within an utterance undergo a series of transitions. For example, a word consists of subword portions of *voiced* (vowels) and *unvoiced* (consonants) sounds. These are characterized by distinctly different statistical properties, which are in turn reflected in transitions of the speech signal from one statistic to another. Handwriting recognition [Chen 95, Vlon 92, Agaz 93, ElYa 99, Aric 02, Ramd 03], texture classification [Chen 95a, Wu 96], blind equalization [Anto 97, Kale 94, Geor 98], musical pattern recognition [Pikr 06] are some other example applications in which the power of HMM modeling has been successfully exploited.

An HMM model is basically a *stochastic finite state automaton*, which generates an observation string, that is, the sequence of observation vectors, $\boldsymbol{x}_1, \boldsymbol{x}_2, \ldots, \boldsymbol{x}_N$. Thus, an HMM model consists of a number of, say K, states and the observation string is produced as a result of successive transitions from one state i to another j. We will adopt the so-called Moore machine model, according to which observations are produced as emissions from the states upon *arrival* (of the transition) at each state.

Figure 9.8 shows a typical diagram of an HMM of three states, where arrows indicate transitions. Such a model could correspond to a short word with three different stationary parts. The model provides information about the successive transitions between the states ($P(i|j), i, j = 1, 2, 3$—temporal modeling of the spoken word) and also about the stationary statistics underlying each state ($p(\boldsymbol{x}|i), i = 1, 2, 3$). This type of HMM model is also known as left to right, because transitions to states with a smaller index are not allowed. Other models also do exist [Rabi 89]. In practice, the states correspond to certain physical characteristics, such as distinct sounds. In speech recognition, the number of states depends on the expected number of such sound phenomena (phonemes)[1] within one word. Actually, a number of states (typically three or four) are used for each phoneme. The average number of observations, resulting from various versions of a spoken word,

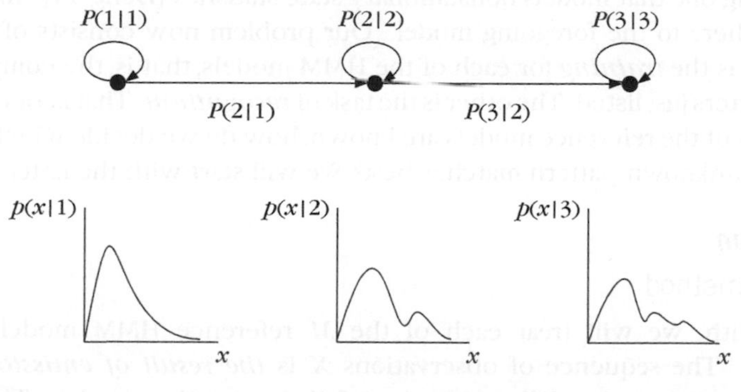

FIGURE 9.8

Model parameters describing a three-state hidden Markov model.

[1] A phoneme is a basic sound unit corresponding to a unique set of articulatory gestures, characterizing the vocal tract articulators for speech sound production [Dell 93].

can also be used as an indication of the number of required states. However, the exact number of states is often the result of experimentation and cannot be determined accurately *a priori*. In blind equalization, the states are associated with the number of clusters formed by the received data [Geor 98]. In character recognition tasks, the states may correspond to line or arc segments in the character [Vlon 92].

In the sequel we will assume that we are given a set of M known reference patterns, and our goal is to *recognize* which one of them an unknown test pattern matches best. This problem was studied in the previous chapter from a template matching (deterministic) perspective. A different (stochastic) path will be taken here. Specifically, we will assume that each known (reference) pattern is described via an HMM model. That is, each of the M patterns is characterized by the following set of parameters:

- The number K_s of the states, $s = 1, 2, \ldots, M$.
- The probability densities $p(\boldsymbol{x}|j), j = 1, 2, \ldots, K_s$, describing the distribution of the observations emitted from state j.
- The transition probabilities $P(i|j), i, j = 1, 2, \ldots, K_s$, among the various states. Some of them can be zero.
- The probabilities $P(i), i = 1, 2, \ldots, K_s$, of the initial state.

Although this is quite a general description of an HMM model, it is worth pointing out that variations are also possible. For example, sometimes the place of the self-transition probability ($P(i|i)$) is taken by the state duration probability distribution, which describes the number of successive stages for which the model stays in state i (Section 9.7). In some other cases, a model for the generation mechanism of the observations is adopted—for example, an autoregressive model [Pori 82] or even a time-varying one that models nonstationary state statistics [Deng 94]. In the sequel we will adhere to the foregoing model. Our problem now consists of two major tasks. One is the *training* for each of the HMM models, that is, the computation of the parameters just listed. The other is the task of *recognition*. That is, once the HMM parameters of the reference models are known, how do we decide which reference model the unknown pattern matches best? We will start with the latter task.

Recognition

Any path method

To start with, we will treat each of the M reference HMM models as a distinct class. The sequence of observations X is *the result of emissions, due to transitions among the different states of the respective model*. The problem then becomes a typical classification task. Given the sequence of N observations $X: \boldsymbol{x}_1, \boldsymbol{x}_2, \ldots, \boldsymbol{x}_N$, resulting from the unknown pattern, decide to which class it belongs. The Bayesian classifier decides in favor of pattern $\mathcal{S}^*$ for which

$$\mathcal{S}^* = \arg\max_{\mathcal{S}} P(\mathcal{S}|X), \quad \text{that is, over all the models} \tag{9.21}$$

and for equiprobable reference models (classes) this is equivalent to

$$S^* = \arg\max_{S} p(X|S) \tag{9.22}$$

where for convenience we have used S to denote the set of parameters describing each HMM model, that is,

$$S = \{P(i|j), p(\boldsymbol{x}|i), P(i), K_s\}$$

For each model S there is more than one possible set of successive state transitions Ω_i, each having probability of occurrence $P(\Omega_i|S)$. Thus, recalling the known rule for probabilities, we can write [Papo 91]

$$p(X|S) = \sum_i p(X, \Omega_i|S) = \sum_i p(X|\Omega_i, S)P(\Omega_i|S) \tag{9.23}$$

In order to find the maximum $p(X|S)$ over all possible models, the quantity in (9.23) has to be computed for each of the M reference models. Its efficient computation can be achieved via an approach similar to the Viterbi algorithm. Indeed, the only difference between (9.23) and (9.6) is that, instead of simply searching for the maximum over all possible state sequences Ω_i, (9.23) requires summing up the respective values for each of them. To this end, let us define as $\alpha(i_k)$ the probability density of the joint event: (a) a path is at state $i_k (i_k \in \{1, 2, \ldots, K_s\})$ at stage k *and* (b) observations $\boldsymbol{x}_1, \boldsymbol{x}_2, \ldots, \boldsymbol{x}_{k-1}$ have been emitted at the previous stages *and* (c) observation $\boldsymbol{x}_k$ is emitted from the state i_k at stage k. From the definition of $\alpha(i_k)$ the following recursive relation is easily understood:

$$\begin{aligned} \alpha(i_{k+1}) &\equiv p(\boldsymbol{x}_1, \ldots, \boldsymbol{x}_{k+1}, i_{k+1}|S) \\ &= \sum_{i_k} \alpha(i_k)P(i_{k+1}|i_k)p(\boldsymbol{x}_{k+1}|i_{k+1}), \quad k = 1, 2, \ldots, N-1 \end{aligned} \tag{9.24}$$

with

$$\alpha(i_1) = P(i_1)p(\boldsymbol{x}_1|i_1)$$

The product $P(i_{k+1}|i_k)p(\boldsymbol{x}_{k+1}|i_{k+1})$ in (9.24) provides the local information for the last transition, and $\alpha(i_k)$ is the information accumulated from the path history up to stage k. As is apparent from its definition, $\alpha(i_k)$ does not depend on the subsequent observations $\boldsymbol{x}_{k+1}, \ldots, \boldsymbol{x}_N$. The definition of a joint probability density function, which depends on *all* available observations, is also possible and will be used later on. To this end, let us define $\beta(i_k)$ as the probability density function of the event: observations $\boldsymbol{x}_{k+1}, \ldots, \boldsymbol{x}_N$ occur at stages $k+1, \ldots, N$, given that at stage k the path is at state i_k. Then after a little thought we conclude that $\beta(i_k)$ obeys the following recursion:

$$\begin{aligned} \beta(i_k) &\equiv p(\boldsymbol{x}_{k+1}, \boldsymbol{x}_{k+2}, \ldots, \boldsymbol{x}_N|i_k, S) \\ &= \sum_{i_{k+1}} \beta(i_{k+1})P(i_{k+1}|i_k)p(\boldsymbol{x}_{k+1}|i_{k+1}), \quad k = N-1, \ldots, 1 \end{aligned} \tag{9.25}$$

where by definition

$$\beta(i_N) = 1, i_N \in \{1, 2, \ldots, K_s\} \tag{9.26}$$

Thus, the probability density of the joint event: (a) a path is at state i_k at stage k *and* (b) $\boldsymbol{x}_1, \ldots, \boldsymbol{x}_N$ have been observed, is given by

$$\begin{aligned}\gamma(i_k) &\equiv p(\boldsymbol{x}_1, \ldots, \boldsymbol{x}_N, i_k|S) \\ &= p(\boldsymbol{x}_1, \ldots, \boldsymbol{x}_k, i_k|S)p(\boldsymbol{x}_{k+1}, \ldots, \boldsymbol{x}_N|i_k, S) \\ &= \alpha(i_k)\beta(i_k)\end{aligned} \tag{9.27}$$

where the assumption about the observations' independence has been employed. Equation (9.27) also justifies the choice $\beta(i_N) = 1, i_N \in \{1, 2, \ldots, K_s\}$.

Let us now return to our original goal of computing the maximum $p(X|S)$. Equation (9.24) suggests that we can write Eq. (9.23) as

$$p(X|S) = \sum_{i_N=1}^{K_s} \alpha(i_N) \tag{9.28}$$

For the computation of (9.28), we need to compute all $\alpha(i_k)$, for $k = 1, 2, \ldots, N$. This is efficiently achieved using the diagram of Figure 9.9. Each node corresponds to a stage k and a state i_k, $i_k = 1, 2, \ldots, K_s$. For each of the nodes the density

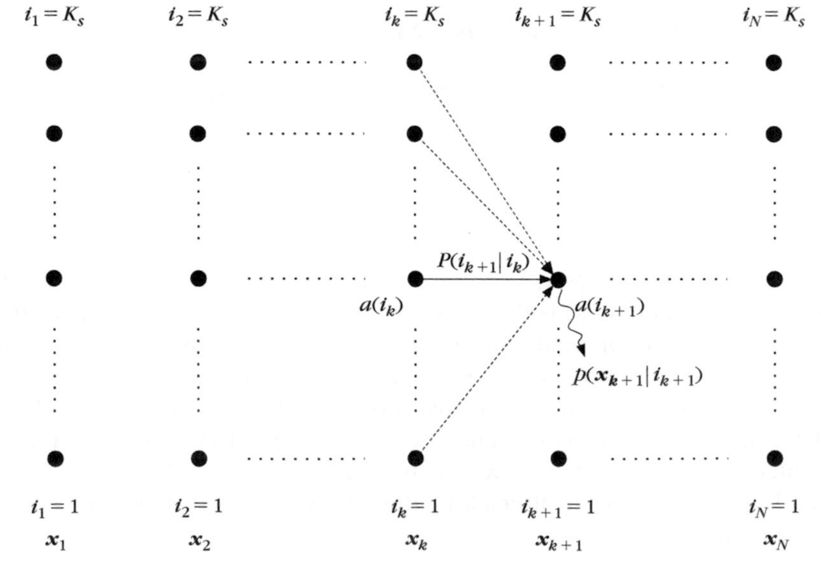

FIGURE 9.9

Diagram showing the computational flow for the any path method.

$\alpha(i_k)$ from Eq. (9.24) is computed. Thus, the number of computations is of the order of NK_s^2. The resulting algorithm is known as the *any path* method, since all paths participate in the final cost. The computation of (9.28) is performed for each of the M models, and the unknown string pattern of the observations $\boldsymbol{x}_1, \boldsymbol{x}_2, \ldots, \boldsymbol{x}_N$ is classified to the reference model S for which $p(X|S)$ becomes maximum.

Best Path Method

An alternative, suboptimal, approach is the so-called *best path* method. According to this method, for a given observation sequence X, we compute the most probable (best) path of states sequence *for each* of the reference models. The task now becomes that of (9.1), and the search for each of the optima can be achieved efficiently via the Viterbi algorithm, with the cost D given as in (9.9),

$$D = \sum_{k=1}^{N} d(i_k, i_{k-1}) \tag{9.29}$$

$$d(i_k, i_{k-1}) = \ln P(i_k|i_{k-1}) + \ln p(\boldsymbol{x}_k|i_k)$$

In other words, for each of the models we compute the maximum of $P(\Omega_i)\, p(X|\Omega_i) = p(\Omega_i, X)$. Hence, the summation in (9.23) is replaced by a maximum operation. The unknown pattern is classified to that reference model S for which the resulting optimal cost D is maximum.

Training

Training is a more difficult task. The states now are not observable, and a direct training approach, such as the one used in Section 9.5, cannot be adopted. The parameters that define each HMM model S can *only* be inferred from the available observations.

One way to achieve this goal is to estimate the unknown parameters, so that the output for each model, (9.29) or (9.28), becomes maximum *for a training set of observations known to belong to the model*. This is an optimization task with a nonlinear cost function, and it is carried out iteratively. To this end, assumptions about the probability density functions $p(\boldsymbol{x}|i)$ are required. The procedure can be simplified if one assumes that the observations $\boldsymbol{x}_k$ can take only discrete values. In such cases probability density functions $p(\boldsymbol{x}|i)$ become probabilities, $P(\boldsymbol{x}|i)$.

Discrete Observation HMM Models

We will assume that the training observation string $\boldsymbol{x}_k, k = 1, 2, \ldots, N$, consists of quantized vectors. In practice, this is achieved via vector quantization techniques, to be discussed later in Chapter 14. Hence, each observation vector can take one only out of L possible distinct values in the l-dimensional space. Thus, observations can be described as integers $r, r = 1, 2, \ldots, L$. The steps for each of the two methods, that is, any path and best path, are as following:

Baum–Welch Reestimation

The "output" quantity in the any path procedure is $p(X|\mathcal{S})$. Thus, estimating the parameters of the model $\mathcal{S}$ so that $p(X|\mathcal{S})$ is a maximum is nothing but a *maximum likelihood* parameter estimation procedure. Before going into the discussion of the iteration steps some definitions are needed.

Definitions

- $\xi_k(i,j,X|\mathcal{S}) \equiv$ the probability of the joint event: (a) a path passes through state i at stage k *and* (b) through state j at the next stage $k+1$ *and* (c) the model generates the available sequence of observations X, given the parameters of the model $\mathcal{S}$.
- $\gamma_k(i|X,\mathcal{S}) \equiv$ the probability of the event: a path passes through state i at stage k *given* the model and the available observation sequence.

From these definitions, it is not difficult to show that

$$\xi_k(i,j) \equiv \xi_k(i,j|X,\mathcal{S}) = \frac{\xi_k(i,j,X|\mathcal{S})}{P(X|\mathcal{S})} \tag{9.30}$$

Mobilizing the definitions in Eqs. (9.24) and (9.25), Eq. (9.30) becomes

$$\xi_k(i,j) = \frac{\alpha(i_k = i)P(j|i)P(\boldsymbol{x}_{k+1}|j)\beta(i_{k+1} = j)}{P(X|\mathcal{S})} \tag{9.31}$$

where $\alpha(i_k = i)$ accounts for the path history terminating at stage k and state i. $\beta(i_{k+1} = j)$ accounts for the future of the path, which at stage $k+1$ is at state j and then evolves *unconstrained* until the end. The product $P(j|i)P(\boldsymbol{x}_{k+1}|j)$ accounts for the local activity at stage k. The other quantity of interest is given by

$$\gamma_k(i) \equiv \gamma_k(i|X,\mathcal{S}) = \frac{\alpha(i_k = i)\beta(i_k = i)}{P(X|\mathcal{S})} \tag{9.32}$$

From the foregoing it is not difficult to see that

- $\sum_{k=1}^{N} \gamma_k(i)$ can be regarded as the expected (over the number of stages) number of times state i occurs, given the model $\mathcal{S}$ and the observation sequence X. When the upper index in the summation is $N-1$, this quantity is the expected number of transitions from state i.
- $\sum_{k=1}^{N-1} \xi_k(i,j)$ can be regarded as the expected number of transitions from state i to state j, given the model and the observation sequence.

The preceding definitions lead us to adopt the following (re)estimation formulas as reasonable estimates of the unknown model parameters.

$$\bar{P}(j|i) = \frac{\sum_{k=1}^{N-1} \xi_k(i,j)}{\sum_{k=1}^{N-1} \gamma_k(i)} \tag{9.33}$$

$$\bar{P}_{\boldsymbol{x}}(r|i) = \frac{\sum_{(k=1 \text{ and } \boldsymbol{x}\to r)}^{N} \gamma_k(i)}{\sum_{k=1}^{N} \gamma_k(i)} \tag{9.34}$$

$$\bar{P}(i) = \gamma_1(i) \tag{9.35}$$

The numerator in (9.34) sums only those of $\gamma_k(i)$ for which the corresponding observation $\boldsymbol{x}_k$ takes the rth discrete value. The iterative algorithm can now be expressed in terms of the following steps:

Iterations

- Initial conditions: Assume initial conditions for the unknown quantities. Compute $P(X|\mathcal{S})$.
- Step 1: From the current estimates of the model parameters reestimate the new model $\bar{\mathcal{S}}$ via Eqs. (9.33) to (9.35).
- Step 2: Compute $P(X \mid \bar{\mathcal{S}})$. If $P(X \mid \bar{\mathcal{S}}) - P(X \mid \mathcal{S}) > \epsilon$ set $\mathcal{S} = \bar{\mathcal{S}}$ and go to step 1. Otherwise stop.

Remarks

- Each iteration improves the model $\bar{\mathcal{S}}$; that is, it is true that $P(X|\bar{\mathcal{S}}) \geq P(X|\mathcal{S})$.
- The algorithm may lead to local maxima; see, for example, [Baum 67, Baum 68, Baum 70]. This is why the algorithm in practice runs a number of times, starting from different initial conditions, in order to find a favorable local maximum for $P(X|\mathcal{S})$. Other computational issues, including parallelism and memory requirements, are treated in [Turi 98].
- The Baum–Welch algorithm is basically an implementation of the EM algorithm, which was introduced in Chapter 2. Indeed, a little thought reveals that the ML estimation of the HMM parameters is a typical ML problem with an incomplete data set, that is, the unobserved states (e.g., [Moon 96, Diga 93]). A generalization of the method that allows multiple observation training sequences is given in [Li 00]. Other, gradient-based, optimizing techniques for the estimation of the unknown parameters have also been suggested and used [Levi 83].
- Practical implementation issues:
  1. *Scaling:* The probabilities $\alpha(i_k), \beta(i_k)$ are obviously less than one, and their values tend to zero very fast as the number of terms in the products (9.24) and (9.25) increases. In practice, the dynamic range of their computed values may exceed the precision range of the computer, so

appropriate scaling is required. A basic procedure is to scale $\alpha(i_k)$ in proportion to the number of stages. If the same scaling factor is used for the $\beta(i_k)$, then on taking their product in the recursions the effect of the scaling cancels out [Levi 83, Rabi 89] (Problem 9.4).

2. *Initial conditions:* This is an omnipresent problem in all iterative optimization algorithms. Usually, the unknown parameters are initialized randomly, subject, of course, to the constraints of the problem. That is, if some transitions are not allowed, the corresponding probabilities are set to zero, and also probabilities must add to one.

3. *Insufficient training data:* Generally, a large amount of training data is necessary to learn the HMM parameters. The observation sequence must be sufficiently long with respect to the number of states of the HMM model. This will guarantee that all state transitions will appear a sufficient number of times, so that the reestimation algorithm learns their respective parameters. If this not the case, a number of techniques have been devised to cope with the issue. For a more detailed treatment the reader may consult [Rabi 89, Dell 93] and the references therein.

Viterbi Reestimation

In the speech literature the algorithm is also known as the *segmental k-means training* algorithm [Rabi 89]. It is related to the best path method.

Definition 1.

- $n_{i|j} \equiv$ *number of transitions from state j to state i.*
- $n_{|j} \equiv$ *number of transitions originated from state j.*
- $n_{i|} \equiv$ *number of transitions terminated at state i.*
- $n(r|i) \equiv$ *number of times observation* $r \in \{1, 2, \ldots, L\}$ *occurs jointly with state i.*

Iterations

- Initial conditions: Assume the initial estimates of the unknown parameters. Obtain the best path and compute D.
- Step 1: From the available best path, reestimate the new model parameters as:

$$\bar{P}(i|j) = \frac{n_{i|j}}{n_{|j}}$$

$$\bar{P}_x(r|i) = \frac{n(r|i)}{n_{i|}}$$

- Step 2: For the new model parameters obtain the best path and compute the corresponding overall cost $\bar{D}$. Compare it with the cost D of the previous iteration. If $\bar{D} - D > \epsilon$ set $D = \bar{D}$ and go to step 1. Otherwise stop.

Symbol $\bar{P}x(r|i)$ is the current iteration estimate of the probability of emitting from state i the rth value from the available palette of the L possible vectors. The preceding algorithm has assumed that the initial state is known; thus no estimation of the respective probabilities is needed. This is, for example, true for left-to-right models, such as the one shown in Figure 9.8. The Viterbi reestimation algorithm can be shown to converge to a proper characterization of the underlying observations [Fu 82, Lee 72].

Continuous Observation HMM

The discrete observation modeling of originally continuous variables suffers from a serious drawback. During the (vector) quantization stage of the signal (e.g., speech segment), a severe loss of information about the original waveform may occur, which can seriously degrade the performance of the recognizer. The alternative is to work with continuous observation modeling, albeit at the expense of higher complexity. This approach requires modeling of the probability densities $p(\boldsymbol{x}|i)$, prior to estimation. Once these have been estimated, the recognition problem evolves along the same lines as with the discrete observation case. The difference exists only in the training task. One way to approach the problem is to assume a parametric model for the probability density function and then use reestimation procedures to compute the unknown model parameters. As we have already discussed in Chapter 2, a very general parameterization of the probability density function is via mixture modeling, that is,

$$p(\boldsymbol{x}|i) = \sum_{m=1}^{L} c_{im} F(\boldsymbol{x}, \boldsymbol{\mu}_{im}, \Sigma_{im}) \tag{9.36}$$

where $F(\cdot,\cdot,\cdot)$ is a density function and $\boldsymbol{\mu}_{im}$, Σ_{im} are the mean vector and the covariance matrix of the mth mixture. We will adhere to Gaussian functions, which are usually employed in practice. The mixture coefficients c_{im} have to satisfy the constraint

$$\sum_{m=1}^{L} c_{im} = 1, \quad 1 \leq i \leq K_s$$

so that

$$\int_{-\infty}^{+\infty} p(\boldsymbol{x}|i)\, d\boldsymbol{x} = 1, \quad 1 \leq i \leq K_s$$

Following arguments similar to those used to reestimate the parameters in the discrete observation HMM case, the following reestimation formulas are obtained [Lipo 82, Juan 85, Juan 86].

$$\bar{c}_{im} = \frac{\sum_{k=1}^{N} \gamma_k(i,m)}{\sum_{k=1}^{N}\sum_{r=1}^{L} \gamma_k(i,r)} \tag{9.37}$$

$$\bar{\boldsymbol{\mu}}_{im} = \frac{\sum_{k=1}^{N} \gamma_k(i,m)\boldsymbol{x}_k}{\sum_{k=1}^{N} \gamma_k(i,m)} \tag{9.38}$$

$$\bar{\Sigma}_{im} = \frac{\sum_{k=1}^{N} \gamma_k(i,m)(\boldsymbol{x}_k - \bar{\boldsymbol{\mu}}_{im})(\boldsymbol{x}_k - \bar{\boldsymbol{\mu}}_{im})^T}{\sum_{k=1}^{N} \gamma_k(i,m)} \tag{9.39}$$

The term $\gamma_k(i,m)$ is the probability density of being at state i and stage k with the mth mixture component accounting for $\boldsymbol{x}_k$, that is,

$$\gamma_k(i,m) = \frac{\alpha(i_k = i)\beta(i_k = i)}{\sum_{i_k=1}^{K_s} \alpha(i_k)\beta(i_k)} \times \frac{c_{im} F(\boldsymbol{x}_k, \boldsymbol{\mu}_{im}, \Sigma_{im})}{\sum_{r=1}^{L} c_{ir} F(\boldsymbol{x}_k, \boldsymbol{\mu}_{ir}, \Sigma_{ir})} \tag{9.40}$$

where c_{im} is the ratio of the expected number of times the system is at state i using the mth mixture component to the overall expected number of times the system is at state i. Similar interpretations can be made for the other formulas too.

When the Viterbi method is employed, reestimation of the parameters is based on averages computed across the best path. For example, for mixture modeling using a single Gaussian ($L = 1$) we get

$$\boldsymbol{\mu}_i = \frac{1}{N_i} \sum_{k=1}^{N} \boldsymbol{x}_k \delta_{ik}$$

$$\Sigma_i = \frac{1}{N_i} \sum_{k=1}^{N} (\boldsymbol{x}_k - \boldsymbol{\mu}_i)(\boldsymbol{x}_k - \boldsymbol{\mu}_i)^T \delta_{ik}$$

where $\delta_{ik} = 1$ for the stages where the path goes through state i and is zero otherwise, and N_i is the respective number of times the path passes through state i.

Remarks

- The algorithms just described estimate the unknown parameters using all the available observations simultaneously. An alternative path, of major practical importance, is to employ adaptive techniques in which new information can be incorporated to adapt an already trained model, without it being necessary to retrain it with all previously used data. It is generally accepted that speaker-dependent recognizers outperform speaker-independent systems, as long as sufficient training data are available. Thus, a long-standing idea is to use speaker-independent recognizers, trained with enough data on a multispeaker platform, and then adapt the model parameters to fit a specific speaker (and/or acoustic environment). This can be achieved by using the minimum number of data from the new speaker. Both batch and sequential schemes have been suggested. Some examples of such learning procedures are given in [Lee 91, Diga 95, Legg 95, Huo 95, Huo 97, Diga 93, Wang 01].
- A drawback of the modeling in (9.36) is that a mixture model is adopted for each of the states. This makes the number of parameters to be estimated rather high. Thus, for a given size of training data, it affects the robustness of the parameter estimation. To alleviate such problems and decrease the number of unknown parameters, so-called tied-mixture densities modeling has been

suggested, where the same Gaussian densities are shared across the mixtures of all the states [Bell 90] or groups of states [Diga 96, Kim 95, Gale 99, Gu 02].

- The Baum–Welch algorithm is an iterative procedure to maximize the likelihood function with respect to the unknown parameters. MAP procedures incorporating prior statistical information have also been proposed, and enhanced performance has been reported [Gauv 94]. An alternative is to optimize with respect to all the unknown parameters, instead of optimizing each HMM model separately, as was the case with ML earlier. The goal of such an optimization approach is to enhance the discrimination capabilities of the models, see, for example, [He 08]. Maximizing the mutual information [Bahl 86] or minimizing the cross-entropy [Ephr 89] and, more recently, the classification error rate by using either a smooth version of it [Juan 92, Juan 97] or the deterministic annealing technique [Rao 01] or controlling the influence of the outliers [Arsl 99] are examples of approaches that enhance performance at the expense of complexity.

  More recently, [Li 04] suggested the use of a deterministic annealing technique that allows one to adapt the number of states during training. This can offer some advantages for those cases where the number of states cannot be accurately predetermined.

- HMM are graphical models and they belong to a class of Bayesian networks (discussed in Chapter 2) known as dynamic Bayesian networks [Neap 04].

9.7 HMM WITH STATE DURATION MODELING

Hidden Markov modeling, as we have approached it so far, falls short of expectations in many cases in practice. Experimental evidence has identified a serious shortcoming associated with the use of the self-transition probabilities, $P(i|i)$, as parameters in the standard HMMs. This is related to the exponential modeling of the state duration probability, $P_i(d)$, that such a modeling implies, where d is the successive number of times the model remains in state i. Indeed, given $P(i|i)$ the probability of a path leaving current state i is equal to $1 - P(i|i)$. Hence, the probability of being in state i for d successive instants (i.e., $d - 1$ self-transitions, and emission of d consecutive observations from state i) is given by

$$P_i(d) = (P(i|i))^{d-1}\,(1 - P(i|i)) \tag{9.41}$$

For many cases (e.g., for a number of audio signals), such an exponential state duration modeling is inappropriate. To alleviate this drawback, it has been suggested to substitute the self-transition probability, $P(i|i)$, by an explicit variable state duration probability $P_i(d)$ in the set of unknown HMM parameters [Ferg 80]. Thus, under this new setting, the set of the unknown parameters defining an HMM consists of

- The number K_s of the states.
- The probability densities $p(\boldsymbol{x}|j)$ (they become probabilities for the case of discrete observation models, i.e., $\boldsymbol{x} \in \{1, 2, \ldots, L\}$).
- The state transition probabilities $P(i|j)$, $i, j = 1, 2, \ldots, K_s$.
- The state duration probabilities, $P_i(d)$, $i = 1, 2, \ldots, K_s$, $1 \leq d \leq D$.
- The probabilities $P(i)$, $i = 1, 2, \ldots, K_s$, of the initial state.

Observe that a maximum allowable state duration D has been adopted. Thus, the model $\mathcal{S}$ can now be written as

$$\mathcal{S} = \{P(i|j), P_i(d), p(\boldsymbol{x}|i), P(i), K_s\}$$

Our goal remains the computation of the maximum of $P(X|\mathcal{S})$ in (9.23). To achieve this in an efficient way, we have to modify the set of auxiliary variables used with the standard HMM so that we can adapt to the needs of the new parameterization. To this end, define $\alpha_k(i)$ to be the probability density of the joint event: (a) a stay at state i ends at stage k *and* (b) observations $\boldsymbol{x}_1, \boldsymbol{x}_2, \ldots, \boldsymbol{x}_k$ have been emitted up to stage k. That is,

$$\alpha_k(i) = p(\boldsymbol{x}_1, \boldsymbol{x}_2, \ldots, \boldsymbol{x}_k, \text{ state } i \text{ ends at stage } k|\mathcal{S}) \tag{9.42}$$

Note the slightly different notation used here, compared to (9.24), to emphasize the different meaning of the involved variables. Since the stay at state i ends at stage k, the next state, at stage $k+1$, can take any value *except* i; that is, $i_{k+1} \neq i$. Furthermore, looking at the path history up to stage k, there are various ways to reach state $i_k = i$. One is to jump to $i_k = i$ from an $i_{k-1} \neq i$, and this suggests that only one symbol is emitted from state i. The probability of this event depends on the value of $P_i(d = 1)$, since we know that the path will depart from state i at $k+1$. The second possibility is the path to be at state i at stage $k-1$ and remain there for two successive stages (i.e., $i_{k-1} = i_k = i$). The probability of such an event is equal to $P_i(d = 2)$. This rationale can be pushed backward $D-1$ steps prior to k (i.e., $i_{k-D+1} = \cdots = i_k = i$), and the probability of this event is given by $P_i(D)$. This, of course, can be applied to all stages prior to k (i.e., $1, 2, \ldots, k-1$). From this discussion, it is not difficult to write the counterpart of recursion (9.24) as

$$\alpha_k(i) = \sum_{(j=1,\, j\neq i)}^{K_s} \sum_{d=1}^{D} \alpha_{k-d}(j) P(i|j) P_i(d) \prod_{m=k-d+1}^{k} p(\boldsymbol{x}_m|i) \tag{9.43}$$

where, once more, statistical independence between observations has been assumed. Initialization of (9.43) requires the following computations as it can easily be understood from the respective definitions:

$$\alpha_1(j) = P(j) P_j(1) p(\boldsymbol{x}_1|j)$$

For $k = 2, 3, \ldots, D$ and $j = 1, 2, \ldots, K_s$,

$$\alpha_k(j) = P(j)P_j(k) \prod_{m=1}^{k} p(\boldsymbol{x}_m|j)$$

$$+ \sum_{(r=1,\, r\neq j)}^{K_s} \sum_{d=1}^{k-1} \alpha_{k-d}(r)P(j|r)P_j(d) \prod_{m=k-d+1}^{k} p(\boldsymbol{x}_m|j)$$

As was the case with the standard form of HMMs during the recognition phase, the desired quantity $p(X|S)$ can now be obtained from

$$p(X|S) = \sum_{i=1}^{K_s} \alpha_N(i) \tag{9.44}$$

which is easily understood from the respective definitions and can efficiently be obtained by the repeated computation of Eq. (9.43) over all ks and is.

For the training phase, in order to derive reestimation formulas for the set of the unknown parameters $(P(i|j), P(i), P_i(d))$ and for a given number of states K_s the following auxiliary variables need to be defined [Rabi 93]. Variable $\alpha_k^*(i)$ is the probability density (probability for the discrete observations case) of the joint event: (a) a path starts its stay at state i at stage $k+1$ *and* (b) observations $\boldsymbol{x}_1, \boldsymbol{x}_2, \ldots, \boldsymbol{x}_k$ have been emitted. That is,

$$\alpha_k^*(i) = p(\boldsymbol{x}_1, \boldsymbol{x}_2, \ldots, \boldsymbol{x}_k, \text{ state } i \text{ starts at stage } k+1|S)$$

From the respective definitions, the following are easily established.

$$\alpha_k^*(i) = \sum_{j=1,\, j\neq i}^{K_s} \alpha_k(j)P(i|j) \tag{9.45}$$

$$\alpha_k(i) = \sum_{d=1}^{D} \alpha_{k-d}^*(i)P_i(d) \prod_{m=k-d+1}^{k} p(\boldsymbol{x}_m|i) \tag{9.46}$$

In addition, let $\beta_k(i)$ be the *conditional* probability density of the event: observations $\boldsymbol{x}_{k+1}, \boldsymbol{x}_{k+2}, \ldots, \boldsymbol{x}_N$ have been observed, given that the path *ends* at state i and at stage k. That is,

$$\beta_k(i) = p(\boldsymbol{x}_{k+1}, \boldsymbol{x}_{k+2}, \ldots, \boldsymbol{x}_N|\text{the path ends its stay at state } i \text{ at stage } k, S)$$

Also, $\beta_k^*(i)$ is the conditional probability density of the event: observations $\boldsymbol{x}_{k+1}, \boldsymbol{x}_{k+2}, \ldots, \boldsymbol{x}_N$ have been observed, given that the path *starts* its stay at state i at stage $k+1$. That is,

$$\beta_k^*(i) = p(\boldsymbol{x}_{k+1}, \boldsymbol{x}_{k+2}, \ldots, \boldsymbol{x}_N|\text{the path starts its stay at state } i \text{ and stage } k+1, S)$$

The previous definitions allow us to write

$$\beta_k(i) = \sum_{j=1,j\neq i}^{K_s} \beta_k^*(j)P(j|i) \tag{9.47}$$

$$\beta_k^*(i) = \sum_{d=1}^{D} \beta_{k+d}(i) P_i(d) \prod_{m=k+1}^{k+d} p(\boldsymbol{x}_m|i) \tag{9.48}$$

Here, via the definitions, the following initial conditions hold (combined with (9.47)):

$$\beta_N(i) = 1, \quad i = 1, 2, \ldots, K_s \tag{9.49}$$

and

$$\beta_k^*(i) = \sum_{d=1}^{N-k} \beta_{k+d}(i) P_i(d) \prod_{k+1}^{k+d} p(\boldsymbol{x}_m|i), \quad k = N-1, \ldots, N-D \tag{9.50}$$

Based on the previous auxiliary variables and the derived relationships, the following reestimation formulas for the set of unknown parameters are obtained for the *discrete observations* case ($\boldsymbol{x}_k \to r \in \{1, 2, \ldots, L\}$).

$$\bar{P}(i) = \frac{P(i)\beta_0^*(i)}{P(X|S)} \tag{9.51}$$

$$\bar{P}(j|i) = \frac{\sum_{k=1}^{N-1} \alpha_k(i) P(j|i) \beta_k^*(j)}{\sum_{j=1}^{K_s} \sum_{k=1}^{N-1} \alpha_k(i) P(j|i) \beta_k^*(j)} \tag{9.52}$$

$$\bar{P}_{\boldsymbol{x}}(r|i) = \frac{\sum_{k=1, \boldsymbol{x}_k \to r}^{N} \left(\sum_{m<k} \alpha_m^*(i)\beta_m^*(i) - \sum_{m<k} \alpha_m(i)\beta_m(i) \right)}{\sum_{r=1}^{L} \sum_{k=1, \boldsymbol{x}_k \to r}^{N} \left(\sum_{m<k} \alpha_m^*(i)\beta_m^*(i) - \sum_{m<k} \alpha_m(i)\beta_m(i) \right)} \tag{9.53}$$

$$\bar{P}_i(d) = \frac{\sum_{k=1}^{N-d} \alpha_k^*(i) P_i(d) \beta_{k+d}(i) \prod_{m=k+1}^{k+d} P(\boldsymbol{x}_m|i)}{\sum_{d=1}^{D} \left(\sum_{k=1}^{N-d} \alpha_k^*(i) P_i(d) \beta_{k+d}(i) \prod_{m=k+1}^{k+d} P(\boldsymbol{x}_m|i) \right)} \tag{9.54}$$

Equation (9.51) is straightforward from the definitions and is an implication of the Bayes theorem. Equation (9.52) is the total number of path transitions from state i to state j along all the stages, divided by the total number of transitions that occur from state i. Equation (9.54) is the ratio of the number of times the path starts its stay at state i with duration d, divided by the number of times state i occurs with any duration.

Equation (9.53) needs a bit more elaboration. The numerator is the number of times observation $\boldsymbol{x}_k \in \{1, 2, \ldots, L\}$ is emitted from state i. To be simultaneously at state i and stage k means that a path can either start its stay at state i at stage k or may have started to be at this state at a previous stage (i.e., at $k-1, k-2, \ldots$) and remain there for a corresponding number of successive instants. However, there is a finite probability for a path to start being at state i at a stage earlier than k but not to survive long enough at this state for us to have the chance to "meet" it there at stage k. The term $\alpha^*_m(i)\beta^*_m(i)$ is the probability that a path starts its stay at state i at some stage $m+1$, and the term $\alpha_{m+1}(i)\beta_{m+1}(i)$ is the probability that a path ends its stay at state i at stage $m+1$. The first summation is the total probability that a path starts its stay at state i at any stage up to k. The second summation is the total probability that a path ends its stay at i at any stage prior to k. Hence, the subtraction of the two summation terms gives the probability of having a path through i at stage k, for the *given observation sequence*. The denominator is the same quantity, but the summation is over all times that state i is visited by a path, regardless of the emitted observation.

Adopting HMM with an explicit state duration probability modeling improves the performance in many recognition tasks compared to the standard HMM. The cost one pays for such an improvement is the increased computational complexity. The storage requirements are increased by an order of D and the computational cost by an order of D^2. Besides it, the state duration model requires D more parameters to be estimated, in addition to those in the standard HMM. This, unavoidably, leads to a demand for longer training sequences to safeguard enough data for the accurate estimation of all unknown parameters. Some of these problems can be minimized by adopting a parametric model for $P_i(d)$, so that the number of unknown parameters is reduced to the number of parameters describing the parametric probability function. To this end, Gaussian, Poisson, and Gamma distribution models have been used [Levi 86, Russ 85]. In [Chie 03] the case of adapting the parameters of the adopted parametric state duration model is treated, to fit nonstationary speech variations for large vocabulary continuous speech recognition.

Best Path Method

For the recognition phase, given a trained HMM and an observation sequence X, the goal now becomes to compute the probability of the most probable path of states sequence. However, this can no longer be achieved by employing the Viterbi algorithm in the form given in Section 9.4. To adapt to the needs of the new parameterization, imposed by the explicit time duration modeling, we have to attack the problem in a slightly different way. For the computation of the best path, up to a node corresponding to stage k and state i, there are now two types of competing paths: (a) paths that end their stay in a state j *different* to i at stage $k-1$ and jump to the node (k,i) and (b) paths that end their stay in a $j \neq i$ state at previous stages, $k-2, \ldots, k-D$, and then jump to state i and remain there for $2, \ldots, D$, time instants, respectively. Let $a_k(i)$ be the optimal cost up to stage k and state i. According to Bellman's principle, for the computation of $a_k(i)$ the following are valid:

- For paths through $(k-1, j)$, $j = 1, 2, \ldots, K_s$, $j \neq i$ and then jumping to i,

$$a_k(i) = a_{k-1}(j)P(i|j)p(\boldsymbol{x}_k|i)P_i(1), \quad j \neq i \tag{9.55}$$

- For paths through nodes $(k-d, j)$, $j \neq i$, $d = 2, \ldots, D$, that jump to i and remain there for d successive instants,

$$\alpha_k(i) = \alpha_{k-d}(j)P(i|j)P_i(d) \prod_{m=k-d+1}^{k} p(\boldsymbol{x}_m|i) \tag{9.56}$$

Thus, the optimal cost associated with node (k, i) results from

$$\alpha_k(i) = \max_{1\leq d\leq D, 1\leq j\leq K_s, j\neq i} [\delta_k(j, d, i)] \tag{9.57}$$

$$\delta_k(j, d, i) = \alpha_{k-d}(j)P(i|j)P_i(d) \prod_{m=k-d+1}^{k} p(\boldsymbol{x}_m|i) \tag{9.58}$$

Equations (9.57) and (9.58) hold for $k > D$. For $k \leq D$, initialization of (9.57) and (9.58) requires the following computations:

$$\alpha_1(i) = P(i)P_i(1)p(\boldsymbol{x}_1|i), \quad i = 1, \ldots, K_s$$

For $k = 2, 3, \ldots, D$,

$$\alpha_k(i) = \max\left\{P(i)P_i(k)\prod_{m=1}^{k} p(\boldsymbol{x}_m|i), \delta_k(j, d, i)\right\}, \quad 1 \leq d < k,\ 1 \leq j \leq K_s,\ j \neq i$$

According to the previous definitions, it turns out that the optimal path is the one ending at state i, where

$$\alpha_N(i) = \arg\max_{1\leq j\leq K_s} \alpha_N(j)$$

Equations (9.57) and (9.58) suggest that for $k > D$ there exist $(K_s \times D - D)$ candidate arguments, $\delta_k(j, d, i)$ for the maximization of each quantity $\alpha_k(i)$.

Reestimation Equations for the Best Path Method

The reestimation formulas for the best path method require maintaining counters to track state transitions, symbol emissions from the individual states, and state durations. For example, the state transition probability $P(i|j)$ is reestimated by counting the number of times the transition from state j to state i appears along the optimal path (computed at the current iteration) and dividing it by the total number of times transitions from state j to any other state are detected. This approach results in the following formulas:

$$\bar{P}(i|j) = \frac{\text{number of transitions from state } j \text{ to state } i}{\text{total number of transitions from state } j}, \quad i \neq j$$

$$\bar{P}_{\boldsymbol{x}}(r|i) = \frac{\text{number of times } \boldsymbol{x} \to r \text{ was emitted from state } i}{\text{total number of observations at state } i}, \quad r = 1, 2, \ldots, L$$

$$\bar{P}_i(d) = \frac{\text{number of times } d \text{ successive observations are emitted from state } i}{\text{total time spent at state } i}$$

A variant of the best path state duration HMM modeling is proposed in [Pikr 06], which has been developed to fit the needs of the music recognition/classification task. To this end, the cost function is modified to account for possible errors that are usually encountered in practice—that is, errors in fundamental frequency estimation or variations among recordings of the same music item due to different instrument players.

Segment Modeling

Although HMM modeling is one of the most powerful and widely used techniques in recognition, it is not without its shortcomings. One of its principal limitations is the required assumption of independence among the observations (conditioned on the state sequence). In fact, this is not truc for most of the cases. Another limitation is the rather weak state duration modeling achieved by standard HMM modeling. To overcome these limitations, a number of schemes have been suggested. Such an example is the state duration HMM modeling. More recently, an cffort was made to present a variety of such schemes in a unified framework, under the notion of *segment modeling*. Here only the basic definitions will be reviewed.

In HMM modeling on the arrival of a transition at a state, *a single* observation (corresponding to a single frame of the original speech samples) is assumed to be emitted and the fundamental observation distribution is *on the frame level*, that is, $p(\boldsymbol{x}|i)$. In contrast, in segment modeling a *segment* X_r^d consisting of d frames, $X_r^d = [\boldsymbol{x}_r, \ldots, \boldsymbol{x}_{r+d-1}]$, is assumed to be emitted upon the arrival at a state. Here d is a random variable itself. The fundamental distribution is now at the segment level, that is, $p(X_r^d|i, d)$. A schematic representation is given in Figure 9.10. The parameters describing a segment model are (a) the number of states; (b) their transition modeling parameters; (c) the joint probability density function for the segment distribution, given the duration d; and (d) the duration probability $P(d|i)$. Training of these parameters follows generalizations of the Baum–Welch and Viterbi schemes. A more detailed treatment of the topic is beyond our scope, and the interested reader may consult, for example, [Oste 96, Russ 97, Gold 99] and the references therein.

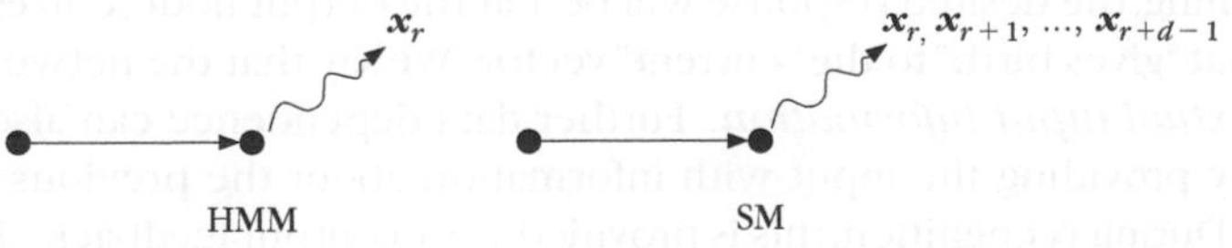

FIGURE 9.10

HMM and segment modeling (SM) for the emission of observations upon arrival of a transition at a state.

9.8 TRAINING MARKOV MODELS VIA NEURAL NETWORKS

The training phase of an HMM-based recognition system is entirely dedicated to the learning of probabilities (and densities). In Chapter 3 we have seen that a *supervised* classifier optimized via certain criteria, such as least squares, can approximate posterior class probabilities. This is the kickoff point for our current concern. States can be treated as classes, and a multilayer perceptron can be used as a nonlinear classifier. The observations feed the input nodes, and the network has as many outputs as the states. Training can be done via the backpropagation algorithm, and the desired responses will be 1 at the true state output and 0 at the others (see Chapter 4). It is now straightforward to see that the outputs of the NN will sufficiently approximate the posterior probabilities $P(i|\boldsymbol{x})$. These can then be changed to $p(\boldsymbol{x}|i)$, as required in the recognition phase of a Markov model-based recognizer, via the Bayes rule

$$p(\boldsymbol{x}|i) = \frac{P(i|\boldsymbol{x})p(\boldsymbol{x})}{P(i)}$$

where the state priors $P(i)$ are determined from their relative occurrence frequencies and $p(\boldsymbol{x})$ is constant for all states during recognition.

In following this procedure, we have made a crucial assumption. That is, the states are treated here as *being observable*, and during training we know the specific state from which each observation originates. For example, in speech recognition this is possible by associating each phoneme in a spoken word with a state. Here lies a disadvantage of this approach, since accurate segmentation of the speech signal is required and the boundaries are not always well defined. This is not the case in the HMM approach, where it is left to the algorithm to decide optimally about the state boundaries. A scheme for unified training of HMM/MLP that avoids the segmentation problem has been suggested [Koni 96]. Let us now turn to the benefits of bringing neural networks into the scene.

We have already mentioned that a major disadvantage of the standard HMM is the assumption of independence among the observations. Using a multilayer perceptron, the underlying statistical dependence can easily be accommodated. Figure 9.11 shows a possible way. Together with the "current" observation vector $\boldsymbol{x}_k$, p "past" as well as p "future" ones appear simultaneously at the input nodes. Thus, the input nodes amount to $(2p + 1)l$, with l being the dimension of the observation vectors. During training, the desired response will be 1 at the output node, corresponding to the state that "gives birth" to the "current" vector. We say that the network is trained with *contextual input information*. Further data dependence can also be accommodated by providing the input with information about the previous state in the sequence. During recognition, this is provided via an output feedback, shown in the figure by the dotted lines [Bourl 90]. Obviously, in such a configuration the output nodes of the network compute the conditional state probabilities $P(i_k|X_{k-p}^{k+p}, i_{k-1})$, where X_{k-p}^{k+p} denotes the contextual input information ranging from $\boldsymbol{x}_{k-p}$ to $\boldsymbol{x}_{k+p}$.

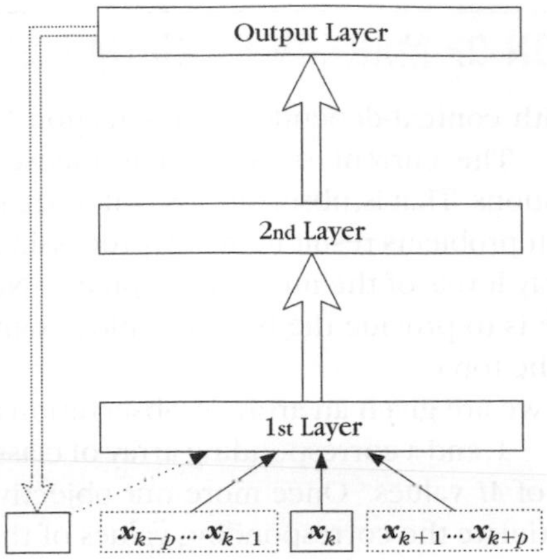

FIGURE 9.11

A multilayer perceptron architecture for training the parameters of a Markov model.

Having these probabilities at our disposal, a number of new "opportunities" open to us. Let us, for example, return to our original goal in (9.1) and treat states as classes. By the chain rule we have

$$P(\Omega_i|X) \equiv P(i_1, i_2, \ldots, i_N|X)$$
$$= P(i_N|i_{N-1}, \ldots, i_1, X) \ldots P(i_{N-1}|i_{N-2}, \ldots, i_1, X)P(i_1|X) \tag{9.59}$$

Taking into account the Markovian property of the state dependence and relaxing a bit the conditional constraint on the observations, this can be written as

$$P(\Omega_i|X) = \prod_k P(i_k|X_{k-p}^{k+p}, i_{k-1}) \tag{9.60}$$

with some appropriate initial conditions. Computing its maximum can easily be done via dynamic programming arguments. A number of other alternatives are also possible; see for example [Bourl 90, Bourl 94, Morg 95] for a more detailed discussion of the topic. In [Pikr 06a] the use of Bayesian networks is suggested in place of neural networks.

Finally, it must be emphasized that in order to obtain good probability estimates the size of the multilayer perceptron must be large enough to have good approximating capabilities. This, of course, requires increased computational resources for the training. Furthermore, the incorporation of the contextual information imposes its own demands on large networks. Another point is that the approximation of probabilities by the network is valid at the global minimum of the minimized cost function, at least in theory. Practical issues affecting the overall performance of such an approach are reported in [Spec 94].

9.9 A DISCUSSION OF MARKOV RANDOM FIELDS

So far, our concern with context-dependent classification has been limited to the one-dimensional case. The current subsection is focused on the related two-dimensional generalizations. That is, observations will be treated as two-dimensional sequences $X(i,j)$. Such problems result in image processing, and observations can be, for example, the gray levels of the image array pixels. No doubt, complications arise, and our aim here is to provide the basic definitions and directions and not a detailed treatment of the topic.

Let us assume that we are given an array of observations $X{:}X(i,j), i = 0, 1, \ldots, N_x - 1, j = 0, 1, \ldots, N_y - 1$, and a corresponding array of classes/states Ω: ω_{ij}, where each ω_{ij} can take one of M values. Once more our objective is, given the array of the observations, to estimate the corresponding values of the state array Ω so that

$$p(X|\Omega)P(\Omega) \text{ is maximum} \tag{9.61}$$

Within the scope of context-dependent classification the values of the elements of Ω will be assumed to be mutually dependent. Furthermore, we will assume that the range of this dependence is limited within a neighborhood. This brings us to the notion of Markov random fields (MRFs) defined in Chapter 7. Thus, for each (i,j) element of the array Ω a respective *neighborhood* $\mathcal{N}_{ij}$ is defined so that

- $\omega_{ij} \notin \mathcal{N}_{ij}$
- $\omega_{ij} \in \mathcal{N}_{kl} \Longleftrightarrow \omega_{kl} \in \mathcal{N}_{ij}$

In words, the (i,j) element does not belong to its own set of neighbors, and if ω_{ij} is a neighbor of ω_{kl}, then ω_{kl} is also a neighbor of ω_{ij}. The Markov property is then defined as

$$P(\omega_{ij}|\bar{\Omega}_{ij}) = P(\omega_{ij}|\mathcal{N}_{ij}) \tag{9.62}$$

where $\bar{\Omega}_{ij}$ includes all the elements of Ω except the (i,j) one. Figure 9.12 gives a typical example of a neighborhood with eight neighbor pixels. Equation (9.62) is a generalization of (9.3). In the one-dimensional case the ordering of the sequence led to the relation (9.4). Unfortunately, this *sequence ordering does not generalize in a natural way* to the two-dimensional case and imposes limitations on the involvement of the computationally elegant dynamic programming techniques [Hans 82].

A seminal paper that had an impact on the use of MRF modeling in image processing and analysis was that of Geman and Geman [Gema 84]. They built upon the important Hammersley–Clifford theorem, which establishes the *equivalence* between Markov random fields and Gibbs distributions [Besa 74]. Thus, we can talk of *Gibbs random fields* (GRFs). A Gibbs conditional probability is of the form

$$P(\omega_{ij}|\mathcal{N}_{ij}) = \frac{1}{Z}\exp\left(-\frac{1}{T}\sum_k F_k(C_k(i,j))\right) \tag{9.63}$$

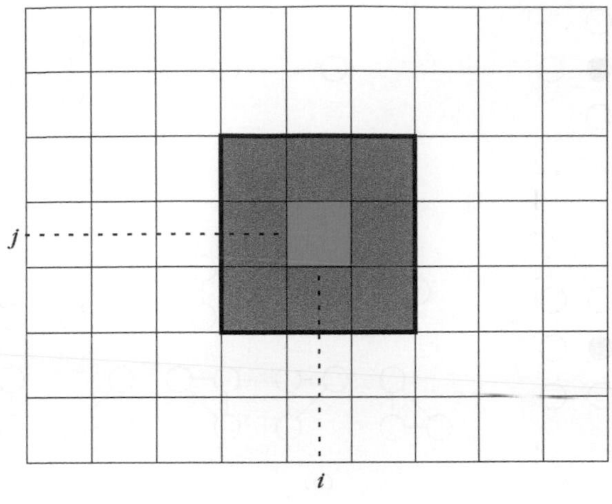

FIGURE 9.12

Example of a neighborhood involving eight neighbors of the (i,j) element (red).

where Z is a normalizing constant so that probabilities sum up to 1, T is a parameter, and $F_k(\cdot)$ are functions of the states of the pixels in the *cliques* $C_k(i,j)$. A clique consists of either a single pixel or a set of pixels, which are neighbors of each other, with respect to the type of the chosen neighborhood. Figure 9.13 shows two cases of neighborhoods and the corresponding sets of cliques. A typical example of the exponent function in (9.63) for the four neighbors case is

$$-\frac{1}{T}\omega_{ij}\left(\alpha_1 + \alpha_2(\omega_{i-1,j} + \omega_{i+1,j}) + \alpha_2(\omega_{i,j-1} + \omega_{i,j+1})\right)$$

where the a_i's are constants.

It turns out that the joint probability $P(\Omega)$ for the Gibbsian model is

$$P(\Omega) = \exp\left(-\frac{U(\Omega)}{T}\right) \tag{9.64}$$

where

$$U(\Omega) = \sum_{i,j}\sum_{k} F_k(C_k(i,j)) \tag{9.65}$$

that is, the sum of the functions over all possible cliques associated with the neighborhood. In many cases, the posterior probability $P(\Omega|X)$, which is to be maximized (i.e., (9.61)) also turns out to be Gibbsian. Such cases result, for example, if the regions in the image are themselves generated by Markov (e.g., Gaussian two-dimensional AR) processes [Deri 86, Chel 85]. *Simulating annealing* techniques can then be employed to obtain the required maximum [Gema 84].

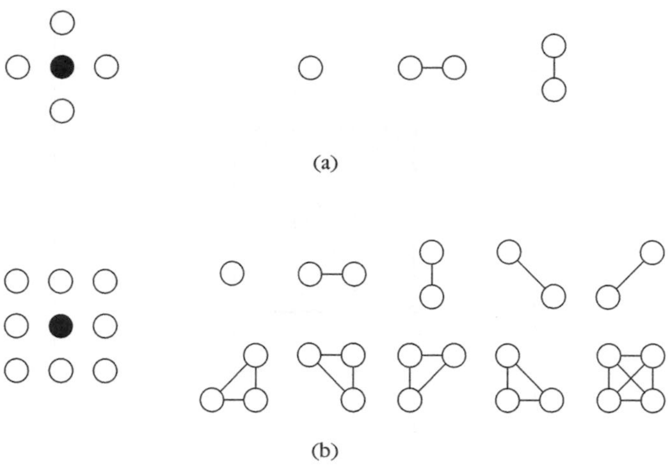

FIGURE 9.13

Two examples of neighborhoods with the corresponding cliques.

Hidden Markov generalizations to the two-dimensional plane have also been considered [Povl 95]. The idea here is to build a *pseudolikelihood function* starting from the local state transition probabilities, using Besag's method for coding the image in mutually independent pixel sets [Besa 74]. An alternative EM formulation of the problem was given in [Zhan 94]. Finally, another direction is the combination of Markov random fields and multiresolution analysis. At the subsampling stage, the Markov property is lost and suitable models are derived for the coarser resolutions. For more details the reader may consult, for example, [Laks 93, Bell 94, Kris 97] and the references therein.

9.10 PROBLEMS

9.1 Assume an HMM model with K states and an observation string X of N continuous observations. Assume that the pdf in each state is described by a Gaussian with known diagonal covariance matrix and unknown mean values. Using the EM algorithm, derive the reestimation recursions.
Hint: Form the complete data set as $Y = (X, \Omega)$, where Ω is the set of the states.

9.2 If the self-transition probability of a state is $P(i|i)$, then the probability of the model being at state i for d successive stages is given by $P_i(d) = (P(i|i))^{d-1}(1-P(i|i))$. Show that the average duration for staying in state i is equal to $\bar{d} = \frac{1}{1-P(i|i)}$.

9.3 In practice, a number Q of different versions of the spoken word are used for training, each resulting in a sequence of observations X_m of length N_m,

$m = 1, 2, \ldots, Q$. Then comment on the following reestimation formulas:

$$\bar{P}(j|i) = \frac{\sum_{m=1}^{Q} \frac{1}{P(X_m|S)} \sum_{k=1}^{N_m-1} \xi_k(i,j,X_m|S)}{\sum_{m=1}^{Q} \frac{1}{P(X_m|S)} \sum_{k=1}^{N_m-1} \alpha^m(i_k = i)\beta^m(i_k = i)}$$

$$\bar{P}_{\boldsymbol{x}}(r|i) = \frac{\sum_{m=1}^{Q} \frac{1}{P(X_m|S)} \sum_{(k=1 \text{ and } x \to r)}^{N_m} \alpha^m(i_k = i)\beta^m(i_k = i)}{\sum_{m=1}^{Q} \frac{1}{P(X_m|S)} \sum_{k=1}^{N_m} \alpha^m(i_k = i)\beta^m(i_k = i)}$$

where superscript m refers to the mth observation sequence, and $\xi(i,j,X|S)$ is defined in (9.30).

9.4 Rederive recursions (9.33) and (9.34) in terms of the scaled versions of α, β

$$\hat{\alpha}(i_k = i) = \frac{1}{c_k}\alpha(i_k = i), \quad \hat{\beta}(i_{k+1} = i) = c_k\beta(i_{k+1} = i)$$

where $c_k = \sum_{i_k=1}^{K_s} \alpha(i_k)$.

9.5 Assume that the HMM models are not equiprobable and let Λ be the set of all the unknown parameters for the M available models. Assume now that a training string X is known to correspond to the model S_r. However, during training, maximization of $P(S_r|X, \Lambda)$ is done with respect to all the parameters and not only those of the specific model. Show that this optimization leads to a maximum ratio between the contribution $p(X|S_r, \Lambda)P(S_r)$ of the correct model and $\sum_{s \neq r} p(X|S_s, \Lambda)P(S_s)$ of the incorrect models. That is, optimization with respect to all the parameters offers maximum discrimination power.

MATLAB PROGRAMS AND EXERCISES

Computer Programs

9.1 *Recognition score for HMMs using the Baum-Welch method.* Write a MATLAB function named *Baum_Welch_Do_HMM* that takes as input: (a) a column vector of initial state probabilities *pi_init*, (b) the transition matrix A, whose (i,j) element is the probability of transition from state i to state j, (c) the matrix of the emission probabilities B, whose (i,j) element is the probability to emit the ith alphabet symbol from state j, and (d) a row vector O, which contains a sequence of the code numbers of discrete symbols. It returns the score produced when the HMM, defined by *pi_init*, A, B, is applied to the sequence of symbols contained in O. Assume that if the alphabet symbols are, say, $s_1, s_2, \ldots, s_q$, the corresponding code numbers are $1, 2, \ldots, q$.

Solution

In order to avoid underflow problems, in the following implementation the score is computed as the log product of scaling factors.

```
function matching_prob=Baum_Welch_Do_HMM(pi_init,A,B,O)
  %Initialization
  T=length(O);
  [M,N]=size(B);
  alpha(:,1)=pi_init .* B(O(1),:)';
  c(1)=1/(sum(alpha(:,1)));
  alpha(:,1)=c(1)*alpha(:,1);
  for t=2:T
    for i=1:N
      alpha(i,t)=sum((alpha(:,t-1).* A(:,i)) * B(O(t),i));
    end
    c(t)=1/(sum(alpha(:,t)));
    alpha(:,t)=c(t)*alpha(:,t);
  end
  matching_prob=-sum(log10(c));
```

9.2 *Viterbi method for Discrete Observation HMMs.* Write a MATLAB function named *Viterbi_Do_HMM* that takes the same inputs as *Baum_Welch_Do_HMM* and returns (a) the best-state sequence and (b) the respective probability produced when the HMM, defined by *pi_init*, *A* and *B*, is applied to *O*, using the Viterbi method.

Solution

In the following implementation, the trellis diagram is constructed first, and then the best state sequence is extracted using the *back_traking* function defined in the computer programs' section of Chapter 8.

```
function [matching_prob,best_path]=Viterbi_Do_HMM...
  (pi_init,A,B,O)
  %Initialization
  T=length(O);
  [M,N]=size(B);
  pi_init(find(pi_init==0))=-inf;
  pi_init(find(pi_init>0))=log10(pi_init(find(pi_init>0)));
  A(find(A==0))=-inf;
  A(find(A>0))=log10(A(find(A>0)));
  B(find(B==0))=-inf;
  B(find(B>0))=log10(B(find(B>0)));
  % First observation
  alpha(:,1)=pi_init + B(O(1),:)';
  pred(:,1)=zeros(N,1);
  % Construct the trellis diagram
  for t=2:T
```

```
    for i=1:N
      temp=alpha(:,t-1)+A(:,i)+B(O(t),i);
      [alpha(i,t),ind]=max(temp);
      pred(i,t)=ind+sqrt(-1)*(t-1);
    end
  end
  [matching_prob,winner_ind]=max(alpha(:,T));
  best_path=back_tracking(pred,winner_ind,T);
```

9.3 ***Viterbi method for Continuous Observation HMMs.*** **Under the hypothesis that the emission pdfs $p(x|i)$ are Gaussians, write a MATLAB function named *Viterbi_Co_HMM*, which takes as inputs: (a) a column vector of initial state probabilities *pi_init*, (b) the transition matrix A, (c) a row vector, m, whose ith element is the mean of the ith Gaussian emission pdf, (d) a row vector, *sigma*, containing the variances of the previous pdfs, and (e) a row vector O, which contains a feature sequence. It returns, (a) the best-state sequence and (b) the respective probability, produced when the HMM, defined by *pi_init*, A, m, *sigma*, is applied to O, using the Viterbi method.**

Solution

```
function [matching_prob,best_path]=Viterbi_Co_HMM...
  (pi_init,A,m,sigma,O)
  %Initialization
  T=length(O);
  [N,N]=size(A);
  pi_init(find(pi_init==0))=-inf;
  pi_init(find(pi_init>0))=log10(pi_init(find(pi_init>0)));
  A(find(A==0))=-inf;
  A(find(A>0))=log10(A(find(A>0)));
  for i=1:N
  alpha(i,1)=pi_init(i)+log10(normpdf(O(1),m(i),sigma(1)));
  end
  pred(:,1)=zeros(N,1);
  % Construction of the trellis diagram
  for t=2:T
    for i=1:N
      temp=alpha(:,t-1)+A(:,i)+log10(normpdf(O(t),m(i),...
        sigma(i)));
      [alpha(i,t),ind]=max(temp);
      pred(i,t)=ind+sqrt(-1)*(t-1);
    end
  end
  [matching_prob,winner_ind]=max(alpha(:,T));
  best_path=back_tracking(pred,winner_ind,T);
```

Computer Experiments

9.1 Two coins are used for a coin-tossing experiment, that is, coin A and coin B. The probability that coin A returns heads is 0.6, and the respective probability for coin B is 0.4. An individual standing behind a curtain decides which coin to toss as follows: the first coin to be tossed is always coin A, the probability that coin A is re-tossed is 0.4, and similarly, the probability that coin B is re-tossed is 0.6. An observer can only have access to the outcome of the experiment, that is, the sequence of heads and tails that is produced. (a) Model the experiment by means of a HMM (i.e., define the vector of the initial state probabilities, the transition matrix and the matrix of the emission probabilities) and (b) use the *Baum_Welch_Do_HMM* function to compute the HMM score for the sequence of observations $\{H, H, T, H, T, T\}$ where H stands for heads and T stands for tails.
Hint: In defining the input sequence of symbols O for *Baum_Welch_Do_HMM* function, use "1" for "H" and "2" for "T".

9.2 For the HMM of the previous experiment, use the *Viterbi_Do_HMM* function to find the best state sequence and respective path probability, for the following observation sequences: $\{H, T, T, T, H\}$ and $\{T, T, T, H, H, H, H\}$.
Hint: In defining the input sequence of symbols O for *Viterbi_Do_HMM* function, use "1" for "H" and "2" for "T".

9.3 Assume that two number generators, Gaussian in nature, operate with mean values 0 and 5, respectively. The values for the respective standard deviations are 1 and 2. The following experiment is carried out behind a curtain: a person tosses a coin to decide which generator will be the first to emit a number. Heads has a probability of 0.4 and stands for generator A. Then the coin is tossed 8 times, and each time the coin decides which generator will emit the next number. An observer has only access to the outcome of the experiment, i.e., to the following sequence of numbers: $\{0.3, 0.4, 0.2, 2.1, 3.2, 5, 5.1, 5.2, 4.9\}$. (a) Model the experiment by means of a HMM that emits continuous observations and (b) use the *Viterbi_Co_HMM* function to compute the best-state sequence and the corresponding probability for the given sequence of numbers.

REFERENCES

[Agaz 93] Agazi O.E., Kuo S.S. "Hidden Markov model based optical character recognition in the presence of deterministic transformations," *Pattern Recognition*, Vol. 26, pp. 1813–1826, 1993.

[Anto 97] Anton-Haro C., Fonollosa J.A.R., Fonollosa J.R. "Blind channel estimation and data detection using HMM," *IEEE Transactions on Signal Processing*, Vol. 45(1), pp. 241–247, 1997.

[Aric 02] Arica N., Yarman-Vural F.T. "Optical character recognition for cursive handwriting," *IEEE Transactions on Pattern Analysis and Machine Intelligence*, Vol. 24(6), pp. 801–813, 2003

[Arsl 99] Arslan L., Hansen J.H.L. "Selective training for hidden Markov models with applications to speech classification," *IEEE Transactions on Speech and Audio Processing*, Vol. 7(1), pp. 46–54, 1999.

[Bahl 86] Bahl L.R., Brown B.F., Desouza P.V. "Maximum mutual information estimation of hidden Markov model parameters for speech recognition," *Proceedings of the IEEE International Conference on Acoustics Speech and Signal Processing*, Vol. 1, pp. 872–875, Japan, 1986.

[Bake 75] Baker J. "The DRAGON system—an overview," *IEEE Transactions on Acoustics Speech and Signal Processing*, Vol. 23(1), pp. 24–29, 1975.

[Baum 67] Baum L.E., Eagon J.A. "An inequality with applications to statistical prediction for functions of Markov processes and to a model for ecology," *Bulletin of the American Mathematical Society*, Vol. 73, pp. 360–362, 1967.

[Baum 70] Baum L.E., Petrie T., Soules G., Weiss N. "A maximization technique occurring in the statistical analysis of probabilistic functions of Markov chains," *Annals of Mathematical Statistics*, Vol. 41, pp. 164–171, 1970.

[Baum 68] Baum L.E., Sell G.R. "Growth functions for transformations of manifolds," *Pacific Journal of Mathematics*, Vol. 27, pp. 211–227, 1968.

[Bell 90] Bellegarda J.R., Nahamoo D. "Tied mixture continuous parameter modeling for speech recognition," *IEEE Transactions on Acoustics Speech and Signal Processing*, Vol. 38(12), pp. 2033–2045, 1990.

[Bell 94] Bello M.G. "A combined Markov random field and wave-packet approach to image segmentation," *IEEE Transactions on Image Processing*, Vol. 3(6), pp. 834–847, 1994.

[Besa 74] Besag J. "Spatial interaction and the statistical analysis of lattice systems," *J. Royal Stat. Soc. B*, Vol. 36(2), pp. 192–236, 1974.

[Bourl 94] Bourland H., Morgan N. *Connectionist Speech Recognition*. Kluwer Academic Publishers, 1994.

[Bourl 90] Bourland H., Wellekens C.J. "Links between Markov models and the multilayer perceptrons," *IEEE Transactions on Pattern Analysis and Machine Intelligence*, Vol. 12(12), pp. 1167–1178, 1990.

[Chel 85] Chellapa R., Kashyap R.L. "Texture synthesis using 2-D noncausal autoregressive models," *IEEE Transactions on Acoustics Speech and Signal Processing*, Vol. 33(1), pp. 194–203, 1985.

[Chen 95a] Chen J.-L., Kundu A. "Unsupervised texture segmentation using multichannel decomposition and hidden Markov models," *IEEE Transactions on Image Processing*, Vol. 4(5), pp. 603–620, 1995.

[Chen 95] Chen M.Y., Kundu A., Srihari S.N. "Variable duration HMM and morphological segmentation for handwritten word recognition," *IEEE Transactions on Image Processing*, Vol. 4(12), pp. 1675–1689, 1995.

[Chie 03] Chien J.-T., Huang C.-H. "Bayesian learning of speech duration models," *IEEE Transactions on Speech and Audio Processing*, Vol. 11(6), pp. 558–567, 2003.

[Dell 93] Deller J., Proakis J., Hansen J. *Discrete Time Processing of Speech Signals*. Macmillan, 1993.

[Deng 94] Deng L., Aksmanovic M. "Speaker-independent phonetic classification using HMM with mixtures of trend functions," *IEEE Transactions on Speech and Audio Processing*, Vol. 5(4), pp. 319–324, 1997.

[Deri 86] Derin H. "Segmentation of textured images using Gibb's random fields," *Computer Vision, Graphics, and Image Processing*, Vol. 35, pp. 72–98, 1986.

[Diga 99] Digalakis V. "Online adaptation of hidden Markov models using incremental estimation algorithms," *IEEE Transactions on Speech and Audio Processing*, Vol. 7(3), pp. 253–261, 1999.

[Diga 95] Digalakis V., Rtischef D., Neumeyer L.G. "Speaker adaptation using constrained estimation of Gaussian mixtures," *IEEE Transaction on Speech and Audio Processing*, Vol. 3(5), pp. 357–366, 1995.

[Diga 96] Digalakis V., Monaco P., Murveit H. "Genones: Generalized mixture tying in continuous HMM model-based speech recognizers," *IEEE Transactions on Speech and Audio Processing*, Vol. 4(4), pp. 281–289, 1996.

[Diga 93] Digalakis V., Rohlicek J.R., Ostendorf M. "ML estimation of a stochastic linear system with the EM algorithm and its application to speech recognition," *IEEE Transactions on Speech and Audio Processing*, Vol. 1(4), pp. 431–441, 1993.

[ElYa 99] El-Yacoubi A., Gilloux M., Sabourin R., Suen C.Y. "An HHM-based approach for off-line unconstrained handwritten word modeling and recognition," *IEEE Transactions on Pattern Analysis and Machine Intelligence*, Vol. 21(8), pp. 752–760, 1999.

[Ephr 89] Ephraim Y., Dembo A., Rabiner L.R. "A minimum discrimination information approach to hidden Markov modelling," *IEEE Transactions on Information Theory*, Vol. 35, pp. 1001–1023, September 1989.

[Ferg 80] Ferguson J. D. "Hiden Markov analysis: An introduction," in *Hidden Markov Models for Speech*, Institute for Defence Analysis, Princeton university, 1980.

[Fu 82] Fu K.S. *Syntactic Pattern Recognition and Applications*, Prentice Hall, 1982.

[Gale 99] Gales M.J.F. "Semitied covariance matrices for hidden Markov models," *IEEE Transactions on Speech and Audio Processing*, Vol. 7(3), pp. 272–281, 1999.

[Gauv 94] Gauvain J.L., Lee C.H. "Maximum a posteriori estimation for multivariate Gaussian mixture observations of Markov chains," *IEEE Transactions on Speech and Audio Processing*, Vol. 2(2), pp. 291–299, 1994.

[Gema 84] Geman S., Geman D. "Stochastic relaxation, Gibbs distributions and the Bayesian restoration of images," *IEEE Transactions on Pattern Analysis and Machine Intelligence*, Vol. 6(6), pp. 721–741, 1984.

[Geor 97] Georgoulakis C., Theodoridis S. "Efficient clustering techniques for channel equalization in hostile environments," *Signal Processing*, Vol. 58, pp. 153–164, 1997.

[Geor 98] Georgoulakis C., Theodoridis S. "Blind equalization for nonlinear channels via hidden Markov modeling," *Proceedings EUSIPCO-98*, Rhodes, Greece, 1998.

[Gold 99] Goldberger J., Burshtein D., Franco H. "Segmental modeling using a continuous mixture of noparametric models," *IEEE Transactions on Speech and Audio Processing*, Vol. 7(3), pp. 262–271, 1999.

[Gu 02] Gu L., Rose K. "Substate tying with combined parameter training and reduction in tied-mixture HMM design," *IEEE Transactions on Speech and Audio Processing*, Vol. 10(3), 2002.

[Hans 82] Hansen F.R., Elliot H. "Image segmentation using simple Markov field models," *Computer Graphics and Image Processing*, Vol. 20, pp. 101–132, 1982.

[He 08] He X., Deng L., Chou W. "Discriminative Learning in Sequential Pattern Recognition—A Unifying Review for Optimization-Oriented Speech Recognition," to appear *IEEE Signal Processing Magazine*, september 2008.

[Huo 95] Huo Q., Chan C., Lee C.H. "Bayesian adaptive learning of the parameters of hidden Markov model for speech recognition," *IEEE Transactions on Speech and Audio Processing*, Vol. 3(5), pp. 334–345, 1995.

[Huo 97] Huo Q., Lee C.H. "On-line adaptive learning of the continuous density HMM based on approximate recursive Bayes estimate," *IEEE Transactions on Speech and Audio Processing*, Vol. 5(2), pp. 161–173, 1997.

[Jeli 76] Jelinek F. "Continuous speech recognition by statistical methods," *Proceedings of the IEEE*, Vol. 64(4), pp. 532–555, 1976.

[Juan 85] Juang B.H. "Maximum likelihood estimation for mixture multivariate stochastic observations of Markov chains," *AT&T System Technical Journal*, Vol. 64, pp. 1235–1249, July–August 1985.

[Juan 97] Juang B.H., Chou W., Lee C.H. "Minimum classification error rate methods for speech recognition," *IEEE Transactions on Speech and Audio Processing*, Vol. 5(3), pp. 257–266, 1997.

[Juan 92] Juang B.H., Katagiri S. "Discriminative learning for minimum error classification," *IEEE Transactions on Signal Processing*, Vol. 40(12), pp. 3043–3054, 1992.

[Juan 86] Juang B.H., Levinson S.E., Sondhi M.M. "Maximum likelihood estimation for multivariate mixture observations of Markov chains," *IEEE Transactions on Information Theory*, Vol. IT-32, pp. 307–309, March 1986.

[Kale 94] Kaleh G.K., Vallet R. "Joint parameter estimation and symbol detection for linear and nonlinear unknown channels," *IEEE Transactions on Communications*, Vol. 42(7), pp. 2406–2414, 1994.

[Kim 95] Kim N.S., Un C.K. "On estimating robust probability distribution in HMM-based speech recognition," *IEEE Transactions on Speech and Audio Processing*, Vol. 3(4), pp. 279–286, 1995.

[Koni 96] Konig Y. "REMAP: Recursive estimation and maximization of a-posteriori probabilities in transition-based speech recognition," Ph.D. thesis, University of California at Berkeley, 1996.

[Kops 03] Kopsinis Y., Theodoridis S. "An efficient low-complexity technique for MLSE equalizers for linear and nonlinear channels," *IEEE Transactions on Signal Processing*, Vol. 51(12), pp. 3236–3249, 2003.

[Kris 97] Krishnamachari S., Chellappa R. "Multiresolution Gauss–Markov random field models for texture segmentation," *IEEE Transactions on Image Processing*, Vol. 6(2), pp. 251–268, 1997.

[Laks 93] Lakshmanan S., Derin H. "Gaussian Markov random fields at multiple resolutions," in *Markov Random Fields: Theory and Applications* (R. Chellappa, ed.), Academic Press, 1993.

[Lee 72] Lee C.H., Fu K.S. "A stochastic syntax analysis procedure and its application to pattern recognition," *IEEE Transactions on Computers*, Vol. 21, pp. 660–666, 1972.

[Lee 91] Lee C.H., Lin C.H., Juang B.H. "A study on speaker adaptation of the parameters of continuous density hidden Markov models," *IEEE Transactions on Signal Processing*, Vol. 39(4), pp. 806–815, 1991.

[Legg 95] Leggetter C.J., Woodland P.C. "Maximum likelihood linear regression for speaker adaptation of continuous density hidden Markov models," *Comput. Speech Lang.*, Vol. 9, pp. 171–185, 1995.

[Levi 86] Levinson S.E. "Continuously variable duration HMMs for automatic speech recognition," *Computer Speech and Language*, Vol. 1, pp. 29–45, March 1986.

[Levi 83] Levinson S.E., Rabiner L.R., Sondhi M.M. "An introduction to the application of the theory of probabilistic functions of a Markov process to automatic speech recognition," *Bell System Technical Journal*, Vol. 62(4), pp. 1035–1074, April 1983.

[Li 04] Li J., Wang J., Zhao Y., Yang Z. "Self adaptive design of hidden Markov models," *Pattern Recognition Letters*, Vol. 25, pp. 197–210, 2004.

[Li 00] Li X., Parizeau M., Plamondon R. "Training hidden Markov models with multiple observations-A combinatorial method," *IEEE Transactions on Pattern Analysis and Machine Intelligence*, Vol. 22(4), pp. 371–377, 2000.

[Lipo 82] Liporace L.A. "Maximum likelihood estimation for multivariate observations of Markov sources," *IEEE Transactions on Information Theory*, Vol. IT-28(5), pp. 729–734, 1982.

[Moon 96] Moon T. "The expectation maximization algorithm," *Signal Processing Magazine*, Vol. 13(6), pp. 47–60, 1996.

[Morg 95] Morgan N. Boulard H. "Continuous speech recognition," *Signal Processing Magazine*, Vol. 12(3), pp. 25–42, 1995.

[Neap 04] Neapolitan R.D. *Learning Bayesian Networks*, Prentice Hall, Gliffs, N.J. 2004.

[Oste 96] Ostendorf M., Digalakis V., Kimball O. "From HMM's to segment models: A unified view of stochastic modeling for speech," *IEEE Transactions on Audio and Speech Processing*, Vol. 4(5), pp. 360–378, 1996.

[Papo 91] Papoulis A. *Probability Random Variables and Stochastic Processes*, 3rd ed., McGraw-Hill 1991.

[Pikr 06] Pikrakis A., Theodoridis S., Kamarotos D. "Classification of musical patterns using variable duration hidden Markov models," *IEEE Transactions on Speech and Audio Processing*, Vol. 14(5), pp. 1795–1807, 2006.

[Pikr06a] Pikrakis A., Gaunakopoulos T., Theodoridis S. "Speech/music discrimination for radio broadcasts using a hybrid HMM-Bayesiay network architecture," *Proceedings, EUSIPCO-Florence*, 2006.

[Pori 82] Poritz A.B. "Linear predictive HMM and the speech signal," *Proceedings of the International Conference on Acoustics, Speech and Signal Processing*, pp. 1291–1294, Paris, 1982.

[Povl 95] Povlow B., Dunn S. "Texture classification using noncausal hidden Markov models," *IEEE Transactions on Pattern Analysis and Machine Intelligence*, Vol. 17(10), pp. 1010–1014, 1995.

[Proa 89] Proakis J. *Digital Communications*, 2nd ed., McGraw-Hill, 1989.

[Rabi 89] Rabiner L. "A tutorial on hidden Markov models and selected applications in speech recognition," *Proceedings of IEEE*, Vol. 77, pp. 257–285, February, 1989.

[Rabi 93] Rabiner L., Juang B.H. *Fundamentals of Speech Recognition*, Prentice Hall, 1993.

[Ramd 03] Ramdane S., Taconet B., Zahour A. "Classification of forms with handwritten fields by planar Markov models," *Pattern Recognition*, Vol. 36, pp. 1045–1060, 2003.

[Rao 01] Rao A.V., Rose K. "Deterministically annealed design of hidden Markov Model speech recognizers," *IEEE Transactions on Speech and Audio Precessing*, Vol. 9(2), pp. 111–127, 2001.

[Russ 97] Russell M., Holmes W. "Linear trajectory segmental HMM's," *IEEE Signal Processing Letters*, Vol. 4(3), pp. 72–75, 1997.

[Russ 85] Russell M.J., Moore R.K. "Explicit modeling of state occupancy in HMMs for automatic speech recognition," *Proceedings of the International Conference on Acoustics, Speech and Signal Processing*, Vol. 1, pp. 5–8, 1985.

[Spec 94] Special issue on neural networks for speech in *IEEE Transactions on Speech and Audio Processing*, Vol. 2(1), 1994.

[Theo 95] Theodoridis S., Cowan C.F.N., See C.M.S. "Schemes for equalization of communication channels with nonlinear impairments," *IEE Proceedings on Communications*, Vol. 142(3), pp. 165–171, 1995.

[Turi 98] Turin W. "Unidirectional and parallel Baum-Welch algorithms," *IEEE Transactions on Speech and Audio Processing*, Vol. 6(6), pp. 516–523, 1998.

[Vite 67] Viterbi A.J. "Error bounds for convolutional codes and an asymptotically optimum decoding algorithm," *IEEE Transactions on Information Theory*, Vol. 13, pp. 260–269, 1967.

[Vlon 92] Vlontzos J.A., Kung S.Y. "Hidden Markov models for character recognition," *IEEE Transactions on Image Processing*, Vol. 1(4), pp. 539–543, 1992.

[Wang 01] Wang S., Zhao Y. "Online Bayesian tree-structured transformation of HMM's with optimal model selection for speaker adaptation," *IEEE Transactions on Speech and Audio Processing*, Vol. 9(6), pp. 663–677, 2001.

[Wu 96] Wu W.R., Wei S.C. "Rotational and gray scale transform invariant texture classification using spiral resampling, subband decomposition, and hidden Markov model," *IEEE Transactions on Image Processing*, Vol. 5(10), pp. 1423–1435, 1996.

[Zhan 94] Zhang J., Modestino J.W., Langan D.A. "Maximum likelihood parameter estimation for unsupervised stochastic model based image segmentation," *IEEE Transactions on Image Processing*, Vol. 3(4), pp. 404–421, 1994.

CHAPTER

Supervised Learning: The Epilogue

10

10.1 INTRODUCTION

This chapter is the last one related to supervised learning, and it is intended to serve three purposes. The first sections focus on the last stage of the design procedure of a classification system. In other words, we assume that an optimal classifier has been designed, based on a selected set of training feature vectors. Our goal now is to evaluate its performance with respect to the probability of classification error associated with the designed system. To this end, methodologies will be developed for the estimation of the classification error probability, using the available, hence finite, set of data. Once the estimated error is considered satisfactory, full evaluation of the system performance is carried out in the real environment for which the system has been designed, such as a hospital for a medical diagnosis system or a factory for an industrial production-oriented system.

It is important to note that the evaluation stage is not cut off from the previous stages of the design procedure. On the contrary, it is an integral part of the procedure. The evaluation of the system's performance will determine whether the designed system complies with the requirements imposed by the specific application and intended use of the system. If this is not the case, the designer may have to reconsider and redesign parts of the system. Furthermore, the misclassification probability can also be used as a performance index, in the feature selection stage, to choose the best features associated with a specific classifier.

The second goal of this chapter is to tie together the various design stages that have been considered separately, so far, in the context of a case study coming from medical ultrasound imaging. Our purpose is to help the reader to get a better feeling, via an example, on how a classification system is built by combining the various design stages. Techniques for feature generation, feature selection, classifier design and system evaluation will be mobilized in order to develop a realistic computer-aided diagnosis medical system to assist a doctor reaching a decision.

In the final sections of the chapter, we will move away from the fully supervised nature of the problem that we have considered so far in the book, and we will allow unlabeled data to enter the scene. As we will see, in certain cases, unlabeled data can offer additional information that can help the designer in cases where the number of labeled data is limited. Semi-supervised learning is gaining in importance in recent years, and it is currently among the hottest research areas. The aim of this chapter is to introduce the reader to the semi-supervised learning basics and to indicate the possible performance improvement that unlabeled data may offer if used properly.

10.2 ERROR-COUNTING APPROACH

Let us consider an M-class classification task. Our objective is to *estimate* the classification error probability by testing the "correct/false" response of an independently designed classifier using a ***finite set*** of N test feature vectors. Let N_i be the vectors in each class, with $\sum_{i=1}^{M} N_i = N$ and P_i the corresponding error probability for class ω_i. ***Assuming independence among the feature vectors***, the probability of k_i vectors from class ω_i being misclassified is given by the binomial distribution

$$\text{prob}\{k_i \textit{ misclassified}\} = \binom{N_i}{k_i} P_i^{k_i}(1 - P_i)^{N_i - k_i} \tag{10.1}$$

In our case the probabilities P_i are not known. An estimate $\hat{P}_i$ results if we maximize (10.1) with respect to P_i. Differentiating and equating to zero result in our familiar estimate

$$\hat{P}_i = \frac{k_i}{N_i} \tag{10.2}$$

Thus, the total error probability estimate is given by

$$\hat{P} = \sum_{i=1}^{M} P(\omega_i)\frac{k_i}{N_i} \tag{10.3}$$

where $P(\omega_i)$ is the occurrence probability of class ω_i. We will now show that $\hat{P}$ is an unbiased estimate of the true error probability. Indeed, from the properties of the binomial distribution (Problem 10.1) we have

$$E[k_i] = N_iP_i \tag{10.4}$$

which leads to

$$E[\hat{P}] = \sum_{i=1}^{M} P(\omega_i)P_i \equiv P \tag{10.5}$$

that is, the true error probability. To compute the respective variance of the estimator, we recall from Problem 10.1 that

$$\sigma_{k_i}^2 = N_iP_i(1 - P_i) \tag{10.6}$$

leading to

$$\sigma_{\hat{P}}^2 = \sum_{i=1}^{M} P^2(\omega_i)\frac{P_i(1-P_i)}{N_i} \tag{10.7}$$

Thus, the error probability estimator in (10.3), which results from simply counting the errors, is unbiased but only asymptotically consistent as $N_i \to \infty$. *Thus, if small data sets are used for testing the performance of a classifier, the resulting estimate may not be reliable.*

In [Guyo 98] the minimum size of the test data set, N, is derived in terms of the true error probability P of the already designed classifier. The goal is to estimate N so that to guarantee, with probability $1-a, 0 \leq a \leq 1$, that P does not exceed the estimated from the test set, $\hat{P}$, by an amount larger than $\epsilon(N,a)$, that is

$$\text{prob}\{P \geq \hat{P} + \epsilon(N,a)\} \leq a \tag{10.8}$$

Let $\epsilon(N,a)$ be expressed as a function of P, that is, $\epsilon(N,a) = \beta P$. An analytical solution for Eq. (10.8) with respect to N is not possible. However, after some approximations certain bounds can be derived. For our purposes, it suffices to consider a simplified formula, which is valid for typical values of a and β ($a = 0.05, \beta = 0.2$),

$$N \approx \frac{100}{P} \tag{10.9}$$

In words, if we want to guarantee, with a risk a of being wrong, that the error probability P will not exceed $\frac{\hat{P}}{1-\beta}$, then N must be of the order given in Eq. (10.9). For $P = 0.01, N = 10{,}000$ and for $P = 0.03, N = 3000$. Note that this result is independent of the number of classes. Furthermore, if the samples in the test data set are not independent, this number must be further increased. Such bounds are also of particular importance, if the objective is to determine the size N of the test data set that provides good confidence in the results, when comparing different classification systems with relatively small differences in their error probabilities.

Although the error-counting approach is by far the most popular one, other techniques have also been suggested in the literature. These techniques estimate the error probability by using smoother versions of the discriminant function(s) realized by the classifier. The error-counting approach can be thought of as an extreme case of a hard limiter, where a 1 or 0 is produced and counted, depending on the discriminant function's response, that is, whether it is false or true, respectively. See, for example, [Raud 91, Brag 04].

10.3 EXPLOITING THE FINITE SIZE OF THE DATA SET

The estimation of the classification error probability presupposes that one has decided upon the data set to which the error counting will be applied. This is not a straightforward task. The set of samples that we have at our disposal is finite, and it has to be utilized for both training and testing. Can we use the same samples for training and testing? If not, what are the alternatives? Depending on the answer to the question, the following methods have been suggested:

- *Resubstitution Method:* The same data set is used, first for training and then for testing. One need not go into mathematical details in order to see that such a procedure is not very fair. Indeed, this is justified by the mathematical analysis. In [Fole 72] the performance of this method was analyzed using normal distributions. The analysis results show that this method provides an *optimistic* estimate of the true error probability. The amount of bias of the resubstitution estimate is a function of the ratio N/l, that is, the data set size and the dimension of the feature space. Furthermore, the variance of the estimate is inversely proportional to the data set size N. In words, ***in order to obtain a reasonably good estimate, N as well as the ratio N/l must be large enough.*** The results from the analysis and the related simulations show that N/l should be at least three and that an upper bound of the variance is $1/8N$. Of course, if this technique is to be used in practice, where the assumptions of the analysis are not valid, experience suggests that the suggested ratio must be even larger [Kana 74]. Once more, the larger the ratio N/l, the more comfortable one feels.

- *Holdout Method:* The available data set is divided into two subsets, one for training and one for testing. The major drawback of this technique is that it reduces the size for both the training and the testing data. Another problem is to decide how many of the N available data will be allocated to the training set and how many to the test set. This is an important issue. In Section 3.5.3 of Chapter 3, we saw that designing a classifier using a finite data set introduces an excess mean error and a variance around it, as different data sets, of the same size, are used for the design. Both of these quantities depend on the size of the training set. In [Raud 91], it is shown that the classification error probability of a classifier, designed using a finite training data set, N, is always higher than the corresponding asymptotic error probability ($N \to \infty$). This excess error decreases as N increases. On the other hand, in our discussion in the previous section we saw that the variance of the error counting depends on the size of the test set, and for small test data sets the estimates can be *unreliable*. Efforts made to optimize the respective sizes of the two sets have not yet led to practical results.

- *Leave-One-Out Method:* This method [Lach 68] alleviates the lack of independence between the training and test sets in the resubstitution method and at the same time frees itself from the dilemma associated with the holdout method. The training is performed using $N-1$ samples, and the test is carried out using the excluded sample. If this is misclassified, an error is counted. This is repeated N times, each time excluding a *different* sample. The total number of errors leads to the estimation of the classification error probability. Thus, training is achieved using, basically, all samples, and at the same time independence between training and test sets is maintained. The major disadvantage of the technique is its high computational complexity. For certain types of classifiers (i.e., linear or quadratic) it turns out that a simple relation exists between the leave-one-out and the resubstitution method ([Fuku 90], Problem 10.2). Thus, in such cases the former estimate is obtained using the latter method with some computationally simple modifications.

The estimates resulting from the holdout and leave-one-out methods turn out to be very similar, for comparable sizes of the test and training sets. Furthermore, it can be shown (Problem 10.3, [Fuku 90]) that the holdout error estimate, for a Bayesian classifier, is an upper bound of the true Bayesian error. In contrast, the resubstitution error estimate is a lower bound of the Bayesian error, confirming our previous comment that it is an optimistic estimate. To gain further insight into these estimates and their relation, let us make the following definitions:

- P_e^N denotes the classification error probability for a classifier designed using a finite set of N training samples.
- $\bar{P}_e^N$ denotes the average $E[P_e^N]$ over all possible training sets of size N.
- P_e is the average asymptotic error as $N \rightarrow \infty$.

It turns out that the holdout and leave-one-out methods (for statistically independent samples) provide an *unbiased* estimate of $\bar{P}_e^N$. In contrast, the resubstitution method provides a biased (underestimated) estimate of $\bar{P}_e^N$. Figure 10.1 shows the trend of a typical plot of $\bar{P}_e^N$ and the average (over all possible sets of size N) resubstitution error as functions of N [Fole 72, Raud 91]. It is readily observed that as the data size N increases, both curves tend to approach the asymptotic P_e.

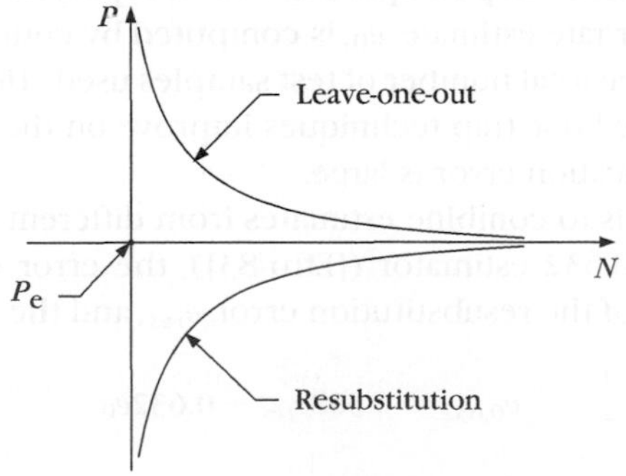

FIGURE 10.1

Plots indicating the general trend of the average resubstitution and leave-one-out error probabilities as functions of the number of training points.

A number of variations and combinations of these basic schemes have also been suggested in the literature. For example, a variation of the leave-one-out method is to leave $k > 1$, instead of one, samples out. The design and test process is repeated for all distinct choices of k samples. References [Kana 74, Raud 91] are two good examples of works discussing various aspects of the topic.

In [Leis 98] a method called *cross-validation with active pattern selection* is proposed, with the goal of reducing the high computational burden required by the leave-one-out method. It is suggested not to leave out (one at a time) all N feature vectors, but only $k < N$. To this end the "good" points of the data set (expected to contribute a 0 to the error) are not tested. Only the k "worst" points are considered. The choice between "good" and "bad" is based on the respective values of the cost function after an initial training. This method exploits the fact that the outputs of the classifier, trained according to the least squares cost function, approximate posterior probabilities, as discussed in Chapter 3. Thus, those feature vectors whose outputs have a large deviation from the desired value (for the true class) are expected to be the ones that contribute to the classification error.

Another set of techniques have been developed around the *bootstrap method* [Efro 79, Hand 86, Jain 87]. A major incentive for the development of these techniques is the variance of the leave-one-out method estimate for small data sets [Efro 83]. According to the "bootstrap" philosophy, new data sets are artificially generated. This is a way to overcome the limited number of available data and create more data in order to better assess the statistical properties of an estimator. Let X be the set of the available data of size N. A *bootstrap design sample set of size* N, X^*, is formed by *random* sampling with *replacement* of the set X. Replacement means that when a sample, say $\boldsymbol{x}_i$, is "copied" to the set X^*, it is not removed from X but is reconsidered in the next sampling. A number of variants have been built upon the bootstrap method. A straightforward one is to design the classifier using a bootstrap sample set and count the errors using the samples from X that do not appear in this bootstrap sample set. This is repeated for different bootstrap sample sets. The error rate estimate, e_0, is computed by counting all the errors and dividing the sum by the total number of test samples used. However, in [Raud 91] it is pointed out that the bootstrap techniques improve on the leave-one-out method only when the classification error is large.

Another direction is to combine estimates from different estimators. For example, in the so-called 0.632 estimator ([Efro 83]), the error estimate is taken as a *convex* combination of the resubstitution error, e_{res}, and the bootstrap error e_0,

$$e_{0.632} = 0.368 e_{res} + 0.632 e_0$$

It has been reported that the 0.632 estimator is particularly effective in cases of small size data sets [Brag 04]. An extension of the 0.632 rule is discussed in [Sima 06] where convex combinations of different estimators are considered and the combining weights are computed via an optimization process.

Confusion Matrix, Recall and Precision

In evaluating the performance of a classification system, the probability of error is sometimes not the only quantity that assesses its performance sufficiently. Let us take for example, an M-class classification task. An important issue is to know whether there are classes that exhibit a higher tendency for confusion. The *confusion matrix* $A = [A(i, j)]$ is defined so that its element $A(i, j)$ is the number of data points whose true class label was i and were classified to class j. From A, one can directly extract the *recall* and *precision* values for each class, along with the overall accuracy:

- *Recall* (R_i). R_i is the percentage of data points with true class label i, which were correctly classified in that class. For example, for a two-class problem, the recall of the first class is calculated as $R_1 = \frac{A(1,1)}{A(1,1)+A(1,2)}$.
- *Precision* (P_i). P_i is the percentage of data points classified as class i, whose true class label is indeed i. Therefore, for the first class in a two-class problem, $P_1 = \frac{A(1,1)}{A(1,1)+A(2,1)}$.
- *Overall Accuracy* (Ac). The overall accuracy, Ac, is the percentage of data that has been correctly classified. Given an M-class problem, Ac is computed from the confusion matrix according to the equation $Ac = \frac{1}{N}\sum_{i=1}^{M} A(i,i)$, where N is the total number of points in the test set.

Take as an example a two-class problem where the test set consists of 130 points from class ω_1 and 150 points from class ω_2. The designed classifier classifies 110 points from ω_1 correctly and 20 points to class ω_2. Also, it classifies 120 points from class ω_2 correctly and 30 points to class ω_1. The confusion matrix for this case is

$$A = \begin{bmatrix} 110 & 20 \\ 30 & 120 \end{bmatrix}$$

The recall for the first class is $R_1 = \frac{110}{130}$ and the precision $P_1 = \frac{110}{140}$. The respective values for the second class are similarly computed. The accuracy is $Ac = \frac{110+120}{130+150}$.

10.4 A CASE STUDY FROM MEDICAL IMAGING

Our goal in this section is to demonstrate the various design stages, discussed in the previous chapters, via a case study borrowed from a real application. It will not come as a surprise to say that focusing on a single example cannot cover all possible design approaches that are followed in practice. However, our aim is to provide a flavor for the newcomer. After all, "perfection is the enemy of the good."

Our chosen application comes from the medical imaging discipline. Our task is to develop a pattern recognition system for the diagnosis of certain liver diseases. Specifically, the system will be presented with *ultrasound* images of the

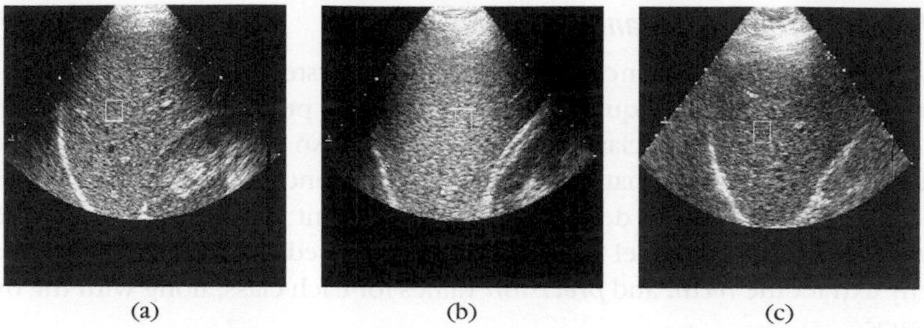

FIGURE 10.2

Ultrasound images corresponding to (a) normal liver, (b) liver with fatty infiltration, and (c) liver with cirrhosis. The square shows the image area on which the analysis was carried out.

liver, and it must be able to recognize *normal* from *abnormal* cases. Abnormal cases correspond to two types of liver diseases, namely, *cirrhosis* and *fatty liver infiltration*. For each case, two different gratings must be recognized, depending on the degree of the disease development [Cavo 97]. Figure 10.2 shows three examples of ultrasound images corresponding to (a) a normal liver, (b) an abnormal liver with fatty infiltration, and (c) an abnormal liver with cirrhosis. It is readily realized that the visual differences between the images are not great. This makes the clinical diagnosis and the diagnostic accuracy very much dependent on the skill of the doctor. Thus, the development of a pattern recognition system can assist the doctor in assessing the case and, together with other clinical findings, reduce the need for invasive techniques (biopsy).

The first stage in the design process involves the close cooperation of the system designer with the specialist, that is, the doctor, in order to establish a "common language" and have the designer *understand* the task and define, in common with the doctor, the goals and requirements of the pattern recognition system. Besides the acceptable error rate, other performance issues come into play, such as complexity, computational time, and cost of the system. The next stage involves various image processing steps, such as image enhancement, in order to assist the system by presenting it only useful information as much as possible. Then things are ripe to begin with the design of the pattern recognition system.

Figure 10.3 outlines the task. There are five possible classes. The pattern recognition system can be designed either around a single classifier, which assigns an unknown image directly to one of the five classes, or around a number of classifiers built on a tree structure philosophy. The latter approach was adopted here. Figure 10.4 illustrates the procedure. A separate classifier was used at each node, and each of them performs a two-class decision. At the first node, the respective classifier decides between normal and abnormal cases. At the second node, images, classified as abnormal, are tested and classified in either the cirrhosis or the fatty liver infiltration class, and so on. The advantage of such a procedure is that we break the problem into a number of simpler ones. It must be stressed, however, that in

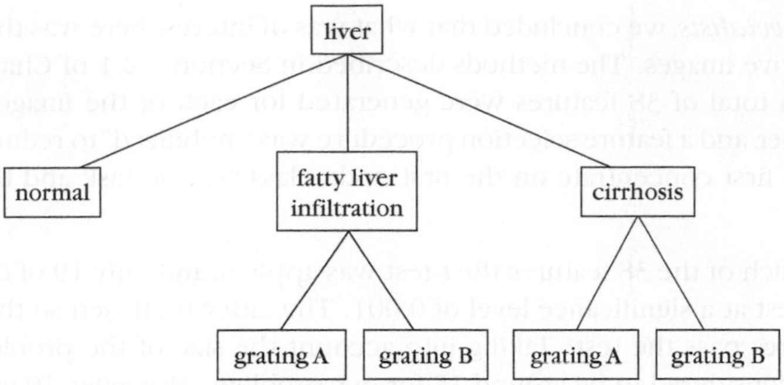

FIGURE 10.3

The classification task.

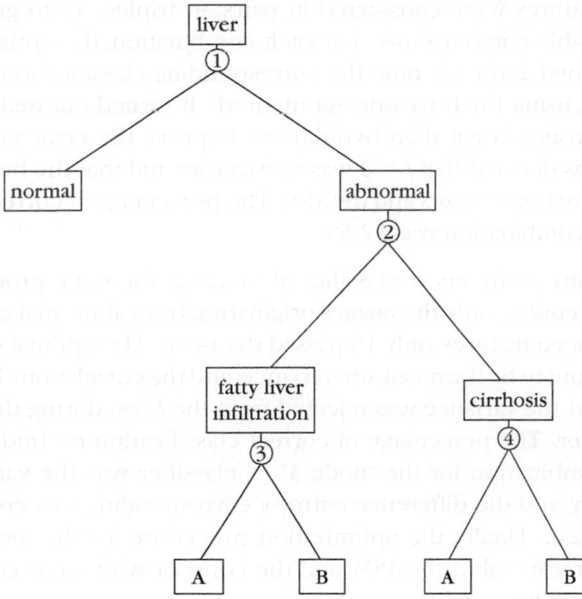

FIGURE 10.4

A tree-structured hierarchy of classifiers.

other applications such a procedure may not be applicable. For the design of the classification system, 150 ultrasound liver images were obtained from a medical center. Fifty of them correspond to normal cases, 55 of them to patients suffering from cirrhosis, and 45 of them to patients suffering from fatty liver infiltration. Three classifiers were adopted for comparison, namely the least squares linear classifier, the minimum Euclidean distance classifier, and the *k*NN for different values of *k*. Each time, the same type of classifier was used for all nodes. *From the discussions* *with the specialists*, we concluded that what was of interest here was the texture of the respective images. The methods described in Section 7.2.1 of Chapter 7 were used, and a total of 38 features were generated for each of the images. This is a large number, and a feature selection procedure was "mobilized" to reduce this number. Let us first concentrate on the first node classification task and the LS linear classifier.

- For each of the 38 features the t-test was applied, and only 19 of them passed the test at a significance level of 0.001. The latter is chosen so that "enough" features pass the test. Taking into account the size of the problem, enough was considered to be around 15 for our problem. However, 19 is still a large number, and a further reduction was considered necessary. For example, 19 is of the same order as 50 (the number of normal patterns), which would lead to poor generalization.

- The 19 features were considered in pairs, in triples, up to groups of seven, in all possible combinations. For each combination, the optimal LS classifier was designed and each time the corresponding classification error rate was estimated, using the leave-one-out method. It turned out that taking the features in groups larger than two did not improve the error rate significantly. Thus, it was decided that $l = 2$ was satisfactory and that the best combination consisted of the *kurtosis* and the ASM. The percentage of correct classification with this combination was 92.5%.

For the design of the linear classifier of "node 2" the same procedure was followed, using, of course, only the images originating from abnormal cases. Of the 38 originally produced features, only 15 passed the t-test. The optimal combination of features turned out to be the mean, the variance, and the correlation. It may be worth pointing out that the variance was rejected from the t-test during the design of the "node 1" classifier. The percentage of correct classification for node 2 was 90.1%. The optimal combination for the "node 3" LS classifier was the variance, entropy, the sum entropy, and the difference entropy corresponding to a correct classification rate of 92.2%. Finally, the optimization procedure for the "node 4" classifier resulted in the mean value, the ASM, and the contrast with a correct classification rate estimate of 83.8%.

Having completed the design with the LS linear classifiers, the same procedure was followed for the Euclidean minimum distance classifier and the kNN classifier. However, in both of these cases the resulting error rate estimates were always higher than the ones obtained with the LS classifier. Thus, the latter one was finally adopted.

Once more, it must be stated that this case study does not and cannot represent the wealth of classification tasks encountered in practice, each with its own specific requirements. We could state, with a touch of exaggeration, of course, that each classification task is like a human being. *Each one has its own personality*! For example, the dimension of our problem was such that it was computationally feasible, with today's technology to follow the procedure described. The feature selection, classifier design, and classification error stages were combined to compute the best combination. This was also a motivation for choosing the specific case study, that is, to demonstrate that the various stages in the design of a classification system are not independent but they can be closely interdependent. However, this may not be possible for a large number of tasks, as, for example, the case of a large multilayer neural network in a high-dimensional feature space. Then the feature selection stage cannot be easily integrated with that of classifier design, and techniques such as those presented in Chapter 5 must be employed. Ideally, what one should aim at is to have a procedure to design the classifiers by minimizing the error probability directly (not the LS, etc.), and at the same time this procedure should be computationally simple (!) to allow also for a search for the optimal feature combination. However, this "utopia" is still quite distant.

10.5 SEMI-SUPERVISED LEARNING

All the methods that we have considered in the book so far have relied on using a set of *labeled* data for the training of an adopted model structure (classifier). The final goal was, always, to design a "machine," which, after the training phase, can predict *reliably* the labels of unseen points. In other words, the scope was to develop a *general rule* based on the *inductive inference* rationale. In such a perspective, the *generalization* performance of the designed classifier was a key issue that has "haunted" every design methodology.

In this section, we are going to relax the design procedure from both "pillars" on which all our methods were so far built; (a) the labeled data set used for the training and (b) our concern about the generalization performance of the developed classifier. Initially, unlabeled data will be brought into the game, and we will investigate the possibility of whether this extra information, in conjunction with the labeled data, can offer performance improvement. Moving on, in a later stage, we will consider cases where the classifier design is not focused on predicting the labels of "future" unseen data points. In contrast, the optimization of a loss function will entirely rely on best serving the needs of a *given set* of unlabeled data, which are at the designer's disposal "now", that is, at the time of the design. The latter concept of designing a classifier is known as *transductive inference* to contrast it to the inductive inference mentioned earlier.

Designing classifiers by exploiting information that resides in both labeled and unlabeled data springs from a fact of life; that is, in many real applications collecting unlabeled data is much easier than the task of labeling them. In a number of cases, the task of labeling is time consuming, and it requires annotation by an expert. Bioinformatics is a field in which unlabeled data is abundant, yet only a relatively small percentage is labeled, as, for example in protein classification tasks. Text classification is another area where unlabeled data is fairly easy to collect while the labeling task requires the involvement of an expert. Annotating music is also a very demanding task, which, in addition, involves a high degree of subjectivity, as, for example, in deciding the genre of a music piece. On the other hand, it is very easy to obtain unlabeled data.

Figures 10.5 and 10.6, inspired by [Sind 06], present two simple examples raising expectations that performance may be boosted by exploiting additional information that resides in an unlabeled data set. In Figure 10.5a we are given two labeled points and one, denoted by "?", whose class is unknown. Based on this limited information, one will readily think that the most sensible decision is to classify the unknown point to the "*" class. In Figure 10.5b, in addition to the previous three points, a set of unlabeled points is shown. Having this more complete picture, one will definitely be tempted to reconsider the previous decision. In this case,

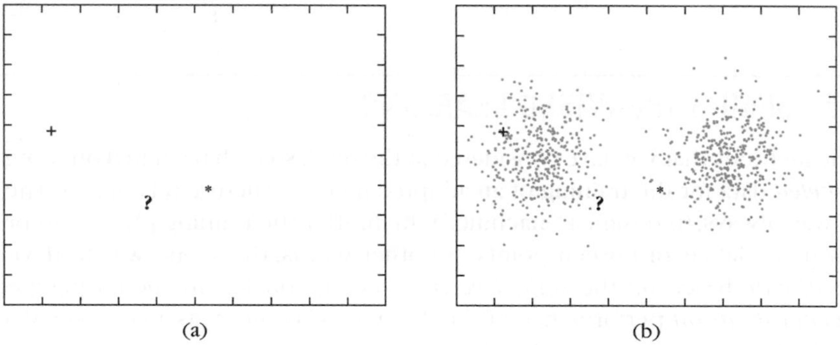

FIGURE 10.5

(a) The unknown point, denoted by "?", is classified in the same class as point "$*$". (b) The setup after a number of unlabeled data have been provided, which leads us to reconsider our previous classification decision.

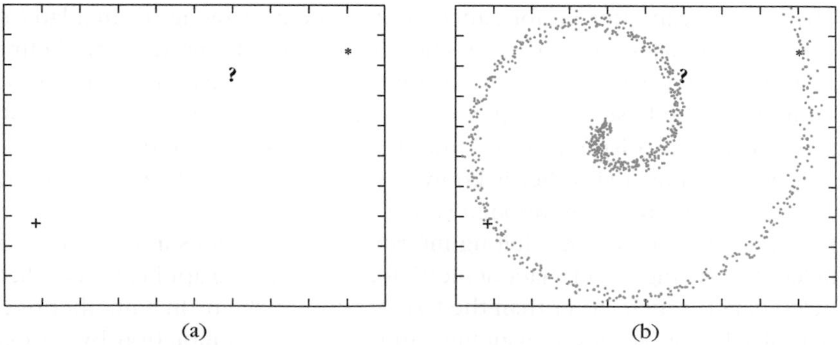

FIGURE 10.6

(a) The unknown point, denoted by "?", is classified in the same class as point "$*$". (b) The setup after a number of unlabeled data have been provided. The latter forces us, again, to reconsider our previous classification decision.

the extra information unveiled by the unlabeled data, and used by our perceptive mechanism, is the *clustered structure* of the data set. Figure 10.6 provides us with a slightly different viewpoint. Once more, we are given the three points, and the same decision as before is drawn (Figure 10.6a). In Figure 10.6b, the unlabeled data reveal a *manifold structure* on which the points reside (see also Section 6.6). The extra piece of information, which is disclosed to us now, is that the unknown point is closer to the "+" than to the "∗" point, if the geodesic, instead of the Euclidean, distance is used. Of course, in both cases, reconsideration of our initial decision is justified only under the following assumptions:

- *Cluster assumption*: If two points are in the same cluster, they are likely to originate from the same class.
- *Manifold assumption*: If the marginal probability distribution, $p(\boldsymbol{x})$, is supported on a manifold, then points lying close to each other on the manifold are more likely to be in the same class. Another way to express this is that the conditional probability, $P(y|\boldsymbol{x})$, of the class label y, is a smooth function of $\boldsymbol{x}$, with respect to the underlying structure of the manifold.

Figure 10.5 illustrates the cluster assumption, and Figure 10.6 the manifold one.

Both assumptions can be seen as particular instances of a more general assumption that covers both classification and regression:

- *Semi-supervised assumption*: If two points are close in a high-density region, then their corresponding outputs should have close values.

In other words, closeness between points is not a decisive factor, if, considered by itself. It has to be considered in the context of the underlying data distribution. This is apparent from the previous two figures. According to the semi-supervised smoothness assumption, if two points are close and linked by a path through a high-density area, they are likely to give closely located outputs. On the other hand, if the path that links them goes through a low-density region, there is no need for the corresponding outputs to be close ([Chap 06a, p. 5]).

Although semi-supervised learning has attracted a lot of interest recently, it is not new as a problem. Semi-supervised learning has been addressed in the statistics community as early as the mid-1960s, for example, [Scud 65]. Transductive learning was introduced by Vapnik and Chervonenkis in the mid-1970s, [Vapn 74]. Over the years, a large number of approaches and algorithms have been proposed, and it is beyond the scope of the present section to cover the area in depth and in its entire breadth. Our goal is to present some of the basic directions that are currently popular and at the same time are based on methods previously addressed in this book.

10.5.1 Generative Models

Generative models are perhaps the oldest semi-supervised methods and they have been used in statistics for many years. The essence behind these methods is to model the class-conditional densities $p(\boldsymbol{x}|y)$, using information provided by both labeled and unlabeled data. Once such a model is available, one can compute the marginal distribution $p(\boldsymbol{x})$

$$p(\boldsymbol{x}) = \sum_y P(y)p(\boldsymbol{x}|y) \tag{10.10}$$

the joint distribution

$$p(y, \boldsymbol{x}) = P(y)p(\boldsymbol{x}|y) \tag{10.11}$$

and finally the quantity that is required by the Bayesian classifier

$$P(y|\boldsymbol{x}) = \frac{P(y)p(\boldsymbol{x}|y)}{\sum_y P(y)p(\boldsymbol{x}|y)} \tag{10.12}$$

The class label is an integer, $y \in \{1, 2, \ldots, M\}$, where M is the number of classes. If $P(y)$ is not known then an estimate of it is used. The above formulas are familiar to us from Chapter 2, but they are repeated here for the sake of completeness.

In the sequel, we adopt a parametric model for the class conditional densities, that is, $p(\boldsymbol{x}|\,y; \boldsymbol{\theta})$. Let also P_y, $y = 1, 2, \ldots, M$, denote the respective estimates of the class priors $P(y)$. Assume that we are given two types of data sets:

- *Unlabeled data*: This data set consists of N_u samples $\boldsymbol{x}_i \in \mathcal{R}^l$, $i = 1, 2, \ldots, N_u$, which are assumed to be independently and identically distributed random vectors drawn from the marginal distribution $p(\boldsymbol{x}; \boldsymbol{\theta}, \boldsymbol{P})$, which is also parameterized in terms of $\boldsymbol{\theta}$, and $\boldsymbol{P} = [P_1, P_2, \ldots, P_M]^T$. The corresponding set is denoted by D_u.

- *Labeled data*: We assume that N_l samples are randomly and independently generated and they are ***subsequently*** labeled by an expert. Let N_y of them be associated with class $y = 1, 2, \ldots, M$, where $N_l = \sum_y N_y$. We adopt the notation $\boldsymbol{z}_{iy}$, $i = 1, 2, \ldots, N_y$, $y = 1, 2, \ldots, M$, to represent the ith sample assigned in the yth class. The set of labeled samples is denoted as $D_l = \{\boldsymbol{z}_{iy}, i = 1, 2, \ldots, N_y, y = 1, 2, \ldots, M\}$. This type of labeling data matches best a number of practical applications. For example, in medical imaging an expert is given a set of images, which have been previously produced, and labels them accordingly. Other "mechanisms" of generating labeled data are also possible, by adopting different assumptions, see [Redn 84, Shas 94, Mill 97, Mill 03].

Our task now is to estimate the set of the unknown parameters, that is, $\boldsymbol{\Theta} \equiv [\boldsymbol{\theta}^T, \boldsymbol{P}^T]^T$ in the mixture model (see Section 2.5.5 of Chapter 2.)

$$p(\boldsymbol{x}; \boldsymbol{\Theta}) = \sum_{y=1}^{M} P_y p(\boldsymbol{x}|y; \boldsymbol{\theta}) \tag{10.13}$$

using the observations in D_u and D_l. For simplicity, in the previous mixture model we have assumed one mixture component per class. This can be relaxed, for example, [Mill 97]. If only D_u was available, then the task would reduce to the mixture modeling task with hidden class (mixture) labels, as discussed in Section 2.5.5.

It is known from statistics and it is readily deduced from the definition of the log-likelihood in Section 2.5, that if the set of observations is the union of two independent sets then the log-likelihood is the sum of the log-likelihoods of the respective sets. In our case the following are valid (see also [Redn 84]):

$$D_u:\quad L_u(\boldsymbol{\Theta}) = \sum_{i=1}^{N_u} \ln p(\boldsymbol{x}_i; \boldsymbol{\Theta}) = \sum_{i=1}^{N_u} \ln \sum_{y=1}^{M} P_y p(\boldsymbol{x}_i|y; \boldsymbol{\theta}) \tag{10.14}$$

$$D_l:\quad L_l(\boldsymbol{\Theta}) = \sum_{y=1}^{M}\sum_{i=1}^{N_y} \ln p(y, \boldsymbol{z}_{iy}; \boldsymbol{\Theta}) + \ln \frac{N_l!}{N_1!N_2!\ldots N_M!}$$

$$= \sum_{y=1}^{M}\sum_{i=1}^{N_y} \ln \left(P_y p(\boldsymbol{z}_{iy}|y; \boldsymbol{\theta})\right) + \ln \frac{N_l!}{N_1!N_2!\ldots N_M!} \tag{10.15}$$

Note that in the case of labeled data the "full" observations of the joint events $(y, \boldsymbol{z}_{iy})$ are made available to us. The second term in the log-likelihood function results from the generalized Bernoulli theorem [Papo 02, p.110]. This is a consequence of the way labeled samples were assumed to occur. Basically, we are given N_l random samples and after the labeling N_1 of them are assigned to class $y = 1$, N_2 of them to class $y = 2$ and so on. However, this term is independent of $\boldsymbol{\theta}$ and $\boldsymbol{P}$ and, in practice, is neglected. The unknown set of parameters $\boldsymbol{\theta}, \boldsymbol{P}$ can now be obtained by maximizing the sum $L_u(\boldsymbol{\Theta}) + L_l(\boldsymbol{\Theta})$ with respect to $\boldsymbol{\theta}$ and $\boldsymbol{P}$. Due to the nature of $L_u(\boldsymbol{\Theta})$ optimization has to be carried out in the framework discussed in Section 2.5.5. The EM algorithm is the most popular alternative toward this end.

In order to get a feeling on how the presence of labeled data affects the results of the EM algorithm, when compared with the case where only unlabeled data are used, let us consider the example of Section 2.5.5, where the conditional densities were assumed to be Gaussians, that is,

$$p(\boldsymbol{x}|y; \boldsymbol{\theta}) = \frac{1}{(2\pi\sigma_y^2)^{l/2}} \exp\left(-\frac{\|\boldsymbol{x} - \boldsymbol{\mu}_y\|^2}{2\sigma_y^2}\right) \tag{10.16}$$

where $\boldsymbol{\theta} = [\boldsymbol{\mu}_1^T, \ldots, \boldsymbol{\mu}_M^T, \sigma_1^2, \ldots, \sigma_M^2]^T$. The E-step now becomes:

E-step:

$$Q(\boldsymbol{\Theta}; \boldsymbol{\Theta}(t)) = \sum_{i=1}^{N_u}\sum_{y=1}^{M} P(y|\boldsymbol{x}_i; \boldsymbol{\Theta}(t)) \ln \left(p(\boldsymbol{x}_i|y; \boldsymbol{\mu}_y, \sigma_y^2)P_y\right)$$

$$+ \sum_{y=1}^{M}\sum_{i=1}^{N_y} \ln \left(p(\boldsymbol{z}_{iy}|y; \boldsymbol{\mu}_y, \sigma_y^2)P_y\right) \tag{10.17}$$

In words, the expectation operation is required for the unlabeled samples only, since for the rest the corresponding labels are known. Using similar steps as in Problem 2.31 and considering both log-likelihood terms, recursions (2.98), (2.99) and (2.100) are modified as follows:

M-step:

$$\mu_y(t+1) = \frac{\sum_{i=1}^{N_u} P(y|\boldsymbol{x}_i;\boldsymbol{\Theta}(t))\boldsymbol{x}_i + \sum_{i=1}^{N_y} \boldsymbol{z}_{iy}}{\sum_{i=1}^{N_u} P(y|\boldsymbol{x}_i;\boldsymbol{\Theta}(t)) + N_y} \tag{10.18}$$

$$\sigma_y^2(t+1) = \frac{\sum_{i=1}^{N_u} P(y|\boldsymbol{x}_i;\boldsymbol{\Theta}(t))\|\boldsymbol{x}_i - \mu_y(t+1)\|^2}{l\left(\sum_{i=1}^{N_u} P(y|\boldsymbol{x}_i;\boldsymbol{\Theta}(t)) + N_y\right)} + \frac{\sum_{i=1}^{N_y} \|\boldsymbol{z}_{iy} - \mu_y(t+1)\|^2}{l\left(\sum_{i=1}^{N_u} P(y|\boldsymbol{x}_i;\boldsymbol{\Theta}(t)) + N_y\right)} \tag{10.19}$$

$$P_y(t+1) = \frac{1}{N_u + N_l}\left(\sum_{i=1}^{N_u} P(y|\boldsymbol{x}_i;\boldsymbol{\Theta}(t)) + N_y\right) \tag{10.20}$$

Remarks

- Provided that the adopted mixture model for the marginal density is correct, the use of unlabeled data is guaranteed to improve performance, for example, [Cast 96]. However, if this is not the case, and the adopted model does not match the characteristics of the true distribution that generates the data, incorporating unlabeled data may actually degrade the performance. This is a very important issue, since in practice it may not be an easy task to have good knowledge about the exact nature of the underlying distribution. This claim has been supported by a number of researchers, and a theoretical justification is provided in [Cohe 04].

- Looking at Eqs. (10.14) and (10.15), we observe that if $N_u >> N_l$, which is usually the case in practice, the unlabeled data term is the dominant one. This is also clear by inspecting the recursion in (10.18)–(10.20). To overcome this, an appropriate weighting of the two log-likelihood terms may be used, for example, [Cord 02, Niga 00]. Another problem associated with the EM algorithm is, as we already know, that it can be trapped in a local maximum. This can also be a source for performance degradation when using unlabeled data. This is treated in [Niga 00].

10.5.2 Graph-Based Methods

In any classification task, the ultimate goal is to predict the class label given the observation $\boldsymbol{x}$. In the generative modeling, the philosophy is to model the "generation" mechanism of the data and also to adopt a model for $p(\boldsymbol{x}|y)$, which then implies all the required information, that is, $p(\boldsymbol{x})$, $p(y, \boldsymbol{x})$, $P(y|\boldsymbol{x})$. However, throughout this book, the majority of the methods we dealt with were developed on a different rationale. If all we need is to infer the class labels, let us model the required information *directly*. As Vapnik stated:

> ***When solving a given problem, try to avoid solving a more general one as an intermediate step.***

For example, if the densities underlying the classes are Gaussians with the same covariance matrix, one need not bother to estimate the covariance parameters; exploiting the fact that the optimum discriminant function is linear can be sufficient to design a good classifier [Vapn 99]. Such techniques are known as ***diagnostic or discriminative*** methods. Linear classifiers, backpropagation neural networks, and support vector machines are typical examples that fall under the diagnostic design methodology. In all these methods, the marginal probability density, $p(\boldsymbol{x})$, was not considered explicitly for the estimation of the corresponding optimal parameters. The obvious question that now arises is whether and how such techniques can benefit from the existence of unlabeled data. The latter "express" themselves via $p(\boldsymbol{x})$. On the other hand, the marginal probability density does not enter into the discriminative models explicitly. The way out comes through ***penalization***, where one forces the solution to respect certain ***general characteristics*** of $p(\boldsymbol{x})$. Typical such characteristics, which have been exploited in semi-supervised learning, are: (a) the clustering structure that may underlie the data distribution and (b) the manifold geometry on which the data might lie. This information can be embedded in the form of ***regularization*** in the optimization of a loss function associated with the classification task (see Section 4.19).

Graph methods fall under the diagnostic design approach, and a number of techniques have been proposed to exploit classification-related information that resides in the data distribution. In order to present the basic rationale behind graph methods, we will focus on a technique that builds around the manifold assumption. This technique also fits nicely with a number of concepts discussed in previous chapters in the book.

As we have already seen in Section 6.7.2, graph methods start with the construction of an undirected graph $G(V, E)$. Each node, v_i, of the graph corresponds to a data point, $\boldsymbol{x}_i$, and the edges connecting nodes, e.g., v_i, v_j, are weighted by a weight $W(i, j)$ that quantifies ***similarity*** between the corresponding points, $\boldsymbol{x}_i$, $\boldsymbol{x}_j$. There was discussed how these weight values can be used to provide information related to the local structure of the underlying manifold—that is, ***the intrinsic geometry*** associated with $p(\boldsymbol{x})$.

Assume that we are given a set of N_l labeled points $\boldsymbol{x}_i$, $i = 1, 2 \ldots, N_l$, and a set of N_u unlabeled points $\boldsymbol{x}_i$, $i = N_l + 1, \ldots, N_l + N_u$. Our kickoff point is Eq. (4.79)

$$\sum_{i=1}^{N_l} \mathcal{L}\left(g(\boldsymbol{x}_i), y_i\right) + \|g\|_H^2 \tag{10.21}$$

where H is used explicitly to denote that the norm of the regularizer is taken in the RKHS space, and we have assumed $\Omega(\cdot)$ to be a square one. The loss function is considered over the labeled data only. In [Belk 04, Sind 06], it is suggested to adding an extra regularization term that *reflects the intrinsic structure* of $p(\boldsymbol{x})$. Using some differential geometry arguments, which are not of interest to us here, it turns out that a quantity that approximately reflects the underlying manifold structure is related to the Laplacian matrix of the graph (see also Section 6.7.2). The proposed in [Belk 04] optimization task is:

$$\arg\min_{g\in H} \frac{1}{N_l}\sum_{i=1}^{N_l} \mathcal{L}\left(g(\boldsymbol{x}_i), y_i\right) + \gamma_H \|g\|_H^2 +$$

$$\frac{\gamma_I}{(N_l+N_u)^2}\sum_{i,j=1}^{N_l+N_u}\left(g(\boldsymbol{x}_i)-g(\boldsymbol{x}_j)\right)^2 W(i,j) \tag{10.22}$$

Observe that two normalizing constants are present in the denominators and account for the number of points contributing to each of the two data terms. The parameters γ_H, γ_I control the relative significance of the two terms in the objective function. Also note that in the last term all points, labeled as well as unlabeled, are considered. For those who do not have "theoretical anxieties," it suffices to understand that the last term in the cost accounts for the local geometry structure. If two points are far apart, the respective $W(i,j)$ is small so their contribution to the cost is negligible. On the other hand, if the points are closely located, $W(i,j)$ is large, and these points have an important "say" in the optimization process. This means that the demand (through the minimization task) of nearby points to be mapped to similar values (i.e., $g(\boldsymbol{x}_i) - g(\boldsymbol{x}_j)$ to be small) will be seriously taken into account. This is basically a smoothness constraint, which is in line with the essence of the manifold assumption stated before. Using similar arguments as in Section 6.7.2, we end up with the following optimization task

$$\arg\min_{g\in H} \frac{1}{N_l}\sum_{i=1}^{N_l} \mathcal{L}\left(g(\boldsymbol{x}_i), y_i\right) + \gamma_H \|g\|_H^2 + \frac{\gamma_I}{(N_l+N_u)^2}\boldsymbol{g}^T L\boldsymbol{g} \tag{10.23}$$

where $\boldsymbol{g} = [g(\boldsymbol{x}_1), g(\boldsymbol{x}_2), \ldots, g(\boldsymbol{x}_{N_l+N_u})]^T$. Recall that the Laplacian matrix is defined as

$$L = D - W$$

where D is the diagonal matrix with elements $D_{ii} = \sum_j^{N_l+N_u} W(i,j)$ and $W = [W(i,j)]$, $i,j = 1, 2, \ldots, N_l + N_u$. A most welcome characteristic of this procedure is that the Representer Theorem, discussed in Section 4.19.1, is still valid and the minimizer of (10.23) admits an expansion

$$g(\boldsymbol{x}) = \sum_{j=1}^{N_l+N_u} a_j K(\boldsymbol{x}, \boldsymbol{x}_j) \tag{10.24}$$

where $K(\cdot,\cdot)$ is the adopted kernel function. Observe that the summation is taken over labeled as well as unlabeled points.

In Section 4.19.1, it was demonstrated how use of the Representer Theorem can facilitate the way the optimal solution is sought. This is also true here. Take as an example the case where the loss function is the least squares one that is, $\mathcal{L}\left(g(\boldsymbol{x}_i), y_i\right) = (y_i - g(\boldsymbol{x}_i))^2$. Then it is easy to show (Problem 10.4) that the coefficients in the expansion (10.24) are given by

$$[a_1, a_2, \ldots a_{N_l+N_u}]^T \equiv \boldsymbol{a} = (J\mathcal{K} + \gamma_H N_l I + \frac{\gamma_I N_l}{(N_l + N_u)^2} L\mathcal{K})^{-1}\boldsymbol{y} \qquad (10.25)$$

where I is the identity matrix, $\boldsymbol{y} = [y_1, y_2, \ldots, y_{N_l}, 0, \ldots, 0]^T$, J the $(N_l + N_u) \times (N_l + N_u)$ diagonal matrix with N_l entries as 1 and the rest 0, i.e., $J = diag(1, 1, \ldots, 1, 0, \ldots, 0)$ and $\mathcal{K} = [K(i,j)]$ is the $(N_l + N_u) \times (N_l + N_u)$ Gram matrix. Combining (10.25) and (10.24) results in the optimum classifier, employing labeled as well as unlabeled data, given by

$$g(\boldsymbol{x}) = \boldsymbol{y}^T (J\mathcal{K} + \gamma_H N_l I + \frac{\gamma_I N_l}{(N_l + N_u)^2} L\mathcal{K})^{-1}\boldsymbol{p} \qquad (10.26)$$

where $\boldsymbol{p} = [K(\boldsymbol{x}, \boldsymbol{x}_1), \ldots, K(\boldsymbol{x}, \boldsymbol{x}_{N_l+N_u})]^T$ (see also Section 4.19.1). The resulting minimizer is known as the *Laplacian regularized kernel least squares* (LRKLS) solution and it can be seen as a generalization of the kernel ridge regressor given in (4.100). Indeed, if $\gamma_I = 0$, unlabeled data do not enter into the game and the last term in the parenthesis becomes zero. Then for $C = \gamma_H N_l$ we obtain the kernel ridge regressor form (Eq. (4.110)). Note that Eq. (10.26) is the result of a scientific evolution process that spans a period spreading over three centuries. It was Gauss in the nineteenth century who solved for the first term in the parenthesis. The second term was added in the mid-1960s, due to the introduction of the notion of regularized optimization ([Tiho 63, Ivan 62, Phil 62]). To our knowledge, ridge regression was introduced in statistics in [Hoer 70]. The kernelized version was developed in mid-1990s following the work of Vapnik and his coworkers, and the Laplacian "edition" was added in the beginning of this century!

We can now summarize the basic steps in computing the LRKLS algorithm:

Laplacian regularized kernel least squares classifier

- Construct a graph using both labeled and unlabeled points. Choose weights $W(i,j)$ as described in Section 6.7.2.
- Choose a kernel function $K(\cdot,\cdot)$ and compute the Gram matrix $\mathcal{K}(i,j)$.
- Compute the Laplacian matrix $L = D - W$.
- Choose γ_H, γ_I.
- Compute $a_1, a_2, \ldots, a_{N_l+N_u}$ from Eq. (10.25).

Given an unknown $\boldsymbol{x}$, compute $g(\boldsymbol{x}) = \sum_{j=1}^{N_l+N_u} a_j K(\boldsymbol{x}, \boldsymbol{x}_j)$. For the two-class classification case the class label, $y \in [+1, -1]$, is obtained by $y = \text{sign}\{g(\boldsymbol{x})\}$.

By changing the cost function and/or the regularization term different algorithms result with different performance trade-offs, for example, [Wu 07, Dela 05, Zhou 04]. Another direction that has been followed within the graph theory framework is what is called *label propagation*, for example, [Zhu 02]. Given a graph, nodes corresponding to the labeled points are assigned their respective class label (e.g., ± 1 for the two-class case) and the unlabeled ones are labeled with a zero. Labels are then propagated in an iterative way through the data set along high-density areas defined by the unlabeled data, until convergence. In [Szum 02] label propagation is achieved by considering Markov random walks on the graph. An interesting point is that the previously discussed two directions to semi-supervised learning, which build on graph theoretic arguments, turn out to be equivalent or, at least, very similar, see, for example, [Beng 06]. Once more, the world is small!

10.5.3 Transductive Support Vector Machines

According to the inductive inference philosophy, one starts from the particular knowledge (training set using labeled data) and then the general rule (the classifier or regressor) is derived, which is subsequently used to predict the labels of *specific* points comprising the test set. In other words, one follows a path

$$\text{particular} \longrightarrow \text{general} \longrightarrow \text{particular}$$

Vapnik and Chervonenkis, pushing the frontiers, questioned whether this is indeed the best path to follow in practice. In cases where the training data set is limited in size, deriving a good general rule becomes a hard task. For such cases, they proposed the *transductive inference* approach, where one follows a "direct" path

$$\text{particular} \longrightarrow \text{particular}$$

In such a way, one may be able to exploit information residing in a *given* test set and obtain improved results. Transductive learning is a special type of semi-supervised learning and the goal is to predict the labels of the points in a *specific* test set by embedding the points of the set, explicitly, in the optimization task. From this perspective, label propagation techniques, discussed before, are also transductive in nature. For a more theoretical treatment of transductive learning, the reader is referred to, for example, [Vapn 06, Derb 03, Joac 02].

In the framework of support vectors machines (see Section 3.7), and for the two class problem, transductive learning is cast as follows. Given the set $D_l = \{\boldsymbol{x}_i, i = 1, 2, \ldots, N_l\}$ of labeled points and the set $D_u = \{\boldsymbol{x}_i, i = N_l + 1, \ldots, N_l + N_u\}$ compute the labels $y_{N_l+1}, \ldots, y_{N_l+N_u}$ of the points in D_u so that the hyperplane that separates the two classes, by taking into consideration *both* labeled and unlabeled points, has maximum margin. The corresponding optimization tasks for the two versions (hard margin and soft margin) become:

Hard margin TSVM

$$\text{minimize} \quad J(y_{N_l+1}, \ldots, y_{N_l+N_u}, \boldsymbol{w}, w_0) = \frac{1}{2}||\boldsymbol{w}||^2 \tag{10.27}$$

$$\text{subject to} \quad y_i(\boldsymbol{w}^T \boldsymbol{x}_i + w_0) \geq 1, \; i = 1, 2, \ldots, N_l \tag{10.28}$$

$$y_i(\boldsymbol{w}^T \boldsymbol{x}_i + w_0) \geq 1, \; i = N_l + 1, \ldots, N_l + N_u \tag{10.29}$$

$$y_i \in \{+1, -1\}, \; i = N_l + 1, \ldots, N_l + N_u \tag{10.30}$$

Soft margin TSVM

$$\text{minimize} \quad J(y_{N_l+1}, \ldots, y_{N_l+N_u}, \boldsymbol{w}, w_0, \boldsymbol{\xi}) = \frac{1}{2}||\boldsymbol{w}||^2 +$$

$$C_l \sum_{i=1}^{N_l} \xi_i + C_u \sum_{i=N_l+1}^{N_l+N_u} \xi_i \tag{10.31}$$

$$\text{subject to} \quad y_i(\boldsymbol{w}^T \boldsymbol{x}_i + w_0) \geq 1 - \xi_i, \; i = 1, 2, \ldots, N_l \tag{10.32}$$

$$y_i(\boldsymbol{w}^T \boldsymbol{x}_i + w_0) \geq 1 - \xi_i, \; i = N_l + 1, \ldots, N_l + N_u \tag{10.33}$$

$$y_i \in \{+1, -1\}, \; i = N_l + 1, \ldots, N_l + N_u \tag{10.34}$$

$$\xi_i \geq 0, \; i = 1, 2, \ldots, N_l + N_u \tag{10.35}$$

where C_l, C_u are user-defined parameters that control the importance of the respective terms in the cost. In [Joac 99, Chap 05] an extra constraint is used that forces the solution to assign unlabeled data to the two classes in roughly the same proportion as that of the labeled ones.

Figure 10.7 shows a simplified example, for the hard margin case, illustrating that the optimal hyperplane, which results if only labeled examples are used (SVM), is different from the one obtained when both labeled and unlabeled data are employed (TSVM). Performing labeling (of the unlabeled samples), so that the margin between the resulting classes is maximized, pushes the decision hyperplane in sparse regions and it is in line with the *clustering* assumption stated in the beginning of this section.

A major difficulty associated with TSVM is that, in contrast to the convex nature of the standard SVM problem, the optimization is over the labels y_i, $i = N_l + 1, \ldots, N_l + N_u$, which are integers in $\{+1, -1\}$ and it is an NP-hard task. To this end, a number of techniques have been suggested. For example, in [DeBi 04] the task is relaxed, and it is solved in the semidefinite programming framework. Algorithms based on coordinate descent searching have been proposed in [Joac 02, Demi 00, Fung 01]. A slight reformulation of the problem is proposed

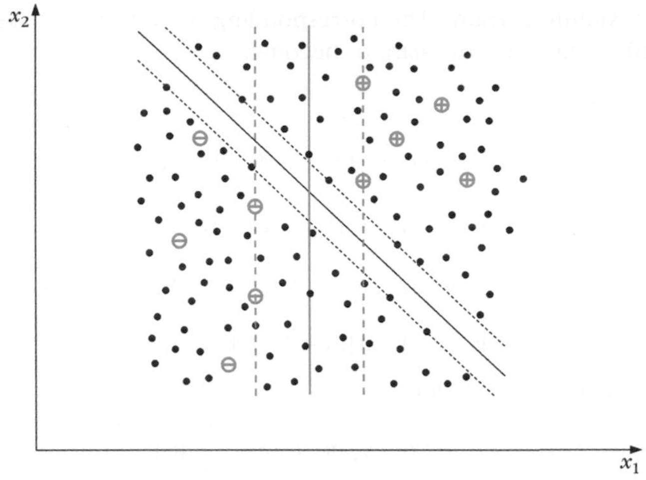

FIGURE 10.7

Red lines correspond to the SVM classifier when only labeled "−" and "+" points are available. The black lines result when the unlabeled data have been considered and have "pushed" the decision hyperplane to an area which is sparse in data.

in [Chap 06]. The constraints associated with the unlabeled data are removed and replaced by $|\boldsymbol{w}^T\boldsymbol{x}_i + w_0| \geq 1 - \xi_i$, $i = N_l + 1, \ldots, N_l + N_u$. Such constraints push the hyperplane away from the unlabeled data, since penalization occurs each time the absolute value becomes less than one. Hence, they are in line with the cluster assumption. This problem formulation has the advantage of removing the combinatorial nature of the problem; yet it remains nonconvex. Strictly speaking, the problem solved in [Chap 06] is not transductive in nature, since one does not try to assign labels to the unlabeled points. All one tries to do is to locate the hyperplane in sparse regions and do not "cut" clusters. This is the reason that such techniques are known as *semi-supervised SVM* or S^3VM (see, e.g., [Benn 98, Sind 06a]). At this point it is interesting to note that the borderline between transductive learning and semi-supervised learning that is inductive in nature is not clearly marked, and it is a topic of ongoing discussion. Take, for example, TSVM. After training, the resulting decision hyperplane can be used for induction to predict the label of an unseen point. We are not going to delve into such issues. A very interesting and quite enlightening discussion on the topic, in a "Platonic dialogue" style, is provided in [Chap 06a, Chapter 25].

Remarks

- Besides the previous semi-supervised methodologies that we have presented, a number of other techniques have been suggested. For example, self-training is simple in concept and a commonly used approach. A classifier is first trained with the labeled data. Then this is applied to the unlabeled data to perform label predictions. Based on a confidence criterion, those of the data that result in confident predictions are added in the labeled training set. The classifier is retrained, and the process is repeated, for example, [Culp 07]. The procedure is similar with that used in the decision feedback equalization (DFE) ([Proa 89]) in communications for more than three decades, in the sense that the classifier uses its own predictions to train itself. Similar in concept is the *co-training* ([Mitc 99, Balc 06, Zhou 07] procedure. However, in this case, the feature set is split into a number of subsets, for example, two, and for each subset a separate classifier is trained using labeled data. The trained classifiers are then used to predict (in their respective feature subspace) labels for the unlabeled points. Each one of the classifiers passes to the other the most confident of the predictions, together with the respective points. The classifiers are then retrained using this new extra information. Splitting the feature set and training the classifiers in different subspaces provides a different complementary "view" of the data points. This reminds us of the classifier combination method where classifiers are trained in different subspaces (Section 4.21). As it was the case there, independence of the two sets is a required assumption.

- A major issue concerning semi-supervised techniques is whether and under what conditions one can obtain enhanced performance compared to training the classifier using labeled data only. A number of papers report that enhanced performance has been obtained. For example, in [Niga 00] a generative approach has been applied to the problem of text classification. It is reported that, using a number of 10,000 unlabeled articles, substantial improvement gains have been attained when the number of labeled documents is small. As the number of labeled data increases from a few tens to a few thousands, the classification accuracies (corresponding to semi-supervised and supervised training) start to converge. In [Chap 06a] a number of semi-supervised techniques were compared using eight benchmark data sets. Some general conclusions are: (a) One must not always expect to obtain improved performance when using unlabeled data. (b) Moreover, the choice of the type of the semi-supervised technique is a crucial issue. The algorithm should "match" the nature of the data set; algorithms that implement the clustering assumption (e.g., TSVM) must be used with data exhibiting a cluster structure, and algorithms that implement the manifold assumption (e.g., Laplacian LS) must be used with data residing on a manifold. So, prior to using a semi-supervised technique, one must have a good "feeling" about the data at hand. This point was also stressed in the remarks given previously, when dealing with the generative methods. It was stated that if the adopted model for the class conditional densities is not correct, then the performance, using unlabeled data, may degrade.

Besides the cases already mentioned before in this section, the performance of semi-supervised techniques has also been tested in the context of other real applications. In [Kasa 04] transductive SVMs were used to recognize promoter sequences in genes. Their results show that TSVM achieve enhanced performance compared to the (inductive) SVM. The news coming from [Krog 04], however, is not that encouraging. In the task of predicting functional properties of proteins, the SVM approach resulted in much better performance compared to TSVM. These results are in line with the comments made before; there is no guarantee for performance improvement when using semi-supervised techniques. Some more samples of applications of semi-supervised learning techniques include [Wang 03] for relevance feedback in image retrieval, [Kock 03] for mail classification, and [Blum 98] for Web mining.

- As it was stated in the beginning of this section, our goal was to present to the reader the main concepts behind semi-supervised learning and in particular to see how techniques that have been used in this book in earlier chapters can be extended to this case. A large number of algorithms and methods are around, and the area is, at the time of writing, the focus of an intense research effort by a number of serious groups worldwide. For deeper and broader information, the interested reader may consult, [Chap 06a, Zhu 07].

- Another type of classification framework was proposed (once more) by Vapnik [Chap 06a, Chapter 24]. It is proposed that an additional data set is used, which is not from the same distribution as the labeled data. In other words, the points of this data set do not belong to either of the classes of interest and are called the Universum. This data set is a form of data-dependent regularization and it encodes prior knowledge related to the problem at hand. The classifier must be trained so that the decision function to result in small values for the points in the Universum; that is, these points are forced to lie close to the decision surface. In [West 06] it is shown that different choices of Universa and loss functions result in some known types of regularizers. Early results reported in [West 06, Sinz 06] indicate that the obtained performance depends on the quality of Universum set. The choice of an appropriate Universum is still an open issue. The results obtained in [Sinz 06] suggest that must be carefully chosen and must contain invariant directions and to be positioned "in between" the two classes.

10.6 PROBLEMS

10.1 Let P be the probability that event A occurs. The probability that event A occurs k times in a sequence of N independent experiments is given by the binomial distribution

$$\binom{N}{k} P^k (1-P)^{N-k}$$

Show that $E[k] = NP$ and $\sigma_k^2 = NP(1-P)$.

10.2 In a two-class problem the classifier to be used is the minimum Euclidean distance one. Assume N_1 samples from class ω_1 and N_2 from class ω_2. Show that the leave-one-out method estimate can be obtained from the resubstitution method, if the distance of $\boldsymbol{x}$ from the class means $d_i(\boldsymbol{x})$, $i = 1, 2$, are modified as $d_i'(\boldsymbol{x}) = (\frac{N_i}{N_i - 1})^2 d_i(\boldsymbol{x})$, if $\boldsymbol{x}$ belongs to class i. Furthermore, show that in this case the leave-one-out method always results in larger error estimates than the resubstitution method.

10.3 Show that for the Bayesian classifier, the estimate provided by the resubstitution method is a lower bound of the true error and that computed from the holdout method is an upper bound.

10.4 Show Eq. (10.25).

REFERENCES

[Balc 06] Balcan M. F., Blum A. "An augmented PAC model for semi-supervised learning," in *Semi-Supervised Learning* (Chapelle O., Schölkopf B., Zien A., eds.), MIT Press, 2006.

[Belk 04] Belkin V., Niyogi P., Sindhwani V. "Manifold regularization: A geometric framework for learning from examples," *Technical Report, TR:2004-06*, Department of Computer Science, University of Chicago, 2004.

[Benn 98] Bennett K., Demiriz A. "Semi-supervised support vector machines," in *Advances in Neural Information Processing Systems*, Vol. 12, 1998.

[Beng 06] Bengio Y., Delalleau O., Le Roux N. "Label propagation and quadratic criterion," in *Semi-Supervised Learning* (Chapelle O., Schölkopf B., Zien A., eds.), MIT Press, 2006.

[Blum 98] Blum A., Mitchell T. "Combining labeled and unlabeled data with co-training," *Proceedings of the 11th Annual Conference on Computational Learning Theory*, pp. 92–100, 1998.

[Brag 04] Braga-Neto U., Dougherty E. "Bolstereol error estimation," *Pattern Recognition*, Vol. 37, pp. 1267–1281, 2004.

[Cast 96] Castelli V., Cover T. "The relative value of labeled and unlabeled samples in pattern recognition with an unknown mixing parameter," *IEEE Transactions on Information Theory*, Vol. 42, pp. 2101–2117, 1996.

[Chap 05] Chapelle O., Zien A. "Semi-supervised classification by low density separation," *Proceedings of the 10th International Workshop on Artificial Intelligence and Statistics*, pp. 57–64, 2005.

[Chap 06] Chapelle O., Chi M., Zien A. "A continuation method for semi-supervised SVMs," *Proceedings of the 23rd International Conference on Machine Learning*, Pittsburgh, PA. 2006.

[Chap 06a] Chapelle O., Schölkopf B., Zien A. *Semi-Supervised Learning*, MIT Press, 2006.

[Cavo 97] Cavouras D., et al. "Computer image analysis of ultrasound images for discriminating and grating liver parenchyma disease employing a hierarchical decision tree scheme and the multilayer perceptron classifier," *Proceedings of Medical Informatics Europe '97*, pp. 517–521, 1997.

[Cohe 04] Cohen I., Cozman F.G., Cirelo M.C., Huang T.S. "Semi-supervised learning of classifiers: Theory, algorithms, and their application to human-computer interaction," *IEEE Transactions on Pattern Analysis and Machine Intelligence*, Vol. 26(12), pp. 1553–1567, 2004.

[Cord 02] Corduneanu A., Jaakola T. "Continuation methods for mixing heterogeneous sources," *Proceedings of 18th Annual Conference on Uncertainty in Artificial Intelligence* (Darwiche A., Friedman N., eds.), Alberta, Canada, Morgan Kaufmann, 2002.

[Culp 07] Culp M., Michailidis G. "An iterative algorithm for extending learners to a semi-supervised setting," *Proceedings of the Joint Statistical Meeting (JSM)*, Salt Lake, Utah, 2007.

[DeBi 04] DeBie T., Christianini N. "Convex methods for transduction ," in *Advances in Neural Information Processing Systems* (Thrun S., Saul L, Schölkopf B., eds.), pp. 73–80, MIT Press, 2004.

[Dela 05] Delalleau O., Bengio Y., Le Roux N. "Efficient non-parametric function induction in semi-supervised learning," *Proceedings of the 10th International Workshop on Artificial Intelligence and Statistics* (Cowell R.G., Ghahramani Z., eds.), pp. 96–103, Barbados, 2005.

[Demi 00] Demiriz A., Bennett K.P. "Optimization approaches to semi-supervised learning," *Applications and Algorithms of Complementarity* (Ferries M.C., Mangasarian O.L., Pang J.S., eds), pp. 121–141, Kluwer, Dordrecht, the Netherlands, 2000.

[Derb 03] Derbeko P., El-Yanif R., Meir R. "Error bounds for transductive learning via compression and clustering," in *Advances in Neural Information Processing Systems*, pp. 1085–1092, MIT Press, 2003.

[Efro 79] Efron B. "Bootstrap methods: Another look at the jackknife," *Annals of Statistics*, Vol. 7, pp. 1–26, 1979.

[Efro 83] Efron B. "Estimating the error rate of a prediction rule: Improvement on cross-validation," *Journal of the American Statistical Association*, Vol. 78, pp. 316–331, 1983.

[Fole 72] Foley D. "Consideration of sample and feature size," *IEEE Transactions on Information Theory*, Vol. 18(5), pp. 618–626, 1972.

[Fung 01] Fung G., Mangasarian O. "Semi-supervised support vector machines for unlabeled data classification," *Optimization Methods and Software*, Vol. 15, pp. 29–44, 2001.

[Fuku 90] Fukunaga K. *Introduction to Statistical Pattern Recognition*, 2nd ed., Academic Press, 1990.

[Guyo 98] Guyon I., Makhoul J., Schwartz R., Vapnik U. "What size test set gives good error rate estimates?" *IEEE Transactions on Pattern Analysis and Machine Intelligence*, Vol. 20(1), pp. 52–64, 1998.

[Hand 86] Hand D.J. "Recent advances in error rate estimation," *Pattern Recognition Letters*, Vol. 5, pp. 335–346, 1986.

[Hoer 70] Hoerl A.E., Kennard R. "Ridge regression: biased estimate for nonorthogonal problems," *Technometrics*, Vol. 12, pp. 55–67, 1970.

[Ivan 62] Ivanov V.V. "On linear problems which are not well-posed," *Soviet Mathematical Docl.*, Vol. 3(4), pp. 981–983, 1962.

[Jain 87] Jain A.K., Dubes R.C., Chen C.C. "Bootstrap techniques for error estimation," *IEEE Transactions on Pattern Analysis and Machine Intelligence*, Vol. 9(9), pp. 628–636, 1987.

[Joac 02] Joachims T. *Learning to Classify Text Using Support Vector Machines*, Kluwer, Dordrecht, the Netherlands, 2002.

[Joac 99] Joachims T. "Transductive inference for text classification using support vector machines," *Proceedings of 16th International Conference on Machine Learning (ICML)*, (Bratko I., Dzeroski S., eds), pp. 200–209, 1999.

[Kana 74] Kanal L. "Patterns in Pattern Recognition," *IEEE Transactions on Information Theory*, Vol. 20(6), pp. 697–722, 1974.

[Kasa 04] Kasabov N., Pang S. "Transductive support vector machines and applications to bioinformatics for promoter recognition," *Neural Information Processing-Letters and Reviews*, Vol. 3(2), pp. 31–38, 2004.

[Kock 03] Kockelkorn M., Lüneburg A., Scheffer T. "Using transduction and multi-view learning to answer emails," *Proceedings of the European Conference on Principles and Practice of Knowledge Discovery in Databases*, pp. 266–277, 2003.

[Krog 04] Krogel M., Scheffer T. "Multirelational learning, text mining and semi-supervised learning for functional genomics," *Machine Learning*, Vol. 57(1/2), pp. 61–81, 2004.

[Lach 68] Lachenbruch P.A., Mickey R.M. "Estimation of error rates in discriminant analysis," *Technometrics*, Vol. 10, pp. 1–11, 1968.

[Leis 98] Leisch F., Jain L.C., Hornik K. "Cross-validation with active pattern selection for neural network classifiers," *IEEE Transactions on Neural Networks*, Vol. 9(1), pp. 35–41, 1998.

[Mill 97] Miller D.J., Uyar H. "A mixture of experts classifier with learning based on both labeled and unlabeled data," *Neural Information Processing Systems*, Vol. 9, pp. 571–577, 1997.

[Mill 03] Miller D.J., Browning J. "A mixture model and EM-algorithm for class discovery, robust classification, and outlier rejection in mixed labeled/unlabeled data sets," *IEEE Transactions on Pattern Analysis and Machine Intelligence*, Vol. 25(11), pp. 1468–1483, 2003.

[Mitc 99] Mitchell T. "The role of unlabeled data in supervised learning," *Proceedings of the 6th International Colloquium on Cognitive Science*, San Sebastian, Spain, 1999.

[Niga 00] Nigam K., McCallum A.K., Thrun S., Mitchell T. "Text classification from labeled and unlabeled documents using EM," *Machine Learning*, Vol. 39, pp. 103–134, 2000.

[Papo 02] Papoulis A., Pillai S.U. *Probability, Random Variables, and Stochastic Processes*, 4th eds, McGraw-Hill, 2002.

[Phil 62] Phillips D.Z. "A technique for numerical solution of certain integral equation of the first kind," *Journal of Association of Computer Machinery (ACM)*, Vol. 9, pp. 84–96.

[Proa 89] Proakis J. *Digital Communications*, McGraw-Hill, 1989.

[Raud 91] Raudys S.J., Jain A.K. "Small size effects in statistical pattern recognition: Recommendations for practitioners," *IEEE Transactions on Pattern Analysis and Machine Intelligence*, Vol. 13(3), pp. 252–264, 1991.

[Redn 84] Redner R.A., Walker H.M. "Mixture densities, maximum likelihood and the EM algorithm," *SIAM Review*, Vol. 26(2), pp. 195–239, 1984.

[Scud 65] Scudder H.J. "Probability of error of some adaptive pattern recognition machines," *IEEE Transactions on Information Theory*, Vol. 11, pp. 363–371, 1965.

[Shas 94] Shashahani B., Landgrebe D. "The effect of unlabeled samples in reducing the small sample size problem and mitigating the Hughes phenomenon," *IEEE Transactions on Geoscience and Remote Sensing*, Vol. 32, pp. 1087–1095, 1994.

[Sima 06] sima C, Dougherty E.R. "Optimal convex error estimators for classification," Pattery Recognist, Vol. 39(9), pp. 1763–1780, 2006.

[Sind 06a] Sindhawi V, Keerthi S., Chapelle O. "Deterministic annealing for semi-supervised kernel machines," *Proceedings of the 23nd nternational Conference on Machine Learning*, 2006.

[Sind 06] Sindhawi V., Belkin M., Niyogi P. "The geometric basis of semi-supervised learning," in *Semi-Supervised Learning* (Chapelle O., Schölkopf B., Zien A., eds.), MIT Press, 2006.

[Sinz 06] Sinz F.H., Chapelle O., Agarwal A., Schölkopf B. "An analysis of inference with the Universum," *Proceedings of the 20th Annual Conference on Neural Information Processing Systems (NIPS)*, MIT Press, Cambridge, Mass., USA, 2008.

[Szum 02] Szummer M., Jaakkola T. "Partially labeled classification with Markov random fields," in *Advances in Neural Information Processing Systems* (Dietterich T.G., Becker S., Ghahramani Z., eds.), MIT Press, 2002.

[Tiho 63] Tikhonov A.N. "On solving ill-posed problems and a method for regularization," *Doklady Akademii Nauk, USSR*, Vol. 153, pp. 501–504, 1963.

[Vapn 74] Vapnik V., Chervonenkis A.Y. *Theory of Pattern Recognition* (in Russian), Nauka, Moskow, 1974.

[Vapn 99] Vapnik V. *The Nature of Statistical Learning Theory*, Springer, 1999.

[Vapn 06] Vapnik V. "Transductive inference and semi-supevised learning," in *Semi-Supervised Learning* (Chapelle O., Schölkopf B., Zien A., eds.), MIT Press, 2006.

[Wang 03] Wang L., Chan K.L., Zhang Z. "Bootstrapping SVM active learning by incorporating unlabeled images for retrieval," *Proceedings of the Conference on Computer Vision and Pattern Recognition*, pp. 629–639, 2003.

[West 06] Weston J., Collobert R., Sinz F., Bottou L., Vapnik V. "Inference with the Universum," *Proceedings of the 23nd International Conference on Machine Learning*, Pittsburgh, PA, 2006.

[Wu 07] Wu M., Schölkopf B. "Transductive classification via local learning regularization," *Proceedings 11th International conference on Artificial Intelligence and Statistics*, San Juan, Puerto Rico, 2007.

[Zhou 04] Zhou D., Bousquet O., Lal T.N., Weston J., Schölkopf "Learning with local and global consistency," in *Advances in Neural Information Processing Systems* (Thrun S., Saul L, Schölkopf B., eds.), pp. 321–328, MIT Press, 2004.

[Zhou 07] Zhou Z.H., Xu J.M. "On the relation between multi-instance learning and semi-supervised learning," *Proceedings of the 24th International Conference on Machine Learning*, Oregon State, 2007.

[Zhu 02] Zhu X., Ghahramani Z. "Learning from labeled and unlabeled data with label propagation," *Technical Report CMU-CALD-02-107*, Carnegie Mellon University, Pittsburgh, PA, 2002.

[Zhu 07] Zhu X. "Semi-supervised learning literature review," *Technical Report, TR 1530*, Computer Science Department, University of Wisconsin-Madison, 2007.

CHAPTER

Clustering: Basic Concepts 11

11.1 INTRODUCTION

All the previous chapters were concerned with supervised classification. In the current and following chapters, we turn to the unsupervised case, where class labeling of the training patterns is not available. Thus, our major concern now is to "reveal" the organization of patterns into "*sensible*" clusters (groups), which will allow us to discover similarities and differences among patterns and to derive useful conclusions about them. This idea is met in many fields, such as the life sciences (biology, zoology), medical sciences (psychiatry, pathology), social sciences (sociology, archaeology), earth sciences (geography, geology), and engineering [Ande 73]. Clustering may be found under different names in different contexts, such as unsupervised learning and learning without a teacher (in pattern recognition), numerical taxonomy (in biology, ecology), typology (in social sciences), and partition (in graph theory). The following example is inspired by biology and gives us a flavor of the problem.

Consider the following animals: sheep, dog, cat (mammals), sparrow, seagull (birds), viper, lizard (reptiles), goldfish, red mullet, blue shark (fish), and frog (amphibians). In order to organize these animals into clusters, we need to define a *clustering criterion*. Thus, if we employ the way these animals bear their progeny as a clustering criterion, the sheep, the dog, the cat, and the blue shark will be assigned to the same cluster, while all the rest will form a second cluster (Figure 11.1a). If the clustering criterion is the existence of lungs, the goldfish, the red mullet, and the blue shark are assigned to the same cluster, while all the other animals are assigned to a second cluster (Figure 11.1b). On the other hand, if the clustering criterion is the environment where the animals live, the sheep, the dog, the cat, the sparrow, the seagull, the viper, and the lizard will form one cluster (animals living outside water); the goldfish, the red mullet, and the blue shark will form a second cluster (animals living only in water); and the frog will form a third cluster by itself, since it may live in the water or out of it (Figure 11.1c). It is worth pointing out that if the existence of a vertebral column is the clustering criterion, all the animals will lie in the same cluster. Finally, we may use composite clustering criteria as

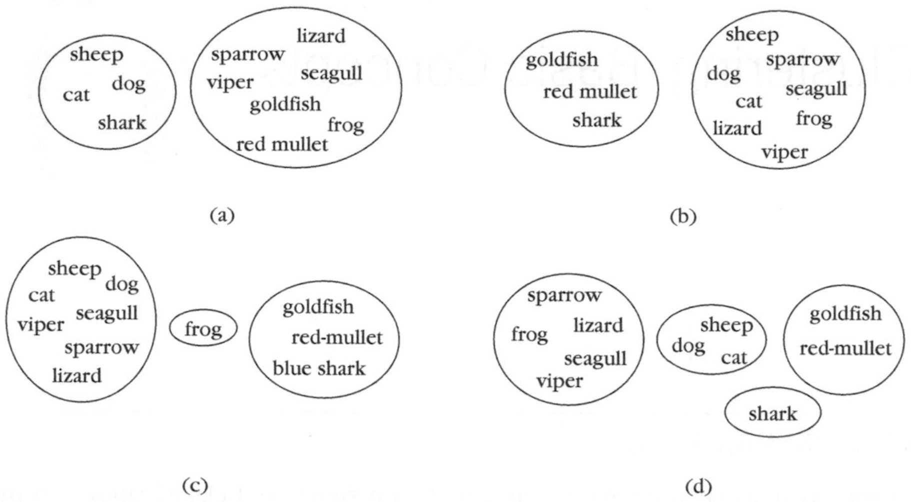

FIGURE 11.1

Resulting clusters if the clustering criterion is (a) the way the animals bear their progeny, (b) the existence of lungs, (c) the environment where the animals live, and (d) the way these animals bear their progeny and the existence of lungs.

well. For example, if the clustering criterion is the way these animals bear their progeny *and* the existence of lungs, we end up with four clusters as shown in Figure 11.1d.

This example shows that the process of assigning objects to clusters may lead to very different results, depending on the specific criterion used for clustering.

Clustering is one of the most primitive mental activities of humans, used to handle the huge amount of information they receive every day. Processing every piece of information as a single entity would be impossible. Thus, humans tend to categorize entities (i.e., objects, persons, events) into clusters. Each cluster is then characterized by the common attributes of the entities it contains. For example, most humans "possess" a cluster "dog." If someone sees a dog sleeping on the grass, he or she will identify it as an entity of the cluster "dog." Thus, the individual will infer that this entity barks even though he or she has never heard this specific entity bark before.

As was the case with supervised learning, we will assume that all patterns are represented in terms of *features*, which form l-dimensional feature vectors.

The basic steps that an expert must follow in order to develop a clustering task are the following:

- *Feature selection*. Features must be properly selected so as to encode as much information as possible concerning the task of interest. Once more, parsimony and, thus, minimum information redundancy among the features is a major goal. As in supervised classification, *preprocessing* of features may be necessary prior to their utilization in subsequent stages. The techniques discussed there are applicable here, too.

- *Proximity measure*. This measure quantifies how "similar" or "dissimilar" two feature vectors are. It is natural to ensure that all selected features contribute equally to the computation of the proximity measure and there are no features that dominate others. This must be taken care of during preprocessing.

- *Clustering criterion*. This criterion depends on the interpretation the expert gives to the term *sensible*, based on the type of clusters that are expected to underlie the data set. For example, a compact cluster of feature vectors in the l-dimensional space, may be sensible according to one criterion, whereas an elongated cluster may be sensible according to another. The clustering criterion may be expressed via a cost function or some other types of rules.

- *Clustering algorithms*. Having adopted a proximity measure and a clustering criterion, this step refers to the choice of a specific algorithmic scheme that unravels the clustering structure of the data set.

- *Validation of the results*. Once the results of the clustering algorithm have been obtained, we have to verify their correctness. This is usually carried out using appropriate tests.

- *Interpretation of the results*. In many cases, the expert in the application field must integrate the results of clustering with other experimental evidence and analysis in order to draw the right conclusions.

In a number of cases, a step known as *clustering tendency* should be involved. This includes various tests that indicate whether or not the available data possess a clustering structure. For example, the data set may be of a completely random nature, thus trying to unravel clusters would be meaningless.

As one may have already suspected, different choices of features, proximity measures, clustering criteria, and clustering algorithms may lead to totally different clustering results. ***Subjectivity is a reality we have to live with from now on.*** To demonstrate this, let us consider the following example. Consider Figure 11.2. How many "sensible" ways of clustering can we obtain for these points? The most "logical" answer seems to be two. The first clustering contains four clusters (surrounded by solid circles). The second clustering contains two clusters (surrounded by dashed lines). Which clustering is "correct"? It seems that there is no definite answer. Both clusterings are valid. The best thing to do is give the results to an expert and let the expert decide about the most sensible one. Thus, the final answer to these questions will be influenced by the expert's knowledge.

The rest of the chapter presents some basic concepts and definitions related to clustering, and it discusses proximity measures that are commonly encountered in various applications.

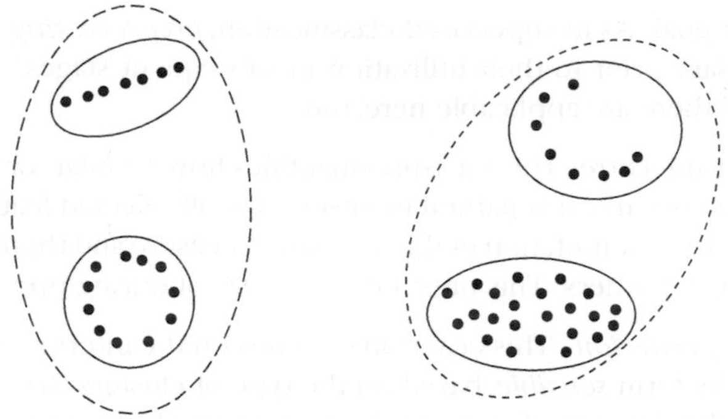

FIGURE 11.2

A coarse clustering of the data results in two clusters, whereas a finer one results in four clusters.

11.1.1 Applications of Cluster Analysis

Clustering is a major tool used in a number of applications. To enrich the list of examples already presented in the introductory chapter of the book, we summarize here four basic directions in which clustering is of use [Ball 71, Ever 01]:

- *Data reduction*. In several cases, the amount of the available data, N, is often very large and as a consequence, its processing becomes very demanding. Cluster analysis can be used in order to group the data into a number of "sensible" clusters, m ($\ll N$), and to process each cluster as a single entity. For example, in data transmission, a representative for each cluster is defined. Then, instead of transmitting the data samples, we transmit a code number corresponding to the representative of the cluster in which each specific sample lies. Thus, data compression is achieved.
- *Hypothesis generation*. In this case we apply cluster analysis to a data set in order to infer some hypotheses concerning the nature of the data. Thus, clustering is used here as a vehicle to suggest hypotheses. These hypotheses must then be verified using other data sets.
- *Hypothesis testing*. In this context, cluster analysis is used for the verification of the validity of a specific hypothesis. Consider, for example, the following hypothesis: "Big companies invest abroad." One way to verify whether this is true is to apply cluster analysis to a large and representative set of companies. Suppose that each company is represented by its size, its activities abroad, and its ability to complete successfully projects on applied research. If, after applying cluster analysis, a cluster is formed that corresponds to companies that are large and have investments abroad (regardless of their ability to complete successfully projects on applied research), then the hypothesis is supported by the cluster analysis.

- *Prediction based on groups*. In this case, we apply cluster analysis to the available data set, and the resulting clusters are characterized based on the characteristics of the patterns by which they are formed. In the sequel, if we are given an unknown pattern, we can determine the cluster to which it is more likely to belong, and we characterize it based on the characterization of the respective cluster. Suppose, for example, that cluster analysis is applied to a data set concerning patients infected by the same disease. This results in a number of clusters of patients, according to their reaction to specific drugs. Then for a new patient, we identify the most appropriate cluster for the patient and, based on it, we decide on his or her medication (e.g., see [Payk 72]).

11.1.2 Types of Features

A feature may take values from a continuous range (subset of $\mathcal{R}$) or from a finite discrete set. If the finite discrete set has only two elements, then the feature is called *binary* or *dichotomous*.

A different categorization of the features is based on the relative significance of the values they take [Jain 88, Spat 80]. We have four categories of features: *nominal*, *ordinal*, *interval-scaled*, and *ratio-scaled*.

The first category, nominal, includes features whose possible values code states. Consider for example a feature that corresponds to the sex of an individual. Its possible values may be 1 for a male and 0 for a female. Clearly, any quantitative comparison between these values is meaningless. The next category, ordinal, includes features whose values can be *meaningfully ordered*. Consider, for example, a feature that characterizes the performance of a student in the pattern recognition course. Suppose that its possible values are 4, 3, 2, 1 and that these correspond to the ratings "excellent," "very good," "good," "not good." Obviously, these values are arranged in a meaningful order. However, the difference between two successive values is of no meaningful quantitative importance.

If, for a specific feature, the difference between two values is meaningful while their ratio is meaningless, then it is an interval-scaled feature. A typical example is the measure of temperature in degrees Celsius. If the temperatures in London and Paris are 5 and 10 degrees Celsius, respectively, then it is meaningful to say that the temperature in Paris is 5 degrees higher than that in London. However, it is meaningless to say that Paris is twice as hot as London.

Finally, if the ratio between two values of a specific feature is meaningful, then this is a ratio-scaled feature, the fourth category. An example of such a feature is weight, since it is meaningful to say that a person who weighs 100 kg is twice as fat as a person whose weight is 50 kg.

By ordering the types of features as nominal, ordinal, interval-scaled, and ratio scaled, one can easily notice that each type of feature possesses all the properties of the types that are before it. For example, an interval-scaled feature has all the properties of the ordinal and nominal types. This information will be of use in Section 11.2.2.

Example 11.1

Suppose that we want to group companies according to their prospects of progress. To this end, we may take into account whether a company is private or public, whether or not the company has activities abroad, its annual budgets for the last, say, three years, its investments, and its rates of change of the budgets and investments. Therefore, each company is represented by a 10×1 vector. The first component of the vector corresponds to a nominal feature, which codes the state "public" or "private." The second component indicates whether or not there are activities abroad. Its possible values are 0, 1, and 2 (discrete range of values), which correspond to "no investments," "poor investments," and "large investments." Clearly, this component corresponds to an ordinal feature. All the remaining features are ratio-scaled.

11.1.3 Definitions of Clustering

The definition of clustering leads directly to the definition of a single "cluster." Many definitions have been proposed over the years (e.g., [John 67, Wall 68, Ever 01]). However, most of these definitions are based on loosely defined terms, such as *similar*, and *alike*, etc., or they are oriented to a specific kind of cluster. As pointed out in [Ever 01], most of these definitions are of vague and of circular nature. This fact reveals the difficulty of having a universally acceptable definition for the term cluster.

In [Ever 01], the vectors are viewed as points in the l-dimensional space, and the clusters are described as "continuous regions of this space containing a relatively high density of points, separated from other high density regions by regions of relatively low density of points." Clusters described in this way are sometimes referred to as *natural clusters*. This definition is closer to our visual perception of clusters in the two- and three-dimensional spaces.

Let us now try to give some definitions for "clustering," which, although they may not be universal, give us an idea of what clustering is. Let X be our data set, that is,

$$X = \{\boldsymbol{x}_1, \boldsymbol{x}_2, \ldots, \boldsymbol{x}_N\}. \tag{11.1}$$

We define as an *m-clustering* of X, $\Re$, the partition of X into m sets (*clusters*), $C_1, \ldots, C_m$, so that the following three conditions are met:

- $C_i \neq \emptyset, i = 1, \ldots, m$
- $\cup_{i=1}^{m} C_i = X$
- $C_i \cap C_j = \emptyset, i \neq j, i, j = 1, \ldots, m$

In addition, the vectors contained in a cluster C_i are "more similar" to each other and "less similar" to the feature vectors of the other clusters. Quantifying the terms *similar* and *dissimilar* depends very much on the types of clusters involved. For example, other *measures* (measuring similarity) are required for compact clusters (e.g., Figure 11.3a), others for elongated clusters (e.g., Figure 11.3b), and different ones for shell-shaped clusters (e.g., Figure 11.3c).

Note that, under the preceding definitions of clustering, each vector belongs to a single cluster. For reasons that will become clear later on, this type of clustering is sometimes called *hard* or *crisp*. An alternative definition is in terms of the *fuzzy sets*, introduced by Zadeh [Zade 65]. A fuzzy clustering of X into m clusters is characterized by m functions u_j where

$$u_j : X \rightarrow [0, 1], \quad j = 1, \ldots, m \tag{11.2}$$

and

$$\sum_{j=1}^{m} u_j(\boldsymbol{x}_i) = 1, \quad i = 1, 2, \ldots, N, \qquad 0 < \sum_{i=1}^{N} u_j(\boldsymbol{x}_i) < N, \quad j = 1, 2, \ldots, m \tag{11.3}$$

These are called *membership functions*. The value of a fuzzy membership function is a mathematical characterization of a set, that is, a cluster in our case, which may not be precisely defined. That is, each vector $\boldsymbol{x}$ belongs to more than one cluster simultaneously "up to some degree," which is quantified by the corresponding value of u_j in the interval [0,1]. Values close to unity show a high "grade of membership" in the corresponding cluster and values close to zero, a low grade of membership. The values of these membership functions are indicative of the structure of the data set, in the sense that if a membership function has close to unity values for two vectors of X, that is, $\boldsymbol{x}_k$, $\boldsymbol{x}_n$, they are considered similar to each other [Wind 82].

The right condition in (11.3) guarantees that there are not trivial cases where clusters exist that do not share any vectors. This is analogous to the condition $C_i \neq \emptyset$ of the aforementioned definition.

The definition of clustering into m distinct sets C_i, given before, can be recovered as a special case of the fuzzy clustering if we define the fuzzy membership functions u_j to take values in $\{0, 1\}$, that is, to be either 1 or 0. In this sense, each data vector belongs exclusively to one cluster and the membership functions are now called *characteristic functions* ([Klir 95]).

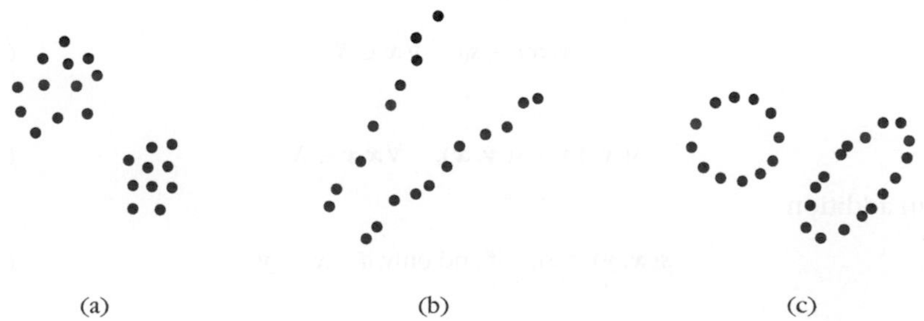

FIGURE 11.3

(a) Compact clusters. (b) Elongated clusters. (c) Spherical and ellipsoidal clusters.

11.2 PROXIMITY MEASURES

11.2.1 Definitions

We begin with definitions concerning measures between vectors, and we will extend them later on to include measures between subsets of the data set X.

A *dissimilarity measure* (DM) d on X is a function.

$$d : X \times X \rightarrow \mathcal{R}$$

where $\mathcal{R}$ is the set of real numbers, such that

$$\exists d_0 \in \mathcal{R} : -\infty < d_0 \leq d(\boldsymbol{x}, \boldsymbol{y}) < +\infty, \quad \forall \boldsymbol{x}, \boldsymbol{y} \in X \tag{11.4}$$

$$d(\boldsymbol{x}, \boldsymbol{x}) = d_0, \quad \forall \boldsymbol{x} \in X \tag{11.5}$$

and

$$d(\boldsymbol{x}, \boldsymbol{y}) = d(\boldsymbol{y}, \boldsymbol{x}), \quad \forall \boldsymbol{x}, \boldsymbol{y} \in X \tag{11.6}$$

If in addition

$$d(\boldsymbol{x}, \boldsymbol{y}) = d_0 \quad \text{if and only if} \quad \boldsymbol{x} = \boldsymbol{y} \tag{11.7}$$

and

$$d(\boldsymbol{x}, \boldsymbol{z}) \leq d(\boldsymbol{x}, \boldsymbol{y}) + d(\boldsymbol{y}, \boldsymbol{z}), \quad \forall \boldsymbol{x}, \boldsymbol{y}, \boldsymbol{z} \in X \tag{11.8}$$

d is called a *metric DM*. Inequality (11.8) is also known as the *triangular inequality*. Finally, equivalence (11.7) indicates that the minimum possible dissimilarity level value d_0 between any two vectors in X is achieved when they are identical. Sometimes we will refer to the dissimilarity level as distance, where the term is not used in its strict mathematical sense.

A *similarity measure* (SM) s on X is defined as

$$s : X \times X \rightarrow \mathcal{R}$$

such that

$$\exists s_0 \in \mathcal{R} : -\infty < s(\boldsymbol{x}, \boldsymbol{y}) \leq s_0 < +\infty, \quad \forall \boldsymbol{x}, \boldsymbol{y} \in X \tag{11.9}$$

$$s(\boldsymbol{x}, \boldsymbol{x}) = s_0, \quad \forall \boldsymbol{x} \in X \tag{11.10}$$

and

$$s(\boldsymbol{x}, \boldsymbol{y}) = s(\boldsymbol{y}, \boldsymbol{x}), \quad \forall \boldsymbol{x}, \boldsymbol{y} \in X \tag{11.11}$$

If in addition

$$s(\boldsymbol{x}, \boldsymbol{y}) = s_0 \quad \text{if and only if} \quad \boldsymbol{x} = \boldsymbol{y} \tag{11.12}$$

and

$$s(\boldsymbol{x}, \boldsymbol{y})s(\boldsymbol{y}, \boldsymbol{z}) \leq [s(\boldsymbol{x}, \boldsymbol{y}) + s(\boldsymbol{y}, \boldsymbol{z})]s(\boldsymbol{x}, \boldsymbol{z}), \quad \forall \boldsymbol{x}, \boldsymbol{y}, \boldsymbol{z} \in X \tag{11.13}$$

s is called a *metric SM*.

Example 11.2

Let us consider the well-known Euclidean distance, d_2

$$d_2(\boldsymbol{x},\boldsymbol{y}) = \sqrt{\sum_{i=1}^{l}(x_i - y_i)^2}$$

where $\boldsymbol{x},\boldsymbol{y} \in X$ and x_i, y_i are the ith coordinates of $\boldsymbol{x}$ and $\boldsymbol{y}$, respectively. This is a dissimilarity measure on X, with $d_0 = 0$; that is, the minimum possible distance between two vectors of X is 0. Moreover, the distance of a vector from itself is equal to 0. Also, it is easy to observe that $d(\boldsymbol{x},\boldsymbol{y}) = d(\boldsymbol{y},\boldsymbol{x})$.

The preceding arguments show that the *Euclidean distance is a dissimilarity measure.* In addition, the Euclidean distance between two vectors takes its minimum value $d_0 = 0$, when the vectors coincide. Finally, it is not difficult to show that the triangular inequality holds for the Euclidean distance (see Problem 11.2). Therefore, the Euclidean distance is a metric dissimilarity measure.

For other measures, the values d_0 (s_0) may be positive or negative.

Not all clustering algorithms, however, are based on proximity measures between vectors. For example, in the hierarchical clustering algorithms[1] one has to compute distances between pairs of sets of vectors of X. In the sequel, we extend the preceding definitions in order to measure "proximity" between subsets of X. Let U be a set containing subsets of X. That is, $D_i \subset X, i = 1, \ldots, k$, and $U = \{D_1, \ldots, D_k\}$. A *proximity measure* $\wp$ *on* U is a function

$$\wp : U \times U \rightarrow \mathcal{R}$$

Equations (11.4)–(11.8) for dissimilarity measures and Eqs. (11.9)–(11.13) for similarity measures can now be repeated with D_i, D_j in the place of $\boldsymbol{x}$ and $\boldsymbol{y}$ and U in the place of X.

Usually, the proximity measures between two sets D_i and D_j are defined in terms of proximity measures between elements of D_i and D_j.

Example 11.3

Let $X = \{\boldsymbol{x}_1, \boldsymbol{x}_2, \boldsymbol{x}_3, \boldsymbol{x}_4, \boldsymbol{x}_5, \boldsymbol{x}_6\}$ and $U = \{\{\boldsymbol{x}_1, \boldsymbol{x}_2\}, \{\boldsymbol{x}_1, \boldsymbol{x}_4\}, \{\boldsymbol{x}_3, \boldsymbol{x}_4, \boldsymbol{x}_5\}, \{\boldsymbol{x}_1, \boldsymbol{x}_2, \boldsymbol{x}_3, \boldsymbol{x}_4, \boldsymbol{x}_5\}\}$. Let us define the following dissimilarity function:

$$d_{\min}^{ss}(D_i,\ D_j) = \min_{\boldsymbol{x} \in D_i,\ \boldsymbol{y} \in D_j} d_2(\boldsymbol{x},\ \boldsymbol{y})$$

where d_2 is the Euclidean distance between two vectors and $D_i,\ D_j \in U$.

The minimum possible value of $d_{\min}^{ss}$ is $d_{\min,0}^{ss} = 0$. Also, $d_{\min}^{ss}(D_i,\ D_i) = 0$, since the Euclidean distance between a vector in D_i and itself is 0. In addition, it is easy to see that the

[1] These algorithms are treated in detail in Chapter 13.

commutative property holds. Thus, this dissimilarity function is a measure. It is not difficult to see that $d^{ss}_{\min}$ is not a metric. Indeed, Eq. (11.7) for subsets of X does not hold in general, since the two sets D_i and D_j may have an element in common. Consider, for example the two sets $\{x_1, x_2\}$ and $\{x_1, x_4\}$ of U. Although they are different, their distance $d^{ss}_{\min}$ is 0, since they both contain x_1.

Intuitively speaking, the preceding definitions show that the DMs are "opposite" to SMs. For example, it is easy to show that if d is a (metric) DM, with $d(x, y) > 0$, $\forall x, y \in X$, then $s = a/d$ with $a > 0$ is a (metric) SM (see Problem 11.1). Also, $d_{\max} - d$ is a (metric) SM, where $d_{\max}$ denotes the maximum value of d among all pairs of elements of X. It is also easy to show that if d is a (metric) DM on a finite set X, such that $d(x, y) > 0$, $\forall x, y \in X$, then so are $-\ln(d_{\max} + k - d)$ and $kd/(1 + d)$, where k is an arbitrary positive constant. On the other hand, if s is a (metric) SM with $s_0 = 1 - \varepsilon$, where ε is a small positive constant, then $1/(1 - s)$ is also a (metric) SM. Similar comments are valid for the similarity and dissimilarity measures between sets $D_i, D_j \in U$.

In the sequel, we will review the most commonly used proximity measures between two points. For each measure of similarity we give a corresponding measure of dissimilarity. We will denote by $b_{\min}$ and $b_{\max}$ the corresponding minimum and maximum values that they take for a finite data set X.

11.2.2 Proximity Measures between Two Points

Real-Valued Vectors

A. Dissimilarity Measures

The most common DMs between real-valued vectors used in practice are:

- The *weighted* l_p metric DMs, that is,

$$d_p(x, y) = \left(\sum_{i=1}^{l} w_i |x_i - y_i|^p\right)^{1/p} \tag{11.14}$$

where x_i, y_i are the ith coordinates of x and y, $i = 1, \ldots, l$, and $w_i \geq 0$ is the ith *weight coefficient*. They are used mainly on real-valued vectors. If $w_i = 1, i = 1, \ldots, l$, we obtain the *unweighted* l_p metric DMs. A well-known representative of the latter category of measures is the *Euclidean distance*, which was introduced in Example 11.2 and is obtained by setting $p = 2$.

The weighted l_2 metric DM can be further generalized as follows:

$$d(x, y) = \sqrt{(x - y)^T B (x - y)} \tag{11.15}$$

where B is a symmetric, positive definite matrix (Appendix B).

This includes the Mahalanobis distance as a special case, and it is also a metric DM.

Special l_p metric DMs that are also encountered in practice are the (weighted) l_1 or *Manhattan norm*,

$$d_1(\boldsymbol{x}, \boldsymbol{y}) = \sum_{i=1}^{l} w_i |x_i - y_i| \tag{11.16}$$

and the (weighted) l_∞ *norm*,

$$d_\infty(\boldsymbol{x}, \boldsymbol{y}) = \max_{1 \leq i \leq l} w_i |x_i - y_i| \tag{11.17}$$

The l_1 and l_∞ norms may be viewed as overestimation and underestimation of the l_2 norm, respectively. Indeed, it can be shown that $d_\infty(\boldsymbol{x}, \boldsymbol{y}) \leq d_2(\boldsymbol{x}, \boldsymbol{y}) \leq d_1(\boldsymbol{x}, \boldsymbol{y})$ (see Problem 11.6). When $l = 1$ all l_p norms coincide.

Based on these DMs, we can define corresponding SMs as $s_p(\boldsymbol{x}, \boldsymbol{y}) = b_{\max} - d_p(\boldsymbol{x}, \boldsymbol{y})$.

- Some additional DMs are the following [Spat 80]:

$$d_G(\boldsymbol{x}, \boldsymbol{y}) = -\log_{10}\left(1 - \frac{1}{l}\sum_{j=1}^{l} \frac{|x_j - y_j|}{b_j - a_j}\right) \tag{11.18}$$

where b_j and a_j are the maximum and the minimum values among the jth features of the N vectors of X, respectively. It can easily be shown that $d_G(\boldsymbol{x}, \boldsymbol{y})$ is a metric DM. Notice that the value of $d_G(\boldsymbol{x}, \boldsymbol{y})$ depends not only on $\boldsymbol{x}$ and $\boldsymbol{y}$ but also on the whole of X. Thus, if $d_G(\boldsymbol{x}, \boldsymbol{y})$ is the distance between two vectors $\boldsymbol{x}$ and $\boldsymbol{y}$ that belong to a set X and $d'_G(\boldsymbol{x}, \boldsymbol{y})$ is the distance between the same two vectors when they belong to a different set X', then, in general, $d_G(\boldsymbol{x}, \boldsymbol{y}) \neq d'_G(\boldsymbol{x}, \boldsymbol{y})$. Another DM is [Spat 80]

$$d_Q(\boldsymbol{x}, \boldsymbol{y}) = \sqrt{\frac{1}{l}\sum_{j=1}^{l}\left(\frac{x_j - y_j}{x_j + y_j}\right)^2} \tag{11.19}$$

Example 11.4

Consider the three-dimensional vectors $\boldsymbol{x} = [0, 1, 2]^T$, $\boldsymbol{y} = [4, 3, 2]^T$. Then, assuming that all w_i's are equal to 1, $d_1(\boldsymbol{x}, \boldsymbol{y}) = 6$, $d_2(\boldsymbol{x}, \boldsymbol{y}) = 2\sqrt{5}$, and $d_\infty(\boldsymbol{x}, \boldsymbol{y}) = 4$. Notice that $d_\infty(\boldsymbol{x}, \boldsymbol{y}) < d_2(\boldsymbol{x}, \boldsymbol{y}) < d_1(\boldsymbol{x}, \boldsymbol{y})$.

Assume now that these vectors belong to a data set X that contains N vectors with maximum values per feature 10, 12, 13 and minimum values per feature 0, 0.5, 1, respectively. Then $d_G(\boldsymbol{x}, \boldsymbol{y}) = 0.0922$. If, on the other hand, $\boldsymbol{x}$ and $\boldsymbol{y}$ belong to an X' with the maximum (minimum) values per feature being 20, 22, 23 (-10, -9.5, -9), respectively, then $d_G(\boldsymbol{x}, \boldsymbol{y}) = 0.0295$.

Finally, $d_Q(\boldsymbol{x}, \boldsymbol{y}) = 0.6455$.

B. Similarity Measures

The most common similarity measures for real-valued vectors used in practice are:

- *The inner product*. It is defined as $s_{inner}(\boldsymbol{x}, \boldsymbol{y}) = \boldsymbol{x}^T\boldsymbol{y} = \sum_{i=1}^{l} x_i y_i$. In most cases, the inner product is used when the vectors $\boldsymbol{x}$ and $\boldsymbol{y}$ are normalized, so that they have the same length a. In these cases, the upper and the lower bounds of s_{inner} are $+a^2$ and $-a^2$, respectively, and $s_{inner}(\boldsymbol{x}, \boldsymbol{y})$ depends exclusively on the angle between $\boldsymbol{x}$ and $\boldsymbol{y}$.

  A corresponding dissimilarity measure for the inner product is $d_{inner}(\boldsymbol{x}, \boldsymbol{y}) = b_{max} - s_{inner}(\boldsymbol{x}, \boldsymbol{y})$.

  Closely related to the inner product is the *cosine similarity measure*, which is defined as

$$s_{cosine}(\boldsymbol{x}, \boldsymbol{y}) = \frac{\boldsymbol{x}^T\boldsymbol{y}}{\|\boldsymbol{x}\|\,\|\boldsymbol{y}\|} \tag{11.20}$$

  where $\|\boldsymbol{x}\| = \sqrt{\sum_{i=1}^{l} x_i^2}$ and $\|\boldsymbol{y}\| = \sqrt{\sum_{i=1}^{l} y_i^2}$ are the lengths of the vectors $\boldsymbol{x}$ and $\boldsymbol{y}$, respectively. This measure is invariant to rotations but not to linear transformations.

- *Pearson's correlation coefficient*. This measure can be expressed as

$$r_{Pearson}(\boldsymbol{x}, \boldsymbol{y}) = \frac{\boldsymbol{x}_d^T\boldsymbol{y}_d}{\|\boldsymbol{x}_d\|\,\|\boldsymbol{y}_d\|} \tag{11.21}$$

  where $\boldsymbol{x}_d = [x_1 - \bar{x}, \ldots, x_l - \bar{x}]^T$ and $\boldsymbol{y}_d = [y_1 - \bar{y}, \ldots, y_l - \bar{y}]^T$, with x_i, y_i being the ith coordinates of $\boldsymbol{x}$ and $\boldsymbol{y}$, respectively, and $\bar{x} = \frac{1}{l}\sum_{i=1}^{l} x_i$, $\bar{y} = \frac{1}{l}\sum_{i=1}^{l} y_i$. Usually, $\boldsymbol{x}_d$ and $\boldsymbol{y}_d$ are called difference vectors. Clearly, $r_{Pearson}(\boldsymbol{x}, \boldsymbol{y})$ takes values between -1 and $+1$. The difference from s_{inner} is that $s_{Pearson}$ does not depend directly on $\boldsymbol{x}$ and $\boldsymbol{y}$ but on their corresponding difference vectors. A related dissimilarity measure can be defined as

$$D(\boldsymbol{x}, \boldsymbol{y}) = \frac{1 - r_{Pearson}(\boldsymbol{x}, \boldsymbol{y})}{2} \tag{11.22}$$

  This takes values in the range [0, 1]. This measure has been used in the analysis of gene-expression data ([Eise 98]).

- Another commonly used SM is the *Tanimoto measure*, which is also known as Tanimoto distance [Tani 58]. It may be used for real- as well as for discrete-valued vectors. It is defined as

$$s_T(\boldsymbol{x}, \boldsymbol{y}) = \frac{\boldsymbol{x}^T\boldsymbol{y}}{\|\boldsymbol{x}\|^2 + \|\boldsymbol{y}\|^2 - \boldsymbol{x}^T\boldsymbol{y}} \tag{11.23}$$

  By adding and subtracting the term $\boldsymbol{x}^T\boldsymbol{y}$ in the denominator of (11.23) and after some algebraic manipulations, we obtain

$$s_T(\boldsymbol{x}, \boldsymbol{y}) = \frac{1}{1 + \frac{(\boldsymbol{x}-\boldsymbol{y})^T(\boldsymbol{x}-\boldsymbol{y})}{\boldsymbol{x}^T\boldsymbol{y}}}$$

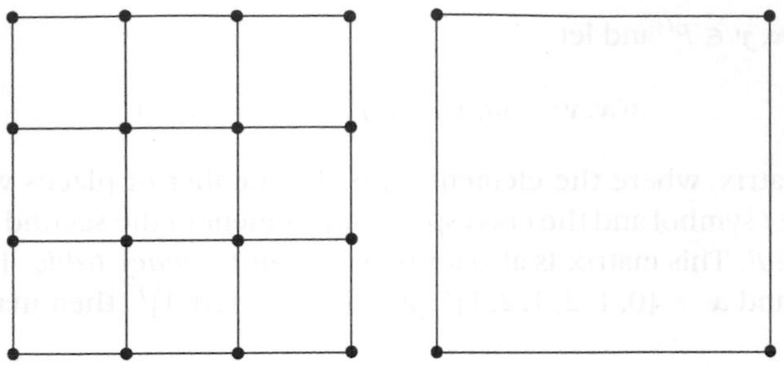

FIGURE 11.4

(a) The $l = 2$ dimensional grid for $k = 4$. (b) The H_2 hypercube (square).

That is, the Tanimoto measure between x and y is inversely proportional to the squared Euclidean distance between x and y divided by their inner product. Intuitively speaking, since the inner product may be considered as a measure of the correlation between x and y, $s_T(x, y)$ is inversely proportional to the squared Euclidean distance between x and y, divided by their correlation.

In the case in which the vectors of X have been normalized to the same length a, the last equation leads to

$$s_T(x, y) = \frac{1}{-1 + 2\frac{a^2}{x^T y}}$$

In this case, s_T is inversely proportional to $a^2/x^T y$. Thus, the more correlated x and y are, the larger the value of s_T.

- Finally, another similarity measure that has been proved useful in certain applications [Fu 93] is the following:

$$s_c(x, y) = 1 - \frac{d_2(x, y)}{\|x\| + \|y\|} \tag{11.24}$$

$s_c(x, y)$ takes its maximum value (1) when $x = y$ and its minimum (0) when $x = -y$.

Discrete-Valued Vectors

We will now consider vectors x whose coordinates belong to the finite set $F = \{0, 1, \ldots, k-1\}$, where k is a positive integer. It is clear that there are exactly k^l vectors $x \in F^l$. One can imagine these vectors as vertices in an l-dimensional grid as depicted in Figure 11.4. When $k = 2$, the grid collapses to the H_l (unit) hypercube.

Consider $\boldsymbol{x}, \boldsymbol{y} \in F^l$ and let

$$A(\boldsymbol{x}, \boldsymbol{y}) = [a_{ij}] \qquad i, j = 0, 1, \ldots, k-1 \tag{11.25}$$

be a $k \times k$ matrix, where the element a_{ij} is the number of places where the first vector has the i symbol and the corresponding element of the second vector has the j symbol, $i, j \in F$. This matrix is also known as a *contingency table*. For example, if $l = 6, k = 3$ and $\boldsymbol{x} = [0, 1, 2, 1, 2, 1]^T, \boldsymbol{y} = [1, 0, 2, 1, 0, 1]^T$, then matrix $A(\boldsymbol{x}, \boldsymbol{y})$ is equal to

$$A(\boldsymbol{x}, \boldsymbol{y}) = \begin{bmatrix} 0 & 1 & 0 \\ 1 & 2 & 0 \\ 1 & 0 & 1 \end{bmatrix}$$

It is easy to verify that

$$\sum_{i=0}^{k-1} \sum_{j=0}^{k-1} a_{ij} = l$$

Most of the proximity measures between two discrete-valued vectors may be expressed as combinations of elements of matrix $A(\boldsymbol{x}, \boldsymbol{y})$.

A. Dissimilarity Measures

- *The Hamming distance* (e.g., [Lipp 87, Gers 92]). It is defined as the number of places where two vectors differ. Using the matrix A, we can define the Hamming distance $d_H(\boldsymbol{x}, \boldsymbol{y})$ as

$$d_H(\boldsymbol{x}, \boldsymbol{y}) = \sum_{i=0}^{k-1} \sum_{j=0, j \neq i}^{k-1} a_{ij} \tag{11.26}$$

that is, the summation of all the off-diagonal elements of A, which indicate the positions where $\boldsymbol{x}$ and $\boldsymbol{y}$ differ.

In the special case in which $k = 2$, the vectors $\boldsymbol{x} \in F^l$ are binary valued and the Hamming distance becomes

$$d_H(\boldsymbol{x}, \boldsymbol{y}) = \sum_{i=1}^{l} (x_i + y_i - 2x_i y_i) = \sum_{i=1}^{l} (x_i - y_i)^2 \tag{11.27}$$

In the case where $\boldsymbol{x} \in F_1^l$, where $F_1 = \{-1, 1\}$, $\boldsymbol{x}$ is called bipolar vector and the Hamming distance is given as

$$d_H(\boldsymbol{x}, \boldsymbol{y}) = 0.5 \left(l - \sum_{i=1}^{l} x_i y_i \right) \tag{11.28}$$

Obviously, a corresponding similarity measure of d_H is $s_H(\boldsymbol{x}, \boldsymbol{y}) = b_{\max} - d_H(\boldsymbol{x}, \boldsymbol{y})$.

- *The l_1 distance*. It is defined as in the case of the continuous-valued vectors, that is,

$$d_1(\boldsymbol{x}, \boldsymbol{y}) = \sum_{i=1}^{l} |x_i - y_i| \tag{11.29}$$

  The l_1 distance and the Hamming distance coincide when binary-valued vectors are considered.

B. Similarity Measures

A widely used similarity measure for discrete-valued vectors is *the Tanimoto measure. It is inspired by the comparison of sets.* If X and Y are two sets and n_X, n_Y, $n_{X \cap Y}$ are the cardinalities (number of elements) of X, Y, and $X \cap Y$, respectively, the Tanimoto measure between two sets X and Y is defined as

$$\frac{n_{X \cap Y}}{n_X + n_Y - n_{X \cap Y}} = \frac{n_{X \cap Y}}{n_{X \cup Y}}$$

In other words, the Tanimoto measure between two sets is the ratio of the number of elements they have in common to the number of all different elements.

We turn now to the Tanimoto measure between two discrete-valued vectors $\boldsymbol{x}$ and $\boldsymbol{y}$. The measure takes into account all pairs of corresponding coordinates of $\boldsymbol{x}$ and $\boldsymbol{y}$, except those whose corresponding coordinates (x_i, y_i) are both 0. This is justified if we have ordinal features and interpret the value of the ith coordinate of, say, $\boldsymbol{y}$ as the degree to which the vector $\boldsymbol{y}$ possesses the ith feature. According to this interpretation, the pairs $(x_i, y_i) = (0, 0)$ are less important than the others. We now define $n_{\boldsymbol{x}} = \sum_{i=1}^{k-1} \sum_{j=0}^{k-1} a_{ij}$ and $n_{\boldsymbol{y}} = \sum_{i=0}^{k-1} \sum_{j=1}^{k-1} a_{ij}$, where a_{ij} are elements of the $A(\boldsymbol{x}, \boldsymbol{y})$ matrix (see Figure 11.5). In words, $n_{\boldsymbol{x}}$ ($n_{\boldsymbol{y}}$) denotes the number of the nonzero coordinates of $\boldsymbol{x}$ ($\boldsymbol{y}$). Then, the Tanimoto measure is defined as

$$s_T(\boldsymbol{x}, \boldsymbol{y}) = \frac{\sum_{i=1}^{k-1} a_{ii}}{n_{\boldsymbol{x}} + n_{\boldsymbol{y}} - \sum_{i=1}^{k-1} \sum_{j=1}^{k-1} a_{ij}} \tag{11.30}$$

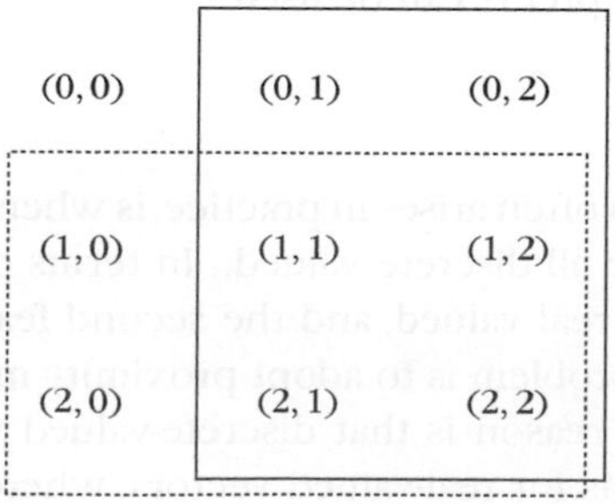

FIGURE 11.5

The elements of a contingency table taken into account for the computation of the Tanimoto measure.

In the special case $k = 2$, this equation results in [Tani 58, Spat 80]

$$s_T(\boldsymbol{x}, \boldsymbol{y}) = \frac{a_{11}}{a_{11} + a_{01} + a_{10}} \tag{11.31}$$

Other similarity functions between $\boldsymbol{x}, \boldsymbol{y} \in F^l$ can be defined using elements of $A(\boldsymbol{x}, \boldsymbol{y})$. Some of them consider only the number of places where the two vectors agree and the corresponding value is not 0, whereas others consider all the places where the two vectors agree. Similarity functions that belong to the first category are

$$\frac{\sum_{i=1}^{k-1} a_{ii}}{l} \quad \text{and} \quad \frac{\sum_{i=1}^{k-1} a_{ii}}{l - a_{00}} \tag{11.32}$$

A representative of the second category is

$$\frac{\sum_{i=0}^{k-1} a_{ii}}{l} \tag{11.33}$$

When dealing with binary-valued vectors (i.e., $k = 2$), probabilistic similarity measures have also been proposed [Good 66, Li 85, Broc 81]. For two binary-valued vectors $\boldsymbol{x}$ and $\boldsymbol{y}$, a measure of this kind, s, is based on the number of positions where $\boldsymbol{x}$ and $\boldsymbol{y}$ agree. The value of $s(\boldsymbol{x}, \boldsymbol{y})$ is then compared with the distances of pairs of randomly chosen vectors, in order to conclude whether $\boldsymbol{x}$ and $\boldsymbol{y}$ are "close" to each other. This task is carried out using statistical tests (see also Chapter 16).

Dynamic Similarity Measures

The proximity measures discussed so far apply to vectors with the same dimension, l. However, in certain applications, such as the comparison of two strings st_1 and st_2 stemming from two different texts, this is not the case. For example, one of the two strings may be shifted with respect to the other. In these cases the preceding proximity measures fail. In such cases, dynamic similarity measures, such as the Edit distance, discussed in Chapter 8, can be used.

Mixed Valued Vectors

An interesting case, which often arises in practice, is when the features of the feature vectors are not all real or all discrete valued. In terms of Example 11.1, the third to the tenth features are real valued, and the second feature is discrete valued. A naive way to attack this problem is to adopt proximity measures (PMs) suitable for real-valued vectors. The reason is that discrete-valued vectors can be accurately compared in terms of PMs for real-valued vectors, whereas the opposite does not lead, in general, to reasonable results. A good PM candidate for such cases is the l_1 distance.

Example 11.5
Consider the vectors $\boldsymbol{x} = [4,\ 1,\ 0.8]^T$ and $\boldsymbol{y} = [1,\ 0,\ 0.4]^T$. Their (unweighted) l_1 and l_2 distances are

$$d_1(\boldsymbol{x}, \boldsymbol{y}) = |4-1| + |1-0| + |0.8-0.4| = 3 + 1 + 0.4 = 4.4$$

and

$$d_2(\boldsymbol{x}, \boldsymbol{y}) = \sqrt{|4-1|^2 + |1-0|^2 + |0.8-0.4|^2} = \sqrt{9 + 1 + 0.16} = 3.187$$

respectively. Notice that in the second case, the difference between the first coordinates of $\boldsymbol{x}$ and $\boldsymbol{y}$ specifies almost exclusively the difference between the two vectors. This is not the case with l_1 distance (see also related comments in Chapter 5, Section 5.2).

Another method that may be employed is to convert the real-valued features to discrete-valued ones, that is, to discretize the real-valued data. To this end, if a feature x_i takes values in the interval $[a, b]$, we may divide this interval into k subintervals. If the value of x_i lies in the rth subinterval, the value $r-1$ will be assigned to it. This strategy leads to discrete-valued vectors, and as a consequence, we may use any of the measures discussed in the previous section.

In [Ande 73] the types nominal, ordinal, and interval-scaled types of features are considered and methods for converting features from one type to another are discussed. These are based on the fact (see Section 11.1.2) that as we move from nominal to interval scaled, we have to impose information on the specific feature, and when we move along the opposite direction, we have to give up information.

A similarity function that deals with mixed valued vectors, without making any conversions to the type of features, is proposed in [Gowe 71]. Let us consider two l-dimensional mixed valued vectors $\boldsymbol{x}_i$ and $\boldsymbol{x}_j$. Then, the similarity function between $\boldsymbol{x}_i$ and $\boldsymbol{x}_j$ is defined as

$$s(\boldsymbol{x}_i, \boldsymbol{x}_j) = \frac{\sum_{q=1}^{l} s_q(\boldsymbol{x}_i, \boldsymbol{x}_j)}{\sum_{q=1}^{l} w_q} \tag{11.34}$$

where $s_q(\boldsymbol{x}_i, \boldsymbol{x}_j)$ is the similarity between the qth coordinates of $\boldsymbol{x}_i$ and $\boldsymbol{x}_j$ and w_q is a weight factor corresponding to the qth coordinate. Specifically, if at least one of the qth coordinates of $\boldsymbol{x}_i$ and $\boldsymbol{x}_j$ is undefined, then $w_q = 0$. Also, if the qth coordinate is a binary variable and it is 0 for both vectors, then $w_q = 0$. In all other cases, w_q is set equal to 1. Finally, if all w_q's are equal to 0 then $s(\boldsymbol{x}_i, \boldsymbol{x}_j)$ is undefined. If the qth coordinates of the two vectors are binary then

$$s_q(\boldsymbol{x}_i, \boldsymbol{x}_j) = \begin{cases} 1, & \text{if } x_{iq} = x_{jq} = 1 \\ 0, & \text{otherwise} \end{cases} \tag{11.35}$$

If the qth coordinates of the two vectors correspond to nominal or ordinal variables, then $s_q(\boldsymbol{x}_i, \boldsymbol{x}_j) = 1$ if x_{iq} and x_{jq} have the same values. Otherwise, $s_q(\boldsymbol{x}_i, \boldsymbol{x}_j) = 0$.

Finally, if the qth coordinates correspond to interval or ratio scaled variables, then

$$s_q(\boldsymbol{x}_i, \boldsymbol{x}_j) = 1 - \frac{|x_{iq} - x_{jq}|}{r_q} \tag{11.36}$$

where r_q is the length of the interval where the values of the qth coordinates lie. One can easily observe that for the case of intervals or ratio-scaled variables, when x_{ik} and x_{jk} coincide, $s_q(\boldsymbol{x}_i, \boldsymbol{x}_j)$ takes its maximum value, which equals 1. On the other hand, if the absolute difference between x_{iq} and x_{jq} equals r_q, then $s_q(\boldsymbol{x}_i, \boldsymbol{x}_j) = 0$. For any other value of $|x_{iq} - x_{jq}|$, $s_q(\boldsymbol{x}_i, \boldsymbol{x}_j)$ lies between 0 and 1.

Example 11.6

Let us consider the following four 5-dimensional feature vectors, each representing a specific company. More specifically, the first three coordinates (features) correspond to their annual budget for the last three years (in millions of dollars), the fourth indicates whether or not there is any activity abroad, and the fifth coordinate corresponds to the number of employees of each company. The last feature is ordinal scaled and takes the values 0 (small number of employees), 1 (medium number of employees), and 2 (large number of employees). The four vectors are

Company	1st bud.	2nd bud.	3rd bud.	Act. abr.	Empl.
1 ($\boldsymbol{x}_1$)	1.2	1.5	1.9	0	1
2 ($\boldsymbol{x}_2$)	0.3	0.4	0.6	0	0
3 ($\boldsymbol{x}_3$)	10	13	15	1	2
4 ($\boldsymbol{x}_4$)	6	6	7	1	1

(11.37)

For the first three coordinates, which are ratio scaled, we have $r_1 = 9.7$, $r_2 = 12.6$, and $r_3 = 14.4$. Let us first compute the similarity between the first two vectors. It is

$$s_1(\boldsymbol{x}_1, \boldsymbol{x}_2) = 1 - |1.2 - 0.3|/9.7 = 0.9072$$

$$s_2(\boldsymbol{x}_1, \boldsymbol{x}_2) = 1 - |1.5 - 0.4|/12.6 = 0.9127$$

$$s_3(\boldsymbol{x}_1, \boldsymbol{x}_2) = 1 - |1.9 - 0.6|/14.4 = 0.9097$$

$$s_4(\boldsymbol{x}_1, \boldsymbol{x}_2) = 0$$

and

$$s_5(\boldsymbol{x}_1, \boldsymbol{x}_2) = 0$$

Also, $w_4 = 0$, while all the other weight factors are equal to 1. Using Eq. (11.34), we finally obtain $s(\boldsymbol{x}_1, \boldsymbol{x}_2) = 0.6824$.

Working in the same way, we find that $s(\boldsymbol{x}_1, \boldsymbol{x}_3) = 0.0541$, $s(\boldsymbol{x}_1, \boldsymbol{x}_4) = 0.5588$, $s(\boldsymbol{x}_2, \boldsymbol{x}_3) = 0$, $s(\boldsymbol{x}_2, \boldsymbol{x}_4) = 0.3047$, $s(\boldsymbol{x}_3, \boldsymbol{x}_4) = 0.4953$.

Fuzzy Measures

In this section, we consider real-valued vectors $\boldsymbol{x}, \boldsymbol{y}$ whose components x_i and y_i belong to the interval $[0, 1], i = 1, \ldots, l$. In contrast to what we have said so far, *the values of x_i are not the outcome of a measuring device*. The closer the x_i to 1 (0), the more likely $\boldsymbol{x}$ possesses (does not possess) the ith feature (characteristic).[2] As x_i approaches 1/2, we become less certain about the possession or not of the ith feature from $\boldsymbol{x}$. When $x_i = 1/2$ we have absolutely no clue whether or not $\boldsymbol{x}$ possesses the ith feature. It is easy to observe that this situation is a generalization of binary logic, where x_i can take only the value 0 or 1 ($\boldsymbol{x}$ possesses a feature or not). In binary logic, there is a certainty about the occurrence of a fact (for example, it will rain or it will not rain). The idea of fuzzy logic is that nothing is happening or not happening with absolute certainty. This is reflected in the values that x_i takes. The binary logic can be viewed as a special case of fuzzy logic where x_i takes only the value 0 or 1.

Next, we will define the similarity between two real-valued variables in [0, 1]. We will approach it as a generalization of the equivalence between two binary variables. The equivalence of two binary variables a and b is given by the following relation:

$$(a \equiv b) = ((NOT\ a)\ AND\ (NOT\ b))\ OR\ (a\ AND\ b) \tag{11.38}$$

Indeed, if $a = b = 0$ (1), the first (second) argument of the *OR* operator is 1. On the other hand if $a = 0$ (1) and $b = 1$ (0), then none of the arguments of the *OR* operator becomes 1.

An interesting observation is that the *AND* (*OR*) operator between two binary variables may be seen as the min (max) operator on them. Also, the *NOT* operation of a binary variable a may be written as $1-a$. In the fuzzy logic context and based on this observation, the logical *AND* is replaced by the operator min, while the logical *OR* is replaced by the operator max. Also, the logical *NOT* on x_i is replaced by $1 - x_i$ [Klir 95]. This suggests that the degree of similarity between two real-valued variables x_i and y_i in [0, 1] may be defined as

$$s(x_i, y_i) = \max(\min(1 - x_i, 1 - y_i), \min(x_i, y_i)) \tag{11.39}$$

Note that this definition includes the special case where x_i and y_i take binary values and results in (11.38).

When we now deal with vectors in the l-dimensional space ($l > 1$), the vector space is the H_l hypercube. In this context, the closer a vector $\boldsymbol{x}$ lies to the center of H_l $(1/2, \ldots, 1/2)$, the greater the amount of uncertainty. That is, in this case we have almost no clue whether $\boldsymbol{x}$ possesses any of the l features. On the other hand, the closer $\boldsymbol{x}$ lies to a vertex of H_l, the less the uncertainty.

Based on similarity s between two variables in [0, 1] given in (11.39), we are now able to define a similarity measure between two vectors. A common similarity

[2] The ideas of this section follow [Zade 73].

measure between two vectors $\boldsymbol{x}$ and $\boldsymbol{y}$ is defined as

$$s_F^q(\boldsymbol{x}, \boldsymbol{y}) = \left(\sum_{i=1}^{l} s(x_i, y_i)^q\right)^{1/q} \tag{11.40}$$

It is easy to verify that the maximum and minimum values of s_F are $l^{1/q}$ and 0, respectively. As $q \to +\infty$, we get $s_F(\boldsymbol{x}, \boldsymbol{y}) = \max_{1 \le i \le l} s(x_i, y_i)$. Also, when $q = 1$, $s_F(\boldsymbol{x}, \boldsymbol{y}) = \sum_{i=1}^{l} s(x_i, y_i)$ (Problem 11.7).

Example 11.7

In this example we consider the case where $l = 3$ and $q = 1$. Under these circumstances, the maximum possible value of s_F is 3. Let us consider the vectors $\boldsymbol{x}_1 = [1, 1, 1]^T$, $\boldsymbol{x}_2 = [0, 0, 1]^T$, $\boldsymbol{x}_3 = [1/2, 1/3, 1/4]^T$, and $\boldsymbol{x}_4 = [1/2, 1/2, 1/2]^T$. If we compute the similarities of these vectors with themselves, we obtain

$$s_F^1(\boldsymbol{x}_1, \boldsymbol{x}_1) = 3\max(\min(1-1, 1-1), \min(1, 1)) = 3$$

and similarly, $s_F^1(\boldsymbol{x}_2, \boldsymbol{x}_2) = 3$, $s_F^1(\boldsymbol{x}_3, \boldsymbol{x}_3) = 1.92$, and $s_F^1(\boldsymbol{x}_4, \boldsymbol{x}_4) = 1.5$. *This is very interesting. The similarity measure of a vector with itself depends not only on the vector but also on its position in the H_l hypercube.* Furthermore, we observe that the greatest similarity value is obtained at the vertices of H_l. As we move toward the center of H_l, the similarity measure between a vector and itself decreases, attaining its minimum value at the center of H_l.

Let us now consider the vectors $\boldsymbol{y}_1 = [3/4, 3/4, 3/4]^T$, $\boldsymbol{y}_2 = [1, 1, 1]^T$, $\boldsymbol{y}_3 = [1/4, 1/4, 1/4]^T$, $\boldsymbol{y}_4 = [1/2, 1/2, 1/2]^T$. Notice that in terms of the Euclidean distance $d_2(\boldsymbol{y}_1, \boldsymbol{y}_2) = d_2(\boldsymbol{y}_3, \boldsymbol{y}_4)$. However, $s_F^1(\boldsymbol{y}_1, \boldsymbol{y}_2) = 2.25$ and $s_F^1(\boldsymbol{y}_3, \boldsymbol{y}_4) = 1.5$. These results suggest that the closer the two vectors to the center of H_l, the less their similarity. On the other hand, the closer the two vectors to a vertex of H_l, the greater their similarity. *That is, the value of $s_F^q(\boldsymbol{x}, \boldsymbol{y})$ depends not only on the relative position of $\boldsymbol{x}$ and $\boldsymbol{y}$ in H_l but also on their closeness to the center of H_l.*

Missing Data

A problem that is commonly met in real-life applications is that of missing data. This means that for some feature vectors we do not know all of their components. This may be a consequence of a failure of the measuring device. Also, in cases such as the one mentioned in Example 11.1, missing data may be the result of a recording error. The following are some commonly used techniques that handle this situation [Snea 73, Dixo 79, Jain 88].

1. Discard all feature vectors that have missing features. This approach may be used when the number of vectors with missing features is small compared to the total number of available feature vectors. If this is not the case, the nature of the problem may be affected.

2. For the ith feature, find its mean value based on the corresponding available values of all feature vectors of X. Then, substitute this value for the vectors where their ith coordinate is not available.

3. For all the pairs of components x_i and y_i of the vectors $\boldsymbol{x}$ and $\boldsymbol{y}$ define b_i as

$$b_i = \begin{cases} 0, & \text{if both } x_i \text{ and } y_i \text{ are available} \\ 1, & \text{otherwise} \end{cases} \tag{11.41}$$

Then, the proximity between $\boldsymbol{x}$ and $\boldsymbol{y}$ is defined as

$$\wp(\boldsymbol{x}, \boldsymbol{y}) = \frac{l}{l - \sum_{i=1}^{l} b_i} \sum_{\text{all } i: b_i = 0} \phi(x_i, y_i) \tag{11.42}$$

where $\phi(x_i, y_i)$ denotes the proximity between the two scalars x_i and y_i. A common choice of ϕ when a dissimilarity measure is involved, is $\phi(x_i, y_i) = |x_i - y_i|$. The rationale behind this approach is simple. Let $[a, b]$ be the interval of the allowable values of $\wp(\boldsymbol{x}, \boldsymbol{y})$. The preceding definition ensures that the proximity measure between $\boldsymbol{x}$ and $\boldsymbol{y}$ spans all $[a, b]$, regardless of the number of unavailable features in both vectors.

4. Find the average proximities $\phi_{avg}(i)$ between all feature vectors in X along all components $i = 1, \ldots, l$. It is clear that for some vectors $\boldsymbol{x}$ the ith component is not available. In that case, the proximities that include x_i are excluded from the computation of $\phi_{avg}(i)$. We define the proximity $\psi(x_i, y_i)$ between the ith components of $\boldsymbol{x}$ and $\boldsymbol{y}$ as $\phi_{avg}(i)$ if at least one of the x_i and y_i is not available, and as $\phi(x_i, y_i)$ if both x_i and y_i are available ($\phi(x_i, y_i)$ may be defined as in the previous case). Then,

$$\wp(\boldsymbol{x}, \boldsymbol{y}) = \sum_{i=1}^{l} \psi(x_i, y_i) \tag{11.43}$$

Example 11.8

Consider the set $X = \{\boldsymbol{x}_1, \boldsymbol{x}_2, \boldsymbol{x}_3, \boldsymbol{x}_4, \boldsymbol{x}_5\}$, where $\boldsymbol{x}_1 = [0, 0]^T$, $\boldsymbol{x}_2 = [1, *]^T$, $\boldsymbol{x}_3 = [0, *]^T$, $\boldsymbol{x}_4 = [2, 2]^T$, $\boldsymbol{x}_5 = [3, 1]^T$. The "$*$" means that the corresponding value is not available.

According to the second technique, we find the average value of the second feature, which is 1, and we substitute it for the "$*$"s. Then, we may use any of the proximity measures defined in the previous sections.

Assume now that we wish to find the distance between $\boldsymbol{x}_1$ and $\boldsymbol{x}_2$ using the third technique. We use the absolute difference as the distance between two scalars. Then $d(\boldsymbol{x}_1, \boldsymbol{x}_2) = \frac{2}{2-1} 1 = 2$. Similarly, $d(\boldsymbol{x}_2, \boldsymbol{x}_3) = \frac{2}{2-1} 1 = 2$.

Finally, if we choose the fourth of the techniques, we must first find the average of the distances between any two values of the second feature. We again use the absolute difference as the distance between two scalars. The distances between any two available values of the second feature are $|0-2| = 2$, $|0-1| = 1$, and $|2-1| = 1$, and the average is $4/3$. Thus, the distance between $\boldsymbol{x}_1$ and $\boldsymbol{x}_2$ is $d(\boldsymbol{x}_1, \boldsymbol{x}_2) = 1 + 4/3 = 7/3$.

11.2.3 Proximity Functions between a Point and a Set

In many clustering schemes, a vector $\boldsymbol{x}$ is assigned to a cluster C taking into account the proximity between $\boldsymbol{x}$ and C, $\wp(\boldsymbol{x}, C)$. There are two general directions for the definition of $\wp(\boldsymbol{x}, C)$. According to the first one, all points of C contribute to $\wp(\boldsymbol{x}, C)$. Typical examples of this case include:

- The *max proximity function:*

$$\wp_{\max}^{ps}(\boldsymbol{x}, C) = \max_{\boldsymbol{y} \in C} \wp(\boldsymbol{x}, \boldsymbol{y}) \tag{11.44}$$

- The *min proximity function:*

$$\wp_{\min}^{ps}(\boldsymbol{x}, C) = \min_{\boldsymbol{y} \in C} \wp(\boldsymbol{x}, \boldsymbol{y}) \tag{11.45}$$

- The *average proximity function:*

$$\wp_{avg}^{ps}(\boldsymbol{x}, C) = \frac{1}{n_C} \sum_{\boldsymbol{y} \in C} \wp(\boldsymbol{x}, \boldsymbol{y}) \tag{11.46}$$

  where n_C is the cardinality of C.

In these definitions, $\wp(\boldsymbol{x}, \boldsymbol{y})$ may be any proximity measure between two points.

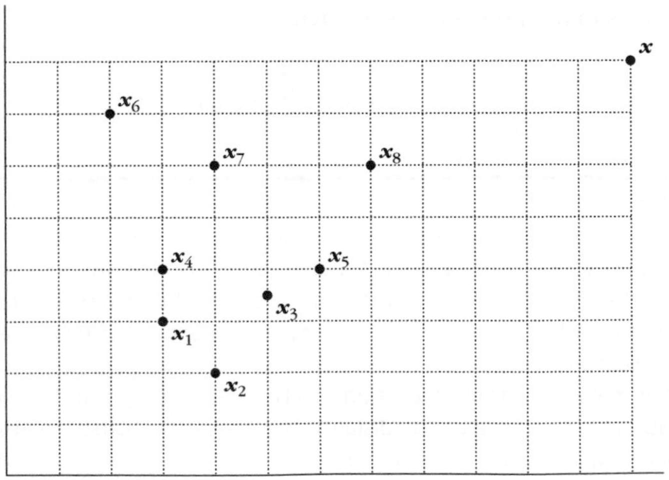

FIGURE 11.6

The setup of Example 11.9.

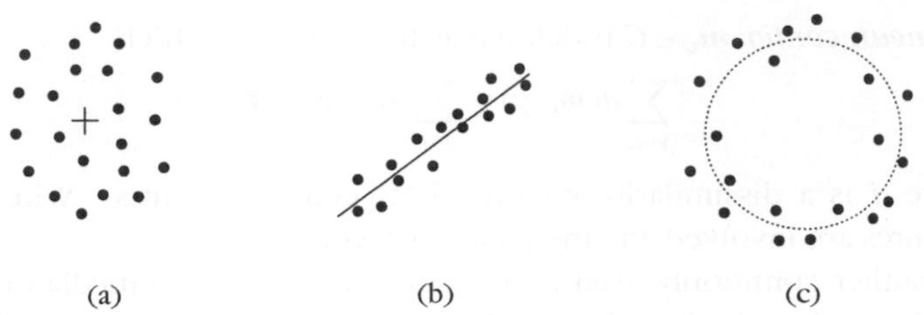

FIGURE 11.7

(a) Compact cluster. (b) Hyperplanar (linear) cluster. (c) Hyperspherical cluster.

Example 11.9

Let $C = \{x_1, x_2, x_3, x_4, x_5, x_6, x_7, x_8\}$, where $x_1 = [1.5, 1.5]^T$, $x_2 = [2, 1]^T$, $x_3 = [2.5, 1.75]^T$, $x_4 = [1.5, 2]^T$, $x_5 = [3, 2]^T$, $x_6 = [1, 3.5]^T$, $x_7 = [2, 3]^T$, $x_8 = [3.5, 3]^T$, and let $x = [6, 4]^T$ (see Figure 11.6). Assume that the Euclidean distance is used to measure the dissimilarity between two points. Then $d^{ps}_{\max}(x, C) = \max_{y \in C} d(x, y) = d(x, x_1) = 5.15$. For the other two distances we have $d^{ps}_{\min}(x, C) = \min_{y \in C} d(x, y) = d(x, x_8) = 2.69$ and $d^{ps}_{avg}(x, C) = \frac{1}{n_C} \sum_{y \in C} d(x, y) = \frac{1}{8} \sum_{i=1}^{8} d(x, x_i) = 4.33$.

According to the second direction, C is equipped with a representative and the proximity between x and C is measured as the proximity between x and the representative of C. Many types of representatives have been used in the literature. Among them, the point, the hyperplane, and the hypersphere are most commonly used.[3] Point representatives are suitable for compact clusters (Figure 11.7a) and hyperplane (hyperspherical) representatives for clusters of linear shape (Figure 11.7b) (hyperspherical shape, Figure 11.7c).

Point Representatives

Typical choices for a point representative of a cluster are:

- The *mean vector* (or *mean point*)

$$m_p = \frac{1}{n_C} \sum_{y \in C} y \tag{11.47}$$

where n_C is the cardinality of C. This is the most common choice when point representatives are employed, and we deal with data of a continuous space. However, it may not work well when we deal with points of a discrete space F^l. This is because it is possible for m_p to lie outside F^l. To cope with this problem, we may use the mean center m_c of C, which is defined next.

[3] In Chapter 14 we discuss the more general family of hyperquadric representatives, which include hyperellipsoids, hyperparabolas, and pairs of hyperplanes.

- The *mean center* $\boldsymbol{m}_c \in C$ is defined as the point for which

$$\sum_{\boldsymbol{y} \in C} d(\boldsymbol{m}_c, \boldsymbol{y}) \leq \sum_{\boldsymbol{y} \in C} d(\boldsymbol{z}, \boldsymbol{y}), \quad \forall \boldsymbol{z} \in C \tag{11.48}$$

where d is a dissimilarity measure between two points. When similarity measures are involved, the inequality is reversed.

Another commonly used point representative is the median center. It is usually employed when the proximity measure between two points is not a metric.

- The *median center* $\boldsymbol{m}_{\text{med}} \in C$ is defined as the point for which

$$\text{med}(d(\boldsymbol{m}_{\text{med}}, \boldsymbol{y})|\boldsymbol{y} \in C) \leq \text{med}(d(\boldsymbol{z}, \boldsymbol{y})|\boldsymbol{y} \in C), \quad \forall \boldsymbol{z} \in C \tag{11.49}$$

where d is a dissimilarity measure between two points. Here med(T), with T being a set of q scalars, is the minimum number in T that is greater than or equal to exactly $[(q+1)/2]$ numbers of T. An algorithmic way to determine med(T) is to list the elements of T in increasing order and to pick the $[(q+1)/2]$ element of that list.

Example 11.10

Let $C = \{\boldsymbol{x}_1, \boldsymbol{x}_2, \boldsymbol{x}_3, \boldsymbol{x}_4, \boldsymbol{x}_5\}$, where $\boldsymbol{x}_1 = [1, 1]^T$, $\boldsymbol{x}_2 = [3, 1]^T$, $\boldsymbol{x}_3 = [1, 2]^T$, $\boldsymbol{x}_4 = [1, 3]^T$, and $\boldsymbol{x}_5 = [3, 3]^T$ (see Figure 11.8). All points lie in the discrete space $\{0, 1, 2, \ldots, 6\}^2$. We use the Euclidean distance to measure the dissimilarity between two vectors in C. The mean point of C is $\boldsymbol{m}_p = [1.8, 2]^T$. It is clear that $\boldsymbol{m}_p$ lies outside the space where the elements of C belong.

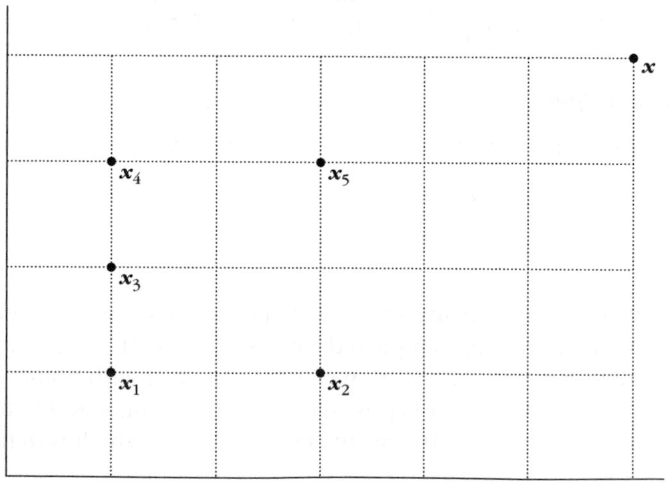

FIGURE 11.8

The setup of Example 11.10.

To find the mean center m_C, we compute, for each point $x_i \in C$, $i = 1, \ldots, 5$, the sum A_i of its distances from all other points of C. The resulting values are $A_1 = 7.83$, $A_2 = 9.06$, $A_3 = 6.47$, $A_4 = 7.83$, $A_5 = 9.06$. The minimum of these values is A_3. Thus, x_3 is the mean center of C.

Finally, for the computation of the median center m_{med} we work as follows. For each vector $x_i \in C$ we form the $n_C \times 1$ dimensional vector T_i of the distances between x_i and each of the vectors of C. Working as indicated, we identify $\text{med}(T_i)$, $i = 1, \ldots, 5$. Thus, $\text{med}(T_1) = \text{med}(T_2) = 2$, $\text{med}(T_3) = 1$, $\text{med}(T_4) = \text{med}(T_5) = 2$. Then we choose $\text{med}(T_j) = \min_{i=1,\ldots,n_C}\{\text{med}(T_i)\} = \text{med}(T_3)$, and we identify x_3 as the median vector of C. In our example, the mean center and the median center coincide. In general, however, this is not the case.

The distances between $x = [6, 4]^T$ and C when the mean point, the mean center, and the median center are used as representatives of C are 4.65, 5.39, and 5.39, respectively.

Hyperplane Representatives

Linear shaped clusters (or hyperplanar in the general case) are often encountered in computer vision applications. This type of cluster cannot be accurately represented by a single point. In such cases we use lines (hyperplanes) as representatives of the clusters (e.g., [Duda 01]).

The general equation of a hyperplane H is

$$\sum_{j=1}^{l} a_j x_j + a_0 = a^T x + a_0 = 0 \tag{11.50}$$

where $x = [x_1, \ldots, x_l]^T$ and $a = [a_1, \ldots, a_l]^T$ is the weight vector of H. The distance of a point x from H is defined as

$$d(x, H) = \min_{z \in H} d(x, z) \tag{11.51}$$

In the case of Euclidean distance between two points and using simple geometric arguments (see Figure 11.9a), we obtain

$$d(x, H) = \frac{|a^T x + a_0|}{\|a\|} \tag{11.52}$$

where $\|a\| = \sqrt{\sum_{j=1}^{l} a_j^2}$.

Hyperspherical Representatives

Clusters of another type are those that are circular (hyperspherical in higher dimensions). These are also frequently encountered in computer vision applications. For such clusters, the ideal representative is a circle (hypersphere).

The general equation of a hypersphere Q is

$$(x - c)^T (x - c) = r^2 \tag{11.53}$$

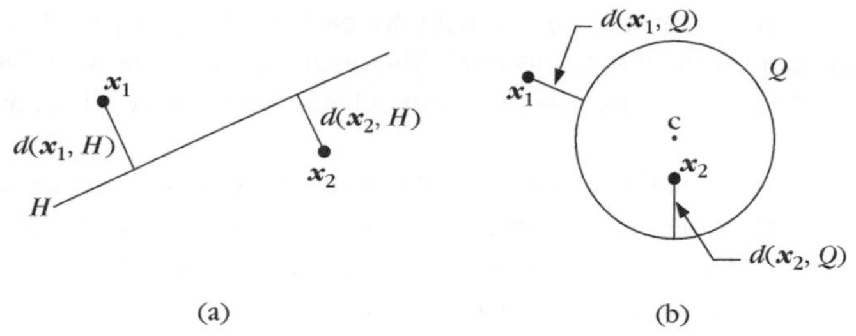

FIGURE 11.9

(a) Distance between a point and a hyperplane. (b) Distance between a point and hypersphere.

where $\boldsymbol{c}$ is the center of the hypersphere and r its radius. The distance from a point $\boldsymbol{x}$ to Q is defined as

$$d(\boldsymbol{x}, Q) = \min_{\boldsymbol{z} \in Q} d(\boldsymbol{x}, \boldsymbol{z}) \tag{11.54}$$

In most of the cases of interest, the Euclidean distance between two points is used in this definition. Figure 11.9b provides geometric insight into this definition. However, other nongeometric distances $d(\boldsymbol{x}, Q)$ have been used in the literature (e.g., [Dave 92, Kris 95, Frig 96]).

11.2.4 Proximity Functions between Two Sets

So far, we have been concerned with proximity measures between points in l-dimensional spaces and proximity functions between points and sets. Our major focus now is on defining proximity functions between sets of points. As we will soon see, some of the clustering algorithms are built upon such information. Most of the proximity functions $\wp^{ss}$ used for the comparison of sets are based on proximity measures, $\wp$, between vectors (see [Duda 01]). If D_i, D_j are two sets of vectors, the most common proximity functions are:

- The *max proximity function*:

$$\wp^{ss}_{\max}(D_i, D_j) = \max_{\boldsymbol{x} \in D_i, \boldsymbol{y} \in D_j} \wp(\boldsymbol{x}, \boldsymbol{y}) \tag{11.55}$$

  It is easy to see that if $\wp$ is a dissimilarity measure, $\wp^{ss}_{\max}$ is not a measure, since it does not satisfy the conditions in Section 11.2.1. $\wp^{ss}_{\max}$ is fully determined by the pair $(\boldsymbol{x}, \boldsymbol{y})$ of the most dissimilar (distant) vectors, with $\boldsymbol{x} \in D_i$ and $\boldsymbol{y} \in D_j$. On the other hand, if $\wp$ is a similarity measure, $\wp^{ss}_{\max}$ is a measure but it is not a metric (see Problem 11.12). In that case $\wp^{ss}_{\max}$ is fully determined by the pair $(\boldsymbol{x}, \boldsymbol{y})$ of the most similar (closest) vectors, with $\boldsymbol{x} \in D_i$ and $\boldsymbol{y} \in D_j$.

- The *min proximity function*:

$$\wp^{ss}_{\min}(D_i, D_j) = \min_{\boldsymbol{x} \in D_i, \boldsymbol{y} \in D_j} \wp(\boldsymbol{x}, \boldsymbol{y}) \tag{11.56}$$

When $\wp$ is a similarity measure, $\wp_{\min}^{ss}$ is not a measure. In this case $\wp_{\min}^{ss}$ is fully determined by the pair $(\boldsymbol{x}, \boldsymbol{y})$ of the most dissimilar (distant) vectors, with $\boldsymbol{x} \in D_i$ and $\boldsymbol{y} \in D_j$. On the other hand, if $\wp$ is a dissimilarity measure, $\wp_{\min}^{ss}$ is a measure, but it is not a metric (see Problem 11.12). In this case $\wp_{\min}^{ss}$ is fully determined by the pair $(\boldsymbol{x}, \boldsymbol{y})$ of the most similar (closest) vectors, with $\boldsymbol{x} \in D_i$ and $\boldsymbol{y} \in D_j$.

- The *average proximity function*:

$$\wp_{\text{avg}}^{ss}(D_i, D_j) = \frac{1}{n_{D_i} n_{D_j}} \sum_{\boldsymbol{x} \in D_i} \sum_{\boldsymbol{y} \in D_j} \wp(\boldsymbol{x}, \boldsymbol{y}) \tag{11.57}$$

where n_{D_i} and n_{D_j} are the cardinalities of D_i and D_j, respectively. It is easily shown that $\wp_{\text{avg}}^{ss}$ is not a measure even though $\wp$ is a measure. In this case, all vectors of both D_i and D_j contribute to the computation of $\wp_{\text{avg}}^{ss}$.

- The *mean proximity function*:

$$\wp_{\text{mean}}^{ss}(D_i, D_j) = \wp(\boldsymbol{m}_{D_i}, \boldsymbol{m}_{D_j}) \tag{11.58}$$

where $\boldsymbol{m}_{D_i}$ is the representative of D_i, $i = 1, 2$. For example, $\boldsymbol{m}_{D_i}$ may be the mean point, the mean center, or the median of D_i. Obviously, this is the proximity function between the representatives of D_i and D_j. It is clear that the mean proximity function is a measure provided that $\wp$ is a measure.

- Another proximity function that will be used later on is based on the mean proximity function and is defined as[4]

$$\wp_e^{ss}(D_i, D_j) = \sqrt{\frac{n_{D_i} n_{D_j}}{n_{D_i} + n_{D_j}}} \wp(\boldsymbol{m}_{D_i}, \boldsymbol{m}_{D_j}) \tag{11.59}$$

where $\boldsymbol{m}_{D_i}$ is defined as in the previous case.

In the last two alternatives we consider only the cases in which D_i's are represented by points. The need for a definition of a proximity function between two sets via their representatives, when the latter are not points, is of limited practical interest.

Example 11.11

(a) Consider the set $D_1 = \{\boldsymbol{x}_1, \boldsymbol{x}_2, \boldsymbol{x}_3, \boldsymbol{x}_4\}$ and $D_2 = \{\boldsymbol{y}_1, \boldsymbol{y}_2, \boldsymbol{y}_3, \boldsymbol{y}_4\}$, with $\boldsymbol{x}_1 = [0, 0]^T$, $\boldsymbol{x}_2 = [0, 2]^T$, $\boldsymbol{x}_3 = [2, 0]^T$, $\boldsymbol{x}_4 = [2, 2]^T$, $\boldsymbol{y}_1 = [-3, 0]^T$, $\boldsymbol{y}_2 = [-5, 0]^T$, $\boldsymbol{y}_3 = [-3, -2]^T$, $\boldsymbol{y}_4 = [-5, -2]^T$. The Euclidean distance is employed as the distance between two vectors. The distances between D_1 and D_2 according to the proximity functions just defined are $d_{\min}^{ss}(D_1, D_2) = 3$, $d_{\max}^{ss}(D_1, D_2) = 8.06$, $d_{\text{avg}}^{ss}(D_1, D_2) = 5.57$, $d_{\text{mean}}^{ss}(D_1, D_2) = 5.39$, $d_e^{ss}(D_1, D_2) = 7.62$.

[4] This definition is a generalization of that given in [Ward 63] (see Chapter 13).

(b) Consider now the set $D_2' = \{z_1, z_2, z_3, z_4\}$, with $z_1 = [1, 1.5]^T$, $z_2 = [1, 0.5]^T$, $z_3 = [0.5, 1]^T$, $z_4 = [1.5, 1]^T$. Notice that the points of D_1 and D_2' lie in two concentric circles centered at $[1, 1]^T$. The radius corresponding to D_1 (D_2') is $\sqrt{2}$ (0.5). The distances between D_1 and D_2' according to the proximity functions are $d_{\min}^{ss}(D_1, D_2') = 1.19$, $d_{\max}^{ss}(D_1, D_2') = 1.80$, $d_{avg}^{ss}(D_1, D_2') = 1.46$, $d_{mean}^{ss}(D_1, D_2') = 0$, $d_e^{ss}(D_1, D_2') = 0$.

Notice that in the last case, in which one of the sets lies in the convex hull of the other, some proximity measures may not be appropriate. For example, the measure based on the distances between the two means of the clusters gives meaningless results. However, this distance is well suited for cases in which the two sets are compact and well separated, especially because of its low computational requirements.

Notice that the proximities between two sets are built on proximities between two points. *Intuitively, one can understand that different choices of proximity functions between sets may lead to totally different clustering results.* Moreover, if we use different proximity measures between points, the same proximity function between sets will lead, in general, to different clustering results. *The only way to achieve proper clustering of the data is by trial and error and, of course, by taking into account the opinion of an expert in the field of application.*

Finally, proximity functions between a vector x and a set D_i may also be derived from the functions defined here, if we set $D_j = \{x\}$.

11.3 PROBLEMS

11.1 Let s be a metric similarity measure on X with $s(x,y) > 0$, $\forall x, y \in X$ and $d(x,y) = a/s(x,y)$, with $a > 0$. Prove that d is a metric dissimilarity measure.

11.2 Prove that the Euclidean distance satisfies the triangular inequality.
Hint: Use the Minkowski inequality, which states that for a positive integer p and two vectors $x = [x_1, \ldots, x_l]^T$ and $y = [y_1, \ldots, y_l]^T$ it holds that

$$\left(\sum_{i=1}^{l} |x_i + y_i|^p\right)^{1/p} \leq \left(\sum_{i=1}^{l} |x_i|^p\right)^{1/p} + \left(\sum_{i=1}^{l} |y_i|^p\right)^{1/p}$$

11.3 Show that:

a. if s is a metric similarity measure on a set X with $s(x,y) \geq 0$, $\forall x, y \in X$, then $s(x,y) + a$ is also a metric similarity measure on X, $\forall a \geq 0$.

b. If d is a metric dissimilarity measure on X, then $d + a$ is also a metric dissimilarity measure on X, $\forall a \geq 0$.

11.4 Let $f : \mathcal{R}^+ \to \mathcal{R}^+$ be a continuous monotonically increasing function such that

$$f(x) + f(y) \geq f(x+y), \quad \forall x, y \in \mathcal{R}^+$$

and let d be a metric dissimilarity measure on a set X with $d_0 \geq 0$. Show that $f(d)$ is also a metric dissimilarity measure on X.

11.5 Let s be a metric similarity measure on a set X, with $s(\boldsymbol{x},\boldsymbol{y}) > 0, \forall \boldsymbol{x},\boldsymbol{y} \in X$ and $f : \mathcal{R}^+ \rightarrow \mathcal{R}^+$ be a continuous monotonically decreasing function such that

$$f(x) + f(y) \geq f\left(\frac{1}{\frac{1}{x} + \frac{1}{y}}\right), \quad \forall x,y \in \mathcal{R}^+$$

Show that $f(s)$ is a metric dissimilarity measure on X.

11.6 Prove that

$$d_\infty(\boldsymbol{x},\boldsymbol{y}) \leq d_2(\boldsymbol{x},\boldsymbol{y}) \leq d_1(\boldsymbol{x},\boldsymbol{y})$$

for any two vectors $\boldsymbol{x}$ and $\boldsymbol{y}$ in X.

11.7 **a.** Prove that the maximum and the minimum values of $s_F(\boldsymbol{x},\boldsymbol{y})$ given in (11.40) are $l^{1/q}$ and 0, respectively.

b. Prove that as $q \rightarrow +\infty$, Eq. (11.40) results in $s_F(\boldsymbol{x},\boldsymbol{y}) = \max_{1 \leq i \leq l} s(x_i, y_i)$.

11.8 Examine whether the similarity functions defined by Eqs. (11.32), (11.33) are metric SMs.

11.9 Let d be a dissimilarity measure on X and $s = d_{\max} - d$ a corresponding similarity measure. Prove that

$$s^{ps}_{\text{avg}}(\boldsymbol{x}, C) = d_{\max} - d^{ps}_{\text{avg}}(\boldsymbol{x}, C), \quad \forall \boldsymbol{x} \in X, \quad C \subset X$$

where s^{ps}_{avg} and d^{ps}_{avg} are defined in terms of s and d, respectively. The definition of $\wp^{ps}_{\text{avg}}$ may be obtained from (11.57), where the first set consists of a single vector.

11.10 Let $\boldsymbol{x},\boldsymbol{y} \in \{0,1\}^l$. Prove that $d_2(\boldsymbol{x},\boldsymbol{y}) = \sqrt{d_{Hamming}(\boldsymbol{x},\boldsymbol{y})}$.

11.11 Consider two points in an l-dimensional space, $\boldsymbol{x} = [x_1,\ldots,x_l]^T$ and $\boldsymbol{y} = [y_1,\ldots,y_l]^T$, and let $|x_i - y_i| = \max_{j=1,\ldots,l}\{|x_j - y_j|\}$. We define the distance $d_n(\boldsymbol{x},\boldsymbol{y})$ as

$$d_n(\boldsymbol{x},\boldsymbol{y}) = |x_i - y_i| + \frac{1}{l - [(l-2)/2]} \sum_{j=1, j \neq i}^{l} |x_j - y_j|$$

This distance has been proposed in [Chau 92] as an approximation of the d_2 (Euclidean) distance.

a. Prove that d_n is a metric.

b. Compare d_n with d_2 in terms of computational complexity.

11.12 Let d and s be a dissimilarity and a similarity measure, respectively. Let $d_{\min}^{ss}$ ($s_{\min}^{ss}$), $d_{\max}^{ss}$ ($s_{\max}^{ss}$), d_{avg}^{ss} (s_{avg}^{ss}), d_{mean}^{ss} (s_{mean}^{ss}) be defined in terms of $d(s)$.

a. Prove that $d_{\min}^{ss}$, d_{mean}^{ss} are measures and $d_{\max}^{ss}$, d_{avg}^{ss} are not.

b. Prove that $s_{\max}^{ss}$, s_{mean}^{ss} are measures while $s_{\min}^{ss}$, s_{avg}^{ss} are not.

11.13 Based on Eqs. (11.55), (11.56), (11.57), and (11.58), derive the corresponding proximity functions between a point and a set. Are these proximity functions measures?

REFERENCES

[Ande 73] Anderberg M.R. *Cluster Analysis for Applications*, Academic Press, 1973.

[Ball 71] Ball G.H. "Classification analysis," Stanford Research Institute, *SRI Project 5533*, 1971.

[Broc 81] Brockett P.L., Haaland P.D., Levine A. "Information theoretic analysis of questionnaire data," *IEEE Transactions on Information Theory*, Vol. 27, pp. 438–445, 1981.

[Chau 92] Chaudhuri D., Murthy C.A., Chaudhuri B.B. "A modified metric to compute distance," *Pattern Recognition*, Vol. 25(7), pp. 667–677, 1992.

[Dave 92] Dave R.N., Bhaswan K. "Adaptive fuzzy c-shells clustering and detection of ellipses," *IEEE Transactions on Neural Networks*, Vol. 3(5), pp. 643–662, 1992.

[Dixo 79] Dixon J.K. "Pattern recognition with partly missing data," *IEEE Transactions on Systems Man and Cybernetics*, Vol. SMC 9, 617–621, 1979.

[Duda 01] Duda R.O., Hart P., Stork D. *Pattern Classification*, 2nd ed., John Wiley & Sons, 2001.

[Eise 98] Eisen M., Spellman P., Brown P., Botstein D. "Cluster analysis and display of genome-wide expression data," *Proceedings of National Academy of Science, USA*, Vol. 95, pp. 14863–14868, 1998.

[Ever 01] Everitt B., Landau S., Leesse M. *Cluster Analysis*, Arnold, 2001.

[Frig 96] Frigui H., Krishnapuram R. "A comparison of fuzzy shell clustering methods for the detection of ellipses," *IEEE Transactions on Fuzzy Systems*, Vol. 4(2), May 1996.

[Fu 93] Fu L., Yang M., Braylan R., Benson N. "Real-time adaptive clustering of flow cytometric data," *Pattern Recognition*, Vol. 26(2), pp. 365–373, 1993.

[Gers 92] Gersho A., Gray R.M. *Vector Quantization and Signal Compression*, Kluwer Academic Publishers, 1992.

[Good 66] Goodall D.W. "A new similarity index based on probability," *Biometrics*, Vol. 22, pp. 882–907, 1966.

[Gowe 67] Gower J.C. "A comparison of some methods of cluster analysis," *Biometrics*, Vol. 23, pp. 623–637, 1967.

[Gowe 71] Gower J.C. "A general coefficient of similarity and some of its properties," *Biometrics*, Vol. 27, pp. 857–872, 1971.

[Gowe 86] Gower J.C., Legendre P. "Metric and Euclidean properties of dissimilarity coefficients," *Journal of Classification*, Vol. 3, pp. 5–48, 1986.

[Hall 67] Hall A.V. "Methods for demonstrating resemblance in taxonomy and ecology," *Nature*, Vol. 214, pp. 830–831, 1967.

[Huba 82] Hubalek Z. "Coefficients of association and similarity based on binary (presence-absence) data—an evaluation," *Biological Review*, Vol. 57, pp. 669–689, 1982.

[Jain 88] Jain A.K., Dubes R.C. *Algorithms for Clustering Data*, Prentice Hall, 1988.

[John 67] Johnson S.C. "Hierarchical clustering schemes," *Psychometrika*, Vol. 32, pp. 241–254, 1967.

[Klir 95] Klir G., Yuan B. *Fuzzy sets and fuzzy logic*, Prentice Hall, 1995.

[Koho 89] Kohonen T. *Self-Organization and Associative Memory*, Springer-Verlag, 1989.

[Kris 95] Krishnapuram R., Frigui H., Nasraoui O. "Fuzzy and possibilistic shell clustering algorithms and their application to boundary detection and surface approximation—Part I," *IEEE Transactions on Fuzzy Systems*, Vol. 3(1), pp. 29–43, February 1995.

[Li 85] Li X., Dubes R.C. "The first stage in two-stage template matching," *IEEE Transactions on Pattern Analysis and Machine Intelligence*, Vol. 7, pp. 700–707, 1985.

[Lipp 87] Lippmann R.P. "An introduction to computing with neural nets," *IEEE ASSP Magazine*, Vol. 4(2), April 1987.

[Payk 72] Paykel E.S. "Depressive typologies and response to amitriptyline," *British Journal of Psychiatry*, Vol. 120, pp. 147–156, 1972.

[Snea 73] Sneath P.H.A., Sokal R.R. *Numerical Taxonomy*, W.H. Freeman & Co., 1973.

[Soka 63] Sokal R.R., Sneath P.H.A. *Principles of Numerical Taxonomy*, W.H. Freeman & Co., 1963.

[Spat 80] Spath H. *Cluster Analysis Algorithms*, Ellis Horwood, 1980.

[Tani 58] Tanimoto T. "An elementary mathematical theory of classification and prediction," *Int. Rpt.*, IBM Corp., 1958.

[Wall 68] Wallace C.S., Boulton D.M. "An information measure for classification," *Computer Journal*, Vol. 11, pp. 185–194, 1968.

[Ward 63] Ward J.H., Jr. "Hierarchical grouping to optimize an objective function," *Journal of the American Statistical Association.*, Vol. 58, pp. 236–244, 1963.

[Wind 82] Windham M.P. "Cluster validity for the fuzzy c-means clustering algorithm," *IEEE Transactions on Pattern Analysis and Machine Intelligence*, Vol. 4(4), pp. 357–363, 1982.

[Zade 65] Zadeh L.A. "Fuzzy sets," *Information and Control*, Vol. 8, pp. 338–353, 1965.

[Zade 73] Zadeh L.A. *IEEE Transactions on Systems Man and Cybernetics SMC-3*, Vol. 28, 1973.

CHAPTER

Clustering Algorithms I: Sequential Algorithms

12

12.1 INTRODUCTION

In the previous chapter, our major focus was on introducing a number of proximity measures. Each of these measures gives a different interpretation of the terms *similar* and *dissimilar*, associated with the types of clusters that our clustering procedure has to reveal. In the current and the following three chapters, the emphasis is on the various clustering algorithmic schemes and criteria that are available to the analyst. As has already been stated, different combinations of a proximity measure and a clustering scheme will lead to different results, which the expert has to interpret.

This chapter begins with a general overview of the various clustering algorithmic schemes and then focuses on one category, known as sequential algorithms.

12.1.1 Number of Possible Clusterings

Given the time and resources, the best way to assign the feature vectors x_i, $i = 1,\ldots,N$, of a set X to clusters would be to identify all possible partitions and to select the most sensible one according to a preselected criterion. However, this is not possible even for moderate values of N. Indeed, let $S(N,m)$ denote the number of all possible clusterings of N vectors into m groups. Remember that, by definition, no cluster is empty. It is clear that the following conditions hold [Spat 80, Jain 88]:

- $S(N,1) = 1$
- $S(N,N) = 1$
- $S(N,m) = 0, \quad \text{for } m > N$

Let L_{N-1}^k be the list containing all possible clusterings of the $N-1$ vectors into k clusters, for $k = m, m-1$. The Nth vector

- Either will be added to one of the clusters of any member of L_{N-1}^m
- Or will form a new cluster to each member of L_{N-1}^{m-1}

Thus, we may write

$$S(N, m) = mS(N-1, m) + S(N-1, m-1) \tag{12.1}$$

The solutions of (12.1) are the so-called Stirling numbers of the second kind (e.g., see [Liu 68]):[1]

$$S(N, m) = \frac{1}{m!} \sum_{i=0}^{m} (-1)^{m-i} \binom{m}{i} i^N \tag{12.2}$$

Example 12.1

Assume that $X = \{\boldsymbol{x}_1, \boldsymbol{x}_2, \boldsymbol{x}_3\}$. We seek to find all possible clusterings of the elements of X in two clusters. It is easy to deduce that

$$\mathcal{L}_2^1 = \{\{\boldsymbol{x}_1, \boldsymbol{x}_2\}\}$$

and

$$\mathcal{L}_2^2 = \{\{\boldsymbol{x}_1\}, \{\boldsymbol{x}_2\}\}$$

Taking into account (12.1), we easily find that $S(3, 2) = 2 \times 1 + = 3$. Indeed, the $\mathcal{L}_3^2$ list is

$$\mathcal{L}_3^2 = \{\{\boldsymbol{x}_1, \boldsymbol{x}_3\}, \{\boldsymbol{x}_2\}\}, \{\{\boldsymbol{x}_1\}, \{\boldsymbol{x}_2, \boldsymbol{x}_3\}\}, \{\{\boldsymbol{x}_1, \boldsymbol{x}_2\}, \{\boldsymbol{x}_3\}\}$$

Especially for $m = 2$, (12.2) becomes

$$S(N, 2) = 2^{N-1} - 1 \tag{12.3}$$

(see Problem 12.1). Some numerical values of (12.2) are [Spat 80]

- $S(15, 3) = 2375101$
- $S(20, 4) = 45232115901$
- $S(25, 8) = 690223721118368580$
- $S(100, 5) \simeq 10^{68}$

It is clear that these calculations are valid for the case in which the number of clusters is fixed. If this is not the case, one has to enumerate all possible clusterings for all possible values of m. From the preceding analysis, it is obvious that evaluating all of them to identify the most sensible one is impractical even for moderate values of N. Indeed, if, for example, one has to evaluate all possible clusterings of 100 objects into five clusters with a computer that evaluates each single clustering in 10^{-12} seconds, the most "sensible" clustering would be available after approximately 10^{48} years!

[1] Compare it with the number of dichotomies in Cover's theorem.

12.2 CATEGORIES OF CLUSTERING ALGORITHMS

Clustering algorithms may be viewed as schemes that provide us with sensible clusterings by considering *only a small fraction of the set containing all possible partitions of X. The result depends on the specific algorithm and the criteria used.* Thus a clustering algorithm is a learning procedure that tries to identify the specific characteristics of the clusters underlying the data set. Clustering algorithms may be divided into the following major categories.

- *Sequential algorithms.* These algorithms produce a single clustering. They are quite straightforward and fast methods. In most of them, all the feature vectors are presented to the algorithm once or a few times (typically no more than five or six times). The final result is, usually, dependent on the order in which the vectors are presented to the algorithm. These schemes tend to produce compact and hyperspherically or hyperellipsoidally shaped clusters, depending on the distance metric used. This category will be studied at the end of this chapter.
- *Hierarchical clustering algorithms.* These schemes are further divided into
  - *Agglomerative algorithms.* These algorithms produce a sequence of clusterings of decreasing number of clusters, m, at each step. The clustering produced at each step results from the previous one by merging two clusters into one. The main representatives of the agglomerative algorithms are the *single and complete link* algorithms. The agglomerative algorithms may be further divided into the following subcategories:
    - Algorithms that stem from the matrix theory
    - Algorithms that stem from graph theory

    These algorithms are appropriate for the recovery of elongated clusters (as is the case with the single link algorithm) and compact clusters (as is the case with the complete link algorithm).
  - *Divisive algorithms.* These algorithms act in the opposite direction; that is, they produce a sequence of clusterings of increasing m at each step. The clustering produced at each step results from the previous one by splitting a single cluster into two.
- *Clustering algorithms based on cost function optimization.* This category contains algorithms in which "sensible" is quantified by a cost function, J, in terms of which a clustering is evaluated. Usually, the number of clusters m is kept fixed. Most of these algorithms use differential calculus concepts and produce successive clusterings while trying to optimize J. They terminate when a local optimum of J is determined. Algorithms of this category are also called *iterative function optimization schemes.* This category includes the following subcategories:

- *Hard or crisp clustering algorithms*, where a vector belongs exclusively to a specific cluster. The assignment of the vectors to individual clusters is carried out optimally, according to the adopted optimality criterion. The most famous algorithm of this category is the Isodata or Lloyd algorithm [Lloy 82, Duda 01].
- *Probabilistic clustering algorithms*, are a special type of hard clustering algorithms that follow Bayesian classification arguments and each vector $\boldsymbol{x}$ is assigned to the cluster C_i for which $P(C_i|\boldsymbol{x})$ (i.e., the *a posteriori* probability) is maximum. These probabilities are estimated via an appropriately defined optimization task.
- *Fuzzy clustering algorithms*, where a vector belongs to a specific cluster up to a certain degree.
- *Possibilistic clustering algorithms*. In this case we measure the possibility for a feature vector $\boldsymbol{x}$ to belong to a cluster C_i.
- *Boundary detection algorithms*. Instead of determining the clusters by the feature vectors themselves, these algorithms adjust iteratively the boundaries of the regions where clusters lie. These algorithms, although they evolve from a cost function optimization philosophy, are different from the above algorithms. All the aforementioned schemes use cluster representatives, and the goal is to locate them in space in an optimal way. In contrast, boundary detection algorithms seek ways of placing optimally boundaries between clusters. This has led us to the decision to treat these algorithms in a separate chapter, together with algorithms to be discussed next.

■ *Other*: This last category contains some special clustering techniques that do not fit nicely in any of the previous categories. These include:

- *Branch and bound clustering algorithms*. These algorithms provide us with *the globally optimal clustering without having to consider all possible clusterings*, for fixed number m of clusters, and for a prespecified criterion that satisfies certain conditions. However, their computational burden is excessive.
- *Genetic clustering algorithms*. These algorithms use an initial population of possible clusterings and iteratively generate new populations, which, in general, contain better clusterings than those of the previous generations, according to a prespecified criterion.
- *Stochastic relaxation methods*. These are methods that guarantee, under certain conditions, convergence in probability to the globally optimum clustering, with respect to a prespecified criterion, at the expense of intensive computations.

It must be pointed out that stochastic relaxation methods (as well as genetic algorithms and branch and bound techniques) are cost function optimization methods. However, each follows a conceptually different approach to the problem compared to the methods of the previous category. This is why we chose to treat them separately.

- *Valley-seeking clustering algorithms.* These algorithms treat the feature vectors as instances of a (multidimensional) random variable $\boldsymbol{x}$. They are based on the commonly accepted assumption that regions of $\boldsymbol{x}$ where many vectors reside correspond to regions of increased values of the respective probability density function (pdf) of $\boldsymbol{x}$. Therefore, the estimation of the pdf may highlight the regions where clusters are formed.
- *Competitive learning algorithms.* These are iterative schemes that do not employ cost functions. They produce several clusterings and they converge to the most "sensible" one, according to a distance metric. Typical representatives of this category are the *basic competitive learning scheme* and the *leaky learning algorithm*.
- *Algorithms based on morphological transformation techniques*. These algorithms use morphological transformations in order to achieve better separation of the involved clusters.
- *Density-based algorithms.* These algorithms view the clusters as regions in the l-dimensional space that are "dense" in data. From this point of view there is an affinity with the valley-seeking algorithms. However, now the approach to the problem is achieved via an alternative route. Algorithmic variants within this family spring from the different way each of them quantifies the term *density*. Because most of them require only a few passes on the data set X (some of them consider the data points only once), they are serious candidates for processing large data sets.
- *Subspace clustering algorithms.* These algorithms are well suited for processing high-dimensional data sets. In some applications the dimension of the feature space can even be of the order of a few thousands. A major problem one has to face is the "curse of dimensionality" and one is forced to equip his/her arsenal with tools tailored for such demanding tasks.
- *Kernel-based methods.* The essence behind these methods is to adopt the "kernel trick," discussed in Chapter 4 in the context of nonlinear support vector machines, to perform a mapping of the original space, X, into a high-dimensional space and to exploit the nonlinear power of this tool.

Advances in database and Internet technologies over the past years have made data collection easier and faster, resulting in large and complex data sets with many patterns and/or dimensions ([Pars 04]). Such very large data sets are met, for example, in *Web mining*, where the goal is to extract knowledge from the Web ([Pier 03]). Two significant branches of this area are *Web content mining* (which aims at the extraction of useful knowledge from the content of Web pages) and *Web usage mining* (which aims at the discovery of interesting patterns of use by analyzing Web usage data). The sizes of web data are, in general, orders of magnitude larger than those encountered in more common clustering applications. Thus, the task of clustering Web pages in order to categorize them according to their content (Web content mining) or to categorize users according to the pages they visit most often (Web usage mining) becomes a very challenging problem. In addition, if in Web content mining each page is represented by a significant number of the words it contains, the dimension of the data space can become very high.

Another typical example of a computational resource-demanding clustering application comes from the area of *bioinformatics*, especially from *DNA microarray analysis*. This is a scientific field of enormous interest and significance that has already attracted a lot of research effort and investment. In such applications, data sets of dimensionality as high as 4000 can be encountered ([Pars 04]).

The need for efficient processing of data sets large in size and/or dimensionality has led to the development of clustering algorithms tailored for such complex tasks. Although many of these algorithms fall under the umbrella of one of the previously mentioned categories, we have chosen to discuss them separately at each related chapter to emphasize their specific focus and characteristics.

Several books—including [Ande 73, Dura 74, Ever 01, Gord 99, Hart 75, Jain 88, Kauf 90, and Spat 80]—are dedicated to the clustering problem. In addition, several survey papers on clustering algorithms have also been written. Specifically, a presentation of the clustering algorithms from a statistical point of view is given in [Jain 99]. In [Hans 97], the clustering problem is presented in a mathematical programming framework. In [Kola 01], applications of clustering algorithms for spatial database systems are discussed. Other survey papers are [Berk 02, Murt 83, Bara 99], and [Xu 05].

In addition, papers dealing with comparative studies among different clustering methods have also appeared in the literature. For example, in [Raub 00] the comparison of five typical clustering algorithms and their relative merits are discussed. Computationally efficient algorithms for large databases are compared in [Wei 00].

Finally, evaluations of different clustering techniques in the context of specific applications have also been conducted. For example, clustering applications for gene-expression data from DNA microarray experiments are discussed in [Jian 04, Made 04], and an experimental evaluation of document clustering techniques is given in [Stei 00].

12.3 SEQUENTIAL CLUSTERING ALGORITHMS

In this section we describe a basic sequential algorithmic scheme, (BSAS), (which is a generalization of that discussed in [Hall 67]), and we also give some variants of it. First, we consider the case where all the vectors are presented to the algorithm *only once*. *The number of clusters is not known a priori in this case*. In fact, new clusters are created as the algorithm evolves.

Let $d(\boldsymbol{x}, C)$ denote the distance (or dissimilarity) between a feature vector $\boldsymbol{x}$ and a cluster C. This may be defined by taking into account either all vectors of C or a representative vector of it (see Chapter 11). *The user-defined parameters required by the algorithmic scheme are the threshold of dissimilarity* Θ *and the maximum allowable number of clusters,* q. The basic idea of the algorithm is the following: As each new vector is considered, it is assigned either to an existing cluster or to a newly created cluster, depending on its distance from the already formed ones. Let m be the number of clusters that the algorithm has created up to now. Then the algorithmic scheme may be stated as:

Basic Sequential Algorithmic Scheme (BSAS)

- $m = 1$
- $C_m = \{\boldsymbol{x}_1\}$
- For $i = 2$ to N
  - Find C_k: $d(\boldsymbol{x}_i, C_k) = \min_{1 \leq j \leq m} d(\boldsymbol{x}_i, C_j)$.
  - If $(d(\boldsymbol{x}_i, C_k) > \Theta)$ AND $(m < q)$ then
    - $m = m + 1$
    - $C_m = \{\boldsymbol{x}_i\}$
  - Else
    - $C_k = C_k \cup \{\boldsymbol{x}_i\}$
    - Where necessary, update representatives[2]
  - End {if}
- End {For}

Different choices of $d(\boldsymbol{x}, C)$ lead to different algorithms, and any of the measures introduced in Chapter 11 can be employed. When C is represented by a single vector, $d(\boldsymbol{x}, C)$ becomes

$$d(\boldsymbol{x}, C) = d(\boldsymbol{x}, \boldsymbol{m}_C) \tag{12.4}$$

[2] This statement is activated in the cases where each cluster is represented by a single vector. For example, if each cluster is represented by its mean vector, this must be updated each time a new vector becomes a member of the cluster.

where m_C is the representative of C. In the case in which the mean vector is used as a representative, the updating may take place in an iterative fashion, that is,

$$m_{C_k}^{new} = \frac{(n_{C_k^{new}} - 1)m_{C_k}^{old} + x}{n_{C_k^{new}}} \tag{12.5}$$

where $n_{C_k^{new}}$ is the cardinality of C_k after the assignment of x to it and $m_{C_k}^{new}$ ($m_{C_k}^{old}$) is the representative of C_k after (before) the assignment of x to it (Problem 12.2).

It is not difficult to realize that the order in which the vectors are presented to the BSAS plays an important role in the clustering results. Different presentation ordering may lead to totally different clustering results, in terms of the number of clusters as well as the clusters themselves (see Problem 12.3).

Another important factor affecting the result of the clustering algorithm is the choice of the threshold Θ. This value directly affects the number of clusters formed by BSAS. If Θ is too small, unnecessary clusters will be created. On the other hand, if Θ is too large a smaller than appropriate number of clusters will be created. In both cases, the number of clusters that best fits the data set is missed.

If the number q of the maximum allowable number of clusters is not constrained, we leave it to the algorithm to "decide" about the appropriate number of clusters. Consider, for example, Figure 12.1, where three compact and well-separated clusters are formed by the points of X. If the maximum allowable number of clusters is set equal to two, the BSAS algorithm will be unable to discover three clusters. Probably, in this case the two rightmost groups of points will form a single cluster. On the other hand, if q is unconstrained, the BSAS algorithm will probably form three clusters (with an appropriate choice of Θ), at least for the case in which the mean vector is used as a representative. However, constraining q becomes necessary when dealing with implementations where the available computational resources are limited. In the next subsection, a simple technique is given for determining the number of clusters.[3]

FIGURE 12.1

Three clusters are formed by the feature vectors. When q is constrained to a value less than 3, the BSAS algorithm will not be able to reveal them.

[3] This problem is also treated in Chapter 16.

Remarks

- The BSAS scheme may be used with similarity instead of dissimilarity measures with appropriate modification; that is, the min operator is replaced by max.

- It turns out that BSAS, with point cluster representatives, favors compact clusters. Thus, it is not recommended if there is strong evidence that other types of clusters are present.

- The BSAS algorithm performs a single pass on the entire data set, X. For each iteration, the distance of the vector currently considered from each of the clusters defined so far is computed. Because the final number of clusters m is expected to be much smaller than N, the time complexity of BSAS is $O(N)$.

- The preceding algorithm is closely related to the algorithm implemented by the ART2 (adaptive resonance theory) neural architecture [Carp 87, Burk 91].

12.3.1 Estimation of the Number of Clusters

In this subsection, a simple method is described for determining the number of clusters (other such methods are discussed in Chapter 16). The method is suitable for BSAS as well as other algorithms, for which the number of clusters is not required as an input parameter. In what follows, BSAS(Θ) denotes the BSAS algorithm with a specific threshold of dissimilarity Θ.

- For $\Theta = a$ to b step c
  - Run s times the algorithm BSAS(Θ), each time presenting the data in a different order.
  - Estimate the number of clusters, m_Θ, as the most frequent number resulting from the s runs of BSAS(Θ).
- Next Θ

The values a and b are the minimum and maximum dissimilarity levels among all pairs of vectors in X, that is, $a = \min_{i,j=1,\ldots,N} d(\boldsymbol{x}_i, \boldsymbol{x}_j)$ and $b = \max_{i,j=1,\ldots,N} d(\boldsymbol{x}_i, \boldsymbol{x}_j)$. The choice of c is directly influenced by the choice of $d(\boldsymbol{x}, C)$. As far as the value of s is concerned, the greater the s, the larger the statistical sample and, thus, the higher the accuracy of the results. In the sequel, we plot the number of clusters m_Θ versus Θ. This plot has a number of flat regions. We estimate the number of clusters as the number that corresponds to the widest flat region. It is expected that at least for the case in which the vectors form well-separated compact clusters, this is the desired number. Let us explain this argument intuitively. Suppose that the data form two compact and well-separated clusters C_1 and C_2. Let the maximum distance between two vectors in C_1 (C_2) be r_1 (r_2) and suppose that $r_1 < r_2$. Also let r ($> r_2$) be the minimum among all distances $d(\boldsymbol{x}_i, \boldsymbol{x}_j)$, with $\boldsymbol{x}_i \in C_1$ and $\boldsymbol{x}_j \in C_2$. It is clear that for $\Theta \in [r_2, r - r_2]$, the number of clusters created by BSAS is 2. In addition, if $r \gg r_2$, the interval has a wide range, and thus it corresponds to a wide flat region in the plot of m_Θ versus Θ. Example 12.2 illustrates the idea.

Example 12.2

Consider two 2-dimensional Gaussian distributions with means $[0, 0]^T$ and $[20, 20]^T$, respectively. The covariance matrices are $\Sigma = 0.5I$ for both distributions, where I is the 2×2 identity matrix. Generate 50 points from each distribution (Figure 12.2a). The number of underlying clusters is 2. The plot resulting from the application of the previously described procedure is shown in Figure 12.2b, with $a = \min_{x_i, x_j \in X} d_2(x_i, x_j)$, $b = \max_{x_i, x_j \in X} d_2(x_i, x_j)$, and $c \simeq 0.3$. It can be seen that the widest flat region corresponds to the number 2, which is the number of underlying clusters.

In the foregoing procedure, we have implicitly assumed that the feature vectors do form clusters. If this is not the case, the method is useless. Methods that deal with the problem of discovering whether any clusters exist are discussed in Chapter 16. Moreover, if the vectors form compact clusters, which are not well separated, the procedure may give unreliable results, since it is unlikely for the plot of m_Θ versus Θ to contain wide flat regions.

In some cases, it may be advisable to consider all the numbers of clusters, m_Θ, that correspond to all flat regions of considerable size in the plot of m_Θ versus Θ. If, for example, we have three clusters and the first two of them lie close to each other and away from the third, the flattest region may occur for $m_\Theta = 2$ and the second flattest for $m_\Theta = 3$. If we discard the second flattest region, we will miss the three-cluster solution (Problem 12.6).

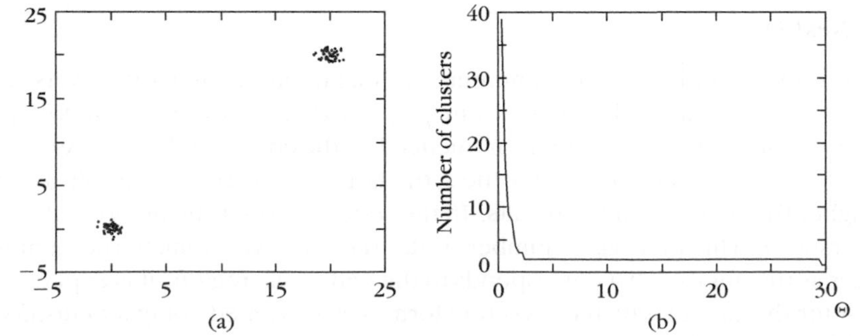

FIGURE 12.2

(a) The data set. (b) The plot of the number of clusters versus Θ. It can be seen that for a wide range of values of Θ, the number of clusters, m, is 2.

12.4 A MODIFICATION OF BSAS

As has already been stated, the basic idea behind BSAS is that each input vector $\boldsymbol{x}$ is assigned to an already created cluster or a new one is formed. Therefore, a decision for the vector $\boldsymbol{x}$ is reached *prior to the final cluster formation*, which is determined after all vectors have been presented. The following refinement of BSAS, which will be called modified BSAS (MBSAS), overcomes this drawback. The cost we pay for it is that the vectors of X have to be presented twice to the algorithm. The algorithmic scheme consists of two phases. The first phase involves the determination of the clusters, via the assignment of *some* of the vectors of X to them. During the second phase, the unassigned vectors are presented for a second time to the algorithm and are assigned to the appropriate cluster. The MBSAS may be written as follows:

Modified Basic Sequential Algorithmic Scheme (MBSAS)

- *Cluster Determination*
- $m = 1$
- $C_m = \{\boldsymbol{x}_1\}$
  - For $i = 2$ to N
  - Find C_k: $d(\boldsymbol{x}_i, C_k) = \min_{1 \leq j \leq m} d(\boldsymbol{x}_i, C_j)$.
  - If $(d(\boldsymbol{x}_i, C_k) > \Theta)$ AND $(m < q)$ then
    - $m = m + 1$
    - $C_m = \{\boldsymbol{x}_i\}$
  - End {if}
- End {For}

Pattern Classification

- For $i = 1$ to N
  - If $\boldsymbol{x}_i$ has not been assigned to a cluster, then
    - Find C_k: $d(\boldsymbol{x}_i, C_k) = \min_{1 \leq j \leq m} d(\boldsymbol{x}_i, C_j)$
    - $C_k = C_k \cup \{\boldsymbol{x}_i\}$
    - Where necessary, update representatives
  - End {if}
- End {For}

The number of clusters is determined in the first phase, and then it is frozen. Thus, the decision taken during the second phase for each vector takes into account all clusters.

When the mean vector of a cluster is used as its representative, the appropriate cluster representative has to be adjusted using Eq. (12.5), after the assignment of each vector in a cluster.

Also, as it was the case with BSAS, MBSAS is sensitive to the order in which the vectors are presented. In addition, because MBSAS performs two passes (one in each phase) on the data set X, it is expected to be slower than BSAS. However, its time complexity is of the same order; that is, $O(N)$.

Finally, it must be stated that, after minor modifications, MBSAS may be used when a similarity measure is employed (see Problem 12.7).

Another algorithm that falls under the MBSAS rationale is the so-called *maxmin* algorithm [Kats 94, Juan 00]. In the MBSAS scheme, a cluster is formed during the first pass, every time the distance of a vector from the already formed clusters is larger than a threshold. In contrast, the max-min algorithm follows a different strategy during the first phase. Let W be the set of all points that have been selected to form clusters, up to the current iteration step. To form a new cluster, we compute the distance of every point in $X - W$ from every point in W. If $\boldsymbol{x} \in X - W$, let $d_{\boldsymbol{x}}$ be the minimum distance of $\boldsymbol{x}$ from all the points in W. This is performed for all points in $X - W$. Then we select the point (say, $\boldsymbol{y}$) whose minimum distance (from the vectors in W) is maximum; that is,

$$d_{\boldsymbol{y}} = \max_{\boldsymbol{x}} d_{\boldsymbol{x}}, \quad \boldsymbol{x} \in X - W$$

If this is greater than a threshold, this vector forms a new cluster. Otherwise, the first phase of the algorithm terminates. It must be emphasized that in contrast to BSAS and MBSAS, the max-min algorithm employs a threshold that is data dependent. During the second pass, points that have not yet been assigned to clusters are assigned to the created clusters as in the MBSAS method. The max-min algorithm, although computationally more demanding than MBSAS, is expected to produce clusterings of better quality.

12.5 A TWO-THRESHOLD SEQUENTIAL SCHEME

As already has been pointed out, the results of BSAS and MBSAS are strongly dependent on the order in which the vectors are presented to the algorithm, as well as on the value of Θ. Improper choice of Θ may lead to meaningless clustering results. One way to overcome these difficulties is to define a "gray" region (see [Trah 89]). This is achieved by employing two thresholds, Θ_1 and $\Theta_2(>\Theta_1)$. If the dissimilarity level $d(\boldsymbol{x}, C)$ of a vector $\boldsymbol{x}$ from its closest cluster C is less than Θ_1, $\boldsymbol{x}$ is assigned to C. If $d(\boldsymbol{x}, C) > \Theta_2$, a new cluster is formed and $\boldsymbol{x}$ is placed in it. Otherwise, if $\Theta_1 \leq d(\boldsymbol{x}, C) \leq \Theta_2$, there exists uncertainty, and the assignment of $\boldsymbol{x}$ to a cluster will take place at a later stage. Let $clas(\boldsymbol{x})$ be a flag that indicates whether $\boldsymbol{x}$ has been classified (1) or not (0). Again, we denote by m the number of clusters that have been formed up to now. In the following, we assume no bounds to the number of clusters (i.e., $q = N$). The algorithmic scheme is:

The Two-Threshold Sequential Algorithmic Scheme (TTSAS)

$m = 0$
$clas(\boldsymbol{x}) = 0, \quad \forall \boldsymbol{x} \in X$
$prev\_change = 0$
$cur\_change = 0$
$exists\_change = 0$

While (there exists at least one feature vector $\boldsymbol{x}$ with $clas(\boldsymbol{x}) = 0$) do

- For $i = 1$ to N
  - if $clas(\boldsymbol{x}_i) = 0$ AND it is the first in the new while loop AND $exists\_change = 0$ then
    - $m = m + 1$
    - $C_m = \{\boldsymbol{x}_i\}$
    - $clas(\boldsymbol{x}_i) = 1$
    - $cur\_change = cur\_change + 1$
  - Else if $clas(\boldsymbol{x}_i) = 0$ then
    - Find $d(\boldsymbol{x}_i, C_k) = \min_{1 \leq j \leq m} d(\boldsymbol{x}_i, C_j)$
    - if $d(\boldsymbol{x}_i, C_k) < \Theta_1$ then
      - $C_k = C_k \cup \{\boldsymbol{x}_i\}$
      - $clas(\boldsymbol{x}_i) = 1$
      - $cur\_change = cur\_change + 1$
    - else if $d(\boldsymbol{x}_i, C_k) > \Theta_2$ then
      - $m = m + 1$
      - $C_m = \{\boldsymbol{x}_i\}$
      - $clas(\boldsymbol{x}_i) = 1$
      - $cur\_change = cur\_change + 1$
    - End {If}
  - Else if $clas(\boldsymbol{x}_i) = 1$ then
    - $cur\_change = cur\_change + 1$
  - End {If}

- End {For}
- *exists_change* = |*cur_change* − *prev_change*|
- *prev_change* = *cur_change*
- *cur_change* = 0

End {While}

The *exists_change* checks whether there exists at least one vector that has been classified at the current pass on X (i.e., the current iteration of the while loop). This is achieved by comparing the number of vectors that have been classified up to the current pass on X, *cur_change*, with the number of vectors that have been classified up to the previous pass on X, *prev_change*. If *exists_change* = 0, that is, no vector has been assigned to a cluster during the last pass on X, the first unclassified vector is used for the formation of a new cluster.

The first *if* condition in the *For* loop ensures that the algorithm terminates after N passes on X (N executions of the *while* loop) at the most. Indeed, this condition forces the first unassigned vector to a new cluster when no vector has been assigned during the last pass on X. This gives a way out to the case in which no vector has been assigned at a given circle.

However, in practice, the number of required passes is much less than N. It should be pointed out that this scheme is almost always at least as expensive as the previous two schemes, because in general it requires at least two passes on X. Moreover, since the assignment of a vector is postponed until enough information becomes available, it turns out that this algorithm is less sensitive to the order of data presentation.

As in the previous case, different choices of the dissimilarity between a vector and a cluster lead to different results. This algorithm also favors compact clusters, when used with point cluster representatives.

Remark

- Note that for all these algorithms no deadlock state occurs. That is, none of the algorithms enters into a state where there exist unassigned vectors that cannot be assigned either to existing clusters or to new ones, regardless of the number of passes of the data to the algorithm. The BSAS and MBSAS algorithms are guaranteed to terminate after a single and after two passes on X, respectively. In TTSAS the deadlock situation is avoided, as we arbitrarily assign the first unassigned vector at the current pass to a new cluster if no assignment of vectors occurred in the previous pass.

Example 12.3

Consider the vectors $\boldsymbol{x}_1 = [2, 5]^T$, $\boldsymbol{x}_2 = [6, 4]^T$, $\boldsymbol{x}_3 = [5, 3]^T$, $\boldsymbol{x}_4 = [2, 2]^T$, $\boldsymbol{x}_5 = [1, 4]^T$, $\boldsymbol{x}_6 = [5, 2]^T$, $\boldsymbol{x}_7 = [3, 3]^T$, and $\boldsymbol{x}_8 = [2, 3]^T$. The distance from a vector $\boldsymbol{x}$ to a cluster C

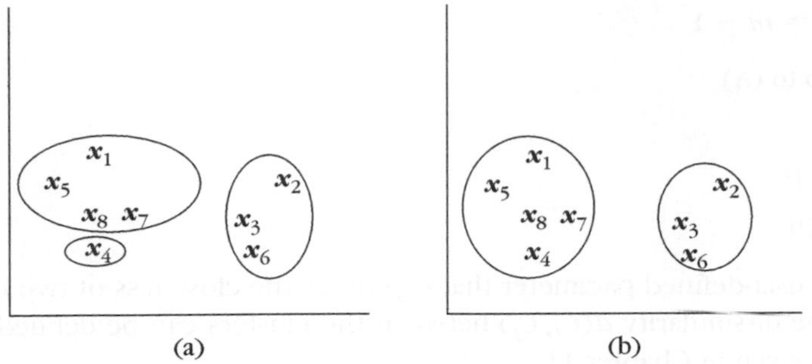

FIGURE 12.3

(a) The clustering produced by the MBSAS. (b) The clustering produced by the TTSAS.

is taken to be the Euclidean distance between $\boldsymbol{x}$ and the mean vector of C. If we present the vectors in the above order to the MBSAS algorithm and we set $\Theta = 2.5$, we obtain three clusters, $C_1 = \{\boldsymbol{x}_1, \boldsymbol{x}_5, \boldsymbol{x}_7, \boldsymbol{x}_8\}$, $C_2 = \{\boldsymbol{x}_2, \boldsymbol{x}_3, \boldsymbol{x}_6\}$, and $C_3 = \{\boldsymbol{x}_4\}$ (see Figure 12.3a).

On the other hand, if we present the vectors in the above order to the TTSAS algorithm, with $\Theta_1 = 2.2$ and $\Theta_2 = 4$, we obtain $C_1 = \{\boldsymbol{x}_1, \boldsymbol{x}_5, \boldsymbol{x}_7, \boldsymbol{x}_8, \boldsymbol{x}_4\}$ and $C_2 = \{\boldsymbol{x}_2, \boldsymbol{x}_3, \boldsymbol{x}_6\}$ (see Figure 12.3b). In this case, all vectors were assigned to clusters during the first pass on X, except $\boldsymbol{x}_4$. This was assigned to cluster C_1 during the second pass on X. At each pass on X, we had at least one vector assignment to a cluster. Thus, no vector is forced to a new cluster arbitrarily.

It is clear that the last algorithm leads to more reasonable results than MBSAS. However, it should be noted that MBSAS also leads to the same clustering if, for example, the vectors are presented with the following order: $\boldsymbol{x}_1, \boldsymbol{x}_2, \boldsymbol{x}_5, \boldsymbol{x}_3, \boldsymbol{x}_8, \boldsymbol{x}_6, \boldsymbol{x}_7, \boldsymbol{x}_4$.

12.6 REFINEMENT STAGES

In all the preceding algorithms, it may happen that two of the formed clusters are very closely located, and it may be desirable to merge them into a single one. Such cases cannot be handled by these algorithms. One way out of this problem is to run the following simple merging procedure, after the termination of the preceding schemes (see [Fu 93]).

Merging procedure

- (A) Find C_i, C_j $(i < j)$ such that $d(C_i, C_j) = \min_{k,r=1,\ldots,m,\, k \neq r} d(C_k, C_r)$
- If $d(C_i, C_j) \leq M_1$ then
  - Merge C_i, C_j to C_i and eliminate C_j.
  - Update the cluster representative of C_i (if cluster representatives are used).
  - Rename the clusters $C_{j+1}, \ldots, C_m$ to $C_j, \ldots, C_{m-1}$, respectively

- $m = m - 1$
- Go to (A)

■ Else
- Stop

■ End {If}

M_1 is a user-defined parameter that quantifies the closeness of two clusters, C_i and C_j. The dissimilarity $d(C_i, C_j)$ between the clusters can be defined using the definitions given in Chapter 11.

The other drawback of the sequential algorithms is their sensitivity to the order of presentation of vectors. Suppose, for example, that in using BSAS, $\boldsymbol{x}_2$ is assigned to the first cluster, C_1, and after the termination of the algorithm four clusters are formed. Then it is possible for $\boldsymbol{x}_2$ to be closer to a cluster different from C_1. However, there is no way for $\boldsymbol{x}_2$ to move to its closest cluster once assigned to another one. A simple way to face this problem is to use the following reassignment procedure:

Reassignment procedure

■ For $i = 1$ to N
- Find C_j such that $d(\boldsymbol{x}_i, C_j) = \min_{k=1,\ldots,m} d(\boldsymbol{x}_i, C_k)$.
- Set $b(i) = j$.

■ End {For}

■ For $j = 1$ to m
- Set $C_j = \{\boldsymbol{x}_i \in X\colon b(i) = j\}$.
- Update the representatives (if used).

■ End {For}

In this procedure, $b(i)$ denotes the closest to $\boldsymbol{x}_i$ cluster. This procedure may be used after the termination of the algorithms or, if the merging procedure is also used, after the termination of the merging procedure.

A variant of the BSAS algorithm combining the two refinement procedures has been proposed in [MacQ 67]. Only the case in which point representatives are used is considered. According to this algorithm, instead of starting with a single cluster, we start with $m > 1$ clusters, each containing one of the first m of the vectors in X. We apply the merging procedure and then we present each of the remaining vectors to the algorithm. After assigning the current vector to a cluster and updating its representative, we run the merging procedure again. If the distance between a vector $\boldsymbol{x}_i$ and its closest cluster is greater than a prespecified threshold, we form a new cluster which contains only $\boldsymbol{x}_i$. Finally, after all vectors have been presented to the algorithm, we run the reassignment procedure once. The merging procedure is applied $N - m + 1$ times. A variant of the algorithm is given in [Ande 73].

A different sequential clustering algorithm that requires a single pass on X is discussed in [Mant 85]. More specifically, it is assumed that the vectors are produced by a mixture of k Gaussian probability densities, $p(\boldsymbol{x}|C_i)$, that is,

$$p(\boldsymbol{x}) = \sum_{j=1}^{k} P(C_j)p(\boldsymbol{x}|C_j; \mu_j, \Sigma_j) \tag{12.6}$$

where μ_j and Σ_j are the mean and the covariance matrix of the jth Gaussian distribution, respectively. Also, $P(C_j)$ is the *a priori* probability for C_j. For convenience, let us assume that all $P(C_j)$'s are equal to each other. The clusters formed by the algorithm are assumed to follow the Gaussian distribution. At the beginning, a single cluster is formed using the first vector. Then, for each newly arrived vector, $\boldsymbol{x}_i$, the mean vector and covariance matrix of each of the m clusters, formed up to now, are appropriately updated and the conditional probabilities $P(C_j|\boldsymbol{x}_i)$ are estimated. If $P(C_q|\boldsymbol{x}_i) = \max_{j=1,\ldots,m} P(C_j|\boldsymbol{x}_i)$ is greater than a prespecifed threshold a, then $\boldsymbol{x}_i$ is assigned to C_q. Otherwise, a new cluster is formed where $\boldsymbol{x}_i$ is assigned. An alternative sequential clustering method that uses statistical tools is presented in [Amad 05].

12.7 NEURAL NETWORK IMPLEMENTATION

In this section, a neural network architecture is introduced and is then used to implement BSAS.

12.7.1 Description of the Architecture

The architecture is shown in Figure 12.4a. It consists of two modules, the matching score generator (MSG) and the MaxNet network (MN).[4]

The first module stores q parameter vectors[5] $\boldsymbol{w}_1, \boldsymbol{w}_2, \ldots, \boldsymbol{w}_q$ of dimension $l \times 1$ and implements a function $f(\boldsymbol{x}, \boldsymbol{w})$, which indicates the similarity between $\boldsymbol{x}$ and $\boldsymbol{w}$. The higher the value of $f(\boldsymbol{x}, \boldsymbol{w})$, the more similar $\boldsymbol{x}$ and $\boldsymbol{w}$ are.

When a vector $\boldsymbol{x}$ is presented to the network, the MSG module outputs a $q \times 1$ vector $\boldsymbol{v}$, with its ith coordinate being equal to $f(\boldsymbol{x}, \boldsymbol{w}_i)$, $i = 1, \ldots, q$.

The second module takes as input the vector $\boldsymbol{v}$ and identifies its maximum coordinate. Its output is a $q \times 1$ vector $\boldsymbol{s}$ with all its components equal to 0 except one that corresponds to the maximum coordinate of $\boldsymbol{v}$. This is set equal to 1. Most of the modules of this type require at least one coordinate of $\boldsymbol{v}$ to be positive.

Different implementations of the MSG can be used, depending on the proximity measure adopted. For example, if the function f is the inner product, the MSG

[4] This is a generalization of the Hamming network proposed in [Lipp 87].
[5] These are also called *exemplar patterns*.

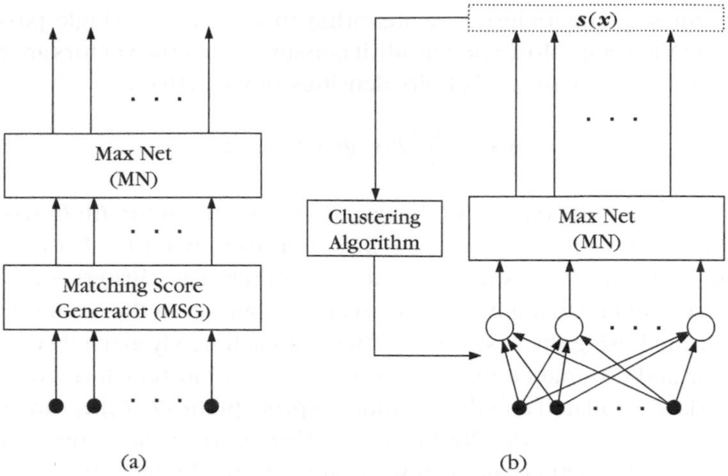

FIGURE 12.4

(a) The neural architecture. (b) Implementation of the BSAS algorithm when each cluster is represented by its mean vector and the Euclidean distance between two vectors is used.

module consists of q *linear* nodes with their threshold being equal to 0. Each of these nodes is associated with a parameter vector $\boldsymbol{w}_i$, and its output is the inner product of the input vector $\boldsymbol{x}$ with $\boldsymbol{w}_i$.

If the Euclidean distance is used, the MSG module also consists of q linear nodes. However, a different setup is required. The weight vector associated with the ith node is $\boldsymbol{w}_i$ and its threshold is set equal to $T_i = \frac{1}{2}(Q - \|\boldsymbol{w}_i\|^2)$, where Q is a positive constant that ensures that at least one of the first layer nodes will output a positive matching score, and $\|\boldsymbol{w}_i\|$ is the Euclidean norm of $\boldsymbol{w}_i$. Thus, the output of the node is

$$f(\boldsymbol{x}, \boldsymbol{w}_i) = \boldsymbol{x}^T \boldsymbol{w}_i + \frac{1}{2}(Q - \|\boldsymbol{w}_i\|^2) \tag{12.7}$$

It is easy to show that $d_2(\boldsymbol{x}, \boldsymbol{w}_i) < d_2(\boldsymbol{x}, \boldsymbol{w}_j)$ is equivalent to $f(\boldsymbol{x}, \boldsymbol{w}_i) > f(\boldsymbol{x}, \boldsymbol{w}_j)$ and thus the output of MSG corresponds to the $\boldsymbol{w}_i$ with the minimum Euclidean distance from $\boldsymbol{x}$ (see Problem 12.8).

The MN module can be implemented via a number of alternatives. One can use either neural network comparators such as the Hamming MaxNet, its generalizations and other feed-forward architectures [Lipp 87, Kout 95, Kout 05, Kout 98] or conventional comparators [Mano 79].

12.7.2 Implementation of the BSAS Algorithm

In this section, we demonstrate how the BSAS algorithm can be mapped to the neural network architecture when (a) each cluster is represented by its mean vector and (b) the Euclidean distance between two vectors is used (see Figure 12.4b). The structure of the Hamming network must also be slightly modified, so that each node in the first layer to has as an extra input the term $-\frac{1}{2}\|\boldsymbol{x}\|^2$. Let $\boldsymbol{w}_i$ and T_i be the weight vector and the threshold of the ith node in the MSG module, respectively. Also let $\boldsymbol{a}$ be a $q \times 1$ vector whose ith component indicates the number of vectors contained in the ith cluster. Also, let $\boldsymbol{s}(\boldsymbol{x})$ be the output of the MN module when the input to the network is $\boldsymbol{x}$. In addition, let t_i be the connection between the ith node of the MSG and its corresponding node in the MN module. Finally, let $sgn(z)$ be the step function that returns 1 if $z > 0$ and 0 otherwise.

The first m of the q $\boldsymbol{w}_i$'s correspond to the representatives of the clusters defined so far by the algorithm. At each iteration step either one of the first m $\boldsymbol{w}_i$'s is updated or a new parameter vector $\boldsymbol{w}_{m+1}$ is employed, whenever a new cluster is created (if $m < q$). The algorithm may be stated as follows.

- Initialization
  - $\boldsymbol{a} = \boldsymbol{0}$
  - $\boldsymbol{w}_i = \boldsymbol{0}, i = 1, \ldots, q$
  - $t_i = 0, i = 1, \ldots, q$
  - $m = 1$
  - For the first vector $\boldsymbol{x}_1$ set
    - $\boldsymbol{w}_1 = \boldsymbol{x}_1$
    - $a_1 = 1$
    - $t_1 = 1$
- Main Phase
  - Repeat
    - Present the next vector $\boldsymbol{x}$ to the network
    - Compute the output vector $\boldsymbol{s}(\boldsymbol{x})$
    - $GATE(\boldsymbol{x}) = AND((1 - \sum_{j=1}^{q}(s_j(\boldsymbol{x}))), sgn(q - m))$
    - $m = m + GATE(\boldsymbol{x})$
    - $a_m = a_m + GATE(\boldsymbol{x})$
    - $\boldsymbol{w}_m = \boldsymbol{w}_m + GATE(\boldsymbol{x})\boldsymbol{x}$
    - $T_m = \Theta - \frac{1}{2}\|\boldsymbol{w}_m\|^2$
    - $t_m = 1$
    - For $j = 1$ to m
      - $a_j = a_j + (1 - GATE(\boldsymbol{x}))s_j(\boldsymbol{x})$
      - $\boldsymbol{w}_j = \boldsymbol{w}_j - (1 - GATE(\boldsymbol{x}))s_j(\boldsymbol{x})(\frac{1}{a_j}(\boldsymbol{w}_j - \boldsymbol{x}))$
      - $T_j = \Theta - \frac{1}{2}\|\boldsymbol{w}_j\|^2$

- ∘ Next j
- Until all vectors have been presented once to the network

Note that only the outputs of the m first nodes of the MSG module are taken into account, because only these correspond to clusters. The outputs of the remaining nodes are not taken into account, since $t_k = 0, k = m+1, \ldots, q$. Assume that a new vector is presented to the network such that $\min_{1 \le j \le m} d(\boldsymbol{x}, \boldsymbol{w}_j) > \Theta$ and $m < q$. Then $GATE(\boldsymbol{x}) = 1$. Therefore, a new cluster is created and the next node is activated in order to represent it. Since $1 - GATE(\boldsymbol{x}) = 0$, the execution of the instructions in the *For* loop does not affect any of the parameters of the network.

Suppose next that $GATE(\boldsymbol{x}) = 0$. This is equivalent to the fact that either $\min_{1 \le j \le m} d(\boldsymbol{x}, \boldsymbol{w}_j) \le \Theta$ or there are no more nodes available to represent additional clusters. Then the execution of the instructions in the For loop results in updating the weight vector and the threshold of the node, k, for which $d(\boldsymbol{x}, \boldsymbol{w}_k) = \min_{1 \le j \le m} d(\boldsymbol{x}, \boldsymbol{w}_j)$. This happens because $s_k(\boldsymbol{x}) = 1$ and $s_j(\boldsymbol{x}) = 0, j = 1, \ldots, q$, $j \neq k$.

12.8 PROBLEMS

12.1 Prove Eq. (12.3) using induction.

12.2 Prove Eq. (12.5).

12.3 This problem aims at the investigation of the effects of the ordering of presentation of the vectors in the BSAS and MBSAS algorithms. Consider the following two-dimensional vectors: $\boldsymbol{x}_1 = [1, 1]^T$, $\boldsymbol{x}_2 = [1, 2]^T$, $\boldsymbol{x}_3 = [2, 2]^T$, $\boldsymbol{x}_4 = [2, 3]^T$, $\boldsymbol{x}_5 = [3, 3]^T$, $\boldsymbol{x}_6 = [3, 4]^T$, $\boldsymbol{x}_7 = [4, 4]^T$, $\boldsymbol{x}_8 = [4, 5]^T$, $\boldsymbol{x}_9 = [5, 5]^T$, $\boldsymbol{x}_{10} = [5, 6]^T$, $\boldsymbol{x}_{11} = [-4, 5]^T$, $\boldsymbol{x}_{12} = [-3, 5]^T$, $\boldsymbol{x}_{13} = [-4, 4]^T$, $\boldsymbol{x}_{14} = [-3, 4]^T$. Also consider the case that each cluster is represented by its mean vector.

a. Run the BSAS and the MBSAS algorithms when the vectors are presented in the given order. Use the Euclidean distance between two vectors and take $\Theta = \sqrt{2}$.

b. Change the order of presentation to $\boldsymbol{x}_1, \boldsymbol{x}_{10}, \boldsymbol{x}_2, \boldsymbol{x}_3, \boldsymbol{x}_4, \boldsymbol{x}_{11}, \boldsymbol{x}_{12}, \boldsymbol{x}_5, \boldsymbol{x}_6, \boldsymbol{x}_7, \boldsymbol{x}_{13}, \boldsymbol{x}_8, \boldsymbol{x}_{14}, \boldsymbol{x}_9$ and rerun the algorithms.

c. Run the algorithms for the following order of presentation: $\boldsymbol{x}_1, \boldsymbol{x}_{10}, \boldsymbol{x}_5, \boldsymbol{x}_2, \boldsymbol{x}_3, \boldsymbol{x}_{11}, \boldsymbol{x}_{12}, \boldsymbol{x}_4, \boldsymbol{x}_6, \boldsymbol{x}_7, \boldsymbol{x}_{13}, \boldsymbol{x}_{14}, \boldsymbol{x}_8, \boldsymbol{x}_9$.

d. Plot the given vectors and discuss the results of these runs.

e. Perform a visual clustering of the data. How many clusters do you claim are formed by the given vectors?

12.4 Consider the setup of Example 12.2. Run BSAS and MBSAS algorithms, with $\Theta = 5$, using the mean vector as representative for each cluster. Discuss the results.

12.5 Consider Figure 12.5. The inner square has side $S_1 = 0.3$, and the sides of the inner and outer square of the outer frame are $S_2 = 1$ and $S_3 = 1.3$, respectively. The inner square contains 50 points that stem from a uniform distribution in the square. Similarly, the outer frame contains 50 points that stem from a uniform distribution in the frame.

a. Perform a visual clustering of the data. How many clusters do you claim are formed by the given points?

b. Consider the case in which each cluster is represented by its mean vector and the Euclidean distance between two vectors is employed. Run BSAS and MBSAS algorithms, with

$$\Theta = \min_{i,j=1,\ldots,100} d(\boldsymbol{x}_i, \boldsymbol{x}_j), \quad \text{to} \quad \max_{i,j=1,\ldots,100} d(\boldsymbol{x}_i, \boldsymbol{x}_j) \text{ with step } 0.2$$

and with random ordering of the data. Give a quantitative explanation for the results. Compare them with the results obtained from the previous problem.

c. Repeat (b) for the case in which d_{min}^{ps} is chosen as the dissimilarity between a vector and a cluster (see Chapter 11).

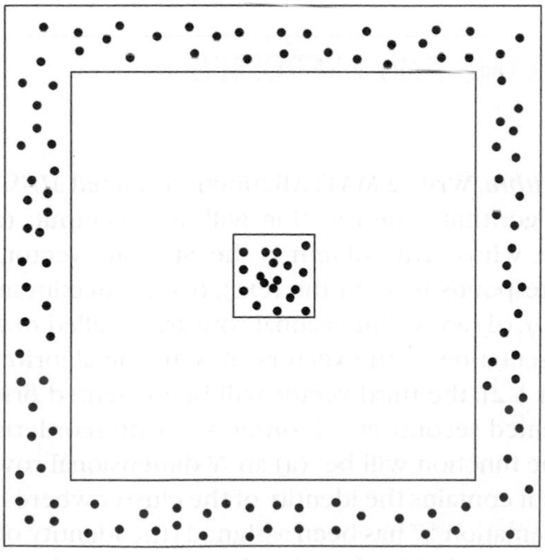

FIGURE 12.5

The setup of Problem 12.5.

12.6 Consider three two-dimensional Gaussian distributions with means $[0,0]^T$, $[6,0]^T$ and $[12,6]^T$, respectively. The covariance matrices for all distributions are equal to the identity matrix I. Generate 30 points from each distribution and let X be the resulting data set. Employ the Euclidean distance and apply the procedure discussed in Section 12.3.1 for the estimation of the number of clusters underlying in X, with $a = \min_{i,j=1,\dots,100} d(\boldsymbol{x}_i, \boldsymbol{x}_j)$, $b = \max_{i,j=1,\dots,100} d(\boldsymbol{x}_i, \boldsymbol{x}_j)$ and $c = 0.3$. Plot m versus Θ and draw your conclusions.

12.7 Let s be a similarity measure between a vector and a cluster. Express the BSAS, MBSAS, and TTSAS algorithms in terms of s.

12.8 Show that when the Euclidean distance between two vectors is in use and the output function of the MSG module is given by Eq. (12.7), the relations $d_2(\boldsymbol{x}, \boldsymbol{w}_1) < d_2(\boldsymbol{x}, \boldsymbol{w}_2)$ and $f(\boldsymbol{x}, \boldsymbol{w}_1) > f(\boldsymbol{x}, \boldsymbol{w}_2)$ are equivalent.

12.9 Describe a neural network implementation similar to the one given in Section 12.7 for the BSAS algorithm when each cluster is represented by the first vector assigned to it.

12.10 The neural network architecture that implements the MBSAS algorithm, if the mean vector is in use, is similar to the one given in Figure 12.4b for the Euclidean distance case. Write the algorithm in a form similar to the one given in Section 12.7 for the MBSAS when the mean vector is in use, and highlight the differences between the two implementations.

MATLAB PROGRAMS AND EXERCISES

Computer Programs

12.1 *MBSAS algorithm.* Write a MATLAB function, named *MBSAS*, that implements the MBSAS algorithm. The function will take as input: (a) an $l \times N$ dimensional matrix, whose ith column is the i-th data vector, (b) the parameter *theta* (it corresponds to Θ in the text), (c) the maximum number of allowable clusters q, (d) an N-dimensional row array, called *order*, that defines the order of presentation of the vectors of X to the algorithm. For example, if *order* = [3 4 1 2], the third vector will be presented first, the fourth vector will be presented second, etc. If *order* = [ ], no reordering takes place. The outputs of the function will be: (a) an N-dimensional row vector *bel*, whose ith component contains the identity of the cluster where the data vector with order of presentation "i" has been assigned (the identity of a cluster is an integer in $\{1, 2, \dots, n\_clust\}$, where n_clust is the number of clusters) and (b) an $l \times n\_clust$ matrix m whose i-th row is the cluster representative of the i-th cluster. Use the Euclidean distance to measure the distance between two vectors.

Solution

In the following code, do not type the asterisks. They will be used later on for reference purposes.

```
function [bel, m]=MBSAS(X,theta,q,order)
  % Ordering the data
  [l,N]=size(X);
  if(length(order)==N)
    X1=[];
    for i=1:N
      X1=[X1 X(:,order(i))];
    end
    X=X1;
    clear X1
  end
  % Cluster determination phase
  n_clust=1; % no. of clusters
  [l,N]=size(X);
  bel=zeros(1,N);
  bel(1)=n_clust;
  m=X(:,1);
  for i=2:N
    [m1,m2]=size(m);
    % Determining the closest cluster representative
    [s1,s2]=min(sqrt(sum((m-X(:,i)*ones(1,m2)).^ 2)));
    if(s1>theta) && (n_clust<q)
      n_clust=n_clust+1;
      bel(i)=n_clust;
      m=[m X(:,i)];
    end(*1)
  end(*2)
  [m1,m2]=size(m);(*3)
  % Pattern classification phase(*4)
  for i=1:N(*5)
    if(bel(i)==0)(*6)
      % Determining the closest cluster representative(*7)
      [s1,s2]=min(sqrt(sum((m-X(:,i)*ones(1,m2)).^ 2)));(*8)
      bel(i)=s2;
      m(:,s2)=((sum(bel==s2)-1)*m(:,s2) +
X(:,i))/sum(bel==s2);
    end
  end
```

12.2 *BSAS algorithm.* Write a MATLAB function, named *BSAS*, that implements the BSAS algorithm. Its inputs and outputs are defined exactly as in the *MBSAS* function.

Solution

In the code given for MBSAS replace the line with (*1) with the command

```
else
```

and remove all the other lines with asterisk.

Computer Experiments

12.1 Consider the data set $X = \{\boldsymbol{x}_1, \boldsymbol{x}_2, \boldsymbol{x}_3, \boldsymbol{x}_4, \boldsymbol{x}_5, \boldsymbol{x}_6, \boldsymbol{x}_7, \boldsymbol{x}_8\}$, where $\boldsymbol{x}_1 = [2, 5]^T$, $\boldsymbol{x}_2 = [8, 4]^T$, $\boldsymbol{x}_3 = [7, 3]^T$ $\boldsymbol{x}_4 = [2, 2]^T$, $\boldsymbol{x}_5 = [1, 4]^T$, $\boldsymbol{x}_6 = [7, 2]^T$, $\boldsymbol{x}_7 = [3, 3]^T$, $\boldsymbol{x}_8 = [2, 3]^T$. Plot the data vectors.

12.2 Run the MBSAS function for $q = 5$ on the above data set for

a. $order = [1, 5, 8, 4, 7, 3, 6, 2], theta = \sqrt{2} + 0.001$

b. $order = [5, 8, 1, 4, 7, 2, 3 6], theta = \sqrt{2} + 0.001$

c. $order = [1, 4, 5, 7, 8, 2, 3, 6], theta = 2.5$

d. $order = [1, 8, 4, 7, 5, 2, 3, 6], theta = 2.5$

e. the same order as in (c) and $theta = 3$

f. the same order as in (d) and $theta = 3$.

Study carefully the results and draw your conclusions.

12.3 Repeat 12.2 for BSAS.

REFERENCES

[Amad 05] Amador J.J. "Sequential clustering by statistical methodology," *Pattern Recognition Letters*, Vol. 26, pp. 2152–2163, 2005.

[Ande 73] Anderberg M.R. *Cluster Analysis for Applications*, Academic Press, 1973.

[Ball 65] Ball G.H. "Data analysis in social sciences," *Proceedings FJCC*, Las Vegas, 1965.

[Bara 99] Baraldi A., Blonda P. "A survey of fuzzy clustering algorithms for pattern recognition, Parts I and II," *IEEE Transactions on Systems, Man and Cybernetics, B. Cybernetics*, Vol. 29(6), pp. 778–801, 1999.

[Bara 99a] Baraldi A., Schenato L. "Soft-to-hard model transition in clustering: a review," Technical Report TR-99-010, 1999.

[Berk 02] Berkhin P. "Survey of clustering data mining techniques," *Technical Report*, Accrue Software Inc., 2002.

[Burk 91] Burke L.I. "Clustering characterization of adaptive reasonance," *Neural Networks*, Vol. 4, pp. 485–491, 1991.

[Carp 87] Carpenter G.A., Grossberg S. "ART2: Self-organization of stable category recognition codes for analog input patterns," *Applied Optics*, Vol. 26, pp. 4919–4930, 1987.

[Duda 01] Duda R.O., Hart P., Stork D. *Pattern Classification*, 2nd ed., John Wiley & Sons, 2001.

[Dura 74] Duran B., Odell P. *Cluster Analysis: A Survey*, Springer-Verlag, Berlin, 1974.

[Ever 01] Everitt B., Landau S., Leesse M. *Cluster Analysis*, Arnold, London, 2001.

[Flor 91] Floreen P. "The convergence of the Hamming memory networks," *IEEE Transactions on Neural Networks*, Vol. 2(4), pp. 449–459, July 1991.

[Fu 93] Fu L., Yang M., Braylan R., Benson N. "Real-time adaptive clustering of flow cytometric data," *Pattern Recognition*, Vol. 26(2), pp. 365–373, 1993.

[Gord 99] Gordon A. *Classification*, 2nd ed., Chapman & Hall, London, 1999.

[Hall 67] Hall A.V. "Methods for demonstrating resemblance in taxonomy and ecology," *Nature*, Vol. 214, pp. 830–831, 1967.

[Hans 97] Hansen P., Jaumard B. "Cluster analysis and mathematical programming," *Mathematical Programming*, Vol. 79, pp. 191–215, 1997.

[Hart 75] Hartigan J. *Clustering Algorithms*, John Wiley & Sons, 1975.

[Jain 88] Jain A.K., Dubes R.C. *Algorithms for Clustering Data*, Prentice Hall, 1988.

[Jain 99] Jain A., Muthy M., Flynn P. "Data clustering: A review," *ACM Computational Surveys*, Vol. 31(3), pp. 264–323, 1999.

[Jian 04] Jiang D., Tang C., Zhang A. "Cluster analysis for gene expression data: A survey," *IEEE Transactions on Knowledge Data Engineering*, Vol. 16(11), pp. 1370–1386, 2004.

[Juan 00] Juan A., Vidal E. "Comparison of four initialization techniques for the *k*-medians clustering algorithm," *Proceedings of Joint IAPR International Workshops SSPR2000 and SPR2000, Lecture Notes in Computer Science*, Vol. 1876, pp. 842–852, Springer-Verlag, Alacant (Spain), September 2000.

[Kats 94] Katsavounidis I., Jay Kuo C.-C., Zhang Z., "A new initialization technique for generalized Lloyd iteration," *IEEE Signal Processing Letters*, Vol. 1(10), pp. 144–146, 1994.

[Kauf 90] Kaufman L., Roussseeuw P. *Finding Groups in Data: An Introduction to Cluster Analysis*. John Wiley & Sons, 1990.

[Kola 01] Kolatch E. "Clustering algorithms for spatial databases: A survey," available at *http://citeseer.nj.nec.com/436843.html*.

[Kout 95] Koutroumbas K. "Hamming neural networks, architecture design and applications," Ph.D. thesis, Department of Informatics, University of Athens, 1995.

[Kout 94] Koutroumbas K., Kalouptsidis N. "Qualitative analysis of the parallel and asynchronous modes of the Hamming network," *IEEE Transactions on Neural Networks*, Vol. 5(3), pp. 380–391, May 1994.

[Kout 98] Koutroumbas K., Kalouptsidis N. "Neural network architectures for selecting the maximum input," *International Journal of Computer Mathematics*, Vol. 68(1–2), 1998.

[Kout 05] Koutroumbas K., Kalouptsidis N., "Generalized Hamming Networks and Applications," *Neural Networks*, Vol. 18, pp. 896–913, 2005.

[Lipp 87] Lippmann R.P. "An introduction to computing with neural nets," *IEEE ASSP Magazine*, Vol. 4(2), April 1987.

[Liu 68] Liu C.L. *Introduction to Combinatorial Mathematics*, McGraw-Hill, 1968.

[Lloy 82] Lloyd S.P. "Least squares quantization in PCM," *IEEE Transactions on Information Theory*, Vol. 28(2), pp. 129–137, March 1982.

[MacQ 67] MacQuenn J.B. "Some methods for classification and analysis of multivariate observations," *Proceedings of the Symposium on Mathematical Statistics and Probability*, 5th ed., Vol. 1, pp. 281–297, AD 669871, University of California Press, Berkeley, 1967.

[Made 04] Madeira S.C., Oliveira A.L. "Biclustering algorithms for biological data analysis: A survey," *IEEE/ACM Transactions on Computational Biology and Bioinformatics*, Vol. 1(1), pp. 24–45, 2004.

[Mano 79] Mano M. *Digital Logic and Computer Design*, Prentice Hall, 1979.

[Mant 85] Mantaras R.L., Aguilar-Martin J. "Self-learning pattern classification using a sequential clustering technique," *Pattern Recognition*, Vol. 18(3/4), pp. 271–277, 1985.

[Murt 83] Murtagh F. "A survey of recent advanced in hierarchical clustering algorithms," *Journal of Computation*, Vol. 26(4), pp. 354–359, 1983.

[Pars 04] Parsons L., Haque E., Liu H. "Subspace clustering for high dimensional data: A review," *ACM SIGKDD Explorations Newsletter*, Vol. 6(1), pp. 90–105, 2004.

[Pier 03] Pierrakos D., Paliouras G., Papatheodorou C., Spyropoulos C.D. "Web usage mining as a tool for personalization: A survey," *User Modelling and User-Adapted Interaction*, Vol. 13(4), pp. 311–372, 2003.

[Raub 00] Rauber A., Paralic J., Pampalk E. "Empirical evaluation of clustering algorithms," *Journal of Inf. Org. Sci.*, Vol. 24(2), pp. 195–209, 2000.

[Sebe 62] Sebestyen G.S. "Pattern recognition by an adaptive process of sample set construction," *IRE Transactions on Information Theory*, Vol. 8(5), pp. S82–S91, 1962.

[Snea 73] Sneath P.H.A., Sokal R.R. *Numerical Taxonomy*, W.H. Freeman, 1973.

[Spat 80] Spath H. *Cluster Analysis Algorithms*, Ellis Horwood, 1980.

[Stei 00] Steinbach M., Karypis G., Kumar V. "A comparison of document clustering techniques," *Technical Report*, 00-034, University of Minnesota, Minneapolis, 2000.

[Trah 89] Trahanias P., Scordalakis E. "An efficient sequential clustering method," *Pattern Recognition*, Vol. 22(4), pp. 449–453, 1989.

[Wei 00] Wei C., Lee Y., Hsu C. "Empirical comparison of fast clustering algorithms for large data sets," *Proceedings of the 33rd Hawaii International Conference on System Sciences*, pp. 1–10, Maui, HI, 2000.

[Xu 05] Xu R., Wunsch D. "Survey of clustering algorithms," *IEEE Transactions on Neural Networks*, Vol. 16(3), pp. 645–678, 2005.

CHAPTER 13

Clustering Algorithms II: Hierarchical Algorithms

13.1 INTRODUCTION

Hierarchical clustering algorithms are of different philosophy from the algorithms described in the previous chapter. Specifically, instead of producing a single clustering, they produce a hierarchy of clusterings. This kind of algorithm is usually found in the social sciences and biological taxonomy (e.g., [El-G 68, Prit 71, Shea 65, McQu 62]). In addition, they have been used in many other fields, including modern biology, medicine, and archaeology (e.g., [Stri 67, Bobe 93, Solo 71, Hods 71]). Applications of the hierarchical algorithms may also be found in computer science and engineering (e.g., [Murt 95, Kank 96]).

Before we describe their basic idea, let us recall that

$$X = \{\boldsymbol{x}_i,\ i = 1, \ldots, N\}$$

is a set of l-dimensional vectors that are to be clustered. Also, recall from Chapter 11 the definition of a clustering

$$\Re = \{C_j,\ j = 1, \ldots, m\}$$

where $C_j \subseteq X$.

A clustering $\Re_1$ containing k clusters is said to be *nested* in the clustering $\Re_2$, which contains $r(<k)$ clusters, if *each* cluster in $\Re_1$ is a subset of a set in $\Re_2$. Note that at least one cluster of $\Re_1$ is a proper subset of $\Re_2$. In this case we write $\Re_1 \sqsubset \Re_2$. For example, the clustering $\Re_1 = \{\{\boldsymbol{x}_1, \boldsymbol{x}_3\}, \{\boldsymbol{x}_4\}, \{\boldsymbol{x}_2, \boldsymbol{x}_5\}\}$ is nested in $\Re_2 = \{\{\boldsymbol{x}_1, \boldsymbol{x}_3, \boldsymbol{x}_4\}, \{\boldsymbol{x}_2, \boldsymbol{x}_5\}\}$. On the other hand, $\Re_1$ is nested neither in $\Re_3 = \{\{\boldsymbol{x}_1, \boldsymbol{x}_4\}, \{\boldsymbol{x}_3\}, \{\boldsymbol{x}_2, \boldsymbol{x}_5\}\}$ nor in $\Re_4 = \{\{\boldsymbol{x}_1, \boldsymbol{x}_2, \boldsymbol{x}_4\}, \{\boldsymbol{x}_3, \boldsymbol{x}_5\}\}$. It is clear that a clustering is not nested to itself.

Hierarchical clustering algorithms produce a *hierarchy of nested clusterings*. More specifically, these algorithms involve N steps, as many as the number of data vectors. At each step t, a new clustering is obtained based on the clustering produced at the previous step $t-1$. There are two main categories of these algorithms, the *agglomerative* and the *divisive hierarchical algorithms*.

The initial clustering $\Re_0$ for the agglomerative algorithms consists of N clusters, each containing a single element of X. At the first step, the clustering $\Re_1$ is produced. It contains $N-1$ sets, such that $\Re_0 \sqsubset \Re_1$. This procedure continues until the final clustering, $\Re_{N-1}$, is obtained, which contains a single set, that is, the set of data, X. Notice that for the hierarchy of the resulting clusterings, we have

$$\Re_0 \sqsubset \Re_1 \sqsubset \cdots \sqsubset \Re_{N-1}$$

The divisive algorithms follow the inverse path. In this case, the initial clustering $\Re_0$ consists of a single set, X. At the first step the clustering $\Re_1$ is produced. It consists of two sets, such that $\Re_1 \sqsubset \Re_0$. This procedure continues until the final clustering $\Re_{N-1}$ is obtained, which contains N sets, each consisting of a single element of X. In this case we have

$$\Re_{N-1} \sqsubset \Re_{N-2} \sqsubset \ldots, \sqsubset \Re_0$$

The next section is devoted to the agglomerative algorithms; the divisive algorithms are discussed briefly in Section 13.4.

13.2 AGGLOMERATIVE ALGORITHMS

Let $g(C_i, C_j)$ be a function defined for all possible pairs of clusters of X. This function measures the proximity between C_i and C_j. Let t denote the current level of hierarchy. Then, the general agglomerative scheme may be stated as follows:

Generalized Agglomerative Scheme (GAS)

- Initialization:
  - Choose $\Re_0 = \{C_i = \{\boldsymbol{x}_i\},\ i = 1, \ldots, N\}$ as the initial clustering.
  - $t = 0$.
- Repeat:
  - $t = t + 1$
  - Among all possible pairs of clusters (C_r, C_s) in $\Re_{t-1}$ find the one, say (C_i, C_j), such that

$$g(C_i, C_j) = \begin{cases} \min_{r,s} g(C_r, C_s), & \text{if } g \text{ is a dissimilarity function} \\ \max_{r,s} g(C_r, C_s), & \text{if } g \text{ is a similarity function} \end{cases} \tag{13.1}$$

  - Define $C_q = C_i \cup C_j$ and produce the new clustering $\Re_t = (\Re_{t-1} - \{C_i, C_j\}) \cup \{C_q\}$.
- Until all vectors lie in a single cluster.

It is clear that this scheme creates a hierarchy of N clusterings, so that each one is nested in all successive clusterings, that is, $\Re_{t_1} \sqsubset \Re_{t_2}$, for $t_1 < t_2$, $t_2 = 1, \ldots, N-1$. Alternatively, we can say that *if two vectors come together into a single cluster at level t of the hierarchy, they will remain in the same cluster for all subsequent clusterings*. This is another way of viewing the nesting property.

A disadvantage of the nesting property is that there is no way to recover from a "poor" clustering that may have occurred in an earlier level of the hierarchy (see [Gowe 67]).[1]

At each level t, there are $N-t$ clusters. Thus, in order to determine the pair of clusters that is going to be merged at the $t+1$ level, $\binom{N-t}{2} \equiv \frac{(N-t)(N-t-1)}{2}$ pairs of clusters have to be considered. Thus, the total number of pairs that have to be examined throughout the whole clustering process is

$$\sum_{t=0}^{N-1} \binom{N-t}{2} = \sum_{k=1}^{N} \binom{k}{2} = \frac{(N-1)N(N+1)}{6}$$

that is, the total number of operations required by an agglomerative scheme is proportional to N^3. However, the exact complexity of the algorithm depends on the definition of g.

13.2.1 Definition of Some Useful Quantities

There are two main categories of agglomerative algorithms. Algorithms of the first category are based on matrix theory concepts, while algorithms of the second one are based on graph theory concepts. Before we enter into their discussion, some definitions are required. The *pattern matrix* $D(X)$ is the $N \times l$ matrix, whose ith row is the (transposed) ith vector of X. The *similarity (dissimilarity) matrix*, $P(X)$, is an $N \times N$ matrix whose (i, j) element equals the similarity $s(\boldsymbol{x}_i, \boldsymbol{x}_j)$ (dissimilarity $d(\boldsymbol{x}_i, \boldsymbol{x}_j)$) between vectors $\boldsymbol{x}_i$ and $\boldsymbol{x}_j$. It is also referred to as the *proximity matrix* to include both cases. In general, P is a symmetric matrix.[2] Moreover, if P is a similarity matrix, its diagonal elements are equal to the maximum value of s. On the other hand, if P is a dissimilarity matrix, its diagonal elements are equal to the minimum value of d. Notice that for a single pattern matrix there exists more than one proximity matrix depending on the choice of the proximity measure $\wp(\boldsymbol{x}_i, \boldsymbol{x}_j)$. However, fixing $\wp(\boldsymbol{x}_i, \boldsymbol{x}_j)$, one can easily observe that for a given pattern matrix there exists an associated single proximity matrix. On the other hand, a proximity matrix may correspond to more than one pattern matrices (see Problem 13.1).

[1] A method that produces hierarchies, which do not, necessarily, possess the nesting property, has been proposed in [Frig 97].

[2] In [Ozaw 83] a hierarchical clustering algorithm, called RANCOR, is discussed, which is based on asymmetric proximity matrices.

Example 13.1

Let $X = \{\boldsymbol{x}_i, i = 1, \ldots, 5\}$, with $\boldsymbol{x}_1 = [1, 1]^T$, $\boldsymbol{x}_2 = [2, 1]^T$, $\boldsymbol{x}_3 = [5, 4]^T$, $\boldsymbol{x}_4 = [6, 5]^T$, and $\boldsymbol{x}_5 = [6.5, 6]^T$. The pattern matrix of X is

$$D(X) = \begin{bmatrix} 1 & 1 \\ 2 & 1 \\ 5 & 4 \\ 6 & 5 \\ 6.5 & 6 \end{bmatrix}$$

and its corresponding dissimilarity matrix, when the Euclidean distance is in use, is

$$P(X) = \begin{bmatrix} 0 & 1 & 5 & 6.4 & 7.4 \\ 1 & 0 & 4.2 & 5.7 & 6.7 \\ 5 & 4.2 & 0 & 1.4 & 2.5 \\ 6.4 & 5.7 & 1.4 & 0 & 1.1 \\ 7.4 & 6.7 & 2.5 & 1.1 & 0 \end{bmatrix}$$

When the Tanimoto measure is used, the similarity matrix of X becomes

$$P'(X) = \begin{bmatrix} 1 & 0.75 & 0.26 & 0.21 & 0.18 \\ 0.75 & 1 & 0.44 & 0.35 & 0.20 \\ 0.26 & 0.44 & 1 & 0.96 & 0.90 \\ 0.21 & 0.35 & 0.96 & 1 & 0.98 \\ 0.18 & 0.20 & 0.90 & 0.98 & 1 \end{bmatrix}$$

Note that in $P(X)$ all diagonal elements are 0, since $d_2(\boldsymbol{x}, \boldsymbol{x}) = 0$, while in $P'(X)$ all diagonal elements are equal to 1, since $s_T(\boldsymbol{x}, \boldsymbol{x}) = 1$.

A *threshold dendrogram*, or simply a *dendrogram*, is an effective means of representing the sequence of clusterings produced by an agglomerative algorithm. To clarify this idea, let us consider again the data set given in Example 13.1. Let us define $g(C_i, C_j)$ as $g(C_i, C_j) = d_{\min}^{ss}(C_i, C_j)$ (see Section 11.2). One may easily see that, in this case, the clustering sequence for X produced by the generalized agglomerative scheme, when the Euclidean distance between two vectors is used, is the one shown in Figure 13.1. At the first step $\boldsymbol{x}_1$ and $\boldsymbol{x}_2$ form a new cluster. At the second step $\boldsymbol{x}_4$ and $\boldsymbol{x}_5$ stick together, forming a single cluster. At the third step $\boldsymbol{x}_3$ joins the cluster $\{\boldsymbol{x}_4, \boldsymbol{x}_5\}$ and, finally, at the fourth step the clusters $\{\boldsymbol{x}_1, \boldsymbol{x}_2\}$ and $\{\boldsymbol{x}_3, \boldsymbol{x}_4, \boldsymbol{x}_5\}$ are merged into a single set, X. The right-hand side of Figure 13.1 shows the corresponding dendrogram. Each step of *the generalized agglomerative sheme (GAS)* corresponds to a level of the dendrogram. *Cutting the dendrogram at a specific level results in a clustering.*

A *proximity dendrogram* is a dendrogram that takes into account the level of proximity where two clusters are merged *for the first time*. When a dissimilarity (similarity) measure is in use, the proximity dendrogram is called a *dissimilarity (similarity) dendrogram*. This tool may be used as an indicator of the natural or

$\{\{x_1\},\{x_2\},\{x_3\},\{x_4\},\{x_5\}\}$

$\{\{x_1, x_2\},\{x_3\},\{x_4\},\{x_5\}\}$

$\{\{x_1, x_2\},\{x_3\},\{x_4, x_5\}\}$

$\{\{x_1, x_2\},\{x_3, x_4, x_5\}\}$

$\{\{x_1, x_2, x_3, x_4, x_5\}\}$

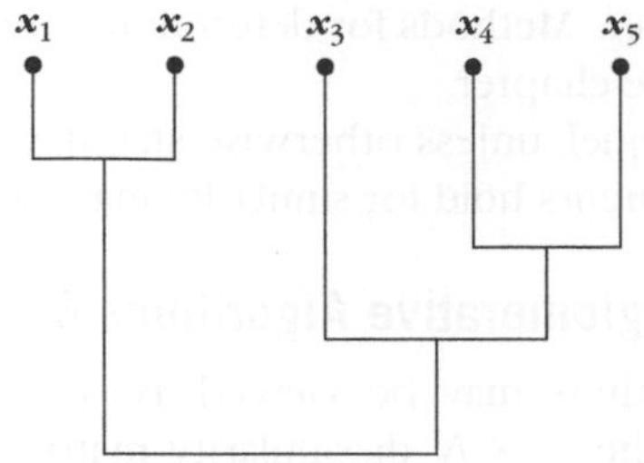

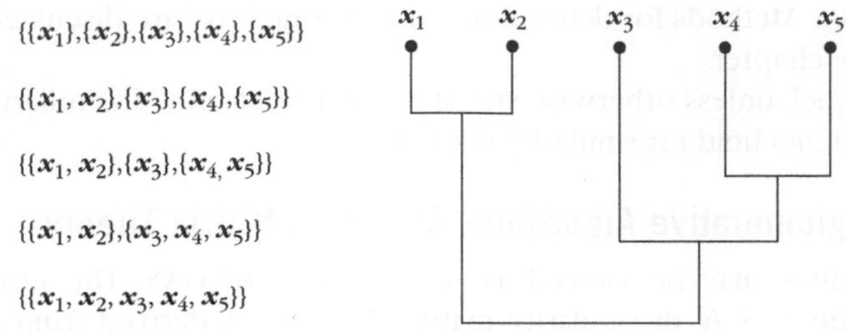

FIGURE 13.1

The clustering hierarchy for X of Example 13.1 and its corresponding dendrogram.

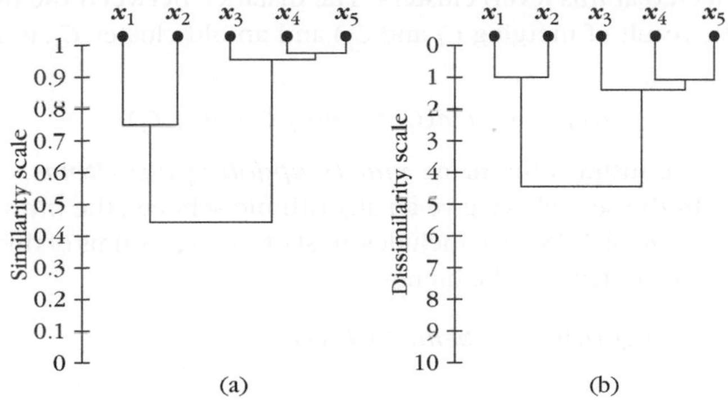

FIGURE 13.2

(a) The proximity (similarity) dendrogram for X using $P'(X)$ from Example 13.1. (b) The proximity (dissimilarity) dendrogram for X using $P(X)$ from Example 13.1.

forced formation of clusters at any level. That is, it may provide a clue about the clustering that best fits the data, as will be explained in Section 13.6. Figure 13.2 shows the similarity and dissimilarity dendrograms for X of Example 13.1 when $P'(X)$ and $P(X)$ are in use, respectively.

Before we proceed to a more detailed discussion of the hierarchical algorithms, an important note is in order. As explained earlier, this kind of algorithm determines a whole hierarchy of clusterings, rather than a single clustering. The determination of the whole dendrogram may be very useful in some applications, such as biological taxonomy (e.g., see [Prit 71]). However, in other applications we are interested only in the specific clustering that best fits the data. If one is willing to use hierarchical algorithms for applications of the latter type, he or she has to decide which clustering of the produced hierarchy is most suitable for the data. Equivalently, one must determine the appropriate level to cut the dendrogram that corresponds to the resulting hierarchy. Similar comments also hold for the divisive algorithms to be discussed later. Methods for determining the cutting level are discussed in the last section of the chapter.

In the sequel, unless otherwise stated, we consider only dissimilarity matrices. Similar arguments hold for similarity matrices.

13.2.2 Agglomerative Algorithms Based on Matrix Theory

These algorithms may be viewed as special cases of GAS. The input in these schemes is the $N \times N$ dissimilarity matrix, $P_0 = P(X)$, derived from X. At each level, t, when two clusters are merged into one, the size of the dissimilarity matrix P_t becomes $(N-t) \times (N-t)$. P_t follows from P_{t-1} by (a) deleting the two rows and columns that correspond to the merged clusters and (b) adding a new row and a new column that contain the distances between the newly formed cluster and the old (unaffected at this level) clusters. The distance between the newly formed cluster C_q (the result of merging C_i and C_j) and an old cluster, C_s, is a function of the form

$$d(C_q, C_s) = f(d(C_i, C_s), d(C_j, C_s), d(C_i, C_j)) \tag{13.2}$$

The procedure justifies the name *matrix updating algorithms*, often used in the literature. In the sequel, we give an algorithmic scheme, the *matrix updating algorithmic scheme (MUAS)*, that includes most of the algorithms of this kind. Again, t denotes the current level of the hierarchy.

Matrix Updating Algorithmic Scheme (MUAS)

- Initialization:
  - $\Re_0 = \{\{\boldsymbol{x}_i\}, i = 1, \ldots, N\}$.
  - $P_0 = P(X)$.
  - $t = 0$
- Repeat:
  - $t = t + 1$
  - Find C_i, C_j such that $d(C_i, C_j) = \min_{r,s=1,\ldots,N,\ r \neq s} d(C_r, C_s)$
  - Merge C_i, C_j into a single cluster C_q and form $\Re_t = (\Re_{t-1} - \{C_i, C_j\}) \cup \{C_q\}$.
  - Define the proximity matrix P_t from P_{t-1} as explained in the text.
- Until $\Re_{N-1}$ clustering is formed, that is, all vectors lie in the same cluster.

Notice that this scheme is in the spirit of the GAS. In [Lanc 67] it is pointed out that a number of distance functions comply with the following update equation:

$$\begin{aligned} d(C_q, C_s) = {} & a_i d(C_i, C_s) + a_j d(C_j, C_s) + b d(C_i, C_j) \\ & + c|d(C_i, C_s) - d(C_j, C_s)| \end{aligned} \tag{13.3}$$

Different values of a_i, a_j, b, and c correspond to different choices of the dissimilarity measure $d(C_i, C_j)$. Equation (13.3) is also a recursive definition of a distance between two clusters, initialized from the distance between the initial point clusters. Another formula, not involving the last term and allowing a_i, a_j, and b to be functions of C_i, C_j, and C_s, is discussed in [Bobe 93]. In the sequel we present algorithms stemming from MUAS and following from Eq. (13.3) for different values of the parameters a_i, a_j, b, c.

The simpler algorithms included in this scheme are:

- The *single link algorithm*. This is obtained from Eq. (13.3) if we set $a_i = 1/2$, $a_j = 1/2, b = 0, c = -1/2$. In this case,

$$d(C_q, C_s) = \min\{d(C_i, C_s), \quad d(C_j, C_s)\} \tag{13.4}$$

  The d_{min}^{ss} measure, defined in Section 11.2, falls under this umbrella.

- The *complete link algorithm*. This follows from Eq. (13.3) if we set $a_i = \frac{1}{2}$, $a_j = \frac{1}{2}, b = 0$ and $c = \frac{1}{2}$. Then we may write[3]

$$d(C_q, C_s) = \max\{d(C_i, C_s), \quad d(C_j, C_s)\}. \tag{13.5}$$

Note that the distance between the merged clusters C_i and C_j does not enter into the above formulas. In the case where a similarity, instead of a dissimilarity, measure is used then (a) for the single link algorithm the operator min should be replaced by max in Eq. (13.4) and (b) for the complete link algorithm the operator max should be replaced by the operator min in Eq. (13.5). To gain a further insight into the behavior of the above algorithms, let us consider the following example.

Example 13.2

Consider the data set shown in Figure 13.3a. The first seven points form an elongated cluster while the remaining four form a rather compact cluster. The numbers on top of the edges connecting the points correspond to the respective (Euclidean) distances between vectors. These distances are also taken to measure the distance between two initial point clusters. Distances that are not shown are assumed to have very large values. Figure 13.3b shows the dendrogram produced by the application of the single link algorithm to this data set. As one can easily observe, the algorithm first recovers the elongated cluster, and the second cluster is recovered at a higher dissimilarity level.

Figure 13.3c shows the dendrogram produced by the complete link algorithm. It is easily noticed that this algorithm proceeds by recovering first compact clusters.

---

[3] Equations (13.4) and (13.5) suggest that merging clusters is a min/max problem for the complete link and a min/min problem for the single link.

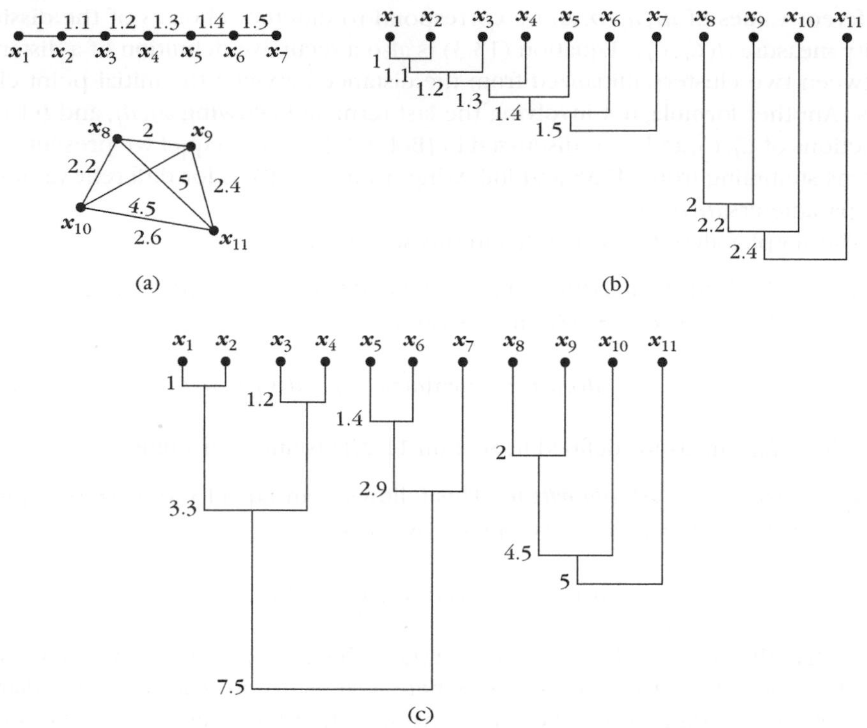

FIGURE 13.3

(a) The data set X. (b) The dissimilarity dendrogram produced by the single link algorithm. (c) The dissimilarity dendrogram produced by the complete link algorithm (the level of the final clustering is not shown).

Remark

- The preceding algorithms are the two extremes of the family described by Eq. (13.3). Indeed, the clusters produced by the single link algorithm are formed at low dissimilarities in the dissimilarity dendrogram. On the other hand, the clusters produced by the complete link algorithm are formed at high dissimilarities in the dissimilarity dendrogram. This happens because in the single link (complete link) algorithm the minimum (maximum) of the distances $d(C_i, C_s)$ and $d(C_j, C_s)$ is used as the distance between $d(C_q, C_s)$. This implies that the single link algorithm has a tendency to favor elongated clusters. This characteristic is also known as the *chaining effect*. On the other hand, the complete link algorithm proceeds by recovering small compact clusters, and it should be preferred if there is evidence that compact clusters underlie X.

The rest of the algorithms, to be discussed next, are compromises between these two extremes.[4]

- The *weighted pair group method average (WPGMA)* algorithm is obtained from Eq. (13.3) if we set $a_i = a_j = \frac{1}{2}, b = 0$, and $c = 0$, that is,

$$d(C_q, C_s) = \frac{1}{2}(d(C_i, C_s) + d(C_j, C_s)) \tag{13.6}$$

  Thus, in this case the distance between the newly formed cluster C_q and an old one C_s is defined as the average of distances between C_i, C_s and C_j, C_s.

- The *unweighted pair group method average (UPGMA)* algorithm is defined if we choose $a_i = \frac{n_i}{n_i+n_j}$, $a_j = \frac{n_j}{n_i+n_j}$, $b = 0, c = 0$, where n_i and n_j are the cardinalities of C_i and C_j, respectively. In this case the distance between C_q and C_s is defined as

$$d(C_q, C_s) = \frac{n_i}{n_i + n_j} d(C_i, C_s) + \frac{n_j}{n_i + n_j} d(C_j, C_s) \tag{13.7}$$

- The *unweighted pair group method centroid (UPGMC)* algorithm results on setting $a_i = \frac{n_i}{n_i+n_j}$, $a_j = \frac{n_j}{n_i+n_j}$, $b = -\frac{n_i n_j}{(n_i+n_j)^2}$, $c = 0$, that is,

$$d_{qs} = \frac{n_i}{n_i + n_j} d_{is} + \frac{n_j}{n_i + n_j} d_{js} - \frac{n_i n_j}{(n_i + n_j)^2} d_{ij} \tag{13.8}$$

  where d_{qs} has been used in place of $d(C_q, C_s)$ for notational simplicity. This algorithm has an interesting interpretation. Let the representatives of the clusters be chosen as the respective means (centroids), that is,

$$\boldsymbol{m}_q = \frac{1}{n_q} \sum_{\boldsymbol{x} \in C_q} \boldsymbol{x} \tag{13.9}$$

  and the dissimilarity to be the squared Euclidean distance between cluster representatives. Then it turns out that this recursive definition of d_{qs} is nothing but the square Euclidean distance between the respective representatives (see Problem 13.2), that is,

$$d_{qs} = \|\boldsymbol{m}_q - \boldsymbol{m}_s\|^2 \tag{13.10}$$

- The *weighted pair group method centroid (WPGMC)* algorithm is obtained if we choose $a_i = a_j = \frac{1}{2}, b = -\frac{1}{4}$, and $c = 0$. That is,

$$d_{qs} = \frac{1}{2} d_{is} + \frac{1}{2} d_{js} - \frac{1}{4} d_{ij} \tag{13.11}$$

[4] The terminology used here follows that given in [Jain 88].

Note that Eq. (13.11) results from (13.8) if the merging clusters have the same number of vectors. Of course, this is not true in general, and the algorithm basically computes the distance between weighted versions of the respective centroids. A notable feature of the WPGMC algorithm is that there are cases where $d_{qs} \leq \min(d_{is}, d_{js})$ (Problem 13.3).

- The *Ward or minimum variance algorithm*. Here, the distance between two clusters C_i and C_j, d'_{ij}, is defined as a weighted version of the squared Euclidean distance of their mean vectors, that is,

$$d'_{ij} = \frac{n_i n_j}{n_i + n_j} d_{ij} \tag{13.12}$$

where $d_{ij} = \|\boldsymbol{m}_i - \boldsymbol{m}_j\|^2$. Thus, in step 2.2 of MUAS we seek the pair of clusters C_i, C_j so that the quantity d'_{ij} is minimum. Furthermore, it can be shown (Problem 13.4) that this distance belongs to the family of Eq. (13.3), and we can write

$$d'_{qs} = \frac{n_i + n_s}{n_i + n_j + n_s} d'_{is} + \frac{n_j + n_s}{n_i + n_j + n_s} d'_{js} - \frac{n_s}{n_i + n_j + n_s} d'_{ij} \tag{13.13}$$

The preceding distance can also be viewed from a different perspective. Let us define

$$e_r^2 = \sum_{\boldsymbol{x} \in C_r} \|\boldsymbol{x} - \boldsymbol{m}_r\|^2$$

as the variance of the rth cluster around its mean and

$$E_t = \sum_{r=1}^{N-t} e_r^2 \tag{13.14}$$

as the total variance of the clusters at the tth level (where $N-t$ clusters are present). We will now show that *Ward's algorithm forms $\Re_{t+1}$ by merging the two clusters that lead to the smallest possible increase of the total variance.* Suppose that clusters C_i and C_j are chosen to be merged into one, say C_q. Let E_{t+1}^{ij} be the total variance after the clusters C_i and C_j are merged in C_q at the $t+1$ level. Then, since all other clusters remain unaffected, the difference $\Delta E_{t+1}^{ij} = E_{t+1}^{ij} - E_t$ is equal to

$$\Delta E_{t+1}^{ij} = e_q^2 - e_i^2 - e_j^2 \tag{13.15}$$

Taking into account that

$$\sum_{\boldsymbol{x} \in C_r} \|\boldsymbol{x} - \boldsymbol{m}_r\|^2 = \sum_{\boldsymbol{x} \in C_r} \|\boldsymbol{x}\|^2 - n_r \|\boldsymbol{m}_r\|^2 \tag{13.16}$$

Eq. (13.15) is written as

$$\Delta E_{t+1}^{ij} = n_i \|\boldsymbol{m}_i\|^2 + n_j \|\boldsymbol{m}_j\|^2 - n_q \|\boldsymbol{m}_q\|^2 \tag{13.17}$$

Using the fact that

$$n_i \boldsymbol{m}_i + n_j \boldsymbol{m}_j = n_q \boldsymbol{m}_q \tag{13.18}$$

Eq. (13.17) becomes

$$\Delta E_{t+1}^{ij} = \frac{n_i n_j}{n_i + n_j} \|\boldsymbol{m}_i - \boldsymbol{m}_j\|^2 = d'_{ij} \tag{13.19}$$

which is the distance minimized by Ward's algorithm. This justifies the name minimum variance.

Example 13.3

Consider the following dissimilarity matrix:

$$P_0 = \begin{bmatrix} 0 & 1 & 2 & 26 & 37 \\ 1 & 0 & 3 & 25 & 36 \\ 2 & 3 & 0 & 16 & 25 \\ 26 & 25 & 16 & 0 & 1.5 \\ 37 & 36 & 25 & 1.5 & 0 \end{bmatrix}$$

where the corresponding squared Euclidean distance is adopted. As one can easily observe, the first three vectors, $\boldsymbol{x}_1$, $\boldsymbol{x}_2$, and $\boldsymbol{x}_3$, are very close to each other and far away from the others. Likewise, $\boldsymbol{x}_4$ and $\boldsymbol{x}_5$ lie very close to each other and far away from the first three vectors. For this problem all seven algorithms discussed before result in the same dendrogram. The only difference is that each clustering is formed at a different dissimilarity level.

Let us first consider the single link algorithm. Since P_0 is symmetric, we consider only the upper diagonal elements. The smallest of these elements equals 1 and occurs at position $(1, 2)$ of P_0. Thus, $\boldsymbol{x}_1$ and $\boldsymbol{x}_2$ come into the same cluster and $\Re_1 = \{\{\boldsymbol{x}_1, \boldsymbol{x}_2\}, \{\boldsymbol{x}_3\}, \{\boldsymbol{x}_4\}, \{\boldsymbol{x}_5\}\}$ is produced. In the sequel, the dissimilarities among the newly formed cluster and the remaining ones have to be computed. This can be achieved via Eq. (13.4). The resulting proximity matrix, P_1, is

$$P_1 = \begin{bmatrix} 0 & 2 & 25 & 36 \\ 2 & 0 & 16 & 25 \\ 25 & 16 & 0 & 1.5 \\ 36 & 25 & 1.5 & 0 \end{bmatrix}$$

Its first row and column correspond to the cluster $\{\boldsymbol{x}_1, \boldsymbol{x}_2\}$. The smallest of the upper diagonal elements of P_1 equals 1.5. This means that at the next stage, the clusters $\{\boldsymbol{x}_4\}$ and $\{\boldsymbol{x}_5\}$ will stick together into a single cluster, producing $\Re_2 = \{\{\boldsymbol{x}_1, \boldsymbol{x}_2\}, \{\boldsymbol{x}_3\}, \{\boldsymbol{x}_4, \boldsymbol{x}_5\}\}$. Employing Eq. (13.4), we obtain

$$P_2 = \begin{bmatrix} 0 & 2 & 25 \\ 2 & 0 & 16 \\ 25 & 16 & 0 \end{bmatrix}$$

where the first row (column) corresponds to $\{\boldsymbol{x}_1, \boldsymbol{x}_2\}$, and the second and third rows (columns) correspond to $\{\boldsymbol{x}_3\}$ and $\{\boldsymbol{x}_4, \boldsymbol{x}_5\}$, respectively. Proceeding as before, at the next stage $\{\boldsymbol{x}_1, \boldsymbol{x}_2\}$

and $\{\boldsymbol{x}_3\}$ will get together in a single cluster and $\Re_3 = \{\{\boldsymbol{x}_1, \boldsymbol{x}_2, \boldsymbol{x}_3\}, \{\boldsymbol{x}_4, \boldsymbol{x}_5\}\}$ is produced. The new proximity matrix, P_3, becomes

$$P_3 = \begin{bmatrix} 0 & 16 \\ 16 & 0 \end{bmatrix}$$

where the first and the second row (column) correspond to $\{\boldsymbol{x}_1, \boldsymbol{x}_2, \boldsymbol{x}_3\}$ and $\{\boldsymbol{x}_4, \boldsymbol{x}_5\}$ clusters, respectively. Finally, $\Re_4 = \{\{\boldsymbol{x}_1, \boldsymbol{x}_2, \boldsymbol{x}_3, \boldsymbol{x}_4, \boldsymbol{x}_5\}\}$ will be formed at dissimilarity level equal to 16.

Working in a similar fashion, we can apply the remaining six algorithms to P_0. Note that in the case of Ward's algorithm, the initial dissimilarity matrix should be $\frac{1}{2}P_0$, due to the definition in Eq. (13.12). However, care must be taken when we apply UPGMA, UPGMC, and Ward's method. In these cases, when a merging takes place the parameters a_i, a_j, b, and c must be properly adjusted. The proximity levels at which each clustering is formed for each algorithm are shown in Table 13.1.

The considered task is a nice problem with two well-defined compact clusters lying away from each other. The preceding example demonstrates that in such "easy" cases all algorithms work satisfactorily (as happens with most of the clustering algorithms proposed in the literature). The particular characteristics of each algorithm are revealed when more demanding situations are faced. Thus, in Example 13.2, we saw the different behaviors of the single link and complete link algorithms. Characteristics of other algorithms, such as the WPGMC and the UPGMC, are discussed next.

13.2.3 Monotonicity and Crossover

Let us consider the following dissimilarity matrix:

$$P = \begin{bmatrix} 0 & 1.8 & 2.4 & 2.3 \\ 1.8 & 0 & 2.5 & 2.7 \\ 2.4 & 2.5 & 0 & 1.2 \\ 2.3 & 2.7 & 1.2 & 0 \end{bmatrix}$$

Table 13.1 The Results Obtained with the Seven Algorithms Discussed when they are Applied to the Proximity Matrix of Example 13.3

	SL	CL	WPGMA	UPGMA	WPGMC	UPGMC	Ward's Algorithm
$\Re_0$	0	0	0	0	0	0	0
$\Re_1$	1	1	1	1	1	1	0.5
$\Re_2$	1.5	1.5	1.5	1.5	1.5	1.5	0.75
$\Re_3$	2	3	2.5	2.5	2.25	2.25	1.5
$\Re_4$	16	37	25.75	27.5	24.69	26.46	31.75

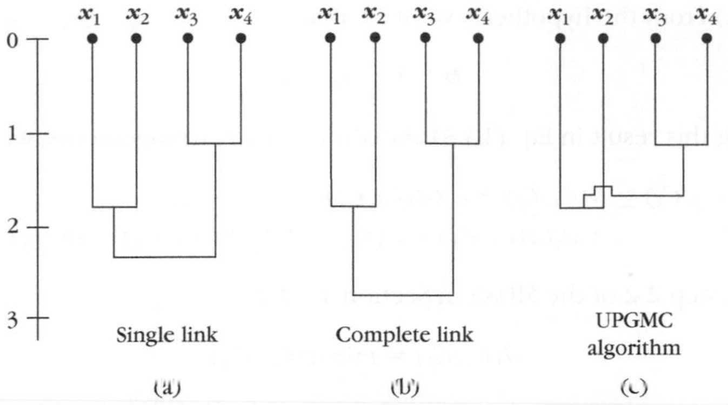

FIGURE 13.4

Dissimilarity dendrograms derived from (a) single link, (b) complete link, and (c) UPGMC and WPGMC when they apply to P. The third dendrogram exhibits the crossover phenomenon.

Application of the single and complete link algorithms to P gives rise to the dissimilarity dendrograms depicted in Figures 13.4a and 13.4b, respectively. Application of the UPGMC and WPGMC algorithms to P results in the same dissimilarity dendrogram, which is shown in Figure 13.4c. In this dendrogram something interesting occurs. The cluster $\{x_1, x_2, x_3, x_4\}$ is formed at a lower dissimilarity level than cluster $\{x_1, x_2\}$. This phenomenon is called *crossover*. More specifically, crossover occurs when a cluster is formed at a lower dissimilarity level than any of its components. The opposite of the crossover is *monotonicity*. Satisfaction of the latter condition implies that each cluster is formed at a higher dissimilarity level than any one of its components. More formally, the monotonicity condition may be stated as follows:

"If clusters C_i and C_j are selected to be merged in cluster C_q, at the tth level of the hierarchy, then the following condition must hold:

$$d(C_q, C_k) \geq d(C_i, C_j)$$

for all $C_k, k \neq i, j, q$."

Monotonicity is a property that is exclusively related to the clustering algorithm and not to the (initial) proximity matrix.

Recall Eq. (13.3) defined in terms of the parameters a_i, a_j, b, and c. In the sequel, a proposition is stated and proved that allows us to decide whether an algorithm satisfies the monotonicity condition.

Proposition 1. *If a_i and a_j are nonnegative, $a_i + a_j + b \geq 1$, and either (a) $c \geq 0$ or (b) $\max\{-a_i, -a_j\} \leq c \leq 0$, then the corresponding clustering method satisfies the monotonicity condition.*

Proof. (a) From the hypothesis we have that

$$b \geq 1 - a_i - a_j$$

Substituting this result in Eq. (13.3) and after some rearrangements, we obtain

$$\begin{aligned} d(C_q, C_s) \geq\ & d(C_i, C_j) + a_i(d(C_i, C_s) - d(C_i, C_j)) \\ & + a_j(d(C_j, C_s) - d(C_i, C_j)) + c|d(C_i, C_s) - d(C_j, C_s)| \end{aligned}$$

Since, from step 2.2 of the MUAS in Section 13.2.2,

$$d(C_i, C_j) = \min_{r,u} d(C_r, C_u)$$

the second and third terms of the last inequality are nonnegative. Moreover, the fourth term of the same inequality is also nonnegative. Therefore, we obtain

$$d(C_q, C_s) \geq d(C_i, C_j)$$

Thus, the monotonicity condition is satisfied.

(b) Let $d(C_i, C_s) \geq d(C_j, C_s)$ (the case where $d(C_i, C_s) < d(C_j, C_s)$ may be treated similarly). As in the previous case,

$$b \geq 1 - a_i - a_j$$

Taking into account this inequality, Eq. (13.3) gives

$$\begin{aligned} d(C_q, C_s) \geq\ & d(C_i, C_j) + a_i(d(C_i, C_s) - d(C_i, C_j)) \\ & + a_j(d(C_j, C_s) - d(C_i, C_j)) + c(d(C_i, C_s) - d(C_j, C_s)) \end{aligned}$$

By adding and subtracting on the right-hand side of this inequality, the term $cd(C_i, C_j)$ and after some manipulations, we obtain

$$\begin{aligned} d(C_q, C_s) \geq\ & (a_j - c)(d(C_j, C_s) - d(C_i, C_j)) + d(C_i, C_j) \\ & + (a_i + c)(d(C_i, C_s) - d(C_i, C_j)) \end{aligned}$$

Since, from the hypothesis, $a_j - c \geq 0$ and

$$d(C_i, C_j) = \min_{r,u} d(C_r, C_u)$$

from step 2.2 of the MUAS we obtain that

$$d(C_q, C_s) \geq d(C_i, C_j) \qquad \square$$

Note that the conditions of Proposition 1 are sufficient but not necessary. This means that algorithms that do not satisfy the premises of this proposition may still satisfy the monotonicity condition. It is easy to note that the single link, the complete link, the UPGMA, the WPGMA, and Ward's algorithm satisfy the premises of

Proposition 1. Thus, all these algorithms satisfy the monotonicity condition. The other two algorithms, the UPGMC and the WPGMC, do not satisfy the monotonicity condition. Moreover, we can construct examples that demonstrate that these two algorithms violate the monotonicity property, as follows from Figure 13.4c. However, it is this does not mean that they always lead to dendrograms with crossovers.

Finally, we note that there have been several criticisms concerning the usefulness of algorithms that do not satisfy the monotonicity condition (e.g., [Will 77, Snea 73]). However, these algorithms may give satisfactory results in the frame of certain applications. Moreover, there is no theoretical guideline suggesting that the algorithms satisfying the monotonicity condition always lead to acceptable results. After all, this ought to be the ultimate criterion for the usefulness of an algorithm (unfortunately, such a criterion does not exist in general).

13.2.4 Implementational Issues

As stated earlier, the computational time of GAS is $O(N^3)$. However, many efficient implementations of these schemes have been proposed in the literature, which reduce the computational time by an order of N. For example, in [Kuri 91] an implementation is discussed, for which the required computational time is reduced to $O(N^2 \log N)$. Also, in [Murt 83, Murt 84, Murt 85], implementations for widely used agglomerative algorithms are discussed that require $O(N^2)$ computational time and either $O(N^2)$ or $O(N)$ storage. Finally, parallel implementations on single instruction multiple data (SIMD) machines are discussed in [Will 89] and [Li 90].

13.2.5 Agglomerative Algorithms Based on Graph Theory

Before we describe the algorithms of this family, let us first recall some basic definitions from graph theory.

Basic Definitions from Graph Theory

A *graph* G is defined as an ordered pair $G = (V, E)$, where $V = \{v_i,\ i = 1, \ldots, N\}$ is a set of *vertices* and E is a set of *edges* connecting some pairs of vertices. An edge, connecting the vertices v_i and v_j, will be denoted either by e_{ij} or by (v_i, v_j). When the ordering of v_i and v_j is of no importance, then we deal with *undirected graphs*. Otherwise, we deal with *directed graphs*. In addition, if no cost is associated with the edges of the graph, we deal with *unweighted graphs*. Otherwise, we deal with *weighted graphs*. In the sequel, we consider graphs where a pair of vertices may be connected by at most one edge. In the framework of clustering, we deal with undirected graphs where each vertex corresponds to a feature vector (or, equivalently, to the pattern represented by the feature vector).

A *path* in G, between vertices v_{i_1} and v_{i_n}, is a sequence of vertices and edges of the form $v_{i_1} e_{i_1 i_2} v_{i_2} \ldots v_{i_{n-1}} e_{i_{n-1} i_n} v_{i_n}$ (Figure 13.5a). Of course, there is no guaranteed, that there will be always exists a path from v_{i_1} to v_{i_n}. If in this path, v_{i_1}

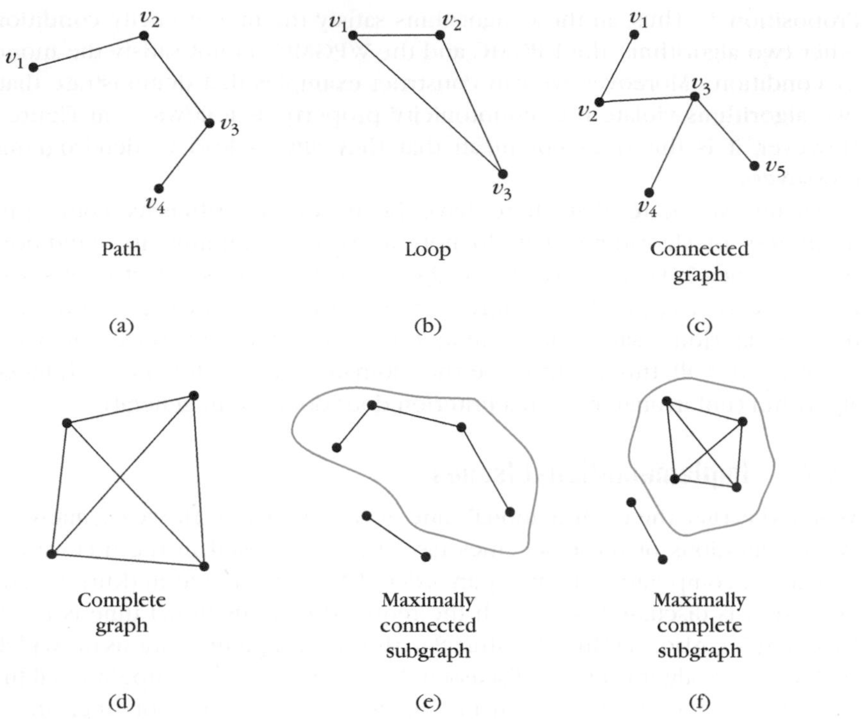

FIGURE 13.5

(a) A path connecting the vertices v_1 and v_4. (b) A loop. (c) A connected graph. (d) A complete graph. (e) Maximally connected subgraph. (f) Maximally complete subgraph.

coincides with v_{i_n}, the path is called a *loop* or *circle* (Figure 13.5b). In the special case in which an edge connects a vertex to itself, we have a *self-loop*.

A *subgraph* $G' = (V', E')$ of G is a graph with $V' \subseteq V$ and $E' \subseteq E_1$, where E_1 is a subset of E, whose edges connect vertices that lie in V'. Clearly, G is a subgraph of itself.

A *subgraph* $G' = (V', E')$ is *connected* if there exists at least one path connecting any pair of vertices in V' (Figure 13.5c). For example, in Figure 13.5c the subgraph with vertices v_1, v_2, v_4, and v_5 is not connected. The subgraph G' is *complete* if every vertex $v_i \in V'$ is connected with every vertex in $V' - \{v_i\}$ (Figure 13.5d).

A *maximally connected subgraph of G* is a connected subgraph G' of G that contains as many vertices of G as possible (Figure 13.5e). A *maximally complete subgraph* is a complete subgraph G' of G that contains as many vertices of G as possible (Figure 13.5f).

A concept that is closely related to the algorithms based on graph theory is that of the *threshold graph*. A threshold graph is an undirected, unweighted graph with

N nodes, each corresponding to a vector of the data set X. In this graph there are no self-loops or multiple edges between any two nodes. Let a be a dissimilarity level. A threshold graph $G(a)$ with N nodes contains an edge between two nodes i and j *if the dissimilarity between the corresponding vectors* $\boldsymbol{x}_i$ *and* $\boldsymbol{x}_j$ *is less than or equal to* a, $i, j = 1, \ldots, N$. Alternatively, we may write

$$(v_i, v_j) \in G(a), \quad \text{if } d(\boldsymbol{x}_i, \boldsymbol{x}_j) \leq a, \quad i, j = 1, \ldots, N \tag{13.20}$$

If similarity measures are used, this definition is altered to

$$(v_i, v_j) \in G(a), \quad \text{if } s(\boldsymbol{x}_i, \boldsymbol{x}_j) \geq a, \quad i, j = 1, \ldots, N$$

A *proximity graph* $G_p(a)$ is a threshold graph $G(a)$, all of whose edges (v_i, v_j) are weighted with the proximity measure between $\boldsymbol{x}_i$ and $\boldsymbol{x}_j$. If a dissimilarity (similarity) measure is used as proximity between two vectors, then the proximity graph is called a *dissimilarity (similarity) graph*. Figure 13.6 shows the threshold and proximity graphs $G(3)$ and $G_p(3)$, $G(5)$ and $G_p(5)$ obtained from the dissimilarity matrix $P(X)$ given in Example 13.1.

The Algorithms

In this section, we discuss agglomerative algorithms based on graph theory concepts. More specifically, we consider graphs, G, of N nodes with each node corresponding to a vector of X. Clusters are formed by connecting nodes together, and this leads to connected subgraphs. Usually, an additional graph property, $h(k)$, must be satisfied by the subgraphs in order to define valid clusters. In this context, the function g involved in GAS is replaced by $g_{h(k)}$, where $h(k)$ is the graph property. Some typical properties that can be adopted are [Jain 88, Ling 72].

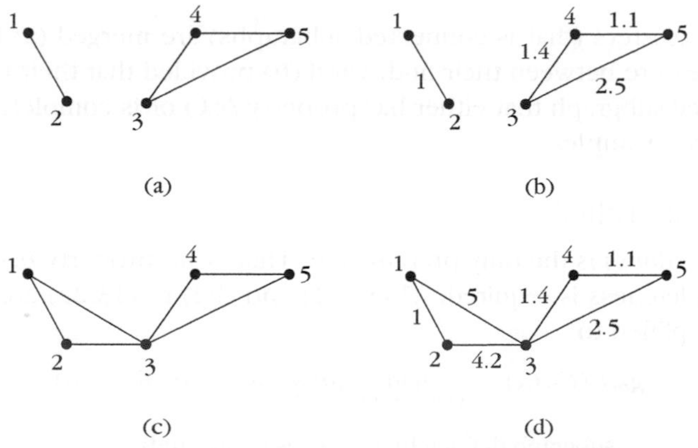

FIGURE 13.6

(a) The threshold graph $G(3)$, (b) the proximity (dissimilarity) graph $G_p(3)$, (c) the threshold graph $G(5)$, (d) the proximity (dissimilarity) graph $G_p(5)$, obtained from the dissimilarity matrix $P(X)$ of Example 13.1.

- *Node connectivity.* The node connectivity of a connected subgraph is the largest integer k such that all pairs of nodes are joined by at least k paths having no nodes in common.
- *Edge connectivity.* The edge connectivity of a connected subgraph is the largest integer k such that all pairs of nodes are joined by at least k paths having no edges in common.
- *Node degree.* The degree of a connected subgraph is the largest integer k such that each node has at least k incident edges.

The general agglomerative scheme in the context of graph theory is known as the *graph theory-based algorithmic scheme (GTAS)*. The algorithm follows exactly the same iteration steps as the generalized agglomerative scheme (GAS), with the exception of step 2.2. This is expressed as

$$g_{h(k)}(C_i, C_j) = \begin{cases} \min_{r,s} g_{h(k)}(C_r, C_s), & \text{for dissimilarity functions} \\ \max_{r,s} g_{h(k)}(C_r, C_s), & \text{for similarity functions} \end{cases} \tag{13.21}$$

The proximity function $g_{h(k)}(C_r, C_s)$ between two clusters is defined in terms of (a) a proximity measure between vectors (that is nodes in the graph) *and* (b) certain constraints imposed by the property $h(k)$ on the subgraphs that are formed. In a more formal way, $g_{h(k)}$ is defined as

$$\begin{aligned} g_{h(k)}(C_r, C_s) &= \min_{\boldsymbol{x}_u \in C_r, \boldsymbol{x}_v \in C_s} \{d(\boldsymbol{x}_u, \boldsymbol{x}_v) \equiv a: \text{ the } G(a) \text{ subgraph} \\ &\qquad \text{defined by } C_r \cup C_s \text{ is (a) connected and either} \\ &\qquad \text{(b1) has the property } h(k) \text{ or (b2) is complete}\}^5 \end{aligned} \tag{13.22}$$

In words, clusters (that is connected subgraphs) are merged (a) based on the proximity measure between their nodes and (b) provided that their merging leads to a connected subgraph that either has property $h(k)$ or is complete. Let us now consider a few examples.

Single Link Algorithm

Here connectedness is the only prerequisite. That is, no property $h(k)$ is imposed and no completeness is required. Thus (b1) and (b2) in (13.22) are ignored and (13.22) is simplified to

$$\begin{aligned} g_{h(k)}(C_r, C_s) = \min_{\boldsymbol{x}_u \in C_r, \boldsymbol{x}_v \in C_s} &\{d(\boldsymbol{x}_u, \boldsymbol{x}_v) \equiv a: \text{ the } G(a) \\ &\text{subgraph defined by } C_r \cup C_s \text{ is connected}\} \end{aligned} \tag{13.23}$$

Let us demonstrate the algorithm via an example.

[5] This means that *all* nodes of $C_r \cup C_s$ participate in the required properties.

Example 13.4

Consider the following dissimilarity matrix:

$$P = \begin{bmatrix} 0 & 1.2 & 3 & 3.7 & 4.2 \\ 1.2 & 0 & 2.5 & 3.2 & 3.9 \\ 3 & 2.5 & 0 & 1.8 & 2.0 \\ 3.7 & 3.2 & 1.8 & 0 & 1.5 \\ 4.2 & 3.9 & 2.0 & 1.5 & 0 \end{bmatrix}$$

The first clustering, $\Re_0$, is the one where each vector of X forms a single cluster (see Figure 13.7). In order to determine the next clustering, $\Re_1$, via the single link algorithm, we need to compute $g_{h(k)}(C_r, C_s)$ for all pairs of the existing clusters. For $\{\boldsymbol{x}_1\}$ and $\{\boldsymbol{x}_2\}$, the value of $g_{h(k)}$ is equal to 1.2, since $\{\boldsymbol{x}_1\} \cup \{\boldsymbol{x}_2\}$ become connected for a first time in $G(1.2)$. Likewise, $g_{h(k)}(\{\boldsymbol{x}_1\}, \{\boldsymbol{x}_3\}) = 3$. The rest of the $g_{h(k)}$ values are computed in a similar fashion. Then, using Eq. (13.21), we find that $g_{h(k)}(\{\boldsymbol{x}_1\}, \{\boldsymbol{x}_2\}) = 1.2$ is the minimum value of $g_{h(k)}$ and thus $\{\boldsymbol{x}_1\}$ and $\{\boldsymbol{x}_2\}$ are merged in order to produce

$$\Re_1 = \{\{\boldsymbol{x}_1, \boldsymbol{x}_2\}, \{\boldsymbol{x}_3\}, \{\boldsymbol{x}_4\}, \{\boldsymbol{x}_5\}\}$$

Following the same procedure, we find that the minimum $g_{h(k)}$ among all pairs of clusters is $g_{h(k)}(\{\boldsymbol{x}_4\}, \{\boldsymbol{x}_5\}) = 1.5$. Thus, $\Re_2$ is given by

$$\Re_2 = \{\{\boldsymbol{x}_1, \boldsymbol{x}_2\}, \{\boldsymbol{x}_3\}, \{\boldsymbol{x}_4, \boldsymbol{x}_5\}\}$$

For the formation of $\Re_3$ we first consider the clusters $\{\boldsymbol{x}_3\}$ and $\{\boldsymbol{x}_4, \boldsymbol{x}_5\}$. In this case $g_{h(k)}(\{\boldsymbol{x}_3\}, \{\boldsymbol{x}_4, \boldsymbol{x}_5\}) = 1.8$, since $\{\boldsymbol{x}_3\} \cup \{\boldsymbol{x}_4, \boldsymbol{x}_5\}$ becomes connected at $G(1.8)$ for a first time. Similarly, we find that $g_{h(k)}(\{\boldsymbol{x}_1, \boldsymbol{x}_2\}, \{\boldsymbol{x}_3\}) = 2.5$, and $g_{h(k)}(\{\boldsymbol{x}_1, \boldsymbol{x}_2\}, \{\boldsymbol{x}_4, \boldsymbol{x}_5\}) = 3.2$. Thus,

$$\Re_3 = \{\{\boldsymbol{x}_1, \boldsymbol{x}_2\}, \{\boldsymbol{x}_3, \boldsymbol{x}_4, \boldsymbol{x}_5\}\}$$

Finally, we find that $g_{h(k)}(\{\boldsymbol{x}_1, \boldsymbol{x}_2\}, \{\boldsymbol{x}_3, \boldsymbol{x}_4, \boldsymbol{x}_5\}) = 2.5$ and $\Re_4$ is formed at this level. Observe that at $G(2.0)$ no clustering is formed.

Remark

- In the single link algorithm no property $h(k)$ is required, and Eq. (13.23) is basically the same as

$$g_{h(k)}(C_r, C_s) = \min_{\boldsymbol{x} \in C_r, \boldsymbol{y} \in C_s} d(\boldsymbol{x}, \boldsymbol{y}) \tag{13.24}$$

  Hence the algorithm is equivalent to its single link counterpart based on matrix theory and both produce the same results (see Problem 13.7).

Complete Link Algorithm

The only prerequisite here is that of completeness; that is, the graph property $h(k)$ is omitted. Since, connectedness is a weaker condition than completeness, subgraphs form valid clusters only if they are complete. Let us demonstrate the algorithms through the case of Example 13.4.

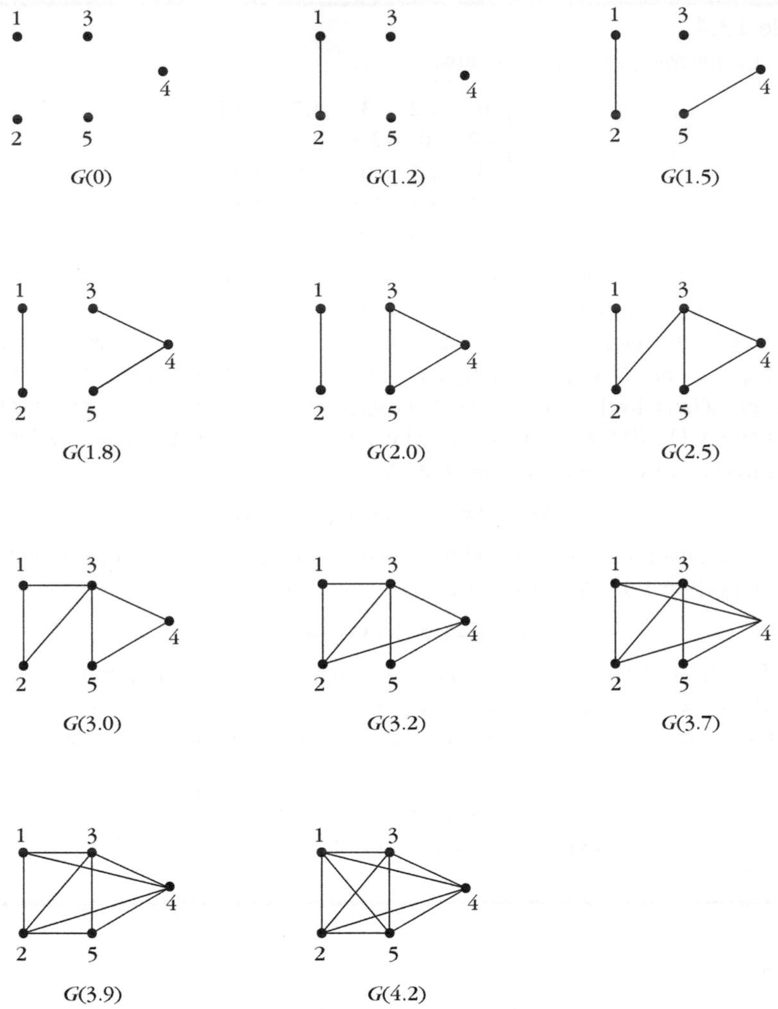

FIGURE 13.7

Threshold graphs derived by the dissimilarity matrix P given in Example 13.4.

Clusterings $\Re_0, \Re_1$, and $\Re_2$ are the same as those produced by the single link algorithm and are formed by $G(0)$, $G(1.2)$, and $G(1.5)$, respectively. Let us derive the $\Re_3$ clustering. It is $g_{h(k)}(\{\boldsymbol{x}_3\}, \{\boldsymbol{x}_4, \boldsymbol{x}_5\}) = 2$, because at $G(2.0)$, $\{\boldsymbol{x}_3\} \cup \{\boldsymbol{x}_4, \boldsymbol{x}_5\}$ becomes complete for the first time. Similarly, $g_{h(k)}(\{\boldsymbol{x}_1, \boldsymbol{x}_2\}, \{\boldsymbol{x}_3\}) = 3$ and $g_{h(k)}(\{\boldsymbol{x}_1, \boldsymbol{x}_2\}, \{\boldsymbol{x}_4, \boldsymbol{x}_5\}) = 4.2$. Thus, the resulting $\Re_3$ clustering is the same as the one obtained by the single link algorithm. The only difference is that it is formed at graph $G(2.0)$ instead of $G(1.8)$, which was the case with the single link algorithm. Finally, the last clustering, $\Re_4$ is defined at $G(4.2)$.

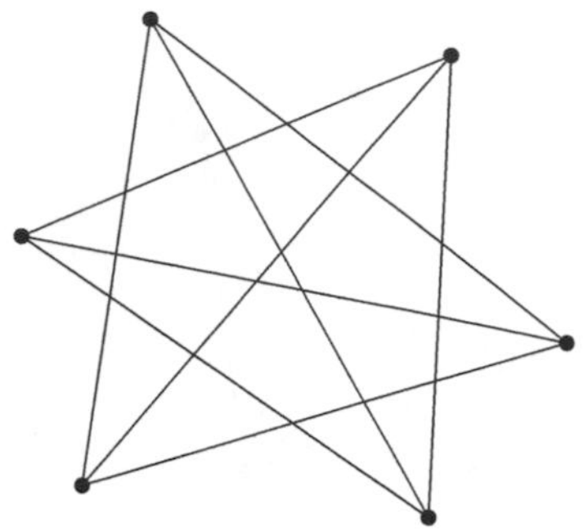

FIGURE 13.8

A graph with node connectivity, edge connectivity, and node degree equal to 3.

Remark

- A little thought suffices to see that Eq. (13.22) for the complete link algorithm is equivalent to

$$g_{h(k)}(C_r, C_s) = \max_{x \in C_r, y \in C_s} d(x, y) \tag{13.25}$$

and, thus, this algorithm is equivalent to its matrix-based counterpart (see Problem 13.8)

The single and the complete link algorithms may be seen as the extreme cases of the GTAS scheme. This is because the criteria adopted for the formation of a new cluster are the weakest possible for the single link and the strongest possible for the complete link algorithm. A variety of algorithms between these two extremes may be obtained if we make different choices for $g_{h(k)}$. From Eq. (13.22) it is clear that this may be achieved by changing the property $h(k)$, where k is a parameter whose meaning depends on the adopted property $h(k)$. For example, in Figure 13.8, the value of k for the properties of node connectivity, edge connectivity, and node degree is 3.

Example 13.5

In this example we demonstrate the operation of the property $h(k)$. Let us consider the following dissimilarity matrix:

$$P(X) = \begin{bmatrix} 0 & 1 & 9 & 18 & 19 & 20 & 21 \\ 1 & 0 & 8 & 13 & 14 & 15 & 16 \\ 9 & 8 & 0 & 17 & 10 & 11 & 12 \\ 18 & 13 & 17 & 0 & 5 & 6 & 7 \\ 19 & 14 & 10 & 5 & 0 & 3 & 4 \\ 20 & 15 & 11 & 6 & 3 & 0 & 2 \\ 21 & 16 & 12 & 7 & 4 & 2 & 0 \end{bmatrix}$$

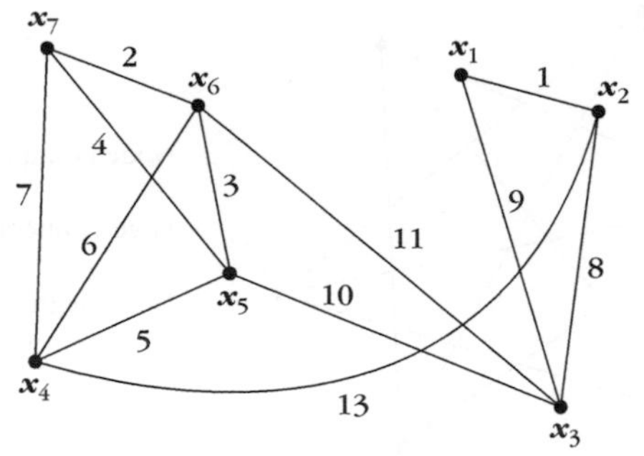

FIGURE 13.9

The proximity graph $G(13)$ derived by the dissimilarity matrix P given in Example 13.5.

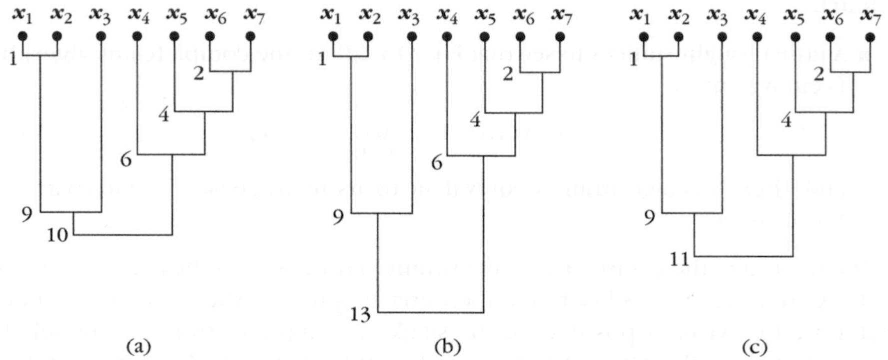

FIGURE 13.10

Dissimilarity dendrograms related to Example 13.5. (a) Dissimilarity dendrogram produced when $h(k)$ is the node degree property, with $k = 2$. (b) Dissimilarity dendrogram produced when $h(k)$ is the node connectivity property, with $k = 2$. (c) Dissimilarity dendrogram produced when $h(k)$ is the edge connectivity property, with $k = 2$.

Figure 13.9 shows the $G(13)$ proximity graph produced by this dissimilarity matrix. Let $h(k)$ be the node degree property with $k = 2$; that is, it is required that each node has at least two incident edges. Then the obtained threshold dendrogram is shown in Figure 13.10a. At dissimilarity level 1, x_1 and x_2 form a single cluster. This happens because $\{x_1\} \cup \{x_2\}$ is complete at $G(1)$, despite the fact that property $h(2)$ is not satisfied (remember the disjunction between conditions (b1) and (b2) in Eq. (13.22)). Similarly, $\{x_6\} \cup \{x_7\}$ forms a cluster at dissimilarity level 2. The next clustering is formed at level 4, since $\{x_5\} \cup \{x_6, x_7\}$ becomes complete in $G(4)$. At level 6, x_4, x_5, x_6, and x_7 lie for the first time in the same cluster. Although this subgraph is not complete, it does satisfy $h(2)$. Finally, at level 9, x_1, x_2, and x_3 come into the same cluster. Note that, although all nodes in the graph have node degree equal to 2, the final clustering will be formed at level 10 because at level 9 the graph is not connected.

Assume now that $h(k)$ is the node connectivity property, with $k = 2$; that is, all pairs of nodes in a connected subgraph are joined by at least two paths having no nodes in common. The dissimilarity dendrogram produced in this case is shown in Figure 13.10b.

Finally, the dissimilarity dendrogram produced when the edge connectivity property with $k = 2$ is employed is shown in Figure 13.10c.

It is not difficult to see that all these properties for $k = 1$ result in the single link algorithm. On the other hand, as k increases, the resulting subgraphs approach completeness.

Example 13.6

Consider again the dissimilarity matrix of the previous example. Assume now that $h(k)$ is the node degree property with $k = 3$. The corresponding dendrogram is shown in Figure 13.11. Comparing the dendrograms of Figures 13.10a and 13.11, we observe that the same clusters in the second case are formed in larger dissimilarity levels.

Clustering Algorithms Based on the Minimum Spanning Tree

A *spanning tree* is a connected graph (containing all the vertices of the graph) and having no loops (that is, there exists only one path connecting any two pairs of

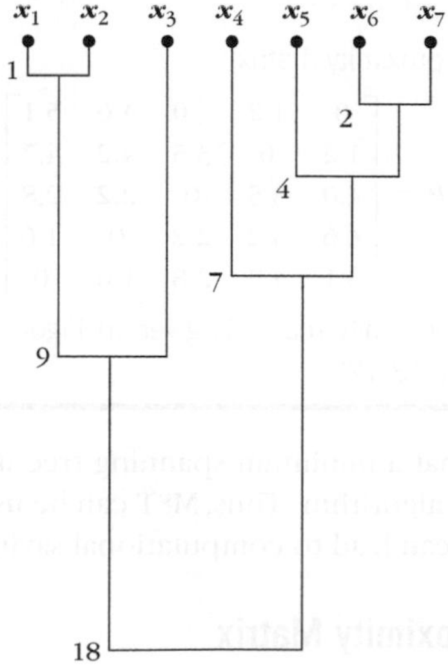

FIGURE 13.11

Threshold dendrogram related to Example 13.6 for the node degree property and $k = 3$.

nodes in the graph). If the edges of the graph are weighted, we define as the *weight of the spanning tree* the sum of the weights of its edges. A *minimum spanning tree (MST)* is a spanning tree with the smallest weight among all spanning trees connecting the nodes of the graph. An MST of a graph may be derived with Prim's algorithm or Kruskal's algorithm (e.g., see [Horo 78]). Note that there may be more than one minimum spanning trees for a given graph. However, when the weights of the edges of G are different from each other, then the MST is unique. In our case, the weights of a graph are derived from the proximity matrix $P(X)$.

Searching for the MST can also be seen as a special case of the GTAS scheme, if we adopt in place of $g_{h(k)}(C_r, C_s)$ the following proximity function:

$$g(C_r, C_s) = \min_{ij}\{w_{ij} : \boldsymbol{x}_i \in C_r,\ \boldsymbol{x}_j \in C_s\} \tag{13.26}$$

where $w_{ij} = d(\boldsymbol{x}_i, \boldsymbol{x}_j)$.

In words, this measure identifies the minimum weight of the MST that connects the subgraphs corresponding to C_r and C_s.

Once the MST has been determined (using any suitable algorithm), we may identify a hierarchy of clusterings as follows: the clusters of the clustering at level t are identified as the *connected* components of G if only the edges of its MST with the smallest t weights are considered. It takes a little thought to see that this hierarchy is identical to the one defined by the single link algorithm, at least for the case in which all the distances between any two vectors of X are different from each other. Thus, this scheme may also be viewed as an alternative implementation of the single link algorithm. The following example demonstrates the operation of this scheme.

Example 13.7

Let us consider the following proximity matrix.

$$P = \begin{bmatrix} 0 & 1.2 & 4.0 & 4.6 & 5.1 \\ 1.2 & 0 & 3.5 & 4.2 & 4.7 \\ 4.0 & 3.5 & 0 & 2.2 & 2.8 \\ 4.6 & 4.2 & 2.2 & 0 & 1.6 \\ 5.1 & 4.7 & 2.8 & 1.6 & 0 \end{bmatrix}$$

The MST derived from this proximity matrix is given in Figure 13.12a. The corresponding dendrogram is given in Figure 13.12b.

It is easy to observe that a minimum spanning tree uniquely specifies the dendrogram of the single link algorithm. Thus, MST can be used as an alternative to the single link algorithm and can lead to computational savings.

13.2.6 Ties in the Proximity Matrix

In cases in which the vectors consist of interval-scaled or ratio-scaled features, the probability of a vector of the data set X being equidistant from two other vectors of X is very small for most practical problems. However, if we deal with ordinal data,

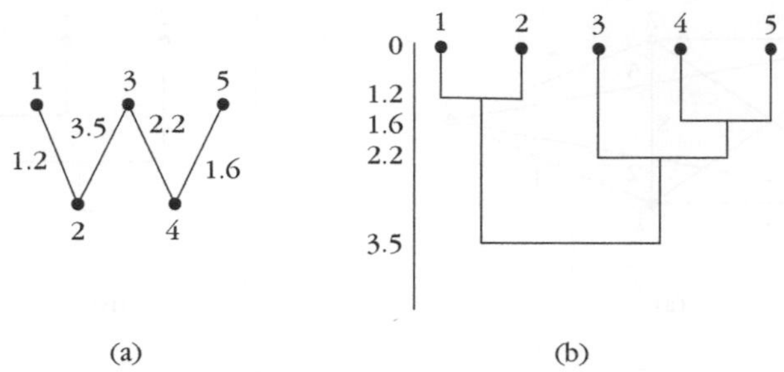

FIGURE 13.12

(a) The minimum spanning tree derived with the dissimilarity matrix given in Example 13.7. (b) The dissimilarity dendrogram obtained with the algorithm based on the MST.

this probability is not negligible. The fact that a vector is equidistant from two other vectors implies that a proximity matrix P will have at least two equal entries in the triangle above its main diagonal (see Example 13.8). It is interesting to see how the hierarchical algorithms behave with such proximity matrices. Let us consider first the family of algorithms based on the graph theory via the following example.

Example 13.8

Consider the following dissimilarity matrix:

$$P = \begin{bmatrix} 0 & 4 & 9 & 6 & 5 \\ 4 & 0 & 3 & 8 & 7 \\ 9 & 3 & 0 & 3 & 2 \\ 6 & 8 & 3 & 0 & 1 \\ 5 & 7 & 2 & 1 & 0 \end{bmatrix}$$

Note that $P(2,3) = P(3,4)$. The corresponding dissimilarity graph $G(9)$ is shown in Figure 13.13a. Figure 13.13b shows the corresponding dissimilarity dendrogram obtained by the single link algorithm. No matter which of the two edges is considered first, the resulting dendrogram remains the same. Figure 13.13c (13.13d) depicts the dendrogram obtained by the complete link algorithm when the $(3,4)$ ($(2,3)$) edge is considered first. Note that the dendrograms of Figures 13.13c and 13.13d are different.

Let us interchange the $P(1,2)$ and $P(2,3)$ entries of P[6] and let P_1 be the new dissimilarity matrix. Figure 13.14a shows the dendrogram obtained by the single link algorithm, and Figure 13.14b depicts the dendrogram obtained by the complete link algorithm. In this case, the complete link algorithm produces the same dendrogram regardless the order in which edges $(1,2)$ and $(3,4)$ are considered.

This example indicates that the single link algorithm leads to the same dendrogram, regardless of how the ties are considered. On the other hand, the complete

[6] Since a dissimilarity matrix is symmetric, $P(2,1)$ and $P(3,2)$ entries are also interchanged.

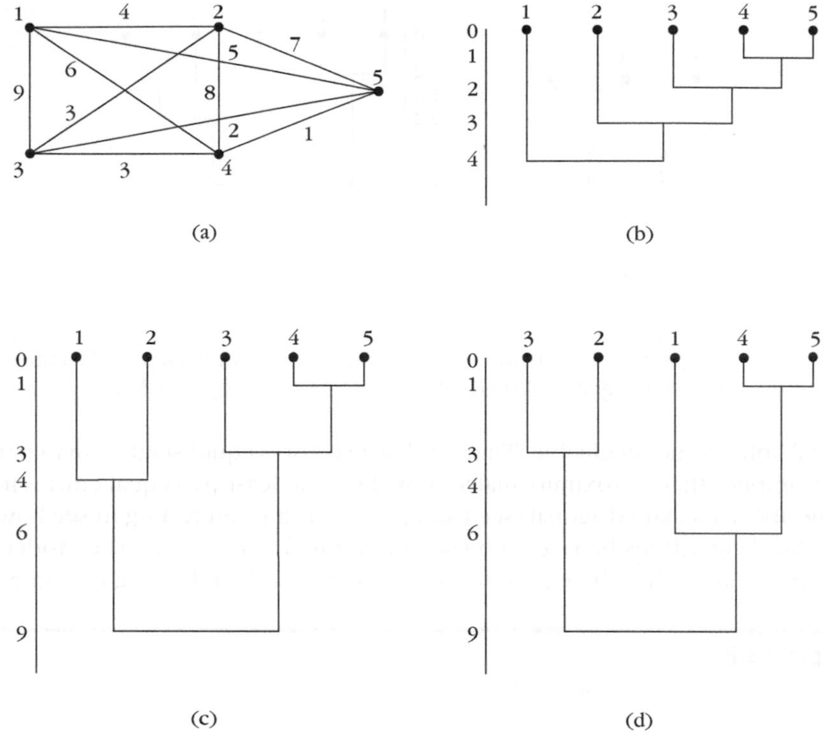

FIGURE 13.13

(a) The dissimilarity graph ($G(9)$) for the dissimilarity matrix given in Example 13.8. (b) The dissimilarity dendrogram obtained by the single link algorithm. (c) The dissimilarity dendrogram obtained by the complete link algorithm when edge $(3, 4)$ is considered first. (d) The dissimilarity dendrogram obtained by the complete link algorithm when edge $(2, 3)$ is considered first.

link algorithm *may* lead to different dendrograms if it follows different ways of considering the ties. The other graph theory-based algorithms, which fall between the single and the complete algorithm, exhibit behavior similar to that of the complete link algorithm (see Problem 13.11).

The same trend is true for the matrix-based algorithms. *Note, however, that in matrix-based schemes ties may appear at a later stage in the proximity matrix* (see Problem 13.12). Thus, as is shown in [Jard 71], the single link algorithm treats the ties in the proximity matrix in the most reliable way; it always leads to the same proximity dendrogram. It seems that every requirement additional to the connectivity property (for graph theory-based algorithms) or to Eq. (13.4) (for matrix theory-based algorithms) produces ambiguity, and the results become sensitive to the order in which ties are processed. From this point of view, the single link algorithm seems to outperform its competitors. This does not mean that all

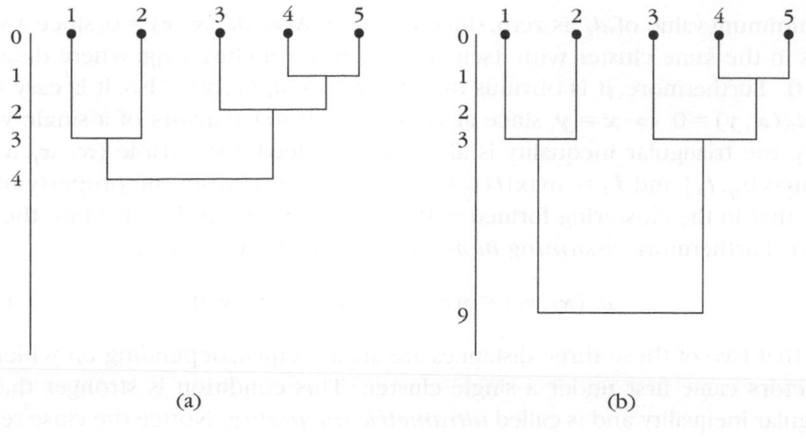

FIGURE 13.14

(a) The dissimilarity dendrogram obtained by the single link algorithm for P_1 given in Example 13.8. (b) The dissimilarity dendrogram obtained by the complete link algorithm for P_1 given in Example 13.8.

the other algorithms are inferior. The algorithm that gives the best results is problem dependent. However, if one decides to use any other algorithm, different from the single link, he or she must treat the possible ties in the proximity matrix very carefully.

13.3 THE COPHENETIC MATRIX

Another quantity associated with the hierarchical algorithms is the *cophenetic matrix*. This will be used as a tool for the validation of clustering hierarchies in Chapter 16.

Let $\Re_{t_{ij}}$ be the clustering at which $\boldsymbol{x}_i$ and $\boldsymbol{x}_j$ are merged in the same cluster *for the first time* (of course, they will remain in the same cluster for all subsequent clusterings). Also let $L(t_{ij})$ be the proximity level at which clustering $\Re_{t_{ij}}$ is defined. We define the *cophenetic distance* between two vectors as

$$d_C(\boldsymbol{x}_i, \boldsymbol{x}_j) = L(t_{ij})$$

In words, the cophenetic distance between two vectors $\boldsymbol{x}_i$ and $\boldsymbol{x}_j$ is defined as the proximity level at which the two vectors are found in the same cluster for the first time.

The cophenetic distance is a metric, *under the assumption of monotonicity*. To prove this, we need to show that the five conditions stated in Chapter 11 are satisfied. Indeed, the first condition holds; that is, $0 \le d_C(\boldsymbol{x}, \boldsymbol{y}) < +\infty$. Note that the minimum value of d_C is zero, since $L(0) = 0$. Also, $d_C(\boldsymbol{x}, \boldsymbol{x}) = 0$, since a vector $\boldsymbol{x}$ lies in the same cluster with itself at the zero level clustering, where dissimilarity is 0. Furthermore, it is obvious that $d_C(\boldsymbol{x}, \boldsymbol{y}) = d_C(\boldsymbol{y}, \boldsymbol{x})$. Also, it is easy to see that $d_C(\boldsymbol{x}, \boldsymbol{y}) = 0 \Leftrightarrow \boldsymbol{x} = \boldsymbol{y}$, since at $L(0)$ each cluster consists of a single vector. Finally, the triangular inequality is also valid. Indeed, for a triple $(\boldsymbol{x}_i, \boldsymbol{x}_j, \boldsymbol{x}_r)$, let $t_1 = \max\{t_{ij}, t_{jr}\}$ and $L_1 = \max\{L(t_{ij}), L(t_{jr})\}$. It is clear, from the property of hierarchy, that in the clustering formed at the t_1 level, $\boldsymbol{x}_i$, $\boldsymbol{x}_j$, and $\boldsymbol{x}_r$ fall into the same cluster. Furthermore, *assuming monotonicity*, $d_C(\boldsymbol{x}_i, \boldsymbol{x}_r) \leq L_1$, or

$$d_C(\boldsymbol{x}_i, \boldsymbol{x}_r) \leq \max\{d_C(\boldsymbol{x}_i, \boldsymbol{x}_j), d_C(\boldsymbol{x}_j, \boldsymbol{x}_r)\} \tag{13.27}$$

Note that two of these three distances are always equal, depending on which pair of vectors came first under a single cluster. This condition is stronger than the triangular inequality and is called *ultrametric inequality*. Notice the close relation of ultrametricity and monotonicity. Monotonicity ensures ultrametricity and, as a consequence, the triangular inequality.

The *cophenetic matrix* is defined as

$$D_C(X) = [d_C(\boldsymbol{x}_i, \ \boldsymbol{x}_j)] = [L(t_{ij})], \quad i,j = 1, \ldots, N$$

It is clear that the cophenetic matrix is symmetric. Moreover, apart from its diagonal elements, it has only $N-1$ distinct entries; that is, it has many ties (duplicate entries). $D_C(X)$ is a special case of dissimilarity matrix, since $d_C(\boldsymbol{x}_i, \boldsymbol{x}_j)$ satisfy the ultrametric inequality.

A hierarchical algorithm can, thus, be viewed as a mapping of the data proximity matrix into a cophenetic matrix.

Example 13.9

Let us consider the dissimilarity dendrogram of Figure 13.2b. The corresponding cophenetic matrix is

$$D_C(X) = \begin{bmatrix} 0 & 1 & 4.2 & 4.2 & 4.2 \\ 1 & 0 & 4.2 & 4.2 & 4.2 \\ 4.2 & 4.2 & 0 & 1.4 & 1.4 \\ 4.2 & 4.2 & 1.4 & 0 & 1.1 \\ 4.2 & 4.2 & 1.4 & 1.1 & 0 \end{bmatrix}$$

13.4 DIVISIVE ALGORITHMS

The divisive algorithms follow the reverse strategy from that of the agglomerative schemes. The first clustering contains a single set, X. At the first step, we search for the best possible partition of X into two clusters. The straightforward method is to consider all possible $2^{N-1}-1$ partitions of X into two sets and to select the optimum, according to a prespecified criterion. This procedure is then applied iteratively to each of the two sets produced in the previous stage. The final clustering consists of N clusters, each containing a single vector of X.

Let us state the general divisive scheme more formally. Here, the tth clustering contains $t+1$ clusters. In the sequel, C_{tj} will denote the jth cluster of the tth clustering $\Re_t$, $t = 0, \ldots, N-1, j = 1, \ldots, t+1$. Let $g(C_i, C_j)$ be a dissimilarity function defined for all possible pairs of clusters. The initial clustering $\Re_0$ contains only the set X, that is, $C_{01} = X$. To determine the next clustering, we consider all possible pairs of clusters that form a partition of X. Among them we choose the pair, denoted by (C_{11}, C_{12}), that maximizes g.[7] These clusters form the next clustering $\Re_1$, that is, $\Re_1 = \{C_{11}, C_{12}\}$. At the next time step, we consider all possible pairs of clusters produced by C_{11} and we choose the one that maximizes g. The same procedure is repeated for C_{12}. Assume now that from the two resulting pairs of clusters, the one originating from C_{11} gives the larger value of g. Let this pair be denoted by (C_{11}^1, C_{11}^2). Then the new clustering, $\Re_2$, consists of C_{11}^1, C_{11}^2, and C_{12}. Relabeling these clusters as C_{21}, C_{22}, C_{23}, respectively, we have $\Re_2 = \{C_{21}, C_{22}, C_{23}\}$. Carrying on in the same way, we form all subsequent clusterings. The general divisive scheme may be stated as follows:

Generalized Divisive Scheme (GDS)

- Initialization
  - Choose $\Re_0 = \{X\}$ as the initial clustering.
  - $t = 0$
- Repeat
  - $t = t + 1$
  - For $i = 1$ to t
    - Among all possible pairs of clusters (C_r, C_s) that form a partition of $C_{t-1,i}$, find the pair $(C_{t-1,i}^1, C_{t-1,i}^2)$ that gives the maximum value for g.
  - Next i
  - From the t pairs defined in the previous step choose the one that maximizes g. Suppose that this is $(C_{t-1,j}^1, C_{t-1,j}^2)$.
  - The new clustering is

$$\Re_t = (\Re_{t-1} - \{C_{t-1,j}\}) \cup \{C_{t-1,j}^1, C_{t-1,j}^2\}$$

  - Relabel the clusters of $\Re_t$.
- Until each vector lies in a single distinct cluster.

[7] We can also use a similarity function. In that case we should choose the pair of clusters that minimizes g.

Different choices of g give rise to different algorithms. One can easily observe that this divisive scheme is computationally very demanding, even for moderate values of N. This is its main drawback, compared with the agglomerative scheme. Thus, if these schemes are to be of any use in practice, some further computational simplifications are required. One possibility is to make compromises and not search for all possible partitions of a cluster. This can be done by ruling out many partitions as not "reasonable," under a prespecified criterion. Examples of such algorithms are discussed in [Gowd 95] and [MacN 64]. The latter scheme is discussed next.

Let C_i be an already formed cluster. Our goal is to split it further, so that the two resulting clusters, C_i^1 and C_i^2, are as "dissimilar" as possible. Initially, we have $C_i^1 = \emptyset$ and $C_i^2 = C_i$. Then, we identify the vector in C_i^2 whose average dissimilarity from the remaining vectors is maximum, and we move it to C_i^1. In the sequel, for each of the remaining $\boldsymbol{x} \in C_i^2$, we compute its average dissimilarity with the vectors of C_i^1, $g(\boldsymbol{x}, C_i^1)$, as well as its average dissimilarity with the rest of the vectors in C_i^2, $g(\boldsymbol{x}, C_i^2 - \{\boldsymbol{x}\})$. If for *every* $\boldsymbol{x} \in C_i^2$, $g(\boldsymbol{x}, C_i^2 - \{\boldsymbol{x}\}) < g(\boldsymbol{x}, C_i^1)$, then we stop. Otherwise, we select the vector $\boldsymbol{x} \in C_i^2$ for which the difference $D(\boldsymbol{x}) = g(\boldsymbol{x}, C_i^2 - \{\boldsymbol{x}\}) - g(\boldsymbol{x}, C_i^1)$ is maximum (among the vectors of C_i^2, $\boldsymbol{x}$ exhibits the maximum dissimilarity with $C_i^2 - \{\boldsymbol{x}\}$ and the maximum similarity to C_i^1) and we move it to C_i^1. The procedure is repeated until the termination criterion is met. This iterative procedure accounts for step 2.2.1 of the "generalized divisive scheme" GDS.

In the preceding algorithm the splitting of the clusters is based on all the features (coordinates) of the feature vectors. Algorithms of this kind are also called *polythetic algorithms*. In fact, all the algorithms that have been or will be considered in this book are polythetic. In contrast, there are divisive algorithms that achieve the division of a cluster based on a single feature at each step. These are the so-called *monothetic algorithms*. Such algorithms are discussed in [Ever 01]. For more details see [Lamb 62, Lamb 66, MacN 65].

A large research effort has been devoted to comparing the performance of a number of the various hierarchical algorithms in the context of different applications. The interested reader may consult, for example, [Bake 74, Hube 74, Kuip 75, Dube 76, Mill 80, Mill 83].

13.5 HIERARCHICAL ALGORITHMS FOR LARGE DATA SETS

As we have seen in Section 13.2 the number of operations for the generalized agglomerative scheme (GAS) is of the order of N^3, and this cannot become less than $O(N^2)$, even if efficient computational schemes are employed. This section is devoted to a special type of hierarchical algorithms that are most appropriate for handling large data sets. As it has been stated elsewhere, the need for such algorithms stems from a number of applications, such as Web mining, bioinformatics, and so on.

The CURE Algorithm

The acronym CURE stands for Clustering Using REpresentatives. The innovative feature of CURE is that it represents each cluster, C, by a set of $k > 1$ representatives, denoted by R_C. By using multiple representatives for each cluster, the CURE algorithm tries to "capture" the shape of each one. However, in order to avoid taking into account irregularities in the border of the cluster, the initially chosen representatives are "pushed" toward the mean of the cluster. This action is also known as "shrinking" in the sense that the volume of space "defined" by the representatives is shrunk toward the mean of the cluster. More specifically, for each C the set R_C is determined as follows:

- Select the point $\boldsymbol{x} \in C$ with the maximum distance from the mean of C and set $R_C = \{\boldsymbol{x}\}$ (the set of representatives).
- For $i = 2$ to $\min\{k, n_C\}^8$
  - Determine $\boldsymbol{y} \in C - R_C$ that lies farthest from the points in R_C and set $R_C = R_C \cup \{\boldsymbol{y}\}$.
- End { For }
- Shrink the points $\boldsymbol{x} \in R_C$ toward the mean $\boldsymbol{m_c}$ in C by a factor a. That is, $\boldsymbol{x} = (1 - a)\boldsymbol{x} + a\boldsymbol{m_c}$, $\forall\ \boldsymbol{x} \in R_C$.

The resulting set R_C is the set of representatives of C. The distance between two clusters C_i and C_j is defined as

$$d(C_i, C_j) = \min_{\boldsymbol{x} \in R_{C_i},\, \boldsymbol{y} \in R_{C_j}} d(\boldsymbol{x}, \boldsymbol{y}) \tag{13.28}$$

Given the previous definitions, the CURE algorithm may be viewed as a special case of the *GAS* scheme. In its original version, the representatives of a cluster, C_q, generated from the agglomeration of two clusters C_i and C_j, are determined by taking into account all the points of C_q and applying the procedure described previously. However, to reduce the time complexity of this procedure, especially for the case of large number of points in each cluster, the representatives of C_q are selected among the $2k$ representatives of C_i and C_j. Such a choice is justified by the fact that the representatives are the most scattered points in each one of the clusters, C_i and C_j. Hence, it is expected that the resulting k representative points for C_q will also be well scattered. Once the final number of m clusters has been established, each point $\boldsymbol{x}$ in X, which is not among the representatives of any of the final clusters, is assigned to the cluster that contains the closest to $\boldsymbol{x}$ representative. The number m of clusters is either supplied to the algorithm by the user, based on his prior knowledge about the structure of the data set, or estimated using methods such as those described in Section 13.6.

[8] n_C denotes the number of points in C.

The worst-case time complexity of CURE can be shown to be $O(N^2 \log_2 N)$ ([Guha 98]). Clearly, this time complexity becomes prohibitive in the case of very large data sets. The technique adopted by the CURE algorithm, in order to reduce the computational complexity, is that of *random sampling*. That is, a sample set X' is created from X, by choosing *randomly* N' out of the N points of X. However, one has to ensure that the probability of missing a cluster of X, due to this sampling, is low. This can be guaranteed if the number of points N' is sufficiently large ([Guha 98]).

Having estimated N', CURE forms a number of $p = N/N'$ sample data sets by successive random sampling. In other words, X is partitioned randomly in p subsets. Let $q > 1$ be a user-defined parameter. Then, the points in each partition are clustered (following the procedure already explained) until N'/q clusters are formed or the distance between the closest pair of clusters to be merged in the next iteration step exceeds a user-defined threshold. Once the clustering procedure applied to each one of the subsets is completed, a second clustering pass on the (at most) $p(N'/q) = N/q$ clusters, obtained from all subsets, is performed. The goal of this pass is to apply the merging procedure described previously to all (at most) N/q clusters so that we end up with the required final number, m, of clusters. To this end, for each of the (at most) N/q clusters k representatives are used. Finally, each point $\boldsymbol{x}$ in the data set, X, that is not used as a representative in any one of the m clusters is assigned to one of them according to the following strategy. First, a random sample of representative points from each of the m clusters is chosen. Then, based on the previous representatives the point $\boldsymbol{x}$ is assigned to the cluster that contains the representative closest to it.

Experiments reported in [Guha 98] show that CURE is sensitive to the choice of the parameters k, N', and a. Specifically, k must be large enough to capture the geometry of each cluster. In addition, N' must be higher than a certain percentage of N (for the reported experiments, N' should be at least 2.5% of N). The choice of the shrinking factor, a, also affects the performance of CURE. For small values of a (small shrinkage), CURE exhibits a behavior similar to the MST algorithm, whereas for large values of a its performance resembles that of algorithms using a single point representative for each cluster. The worst-case execution time for CURE increases quadratically with the sample size N', that is, $O(N'^2 \log_2 N')$ ([Guha 98]).

Remarks

- The algorithm exhibits low sensitivity with respect to outliers within the clusters. The reason is that shrinking the scattered points toward the mean "dampens" the adverse effects due to outliers ([Guha 98]).
- The problem of outliers that form clusters by themselves is faced by CURE as follows. Because the outliers usually form small clusters, a few stages prior to the termination of the algorithm a check for clusters containing very few points takes place. These clusters are likely to consist of outliers, and they are removed.

- If N/q is chosen to be sufficiently large compared to the final number (m) of clusters in X, it is expected that the points in a subset X' that are merged to the same cluster during the application of CURE on X', will also belong to the same cluster (as if the entire data set, X, were taken into account). In other words, it is expected that the quality of the final clustering obtained by CURE will not be significantly affected by the partitioning of X.
- For $a=1$, all representatives diminish to the mean of the cluster.
- The CURE algorithm can reveal clusters with nonspherical or elongated shapes, as well as clusters of wide variance in size.
- In [Guha 98], the task of the efficient implementation of the algorithm is discussed using the heap and the $k-d$ tree data structures (see also [Corm 90] and [Same 89]).

The ROCK Algorithm

The RObust Clustering using linKs (ROCK) algorithm is best suited for nominal (categorical) features. For this type of data, it makes no sense to choose the mean of a cluster as a representative. In addition, the proximity between two feature vectors whose coordinates stem from a discrete data set cannot be adequately quantified by any of the l_p distances. ROCK introduces the idea of ***links***, in place of distances, to merge clusters.

Before we proceed further, some definitions are in order. Two points $\boldsymbol{x}, \boldsymbol{y} \in X$ are considered *neighbors* if $s(\boldsymbol{x}, \boldsymbol{y}) \geq \theta$, where $s(\cdot)$ is an appropriately chosen *similarity* function and θ is a user-defined parameter, which defines the similarity level according to which two points can be considered "similar." Let $link(\boldsymbol{x}, \boldsymbol{y})$ be the number of *common* neighbors between $\boldsymbol{x}$ and $\boldsymbol{y}$. Consider the graph whose vertices are the points in X and whose edges connect points that are neighbors. Then, it can easily be determined that $link(\boldsymbol{x}, \boldsymbol{y})$ may be viewed as the number of distinct paths of length 2 connecting $\boldsymbol{x}$ and $\boldsymbol{y}$.

Assume now that there exists a function $f(\theta)<1$ that is dependent on the data set as well as on the type of clusters we are interested in, with the following property: each point assigned to a cluster C_i has approximately $n_i^{f(\theta)}$ neighbors in C_i, where n_i is the number of points in C_i. Assuming that cluster C_i is large enough and that points outside C_i result in a very small number of links to the points of C_i, each point in C_i contributes approximately to $n_i^{2f(\theta)}$ ***links***. Thus, the expected total number of ***links*** among all pairs in C_i is $n_i^{1+2f(\theta)}$ (Problem 13.15).

The "closeness" between two clusters is assessed by the function

$$g(C_i, C_j) = \frac{link(C_i, C_j)}{(n_i + n_j)^{1+2f(\theta)} - n_i^{1+2f(\theta)} - n_j^{1+2f(\theta)}} \tag{13.29}$$

where $link(C_i, C_j) = \sum_{\boldsymbol{x} \in C_i, \boldsymbol{y} \in C_j} link(\boldsymbol{x}, \boldsymbol{y})$. Note that the denominator in this fraction is the expected total number of links *between* the two clusters. Clearly, the larger the value of $g(\cdot)$ the more similar the clusters C_i and C_j. The number of clusters, m, is a user-defined parameter. At each iteration, the clusters with maximum $g(\cdot)$ are merged. The procedure stops when the number of clusters formed becomes equal to the desired number of clusters, m, or the number of *links* between every pair of clusters in a clustering $\Re_t$ is 0.

In the case where the data set X is very large (and in order to improve the execution time), ROCK is applied on a reduced data set, X', stemming from X via random sampling, whose size N' is estimated as in the CURE algorithm. After the clusters in the sample data set X' have been identified, the points in X that were not selected in the sample subset are assigned to a cluster via the following strategy. First, a subset L_i of points from each cluster C_i is selected. In the sequel, for each $\boldsymbol{z}$ in $X - X'$ and each cluster C_i, the number N_i of its neighbors among the L_i points is determined. Then $\boldsymbol{z}$ is assigned to the cluster C_i for which the quantity $N_i/(n_{L_i} + 1)^{f(\theta)}$ is maximum, where n_{L_i} is the number of points in L_i. Note that the denominator in this expression is the expected number of neighbors of $\boldsymbol{z}$ in $L_i \cup \{\boldsymbol{z}\}$.

Remarks

- In [Guha 00], it is proposed using $f(\theta) = (1 - \theta)/(1 + \theta)$, with $\theta < 1$.
- The algorithm makes a rather strong hypothesis about the existence of the function $f(\theta)$. In other words, it poses a constraint for each cluster in the data set, which may lead to poor results if the clusters in the data set do not satisfy this hypothesis. However, in the experimental cases discussed in [Guha 00] the clustering results were satisfactory for the choice of $f(\theta)$.
- For large values of N, the worst-case time complexity of ROCK is similar to that of CURE.

In [Dutt 05] it is proved that, under certain conditions, the clusters produced by the ROCK algorithm are the connected components of a certain graph, called *link graph* whose vertices correspond to the data points. Based on the above result a quick version of ROCK, called QROCK, is described, which simply identifies the connected components of the link graph.

The Chameleon Algorithm

The algorithms CURE and ROCK, described previously, are based on "static" modeling of the clusters. Specifically, CURE models each cluster by the same number, k, of representatives, whereas ROCK poses constraints on the clusters through $f(\theta)$. Clearly, these algorithms may fail to unravel the clustering structure of the data set in cases where the individual clusters do not obey the adopted model, or when noise is present. In the sequel, another hierarchical clustering algorithm, known as *Chameleon*, is presented. The algorithm is capable of recovering clusters of various shapes and sizes.

To quantify the similarity between two clusters, we define the concepts of *relative interconnectivity* and *relative closeness*. Both of these quantities are defined in terms of graph theory concepts. Specifically, a graph $G = (V, E)$ is constructed so that each data point corresponds to a vertex in V and E contains edges among vertices. The weight of each edge is set equal to the *similarity* between the points associated with the connected vertices.

Before we proceed further, the following definitions are in order. Let C be the set of points corresponding to a subset of V. Assume that C is partitioned into two nonempty subsets C_1 and C_2 ($C_1 \cup C_2 = C$). The subset of the edges E' of E that connect C_1 and C_2 form the *edge cut set*. If the sum of the weights of the edge cut set, E', corresponding to the (C_1, C_2) partition of C is minimum among all edge cut sets resulting from all possible partitions of C into two sets (excluding the empty set cases), E' is called the *minimum cut set* of C. If, now, C_1 and C_2 are constrained to be of approximately equal size, then the minimum sum E' (over all possible partitions of approximately equal size) is known as the *minimum cut bisector* of C. For example, suppose that C is represented by the vertices v_1, v_2, v_3, v_4, and v_5 in Figure 13.15. The edges of the respective graph G are e_{12}, e_{23}, e_{34}, e_{45}, and e_{51}, with weights 0.1, 0.2, 0.3, 0.4, and 0.5, respectively. Then, the sums of the weights of the edges of the sets $\{e_{51}, e_{34}\}$, $\{e_{12}, e_{23}\}$, and $\{e_{12}, e_{34}\}$ are 0.8, 0.3, and 0.4, respectively. The second edge cut set corresponds to the minimum cut set of C, whereas the third corresponds to the minimum cut bisector of C.

Relative Interconnectivity

Let E_{ij} be the set of edges connecting points in C_i with points in C_j, and E_i be the set of edges that corresponds to the minimum cut bisector of C_i. The *absolute interconnectivity*, $|E_{ij}|$, between two clusters C_i and C_j, is defined as the sum of the weights of the edges in E_{ij}. Equivalently, this is the edge cut set associated with the partition of $C_i \cup C_j$ into C_i and C_j. The *internal interconnectivity* $|E_i|$ of a cluster C_i is defined as the sum of the weights of its minimum cut bisector E_i. The relative

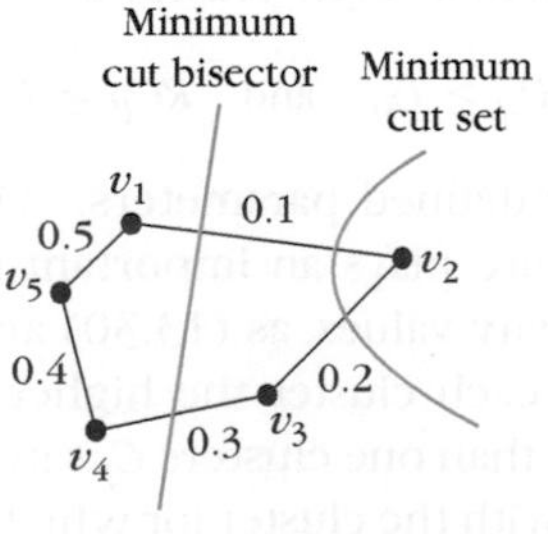

FIGURE 13.15

Partitions leading to the minimum cut set and the minimum cut bisector of the set C containing five vertices (see text for explanation).

interconnectivity between two clusters, C_i and C_j, is defined as

$$RI_{ij} = \frac{|E_{ij}|}{\frac{|E_i|+|E_j|}{2}} \tag{13.30}$$

Relative Closeness

Let S_{ij} be the average weight of the edges in the set E_{ij}, and S_i be the average weight of the edges in E_i. Then the relative closeness between two clusters C_i and C_j is defined as

$$RC_{ij} = \frac{S_{ij}}{\frac{n_i}{n_i+n_j}S_i + \frac{n_j}{n_i+n_j}S_j} \tag{13.31}$$

where n_i and n_j are the number of points in C_i and C_j, respectively.

Having defined the previous quantities, we proceed now to description of the algorithm. This algorithm, unlike most hierarchical algorithms that are either agglomerative or divisive, enjoys both of these characteristics. Initially, a k-nearest neighbor graph is created. More specifically, each vertex of the graph corresponds to a feature vector, and an edge is added between two vertices if *at least* one of the corresponding points is among the k-nearest neighbors of the other (typically k takes values in the range from 5 to 20). Note that if a point $\boldsymbol{x}$ is among the k nearest neighbors of a point $\boldsymbol{y}$, this does not necessarily mean that $\boldsymbol{y}$ is among the k nearest neighbors of $\boldsymbol{x}$.

The first phase of the algorithm is the divisive phase. Initially, all points belong to a single cluster. This cluster is partitioned into two clusters so that the sum of the weights of the edge cut set between the resulting clusters is minimized and each of the resulting clusters contains at least 25% of the vertices of the initial cluster. Then, at each step the largest cluster is selected and is partitioned as indicated previously. This procedure terminates when all obtained clusters, at a given level, contain fewer than q points (a user-defined parameter, typically chosen in the range 1 to 5% of the total number of points in X). In the sequel, the agglomerative phase is applied on the set of clusters that have resulted from the previous phase. Specifically, two clusters C_i and C_j are merged to a single cluster if

$$RI_{ij} \geq T_{RI} \quad \text{and} \quad RC_{ij} \geq T_{RC} \tag{13.32}$$

where T_{RI} and T_{RC} are user-defined parameters. Observe that for merging two clusters their internal structure plays an important role through their respective S and internal interconnectivity values, as (13.30) and (13.31) suggest. The more similar the elements within each cluster the higher "their resistance" in merging with another cluster. If more than one clusters C_j satisfy both conditions for a given cluster C_i, then C_i is merged with the cluster for which $|E_{ij}|$ is the highest among the candidates C_js. In addition, unlike most agglomerative algorithms, it is permissible to merge more than one pair of clusters at a given level of hierarchy (provided, of course, that these pairs satisfy the condition (13.32)).

A different rule that may be employed for the choice of the clusters to be merged is the following: merge those clusters that maximize the quantity

$$RI_{ij}RC_{ij}^{a} \tag{13.33}$$

where a gives the relative importance between RI and RC. Typically, a is chosen between 1.5 and 3.

Example 13.10

To gain a better understanding of the rationale behind the Chameleon algorithm, let's consider the overly-simplistic four-cluster data set depicted in Figure 13.16. We assume that $k = 2$. The similarities between the connected pairs of points for C_1, C_2, C_3, and C_4 are as shown in the figure (the weights between the connected pairs of points in C_3 and C_4 not shown in the figure are all equal to 0.9). In addition, the similarity between the closest points between clusters C_1 and C_2 is 0.4, whereas the similarity between the closest points between clusters C_3 and C_4 is 0.6. Note that although C_3 and C_4 lie closer to each other than C_1 and C_2, each exhibits a significantly higher degree of internal interconnectivity, compared to each of the C_1 and C_2.

For the previous clusters we have $|E_1| = 0.48$, $|E_2| = 0.48$, $|E_3| = 0.9 + 0.55 = 1.45$, $|E_4| = 1.45$, $|S_1| = 0.48$, $|S_2| = 0.48$, $|S_3| = 1.45/2 = 0.725$, and $|S_4| = 0.725$. In addition, $|E_{12}| = 0.4$, $|E_{34}| = 0.6$, $|S_{12}| = 0.4$, and $|S_{34}| = 0.6$. Taking into account the definitions of RI and RC, we have $RI_{12} = 0.833$, $RI_{34} = 0.414$ and $RC_{12} = 0.833$, $RC_{34} = 0.828$. We see that both RI and RC favor the merging of C_1 and C_2 against the merging of C_3 and C_4.

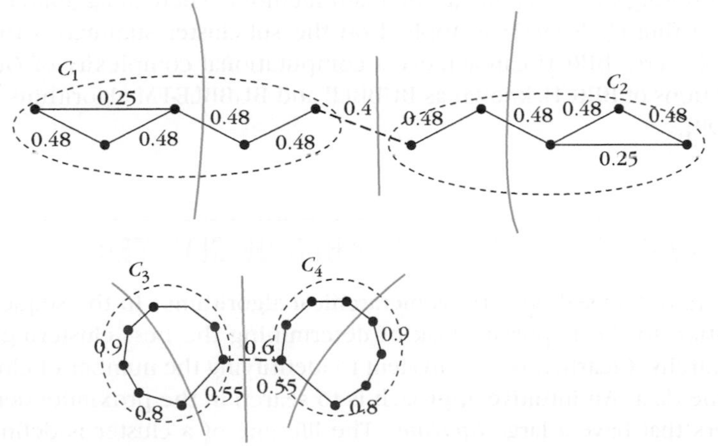

FIGURE 13.16

The high degree of internal interconnectivity of C_3 and C_4 favors the merging of C_1 with C_2, although the neighbors between C_1 and C_2 are less similar than the neighbors between C_3 and C_4.

Thus, Chameleon merges C_1 and C_2, which is in agreement with intuition. Note that the MST algorithm would merge the clusters C_3 and C_4.

Remarks

- This algorithm requires user-defined parameters (k, q and T_{RI}, T_{RC} for the rule given in (13.32) or a for the rule given in (13.33)). Experiments reported in [Kary 99] show that Chameleon is not very sensitive to the choice of the parameters k, q, and a.
- Chameleon requires large data sets in order to have more accurate estimates for $|E_{ij}|$, $|E_i|$, S_{ij}, and S_i. Thus, the algorithm is well suited for the applications where the volume of the available data is large.
- For large N, the worst-case time complexity of the algorithm is $O(N(\log_2 N + m))$, where m is the number of clusters formed after completion of the first (divisive) phase of the algorithm.

An alternative clustering method that employs agglomerative algorithms and is suitable for processing very large data sets is the so-called BIRCH method (Balanced Iterative Reducing and Clustering using Hierarchies) ([Zhan 96]). This method has been designed so as to minimize the number of I/O operations (that is, the exchange of information between the main memory and the secondary memory where the data set is stored). It performs a preclustering of data and stores a compact summary for each generated subcluster in a specific data structure called $CF - tree$. More specifically, it generates the maximum number of subclusters such that the resulting $CF - tree$ fits in the main memory. Then an agglomerative clustering algorithm ([Olso 93]) is applied on the subcluster summaries to produce the final clusters. BIRCH can achieve a computational complexity of $O(N)$. Two generalizations of BIRCH, known as BUBBLE and BUBBLE-FM algorithms, are given in [Gant 99].

13.6 CHOICE OF THE BEST NUMBER OF CLUSTERS

So far, we have focused on various hierarchical algorithms. In the sequel we turn our attention to the important task of determining the best clustering within a given hierarchy. Clearly, this is equivalent to identifying the number of clusters that best fits the data. An intuitive approach is to search in the proximity dendrogram for clusters that have a large *lifetime*. The lifetime of a cluster is defined as the absolute value of the difference between the proximity level at which it is created and the proximity level at which it is absorbed into a larger cluster. For example, the dendrogram of Figure 13.17a suggests that two major clusters are present and that of Figure 13.17b suggests only one. In [Ever 01], experiments are conducted to

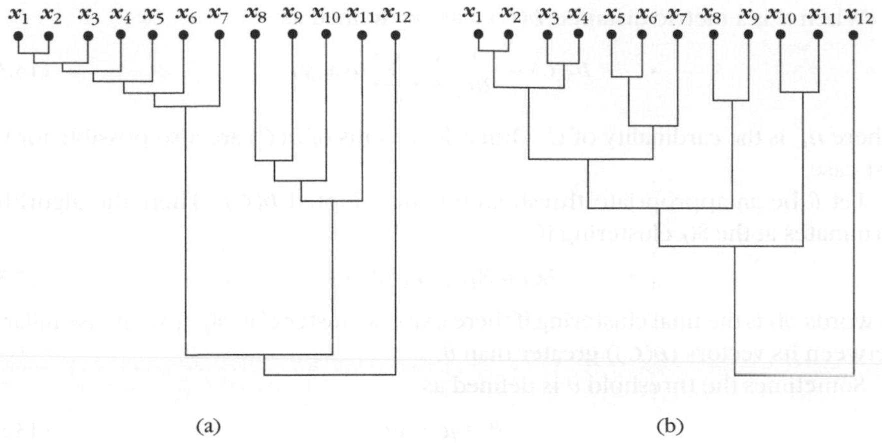

FIGURE 13.17

(a) A dendrogram that suggests that there are two major clusters in the data set. (b) A dendrogram indicating that there is a single major cluster in the data set.

assess the behavior of various agglomerative algorithms when (a) a single compact cluster and (b) two compact clusters are formed by the vectors of X. However, human subjectivity is required to reach conclusions.

Many formal methods that may be used in cooperation with both hierarchical and nonhierarchical algorithms for identifying the best number of clusters for the data at hand have been proposed (e.g., [Cali 74, Duda 01, Hube 76]).

A comparison of many such methods is given in [Mill 85]. In the sequel, we discuss two methods, proposed in [Bobe 93] for identifying the clustering that best fits the data that are appropriate for agglomerative algorithms. The clustering algorithm does not necessarily produce the whole hierarchy of N clusterings, but it terminates when the clustering that best fits the data has been achieved, according to a criterion.

Method I

This is an extrinsic method, in the sense that it requires determination of the value of a specific parameter by the user. It involves the definition of a function $h(C)$ that measures the dissimilarity between the vectors of the same cluster C. That is, we can view it as a "self-similarity" measure. For example, $h(C)$ may be defined as

$$h_1(C) = \max\{d(\boldsymbol{x},\boldsymbol{y}), \boldsymbol{x},\boldsymbol{y} \in C\} \tag{13.34}$$

or

$$h_2(C) = \text{med}\{d(\boldsymbol{x},\boldsymbol{y}), \boldsymbol{x},\boldsymbol{y} \in C\} \tag{13.35}$$

(see Figure 13.18a).

When d is a metric distance, $h(C)$ may be defined as

$$h_3(C) = \frac{1}{2n_C} \sum_{x \in C} \sum_{y \in C} d(x, y) \tag{13.36}$$

where n_C is the cardinality of C. Other definitions of $h(C)$ are also possible for the last case.

Let θ be an appropriate threshold for the adopted $h(C)$. Then, the algorithm terminates at the $\Re_t$ clustering if

$$\exists C_j \in \Re_{t+1} : h(C_j) > \theta \tag{13.37}$$

In words, $\Re_t$ is the final clustering if there exists a cluster C in $\Re_{t+1}$ with dissimilarity between its vectors ($h(C)$) greater than θ.

Sometimes the threshold θ is defined as

$$\theta = \mu + \lambda\sigma \tag{13.38}$$

where μ is the average distance between any two vectors in X and σ is its variance. The parameter λ is a user-defined parameter. Thus, the need for specifying an appropriate value for θ is transferred to the choice of λ. However, λ may be estimated more reasonably than θ.

Method II

This is an intrinsic method; that is, in this case only the structure of the data set X is taken into account. According to this method, the final clustering $\Re_t$ *must* satisfy the following relation:

$$d_{\min}^{ss}(C_i, C_j) > \max\{h(C_i), h(C_j)\}, \quad \forall C_i, C_j \in \Re_t \tag{13.39}$$

where $d_{\min}^{ss}$ is defined in Chapter 11. In words, in the final clustering, the dissimilarity between every pair of clusters is larger than the "self-similarity" of each of them (see Figure 13.18b). Note that this is only a necessary condition.

Finally, it must be stated that all these methods are based on heuristic arguments, and they are indicative only of the best clustering.

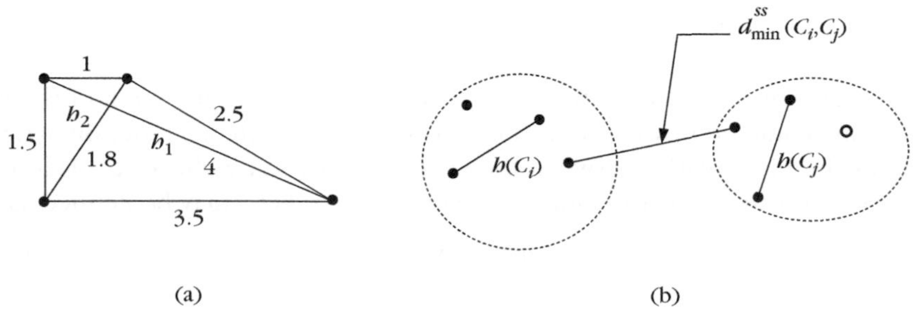

FIGURE 13.18

(a) Examples of "self-similarity" measures. (b) Illustration of the termination condition for method II.

13.7 PROBLEMS

13.1 Consider the Euclidean distance as the proximity measure between two vectors. Consider only ratio-scaled vectors. Prove that:

a. A pattern matrix uniquely identifies the corresponding proximity matrix.

b. A proximity matrix does not identify a pattern matrix uniquely. Furthermore, there are proximity matrices which do not correspond to any pattern matrix.

Hint: (b) Consider, for example, the translations of the points of the data set X.

13.2 Derive Eq. (13.10) from Eq. (13.8).
Hint: Make use of the following identities:

$$n_3 \boldsymbol{m}_3 = n_1 \boldsymbol{m}_1 + n_2 \boldsymbol{m}_2 \tag{13.40}$$

and

$$n_1 \|\boldsymbol{m}_1 - \boldsymbol{m}_3\|^2 + n_2 \|\boldsymbol{m}_2 - \boldsymbol{m}_3\|^2 = \frac{n_1 n_2}{n_1 + n_2} \|\boldsymbol{m}_1 - \boldsymbol{m}_2\|^2 \tag{13.41}$$

where C_1 and C_2 are any two clusters and $C_3 = C_1 \cup C_2$.

13.3 Show that for the WPGMC algorithm there are cases where $d_{qs} \leq \min(d_{is}, d_{js})$.

13.4 Prove

$$d'_{qs} = \frac{n_q n_s}{n_q + n_s} d_{qs} \tag{13.42}$$

Hint: Multiply both sides of

$$\|\boldsymbol{m}_q - \boldsymbol{m}_s\|^2 = \frac{n_i}{n_i + n_j} d_{is} + \frac{n_j}{n_i + n_j} d_{js} - \frac{n_i n_j}{(n_i + n_j)^2} d_{ij}$$

by $(n_i + n_j)n_s/(n_i + n_j + n_s)$ (This equation holds from Problem 13.2).

13.5 **a.** Prove Eq. (13.16).

b. Complete the proof of Eq. (13.19).

Hint: Take the squares of both sides of Eq. (13.18).

13.6 Consider the proximity matrix given in Example 13.5. Find the proximity dendrograms derived by the GTAS algorithm when $h(k)$ is (a) the node connectivity property and (b) the edge connectivity property, with $k = 3$.

13.7 Prove that the distance between two clusters C_r and C_s, $d(C_r, C_s)$, which are at the same level of the hierarchy produced by the single link algorithm, given by Eq. (13.4), may be written as

$$d(C_r, C_s) = \min_{x \in C_r, y \in C_s} d(x, y) \tag{13.43}$$

That is, the single link algorithms based on matrix and graph theory are equivalent.

Hint: Proceed by induction on the level of hierarchy t. Take into account that the clusterings $\Re_t$ and $\Re_{t+1}$ have $N - t - 2$ common clusters.

13.8 Prove that the distance between two clusters C_r and C_s, $d(C_r, C_s)$, which are at the same level of the hierarchy produced by the complete link algorithm, given by Eq. (13.5), may be written as

$$d(C_r, C_s) = \max_{x \in C_r, y \in C_s} d(x, y) \tag{13.44}$$

That is, the complete link algorithms based on matrix and graph theory are equivalent.

Hint: Take into account the hints of the previous problem.

13.9 State and prove the propositions of the previous two problems when similarity measures between two vectors are used.

13.10 Consider the following proximity matrix:

$$P = \begin{bmatrix} 0 & 4 & 9 & 6 & 5 \\ 4 & 0 & 1 & 8 & 7 \\ 9 & 1 & 0 & 3 & 2 \\ 6 & 8 & 3 & 0 & 1 \\ 5 & 7 & 2 & 1 & 0 \end{bmatrix}$$

Apply the single and complete link algorithms to P and comment on the resulting dendrograms.

13.11 Consider the dissimilarity matrix P given in Example 13.5. Let us change $P(3, 4)$ to 6 ($P(4, 6)$ is also equal to 6). Also let $h(k)$ be the node degree property with $k = 2$. Run the corresponding graph theory algorithm when (a) the edge (3, 4) is considered first and (b) the edge (4, 6) is considered first. Comment on the resulting dendrograms.

13.12 Consider the following dissimilarity matrix:

$$P = \begin{bmatrix} 0 & 4 & 9 & 6 & 5 \\ 4 & 0 & 3 & 8 & 7 \\ 9 & 3 & 0 & 3 & 2 \\ 6 & 8 & 3 & 0 & 1 \\ 5 & 7 & 2 & 1 & 0 \end{bmatrix}$$

a. Determine all possible dendrograms resulting from application of the single and the complete link algorithms to P and comment on the results.

b. Set $P(3,4) = 4$, $P(1,2) = 10$, and let P_1 be the new proximity matrix. Note that P_1 contains no ties. Determine all possible dendrograms resulting from the application of the UPGMA algorithm to P_1.

13.13 Consider the general divisive scheme. Assume that at step 2.2.1 of the algorithm the cluster $C^1_{t-1,t}$ consists of a single vector, for $i = 1, \ldots, t$, $t = 1, \ldots, N$. Compute the number of pairs of clusters that have to be examined during the whole clustering process. Discuss the merits and the disadvantages of this scheme.

13.14 Does the alternative divisive algorithm discussed in Section 13.4 guarantee the determination of the optimum possible clustering at each level? Is it reasonable to extend the answer to this question to other such divisive algorithms?

13.15 In the ROCK algorithm prove that the expected total number of links among all pairs of points in a cluster, C, is $n^{1+2f(\theta)}$, where n is the cardinality of C.

13.16 Explain the physical meaning of Eq. (13.38).

MATLAB PROGRAMS AND EXERCISES

Computer Programs

13.1 *Convert matrix to vector.* Write a MATLAB function named *convert_prox_mat* that takes as input an $N \times N$ dimensional proximity matrix *prox_mat* and returns an $N(N-1)/2$ dimensional row vector *proc_vec* that contains the upper diagonal elements of *prox_mat* in the following order: $(1,2)$, $(1,3)$, $\ldots (1,N)$, $(2,3)$, $\ldots (2,N)$, $\ldots (N-1,N)$.

Solution

```
function prox_vec=convert_prox_mat(prox_mat)
  [N,N]=size(prox_mat);
  prox_vec=[];
  for i=1:N-1
    prox_vec=[prox_vec prox_mat(i,i+1:N)];
  end
```

13.2 *Single link, Complete link algorithms.* Write MATLAB code for the single link and complete link algorithms.

Solution

Just type

```
Z=linkage(prox_vec,'single')
```

for the single link algorithm and

```
Z=linkage(prox_vec,'complete')
```

for the complete link algorithm. The function *linkage* is a built-in MATLAB function that takes as inputs (a) a proximity vector in the form described before in 13.1 and (b) the type of the agglomerative algorithm that will be used. It returns an $(N-1) \times 3$ matrix Z, each line of which corresponds to a clustering of the hierarchy. The first two elements of the ith line of Z contain the indices of the objects that were linked at this level to form a new cluster. The index value $N + i$ is assigned to the new cluster. If, for example, $N = 7$ and at the first level the elements 1 and 4 are merged, the cluster consisting of these elements will be represented by the integer $8(= 7 + 1)$. If cluster 8 appears at a later row, this means that this cluster is being combined with another cluster in order to form a larger cluster. Finally, the third element of the ith line of Z contains the dissimilarity between the clusters that have been merged at this level.

13.3 Generate the dendrogram corresponding to the output of an agglomerative algorithm.

Solution

Just type

```
H=dendrogram(Z);
```

The *dendrogram* is a built-in MATLAB function that takes as input the output of the *linkage* function and generates the corresponding dendrogram.

Computer Experiments

13.1 Consider the following dissimilarity matrix

prox_mat =

0.0	2.0	4.2	6.6	9.2	12.0	15.0	300	340	420
2.0	0.0	2.2	4.6	7.2	10.0	13.0	280	320	400
4.2	2.2	0.0	2.4	5.0	7.8	10.8	270	310	390
6.6	4.6	2.4	0.0	2.6	5.4	8.4	260	300	380
9.2	7.2	5.0	2.6	0.0	2.8	5.8	262	296	388
12.0	10.0	7.8	5.4	2.8	0.0	3.0	316	280	414
15.0	13.0	10.8	8.4	5.8	3.0	0.0	380	326	470
300	280	270	260	262	316	380	0.0	4.0	4.4
340	320	310	300	296	280	326	4.0	0.0	9.0
420	400	390	380	388	414	470	4.4	9.0	0.0

a. Apply the single link and the complete link algorithms on *prox_mat* and draw the corresponding dissimilarity dendrograms.

b. Comment on the way the successive clusterings are created in each of the two algorithms.

c. Compare the dissimilarity thresholds where clusters are merged for both single link and complete link algorithms.

d. Can you draw any conclusion about the most natural clustering for the above data set by considering the dissimilarity thresholds where the clusters are merged for both algorithms?

REFERENCES

[Ande 73] Anderberg M.R. *Cluster Analysis for Applications*, Academic Press, 1973.

[Bake 74] Baker F.B. "Stability of two hierarchical grouping techniques. Case 1. Sensitivity to data errors," *J. Am. Statist. Assoc.*, Vol. 69, pp. 440–445, 1974.

[Bobe 93] Boberg J., Salakoski T. "General formulation and evaluation of agglomerative clustering methods with metric and non-metric distances," *Pattern Recognition*, Vol. 26(9), pp. 1395–1406, 1993.

[Cali 74] Calinski R.B., Harabasz J. "A dendrite method for cluster analysis," *Communications in Statistics*, Vol. 3, pp. 1–27, 1974.

[Corm 90] Cormen T.H., Leiserson C.E., Rivest R.L. *Introduction to Algorithms*, MIT Press, 1990.

[Day 84] Day W.H.E., Edelsbrunner H. "Efficient algorithms for agglomerative hierarchical clustering methods," *Journal of Classification*, Vol. 1(1), pp. 7–24, 1984.

[Dube 76] Dubes R., Jain A.K. "Clustering techniques: The user's dilemma," *Pattern Recognition*, Vol. 8, pp. 247–260, 1976.

[Dube 87] Dubes R. "How many clusters are best?—An experiment," *Pattern Recognition*, Vol. 20(6), pp. 645–663, 1987.

[Duda 01] Duda R., Hart P., Stork D. *Pattern Classification*, 2nd ed., John Wiley & Sons, 2001.

[Dutt 05] Dutta M., Mahanta A.K., Pujari A.K. "QROCK: A quick version of the ROCK algorithm for clustering catergorical data," *Pattern Recognition Letters*, Vol. 26, pp. 2364–2373, 2005.

[El-G 68] El-Gazzar A., Watson L., Williams W.T., Lance G. "The taxonomy of Salvia: A test of two radically different numerical methods," *J. Linn. Soc. (Bot.)*, Vol. 60, pp. 237–250, 1968.

[Ever 01] Everitt B., Landau S., Leese M. *Cluster Analysis*, Arnold, 2001.

[Frig 97] Frigui H., Krishnapuram R. "Clustering by competitive agglomeration," *Pattern Recognition*, Vol. 30(7), pp. 1109–1119, 1997.

[Gant 99] Ganti V., Ramakrishnan R., Gehrke J., Powell A., French J. "Clustering large datasets in arbitrary metric spaces," *Proceedings 15th International Conference on Data Engineering*, pp. 502–511, 1999.

[Gowd 95] Gowda Chidananda K., Ravi T.V. "Divisive clustering of symbolic objects using the concepts of both similarity and dissimilarity," *Pattern Recognition*, Vol. 28(8), pp. 1277–1282, 1995.

[Gowe 67] Gower J.C. "A comparison of some methods of cluster analysis," *Biometrics*, Vol. 23, pp. 623–628, 1967.

[Guha 98] Guha S., Rastogi R., Shim K. "CURE: An efficient clustering algorithm for large databases," *Proceedings of the ACM SIGMOD Conference on Management of Data*, pp. 73–84, 1998.

[Guha 00] Guha S., Rastogi R., Shim K. "ROCK: A robust clustering algorithm for categorical attributes," *Information Systems*, Vol. 25, No. 5, pp. 345–366, 2000.

[Hods 71] Hodson F.R. "Numerical typology and prehistoric archaeology," in *Mathematics in Archaeological and Historical Sciences* (Hodson F.R., Kendell D.G., Tautu P.A., eds.), University Press, Edinburgh, 1971.

[Horo 78] Horowitz E., Sahni S. *Fundamentals of Computer Algorithms*, Computer Science Press, 1978.

[Hube 74] Hubert L.J. "Approximate evaluation techniques for the single link and complete link hierarchical clustering procedures," *J. Am.Statist. Assoc.*, Vol. 69, pp. 698–704, 1974.

[Hube 76] Hubert L.J., Levin J.R. "A general statistical framework for assessing categorical clustering in free recall," *Psychological Bulletin*, Vol. 83, pp. 1072–1080, 1976.

[Jain 88] Jain A.K., Dubes R.C. *Algorithms for Clustering Data*, Prentice Hall, 1988.

[Jard 71] Jardine N., Sibson R. *Numerical Taxonomy*, John Wiley & Sons, 1971.

[Kank 96] Kankanhalli M.S., Mehtre B.M., Wu J.K. "Cluster-based color matching for image retrieval," *Pattern Recognition*, Vol. 29(4), pp. 701–708, 1996.

[Kary 99] Karypis G., Han E., Kumar V. "Chameleon: Hierarchical clustering using dynamic modeling," *IEEE Computer*, Vol. 32, No. 8, pp. 68–75, 1999.

[Kuip 75] Kuiper F.K., Fisher L. "A Monte Carlo comparison of six clustering procedures," *Biometrics*, Vol. 31, pp. 777–783, 1975.

[Kuri 91] Kurita T. "An efficient agglomerative clustering algorithm using a heap," *Pattern Recognition*, Vol. 24, pp. 205–209, 1991.

[Lamb 62] Lambert J.M., Williams W.T. "Multivariate methods in plant technology, IV. Nodal analysis," *J. Ecol.*, Vol. 50, pp. 775–802, 1962.

[Lamb 66] Lambert J.M., Williams W.T. "Multivariate methods in plant technology, IV. Comparison of information analysis and association analysis," *J. Ecol.*, Vol. 54, pp. 635–664, 1966.

[Lanc 67] Lance G.N., Williams W.T. "A general theory of classificatory sorting strategies: II. Clustering systems," *Computer Journal*, Vol. 10, pp. 271–277, 1967.

[Li 90] Li X. "Parallel algorithms for hierarchical clustering and cluster validity," *IEEE Transactions on Pattern Analysis and Machine Intelligence*, Vol. 12(11), pp. 1088–1092, 1990.

[Ling 72] Ling R.F. "On the theory and construction of k-clusters," *Computer Journal*, Vol. 15, pp. 326–332, 1972.

[Liu 68] Liu C.L. *Introduction to Combinatorial Mathematics*, McGraw-Hill, 1968.

[MacN 64] MacNaughton-Smith P., Williams W.T., Dale M.B., Mockett L.G. "Dissimilarity analysis," *Nature*, Vol. 202, pp. 1034–1035, 1964.

[MacN 65] MacNaughton-Smith P. "Some statistical and other numerical techniques for classifying individuals," *Home Office Research Unit Report No.* 6, London: H.M.S.O., 1965.

[Marr 71] Marriot F.H.C. "Practical problems in a method of cluster analysis," *Biometrics*, Vol. 27, pp. 501-514, 1971.

[McQu 62] McQuitty L.L. "Multiple hierarchical classification of institutions and persons with reference to union-management relations and psychological well-being," *Educ. Psychol. Measur.*, Vol. 23, pp. 55-61, 1962.

[Mill 80] Milligan G.W. "An examination of the effect of six types of error perturbation on fifteen clustering algorithms," *Psychometrika*, Vol. 45, pp. 325-342, 1980.

[Mill 85] Milligan G.W., Cooper M.C. "An examination of procedures for determining the number of clusters in a data set," *Psychometrika*, Vol. 50(2), pp. 159-179, 1985.

[Mill 83] Milligan G.W., Soon S.C., Sokol L.M. "The effect of cluster size, dimensionality and the number of clusters on recovery of true cluster structure," *IEEE Transactions on Pattern Analysis and Machine Intelligence*, Vol. 5(1), January 1983.

[Murt 83] Murtagh F. "A survey of recent advances in hierarchical clustering algorithms," *Journal of Computation*, Vol. 26, pp. 354-359, 1983.

[Murt 84] Murtagh F. "Complexities of hierarchic clustering algorithms: State of the art," *Computational Statistics Quarterly*, Vol. 1(2), pp. 101-113, 1984.

[Murt 85] Murtagh F. *Multidimensional clustering algorithms*, Physica-Verlag, COMPSTAT Lectures, Vol. 4, Vienna, 1985.

[Murt 95] Murthy M.N., Jain A.K. "Knowledge-based clustering scheme for collection, management and retrieval of library books," *Pattern Recognition*, Vol. 28(7), pp. 949-963, 1995.

[Olso 93] Olson C.F. "Parallel algorithms for hierarchical clustering," Technical Report, University of California at Berkeley, December. 1993.

[Ozaw 83] Ozawa K. "Classic: A hierarchical clustering algorithm based on asymmetric similarities," *Pattern Recognition*, Vol. 16(2), pp. 201-211, 1983.

[Prit 71] Pritchard N.M., Anderson A.J.B. "Observations on the use of cluster analysis in botany with an ecological example," *J. Ecol.*, Vol. 59, pp. 727-747, 1971.

[Rolp 73] Rolph F.J. "Algorithm 76: Hierarchical clustering using the minimum spanning tree," *Journal of Computation*, Vol. 16, pp. 93-95, 1973.

[Rolp 78] Rolph F.J. "A probabilistic minimum spanning tree algorithm," *Information Processing Letters*, Vol. 7, pp. 44-48, 1978.

[Same 89] Samet H. *The Design and Analysis of Spatial Data Structures*. Addison-Wesley, Boston, 1989.

[Shea 65] Sheals J.G. "An application of computer techniques to Acrine taxonomy: A preliminary examination with species of the Hypoaspio-Androlaelaps complex Acarina," *Proc. Linn. Soc. Lond.*, Vol. 176, pp. 11-21, 1965.

[Snea 73] Sneath P.H.A., Sokal R.R. *Numerical Taxonomy*, W.H. Freeman & Co., 1973.

[Solo 71] Solomon H. "Numerical taxonomy," in *Mathematics in the Archaeological and Historical Sciences* (Hobson F.R., Kendall D.G., Tautu P.A., eds.), University Press, 1971.

[Stri 67] Stringer P. "Cluster analysis of non-verbal judgement of facial expressions," *Br. J. Math. Statist. Psychol.*, Vol. 20, pp. 71-79, 1967.

[Will 89] Willett P. "Efficiency of hierarchic agglomerative clustering using the ICL distributed array processor," *Journal of Documentation*, Vol. 45, pp. 1–45, 1989.

[Will 77] Williams W.T., Lance G.N. "Hierarchical classificatory methods," in *Statistical Methods for Digital Computers* (Enslein K., Ralston A., Wilf H.S., eds.), John Wiley & Sons, 1977.

[Zhan 96] Zhang T., Ramakrishnan R., Livny M. "BIRCH: An efficient data clustering method for very large databases," *Proceedings of the ACM SIGMOD Conference on Management of Data*, pp. 103–114, Montreal, Canada, 1996.

CHAPTER

Clustering Algorithms III: Schemes Based on Function Optimization

14

14.1 INTRODUCTION

One of the most commonly used families of clustering schemes relies on the optimization of a cost function J using differential calculus techniques (e.g., see [Duda 01, Bezd 80, Bobr 91, Kris 95a, Kris 95b]). The cost J is a function of the vectors of the data set X and it is parameterized in terms of an unknown parameter vector, $\boldsymbol{\theta}$. For most of the schemes of the family, the number of clusters, m, is assumed to be known.

Our goal is the estimation of $\boldsymbol{\theta}$ that characterizes best the clusters underlying X. The parameter vector $\boldsymbol{\theta}$ is strongly dependent on the shape of the clusters. For example, for compact clusters (see Figure 14.1a), it is reasonable to adopt as parameters a set of m points, $\boldsymbol{m}_i$, in the l-dimensional space, each corresponding to a cluster—thus, $\boldsymbol{\theta} = [\boldsymbol{m}_1^T, \boldsymbol{m}_2^T, \ldots, \boldsymbol{m}_m^T]^T$. On the other hand, if ring-shaped clusters are expected (see Figure 14.1b), it is reasonable to use m hyperspheres $C(\boldsymbol{c}_i, r_i)$, $i = 1, \ldots, m$, as representatives, where $\boldsymbol{c}_i$ and r_i are the center and the radius of the ith hypersphere, respectively. In this case, $\boldsymbol{\theta} = [\boldsymbol{c}_1^T, r_1, \boldsymbol{c}_2^T, r_2, \ldots, \boldsymbol{c}_m^T, r_m]^T$.

Spherical or, in general, shell-shaped clusters[1] are encountered in many robot vision applications. The basic problem here is the identification of objects (patterns) lying in a *scene* (which is a region in the three-dimensional space), and the estimation of their relative positions, using a single or several *images* (two-dimensional projections of the scene). An important task of this problem is the identification of the boundaries of the objects in the image. Given an image (see, e.g., Figure 14.2a), we may identify the pixels that constitute the boundary of the objects using appropriate operators (see, e.g., [Horn 86, Kare 94]) (see Figure 14.2b). Then, the boundaries of the objects may be considered as shell-shaped or linear-shaped clusters and clustering algorithms may be mobilized in order to recover their exact shape and location in the image. In fact, clustering techniques have exhibited

[1] These may be hyperellipsoids, hyperparabolas, etc.

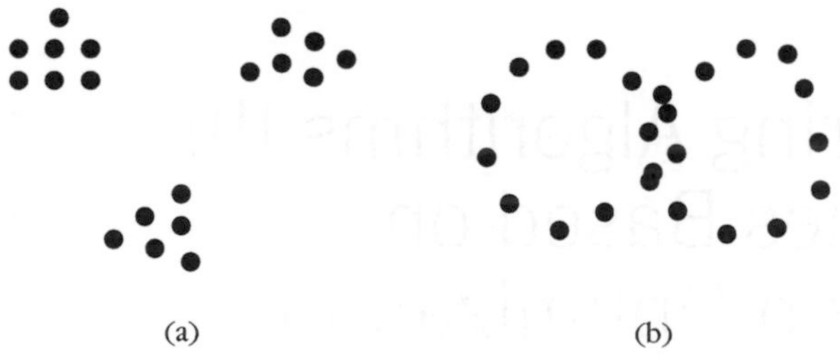

FIGURE 14.1

(a) Compact clusters. (b) Spherical clusters.

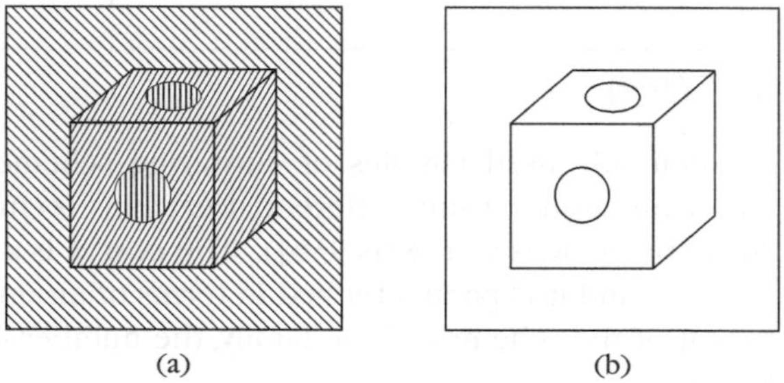

FIGURE 14.2

(a) The original image of a scene. (b) The image after the application of appropriate operators.

satisfactory results at least when the boundaries are known to be shell shaped (e.g., [Kris 95a]).

A distinct characteristic of most of the algorithms of this chapter, compared with the algorithms of the previous chapter, is that the cluster representatives are computed using *all* the available vectors of X, and not only the vectors that have been assigned to the respective cluster. We will focus on four major categories of algorithms: the mixture decomposition, the fuzzy, the possibilistic and the hard clustering algorithms. In the first, the cost function is constructed on the basis of random vectors, and assignment to clusters follows probabilistic arguments, in the spirit of the Bayesian classification. The conditional probabilities here result from the optimization process. In the fuzzy approach a proximity function between a vector and a cluster is defined, and the "grade of membership" of a vector in a cluster is provided by the set of membership functions. As is always the case with fuzzy approaches, the values of the membership functions of a vector in the various clusters are interrelated. This constraint is removed in the case of the possibilistic approach. Finally, hard clustering may be viewed as a special case of the fuzzy clustering approach, where each vector belongs exclusively to a cluster. This category includes the celebrated *k-means* clustering algorithm.

14.2 MIXTURE DECOMPOSITION SCHEMES

The basic reasoning behind this algorithmic family springs from our familiar Bayesian philosophy. We assume that there are m clusters, $C_j, j = 1, \ldots, m$, underlying the data set.[2] Each vector $\boldsymbol{x}_i$, $i = 1, \ldots, N$, belongs to a cluster C_j with probability $P(C_j|\boldsymbol{x}_i)$. A vector $\boldsymbol{x}_i$ is appointed to the cluster C_j if

$$P(C_j|\boldsymbol{x}_i) > P(C_k|\boldsymbol{x}_i), \quad k = 1, \ldots, m, k \neq j$$

The differences from the classification task of Chapter 2 are that (a) no training data with known cluster labeling are available and (b) the *a priori* probabilities $P(C_j) \equiv P_j$ are not known either. Thus, although the goal is the same, the tools have to be different. Basically, this is a typical task with an incomplete training data set. We are missing the corresponding cluster labeling information for each data point $\boldsymbol{x}_i$. Thus, the task fits nicely in the framework introduced in Section 2.5.5.

From Eq. (2.95) and adopting the notation for the needs of the current chapter we have

$$Q(\boldsymbol{\Theta}; \boldsymbol{\Theta}(t)) = \sum_{i=1}^{N} \sum_{j=1}^{m} P(C_j|x_i; \boldsymbol{\Theta}(t)) \ln\left(p(\boldsymbol{x}_i|C_j; \boldsymbol{\theta})P_j\right), \tag{14.1}$$

where $\boldsymbol{\theta} = [\boldsymbol{\theta}_1^T, \ldots, \boldsymbol{\theta}_m^T]^T$, with $\boldsymbol{\theta}_j$ being the parameter vector corresponding to the j-th cluster, $\boldsymbol{P} = [P_1, \ldots, P_m]^T$, with P_j being the *a priori* probability for the jth cluster and $\boldsymbol{\Theta} = [\boldsymbol{\theta}^T, \boldsymbol{P}^T]^T$. The above equation results from application of the E-step of the EM algorithm. The M-step of the algorithm is

$$\boldsymbol{\Theta}(t+1) = arg \max_{\boldsymbol{\Theta}} Q(\boldsymbol{\Theta}; \boldsymbol{\Theta}(t)). \tag{14.2}$$

Assuming that all pairs of $\boldsymbol{\theta}_k, \boldsymbol{\theta}_j$'s are functionally independent, that is, no $\boldsymbol{\theta}_k$ gives any information about $\boldsymbol{\theta}_j (j \neq i)$, we estimate $\boldsymbol{\theta}_j$ from Eq. (14.2) as follows:

$$\sum_{i=1}^{N} \sum_{j=1}^{m} P(C_j|\boldsymbol{x}_i; \boldsymbol{\Theta}(t)) \frac{\partial}{\partial \boldsymbol{\theta}_j} \ln p(\boldsymbol{x}_i|C_j; \boldsymbol{\theta}_j) = 0 \tag{14.3}$$

Maximization with respect to $\boldsymbol{P}$ is a constraint optimization problem since

$$P_k \geq 0, \quad k = 1, \ldots, m, \quad \text{and} \quad \sum_{k=1}^{m} P_k = 1 \tag{14.4}$$

The corresponding Lagrangian function is

$$Q(\boldsymbol{P}, \lambda) = Q(\boldsymbol{\Theta}; \boldsymbol{\Theta}(t)) - \lambda \left(\sum_{k=1}^{m} P_k - 1 \right) \tag{14.5}$$

[2] Recall that the number m is assumed to be known.

Taking the partial derivative of $\mathcal{Q}(\boldsymbol{P}, \lambda)$ with respect to P_j and setting it equal to 0, and after some algebra we obtain

$$P_j = \frac{1}{\lambda} \sum_{i=1}^{N} P(C_j|\boldsymbol{x}_i; \boldsymbol{\Theta}(t)), \quad j = 1, \ldots, m \tag{14.6}$$

Substituting the above equations into Eq. (14.4), we obtain

$$\lambda = \sum_{i=1}^{N} \sum_{j=1}^{m} P(C_j|\boldsymbol{x}_i; \boldsymbol{\Theta}(t)) = N \tag{14.7}$$

Thus, Eq. (14.6) gives

$$P_j = \frac{1}{N} \sum_{i=1}^{N} P(C_j|\boldsymbol{x}_i; \boldsymbol{\Theta}(t)) \quad j = 1, \ldots, m \tag{14.8}$$

Taking into account Eqs. (14.3), (14.8), and (2.87), the EM algorithm for this case may be stated as

Generalized Mixture Decomposition Algorithmic Scheme (GMDAS)

- Choose initial estimates, $\boldsymbol{\theta} = \boldsymbol{\theta}(0)$ and $\boldsymbol{P} = \boldsymbol{P}(0)$.[3]
- $t = 0$
- Repeat
  - Compute

$$P(C_j|\boldsymbol{x}_i; \boldsymbol{\Theta}(t)) = \frac{p(\boldsymbol{x}_i|C_j; \boldsymbol{\theta}_j(t))P_j(t)}{\sum_{k=1}^{m} p(\boldsymbol{x}_i|C_k, \boldsymbol{\theta}_k(t))P_k(t)} \tag{14.9}$$

    $i = 1, \ldots, N, \quad j = 1, \ldots, m.$

  - Set $\boldsymbol{\theta}_j(t+1)$ equal to the solution of the equation

$$\sum_{i=1}^{N} \sum_{j=1}^{m} P(C_j|\boldsymbol{x}_i; \boldsymbol{\Theta}(t)) \frac{\partial}{\partial \boldsymbol{\theta}_j} \ln p(\boldsymbol{x}_i|C_j; \boldsymbol{\theta}_j) = \boldsymbol{0} \tag{14.10}$$

    with respect to $\boldsymbol{\theta}_j$, for $j = 1, \ldots, m$.

  - Set

$$P_j(t+1) = \frac{1}{N} \sum_{i=1}^{N} P(C_j|\boldsymbol{x}_i; \boldsymbol{\Theta}(t)), \quad j = 1, \ldots, m \tag{14.11}$$

  - $t = t + 1$
- Until convergence, with respect to $\boldsymbol{\Theta}$, is achieved.

[3] Initial conditions must satisfy the constraints.

A suitable termination criterion for the algorithm is the following:

$$\|\Theta(t+1) - \Theta(t)\| < \varepsilon$$

where $\|\cdot\|$ is an appropriate vector norm and ε is a "small" user-defined constant. This scheme is guaranteed to converge to a global or a local maximum of the log-likelihood function. However, even if a local maximum solution is reached, it may still capture satisfactorily the underlying clustering structure of X.

Once the algorithm has converged, vectors are assigned to clusters according to the final estimates $P(C_j|\boldsymbol{x}_i)$ of Eq. (14.9). Hence, the task now becomes a typical Bayesian classification problem, if we treat each cluster as a separate class.

14.2.1 Compact and Hyperellipsoidal Clusters

In this section, we focus our attention on the case in which the vectors of X form compact clusters. A distribution that is suitable for clusters of this scheme is the normal distribution, that is,

$$p(\boldsymbol{x}|C_j;\boldsymbol{\theta}_j) = \frac{1}{(2\pi)^{l/2}|\Sigma_j|^{1/2}} \exp\left(-\frac{1}{2}(\boldsymbol{x}-\boldsymbol{\mu}_j)^T \Sigma_j^{-1}(\boldsymbol{x}-\boldsymbol{\mu}_j)\right), \quad j = 1,\ldots,m \tag{14.12}$$

or

$$\ln p(\boldsymbol{x}|C_j;\boldsymbol{\theta}_j) = \ln\frac{|\Sigma_j|^{-1/2}}{(2\pi)^{l/2}} - \frac{1}{2}(\boldsymbol{x}-\boldsymbol{\mu}_j)^T \Sigma_j^{-1}(\boldsymbol{x}-\boldsymbol{\mu}_j), \quad j = 1,\ldots,m \tag{14.13}$$

In this case, each vector $\boldsymbol{\theta}_j$ consists of the l parameters of the mean $\boldsymbol{\mu}_j$ and the $l(l+1)/2$ independent parameters of Σ_j. A parameter reduction may be obtained by assuming that the covariance matrices are diagonal. If this assumption is too strict, another commonly used assumption is that all covariance matrices are equal. In the former case $\boldsymbol{\theta}$ consists of $2ml$ parameters, while in the latter it consists of $ml + l(l+1)/2$ parameters.

Combining Eq. (14.12) and Eq. (14.9) results in

$$P(C_j|\boldsymbol{x};\Theta(t)) = \frac{|\Sigma_j(t)|^{-1/2}\exp\left(-\frac{1}{2}(\boldsymbol{x}-\boldsymbol{\mu}_j(t))^T\Sigma_j^{-1}(t)(\boldsymbol{x}-\boldsymbol{\mu}_j(t))\right)P_j(t)}{\sum_{k=1}^{m}|\Sigma_k(t)|^{-1/2}\exp\left(-\frac{1}{2}(\boldsymbol{x}-\boldsymbol{\mu}_k(t))^T\Sigma_k^{-1}(t)(\boldsymbol{x}-\boldsymbol{\mu}_k(t))\right)P_k(t)} \tag{14.14}$$

In the sequel, we consider the problem in its most general form; that is, we assume that all the means $\boldsymbol{\mu}_j$ and the covariance matrices Σ_j are unknown. We also assume that, in general, all Σ_j's are different from each other. Following an approach similar to the one described in Chapter 2, the updating equations for $\boldsymbol{\mu}_j$'s and Σ_j's from the M-step are

$$\mu_j(t+1) = \frac{\sum_{k=1}^{N} P(C_j|\boldsymbol{x}_k;\boldsymbol{\Theta}(t))\boldsymbol{x}_k}{\sum_{k=1}^{N} P(C_j|\boldsymbol{x}_k;\boldsymbol{\Theta}(t))} \tag{14.15}$$

and

$$\Sigma_j(t+1) = \frac{\sum_{k=1}^{N} P(C_j|\boldsymbol{x}_k;\boldsymbol{\Theta}(t))(\boldsymbol{x}_k-\mu_j(t))(\boldsymbol{x}_k-\mu_j(t))^T}{\sum_{k=1}^{N} P(C_j|\boldsymbol{x}_k;\boldsymbol{\Theta}(t))} \tag{14.16}$$

$j = 1, \ldots, m$.

Thus, in the Gaussian case these two equations replace Eq. (14.10), and Eq. (14.14) replaces Eq. (14.9) in the corresponding steps of the GMDAS algorithm.

Remark

- Notice that this scheme is computationally very demanding, because at each iteration step the inverses of m covariance matrices are required for the computation of $P(C_j|\boldsymbol{x}_i;\boldsymbol{\Theta}(t))$. As stated earlier, one way to relax this demand is to assume that all covariance matrices are diagonal or that all are equal to each other. In the latter case, only one inversion is required at each iteration step.

Example 14.1

(a) Consider three 2-dimensional normal distributions with means $\mu_1 = [1,1]^T$, $\mu_2 = [3.5, 3.5]^T$, $\mu_3 = [6,1]^T$ and covariance matrices

$$\Sigma_1 = \begin{bmatrix} 1 & -0.3 \\ -0.3 & 1 \end{bmatrix}, \quad \Sigma_2 = \begin{bmatrix} 1 & 0.3 \\ 0.3 & 1 \end{bmatrix}, \quad \Sigma_3 = \begin{bmatrix} 1 & 0.7 \\ 0.7 & 1 \end{bmatrix}$$

respectively.

A group of 100 vectors is generated from each distribution. These groups constitute the data set X. Figure 14.3a is a plot of the generated data.

We initialize $P_j = 1/3$, $j = 1, 2, 3$. Also, we set $\mu_i(0) = \mu_i + \boldsymbol{y}_i$, where $\boldsymbol{y}_i$ is an 2×1 vector with random coordinates, uniformly distributed in the interval $[-1,1]^T$. Similarly, we define $\Sigma_i(0)$, $i = 1, 2, 3$. We set $\varepsilon = 0.01$. Using these initial conditions, the GMDAS for Gaussian pdf's terminates after 38 iterations. The final parameter estimates obtained are $\boldsymbol{P}' = [0.35, 0.31, 0.34]^T$, $\mu_1' = [1.28, 1.16]^T$, $\mu_2' = [3.49, 3.68]^T$, $\mu_3' = [5.96, 0.84]^T$, and

$$\Sigma_1' = \begin{bmatrix} 1.45 & 0.01 \\ 0.01 & 0.57 \end{bmatrix}, \quad \Sigma_2' = \begin{bmatrix} 0.62 & 0.09 \\ 0.09 & 0.74 \end{bmatrix}, \quad \Sigma_3' = \begin{bmatrix} 0.30 & 0.0024 \\ 0.0024 & 1.94 \end{bmatrix}$$

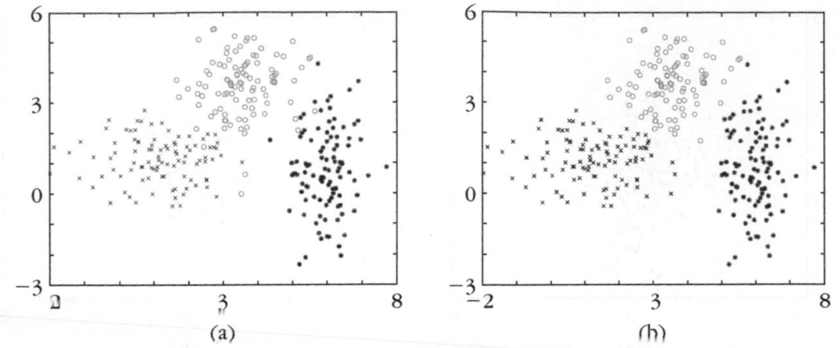

FIGURE 14.3

(a) A data set that consists of three groups of points. (b) The results from the application of GMDAS when normal mixtures are used.

For comparison, the sample mean values are $\hat{\boldsymbol{\mu}}_1 = [1.16, 1.13]^T$, $\hat{\boldsymbol{\mu}}_2 = [3.54, 3.56]^T$, $\hat{\boldsymbol{\mu}}_3 = [5.97, 0.76]^T$, respectively. Also, the sample covariance matrices are

$$\hat{\Sigma}_1 = \begin{bmatrix} 1.27 & -0.03 \\ -0.03 & 0.52 \end{bmatrix}, \quad \hat{\Sigma}_2 = \begin{bmatrix} 0.70 & 0.07 \\ 0.07 & 0.98 \end{bmatrix}, \quad \hat{\Sigma}_3 = \begin{bmatrix} 0.32 & 0.05 \\ 0.05 & 1.81 \end{bmatrix}$$

respectively.

As we can see, the final estimates of the algorithm are close enough to the means and the covariance matrices of the three groups of vectors.

Once the unknown parameters of the model have been estimated, the data vectors are assigned to clusters according to the estimated values of $P(C_j|\boldsymbol{x}_i)$. Figure 14.3b shows the assignment of points to the three clusters, which is in close agreement with the original structure. A way to assess the performance of the resulting model estimates is via the so-called *confusion matrix*. This is a matrix A whose (i, j) element is the number of vectors that originate from the ith distribution and are assigned to the jth cluster.[4] For our example this is

$$A_1 = \begin{bmatrix} 99 & 0 & 1 \\ 0 & 100 & 0 \\ 3 & 4 & 93 \end{bmatrix}$$

This example indicates that 99% of the data from the first distribution are assigned to the same cluster (the first one). Similarly, all the data from the second distribution are assigned to the same cluster (the second one) and, finally, 93% of the data from the third distribution are assigned to the same cluster (the third one).

[4] It should be noted here that in real clustering applications the confusion matrix cannot be defined, since we do not know *a priori* the cluster where each feature vector belongs. However, we may use it in artificial experiments such as this one, in order to evaluate the performance of the clustering algorithms.

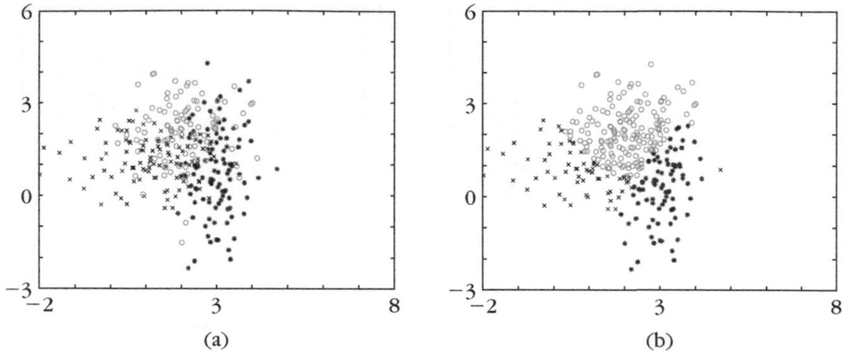

FIGURE 14.4

(a) The data set, which consists of three overlapping groups of points. (b) The results of the GMDAS when Gaussian mixtures are used.

(b) Let us now consider the case in which the three normal distributions are located closer, for example, $\boldsymbol{\mu}_1 = [1, 1]^T$, $\boldsymbol{\mu}_2 = [2, 2]^T$, $\boldsymbol{\mu}_3 = [3, 1]^T$, and with the same covariance matrices as before (Figure 14.4). We initialize $\boldsymbol{\mu}_i$ and Σ_i, $i = 1, 2, 3$, as in the previous case and run the GMDAS for Gaussian pdf's. The confusion matrix for this case is

$$A_2 = \begin{bmatrix} 85 & 4 & 11 \\ 35 & 56 & 9 \\ 26 & 0 & 74 \end{bmatrix}$$

As expected, each one of the obtained clusters contains a significant percentage of points from more than one distribution.

Example 14.2

The data set X, which is depicted in Figure 14.5a, consists of two intersecting ring-shaped clusters. Each cluster consists of 500 points. We run the GMDAS with Gaussians and $m = 2$ and $\varepsilon = 0.01$. The algorithm terminates after 72 iterations, and the results are shown in Figure 14.5b. *As expected, the algorithm fails to recover the underlying clustering structure of X, because it seeks compact clusters. Generally speaking, GMDAS using Gaussians reveals clusters that are as compact as possible, even though the clusters underlying X may have a different shape.* Even worse, it will identify clusters in X even though there is no clustering structure in it.[5]

[5] In general, before we apply any clustering algorithm to identify clusters contained in X, we should first check whether *there exists* any clustering structure in X. This procedure refers to *clustering tendency* and is considered in Chapter 16.

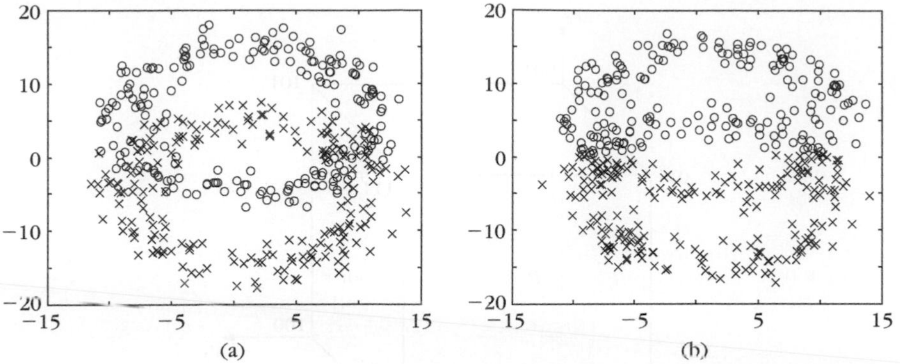

FIGURE 14.5

(a) A data set that consists of ring-shaped intersecting clusters. (b) The results from the application of GMDAS when Gaussian mixtures are used.

In [Zhua 96], the case of Gaussian pdf's, contaminated by unknown outlier distributions, $h(\boldsymbol{x}_i|C_j)$, is considered. In this case, we can write $p(\boldsymbol{x}|C_j) = (1-\varepsilon_j)G(\boldsymbol{x}|C_j) + \varepsilon_j h(\boldsymbol{x}|C_j)$, where ε_j is the level of contamination and $G(\boldsymbol{x}|C_j)$ is the jth Gaussian distribution. Under the assumption that all $h(\boldsymbol{x}_i|C_j)$ are constant, that is, $h(\boldsymbol{x}_i|C_j) = c_j$, $i = 1,\ldots,N$, $p(\boldsymbol{x}|C_j)$ may be written as $p(\boldsymbol{x}|C_j) = (1-\varepsilon_j)G(\boldsymbol{x}|C_j) + \varepsilon_j c_j$. Then we may use the preceding methodology in order to identify the mean and the covariance matrices of the normal distributions $G(\boldsymbol{x}|C_j)$ as well as the values of ε_j and c_j.

In [Figu 02] an alternative mixture decomposition scheme is proposed, which does not demand *a priori* knowledge of m and, in addition, it does not require careful initialization.

14.2.2 A Geometrical Interpretation

As mentioned earlier, the conditional probability, $P(C_j|\boldsymbol{x}_i)$, indicates how likely it is that $\boldsymbol{x}_i \in X$ belongs to C_j, $i = 1,\ldots,N$, subject, of course, to the constraint

$$\sum_{j=1}^{m} P(C_j|\boldsymbol{x}_i) = 1 \tag{14.17}$$

This may be viewed as the equation of an $(m-1)$-dimensional hyperplane. For notational purposes, let $P(C_j|\boldsymbol{x}_i) \equiv y_j$, $j = 1,\ldots,m$. Then Eq. (14.17) may be written as

$$\boldsymbol{a}^T\boldsymbol{y} = 1 \tag{14.18}$$

where $\boldsymbol{y}^T = [y_1,\ldots,y_m]$ and $\boldsymbol{a}^T = [1,1,\ldots,1]$. That is, $\boldsymbol{y}$ is allowed to move on the hyperplane defined by the previous equation. In addition, since $0 \le y_j \le 1$, $j = 1,\ldots,m$, $\boldsymbol{y}$ lies also inside the unit hypercube (see Figure 14.6).

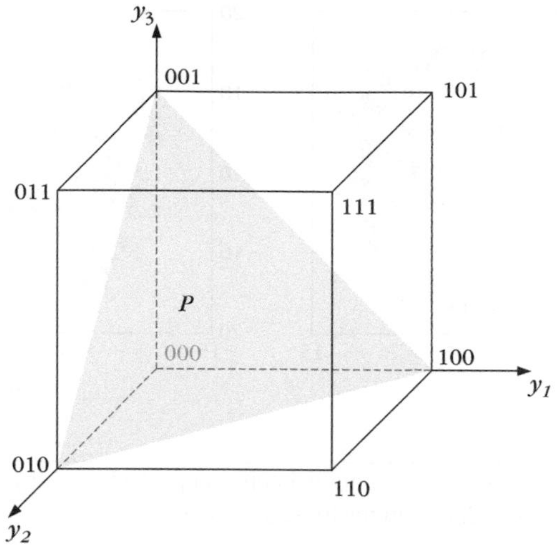

FIGURE 14.6

The hypercube for $m = 3$. The point $\boldsymbol{y}$ is allowed to move only on the shaded region of $\boldsymbol{P}$.

This interpretation allows us to derive some useful conclusions for the so-called *noisy feature vectors* or *outliers*. Let $\boldsymbol{x}_i$ be such a vector. Since Eq. (14.17) holds for $\boldsymbol{x}_i$, at least one of the y_j's, $j = 1, \ldots, m$, is significant (it lies in the interval $[1/m, 1]$). Thus, $\boldsymbol{x}_i$ will affect, at least, the estimates for the corresponding cluster C_j, through Eqs. (14.9), (14.10), and (14.11), and this makes GMDAS sensitive to outliers. The following example clarifies this idea further.

Example 14.3

Consider the data set X shown in Figure 14.7. It consists of 22 vectors. The first (next) 10 vectors are drawn from the normal distribution with mean $\boldsymbol{\mu}_1 = [0, 0]^T$ ($\boldsymbol{\mu}_2 = [4.5, 4.5]^T$) and covariance matrix $\Sigma_1 = I$ ($\Sigma_2 = I$), where I is the 2×2 identity matrix. The last two points are $\boldsymbol{x}_{21} = [-6, 5]^T$ and $\boldsymbol{x}_{22} = [11, 0]^T$, respectively. We run the GMDAS for Gaussian pdf's on X. The estimates of $\boldsymbol{P}$, $\boldsymbol{\mu}_j$, and Σ_j, $j = 1, 2$, obtained after five iterations, are

$$\boldsymbol{P}' = [0.5, 0.5]^T, \quad \boldsymbol{\mu}_1' = [-0.58, 0.35]^T, \quad \boldsymbol{\mu}_2' = [4.98, 4.00]^T$$

$$\Sigma_1' = \begin{bmatrix} 4.96 & -2.01 \\ -2.01 & 2.63 \end{bmatrix}, \quad \Sigma_2' = \begin{bmatrix} 3.40 & -2.53 \\ -2.53 & 3.27 \end{bmatrix}$$

The resulting values of $P(C_j|\boldsymbol{x}_i)$, $j = 1, 2$, $i = 1, \ldots, 22$, are shown in Table 14.1. Although $\boldsymbol{x}_{21}$ and $\boldsymbol{x}_{22}$ may be considered as outliers, since they lie away from the two clusters, we get that $P(C_1|\boldsymbol{x}_{21}) = 1$ and $P(C_2|\boldsymbol{x}_{22}) = 1$, due to the constraint $\sum_{j=1}^{2} P(C_j|\boldsymbol{x}_i) = 1$. This

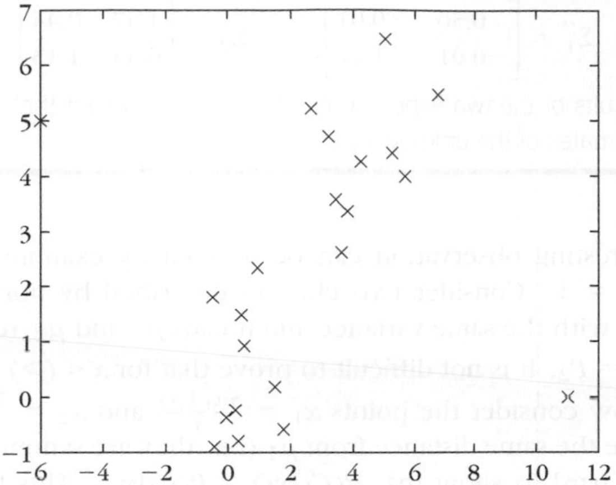

FIGURE 14.7

The data set for Example 14.3.

Table 14.1 The Resulting *a Posteriori* Probabilities for the Data Set of Example 14.3

feat. vec.	$P(C_1\|x)$	$P(C_2\|x)$	feat. vec.	$P(C_1\|x)$	$P(C_2\|x)$
x_1	0	1	x_{12}	1	0
x_2	0	1	x_{13}	1	0
x_3	0	1	x_{14}	1	0
x_4	0	1	x_{15}	1	0
x_5	0	1	x_{16}	1	0
x_6	0	1	x_{17}	1	0
x_7	0	1	x_{18}	1	0
x_8	0	1	x_{19}	0.99	0.01
x_9	0	1	x_{20}	1	0
x_{10}	0	1	x_{21}	0	1
x_{11}	1	0	x_{22}	1	0

implies that these points have a nonnegligible impact on μ_1, μ_2, Σ_1, and Σ_2. Indeed, if we run GMDAS for Gaussian pdf's on $X_1 = \{x_i : i = 1, \ldots, 20\}$ (i.e., we exclude the last two points), using the same initial conditions, we obtain after five iterations:

$$P'' = [0.5, 0.5]^T, \quad \mu_1'' = [-0.03, -0.12]^T, \quad \mu_2'' = [4.37, 4.40]^T$$

$$\Sigma_1'' = \begin{bmatrix} 0.50 & -0.01 \\ -0.01 & 1.22 \end{bmatrix}, \quad \Sigma_2'' = \begin{bmatrix} 1.47 & 0.44 \\ 0.44 & 1.13 \end{bmatrix}$$

Comparing the results of the two experiments, it is easily observed that the last setup gives more accurate estimates of the unknown parameters.

Another interesting observation can be derived by examining the following situation. Let $l = 1$. Consider two clusters described by normal distributions $p(x|C_j), j = 1, 2$, with the same variance and means μ_1 and μ_2, respectively ($\mu_1 < \mu_2$). Also let $P_1 = P_2$. It is not difficult to prove that for $x < (>) \frac{\mu_1+\mu_2}{2}$, $P(C_1|x) > (<) P(C_2|x)$. Now consider the points $x_1 = \frac{3\mu_1+\mu_2}{4}$ and $x_2 = \frac{5\mu_1-\mu_2}{4}$. Although these points have the same distance from μ_1 (i.e., they are symmetric with respect to μ_1), it is not hard to show that $P(C_1|x_1) > P(C_1|x_2)$. This happens because $P(C_1|x)$ and $P(C_2|x)$ are *related* through Eq. (14.17). Thus, the probability of having x in one cluster is affected by the probability of belonging to the other. We will soon see that we can free ourselves from such an interrelation.

14.3 FUZZY CLUSTERING ALGORITHMS

One of the difficulties associated with the previously discussed probabilistic algorithms is the involvement of the pdf's, for which a suitable model has to be assumed. In addition, it is not easy to handle cases where the clusters are not compact but are shell shaped. A family of clustering algorithms that emancipates itself from such constraints is that of fuzzy clustering algorithms. These schemes have been the subject of intensive research during the past three decades. The major point that differentiates the two approaches is that in the fuzzy schemes a vector *belongs simultaneously* to more than one cluster, whereas in the probabilistic schemes, each vector belongs *exclusively* to a single cluster.

As already discussed in Chapter 11, a *fuzzy m-clustering of X* is defined by a set of functions $u_j : X \to A, j = 1, \ldots, m$, where $A = [0, 1]$.

In the case where $A = \{0, 1\}$, a *hard m-clustering of X* is defined. In this case, each vector belongs exclusively to a single cluster.

As in the previous section, it is assumed that the number of clusters as well as their shape is known *a priori*. The shape of the clusters is characterized by the adopted set of parameters. For example, if we deal with compact clusters, a point representative is used to represent each cluster; that is, each cluster is represented by l parameters. On the other hand, if we deal with noncompact but, say, hyperspherical clusters, a hypersphere is used as a representative of each cluster. In this case, each cluster is represented by $l + 1$ parameters (l for the center of the hypersphere and 1 for its radius).

In the sequel we use the following notation: $\boldsymbol{\theta}_j$ is the parameterized representative of the jth cluster, $\boldsymbol{\theta} \equiv [\boldsymbol{\theta}_1^T, \ldots, \boldsymbol{\theta}_m^T]^T$, U is an $N \times m$ matrix whose (i,j)

element equals $u_j(\boldsymbol{x}_i)$, $d(\boldsymbol{x}_i, \boldsymbol{\theta}_j)$ is the dissimilarity between $\boldsymbol{x}_i$ and $\boldsymbol{\theta}_j$, and $q(>1)$ is a parameter called a *fuzzifier*. The role of the latter will be clarified shortly. Most of the well-known fuzzy clustering algorithms are those derived by minimizing a cost function of the form

$$J_q(\boldsymbol{\theta}, U) = \sum_{i=1}^{N}\sum_{j=1}^{m} u_{ij}^q d(\boldsymbol{x}_i, \boldsymbol{\theta}_j) \tag{14.19}$$

with respect to $\boldsymbol{\theta}$ and U, subject to the constraints

$$\sum_{j=1}^{m} u_{ij} = 1, \quad i = 1, \ldots, N \tag{14.20}$$

where

$$u_{ij} \in [0, 1], \quad i = 1, \ldots, N, \quad j = 1, \ldots, m,$$

$$0 < \sum_{i=1}^{N} u_{ij} < N, \quad j = 1, 2, \ldots, m \tag{14.21}$$

In other words, the grade of membership of $\boldsymbol{x}_i$ in the jth cluster is related to the grade of membership of $\boldsymbol{x}_i$ to the rest $m - 1$ clusters through Eq. (14.20). Different values of q in Eq. (14.19) bias $J_q(\boldsymbol{\theta}, U)$ toward either the fuzzy or the hard clusterings. More specifically, for fixed $\boldsymbol{\theta}$, *if $q = 1$, no fuzzy clustering is better than the best hard clustering in terms of $J_q(\boldsymbol{\theta}, U)$. However, if $q > 1$, there are cases in which fuzzy clusterings lead to lower values of $J_q(\boldsymbol{\theta}, U)$ than the best hard clustering.* Let us clarify these ideas further using the following example.

Example 14.4

Let $X = \{\boldsymbol{x}_1, \boldsymbol{x}_2, \boldsymbol{x}_3, \boldsymbol{x}_4\}$, where $\boldsymbol{x}_1 = [0, 0]^T$, $\boldsymbol{x}_2 = [2, 0]^T$, $\boldsymbol{x}_3 = [0, 3]^T$, $\boldsymbol{x}_4 = [2, 3]^T$. Let $\boldsymbol{\theta}_1 = [1, 0]^T$, $\boldsymbol{\theta}_2 = [1, 3]^T$ be the cluster representatives. Suppose also that the Euclidean distance between a vector and a representative is in use. The hard two-cluster clustering that minimizes $J_q(\boldsymbol{\theta}, U)$, for the above choice of $\boldsymbol{\theta}_1, \boldsymbol{\theta}_2$, can be represented by

$$U_{hard} = \begin{bmatrix} 1 & 0 \\ 1 & 0 \\ 0 & 1 \\ 0 & 1 \end{bmatrix}$$

The value of $J_q(\boldsymbol{\theta}, U)$ in this case Eq. (14.19) is $J_q^{hard}(\boldsymbol{\theta}, U) = 4$. Obviously, hard clusterings do not depend on q.

Assume now that $q = 1$ and u_{ij}'s are between 0 and 1. Then the value of the cost function becomes

$$J_1^{fuzzy}(\boldsymbol{\theta}, U) = \sum_{i=1}^{2}(u_{i1} + u_{i2}\sqrt{10}) + \sum_{i=3}^{4}(u_{i1}\sqrt{10} + u_{i2})$$

Since for each $\boldsymbol{x}_i$ both u_{i1} and u_{i2} are positive and $u_{i1} + u_{i2} = 1$, it easily follows that $J_1^{fuzzy}(\boldsymbol{\theta}, U) > 4$. Thus, the hard clustering always results in better values of $J_q^{fuzzy}(\boldsymbol{\theta}, U)$, compared with their fuzzy counterparts, when $q = 1$.

Assume now that $q = 2$. The reader should easily verify that when $u_{i2} \in [0, 0.48]$ for $i = 1, 2$ and $u_{i1} \in [0, 0.48]$ for $i = 3, 4$, and, of course, $u_{i1} = 1 - u_{i2}$, for each $\boldsymbol{x}_i$, then the value of $J_2^{fuzzy}(\boldsymbol{\theta}, U)$ is less than 4 (see Problem 14.7). Thus, in this case fuzzy clusterings are favored over hard ones.

Finally, let $q = 3$. In this case, it is easily verified that when $u_{i2} \in [0, 0.67]$ for $i = 1, 2$ and $u_{i1} \in [0, 0.67]$ for $i = 3, 4$ and $u_{i1} = 1 - u_{i2}$, for each $\boldsymbol{x}_i$, then the value of $J_3^{fuzzy}(\boldsymbol{\theta}, U)$ is also less than 4.

Minimization of $J_q(\theta, U)$

We first assume that no $\boldsymbol{x}_i$ coincides with any of the representatives. More formally, for an $\boldsymbol{x}_i$ let Z_i be the set that contains the indices of the representatives $\boldsymbol{\theta}_j$ for which $d(\boldsymbol{x}_i, \boldsymbol{\theta}_j) = 0$. According to our assumption, $Z_i = \emptyset$, for all i. In the sequel, for ease of notation, we drop the index q from $J_q(\boldsymbol{\theta}, U)$. Let us consider first U. Minimization of $J_q(\boldsymbol{\theta}, U)$ with respect to U, subject to the constraint (14.20), leads to the following Lagrangian function:

$$\mathcal{J}(\boldsymbol{\theta}, U) = \sum_{i=1}^{N}\sum_{j=1}^{m} u_{ij}^q d(\boldsymbol{x}_i, \boldsymbol{\theta}_j) - \sum_{i=1}^{N} \lambda_i \left(\sum_{j=1}^{m} u_{ij} - 1\right) \tag{14.22}$$

The partial derivative of $\mathcal{J}(\boldsymbol{\theta}, U)$ with respect to u_{rs} is

$$\frac{\partial \mathcal{J}(\boldsymbol{\theta}, U)}{\partial u_{rs}} = q u_{rs}^{q-1} d(\boldsymbol{x}_r, \boldsymbol{\theta}_s) - \lambda_r \tag{14.23}$$

Setting $\partial \mathcal{J}(\boldsymbol{\theta}, U)/\partial u_{rs}$ equal to 0 and solving with respect to u_{rs}, we obtain

$$u_{rs} = \left(\frac{\lambda_r}{q d(\boldsymbol{x}_r, \boldsymbol{\theta}_s)}\right)^{\frac{1}{q-1}}, \quad s = 1, \ldots, m \tag{14.24}$$

Substituting u_{rs} from the previous equation in the constraint equation $\sum_{j=1}^{m} u_{rj} = 1$, we obtain

$$\sum_{j=1}^{m} \left(\frac{\lambda_r}{q d(\boldsymbol{x}_r, \boldsymbol{\theta}_j)}\right)^{\frac{1}{q-1}} = 1$$

or

$$\lambda_r = \frac{q}{\left(\sum_{j=1}^{m} \left(\frac{1}{d(\boldsymbol{x}_r, \boldsymbol{\theta}_j)}\right)^{\frac{1}{q-1}}\right)^{q-1}} \tag{14.25}$$

Combining Eq. (14.25) with (14.24) and using a bit of algebra, we obtain

$$u_{rs} = \frac{1}{\sum_{j=1}^{m}\left(\frac{d(\boldsymbol{x}_r,\boldsymbol{\theta}_s)}{d(\boldsymbol{x}_r,\boldsymbol{\theta}_j)}\right)^{\frac{1}{q-1}}} \tag{14.26}$$

$r = 1, \ldots, N, s = 1, \ldots, m$.

Now consider the parameter vector $\boldsymbol{\theta}_j$. Taking the gradient of $J(\boldsymbol{\theta}, U)$ with respect to $\boldsymbol{\theta}_j$ and setting it equal to zero, we obtain

$$\frac{\partial J(\boldsymbol{\theta}, U)}{\partial \boldsymbol{\theta}_j} = \sum_{i=1}^{N} u_{ij}^q \frac{\partial d(\boldsymbol{x}_i, \boldsymbol{\theta}_j)}{\partial \boldsymbol{\theta}_j} = \boldsymbol{0}, \quad j = 1, \ldots, m \tag{14.27}$$

Equations (14.26) and (14.27) are coupled and, in general, cannot give closed-form solutions. One way to proceed is to employ the following iterative algorithmic scheme, in order to obtain estimates for U and $\boldsymbol{\theta}$.

Generalized Fuzzy Algorithmic Scheme (GFAS)

- Choose $\boldsymbol{\theta}_j(0)$ as initial estimates for $\boldsymbol{\theta}_j, j = 1, \ldots, m$.
- $t = 0$
- Repeat
  - For $i = 1$ to N
    - For $j = 1$ to m
      - $u_{ij}(t) = \frac{1}{\sum_{k=1}^{m}\left(\frac{d(\boldsymbol{x}_i,\boldsymbol{\theta}_j(t))}{d(\boldsymbol{x}_i,\boldsymbol{\theta}_k(t))}\right)^{\frac{1}{q-1}}}$
    - End {For-j}
  - End {For-i}
  - $t = t + 1$
  - For $j = 1$ to m
    - *Parameter updating:* Solve

$$\sum_{i=1}^{N} u_{ij}^q(t-1) \frac{\partial d(\boldsymbol{x}_i, \boldsymbol{\theta}_j)}{\partial \boldsymbol{\theta}_j} = \boldsymbol{0} \tag{14.28}$$

      with respect to $\boldsymbol{\theta}_j$ and set $\boldsymbol{\theta}_j(t)$ equal to this solution.
  - End {For-j}
- Until a termination criterion is met.

As the termination criterion we may employ $\|\boldsymbol{\theta}(t) - \boldsymbol{\theta}(t-1)\| < \varepsilon$, where $\|\cdot\|$ is any vector norm and ε is a "small" user-defined constant.

Remarks

- If, for a given $\boldsymbol{x}_i$, $Z_i \neq \emptyset$, we arbitrarily choose u_{ij}'s, with $j \in Z_i$, such that $\sum_{j \in Z_i} u_{ij} = 1$ and $u_{ij} = 0$, for $j \notin Z_i$. That is, $\boldsymbol{x}_i$ is shared arbitrarily among the clusters whose representatives coincide with $\boldsymbol{x}_i$, subject to the constraint (14.20). In the case in which $\boldsymbol{x}_i$ coincides with a single representative, say $\boldsymbol{\theta}_j$, the condition becomes $u_{ij} = 1$, and $u_{ik} = 0, k \neq j$.
- The algorithmic scheme may also be initialized from $U(0)$ instead of $\boldsymbol{\theta}_j(0)$, $j = 1, \ldots, m$, and start iterations with computing $\boldsymbol{\theta}_j$ first.
- The above iterative algorithmic scheme is also known as the *alternating optimization (AO) scheme*, since at each iteration step U is updated for fixed $\boldsymbol{\theta}$, and then $\boldsymbol{\theta}$ is updated for fixed U ([Bezd 95, Hopp 99]).

In the sequel the algorithm is specialized to three commonly encountered cases.

14.3.1 Point Representatives

In the case of compact clusters, a point representative is used for each cluster; that is, $\boldsymbol{\theta}_j$ consists of l parameters. In this case, the dissimilarity $d(\boldsymbol{x}_i, \boldsymbol{\theta}_j)$ may be any distance measure between two points. Two common choices for $d(\boldsymbol{x}_i, \boldsymbol{\theta}_j)$ are (see also Chapter 11)

$$d(\boldsymbol{x}_i, \boldsymbol{\theta}_j) = (\boldsymbol{x}_i - \boldsymbol{\theta}_j)^T A (\boldsymbol{x}_i - \boldsymbol{\theta}_j) \tag{14.29}$$

where A is a symmetric, positive definite matrix, and the *Minkowski distance*,

$$d(\boldsymbol{x}_i, \boldsymbol{\theta}_j) = \left(\sum_{k=1}^{l} |x_{ik} - \theta_{jk}|^p \right)^{\frac{1}{p}} \tag{14.30}$$

where p is a positive integer and x_{ik}, θ_{jk} are the kth coordinates of $\boldsymbol{x}_i$ and $\boldsymbol{\theta}_j$, respectively. Let us now see the specific form of GFAS under these choices.

- When the distance given in Eq. (14.29) is in use, we have

$$\frac{\partial d(\boldsymbol{x}_i, \boldsymbol{\theta}_j)}{\partial \boldsymbol{\theta}_j} = 2A(\boldsymbol{\theta}_j - \boldsymbol{x}_i) \tag{14.31}$$

Substituting Eq. (14.31) into Eq. (14.28), we obtain

$$\sum_{i=1}^{N} u_{ij}^q (t-1) 2A(\boldsymbol{\theta}_j - \boldsymbol{x}_i) = \mathbf{0}$$

Since A is positive definite, it is invertible. Premultiplying both sides of this equation with A^{-1} and after some simple algebra, we obtain

$$\boldsymbol{\theta}_j(t) = \frac{\sum_{i=1}^{N} u_{ij}^q(t-1)\boldsymbol{x}_i}{\sum_{i=1}^{N} u_{ij}^q(t-1)} \tag{14.32}$$

The resulting algorithm is also known as *Fuzzy c-Means (FCM)*[6] or *Fuzzy k-means algorithm* and has been discussed extensively in the literature (e.g. [Bezd 80, Cann 86, Hath 86, Hath 89, Isma 86]).

- Let us now examine the case in which Minkowski distances are in use. In the sequel, we consider only the case where p is even and $p < +\infty$. In this case, we can guarantee the differentiability of $d(\boldsymbol{x}_i, \boldsymbol{\theta}_j)$ with respect to $\boldsymbol{\theta}_j$. Equation (14.30) then gives

$$\frac{\partial d(\boldsymbol{x}_i, \boldsymbol{\theta}_j)}{\partial \theta_{jr}} = \frac{(\theta_{jr} - x_{ir})^{p-1}}{\left(\sum_{k=1}^{l} |x_{ik} - \theta_{jk}|^p\right)^{1-\frac{1}{p}}}, \quad r = 1, \ldots, l \tag{14.33}$$

Substituting Eq. (14.33) into Eq. (14.28), we obtain

$$\sum_{i=1}^{N} u_{ij}^q(t-1) \frac{(\theta_{jr} - x_{ir})^{p-1}}{\left(\sum_{k=1}^{l} |x_{ik} - \theta_{jk}|^p\right)^{1-\frac{1}{p}}} = 0, \quad r = 1, \ldots, l \tag{14.34}$$

Hence, we end up with a system of l nonlinear equations and l unknowns, that is, the coordinates of $\boldsymbol{\theta}_j$. This can be solved by an iterative technique, such as the Gauss–Newton or the Levenberg-Marquardt (L-M) method (e.g., [Luen 84]).

The resulting algorithms are also known as *pFCM*, where p indicates the employed Minkowski distance ([Bobr 91]).

In the iterative technique, the initial estimates at step t can be the estimates obtained from the previous iteration step $t-1$.

Example 14.5

(a) Consider the setup of Example 14.1(a). We run the GFAS first for the distance defined in Eq. (14.29), when (i) A is the identity 2×2 matrix, and (ii) $A = \begin{bmatrix} 2 & 1.5 \\ 1.5 & 2 \end{bmatrix}$, and (iii) the Minkowski distance with $p = 4$ is used. The algorithm is initialized as in the Example 14.1, with $\boldsymbol{\theta}_j$ in the place of $\boldsymbol{\mu}_j$. The fuzzifier q was set equal to 2.

The estimates for $\boldsymbol{\theta}_1$, $\boldsymbol{\theta}_2$, and $\boldsymbol{\theta}_3$ are $\boldsymbol{\theta}_1 = [1.37, 0.71]^T$, $\boldsymbol{\theta}_2 = [3.14, 3.12]^T$, and $\boldsymbol{\theta}_3 = [5.08, 1.21]^T$ for case (i), $\boldsymbol{\theta}_1 = [1.47, 0.56]^T$, $\boldsymbol{\theta}_2 = [3.54, 1.97]^T$, and $\boldsymbol{\theta}_3 = [5.21, 2.97]^T$ for

[6] A variant of the FCM tailored to a specific medical application is discussed in [Siya 05].

case (ii), and $\boldsymbol{\theta}_1 = [1.13, 0.74]^T$, $\boldsymbol{\theta}_2 = [2.99, 3.16]^T$, and $\boldsymbol{\theta}_3 = [5.21, 3.16]^T$ for case (iii). The corresponding confusion matrices (see Example 14.1) are

$$A_i = \begin{bmatrix} 98 & 2 & 0 \\ 14 & 84 & 2 \\ 11 & 0 & 89 \end{bmatrix}, \quad A_{ii} = \begin{bmatrix} 63 & 11 & 26 \\ 5 & 95 & 0 \\ 39 & 23 & 38 \end{bmatrix}, \quad A_{iii} = \begin{bmatrix} 96 & 0 & 4 \\ 11 & 89 & 0 \\ 13 & 2 & 85 \end{bmatrix}$$

Notice that in the cases of A_i and A_{iii}, almost all vectors from the same distribution are assigned to the same cluster. Note that for the construction of the confusion matrices we took the liberty to assign each point $\boldsymbol{x}_i$ to the cluster, for which the respective u_{ij} has the maximum value.

(b) Let us now consider the setup of Example 14.1(b). We run the GFAS algorithm for the three cases described in (a). The estimates for $\boldsymbol{\theta}_1$, $\boldsymbol{\theta}_2$, and $\boldsymbol{\theta}_3$ are $\boldsymbol{\theta}_1 = [1.60, 0.12]^T$, $\boldsymbol{\theta}_2 = [1.15, 1.67]^T$, and $\boldsymbol{\theta}_3 = [3.37, 2.10]^T$ for case (i), $\boldsymbol{\theta}_1 = [1.01, 0.38]^T$, $\boldsymbol{\theta}_2 = [2.25, 1.49]^T$, $\boldsymbol{\theta}_3 = [3.75, 2.68]^T$ for case (ii), and $\boldsymbol{\theta}_1 = [1.50, -0.13]^T$, $\boldsymbol{\theta}_2 = [1.25, 1.77]^T$, $\boldsymbol{\theta}_3 = [3.54, 1.74]^T$ for case (iii). The corresponding confusion matrices are

$$A_i = \begin{bmatrix} 51 & 46 & 3 \\ 14 & 47 & 39 \\ 43 & 0 & 57 \end{bmatrix}, \quad A_{ii} = \begin{bmatrix} 79 & 21 & 0 \\ 19 & 58 & 23 \\ 28 & 41 & 31 \end{bmatrix}, \quad A_{iii} = \begin{bmatrix} 51 & 3 & 46 \\ 37 & 62 & 1 \\ 11 & 36 & 53 \end{bmatrix}$$

Let us now comment on these results. First, as expected, the closer the clusters are, the worse the performance of all the algorithms. Also, when the distance given in Eq. (14.29) is employed, the choice of the matrix A is critical. For the case of our example, when $A = I$, the GFAS identifies almost perfectly the clusters underlying X, when they are not too close to each other. The same holds true for the Minkowski distance with $p = 4$.

Remarks

- The choice of the fuzzifier q in significant for the fuzzy clustering algorithms. Especially for the FCM, heuristic guidelines for the choice of q are given in [Bezd 81], while in [Gao 00] a method for selecting q based on fuzzy decision theory concepts is discussed.
- Several generalized FCM schemes have been proposed in the literature. These are derived from the minimization of cost functions that result from the basic one given in eq. (14.19) by adding suitable terms (see e.g. [Yang 93, Lin 96, Pedr 96, Ozde 02, Yu 03]).
- Kernelized versions of the FCM are discussed in [Chia 03, Shen 06, Zeyu 01, Zhan 03, Zhou 04]. Also, a comparative study of the kernelized verisons of FCM and the FCM itself is reported in [Grav 07].

14.3.2 Quadric Surfaces as Representatives

In this section we consider the case of clusters of quadric shape, such as hyperellipsoids and hyperparaboloids. In the sequel, we present four algorithms of this type, out of a large number that have appeared in the literature.

Our first concern is to define the distance between a point and a quadric surface, as Eq. (14.19) demands. The next section is devoted to the definition and the physical explanation of some commonly used distances of this kind.

Distances between a Point and a Quadric Surface

In this section we introduce definitions in addition to those discussed in Chapter 11 concerning the distance between a point and a quadric surface.

We recall that the general equation of a quadric surface, Q, is

$$\boldsymbol{x}^T A \boldsymbol{x} + \boldsymbol{b}^T \boldsymbol{x} + c = 0 \tag{14.35}$$

where A is an $l \times l$ symmetric matrix, $\boldsymbol{b}$ is an $l \times 1$ vector, c is a scalar, and $\boldsymbol{x} = [x_1, \ldots, x_l]^T$. The $A, \boldsymbol{b}$ and c quantities are the parameters defining Q. For various choices of these quantities we obtain hyperellipses, hyperparabolas, and so on. An alternative to the Eq. (14.35) form is easily verified to be (see Problem 14.8)

$$\boldsymbol{q}^T \boldsymbol{p} = 0 \tag{14.36}$$

where

$$\boldsymbol{q} = [\overbrace{x_1^2, x_2^2, \ldots, x_l^2}^{l}, \overbrace{x_1x_2, \ldots, x_{l-1}x_l}^{l(l-1)/2}, \overbrace{x_1, x_2, \ldots, x_l, 1}^{l+1}]^T \tag{14.37}$$

and

$$\boldsymbol{p} = [p_1, p_2, \ldots, p_l, p_{l+1}, \ldots, p_r, p_{r+1}, \ldots, p_s]^T \tag{14.38}$$

with $r = l(l+1)/2$ and $s = r + l + 1$. Vector $\boldsymbol{p}$ is easily derived from A, $\boldsymbol{b}$, and c so that Eq. (14.36) is satisfied.

Algebraic Distance

The *squared algebraic distance* between a point $\boldsymbol{x}$ and a quadric surface Q is defined as

$$d_a^2(\boldsymbol{x}, Q) = (\boldsymbol{x}^T A \boldsymbol{x} + \boldsymbol{b}^T \boldsymbol{x} + c)^2 \tag{14.39}$$

Using the alternative formulation in Eq. (14.36), $d_a^2(\boldsymbol{x}, Q)$ can be written as

$$d_a^2(\boldsymbol{x}, Q) = \boldsymbol{p}^T M \boldsymbol{p} \tag{14.40}$$

where $M = \boldsymbol{q}\boldsymbol{q}^T$. The algebraic distance could be seen as a generalization of the distance of a point from a hyperplane (see Chapter 11). Its physical meaning will become clear later on. For the derivation of the GFAS algorithm, based on the squared algebraic distance, it is more convenient to use the last formulation in (14.40).

Perpendicular Distance

Another distance between a point $\boldsymbol{x}$ and a quadric surface Q is the *squared perpendicular distance* defined as

$$d_p^2(\boldsymbol{x}, Q) = \min_{\boldsymbol{z}} \|\boldsymbol{x} - \boldsymbol{z}\|^2 \tag{14.41}$$

subject to the constraint that

$$\boldsymbol{z}^T A \boldsymbol{z} + \boldsymbol{b}^T \boldsymbol{z} + c = 0 \tag{14.42}$$

In words, this definition states that the distance between $\boldsymbol{x}$ and Q is defined as the squared Euclidean distance between $\boldsymbol{x}$ and the point $\boldsymbol{z}$ of Q closest to $\boldsymbol{x}$. $d_p(\boldsymbol{x}, Q)$ is the length of the perpendicular line segment from $\boldsymbol{x}$ to Q. Although this definition seems to be the most reasonable one from an intuitive point of view, its computation is not straightforward. More precisely, it involves Lagrangian formalization. Specifically, we define

$$\mathcal{D}(\boldsymbol{x}, Q) = \|\boldsymbol{x} - \boldsymbol{z}\|^2 - \lambda(\boldsymbol{z}^T A \boldsymbol{z} + \boldsymbol{b}^T \boldsymbol{z} + c) \tag{14.43}$$

Taking into account that A is symmetric, the gradient of $\mathcal{D}(\boldsymbol{x}, Q)$ with respect to $\boldsymbol{z}$ is

$$\frac{\partial \mathcal{D}(\boldsymbol{x}, Q)}{\partial \boldsymbol{z}} = 2(\boldsymbol{x} - \boldsymbol{z}) - 2\lambda A \boldsymbol{z} - \lambda \boldsymbol{b}$$

Setting $\partial \mathcal{D}(\boldsymbol{x}, Q)/\partial \boldsymbol{z}$ equal to $\mathbf{0}$ and after some algebra, we obtain

$$\boldsymbol{z} = \frac{1}{2}(I + \lambda A)^{-1}(2\boldsymbol{x} - \lambda \boldsymbol{b}) \tag{14.44}$$

To compute λ, we substitute $\boldsymbol{z}$ in Eq. (14.42), and we obtain a polynomial of λ of degree $2l$. For each of the real roots, λ_k, of this polynomial, we determine the corresponding $\boldsymbol{z}_k$. Then, $d_p(\boldsymbol{x}, Q)$ is defined as

$$d_p^2(\boldsymbol{x}, Q) = \min_{\boldsymbol{z}_k} \|\boldsymbol{x} - \boldsymbol{z}_k\|^2$$

Radial Distance

This distance is suitable when Q is a hyperellipsoidal. Then Eq. (14.35) can be brought into the form

$$(\boldsymbol{x} - \mathbf{c})^T A(\boldsymbol{x} - \mathbf{c}) = 1 \tag{14.45}$$

where $\mathbf{c}$ is the center of the ellipse and A is a symmetric positive definite matrix,[7] which determines the major and the minor axis of the ellipse as well as its orientation.

[7] Obviously, this matrix is in general different (yet related) from the A matrix used in Eq. (14.35). We use the same symbol for notational convenience.

The *squared radial distance* [Frig 96] between a point $\boldsymbol{x}$ and Q is defined as

$$d_r^2(\boldsymbol{x}, Q) = \|\boldsymbol{x} - \boldsymbol{z}\|^2 \tag{14.46}$$

subject to the constraints

$$(\boldsymbol{z} - \mathbf{c})^T A(\boldsymbol{z} - \mathbf{c}) = 1 \tag{14.47}$$

and

$$(\boldsymbol{z} - \mathbf{c}) = a(\boldsymbol{x} - \mathbf{c}) \tag{14.48}$$

In words, we first determine the intersection point, $\boldsymbol{z}$, between the line segment $\boldsymbol{x} - \mathbf{c}$ and Q, and then we compute $d_r(\boldsymbol{x}, Q)$ as the squared Euclidean distance between $\boldsymbol{x}$ and $\boldsymbol{z}$ (see Figure 14.8).

Normalized Radial Distance

The *squared normalized radial distance* between $\boldsymbol{x}$ and Q is also appropriate for hyperellipsoids and is defined as

$$d_{nr}^2(\boldsymbol{x}, Q) = \left(\left((\boldsymbol{x} - \mathbf{c})^T A(\boldsymbol{x} - \mathbf{c})\right)^{1/2} - 1\right)^2 \tag{14.49}$$

It can be shown (Problem 14.10) that

$$d_r^2(\boldsymbol{x}, Q) = d_{nr}^2(\boldsymbol{x}, Q)\|\boldsymbol{z} - \mathbf{c}\|^2 \tag{14.50}$$

where $\boldsymbol{z}$ is the intersection of the line segment $\boldsymbol{x} - \mathbf{c}$ with Q. This justifies the term "*normalized*."

The following examples give some insight into the distances that have been defined.

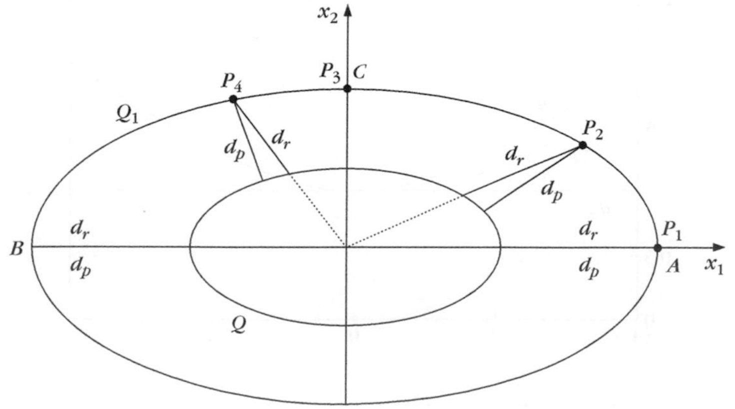

FIGURE 14.8

Graphical representation of the perpendicular and radial distances.

Example 14.6

Consider an ellipse Q centered at $\mathbf{c} = [0,0]^T$, with

$$A = \begin{bmatrix} 0.25 & 0 \\ 0 & 1 \end{bmatrix}$$

and an ellipse Q_1 centered at $\mathbf{c} = [0,0]^T$, with

$$A_1 = \begin{bmatrix} 1/16 & 0 \\ 0 & 1/4 \end{bmatrix}$$

Let $P(x_1, x_2)$ be a point in Q_1 moving from $A(4,0)$ to $B(-4,0)$ and always having its x_2 coordinate positive (Figure 14.8). Figure 14.9 illustrates how the four distances vary as P moves from A to B. One can easily observe that d_a and d_{nr} do not vary as P moves. This means that all points lying on an ellipse sharing the same center as Q and, having the same orientation as it, have the same d_a and d_{nr} distances from Q. However, this is not the case with the other two distances. Figure 14.8 shows graphically the d_p and d_r distances for various instances of P. As expected, the closer P is to the point $C(2,0)$, the smaller the d_p and d_r distances. Also, Figure 14.8 indicates that d_r can be used as an approximation of d_p, since, as we saw earlier, it is hard to compute d_p. However, it should be recalled that d_p is applicable when general quadric surfaces are considered, whereas d_r is used only when hyperellipsoids are considered.

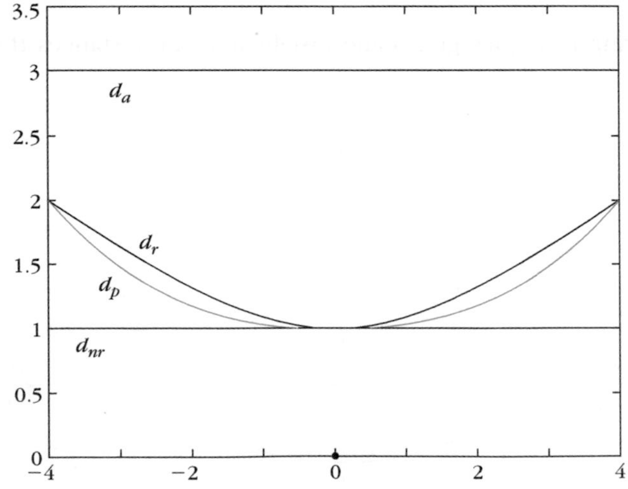

FIGURE 14.9

Variation of the distances d_p, d_a, d_{nr}, and d_r as P moves from $A(4,0)$ to $B(-4,0)$, with its x_2 coordinate being positive. The horizontal axis corresponds to the x_1 coordinate of the various points considered.

Example 14.7

Consider the two ellipses Q_1 and Q_2 shown in Figure 14.10, with equations

$$(\boldsymbol{x} - \boldsymbol{c}_j)^T A_j(\boldsymbol{x} - \boldsymbol{c}_j) = 1, \quad j = 1, 2$$

where $\boldsymbol{c}_1 = [0, 0]^T$, $\boldsymbol{c}_2 = [8,\ 0]^T$ and

$$A_1 = \begin{bmatrix} 1/16 & 0 \\ 0 & 1/4 \end{bmatrix}, \qquad A_2 = \begin{bmatrix} 1/4 & 0 \\ 0 & 1 \end{bmatrix}$$

Also consider the points $A(5, 0)$, $B(3, 0)$, $C(0, 2)$, and $D(5.25, 1.45)$. The distances d_a, d_p, d_{nr}, d_r between each of these points and Q_1 and Q_2 are shown in Table 14.2. From this table we observe that the d_p distances from A, B, and C to Q_1 are equal. Moreover, as expected, the d_r distance is always greater than or equal to d_p (when the equality holds?). Also, d_p is unbiased toward the size of the ellipses ($d_p(A, Q_1) = d_p(A, Q_2)$). Finally, d_a and d_{nr} are biased toward larger ellipses ($d_a(A, Q_1) < d_a(A, Q_2)$ and $d_{nr}(A, Q_1) < d_{nr}(A, Q_2)$).

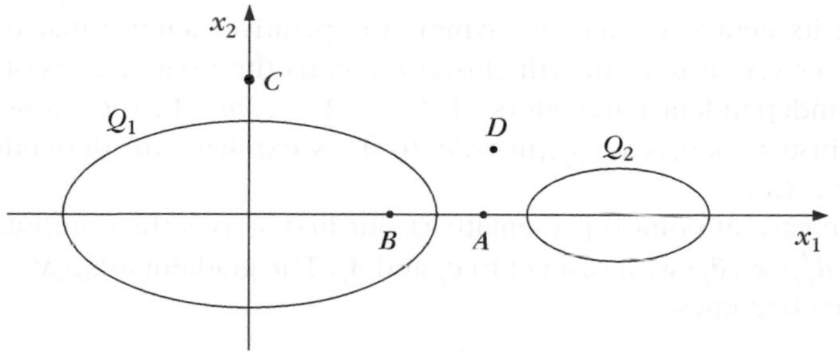

FIGURE 14.10

The setup of Example 14.7.

Table 14.2 Comparison of the various distances between points and hyperellipsoids

	d_a		d_p		d_{nr}		d_r	
	Q_1	Q_2	Q_1	Q_2	Q_1	Q_2	Q_1	Q_2
A	0.32	1.56	1	1	0.06	0.25	1	1
B	0.19	27.56	1	9	0.06	2.25	1	9
C	1.56	576	1	44.32	0.25	16	1	46.72
D	1.56	9.00	2.78	1.93	0.25	1	3.30	2.42

In the sequel, we derive some well-known algorithms suitable for shell-shaped clusters. These algorithms are usually called *fuzzy shell clustering algorithms*, and the representatives of the clusters are (in most cases) hyperquadrics.

Fuzzy Shell Clustering Algorithms

The first two algorithms that are examined are suitable for hyperellipsoid-shaped clusters. The first of them [Dave 92a, Dave 92b] is called the *adaptive fuzzy C-shells (AFCS) clustering algorithm*, and the second one is known as the *fuzzy C ellipsoidal shells (FCES) algorithm* [Kris 95a].

The Adaptive Fuzzy C-Shells (AFCS) Algorithm

The AFCS uses the squared distance d_{nr} between a point and a hyperellipsoidal (Eq. 14.49). Thus, Eq. (14.19) becomes

$$J_{nr}(\boldsymbol{\theta}, U) = \sum_{i=1}^{N}\sum_{j=1}^{m} u_{ij}^{q} d_{nr}^{2}(\boldsymbol{x}_i, Q_j) \tag{14.51}$$

It is clear that in this case the parameters used to identify a representative (an ellipse) are its center, $\boldsymbol{c}_j$, and the symmetric, positive definite matrix, A_j. Thus, the parameter vector $\boldsymbol{\theta}_j$ of the jth cluster contains the l parameters of $\boldsymbol{c}_j$ plus the $l(l+1)/2$ independent parameters of A_j, $j = 1, \ldots, m$. In the sequel, we write $d_{nr}(\boldsymbol{x}_i, \boldsymbol{\theta}_j)$ instead of $d_{nr}(\boldsymbol{x}_i, Q_j)$, in order to show explicitly the dependence on the parameter vector.

As in the case of point representatives, our first step is the computation of the gradient of $d_{nr}^2(\boldsymbol{x}_i, \boldsymbol{\theta}_j)$ with respect to $\boldsymbol{c}_j$ and A_j. The gradient $\partial d_{nr}^2(\boldsymbol{x}_i, \boldsymbol{\theta}_j)/\partial \boldsymbol{c}_j$ after some algebra becomes

$$\frac{\partial d_{nr}^2(\boldsymbol{x}_i, \boldsymbol{\theta}_j)}{\partial \boldsymbol{c}_j} = -2\frac{d_{nr}(\boldsymbol{x}_i, \boldsymbol{\theta}_j)}{\phi(\boldsymbol{x}_i, \boldsymbol{\theta}_j)} A_j(\boldsymbol{x}_i - \boldsymbol{c}_j) \tag{14.52}$$

where

$$\phi^2(\boldsymbol{x}_i, \boldsymbol{\theta}_j) = (\boldsymbol{x}_i - \boldsymbol{c}_j)^T A_j (\boldsymbol{x}_i - \boldsymbol{c}_j) \tag{14.53}$$

Let a_{rs}^j be the (r, s) element of A_j and x_{ir}, c_{jr} the rth coordinates of $\boldsymbol{x}_i$ and $\boldsymbol{c}_j$, respectively. Then, the partial derivative of $d_{nr}^2(\boldsymbol{x}_i, \boldsymbol{\theta}_j)$ with respect to a_{rs}^j, after some elementary manipulations, becomes

$$\frac{\partial d_{nr}^2(\boldsymbol{x}_i, \boldsymbol{\theta}_j)}{\partial a_{rs}^j} = \frac{d_{nr}(\boldsymbol{x}_i, \boldsymbol{\theta}_j)}{\phi(\boldsymbol{x}_i, \boldsymbol{\theta}_j)}(x_{ir} - c_{jr})(x_{is} - c_{js})$$

Thus,

$$\frac{\partial d_{nr}^2(\boldsymbol{x}_i, \boldsymbol{\theta}_j)}{\partial A_j} = \frac{d_{nr}(\boldsymbol{x}_i, \boldsymbol{\theta}_j)}{\phi(\boldsymbol{x}_i, \boldsymbol{\theta}_j)}(\boldsymbol{x}_i - \boldsymbol{c}_j)(\boldsymbol{x}_i - \boldsymbol{c}_j)^T \tag{14.54}$$

Substituting Eqs. (14.52) and (14.54) in (14.28), and after some minor manipulations, the parameter updating part of GFAS becomes

- Parameter updating:
  - Solve with respect to c_j and A_j the following equations.

$$\sum_{i=1}^{N} u_{ij}^q(t-1)\frac{d_{nr}(x_i, \theta_j)}{\phi(x_i, \theta_j)}(x_i - c_j) = 0$$

and

$$\sum_{i=1}^{N} u_{ij}^q(t-1)\frac{d_{nr}(x_i, \theta_j)}{\phi(x_i, \theta_j)}(x_i - c_j)(x_i - c_j)^T = O$$

where

$$\phi^2(x_i, \theta_j) = (x_i - c_j)^T A_j (x_i - c_j)$$

and

$$d_{nr}^2(x_i, \theta_j) = (\phi(x_i, \theta_j) - 1)^2$$

  - Set $c_j(t)$ and $A_j(t), j = 1, \ldots, m$, equal to the resulting solutions.
- End parameter updating.

Once more, the above system of equations can be solved by employing iterative techniques.

Variants of the algorithm, imposing certain constraints on $A_j, j = 1, \ldots, m$, have also been proposed in [Dave 92a] and [Dave 92b].

Example 14.8

Consider the three ellipses in Figure 14.11a, with centers $c_1 = [0, 0]^T$, $c_2 = [8, 0]^T$, and $c_3 = [1, 1]^T$, respectively. The corresponding matrices that specify their major and minor axes, as well as their orientation, are

$$A_1 = \begin{bmatrix} \frac{1}{16} & 0 \\ 0 & \frac{1}{4} \end{bmatrix}, \quad A_2 = \begin{bmatrix} \frac{1}{4} & 0 \\ 0 & 1 \end{bmatrix}, \quad A_3 = \begin{bmatrix} \frac{1}{8} & 0 \\ 0 & \frac{1}{4} \end{bmatrix}$$

respectively. We generate 100 points, x_i, from each ellipse and we add to each of these points a random vector whose coordinates stem from the uniform distribution in $[-0.5, 0.5]$. The initial values for the c_i's and the A_i's, $i = 1, 2, 3$, are $c_i(0) = c_i + z, i = 1, 2, 3$, with z taken to be $z = [0.3, 0.3]^T$ and $A_i(0) = A_i + Z$, $i = 1, 2, 3$, where all the elements of Z are equal to 0.2. The fuzzifier q was also set equal to 2. Application of the AFCS algorithm to this data set gives, after four iterations, the results shown in Figure 14.11b. Thus, the algorithm has identified the ellipses to a good approximation.

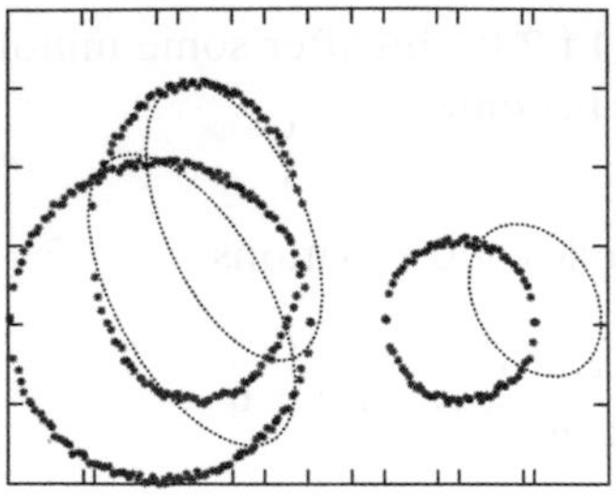

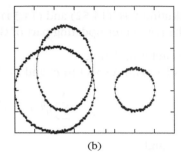

FIGURE 14.11

The setup of Example 14.8. Thick dots represent the points of the data set. Thin dots represent (a) the initial estimates and (b) the final estimates of the ellipses.

The Fuzzy C Ellipsoidal Shells (FCES) Algorithm

This algorithm uses the squared radial distance between a point and a hyperellipsoidal. Equation (14.19) now becomes

$$J_r(\boldsymbol{\theta}, U) = \sum_{i=1}^{N}\sum_{j=1}^{m} u_{ij}^q d_r^2(\boldsymbol{x}_i, \boldsymbol{\theta}_j) \tag{14.55}$$

Defining the $\boldsymbol{\theta}_j$'s as in the previous case and carrying out the steps followed for the derivation of the AFCS, we end up with the following equations for $\boldsymbol{c}_j$ and A_j (see Problem 14.11):

$$\sum_{i=1}^{N} u_{ij}^q(t-1)\left[\frac{\|\boldsymbol{x}_i - \boldsymbol{c}_j\|^2(1-\phi(\boldsymbol{x}_i, \boldsymbol{\theta}_j))}{\phi^4(\boldsymbol{x}_i, \boldsymbol{\theta}_j)}A_j - \left(1 - \frac{1}{\phi(\boldsymbol{x}_i, \boldsymbol{\theta}_j)}\right)^2 I\right](\boldsymbol{x}_i - \boldsymbol{c}_j) = \mathbf{0} \tag{14.56}$$

and

$$\sum_{i=1}^{N} u_{ij}^q(t-1)\frac{\phi(\boldsymbol{x}_i, \boldsymbol{\theta}_j) - 1}{\phi^4(\boldsymbol{x}_i, \boldsymbol{\theta}_j)}\|\boldsymbol{x}_i - \boldsymbol{c}_j\|^2(\boldsymbol{x}_i - \boldsymbol{c}_j)(\boldsymbol{x}_i - \boldsymbol{c}_j)^T = O \tag{14.57}$$

where $\phi(\boldsymbol{x}_i, \boldsymbol{\theta}_j)$ is defined as in the case of the AFCS algorithm.

The following two algorithms are proposed in [Kris 95a] and [Frig 96]. In contrast to the previous algorithms, they may fit quadrics of any shape to the data set. They are called the *fuzzy C quadric shells (FCQS) algorithm* and *modified fuzzy C quadric shells (MFCQS) algorithm*, respectively.

Fuzzy C Quadric Shells (FCQS) Algorithm

The FCQS algorithm is suitable for recovering general hyperquadric shapes. It uses the squared algebraic distance. Equation (14.19) now becomes

$$J_a(\boldsymbol{\theta}, U) = \sum_{i=1}^{N}\sum_{j=1}^{m} u_{ij}^q d_a^2(\boldsymbol{x}_i, \boldsymbol{\theta}_j) = \sum_{i=1}^{N}\sum_{j=1}^{m} u_{ij}^q \boldsymbol{p}_j^T M_i \boldsymbol{p}_j \tag{14.58}$$

where $\boldsymbol{p}_j$ is defined in Eq. (14.38) and $M_i = \boldsymbol{q}_i\boldsymbol{q}_i^T$, with $\boldsymbol{q}_i$ defined in Eq. (14.37).

We recall that $\boldsymbol{p}_j$ incorporates all the parameters of the jth quadric surface (see Eq. 14.40), that is, $\boldsymbol{\theta}_j = \boldsymbol{p}_j$. Direct minimization of $J_a(\boldsymbol{\theta}, U)$ with respect to $\boldsymbol{p}_j$ would lead to the trivial zero solution for $\boldsymbol{p}_j$. Thus, constraints on $\boldsymbol{p}_j$ must be imposed, and a number of those have been proposed in the literature. Different constraints lead to different algorithms. Examples of such constraints are [Kris 95a] (i) $\|\boldsymbol{p}_j\|^2 = 1$, (ii) $\sum_{k=1}^{r+l} p_{jk}^2 = 1$, (iii) $p_{j1} = 1$, (iv) $p_{js}^2 = 1$, and (v) $\|\sum_{k=1}^{l} p_{jk}^2 + 0.5\sum_{k=l+1}^{r} p_{jk}^2\|^2 = 1$ (Problem 14.12). Each of these constraints has its advantages and disadvantages. For example, constraints (i) and (ii) [Gnan 77, Pato 70] do not preserve the invariance under translation and rotation of d_a. However, they are able to identify planar clusters. Also, constraint (iii) [Chen 89] precludes linear clusters and can lead to poor results if the points in X are approximately coplanar.

Modified Fuzzy C Quadric Shells (MFCQS) Algorithm

A different C-shells quadric algorithm is obtained if we employ the squared perpendicular distance d_p between a point and a quadric surface. However, because of the difficulty of its estimation, the resulting problem is much more difficult than those examined before. In this case, due to the complex nature of d_p, minimization of the $J_p(\boldsymbol{\theta}, U)$ with respect to the parameter vector $\boldsymbol{\theta}_j$ becomes very complex [Kris 95a].

One way to simplify things is to use the following alternative scheme. For the computation of u_{ij}'s the perpendicular distance d_p is used, and for the estimation of the parameters $\boldsymbol{\theta}_j, j = 1, \ldots, m$, the updating scheme of FCQS is employed (recall that in FCQS, $\boldsymbol{\theta}_i = \boldsymbol{p}_i$). In other words, the grade of membership of a vector $\boldsymbol{x}_i$ in a cluster is determined using the perpendicular distance, and the updating of the parameters of the representatives is carried out using the parameter updating part of the FCQS algorithm. However, this simplification implies that the algebraic and the perpendicular distances should be close to each other (see also Problem 14.13). This modification leads to the so called modified FCQS (MFCQS) algorithm.

Another algorithm, discussed in [Kris 95a] and [Frig 96], is the fuzzy C planoquadric shells (FCPQS) algorithm. This algorithm uses a first-order approximation of the algebraic distance, and it is derived, as are all the others, by taking derivatives of the resulting cost function with respect to the parameter vector, $\boldsymbol{\theta}_j$, and setting them equal to zero.

Finally, fuzzy clustering algorithms that are able to detect spherical clusters are discussed in [Dave 92a, Kris 92a, Kris 92b, Man 94]. However, most of these may be viewed as special cases of the algorithms developed to fit ellipses.

14.3.3 Hyperplane Representatives

In this section, we discuss algorithms that suitable for the recovery of hyperplanar clusters [Kris 92a]. Algorithms of this kind can be applied to the surface-fitting problem, which is one of the most important tasks in computer vision. In this problem, the surfaces of an object depicted in the image are approximated by planar surfaces. Successful identification of the surfaces is a prerequisite for the identification of the objects depicted in an image.

Some of these algorithms, such as the fuzzy c-varieties (FCV) algorithm [Ande 85], are based on minimization of the distances of the vectors in X from hyperplanes (see Chapter 11). However, FCV tends to recover very long clusters, and thus, collinear distinct clusters may be merged into a single cluster.

In this section we describe an algorithm, known as the Gustafson–Kessel (G-K) algorithm (see, e.g., [Kris 92a, Kris 99a]). According to this algorithm, planar clusters are represented by centers $\boldsymbol{c}_j$ and covariance matrices Σ_j. Defining $\boldsymbol{\theta}_j$ as in previous cases, we define the squared distance between a vector $\boldsymbol{x}$ and the jth cluster as the scaled Mahalanobis distance

$$d_{GK}^2(\boldsymbol{x}, \boldsymbol{\theta}_j) = |\Sigma_j|^{1/l}(\boldsymbol{x} - \boldsymbol{c}_j)^T \Sigma_j^{-1}(\boldsymbol{x} - \boldsymbol{c}_j) \tag{14.59}$$

Let us now gain some insight into the behavior of this distance. A well-known property that characterizes the distance d_H of a point from a hyperplane, as defined in Chapter 11, is that all points lying on a hyperplane H_1 parallel to a given hyperplane H, have the same d_H distance from H. This will be our starting point for the investigation of d_{GK}^2.

Example 14.9

Consider the setup of Figure 14.12a, where a single cluster C is present, and let $\boldsymbol{\theta}$ be its parameter vector. The points of C are of the form $[x_{i1}, x_{i2}]^T$ where $x_{i1} = -2 + 0.1i$, $i = 0, 1, 2, \ldots, 40$, and the corresponding x_{i2}'s are random numbers following the uniform distribution in $[-0.1, 0.1]$.

Consider also the points of the line segment u connecting the points $(-2, 2)$ and $(2, 2)$. Figure 14.12b shows the distances $d_{GK}(\boldsymbol{x}, \boldsymbol{\theta})$ of the points $\boldsymbol{x} \in u$ from C. As we can see, all these distances are almost the same. Indeed, the relative difference $(d_{\max} - d_{\min})/d_{\max}$ between the maximum $d_{\max}$ and the minimum $d_{\min}$ values is approximately equal to 0.02.

Now consider the larger line segment $v_1(v_2)$ that connects $(-8, 2)((-8, -2))$ and $(8, 2)((8, -2))$. The distances $d_{GK}(\boldsymbol{x}, \boldsymbol{\theta})$ of the points $\boldsymbol{x} \in v_1(v_2)$ from C are shown in Figure 14.12b. Note that although we have larger variations compared to the previous case, they still remain relatively small (the relative difference $(d_{\max} - d_{\min})/d_{\max}$ is approximately 0.12).

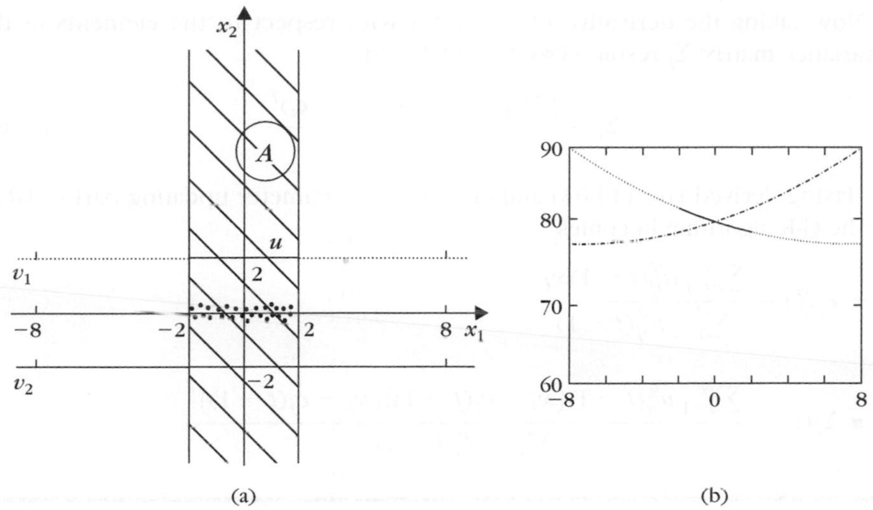

FIGURE 14.12

(a) The setup of the Example 14.9. (b) The solid line corresponds to the distances of the points of $\boldsymbol{u}$ from C. The dashed line corresponds to the distances of the points of the line segment v_1 from C (the solid line is part of the dashed line). Also, the dash-dotted line corresponds to the distances of the points of the line segment v_2 from C.

The G-K algorithm can be derived via the minimization of

$$J_{GK}(\boldsymbol{\theta}, U) = \sum_{i=1}^{N}\sum_{j=1}^{m} u_{ij}^q d_{GK}^2(\boldsymbol{x}_i, \boldsymbol{\theta}_j) \tag{14.60}$$

Taking the gradient of $J_{GK}(\boldsymbol{\theta}, U)$ with respect to $\boldsymbol{c}_j$, we obtain

$$\frac{\partial J_{GK}(\boldsymbol{\theta}, U)}{\partial \boldsymbol{c}_j} = \sum_{i=1}^{N} u_{ij}^q \frac{\partial d_{GK}^2(\boldsymbol{x}_i, \boldsymbol{\theta}_j)}{\partial \boldsymbol{c}_j} \tag{14.61}$$

The gradient of the distance, after a bit of algebra, becomes

$$\frac{\partial d_{GK}^2(\boldsymbol{x}_i, \boldsymbol{\theta}_j)}{\partial \boldsymbol{c}_j} = -2|\Sigma_j|^{1/l}\Sigma_j^{-1}(\boldsymbol{x}_i - \boldsymbol{c}_j) \tag{14.62}$$

Substituting $\partial d^2(\boldsymbol{x}_i, \boldsymbol{\theta}_j)/\partial \boldsymbol{c}_j$ from Eq. (14.62) into (14.61) and setting $\partial J_{GK}(\boldsymbol{\theta}, U)/\partial \boldsymbol{c}_j$ equal to zero, we obtain[8]

$$\boldsymbol{c}_j = \frac{\sum_{i=1}^{N} u_{ij}^q \boldsymbol{x}_i}{\sum_{i=1}^{N} u_{ij}^q} \tag{14.63}$$

[8] We also make the mild assumption that the covariance matrix is invertible.

Now taking the derivative of $J_{GK}(\boldsymbol{\theta}, U)$ with respect to the elements of the covariance matrix, Σ_j results (Problem 14.16) in

$$\Sigma_j = \frac{\sum_{i=1}^{N} u_{ij}^q (\boldsymbol{x}_i - \boldsymbol{c}_j)(\boldsymbol{x}_i - \boldsymbol{c}_j)^T}{\sum_{i=1}^{N} u_{ij}^q} \tag{14.64}$$

Having derived Eqs. (14.63) and (14.64), the parameter updating part of GFAS for the G-K algorithm becomes

- $\boldsymbol{c}_j(t) = \dfrac{\sum_{i=1}^{N} u_{ij}^q (t-1)\boldsymbol{x}_i}{\sum_{i=1}^{N} u_{ij}^q (t-1)}$

- $\Sigma_j(t) = \dfrac{\sum_{i=1}^{N} u_{ij}^q (t-1)\big(\boldsymbol{x}_i - \boldsymbol{c}_j(t-1)\big)\big(\boldsymbol{x}_i - \boldsymbol{c}_j(t-1)\big)^T}{\sum_{i=1}^{N} u_{ij}^q (t-1)}$

Example 14.10

(a) Consider Figure 14.13a. It consists of three linear clusters. Each cluster contains 41 points. The points of the first cluster lie around the line $x_2 = x_1 + 1$, while the points of the second and the third clusters lie around the lines $x_2 = 0$ and $x_2 = -x_1 + 1$, respectively. The $\boldsymbol{c}_j$'s, $j = 1, 2, 3$, are randomly initialized and the threshold of the termination criterion, ε, is set to 0.01. The G-K algorithm converges after 26 iterations. As shown in Figure 14.13b, the G-K identifies correctly the clusters underlying X.

(b) Now consider Figure 14.14a. We also have three clusters, each consisting of 41 points. The first and the third clusters are the same as in Figure 14.13a, while the points of the second cluster lie around the line $x_2 = 0.5$. Note that in this case the three intersection points between

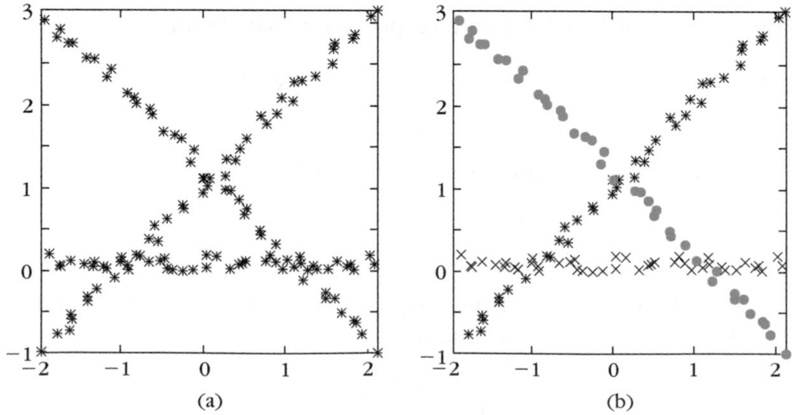

FIGURE 14.13

(a) The data set X for Example 14.10(a). (b) The results of the G-K algorithm.

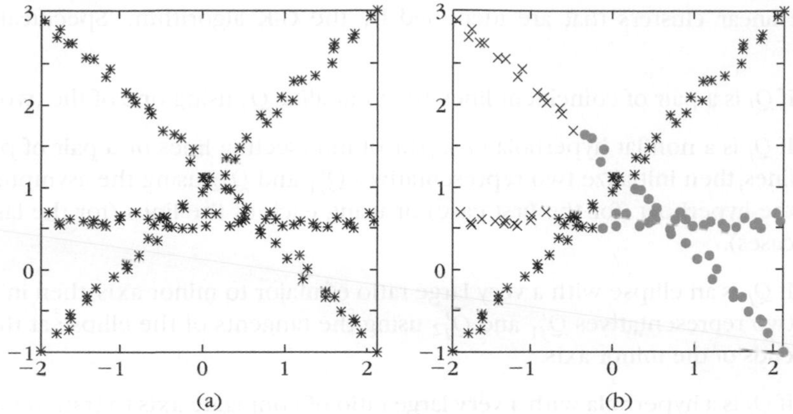

FIGURE 14.14

(a) The data set X for Example 14.10(b). (b) The results of the G-K algorithm.

any pair of lines lie very close to each other. The G-K algorithm terminates after 38 iterations. The results obtained are shown in Figure 14.14b. In this case, the G-K algorithm fails to identify the clusters correctly.

14.3.4 Combining Quadric and Hyperplane Representatives

In this section, we assume that $l = 2$. Consider the case in which X contains quadric-shaped clusters as well as linear clusters. How can we accurately identify both kinds of clusters? If we run an algorithm that fits quadric curves to the clusters, the linear clusters will not be properly represented. On the other hand, if we run an algorithm that fits lines to the clusters, the ellipsoidally and hyperbolically shaped clusters will be poorly represented. A way out of this problem is discussed in [Kris 95a]. The idea is to run the FCQS algorithm first on the whole data set X. This algorithm can be used to detect linear clusters, even though the adopted constraints force all representatives to be of second degree. This happens since, in practice, FCQS fits a pair of coincident lines for a single line, a hyperbola for two intersecting lines and a very "flat" hyperbola or a very elongated ellipse or a pair of lines for two parallel lines [Kris 95a]. The identification of "extreme" quadric curves (i.e., extremely elongated ellipses, "flat" hyperbolas, a set of lines) after the termination of the algorithm is a strong indication that X contains linear clusters. In order to represent these extreme clusters more accurately, we can run the G-K algorithm on the set X', which contains only the vectors that belong to them (with a high grade of membership). However, different actions have to be carried out depending on the shape of each extreme quadric curve. Let Q_j be the representative curve of the jth cluster, $j = 1, \ldots, m$, identified by FCQS and Q_j''s the representative curves of the linear clusters that are identified by the G-K algorithm. Specifically, we have

- If Q_j is a pair of coincident lines, then initialize Q'_j using one of the two lines.
- If Q_j is a nonflat hyperbola *or* a pair of intersecting lines *or* a pair of parallel lines, then initialize two representatives Q'_{j1} and Q'_{j2} using the asymptotes of the hyperbola (for the first case) or using each of the lines (for the last two cases).
- If Q_j is an ellipse with a very large ratio of major to minor axis, then initialize two representatives Q'_{j1} and Q'_{j2} using the tangents of the ellipse at the two ends of the minor axis.
- If Q_j is a hyperbola with a very large ratio of conjugate axis to transverse axis, then initialize two representatives Q'_{j1} and Q'_{j2} using the two tangents of the hyperbola at its two vertices.

Since the initialization of the Q'_j representatives is very good, it is expected that the G-K algorithm will converge in a few iterations to a satisfactory solution.

14.3.5 A Geometrical Interpretation

Arguments similar to those given in Section 14.2.2 can also be repeated here. Now u_{ij} takes the place of $P(C_j|\boldsymbol{x}_i)$. The constraint equation in this case is

$$\sum_{j=1}^{m} u_{ij} = 1, \quad i = 1, \ldots, N. \tag{14.65}$$

The vector $\boldsymbol{y}$ associated with vector $\boldsymbol{x}_i$ becomes $\boldsymbol{y} = [u_{i1}, u_{i2}, \ldots, u_{im}]$ and it is also restricted on the hyperplane defined by the constraint (14.65) in the H_m hypercube. If we carry out the experiments discussed in Section 14.2.2, we will draw similar conclusions with respect to the effect of the outliers on the performance of the fuzzy algorithms.

In [Mena 00] an algorithm called *fuzzy c + 2 means* is introduced. This is an extension of GFAS for point representatives, where the outliers as well as the points that lie near the cluster boundaries are treated so as to control their effect on the estimates of the cluster representatives.

14.3.6 Convergence Aspects of the Fuzzy Clustering Algorithms

Although fuzzy clustering algorithms are obtained by minimizing a cost function of the form of Eq. (14.19), little is known about their convergence behavior. More specifically, it has been proved [Bezd 80, Hath 86], using the global convergence theorem of Zangwill [Luen 84], that when a Mahalanobis distance is used (or other distances satisfying certain conditions discussed in [Bezd 80]), the iteration sequence produced by the fuzzy c-means (FCM) algorithm either converges to a stationary point of the cost function in a finite number of iteration steps or it has at

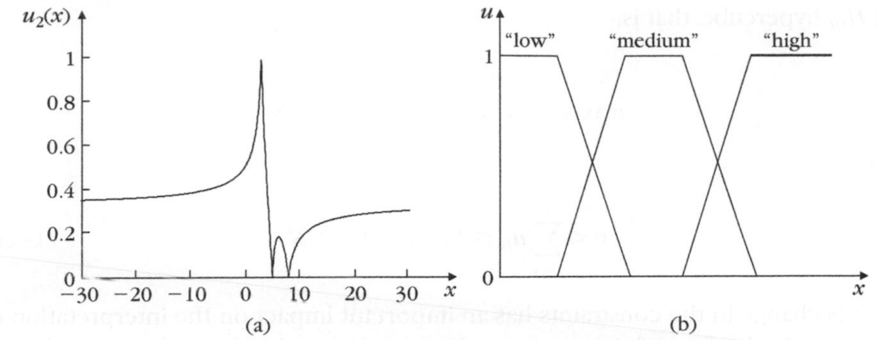

FIGURE 14.15

(a) The membership function $u_2(x)$ given by Eq. (14.26), for the one-dimensional case, with $\theta_1 = 5$, $\theta_2 = 3$, $\theta_3 = 8$, $q = 2$ and $d(x, \theta_i) = |x - \theta_i|$. (b) Examples of membership functions characterizing "low,", "medium," and "high" for a specific quantity.

least one subsequence that converges to a stationary point of the cost function. This point may be a local (or global) optimum or a saddle point. Tests for the identification of the nature of the convergence point are discussed in [Isma 86, Hath 86, Kim 88]. More recently in [Grol 05] it is shown that sequence produced by the FCM converges to a stationary point of the cost function. Issues concerning numerical convergence aspects of the FCM algorithms are discussed in [Bezd 92].

14.3.7 Alternating Cluster Estimation

It is not difficult to notice that the membership functions $u_j(\boldsymbol{x}_i)$, associated with the u_{ij}'s used in GFAS (Eq. (14.26)), are neither convex nor monotonous (see, for example, Figure 14.15a). However, in fuzzy rule-based systems convexity of the membership functions is an important requirement. For example, linguistic characterizations such as "low," "medium," or "high" require convex membership functions of the form shown in Figure 14.15b. In such cases it may be preferable to adopt a specific membership function and use the alternating updating philosophy used in GFAS to estimate u_{ij}'s and $\boldsymbol{\theta}_j$. The resulting algorithmic scheme is known as *alternating cluster estimation (ACE)* ([Runk 99, Hopp 99]) and GFAS may be viewed as a special case of it. Obviously, in this case, the solution obtained is not necessarily related to an optimizing criterion.

14.4 POSSIBILISTIC CLUSTERING

The algorithms of this section are relaxed from constraints such as in (14.17) and (14.65) [Kris 93, Kris 96]. Speaking in the terms of Section 14.3.5, this means that the vector $\boldsymbol{y}$, with coordinates the u_{ij}'s, will be allowed to move anywhere in the H_m hypercube, that is,

$$u_{ij} \in [0, 1]$$
$$\max_{j=1,\dots,m} u_{ij} > 0, \quad i = 1, \dots, N$$

and

$$0 < \sum_{i=1}^{N} u_{ij} \leq N, \quad i = 1, \dots, N \tag{14.66}$$

This change in the constraints has an important impact on the interpretation of the u_{ij}'s. In the fuzzy framework, u_{ij} denotes the grade of membership of $\boldsymbol{x}_i$ in the jth cluster. Here, u_{ij} may be interpreted as the degree of compatibility of $\boldsymbol{x}_i$ with the jth cluster representative, or, following [Zade 78], the possibility that $\boldsymbol{x}_i$ belongs to the jth cluster. Note that *the possibility that $\boldsymbol{x}_i$ belongs to the jth cluster depends exclusively on $\boldsymbol{x}_i$ and the cluster representative of the jth cluster; that is, it is independent of the possibilities that $\boldsymbol{x}_i$ belongs to any other cluster.*

For convenience, let us recall here that the cost function to be minimized is

$$J(\boldsymbol{\theta}, U) = \sum_{i=1}^{N}\sum_{j=1}^{m} u_{ij}^q d(\boldsymbol{x}_i, \boldsymbol{\theta}_j) \tag{14.67}$$

Obviously, direct minimization with respect to U will lead to the trivial zero solution. In order to avoid this situation, we must insert an additional term in $J(\boldsymbol{\theta}, U)$. This term, $f(U)$, will be a function of u_{ij}'s only. Motivated by the discussion in Section 14.2.2, it will be chosen in such a way so as to minimize the effects of outliers. As will become apparent soon, one such choice of $f(U)$ is

$$f(U) = \sum_{j=1}^{m} \eta_j \sum_{i=1}^{N} (1 - u_{ij})^q \tag{14.68}$$

Then, the cost function becomes

$$J(\boldsymbol{\theta}, U) = \sum_{i=1}^{N}\sum_{j=1}^{m} u_{ij}^q d(\boldsymbol{x}_i, \boldsymbol{\theta}_j) + \sum_{j=1}^{m} \eta_j \sum_{i=1}^{N} (1 - u_{ij})^q \tag{14.69}$$

where η_j are suitably chosen positive constants.

The minimum of $J(\boldsymbol{\theta}, U)$, with respect to u_{ij}, is obtained by

$$\frac{\partial J(\boldsymbol{\theta}, U)}{\partial u_{ij}} = q u_{ij}^{q-1} d(\boldsymbol{x}_i, \boldsymbol{\theta}_j) - q\eta_j (1 - u_{ij})^{q-1} = 0$$

or

$$u_{ij} = \frac{1}{1 + \left(\frac{d(\boldsymbol{x}_i, \boldsymbol{\theta}_j)}{\eta_j}\right)^{\frac{1}{q-1}}} \tag{14.70}$$

In words, u_{ij} is inversely proportional to the dissimilarity between $\boldsymbol{x}_i$, and the representative of the jth cluster. Loosely speaking, u_{ij} denotes the degree to which the representative of the jth cluster *should be "stretched"* in order to match $\boldsymbol{x}_i$. Large (small) values of u_{ij} indicate little (large) stretch for the jth representative. *It is clear that for a specific vector $\boldsymbol{x}_i$, this "stretching" action can be carried out independently for each cluster.*

The meaning of the second term in Eq. (14.69) is clearer now. Its effect is to minimize the influence of outliers in the estimation of the $\boldsymbol{\theta}_j$'s. Indeed, large dissimilarity levels correspond to small u_{ij}'s and they have little effect on the first term in the cost function, which controls the estimation of $\boldsymbol{\theta}_j$'s.

Since the second term does not involve the representatives of the clusters, one may easily conclude that in possibilistic clustering schemes, the updating of the parameters of each cluster is carried out in exactly the same way as in the case of their fuzzy counterparts.

Generalized Possibilistic Algorithmic Scheme (GPAS)

- Fix $\eta_j, j = 1, \ldots, m$.
- Choose $\boldsymbol{\theta}_j(0)$ as the initial estimates of $\boldsymbol{\theta}_j, j = 1, \ldots, m$.
- $t = 0$.
- Repeat
  - For $i = 1$ to N
    - For $j = 1$ to m

$$u_{ij}(t) = \frac{1}{1 + \left(\frac{d(\boldsymbol{x}_i, \boldsymbol{\theta}_j(t))}{\eta_j}\right)^{\frac{1}{q-1}}}$$

    - End {For-j}
  - End {For-i}
  - $t = t + 1$
  - For $j = 1$ to m
    - *Parameter updating*: Solve

$$\sum_{i=1}^{N} u_{ij}^q(t-1) \frac{\partial d(\boldsymbol{x}_i, \boldsymbol{\theta}_j)}{\partial \boldsymbol{\theta}_j} = \mathbf{0} \qquad (14.71)$$

      with respect to $\boldsymbol{\theta}_j$ and set $\boldsymbol{\theta}_j(t)$ equal to the computed solution.
  - End {For-j}.
- Until a termination criterion is met.

As usual, we may employ $\|\boldsymbol{\theta}(t) - \boldsymbol{\theta}(t-1)\| < \varepsilon$ as a termination criterion. Based on the preceding generalized scheme, for each of the fuzzy clustering algorithms, defined in the previous section, we can derive a corresponding possibilistic one.

An interesting observation is that, since for each vector $\boldsymbol{x}_i$, u_{ij}'s, $j = 1, \ldots, m$, are independent of each other, we can write $J(\boldsymbol{\theta}, U)$ as

$$J(\boldsymbol{\theta}, U) = \sum_{j=1}^{m} J_j$$

where

$$J_j = \sum_{i=1}^{N} u_{ij}^q d(\boldsymbol{x}_i, \boldsymbol{\theta}_j) + \eta_j \sum_{i=1}^{N} (1 - u_{ij})^q \tag{14.72}$$

Each J_j corresponds to a different cluster and the minimization of $J(\boldsymbol{\theta}, U)$ with respect to the u_{ij}'s can be carried out separately for each J_j.

The value of η_j determines the relative significance of the two terms in (14.72) and it is related to the size and "shape" of the jth cluster, $j = 1, \ldots, m$. More specifically, as can be seen from Figure 14.16, η_j determines the dissimilarity level between a vector $\boldsymbol{x}_i$ and the representative $\boldsymbol{\theta}_j$ at which u_{ij} becomes equal to 0.5. Thus, η_j determines the influence of a specific point on the estimation of the jth cluster representative.

In general, the size of η_j is assumed constant during the execution of the algorithm. One way to estimate its value, under the assumption that X does not contain many outliers, is to run the generalized fuzzy algorithmic scheme (GFAS) and after its convergence to estimate η_j as [Kris 96]

$$\eta_j = \frac{\sum_{i=1}^{N} u_{ij}^q d(\boldsymbol{x}_i, \boldsymbol{\theta}_j)}{\sum_{i=1}^{N} u_{ij}^q} \tag{14.73}$$

or

$$\eta_j = \frac{\sum_{u_{ij} > a} d(\boldsymbol{x}_i, \boldsymbol{\theta}_j)}{\sum_{u_{ij} > a} 1} \tag{14.74}$$

where a is an appropriate threshold. In words, η_j is defined as a weighted average of the dissimilarities between the vectors $\boldsymbol{x}_i$ and $\boldsymbol{\theta}_j$. Once η_j's have been fixed, the GPAS algorithm can be applied.

In Figure 14.16, u_{ij} versus $d(\boldsymbol{x}_i, \boldsymbol{\theta}_j)/\eta_j$ is plotted for various choices of q (see Eq. (14.70)). From this diagram, it can be seen that q determines the rate of decrease of u_{ij} with respect to $d(\boldsymbol{x}_i, \boldsymbol{\theta}_j)$. For $q = 1$, all points $\boldsymbol{x}_i$ with $d(\boldsymbol{x}_i, \boldsymbol{\theta}_j) > \eta_j$ have $u_{ij} = 0$. On the other hand, as $q \rightarrow +\infty$, u_{ij} tends to a constant and all the vectors of X contribute equally to the estimation of the representative of the jth cluster.

It is worth noting here that q has different meanings in the possibilistic and the fuzzy framework. In the first case, high values of q imply almost equal contributions of *all* feature vectors to *all* clusters, whereas in the second case, high values of q imply increased *sharing* of the vectors among all clusters [Kris 96]. This implies that, in general, different values of q are required to provide satisfactory results for the two cases.

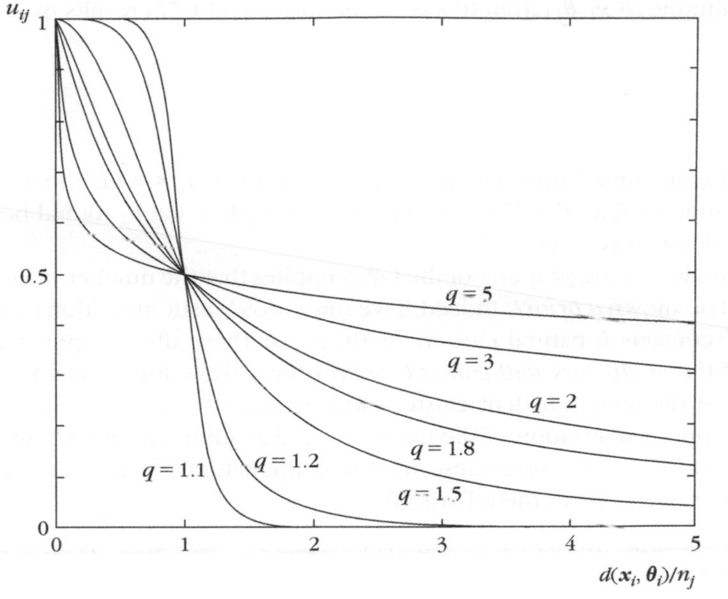

FIGURE 14.16

Plots of the membership function for various values of q.

14.4.1 The Mode-Seeking Property

The generalized mixture decomposition algorithmic scheme (GMDAS) and the generalized fuzzy algorithmic scheme (GFAS) are *partition algorithmic schemes*—that is, schemes that always end up with the predetermined number of clusters m, no matter how many "naturally formed" clusters underlie X. If, for example, the data set X contains two clusters and we run GMDAS or GFAS with $m = 3$, these algorithms will split at least one natural cluster and will end up with three clusters.

This is not the case however with the generalized possibilistic algorithmic scheme (GPAS). Algorithms of this kind are known as *mode-seeking algorithms*—that is, algorithms searching for dense regions of vectors in X.[9] In order to see this, let us consider again the individual functions J_j. Solving Eq. (14.70) with respect to $d(\boldsymbol{x}_i, \boldsymbol{\theta}_j)$, we obtain

$$d(\boldsymbol{x}_i, \boldsymbol{\theta}_j) = \eta_j \left(\frac{1 - u_{ij}}{u_{ij}} \right)^{q-1}$$

[9] Such algorithms are also considered in Chapter 15.

Substituting $d(\boldsymbol{x}_i, \boldsymbol{\theta}_j)$ from this equation into Eq. (14.72) results in

$$J_j = \eta_j \sum_{i=1}^{N} (1 - u_{ij})^{q-1} \tag{14.75}$$

For fixed η_j, minimization of J_j requires maximization of u_{ij}'s, which, in turn, requires minimization of $d(\boldsymbol{x}_i, \boldsymbol{\theta}_j)$. The last requirement implies that $\boldsymbol{\theta}_j$ should be placed in a region dense in vectors of X.

The mode-seeking property of the GPAS implies that the number of clusters in X need not be known *a priori*. Indeed, if we run a possibilistic algorithm for m clusters while X contains k natural clusters, with $m > k$, then, after proper initialization, *some of the m clusters will coincide with others* [Kris 96]. It is hoped that the number of the noncoincident clusters will be equal to k. If, on the other hand, $m < k$, proper initialization of the possibilistic algorithm will potentially lead to m different clusters. Of course, these are not all the natural clusters formed in X, but at least they are some of them [Kris 96].

Example 14.11

This example demonstrates the mode-seeking property. Consider three two-dimensional Gaussian distributions with means $\boldsymbol{\mu}_1 = [1, 1]^T, \boldsymbol{\mu}_2 = [6, 1]^T, \boldsymbol{\mu}_3 = [6, 6]^T$ and covariance matrices $\Sigma_j = I, j = 1, 2, 3$. One hundred vectors are generated from each distribution. These constitute the data set X. We set $q = 1.5$ and, finally, we employ the squared Euclidean distance. It is not difficult to realize that under the above choice, Eq. (14.71) gives

$$\boldsymbol{\theta}_j(t) = \frac{\sum_{i=1}^{N} u_{ij}^q(t-1)\boldsymbol{x}_i}{\sum_{i=1}^{N} u_{ij}^q(t-1)} \tag{14.76}$$

(a) Let $m = 3$. The initial estimates of $\boldsymbol{\theta}_j$'s (which, in this case, are vectors in the two-dimensional space) in GPAS are $\boldsymbol{\theta}_j(0) = \boldsymbol{\mu}_j + \boldsymbol{z}_j, j = 1, 2, 3$, where the $\boldsymbol{z}_j$'s are two-dimensional vectors whose components are drawn from the uniform distribution in $[-2, 2]$. Also, we set $\eta_j = 1.5, j = 1, 2, 3$. Application of the GPAS causes the movement of each one of the $\boldsymbol{\theta}_j$'s toward the mean of each distribution (i.e., toward dense regions). Indeed, the final estimates for $\boldsymbol{\theta}_j$'s obtained after 12 iterations, are $\boldsymbol{\theta}_1 = [0.93, 0.60]^T, \boldsymbol{\theta}_2 = [5.88, 1.12]^T$, and $\boldsymbol{\theta}_3 = [6.25, 5.86]^T$, which compare very favorably to $\boldsymbol{\mu}_j$'s.

(b) Let $m = 4$. In this case, $\boldsymbol{\theta}_j$'s, $j = 1, 2, 3$ are initialized as in the previous example, while $\boldsymbol{\theta}_4$ is initialized as $\boldsymbol{\mu}_1 + \boldsymbol{z}_4$. Application of GPAS in this case causes the movement of $\boldsymbol{\theta}_1$ and $\boldsymbol{\theta}_4$ toward the dense region that corresponds to the first distribution. Also, $\boldsymbol{\theta}_2$ and $\boldsymbol{\theta}_3$ move toward the dense regions that correspond to the second and the third distribution, respectively. The resulting values for $\boldsymbol{\theta}_j$'s, obtained after 12 iterations, are $\boldsymbol{\theta}_1 = [0.93, 0.60]^T, \boldsymbol{\theta}_2 = [5.88, 1.12]^T, \boldsymbol{\theta}_3 = [6.25, 5.86]^T$, and $\boldsymbol{\theta}_4 = [0.94, 0.60]^T$.

(c) Let $m = 2$. We initialize $\boldsymbol{\theta}_1$ and $\boldsymbol{\theta}_2$ as in (a). Application of the GPAS algorithm causes the movement of $\boldsymbol{\theta}_1$ and $\boldsymbol{\theta}_2$ toward the dense regions corresponding to first and the second distribution, respectively. The resulting values for $\boldsymbol{\theta}_j$'s, obtained after 11 iterations, are $\boldsymbol{\theta}_1 = [0.93, 0.60]^T$ and $\boldsymbol{\theta}_2 = [5.88, 1.12]^T$.

14.4.2 An Alternative Possibilistic Scheme

An alternative possibilistic algorithm may be derived from the function [Kris 96]

$$J_1(\boldsymbol{\theta}, U) = \sum_{i=1}^{N}\sum_{j=1}^{m} u_{ij} d(\boldsymbol{x}_i, \boldsymbol{\theta}_j) + \sum_{j=1}^{m} \eta_j \sum_{i=1}^{N} (u_{ij} \ln u_{ij} - u_{ij}) \tag{14.77}$$

Note that q is not involved in the definition of $J_1(\boldsymbol{\theta}, U)$. Also, in this case the second term is negative. Setting the partial derivative of $J_1(\boldsymbol{\theta}, U)$ with respect to u_{ij} equal to 0 and solving for u_{ij}, we obtain the following necessary condition for each u_{ij} to be a minimum of $J_1(\boldsymbol{\theta}, U)$:

$$u_{ij} = \exp\left(-\frac{d(\boldsymbol{x}_i, \boldsymbol{\theta}_j)}{\eta_j}\right) \tag{14.78}$$

Hence, u_{ij} decreases more rapidly with $d(\boldsymbol{x}_i, \boldsymbol{\theta}_j)$ than in the previous case (Eq. 14.70). Let us consider a point $\boldsymbol{x}_i$ and a cluster representative $\boldsymbol{\theta}_j$. For the same distance d, (14.78) leads to smaller values of u_{ij} than those derived from (14.70). This means that increased "stretching" is demanded for the former case. *This is an indication that this algorithmic scheme may be used when the clusters are expected to lie close to each other.*

14.5 HARD CLUSTERING ALGORITHMS

In this section we return to the world where each vector belongs *exclusively* to a single cluster. This is why such schemes are called *hard* or *crisp* clustering algorithms. *It turns out that some of the most well-known and widely used clustering algorithms fall into this category.* Our starting point is the assumption that the membership coefficients u_{ij} are either 1 or 0. Moreover, they are 1 for one cluster, C_j, and zero for all the others, $C_k, k \neq j$, that is,

$$u_{ij} \in \{0, 1\}, \qquad j = 1, \ldots, m \tag{14.79}$$

and

$$\sum_{j=1}^{m} u_{ij} = 1 \tag{14.80}$$

This situation may be seen as a special case of the fuzzy algorithmic schemes. However, the cost function

$$J(\boldsymbol{\theta}, U) = \sum_{i=1}^{N}\sum_{j=1}^{m} u_{ij} d(\boldsymbol{x}_i, \boldsymbol{\theta}_j) \tag{14.81}$$

is *no longer differentiable with respect to* $\boldsymbol{\theta}_j$. Despite that, the general framework of the generalized fuzzy algorithmic schemes, can be adopted for the special case of hard clustering. Such schemes have been used extensively in practice (e.g., [Duda 01]).

Let us fix $\boldsymbol{\theta}_j, j = 1, \ldots, m$. Since for each vector $\boldsymbol{x}_i$ only one u_{ij} is 1 and all the others are 0, it is straightforward to see that $J(\boldsymbol{\theta}, U)$ in Eq. (14.81) is minimized if we assign each $\boldsymbol{x}_i$ to its closest cluster, that is,

$$u_{ij} = \begin{cases} 1, & \text{If } d(\boldsymbol{x}_i, \boldsymbol{\theta}_j) = \min_{k=1,\ldots,m} d(\boldsymbol{x}_i, \boldsymbol{\theta}_k) \\ 0, & \text{otherwise} \end{cases} \quad i = 1, \ldots, N \tag{14.82}$$

Let us now fix u_{ij}s. Working as in the fuzzy algorithms case, the updating equations of the parameter vectors, $\boldsymbol{\theta}_j$, of the clusters are

$$\sum_{i=1}^{N} u_{ij} \frac{\partial d(\boldsymbol{x}_i, \boldsymbol{\theta}_j)}{\partial \boldsymbol{\theta}_j} = \mathbf{0}, \quad j = 1, \ldots, m \tag{14.83}$$

Having derived Eqs. (14.82) and (14.83), we are now in a position to write down the generalized hard clustering algorithmic scheme

Generalized Hard Algorithmic Scheme (GHAS)

- Choose $\boldsymbol{\theta}_j(0)$ as initial estimates for $\boldsymbol{\theta}_j, j = 1, \ldots, m$.
- $t = 0$
- Repeat
  - For $i = 1$ to N
    - For $j = 1$ to m
      - — *Determination of the partition:*[10]

$$u_{ij}(t) = \begin{cases} 1, & \text{if } d(\boldsymbol{x}_i, \boldsymbol{\theta}_j(t)) = \min_{k=1,\ldots,m} d(\boldsymbol{x}_i, \boldsymbol{\theta}_k(t)) \\ 0, & \text{otherwise,} \end{cases}$$

    - End {For-j}
  - End {For-i}
  - $t = t + 1$
  - For $j = 1$ to m
    - *Parameter updating*: Solve

$$\sum_{i=1}^{N} u_{ij}(t-1) \frac{\partial d(\boldsymbol{x}_i, \boldsymbol{\theta}_j)}{\partial \boldsymbol{\theta}_j} = \mathbf{0} \tag{14.84}$$

[10] In the case in which two or more minima occur, an arbitrary choice is made.

with respect to θ_j and set $\theta_j(t)$ equal to the computed solution.

- End {For-j}.

■ Until a termination criterion is met.

Note that in the update of each θ_j, only the vectors x_i closest to it (i.e., those x_i's for which $u_{ij}(t-1) = 1$) are used. As usual, the termination criterion $\|\theta(t) - \theta(t-1)\| < \varepsilon$ can be used. Alternatively, GHAS may terminate if U remains unchanged for two successive iterations.

Each hard clustering algorithm has its corresponding fuzzy clustering algorithm. As with the fuzzy clustering algorithms, we may obtain hard clustering algorithms when θ_js represent points, quadric surfaces, or hyperplanes. The updating equations for the parameter vectors θ_j in the hard clustering algorithms are obtained from their fuzzy counterparts if we set $q = 1$.

Remarks

- Hard clustering algorithms are not as robust as fuzzy clustering algorithms when other than point representatives are employed. If, for example, hyperplane representatives are used and the G-K algorithm is adopted, we must have an adequate number of vectors N from all underlying clusters in order to avoid degenerate cases where Σ_j is not invertible [Kris 92a].

- The determination of the partition part of the algorithms optimizes $J(\theta, U)$ with respect to U given a set of representatives θ_j. On the other hand, the parameter updating phase optimizes $J(\theta, U)$ with respect to θ given a specific partition. Note that this procedure does not necessarily lead to a (local) optimum of $J(\theta, U)$.

14.5.1 The Isodata or k-Means or c-Means Algorithm

This is one of the most popular and well-known clustering algorithms [Duda 01, Ball 67, Lloy 82]. It can be viewed as a special case of the generalized hard clustering algorithmic scheme when point representatives are used and the *squared* Euclidean distance is adopted to measure the dissimilarity between vectors x_i and cluster representatives θ_j. Before we state the algorithm explicitly, some further comments may be of interest. For this case Eq. (14.81) becomes

$$J(\theta, U) = \sum_{i=1}^{N}\sum_{j=1}^{m} u_{ij}\|x_i - \theta_j\|^2 \tag{14.85}$$

This is nothing but the trace of the within scatter matrix S_w, defined in Chapter 5. That is,

$$J(\theta, U) = \text{trace}\{S_w\} \tag{14.86}$$

For the above choice of distance, Eq. (14.83) gives that θ_j is the mean vector of the jth cluster. *Applying the generalized hard algorithmic scheme for this specific choice, it turns out that the algorithm converges to a minimum of the cost function.* In other words, the isodata algorithm recovers clusters that are as compact as possible. It must be emphasized however, that this convergence result is not valid for other distances, including the Euclidean distance. For example, when Minkowski distances are used, the algorithm converges but not necessarily to a minimum of the corresponding cost function [Seli 84a].

The Isodata or k-Means or c-Means Algorithm

- Choose arbitrary initial estimates $\theta_j(0)$ for the θ_j's, $j = 1, \ldots, m$.
- Repeat
  - For $i = 1$ to N
    - Determine the closest representative, say θ_j, for x_i.
    - Set $b(i) = j$.
  - End {For}
  - For $j = 1$ to m
    - Parameter updating: Determine θ_j as the mean of the vectors $x_i \in X$ with $b(i) = j$.
  - End {For}.
- Until no change in θ_j's occurs between two successive iterations.

As with all the algorithms that use point representatives, isodata is suitable for recovering compact clusters. A sequential version of the k-means (see, for example, [Pena 99]) results if the updating of the representatives takes place immediately after determining the representative that lies closest to the currently considered vector x_i. Clearly, the result of this version of the algorithm is dependent on the order in which the vectors are considered. A version of the k-means algorithm, where in each cluster C_i the number of vectors is constrained *a priori* to be n_i, is proposed in [Ng 00].

Example 14.12

(a) Consider the setup of Example 14.1(a). In this case θ_js correspond to the μ_js. We set $m = 3$ and we initialize randomly θ_js. After convergence, the isodata algorithm identifies successfully the underlying clusters in X, as indicated by the corresponding confusion matrix,

$$A = \begin{bmatrix} 94 & 3 & 3 \\ 0 & 100 & 0 \\ 9 & 0 & 91 \end{bmatrix}.$$

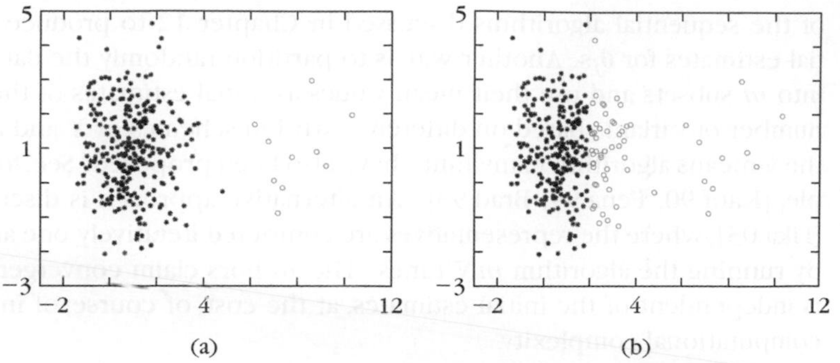

FIGURE 14.17

(a) The data set. (b) The results of the isodata algorithm.

The resulting values of θ_js are $\theta_1 = [1.19, 1.16]^T$, $\theta_2 = [3.76, 3.63]^T$ and $\theta_3 = [5.93, 0.55]^T$.

(b) Let us now consider two 2-dimensional Gaussian distributions with means $\mu_1 = [1, 1]^T$ and $\mu_2 = [8, 1]^T$ and covariance matrices $\Sigma_1 = 1.5I$ and $\Sigma_2 = I$, respectively. We generate 300 points from the first distribution and 10 points from the second distribution in order to form the data set X (Figure 14.17a). Also, we set $m = 2$ and we initialize randomly θ_1 and θ_2. After convergence, we observe that the large group has been split into two parts and, in addition, the right part joins the vectors of the second distribution in the same cluster (Figure 14.17b). Specifically, the results of the algorithm are $\theta_1 = [0.54, 0.94]^T$ and $\theta_2 = [3.53, 0.99]^T$ and 61 vectors from the first distribution are assigned to the same cluster with the ten vectors of the second distribution. The above situation reveals a weakness of the algorithm to deal accurately with clusters having significantly different sizes.

A major advantage of the k-means algorithm is its computational simplicity, which makes it an attractive candidate for a variety of applications. Its time complexity is $O(Nmq)$, where q is the number of iterations required for convergence. Because in practice m and q are significantly less than N, k-means becomes eligible for processing large data sets. Furthermore, its conceptual simplicity has been a source of inspiration to many authors, who have proposed a number of variants in order to remedy drawbacks associated with the algorithm. Some of them are summarized in the following

- As all optimization schemes considered in this chapter, the k-means algorithm cannot guarantee convergence to the global minimum of $J(\theta, U)$. Equivalently, different initial partitions may lead k-means to produce different final clusterings, each corresponding to a different local minimum of $J(\theta, U)$. To minimize or even overcome this drawback, a number of strategies have been suggested.
  - Instead of initializing θ_js by m randomly chosen points (some suggest random initialization with points drawn from X, [Forg 65]), one can use any

of the sequential algorithms discussed in Chapter 12 to produce the initial estimates for θ_js. Another way is to partition randomly the data set, X, into m subsets and use their mean values as initial estimates of the θ_js. A number of variants based on different partition schemes of X and running the k-means algorithm many times have also been proposed. See, for example, [Kauf 90, Pena 99, Brad 98]. An alternative approach is discussed in [Lika 03], where the representatives are computed iteratively, one at a time, by running the algorithm mN times. The authors claim convergence that is independent of the initial estimates, at the cost, of course, of increased computational complexity.

- Another path is to adopt tools from stochastic optimization techniques, such as simulated annealing and genetic algorithms (see also Chapter 15), in that such techniques guarantee, in probability, the computation of the global minimum of $J(\theta, U)$ at the cost of excessive computations. Extensions of the k-means in this spirit are discussed in [Kris 99] and [Pata 01].

- Although computing the optimal partition for the k-means, as well as the k-medoids algorithm to be discussed next, is an NP-hard problem, recent theoretical work shows that it is possible to find solutions that are provably good approximations. In addition, this can be achieved via reasonably efficient techniques, see, for example, [Indy 99, Kuma 04, Kanu 04].

■ The number of clusters m in the data set, X, is required as an input parameter to the algorithm. Clearly, a poor estimate of m will prevent the algorithm to unravel the underlying clustering structure in X. To this end, a number of variants of the k-means algorithm have been suggested, employing various splitting, merging, and discarding operations among the resulting clusters [Ande 73] (based on suitably chosen user-defined parameters). No doubt, such *ad hoc* techniques are no more the result of an optimization process.

An alternative method for estimating m is to apply the procedure described in Section 12.3, using a sequential clustering algorithm.

■ k-means is sensitive to outliers and noise. The outliers, being points in X, are necessarily assigned to one of the clusters. Thus, they influence the respective means and, as a consequence, the final clustering. Taking into account that in general small clusters are likely to be formed by outliers, a version of the algorithm given in [Ball 67] deals with outliers by simply discarding "small" clusters.

In addition, this drawback of the k-means gave rise to the so-called *k-medoids algorithms* (see next section), where each cluster is represented by one of its points. This way of representing clusters is less sensitive to outliers, at the cost of increased computational complexity.

- *k*-means is generally applicable to data sets with continuous valued feature vectors, and in principle it is not suitable for data with nominal (categorical) coordinates. Variants of the *k*-means that can deal with data sets consisting of data stemming from a finite discrete-valued domain are discussed in [Huan 98, Gupa 99]. The *k-medoids* algorithms, discussed next, are another possibility.
- As it is currently the trend kernelized versions of the *k*-means algorithm have also been proposed, see, for example, [Scho 98, Giro 02].

Other advances related to the *k*-means and other square-error-based clustering algorithms can be found in [Hans 01, Kanu 00, Pata 02, Su 01, Wags 01].

14.5.2 *k*-Medoids Algorithms

In the *k*-means algorithm described in the previous section, each cluster is represented by the mean of its vectors. In the *k-medoids* methods, discussed in this section, each cluster is represented by a vector selected *among the elements* of X, and we will refer to it as the *medoid*. Apart from its medoid, each cluster contains all vectors in X that (a) are not used as medoids in other clusters and (b) lie closer to its medoid than to the medoids representing the other clusters. Let Θ be the set of medoids for all clusters. We will denote by I_Θ the set of indices of the points in X that constitute Θ, and by $I_{X-\Theta}$ the set of indices of the points that are not medoids. Thus, if for example $\Theta = \{\boldsymbol{x}_1, \boldsymbol{x}_5, \boldsymbol{x}_{13}\}$ is the set of medoids for a three-cluster case then $I_\Theta = \{1, 5, 13\}$. The quality of the clustering associated with a given set Θ of medoids is assessed through the cost function

$$J(\Theta, U) = \sum_{i \in I_{X-\Theta}} \sum_{j \in I_\Theta} u_{ij} d(\boldsymbol{x}_i, \boldsymbol{x}_j) \tag{14.87}$$

and

$$u_{ij} = \begin{cases} 1, & \text{if } d(\boldsymbol{x}_i, \boldsymbol{x}_j) = \min_{q \in I_\Theta} d(\boldsymbol{x}_i, \boldsymbol{x}_q) \\ 0, & \text{otherwise} \end{cases} \qquad i = 1, \ldots, N \tag{14.88}$$

Thus, obtaining the set of medoids Θ that best represents the data set, X, is equivalent to minimizing $J(\Theta, U)$. Note that Eqs. (14.87) and (14.81) are almost identical. The only difference is that $\boldsymbol{\theta}_j$ in Eq. (14.81) is replaced by $\boldsymbol{x}_j$, in that each cluster is now represented by a vector in X.

Representing clusters using medoids has two advantages over the *k*-means algorithm. First, it can be used with data sets originating from either continuous or discrete domains, whereas *k*-means is suited only for the case of continuous domains because in a discrete domain application the mean of a subset of the data vectors is not necessarily a point lying in the domain (see Figure 14.18a). Second, *k*-medoids algorithms tend to be less sensitive to outliers compared to the *k*-means algorithm (see Figure 14.18b). However, it should be noted that the mean of a cluster has a clear geometrical and statistical meaning, which is not necessarily the case with the

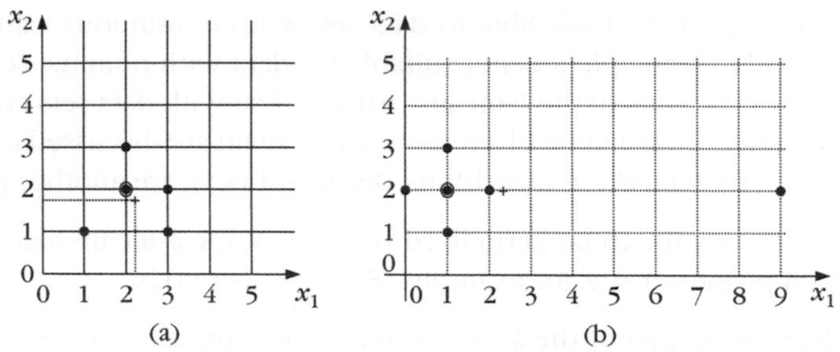

FIGURE 14.18

(a) The five-point two-dimensional set stems from the discrete domain $\mathcal{D} = \{1, 2, 3, 4, \ldots\} \times \{1, 2, 3, 4, \ldots\}$. Its medoid is the circled point. The mean of the vectors of the set is denoted by "+" and does not belong to $\mathcal{D}$. (b) In the six-point two-dimensional set, the point (9, 2) can be considered an outlier. Clearly, the outlier affects significantly the position of the mean of the set, whereas it has no affect on the position of its medoid.

medoids. In addition, the algorithms for the estimation of the best set of medoids are computationally more demanding compared to the *k*-means algorithm.

In the sequel, we describe three *k*-medoids algorithms: PAM (Partitioning Around Medoids), CLARA (Clustering LARge Applications), and CLARANS (Clustering Large Applications based on RANdomized Search). Note that the last two algorithms are inspired by PAM but are suitable for dealing with large data sets more efficiently than PAM.

The PAM Algorithm

To determine the set Θ of the m medoids that best represent the data set, PAM uses a function optimization procedure that minimizes $J(\Theta, U)$, subject to the constraint that the representatives of the clusters are themselves elements of X. Before proceeding any further, some definitions are in order. Two sets of medoids Θ and Θ', each consisting of m elements, are called ***neighbors*** if they share $m - 1$ elements. Clearly, the number of neighbors a set $\Theta \subset X$ with m elements can have is $m(N - m)$. Also, let Θ_{ij} denote the neighbor of Θ that results if $\boldsymbol{x}_j$, $j \in I_{X-\Theta}$ replaces $\boldsymbol{x}_i$, $i \in I_\Theta$. Finally, let $\Delta J_{ij} = J(\Theta_{ij}, U_{ij}) - J(\Theta, U)$.

PAM starts with a set Θ of m medoids, which are *randomly* selected out of X. Then, among all $m(N - m)$ neighbors, Θ_{ij}, $i \in I_\Theta$, $j \in I_{X-\Theta}$, of the set Θ, we select Θ_{qr}, $q \in I_\Theta$, $r \in I_{X-\Theta}$, with $\Delta J_{qr} = \min_{ij} \Delta J_{ij}$. If ΔJ_{qr} is negative, then Θ is replaced by Θ_{qr} and the same procedure is repeated. Otherwise, if $\Delta J_{qr} \geq 0$ the algorithm has reached a local minimum and terminates. Once the set Θ that best represents the data has been determined, each $\boldsymbol{x} \in X - \Theta$ is assigned to the cluster represented by the closest to it medoid.

Let us focus now on the computation of ΔJ_{ij}. This quantity may be written as

$$\Delta J_{ij} = \sum_{h \in I_{X-\Theta}} C_{hij} \tag{14.89}$$

where C_{hij} is the difference in J resulting from the (possible) assignment of the vector $\boldsymbol{x}_h \in X - \Theta$ from the cluster it currently belongs to another, as a consequence of the replacement of $\boldsymbol{x}_i \in \Theta$ by $\boldsymbol{x}_j \in X - \Theta$. For the computation of C_{hij} we consider the following four cases.

- Suppose that $\boldsymbol{x}_h$ belongs to the cluster represented by $\boldsymbol{x}_i$. Also, let $\boldsymbol{x}_{h2} \in \Theta$ denote the second closest to $\boldsymbol{x}_h$ representative. If $d(\boldsymbol{x}_h, \boldsymbol{x}_j) \geq d(\boldsymbol{x}_h, \boldsymbol{x}_{h2})$ (see Figure 14.19a), then after the replacement of $\boldsymbol{x}_i$ by $\boldsymbol{x}_j$ in Θ, $\boldsymbol{x}_h$ will now be represented by $\boldsymbol{x}_{h2}$. Thus,

$$C_{hij} = d(\boldsymbol{x}_h, \boldsymbol{x}_{h2}) - d(\boldsymbol{x}_h, \boldsymbol{x}_i) \geq 0 \tag{14.90}$$

The equality corresponds to the case of a tie, that is, $d(\boldsymbol{x}_h, \boldsymbol{x}_{h2}) = d(\boldsymbol{x}_h, \boldsymbol{x}_i)$.

- Suppose again that $\boldsymbol{x}_h$ belongs to the cluster represented by $\boldsymbol{x}_i$ and let $\boldsymbol{x}_{h2}$ denote the second closest to $\boldsymbol{x}_h$ representative. If $d(\boldsymbol{x}_h, \boldsymbol{x}_j) \leq d(\boldsymbol{x}_h, \boldsymbol{x}_{h2})$ (see Figure 14.19b–c), then after the replacement of $\boldsymbol{x}_i$ by $\boldsymbol{x}_j$ in Θ, $\boldsymbol{x}_h$ will now be represented by $\boldsymbol{x}_j$. Thus,

$$C_{hij} = d(\boldsymbol{x}_h, \boldsymbol{x}_j) - d(\boldsymbol{x}_h, \boldsymbol{x}_i) \tag{14.91}$$

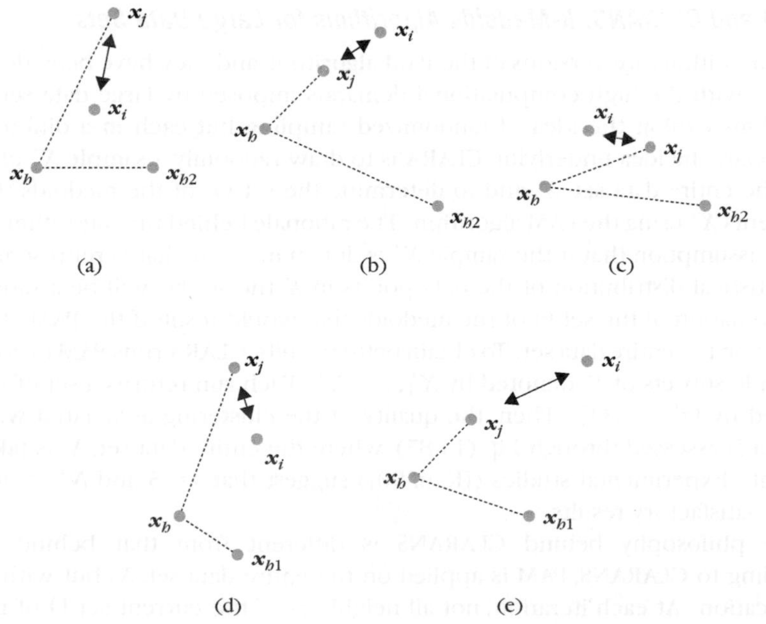

FIGURE 14.19

Different cases encountered in the computation of $C_{\hat{h}ij}$: (a) $C_{\hat{h}ij} > 0$, (b) $C_{\hat{h}ij} < 0$, (c) $C_{\hat{h}ij} > 0$, (d) $C_{\hat{h}ij} = 0$, (e) $C_{\hat{h}ij} < 0$.

In this case, C_{hij} may be either negative, zero, or positive (e.g., Figure 14.19b–c).

- Suppose now that $\boldsymbol{x}_h$ is not represented by $\boldsymbol{x}_i$ and let $\boldsymbol{x}_{h1}$ be the closest to $\boldsymbol{x}_h$ medoid. If $d(\boldsymbol{x}_h, \boldsymbol{x}_{h1}) \leq d(\boldsymbol{x}_h, \boldsymbol{x}_j)$ (see Figure 14.19d), then $\boldsymbol{x}_h$ will continue to be represented by $\boldsymbol{x}_{h1}$. Thus,

$$C_{hij} = 0 \tag{14.92}$$

- Finally, suppose that $\boldsymbol{x}_h$ is not represented by $\boldsymbol{x}_i$ and let $\boldsymbol{x}_{h1}$ be the closest to $\boldsymbol{x}_h$ medoid. If $d(\boldsymbol{x}_h, \boldsymbol{x}_{h1}) > d(\boldsymbol{x}_h, \boldsymbol{x}_j)$ (see Figure 14.19e), then

$$C_{hij} = d(\boldsymbol{x}_h, \boldsymbol{x}_j) - d(\boldsymbol{x}_h, \boldsymbol{x}_{h1}) < 0 \tag{14.93}$$

Experimental results ([Kauf 90]) show that PAM works satisfactorily for relatively small data sets. However, it becomes inefficient for large data sets because its time complexity per iteration increases quadratically with respect to N. This is easily verified because at each iteration the term ΔJ_{ij} for $m(N-m)$ pairs of vectors has to be calculated. In addition, for the computation of a single ΔJ_{ij}, $N-m$ of C_{hij} terms have to be considered (see Eq. (14.89)). Thus, the total complexity of the algorithm per iteration amounts to $O(m(N-m)^2)$.

CLARA and CLARANS: k-Medoids Algorithms for Large Data Sets

These algorithms are versions of the PAM algorithm, and they have been developed to cope with the high computational demands imposed by large data sets. Both algorithms exploit the idea of randomized sampling but each in a different way. Specifically, the idea underlying CLARA is to draw randomly a sample X' of size N' from the entire data set, X, and to determine the set Θ' of the medoids that best represents X' using the PAM algorithm. The rationale behind this algorithm is based on the assumption that if the sample X' is drawn in a way that is representative of the statistical distribution of the data points in X the set Θ' will be a satisfactory approximation of the set Θ of the medoids that would result if the PAM algorithm was run on the entire data set. To obtain better results, CLARA runs PAM on a number of sample subsets of X, denoted by $X'_1, \ldots, X'_s$. Each run returns a set of medoids denoted by $\Theta'_1, \ldots, \Theta'_s$. Then, the quality of the clustering associated with each of them is assessed through Eq. (14.87), where the entire data set, X, is taken into account. Experimental studies ([Kauf 90]) suggest that $s=5$ and $N' = 40 + 2m$ lead to satisfactory results.

The philosophy behind CLARANS is different from that behind CLARA. According to CLARANS, PAM is applied on the entire data set, X, but with a slight modification. At each iteration, not all neighbors of the current set Θ of medoids are considered. Instead, only a randomly selected fraction $q < m(N-m)$ of them is utilized. The selected neighbors are considered in a sequential manner and if the currently considered neighbor Θ_{ij} of Θ is better than Θ (in terms of J) then Θ is replaced by Θ_{ij} and the procedure is repeated. When none of the q selected neighbors of Θ is better than Θ, in terms of J, then Θ is considered to be "local minimum."[11] Then CLARANS starts from another arbitrarily chosen Θ, and the same procedure is followed in order to obtain a different "local minimum." This is repeated for a predetermined number of times, s, and the algorithm outputs the best among the s "local minima." In the sequel, based on this set of medoids each point $\boldsymbol{x} \in X - \Theta$ is assigned to the cluster whose representative is closest to $\boldsymbol{x}$.

Remarks

- The performance of CLARANS depends on the two parameters q and s. As q gets closer to $m(N - m)$, CLARANS approaches PAM and the time complexity increases. As suggested in [Ng 94a], a typical value for s is 2, whereas for q it is suggested to be chosen as the maximum between $0.12m\,(N - m)$ and 250.
- The CLARANS algorithm can also be described in terms of graph theory concepts ([Ng 94]).
- CLARANS unravels better-quality clusterings than CLARA. On the other hand, in some cases CLARA runs significantly faster than CLARANS ([Ng 94]). It must be pointed out that CLARANS retains its quadratic computational nature and is thus not appropriate for very large data sets.

14.6 VECTOR QUANTIZATION

An area that has close affinity with clustering is that of vector quantization (VQ), and it has been the focus of intensive research effort over the past years (e.g., [Gray 84, Gers 92]). Vector quantization techniques are used mainly for data compression, which is a prerequisite for achieving better computer storage utilization and better bandwidth utilization (in communications). Let T be the set of all possible vectors for the problem at hand. The task of VQ may be stated in the general case in which T is a continuous subset of $\mathcal{R}^l$. The idea is rather simple. We separate T into m distinct regions R_j that exhaust T, and we represent each of them with an l-dimensional vector, the so-called *code vector* or *reproduction vector*, $\boldsymbol{\theta}_j, j = 1, \ldots, m$. In the sequel, given an $\boldsymbol{x} \in T$, we determine the region where it belongs, say R_j, and we adopt the corresponding representative $\boldsymbol{\theta}_j$, instead of $\boldsymbol{x}$, for further use, that is, storage or transmission. This is obviously associated with some information loss, which is known as *distortion*. The major goal in VQ is to define the regions R_j and their representatives $\boldsymbol{\theta}_j$ so that distortion is minimized.

After stating the general idea, let us proceed now to some more formal definitions. A *vector quantizer* Q of dimension l and size m is a mapping of T to a finite set C,

[11] Note that Θ may not actually be a local minimum because it may have a nonselected neighbor that gives lower value of J.

which is called the *reproduction set* and contains m output reproduction points, the code vectors or *code words*. Thus

$$Q : T \rightarrow C$$

where $C = \{\boldsymbol{\theta}_1, \boldsymbol{\theta}_2, \ldots, \boldsymbol{\theta}_m\}$ with $\boldsymbol{\theta}_i \in T$. Each code vector $\boldsymbol{\theta}_i$ represents a specific region R_i of the vector space.

The next question that naturally arises is how one can select the code vectors $\boldsymbol{\theta}_j$ in such a way as to achieve the least possible distortion. A usual approach is to optimize an appropriate criterion function, which in this framework is also known as a *distortion function*, with respect to $\boldsymbol{\theta}_j$'s. Let $\boldsymbol{x}$ be a random vector that models T and $p(\boldsymbol{x})$ its corresponding pdf.

A commonly used distortion criterion is the *average expected quantization error*, which is defined as

$$D(Q) = \sum_{j=1}^{m} D_j(Q) \tag{14.94}$$

where

$$D_j(Q) = \int_{R_j} d(\boldsymbol{x}, \boldsymbol{\theta}_j) p(\boldsymbol{x})\, d\boldsymbol{x} \tag{14.95}$$

$D_j(Q)$ is known as the average quantization error for region R_j. The quantity d is a distance measure, for example, Euclidean, and it is also referred to as a *distortion measure*.

When a finite number of samples, $\boldsymbol{x}_1, \boldsymbol{x}_2, \ldots, \boldsymbol{x}_N$, of $\boldsymbol{x}$ is available, the distortion criterion becomes

$$D(Q) = \sum_{i=1}^{N} d(\boldsymbol{x}_i, Q(\boldsymbol{x}_i)) P(\boldsymbol{x}_i) \tag{14.96}$$

where $Q(\boldsymbol{x}_i) \in C$ is the code vector that represents $\boldsymbol{x}_i$ and $P(\boldsymbol{x}_i)(>0)$ $i = 1, \ldots, N$, the respective probabilities.

In [Gers 92] it is shown that the following conditions are *necessary* for a given quantizer to be optimal. The first refers to the *encoder* part of the vector quantizer, that is, the optimal way in which T is partitioned in the regions $R_j, j = 1, \ldots, m$, given the code vectors $\boldsymbol{\theta}_j$. It is known as the *nearest neighbor condition*, and it states that

- *For fixed* C,

$$Q(\boldsymbol{x}) = \boldsymbol{\theta}_j \quad \text{only if } d(\boldsymbol{x}, \boldsymbol{\theta}_j) \leq d(\boldsymbol{x}, \boldsymbol{\theta}_k), \quad k = 1, \ldots, m, k \neq j$$

The second condition refers to the *decoder* part of the VQ, that is, the optimal way the code words $\boldsymbol{\theta}_j$ are chosen, given the partition regions $R_j, j = 1, \ldots, m$. It is known as the *centroid condition*, and it is stated as

- *For a fixed partition $R_1, R_2, \ldots, R_m$, each $\boldsymbol{\theta}_j$ is chosen such that*

$$\int_{R_j} d(\boldsymbol{x}, \boldsymbol{\theta}_j) p(\boldsymbol{x})\, d\boldsymbol{x} = \min_{\boldsymbol{y}} \int_{R_j} d(\boldsymbol{x}, \boldsymbol{y}) p(\boldsymbol{x})\, d\boldsymbol{x}$$ [12]

In the case that T is finite, the integrals are replaced with summations. One way to compute the code vectors of the set C is to start with an arbitrary initial estimate of the code vectors and to iteratively apply the nearest neighbor condition and the centroid condition, interchangeably, until a termination criterion is satisfied.[13] This is the well-known Lloyd's algorithm [Lloy 82].[14] Note that if $P(\boldsymbol{x}_i) = 1/N$, $\forall \boldsymbol{x}_i \in T$, Lloyd's algorithm coincides with the generalized hard clustering algorithmic scheme. This is not surprising. Both algorithms try to place optimally point representatives in space. Note, however, that despite algorithmic similarities, the two tasks have different goals. The goal of VQ is to place points in space in a way that is representative of the data distribution. On the other hand, clustering focuses on revealing the underlying clusters in X, if they exist.

Finally, it is worth pointing out that many other models for vector quantization have been proposed in the literature. For example, hierarchical and fuzzy vector quantizers are discussed in [Lutt 89] and [Kara 96], respectively.

APPENDIX

Derivation of μ_j and Σ_j for the EM Algorithm (Section 14.2)

Equations (14.3) and (14.13) for $\boldsymbol{\mu}_j$ lead to

$$\boldsymbol{\mu}_j = \frac{\sum_{k=1}^{N} P(C_j|\boldsymbol{x}_k; \boldsymbol{\Theta}(t))\boldsymbol{x}_k}{\sum_{k=1}^{N} P(C_j|\boldsymbol{x}_k; \boldsymbol{\Theta}(t))} \tag{14.97}$$

for $j = 1, \ldots, m$.

Let us now turn our attention to Σ_j. Let σ_{rs} be the (r, s) element of Σ_j^{-1}. Then Eq. (14.13) gives

$$\frac{\partial}{\partial \sigma_{rs}} \ln p(\boldsymbol{x}|C_j; \boldsymbol{\theta}_j) = \frac{1}{2}|\Sigma_j| \frac{\partial}{\partial \sigma_{rs}} |\Sigma_j^{-1}| - \frac{1}{2}(x_r - \mu_{jr})(x_s - \mu_{js})$$

or

$$\frac{\partial}{\partial \sigma_{rs}} \ln p(\boldsymbol{x}|C_j; \boldsymbol{\theta}_j) = \frac{1}{2}|\Sigma_j|\sigma_{rs} - \frac{1}{2}(x_r - \mu_{jr})(x_s - \mu_{js})$$

[12] An additional optimality condition, given in [Gers 92], is that no vector in T is equidistant from two (or more) code vectors.
[13] One such criterion is to have the same values for all $\boldsymbol{\theta}_j$'s, $j = 1, \ldots, m$, for two successive iterations.
[14] For the special case in which the squared Euclidean distance is considered, the centroid condition becomes $\boldsymbol{\theta}_j = (1/n_j) \sum_{\boldsymbol{x} \in R_j} \boldsymbol{x}$, where n_j is the number of vectors that lie in R_j. The corresponding algorithm is the isodata algorithm, which in this framework is also called the LBG algorithm [Lind 80].

where $\boldsymbol{\sigma}_{rs}$ is the cofactor of σ_{rs}[15] and x_r, μ_{jr} (x_s, μ_{js}) are the rth (sth) coordinates of $\boldsymbol{x}$ and $\boldsymbol{\mu}_j$, respectively. Thus,

$$\frac{\partial \ln p(\boldsymbol{x}|C_j;\boldsymbol{\theta}_j)}{\partial \Sigma_j^{-1}} = \frac{1}{2}|\Sigma_j|\boldsymbol{\sigma} - \frac{1}{2}(\boldsymbol{x}-\boldsymbol{\mu}_j)(\boldsymbol{x}-\boldsymbol{\mu}_j)^T \tag{14.98}$$

where $\boldsymbol{\sigma}$ is the matrix of the cofactors of Σ_j^{-1}. Since, in general, $|\Sigma^{-1}| \neq 0$, the following identity holds from linear algebra:

$$\Sigma_j^{-1}\boldsymbol{\sigma}^T = |\Sigma_j^{-1}|I$$

Premultiplying both sides of this equation by Σ_j and noting that σ is a symmetric matrix, we obtain

$$\boldsymbol{\sigma} = |\Sigma_j^{-1}|\Sigma_j$$

Substituting the last result into Eq. (14.98), we obtain

$$\frac{\partial \ln p(\boldsymbol{x}|C_j;\boldsymbol{\theta}_j)}{\partial \Sigma_j^{-1}} = \frac{1}{2}\Sigma_j - \frac{1}{2}(\boldsymbol{x}-\boldsymbol{\mu}_j)(\boldsymbol{x}-\boldsymbol{\mu}_j)^T \tag{14.99}$$

Substituting this result into Eq. (14.3) and after some manipulations, we finally obtain

$$\Sigma_j = \frac{\sum_{k=1}^{N} P(C_j|\boldsymbol{x}_k;\boldsymbol{\Theta}(t))(\boldsymbol{x}_k-\boldsymbol{\mu}_j)(\boldsymbol{x}_k-\boldsymbol{\mu}_j)^T}{\sum_{k=1}^{N} P(C_j|\boldsymbol{x}_k;\boldsymbol{\Theta}(t))} \tag{14.100}$$

for $j = 1, \ldots, m$.

14.7 PROBLEMS

14.1 Consider the case in which there exist m clusters in X, which are characterized by normal distributions of unknown means and known covariance matrices; that is, the parameter vector $\boldsymbol{\theta}$ consists only of the parameters of $\boldsymbol{\mu}_j$, $j = 1, \ldots, m$. State the corresponding generalized mixture decomposition algorithmic scheme (GMDAS).

14.2 Consider the case that there exist m clusters in X which are described by normal distributions of unknown means and unknown diagonal covariance matrices. Derive the corresponding GMDAS.

14.3 Consider the case that there exist m clusters in X which are described by normal distributions. Derive the maximum likelihood estimates of the means $\boldsymbol{\mu}_j$ and covariance matrices Σ_j when:

a. the means and the covariance matrices are unknown and

[15] Recall that the cofactor of the element a_{ij} of a matrix A is the determinant of the matrix that results from A if we delete its ith row and its jth column.

b. the means and the covariance matrices are unknown but $\Sigma_j = \Sigma$, $j = 1, \ldots, m$.

Compare the results of (a) with those of Section 14.2.1.

14.4 Consider the data set $X = \{\boldsymbol{x}_i \in \mathcal{R}^2, i = 1, \ldots, 16\}$, where $\boldsymbol{x}_1 = [2, 0]^T$, $\boldsymbol{x}_2 = [\sqrt{2}, \sqrt{2}]^T$, $\boldsymbol{x}_3 = [0, 2]^T$, $\boldsymbol{x}_4 = [-\sqrt{2}, \sqrt{2}]^T$, $\boldsymbol{x}_5 = [-2, 0]^T$, $\boldsymbol{x}_6 = [-\sqrt{2}, -\sqrt{2}]^T$, $\boldsymbol{x}_7 = [0, -2]^T$, $\boldsymbol{x}_8 = [\sqrt{2}, -\sqrt{2}]^T$. The remaining points $\boldsymbol{x}_i, i = 9, \ldots, 16$, are obtained from the first eight points as follows. The first coordinate of $\boldsymbol{x}_i, i = 9, \ldots, 16$, equals the first coordinate of $\boldsymbol{x}_{i-8}$ plus 6, while the second coordinate of $\boldsymbol{x}_i$, $i = 9, \ldots, 16$, equals the second coordinate of $\boldsymbol{x}_{i-8}$.

a. Run the GMDAS, with Gaussian pdf's, to obtain estimates of $\boldsymbol{\mu}_j$, and Σ_j, $j = 1, 2$.

b. Does the algorithm determine the clusters that underlie X correctly? Justify your answer.

Hint: In the rest problems, where possible, one may use the MATLAB codes given in the Computer Programs section of this chapter.

14.5 Consider the setup of Problem 14.4, with the difference that the points $\boldsymbol{x}_i$, $i = 9, \ldots, 16$, are derived as follows. The first coordinate of $\boldsymbol{x}_i, i = 9, \ldots, 16$, equals the first coordinate of $\boldsymbol{x}_{i-8}$ plus 2, while the second coordinate of $\boldsymbol{x}_i$, $i = 9, \ldots, 16$, equals to the second coordinate of $\boldsymbol{x}_{i-8}$.

a. Run the GMDAS, with Gaussian pdf's, to obtain estimates for $\boldsymbol{\mu}_j$, and Σ_j, $j = 1, 2$.

b. Does the algorithm determine the clusters that underlie X accurately? Justify your answer.

c. Compare the results obtained in this and the previous problem.

14.6 Consider four two-dimensional distributions with means $\boldsymbol{\mu}_1 = [0, 0]^T$, $\boldsymbol{\mu}_2 = [2, 2]^T$, $\boldsymbol{\mu}_3 = [4, 0]^T$, $\boldsymbol{\mu}_4 = [7, 0]^T$, respectively, and covariance matrices

$$\Sigma_1 = \begin{bmatrix} 1 & 0.3 \\ 0.3 & 1 \end{bmatrix}, \quad \Sigma_2 = \begin{bmatrix} 1 & 0 \\ 0 & 1 \end{bmatrix}$$

$$\Sigma_3 = \begin{bmatrix} 1 & -0.5 \\ -0.5 & 1 \end{bmatrix}, \quad \Sigma_4 = \begin{bmatrix} 1 & 0.5 \\ 0.5 & 1 \end{bmatrix}$$

respectively. Draw 80 points from each distribution and let X be the set that contains these 320 points. Initialize $\boldsymbol{\mu}_i$ and Σ_i, $i = 1, \ldots, 4$, as in Example 14.1. Set $m = 4$, $\varepsilon = 0.01$ and run the GMDAS, with Gaussian pdf's.

a. What are the estimates of $\boldsymbol{\mu}_j, j = 1, \ldots, 4$, and $\Sigma_j, j = 1, \ldots, 4$?

b. Assign each feature vector $\boldsymbol{x} \in X$ to a cluster C_j according to the Bayes decision rule.

c. Derive the respective confusion matrix.

d. Run the algorithm for $m = 3$ and $m = 2$ and repeat steps (a) and (b). Discuss the results.

14.7 In the framework of Example 14.4, prove that for $m = 2, q \in \{2, 3\}$ and for fixed $\boldsymbol{\theta}$, there are cases where the fuzzy clusterings are favored against the hard ones.

14.8 Find the relation between $\boldsymbol{p}$ and $A, \boldsymbol{b}$, and c so that Eqs. (14.35) and (14.36) are equivalent.
Hint: Consider each coordinate of $\boldsymbol{p}$ separately.

14.9 Let $l = 2$. Prove that the substitution of $\boldsymbol{z}$, as given by Eq. (14.44), into Eq. (14.42) leads to a fourth-degree polynomial, with respect to λ.

14.10 Prove Eq. (14.50).

14.11 **a.** Derive Eqs. (14.56) and (14.57) for the fuzzy C ellipsoidal shells (FCES) algorithm.

b. Write the parameter determination part of the fuzzy C ellipsoidal shells (FCES) algorithm.

14.12 Derive the fuzzy C quadric shells (FCQs) algorithm by minimizing Eq. (14.58) under constraint (v).

14.13 **a.** State explicitly the modified fuzzy C quadric shells (MFCQS) algorithm.

b. Under what general conditions are the algebraic and the perpendicular distances close to each other?

14.14 What is the relation between the perpendicular and the radial distance between a point $\boldsymbol{x}$ and a hyperspherical cluster?

14.15 **a.** Derive the AFCS algorithm [Dave 92b] for the case that spherical clusters are to be recovered. The distance between a point $\boldsymbol{x}$ and a hypersphere Q with center $\boldsymbol{c}$ and radius r is

$$d^2(\boldsymbol{x}, Q) = (\|\boldsymbol{x} - \boldsymbol{c}\| - r)^2$$

b. Derive the fuzzy C spherical shells (FCSS) algorithm [Kris 92b] for the case that spherical clusters are to be identified.

14.16 Prove Eq. (14.64).

14.17 Derive the possibilistic algorithm obtained via minimization of the function J_1 given in Eq. (14.77).

14.18 Plot u_{ij} versus $d(\boldsymbol{x}_i, \boldsymbol{\theta}_j)/\eta_j$, using Eq. (14.78). Compare this plot with the one shown in Figure 14.15.

14.19 Compare the isodata algorithm with the variant of the BSAS proposed in MACQ 67 and outlined in Section 12.6.

MATLAB PROGRAMS AND EXCERCISES

Computer Programs

14.1 *GMDAS algorithm.* Write a MATLAB function named *GMDAS* that implements the GMDAS algorithm when normal distributions are adopted for the representation of the clusters. The function takes as input (a) an $l \times N$ dimensional matrix X whose columns are the data vectors, (b) an $l \times m$ dimensional matrix mv whose ith column contains an initial estimate of the mean of the ith normal distribution, (c) an $l \times l \times m$ dimensional matrix mc, whose ith two-dimensional $l \times l$ component contains an initial estimate of the covariance matrix of the ith normal distribution, (d) a parameter e used in the termination condition of the algorithm, which is $||\boldsymbol{\Theta}(t) - \boldsymbol{\Theta}(t-1)|| < e$, (e) the maximum number of allowable iterations, *maxiter*, (f) a seed *sed* for the *rand* MATLAB function. The output consists of (a) an m dimensional row vector ap with the a priori probabilities, (b) an $N \times m$ dimensional matrix cp, whose ith row contains the conditional probabilities $P(C_j|\boldsymbol{x}_i), j = 1, \ldots, m$, (c)-(d) the final estimates mv and mc of the mean values and covariance matrices of the normal distributions, respectively, (e) the number of iterations required for convergence, (f) the vector *diffvec* that contains the differences between successive values of $\boldsymbol{\Theta}$, during the training phase.

Solution

For an implementation of this function see in http://www.di.uoa.gr/~stpatrec.

14.2 *Random initialization.* Write a MATLAB function named *rand_vec*, that selects randomly m vectors in the range of values of a given data set. The function takes as input (a) an $l \times N$ dimensional matrix X, whose columns are the data vectors, (b) the number m of column vectors that will be produced, (c) a seed (integer) for the initialization of the *rand* MATLAB function. It returns an $l \times m$ dimensional matrix consisting of the randomly selected column vectors, which will be used for initialization purposes.

Solution

```
function w=rand_vec(X,m,sed)
  rand('seed',sed)
  mini=min(X');
  maxi=max(X');
  w=rand(size(X,1),m);
  for i=1:m
    w(:,i)=w(:,i).*(maxi'-mini')+mini';
  end
```

14.3 ***k-means algorithm.*** Write a MATLAB function, named *k_means*, that implements the *k*-means algorithm. The function takes as input (a) an $l \times N$ dimensional matrix X, each column of which is a data vector, (b) an $l \times m$ dimensional matrix w, the ith column of which is the initial estimate of the ith representative. The output consists of (a) a matrix w similar to the previous one, which contains the final estimates of the representatives and (b) an N-dimensional row vector whose ith element contains the identity number of the cluster where the ith vector belongs (an integer in the set $\{1, 2, \ldots, m\}$).

Solution

```
function [w,bel]=k_means(X,w)
  [l,N]=size(X);
  [l,m]=size(w);
  e=1;
  iter=0;
  while(e~=0)
     iter=iter+1;
     w_old=w;
     dist_all=[];
     for j=1:m
       dist=sum(((ones(N,1)*w(:,j)'-X').^2)');
       dist_all=[dist_all; dist];
     end
     [q1,bel]=min(dist_all);
     for j=1:m
       if(sum(bel==j)~=0)
          w(:,j)=sum(X'.*((bel==j)'*ones(1,l))) / sum(bel==j);
       end
     end
     e=sum(abs(w-w_old));
  end
```

Computer Experiments

14.1 **a.** Generate $q = 50$ two-dimensional vectors from three normal distributions with mean values $[1,\ 1]^T$, $[5,\ 5]^T$, $[9,\ 1]^T$ and covariance matrices $\begin{bmatrix} 1 & 0.4 \\ 0.4 & 1 \end{bmatrix}$, $\begin{bmatrix} 1 & -0.6 \\ -0.6 & 1 \end{bmatrix}$, $\begin{bmatrix} 1 & 0 \\ 0 & 1 \end{bmatrix}$, respectively. Form the 2×150 dimensional matrix X, whose columns are the data vectors produced before.

b. Run the *GMDAS* algorithm on X setting $e = 0.01$, $maxiter = 300$, $sed = 110$ and initializing randomly the mv and mc using the *rand* MATLAB function.

c. Compute the sample mean and the sample covariance matrix for the vectors from each distribution and compare them with the corresponding estimates produced by the algorithm.

d. Comment on the conditional probabilities for each vector.

e. Repeat (b)–(d) five times for different initial estimates for mv and mc.

Hint: Assuming that the first group of q vectors in X is generated from the first distribution, the second group of q vectors in X is generated from the second distribution and so on, the sample mean and sample covariance matrix of the ith group are computed via $sum\ (X(:, (i-1)*q+1 : i*q)')'/q$ and $cov(X(:, (i-1)*q+1 : i*q)')$, respectively.

14.2 Repeat 14.1 when the mean values of the normal distributions are $[1,\ 1]^T$, $[3.5,\ 3.5]^T$, $[6,\ 1]^T$.

14.3 a. Repeat 14.1 when the mean values of the normal distributions are $[1,\ 1]^T$, $[2,\ 2]^T$, $[3,\ 1]^T$.

b. Compare the results obtained in (a) with those obtained in 14.1 and 14.2 and draw your conclusions.

14.4 a. Generate 100 two-dimensional vectors from each one of the three normal distributions with mean values $m1 = [2,\ 2]^T$, $m2 = [6,\ 6]^T$, $m3 = [10,\ 2]^T$ and covariance matrices $S1 = S2 = S3 = 0.5 * I$. Form the 2×300 dimensional matrix X using as columns the vectors generated previously from the three distributions.

b. Run the *k*-means algorithm on X for $m = 2, 3, 4$ representatives using 10 different (randomly selected) initial conditions for the representatives, for each value of m.

c. Comment on the results.

Hint: Use the *rand_vec* function for the initialization of the representatives.

14.5 In the data set of the previous experiment apply the *k*-means algorithm for $m = 3$ representatives, initializing the representatives to the vectors $[-100,\ -100]^T$, $[4.5,\ 6.5]^T$, $[3.5,\ 5.5]^T$. Comment on the results.

14.6 a. Generate 400 two-dimensional vectors from the normal distribution with mean $[0,\ 0]^T$ and covariance matrix $1.5 * I$ and another 15 two-dimensional vectors from the normal distribution with mean $[7,\ 0]^T$ and covariance matrix I. Form the 2×415 dimensional matrix X using as columns the vectors generated previously from both distributions.

b. Run the *k*-means algorithm on X for $m = 2$ representatives using 10 different randomly selected initial conditions for the representatives.

c. Comment on the results.
Hint: Use the *rand_vec* function for the initialization of the representatives.

14.7 **a.** Generate 20 two-dimensional vectors from each one of the two normal distributions with mean values $m1 = [0, 0]^T$, $m2 = [6, 6]^T$ and covariance matrices $S1 = S2 = 0.5 * I$. Form the 2×40 dimensional matrix using as columns the vectors generated previously from the two distributions.

b. Run the fuzzy *c*-means (FCM) algorithm on the above data set with $m = 2$ representatives, initialized randomly. Comment on the grade of memberships of the data vectors in the two resulting clusters.
Hint: Just type

```
[w,U,obj_fun]=fcm(X,m)
```

This function returns (a) the cluster representatives in the rows of w, (b) the grade of membership of each vector to each cluster in matrix U, and (c) the values of the objective function during iterations.

14.8 Repeat the previous experiment when $S1 = S2 = 6 * I$.

14.9 Run the FCM algorithm for $m = 3$ representatives on the data sets produced in 14.7 and 14.8 and comment on the results.

14.10 Run the *k*-means algorithm on the data sets of experiments 14.7 and 14.8 for $m = 2$ randomly initialized representatives and compare the final values of the representatives with those produced by the FCM algorithm on these data sets. Comment on the results.

REFERENCES

[Ande 73] Anderberg M.R. *Cluster analysis for applications*, Academic Press, 1973.

[Ande 85] Anderson I., Bezdek J.C. "An application of the c-varieties clustering algorithms to polygonal curve fitting," *IEEE Transactions on Systems Man and Cybernetics*, Vol. 15, pp. 637–639, 1985.

[Ball 67] Ball G.H., Hall D.J. "A clustering technique for summarizing multivariate data," *Beh-avioral Science*, Vol. 12, pp. 153–155, March 1967.

[Barn 96] Barni M., Cappellini V., Mecocci A. "Comments on 'A possibilistic approach to clustering'," *IEEE Transactions on Fuzzy Systems*, Vol. 4(3), pp. 393–396, August 1996.

[Berk 02] Berkhin P. "Survey of clustering data mining techniques," *Technical Report*, Accrue Software Inc., 2002.

[Bezd 80] Bezdek J.C. "A convergence theorem for the fuzzy Isodata clustering algorithms," *IEEE Transactions on Pattern Analysis and Machine Intelligence*, Vol. 2(1), pp. 1–8, 1980.

[Bezd 81] Bezdek J.C., *Pattern Recognition with Fuzzy Objective Function Algorithms*, Plenum, 1981.

[Bezd 92] Bezdek J.C., Hathaway R.J. "Numerical convergence and interpretation of the fuzzy c-shells clustering algorithm," *IEEE Transactions on Neural Networks*, Vol. 3(5), pp. 787-793, September 1992.

[Bezd 95] Bezdek J.C., Hathaway R.J., Pal N.R., "Norm-induced shell prototypes (NISP) clus- tering," *Neural, Parallel and Scientific Computations*, Vol. 3, pp. 431-450, 1995.

[Bobr 91] Bobrowski L., Bejdek J.C. "c-Means clustering with l_1 and l_∞ norms," *IEEE Transactions on Systems Man and Cybernetics*, Vol. 21(3), pp. 545-554, May/June 1991.

[Brad 98] Bradley P., Fayyad U. "Refining initial points for *K*-means clustering," *Proceedings of the 15th International Conference on Machine Learning*, pp. 91-99, 1998.

[Cann 86] Cannon R.L., Dave J.V., Bezdek J.C. "Efficient implementation of the fuzzy c-means clustering algorithms," *IEEE Transactions on PAMI*, Vol. 8(2), pp. 248-255, March 1986.

[Chen 89] Chen D.S. "A data-driven intermediate level feature extraction algorithm," *IEEE Transactions on PAMI*, Vol. 11(7), pp. 749-758, July 1989.

[Chia 03] Chiang J.H., Hao P.Y., "A new kernel-based fuzzy clustering approach: support vector clustering with cell growing," *IEEE Transactions of Fuzzy Systems*, Vol. 11(4), pp. 518-527, 2003.

[Dave 92a] Dave R.N., Bhaswan K. "Adaptive fuzzy c-shells clustering and detection of ellipses," *IEEE Transactions on Neural Networks*, Vol. 3(5), pp. 643-662, 1992.

[Dave 92b] Dave R.N. "Generalized fuzzy c-shells clustering and detection of circular and elliptical boundaries," *Pattern Recognition*, Vol. 25(7), pp. 713-721, 1992.

[Duda 01] Duda R.O., Hart P.E., Stork D. *Pattern Classification*, John Wiley & Sons, 2001.

[Figu 02] Figueiredo M., Jain A.K. "Unsupervised learning of finite mixture models," *IEEE Transactions on Pattern Analysis and Machine Intelligence*, Vol. 24(3), pp. 381-396, 2002.

[Forg 65] Forgy E. "Cluster analysis of multivariate data: Efficiency vs. interpretability of classifications," *Biometrics*, Vol. 21, pp. 768-780, 1965.

[Frig 96] Frigui H., Krishnapuram R. "A comparison of fuzzy shell clustering methods for the detection of ellipses," *IEEE Transactions on Fuzzy Systems*, Vol. 4(2), pp. 193-199, May 1996.

[Gao 00] Gao X., Li J., Xie W., "Parameter optimization in FCM clustering algorithms," *Proc. of the Int. Conf. on Signal Processing (ICSP) 2000*, pp. 1457-1461, 2000.

[Gers 79] Gersho A. "Asymptotically optimal block quantization," *IEEE Transactions on Information Theory*, Vol. 25(4), pp. 373-380, 1979.

[Gers 92] Gersho A., Gray R.M. *Vector Quantization and Signal Compression*, Kluwer Publishers, 1992.

[Giro 02] Girolami M. "Mercer kernel based clustering in feature space," *IEEE Transactions on Neural Networks*, Vol. 13(3), pp. 780-784, 2002.

[Gnan 77] Gnanadesikan R. *Methods for Statistical Data Analysis of Multivariate Observations*, John Wiley & Sons, 1977.

[Grav 07] Graves D., Pedrycz W., "Fuzzy c-means, Gustafson-Kessel FCM, and kernel-based FCM. A comparative study," in *Analysis and Design on Intelligent Systems using Soft Computing Techniques*, eds. Mellin P. *et al.*, Springer, pp. 140-149, 2007.

[Gray 84] Gray R.M. "Vector quantization," *IEEE ASSP Magazine*, Vol. 1, pp. 4-29, 1984.

[Grol 05] Groll L., Jakel J., "An new convergence proof of fuzzy c-means", *IEEE Transactions on Fuzzy Systems*, Vol. 13(5), pp. 717-720, 2005.

[Gupa 99] Gupata S., Rao K., Bhatnagar V. "*K*-means clustering algorithm for categorical attributes," *Proceedings of the 1st International Conference on Data Warehousing and Knowledge Discovery*, pp. 203–208, Florence, Italy, 1999.

[Hans 01] Hansen P., Mladenovic. "*J*-means: A new local search heuristic for minimum sum of squares clustering." *Pattern Recognition*, Vol. 34, pp. 405–413, 2001.

[Hath 86] Hathaway R.J., Bezdek J.C. "Local convergence of the fuzzy c-means algorithms," *Pattern Recognition* Vol. 19(6), pp. 477–480, 1986.

[Hath 89] Hathaway R.J., Davenport J.W., Bezdek J.C. "Relational duals of the c-means clustering algorithms," *Pattern Recognition*, Vol. 22(2), pp. 205–212, 1989.

[Hath 93] Hathaway R.J., Bezdek J.C. "Switching regression models and fuzzy clustering," *IEEE Transactions on Fuzzy Systems*, Vol. 1(3), pp. 195–204, August 1993.

[Hath 95] Hathaway R.J., Bezdek J.C. "Optimization of clustering criteria by reformulation," *IEEE Transactions on Fuzzy Systems*, Vol. 3(2), pp. 241–245, 1995.

[Hopp 99] Hoppner F., Klawonn F., Kruse R., Runkler T. *Fuzzy Cluster Analysis*, John Wiley & Sons, 1999.

[Horn 86] Horn B.K.P. *Robot Vision*, MIT Press, 1986.

[Huan 98] Huang Z. "Extensions to the *K*-means algorithm for clustering large data sets with categorical values," *Data Mining Knowledge Discovery*, Vol. 2, pp. 283–304, 1998.

[Indy 99] Indyk P. "A sublinear time approximation scheme for clustering in metric spaces," *Foundations of Computer Science (FOCS)*, pp. 154–159, 1999.

[Isma 86] Ismail M.A., Selim S.Z. "Fuzzy c-means: Optimality of solutions and effective termination of the algorithm," *Pattern Recognition*, Vol. 19(6), pp. 481–485, 1986.

[Kanu 00] Kanungo T., Mount D.M., Netanyahu N., Piatko C., Silverman R., Wu A. "An efficient *k*-means clustering algorithm: Analysis and implementation," *IEEE Transactions on Pattern Analysis and Machine Intelligence*, Vol. 24(7), pp. 881–892, 2000.

[Kanu 04] Kanungo T., Mount D.M., Netanyahu N., Piatko C., Silverman R., Wu A. "A local search approximation algorithm for *k*-means clustering," *Computational Geometry*, Vol. 28(2–3), pp. 89–112, 2004.

[Kara 96] Karayiannis N.B., Pai P.-I. "Fuzzy algorithms for learning vector quantization," *IEEE Transactions on Neural Networks*, Vol. 7(5), pp. 1196–1211, September 1996.

[Kare 94] Karen D., Cooper D., Subrahmonia J. "Describing complicated objects by implicit polynomials," *IEEE Transactions on Pattern Analysis and Machine Intelligence*, Vol. 16(1), pp. 38–53, 1994.

[Kauf 90] Kaufman L., Rousseeuw P. *Finding groups in data: An introduction to cluster analysis*. John Wiley & Sons, 1990.

[Kim 88] Kim T., Bezdek J.C., Hathaway R.J. "Optimality tests for fixed points of the fuzzy c-means algorithm," *Pattern Recognition*, Vol. 21(6), pp. 651–663, 1988.

[Kris 92a] Krishnapuram R., Freg C.-P. "Fitting an unknown number of lines and planes to image data through compatible cluster merging," *Pattern Recognition*, Vol. 25(4), pp. 385–400, 1992.

[Kris 92b] Krishnapuram R., Nasraoui O., Frigui H. "The fuzzy c spherical shells algorithm: A new approach," *IEEE Transactions on Neural Networks*, Vol. 3(5), pp. 663–671, 1992.

[Kris 93] Krishnapuram R., Keller J.M. "A possibilistic approach to clustering," *IEEE Transactions on Fuzzy Systems*, Vol. 1(2), pp. 98–110, May 1993.

[Kris 95a] Krishnapuram R., Frigui H., Nasraoui O. "Fuzzy and possibilistic shell clustering algorithms and their application to boundary detection and surface approximation—Part I," *IEEE Transactions on Fuzzy Systems*, Vol. 3(1), pp. 29–43, February 1995.

[Kris 95b] Krishnapuram R., Frigui H., Nasraoui O. "Fuzzy and possibilistic shell clustering algorithms and their application to boundary detection and surface approximation—Part II," *IEEE Transactions on Fuzzy Systems*, Vol. 3(1), pp. 44–60, February 1995.

[Kris 96] Krishnapuram R., Keller J.M. "The possibilistic c-means algorithm: Insights and recommendations," *IEEE Transactions on Fuzzy Systems*, Vol. 4(3), pp. 385–393, August 1996.

[Kris 99] Krishna K., Muthy M. "Genetic *k*-means algorithm," *IEEE Transactions on Systems, Man and Cybernetics*, Vol. 29(3), pp. 433–439, 1999.

[Kris 99a] Krishnapuram R., Kim J., "A note on the Gustafson-Kessel and adaptive fuzzy clustering algorithms," *IEEE Transactions on Fuzzy Systems*, Vol. 7(4), pp. 453–461, 1999.

[Kuma 04] Kumar A., Sabharwal Y., Sen S. "A simple linear time (1+)-approximation algorithm for *k*-means clustering in any dimension," *Foundations of Computer Science (FOCS)*, pp. 454–462, 2004.

[Lika 03] Likas A., Vlassis N., Verbeek J. "The global *K*-means clustering algorithm," *Pattern Recognition*, Vol. 36(2), pp. 451–461, 2003.

[Lin 96] Lin J.S., Cheng K.S., Mao C.W., "Segmentation of multispectral magnetic resonance image using penalized fuzzy competitive learning network," *Computers and Biomedical Research*, Vol. 29, pp. 314–326, 1996.

[Lind 80] Linde Y., Buzo A., Gray R.M. "An algorithm for vector quantizer design," *IEEE Transactions on Communications*, Vol. 28, pp. 84–95, 1980.

[Lloy 82] Lloyd S.P. "Least squares quantization in PCM," *IEEE Transactions on Information Theory*, Vol. 28(2), pp. 129–137 March 1982.

[Luen 84] Luenberger D.G. *Linear and Nonlinear Programming*, Addison Wesley, 1984.

[Lutt 89] Luttrell S.P. "Hierarchical vector quantization," *IEE Proceedings (London)*, Vol. 136 (Part I), pp. 405–413, 1989.

[MacQ 67] MacQueen J.B. "Some methods for classification and analysis of multivariate observations," *Proceedings of the Symposium on Mathematical Statistics and Probability, 5th Berkeley*, Vol. 1, pp. 281–297, AD 669871, University of California Press, 1967.

[Man 94] Man Y., Gath I. "Detection and separation of ring-shaped clusters using fuzzy clustering," *IEEE Transactions on PAMI*, Vol. 16(8), pp. 855–861, August 1994.

[Mena 00] Menard M., Demko C., Loonis P. "The fuzzy c+2 means: solving the ambiguity rejection in clustering," *Pattern Recognition*, Vol. 33, pp. 1219–1237, 2000.

[Ng 94] Ng R., Han J. "Efficient and effective clustering methods for spatial data mining." *Proceedings of the 20th Conference on VLDB*, pp. 144–155, Santiago, Chile, 1994.

[Ng 94a] Ng R., Han J. "Efficient and effective clustering methods for spatial data mining." *Technical Report 94-13*, University of British Columbia.

[Ng 00] Ng M. K. "A note on constrained k-means algorithms," *Pattern Recognition*, Vol. 33, pp. 515–519, 2000.

[Ozde 01] Özdemir D., Akarun L., "Fuzzy algorithms for combined quantization and dithering," *IEEE Transactions on Image Processing*, Vol. 10(6), pp. 923–931, 2001.

[Ozde 02] Özdemir D., Akarun L., "A fuzzy algorithm for color quantization and images," *Pattern Recognition*, Vol. 35, pp. 1785–1791, 2002.

[Pata 01] Patane G., Russo M. "The enhanced-LBG algorithm," *Neural Networks*, Vol. 14(9), pp. 1219–1237, 2001.

[Pata 02] Patane G., Russo M. "Fully automatic clustering system," *IEEE Transactions on Neural Networks*, Vol. 13(6), pp. 1285–1298, 2002.

[Pato 70] Paton K. "Conic sections in chromosome analysis," *Pattern Recognition*, Vol. 2(1), pp. 39–51, January 1970.

[Pedr 96] Pedrycz W., "Conditional fuzzy c-means," *Pattern Recognition Letters*, Vol. 17, pp. 625–632, 1996.

[Pena 99] Pena J., Lozano J., Larranaga P. "An empirical comparison of four initialization methods for the *k*-means algorithm," *Pattern Recognition Letters*, Vol. 20, pp. 1027–1040, 1999.

[Runk 99] Runkler T.A., Bezdek J.C. "Alternating cluster estimation: A new tool for clustering and function approximation," *IEEE Trans. on Fuzzy Systems*, Vol. 7, No. 4, pp. 377–393, August 1999.

[Sabi 87] Sabin M.J. "Convergence and consistency of fuzzy c-means/Isodata algorithms," *IEEE Transactions on PAMI*, Vol. 9(5), pp. 661–668, September 1987.

[Scho 98] Schölkopf B., Smola A.J., Müller "Nonlinear component analysis as a kernel eigenvalue problem," *Neural Computation*, Vol. 10(5), pp. 1299–1319, 1998.

[Seli 84a] Selim S.Z., Ismail M.A. "K-means type algorithms: A generalized convergence theorem and characterization of local optimality, "*IEEE Transactions on PAMI*, Vol. 6(1), pp. 81–87, 1984.

[Seli 84b] Selim S.Z., Ismail M.A. "Soft clustering of multidimensional data: A semifuzzy approach," *Pattern Recognition*, Vol. 17(5), pp. 559–568, 1984.

[Seli 86] Selim, S.Z., Ismail, M.A. "On the local optimality of the fuzzy Isodata clustering algorithm," *IEEE Transactions on PAMI*, Vol. 8(2), pp. 284–288, March 1986.

[Shen 06] Shen H., Yang J., Wang S., Liu X., "Attribute weighted Mercer kernel-based fuzzy clustering algorithm for general non-spherical data sets," *Soft Computing*, Vol. 10(11), pp. 1061–1073, 2006.

[Siya 05] Siyal M.Y., Yu L., "An intelligent modified fuzzy c-means based algorithm for bias estimation and segmentation of brain MRI," *Pattern Recognition Letters*, Vol. 26(13), pp. 2052–2062, 2005.

[Spra 66] Spragins J. "Learning without a teacher," *IEEE Transactions on Information Theory*, Vol. IT-12, pp. 223–230, April 1966.

[Su 01] Su M., Chou C. "A modified version of the *K*-means algorithm with a distance based on cluster symmetry," *IEEE Transactions on Pattern Analysis and Machine Intelligence*, Vol. 23(6), pp. 674–680, 2001.

[Wags 01] Wagstaff K., Rogers S., Schroedl S. "Constrained *k*-means clustering with background knowledge," *Proceedings of the 8th International Conference on Machine Learning*, pp. 577–584, 2001.

[Wei 94] Wei W., Mendel J.M. "Optimality tests for the fuzzy c-means algorithm," *Pattern Recognition*, Vol. 27(11), pp. 1567–1573, 1994.

[Yama 80] Yamada Y., Tazaki S., Gray R.M. "Asymptotic performance of block quantizers with difference distortion measures," *IEEE Transactions on Information Theory*, Vol. 26(1), pp. 6–14, 1980.

[Yang 93] Yang M.S., "On a class of fuzzy classification maximum likelihood procedures," *Fuzzy Sets and Systems*, Vol. 57, pp. 365–375, 1993.

[Yu 03] Yu J., Yang M., "A study on a generalized FCM," in *Rough Sets, Fuzzy Sets, Data Mining, and Granular Computing*, eds. Wang G. *et al*, pp. 390–393, Springer, 2003.

[Zade 78] Zadeh L.A. "Fuzzy sets as a basis for a theory of possibility," *Fuzzy Sets and Systems*, Vol. 1, pp. 3–28, 1978.

[Zeyu 01] Zeyu L., Shiwei T., Jing X., Jun J., "Modified FCM clustering based on kernel mapping," *Proc. of Int. Society of Optical Engineering*, Vol. 4554 pp. 241–245, 2001.

[Zhan 03] Zhang D.Q., Chen S.C., "Clustering incomplete data using kernel-based fuzzy c-means algorithm," *Neural Processing Letters*, Vol. 18(3), pp. 155–162, 2003.

[Zhou 04] Zhou S., Gan J., "Mercer kernel fuzzy c-means algorithm and prototypes of clusters," *Proc. Cong. on Int. Data Engineering and Automated Learning*, pp. 613–618, 2004.

[Zhua 96] Zhuang X., Huang Y., Palaniappan K., Zhao Y. "Gaussian mixture density modelling, decomposition and applications," *IEEE Transactions on Image Processing*, Vol. 5, pp. 1293–1302, September 1996.

CHAPTER 15

Clustering Algorithms IV

15.1 INTRODUCTION

The clustering algorithms presented in the previous two chapters evolved along two distinct major philosophies. The current chapter presents categories of algorithms that cannot be included in either of the previous two families, and they stem from various ideas. The first one includes clustering algorithms based on graph theory concepts, such as the minimum spanning tree, the directed tree and spectral clustering. The second category includes competitive learning algorithms. The third category includes branch and bound algorithms. These schemes guarantee to provide globally optimal clustering, in terms of a prespecified optimality criterion, at the cost of increased computational requirements. The fourth category contains algorithms that are based on morphological transformations. These have been inspired by the corresponding methods used in signal and image processing. The fifth category contains algorithms that are not based on cluster representatives but, instead, seek to place boundaries between clusters. Algorithms of the sixth category treat clusters as dense in data regions of the feature space separated by regions sparse in data. Alternatively, clusters may be viewed as peaks of the pdf, underlying the data in X, separated by valleys. The seventh category includes additional algorithms that are based on function optimization, such as simulated annealing and deterministic annealing. The difference from the algorithms presented in Chapter 14 is that the optimizing methods used in this chapter do not involve differential calculus concepts. In addition, this category also includes genetic algorithms modified suitably for clustering tasks. Finally, the eighth category includes algorithms that combine clusterings in order to produce a final (hopefully more accurate) one.

15.2 CLUSTERING ALGORITHMS BASED ON GRAPH THEORY

The algorithms of this family are capable of detecting clusters of various shapes, at least for the case in which they are well separated. Detection of clusters of various shapes is a feature that is shared by only a few other clustering algorithms.

15.2.1 Minimum Spanning Tree Algorithms

The first algorithm is based on the idea of the minimum spanning tree (MST) (Chapter 13) and is motivated by the way human perception works [Zahn 71]. More precisely, humans organize information with the most economical encoding [Hoch 64]. For example, the more likely way for a human to organize the points in Figure 15.1 is in four groups (clusters).

Let us consider the complete graph G, each node of which corresponds to a point of X. The weight of an edge $e = (\boldsymbol{x}_i, \boldsymbol{x}_j)$, w_e, connecting two nodes $\boldsymbol{x}_i$ and $\boldsymbol{x}_j$, is set equal to the distance $d(\boldsymbol{x}_i, \boldsymbol{x}_j)$ of the corresponding points in the feature space. Also, we say that two edges e_1 and e_2 are *k steps away* from each other if the minimum path that connects a vertex of e_1 and a vertex of e_2 has length equal to $k - 1$, that is, contains $k - 1$ edges.

The idea of the algorithm is the following: ***determine the minimum spanning tree of G and then remove the edges that are "unusually" large compared with their neighboring edges.*** These edges are called *inconsistent*, and it is expected that they connect points from different clusters. Next, we discuss a way to determine inconsistent edges. For each edge e, we consider all the edges, e_i, that lie k steps away from it, at the most, and we compute the mean, m_e, and the standard deviation, σ_e, of their weights. If w_e lies more than q (typically $q = 2$) standard deviations (σ_e) away from m_e, then we consider e as inconsistent. From this, it is clear that the characterization of an edge as inconsistent is somewhat subjective and depends on k and q, which are preselected.

Example 15.1

Consider Figure 15.2. Let $k = 2$ and $q = 3$. The edges lying two steps at the most from e_0 are e_i, $i = 1, \ldots, 10$. The mean m_{e_0} and the standard deviation σ_{e_0}, corresponding to e_0 are 2.3 and 0.95, respectively. Thus w_{e_0} lies 15.5 standard deviations (σ_{e_0}) away from m_{e_0}. Therefore, e_0 is an inconsistent edge.

Let us now consider the edge e_{11}. Working as before, we find that $m_{e_{11}}$ and $\sigma_{e_{11}}$ are 2.5 and 2.12, respectively. Thus $w_{e_{11}}$ is 0.24 standard deviations ($\sigma_{e_{11}}$) away from $m_{e_{11}}$. Therefore, e_{11} is not an inconsistent edge.

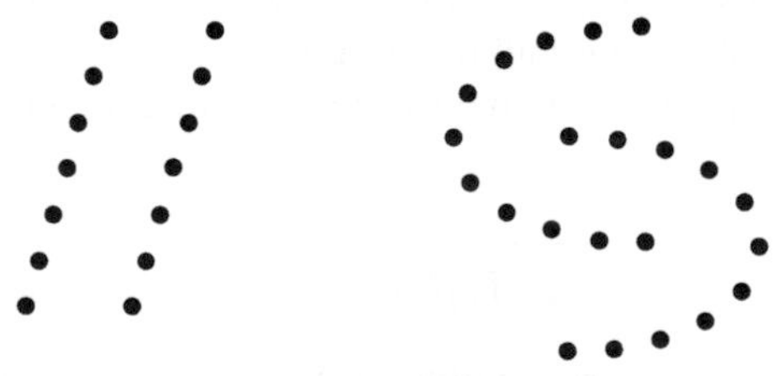

FIGURE 15.1

An arrangement of clusters.

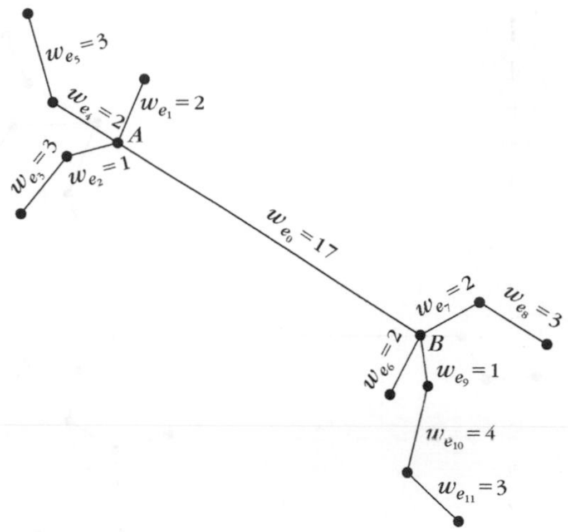

FIGURE 15.2

The minimum spanning tree of a graph. The edge e_0 is inconsistent, while e_{11} is consistent.

After these definitions, the MST clustering algorithm may be stated as follows.

The MST Clustering Algorithm

- Construct a complete graph G such that:
  - its vertices correspond to the vectors of X.
  - $w_{(x_i,x_j)} = d(x_i, x_j), i, j = 1, \ldots, N, i \neq j$.
- Determine the MST of G.
- Identify the inconsistent edges of the MST.
- The clusters are the connected components of the MST after the removal of the inconsistent edges.

This algorithm works satisfactorily for many cases where the clusters are well separated. However, this is not a panacea. Let us consider for example Figure 15.3. The edge AB has a very large neighboring edge (BC), which increases m_{AB} and σ_{AB}. Thus, AB may not be characterized as inconsistent and, as a consequence, the vectors from regions R_1 and R_2 are considered as members of the same cluster [Jarv 78].

Some suggestions for the use of the MST algorithm for the cases where we have touching clusters (Figure 15.4a), as well as for the case where the clusters have different densities (Figure 15.4b), are discussed in [Zahn 71]. However, they implicitly require knowledge of the shape of the clusters.

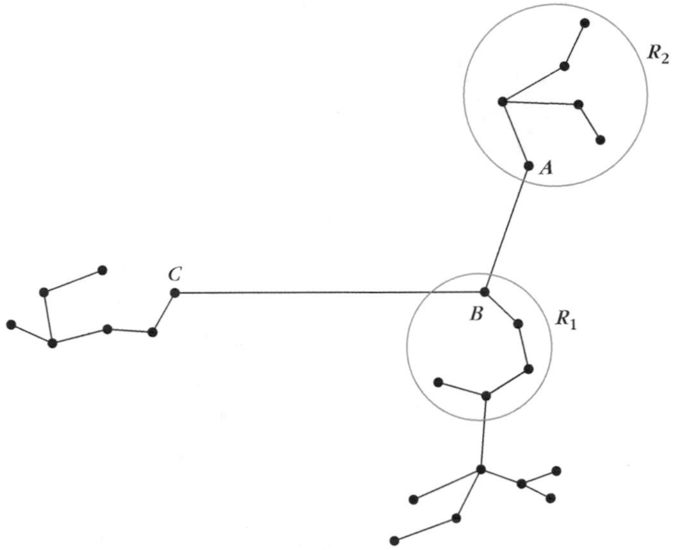

FIGURE 15.3

The MST clustering algorithm will assign the vectors of the regions R_1 and R_2 to the same cluster.

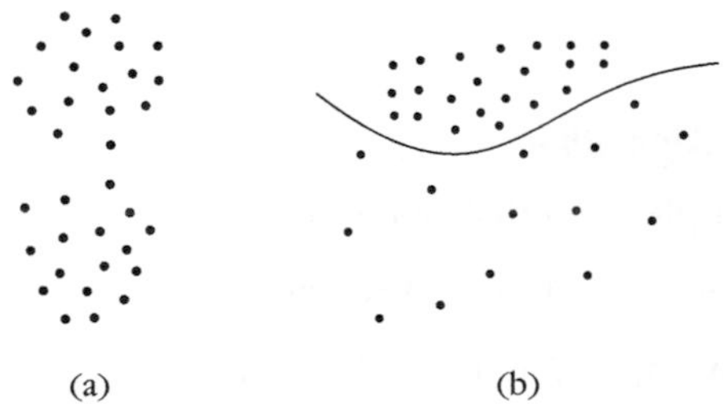

FIGURE 15.4

(a) Touching clusters. (b) Clusters with different densities.

Remark

- Note that this algorithm does not depend on the order in which data are considered by the algorithm and, also, no initial conditions are required, as is the case with the algorithms of Chapter 14.

15.2.2 Algorithms Based on Regions of Influence

An extension of the MST involves regions of influence for each pair of vectors of X. This idea was used by many researchers (e.g., [Tous 80, Gabr 69, Urqu 82]) in order to overcome the problems associated with the MST algorithms.

Let us consider two distinct vectors, $\mathbf{x}_i, \mathbf{x}_j \in X$. Their region of influence is defined as

$$R(\mathbf{x}_i, \mathbf{x}_j) = \{\mathbf{x}: cond(d(\mathbf{x}, \mathbf{x}_i), d(\mathbf{x}, \mathbf{x}_j), d(\mathbf{x}_i, \mathbf{x}_j)), \mathbf{x}_i \neq \mathbf{x}_j\} \tag{15.1}$$

where $cond(d(\mathbf{x}, \mathbf{x}_i), d(\mathbf{x}, \mathbf{x}_j), d(\mathbf{x}_i, \mathbf{x}_j))$ is a condition among the distances $d(\mathbf{x}, \mathbf{x}_i)$, $d(\mathbf{x}, \mathbf{x}_j)$, and $d(\mathbf{x}_i, \mathbf{x}_j)$. Different choices of *cond* give rise to different shapes of regions of influence. Typical choices of *cond*, proposed in [Tous 80] and [Gabr 69], are

$$\max\{d(\mathbf{x}, \mathbf{x}_i), d(\mathbf{x}, \mathbf{x}_j)\} < d(\mathbf{x}_i, \mathbf{x}_j) \tag{15.2}$$

and

$$d^2(\mathbf{x}, \mathbf{x}_i) + d^2(\mathbf{x}, \mathbf{x}_j) < d^2(\mathbf{x}_i, \mathbf{x}_j) \tag{15.3}$$

respectively. Also, in [Urqu 82], the following two alternatives are proposed:

$$(d^2(\mathbf{x}, \mathbf{x}_i) + d^2(\mathbf{x}, \mathbf{x}_j) < d^2(\mathbf{x}_i, \mathbf{x}_j))\ OR$$
$$(\sigma \min\{d(\mathbf{x}, \mathbf{x}_i), d(\mathbf{x}, \mathbf{x}_j)\} < d(\mathbf{x}_i, \mathbf{x}_j)) \tag{15.4}$$

and

$$(\max\{d(\mathbf{x}, \mathbf{x}_i), d(\mathbf{x}, \mathbf{x}_j)\} < d(\mathbf{x}_i, \mathbf{x}_j))\ OR$$
$$(\sigma \min\{d(\mathbf{x}, \mathbf{x}_i), d(\mathbf{x}, \mathbf{x}_j)\} < d(\mathbf{x}_i, \mathbf{x}_j)) \tag{15.5}$$

where σ is a factor called ***relative edge consistency***. This factor affects the size of the region of influence defined by $\mathbf{x}_i$ and $\mathbf{x}_j$. The shapes of these regions are shown in Figure 15.5. Other choices of *cond* are also possible. An algorithm based on the idea of the regions of influence is described next.

Algorithm Based on Regions of Influence

- For $i = 1$ to N
  - For $j = i + 1$ to N

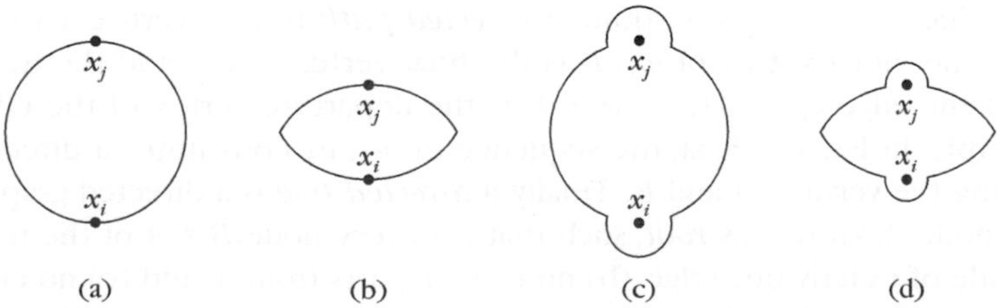

FIGURE 15.5

The shapes of the regions defined by (a) condition (15.2), (b) condition (15.3), (c) condition (15.4), and (d) condition (15.5).

    - Determine the region of influence $R(\boldsymbol{x}_i, \boldsymbol{x}_j)$.
    - If $R(\boldsymbol{x}_i, \boldsymbol{x}_j) \cap (X - \{\boldsymbol{x}_i, \boldsymbol{x}_j\}) = \emptyset$ then
      - — Add the edge connecting $\boldsymbol{x}_i, \boldsymbol{x}_j$
    - End {If }
  - End {For}
- End {For}
- Determine the connected components of the resulted graph and identify them as clusters.

In words, the edge between $\boldsymbol{x}_i$ and $\boldsymbol{x}_j$ is added if no other vector of X lies in $R(\boldsymbol{x}_i, \boldsymbol{x}_j)$. This is because when $\boldsymbol{x}_i$ and $\boldsymbol{x}_j$ are closely located, it is expected that no other points of X will be in $R(\boldsymbol{x}_i, \boldsymbol{x}_j)$. The opposite is obviously true for points located further away.

The algorithm is insensitive to the order in which the pairs of vectors are considered. Also, for the last two choices of *cond*, the value of σ must be chosen *a priori*. The graphs produced by these algorithms when (15.2) and (15.3) are used are also called *relative neighborhood graphs (RNGs)* and *Gabriel graphs (GGs)*, respectively.

These techniques avoid situations such as the one shown in Figure 15.3. Moreover, several results are given in [Urqu 82] showing the superior performance for the last two choices of *cond* compared with the first two. Also, in [Urqu 82] it is shown how the idea of the regions of influence may be used to give rise to hierarchical algorithms. Finally, in [Ozbo 95], empirically defined regions of influence are used, that exhibit satisfactory behavior.

15.2.3 Algorithms Based on Directed Trees

An alternative clustering scheme, based on the idea of directed trees, is proposed in [Koon 76]. Before we proceed, let us give some definitions. We recall that a *directed graph* is a graph whose edges are directed (see Figure 15.6a). We say that a set of edges $e_{i_1}, \ldots, e_{i_q}$ constitute a *directed path* from a vertex A to a vertex B, if A is the initial vertex of e_{i_1}, B is the final vertex of e_{i_q}, and the destination vertex of the edge $e_{i_j}, j = 1, \ldots, q-1$, is the departure vertex of the edge $e_{i_{j+1}}$. For example, in Figure 15.6a, the sequence e_1, e_2, e_3 constitutes a directed path connecting the vertices A and B. Finally, a *directed tree* is a directed graph with a specific node A, known as *root*, such that (a) every node $B \neq A$ of the tree is the initial node of exactly one edge, (b) no edge departs from A, and (c) no circles are encountered, that is, there is no directed path from a node to itself (see Figure 15.6b).

The idea of the algorithm is the identification of directed trees in a graph, corresponding to the points of X, so that each of them corresponds to a cluster.

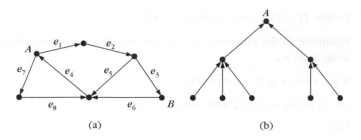

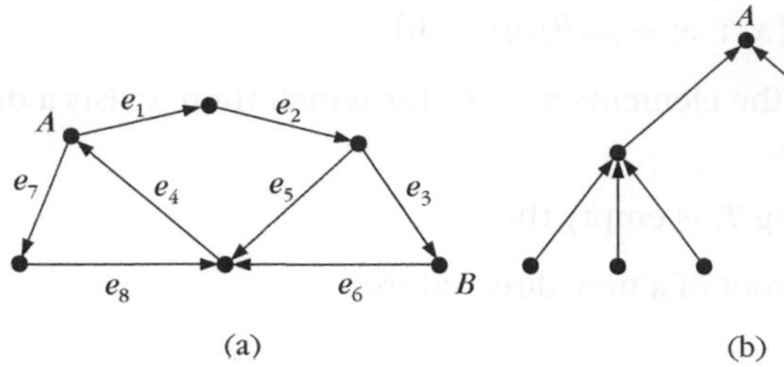

FIGURE 15.6

(a) A directed graph. (b) A directed tree.

The vectors of X are processed sequentially. For each point x_i, we define its neighborhood as

$$\rho_i(\theta) = \{x_j \in X: d(x_i, x_j) \leq \theta, x_j \neq x_i\} \tag{15.6}$$

where θ determines the size of the neighborhood and $d(x_i, x_j)$ is the distance between the corresponding vectors of X. Let $n_i = |\rho_i(\theta)|$ be the number of points of X lying in $\rho_i(\theta)$. Finally, let $g_{ij} = (n_j - n_i)/d(x_i, x_j)$. This quantity will be used to determine the position of the point x_i in a directed tree. After these definitions, the clustering algorithm may be stated as follows.

Clustering Algorithm Based on Directed Trees

Set θ to a specific value.
Determine $n_i, i = 1, \ldots, N$.
Compute $g_{ij}, i, j = 1, \ldots, N, i \neq j$.
For $i = 1$ to N

- if $n_i = 0$ then
  - x_i is the root of a new directed tree.
- else
  - Determine x_r such that $g_{ir} = \max_{x_j \in \rho_i(\theta)} g_{ij}$.
  - If $g_{ir} < 0$ then
    - x_i is the root of a new directed tree.
  - Else if $g_{ir} > 0$ then
    - x_r is the parent of x_i.[1]
  - Else if $g_{ir} = 0$ then

[1] We say that x_r is the parent of x_i if there exists a directed edge from x_i to x_r.

    ○ Define $T_i = \{\boldsymbol{x}_j : \boldsymbol{x}_j \in \rho_i(\theta), g_{ij} = 0\}$.
    ○ Eliminate all the elements $\boldsymbol{x}_j \in T_i$, for which there exists a directed path from $\boldsymbol{x}_j$ to $\boldsymbol{x}_i$.
    ○ If the resulting T_i is empty then
        — $\boldsymbol{x}_i$ is the root of a new directed tree.
    ○ Else
        — The parent of $\boldsymbol{x}_i$ is $\boldsymbol{x}_q$ such that $d(\boldsymbol{x}_i, \boldsymbol{x}_q) = \min_{\boldsymbol{x}_s \in T_i} d(\boldsymbol{x}_i, \boldsymbol{x}_s)$.
    ○ End {if}
  • End {if}
■ End {if}

End {for}

The directed trees formed by these steps identify the clusters.

It is clear from the preceding algorithm that the root, say $\boldsymbol{x}_i$, of a directed tree has the largest n_i among the points lying in $\rho_i(\theta)$. That is, among the points lying in $\rho_i(\theta)$, $\boldsymbol{x}_i$ is the point with the most dense neighborhood. It should be pointed out that the branch that handles the case in which $g_{ir} = 0$ ensures that no circles will occur. Also, this algorithm is sensitive to the order in which the vectors are processed. Finally, it can be shown that for proper values of θ and large N this scheme behaves as a mode-seeking algorithm [Koon 76].

Example 15.2

Consider Figure 15.7. The size of the edge of the grid is 1 and the diagonal of a small rectangle equals $\sqrt{2}$. Also, let $X = \{\boldsymbol{x}_i, i = 1, \ldots, 11\}$. It is clear that the vectors of X form two well-separated clusters. Let $\theta = 1.1$. Applying the preceding algorithm on X, we determine the two directed trees shown in Figure 15.7. However, if we present $\boldsymbol{x}_5$ before $\boldsymbol{x}_4$, the left-directed tree will have a different root. Nevertheless, the final results of the algorithm remain the same in this (rather easy) case.

15.2.4 Spectral Clustering

Spectral clustering is a class of graph-based techniques that unravel the structural properties of a graph using information conveyed by the spectral decomposition (eigendecomposition) of an associated matrix. The elements of this matrix code the underlying similarities among the nodes (data points) of the graph. Spectral clustering algorithms have attracted a lot of interest over the last years. Their high popularity springs from their improved performance in a number of applications, where several classical techniques fail and also from a number of interesting related theoretical issues. Among the earlier works on spectral clustering are [Scot 90] and [Hage 92].

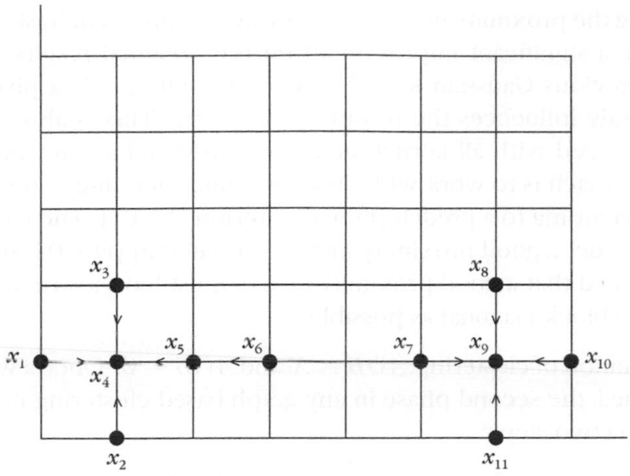

FIGURE 15.7

The setup of Example 15.2.

In this book, we will focus on the simplest task of bi-partitioning a given data set, X, into two clusters, A and B. Generalizations will be discussed later on. Let $X = \{\boldsymbol{x}_1, \boldsymbol{x}_2, \ldots, \boldsymbol{x}_N\} \subset \mathcal{R}^l$. In graph-based clustering methods the following steps are in order:

- Construct a graph $G(V, E)$, where each point of the graph corresponds to a point $\boldsymbol{x}_i$, $i = 1, 2, \ldots, N$, of X. We will further assume that G is undirected and connected (Section 13.2.5).
- Weigh each one of the edges of the graph, e_{ij}, by a weight $W(i,j)$ that measures the similarity between the respective nodes, v_i, v_j in G^2. The set of weights defines the proximity (sometimes called affinity) $N \times N$ matrix W with elements

$$W \equiv [W(i,j)], \quad i,j = 1, 2, \ldots, N$$

The proximity matrix is assumed to be symmetric, that is, $W(i,j) = W(j,i)$. The choice of the weights is up to the user and it is a problem dependent task. A common choice is

$$W(i,j) = \begin{cases} \exp\left(-\frac{||\boldsymbol{x}_i - \boldsymbol{x}_j||^2}{2\sigma^2}\right), & \text{if } ||\boldsymbol{x}_i - \boldsymbol{x}_j|| < \epsilon \\ 0, & \text{otherwise} \end{cases}$$

where ϵ is a user-defined constant and $||\cdot||$ is the Euclidean norm.

[2] For notational convenience in some places we use i instead of v_i.

> Choosing the proximity matrix is not always an "innocent" task. A right choice can have a significant improvement on the obtained results. For example, in the previous Gaussian kernel case, determining σ is a pivotal issue that significantly influences the resulting clustering. This is also a problem that we have faced with all kernel methods considered in previous chapters. A naive approach is to work with different values of σ and choose the one that is best according to a predetermined criterion [Ng 01]. The issue of how one can construct a good proximity matrix is treated in [Fisc 05, Weis 99]. There is also stated that a good proximity matrix must have a structure which is as close to a block diagonal as possible.

By the definition of clustering, $A \cup B = X$ and $A \cap B = \emptyset$. Once a weighted graph has been formed, the second phase in any graph-based clustering method consists of the following two steps:

- Choose an appropriate clustering criterion for the partitioning of the graph.
- Adopt an efficient algorithmic scheme to perform the partitioning, in accordance with the previously adopted clustering criterion.

A commonly used clustering criterion is the so called *cut* [Wu 93]. If A and B are the resulting clusters, the associated *cut* is defined as

$$cut(A,B) = \sum_{i \in A,\, j \in B} W(i,j) \tag{15.7}$$

Selecting A and B so that the respective $cut(A,B)$ is minimized means that the set of edges, connecting nodes in A with nodes in B, have the minimum sum of weights, that is, points in A and B have the least similarity compared to any other bi-partitioning. However, this simple criterion turns out to form clusters of small size of isolated points (least similar with the rest of the nodes). This is illustrated in Figure 15.8. The minimum *cut* criterion would result in the two clusters separated by the dotted line, although the partition by the full line seems to be a more natural partitioning.

To remedy the previous drawback, the *normalized cut* criterion has been suggested in [Shi 00]. This is one of the most popular criteria used in spectral clustering.

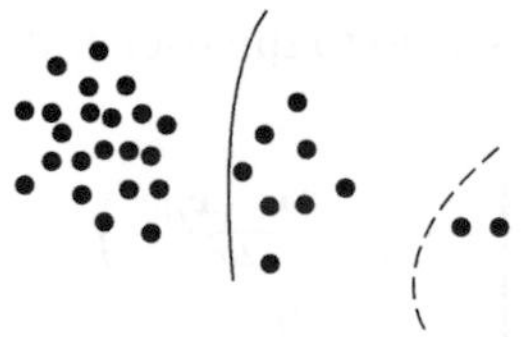

FIGURE 15.8

The cut criterion has the tendency to form small clusters of isolated points, as for example the two points separated by the dotted line. A more natural clustering for this case results by the full line.

The driving force behind this criterion is to minimize the *cut* and at the same time trying to keep the sizes of the clusters large. To this end, for each node, $v_i \in V$ in the graph G, define the index

$$D_{ii} = \sum_{j \in V} W(i,j) \tag{15.8}$$

This is an index measuring the "significance" of a node, $v_i, i = 1, 2, \ldots, N$. The higher the value of D_{ii} is, the higher the similarity of the respective ith node with respect to the rest of the nodes. A low D_{ii} value indicates an isolated (remote) point. Given a cluster A, a measure of the total significance of the points in A is given by the following index

$$V(A) = \sum_{i \in A} D_{ii} = \sum_{i \in A,\, j \in V} W(i,j) \tag{15.9}$$

where $V(A)$ is sometimes known as the *volume* or the *degree* of A. It is obvious that small and isolated clusters will have a small $V(\cdot)$. The *normalized cut* between two clusters A, B is defined as

$$Ncut(A,B) = \frac{cut(A,B)}{V(A)} + \frac{cut(A,B)}{V(B)} \tag{15.10}$$

It is easy to see that a small cluster, for example, A, will result in a large value (close to one) for the previous ratio, since in such a case $cut(A, B)$ will be a large percentage of $V(A)$.

Minimization of the $Ncut(A, B)$ turns out to be an NP-hard task. To alleviate this difficulty, we will reformulate the problem to bring it in a form that allows an efficient approximate solution. To this end define ([Belk 03])

$$y_i = \begin{cases} \frac{1}{V(A)}, & \text{if } i \in A \\ -\frac{1}{V(B)}, & \text{if } i \in B \end{cases} \tag{15.11}$$

$$y = [y_1, y_2, \ldots, y_N]^T$$

In words, each y_i can be thought of as a cluster indicator of the corresponding point $\boldsymbol{x}_i$, $i = 1, 2, \ldots, N$. Taking into account the definitions in (15.11), it is straightforward to see that (see also Section 6.7)

$$y^T L y = \frac{1}{2} \sum_{i \in V} \sum_{j \in V} (y_i - y_j)^2 W(i,j)$$

$$= \sum_{i \in A} \sum_{j \in B} \left(\frac{1}{V(A)} + \frac{1}{V(B)} \right)^2 cut(A,B)$$

$$\propto \left(\frac{1}{V(A)} + \frac{1}{V(B)} \right)^2 cut(A,B) \tag{15.12}$$

since the contribution of $y_i - y_j$ is zero for points in the same cluster. The symbol $\propto$ denotes proportionality and

$$L = D - W, \; D \equiv \text{diag}\{D_{ii}\}$$

is the Laplacian matrix of the graph (Section 6.7). Matrix D is diagonal with the elements D_{ii} along the main diagonal. In addition we have that

$$y^T D y = \sum_{i \in A} y_i^2 D_{ii} + \sum_{j \in B} y_j^2 D_{jj} = \frac{1}{V(A)^2} V(A) + \frac{1}{V(B)^2} V(B)$$
$$= \frac{1}{V(A)} + \frac{1}{V(B)} \tag{15.13}$$

Combining (15.12) and (15.13), it turns out that minimizing $Ncut(A, B)$ is equivalent with minimizing

$$J = \frac{y^T L y}{y^T D y} \tag{15.14}$$

subject to the constraint that $y_i \in \{\frac{1}{V(A)}, -\frac{1}{V(B)}\}$. Furthermore, direct substitution of the definitions of the involved quantities results in

$$y^T D \mathbf{1} = 0 \tag{15.15}$$

where $\mathbf{1}$ is the N-dimensional vector with all its elements being equal to 1. In order to overcome the NP-hard nature of the original task we will solve, instead, the relaxed problem of minimizing (15.14) subject to the constraint in (15.15). The unknown variables y_i, $i = 1, 2, \ldots, N$, are now assumed to lie on the real axis. We are already very close to a well known optimization task. Define

$$z \equiv D^{1/2} y$$

Then (15.14) becomes

$$J = \frac{z^T \tilde{L} z}{z^T z} \tag{15.16}$$

and the constraint in (15.15)

$$z^T D^{1/2} \mathbf{1} = 0 \tag{15.17}$$

where $\tilde{L} \equiv D^{-1/2} L D^{-1/2}$ and it is known as the *normalized graph Laplacian* matrix. It can easily be shown that $\tilde{L}$ has the following properties (see also, Section 6.7)

- It is symmetric and nonnegative definite. Thus, all its eigenvalues are nonnegative and the corresponding eigenvectors are orthogonal to each other (see Appendix B).

- It can easily be checked out that $D^{1/2}\mathbf{1}$ is an eigenvector corresponding to a zero eigenvalue, that is,

$$\tilde{L}D^{1/2}\mathbf{1} = 0$$

Obviously, the zero eigenvalue is the smallest one of $\tilde{L}$, due to its nonnegative definite nature. As it was stated in Section 6.7, if the graph is connected then there is one eigenvector associated with the zero eigenvalue. This is an assumption which is adopted here.

We have by now all the ingredients to perform the final optimization. Observe that the ratio in (15.16) is the celebrated Rayleigh quotient. Recall from linear algebra, for example, [Golu 89], that

- The smallest value of the quotient, with respect to $\boldsymbol{z}$, is equal to the smallest eigenvalue of $\tilde{L}$ and it occurs for $\boldsymbol{z}$ equal to the eigenvector corresponding to this (smallest) eigenvalue.
- If we constraint the solution to be orthogonal to all eigenvectors associated with the j smallest eigenvalues, $\lambda_0 \leq \lambda_1 \leq \cdots \leq \lambda_{j-1}$, the Rayleigh quotient is minimized by the eigenvector corresponding to the next smallest eigenvalue, λ_j, and the minimum value is equal to λ_j.

Taking into account a) the orthogonality condition in the constraint (15.17) and b) the fact that $D^{1/2}\mathbf{1}$ is the eigenvector corresponding to the smallest eigenvalue $\lambda_0 = 0$, we end up that:

The optimal solution vector $\boldsymbol{z}$ minimizing the Rayleigh quotient in (15.16), subject to the constraint (15.17), is the eigenvector corresponding to the second smallest eigenvalue of $\tilde{L}$.

We are now ready to summarize the steps for our spectral clustering algorithm.

1. Given a set of points, $\boldsymbol{x}_1, \boldsymbol{x}_2, \ldots, \boldsymbol{x}_N$, set up the weighted graph $G(V, E)$. Form the proximity matrix W by adopting a similarity rule.
2. Form the matrices $D, L = D - W$ and $\tilde{L}$. Perform the eigenanalysis

$$\tilde{L}\boldsymbol{z} = \lambda \boldsymbol{z}$$

   of the normalized Laplacian matrix $\tilde{L}$. Compute the eigenvector $\boldsymbol{z}_1$ corresponding to the second smallest eigenvalue λ_1. Compute the vector

$$\boldsymbol{y} = D^{-1/2}\boldsymbol{z}_1$$

3. Discretize the components of $\boldsymbol{y}$ according to a threshold value.

The final step is necessary since the components of the obtained solution are real-valued and our required solution is discrete. To this goal, different techniques can be applied. For example, the threshold can be taken to be equal to zero. Another choice is to adopt the median value of the components of the optimum eigenvector. An alternative approach would be to select the threshold value that results in the minimum *cut* value.

The eigenanalysis of an $N \times N$ matrix, using a general purpose solver, amounts to $O(N^3)$ operations. Thus, for large number of data points, this may be prohibitive for some applications. However, for most of the practical applications, the resulting graph is only locally connected, and the proximity matrix is *sparse*. Moreover, only the smallest eigenvalues/eigenvectors are required and also the accuracy is not of major issue, since the solution is to be discretized. In such a setting, the efficient Lanczos algorithm (e.g., [Golu 89]) can be mobilized and the computational requirements drop down to approximately $O(N^{3/2})$.

So far, the partition of a data set into two clusters has been considered. If more clusters are expected, the scheme can be used in a hierarchical mode, where, at each step, each one of the resulting clusters is partitioned further into two clusters. This is continued until a prespecified criterion is satisfied. In [Shi 00] it is suggested that the third smallest eigenvalue can be used to sub-partition the first two clusters and so on. However, this procedure tends to become unreliable due to approximation errors.

In our discussion, so far, we focused on a specific clustering criterion, that is, the normalized cut, in order to present the basic philosophy behind the spectral clustering techniques. No doubt, a number of other criteria have been proposed in the related literature. For example, the ratio cut ([Chan 94]) is defined as

$$Rcut(A, B) = \frac{cut(A, B)}{|A|} + \frac{cut(A, B)}{|B|}$$

In [Kann 00] the Cheeger constant is used as the partition criterion, that is,

$$h_G = \frac{cut(A, B)}{\min(V(A), V(B))}$$

For each criterion, a different eigendecomposition problem results, each with its advantages and drawbacks. In [Verm 03], a review and a comparative study of a number of popular spectral clustering algorithms is presented. A comparative and insightful study of a number of spectral clustering algorithms is provided in [Weis 99].

The literature on spectral clustering is large and it is beyond the purpose of this section to cover it in detail. Besides the criteria mentioned before, other approaches to spectral clustering have also been proposed. For example, in [Meil 00] the pairwise similarities are interpreted as edge flows in a Markov random walk leading to a probabilistic interpretation of spectral clustering. In [Xian 08], the issues on how to determine the number of clusters and how to deal with noisy and sparse data are considered. They tackle this problem via a data-driven approach that selects the most relevant eigenvectors, which provide information about the natural grouping of the data. Also, in [Jens 04] a spectral clustering algorithm based on an information theoretic framework is discussed. For more information, besides the references given before, the interested reader may consult, for example, [Chun 97, vonL 07].

Spectral clustering has been used in a number of applications such as image segmentation and motion tracking [Shi 00, Qiu 07], circuit layout [Chan 94], gene expression [Kann 00], machine learning [Ng 01], load balancing [Hend 93].

Remarks

- For those of the readers who have also covered Section 6.7, it is obvious to observe the close resemblance between spectral clustering and the dimensionality reduction methods that preserve locality (Laplacian eigenmaps and LLE). Attempting to preserve neighborhood information, while projecting in the low dimensional subspace, may be interpreted as imposing a "soft" clustering on the data, [Belk 03]. In spectral clustering, one can look at each y_i as the nonlinear mapping onto the real axis of the data point, $\boldsymbol{x}_i$, $i = 1, 2, \ldots, N$, and a hard clustering is obtained after discretization. Moreover, as Eq. (15.12) suggests, the cluster indicators y_i are "forced" to take similar values for closely located points. This is a consequence of the minimization task in Eq. (15.14) and of the fact that far away points result in small or zero values of the weights.
- A very interesting result, that ties the "old" and the "new", is the establishment of the mathematical equivalence between a weighted form of the kernelized k-means objective and a general weighted graph clustering objective. The normalized cut and the ratio cut objectives fall under this category. Thus, a kernelized version of the classical k-means algorithm can be employed to solve the task instead of a matrix eigendecomposition. This may have certain computational advantages, especially for large problems. However, spectral decomposition computes the global optimal, in contrast to the k-means algorithm that may be trapped in local minima. Such issues and a novel algorithm, exploiting this equivalence, are discussed in, for example, [Dill 07, Zass-05].

Example 15.3

In the example shown in Figure 15.9a, two concentric circularly shaped clusters are shown. The first cluster is spread around the circle of radius equal to 3 centered at (0, 0), while the second cluster is spread around the circle of radius equal to 6 and also centered at (0, 0). The spectral clustering algorithm with $\sigma^2 = 2$ and $\varepsilon = 2$, using the normalized cut criterion, is applied on the previous data set and the results are shown in Figure 15.9b. Clearly, the algorithm identifies successfully the two clusters. On the contrary, the k-means algorithm fails to identify the clusters successfully, as shown in Figure 15.9c. Recall that the k-means algorithm has the tendency of recovering compact clusters.

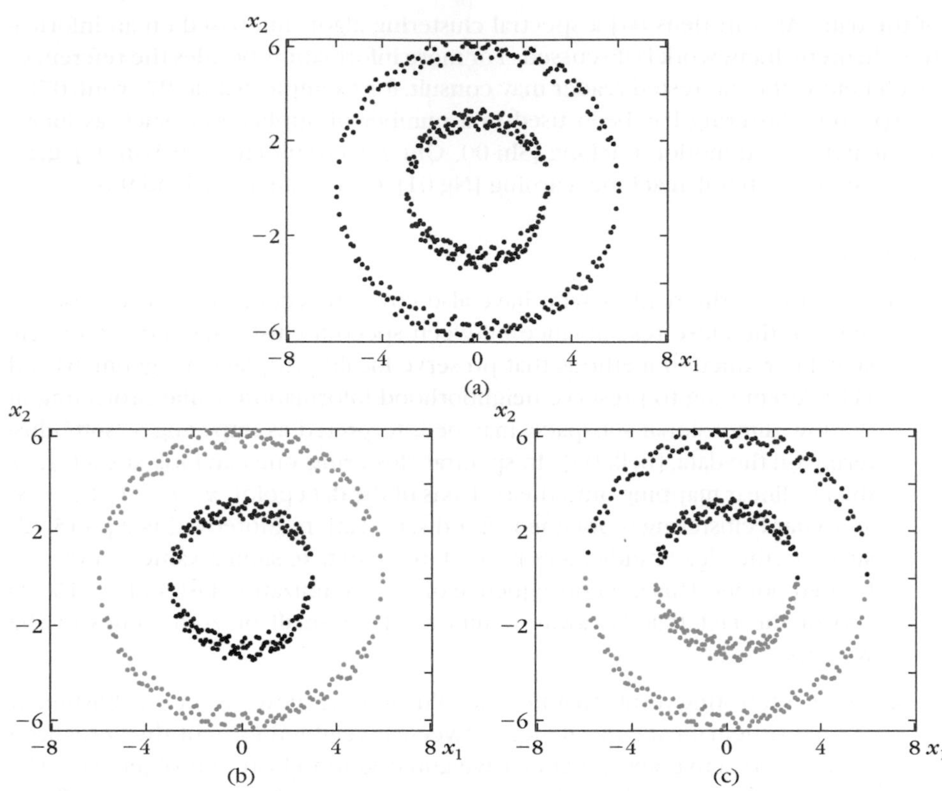

FIGURE 15.9

(a) The data set. (b) The two clusters (denoted by different colors) obtained by the spectral clustering algorithm. (c) The two clusters obtained by the k-means algorithm.

15.3 COMPETITIVE LEARNING ALGORITHMS

These algorithms employ a set of representatives w_j, $j = 1, \ldots, m$.[3] *Their goal is to move each of them to regions of the vector space that are "dense" in vectors of X.* The representatives are updated each time a new vector $x \in X$ is presented to the algorithm. Algorithms of this type are called *pattern mode algorithms*. This is a point of differentiation from the hard clustering algorithms discussed in Chapter 14. There the updating of the representatives takes place after the presentation of *all* the vectors of X to the algorithm. Algorithms of the latter kind are also called *batch mode algorithms*. It must be emphasized that competitive learning algorithms do not necessarily stem from the optimization of a cost function.

[3] We use w_j here instead of θ_j to comply with the notation usually adopted for this type of schemes.

The general idea is very simple. When a vector $\boldsymbol{x}$ is presented to the algorithm, all representatives *compete* with each other. The winner of this competition is the representative that lies closer (according to some distance measure) to $\boldsymbol{x}$. Then, the winner is updated so as to move toward $\boldsymbol{x}$, while the losers either remain unchanged or are updated toward $\boldsymbol{x}$ but at a much slower rate.

Although, in most of the cases, $\boldsymbol{w}_j$'s are points in the l-dimensional space, other choices are also possible. For example, the representatives may be hyperplanes [Likh 97]. In the sequel, we consider only the case in which $\boldsymbol{w}_j$'s are points in the l-dimensional space.

Let t be the current iteration and t_{max} the maximum allowable number of iterations. Also, let m be the current number, m_{init} the initial number, and m_{max} the maximum allowable number of clusters (representatives). Then, a generalized competitive learning scheme (GCLS) may be stated as follows.

Generalized Competitive Learning Scheme (GCLS)

- $t = 0$
- $m = m_{init}$
- (A) Initialize any other necessary parameters (depending on the specific scheme).
- Repeat
  - $t = t + 1$
  - Present the next $\boldsymbol{x} \in X$ to the algorithm.
  - (B) Determine the winning representative $\boldsymbol{w}_j$.
  - (C) If (($\boldsymbol{x}$ is not "similar" to $\boldsymbol{w}_j$) OR (other condition)) AND ($m < m_{max}$) then
    - $m = m + 1$
    - $\boldsymbol{w}_m = \boldsymbol{x}$
  - Else
    - (D) *Parameter updating*

$$\boldsymbol{w}_j(t) = \begin{cases} \boldsymbol{w}_j(t-1) + \eta h(\boldsymbol{x}, \boldsymbol{w}_j(t-1)), & \text{if } \boldsymbol{w}_j \text{ is the winner} \\ \boldsymbol{w}_j(t-1) + \eta' h(\boldsymbol{x}, \boldsymbol{w}_j(t-1)), & \text{otherwise} \end{cases} \tag{15.18}$$

  - End
- (E) Until (convergence has occurred) OR ($t > t_{max}$)
- Identify the clusters represented by $\boldsymbol{w}_j$'s, by assigning each vector, $\boldsymbol{x} \in X$, to the cluster that corresponds to the representative closest to $\boldsymbol{x}$.

The function $h(\boldsymbol{x}, \boldsymbol{w}_i)$ is an appropriately defined function. Also, η and η' are the *learning rates* controlling the update of the winner and the losers, respectively.

The parameter η' may be different for different losers. The similarity between a vector $\boldsymbol{x}$ and a representative $\boldsymbol{w}_j$ may be characterized by a threshold of similarity Θ; that is, if $d(\boldsymbol{x}, \boldsymbol{w}_j) > \Theta$, for some distance measure, $\boldsymbol{x}$ and $\boldsymbol{w}_j$ are considered as dissimilar and $\boldsymbol{w}_j$ cannot be used to represent $\boldsymbol{x}$ accurately. It is clear that improper choice of the value of Θ may lead to misleading results.

Termination of the algorithm is achieved via our familiar criterion $\|\boldsymbol{W}(t) - \boldsymbol{W}(t-1)\| < \varepsilon$, where $\boldsymbol{W} = [\boldsymbol{w}_1^T, \ldots, \boldsymbol{w}_m^T]^T$.

With appropriate choices of the parts (A), (B), (C), and (D), most of the competitive learning algorithms may be viewed as special cases of the GCLS. In the sequel, unless otherwise stated, we use the Euclidean distance, although other distances may also be used.

15.3.1 Basic Competitive Learning Algorithm

In this algorithm $m = m_{\text{init}} = m_{\max}$; that is, the number of representatives is constant. Thus, condition (C) is never satisfied. Also, no other parameters are necessary, so part (A) is omitted. The determination of the winning representative (part (B)) is carried out using the following rule.

- The representative $\boldsymbol{w}_j$ is the winner on $\boldsymbol{x}$ if

$$d(\boldsymbol{x}, \boldsymbol{w}_j) = \min_{k=1,\ldots,m} d(\boldsymbol{x}, \boldsymbol{w}_k)$$

Besides the Euclidean distance, other distances may also be used, depending on the application at hand. For example, in [Ahal 90], the Itakura-Saito distortion is proposed when dealing with speech coding applications in the clustering framework. Moreover, similarity measures may also be used (see, e.g., [Fu 93]). In this case, the min operator in the preceding relation is replaced by the max operator. Finally, the updating of the representatives (part (D)) is carried out as follows:

$$\boldsymbol{w}_j(t) = \begin{cases} \boldsymbol{w}_j(t-1) + \eta(\boldsymbol{x} - \boldsymbol{w}_j(t-1)), & \text{if } \boldsymbol{w}_j \text{ is the winner} \\ \boldsymbol{w}_j(t-1), & \text{otherwise} \end{cases} \tag{15.19}$$

where η is the learning rate and takes values in $[0, 1]$. According to this algorithm, the losers remain unchanged. On the other hand, the winner $\boldsymbol{w}_j(t)$ moves toward $\boldsymbol{x}$. The size of the movement depends on η. In the extreme case where $\eta = 0$, no updating takes place. On the other hand, if $\eta = 1$, the winning representative is placed on $\boldsymbol{x}$. For all other choices of η, the new value of the winner lies in the line segment formed by $\boldsymbol{w}_j(t-1)$ and $\boldsymbol{x}$.

It is clear that this algorithm requires an accurate determination of the number of representatives; that is, knowledge of the number of clusters is required. Another related problem that may arise is associated with the initialization of $\boldsymbol{w}_j$'s. If a representative is initialized far away from its closest vector in X,[4] it will never win.

[4] More specifically, if it lies far away from the convex hull defined by the vectors of X.

Thus, the vectors of X will be represented by the remaining representatives. An easy way to avoid this situation is to initialize all representatives using vectors of X.

In the special case in which the vectors are always presented in the same order, that is, $x_1, x_2, \ldots, x_N, x_1, x_2, \ldots, x_N, \ldots$, and under the assumption that after an iteration t_0 each representative wins on the same vectors, which is reasonable at least for the case where well-separated clusters are formed by the vectors of X, it can be shown that each representative converges to a weighted mean of the vectors it represents [Kout 95].

This algorithm has also been studied for a variable learning rate (e.g., [Likh 97]). Typical constraints for $\eta(t)$ in this case are:

- $\eta(t)$ is a positive decreasing sequence and $\eta(t) \to 0$.
- $\sum_{t=0}^{\infty} \eta(t) = \infty$.
- $\sum_{t=0}^{\infty} \eta^r(t) < +\infty$, for $r > 1$.

Note that these constraints are very similar to those required by the Robbins–Monro schemes, discussed in Section 3.4.2. This is not a coincidence. Let us consider for example the trivial case of a single representative ($m = 1$). If $\eta = \eta(t)$, the updating equation may be seen as the Robbins–Monro iteration for solving the problem

$$E[h(x, w)] = 0$$

where $h(x, w) = x - w$.

Finally, competitive learning algorithms for binary-valued vectors are discussed in [Rume 86, Mals 73].

15.3.2 Leaky Learning Algorithm

This algorithm is the same as the basic competitive learning algorithm except for the updating equation of the representatives, which is

$$w_j(t) = \begin{cases} w_j(t-1) + \eta_w(x - w_j(t-1)), & \text{if } w_j \text{ is the winner} \\ w_j(t-1) + \eta_l(x - w_j(t-1)), & \text{if } w_j \text{ is a loser} \end{cases} \tag{15.20}$$

where η_w and η_l are the learning rates in [0, 1] and $\eta_w \gg \eta_l$. The basic competitive learning scheme may be viewed as a special case of the leaky learning scheme, for $\eta_l = 0$. In the general case where η_w and η_l are both positive, all representatives move toward x. However, the losers move toward x at a much slower rate than the winner.

This algorithm does not suffer from the problem of poor initialization of the representatives. This is because the second branch of (15.20) ensures that even if some representatives are initialized away from their closest vectors of X, they will eventually come closer to the region where the vectors of X are located.

An algorithm in the same spirit is the *neural-gas* algorithm. However, in this case $n_w = \varepsilon$ and $\eta_l = \varepsilon g(k_j(x, w_j(t-1)))$, where $\varepsilon \in [0, 1]$ is the step size of the updating, $k_j(x, w_j(t-1))$ is the number of representatives that lie closer to x than

$w_j(t-1)$ and $g(\cdot)$ is a function that takes the value 1 for $k_j(x, w_j(t-1)) = 0$ and decays to zero as $k_j(x, w_j(t-1))$ increases. It is worth noting that this algorithm results from the optimization of a cost function via a gradient descent technique ([Mart 93]).

15.3.3 Conscientious Competitive Learning Algorithms

Another way to utilize the representative power of w_j's is to discourage a representative from winning if it has won many times in the past. This is achieved by assigning a "conscience" to each representative. Several models of conscience have been proposed in the literature (e.g., [Gros 76a, Gros 76b, Gros 87, Hech 88, Chen 94, Uchi 94]).

Perhaps the simplest way to implement this idea is to equip each representative, $w_j, j = 1, \ldots, m$, with a counter f_j, that counts the times that w_j wins. One way to proceed is the following [Ahal 90]. At the initialization stage (part (A)) we set $f_j = 1, j = 1, \ldots, m$. We define

$$d^*(x, w_j) = d(x, w_j) f_j$$

and part (B) becomes the following:

- The representative w_j is the winner on x if

$$d^*(x, w_j) = \min_{k=1,\ldots,m} d^*(x, w_k)$$

- $f_j(t) = f_j(t-1) + 1$.

This setup ensures that the distance is penalized to discourage representatives that have won many times. The parts (C) and (D) are the same as their corresponding parts of the basic competitive learning algorithm, and also $m = m_{\text{init}} = m_{\max}$.

An alternative way is to utilize f_j via the equation [Chou 97]

$$f_j = f_j + d(x, w_j)$$

Other schemes of this kind may be found in [Ueda 94, Zhu 94, Butl 96, Chou 97].

A different approach is followed in [Chen 94]. Here, in part (A), all f_j's are initialized to $1/m$. We define $d^*(x, w_j)$ as

$$d^*(x, w_j) = d(x, w_j) + \gamma(f_j - 1/m)$$

where γ is a conscience parameter. Letting $z_j(x)$ be 1 if w_j wins on x and 0 otherwise, part (B) of the algorithm becomes

- The representative w_j is the winner on x if

$$d^*(x, w_j) = \min_{k=1,\ldots,m} d^*(x, w_k)$$

- $f_j(t) = f_j(t-1) + \varepsilon(z_j(x) - f_j(t-1))$

where $0 < \varepsilon \ll 1$. As we can easily observe, f_j increases for the winner and, in contrast to the previous case, decreases for the losers. Guidelines for the choice of the appropriate values of ε and γ, as well as a version of the algorithm where the value of γ is adaptively adjusted, are discussed in [Chen 94].

15.3.4 Competitive Learning–Like Algorithms Associated with Cost Functions

The basic philosophy behind the competitive learning schemes is to move representatives toward their closest points. If we want to express this in terms of a cost function, then a possible way is the following. Let us consider the cost function

$$J(\boldsymbol{W}) = \frac{1}{2m}\sum_{i=1}^{N}\sum_{j=1}^{m} z_j(\boldsymbol{x}_i)\|\boldsymbol{x}_i - \boldsymbol{w}_j\|^2 \tag{15.21}$$

where $\boldsymbol{W} = [\boldsymbol{w}_1^T, \ldots, \boldsymbol{w}_m^T]^T$ and $z_j(\boldsymbol{x}) = 1$, if $\boldsymbol{w}_j$ lies closer to $\boldsymbol{x}$, and 0 otherwise. This is basically the cost function associated with the isodata algorithm (Chapter 14), and it is not differentiable, due to the presence of $z_j(\boldsymbol{x})$. One way to overcome the problem of differentiability of $J(\boldsymbol{W})$ is to smooth $z_j(\boldsymbol{x})$. This implies that the concept of the competition is abandoned. Instead, each representative is updated in proportion to its distance from $\boldsymbol{x}$.

One way to smooth $z_j(\boldsymbol{x})$ is to redefine it as

$$z_j(\boldsymbol{x}) = \frac{\|\boldsymbol{x} - \boldsymbol{w}_j\|^{-2}}{\sum_{r=1}^{m}\|\boldsymbol{x} - \boldsymbol{w}_r\|^{-2}}, \quad j = 1, \ldots, m \tag{15.22}$$

where $\|\cdot\|$ is the Euclidean distance between two vectors. Clearly, $z_j(\boldsymbol{x})$ is no longer strictly equal to 0 or 1 but lies in $[0, 1]$. More specifically, the closer the $\boldsymbol{w}_j$ to $\boldsymbol{x}$, the larger the $z_j(\boldsymbol{x})$.

Using the preceding definition and after some rearrangements, $J(\boldsymbol{W})$ becomes

$$J(\boldsymbol{W}) = \frac{1}{2}\sum_{i=1}^{N}\left(\sum_{j=1}^{m}\|\boldsymbol{x}_i - \boldsymbol{w}_j\|^{-2}\right)^{-1} \tag{15.23}$$

The gradient of $J(\boldsymbol{W})$ with respect to $\boldsymbol{w}_k$, $\partial J/\partial \boldsymbol{w}_k$, after some algebra, becomes

$$\frac{\partial J}{\partial \boldsymbol{w}_k} = -\sum_{i=1}^{N} z_k^2(\boldsymbol{x}_i)(\boldsymbol{x}_i - \boldsymbol{w}_k) \tag{15.24}$$

In the context of the gradient descent algorithms and using the "instantaneous" value of the gradient, as it was the case with the backpropagation algorithm, the following updating algorithm results.

$$\boldsymbol{w}_k(t) = \boldsymbol{w}_k(t-1) + \eta(t) z_k^2(\boldsymbol{x})(\boldsymbol{x} - \boldsymbol{w}_k(t-1)), \quad k = 1, \ldots, m \tag{15.25}$$

where $\boldsymbol{x}$ is the vector currently presented to the algorithm.

Notice that in this scheme *all* representatives are updated in proportion to their distance from $\boldsymbol{x}$. Thus, by smoothing $z_k(\boldsymbol{x})$, we end up with algorithms that are competitive in a wider sense, for which general tools may apply for the establishment of their convergence properties.

Alternative choices of $z_j(\boldsymbol{x})$ and $J(W)$, leading to more general algorithmic schemes, are given in [Masu 93].

15.3.5 Self-Organizing Maps

So far, we have implicitly assumed that the representatives, $\boldsymbol{w} \in \mathcal{R}^l$, are not interrelated. We will now remove this assumption. Furthermore, the representatives will be "forced" to be *topologically ordered* in the one-dimensional or two-dimensional space. In other words, each $\boldsymbol{w}$ is parameterized in terms of an integer pair (for the two-dimensional case) (i, j), $i = 1, 2, \ldots, I$, $j = 1, 2, \ldots, J$, where IJ is the number of representatives. In this way, a *grid* of nodes is defined. The goal in this section is to place the representatives so that data points that lie close in the $\mathcal{R}^l$ space to be represented by representatives that lie close in the grid. Alternatively, the one-dimensional or two-dimensional grid can be seen as a *map* where we require that different "dense" in data regions, in the data space, are mapped to different regions in the map.

The concept of the topological ordering implies the adoption of a *topological distance* between two representatives. For example, if (i_1, j_1) and (i_2, j_2) denote the positions of two representatives in a (two-dimensional) grid of nodes, their topological distance may be defined as the l_1 distance (Section 11.2.2) between the two integer pairs. In this respect, in Figure 15.10a, $\boldsymbol{w}_r$ and $\boldsymbol{w}_q$ are topologically close to each other in the two-dimensional grid, while $\boldsymbol{w}_s$ is far from both of them. Figure 15.10b shows the case of an one-dimensional grid. In the sequel, we define a *topological neighborhood* Q_j for each representative $\boldsymbol{w}_j$, which consists of representatives that are close to $\boldsymbol{w}_j$ in terms of the topological distance. Typical topological neighborhood shapes are shown in Figures 15.11a and 15.11b for the two-dimensional and the one-dimensional cases, respectively. However, for the two-dimensional case, other shapes, such as hexagonal or rhombic, may also be employed.

As it was the case with the algorithms discussed in previous sections, at each iteration step, t, a single data vector, $\boldsymbol{x}$, is presented to the algorithm. However now, when a representative $\boldsymbol{w}_j$ wins on a given data vector $\boldsymbol{x}$, *all* the representatives *inside* its respective neighborhood $Q_j(t)$ are also updated (move toward $\boldsymbol{x}$). Note that the neighborhood is allowed to change with the iteration step. After a random initialization of the representatives, it is expected that, at the beginning of the training, topologically close representatives may win on data points that may not, necessarily, lie close in the data space. However, as the training evolves, this phenomenon decays and *after convergence, topologically neighboring representatives on the grid win on vectors lying in the same region in the data space. In contrast, topologically distant representatives win on data vectors that lie in*

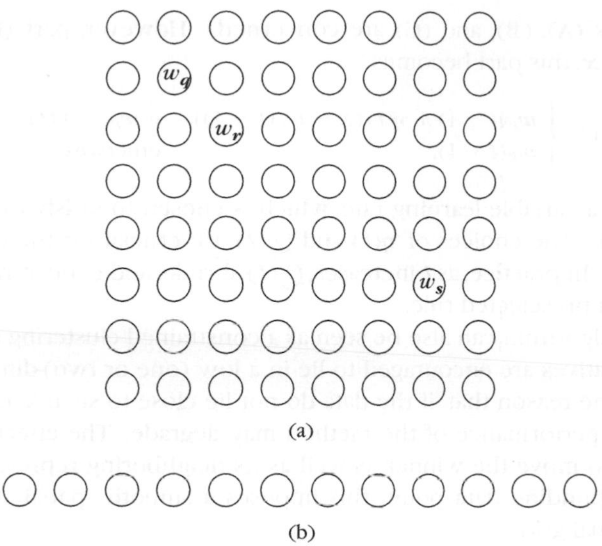

FIGURE 15.10

(a) An 8×8 two-dimensional grid of topologically ordered representatives. Adopting l_1 as the topological distance between two representatives, $\boldsymbol{w}_r$ and $\boldsymbol{w}_q$ are topologically close to each other, while $\boldsymbol{w}_s$ is distant from both of them. (b) A 1×12 one-dimensional grid.

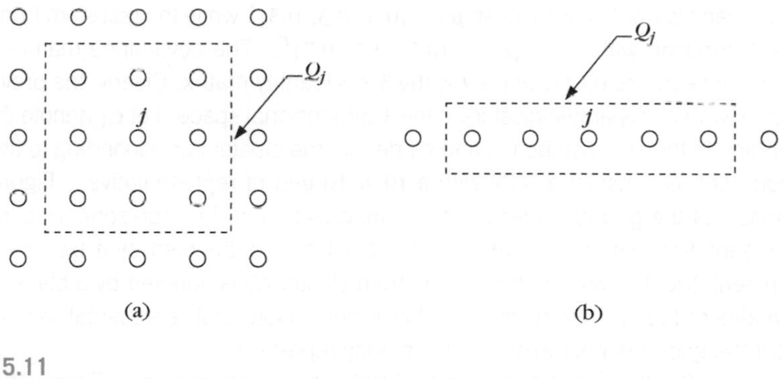

FIGURE 15.11

(a) A square-shaped 3×3 neighborhood. (b) A squared-shaped 5×1 neighborhood.

different regions in the data space. **In other words,** ***each dense in data region in the data space is represented by a set of topologically neighboring representatives.***

The previous described approach leads to the celebrated *Kohonen self-organizing mapping (SOM) scheme* [Koho 89, Koho 95, Kask 98, Diam 07] and it has extensively been used for applications such as data visualization. In its simplest form, SOM may be viewed as a special case of the generalized competitive learning scheme (GCLS). Precisely, it is the same as the basic competitive learning algorithm

as far as parts (A), (B), and (C) are concerned. However, part (D) is different. If w_j wins on x, this part becomes

$$w_k(t) = \begin{cases} w_k(t-1) + \eta(t)(x - w_k(t-1)), & \text{if } w_k \in Q_j(t) \\ w_k(t-1), & \text{otherwise} \end{cases} \tag{15.26}$$

where $\eta(t)$ is a variable learning rate, which is chosen to satisfy the conditions in Section 15.3.1. The choices of $\eta(t)$ and $Q_j(t)$ are crucial for the convergence of the algorithm. In practice, as t increases, $Q_j(t)$ shrinks and concentrates around w_j, according to a preselected rule.

The SOM algorithm can also be seen as a constrained clustering scheme, where the representatives are encouraged to lie in a low (one or two)-dimensional manifold. This is the reason that, if the data do not lie close to such a low dimensional manifold, the performance of the method may degrade. The effect of the update in (15.26) is to move the winner as well as its neighboring representatives closer to the corresponding data point; this imposes a smooth spatial structure in the low-dimensional grid.

A suboptimal method for selecting the winner representative that may lead to computational savings is presented in [Vish 00]. Kernelized versions of SOM have appeared in [Inok 04, Macd 00].

Example 15.4

Let X be a data set consisting of 200 3-dimensional data points. The first 100 of them stem from a Gaussian distribution with mean $\mu_1 = [0.3, 0.3, 0.3]^T$ while the rest stem from another Gaussian distribution with mean $\mu_2 = [0.7, 0.7, 0.7]^T$. The covariance matrices of both distributions are equal to $0.01I$, where I is the 3×3 identity matrix. Clearly, the previous data vectors form two well separated clusters in the 3-dimensional space. Let C_1 denote the cluster corresponding to the first distribution and C_2 denote the cluster corresponding to the second distribution. Let us consider a SOM with a 10×10 grid of representatives. Figure 15.12a is a "snapsot" of the grid just after a random initialization of the representatives in $[0, 1]^3$. Each representative that wins on vectors from cluster C_1 is denoted by a blue circle, while each representative that wins on the vectors from cluster C_2 is denoted by a black circle. All the rest nodes of the grid are denoted by black dots. Note that representatives are spread throughout the grid, irrespective of the cluster they represent.

The results after the convergence of the SOM scheme are shown in Figure 15.12b; the representatives are concentrated in two distinctly different regions of the grid, depending on the cluster they represent.

15.3.6 Supervised Learning Vector Quantization

A supervised variant of the competitive schemes has been suggested and extensively used in the context of VQ [Koho 89, Kosk 92]. In this case, each cluster is treated as a class and the available vectors have known class labels. In this framework, let m be the number of classes. Supervised VQ *uses a set of m representatives, one*

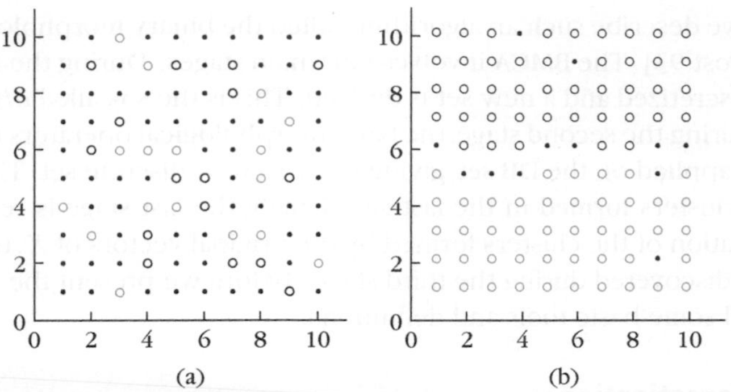

FIGURE 15.12

(a) The state of the map after the first iteration step for the case of the Example 15.4. The representatives of both clusters (black and red circles respectivley) are spread throughout the grid. (b) The state of the grid after the completion of the training. Now the clusters are represented by neighboring representatives which occupy two clearly distinct regions in the grid. Black dots are points where no representatives have been allocated by the algorithm.

for each class, and tries to place them in such a way that each class is "optimally" represented. The simplest version of the supervised VQ (also called LVQ1 [Tsyp 73]) may be derived from generalized competitive learning schemes by keeping parts (A),(B), and (C) the same as in the basic competitive learning scheme and modifying part (D) to

$$w_j(t) = \begin{cases} w_j(t-1) + \eta(t)(x - w_j(t-1)), & \text{if } w_j \text{ correctly wins on } x \\ w_j(t-1) - \eta(t)(x - w_j(t-1)), & \text{if } w_j \text{ wrongly wins on } x \\ w_j(t-1), & \text{otherwise} \end{cases} \quad (15.27)$$

It is clear that the information related to the known class labels determines the direction in which w_j is moved. Specifically, we move w_j (a) toward x if w_j wins and x belongs to the jth class and (b) away from x if w_j wins and x does not belong to the jth class. In addition, all other representatives remain unaltered. Such algorithms have been used in speech recognition and OCR applications.

A variant of this scheme, where more than one representative is used to represent each class, is discussed in [Koho 89].

15.4 BINARY MORPHOLOGY CLUSTERING ALGORITHMS (BMCAs)

Algorithms of this type are suitable for cases in which clusters are not properly represented by a single representative ([Post 93, Mora 00]). The idea here is to map X to a discrete set S that facilitates the clustering procedure and then use the identified clusters in S as a guide for the identification of the clusters in X. In the sequel, we describe such an algorithm, called the binary morphology clustering algorithm [Post 93]. The BMCA involves four main stages. During the first stage the data set is discretized and a new set is derived. This is the so-called *discrete binary (DB) set*. During the second stage, the basic morphological operators (opening and closing) are applied on the DB set, giving rise to a new discrete set. The third stage reveals the clusters formed in the last set. Finally, the last stage is responsible for the identification of the clusters formed by the original vectors of X, using as guide the clusters discovered during the third stage. Before we present the algorithm, let us first recall some basic tools and definitions.

15.4.1 Discretization

During the first step of the discretization stage, we normalize the vectors $\boldsymbol{x} \in X$ so that all their coordinates lie in the range $[0, r-1]$, where r is a user-defined parameter. This is achieved via the following transformation:

$$y_{ij} = \frac{x_{ij} - \min_{q=1,\ldots,N} x_{qj}}{\max_{q=1,\ldots,N} x_{qj} - \min_{q=1,\ldots,N} x_{qj}}(r-1), \quad i = 1,\ldots,N, j = 1,\ldots,l \tag{15.28}$$

where x_{ij} denotes coordinate j of vector i. Let us denote the resulting set by X', that is,

$$X' = \{\boldsymbol{y}_i \in [0, r-1]^l, i = 1,\ldots,N\}$$

Next, we discretize $[0, r-1]^l$ into r^l hypercubes. This is achieved by segmenting the $[0, r-1]$ interval, for each coordinate, into r subintervals (see Figure 15.13). Each hypercube is identified by the coordinates of its lower left corner. The parameter r defines the "resolution" of $[0, r-1]^l$.

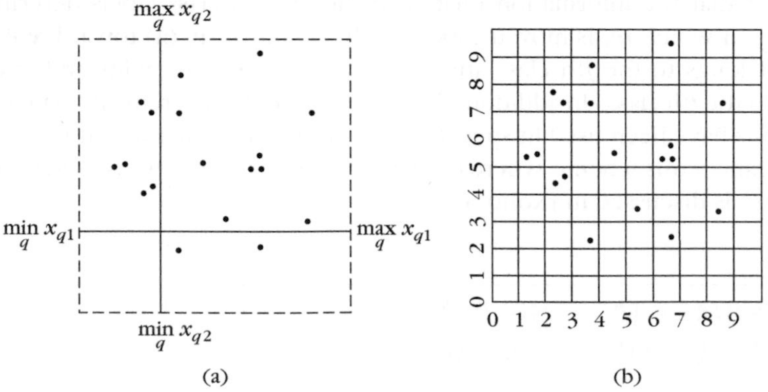

FIGURE 15.13

(a) The original data set X. (b) Normalization of the data set X in $[0, r-1]^l$, with $r = 10$, and discretization. The nonempty hypercubes define the discrete binary set.

In the sequel, we identify the hypercube where each vector $\boldsymbol{y}_i \in X'$ lies. This can be accomplished by simply taking the integer part of each coordinate of $\boldsymbol{y}_i$. The resulting vector, $\boldsymbol{z}_i$, will be the identity label of a hypercube in the defined grid. More specifically,

$$z_{ij} = [y_{ij}], \quad i = 1, \ldots, N, \quad j = 1, \ldots, l$$

where $[x]$ denotes the integer part of x. Let S be the set containing the new vectors $\boldsymbol{z}_i$, after removing all duplicates. Thus, each element of S corresponds to a nonempty hypercube, and S is the discrete binary set.

15.4.2 Morphological Operations

These operations are applied only to sets with discrete-valued vectors. The simplest operations of this kind are *dilation* and *erosion*. Based on these two operations, *opening* and *closing* operations are defined.

Let Y and T be subsets of Z^l, where Z is the set of integers and $\boldsymbol{s}$ a vector in Z^l. The *translation of Y by* $\boldsymbol{s}$ is defined as

$$Y_{\boldsymbol{s}} = \{\boldsymbol{t} \in Z^l : \boldsymbol{t} = \boldsymbol{x} + \boldsymbol{s}, \ \boldsymbol{x} \in Y\} \tag{15.29}$$

Definition 1. *The dilation of Y by T, denoted by $Y \oplus T$, is defined as*

$$Y \oplus T = \{\boldsymbol{e} \in Z^l : \boldsymbol{e} = \boldsymbol{x} + \boldsymbol{s}, \boldsymbol{x} \in Y, \ \boldsymbol{s} \in T\} \tag{15.30}$$

Equivalently, the set $Y \oplus T$ is determined by translating Y by all elements of T and taking the union of the resulting sets [Gonz 93].

Definition 2. *The erosion of Y by T is denoted by $Y \ominus T$ and is defined as*

$$Y \ominus T = \{\boldsymbol{f} \in Z^l : \boldsymbol{x} = \boldsymbol{f} + \boldsymbol{s}, \boldsymbol{x} \in Y, \ \forall \, \boldsymbol{s} \in T\} \tag{15.31}$$

Equivalently, the set $Y \ominus T$ is determined by translating Y by all elements of T and taking the intersection of the resulting sets [Gonz 93].

In both of these cases, T is called the *structuring element*. Usually, it has a hypercubical shape (see Figure 15.14) but other choices, such as hyperspherical shape, are also possible.

Example 15.5

Let us consider a two-dimensional normal density function with mean $\boldsymbol{\mu} = [0, 0]^T$ and covariance matrix $\Sigma = 3I$, where I is the 2×2 identity matrix. Let X be a set containing 200 vectors stemming from this distribution (Figure 15.15a). We apply the discretization process on X, with $r = 20$, in order to obtain the corresponding discrete binary set, which is denoted by Y (Figure 15.15b). Let T be the 3×3 structuring element shown in Figure 15.14a, which consists of the points

$$\begin{aligned} T &= \{\boldsymbol{t}_1, \boldsymbol{t}_2, \ldots, \boldsymbol{t}_9\} \\ &= \{(-1,-1), (-1,0), (-1,1), (0,-1), (0,0), (0,1), (1,-1), (1,0), (1,1)\} \end{aligned}$$

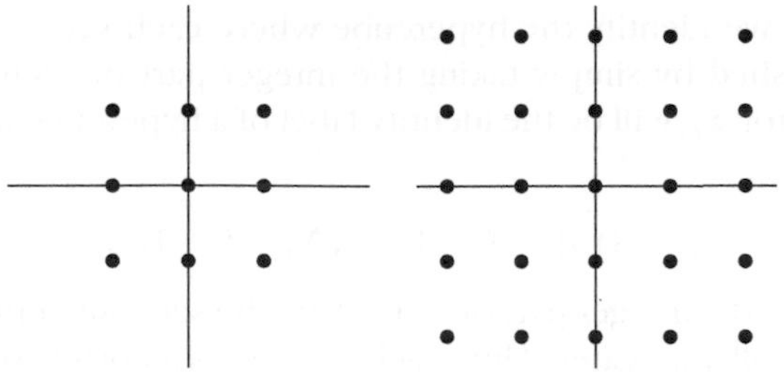

FIGURE 15.14

(a) A squared 3×3 structuring element. (b) A squared 5×5 structuring element.

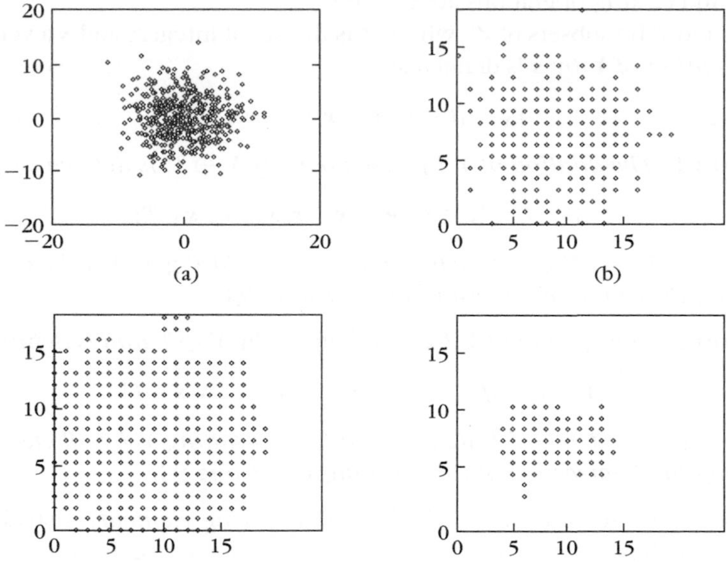

FIGURE 15.15

(a) The original data set X given in Example 15.5. (b) The discrete binary set Y derived from X. (c) The set Y dilated by T. (d) The set Y eroded by T.

In order to derive the dilation of Y by T, we compute the sets Y_i, $i = 1, \ldots, 9$, produced by the translation of Y by each element t_i of T, $i = 1, \ldots, 9$, and in the sequel, we take the union of all Y_i's. This is the dilation of Y by T (Figure 15.15c). The erosion of Y by T is computed by taking the intersection of all Y_i's defined above. The result of this operation is shown in Figure 15.15d.

Opening and *closing* are two additional basic operations that are defined in terms of the dilation and the erosion.

Definition 3. *The opening of Y by T is denoted by Y_T and is defined as*

$$Y_T = (Y \ominus T) \oplus T \tag{15.32}$$

that is, the opening of Y by T is the erosion of Y by T followed by the dilation of the resulting $Y \ominus T$ by T.

Definition 4. *The closing of Y by T is denoted by Y^T and is defined as*

$$Y^T = (Y \oplus T) \ominus T \tag{15.33}$$

that is, the closing of Y by T is the dilation of Y by T followed by the erosion of the resulting $Y \oplus T$ by T.

In general, Y is different from Y_T and Y^T. Note that the opening operation smooths out the boundary of Y by discarding irrelevant details of it. On the other hand, the closing operation fills the gaps in the set Y. These observations show that *opening and closing tend to produce new sets with simpler shapes than the original ones.* As pointed out in [Post 93], *"Opening and closing seem to be very effective to eliminate isolated groups of set points and holes, provided that these details do not exceed the size of the structuring element."* The following example shows how the opening and closing operations work.

Example 15.6

Let us consider the discrete binary set Y (see Figure 15.15b) and the structuring element T of Example 15.5. We derive first the opening of Y by T. The result is shown in Figure 15.16a. As one can observe, the resulting set retains the basic shape of Y, while irrelevant details

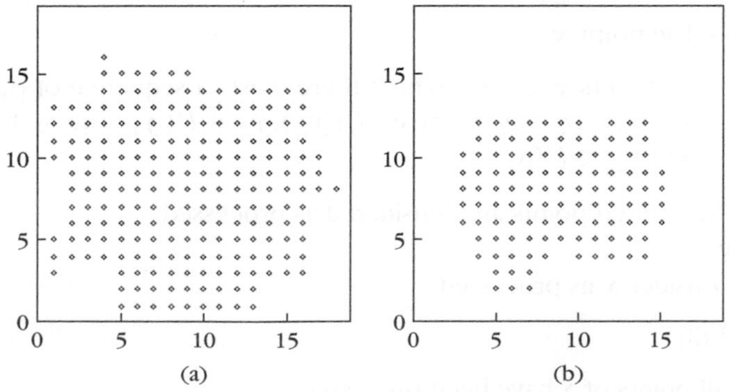

FIGURE 15.16

(a) The set Y given in Example 15.5, opened by T. (b) The set Y given in Example 15.5, closed by T.

of the boundary of Y have been discarded. The action of closing Y by T is shown in Figure 15.16b.

The above arguments indicate that the structuring element T plays an important role in the outcome of the above operations. Unfortunately, there are no general guidelines for choosing the appropriate T.

15.4.3 Determination of the Clusters in a Discrete Binary Set

We begin with a description of a rather simple algorithm suitable for clusters formed by the points of a discrete-valued data set $S \subset \{0, 1, \ldots, r-1\}^l$. Let us first define the neighborhood, $V(\boldsymbol{x})$, of a point $\boldsymbol{x} \in S$ as

$$V(\boldsymbol{x}) = \{\boldsymbol{y} \in S - \{\boldsymbol{x}\}: d(\boldsymbol{x}, \boldsymbol{y}) \leq d_q\}$$

where d may be any distance measure between two points (see Chapter 11) and d_q is a distance threshold. Also, let θ be a threshold of the density of the neighborhood $V(\boldsymbol{x})$ of a point $\boldsymbol{x}$. That is, if $V(\boldsymbol{x})$ contains at least θ points of S, it is considered "dense." These are user-defined parameters.

Let $U(\boldsymbol{x})$ be the immediate neighborhood of $\boldsymbol{x}$, that is, the set that contains all points lying at a (Euclidean) distance less than or equal to $\sqrt{l}$ from $\boldsymbol{x}$.

Cluster Detection Algorithm for Discrete-Valued Sets (CDADV)

- Initially no vector is considered as processed.
- Repeat
  - Choose a nonprocessed point $\boldsymbol{x}$ of S.
  - Determine the neighborhood $V(\boldsymbol{x})$.
  - If $V(\boldsymbol{x})$ contains at least θ points then
    - Create a new cluster that includes:
      - The point $\boldsymbol{x}$
      - All points $\boldsymbol{y} \in S$ for which there exists a sequence of points $\boldsymbol{y}_{j_s} \in S$, $s = 1, \ldots, q_y$, such that $\boldsymbol{y} \in U(\boldsymbol{y}_{j_1}), \boldsymbol{y}_{j_s} \in U(\boldsymbol{y}_{j_{s+1}}), s = 1, \ldots, q_y - 1$, and $\boldsymbol{y}_{j_{q_y}} \in U(\boldsymbol{x})$.
    - The defined points are considered as processed.
  - Else
    - Consider $\boldsymbol{x}$ as processed
  - End {if}
- Until all points of S have been processed.

Example 15.7

Consider the setup of Figure 15.17a. We choose $d_q = \sqrt{2}$ and $\theta = 1$ (in this case $V(\boldsymbol{x}) \equiv U(\boldsymbol{x})$). Also, the sides of the squares depicted in Figure 15.17a are of unit length. The CDADV

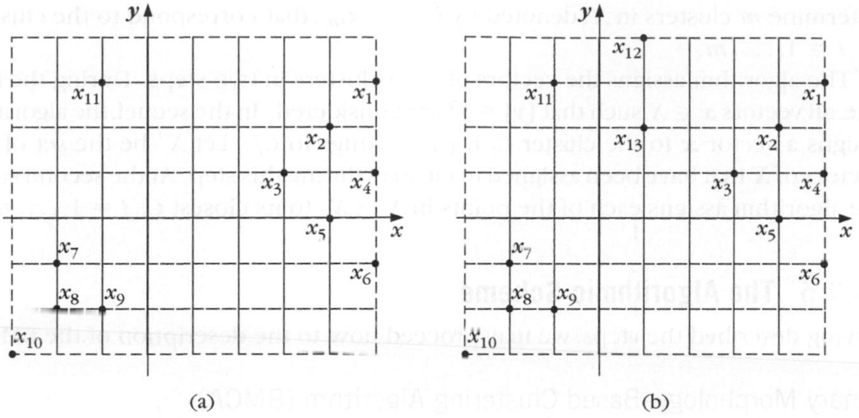

FIGURE 15.17

(a) The setup of Example 15.7. (b) A data set containing outliers.

considers first $\boldsymbol{x}_1$. Since it is "dense" (i.e., its neighborhood contains at least one point of S apart from $\boldsymbol{x}_1$), a new cluster is created. $\boldsymbol{x}_2$ also belongs to this cluster because $\boldsymbol{x}_2 \in U(\boldsymbol{x}_1)$. Moreover, $\boldsymbol{x}_3$, $\boldsymbol{x}_4$ belong to this cluster because $\boldsymbol{x}_3, \boldsymbol{x}_4 \in U(\boldsymbol{x}_2)$ and $\boldsymbol{x}_2 \in U(\boldsymbol{x}_1)$. In addition, $\boldsymbol{x}_5$ belongs to this cluster because $\boldsymbol{x}_5 \in U(\boldsymbol{x}_3)$, $\boldsymbol{x}_3 \in U(\boldsymbol{x}_2)$, and $\boldsymbol{x}_2 \in U(\boldsymbol{x}_1)$. Finally, since $\boldsymbol{x}_6 \in U(\boldsymbol{x}_5)$, $\boldsymbol{x}_5 \in U(\boldsymbol{x}_4)$, $\boldsymbol{x}_4 \in U(\boldsymbol{x}_2)$, and $\boldsymbol{x}_2 \in U(\boldsymbol{x}_1)$, $\boldsymbol{x}_6$ also belongs to this cluster. Working similarly, we find that $\boldsymbol{x}_7$, $\boldsymbol{x}_8$, $\boldsymbol{x}_9$, and $\boldsymbol{x}_{10}$ form a second cluster, while no action is taken for $\boldsymbol{x}_{11}$.

In the preceding scheme, all points of S are processed by the algorithm, regardless of their density. In fact, CDADV works well when the points of S form well-separated clusters. However, if this is not the case, for example, when we have a small number of outliers in S lying between its clusters (Figure 15.17b), we may proceed as follows. As it is expected, the neighborhood of an outlier is rather "sparse," and we first define a lower threshold θ_1 ($\leq \theta$) for the density of the neighborhood of a point and we consider the set S' of all points of S whose neighborhoods have density at least θ_1. That is, we exclude the outliers from S'. Then we run the CDADV on S' using θ, and, after its completion, we assign each vector of $S - S'$ to the cluster where its nearest point in S' belongs. Note that the distance function between a point and a cluster that we employ for this stage should not involve cluster representatives, since S is a discrete-valued set.

15.4.4 Assignment of Feature Vectors to Clusters

This subsection deals with the final stage of the algorithmic procedure. Let us recall that S is the discrete-valued set obtained from X, after applying the opening and closing transformations. Let $C'_1, \ldots, C'_m$ be the clusters formed in S, determined by the previously discussed CDADV algorithm. The aim of the current task is to determine m clusters in X, denoted by $C_1, \ldots, C_m$, that correspond to the clusters $C_i', i = 1, \ldots, m$.

The algorithm assigns the vectors of X to clusters in two steps. During the first one, all vectors $\boldsymbol{x} \in X$ such that $[\boldsymbol{y}] \in S$[5] are considered. In the sequel, the algorithm assigns a vector $\boldsymbol{x}$ to the cluster C_i if $[\boldsymbol{y}]$ belongs to C_i'. Let X' be the set of the vectors of X that have been assigned to clusters during this step. At the second stage, the algorithm assigns each of the points in $X - X'$ to its closest $C_j, j = 1, \ldots, m$.

15.4.5 The Algorithmic Scheme

Having described the steps, we may proceed now to the description of the BMCA.

Binary Morphology–Based Clustering Algorithm (BMCA)

- *1st stage.* Discretize the data set X and let S be the resulting discrete binary set.
- *2nd stage.*
  - (a) Apply the opening transformation on S using a preselected structuring element T, to obtain S_T.
  - (b) Apply the closing transformation on S_T using T. Let $S_1 = (S_T)^T$ be the set obtained.
- *3rd stage.* Determine the underlying clusters of S_1 using the CDADV algorithm.
- *4th stage.* Based on the clusters formed in S_1, determine the underlying clusters of X.

It should be noted here that different choices of morphological operators can be used at the second stage of the algorithm. Thus, for example, one may use either the opening or the closing operator or both of them in the reverse order.

BMCA is sensitive to the parameter r and the structuring element T. These parameters may cause overestimation or underestimation of the true number of clusters underlying X. However, it is expected that when X contains clusters, their number remains unchanged for a significant range of values of the parameters involved (a similar situation has been met earlier in Chapter 12). Based on this assumption, we run the first three stages of the algorithm for various values of r and different T (for simplicity we may assume that T has a hypercubic scheme and, thus, its only parameter that is subject to change is the length of its side a). Then we plot the number of the resulting clusters versus r and a and we consider the widest area in the (r, a) plane, for which the number of clusters remains unchanged. The final values for r and a are chosen to be those corresponding to the middle point of the

---

[5] By $[\boldsymbol{y}]$ we denote the l-dimensional vector whose ith coordinate is the integer part of the ith coordinate of the l-dimensional vector $\boldsymbol{y}$.

above area. Using these values, we run the BMCA algorithmic scheme in order to determine the clusters of X.

A major drawback of the procedure is that it requires intensive computations, since many combinations of the values for r and a have to be considered. A way to reduce the required computations is to fix one of the two parameters to a reasonable value (if this is possible) and to apply the procedure only to the other parameter.

An important observation is that when the underlying clusters of X are compact and well separated, algorithms such as the isodata give better results than BMCA (see Problem 15.6). However, the situation is reversed when this is not the case. Let us consider the following example.

Example 15.8

Let X be a data set consisting of 1000 vectors. The first 500 of them, (x_{i_1}, x_{i_2}), are defined as

$$x_{i_1} = (i-1)\frac{2s}{500}$$

$$x_{i_2} = \sqrt{s^2 - x_{i_1}^2} + z_i$$

where $s = 10$ and z_i stems from a Gaussian distribution with zero mean and unit variance, $i = 1, \ldots, 500$. Similarly, the remaining vectors (x_{i_1}, x_{i_2}) are defined as

$$x_{i_1} = b_1 + (i-501)\frac{2s}{500}$$

$$x_{i_2} = b_2 + \sqrt{s^2 - (x_{i_1} - b_1)^2} + z_i'$$

where $b_1 = -10$, $b_2 = 3$, $s = 10$, and z_i' is normally distributed with zero mean and unit variance, $i = 501, \ldots, 1000$. It is not difficult to realize that the first 500 feature vectors spread around the upper half-circle with radius 10 centered at $(0, 0)$. Similarly, the rest of the 500 vectors spread around the lower half-circle with radius 10, centered at the point $(-10, 3)$ (see Figure 15.18a).

Clearly, two clusters are formed in X and each of them cannot be represented satisfactorily by a single point representative. In our simulations we use the 3×3 structuring element defined in Example 15.5 and $r = 25$. Also, the Euclidean distance between two vectors is adopted. The discrete binary set S, resulting from the discretization process, is shown in Figure 15.18b. Figure 15.18c shows the result of the opening of S by T, and Figure 15.18d shows the result of the closing of S_T by T. The two clusters involved in the last set are well separated. Application of the third stage of the BMCA algorithmic scheme reveals these two clusters. Finally, application of the fourth stage of the algorithm determines the clusters formed in X. The results obtained are excellent. Only two of the first 500 vectors were misclassified, and only 1 of the remaining 500 vectors was misclassified. Thus, 99.7% of the vectors of X were correctly classified. In contrast, the results obtained with the isodata algorithm were much inferior to these results.

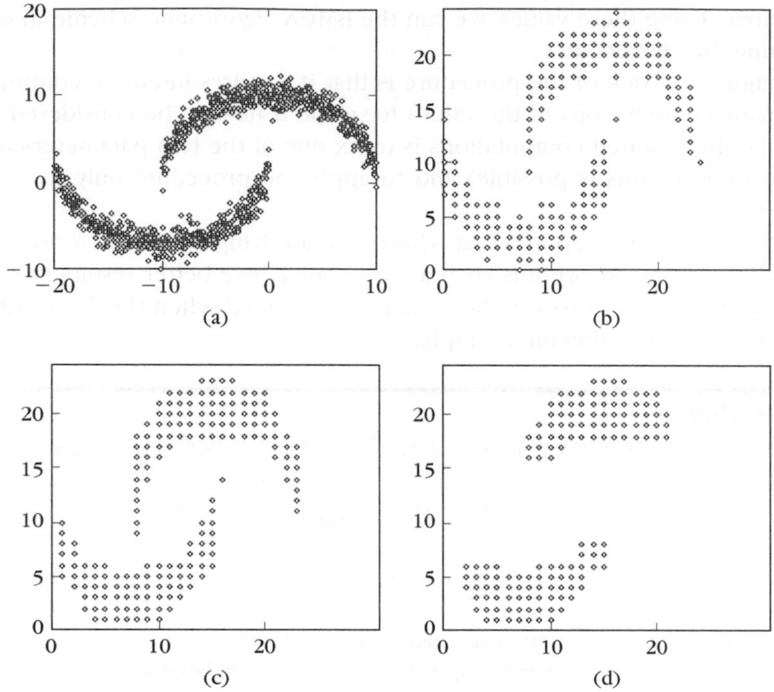

FIGURE 15.18

(a) The original data set X. (b) The set S resulting from the discretization of X. (c) The set resulting from opening of S by T. (d) The set resulting from closing of $S_{T'}$ by T.

15.5 BOUNDARY DETECTION ALGORITHMS

Most of the algorithms discussed so far determine clusters based on either the distance between vectors and clusters or the distance between clusters. In this section, a different rationale is discussed. Clusters are formed via the estimation of the boundary surfaces that separate them [Atiy 90]. This approach is well suited when the underlying clusters are compact. The idea is rather simple. The compact clusters are viewed as dense regions, in the l-dimensional space, separated by regions sparse in data vectors. Therefore, it suffices to begin with an initial estimate of the boundary and move it iteratively to regions that are sparse in vectors.

Let us consider first the case in which two clusters are present. Let $g(\boldsymbol{x}; \boldsymbol{\theta})$ be the function describing the decision boundary between the two clusters, where $\boldsymbol{\theta}$ is the unknown parameter vector describing the surface. If, for a specific $\boldsymbol{x}, g(\boldsymbol{x}; \boldsymbol{\theta}) > 0$, then $\boldsymbol{x}$ belongs to the first cluster, denoted by C^+. Otherwise, $\boldsymbol{x}$ belongs to the second cluster, denoted by C^-. The goal is to determine the unknown parameter vector $\boldsymbol{\theta}$. The situation looks similar to the supervised case where we identify the decision boundary between classes, utilizing the labeling information of the feature vectors. However, no such information is available here. In the present case, *the adjustment of $\boldsymbol{\theta}$ relies exclusively on the distances of the vectors of X from the decision boundary.*

To this end, we define a cost function J, whose maximization will lead to locally optimal values for $\boldsymbol{\theta}$. Let J be defined as

$$J(\boldsymbol{\theta}) = \frac{1}{N}\sum_{i=1}^{N} f^2\left(g(\boldsymbol{x}_i;\boldsymbol{\theta})\right) - \left(\frac{1}{N}\sum_{i=1}^{N} f(g(\boldsymbol{x}_i;\boldsymbol{\theta}))\right)^{2q} \tag{15.34}$$

where q is a constant positive integer and $f(x)$ is a monotonically increasing symmetric squashing function with

$$\lim_{x\to+\infty} f(x) = 1, \qquad \lim_{x\to-\infty} f(x) = -1, \quad \text{and} \quad f(0) = 0 \tag{15.35}$$

A common choice for such a function is the hyperbolic tangent

$$f(x) = \frac{1 - e^{-x}}{1 + e^{-x}}$$

Each of the two terms in (15.34) has a maximum value of 1. Also, $J(\boldsymbol{\theta})$ is nonnegative since

$$J(\boldsymbol{\theta}) \geq \frac{1}{N}\sum_{i=1}^{N}\left(f(g(\boldsymbol{x}_i;\boldsymbol{\theta})) - \frac{1}{N}\sum_{k=1}^{N} f(g(\boldsymbol{x}_k;\boldsymbol{\theta}))\right)^2 \geq 0 \tag{15.36}$$

One can easily observe that the first term in Eq. (15.34) is maximized when all $\boldsymbol{x} \in X$ lie away from the boundary. In this case $f^2(g(\boldsymbol{x}_i;\boldsymbol{\theta})) \to 1$ and the first term attains values close to 1. However, this argument holds also true if *all* vectors of X lie on the same side of the boundary and away from it. The role of the second term is to discourage such trivial solutions. Indeed, in such cases $(\frac{1}{N}\sum_{i=1}^{N} f(g(\boldsymbol{x}_i;\boldsymbol{\theta})))^{2q} \to 1$ and therefore $J(\boldsymbol{\theta})$ approaches zero, its minimum value. The role of q is to control the impact of the second term on the cost function J.

Let us now consider an intermediate case in which the boundary lies between two dense regions. In such cases, the contribution of the second term to J is small. Let us demonstrate it via a simplified example. Assume that the decision surface is a hyperplane H and that at the positive (negative) side of H we have k points lying at distance a $(-a)$ away from it. Then it is not difficult to show that the second term becomes zero while the first equals $f^2(a\|\boldsymbol{\theta}\|)$.

In the sequel we adopt a steepest ascent scheme in order to determine the optimal value for $\boldsymbol{\theta}$. Let θ_j be a coordinate of $\boldsymbol{\theta}$. Then

$$\theta_j(t+1) = \theta_j(t) + \mu \left.\frac{\partial J(\boldsymbol{\theta})}{\partial \theta_j}\right|_{\theta_j=\theta_j(t)}$$

or

$$\theta_j(t+1) = \theta_j(t) + \mu \left(\frac{2}{N} \sum_{i=1}^{N} f(g(\boldsymbol{x}_i;\boldsymbol{\theta})) \frac{\partial f(g(\boldsymbol{x}_i;\boldsymbol{\theta}))}{\partial g(\boldsymbol{x}_i;\boldsymbol{\theta})} \frac{\partial g(\boldsymbol{x}_i;\boldsymbol{\theta})}{\partial \theta_j} \right.$$
$$\left. - \frac{2q}{N} \left(\frac{1}{N} \sum_{i=1}^{N} f(g(\boldsymbol{x}_i;\boldsymbol{\theta})) \right)^{2q-1} \sum_{i=1}^{N} \frac{\partial f(g(\boldsymbol{x}_i;\boldsymbol{\theta}))}{\partial g(\boldsymbol{x}_i;\boldsymbol{\theta})} \frac{\partial g(\boldsymbol{x}_i;\boldsymbol{\theta})}{\partial \theta_j} \right) \Bigg|_{\theta_j = \theta_j(t)} \tag{15.37}$$

For the simple case where $g(\boldsymbol{x};\boldsymbol{\theta})$ is a hyperplane we can write

$$g(\boldsymbol{x};\boldsymbol{\theta}) = \boldsymbol{w}^T \boldsymbol{x} + w_0$$

where $\boldsymbol{\theta} \equiv [\boldsymbol{w}\ w_0]^T$. The updating equation for the parameters follows directly from Eq. (15.37) if we notice that

$$\frac{\partial g(\boldsymbol{x};\boldsymbol{\theta})}{\partial w_j} = \begin{cases} x_j, & j = 1,\ldots,l \\ 1, & j = 0 \end{cases}$$

The resulting algorithm is rather simple in its formulation and may be stated as follows

Boundary Detection Algorithm (BDA)

- Choose an initial value $\boldsymbol{\theta}(0)$ for the parameter vector.
- Compute $J(\boldsymbol{\theta}(0))$ using Eq. (15.34).
- $t = 0$.
- Repeat
  - $t = t + 1$
  - Compute $\boldsymbol{\theta}(t)$ using Eq. (15.37).
  - Compute $J(\boldsymbol{\theta}(t))$ using Eq. (15.34).
- Until $\left| \frac{J(\boldsymbol{\theta}(t+1)) - J(\boldsymbol{\theta}(t))}{J(\boldsymbol{\theta}(t))} \right| < \varepsilon$.

We note here that the coordinates of $\boldsymbol{\theta}$ should not grow in an unbounded way. In the case that $g(\boldsymbol{x};\boldsymbol{\theta})$ corresponds to a hyperplane, a bounded condition for the coordinates of $\boldsymbol{\theta}$ could be $\|\boldsymbol{\theta}\| \le a$, where a is a user-defined parameter.

Let us now consider the case in which more than two clusters underlie X. In this case, we follow a hierarchical procedure. First we divide X into two clusters X^+ and X^- using the boundary detection algorithm. Then, using the algorithm again, we further divide X^+ (X^-) and obtain X^{+-} and X^{++} (X^{--} and X^{-+}). This procedure is then applied iteratively to the resulting clusters until a specific termination criterion is met. This procedure reminds us of the neural network design discussed in Chapter 4 [Atiy 90].

One can easily observe that if no sparse regions exist in a formed cluster C, the division of C will result in a low value of $J(\boldsymbol{\theta})$. Thus, an appropriate criterion that may be used to check whether C contains more clusters is

$$J(\boldsymbol{\theta}) \leq b$$

where b is a user-defined threshold. If this happens, the division of C is not acceptable. Otherwise, the division of C is accepted and we proceed with C^+ and C^-. It is clear that the smaller the value of b, the more clusters will be defined. On the other hand, higher values of b result in the acceptance of fewer clusters with well-defined borders among them. Thus, b should be chosen with care.

A different algorithm, called *OptiGrid*, that separates clusters by applying a suitably defined grid of hyperplanes on the feature space, is discussed in [Hinn 99].

15.6 VALLEY-SEEKING CLUSTERING ALGORITHMS

The method discussed here is in the same spirit as that of the previous section. Let $p(\boldsymbol{x})$ be the density function describing the distribution of the vectors in X. An alternative way to attack the clustering problem is to view the clusters as peaks of $p(\boldsymbol{x})$ separated by valleys. Inspired by this consideration, one can search to identify such valleys, and try to move and place the borders of the clusters in these valleys.

In the sequel, we discuss an iterative and computationally effective algorithm based on this idea [Fuku 90]. Once more, let $V(\boldsymbol{x})$ be the local region of $\boldsymbol{x}$, that is,

$$V(\boldsymbol{x}) = \{\boldsymbol{y} \in X - \{\boldsymbol{x}\}: d(\boldsymbol{x},\boldsymbol{y}) \leq a\} \tag{15.38}$$

where a is a user-defined parameter. The distance $d(\boldsymbol{x},\boldsymbol{y})$ can be taken to be

$$d(\boldsymbol{x},\boldsymbol{y}) = (\boldsymbol{y}-\boldsymbol{x})^T A(\boldsymbol{y}-\boldsymbol{x}) \tag{15.39}$$

where A is a symmetric positive definite matrix. Also, let k_j^i denote the number of vectors of the j cluster that belong to $V(\boldsymbol{x}_i)$, excluding $\boldsymbol{x}_i$. Also, let $c_i \in \{1,\ldots,m\}$ denote the cluster to which $\boldsymbol{x}_i$ belongs, $i = 1,\ldots,N$. Then the algorithm is stated as follows.

Valley-Seeking Algorithm

- Fix a.
- Fix the number of clusters m.
- Define an initial clustering of X.
- Repeat
  - For $i = 1$ to N

- Find j: $k_j^i = \max_{q=1,\ldots,m} k_q^i$ [6]
- Set $c_i = j$

- End {For}
- For $i = 1$ to N
  - Assign x_i to cluster C_{c_i}.
- End {For}

- Until no reclustering of vectors occurs.

Remarks

- Observe that the preceding algorithm has close similarities to the Parzen windows method, for pdf estimation, discussed in Chapter 2. Indeed, all it does is to move a window $d(x, y) \leq a$ at x and to count points from different clusters. Then it assigns x to the cluster with the larger number of points in the window. That is, to the cluster with the highest local pdf density. This is equivalent with moving the boundary away from the "winning" cluster.
- The preceding is a mode seeking algorithm. That is, if more than enough clusters are initially appointed, some of them may be empty.

Example 15.9

(a) Consider Figure 15.19a. X consists of the following 10 vectors: $x_1 = [0, 1]^T$, $x_2 = [1, 0]^T$, $x_3 = [1, 2]^T$, $x_4 = [2, 1]^T$, $x_5 = [1, 1]^T$, $x_6 = [5, 1]^T$, $x_7 = [6, 0]^T$, $x_8 = [6, 2]^T$, $x_9 = [7, 1]^T$, $x_{10} = [6, 1]^T$. The squared Euclidean distance is employed.

The initial clustering consists of two clusters as shown in Figure 15.19a. Also, a decision line, b_1, separating the two clusters is shown (Figure 15.19a). Let $a = 1.415$.[7] After the first iteration of the algorithm, x_4 is assigned to the cluster denoted by "x." This is equivalent to moving the decision curve separating the two clusters to the valley between the two high density areas.

(b) Now consider Figure 15.19b. The set X remains unchanged. Also, the initial clustering remains the same except that x_6 is assigned to the cluster denoted by "x." Three curves, b_1, b_2, and b_3, can now be used to separate the clusters. After the first iteration of the algorithm, x_4 is assigned to the cluster denoted "x" and x_6 is assigned to the cluster denoted by "o." Equivalently, b_1, b_2 can be thought to move to the valley between the two peaks in the place of b_3.

---

[6] If ties occur, that is, more than one maximum are encountered, we choose the one with the smallest index.

[7] This number is slightly greater than $\sqrt{2}$.

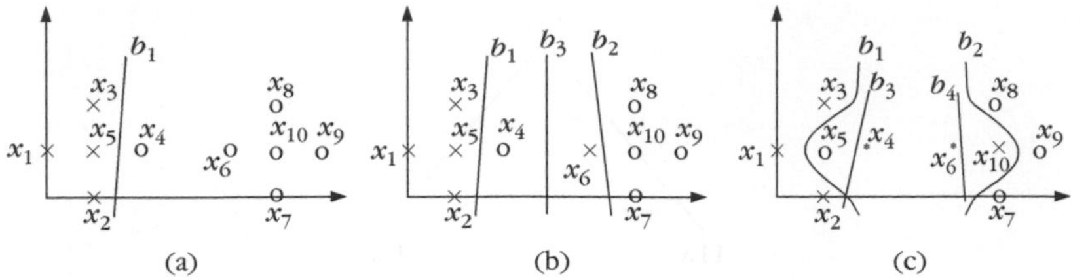

FIGURE 15.19

(a) The setup of Example 15.9(a). (b) The setup of Example 15.9(b). (c) The setup of Example 15.9(c).

(c) Finally, let us consider Figure 15.19c. Again X remains unchanged. However, the initial clustering consists of three clusters, denoted by "x," "o," and "$*$." Also, the initial clustering is the same as that of (b) except that $\boldsymbol{x}_4$ and $\boldsymbol{x}_6$ are assigned to the cluster denoted by "$*$," $\boldsymbol{x}_5$ is assigned to the cluster denoted by "o," and, finally, $\boldsymbol{x}_{10}$ is assigned to the cluster denoted by "x." The decision surfaces consist of the curves b_i, $i = 1, \ldots, 4$. After the first iteration of the algorithm, $\boldsymbol{x}_4$ and $\boldsymbol{x}_5$ are assigned to the cluster denoted by "x." Likewise, $\boldsymbol{x}_6$ and $\boldsymbol{x}_{10}$ are assigned to the cluster denoted by "o." Finally, the rest vectors remain in the clusters where they were initially assigned. The important point here is that, although we initially considered three clusters, the algorithm ends up with two. This is because only two peaks are present. Moreover, in all cases, the decision surface is moved to the valley between the two peaks.

It should be emphasized here that the algorithm is sensitive to the value of the parameter a. Thus, one should run the algorithm several times for different values of a and interpret the results very carefully.

An alternative algorithm based on similar ideas is discussed in [Fuku 90]. It identifies the underlying clusters of X by moving the $\boldsymbol{x}_i \in X$ toward the direction of $\partial p(\boldsymbol{x})/\partial \boldsymbol{x}$, computed at $\boldsymbol{x}_i$, by $\eta \partial p(\boldsymbol{x})/\partial \boldsymbol{x}$, where η is a user-defined parameter. Iterating this procedure, points of the same cluster converge towards the same point in space (a method for the estimation of the gradient of $p(\boldsymbol{x})$ is given in [Fuku 90]). Finally, other related algorithms can be found in [Touz 88, Chow 97].

15.7 CLUSTERING VIA COST OPTIMIZATION (REVISITED)

In this section we present four optimization methods that have been used successfully in many fields of application.

15.7.1 Branch and Bound Clustering Algorithms

As already stated in Chapter 5, *branch and bound* methods compute the globally optimal solution to combinatorial problems, according to a prespecified criterion

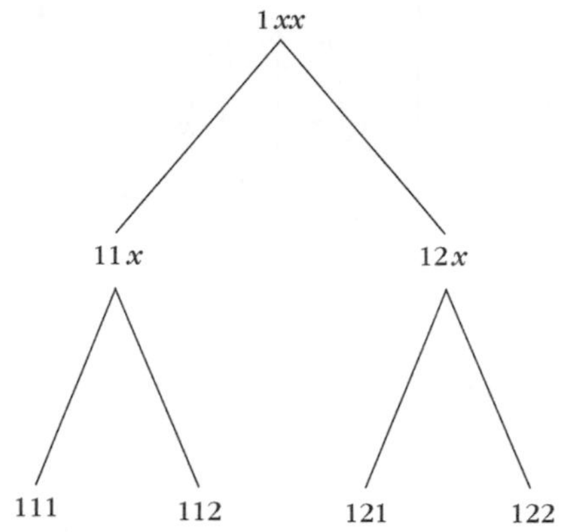

FIGURE 15.20

The classification tree corresponding to the grouping of three vectors in two clusters.

(cost) function J, overcoming the need for exhaustive search.[8] They are applicable to monotonic criteria. That is, if k vectors of X have been assigned to clusters, the assignment of an extra vector to a cluster does not decrease the value of J.

We will first attempt to gain some insight into these methods by considering an example. Let us assume that our goal is to find the best way (with respect to a criterion J) in which three vectors can be clustered in two clusters. To this end, we construct the *classification tree* of Figure 15.20. Each node is characterized by a string of three symbols, namely 1, 2, and x. For example, the string "122" means that the first vector is assigned to cluster 1 while the other two are assigned to cluster 2. Also, the string "1xx" means that the first vector is assigned to cluster 1 while the other two remain unassigned. The first vector is always assigned to the first cluster. Note that each leaf corresponds to an actual clustering and there are as many leaves as the possible clusterings of the three vectors in two clusters. All the other nodes correspond to the so-called *partial clusterings*, that is, to clusterings where not all the vectors of X have been assigned yet to a cluster.

We are now ready to see where the computational saving comes from. Let us assume that at an iteration step of the algorithm, the best computed value for the criterion J is B. Then, if at a node the corresponding value of J is greater than B, *no further search is performed for all subsequent descendants springing from this node.* This is because of the monotonicity of the criterion, which ensures that all descendants will result in values of J no less than B.

More formally, let $\mathcal{C}_r = [c_1, \ldots, c_r]$, $1 \leq r \leq N$, denote a partial clustering, where $c_i \in \{1, 2, \ldots, m\}$, $c_i = j$ if the vector $\boldsymbol{x}_i$ belongs to cluster C_j, and the vectors $\boldsymbol{x}_{r+1}, \ldots, \boldsymbol{x}_N$ are not yet assigned to any cluster.

[8] In the sequel we consider only the minimization problem.

In the sequel, we focus on compact clusters and we give a simple branch and bound algorithm [Koon 75]. We assume that the number of clusters, m, is fixed. The criterion function employed is defined as

$$J(\mathcal{C}_r) = \sum_{i=1}^{r} \|\boldsymbol{x}_i - \boldsymbol{m}_{c_i}(\mathcal{C}_r)\|^2 \tag{15.40}$$

where $\boldsymbol{m}_{c_i}$ is the mean vector of the cluster C_{c_i}, that is,

$$\boldsymbol{m}_j(\mathcal{C}_r) = \frac{1}{n_j(\mathcal{C}_r)} \sum_{\{q=1,\ldots,r: c_q=j\}} \boldsymbol{x}_q, \quad j = 1, \ldots, m \tag{15.41}$$

with $n_j(\mathcal{C}_r)$ being the number of vectors $\boldsymbol{x} \in \{\boldsymbol{x}_1, \ldots, \boldsymbol{x}_r\}$ that belong to cluster C_j. Note that the computation of the mean vectors of the clusters takes into account only the first r vectors. We assume that $J(\mathcal{C}_1) = 0$. One can easily verify that

$$J(\mathcal{C}_{r+1}) = J(\mathcal{C}_r) + \Delta J(\mathcal{C}_r) \tag{15.42}$$

with $\Delta J(\mathcal{C}_r) \geq 0$. In words, $\Delta J(\mathcal{C}_r)$ denotes the increase in the value of J when the next vector is assigned to a cluster. More precisely, assuming that the $r+1$ vector is assigned to the cluster C_j, it can be shown that

$$\Delta J(\mathcal{C}_r) = \frac{n_j(\mathcal{C}_r)}{n_j(\mathcal{C}_r) + 1} \|\boldsymbol{x}_{r+1} - \boldsymbol{m}_j(\mathcal{C}_r)\|^2 \tag{15.43}$$

Let $\mathcal{C}_N^* = [c_1^*, \ldots, c_N^*]$ denote the optimal clustering. In the sequel, the index r denotes the vector that is currently considered for cluster assignment. Then the algorithm may be stated as follows.

Branch and Bound Clustering (BBC) Algorithm

$r = 1$
$B = +\infty$
While $r \neq 0$ do

- If ($J(\mathcal{C}_r) < B$) AND ($r < N$) then
  - $r = r + 1$
  - Among all possible assignments, c_r, of $\boldsymbol{x}_r$ that have not been tested yet, choose the one that minimizes the value of $\Delta J(\mathcal{C}_r)$.[9]
- End {If}
- If ($J(\mathcal{C}_r) < B$) AND ($r = N$) then
  - $\mathcal{C}_N^* = \mathcal{C}_N$
  - $B = J(\mathcal{C}_N)$
- End {If}

[9] If more than one c_r's minimize $\Delta J(\mathcal{C}_r)$, choose the smallest one.

- If (($J(\mathcal{C}_r) \leq B$) AND ($r = N$)) OR ($J(\mathcal{C}_r) > B$) then
  - (A) $r = r - 1$
  - If ($r = 0$) then
    - Stop
  - Else
    - If all possible clusterings that branch from this node have been exhausted for the rth vector then
      - Go to (A)
    - Else
      - $r = r + 1$
      - Among all possible c_r's that have not been tested yet, choose another path, the one that minimizes the value of $\Delta J(\mathcal{C}_r)$.
    - End {If}
  - End {If}
- End {If}

End {While}

The algorithm starts from the initial node of the tree and goes down until either (i) a leaf or (ii) a node q with cost function value greater than B is encountered. In case (i) if the cost for the clustering that corresponds to that leaf is less than B, then this cost becomes the new bound B, and the clustering is the best clustering found so far. In case (ii) all subsequent clusterings branching from q are not considered any further and we say that they are exhausted. The algorithm, then, backtracks to the parent node of q in order to span a different path. If all paths branching from the parent of q have already been considered, we move to the grandparent of q. The algorithm terminates when all possible paths have been considered explicitly or implicitly (via the aforementioned case (ii)). Clearly, in the beginning, the BBC algorithm spans first a whole path from the initial node of the tree down to a leaf. The cost function of the clustering corresponding to that leaf is the new value of B.

It is clear that the tighter the upper bound B, the more paths are rejected without explicit consideration. Variations of this algorithm that use better estimates of B are discussed in [Koon 75]. Moreover, the substitution of the values $J(\mathcal{C}_r)$ with tighter lower bounds of the optimal value of J is also suggested in [Koon 75]. This leads to the rejection of many more clusterings without considering them explicitly.

The major disadvantage of this algorithm is the excessive amount of computational time it requires even for moderate values of N.[10] In addition, this time is unpredictable. One way to face this problem is to run the algorithm for a prechosen time and use the best clustering found so far. It is clear that in this

[10] This is a common feature for the methods performing global optimization.

case, the algorithm can no longer guarantee the determination of the globally optimal clustering. Versions of the branch and bound algorithm that achieve some computational savings by reducing redundant J evaluations have appeared in [Yu 93, Chen 95].

15.7.2 Simulated Annealing

This is a global optimization algorithm. More specifically, under certain conditions, it guarantees, in probability, the computation of the globally optimal solution of the problem at hand via the minimization of a cost function J. This algorithm has been proposed by Kirkpatrick *et al.* [Kirk 83] (see also [Laar 87]) and it is inspired by the problem of condensed matter in physics.[11] In contrast to the algorithms that allow corrections of the unknown parameters only to directions that reduce the cost function J, simulated annealing allows moves that, temporarily, may increase the value of J. The rationale is that, by allowing such moves, we may escape from the region of attraction of a local minimum.

A very important parameter of this method is the so-called temperature T, which is the analog of the temperature in physical systems. The algorithm starts with a high temperature, which is reduced gradually. A *sweep* is the time that has to be spent at a given temperature so that the system can enter the "thermal equilibrium" state. Let T_{max} and C_{init} denote the initial value of the temperature, T, and the initial clustering, respectively. Also, C denotes the current clustering and t the current sweep. The general scheme of simulated annealing, in the clustering context, is the following.

Simulated Annealing for Clustering

- Set $T = T_{max}$ and $C = C_{init}$.
- $t = 0$.
- Repeat
  - $t = t + 1$
  - Repeat
    - Compute $J(C)$.
    - Produce a new clustering, C', by assigning a randomly chosen vector from X to a different cluster.
    - Compute $J(C')$.
    - If $\Delta J = J(C') - J(C) < 0$ then
      - (A) $C = C'$

[11] Also, it shares many common features with the Metropolis algorithm [Metr 53].

- Else
  - (B) $C = C'$, with probability $P(\Delta J) = e^{-\Delta J/T}$.
- End if

- Until an equilibrium state is reached at this temperature.
- $T = f(T_{\max}, t)$

- Until a predetermined value T_{min} for T is reached.

It is clear that high values of the temperature imply that almost all movements of vectors between clusters are allowed, since, as $T \to +\infty$, $P(\Delta J) \simeq 1$. On the other hand, for low values of T, fewer moves of the (B) type are allowed and, finally, as $T \to 0$, the probability of such moves tends to zero. Thus, as T is lowered, it becomes more probable that clusterings that correspond to lower values of J will be reached. On the other hand, by keeping T positive, we ensure a nonzero probability for escaping from a local minimum.

A difficulty with this algorithm is the determination of the equilibrium state at a specific temperature. One heuristic rule for this case is to consider that the equilibrium state has been reached if for k successive random reassignments of patterns, C remains unchanged (typically k is of the order of a few thousand). Further discussion of this direction is provided in [Klei 89]. Also, another crucial point is the schedule for lowering T. It has been shown that if

$$T = T_{max}/\ln(1 + t) \tag{15.44}$$

this scheme converges to the global minimum with probability 1 [Gema 84]. However, this schedule is too slow. A faster schedule for lowering T is discussed in [Szu 86]. Despite this, the main disadvantage of this algorithm remains the vast amount of computations required.

Finally, in [Al-S 93] simulated annealing is used in terms of fuzzy clustering. Experiments with simulated annealing in clustering problems are presented in [Klei 89, Brow 92].

15.7.3 Deterministic Annealing

This is a hybrid parametric scheme combining the advantages of simulated annealing and the deterministic clustering methods. In contrast to simulated annealing, where successive clusterings are obtained by randomly disturbing the current one, no random disturbances occur here. On the other hand, the cost function is changed slightly, in order to accommodate the parameter $\beta = 1/T$, where T is defined as in the simulated annealing methods.[12] In contrast to simulated annealing, deterministic annealing is the counterpart of the phase transition phenomenon that is observed when the temperature of a material changes [Rose 91].

[12] We choose to work with β instead of T, because this notation is generally used for this algorithm.

In this framework, a set of representatives $\boldsymbol{w}_j, j = 1, \ldots, m$ (m is fixed), is adopted and our goal is to locate them in appropriate positions so that a distortion function is minimized. To this end, the following "effective" cost function J is constructed [Rose 91]:

$$J = -\frac{1}{\beta} \sum_{i=1}^{N} \ln \left(\sum_{j=1}^{m} e^{-\beta d(\boldsymbol{x}_i, \boldsymbol{w}_j)} \right) \tag{15.45}$$

where m is the number of clusters. Differentiating J with respect to the representative $\boldsymbol{w}_r$ and setting it equal to 0, we obtain

$$\frac{\partial J}{\partial \boldsymbol{w}_r} = \sum_{i=1}^{N} \left(\frac{e^{-\beta d(\boldsymbol{x}_i, \boldsymbol{w}_r)}}{\sum_{j=1}^{m} e^{-\beta d(\boldsymbol{x}_i, \boldsymbol{w}_j)}} \right) \frac{\partial d(\boldsymbol{x}_i, \boldsymbol{w}_r)}{\partial \boldsymbol{w}_r} = 0 \tag{15.46}$$

It is clear that the ratio in parentheses takes values in [0, 1] and, in addition, all these terms for a specific $\boldsymbol{x}_i$ sum up to 1. Thus, it may be interpreted as the probability, P_{ir}, that $\boldsymbol{x}_i$ belongs to $C_r, r = 1, \ldots, m$. Then Eq. (15.46) can be written as

$$\sum_{i=1}^{N} P_{ir} \frac{\partial d(\boldsymbol{x}_i, \boldsymbol{w}_r)}{\partial \boldsymbol{w}_r} = 0 \tag{15.47}$$

In the sequel, we assume that $d(\boldsymbol{x}, \boldsymbol{w})$ is a convex function of $\boldsymbol{w}$ for fixed $\boldsymbol{x}$. Note that for $\beta = 0$, all P_{ij}'s are equal to $1/m$, for all $\boldsymbol{x}_i$'s, $i = 1, \ldots, N$. Thus, in this case Eq. (15.47) becomes

$$\sum_{i=1}^{N} \frac{\partial d(\boldsymbol{x}_i, \boldsymbol{w}_r)}{\partial \boldsymbol{w}_r} = 0 \tag{15.48}$$

Since $d(\boldsymbol{x}, \boldsymbol{w})$ is a convex function, $\sum_{i=1}^{N} d(\boldsymbol{x}_i, \boldsymbol{w}_r)$ is also a convex function and, thus, it has a unique minimum, which may be captured by any gradient descent scheme. Thus, in this case, *all the resulting representatives coincide with this unique global minimum.* That is, all data belong to a single cluster. As the value of β increases, it reaches a critical value where a phase transition occurs (alternatively, the probabilities P_{ir} "depart sufficiently" from the uniform model); that is, the clusters are no longer optimally represented by a single representative. Thus, the representatives split up in order to provide an optimal representation of the data set at the new phase. Further increase of β causes a new phase transition and the available representatives are further split up. By choosing m to be greater than the "actual" number of clusters, we ensure the ability of the algorithm to represent the data set properly. In the worst case, some of the representatives will coincide.

Note that as β increases, the probabilities P_{ij} depart from the uniform model and approach the hard clustering model; that is, for all $\boldsymbol{x}_i$, $P_{ir} \simeq 1$ for some r, and $P_{ij} \simeq 0$ for $j \neq r$.

Thus, the requirement for each vector to be assigned to a specific cluster with probability close to unity may serve as a termination criterion for the algorithm.

Schedules for the increase of β are discussed in [Rose 91]. Although simulation results show satisfactory performance of the algorithm, it is not guaranteed that it reaches the globally optimum clustering. Other applications of deterministic annealing to clustering are discussed in [Hofm 97, Beni 94].

15.7.4 Clustering Using Genetic Algorithms

Genetic algorithms have been inspired by the natural selection mechanism introduced by Darwin. They apply certain operators to a population of solutions of the problem at hand, in such a way that the new population is improved compared with the previous one according to a prespecified criterion function J. This procedure is applied for a preselected number of iterations and the output of the algorithm is the best solution found in the last population or, in some cases, the best solution found during the evolution of the algorithm.

In general, the solutions of the problem at hand are coded[13] and the operators are applied to the coded versions of the solutions. The way the solutions are coded plays an important role in the performance of a genetic algorithm. Inappropriate coding may lead to poor performance.

The operators used by genetic algorithms simulate the way natural selection is carried out. The most well-known operators used are the *reproduction*, *crossover*, and *mutation* operators applied in that order to the current population. The reproduction operator ensures that, in probability, the better (worse) a solution in the current population is, the more (less) replicates it has in the next population. The crossover operator, which is applied to the temporary population produced after the application of the reproduction operator, selects pairs of solutions randomly, splits them at a random position, and exchanges their second parts. Finally, the mutation operator, which is applied after the application of the reproduction and crossover operators, selects randomly an element of a solution and alters it with some probability. The last operator may be viewed as a way out of getting stuck in local minima. Apart from these three operators, many others have been proposed in the literature (e.g., [Mich 94]).

Besides the coding of the solutions, other parameters, such as the number of solutions in a population, p,[14] the probability with which we select two solutions for crossover, and the probability with which an element of a solution is mutated, play very important roles in the performance of the algorithm.

Several genetic algorithms with application to clustering have been proposed (e.g., [Bhan 91, Andr 94, Sche 97, Maul 00, Tsen 00]). In the sequel, we briefly discuss a simple parametric one that is suitable for hard clustering. We assume that the number of clusters, m, is fixed. As stated before, the first thing we have to decide

[13] Binary representations as well as more general ones are possible.
[14] This may be fixed or varied.

is how to code a solution. A simple (but not unique) way to achieve this is to use the representatives in order to form the following string:

$$[\boldsymbol{w}_1, \boldsymbol{w}_2, \ldots, \boldsymbol{w}_m] \tag{15.49}$$

or, in more detail,

$$[w_{11}, w_{12}, \ldots, w_{1l}, w_{21}, w_{22}, \ldots, w_{2l}, \ldots, w_{m1}, w_{m2}, \ldots, w_{ml}] \tag{15.50}$$

The cost function we use is

$$J = \sum_{i=1}^{N} u_{ij} d(\boldsymbol{x}_i, \boldsymbol{w}_j) \tag{15.51}$$

where

$$u_{ij} = \begin{cases} 1, & d(\boldsymbol{x}_i, \boldsymbol{w}_j) = \min_{k=1,\ldots,m} d(\boldsymbol{x}_i, \boldsymbol{w}_k) \\ 0, & \text{otherwise} \end{cases} \quad i = 1, \ldots, N \tag{15.52}$$

The allowable cut points for the crossover operator are between different representatives. Also, in this case the mutation operator selects randomly a coordinate of a vector of a solution and decides randomly to add a small random number to it.

An alternative to this algorithm is the following. Before we apply the reproduction operator to the current population, we run the hard clustering algorithm, described in Chapter 14, p times, each time using a different solution of the current population as the initial state. The p solutions produced after the convergence of the hard clustering algorithm constitute the population to which the reproduction operator will be applied. It is expected that this modification will give better results, as it is likely that the resulting p solutions are local minima of the cost function. This modified algorithm has been reported to give satisfactory results in a color image quantization application, [Sche 97].

15.8 KERNEL CLUSTERING METHODS

The minimal-enclosure (hyper)sphere (i.e., the sphere with the minimum volume enclosing all points in a vector space X) was briefly discussed in Chapter 5 (Problem 5.20). In [Tax 99], this problem was considered in a more relaxed framework, allowing some of the points of the data set to lie outside the volume of the sphere. This viewpoint makes the problem less sensitive to outliers. The optimization task now becomes similar to the soft-margin SVM and can be casted as

$$\text{minimize} \quad r^2 + C \sum_{i=1}^{N} \xi_i \tag{15.53}$$

$$\text{subject to} \quad ||\boldsymbol{x} - \mathbf{c}||^2 \leq r^2 + \xi_i, \; i = 1, 2, \ldots, N \tag{15.54}$$

$$\xi_i \geq 0, \quad i = 1, 2, \ldots, N \tag{15.55}$$

In other words, the radius r of the sphere, centered at $\mathbf{c}$, enclosing the data points of X, is minimized. However, some points in X are allowed to lie outside it ($\xi > 0$), but at the same time the number of these points must be as small as possible (due to the second term in Equation (15.53)). The Lagrangian of the above constrained problem (Chapter 3 and Appendix C) is given by

$$\mathcal{L}(r, \mathbf{c}, \xi, \mu, \lambda) = r^2 + C\sum_{i=1}^{N} \xi_i - \sum_{i=1}^{N} \mu_i \xi_i - \sum_{i=1}^{N} \lambda_i \left(r^2 + \xi_i - ||\boldsymbol{x}_i - \mathbf{c}||^2\right) \tag{15.56}$$

Taking the derivatives of the above and equating to zero, the Wolfe dual form results in

$$\max_{\lambda} \left(\sum_{i=1}^{N} \lambda_i \boldsymbol{x}_i^T \boldsymbol{x}_i - \sum_{i=1}^{N}\sum_{j=1}^{N} \lambda_i \lambda_j \boldsymbol{x}_i^T \boldsymbol{x}_j \right) \tag{15.57}$$

$$\text{subject to } 0 \leq \lambda_i \leq C, \; i = 1, 2, \ldots, N \tag{15.58}$$

$$\sum_{i=1}^{N} \lambda_i = 1 \tag{15.59}$$

and the KKT conditions are

$$\mu_i \xi_i = 0 \tag{15.60}$$

$$\lambda_i \left[r^2 + \xi_i - ||\boldsymbol{x}_i - \mathbf{c}||^2\right] = 0 \tag{15.61}$$

$$\mathbf{c} = \sum_{i=1}^{N} \lambda_i \boldsymbol{x}_i \tag{15.62}$$

$$\lambda_i = C - \mu_i, \; i = 1, 2, \ldots, N \tag{15.63}$$

From these conditions the following remarks are easily deduced.

- Only points with $\lambda_i \neq 0$ contribute to the definition of the center of the optimal sphere (Eq. (15.62)). These points are known as *support vectors*.
- Points with $\xi_i > 0$ correspond to $\mu_i = 0$ (Eq. (15.60)), which leads to $\lambda_i = C$ (Eq. (15.63)) and, as (15.61) suggests, these points lie outside the sphere. We will refer to these points as *bounded support vectors*.
- Points with $0 < \lambda_i < C$ have corresponding $\mu_i > 0$ leading to $\xi_i = 0$ (Eq. (15.60)) and, as Equation (15.61) suggests, these points lie on the sphere.
- Points with $\lambda_i = 0$ correspond to $\xi_i = 0$. As (15.61) suggests, all points lying inside the sphere satisfy, necessarily, these two conditions.

As we already know from the SVM theory, exposed in Chapter 4, all we have said so far is still valid if the input space, X, is mapped into a high-dimensional Hilbert space, H, via the "kernel trick." The task now becomes

$$\max_{\lambda} \left(\sum_{i=1}^{N} \lambda_i K(\boldsymbol{x}_i, \boldsymbol{x}_i) - \sum_{i=1}^{N} \sum_{j=1}^{N} \lambda_i \lambda_j K(\boldsymbol{x}_i, \boldsymbol{x}_j) \right) \tag{15.64}$$

$$\text{subject to } 0 \le \lambda_i \le C, \; i = 1, 2, \ldots, N \tag{15.65}$$

$$\sum_{i=1}^{N} \lambda_i = 1 \tag{15.66}$$

Maximization of the previous leads to the computation of the optimal Lagrange multipliers.

In [Ben 01] a methodology is presented that exploits the minimal enclosure sphere problem to unravel the clustering structure underlying the data set X. Let us denote the (implicit) mapping, induced by the adopted kernel function, from the original X to the high-dimensional space H as

$$\boldsymbol{x} \in X \longrightarrow \phi(\boldsymbol{x}) \in H \tag{15.67}$$

The distance of $\phi(\boldsymbol{x})$ from the center of the optimal sphere $\boldsymbol{c} = \sum_i \lambda_i \phi(\boldsymbol{x}_i)$ (as Equation (15.62) suggests) is equal to

$$r^2(\boldsymbol{x}) \equiv ||\phi(\boldsymbol{x}) - \boldsymbol{c}||^2 = \phi^T(\boldsymbol{x})\phi(\boldsymbol{x}) \tag{15.68}$$

$$+ \boldsymbol{c}^T \boldsymbol{c} - 2\phi^T(\boldsymbol{x})\boldsymbol{c} \tag{15.69}$$

or

$$r^2(\boldsymbol{x}) = K(\boldsymbol{x}, \boldsymbol{x}) + \sum_{i=1}^{N} \sum_{j=1}^{N} \lambda_i \lambda_j K(\boldsymbol{x}_i, \boldsymbol{x}_j) \tag{15.70}$$

$$- 2 \sum_{i=1}^{N} \lambda_i K(\boldsymbol{x}, \boldsymbol{x}_i) \tag{15.71}$$

Obviously, for any of the support vectors, $\boldsymbol{x}_i$, this function is equal to the radius of the optimum sphere, that is,

$$r^2(\boldsymbol{x}_i) = r^2 \tag{15.72}$$

The contours formed in the original vector space, X, defined by the points

$$\{\boldsymbol{x} : r(\boldsymbol{x}) = r\} \tag{15.73}$$

are *interpreted* as forming cluster boundaries. It is apparent that the shapes of these contours are heavily dependent on the specific kernel function that has been adopted. In view of what we have already said, the vectors in X whose images in H

are support vectors ($0<\lambda_i<C$) lie *on* these contours, points with images inside the sphere ($\lambda_i = 0$) lie *inside* these, contours, and points whose images lie outside the sphere ($\lambda_i = C$) lie outside these contours. A schematic representation is shown in Figure 15.21.

Contours, of course, do not suffice to define clusters. In [Ben 01] a geometric approach is suggested to differentiate points of the same and of different clusters. If the line segment, joining two points, does not cross a contour in X, these points reside in the same cluster. However, if a line segment crosses a contour it means that there are points on it (e.g., $\boldsymbol{y}$ in Figure 15.21), whose images in H lie outside the optimal sphere. Therefore, in order to check whether a line segment crosses a contour it suffices to detect some points on it with the property $r(\boldsymbol{y})>r$. This leads to the definition of the *adjacency* matrix, A, with elements A_{ij}, referring to the pair of points $\boldsymbol{x}_i,\ \boldsymbol{x}_j \in X$, whose images lie in or on the sphere in H

$$A_{ij} = \begin{cases} 1 & \text{no points, } \boldsymbol{y}\text{, on the respective segment exist such that } r(\boldsymbol{y}) > r \\ 0 & \text{otherwise} \end{cases} \tag{15.74}$$

Clusters are now defined as the ***connected components*** of the graph induced by A. An extension of the previous that combines the concept of fuzzy membership is discussed in [Chia 03]. In [Wang 07] a variant of the method is presented, which utilizes the minimal-enclosure hyperellipsoid instead of a hypersphere.

The other direction that has been followed, in order to exploit the power springing from the nonlinear nature of the "kernel trick," is of the same nature as the kernel-based principle component analysis (PCA), discussed in Chapter 6. To this end, any clustering approach that can be casted, throughout, in terms of inner product computations is a candidate to accommodate the "kernel trick." The classical k-Means algorithm, as well as some of its variants, are typical examples that have been proposed and used. See, for example, [Scho 98, Cama 05, Giro 02]. This is because the Euclidean distance, which is in one way or another at the heart of these

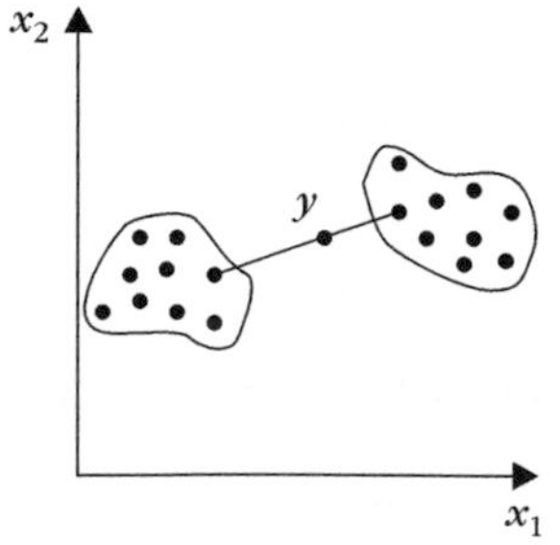

FIGURE 15.21

Point $\boldsymbol{y}$ on the line segment, crossing the contours, has an image outside the minimal enclosure sphere.

algorithms, is itself an inner product and all involved computations can be expressed in terms of inner products.

The advantage of the kernel-based methods lies in the fact that they can reveal clusters of arbitrary shapes, due to the nonlinear nature of the mapping. On the other hand, the results are very sensitive to the choice of the specific kernel function as well as of its defining parameters. Moreover, the required computational demands pose an obstacle for the use of these methods for large data sets

15.9 DENSITY-BASED ALGORITHMS FOR LARGE DATA SETS

In this framework, clusters are considered as regions in the l-dimensional space that are "dense" in points of X (Note the close agreement between this way of viewing clusters and the definition of clusters given in [Ever 01]). Most of the density-based algorithms do not impose any restrictions to the shape of the resulting clusters. Thus, these algorithms have the ability to recover arbitrarily shaped clusters. In addition, they are able to handle efficiently the outliers. Moreover, the time complexity of these algorithms is lower than $O(N^2)$, which makes them eligible for processing large data sets.

Typical density-based algorithms are the DBSCAN ([Este 96]), the DBCLASD ([Xu 98]), and the DENCLUE ([Hinn 98]). Although these algorithms share the same basic philosophy, they differ in the way the density is quantified.

15.9.1 The DBSCAN Algorithm

The "density" as it is considered in DBSCAN (Density-Based Spatial Clustering of Applications with Noise) around a point $\boldsymbol{x}$ is estimated as the number of points in X that fall inside a certain region in the l-dimensional space surrounding $\boldsymbol{x}$. In the sequel, we will consider this region to be a hypersphere $V_\varepsilon(\boldsymbol{x})$ centered at $\boldsymbol{x}$, whose radius ε is a user-defined parameter. In addition, let $N_\varepsilon(\boldsymbol{x})$ denote the number of points of X lying in $V_\varepsilon(\boldsymbol{x})$. An additional user-defined parameter is the minimum number of points, q, that must be contained in $V_\varepsilon(\boldsymbol{x})$, in order for $\boldsymbol{x}$ to be considered an "interior" point of a cluster. Before we proceed, the following definitions are in order.

Definition 5. *A point $\boldsymbol{y}$ is* directly density reachable *from a point $\boldsymbol{x}$ (see Figure 15.22a) if*

(i) $\boldsymbol{y} \in V_\varepsilon(\boldsymbol{x})$ and

(ii) $N_\varepsilon(\boldsymbol{x}) \geq q$.

Definition 6. *A point $\boldsymbol{y}$ is* density reachable *from a point $\boldsymbol{x}$ in X if there is a sequence of points $\boldsymbol{x}_1, \boldsymbol{x}_2, \ldots, \boldsymbol{x}_p \in X$, with $\boldsymbol{x}_1 = \boldsymbol{x}$, $\boldsymbol{x}_p = \boldsymbol{y}$, such that $\boldsymbol{x}_{i+1}$ is directly density reachable from $\boldsymbol{x}_i$ (see Figure 15.22b).*

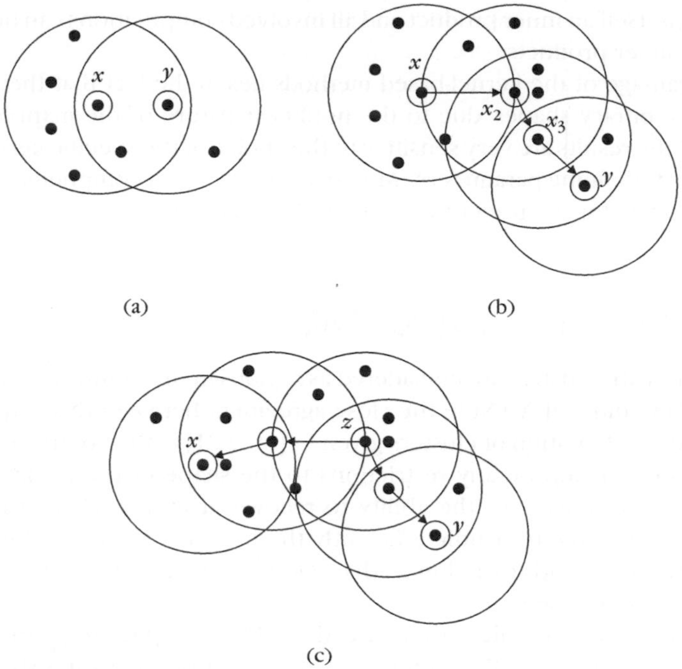

FIGURE 15.22

Assuming that $q = 5$, (a) $\boldsymbol{y}$ is directly density reachable from $\boldsymbol{x}$, but not vice versa, (b) $\boldsymbol{y}$ is density reachable from $\boldsymbol{x}$, but not vice versa, and (c) $\boldsymbol{x}$ and $\boldsymbol{y}$ are density connected (in addition, $\boldsymbol{y}$ is density reachable from $\boldsymbol{x}$, but not vice versa).

It is easy to note that if $\boldsymbol{x}$ and $\boldsymbol{y}$ in the previous definition are "dense enough" (that is, $N_\varepsilon(\boldsymbol{x}) \geq q$ and $N_\varepsilon(\boldsymbol{y}) \geq q$), the density reachability property is symmetric. On the other hand, symmetry is lost in the case where either of the points, $\boldsymbol{x}$ and $\boldsymbol{y}$, has less than q points in its neighborhood. However, in this case there must be a third point $\boldsymbol{z} \in X$ such that both $\boldsymbol{x}$ and $\boldsymbol{y}$ are density reachable from $\boldsymbol{z}$. This leads to the definition of the *density-connectivity*.

Definition 7. *A point* $\boldsymbol{x}$ *is* density connected *to a point* $\boldsymbol{y} \in X$ *if there exists* $\boldsymbol{z} \in X$ *such that both* $\boldsymbol{x}$ *and* $\boldsymbol{y}$ *are density reachable from* $\boldsymbol{z}$ *(see Figure 15.22c).*

After the previous definitions, a *cluster* C in the DBSCAN framework is defined as a nonempty subset of X satisfying the following conditions:

(i) If $\boldsymbol{x}$ belongs to C and $\boldsymbol{y} \in X$ is density reachable from $\boldsymbol{x}$, then $\boldsymbol{y} \in C$.

(ii) For each pair $(\boldsymbol{x}, \boldsymbol{y}) \in C$, $\boldsymbol{x}$ and $\boldsymbol{y}$ are density connected.

Let $C_1, \ldots, C_m$ be the clusters in X. Then, the set of points that are not contained in any of the clusters $C_1, \ldots, C_m$ is known as *noise*.

In the sequel, we define a point $\boldsymbol{x}$ as a *core point* if it has at least q points in its neighborhood. Otherwise, $\boldsymbol{x}$ is said to be a *noncore point*. A noncore point may be either a *border point* of a cluster (that is, density reachable from a core point) or a *noise point* (that is, not density reachable from other points in X). Having established the previous definitions, the following two propositions hold true ([Este 96]).

Proposition 1. If $\boldsymbol{x}$ is a core point and D is the set of points in X that are density reachable from $\boldsymbol{x}$, then D is a cluster.

Proposition 2. If C is a cluster and $\boldsymbol{x}$ a core point in C, then C equals to the set of the points $\boldsymbol{y} \in X$ that are density reachable from $\boldsymbol{x}$.

These two propositions imply that *a cluster is uniquely determined by any of its core points*. Keeping in mind the last conclusion, we proceed now with the description of the DBSCAN algorithm. Let X_{un} be the set of points in X that have not been considered yet, and let m denote the number of clusters.

DBSCAN Algorithm

- Set $X_{un} = X$
- Set $m = 0$
- While $X_{un} \neq \emptyset$ do
  - Arbitrarily select a $\boldsymbol{x} \in X_{un}$.
  - If $\boldsymbol{x}$ is a noncore point then
    - Mark $\boldsymbol{x}$ as noise point.
    - $X_{un} = X_{un} - \{\boldsymbol{x}\}$
  - If $\boldsymbol{x}$ is a core point then
    - $m = m + 1$
    - Determine all density-reachable points in X from $\boldsymbol{x}$.
    - Assign $\boldsymbol{x}$ and the previous points to the cluster C_m. The border points that may have been marked as noise are also assigned to C_m.
    - $X_{un} = X_{un} - C_m$
  - End { if }
- End { while }

A delicate point of the algorithm is the following. Suppose that a border point $\boldsymbol{y}$ of a cluster C has currently been selected by the algorithm. This point will be marked as a noise point, through the first branch of the *if* statement of the algorithm. Note, however, that when, later on, a core point $\boldsymbol{x}$ of C will be considered from which $\boldsymbol{y}$ is density reachable, then $\boldsymbol{y}$ will be identified as a density-reachable point from $\boldsymbol{x}$ and will be assigned to C. On the other hand, if $\boldsymbol{y}$ is a noise point it will be marked as such and because it is not density reachable by any of the core points in X its "noise" label will remain unaltered.

Remarks

- The results of the algorithm are greatly influenced by the choice of ε and q. Different values of the parameters may lead to totally different results. One should select these parameters so that the algorithm is able to detect the least "dense" cluster. In practice, one has to experiment with several values for ε and q in order to identify their "best" combination for the data set at hand.
- Implementation of the algorithm by adopting the R^*-tree data structure can achieve time complexity of $O(N \log_2 N)$ for low-dimensional data sets ([Berk 02]).
- The DBSCAN is not appropriate for cases where the clusters in X exhibit significant differences in density, and it is not well suited for high-dimensional data ([Xu 05]).
- It is also worth noting the close resemblance of the previous algorithm with the CDADV algorithm described in Section 15.4.3. However, the latter is suitable for discrete-valued vectors.

An extension of DBSCAN that overcomes the necessity of choosing carefully the parameters ε and q is the OPTICS (Ordering Points To Identify the Clustering Structure) algorithm ([Anke 99]). This generates a density-based cluster ordering, representing the intrinsic hierarchical cluster structure of the data set in a comprehensible form ([Bohm 00]). Experiments indicate that the run time of OPTICS is roughly equal to 1.6 of the runtime required by DBSCAN ([Berk 02]). On the other hand, in practice one has to run DBSCAN more than one time for different values of ε and q. Another extension of DBSCAN is given in [Sand 98].

15.9.2 The DBCLASD Algorithm

An alternative to the DBSCAN approach is followed by the DBCLASD (Distribution-Based Clustering of LArge Spatial Databases) algorithm ([Xu 98]). In this case, the distance d of a point $\boldsymbol{x} \in X$ to its nearest neighbor is considered a *random variable* and the "density" is quantified in terms of probability distribution of d. In addition, it is assumed that the points in each cluster are uniformly distributed. However, not all points in X are assumed to be uniformly distributed. Based on this assumption, the distribution of d can be derived analytically ([Xu 98]). In the framework of DBCLASD, a *cluster* C is defined as a nonempty subset of X with the following properties.

(a) The distribution of the distances, d, between points in C and their nearest neighbors is in agreement with the distribution derived theoretically, within some confidence interval. (This is carried out by using the χ^2 test.)

(b) It is the *maximal set* with the previous property. That is, the insertion of additional points neighboring to points in C will cause (a) not to hold anymore.

(c) It is *connected*. Having applied a grid of cubes on the feature space, this property implies that for any pair of points $(\boldsymbol{x}, \boldsymbol{y})$ from C there exists a path of adjacent cubes that contains at least one point in C that connects $\boldsymbol{x}$ and $\boldsymbol{y}$.

In this algorithm, the points are considered in a sequential manner. A point that has been assigned to a cluster may be reassigned to another cluster at a later stage of the algorithm. In addition, some points are not assigned to any of the clusters determined so far, but they are tested again at a later stage.

Among the merits of the algorithm is that it is able to determine arbitrarily shaped clusters of various densities in X, and it requires no parameter definition. Experimental results given in [Xu 98] indicate that the runtime of DBCLASD is roughly twice the runtime of DBSCAN, whereas DBCLASD outperforms CLARANS by a factor of at least 60.

15.9.3 The DENCLUE Algorithm

In the previously described methods, "density" was defined based on the distance between neighboring points $\boldsymbol{x} \in X$, in either a deterministic (DBSCAN) or a probabilistic setting (DBCLASD). In the following, "density" is quantified via a different route. Specifically, for each point $\boldsymbol{y} \in X$ a so-called *influence function* $f^{\boldsymbol{y}}(\boldsymbol{x}) \geq 0$ is defined, which decreases to zero as $\boldsymbol{x}$ "moves away" from $\boldsymbol{y}$. Typical examples of $f^{\boldsymbol{y}}(\boldsymbol{x})$ include

$$f^{\boldsymbol{y}}(\boldsymbol{x}) = \begin{cases} 1, & \text{if } d(\boldsymbol{x}, \boldsymbol{y}) < \sigma \\ 0, & \text{otherwise} \end{cases} \tag{15.75}$$

and

$$f^{\boldsymbol{y}}(\boldsymbol{x}) = e^{-\frac{d(\boldsymbol{x},\boldsymbol{y})^2}{2\sigma^2}} \tag{15.76}$$

where $d(\boldsymbol{x}, \boldsymbol{y})$ denotes the distance between $\boldsymbol{x}$ and $\boldsymbol{y}$ (this may be the Euclidean distance or any other dissimilarity measure) and σ is a user-defined parameter. Then, the *density function* based on X is defined as

$$f^X(\boldsymbol{x}) = \sum_{i=1}^{N} f^{\boldsymbol{x}_i}(\boldsymbol{x}) \tag{15.77}$$

Note the similarity of the previous with the Parzen windows approximation of the density function $p(\boldsymbol{x})$, discussed in Chapter 2. The goal here is to identify all "significant" local maxima, $\boldsymbol{x}_j^*, j = 1, \ldots, m$, of $f^X(\boldsymbol{x})$ and then to create a cluster C_j for each $\boldsymbol{x}_j^*$ and to assign to C_j all the points of X that lie in the *region of attraction* of $\boldsymbol{x}_j^*$. The region of attraction of $\boldsymbol{x}_j^*$ is defined as the set of points $\boldsymbol{x} \in \mathcal{R}^l$ such that if a "hill-climbing" method (a hill-climbing method aims at determining the local maxima of a function; a typical example is the steepest ascent method, see Appendix C) is applied, initialized by $\boldsymbol{x}$, it will terminate arbitrarily close to $\boldsymbol{x}_j^*$.

As can be seen from Eqs. (15.75) and (15.76), σ quantifies the influence of a specific data point of X in the $\mathcal{R}^l$ space. In addition, a parameter ξ is required in order to quantify the significance of a local maximum. Thus, a local maximum is considered significant if $f^X(\boldsymbol{x}_j^*) \geq \xi$. Since by definition $f^{\boldsymbol{x}_i}(\boldsymbol{x})$ decreases as $\boldsymbol{x}$ moves away from $\boldsymbol{x}_i$, it is expected that $f^X(\boldsymbol{x})$ can be approximated satisfactorily by

$$\hat{f}^X(\boldsymbol{x}) = \sum_{\boldsymbol{x}_i \in Y(\boldsymbol{x})} f^{\boldsymbol{x}_i}(\boldsymbol{x}) \tag{15.78}$$

where $Y(\boldsymbol{x})$ is the set of points in X that lie "close" to $\boldsymbol{x}$.

A well-known representative that evolves around the previous methodology is the DENCLUE (DENsity-based CLUstEring) algorithm ([Hinn 98]). Since the "significant" local maxima are expected to be located in regions "dense" in data (and in order to reduce time complexity), DENCLUE applies first a *preclustering step* to determine regions that are dense in points of X. To this end, an l-dimensional grid of edge-length 2σ is applied on the feature space that contains X and the set D_p of the (hyper)cubes that contain at least one point of X is determined. Then, the subset D_{sp} of D_p that contains the highly populated cubes of D_p (a high populated cube contains at least $\xi_c > 1$ points of X, where ξ_c is a user-defined parameter) is determined. For each highly populated cube, c, a *connection* is defined with all neighboring cubes c_j in D_p for which $d(\boldsymbol{m}_c, \boldsymbol{m}_{c_j})$ is no greater than 4σ, where $\boldsymbol{m}_c$ and $\boldsymbol{m}_{c_j}$ are the mean values of the points in the respective cubes.

Once the previous step has been completed, DENCLUE proceeds as follows. First, it considers the set D_r of the highly populated cubes and the cubes that have at least one connection with a highly populated cube. The search for local maxima is constrained within the cubes in D_r. Then, for each point $\boldsymbol{x}$ in a cube $c \in D_r$ its neighboring points that influence it are considered. Specifically, $Y(\boldsymbol{x})$ contains all points that belong to cubes c_j in D_r such that the means of c_js lie at a distance less than $\lambda\sigma$ from $\boldsymbol{x}$ (typically $\lambda = 4$). Then, a hill climbing method ([Hinn 98]) starting from $\boldsymbol{x}$ is applied. The local maximum $\boldsymbol{x}^*$ to which the method converges is then tested to see if it is a significant local maximum (i.e., if $\hat{f}^X(\boldsymbol{x}^*) \geq \xi$). If it is not, no additional action is taken. Otherwise, if a cluster associated to $\boldsymbol{x}^*$ has not been created yet it is created now and $\boldsymbol{x}$ is assigned to it. In addition, all points of X for which the hill-climbing method leads to $\boldsymbol{x}^*$ will be assigned to the same cluster. Shortcuts that allow the assignment of points to clusters based on certain conditions to be satisfied, without having to apply the hill-climbing procedure, are also discussed in [Hinn 98].

The method, like all density-based methods, is able to detect arbitrarily shaped clusters ([Hinn 98]). In addition, DENCLUE deals with noise very satisfactorily. A procedure for selecting appropriate values for the parameters σ and ξ is discussed in ([Hinn 98]). The worst-case time complexity of DENCLUE is $O(N \log_2 N)$; that is, of the same order as the time complexity of DBSCAN. However, experimental results reported in [Hinn 98] indicate that DENCLUE can be significantly faster than DBSCAN, and the reported experiments indicate that the *average* time complexity for DENCLUE is $O(\log_2 N)$. This is a consequence of the fact that only a small fraction of the total number of points in X are considered for determining the clusters. It has also been pointed out that the algorithm works efficiently with high-dimensional data, and it has been applied to a molecular biology experiment ([Hinn 98]).

A very recent density-based clustering algorithm, named ADACLUS, is discussed in [Noso 08]. It is able to discover clusters of various shapes and densities and to detect boundaries of the clusters. Also, it is robust to noise and it does not oblige the user to define the values of certain parameters. Finally, it has low computational requirements, which is very important in the processing of large data sets.

15.10 CLUSTERING ALGORITHMS FOR HIGH-DIMENSIONAL DATA SETS

As we have already discussed at several points in this book, concerning algorithms that have been designed to meet the needs of large data sets, special data structures for indexing data such as *R*-trees, $k-d$ trees, and so on have been employed in order to facilitate access to the data. As it is pointed out in [Berk 02], the performance of most of these data structures degrades to the level of sequential search for $l > 20$. Thus, one can consider the value of 20 as a "lower bound" that quantifies high dimensionality. In fact, there are applications such as bioinformatics or web mining in which the dimensionality of the feature space can be as high as a few thousands.

Most of the algorithms discussed so far consider all dimensions of the feature space simultaneously, trying to utilize as much of the available information as possible. However, this approach may turn to be problematic in high-dimensional spaces. One source of problems is the "curse of dimensionality" which besides the complexity issues discussed in Section 2.5.6 now shows another of its "faces." Having fixed the number of data points, N, as the dimensionality of the feature space increases, the points are spread out in the space. This can easily be verified by considering two points in the two-dimensional space. Then add a third dimension to each of them and compare the resulting Euclidean distances for both cases. As the points spread out in very high-dimensional spaces they become almost equidistant. Clearly, in such a case the terms *similarity* and *dissimilarity* between two points become increasingly meaningless ([Pars 04]). A second source of problems is that often in very high-dimensional spaces only a small fraction of the features contributes to the formation of each cluster. In other words, clusters can be identified in subspaces of the original feature space. In terms of Figure 15.23a, the clusters C_1 and C_2 are the result of the concentration of the values of x_2 within two small intervals, whereas the values along x_1 do not show such a preference. In a similar manner, the clusters in Figure 15.23b are due to data concentrations in the (x_1, x_2) subspace. Observe that these clusters could be identified by projecting the points in (x_1, x_2).

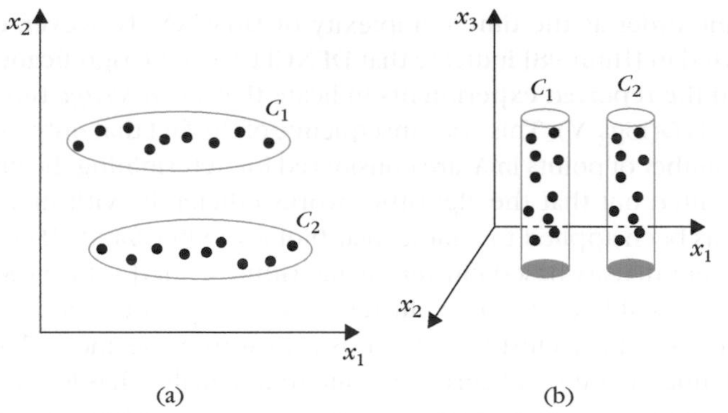

FIGURE 15.23

(a) The clusters C_1 and C_2 are the result of concentrations of the projections of the data points along x_2. (b) The clusters in this three-dimensional case are the result of concentrations of the projections of the data points in the (x_1, x_2) subspace.

Clearly, a way out of this situation is to work on subspaces of dimension lower than l. In the sequel we discuss two main approaches to this direction: the *dimensionality reduction clustering approach* and the *subspace clustering approach*.

15.10.1 Dimensionality Reduction Clustering Approach

The general idea of this approach is to identify an l'-dimensional space $H_{l'}$ ($l' < l$), project the data points in X onto it, and apply a clustering algorithm on the projections of the points of X into $H_{l'}$. For the identification of $H_{l'}$ one may use (a) *feature generation methods*, (b) *feature selection methods*, and (c) *random projections*. In the sequel, we "touch" briefly the first two methodologies, in that they build upon the techniques treated in more detail in Chapters 5 and 6, and continue to pursue the random projection method.

Feature generation methods, such as the principal component analysis (PCA) and the singular value decomposition (SVD), generally preserve the distances between the points in the high-dimensional space when these are mapped to the lower-dimensional space. These methods are very useful in producing compact representations of the original high-dimensional feature space. Algorithms that adopt feature generation methods are discussed in, for example, [Deer 90, Ding 99] and [Ding 02]. A notable characteristic of the latter work is that feature generation is integrated with the clustering algorithm (*k*-means or *EM*) and as a consequence applies iteratively as the clustering algorithm evolves. Feature generation methods can be proved useful in cases where a significant number of features contributes to the identification of the clusters. In addition to PCA or SVD, other techniques used for dimensionality reduction can also be employed, such as nonlinear PCA, ICA., and so on (see Chapter 6).

An alternative approach is to employ feature selection methods, in order to identify those features that are the main contributors to the formation of the clusters.

However, the feature selection methods proposed for the supervised case are no longer suitable in the clustering framework. In general, the criteria used to evaluate the "goodness" of a specific subset of features follows either the *wrapper model* or the *filter model* ([Koha 97]). According to the former, the clustering algorithm, $\mathcal{C}$, is first chosen and a subset of features, $\mathcal{F}_i$, is evaluated through the results obtained from the application of $\mathcal{C}$ on the data set X, where for each point only the features in $\mathcal{F}_i$ are taken into account. According to the latter, the evaluation of a subset of features is carried out using intrinsic properties of the data, prior to the application of the clustering algorithm. Feature selection methods are discussed in [Blum 97, Liu 98, Pena 01, Yu 03]. The feature selection approach is useful when all clusters lie in the same subspace of the feature space.

Clustering Using Random Projections

Unlike the previously cited dimensionality reduction methods, where $H_{l'}$ is identified deterministically, in the present case $H_{l'}$ will be identified in a random manner. Noting that a projection from an l-dimensional space, H_l, to an l'-dimensional space $H_{l'}$ ($l' < l$) is uniquely defined via an $l' \times l$ *projection matrix* A, the issues to be addressed here are: (a) the proper estimate of l' and (b) the definition of the projection matrix. In [Achl 01, Dasg 99] estimates of l' are given that guarantee (with a certain probability) that the distances between the points of the data set X, in the original feature space H_l, will be preserved up to a factor $1 \pm \varepsilon$ (where $\varepsilon > 0$ is an arbitrarily chosen constant) after the projection of X to a randomly chosen l'-dimensional space $H_{l'}$, whose projection matrix is constructed by following certain simple probabilistic rules. More specifically, in [Dasg 99] l' is shown to be bounded below by $4(\varepsilon^2/2 - \varepsilon^3/3)^{-1} \ln N$. (Note, however, that the choice for l' does not guarantee, even in probability that the separation of clusters is preserved in the general case of arbitrarily shaped clusters. This problem is studied in [Dasg 00] for the special case where the clusters stem from Gaussian distributions.)

As far as A is concerned, one way to construct it is to set each of its entries equal to a value stemming from an i.i.d. zero-mean unit variance Gaussian distribution and then to normalize each row to the unit length ([Fern 03]). Another way is to set each entry of A equal to -1 or $+1$, with probability 0.5. Still another way to generate A is to set each of its entries equal to $+\sqrt{3}$ with probability 1/6, $-\sqrt{3}$ with probability 1/6, or 0 with probability 2/3 ([Achl 01]).

After the definition of the projection matrix A, one proceeds by projecting the points of X into $H_{l'}$ and performing a clustering algorithm on the projections of the points of X into $H_{l'}$. However, a significant problem may arise here. Different random projections may lead to totally different clustering results. One way to cope with this problem is to perform several random projections, apply a clustering algorithm on the projections of X to each of them, and combine the clustering results in order to produce the final clustering.

A method in this spirit is discussed in the sequel ([Fern 03]). First, the dimension l' of the lower-dimensional space is selected and r different projection matrices, $A_1, \ldots, A_r$, are generated using the first of the three ways described previously. Then, the Generalized Mixture Decomposition Algorithmic Scheme (GMDAS) for compact and hyperellipsoidal clusters (Section 14.2) is applied on each one of the r random projections of X. Let $P(C_j^s|\boldsymbol{x}_i)$ denote the probability that $\boldsymbol{x}_i$ belongs to the j-th cluster in the s-th projection (C_j^s) after the execution of GMDAS on each projection of X.

For each projection, a similarity matrix P^s is created, whose (i,j) element is defined as

$$P_{ij}^s = \sum_{q=1}^{m_s} P(C_q^s|\boldsymbol{x}_i)P(C_q^s|\boldsymbol{x}_j) \tag{15.79}$$

where m_s is the number of clusters in the s-th projection. Actually, P_{ij}^s is the probability that $\boldsymbol{x}_i$ and $\boldsymbol{x}_j$ belong to the same cluster at the s-th projection. Then, the average proximity matrix P is defined, so that its (i,j) element is equal to the average of P_{ij}^s, $s = 1,\ldots,r$.

In the sequel, the Generalized Agglomerative Scheme (GAS) (Section 13.2) is employed in order to identify the final clusters. The similarity between two clusters C_i and C_j is defined as $\min_{\boldsymbol{x}_u \in C_i,\ \boldsymbol{x}_v \in C_j} P_{uv}$ (actually, this choice complies with the philosophy of the complete link algorithm).

For the estimation of the number of clusters m underlying X, it is proposed to let GAS run until all points of X are agglomerated to a single cluster. Then, the similarity between the closest pair of clusters, determined in each iteration of GAS, is plotted versus the number of iterations. The most abrupt decrease in the plot is determined and m is set equal to the value corresponding to this decrease. Results reported in [Fern 03] show that in principle this method produces better clusters and is more robust than the principal component analysis method followed by the EM algorithm. The complexity of this algorithm is controlled by the GAS algorithm and is larger than $O(N^2)$.

15.10.2 Subspace Clustering Approach

The general strategy followed by the dimensionality reduction methods was to project the data set into a lower-dimensional space and to apply a clustering algorithm on the projections of the data points in this space. This strategy copes well with the problems rising from the "curse of dimensionality," as stated previously. However, these methods cannot deal well with the cases where (a) a small number of features contributes to the identification of clusters and (b) different clusters reside in different subspaces of the original feature space. The latter situation is depicted in Figure 15.24.

A way to overcome these shortcomings is to develop special types of clustering algorithms that are able to search for clusters within the *various subspaces of the feature space*. In other words, these algorithms will reveal the clusters as well as the subspaces where they reside. Algorithms of this type are known as *subspace clustering algorithms* (SCAs). In contrast to all of the algorithms considered so far in the book, SCAs allow us to seek for clusters in different subspaces of the original feature space. SCAs algorithms can be divided into two categories: (a) grid-based

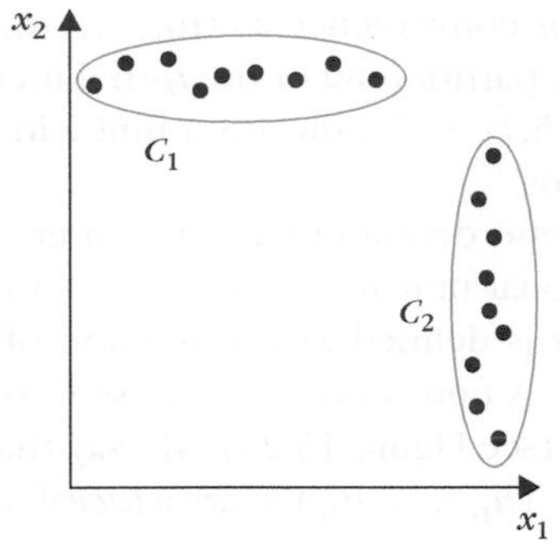

FIGURE 15.24

By projecting the data points in both x_1 and x_2 directions, it is readily seen that cluster C_1 lies in the x_2 subspace, whereas cluster C_2 lies in x_1.

SCAs and (b) point-based SCAs. In the sequel, we explore the basic philosophy behind the algorithms in each of the two categories and focus on some typical representatives.

Grid-based Subspace Clustering Algorithms (GBSCAs)

The main strategy adopted by these algorithms consists of the following steps: (a) identify the subspaces of the feature space that are *likely* to contain clusters, (b) determine the clusters lying in each of these subspaces, and (c) obtain descriptions of the resulting clusters.

The algorithms of this family apply an l-dimensional grid on the feature space and identify the subspaces that are likely to contain clusters, based on the k-dimensional units (boxes) ($k \leq l$) defined by the grid. However, the consideration of all possible subspaces becomes infeasible, especially when high-dimensional data sets are considered. To solve this problem, the algorithms establish certain criteria that are indicative of the presence of clusters in a subspace. These criteria must comply with the so-called *downward closure property*, which states that if a criterion is satisfied in a k-dimensional space, it is also satisfied in all of its $(k-1)$-dimensional subspaces. This allows the identification of the subspaces in an iterative bottom-up fashion, from lower to higher dimensional subspaces.

Having established the subspaces via the previously described procedure, the algorithms seek for clusters in each of them. Clusters are identified as maximally connected components of units in each subspace. In the sequel, we describe in more detail the CLIQUE and ENCLUS algorithms, which are among the most popular representatives of this family.

The CLIQUE (CLustering In QUEst) Algorithm

CLIQUE ([Agra 98]) partitions the feature space by applying an l-dimensional grid on it of edge size ξ (a user-defined parameter). Each unit u is written as $u_{t_1} \times \ldots \times u_{t_k}$

$(t_1 < \ldots < t_k, k \le l)$, or for convenience as $(u_{t_1}, \ldots, u_{t_k})$, where $u_{t_i} = [a_{t_i}, b_{t_i})$ is a right-open interval in the partitioning of the t_i-th dimension of the feature space. For example, $t_1 = 2, t_2 = 5, t_3 = 7$ indicates a unit lying in the subspace spanned by x_2, x_5, and x_7 dimensions.

Before we proceed, some definitions are in order. We say that a point $\boldsymbol{x}$ is *contained* in a k-dimensional unit $u = (u_{t_1}, \ldots, u_{t_k})$ if $a_{t_i} \le x_{t_i} < b_{t_i}$ for all t_i. The ***selectivity*** of a unit u is defined as the fraction of the total number of data points (N) contained in u. A unit u is called *dense* if its selectivity is greater than a user-defined threshold τ (see Figure 15.25). We say that two k-dimensional units, $u = (u_{t_1}, \ldots, u_{t_k})$ and $u' = (u'_{t_1}, \ldots, u'_{t_k})$, share a *face* if there are $(k-1)$ dimensions (e.g., $x_{t_1} \ldots, x_{t_{k-1}}$), such that $u_{t_j} = u'_{t_j}$, $j = 1, 2, \ldots, k-1$, and either $a_{t_k} = b'_{t_k}$ or $b_{t_k} = a'_{t_k}$. For example, units u_{12} and u_{22} in Figure 15.25 share a one-dimensional face. Two k-dimensional units u_1 and u_2 are said to be ***directly connected*** if they have in common a $(k-1)$-dimensional face. In addition, two k-dimensional units are said to be *connected* if there exists a sequence of k-dimensional units $v_1, \ldots, v_s$, with $v_1 = u_1$ and $v_s = u_2$, such that each pair (v_i, v_{i+1}) of units is directly connected. Finally, in the present framework, a *cluster* is defined as a maximal set of connected dense units in k dimensions.

We proceed now with the description of the algorithm. CLIQUE consists of three main stages. In the first stage identification of the subspaces that contain clusters takes place. This is carried out by first identifying all k-dimensional dense units $(1 \le k \le l)$ and then selecting those subspaces that contain dense units.

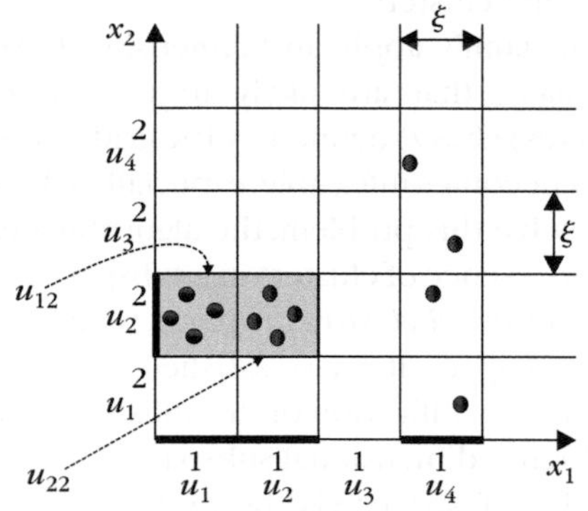

FIGURE 15.25

A two-dimensional grid of lines of edge size ξ is applied in the two-dimensional feature space, defining two-dimensional and one-dimensional boxes (units). In the figure, u_i^q denotes the i-th one-dimensional unit along x_q and u_{ij} denotes the two-dimensional unit resulting from the Cartesian product of the i-th and j-th intervals along x_1 and x_2, respectively. In addition, let $\tau = 3$. Then u_1^1, u_2^1, u_4^1, u_2^2 are one-dimensional dense units, each containing 4, 4, 4, and 9 points, respectively, whereas u_{12} and u_{22} are two-dimensional dense units, each containing 4 points.

(In other words, the criterion a subspace has to satisfy to be selected is to have at least one dense unit.) We proceed by identifying dense units in a bottom-up fashion from lower to higher dimensionality. First, all one-dimensional dense units are identified, along each dimension of the feature space. At each step, the set D_k of the k-dimensional dense units is determined based on the set D_{k-1} of the $(k-1)$-dimensional dense units. To this end, a k-dimensional dense unit u in D_k results from two $(k-1)$-dimensional dense units in D_{k-1} that share a $(k-2)$-dimensional face *and at the same time*, all $(k-1)$-dimensional projections must belong to D_{k-1}. Clearly, this procedure can be terminated to a dimension less than l. Having identified the dense units, the subspaces that contain at least one such unit can in turn be determined.

The rationale behind this approach relies on the *downward closure property* of the density, which states that: "if there is a dense unit u in a k-dimensional space, there are also dense units in the projections of u in all $(k-1)$-dimensional subspaces of the k-dimensional space" (Note how the downward closure property appears in Figure 15.25) [Pars 04].

This procedure may lead to an increasing number of dense units in all subspaces (especially for large l), which increases significantly the required computational time. This problem can be overcome as follows. After the completion of the procedure, we consider all subspaces that contain dense units and sort *the subspaces* according to their *coverage*; that is, the fraction of the number of points of the original data set they contain. Then, subspaces with large coverage are selected and the rest are pruned. The threshold under which a coverage is considered "low" is determined by the optimization of a suitably defined Minimum Description Length criterion function ([Agra 98]).

During the second stage, CLIQUE identifies the clusters. The input to this stage are the subspaces with high coverage as determined by the previous step. *The clusters are formed in each subspace separately*. Specifically, for each selected subspace the dense units in it are considered. One such unit is randomly picked, and all of the dense units connected to it are identified and we assign all of them to a new cluster C. If there are other dense units left in the current subspace, the same procedure is applied to form a second cluster, and so on.

Finally, the goal of the third stage is to derive a *minimal cluster description* for each cluster; that is, to express each cluster as the minimum possible union of hyperrectangular regions (see Figure 15.26a). This stage consists of two phases. During the first phase, we pick randomly a (dense) unit of a cluster C formed by the previous stage, and we grow it in both directions along a dimension, trying to cover as many units in C as possible. Then the unit is grown along the second direction (as in the previous case) and we continue until all dimensions have been considered once. If there are uncovered units, we repeat the procedure starting from a new uncovered point. For example, the description in Figure 15.26a may result by selecting u_1 and growing along x_1 and then along x_2, producing A. In a similar way, B is produced starting from the (uncovered by A) unit u_2. In the second phase we consider all covers produced for each cluster and remove those whose

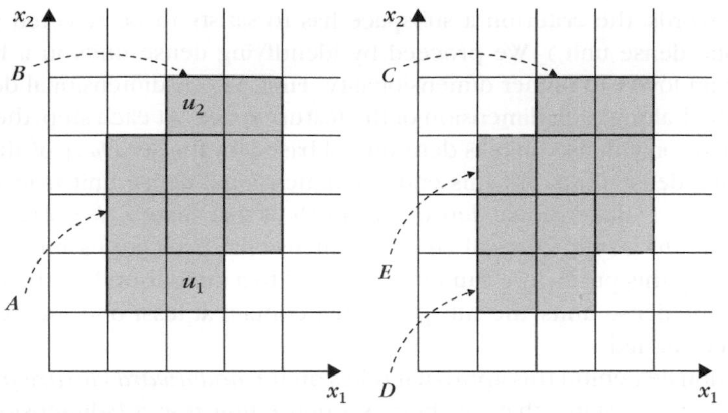

FIGURE 15.26

The minimal cluster description of the shaded region is $A \cup B$, shown in (a), whereas a nonminimal cluster description of the same region ($C \cup D \cup E$) is shown in (b).

units are covered by at least another cover. Example 15.10 will help us gain more insight on how CLIQUE works.

Example 15.10

Consider the two-dimensional data set shown in Figure 15.27, where the two-dimensional grid applied is also shown. By u_i^q, we denote the i-th one-dimensional unit along the q-th dimension, whereas by u_{ij} we denote the two-dimensional unit which results from the Cartesian product of the i-th unit along the first direction (x_1) times the j-th unit along the second direction (x_2). Assume that $\xi = 1$, which implies that $u_i^q = [i-1,\ i)$, and that $\tau = 8\%$ (since we have 69 points in total, each unit with more than 5 points is considered to be dense). In addition, the points in units u_{48}, u_{58}, u_{75}, u_{76}, u_{83}, and u_{93} are collinear in the direction of one of the two axes. This implies that, for example, u_{48} contributes to u_8^2 a single point. Similar observations hold for the rest of the units. Applying the first stage of the CLIQUE algorithm, we identify the set D_1 of the one-dimensional dense units, which equals

$$D_1 = \{u_2^1,\ u_3^1,\ u_4^1,\ u_5^1,\ u_8^1,\ u_9^1,\ u_1^2,\ u_2^2,\ u_3^2,\ u_5^2,\ u_6^2\}$$

Based on D_1, D_2 is then determined and it equals

$$D_2 = \{u_{21},\ u_{22},\ u_{32},\ u_{33},\ u_{83},\ u_{93}\}$$

Note that although each of the u_{48}, u_{58}, u_{75}, u_{76} contains more than 5 points they are not included in D_2, since each of them has one one-dimensional projection outside D_1 (for example, for u_{75}, u_5^2 belongs to D_1, wheras u_7^1 is not included in D_1). Furthermore, although it seems unnatural for u_{83} and u_{93} to be included in D_2 they are included, since u_3^2 is dense. However, u_3^2 is characterized as dense not because of the projections of the points in u_{83} and

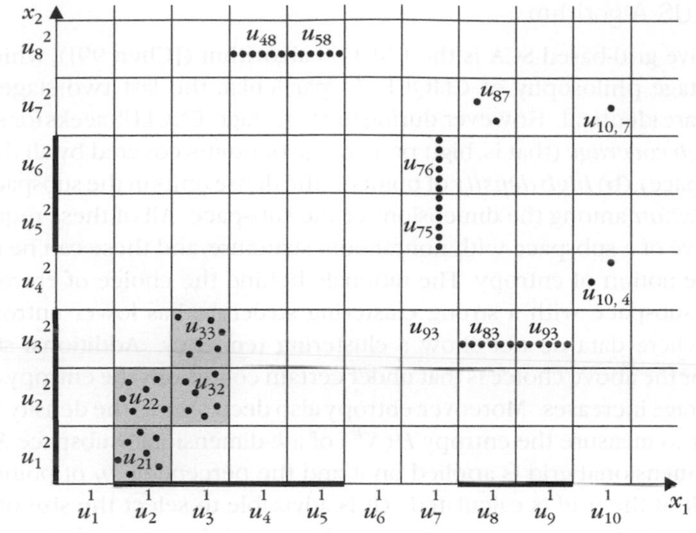

FIGURE 15.27

The set up of Example 15.10.

u_{93} but because of the projections of the points in u_{33}. Skipping the pruning step associated with the coverage criterion, the second stage of the algorithm ends up with the one-dimensional clusters $C_1 = \{u_2^1, u_3^1, u_4^1, u_5^1\}$, $C_2 = \{u_8^1, u_9^1\}$, $C_3 = \{u_1^2, u_2^2, u_3^2\}$, $C_4 = \{u_5^2, u_6^2\}$ and the two-dimensional clusters $C_5 = \{u_{21}, u_{22}, u_{32}, u_{33}\}$, $C_6 = \{u_{83}, u_{93}\}$.
Finally, the representation of the clusters after the completion of the third stage is as follows: $C_1 = \{(x_1) : 1 \le x_1 < 5\}$, $C_2 = \{(x_1) : 7 \le x_1 < 9\}$, $C_3 = \{(x_2) : 0 \le x_2 < 3\}$, $C_4 = \{(x_2) : 4 \le x_2 < 6\}$, $C_5 = \{(x_1, x_2) : 1 \le x_1 < 2,\ 0 \le x_2 < 2\} \cup \{(x_1, x_2) : 2 \le x_1 < 3,\ 1 \le x_2 < 3\}$, $C_6 = \{(x_1, x_2) : 7 \le x_1 < 9,\ 2 \le x_2 < 3\}$. Note that C_2 and C_6 are essentially the same cluster, which is reported twice by the algorithm.

Among the performance features of the CLIQUE algorithm is that it automatically determines the subspaces of the original feature space where high-density clusters exist. In addition, it is insensitive to the order the data are presented to the algorithm and does not impose any *data* distribution hypothesis on the data set. Also, it scales linearly with N but scales exponentially with l. In addition, the accuracy of the determined clusters may be degraded because the clusters are not given in terms of the points in X but as unions of dense units. Moreover, the performance of the algorithm is heavily dependent on the choices of ξ and τ, and it is not obvious how they must be selected in practice. Also, there is a large overlap among the reported clusters because for each dense unit (the structuring element of a cluster) all of its projections are also dense and contribute to clusters in lower-dimensional subspaces. Finally, there is a risk of losing small but meaningful clusters after the pruning of subspaces based on their coverage.

The ENCLUS Algorithm

An alternative grid-based SCA is the ENCLUS algorithm ([Chen 99]), which adopts the three-stage philosophy of CLIQUE. In particular, the last two stages of both algorithms are identical. However, during the first stage, ENCLUS seeks for subspaces with (a) *high coverage* (that is, high percentage of points covered by all dense units of the subspace), (b) *high density* of points in the dense units in the subspace, and (c) *high correlation* among the dimensions of the subspace. All of these requirements are indicative of a subspace with nonrandom structure, and these can be described utilizing the notion of entropy. The rationale behind the choice of entropy is that typically a subspace with a strong clustering structure has lower entropy than a subspace where data do not show a clustering tendency. Additional supportive evidence for the above choice is that under certain conditions the entropy decreases as the coverage increases. Moreover, entropy also decreases as the density increases.

In order to measure the entropy $H(X^k)$ of a k-dimensional subspace X^k ($k \leq l$) of X, a k-dimensional grid is applied on it and the percentage p_i of points that fall in each unit of the grid is calculated. (It is advisable to select the size of the edge of the grid such that each unit contains at least a minimum number of points. In [Devo 95] this minimum value is taken to be equal to 35.) Then, $H(X^k)$ is defined as

$$H(X^k) = -\sum_{i=1}^{n} p_i \log_2 p_i \tag{15.80}$$

where n is the total number of units. Also, an additional entropy-based measure, called *interest*, which quantifies the degree of correlation among the dimensions of a subspace, is defined as

$$\text{interest}\,(X^k) = \sum_{j=1}^{k} H(x_{t_j}^k) - H(X^k) \tag{15.81}$$

where $t_1 < \ldots < t_k$ ($k \leq l$) with $x_{t_j}^k$ denoting the j-th dimension of the X^k subspace. The higher the interest the stronger the correlation among the dimensions of X^k. Note that the minimum value of interest is 0, which is achieved when $x_{t_j}^k$ are *independent* from each other. (By definition, $x_{t_j}^k$s are *independent* if and only if $H(X^k) = \sum_{j=1}^{k} H(x_{t_j}^k)$ [Chen 99].)

A subspace with "low" entropy (below a user-defined threshold ω) is considered to have *good clustering*. A *significant* subspace is a subspace with good clustering *and* interest above a user-defined threshold ε. Following these necessary definitions, we are now ready to outline the way ENCLUS identifies significant subspaces. (In other words, a subspace is selected provided it satisfies the "low entropy" and "high interest" criteria.) Let B_k denote the set of significant k-dimensional subspaces and D_k the set of subspaces having good clustering but low interest. As in CLIQUE, a bottom-up strategy in the consideration of the subspaces is adopted. First, the set A_1 of all one-dimensional subspaces is considered and each is examined if it is significant or has a good clustering but is of low interest (below ε). In the former case, the subspace at hand is assigned to B_1, whereas the latter is assigned to D_1. In an iterative procedure, having defined B_k and D_k we determine the set A_{k+1} as follows: a $(k+1)$-dimensional subspace, $\mathcal{E}$, is assigned to A_{k+1} if (a) it is the result of the union of two k-dimensional subspaces in D_k, sharing $(k-1)$ dimensions, and (b) all k-dimensional projections of $\mathcal{E}$ belong to D_k. If A_{k+1} is nonempty, the procedure repeated. Otherwise, the procedure terminates and returns all significant subspaces found; that is, $B_1 \cup \ldots \cup B_k$.

The rationale behind this approach relies on the *downward closure property* of entropy, which states that if a k-dimensional space is of low entropy all of its $(k-1)$-dimensional subspaces are also of low entropy ([Chen 99]).

A variant of this scheme, based on a different criterion for measuring the correlation among the different dimensions of a subspace, is discussed in [Chen 99]. ENCLUS shares most of the features of CLIQUE; that is, insensitivity in the order of consideration of the data, the utility to unravel (in principle) arbitrarily shaped clusters, overlap among reported clusters, and linear dependence of the computational time on N. Also, the performance of ENCLUS is heavily dependent on the choice of the edge of the grid as well as on the parameters ω and ε.

Another algorithm of the grid-based SCA family is the so-called MAFIA algorithm (Merging of Adaptive Finite IntervAls) ([Goil 99]). In contrast to both algorithms discussed previously, where the grid applied on a subspace is static (that is, the edge size ξ is fixed for all dimensions), in MAFIA the grid is adaptively adjusted to match the distribution of the data. More specifically, the algorithm divides each dimension x_i of the feature space into (one-dimensional) windows of small size d. Then, it projects the points of the data set on each dimension, determines the projections that lie in each window, and sets the value of the corresponding window equal to the maximum among the projections lying in the window. In the sequel, it considers each dimension separately, scans its windows from left to right, and merges two adjacent windows if their values happen to be within a user-defined threshold β (a typical value for β is 20%). In the case where all windows in x_i are merged to a single one, which implies that the data are uniformly distributed along x_i, MAFIA applies on this a grid of fixed edge size. The unions of the windows resulting from this procedure are the one-dimensional units. Once these have been established, all other k-dimensional units can be defined $(1 \leq k \leq l)$, and the algorithm proceeds exactly as CLIQUE.

Parallel versions of the MAFIA algorithm are discussed in [Goil 99]. As it was the case with the CLIQUE and ENCLUS, MAFIA is sensitive on the choice of the related parameters. The required computational time increases linearly with N and performs better than the other two algorithms of this family, as far as the dimensionality is concerned. However, the computational time still grows exponentially as l increases ([Pars 04]). Finally, considering its parallel implementation as well as some other improvements, it is reported that MAFIA performs much faster than CLIQUE. Other algorithms that exploit the adaptive partition of each dimension are the *cell-based clustering method (CBF)* ([Chan 02]) and the *CLTree algorithm* ([Liu 00]).

Point-based Subspace Clustering Algorithms (PBSCA)

According to the philosophy of grid-based algorithms, clusters are defined as unions of dense units. In addition, a data point may "contribute" to more than one cluster in different subspaces through its projections. Moreover, the identification of clusters takes place after the establishment of the appropriate subspaces. In the present framework, a different philosophy is adopted. The clusters are defined in terms of data points and each data point contributes to a single cluster. Furthermore, the clusters as well as the subspaces in which they live are simultaneously determined in an iterative fashion. Two typical representatives of this category, the PROCLUS and ORCLUS, are described next.

The PROCLUS Algorithm

This algorithm ([Agga 99]) borrows concepts from the k-medoids algorithmic family, described in Chapter 14. Its main idea is to generate an initial set of medoids and to iterate until an "optimum" set of medoids results. However, at each iteration a special treatment is required in order to determine the subspace where each cluster resides. The number of clusters, m, as well as the *average dimensionality*, s, of the subspaces where the clusters lie, are given as inputs to the algorithm. PROCLUS consists of three main stages, namely, the *initialization stage*, the *iterative stage*, and the *refinement stage*.

In the initialization stage a sample X' of size am is generated via a random selection from the entire data set (a is a constant positive integer). Then, a subset X'' of X' consisting of bm points ($b < a$) is selected such that each of its points lies as far as possible from the remaining points in it. The latter set is likely to contain points from all "physical" clusters underlying X. Then, a set Θ with a corresponding index set I_Θ (for the notation see related section in Chapter 14) containing m randomly selected elements of X'' is formed. Its elements are taken as the initial estimates of the medoids of the m clusters. The iterative stage of the algorithm is outlined in the following in algorithmic form.

- Set $cost = \infty$
- $iter = 0$
- Repeat
  - $iter = iter + 1$
  - (A) For each $i \in I_\Theta$ determine the set of dimensions D_i of the subspace where cluster C_i lives.
  - (B) For each $i \in I_\Theta$ determine the corresponding cluster C_i.
  - (C) Compute the cost $J(\Theta)$ associated with Θ.

- if $J(\Theta) < cost$ then
  - $\Theta_{best} = \Theta$
  - $cost = J(\Theta_{best})$
- End { if }
- (D) Determine the "bad" medoids of Θ_{best}.
- Set Θ equal to Θ_{best} and replace its bad medoids with randomly selected points from X''.

■ Until a termination condition is satisfied.

In the sequel we describe in more detail the steps (A) through (D) of the algorithm.

■ *(A) Determination of the cluster subspaces:* For each $\boldsymbol{x}_i, i \in I_\Theta$, the minimum among its distances from all other medoids in Θ is computed; that is, $\delta_i = \min_{r \in I_\Theta - \{i\}} d(\boldsymbol{x}_i, \boldsymbol{x}_r)$. Then, for each $\boldsymbol{x}_i$, $i \in I_\Theta$, the set L_i containing the points of X that lie in the sphere of radius δ_i, centered at $\boldsymbol{x}_i$, is determined. Along each dimension, j, of the feature space, the average distance between each $\boldsymbol{x} \in L_i$ and $\boldsymbol{x}_i$, $i \in I_\Theta$ is calculated (i.e., $d_{ij} = \sum_{\boldsymbol{x} \in L_i} |x_j - x_{ij}|/|L_i|$), where $|L_i|$ is the cardinality of L_i. In the sequel, for each $\boldsymbol{x}_i$, $i \in I_\Theta$, the average distance and standard deviation along all dimensions is computed; that is, $e_i = (\sum_{j=1}^{l} d_{ij})/l$ and $\sigma_i = \sqrt{\sum_{j=1}^{l}(d_{ij} - e_i)^2/(l-1)}$.

Then, for each medoid $\boldsymbol{x}_i$, $i \in I_\Theta$ and for each dimension of the feature space, the value $z_{ij} = (d_{ij} - e_i)/\sigma_i$ is computed. Clearly, the smaller the value of z_{ij} the more concentrated the points in L_i around $\boldsymbol{x}_i$ along the j-th dimension are. Then the D_is are determined based exclusively on the z_{ij}s values, under the following conditions: (a) each D_i has at least two dimensions and (b) the total number of dimensions contained in all D_is is equal to sm, where s is the (user-defined) average dimensionality of the cluster subspaces. Specifically, for each medoid $\boldsymbol{x}_i$, $i \in I_\Theta$ the two dimensions with the smallest z_{ij} values among the z_{iq}s, $q = 1, \ldots, l$, are assigned to D_i. Then, all the rest $m(l-2)$ z_{ij} values, for all medoids, are simultaneously considered, the $m(s-2)$ lowest of them are identified, and the corresponding dimensions are assigned to the appropriate D_i. If, for example, z_{34} is among the $m(s-2)$ lowest values, the fourth dimension will be assigned to D_3.

■ *(B) Determination of the clusters:* For each data point, its *Manhattan segmental distance* from each medoid $\boldsymbol{x}_i$, $i \in I_\Theta$ is computed; that is,

$$d_{D_i}(\boldsymbol{x}, \boldsymbol{x}_i) = \frac{\sum_{j \in D_i} |x_j - x_{ij}|}{|D_i|} \tag{15.82}$$

Note that in $d_{D_i}(\boldsymbol{x}, \boldsymbol{x}_i)$ only the coordinates in the dimensions that belong to D_i are taken into account. In addition, this distance is defined as an average and not as a summation. Then $\boldsymbol{x}$ is assigned to the cluster whose medoid lies closer to it.

- *(C) Computation of $J(\Theta)$:* $J(\Theta)$ is defined as the average Manhattan segmental distance between the points of X and the *means* of the clusters to which they belong; that is,

$$J(\Theta) = \frac{1}{N} \sum_{i=1}^{m} \sum_{\boldsymbol{x} \in C_i} d_{D_i}(\boldsymbol{x}, \boldsymbol{m}_i) \tag{15.83}$$

where $\boldsymbol{m}_i$ is the mean of the vectors in C_i.

- *(D) Determination of "bad" medoids:* A medoid is considered to be a "bad" one if (a) its corresponding cluster has the least number of points and (b) its corresponding cluster has less than $(N/m)q$ points, where q is a user-defined constant (typically $q = 0.1$). The rationale behind this rule is that a medoid whose corresponding cluster is of small size is likely to be an outlier or to belong to a "physical" cluster that contains at least one of the other medoids in Θ.

Finally, in the refinement stage for the set Θ_{best} that has been determined during the iteration stage the sets D_i are recomputed by applying the (A) step of the iteration stage on the C_is resulting from the iteration stage, rather than the L_is. Once the new D_is have been established, the C_is are recomputed based on the new D_is.

Like many other algorithms of this algorithmic class, PROCLUS is biased toward hyperspherically shaped clusters. In addition, the cluster subspaces must be of similar size since the average subspace dimensionality is required as input to the algorithm. Also, special care is required during the initialization stage in order to get representative points (in the set X'') from all the ("physical") clusters underlying the data set X. Otherwise, some clusters will not be recovered. In general, PROCLUS is sensitive to the input parameters it requires, which are not always easy to determine. On the other hand, PROCLUS is somewhat faster than CLIQUE on large data sets ([Pars 04]) and the required computational effort increases linearly with the dimension of the feature space, l ([Agga 99]).

The ORCLUS Algorithm

This is a point-based SCA algorithm of an agglomerative hierarchical nature ([Agga 00]). However, unlike the classical agglomerative hierarchical algorithms apart from reducing the number of clusters at each iteration (down to a user-defined value m) ORCLUS also successively reduces the dimensionality of the subspaces, where the clusters lie, down to a (user-defined) dimensionality l. (Only for this section, we denote the dimensionality of the feature space with l_0 instead of l. With l we denote the user-defined value to which the dimensionality of the cluster subspaces will gradually converge.) At each iteration of the algorithm the number of clusters as well as the dimensionality of the subspace of each cluster are reduced by user-defined factors $a < 1$ (a typical value for a is 0.5) and $b < 1$, respectively. However, a and b must be chosen so that the reduction of the initial number of clusters m_0, down to m, and the reduction of the initial dimensionality l_0 (of the original feature space), down to l, to be achieved in the same number of iterations. Assuming that t is the total number of iterations, we have that $m = a^t m_0$ and $l = b^t l_0$, which implies that a and b are related via the condition

$$\frac{\ln(m/m_0)}{\ln(l/l_0)} = \frac{\ln a}{\ln b} \tag{15.84}$$

In addition, the subspace where each cluster lies is represented by a set of vectors, which are not necessarily parallel to the axes of the original feature space. Specifically, the set of vectors $\mathcal{E}_i$, defining the "best" q-dimensional subspace for the cluster C_i, is chosen as the subspace where the points of C_i exhibit the least spread (highest concentration). Hence, $\mathcal{E}_i$ consists of the eigenvectors of the $l_0 \times l_0$ covariance matrix Σ_i of the points in C_i that correspond to the q smallest eigenvalues of Σ_i (see also Section 6.3). The sum of these eigenvalues is the (projected) *energy* of the cluster C_i in $\mathcal{E}_i$, denoted by $E(C_i, \mathcal{E}_i)$. The ORCLUS algorithm is summarized as follows.

- Generate a set S_0 consisting of $m_0 (> m)$ points selected randomly from X.
- Set $m_c = m_0, l_c = l_0$ and $S_c = S_0$.
- Set $\mathcal{E}_i$ equal to the set of vectors defining the original feature space, for $i = 1, \ldots, m_c$.
- Set $a = 0.5$ and compute b by solving Equation (15.84).
- While $m_c > m$ do
  - For each i, $i = 1, \ldots, m_c$, define the C_i cluster as the one containing all points in X that lie closer to the i-th element of S_c. (In this case, the distance between two points is computed in the $\mathcal{E}_i$ subspace.)
  - For each $i, i = 1, \ldots, m_c$, define $\mathcal{E}_i$ as the set of eigenvectors corresponding to the l_c smallest eigenvalues of the $l_0 \times l_0$ covariance matrix of C_i.
  - Set $m_{new} = \max\{m, am_c\}$ and $l_{new} = \max\{l, bl_c\}$.
  - For each pair $(C_i, C_j), i, j = 1, \ldots, m_c, i < j$ determine $\mathcal{E}_{ij}$ for the $C_i \cup C_j$ as well as $E(C_i \cup C_j, \mathcal{E}_{ij})$.
  - While $m_c > m_{new}$ do
    - Determine $E(C_u \cup C_v, \mathcal{E}_{uv}) = \min_{i,j=1,\ldots,m_c,\ i \neq j} E(C_i \cup C_j, \mathcal{E}_{ij})$ and merge C_u and C_v to $C_r = C_u \cup C_v$.
    - Recompute the necessary $E(C_i \cup C_r, \mathcal{E}_{ir})$s in light of the previous merging.
    - $m_c = m_c - 1$

- End { While }. (Note that during this *While* loop the subspace dimension remains unchanged.)

- $m_c = m_{\text{new}}$
- $l_c = l_{\text{new}}$
- Set S_c equal to the means of the m_{new} clusters formed by the previous *While* loop.
- End { While }

As PROCLUS, ORCLUS is biased toward hyperspherical clusters, due to the fact that the mean of a cluster is used as its representative. In addition, the required computational time is $O(m_0^3 + m_0 N l_0 + m_0^2 l_0^3)$. That is, it increases linearly with N and cubically with l_0. It is worth noting that although increasing the value of m_0 increases the computational time this may improve the quality of the final clustering. Furthermore, note that the subspaces of all clusters are restricted to the same dimensionality. Criteria for choosing a proper value for l as well as extensions of the algorithm that are able to handle outliers are discussed in [Agga 00]. Finally, random sampling techniques may be used to reduce the computational time. Other point-based subspace clustering algorithms are discussed in [Frie 02, Woo 02], and [Yang 02].

Remarks

- In GBSCAs, the clusters are represented as unions of dense units (which is a rather "rough" description), whereas in PBSCAs the clusters are represented in terms of data points (exact description). Moreover, in GBSCAs each point may contribute to more than one cluster in different subspaces through its projections, whereas in the PBSCAs each point contributes to a single cluster.
- In GBSCAs the identification of clusters is carried out only after the appropriate subspaces have been determined. In contrast, in PBSCAs the clusters as well as the subspaces where they lie are simultaneously determined in an iterative fashion.
- The GBSCAs are able, in principle, to unravel arbitrarily shaped clusters, whereas several PBSCAs are biased toward hyperellepsoidal clusters.
- The computational time required by most of the GBSCAs and PBSCAs scales linearly with N, the number of points. (For PROCLUS this has been established experimentally [Agga 99].)
- The computational time required by the described GBSCAs increases exponentially with the dimensionality of the feature space l, whereas in PBSCAs it exhibits a polynomial dependence.
- In the GBSCAs there are no restrictions concerning the dimensionality of the subspaces, wheras the PBSCAs pose constraints on it.

- In GBSCAs there exists a large overlap in the resulting clusters. On the contrary, most of the PBSCAs produce disjoint clusters.
- Both GBSCAs and PBSCAs are sensitive to the choice of the involved user-defined parameters.

15.11 OTHER CLUSTERING ALGORITHMS

A clustering algorithm that is based on the so-called *tabu search method* is presented in [Al-S 95]. Its initial state is an arbitrarily chosen clustering. The algorithm proceeds as follows. Based on the current state of the algorithm, a set of candidate clusterings, A, is created. The next state is chosen to be the "best" element of A, according to some criterion function. Certain criteria are used to prevent the algorithm from returning to recently visited states. The procedure is repeated for a prespecified number of iterations. Preliminary results reported in [Al-S 95] show that this algorithm compares favorably with the hard clustering and the simulated annealing algorithms. A recent tabu search based heuristic scheme for clustering is presented in [Sung 00].

A method that directly relates clusters to peak values of the pdf has been suggested in [Tou 74], where the estimation of the pdf is achieved via Parzen windows. A related method is the *mountain method* (see, e.g., [Dave 97]). The idea is to assign to each vector, $\boldsymbol{x}$, an energy source. The generated potential has a peak at $\boldsymbol{x}$ and rapidly decays as we move away from it. The total potential function at a specific point in the vector space is the summation of the potentials produced by all the vectors of X. We compute the value of this function at each data point and form the array $\boldsymbol{v}$ of the N resulting values. The maximum one, which corresponds to the highest peak, is identified and the corresponding vector is considered as the representative of the first cluster. Then we remove the largest value of $\boldsymbol{v}$ and update the rest of the components appropriately. This procedure is repeated until a specific termination criterion is met.

An algorithm that combines ideas from both the fuzzy clustering schemes and the agglomerative algorithms (Chapter 13) is discussed in [Frig 97]. This scheme produces clusterings that minimize the following cost function

$$J(\boldsymbol{\theta}, U) = \sum_{i=1}^{N}\sum_{j=1}^{m} u_{ij}^2 d(\boldsymbol{x}_i, C_j) - a\sum_{j=1}^{m}\left[\sum_{i=1}^{N} u_{ij}^2\right] \tag{15.85}$$

where now m varies. Clearly, the first term in the above equation is minimized when $m = N$, and the second term is maximized when $m = 1$. It is worth mentioning that the hierarchies of clusterings produced do not necessarily possess the nested property.

Besides the preceding algorithms, many other clustering algorithms based on very different ideas have been proposed. For example, in [Matt 91] a scheme that does not use the concept of the distance between vectors is proposed. Also, in

[Kodr 88] a clustering technique based on a conceptual distance is presented. Clustering approaches that borrow concepts from gravity theory are discussed in [Oyan 01, Chen 05]. Clustering algorithms, suitable for discrete-valued feature vectors, that construct classification trees are discussed in [Fish 87, Bisw 98]. A graph theory-based algorithm that uses a probabilistic framework is discussed in [Ben 99].

Another technique combining supervised and unsupervised methods has been proposed in [Pedr 97]. The latter may be useful in applications in which only a fraction of the data have a class label for the training.

In [Robe 00], a different clustering method is discussed. The clustering problem is stated in information theoretic terms and it is shown that minimization of the entropy can be used in order to estimate the clustering structure underlying the data set X. This is equivalent to obtaining the structure associated with maximum certainty. Another algorithm based on information theoretic criteria has been suggested in [Goks 02]. It is a valley-seeking algorithm and builds upon Renyi's entropy estimator.

Another interesting clustering algorithm that utilizes the wavelet transform (see Section 6.13) is the so-called *WaveCluster* algorithm ([Shei 98]). The method applies, first, an l-dimensional grid on the feature space by dividing each dimension in r intervals and it determines the data points contained in each unit (box), M_i, of the grid. In the sequel, an l-dimensional signal is generated by representing each unit by the number of points it contains. Then, it applies the l-dimensional wavelet transform on the units (points) M_i of the grid (the multidimensional wavelet transform is actually a generalization of the two-dimensional wavelet transform discussed in Section 6.14) and produces a new set of units T_j in the transformed space at various resolutions. Then, at each resolution it determines the clusters in the transformed space as connected components of T_j units. In the sequel, based on the correspondence between M_is and T_js the algorithm assigns the points contained in each M_i to the appropriate cluster. Among the advantages of WaveCluster is the efficient handling of outliers, its insensitivity to the order of the presentation of the points to the algorithm, its ability to determine arbitrarily shaped clusters, and the fact that it does not require as input the exact number of the clusters. Its computational complexity is $O(N + r^l)$. Thus, the algorithm is best suited for large data sets of relatively low dimensionality.

In recent years there has been an increasing interest in clustering sequential data. DNA sequencing and web data mining are examples of two typical applications of this type. Techniques for clustering this type of data are based on tools discussed in Chapters 8 and 9, with the Edit distance and the HMM being among the most popular. A major problem associated with these applications is the very long length of the sequences to be matched, which renders most of the classical algorithms computationally infeasible. To overcome such difficulties, a number of algorithmic schemes have been suggested adopting various heuristics. Some well-known schemes include BLAST [Alts 97] and FASTA [Pear 88].

Reviews of such clustering methods and applications are discussed in, for example, [Durb 98, Gusf 97, Beng 99, Mill 01, Liew 05].

Constraint clustering

In many cases, the use of labeled data is critical for the success of the clustering process. A subset of labeled data also makes the evaluation of the resulting clustering more accurate. Consequently, several clustering approches have been introduced that use both labeled and unlabeled data, as it was the case with the semi-supervised learning treated in chapter 10.

Most of these approaches rely on getting input from the user in terms of constraints on feasible clusterings. The simplest constraints are placed on pairs of patterns and either force both patterns to be in the same cluster (*Must-Link constraint*), or force them to be in different clusters (*Cannot-Link constraint*). Initial efforts concentrated on modifying existing algorithms, such as the *k*-means or hierarchical algorithms, to operate with constraints ([Wags 01, Davi 05]). More recent approaches use the constraints imposed by the user to learn distance measures that conform to the user expectation, as this is expressed by the given constraints ([Basu 04, Xing 02, Kuli 05, Halk 08]).

In [Bene 00, Bane 02] two modifications of the *k*-means algorithm are proposed that deal with the case where the number of points in each cluster is bounded below. Also, an approach that builds balanced clusters is discussed in [Stre 00]. In [Tung 01] the problem of clustering two-dimensional data in the presence of *obstacles* in the feature space is considered. In this case, the distance between two points is defined as the shortest path from one point to the other, taking into account the obstacles in the feature space.

15.12 COMBINATION OF CLUSTERINGS

Throughout the second part of the book, we discussed algorithms that produce a single clustering (or a hierarchy of clusterings) for a given data set $X = \{\boldsymbol{x}_1, \boldsymbol{x}_2, \ldots, \boldsymbol{x}_N\}$, $\boldsymbol{x}_i \in \mathcal{R}^l$, $i = 1, 2, \ldots, N$. In this section, the situation is different. A (final) clustering is obtained based on a set of n different clusterings of X. Specifically, the aim here is two-fold: (a) the production of an appropriate set of clusterings, $\mathcal{E}$, for the data set, X, called *ensemble* of clusterings, and (b) the combination of the clusterings of $\mathcal{E}$ to produce a final clustering, called *consensus clustering*. The main motivation for considering such techniques is that it is expected that the resulting consensus clustering, based on $\mathcal{E}$, will model the data better that any single clustering.

Various techniques have been proposed in the above spirit but none of them seems to clearly outperform the rest ([Topc 05]). In the sequel, after giving some necessary definitions, we consider separately the two previously stated goals and we focus on the most representative methods for each case.

Let $\mathcal{E} = \{\mathcal{R}_1, \mathcal{R}_2, \ldots, \mathcal{R}_n\}$ be a set of clusterings of the data set X, where $\mathcal{R}_i = \{C_i^1, C_i^2, \ldots, C_i^{m_i}\}$, with m_i being the number of clusters in the $\mathcal{R}_i$ clustering. The superscript j in C_i^j denotes the *label* of the corresponding cluster in the $\mathcal{R}_i$ clustering. Alternatively, each clustering, $\mathcal{R}_i$, can be represented by an N dimensional row vector, $\boldsymbol{y}_i$, called *label vector*, whose k-th element $y_i(k)$ contains the cluster label of the k-th data point. If for example $\mathcal{R}_i = \{C_i^1, C_i^2, C_i^3\} = \{\{\boldsymbol{x}_1, \boldsymbol{x}_2, \boldsymbol{x}_6, \boldsymbol{x}_{10}\}, \{\boldsymbol{x}_3, \boldsymbol{x}_4, \boldsymbol{x}_7\}, \{\boldsymbol{x}_5, \boldsymbol{x}_8, \boldsymbol{x}_9\}\}$, then $\boldsymbol{y}_i = [1, 1, 2, 2, 3, 1, 2, 3, 3, 1]$. Clearly, both $\mathcal{R}_i$ and $\boldsymbol{y}_i$ are equivalent representations of the same clustering.

A. Generation of an ensemble of clusterings

Generation of each clustering $\mathcal{R}_i$ of $\mathcal{E}$ involves the following two steps: (a) the choice of a subspace to project the data points in X and (b) the application of a clustering algorithm on this subspace. Note that, in general, the $\mathcal{R}_i$s are not constrained to have the same number of clusters. Unless otherwise stated, this assumption has been adopted here.

In analogy with the schemes for combining classifiers (Section 4.21), it is desirable for $\mathcal{E}$ to contain clusterings that are as "independent" as possible. To this end, the following general directions are followed:

- ***All data, all features.*** In this case, all l features and all N data points are used. The n different clusterings, $\mathcal{R}_i$, $i = 1, 2, \ldots, n$, result by employing either different clustering algorithms or the same clustering algorithm with different parameters (e.g., in the case of the k-means algorithm, different initial conditions or different distance measures can be used) (see, e.g., [Fred 05]). This method of generating the $\mathcal{R}_i$s is also called *robust centralized clustering* ([Stre 02]).

- ***All data, some features.*** In this case, all the data points of X are considered. A number of n sets, $X_i, i = 1, \ldots, n$, are formed from X, where in each one of them the data points are represented (a) either by selecting a subset of features or (b) by projecting onto a randomly chosen lower dimensional space, see for example, [Fern 03, Topc 05] (see Section 15.10.1). Clearly, the cardinality of each X_i is N. In the sequel, eihter the same algorithm (also called *base algorithm*) with the same parameters (e.g., [Fern 04]) or different algorithms (e.g., [Stre 02]) are applied on the X_is in order to produce the respective clusterings $\mathcal{R}_i$s. This method of generating the $\mathcal{R}_i$s is also called *feature-distributed clustering* ([Stre 02]).

- ***Some data, all features.*** In this case, techniques like bootstrapping or sampling are applied on X, in order to produce data sets $X_i, i = 1, \ldots, n$, on which (usually) the same clustering algorithm is applied to produce the respective n clusterings of $\mathcal{E}$. The points of X, which have not been selected to participate in X_i, may be assigned to their nearest cluster in $\mathcal{R}_i$. When X is a high dimensional data set, its points may first be projected to a lower dimensional space (using, e.g., PCA), forming a lower dimensional data set X'. Then the method that generates the X_is is applied on the reduced dimensionality data set X' (e.g., [Fern 04]).

B. Combination of clusterings

Having generated the clustering ensemble, $\mathcal{E} = \{\mathcal{R}_1, \mathcal{R}_2, \ldots, \mathcal{R}_n\}$, the next step is to combine the $\mathcal{R}_i$s in order to produce the final (consensus) clustering $\mathcal{F} = \{F_1, \ldots, F_m\}$.

Among the various combination techniques that have been proposed, several of them make use of the so-called *co-association matrix* $\mathcal{C}$. This is an $N \times N$ dimensional matrix, whose (i, j) element $c(i, j)$ equals to n_{ij}/n, where n_{ij} is the number of times the ith and the jth points of X have been assigned to the same cluster, among the n clusterings of $\mathcal{E}$. Note that $c(i, j) \in [0, 1]$. Clearly, large values of $c(i, j)$ imply that the ith and jth points of X are likely to be similar to each other.

In the sequel, a brief description is given for the most popular methods for combining clusterings.

- *Methods based on the co-association matrix.* The co-association matrix $\mathcal{C}$ is computed and then, using $\mathcal{C}$ as a similarity matrix, the single link algorithm is applied. From the resulting dendrogram, the clustering with the larger lifetime (see Section 13.6) is selected as the final clustering ([Fred 05]). Other hierarchical clustering algorithms, such as the complete link and the average link, can also be applied on $\mathcal{C}$ (e.g., [Topc 05]).

  In general, these methods require a large number of clusterings n, in order to estimate more reliably the elements of $\mathcal{C}$.

- *Graph-based methods.* In the sequel, we discuss three different formulations in this framework.
  - *Instance-based graph formulation (IBGF).* A fully connected (complete) graph $G(V, E)$ is constructed, where *each vertex of* V *corresponds to a data point* and each edge e_{ij} is weighted by $c(i, j)$ (the (i, j) element of the co-association matrix). Then the graph is partitioned into m disjoint subsets of vertices $V_1, \ldots, V_m$ such that: (i) the sum of the weights of the edges that connect vertices from different subsets is minimized and (ii) the V_is are approximately of the same size (note, however, that (ii) is not always required). To this end, different optimization criteria are used, such as the *normalized cut criterion* ([Shi 00]) and the *ratio-cut criterion* ([Hage 98]) (see also Section 15.2.4 on spectral clustering).
  - *Cluster-based graph formulation (CBGF).* Let

$$\mathcal{E}' = \{C_1^1, \ldots, C_1^{m_1}, C_2^1, \ldots, C_2^{m_2}, \ldots, C_n^1, \ldots, C_n^{m_n}\}$$

be the set of all the clusters contained in the n clusterings of $\mathcal{E}$ and let $t = \sum_{i=1}^{n} m_i$ be its cardinality. In this case, a graph $G = (V, E)$ is constructed where *each vertex of* V *corresponds to a cluster in* $\mathcal{E}'$. Also, the weight w_{ij}

of the edge connecting the vertices associated with the clusters $C_i, C_j \in \mathcal{E}'$ is defined as the *Jaccard measure*, which is the ratio of the number of the data points in $C_i \cap C_j$ to the number of the data points in $C_i \cup C_j$, that is, $w_{ij} = \frac{|C_i \cap C_j|}{|C_i \cup C_j|}$. Then, an m-partition $\mathcal{P} = \{P_1, \ldots, P_m\}$ of the graph is obtained, under the constraints (i) and (ii) used in the previous case (Note that each P_i corresponds to a set of clusters). Ultimately, the final clustering $\mathcal{F} = \{F_1, \ldots, F_m\}$ is obtained as follows: For each data point $\boldsymbol{x} \in X$ we count the number of its occurrences in the clusters contained in $P_i, i = 1, \ldots, m$. Then $\boldsymbol{x}$ is assigned to F_i, if it is more frequently met in P_i ([Fern 04]).

- *Hybrid bipartite graph formulation (HBGF).* Let $\mathcal{E}'$ and t be defined as in CBFG. A graph $G = (V, E)$ is constructed, however, in this case, *data points as well as clusters in $\mathcal{E}'$ are represented by vertices.* Thus, V contains a total of $N + t$ vertices. The weight w_{ij} between the v_i and v_j vertices of the graph equals to 0 if either both vertices correspond to data points or both correspond to clusters. Otherwise, if v_j corresponds to a cluster and v_i to a point *and in addition* the point belongs to this cluster, then $w_{ij} = w_{ji} = 1$. In all other cases we set $w_{ij} = 0$. A graph clustering approach (e.g., spectral clustering) is then applied to the previous graph, leading to the consensus (final) clustering of the data in X.

  Consider, for example, the following case: $\mathcal{E} = \{\mathcal{R}_1, \mathcal{R}_2\}$, where $\mathcal{R}_1 = \{C_1^1, C_1^2\} = \{\{\boldsymbol{x}_1, \ldots, \boldsymbol{x}_6\}, \{\boldsymbol{x}_7, \boldsymbol{x}_8, \boldsymbol{x}_9\}\}$ and $\mathcal{R}_2 = \{C_2^1, C_2^2\} = \{\{\boldsymbol{x}_1, \ldots, \boldsymbol{x}_5\}, \{\boldsymbol{x}_6, \ldots, \boldsymbol{x}_9\}\}$. The corresponding graph, for this case, is shown in Figure 15.28. The line shows a bi-partition of the graph which implies that the final clustering is $\{\{\boldsymbol{x}_1, \ldots, \boldsymbol{x}_5\}, \{\boldsymbol{x}_6, \ldots, \boldsymbol{x}_9\}\}$.

In general, IBGF exhibits higher computational complexity compared to CBGF and HBGF. This happens because in IBGF a fully connected graph is generated and the graph partitioning problem is of size $O(N^2)$. On the other hand, in CBGF the size of the problem is $O(t^2)$, where (usually) $t \ll N$, while in HBGF the size of the problem is $O(nN)$.

Experimental results ([Fern 04]) suggest that HBGF achieves comparable or better performance than IBGF, CBGF. Also, HBGF consistently improves over the average performance of the members of the clustering ensemble $\mathcal{E}$ (this is measured in terms of data sets, where the true clusters are known). In addition, all methods perform, in general, better as n increases.

Another graph-based method is discussed in [Ayad 03]. A graph $G = (V, E)$ is constructed where each vertex of V corresponds to a data point of X and the co-association matrix is used to define the *nearest neighbor vertices* of a given vertex. In the sequel (a) each edge between two vertices v_i and v_j is weighted accordingly, depending on the number of their common nearest neighbors, n_{ij}, provided that n_{ij} exceeds a certain threshold (otherwise, the edge weight is set to 0) and (b) each vertex is weighted according to the number of its nearest neighbor vertices that shares with other vertices. Having

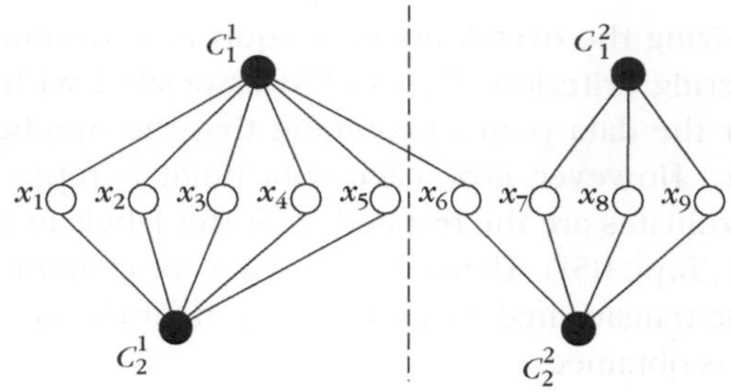

FIGURE 15.28

The graph constructed by HBGF for an ensemble of two clusterings, each one containing two clusters. Only edges associated with nonzero weights are shown. The dashed line cuts the graph and it defines the ensemble clustering (see text for more explanation).

defined G, a graph partition algorithm is employed to determine the final clustering of X, subject to some constraints imposed on the weights of the vertices.

- *Cost function optimization methods.* In this framework, the ensemble clustering $\mathcal{F} = \{F_1, \ldots, F_m\}$ is obtained via cost optimization techniques. In the sequel, three such methods are described.
  - *Utility function optimization.* According to this method, $\mathcal{F}$, also called *median clustering*, is defined as the clustering that "summarizes" best the clusterings of $\mathcal{E}$, according to the *utility function criterion* ([Fish 87, Topc 05]). The latter measures the quality of a candidate median clustering $\mathcal{F}$ against some other clustering $\mathcal{R}_i$ and it is defined as

$$U(\mathcal{F}, \mathcal{R}_i) = \sum_{r=1}^{m} P(F_r) \sum_{j=1}^{m_i} P(C_i^j|F_r)^2 - \sum_{j=1}^{m_i} P(C_i^j)^2 \tag{15.86}$$

where $P(F_r) = |F_r|/N, P(C_i^j) = |C_i^j|/N$ and $P(C_i^j|F_r) = |C_i^j \cap F_r|/|F_r|$, with $|A|$ denoting the cardinality of the set A. $U(\mathcal{F}, \mathcal{R}_i)$ measures the agreement between the two clusterings $\mathcal{R}_i$ and $\mathcal{F}$. It turns out that $U(\mathcal{F}, \mathcal{R}_i)$ achieves its maximum value when $\mathcal{F}$ and $\mathcal{R}_i$ are identical (in this case, the probabilities $P(C_i^j|F_r)$ are either one or zero), while as $\mathcal{F}$ and $\mathcal{R}_i$ deviate from each other $U(\mathcal{F}, \mathcal{R}_i)$ decreases.

The *overall utility* of $\mathcal{F}$ on $\mathcal{E}$ is defined as

$$U(\mathcal{F}, \mathcal{E}) = \sum_{i=1}^{n} U(\mathcal{F}, \mathcal{R}_i) \tag{15.87}$$

The best summary of the clusterings of $\mathcal{E}$ (the median clustering) follows from the maximization of the cost given in (15.87). In [Mirk 01], it is shown that maximizing the overall utility is equivalent to minimizing the square error clustering criterion (Eq. (14.85) associated with the k-means algorithm) over the data points (assuming that the number of clusters m in $\mathcal{F}$ is fixed). However, now, each data point is represented by a vector whose coordinates are the respective cluster labels in the n clusterings of $\mathcal{E}$ ([Mirk 01, Topc 05]). Hence by applying, for example, the k-means algorithm on the transformed data set a solution for the maximization problem of $U(\mathcal{F}, \mathcal{E})$ is obtained.

- *Normalized mutual information (NMI) criterion.* This criterion has been inspired from the information theory, where the statistical information shared between two distributions is assessed by the *mutual information measure*. In the context of clustering, NMI is a measure analogous to the mutual information between two distributions and it is defined as

$$NMI(\mathcal{R}_1, \mathcal{R}_2) = \frac{2}{N} \sum_{q=1}^{m_1} \sum_{r=1}^{m_2} n_q^r \log_{m_1 m_2} \left(\frac{n_q^r N}{n_q n_r} \right) \tag{15.88}$$

where m_1, m_2 are the number of clusters in $\mathcal{R}_1$ and $\mathcal{R}_2$, respectively, $n_q^r = |C_1^q \cap C_2^r|$, $n_q = |C_1^q|$, $n_r = |C_2^r|$, where $|A|$ denotes the cardinality of a set A. Let us define the *average normalized mutual information criterion (ANMI)* between a clustering $\mathcal{F}$ and an ensemble of clusterings $\mathcal{E}$ as follows

$$ANMI(\mathcal{F}, \mathcal{E}) = \frac{1}{n} \sum_{i=1}^{n} NMI(\mathcal{F}, \mathcal{R}_i) \tag{15.89}$$

Then the ensemble clustering $\mathcal{F}$ is obtained as the one that maximizes the previous criterion ([Stre 02]).

It is worth noting that NMI may also be used to assess the performance of a clustering algorithm when it is performed on a data set where the labels of the data points are known (e.g. [Fred 05]).

- *Mixture model formulation.* Such a formulation is discussed in [Topc 05]. In this framework, each data point $\boldsymbol{x}_i \in X$ is represented by a new n-dimensional vector, $\boldsymbol{x}_i'$, whose jth component equals to the cluster label of $\boldsymbol{x}_i$ in the jth clustering. Let $X' = \{\boldsymbol{x}_1', \ldots, \boldsymbol{x}_N'\}$. Recall that $\boldsymbol{y}_i$ is the label vector for clustering $\mathcal{R}_i$; the relation between the $\boldsymbol{x}_i$s, $\boldsymbol{y}_i$s and $\boldsymbol{x}_i'$s is schematically given below:

$$\begin{array}{ccccccccc}
 & & & y_1 & \cdots & y_n & & \\
\hline
\boldsymbol{x}_1 & \rightarrow & [& y_1(1) & \ldots & y_n(1) &] & \equiv \boldsymbol{x}_1' \\
\boldsymbol{x}_2 & \rightarrow & [& y_1(2) & \ldots & y_n(2) &] & \equiv \boldsymbol{x}_2' \\
\vdots & \rightarrow & & & \vdots & & & \\
\boldsymbol{x}_N & \rightarrow & [& y_1(N) & \ldots & y_n(N) &] & \equiv \boldsymbol{x}_N'
\end{array}$$

The goal, now, is to determine the (unknown) cluster labels of the data in the final clustering $\mathcal{F} = \{F_1, \ldots, F_m\}$. To this end, a finite mixture model of probability functions is defined

$$P(\boldsymbol{x}'; \boldsymbol{\Theta}) = \sum_{q=1}^{m} P_q P(\boldsymbol{x}' | F_q; \boldsymbol{\theta}_q) \tag{15.90}$$

where m is the number of clusters in the consensus clustering $\mathcal{F}$, P_q is the a priori probability of the qth cluster in $\mathcal{F}$ and $P(\boldsymbol{x}' | F_q; \boldsymbol{\theta}_q)$ is the probability function describing the cluster F_q, which is parametrized by the vector $\boldsymbol{\theta}_q$. Finally, $\boldsymbol{\Theta} = \{P_1, \ldots, P_m, \boldsymbol{\theta}_1, \ldots, \boldsymbol{\theta}_m\}$ is the set of all the parameters of the model.

Assuming statistical independence among the components of $\boldsymbol{x}'$, it follows that

$$P(\boldsymbol{x}' | F_q; \boldsymbol{\theta}_q) = \Pi_{j=1}^{n} P^j(x_j' | F_q; \theta_q^j) \tag{15.91}$$

Since the values of x_j' are nominal (integers) it seems natural to model the respective probabilities via a multinomial distribution, that is,

$$P^j(x_j' | F_q; \theta_q^j) = \Pi_{r=1}^{m_j} \theta_{jq}(r)^{\delta(x_j', r)} \tag{15.92}$$

where $\delta(a, b) = 1$, if $a = b$ and 0 otherwise. Keeping in mind the definition of $\boldsymbol{x}'$, $\theta_{jq}(r)$ models the probability that the data point $\boldsymbol{x}$ belongs to the rth cluster in the jth clustering $\mathcal{R}_j$ of $\mathcal{E}$, given that $\boldsymbol{x}$ belongs to cluster F_q of $\mathcal{F}$.

In the sequel, the EM algorithm can be employed to derive the estimates for the parameters of $\boldsymbol{\Theta}$ (see [Topc 05]). Experimental results ([Topc 05]) show that the EM algorithm converges fast and its performance is slightly better than that of graph-based methods for small n. As n increases, the performance of EM is expected to degrade, due to the increase of the parameters that have to be estimated. An alternative EM formulation of the problem is discussed in [Lang 05].

An interesting variation of the clustering combination problem is discussed in [Topc 04]. In this case, the sets X_i, $i = 1, \ldots, n$, used for the generation of the ensemble $\mathcal{E}$, are generated *sequentially* via sampling of X, where now *the sampling probabilities for the data points differ from each other*. Specifically, for each X_i (a) the clustering $\mathcal{R}_i$ is produced (b) the sampling probabilities for the data points in X are re-estimated according to a rule and (c) the next set X_{i+1} performs sampling on X based on the updated probabilities. The essence of this method is to assign higher sampling probabilities to points of X, which lie in regions where overlap between clusters is encountered. In this way, the focus of the X_is to be formed at the later stages will be on those overlapping regions. Experimental results reported in [Topc 04] suggest that this technique improves (even slightly) the results of other methods, where the X_is are drawn independently from each other. Other techniques are discussed in, for example, [Law 04, Qian 00, Fred 05] where the case of combining different objective functions is also considered.

Remarks

- In comparing two clusterings usually one faces the problem of determining the optimal correspondence between the two clusterings. For example the clusterings with cluster vectors $y_1 = [1,\ 1,\ 1,\ 1,\ 2,\ 2,\ 2]$ and $y_2 = [2,\ 2,\ 2,\ 2,\ 1,\ 1,\ 1]$ define exactly the same clustering on the data set, however, the same cluster in the above clusterings is denoted with different labels. This problem is usually faced using the so called *Hungarian method* (see, e.g., [Papa 82]).
- The problem of combining various clusterings is in close affinity with the so-called *distributed clustering*, which has attracted significant attention due to the increasing size of the current databases ([Kots 04]). In distributed clustering, the objects to be clustered and/or their features reside in different (local) sites in one of the following ways: (a) each site has access to a specific subset of features from *all* objects (*feature distributed clustering*), (b) each site stores *all* the available features for *some* objects (*object distributed clustering*), (c) each site stores *some* features for *some* objects (*feature/object distributed clustering*). Instead of transmitting all of them to a single site and perform a standard clustering algorithm, the data are clustered independently on the different local sites and then the clustering results are moved to a single site, where they contribute to the generation of a final clustering.

15.13 PROBLEMS

15.1 Consider the set $X = \{\boldsymbol{x}_i, i = 1, \ldots, 7\}$, where $\boldsymbol{x}_1 = [1, 1]^T$, $\boldsymbol{x}_2 = [1,\ 2]^T$, $\boldsymbol{x}_3 = [2, 1]^T, \boldsymbol{x}_4 = [3,\ 1]^T, \boldsymbol{x}_5 = [6, 1]^T, \boldsymbol{x}_6 = [7,\ 1]^T, \boldsymbol{x}_7 = [6, 2]^T$.

a. Determine the value of q (Section 15.2.1) for which the MST clustering algorithm gives two clusters.

b. Apply the algorithms that are based on the idea of regions of influence when these regions are defined by Eqs. (15.2)–(15.5).

c. Run the directed tree-based clustering algorithm and determine the values of θ for which it gives two clusters.

15.2 Consider the basic competitive learning algorithm with $\eta = 0.2$. Let $X = \{x_i, i = 1, \ldots, 4\}$, where $x_1 = -3, x_2 = -2, x_3 = 2, x_4 = 3$. Also, let $m = 2$ and $w_1 = -1$ and $w_2 = 1$. Assume that the vectors of X are presented to the algorithm in the same order, $x_1, x_2, x_3 x_4, x_1, x_2, x_3, x_4, \ldots$. Let us call the time required for the consideration of the feature vectors of X once the *updating circle*.

a. Show that w_1 (w_2) always wins on the first (last) two feature vectors when the squared Euclidean distance is in use.

b. Will w_1 and w_2 converge to the values -2.5 and 2.5, respectively, after an infinite number of updating circles? Give intuitive arguments.

15.3 What would the behavior of the leaky learning algorithm be if $\eta_w = \eta_l$?

15.4 *von der Malsburg learning rule.* Assume that the data set X consists of N binary-valued feature vectors, and for each of the m available representatives, $\boldsymbol{w}_j$, we have $\sum_{k=1}^{l} w_{jk} = 1$, where w_{jk} is the kth coordinate of $\boldsymbol{w}_j$. The rule may be stated as follows:

- Present an input vector $\boldsymbol{x} \in X$.
- Determine the winner, $\boldsymbol{w}_j$, of the competition for $\boldsymbol{x}$.
- Update the representatives

$$w_{jk}^{\text{new}} = \begin{cases} w_{jk} + \eta\left(\frac{x_k}{n_x} - w_{jk}\right), & \text{if } \boldsymbol{w}_j \text{ wins on } \boldsymbol{x} \\ w_{jk}, & \text{if } \boldsymbol{w}_j \text{ loses on } \boldsymbol{x} \end{cases} \tag{15.93}$$

where x_k is the kth coordinate of $\boldsymbol{x}$. In the last equation $n_x = \sum_{k=1}^{l} x_k$, that is, it is equal to the number of 1's contained in $\boldsymbol{x}$, and η, the learning rate, takes values in $[0, 1]$. The updating rule may be stated in words as follows: "If a representative wins, each of its coordinates gives up some proportion η that in the sequel is equally distributed among the coordinates w_{ji} that correspond to $x_k = 1$. All the remaining representatives do not change."

a. Verify that this statement is equivalent to the updating rule given by Eq. (15.93).

b. Prove that $\sum_{k=1}^{l} w_{jk}^{\text{new}} = 1, j = 1, \ldots, m$.

15.5 Prove Eq. (15.43).

15.6 Consider three 2-dimensional Gaussian probability density functions with means $\boldsymbol{\mu}_1 = [-1, -1]^T$, $\boldsymbol{\mu}_2 = [6, 3]^T$, $\boldsymbol{\mu}_3 = [-0.7, 7]^T$ and covariance matrices $\Sigma_1 = \Sigma_2 = \Sigma_3 = 2I$, respectively, where I is the 2×2 identity matrix. Draw 200 points from each distribution and form a data set X with the resulting 600 points.

a. Run the Binary Morphology Clustering Algorithm (BMCA) when T is the 3×3 square structuring element given in Example 15.5. For each case use different values of r and proceed to the third stage of the algorithm with the best one of them.

b. Run the Generalized Hard Clustering Scheme (GHAS) algorithm from Chapter 14, using the squared Euclidean distance as the dissimilarity function between two vectors, for the optimum number of clusters derived by the previous procedure. Compare the results of the two algorithms.

15.7 For the data set X given in Example 15.8, run the isodata algorithm assuming that the number of clusters is 2. Compare the results obtained in Example 15.8 and those obtained with isodata. Give a qualitative explanation of the differences that may be observed.

15.8 Verify Eqs. (15.36) and (15.37).

15.9 Derive the updating equation for the coordinates of the parameter vector $\boldsymbol{\theta}$ when $g(\boldsymbol{x}; \boldsymbol{\theta})$, defined in Section 15.6, is a quadratic function of $\boldsymbol{\theta}$.
Hint: In this case $g(\boldsymbol{x}; \boldsymbol{\theta}) = w_0 + \sum_{i=1}^{l} w_i x_i + \sum_{s=1}^{l} \sum_{r=1}^{l} w_{sr} x_s x_r$.

15.10 Consider two 2-dimensional Gaussian distributions with means $\boldsymbol{\mu}_1 = [0, 0]^T$ and $\boldsymbol{\mu}_2 = [3, 3]^T$ and covariance matrices $\Sigma_1 = I$ and $\Sigma_2 = 1.5I$, respectively, where I is the 2×2 identity matrix. Create a data set X such that 100 feature vectors stem from the first and another 100 feature vectors stem from the second Gaussian distribution.

a. Run the Boundary Detection Algorithm (BDA) algorithm on X assuming that the decision boundary is a hyperplane. Use the hyperbolic tangent as f.

b. Run the BDA algorithm on the X^+ and X^- stemming from the previous running. Comment on the results.

15.11 **a.** What is the shape of $V(\boldsymbol{x})$, defined by Eq. (15.38) when $d(\boldsymbol{x}, \boldsymbol{y})$ is given as in Eq. (15.39)?

b. How should $d(\boldsymbol{x}, \boldsymbol{y})$ be defined in order to have a hypercubical shape for $V(\boldsymbol{x})$?

c. Does the shape of $V(\boldsymbol{x})$ affects the behavior of the valley-seeking clustering algorithm? Give an example.

15.12 Consider the set $X = \{\boldsymbol{x}_i, i = 1, \ldots, 10\}$, where $\boldsymbol{x}_1 = [0, 1]^T$, $\boldsymbol{x}_2 = [0, 2]^T$, $\boldsymbol{x}_3 = [0, 3]^T$, $\boldsymbol{x}_4 = [0, 4]^T$, $\boldsymbol{x}_5 = [1, 1]^T$, $\boldsymbol{x}_6 = [1, 2]^T$, $\boldsymbol{x}_7 = [1, 3]^T$, $\boldsymbol{x}_8 = [2, 1]^T$, $\boldsymbol{x}_9 = [2, 2]^T$, $\boldsymbol{x}_{10} = [2, 3]^T$. Initially, the first six of them belong to cluster C_1 and the next four belong to cluster C_2. Apply the valley-seeking algorithm to X and comment on the results.

15.13 If $T_{max} = 5$ and $T_{min} = 0.5$, estimate the number of sweeps required with the simulated annealing algorithm in order to determine (in probability) the clustering with the globally minimum value of J.
Hint: Use Eq. (15.44).

15.14 Modify the deterministic annealing algorithm so that the number of representatives is not fixed *a priori* but increases as β increases.

15.15 Consider the function

$$J = \sum_{i=1}^{N} d(\boldsymbol{x}_i, C_{\boldsymbol{x}_i}) \tag{15.94}$$

where $C_{\boldsymbol{x}_i}$ is the cluster to which $\boldsymbol{x}_i$ belongs and $d(\boldsymbol{x}_i, C_{\boldsymbol{x}_i})$ is a distance between a point and a set using no representative for the set (e.g., Chapter 11). Propose a coding of the solutions for a genetic algorithm that uses this function. Discuss the merits and the disadvantages of the proposed coding.

MATLAB PROGRAMS AND EXERCISES

Computer Programs

15.1 *Competitive learning.* Write a MATLAB function named *leaky_learn* that implements the leaky learning algorithm. This function takes as inputs: (a) an $l \times N$ dimensional matrix X each column of which is a data vector, (b) an $l \times m$ dimensional matrix w whose columns contain initial estimates of the m representatives, (c) the learning rate for the winner unit, gw, (d) the learning rate for the rest of the units, gl, (e) the maximum number of iterations, (f) the parameter e, which is used in the termination condition of the algorithm ($||w(t) - w(t-1)|| < e$). The output consists of: (a) the vector w whose columns are the final estimates of the representatives and (b) an N dimensional row vector *bel*, the *i*-th element of which contains the cluster to which the *i*-th vector belongs. The Euclidean distance is used to measure the distance between two vectors and the vectors are presented always in the same order to the algorithm, until convergence.

Solution

```
function [w,bel]=leaky_learn(X,w,gw,gl,maxiter,e)
  [l,N]=size(X);
  [l,m]=size(w);
  diff=e+1;
  iter=0;
  while(diff>e)&(iter<=maxiter)
    iter=iter+1;
    wold=w;
    for i=1:N
      %Computation of the distances
      dist=sum((X(:,i)*ones(1,m)-w).^ 2);
      [mval,mind]=min(dist);
      %Updating the representatives
      w=w+gl*(X(:,i)*ones(1,m)-w);
```

```
      w(:,mind)=w(:,mind)+(gw-gl)*(X(:,i)-w(:,mind));
    end
    diff=sum(sum(abs(w-wold)))
  end
  %Assigning vectors to clusters
  bel=zeros(1,N);
  for i=1:N
    dist=sum((X(:,i)*ones(1,m)-w).^ 2);
    [mval,mind]=min(dist);
    bel(i)=mind;
  end
```

15.2 *Self-Organizing Map.* Write a MATLAB function named *som_experi* that implements the Self-Organizing Map. More specifically, this function takes as input (a) an $l \times N$ dimensional matrix X each column of which is a data vector, (b) the cluster label (a positive integer) where each vector belongs (this is used only in the plot of the results), (c) the number of iterations, *iter*, the algorithm will perform, (d) the size *side* of the side of the two-dimensional grid (only squared grids are considered). The function returns: (a) a matrix w each column of which contains the final estimates of the representatives and (b) a plot of the grid after the convergence of SOM, where the representatives of the same cluster are denoted by the same color.

Solution

The following function can plot a map for up to 4 different clusters.

```
function w=som_experi(X,y,iter,side)
  [l,N]=size(X);
  p=[side side];
  q=side^2; %Number of representatives
  minmax=[];
  for i=1:l
    minmax=[minmax; min(X(i,:)) max(X(i,:))];
  end
  % Defining and training the SOM
  net=newsom(minmax,p,'gridtop', 'mandist');
  net.trainParam.epochs=iter;
  net.trainParam.show=50;
  net = train(net,X);
  % Check if representatives represent data points
  % of a cluster.
  w=net.iw1';
  repr=zeros(1,q);
  map=zeros(side,side);
```

```
for i=1:N
   [s1,s2]=min(sum((X(:,i)*ones(1,q)-w).^2));
   repr(s2)=y(i);
end
% Creation of the map
for i=1:q
   i1=fix(i/side)+(mod(i,side)>0);
   i2=mod(i,side)+side*(mod(i,side)==0);
   map(i1,i2)=repr(i);
end
% Plot of the map. Up to four clusters can be plotted
if(max(y)<=4)
   figure(1), hold on
   palet=['k.';'ro'; 'ko'; 'go'; 'bo'];
   for i=1:side
      for j=1:side
         figure(1), plot(j,i,palet(map(i,j)+1,:))
      end
   end
end
```

Computer Experiments

15.1 **a.** Generate three data sets each one consisting of $q = 100$ two dimensional vectors stemming from normal distributions with means $[1, 1]^T$, $[5, 5]^T$, $[9, 1]^T$ and covariance matrices all equal to the 2×2 identity matrix I. Form the matrix X using as columns the data vectors from these data sets.

b. Initialize randomly $m = 3$ representatives using the *rand_vec* function (see in the "computer programs" section of chapter 14) and apply the leaky learning algorithm with $gw = 0.1$, $gl = 0.001$, $maxiter = 200$ and $e = 0.001$.

c. Repeat (b) with $gl = 0$ (basic competitive learning algorithm), initializing the representatives from the same values as in (b).

d. Comment on the results.

15.2 **a.** Repeat 1(b) and 1(c) using as initial estimates for the representatives the following: $[-100, -100]^T$, $[3.5, 4.5]^T$, $[2.5, 3]^T$. Comment on the results.

b. Taking into account the comments of 1 and 2(a) can you propose a combination of the leaky learning algorithm and the basic competitive learning algorithm that takes advantage of the merits of both methods?

15.3 **a.** Consider three 3-dimensional Gaussian distributions with means $[0.3, 0.3, 0.3]^T$, $[0.7, 0.7, 0.7]^T$ and $[0.3, 0.7, 0.3]^T$ and covariance matrices equal to $0.01I$, where I is the 3×3 identity matrix. Generate 100 data vectors from each distribution and let X be the resulting data set containing the above vectors (300 in total).

b. Apply the Self-Organizing Map (SOM) on the previous data set using a squared two-dimensional grid of size 10×10. Let the algorithm run for 300 iterations.

c. Repeat (b) for grid sizes 6×6 and 15×15.

d. Comment on the results.

REFERENCES

[Achl 01] Achlioptas D. "Database-friendly random projections," *Symposium on Principles of Database Systems (PODS)*, pp. 274–281, 2001.

[Agga 99] Aggarwal C.C., Wolf J.L., Yu P.S., Procopiuc C., Park J.S. "Fast algorithms for projected clustering," *Proceedings of the 1999 ACM SIGMOD International Conference on Management of Data*, pp. 61–72, 1999.

[Agga 00] Aggarwal C.C., Yu P.S. "Finding generalized projected clusters in high dimensional spaces," *Proceedings of the 2000 ACM SIGMOD International Conference on Management and Data*, pp. 70–81, 2000.

[Agra 98] Agrawal R., Gehrke J., Gunopoulos D., Raghavan P. "Automatic subspace clustering of high dimensional data for data mining applications," *Proceedings of the 1998 ACM SIGMOD International Conference on Management of Data*, pp. 94–105, 1998.

[Ahal 90] Ahalt S.C., Krishnamurthy A.K., Chen P., Melton D.E. "Competitive learning algorithms for vector quantization," *Neural Networks*, Vol. 3, pp. 277–290, 1990.

[Al-S 95] Al-Sultan K.S. "A tabu search to the clustering problem," *Pattern Recognition*, Vol. 28(9), pp. 1443–1451, 1995.

[Al-S 93] Al-Sultan K.S., Selim S.Z. "A global algorithm for the fuzzy clustering problem," *Pattern Recognition*, Vol. 26(9), pp. 1357–1361, 1993.

[Alts 97] Altschul S.F., Madden T.L., Schaffer A.A., Zhang J., Zhang Z., Miller W., Lipman D.J. "Gapped-BLAST and PSI-BLAST: A new generation of protein database search programs," *Nucleic Acids Research*, Vol. 25, pp. 3389–3402, 1997.

[Andr 94] Andrey P., Tarroux P. "Unsupervised image segmentation using a distributed genetic algorithm," *Pattern Recognition*, Vol. 27(5), pp. 659–673, 1994.

[Anke 99] Ankerst M., Breunig M., Kriegel H.-P., Sander J. "OPTICS: Ordering points to identify clustering structure," *Proceedings of the ACM SIGMOD Conference*, pp. 49–60, Philadelphia, PA, 1999.

[Atiy 90] Atiya A.F. "An unsupervised learning technique for artificial neural networks," *Neural Networks*, Vol. 3, pp. 707–711, 1990.

[Ayad 03] Ayad H., Kamel M. "Finding natural clusters using multi-clusterer combiner based one shared nearest neighbors," *Multiple Classifier Systems: 4th International Workshop*, 2003.

[Bane 02] Banerjee A., Ghosh J. "On scaling up balanced clustering algorithms," *Proceedings of the 2nd SIAM International Conference on Data Mining*, pp. 333–349, Arlington, VA, 2002.

[Banz 90] Banzhaf W., Haken H. "Learning in a competitive network," *Neural Networks*, Vol. 3, pp. 423–435, 1990.

[Basu 04] Basu S., Bilenko M., Mooney R.J. "A probabilistic framework for semi-supervised clustering," *International Conference on Knowledge Discovery and Data Mining*, pp. 59–68, 2004.

[Belk 03] Belkin M., Niyogi P. "Laplacian eigenmaps for dimensionality reduction and data representation," *Neural Computation*, Vol. 15(6), pp. 1373–1396, 2003.

[Ben 99] Ben-Dor A., Shamir R., Yakhimi Z. "Clustering gene expression patterns," *Journal of Computational Biology*, Vol. 6, pp. 281–297, 1999.

[Ben 01] Ben-Hur A., Horn D., Siegelmann H.T., Vapnik V. "Support vector clustering," *Journal of Machine Learning Research*, Vol. 2, pp. 125–137, 2001.

[Bene 00] Bennett K.P., Bradley P.S., Demiriz A. "Constraint *k*-means clustering," *Technical Report MSR-TR-2000-65*, Microsoft Research, Redmond, CA, 2000.

[Beng 99] Bengio Y. "Markovian models for sequential data," *Neural Computation Survey*, Vol. 2, pp. 129–162, 1999.

[Beni 94] Beni G., Liu X. "A least biased fuzzy clustering method," *IEEE Transactions on Pattern Analysis and Machine Intelligence*, Vol. 16, pp. 954–960, September 1994.

[Benv 87] Benveniste A., Metivier M., Priouret P. *Adaptive Algorithms and Stochastic Approximation*, Springer-Verlag, 1987.

[Berk 02] Berkhin P. "Survey of clustering data mining techniques," Technical report, Accrue Software, San Jose, CA, 2002.

[Bhan 91] Bhanu B., Lee S., Ming J. "Self-optimizing image segmentation system using a genetic algorithm," *Proceedings, Fourth International Conference on Genetic Algorithms*, pp. 362–369, 1991.

[Bisw 98] Biswas G., Weinberg J.B., Fisher D.H. "ITERATE: A conceptual clustering algorithm for data mining," *IEEE Transactions on Systems, Man and Cybernetics, Part C*, Vol. 28(2), pp. 100–111, 1998.

[Blum 97] Blum A., Langley P. "Selection of relevant features and examples in machine learning," *Artificial Intelligence*, Vol. 97, pp. 245–271, 1997.

[Bohm 00] Bohm C., Braunmuller B., Breunig M., Kriegel H.P. "High performance clustering based on the similarity join," *Proceedings of the 9th International Conference on Information and Knowledge Management, CIKN*, pp. 298–313, Washington, DC, 2000.

[Brow 92] Brown D.E., Huntley C.L. "A practical application of simulated annealing to clustering," *Pattern Recognition*, Vol. 25(4), pp. 401–412, 1992.

[Burr 91] Burrascano P. "Learning vector quantization for the probabilistic neural network," *IEEE Transactions on Neural Networks*, Vol. 2(4), pp. 458–461, 1991.

[Butl 96] Butler D., Jiang J. "Distortion equalized fuzzy competitive learning for image data vector quantization," *Proceedings of ICAPPS'96*, pp. 3390–3393, 1996.

[Cama 05] Camastra F. "A novel kernel method for clustering," *IEEE Transactions on Pattern Analysis and Machine Intelligence*, Vol. 27(5), pp. 801–805, 2005.

[Chan 02] Chang J.-W., Jin D.-S. "A new cell-based clustering method for large high-dimensional data in data mining applications," *Proceedings of 2002 ACM Symposium on Applied Computing*, pp. 503–507, 2002.

[Chan 94] Chan P., Schlag M., Zien J. "Spectral *k*-way ratio cut partitioning," *IEEE Transactions on Computer Aided Design of Integrated Circuits and Systems*, Vol. 13, pp.1088–1096, 1994.

[Chen 94] Chen L., Chang S. "An adaptive conscientious competitive learning algorithm and its applications," *Pattern Recognition*, Vol. 27(12), pp. 1787–1813, 1994.

[Chen 99] Cheng C.-H., Fu A.W., Zhang Y. "Entropy-based subspace clustering for mining numerical data," *Proceedings of the Fifth ACM SIGKDD International Conference on Knowledge Discovery and Data Mining*, pp. 84–93, 1999.

[Chen 03] Chen X. "An improved branch and bound algorithm for feature selection," *Pattern Recognition Letters*, Vol. 24, pp. 1925–1933, 2003.

[Chen 05] Chen C.-Y., Hwang S.-C., Oyang Y.-J., "A statistics-based approach to control the quality of subclusters in incremental gravitational clustering", *Pattern Recognition*, Vol. 38, pp. 2256–2269, 2005.

[Chia 03] Chiang J., Hao P. "A new kernel-based fuzzy clustering approach: Support vector clustering with cell growing," *IEEE Transactions on Fuzzy Systems*, Vol. 11(4), pp. 518–527, 2003.

[Chou 97] Chou C.S., Siu W. "Distortion sensitive competitive learning for vector quantizer design," *IEEE Proc.*, pp. 3405–3408, 1997.

[Chow 97] Chowdhury N., Murthy C.A. "Minimal spanning tree based clustering technique: Relationship with Bayes classifier," *Pattern Recognition*, Vol. 30(11), pp. 1919–1929, 1997.

[Chun 97] Chung F.R.K. *Spectral Graph theory*, American Mathematical Society, 1997.

[Davi 05] Davidson I., Ravi I.I. "Agglomerative hierarchical clustering with constraints: theoretical and empirical results," *9th European Conference on Principles and Practice of Knowledge Discovery in Databases (PKDD)*, pp. 59–70, 2005.

[Dasg 99] Dasgupta S., Gupta A. "An elementary proof of the Johnson–Lindenstrauss lemma," Technical Report TR-99-006, International Computer Science Institute, Berkeley, California, 1999.

[Dasg 99a] Dasgupta S. "Learning mixtures of Gaussians," *IEEE Symposium on Foundations of Computer Science (FOCS)*, 1999.

[Dasg 00] Dasgupta S. "Experiments with random projections," *Proceedings of the 16th Conference of Uncertainty in Artificial Intelligence (UAI)*, 2000.

[Dave 97] Dave R.N., Krishnapuram R. "Robust clustering methods: A unified view," *IEEE Transactions on Fuzzy Systems*, Vol. 5(2), pp. 270–293, May 1997.

[Deer 90] Deerwester S., Dumais S.T., Landauer T.K., Furnas G.W., Harshman R.A. "Indexing by latent semantic analysis," *Jouranl of American Society of Information Sciences*, Vol. 41, pp. 391–407, 1990.

[Dela 80] Delattre M., Hansen P. "Bicriterion cluster analysis," *IEEE Transactions on Pattern Analysis and Machine Intelligence*, Vol. 2(4), pp. 277–291, July 1980.

[Devo 95] Devorer J.L. *Probability and Statistics for Engineering and the Sciences*, (4th ed.), Duxbury Press, 1995.

[Diam 07] Diamantaras K. *Artificial Neural Networks*, Klidarithmos, 2007, (in Greek).

[Dill 07] Dillon I., Guan Y., Kullis B. "Weighted graph cuts without eigenvectors: A multilevel approach," *IEEE Transactions on Pattern Analysis and Machine Intelligence*, Vol. 29(11), pp. 1945–1957, 2007.

[Ding 99] Ding C.H.Q. "A similarity-based probability model for latent semantic indexing," *Proceedings of the 22th ACM SIGIR Conference*, pp. 59–65, 1999.

[Ding 02] Ding C.H.Q., He X., Zha H., Simon H.D. "Adaptive dimension reduction for clustering high dimensional data," *Proceedings of the 2nd IEEE International Conference on Data Mining*, pp. 147–154, 2002.

[Durb 98] Durbin R., Eddy S., Krogh A., Mitchison G. *Biological Sequence Analysis: Probabilistic Models of Proteins and Nucleic Acids*, Cambridge University Press, UK, 1998.

[Este 96] Ester M., Kriegel H.-P., Sander J., Xu X. "A density-based algorithm for discovering clusters in large spatial databases with noise," *Proceedings of the 2nd International Conference on Knowledge Discovery and Data Mining*, pp. 226–231, Portland, OR, 1996.

[Ever 01] Everitt B., Landau S., Leese M. *Cluster Analysis*, Arnold, 2001.

[Fern 04] Fern X.Z., Brodley C.E. "Solving ensemble problems by bipartite graph partitioning," *Proceedings of the 21th International Conference on Machine Learning*, Banff, Canada, 2004.

[Fern 03] Fern X.Z., Brodley C.E. "Random projection for high dimensional data clustering: A cluster ensemble approach," *Proceedings of the 20th International Conference on Machine Learning*, 2003.

[Fisc 05] Fischer I., Poland J. "Amplifying the block matrix structure for spectral clustering," *Technical Report No. IDSIA-03-05*, Instituto Dalle Molle di Studi sull' Intelligenza Artificialle (IDSIA), 2005.

[Fish 87] Fisher D. "Knowledge acquisition via incremental conceptual clustering," *Machine Learning*, Vol. 2, pp. 139–172, 1987.

[Fred 05] Fred A., Jain A.K., "Combining multiple clustering using evidence accumulation," *IEEE Transactions on Pattern Analysis and Machine Intelligence*, Vol. 27(6), pp. 835–850, 2005.

[Fric 02] Friedman J.H., Meulman "Clustering objects on subsets of attributes," *http://citeseer.nj.nec.com/friedman02clustering.html*, 2002.

[Frig 97] Frigui H., Krishnapuram R. "Clustering by competitive agglomeration," *Pattern Recognition*, Vol. 30(7), pp. 1109–1119, 1997.

[Fu 93] Fu L., Yang M., Braylan R., Benson N. "Real-time adaptive clustering of flow cytometric data," *Pattern Recognition*, Vol. 26(2), pp. 365–373, 1993.

[Fuku 90] Fukunaga K. *Introduction to Statistical Pattern Recognition*, 2nd ed., Academic Press, 1990.

[Gabr 69] Gabriel K.R., Sokal R.R. "A new statistical approach to geographic variation analysis," *Syst. Zool.* Vol. 18, pp. 259–278, 1969.

[Gan 04] Gan G., Wu J. "Subspace clustering for high dimensional categorical data," *ACM SIGKDD Explorations Newsletter*, Vol. 6(2), pp. 87–94, 2004.

[Gema 84] Geman S., Geman D. "Stochastic relaxation, Gibbs distribution and Bayesian restoration of images," *IEEE Transactions on Pattern Analysis and Machine Intelligence*, Vol. 6, pp. 721–741, 1984.

[Gers 79] Gersho A. "Asumptotically optimal block quantization," *IEEE Transactions on Information Theory*, Vol. 25(4), pp. 373–380, 1979.

[Gers 92] Gersho A., Gray R.M. *Vector Quantization and Signal Compression*, Kluwer Academic, 1992.

[Giro 02] Girolami M. "Mercer kernel-based clustering in feature space," *IEEE Transactions on Neural Networks*, Vol. 13(2), pp. 780–784, 2002.

[Goil 99] Goil S., Nagesh H., Choudhary A. "Mafia: Efficient and scalable subspace clustering for very large data sets," Technical Report CPDC-TR-9906-010, Northwestern University, June 1999.

[Goks 02] Goksay E., Principe J.C. "Information theoretic clustering," *IEEE Transactions on Pattern Analysis and Machine Intelligence*, Vol. 24(2), pp. 158–171, 2002.

[Golu 89] Golub G.H., Van Loan C.F. *Matrix Computations*, John Hopkins Press, 1989.

[Gonz 93] Gonzalez R.C., Woods R.E. *Digital Image Processing*, Addison Wesley, 1993.

[Gray 84] Gray R.M. "Vector quantization," *IEEE ASSP Magazine*, pp. 4–29, April 1984.

[Gros 76a] Grossberg S. "Adaptive pattern classification and universal recoding: I. Parallel development and coding of neural feature detectors," *Biological Cybernetics*, Vol. 23, pp. 121–134, 1976.

[Gros 76b] Grossberg S. "Adaptive pattern classification and universal recoding: II. Feedback, expectation, olfaction, illusions," *Biological Cybernetics*, Vol. 23, pp. 187–202, 1976.

[Gros 87] Grossberg S. "Competitive learning: From interactive activation to adaptive resonance," *Cognitive Science*, Vol. 11, pp. 23–63, 1987.

[Gusf 97] Gusfield D. *Algorithms on Strings, Trees and Sequences: Computer Science and Computational Biology*, Cambridge University Press, UK, 1997.

[Halk 08] Halkidi M., Gunopulos D., Vazirgiannis M., Kumar N., Domeniconi C. "A clustering framework based on subjective and objective validity criteria," *ACM Transactions on Knowledge Discovery from Data*, Vol. 1(4), 2008.

[Hage 98] Hagen L., Kahng A. "New spectral methods for ratio cut partitioning and clustering," *IEEE Transactions on Computer-Aided-Design of Integrated Circuits and Systems*, Vol. 11, pp. 1074–1085, 1998.

[Hage 92] Hagen L.W., Kahng A.B. "New spectral methods for ratio cut partitioning and clustering," *IEEE Transactions on Computer Aided Design of Integrated Circuits and Systems*, Vol. 11(9), pp. 1074–1085, 1992.

[Hech 88] Hecht-Nielsen R. "Applications of counter-propagation networks," *Neural Networks*, Vol. 1(2), pp. 131–141, 1988.

[Hend 93] Hendrickson B., Leland R. "Multidimensional spectral load balancing," *Proceedings 4th SIAM Conference on Parallel Processing*, pp. 953–961, 1993.

[Hinn 98] Hinneburg A., Keim D. "An efficient approach to clustering large multimedia databases with noise," *Proceedings of the 4th ACM SIGKDD*, pp. 58–65, New York, NY, 1998.

[Hinn 99] Hinneburg A., Keim D.A. "Optimal grid-clustering: Towards breaking the curse of dimensionality in high-dimensional clustering," *Proceedings of the 25th Conference on Very Large Databases*, Edinburgh, Scotland, 1999.

[Hoch 64] Hochberg J.E. *Perception*, Prentice-Hall, 1964.

[Hofm 97] Hofmann T., Buchmann J.M. "Pairwise data clustering by deterministic annealing," *IEEE Transactions on Pattern Analysis and Machine Intelligence*, Vol. 19(1), pp. 1–14, 1997.

[Inok 04] Inokuchi R., Miyamoto S. "LVQ clustering and SOM using a kernel function," *Proceedings of the IEEE International Conference on Fuzzy Systems*, Vol. 3, pp. 1497–1500, 2004.

[Jarv 78] Jarvis R.A. "Shared nearest neighbor maximal spanning trees for cluster analysis," *Proceedings, 4th Joint Conference on Pattern Recognition*, Kyoto, Japan, pp. 308–313, 1978.

[Jayn 82] Jaynes E.T. "On the rationale of maximum-entropy methods," *Proc. IEEE*, Vol. 70(9), pp. 939–952, September 1982.

[Jens 04] Jenssen R., Eltoft T., Principe J.C. "Information theoretic spectral clustering," *Proceedings of the International Joint Conference on Neural Networks*, pp. 111-116, 2004.

[Kann 00] Kannan R., Vempala S., Vetta A. "On clusterings- good, bad and spectral," *41st Annual Symposium on Foundations of Computer Science, FOCS*, pp. 367–377, Redondo Beach, California, USA, 2000.

[Kara 96] Karayiannis N.B., Pai P. "Fuzzy algorithms for learning vector quantization," *IEEE Transactions on Neural Networks*, Vol. 7(5), pp. 1196–1211, 1996.

[Kask 98] Kaski S., Kangas J., Kohonen T. "Bibliography of SOM papers: 1981–1997," *Neural Computing Reviews*, Vol. 1, pp. 102–350, 1998.

[Kirk 83] Kirkpatrick S., Gelatt C.D. Jr., Vecchi M.P. "Optimization by simulated annealing," *Science*, Vol. 220, pp. 671–680, 1983.

[Klei 89] Klein R.W., Dubes R.C. "Experiments in projection and clustering by simulated annealing," *Pattern Recognition*, Vol. 22(2), pp. 213–220, 1989.

[Kodr 88] Kodratoff Y., Tecuci G. "Learning based on conceptual distance," *IEEE Transactions on Pattern Analysis and Machine Intelligence*, Vol. 10(6), pp. 897–909, 1988.

[Koha 97] Kohavi R., John G. "Wrappers for feature subset selection," *Artificial Intelligence*, Vol. 97(1-2), pp. 273–324, 1997.

[Koho 89] Kohonen T. *Self-Organization and Associative Memory*, 2nd ed., Springer-Verlag, 1989.

[Koho 95] Kohonen T. *Self-Organizing Maps*, Springer-Verlag, 1995.

[Koon 75] Koontz W.L.G., Narendra P.M., Fukunaga K. "A branch and bound clustering algorithm," *IEEE Transactions on Computers*, Vol. 24(9), pp. 908–914, September 1975.

[Koon 76] Koontz W.L.G., Narendra P.M., Fukunaga K. "A graph-theoretic approach to nonparametric cluster analysis," *IEEE Transactions on Computers*, Vol. 25(9), pp. 936–944, September 1976.

[Kosk 91] Kosko B. "Stochastic competitive learning," *IEEE Transactions on Neural Networks*, Vol. 2(5), pp. 522–529, 1991.

[Kosk 92] Kosko B. *Neural Networks for Signal Processing*, Prentice Hall, 1992.

[Kotr 92] Kotropoulos C., Auge E., Pitas I. "Two-layer learning vector quantizer for color image quantization," *Signal Processing VI*, pp. 1177–1180, 1992.

[Kots 04] Kotsiantis S.B., Pintelas P.E. "Recent advances in clustering: a brief survey," *WSEAS Transactions on Information Science and Applications*, Vol. 1(1), pp. 73–81, 2004.

[Kout 95] Koutroumbas K. "Hamming neural networks, architecture design and applications," Ph.D. dissertation, Department of Informatics, University of Athens, 1995 (in Greek).

[Kuli 05] Kulis B., Basu S., Dhilon I.S., Mooney R.J., "Semi-supervised graph clustering: a kernel approach," *International Conference on Machine Learning*, pp. 457–464, 2005.

[Laar 87] van Laarhoven P.J.M., Aarts E.H.L. *Simulated Annealing: Theory and Applications*, Reidel, Hingham, MA, 1987.

[Lang 05] Lange T., Buhmann J.M. "Combining partitions by probabilistic label aggregation," *Proceedings of the 11th ACM SIGKDD International Conference on Knowledge discovery in data mining*, pp. 147-155, 2005.

[Law 04] Law M., Topchy A., Jain A.K., "Multiobjective data clustering," *Proceedings of the IEEE Computer Society Conference on Computer Vision and Pattern Recognition*, Vol. 2, pp. 424-430, 2004.

[Liew 05] Liew A.W.C., Yan H., Yang M. "Pattern recognition techniques for the emerging field of bioinformatics: A review," *Pattern Recognition*, Vol. 38, pp. 2055-2073, 2005.

[Likh 97] Likhovidov V. "Variational approach to unsupervised learning algorithms of neural networks," *Neural Networks*, Vol. 10(2), pp. 273-289, 1997.

[Lind 80] Linde Y., Buzo A., Gray R.M. "An algorithm for vector quantizer design," *IEEE Transactions on Communications*, Vol. 28(1), pp. 84-95, 1980.

[Liu 98] Liu H., Motoda H. *Feature Selection for Knowledge Discovery and Data Mining*, Kluwer Academic Publishers, 1998.

[Liu 00] Liu B., Xia Y., Yu P.S. "Clustering through decision tree construction," *Proceedings of the ninth International Conference on Information and Knowledge Management*, pp. 20-29, 2000.

[Luen 84] Luenberger D.G. *Linear and Nonlinear Programming*, Addison Wesley, 1984.

[Macd 00] Macdonald D., Fyfe C. "The kernel self- organizing map," *Proceedings of the fourth International Conference on Knowledge-based Intelligent Engineering Systems and Allied Technologies*, Vol. 1, pp. 317-320, 2000.

[Mals 73] von der Maslburg. "Self-organization sensitive cells in the striate cortex," *Kybernetic*, Vol. 14, pp. 85-100, 1973.

[Mara 80] Maragos P., Schafer R.W. "Morphological systems for multidimensional signal processing," *Proc. IEEE*, Vol. 78(4), pp. 690-710, April 1980.

[Mart 93] Martinetz T.M., Berkovich S.G., Schulten K.J. "Neural-gas network for vector quantization and its application to time-series prediction," *IEEE Transactions on Neural Networks*, Vol. 4(4), pp. 558-569, July 1993.

[Masu 93] Masuda T. "Model of competitive learning based upon a generalized energy function," *Neural Networks*, Vol. 6, pp. 1095-1103, 1993.

[Matt 91] Matthews G., Hearne J. "Clustering without a metric," *IEEE Transactions on Pattern Analysis and Machine Intelligence*, Vol. 13(2), pp. 175-184, 1991.

[Maul 00] Maulik U., Bandyopadhyay S. "Genetic algorithm-based clustering technique," *Pattern Recognition*, Vol. 33, pp. 1455-1465, 2000.

[Meil 00] Meilă M., Shi J. "A random walk view of spectral segmentation," *Proceedings Neural Information Processing Conference*, pp. 873-879, 2000.

[Metr 53] Metropolis N., Rosenbluth A.W., Rosenbluth M.N., Teller A.H., Teller E. "Equations of state calculations by fast computing machines," *Journal of Chemical Physics*, Vol. 21, pp. 1087-1092, 1953.

[Mich 94] Michalevitz Z. *Genetic Algorithms + Data Structures = Evolutionary Programming*, 2nd ed., Springer-Verlag, 1994.

[Mill 01] Miller W. "Comparison of genomic DNA sequences: Solved and unsolved problems," *Bioinformatics*, Vol. 17, pp. 391-397, 2001.

[Mirk 01] Mirkin B. "Reinterpreting the category utility function," *Machine Learning*, Vol. 45(2), pp. 219–228, 2001.

[Mora 00] Morales E., Shih F.Y. "Wavelet coefficients clustering using morphological operations and pruned quadtrees," *Pattern Recognition*, Vol. 33, pp. 1611–1620, 2000.

[Ng 01] Ng A.Y., Jordan M., Weiss Y. "On spectral clustering analysis and an algorithm," *Proceedings 14th Conference on Advances in Neural Information Processing Systems*, 2001.

[Noso 08] Nosovskiy G.V., Liu D., Sourina O. "Automatic clustering and boundary detection algorithm based on adaptive influence function," *Pattern Recognition*, Vol. 41, pp. 2757–2776, 2008.

[Nyec 92] Nyeck A., Mokhtari H., Tosser-Roussey A. "An improved fast adaptive search algorithm for vector quantization by progressive codebook arrangement," *Pattern Recognition*, Vol. 25(8), pp. 799–802, 1992.

[Oyan 01] Oyang Y.-J., Chen C.-Y., Yang T.-W. "A study on the hierarchical data clustering algorithm based on gravity theory," *Lecture Notes in Artificial Intelligence: Principles of Data Mining and Knowledge Discovery*, Vol. 2168, pp. 350–361, Springer, 2001.

[Ozbo 95] Osbourn G.C., Martinez R.F. "Empirically defined regions of influence for cluster analysis," *Pattern Recognition*, Vol. 28(11), pp. 1793–1806, 1995.

[Papa 82] Papadimitriou C.H., Steiglitz K. *Combinatorial Optimization: algorithms and complexity*, Prentice Hall, 1982.

[Pars 04] Parsons L., Haque E., Liu H. "Subspace clustering for high dimensional data: A review," *ACM SIGKDD Explorations Newsletter*, Vol. 6(1), pp. 90–105, 2004.

[Pear 88] Pearson W. "Improved tools for biological sequence comparison," *Proceedings National Academy of Sciences*, Vol. 85, pp. 2444–2448, 1988.

[Pedr 97] Pedrycz W., Waletzky J. "Neural-network front ends in unsupervised learning," *IEEE Transactions on Neural Networks*, Vol. 8(2), pp. 390–401, March 1997.

[Pena 01] Pena J.M., Lozano J.A., Larranaga P., Inza I. "Dimensionality reduction in unsupervised learning of conditional gaussian networks," *IEEE Transactions on Pattern Analysis and Machine Intelligence*, Vol. 23(6), pp. 590–603, 2001.

[Post 93] Postaire J.G., Zhang R.D., Lecocq-Botte C. "Cluster analysis by binary morphology," *IEEE Transactions on Pattern Analysis and Machine Intelligence*, Vol. 15(2), pp. 170–180, 1993.

[Proc 02] Procopiuc C.M., Jones M., Agarwal P.K., Murali T.M. "A Monte-Carlo algorithm for fast projective clustering," *Proceedings of the 2002 ACM SIGMOD International Conference on Management of Data*, pp. 418–427, 2002.

[Qian 00] Qian Y., Suen C. "Clustering combination method," *15th International Conference on Pattern Recognition (ICPR00)*, Vol. 2, pp. 732–735, 2000.

[Qiu 07] Qiu H., Hancock E.R. "Clustering and embedding using commute times," *IEEE Transactions on Pattern Analysis and Machine Intelligence*, Vol. 29(11), pp. 1873–1890, 2007.

[Robe 00] Roberts S.J., Everson R., Rezek I. "Maximum certainty data partitioning," *Pattern Recognition*, pp. 833–839, 2000.

[Rose 91] Rose K. "Deterministic annealing, clustering and optimization," Ph.D. dissertation, California Institute of Technology, 1991.

[Rose 93] Rose K., Gurewitz E., Fox G.C. "Constrained clustering as an optimization method," *IEEE Transactions on Pattern Analysis and Machine Intelligence*, Vol. 15(8), pp. 785–794, 1993.

[Rume 86] Rumelhart D.E., Zipser D. "Feature discovery by competitive learning," *Cognitive Science*, Vol. 9, pp. 75–112, 1986.

[Rume 86] Rumelhart D.E., McLelland J.L. *Parallel Distributed Processing*, Cambridge, MA: MIT Press, 1986.

[Sand 98] Sander J., Ester M., Kriegel H.-P., Xu X. "Density based clustering in spatial databases: The algorithm GDBSCAN and its applications," *Data Mining and Knowledge Discovery*, Vol. 2(2), pp. 169–194, 1998.

[Sche 97] Scheunders P. "A genetic c-means clustering algorithm applied to color image quantization," *Pattern Recognition*, Vol. 30(6), pp. 859–866, 1997.

[Scho 98] Schölkopf B., Smola A., Müller K.R. "Nonlinear component analysis as a kernel eigenvalue problem," *Neural Computation*, Vol. 10(5), pp. 1299–1319, 1998.

[Scot 90] Scott G., Longuet-Higgins H. "Feature grouping by relocalization of eigenvectors of the proximity matrix," *Proceedings British Machine Vision Conference*, pp. 103–108, 1990.

[Shei 98] Sheikholeslami G., Chatterjee S., Zhang A. "WaveCluster: A multi-resolution clustering approach for very large spatial databases," *Proceedings of the 24th Conference on Very Large Databases*, New York, 1998.

[Shi 00] Shi J., Malik J. "Normalized cuts and image segmentation," *IEEE Transactions on Pattern Analysis and Machine Intelligence*, Vol. 22(8), pp.888–905, 2000.

[Stre 02] Strehl A., Ghosh J. "Cluster ensembles - a knowledge reuse framework for combining multiple partitions," *Journal of Machine Learning Research*, Vol. 3, pp. 583–617, 2002.

[Stre 00] Strehl A., Ghosh J. "A scalable approach to balanced, high-dimensional clustering of market baskets," *Proceedings of the 17th International conference on High Performance Computing*, pp. 525–536, Springer LNCS, Bangalore, India, 2000.

[Sung 00] Sung C.S., Jin H.W., "A tabu-search-based heuristic for clustering," *Pattern Recognition*, Vol. 33, pp. 849–858, 2000.

[Szu 86] Szu H. "Fast simulated annealing," in *Neural Networks for Computing* (Denker J.S., ed.), American Institute of Physics, 1986.

[Tax 99] Tax D.M.J., Duin R.P.W. "Support vector domain description," *Pattern Recognition Letters*, Vol. 20, pp. 1191–1199, 1999.

[Topc 05] Topchy A., Jain A.K., Punch W. "Clustering ensembles: Models of consensus and weak partitions," *IEEE Transactions on Pattern Analysis and Machine Intelligence*, Vol. 27(12), pp. 1866–1881, 2005.

[Topc 04] Topchy A., Minaei B., Jain A.K., Punch W. "Adaptive clustering ensembles," *Proceedings of the International Conference on Pattern Recognition (ICPR)*, U.K., August 23–26, 2004.

[Tou 74] Tou J.T., Gonzales R.C. *Pattern Recognition Principles*, Addison-Wesley, 1974.

[Tous 80] Toussaint G.T. "The relative neighborhood graph of a finite planar set," *Pattern Recognition*, Vol. 12, pp. 261–268, 1980.

[Touz 88] Touzani A., Postaire J.G. "Mode detection by relaxation," *IEEE Transactions on Pattern Analysis and Machine Intelligence*, Vol. 10(6), pp. 970–977, 1988.

[Tsen 00] Tseng L.Y., Yang S.B., "A genetic clustering algorithm for data with non-spherical-shape clusters," *Pattern Recognition*, Vol. 33, pp. 1251–1259, 2000.

[Tsyp 73] Tsypkin Y.Z. *Foundations of the Theory of Learning Systems*, Academic Press, 1973.

[Tung 01] Tung A.K.H., Ng R.T., Lakshmanan L.V.S., Han J. "Constraint-based clustering in large databases," *Proceedings of the 8th International Conference on Database Theory*, London, 2001.

[Uchi 94] Uchiyama T., Arbib M.A. "An algorithm for competitive learning in clustering problems," *Pattern Recognition*, Vol. 27(10), pp. 1415–1421, 1994.

[Ueda 94] Ueda N., Nakano R. "A new competitive learning approach based on an equidistortion principle for designing optimal vector quantizers," *Neural Networks*, Vol. 7(8), pp. 1211–1227, 1994.

[Urqu 82] Urquhart R. "Graph theoretical clustering based on limited neighborhood sets," *Pattern Recognition*, Vol. 15(3), pp. 173–187, 1982.

[Verm 03] Verma D., Meilă M. "A comparison of spectral clustering algorithms," *Technical Report, UW-CSE-03-05-01*, University of Washington, Seattle, CSE Department, 2003.

[Vish 00] Vishwanathan S.V.N., Murty M.N., "Kohonen's SOM with cashe," *Pattern Recognition*, Vol. 33, pp. 1927–1929, 2000.

[vonL 07] von Luxburg U. "A Tutorial on Spectral Clustering," *Statistics and Computing*, Vol. 17(4), 2007.

[Wags 01] Wagstaff K., Cardie C., Rogers S., Schrodl S. "Constraint k-means clustering with background knowledge," *International Conference on Machine Learning*, pp. 577–584, 2001.

[Weis 99] Weiss Y. "Segmentation using eigenvectors: A unifying view," *Proceedings 7th IEEE International Conference on Computer Vision*, pp. 975–982, 1999.

[Wang 07] Wang D., Shi L., Yeung D.S., Tsang E.C.C., Heng P.A. "Ellipsoidal support vector clustering for functional MRI analysis," *Pattern Recognition*, Vol. 40(10), pp. 2685–2695, 2007.

[Woo 02] Woo K.-G., Lee J.-H. "FINDIT: A fast and intelligent subspace clustering algorithm using dimension voting," Ph.D. Thesis, Korea Advanced Institute of Science and Technology, Taejon, Korea, 2002.

[Wu 93] Wu Z., Leahy R. "An optimal graph theoretic approach to data clustering: Theory and its applications to image segmentation," *IEEE Transactions on Pattern Analysis and Machine Intelligence*, Vol. 15(11), pp. 1101–1113, 1993.

[Xian 08] Xiang T., Gong S. "Spectral clustering with eigenvalue selection," *Pattern Recognition*, Vol. 41(3), pp. 1012–1029, 2008.

[Xie 93] Xie Q., Laszlo A., Ward R.K. "Vector quantization technique for nonparametric classifier design," *IEEE Transactions on Pattern Analysis and Machine Intelligence*, Vol. 15(12), pp. 1326–1330, 1993.

[Xing 02] Xing E.P., Ng A.Y., Jordan M.I., Russell S.J. "Distance metric learning with application to clustering with side-information," *International Conference on Neural Information Processing Systems*, pp. 505–512, 2002.

[Xu 98] Xu X., Ester M., Kriegel H.P., Sander J. "A distribution-based clustering algorithm for mining in large spatial databases," *Proceedings of the 14th ICDE*, pp. 324–331, Orlando, FL, 1998.

[Xu 05] Xu R., Wunsch D. II "Survey of clustering algorithms," *IEEE Transactions on Neural Networks*, Vol. 16(3), pp. 645–677, 2005.

[Yama 80] Yamada Y., Tazaki S., Gray R.M. "Asymptotic performance of block quantizers with difference distortion measures," *IEEE Transactions on Information Theory*, Vol. 26(1), pp. 6–14, 1980.

[Yang 02] Yang J., Wang W., Wang H., Yu P. "δ-clusters: Capturing subspace correlation in a large data set," *Proceedings of the 18th International Conference on Data Engineering*, pp. 517–528, 2002.

[Yu 93] Yu B., Yuan B. "A more efficient branch and bound algorithm for feature selection," *Pattern Recognition*, Vol. 26, pp. 883–889, 1993.

[Yu 03] Yu L., Liu H. "Feature selection for high-dimensional data: A fast correlation-based filter solution," *Proceedings of the 20th International Conference on Machine Learning*, pp. 856–863, 2003.

[Zahn 71] Zahn C.T. "Graph-theoretical methods for detecting and describing gestalt clusters," *IEEE Transactions on Computers*, Vol. 20(1), pp. 68–86, January 1971.

[Zass-05] Zass R., Shashua A. "A unifying approach to hard and probabilistic clustering," *Proceedings of the 10th IEEE International Conference on Computer Vision, ICCV*, pp. 294–301, 2005.

[Zhu 94] Zhu C., Li L., He Z., Wang J. "A new competitive learning learning algorithm for vector quantization," *Proceedings of ICASSP'94*, pp. 557–560, 1994.

CHAPTER

Cluster Validity 16

16.1 INTRODUCTION

A common characteristic of the majority of the clustering algorithms, discussed in the previous chapters, is that they *impose* a clustering structure on the data set X, even though X may not possess such a structure. In the latter case, the results produced after the application of a clustering algorithm on X are not indicative of the structure of X. In other words, *cluster analysis is not a panacea*. That is, we must have an indication that the vectors of X form clusters before we apply a clustering algorithm. The problem of verifying whether X possesses a clustering structure, without identifying it explicitly, is known as *clustering tendency* and is discussed at the end of the chapter.

Let us now assume that X possesses a clustering structure and we want to unravel it. A different kind of problem is encountered now. Recall that all the clustering algorithms require knowledge of the values of specific parameters and, in addition, some of them impose restrictions on the shape of the clusters (e.g., compact, hyper-ellipsoidal). As already shown in the previous chapters, poor estimation of these parameters and inappropriate restrictions on the shape of the clusters (wherever such restrictions are required) may lead to incorrect conclusions about the clustering structure of X. Thus, the need for further evaluation of the results of a clustering algorithm is apparent.

In this chapter, we discuss methods suitable for quantitative evaluation of the results of a clustering algorithm. This task is known under the general term *cluster validity*. However, it must be emphasized that the results obtained by these methods are *only* tools at the disposal of the expert in order to evaluate the resulting clustering.

Let $\mathcal{C}$ denote the clustering structure resulting from the application of a clustering algorithm on X. This may be a hierarchy of clusterings, as is the case with the hierarchical algorithms, or a single clustering, as happens with all the other algorithms discussed in the previous chapters. Cluster validity can be approached in three possible directions. First, we may evaluate $\mathcal{C}$ in terms of an independently drawn structure, which is imposed on X *a priori* and reflects our intuition about the clustering structure of X. The criteria used for the evaluation of this kind are called *external criteria*. In addition, external criteria may be used to measure the degree to which the available data confirm a prespecified structure, without applying any clustering algorithm to X. Second, we may evaluate C in terms of quantities that involve the vectors of X themselves, for example, the proximity matrix. The criteria used for this kind of evaluation are called *internal criteria*. Finally, we may evaluate C by comparing it with other clustering structures, resulting from the application of the same clustering algorithm, but with different parameter values, or of other clustering algorithms to X. Criteria of this kind are called *relative criteria*.

The cluster validation methods based on external or internal criteria rely on statistical hypothesis testing, which was introduced in Chapter 5. The following section contains some additional definitions to be used in this chapter.

16.2 HYPOTHESIS TESTING REVISITED

Let H_0 and H_1 be the null and alternative hypotheses, respectively,

$$H_1 : \boldsymbol{\theta} \neq \boldsymbol{\theta}_0$$

$$H_0 : \boldsymbol{\theta} = \boldsymbol{\theta}_0$$

Also let $\bar{D}_\rho$ be the critical interval corresponding to significance level ρ of a test statistic q, and Θ_1 the set of all possible values that $\boldsymbol{\theta}$ may take under hypothesis H_1. The *power function* of the test is defined as

$$W(\boldsymbol{\theta}) = P(q \in \bar{D}_\rho | \boldsymbol{\theta} \in \Theta_1) \tag{16.1}$$

For a specific $\boldsymbol{\theta} \in \Theta_1$, $W(\boldsymbol{\theta})$ is known as the *test power under the alternative* $\boldsymbol{\theta}$. In words, $W(\boldsymbol{\theta})$ is the probability that q lies in the critical region when the value of the parameter vector is $\boldsymbol{\theta}$. This is the probability of making the correct decision when H_0 is rejected. The power function can be used for the comparison of two different statistical tests. The test whose power under the alternative hypotheses is greater is always preferred.

There are two types of errors associated with a statistical test.

- Suppose that H_0 is true. If $q(\boldsymbol{x}) \in \bar{D}_\rho$, H_0 will be rejected even if it is true. This is called a type I error. The probability of such an error is ρ. The probability of accepting H_0 when it is true is $1 - \rho$.
- Suppose that H_0 is false. If $q(\boldsymbol{x}) \notin \bar{D}_\rho$, H_0 will be accepted even if it is false. This is called a type II error. The probability of such an error is $1 - W(\boldsymbol{\theta})$, and it depends on the specific value of $\boldsymbol{\theta}$.

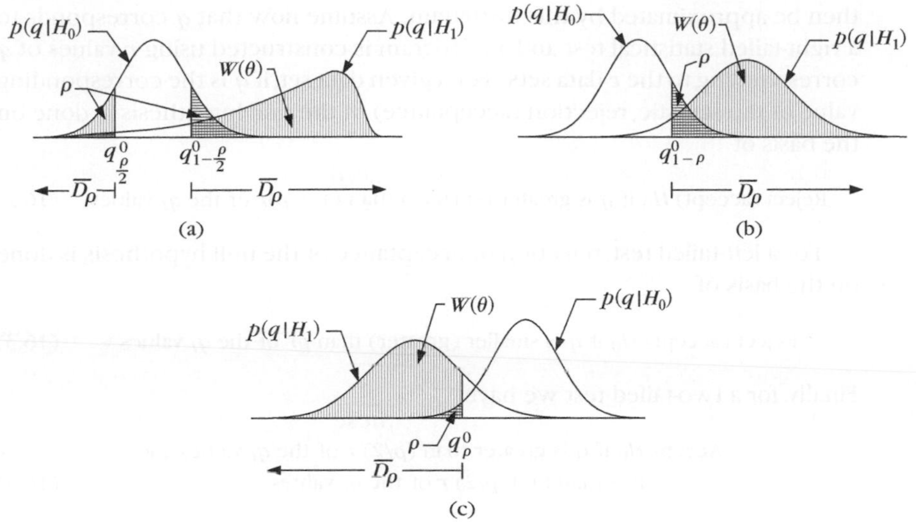

FIGURE 16.1

Critical regions of (a) A two-tailed test, (b) A right-tailed test, and (c) A left-tailed test. q^0_a is the a percentile of q under H_0.[1]

In practice, the final decision to reject or accept H_0 is based partially on the preceding statements as well as on other factors, such as the cost of a wrong decision. Thus, the terms "accept" and "reject" H_0 must be interpreted accordingly.

The probability density function (pdf) of the statistic q, under H_0, for most of the statistics used in practice has a single maximum and the $\bar{D}_\rho$ region is either a half-line or the union of two half-lines. These assumptions have also been adopted here. Figure 16.1 shows the three possible cases for $\bar{D}_\rho$. In the first case, $\bar{D}_\rho$ is the union of two half-lines. Such a test is known as a *two-tailed statistical test*. The other two tests are called *one-tailed* statistical tests, because $\bar{D}_\rho$ consists of a single half-line. Figure 16.1a is an example of a two-tailed statistical test[2] and Figures 16.1b and 16.1c are examples of a right- and a left-tailed test, respectively.

In many practical cases the exact form of the pdf of a statistic q, under a given hypothesis, is not available and it is difficult to obtain. In the sequel, we discuss two methods for estimating pdf's via simulations.

- *Monte Carlo techniques* [Shre 64, Sobo 84] rely on simulating the process at hand using a sufficient number of computer-generated data. For each of the, say r, data sets, X_i, we compute the value of q, denoted by q_i, and then we construct the corresponding histogram of these values. The unknown pdf can

[1] The a percentile of q is the smallest number q_a such that $a = P(q \le q_a)$.
[2] More general versions of a two-tailed statistical test are also possible (e.g., [Papo 91]).

then be approximated by this histogram. Assume now that q corresponds to a right-tailed statistical test and a histogram is constructed using r values of q corresponding to the r data sets. For a given data set, if q is the corresponding value of the statistic, rejection (acceptance) of the null hypothesis is done on the basis of

$$\text{Reject (accept) } H_0 \text{ if } q \text{ is greater (smaller) than } (1-\rho)r \text{ of the } q_i \text{ values} \quad (16.2)$$

For a left-tailed test, rejection or acceptance of the null hypothesis is done on the basis of

$$\text{Reject (accept) } H_0 \text{ if } q \text{ is smaller (greater) than } \rho r \text{ of the } q_i \text{ values} \quad (16.3)$$

Finally, for a two-tailed test we have

$$\text{Accept } H_0 \text{ if q is greater than } (\rho/2)\, r \text{ of the } q_i \text{ values and less than } (1-\rho/2)\, r \text{ of the } q_i \text{ values} \quad (16.4)$$

- *Bootstrapping techniques* constitute an alternative way to cope with a limited amount of data. The idea here is to parameterize the unknown pdf in terms of an unknown parameter. To cope with the limited amount of data and in order to improve the accuracy of the estimate of the unknown pdf parameter, several "fake" data sets $X_1, \ldots, X_r$ are created by sampling X with replacement, as discussed in Chapter 10.

  Typically, good estimates are obtained if r is between 100 and 200. For a more detailed discussion and applications of the bootstrapping techniques see, for example, [Diac 83, Efro 79, Jain 87a, Jain 87b].

16.3 HYPOTHESIS TESTING IN CLUSTER VALIDITY

In this framework, the null hypothesis H_0 will be expressed in a slightly different way. This is because our major concern is not to test a parameter against a specific value. In contrast, our concern here is to test whether the data of X possess a "random" structure or not. *Thus, in this case, the null hypothesis H_0 should be a statement of randomness concerning the structure of X.* Thus, our goal is now twofold.

- First, we must generate a reference *data population under the random hypothesis*, that is, a data population that models a random structure.
- Second, we must define an appropriate statistic, whose values are indicative of the structure of a data set, and compare the value that results from our data set X against the value obtained from the reference (random) population.

There are three different ways to generate the reference population under the null (randomness) hypothesis, each being appropriate for different situations.

- ***Random position hypothesis.*** This hypothesis is appropriate for ratio data. It *requires* that *"All the arrangements of N vectors in a specific region of the l-dimensional space are equally likely to occur."* Such regions may be the H_l hypercube or the l-dimensional hypersphere. One way to produce such an arrangement is to insert each point randomly in this region of the l-dimensional space, according to the uniform distribution. The random position hypothesis can be used with either external or internal criteria.
  - *Internal criteria.* In this case, the statistic q is defined so as to measure the degree to which a clustering structure, produced by a clustering algorithm, matches the proximity matrix of the corresponding data set. Let X_i be a set of N vectors generated according to the random position hypothesis and P_i be the corresponding proximity matrix. In the sequel, we apply the same clustering algorithm to each X_i and to our data set X and let C_i and C be the resulting clustering structures, respectively. For each case, the value of the statistic q is computed. The random hypothesis, H_0, is then rejected if the value q, resulting from X lies in the critical interval $\bar{D}_\rho$ of the statistic pdf of the reference population (i.e., under H_0), that is, if q is unusually small or large.
  - *External criteria.* The statistic q is defined so as to measure the degree of correspondence between a *prespecified structure* $\mathcal{P}$ imposed on X and the clustering that results after the application of a specific clustering algorithm to X. Then, the value of q corresponding to the clustering C resulting from the data set X is tested against the q_i's, corresponding to the clusterings resulting from the reference population generated under the random position hypothesis. Once more, the random hypothesis is rejected if q is unusually large or small.

- ***Random graph hypothesis.*** It is usually adopted when only internal information (i.e., information that concerns only the vectors themselves or their relationships) is available. It is appropriate when ordinal proximities between vectors are used. Before we proceed, let us define the ordinal, or rank order, $N \times N$ matrix A as a symmetric matrix with zero diagonal elements (provided that dissimilarity measures are used) and with its upper diagonal elements being integers in the range $[1, N(N-1)/2]$. The entry $A(i,j)$ of A provides only qualitative information about the dissimilarity between the corresponding vectors $\boldsymbol{x}_i$ and $\boldsymbol{x}_j$. If, for example, $A(2,3) = 3$ and $A(2,5) = 5$, we can only conclude that $\boldsymbol{x}_2$ is more similar to $\boldsymbol{x}_3$ than $\boldsymbol{x}_5$. That is, in this context, comparing dissimilarities is meaningless (recall the comments made in Chapter 11, concerning ordinal type data).

  Let A_i be an $N \times N$ rank order proximity matrix with no ties; that is, all entries in the upper diagonal are different from each other. Under the random graph hypothesis, the reference population consists of such matrices A_i each one generated by inserting randomly the integers in the range

$[1, N(N-1)/2]$, in its upper diagonal entries. Let P be the ordinal proximity matrix associated with the given data set X and $\mathcal{C}$ be the clustering structure produced by the application of a specific algorithm to P. Finally, let $\mathcal{C}_i$ be the clustering structure produced when the same algorithm is applied to A_i. We may now proceed as in the previous case and define a statistic q that measures the agreement between a rank order (proximity) matrix and the corresponding clustering structure. If the value of q, corresponding to P and $\mathcal{C}$, is unusually large or small, the random hypothesis is rejected.

It must be emphasized that the random graph hypothesis is not appropriate for ratio-scaled data. Let us take, for example, the case where the Euclidean distance is in use and $l \leq N-2$ and consider the points $x_1 = 0$, $x_2 = 1$, $x_3 = 3$ on the real line. It is clear that the distance between x_1 and x_3 cannot be smaller than the distance between x_2 and x_3. That is, the matrix $A = \begin{bmatrix} 0 & 2 & 1 \\ 2 & 0 & 3 \\ 1 & 3 & 0 \end{bmatrix}$ is not a valid proximity matrix for these ratio-scaled data.

- *Random label hypothesis.* Let us consider all possible partitions, $\mathcal{P}'$, of X into m groups. Each partition may be defined in terms of a mapping g from X to $\{1, \ldots, m\}$. The random label hypothesis assumes that *all* possible mappings are *equally likely*. The statistic q can be defined so as to measure the degree to which information inherent in the data set X, such as the proximity matrix P, matches a specific partition. The statistic q is then used to test the degree of match between P and an externally imposed partition $\mathcal{P}$, against the q_i's corresponding to the random partitions generated under the random label hypothesis. Once more, H_0 is then rejected if q is unusually large or small.

In the sequel, we give a number of statistic indices appropriate for external and, then, for internal criteria.

16.3.1 External Criteria

External criteria are used either (a) for the comparison of a clustering structure $\mathcal{C}$, produced by a clustering algorithm, with a partition $\mathcal{P}$ of X drawn independently from $\mathcal{C}$ or (b) for measuring the degree of agreement between a predetermined partition $\mathcal{P}$ and the proximity matrix of X, P.

Comparison of $\mathcal{P}$ with a Clustering $\mathcal{C}$

In this case, $\mathcal{C}$ may be either a specific hierarchy of clusterings or a specific clustering. The latter may be produced either by cutting the dendrogram produced by a hierarchical algorithm at a given level (see Chapter 13) or by any other algorithm discussed in the previous chapters. However, a prespecified hierarchy of partitions is rarely available in practice. Thus, the problem of validating hierarchies of clusterings is of limited practical interest.

In the sequel, we consider the validation task concerning a clustering, $\mathcal{C}$, resulting from a specific clustering algorithm, in terms of an independently drawn partition $\mathcal{P}$ of X. Let $\mathcal{C} = \{C_1, \ldots, C_m\}$ and $\mathcal{P} = \{P_1, \ldots, P_s\}$. Note that the number of clusters in $\mathcal{C}$ need not be the same as the number of groups in $\mathcal{P}$. Our goal is to define appropriate statistical indices to be used for the hypothesis test.

Let n_{ij} denote the number of vectors that belong to C_i and P_j simultaneously. Also let $n_i^C = \sum_{j=1}^{s} n_{ij}$; that is, n_i^C is the number of vectors that belong to C_i. Similarly, we define the number of vectors that belong to P_j as $n_j^P = \sum_{i=1}^{m} n_{ij}$.

Consider a pair of vectors $(\boldsymbol{x}_v, \boldsymbol{x}_u)$. We refer to it as (a) *SS* if both vectors belong to the same cluster in $\mathcal{C}$ and to the same group in $\mathcal{P}$, (b) *DD* if both vectors belong to different clusters in $\mathcal{C}$ and to different groups in $\mathcal{P}$, (c) *SD* if the vectors belong to the same cluster in $\mathcal{C}$ and to different groups in $\mathcal{P}$, and (d) *DS* if the vectors belong to different clusters in $\mathcal{C}$ and to the same group in $\mathcal{P}$. Let a, b, c, and d be the number of *SS*, *SD*, *DS*, and *DD* pairs of vectors of X, respectively. Then $a + b + c + d = M$, where M is the total number of possible pairs in X, that is, $M = N(N-1)/2$.

Example 16.1

Let $X = \{\boldsymbol{x}_i, i = 1, \ldots, 6\}$, $\mathcal{C} = \{\{\boldsymbol{x}_1, \boldsymbol{x}_2, \boldsymbol{x}_3\}, \{\boldsymbol{x}_4, \boldsymbol{x}_5\}, \{\boldsymbol{x}_6\}\}$, and $\mathcal{P} = \{\{\boldsymbol{x}_1, \boldsymbol{x}_2, \boldsymbol{x}_3\}, \{\boldsymbol{x}_4, \boldsymbol{x}_5, \boldsymbol{x}_6\}\}$. The following table shows the type of all pairs of vectors in X.

	$\boldsymbol{x}_1$	$\boldsymbol{x}_2$	$\boldsymbol{x}_3$	$\boldsymbol{x}_4$	$\boldsymbol{x}_5$	$\boldsymbol{x}_6$
$\boldsymbol{x}_1$		SS	SS	DD	DD	DD
$\boldsymbol{x}_2$			SS	DD	DD	DD
$\boldsymbol{x}_3$				DD	DD	DD
$\boldsymbol{x}_4$					SS	DS
$\boldsymbol{x}_5$						DS
$\boldsymbol{x}_6$						

From this table we obtain $a = 4$, $b = 0$, $c = 2$, and $d = 9$.

Let $m_1 = a + b$ be the number of pairs of vectors that belong to the same cluster in $\mathcal{C}$ and $m_2 = a + c$ be the number of pairs of vectors that belong to the same group in $\mathcal{P}$. Using the preceding definitions, we can define statistical indices (statistics) in order to measure the degree to which $\mathcal{C}$ matches $\mathcal{P}$. Such statistical indices are the following:

- Rand statistic

$$R = (a + d)/M \tag{16.5}$$

- Jaccard coefficient

$$J = a/(a + b + c) \tag{16.6}$$

- Fowlkes and Mallows index

$$FM = a/\sqrt{m_1 m_2} = \sqrt{\frac{a}{a+b}\frac{a}{a+c}} \tag{16.7}$$

The term $a + d$ is the number of *SS* pairs of vectors plus the number of *DD* pairs. Thus, the Rand statistic measures the fraction of the total number of pairs that are either *SS* or *DD*. The Jaccard coefficient follows the same philosophy as the Rand statistic, except that it excludes d. The values of these two statistics are between 0 and 1. However, a prerequisite for achieving the maximum value is to have $m = s$, which, in general, is not always the case.

For all the above defined indices, it is clear that the larger their value, the higher the agreement between $\mathcal{C}$ and $\mathcal{P}$, that is, ***all the corresponding statistical tests are right tailed***.

Another very popular statistic that is, frequently used in conjunction with external criteria is Hubert's Γ statistic (e.g., [Hube 76, Mant 67, Bart 62]). It measures the correlation between two matrices, X and Y, of dimension $N \times N$, drawn independently of each other. For symmetric matrices this can be written as

- Hubert's Γ statistic

$$\Gamma = (1/M)\sum_{i=1}^{N-1}\sum_{j=i+1}^{N} X(i,j)Y(i,j) \tag{16.8}$$

where $X(i,j)$ and $Y(i,j)$ are the (i,j) elements of the matrices X and Y, respectively. High values of Γ indicate close agreement between X and Y. The normalized version of the Γ statistic, denoted by $\hat{\Gamma}$, is also used.

- Normalized Γ statistic

$$\hat{\Gamma} = \frac{(1/M)\sum_{i=1}^{N-1}\sum_{j=i+1}^{N}(X(i,j)-\mu_X)(Y(i,j)-\mu_Y)}{\sigma_X\sigma_Y} \tag{16.9}$$

where μ_X, μ_Y, σ_X^2, and σ_Y^2 are the respective means and variances, that is, $\mu_X = (1/M)\sum_{i=1}^{N-1}\sum_{j=i+1}^{N} X(i,j)$, $\sigma_X^2 = (1/M)\sum_{i=1}^{N-1}\sum_{j=i+1}^{N} X(i,j)^2 - \mu_X^2$ (similarly we define μ_Y and σ_Y^2). The values of $\hat{\Gamma}$ are between -1 and 1.

Let us set $X(i,j)$ equal to 1 if $\boldsymbol{x}_i$ and $\boldsymbol{x}_j$ belong to the same cluster in $\mathcal{C}$ and 0 otherwise, and $Y(i,j)$ equal to 1 if $\boldsymbol{x}_i$ and $\boldsymbol{x}_j$ belong to the same group in $\mathcal{P}$ and 0 otherwise. It can then be shown (see Problem 16.2) that in this case the $\hat{\Gamma}$ statistic becomes equal to

$$\hat{\Gamma} = (Ma - m_1 m_2)/\sqrt{m_1 m_2 (M - m_1)(M - m_2)} \tag{16.10}$$

Unusually large absolute values of $\Gamma(\hat{\Gamma})$ suggest that $\mathcal{C}$ and $\mathcal{P}$ agree with each other.

As almost always happens in practice, the exact computation of the pdf of all these indices, under the null hypothesis, is very difficult. Thus, we use Monte Carlo techniques for their estimation. In the sequel, we discuss such a procedure, which is based on the random position hypothesis. Data are assumed to be ratio scaled.

- For $i = 1$ to r
  - Generate a data set X_i of N vectors in the area of interest of X, so that the vectors are uniformly distributed in it.
  - Assign each vector $y_j^i \in X_i$ to the group where the $x_j \in X$ belongs, according to the structure imposed by $\mathcal{P}$.
  - Run the same clustering algorithm, used for obtaining $\mathcal{C}$, on X_i and let $\mathcal{C}_i$ be the resulting clustering.
  - Compute the value $q(\mathcal{C}_i)$ of the corresponding statistical index q for $\mathcal{P}$ and $\mathcal{C}_i$.
- End {For}
- Create the histogram of $q(\mathcal{C}_i)$'s.

The following example demonstrates how this methodology can be used in practice.

Example 16.2

(a) Consider a data set X of 100 vectors in the H_3 hypercube. The data are generated to form four groups, each consisted of 25 vectors. Each group is generated by a normal distribution. The first group of 25 vectors of X is generated from the first distribution while the second, third, and fourth groups of 25 vectors are generated from the second, the third, and the fourth distribution, respectively. The covariance matrices of all distributions are equal to $0.2I$, where I is the 3×3 identity matrix. The mean vectors for the four distributions are $[0.2, 0.2, 0.2]^T$, $[0.5, 0.2, 0.8]^T$, $[0.5, 0.8, 0.2]^T$, and $[0.8, 0.8, 0.8]^T$, respectively. If a distribution generates a vector that is, outside the unit hypercube, it is ignored and replaced by another that lies inside H_3. It is not difficult to realize that the points of X form four compact and well-separated clusters.

We assume that the external information is: "The vectors of X belong to four different groups P_1, P_2, P_3, and P_4, such that P_1 contains the first 25 vectors of X and P_2, P_3, and P_4 contain the second, third, and fourth groups of 25 vectors of X, respectively."

We run the isodata algorithm for $m = 4$ and let $\mathcal{C}$ be the resulting clustering. We compute the values of the Rand, R, the Jaccard, J, the Fowlkes and Mallows, FM, and the $\hat{\Gamma}$ statistics for $\mathcal{C}$ and $\mathcal{P}$. These are 0.91, 0.68, 0.81, and 0.75, respectively. Next, we estimate the distribution of these statistics using the procedure described before. Specifically, 100 data sets X_i, $i = 1, \ldots, 100$, are generated, each of them consisting of 100 randomly selected vectors in H_3, following the uniform distribution. According to the $\mathcal{P}$ defined earlier, we assign the first 25 of them to P_1 and the second, third, and fourth groups of 25 vectors to P_2, P_3, and P_4, respectively. For each X_i we run the isodata algorithm for $m = 4$ and we produce the clustering $\mathcal{C}_i$, $i = 1, \ldots, 100$. Then we compute the values of the four statistics, R_i, J_i, FM_i, and $\hat{\Gamma}_i$ for each $\mathcal{C}_i$ and $\mathcal{P}$, $i = 1, \ldots, 100$. We set the significance level at $\rho = 0.05$.

Then, in terms of a given statistic, we accept or reject the null hypothesis (i.e., the random hypothesis) according to the conditions given in Section 16.2. In our case R is greater than all R_i's. Similarly, J, FM, and $\hat{\Gamma}$ are greater than all J_i's, FM_i's, and $\hat{\Gamma}_i$'s, respectively. Thus, all statistics reject the null hypothesis at significance level $\rho = 0.05$.

(b) Now let X' be a data set constructed as X, but with the covariance matrices of the normal distributions equal to $0.6I$. In this case, the vectors of X form weak clusters, that is, clusters that exhibit "large" spread around their mean vector. The values of the four statistics in this case are $R = 0.64$, $J = 0.15$, $FM = 0.27$, and $\hat{\Gamma} = 0.03$. R is greater than 99 of the R_i's. Similarly, J, FM, and $\hat{\Gamma}$ are greater than 94 J_i's, 94 FM_i's, and 98 $\hat{\Gamma}_i$'s, respectively. Thus, according to the Rand and $\hat{\Gamma}$ statistics, the null hypothesis is rejected at significance level $\rho = 0.05$. However, this is not the case for the other two indices.

This situation illustrates the fact that different statistics may lead to different conclusions when no clear-cut situations are considered (see also comparative studies in [Mill 80, Mill 83, Mill 85]).

(c) Let us now construct X'' by selecting the covariance matrices equal to $0.8I$. In this case, the vectors of X'' are so dispersed that, practically, X'' does not exhibit any clustering structure. The values of the four statistics in this case are $R = 0.63$, $J = 0.14$, $FM = 0.25$, and $\hat{\Gamma} = -0.01$. Specifically, R is greater than 62, from the total of 100, R_i's. Similarly, J, FM, and $\hat{\Gamma}$ are greater than 48 R_i's, 48 J_i's, and 55 $\hat{\Gamma}_i$'s, respectively. Thus, according to all statistics, the null hypothesis is not rejected at significance level $\rho = 0.05$.

Remark

- For each of these statistics, q, there exists a corresponding "corrected" statistic q', which is a normalized version of q and is defined as

$$q' = \frac{q - E(q)}{\max(q) - E(q)} \tag{16.11}$$

  where $\max(q)$ is the maximum possible value of q and $E(q)$ is the mean value of q, under the null hypothesis. Its values are between 0 and 1. The maximum value is always achievable when a perfect match between $\mathcal{C}$ and $\mathcal{P}$ occurs and the minimum if $\mathcal{C}$ and $\mathcal{P}$ have been chosen by chance. The problem encountered here is the computation of $E(q)$ and $\max(q)$. This problem is attacked in [Hube 85], for the Rand statistic, under the assumption that the maximum value of the Rand statistic is 1. The same problem for the Fowlkes–Mallows index is treated in [Fowl 83].

Assessing the Agreement between $\mathcal{P}$ and Proximity Matrix P

In this section, we show that the Γ statistic can be used to measure the degree to which the proximity matrix P of X matches a partition $\mathcal{P}$, which is imposed *a priori* on X. Recall that $\mathcal{P}$ may be viewed as a mapping g of X to $\{1, \ldots, m\}$. Let us consider the matrix Y whose (i,j) element, $Y(i,j)$, is defined as follows:

$$Y(i,j) = \begin{cases} 1, & \text{if } g(\boldsymbol{x}_i) \neq g(\boldsymbol{x}_j) \\ 0, & \text{otherwise} \end{cases} \tag{16.12}$$

for $i,j = 1,\ldots,N$. It is clear that Y is symmetric. Then, the Γ (or $\hat{\Gamma}$) statistic is applied to the proximity matrix P and Y. Its value is a measure of the degree to which Y matches P.

In order to estimate the pdf of Γ (or $\hat{\Gamma}$) under the *random label hypothesis*, we produce, say, r mappings g_i, $i = 1,\ldots,r$.[3] For each of them we form the corresponding Y_i matrix and we apply the Γ (or $\hat{\Gamma}$) statistic to P and Y_i, $i = 1,\ldots,r$. Then we proceed as usual for the acceptance or rejection of the random label hypothesis.

Example 16.3

We consider a data set X of 64 two-dimensional vectors. The first 16 of them spring out of a normal distribution with mean $[0.2, 0.2]^T$, and the remaining three groups of 16 vectors stem from three normal distributions with means $[0.2, 0.8]^T$, $[0.8, 0.2]^T$, and $[0.8, 0.8]^T$, respectively. The covariance matrices of all distributions are equal to $0.15I$. Let P be the proximity matrix of X when the squared Euclidean distance is in use. Also, we set the significance level at $\rho = 0.05$.

(a) Let $\mathcal{P} = \{P_1, P_2, P_3, P_4\}$. Suppose that the first set of 16 vectors is assigned to P_1, the second is assigned to P_2, the third is assigned to P_3, and the last to P_4. Based on this information, we form Y as described before and we compute the value of $\hat{\Gamma}$ for P and Y, which is found to be 0.77. Then we generate random partitions $\mathcal{P}_i$, $i = 1,\ldots,100$, we form the corresponding matrices Y_i, and we compute the values $\hat{\Gamma}_i$ between P and each of the Y_i's. It turns out that $\hat{\Gamma}$ is greater than all of these values. Thus, the null hypothesis is rejected at significance level ρ.

(b) Assume now that the external information $\mathcal{P}$ assigns randomly 16 vectors of X to each P_i. It is clear that the external information does not agree with the underlying structure of X. If we apply the same procedure as before, we find that $\hat{\Gamma} = -0.01$, which is less than 70 values of $\hat{\Gamma}_i$. Thus, the randomness hypothesis is accepted.

16.3.2 Internal Criteria

Our aim here is to verify whether the clustering structure produced by a clustering algorithm fits the data, using only information inherent in the data. In the sequel, unless otherwise stated, we consider the case in which the data are represented by their proximity matrix. Two cases are considered: (a) the clustering structure is a hierarchy of clusterings and (b) the clustering structure consists of a single clustering.

Validation of Hierarchies of Clusterings

We recall that the dendrogram produced by a hierarchical clustering algorithm may be represented by the respective cophenetic matrix, P_c. *We will define statistical indices that measure the degree of agreement between the cophenetic matrix,*

[3] Typically, $r = 100$.

P_c, produced by a specific hierarchical clustering algorithm, with the proximity matrix P of X. Because both matrices are symmetric and have their diagonal elements equal to 0,[4] we consider only the $M \equiv N(N-1)/2$ upper diagonal elements of P_c and P. Let d_{ij} and c_{ij} be the (i,j) element of P and P_c, respectively.

The first index, known as the *cophenetic correlation coefficient* (*CPCC*) measures the correlation between P_c and P and is used when the matrices are interval or ratio scaled. It is defined as

$$CPCC = \frac{(1/M)\sum_{i=1}^{N-1}\sum_{j=i+1}^{N} d_{ij}c_{ij} - \mu_P\mu_C}{\sqrt{\left((1/M)\sum_{i=1}^{N-1}\sum_{j=i+1}^{N} d_{ij}^2 - \mu_P^2\right)\left((1/M)\sum_{i=1}^{N-1}\sum_{j=i+1}^{N} c_{ij}^2 - \mu_C^2\right)}} \tag{16.13}$$

where the corresponding mean values are defined as in Eq. (16.9). It can be shown that the values of the *CPCC* are between -1 and 1 (see Problem 16.4). The closer the *CPCC* index to 1, the better the agreement between the cophenetic and the proximity matrix. The *CPCC* statistic has been studied by various researchers (see, e.g., [Rolp 68, Rolp 70, Farr 69]). The major difficulty associated with it is that it depends on many parameters of the problem, such as the size of X, the clustering algorithm used and the employed proximity measure. Hence, the exact computation of its pdf under H_0 is very difficult. Once more, one is forced to use Monte Carlo techniques for the estimation of its distribution, under H_0. According to the random position hypothesis, we generate r sets X_i, whose vectors are randomly distributed according to the uniform distribution, and we apply to each X_i the same hierarchical algorithm that has produced P_c. Then, we compute *CPCC* for the proximity matrix of X_i, P_i, and the resulting cophenetic matrix, P_{c_i} and we construct the corresponding histogram.

Interestingly enough, in [Rolp 70], it is stated that even high values of *CPCC* (near 0.9) should be handled with caution when the unweighted pair group method average (UPGMA) algorithm is in use (Chapter 13), as there are cases for which even such large values cannot guarantee close agreement between the cophenetic and the proximity matrix.

Another statistical index, which is suitable for cases in which P_c and P are ordinally scaled, is the γ statistic, which is described in the sequel. Let v_p and v_c be two vectors of dimension $N(N-1)/2$, each containing the upper diagonal elements of P and P_c, respectively, ordered by rows. Let (v_{p_i}, v_{p_j}) and (v_{c_i}, v_{c_j}) be two pairs of elements of v_p and v_c, respectively. The following definitions are in order.

A set of pairs $\{(v_{p_i}, v_{p_j}), (v_{c_i}, v_{c_j})\}$ is called

- concordant if

$$\left((v_{p_i} < v_{c_i})\ \&\ (v_{p_j} < v_{c_j})\right)\ or\ \left((v_{p_i} > v_{c_i})\ \&\ (v_{p_j} > v_{c_j})\right)$$

[4] This implies that we use a dissimilarity measure.

- discordant if

$$((v_{p_i} < v_{c_i}) \& (v_{p_j} > v_{c_j})) \ or \ ((v_{p_i} > v_{c_i}) \& (v_{p_j} < v_{c_j}))$$

Finally, a set of pairs is neither concordant nor discordant if $v_{p_i} = v_{c_i}$ or $v_{p_j} = v_{c_j}$. Let S_+ and S_- be the numbers of the concordant and discordant pairs, respectively. Then γ is defined as

$$\gamma = \frac{S_+ - S_-}{S_+ + S_-} \tag{16.14}$$

The γ statistic takes values between -1 and 1.

Example 16.4

Let $v_P = [3, 2, 1, 5, 2, 6]^T$ and $v_C = [2, 3, 5, 1, 6, 4]^T$. For all possible 16 pairs of pairs we have

Index	v_P	v_C		Index	v_P	v_C	
(1, 2)	(3, 2)	(2, 3)	dis.	(2, 6)	(2, 6)	(3, 4)	dis.
(1, 3)	(3, 1)	(2, 5)	dis.	(3, 4)	(1, 5)	(5, 1)	dis.
(1, 4)	(3, 5)	(2, 1)	con.	(3, 5)	(1, 2)	(5, 6)	con.
(1, 5)	(3, 2)	(2, 6)	dis.	(3, 6)	(1, 6)	(5, 4)	dis.
(1, 6)	(3, 6)	(2, 4)	con.	(4, 5)	(5, 2)	(1, 6)	dis.
(2, 3)	(2, 1)	(3, 5)	con.	(4, 6)	(5, 6)	(1, 4)	con.
(2, 4)	(2, 5)	(3, 1)	dis.	(5, 6)	(2, 6)	(6, 4)	dis.
(2, 5)	(2, 2)	(3, 6)	con.				

Thus, $S_+ = 6$, $S_- = 9$ and $\gamma = -1/5 = -0.2$.

The γ statistic depends on all the factors of the problem at hand and, as a consequence, the estimate of its pdf under the randomness hypothesis (H_0) is also difficult to derive. Thus, one has to use Monte Carlo techniques once again for the estimation of the pdf of γ under H_0. In this case, the random graph hypothesis is used. Specifically, we produce r random rank order proximity matrices P_i, with no ties, and we run the algorithm that produced P_c on each of them. Then we compute the value of γ for each P_i and its corresponding cophenetic matrix P_{c_i} and we form the histogram for the values of γ.

Remarks

- It has been conjectured [Hube 74] that when the single and the complete link algorithms are used, the statistic $N\gamma - a \ln N$ follows (approximately) the standard normal distribution. The constant a is set equal to 1.1 (1.8) when the single (complete) link algorithm is used. If we adopt this conjecture, it relieves us of the computational burden of the Monte Carlo method.
- The γ statistic may also be used to compare the results for two different hierarchies of clusterings resulting from two different clustering algorithms. (e.g., [Bake 74, Hube 74], Problem 16.5).

Another measure that is, suitable for ordinal-scaled P and P_c is Kudall's τ statistic [Cunn 72], which is defined as

$$\tau = \frac{S_+ - S_-}{N(N-1)/2} \tag{16.15}$$

The difference from the γ statistic is that the denominator here extends to all sets of pairs, whereas in the case of the γ statistic the sets of pairs that are neither concordant nor discordant are excluded.

Validation of Individual Clusterings

Our goal here is to investigate whether a given clustering $\mathcal{C}$, consisting of m clusters, matches information that is, inherent in the data set X. In the sequel, we show that the Γ (or $\hat{\Gamma}$) statistic can be used in order to achieve this goal. Once again, we use the proximity matrix P as a measure representing the structural information inherent in the data. The (i,j) element of the matrix Y is defined as

$$Y(i,j) = \begin{cases} 1, & \text{if } \boldsymbol{x}_i \text{ and } \boldsymbol{x}_j \text{ belong to different clusters} \\ 0, & \text{otherwise} \end{cases} \tag{16.16}$$

for $i,j = 1,\ldots,N$. It is clear that Y is symmetric. Then the Γ (or $\hat{\Gamma}$) statistic is applied to P and Y. Its value is a measure of the degree of correspondence between P and Y.

The *random position hypothesis* is employed. For each of the resulting random data sets X_i, the proximity matrix P_i is computed. Then we apply, to each of them, the clustering algorithm used to produce $\mathcal{C}$. Let $\mathcal{C}_i$, $i = 1,\ldots,r$, be the resulting clusterings of m clusters. We compute Y_i and Γ_i. Finally, we decide for the rejection or acceptance of the null hypothesis at a given significance level ρ according to the conditions given in Section 16.2.

Example 16.5

Consider a data set X of 100 vectors in the H_2 hypercube. The vectors are generated to form four groups, each of 25 vectors. Each group is generated by a normal distribution. The corresponding covariance matrices are all equal to $0.1I$ and the mean vectors are $[0.2, 0.2]^T$, $[0.8, 0.2]^T$, $[0.2, 0.8]^T$, $[0.8, 0.8]^T$, respectively. We apply the isodata algorithm and let $\mathcal{C}$ be the resulting clustering. Computing the corresponding matrices Y and P, we obtain $\hat{\Gamma} = 0.5704$. Then we generate 100 data sets, X_i, whose vectors are randomly distributed in H_2, following the uniform distribution. The isodata algorithm is applied to each of them, and let $\mathcal{C}_i$, $i = 1,\ldots,100$, be the resulting clusterings. Computing Y_i and P_i associated with the resulting clusterings for each X_i, it turns out that 99 of the corresponding $\hat{\Gamma}_i$ values are smaller than $\hat{\Gamma}$. Thus, the null hypothesis is rejected at significance level $\rho = 0.05$.

Repeating the experiment but with covariance matrices equal to $0.2I$, we find that $\hat{\Gamma}$ is greater than 86 of 100 $\hat{\Gamma}_i$ values. Thus, the null hypothesis is not rejected at significance level $\rho = 0.05$.

16.4 RELATIVE CRITERIA

So far, clustering validation has been performed on the basis of statistical tests. A major drawback of most of these techniques is their high computational demands, due to the required Monte Carlo methodology. In this section, a different approach is discussed that does not involve statistical tests. To this end, *a set of clusterings is considered and the goal is to choose the best one according to a prespecified criterion*. More specifically, let $\mathcal{A}$ be the set of parameters associated with a specific algorithm. For example, for the algorithms of Chapter 14, $\mathcal{A}$ contains the number of clusters, m, as well as the initial estimates of the parameter vectors associated with each cluster. The problem can be stated as follows:

"Among the clusterings produced by a specific clustering algorithm, for different values of the parameters in $\mathcal{A}$, choose the one that best fits the data set X."

We consider the following cases:

- *$\mathcal{A}$ does not contain the number of clusters, m*, as a parameter (such as the algorithms based on graph theory, the morphological clustering algorithm and the boundary detection algorithms).

  The choice of the "best" parameter values for this type of algorithm is based on the assumption that *if X possesses a clustering structure, this structure is captured for a "wide" range of values of the parameters* in $\mathcal{A}$ (e.g., [Post 93]). Based on this assumption, we proceed as follows. We run the algorithm for a wide range of values of its parameters and we choose the widest range for which, m, remains constant (typically $m \ll N$). Then we choose as appropriate values of the parameters of $\mathcal{A}$ the values that correspond to the middle of this range. Note that, implicitly, this procedure also identifies the number of clusters that underlie X.

Example 16.6

(a) We consider a data set X, consisted of three groups of 100 two-dimensional vectors. These groups are formed from normal distributions with means $[0, 0]^T$, $[8, 4]^T$, and $[8, 0]^T$, respectively, and covariance matrices equal to $1.5I$. As one can easily observe in Figure 16.2a, the three groups form three compact and well-separated clusters. We run the binary morphology clustering algorithm (BMCA), using the 3×3 structuring element (Figure 15.10a), with the resolution parameter r ranging from 1 to 77 and we plot the number of clusters versus r (Figure 16.2b). We observe that for any value of r between 37 and 67, the number of clusters remains constant and equal to 3. Taking into account that this range of values is the largest one, we choose $r = 52$, and we conclude that our data form three clusters.

(b) Generate another data set, as before, but with the covariance matrices equal to $2.5I$. This data set is depicted in Figure 16.3a. We observe that in this case the three groups are so dispersed that they practically cannot be distinguished from each other. We run BMCA once again, using the 3×3 structuring element, for r ranging from 1 to 77, with step 1, and we

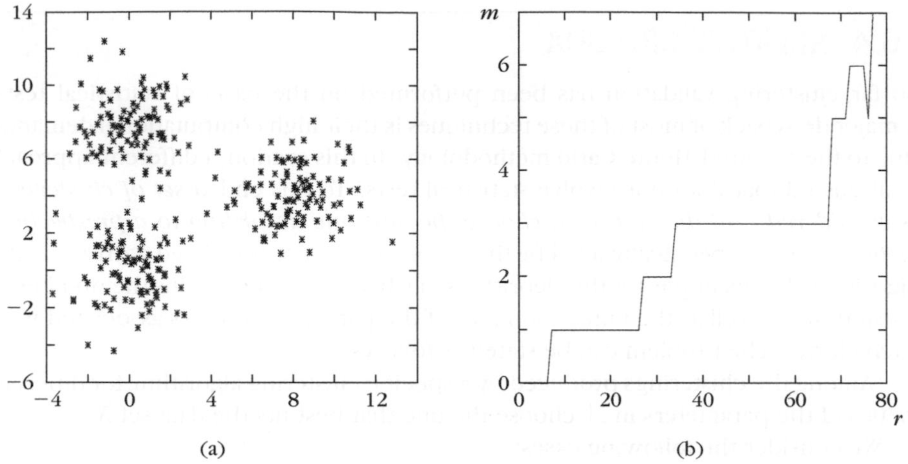

FIGURE 16.2

(a) Three well-separated clusters. (b) The plot of the number of clusters m versus the resolution parameter r, using the binary morphology clustering algorithm (BMCA).

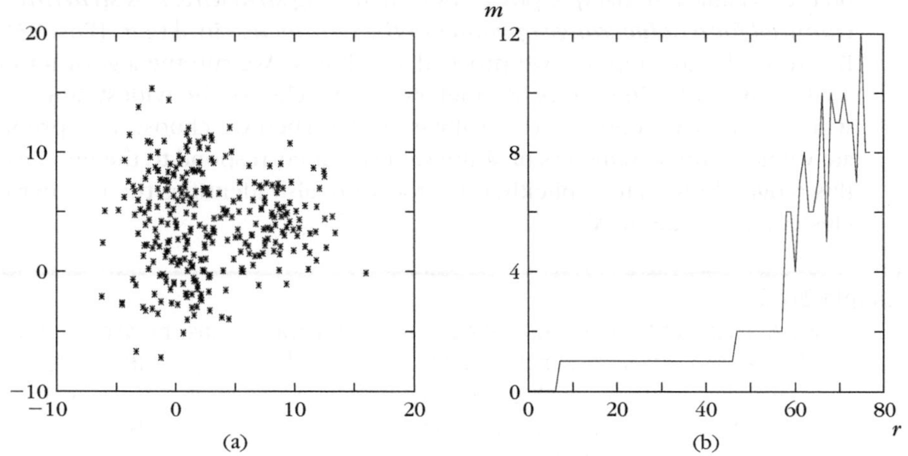

FIGURE 16.3

(a) Three overlapped clusters. (b) The plot of m versus r.

plot the number of clusters versus r (see Figure 16.3b). In this case, for $r = 7, \ldots, 46$, the number of clusters remains constant and the corresponding value of m is 1.

- ***A contains m as a parameter*** (such as the fuzzy and hard clustering algorithms discussed in Chapter 14). For this case, a different procedure is followed. We first select a suitable performance index q. The "best" clustering is identified, in terms of q, via the following procedure. We run the clustering algorithm at hand for all values of m between a minimum $m_{\min}$ and a maximum $m_{\max}$, where $m_{\min}$ and $m_{\max}$ are chosen *a priori*. For each value of m, we run the algorithm r times, using different sets of values for the other parameters of $\mathcal{A}$.[5] Then we plot the best values of q, obtained for each m, versus m and we seek the maximum or the minimum of this plot, according to whether large or small values of q indicate good clusterings. This procedure works well if q exhibits no trend with respect to m. However, as we will see, several of the commonly used indices q exhibit an increasing (decreasing) trend as m increases. Thus, locating the maximum (minimum) versus m is no longer indicative of a good clustering. For indices that exhibit such behavior, in the range $[m_{\min}, m_{\max}]$, we search for values of m at which a significant local change in the value of q occurs. This change appears in the plot as a significant "knee." *The presence of such a knee is an indication of the number of clusters underlying X. On the other hand, the absence of such a knee may be an indication that X possesses no clustering structure.*

Another source of complication, associated with many of the indices used in this framework is that their behavior depends on many other factors such as the number of vectors in X and their dimensionality. The situation is demonstrated via the following example.

Example 16.7

(a) In this example, we consider 16 different data sets with different numbers of vectors and dimensionalities. Specifically, we consider four 2-dimensional data sets of 50, 100, 150, and 200 vectors; four 4-dimensional data sets of 50, 100, 150, and 200 vectors; four 6-dimensional data sets of 50, 100, 150, and 200 vectors; and four 8-dimensional data sets of 50, 100, 150, and 200 vectors. The vectors of the data sets lie in the H_i hypercube, $i = 2, 4, 6, 8$, respectively. All the data sets contain four compact and well-separated clusters. All the clusters stem from normal distributions with means $[\overbrace{0.2, \ldots, 0.2}^{i}]^T$, $[\overbrace{0.2, \ldots, 0.2}^{i/2}, \overbrace{0.8, \ldots, 0.8}^{i/2}]^T$, $[\overbrace{0.8, \ldots, 0.8}^{i/2}, \overbrace{0.2, \ldots, 0.2}^{i/2}]^T$, $[\overbrace{0.8, \ldots, 0.8}^{i}]^T$, where i is the dimensionality, and covariance matrices $0.2I_i$, where I_i is the $i \times i$ identity matrix. For each of these data sets we run the isodata algorithm for $m = 1, \ldots, 10$, and we compute the corresponding values of the cost function J. For this case, only a single run is performed for each m. In Figure 16.4 we plot J versus the number of clusters, m, for the cases of 50, 100, 150, and 200 vectors and for different dimensionalities.

One can easily notice that the higher the dimensionality, the sharper the knee at $m = 4$. Moreover, as the size of the data set increases, the knee at $m = 4$ becomes sharper, even at lower dimensionalities. Rules for automatic identification of a knee are discussed in [Dube 87a].

[5] For example, if the k-means algorithm is used, we run it using different initial conditions.

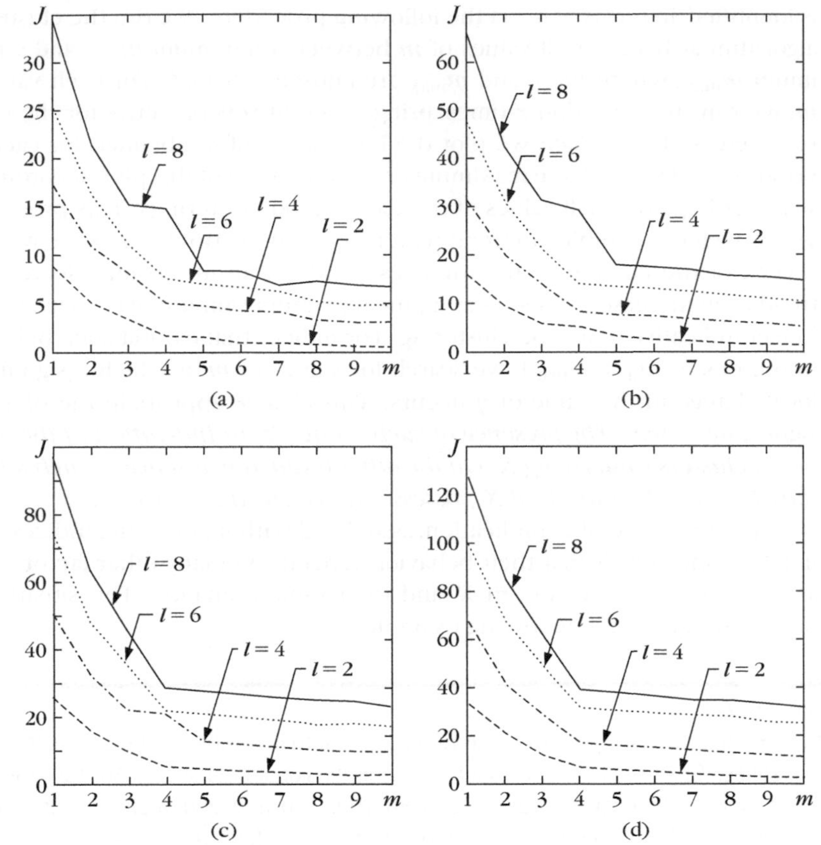

FIGURE 16.4

Plots of J versus m for (a) $N = 50$, (b) $N = 100$, (c) $N = 150$, (d) $N = 200$, for clustered data.

(b) We again construct 16 data sets, but now the vectors in each of them are randomly distributed in the H_l hypercube, according to the uniform distribution. If we carry out the same procedure as before, we see in Figure 16.5 that there are no sharp knees in the plots. Thus, the absence of sharp knees in the plots *may be* an indication of the absence of clustering structure.

16.4.1 Hard Clustering

In this section we discuss indices that are suitable for hard clusterings. In the sequel unless otherwise stated, we consider only the case of compact clusters.

- *The modified Hubert* Γ *statistic.* Let $c_i = k$ if the vector $\boldsymbol{x}_i$ belongs to cluster C_k. Also let Q be the $N \times N$ matrix whose (i,j) element, $Q(i,j)$, is equal to the distance $d(\boldsymbol{w}_{c_i}, \boldsymbol{w}_{c_j})$ between the representatives of the clusters where $\boldsymbol{x}_i$

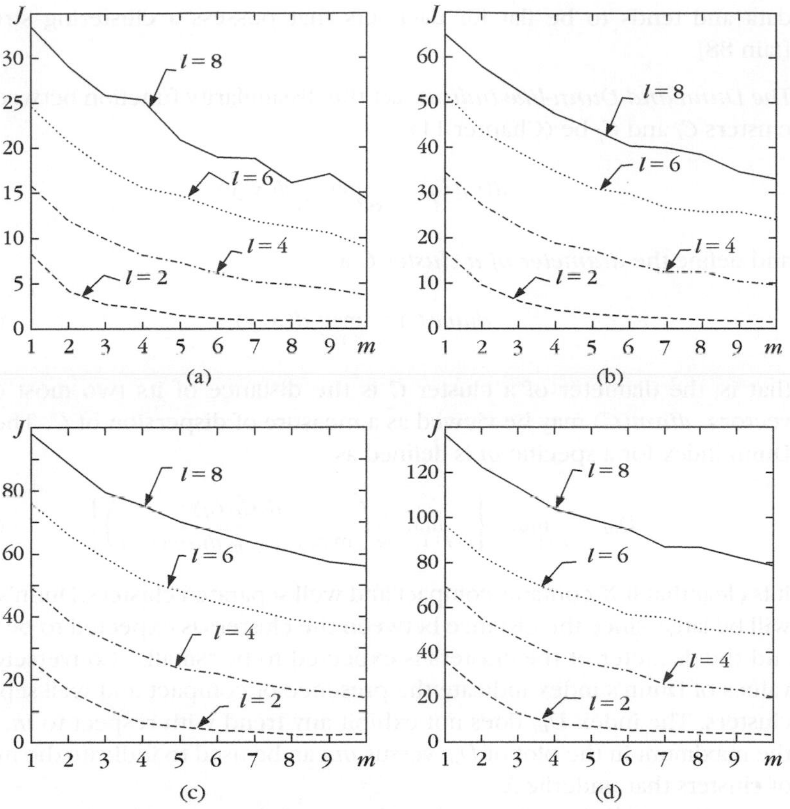

FIGURE 16.5

Plots of J versus m for (a) $N = 50$, (b) $N = 100$, (c) $N = 150$, (d) $N = 200$, for random data.

and x_j belong. The modified Hubert Γ statistic is defined as in Eq. (16.8) and it is applied to the proximity matrix P of the data set X and the matrix Q (of course, the same distance measure must be used for both P and Q). Similarly, we can define the normalized modified Hubert Γ statistic. It is clear that if $d(w_{c_i}, w_{c_j})$ is close to $d(x_i, x_j)$, for $i,j=1,\ldots,N$, that is, when compact clusters are encountered in X, P and Q will be in close agreement and the values of Γ and $\hat{\Gamma}$ will be high. Conversely, high values of $\Gamma(\hat{\Gamma})$ indicate the existence of compact clusters. If the opposite is true, the values of the modified Γ and $\hat{\Gamma}$ indices are expected to be low. Thus, in the plot of $\hat{\Gamma}$ versus m, we seek a significant knee that corresponds to a significant increase of $\hat{\Gamma}$. The value of m at which this knee occurs indicates the number of clusters tha underlie X.

For $m = 1$ and $m = N$ the index is not defined. Also, this index tends to increase as m increases toward N (see Problem 16.6) for random data and tends to be flat for data sets that possess a clustering structure [Jain 88].

- *The Dunn and Dunn-like indices.* Let the dissimilarity function between two clusters C_i and C_j be (Chapter 11)

$$d(C_i, C_j) = \min_{x \in C_i, y \in C_j} d(x, y) \tag{16.17}$$

and define the *diameter of a cluster C* as

$$diam(C) = \max_{x, y \in C} d(x, y) \tag{16.18}$$

that is, the diameter of a cluster C is the distance of its two most distant vectors. $diam(C)$ may be viewed as a measure of dispersion of C. Then, the Dunn index for a specific m is defined as

$$D_m = \min_{i=1,\ldots,m} \left\{ \min_{j=i+1,\ldots,m} \left(\frac{d(C_i, C_j)}{\max_{k=1,\ldots,m} diam(C_k)} \right) \right\} \tag{16.19}$$

It is clear that if X contains compact and well-separated clusters, Dunn's index will be large, since the distance between the clusters is expected to be "large" and the diameter of the clusters is expected to be "small." Conversely, large values of Dunn's index indicate the presence of compact and well-separated clusters. The index D_m does not exhibit any trend with respect to m, hence the maximum in the plot of D_m versus m can be used to indicate the number of clusters that underlie X.

In [Dunn 74], it is shown that if $D_m > 1$ for a specific clustering, then this clustering contains compact and well-separated clusters.

A disadvantage of the Dunn index is the considerable amount of time required for its computation (see Problem 16.7). Moreover, Dunn's index is sensitive to the presence of noisy vectors in X, because these are likely to increase the value of the denominator of Eq. (16.19).

In [Pal 97] three Dunn-like indices are proposed that are more robust to the presence of noisy vectors. Furthermore, preliminary simulation results show that they may be used for cases in which shell-shaped clusters underlie X. These three indices are based on the concepts of the minimum spanning tree (MST), the relative neighborhood graph (RNG), and the Gabriel graph (GG), discussed in Chapter 15. Let us consider explicitly the index based on the MST concept. The other two are defined using similar arguments.

Consider a cluster C_i and the *complete* graph G_i having vertices that correspond to the vectors of C_i. The weight, w_e, of an edge, e, of this graph equals the distance between its two end points, x and y, that is, $w_e = d(x, y)$. Let E_i^{MST} be the set of edges of the MST of G_i and let e_i^{MST} be the edge in E_i^{MST} with the maximum weight. Then the diameter of C_i, diam_i^{MST}, is defined as the weight of e_i^{MST} (see Figure 16.6).

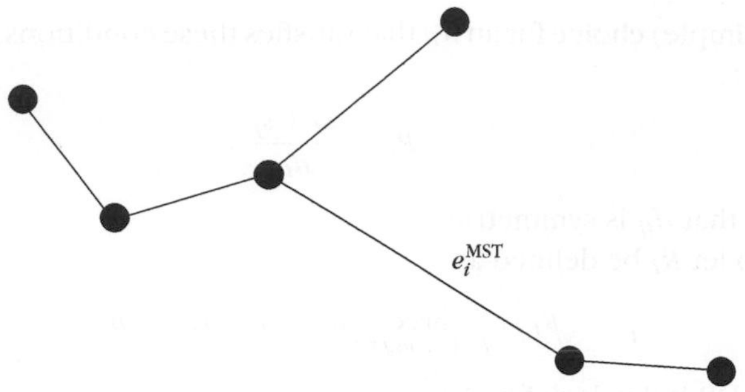

FIGURE 16.6

A minimum spanning tree.

The dissimilarity between two clusters is defined as the distance of their mean vectors, $d(C_i, C_j) = d(\boldsymbol{m}_i, \boldsymbol{m}_j)$. Then, the Dunn-like index, based on the concept of the MST, is defined as

$$D_m^{MST} = \min_{i=1,\dots,m}\left\{\min_{j=i+1,\dots,m}\left(\frac{d(C_i, C_j)}{\max_{k=1,\dots,m} diam_k^{MST}}\right)\right\} \tag{16.20}$$

The maximum in the plot of D_m^{MST} versus m indicates the underlying number of clusters in X. Similar arguments are followed to define Dunn-like indices for GG and RNG graphs (see Problem 16.8).

- *The Davies–Bouldin (DB) and DB-like indices.* Let s_i be a measure of dispersion of a cluster C_i (i.e., a measure of its spread around its mean vector) and $d(C_i, C_j) \equiv d_{ij}$ the dissimilarity between two clusters, using an appropriate dissimilarity measure. Based on these, a similarity index R_{ij} between C_i and C_j is defined to satisfy the following conditions [Davi 79]:

  (C1) $R_{ij} \geq 0$.
  (C2) $R_{ij} = R_{ji}$.
  (C3) If $s_i = 0$ and $s_j = 0$ then $R_{ij} = 0$.
  (C4) If $s_j > s_k$ and $d_{ij} = d_{ik}$ then $R_{ij} > R_{ik}$.
  (C5) If $s_j = s_k$ and $d_{ij} < d_{ik}$ then $R_{ij} > R_{ik}$.

  These conditions state that R_{ij} is nonnegative and symmetric. If both clusters, C_i and C_j, collapse to a single point, then $R_{ij} = 0$. A cluster C_i with the same distance from two other clusters, C_j, C_k, is more similar to the cluster with the largest dispersion (condition (C4)). For the case of equal dispersions and different dissimilarity levels, the cluster C_i is more similar to the closer of the two (condition (C5)).

A (simple) choice for an R_{ij} that satisfies these conditions is the following [Davi 79]:

$$R_{ij} = \frac{s_i + s_j}{d_{ij}} \tag{16.21}$$

provided that d_{ij} is symmetric.

Also let R_i be defined as

$$R_i = \max_{j=1,\dots,m, j\neq i} R_{ij}, \quad i = 1,\dots,m \tag{16.22}$$

Then the DB index is defined as

$$DB_m = \frac{1}{m}\sum_{i=1}^{m} R_i \tag{16.23}$$

That is, DB_m is the average similarity between each cluster $C_i, i = 1,\dots,m$, and its most similar one. As it is desirable for the clusters to have the minimum possible similarity to each other, we seek clusterings that minimize DB. On the other hand, small values of DB are indicative of the presence of compact and well-separated clusters. The DB_m index exhibits no trends with respect to m [Davi 79], thus we seek the minimum value of DB_m, in the plot of DB_m versus m.

In [Davi 79], the dissimilarity $d(C_i, C_j)$ between two clusters is defined as

$$d_{ij} = \|\boldsymbol{w}_i - \boldsymbol{w}_j\|_q = \left(\sum_{k=1}^{l} |w_{ik} - w_{jk}|^q\right)^{1/q} \tag{16.24}$$

Also, the dispersion of a cluster C_i is defined as

$$s_i = \left(\frac{1}{n_i}\sum_{\boldsymbol{x}\in C_i} \|\boldsymbol{x} - \boldsymbol{w}_i\|^r\right)^{1/r} \tag{16.25}$$

where n_i is the number of vectors in C_i. (Compare this definition with that of the diameter of a cluster defined earlier.)

In [Pal 97] three variants of the DB index, based again on the MST, RNG, and GG concepts, are proposed. We focus on the MST case. Let s_i^{MST} be the $\text{diam}_i^{\text{MST}}$, as defined in the Dunn-like index, and let d_{ij} be the distance between the mean vectors of C_i, C_j. Then, we define

$$R_{ij}^{\text{MST}} = \frac{s_i^{\text{MST}} + s_j^{\text{MST}}}{d_{ij}} \tag{16.26}$$

It is easy to show that R_{ij}^{MST} satisfies the conditions (C1)–(C5) (see Problem 16.10). Defining $R_i^{\text{MST}} = \max_{j=1,\dots,m, j\neq i} R_{ij}^{\text{MST}}$, the MST DB index is defined as

$$DB_m^{\text{MST}} = \frac{1}{m}\sum_{i=1}^{m} R_i^{\text{MST}} \tag{16.27}$$

The minimum in the plot of DB_m^{MST} versus m is an indication of the number of clusters that underlie X.

Using arguments similar to these, we may define DB_m^{RNG} and DB_m^{GG}.

- ***The silhouette index*** ([Kauf 90]). Let C_{c_i} denote the cluster where $\boldsymbol{x}_i \in X$ belongs, $i = 1, \ldots, N$. For each $\boldsymbol{x}_i$ let a_i be the average distance between $\boldsymbol{x}_i$ and the rest of the elements of C_{c_i}, that is,

$$a_i = d_{\text{avg}}^{\text{ps}}(\boldsymbol{x}_i, C_{c_i} - \{\boldsymbol{x}_i\})$$

where $d_{\text{avg}}^{\text{ps}}(\cdot, \cdot)$ denotes the average distance measure between a point and a set (see Section 11.2.1). Let also b_i be the average distance between $\boldsymbol{x}_i$ and its closest cluster excluding C_{c_i}, that is,

$$b_i = \min_{k=1,\ldots,m, k \neq c_i} d_{\text{avg}}^{\text{ps}}(\boldsymbol{x}_i, C_k)$$

Then the *silhouette width* of $\boldsymbol{x}_i$ is defined as

$$s_i = \frac{b_i - a_i}{\max(b_i, a_i)} \tag{16.28}$$

It is not difficult to see that $-1 \le s_i \le 1$. Values of s_i close to 1 imply that the distance of $\boldsymbol{x}_i$ from the cluster where it belongs (C_{c_i}) is significantly less than the distance between $\boldsymbol{x}_i$ and its nearest cluster excluding C_{c_i}. This is an indication that $\boldsymbol{x}_i$ is well clustered. On the other hand, values of s_i close to -1 imply that the distance between $\boldsymbol{x}_i$ and C_{c_i} is significantly higher than the distance between $\boldsymbol{x}_i$ and its nearest cluster excluding C_{c_i}. This is an indication that $\boldsymbol{x}_i$ is not well clustered. Finally, values of s_i close to 0 indicate that $\boldsymbol{x}_i$ lies close to the border between the two clusters.

Based on the definition of s_i, the *silhouette of the cluster* $C_j, j = 1, \ldots, m$, is defined as

$$S_j = \frac{1}{n_j} \sum_{i: \boldsymbol{x}_i \in C_j} s_i \tag{16.29}$$

where n_j is the cardinality of C_j, and the *global silhouette index* is defined as

$$S_m = \frac{1}{m} \sum_{j=1}^{m} S_j \tag{16.30}$$

Clearly, $S_m \in [-1, \ 1]$. In addition, the higher the value of S_m, the better the corresponding clustering is. Therefore, the maximum in the plot of S_m versus m is taken to indicate the underlying number of clusters in X.

- ***The Gap statistic*** ([Tibs 01]). Let D_q denote the sum of the distances between all pairs of patterns in cluster C_q, that is,

$$D_q = \sum_{\boldsymbol{x}_i \in C_q} \sum_{\boldsymbol{x}_j \in C_q} d(\boldsymbol{x}_i, \boldsymbol{x}_j)$$

and let

$$W_m = \sum_{q=1}^{m} \frac{1}{2n_q} D_q \tag{16.31}$$

Clearly, a low value of W_m indicates a clustering of compact clusters.

The idea here is to compare the curve of $\log W_m$ versus m with the corresponding curve obtained from data uniformly distributed within a hyperrectangle that contains the data points of X [Hast 01] (see also [Tibs 01] for a more formal discussion on this issue). To this end, and for each m, n data sets $X_m^r, r = 1, \ldots, n$, are generated, as indicated before, and the average (in theory the expectation) $E_n(\log(W_m^r))$ over the $\log(W_m^r)$s of the corresponding X_m^rs is computed. Then the value of m for which $\log(W_m)$ falls the farthest below the reference curve formed by $E_n(\log(W_m^r))$ is taken to indicate the number of clusters in X. This is formalized via the so-called *Gap statistic*, which is defined as

$$\text{Gap}_n(m) = E_n(\log(W_m^r)) - \log(W_m) \tag{16.32}$$

The estimate of the number of clusters in X is taken to be the value that maximizes $\text{Gap}_n(m)$ (within some tolerance).

For the computational implementation of the Gap statistic we proceed as follows:

- For each value of m in $[m_{\min}, m_{\max}]$ do
  - Cluster the data set X and compute $\log(W_m)$
  - Generate n reference data sets and compute the Gap statistic via Eq. (16.32).
  - Define $s_m = sd_m\sqrt{1 + 1/n}$, where sd_m is the standard deviation of the $\log(W_m^r)$s around their average value.
- Choose the smallest m for which $\text{Gap}_n(m) \geq \text{Gap}_n(m+1) - s_{m+1}$

The Gap statistic can be used with any distance measure between points. In addition, it works well for the case where the data of X form a single cluster. Experimental results ([Tibs 01]) show that the Gap statistic outperforms several other indices. However, when the data are concentrated on a subspace of $\mathcal{R}^l$, the method generating the X_m^rs, as described before, degrades the performance of the Gap statistic.

- *Information Theory based criteria.* A different philosophy that may be used for the estimation of the number of clusters m relies on the determination of a model that best fits the available data, without having any knowledge of their true distribution (see, for example, [Lu 00]).

  Let us define the following criterion function:

$$C(\boldsymbol{\theta}, K) = -2L(\boldsymbol{\theta}) + \phi(K) \tag{16.33}$$

where $\boldsymbol{\theta}$ is the parameter vector of the model, $L(\boldsymbol{\theta})$, is the log-likelihood function (see Eq. (2.58)), K is the order of the model, that is, the dimensionality of $\boldsymbol{\theta}$, and ϕ is an increasing function of K. Typical choices of ϕ are $\phi(K) = 2K$ (Akaike Information Criterion, AIC [Akai 85]), $\phi(K) = \frac{2KN}{N-K-1}$ (Consistent AIC [Hurv 89]), $\phi(K) = K \ln N$ (Minimum Description Length (MDL) Criterion [Riss 78, Riss 89] and Bayesian Information Criterion (BIC) [Schw 76, Fral 98]). Note that K is a strictly increasing function of the number of clusters, m, since the higher the m, the larger the dimensionality of $\boldsymbol{\theta}$. For example, in the case where $p(\boldsymbol{x}; \boldsymbol{\theta})$ is a weighted summation of m l-dimensional Gaussian distributions, each one corresponding to a cluster, $\boldsymbol{\theta}$ consists of the ml parameters associated with the mean values of the distributions, plus $m\frac{l(l+1)}{2}$ parameters associated with the covariance matrices of the distributions plus the $m - 1$ weighting parameters. Thus, $K = (l + \frac{l(l+1)}{2} + 1)m - 1$. In words, K is an increasing linear function of m.

The aim is to minimize C with respect to $\boldsymbol{\theta}$ and K. We proceed as follows. First, the set of candidate models is fixed, involving models of similar structure but of different orders. Let $m \in [m_{\min}, m_{\max}]$ for the models of the above set. Then for each value of $m_i \in [m_{\min}, m_{\max}]$, we optimize $C(\boldsymbol{\theta}, m_i)$ with respect to $\boldsymbol{\theta}$, that is, we determine the maximum likelihood estimation $\boldsymbol{\theta}_i$. Then, among all pairs $(\boldsymbol{\theta}_i, m_i)$, we choose the one, say $(\boldsymbol{\theta}_j, m_j)$, that minimizes C. Thus, the estimated number of clusters is m_j. In the case where it is desirable to choose the best among models of different structure, we first identify all the subsets, each one containing similar models of differing order. Then, we determine the best model of each subset as described above. Finally, among these models, we select the one that leads to the minimum value of C.

Other indices suitable for hard clusterings have also been proposed. For example, in [Mill 80] and [Mill 85] many indices of this kind are tested on specific data sets (see, also, [Gord 99]). Also, in [Kirl 00] two new indices are presented and their relation to the method discussed in Section 12.3, for estimating the number of clusters, is investigated. Additional indices may be found in [Shar 96, Halk 00]. In [Halk 01] an index that takes into account the density of the clusters is proposed. An evaluation of several indices may be found in [Halk 02a, Halk 02b]. Finally, in [Bout 04] a number of validity indices suitable for graph partitioning is considered.

16.4.2 Fuzzy Clustering

In this section we consider indices suitable for fuzzy clustering. In this context, we seek clusterings that are not very fuzzy, that is, those whose clusters exhibit small overlap. In other words, we seek clusterings where most of the vectors of X exhibit high grade of membership in only one cluster. Recall that a fuzzy clustering is defined by the $N \times m$ matrix $U = [u_{ij}]$, where u_{ij} denotes the grade of membership of the vector $\boldsymbol{x}_i$ in the j-th cluster. Also, let $W = \{\boldsymbol{w}_j, j = 1, \ldots, m\}$ be the set of the cluster representatives.

The strategy followed for the hard clustering case is also adopted here. That is, we define an appropriate index q (not to be confused with the fuzzifier) and we search for the minimum or the maximum in the plot of q versus m. In the case where q exhibits a trend with respect to m in the range $[m_{\min}, m_{\max}]$, we seek a significant knee of decrease or increase of q.

Indices for Clusters with Point Representatives

A. Indices that Involve Only U

One such index is the *partition coefficient* [Bezd 74], which is defined as

$$PC = \frac{1}{N}\sum_{i=1}^{N}\sum_{j=1}^{m} u_{ij}^2 \tag{16.34}$$

where u_{ij}'s are the values obtained after the convergence of the adopted fuzzy clustering algorithm.

The range of values for PC is $[1/m, 1]$. This index is computed for values of m greater than 1, since for $m = 1$, it is trivially equal to 1. The closer to unity the PC, the harder the clustering is or, alternatively, the smaller the "sharing" of the vectors in X among different clusters. The lowest value of PC is obtained when all u_{ij}'s are equal, that is, $u_{ij} = 1/m, j = 1, \ldots, m, i = 1, \ldots, N$. Thus, the closer the value of PC to $1/m$, the fuzzier the clustering. A value close to $1/m$ indicates that either X possesses no clustering structure or the adopted clustering algorithm failed to unravel it [Pal 95].

Another index of this category is the *partition entropy coefficient* [Bezd 75], which is defined as

$$PE = -\frac{1}{N}\sum_{i=1}^{N}\sum_{j=1}^{m}(u_{ij}\log_a u_{ij}) \tag{16.35}$$

where a is the base of the logarithm. This index is also computed for values of m greater than 1. Its minimum value equals 0 and its maximum $\log_a m$. The closer the value of PE to 0, the harder the clustering is. On the other hand, the closer the value of PE to $\log_a m$, the fuzzier the clustering is. As in the previous case, values close to $\log_a m$ indicate the absence of any clustering structure in X or the inability of the clustering algorithm to reveal it [Pal 95].

Both of these indices measure the amount of "overlap" among clusters, without utilizing any additional information concerning the positions of the data vectors and the cluster representatives in space.

A disadvantage of both PC and PE indices is that they exhibit a dependence on m with a trend to increase or decrease, respectively, as m increases. Thus, one seeks significant knees of increase (for PC) or decrease (for PE) in the plot of the indices versus m. Moreover, they are also sensitive to the fuzzifier q. It can be shown (Problem 16.13) that as $q \to 1^+$,[6] both PC and PE give the same values for all m's;

[6] This notation means that q tends to 1 from the right.

that is, they are unable to discriminate between different values of m. On the other hand, as $q \to \infty$, both PC and PE exhibit the most significant knee at $m = 2$ (see Problem 16.13). The behavior of PC and PE is illustrated via the following example.

Example 16.8

Let X be a data set that consists of three groups of two-dimensional vectors, each containing 100 vectors. The groups stem from normal distributions with means $[1, 1]^T$, $[4, 4]^T$, $[7, 1]^T$, respectively (see Figure 16.7a). All covariance matrices are equal to the identity 2×2 matrix. We run the fuzzy c-means algorithm for $q = 1.5, 2, 3, 5$ and $m = 1, \ldots, 10$. Figure 16.7b shows the behavior of the PC index. One can observe that for $q = 1.5$ and $q = 2$, the corresponding plots exhibit a significant knee at $m = 3$, which is the correct number of (the natural) clusters. The plots for $q = 3$ and $q = 5$ coincide. This implies that no significant change in the behavior of the index is expected for $q \geq 3$. Moreover, no peak is encountered; that is, no conjecture can be made for the number of clusters. Also, notice the general decreasing trend as m increases.

Figure 16.7c shows the behavior of the PE index. The plots corresponding to $q = 1.5$ and $q = 2$ exhibit a significant knee at $m = 3$. Also, as in the previous case, no significant change in the behavior of the index is expected for $q \geq 3$, and no minimum is encountered for $q \geq 3$. Finally, PE exhibits an increasing trend as m increases.

Other indices of this kind have also been proposed in the literature (e.g., [Wind 81]). We consider next indices that involve X, U, and W.

B. Indices Involving W, U, and the Data Set X

Let us define the so-called *cluster variation* as $\sigma_j^q = \sum_{i=1}^N u_{ij}^q \|x_i - w_j\|^2$ (compare this with the dispersion used in the DB index) and the *total variation* as $\sigma_q = \sum_{j=1}^m \sigma_j^q$. The parameter σ_q may be viewed as a measure of compactness of the specific clustering. Also let $d_{\min} = \min_{i,j=1,\ldots,m, i\neq j} \|w_i - w_j\|^2$ be a measure of

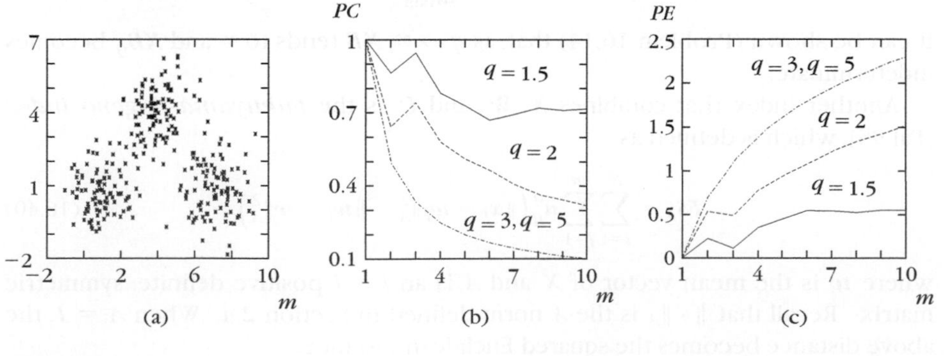

FIGURE 16.7

(a) The data set. (b) The plot of PC versus m. (c) The plot of PE versus m.

separability of the clusters in X, where w_j denotes the representative of the j-th cluster $j = 1, \ldots, m$. Then the *Xie-Beni index*, which is also called the *compactness and separation validity function*, is defined as

$$XB = \frac{\sigma_2/N}{d_{\min}} \tag{16.36}$$

This index is usually employed for the validation of clusterings produced by the fuzzy c-means algorithm when the Euclidean distance is in use. Note that despite the fact that the fuzzifier q in the fuzzy c-means may have any value greater than 1, *in the XB index the value of q involved in σ_q is restricted to* 2.

It is clear that for compact and well-separated clusters, small values of XB are expected. On the other hand, small values of XB indicate compact and well-separated clusters. As stated in [Xie 91], the XB index decreases monotonically as m gets very close to N. One way to handle this problem is to determine the starting point, $m_{\max}$, of the monotonicity behavior and to search for the minimum value of XB in the range $[2, m_{\max}]$.

Let

$$J_q = \sum_{i=1}^{N} \sum_{j=1}^{m} u_{ij}^q \|x_i - w_j\|^2 \tag{16.37}$$

(recall that this is the cost function minimized by the fuzzy c-means clustering algorithm when the squared Euclidean distance is in use). Then XB may be written in terms of J_2 as

$$XB = \frac{J_2}{N d_{\min}} \tag{16.38}$$

Thus, minimization of XB implies minimization of J_2.

Removal of the constraint $q = 2$, used in the definition of XB, allows the definition of the *generalized XB index* as

$$XB_q = \frac{\sigma_q}{N d_{\min}} \tag{16.39}$$

It can be shown (Problem 16.14) that, as $q \to \infty$, XB tends to ∞ and XB_q becomes indeterminate.

Another index that combines X, W, and U is the *Fukuyama-Sugeno index* [Pal 95], which is defined as

$$FS_q = \sum_{i=1}^{N} \sum_{j=1}^{m} u_{ij}^q \left(\|x_i - w_j\|_A^2 - \|w_j - w\|_A^2\right) \tag{16.40}$$

where w is the mean vector of X and A is an $l \times l$ positive definite, symmetric matrix. Recall that $\|\cdot\|_A$ is the A norm defined in Section 2.4. When $A = I$, the above distance becomes the squared Euclidean distance.

The first of the two terms in the parenthesis measures the compactness of the clusters, and the second one measures the distance of the cluster representatives,

$\boldsymbol{w}_i$, from the overall mean vector $\boldsymbol{w}$. It is clear that for compact and well-separated clusters we expect small values for FS_q. Furthermore, small values of FS_q are indicative of compact and well-separated clusters. As far as the limiting behavior of FS_q is concerned, it can be shown (Problem 16.15) that (a) as $q \to 1^+$, FS_q behaves like $tr(S_w)$, the trace of the within scatter matrix (see Chapter 5), and (b) as $q \to \infty$, FS_q tends to 0.

In [Gath 89], three additional indices are proposed that are based on the concepts of hypervolume and density. Let us define the *fuzzy covariance matrix* of the j-th cluster as

$$\Sigma_j = \frac{\sum_{i=1}^{N} u_{ij}^q (\boldsymbol{x}_i - \boldsymbol{w}_j)(\boldsymbol{x}_i - \boldsymbol{w}_j)^T}{\sum_{i=1}^{N} u_{ij}^q} \tag{16.41}$$

The *fuzzy hypervolume* of the j-th cluster is defined as

$$V_j = |\Sigma_j|^{1/2} \tag{16.42}$$

where $|\Sigma_j|$ is the determinant of Σ_j. Note that this is a measure of compactness of the j-th cluster. The smaller the value of V_j, the more "compact" the j-th cluster is.

The *total fuzzy hypervolume* is defined as

$$FH = \sum_{j=1}^{m} V_j \tag{16.43}$$

Small values of FH indicate the existence of compact clusters.

Let $X_j = \{\boldsymbol{x} \in X: (\boldsymbol{x} - \boldsymbol{w}_j)^T \Sigma_j^{-1} (\boldsymbol{x} - \boldsymbol{w}_j) < 1\}$; that is, X_j contains all the vectors in X that are within a prespecified (small) region around $\boldsymbol{w}_j$. Also let $S_j = \sum_{\boldsymbol{x}_i \in X_j} u_{ij}$ be the so-called sum of central members of the jth cluster. The quantity S_j/V_j is called the *fuzzy density* of the jth cluster. Then the *average partition density* is defined as

$$PA = \frac{1}{m} \sum_{j=1}^{m} \frac{S_j}{V_j} \tag{16.44}$$

A different measure is the *partition density* index and it is defined as

$$PD = \frac{S}{FH} \tag{16.45}$$

where $S = \sum_{j=1}^{m} S_j$.

"Compact" clusters lead to large values of PA and PD, and vice versa, large values of PA and PD are indications of "compact" clusters.

Note that all these indices, except PE and PC, may be used in the framework of hard clustering as well, by defining

$$u_{ij} = \begin{cases} 1, & \text{if } d(\boldsymbol{x}_i, C_j) = \min_{k=1,\ldots,m} d(\boldsymbol{x}_i, C_k) \\ 0, & \text{otherwise} \end{cases} \quad i = 1, \ldots, N \tag{16.46}$$

Additional indices for fuzzy clustering validation are discussed in [Boug 04, Sent 07].

Indices for Shell-Shaped Clusters

Let us now focus on the case of shell-shaped clusters (see Chapter 14). The *PE* and *PC* indices, discussed previously, may also be used in this case, since they involve no information concerning the geometrical characteristics of X.[7] However, the rest of the previously discussed indices need to be modified accordingly. Here, the representatives of each cluster, are shell shaped and they are denoted by β_j. The parameter vector $\boldsymbol{\theta}_j$ contains all the necessary parameters for the identification of β_j. For a vector $\boldsymbol{x}_i$, we define its distance from β_j in terms of

$$\tau_{ij} = \boldsymbol{x}_i - \boldsymbol{x}_j^i \tag{16.47}$$

where $\boldsymbol{x}_j^i$ is the point on β_j that is, closer to $\boldsymbol{x}_i$. It is not difficult to show (Problem 16.16) that for clusters of spherical shape, where $\boldsymbol{\theta}_j = (\boldsymbol{c}_j, r_j)$, with $\boldsymbol{c}_j$ being the center and r_j being the radius of the corresponding sphere,

$$\tau_{ij} = (\boldsymbol{x}_i - \boldsymbol{c}_j) - r_j \frac{\boldsymbol{x}_i - \boldsymbol{c}_j}{\|\boldsymbol{x}_i - \boldsymbol{c}_j\|} \tag{16.48}$$

However, for general types of shells, computation of the τ_{ij}'s is not always an easy task. In such cases we resort to approximations of $\boldsymbol{x}_j^i$ ([Kris 95a]).

The *fuzzy shell covariance matrix* for the j-th cluster is defined in accordance with Eq. (16.41) as

$$\Sigma_j^S = \frac{\sum_{i=1}^N u_{ij}^q \tau_{ij} \tau_{ij}^T}{\sum_{i=1}^N u_{ij}^q} \tag{16.49}$$

Then the *shell hypervolume* of a cluster is defined as

$$V_j^S = |\Sigma_j^S|^{1/2} \tag{16.50}$$

Let us define $X_j^S = \{\boldsymbol{x}_i : \tau_{ij}^T (\Sigma_j^S)^{-1} \tau_{ij} < 1\}$ and $S_j^S = \sum_{\boldsymbol{x}_i \in X_j^S} u_{ij}$. Then, the *fuzzy shell density*, the *average partition shell density*, and the *shell partition density* are defined as before.

Finally, another measure suitable for shell-shaped clusters is the *total fuzzy average shell thickness*, T^S [Kris 95b], which is defined as

$$T^S = \sum_{j=1}^m T_j^S \tag{16.51}$$

where T_j^S is the so-called *fuzzy average shell thickness* of the jth cluster, defined as

$$T_j^S = \frac{\sum_{i=1}^N u_{ij}^q \|\tau_{ij}\|^2}{\sum_{i=1}^N u_{ij}^q} \tag{16.52}$$

---

[7] Such characteristics may concern the shape of the clusters, the position of the representatives, etc.

It is clear that the "thicker" the clusters, the smaller the value of T^S. Furthermore, small values of T^S indicate "thick" clusters. However, T^S tends to decrease monotonically as the number of clusters increases.

The comments made for the total fuzzy hypervolume, the average partition density and the partition density indices are also valid here.

Note that PA^S, PD^S, and T^S *can be thought as measures of the density of the clusters formed by the vectors of X around their representatives.*

A few other indices of this kind have also been proposed and discussed in [Dave 90, Kris 93]. A detailed overview of objective structural validity criteria is given in [Halk 02a, Halk 02b]. A general comment applied to all these indices is that they are sensitive to the size and the density of the points in the clusters.

Finally, using Eq. (16.46), we obtain the shell density, the average partition shell density, and the shell partition density for the hard clustering case.

Remarks

- An alternative way of determining the number of clusters underlying in the data set X, is to perform the so called *progressive clustering method* (e.g., [Kris 95b]). According to this method we run first the clustering algorithm at hand for an overspecified number of clusters, m. Then, we remove spurious clusters, we merge compatible clusters and we identify the "good" clusters. Let k be the number of spurious and "good" clusters defined above. Then, we temporarily remove the vectors contained in the above clusters from the data set and we apply the algorithm on the reduced data set for $m-k$ clusters. This procedure is repeated until no "good" clusters can be removed anymore or no vectors are left in the data set. The output of the above method is the set of the "good" clusters determined above.

  The advantage of this method is that, in general, it is not necessary to run the clustering algorithm for all values of m in a prespecified range. Also, this method is less influenced by the presence of noise. However, one must establish criteria concerning the merging and the removing operations involved in the above method as well as criteria for the identification of "good" clusters.

- A different philosophy for the determination of the number of clusters underlying in the data set X employs the idea of information criteria (IC), such as Akaike and the Minimum Description Length (MDL) criteria (see, e.g., [Sclo 87, Lang 98]).

16.5 VALIDITY OF INDIVIDUAL CLUSTERS

There are two cases in which individual cluster validity may be of interest. One is when we want to test whether a given subset of X forms a "good" cluster. "Good" in this case is interpreted in terms of compactness, with respect to its own data, and isolation with respect to the other vectors of X. The other case concerns the validation of a cluster resulting from the application of a clustering algorithm. To this end, both external and internal criteria may be used.

16.5.1 External Criteria

In this section we consider hard clusters and ordinal-type proximity matrices [Bail 82, Jain 88]. The goal is to test whether a given subset of X forms a compact and well-separated cluster. In [Bail 82], two indices are defined, one for compactness and one for isolation. Both are based on graph theory concepts. However, some necessary definitions must first be provided.

Let us consider the proximity graph $G(p)$, with p ($<N(N-1)/2$) edges, whose vertices correspond to the N vectors of X and whose edges correspond to the p smallest entries of the upper diagonal of the proximity matrix of X, P. In other words, a pair of vertices $\boldsymbol{x}_i$ and $\boldsymbol{x}_j$ is connected with an edge if the dissimilarity $d(\boldsymbol{x}_i, \boldsymbol{x}_j)$ is among the p smallest dissimilarity values of all possible pairs of vectors in X (see Chapter 13). Also, let C be a predetermined subset of X, with k vectors. Our goal is to determine whether C is a good cluster. For the $G(p)$ and the given P, we define the sets $A_{\text{in}}(p)$, $A_{\text{out}}(p)$, and $A_{\text{bet}}(p)$ as follows: (a) $A_{\text{in}}(p)$ is the set of edges whose end points are vectors in C, (b) $A_{\text{out}}(p)$ is the set of edges whose end points are vectors in $X-C$, and (c) $A_{\text{bet}}(p)$ is the set of edges that connect vectors in C with vectors in $X-C$.

For a given $G(p)$, let $q_C(p)$ be the number of edges connecting vertices in C with vertices in $X-C$ and $r_C(p)$ be the set of edges connecting vertices in C. Clearly, these indices depend on p. It is easy to see that low values of $q_C(p)$ indicate a well-isolated cluster, and large values of $r_C(p)$ indicate a compact cluster. In order to extract conclusions about the compactness and isolation of C, we consider the behavior of these indices with respect to p. To this end, we plot the indices versus p. It is expected that an isolated and compact cluster will exhibit low values for $q_C(p)$ and high values for $r_C(p)$, for a "wide" range of values of p. The size of this range is application dependent.

A drawback of these indices is that they do not provide information with respect to a random population. To overcome this, an extension of the indices within the probabilistic framework is discussed in [Bail 82].

16.5.2 Internal Criteria

The aim here is to validate a single cluster that results from a clustering algorithm using only the information residing in the proximity matrix, P, of X.

- *Hard clustering case*
  - *Ordinal proximity matrices.* A method for the evaluation of a cluster is given in [Ling 72] and [Ling 73]. This method is well suited for hierarchies of clusterings, produced by a hierarchical clustering algorithm. It relies on

the lifetime, $L(C)$, of a cluster C, which is given by $L(C) = d_a(C) - d_f(C)$ where $d_f(C)$ is the level of hierarchy where C is formed and $d_a(C)$ is the level where C is absorbed in a larger cluster. The statistical index used is the so-called Ling index, which is defined as the probability of the lifetime of a randomly selected cluster exceeding $L(C)$. Finally, other methods in this category are the so-called *best case method* [Bail 82] and the *CM (clustering method) reachable method* [Bake 76].

- *Ratio-scaled proximity matrices.* In this case, we may adopt the hard hypervolume and the hard density (Section 16.4.2) when we seek compact clusters and the hard shell hypervolume and hard shell density when shell-shaped clusters are considered. Here, an empirically established threshold, ε, is used and, according to whether the value of the index is greater or less than ε, C is characterized as a "good" cluster or not.

- *Fuzzy clustering case.* We first focus on shell-shaped clusters. In this context, "good" clusters are those that are *"compact" around their shell representatives.* In this framework, one can use the shell hypervolume, the shell density indices, and the fuzzy average shell thickness, defined in Section 16.4. Based on these indices, a cluster is characterized as a good one according to whether the value of the corresponding index is greater or less than a prespecified threshold ε.

All these indices do not take into account the fact that shell clusters lie in subspaces of the vector space [Kris 95b]. A criterion that takes this observation into account is the *surface density* criterion. We present the two-dimensional case. The criterion measures the number of points in a cluster per unit curve length. Let us define X' as the set of the vectors in C that lie at a distance smaller than or equal to $\tau_{\max}$ from the shell representative β of C and let $S = \sum_{j:\boldsymbol{x}_j \in X'} u_j$. Then, the surface density δ of a cluster C is defined as

$$\delta = \frac{S}{2\pi r_{\text{eff}}} \tag{16.53}$$

where r_{eff} is defined as

$$r_{\text{eff}} = \sqrt{\text{trace}\{\Sigma\}} \tag{16.54}$$

where Σ is given in Eq. (16.49) and $tr(\Sigma)$ is the trace of Σ. The quantity $2\pi r_{\text{eff}}$ may be viewed as an estimate of the arc length of C (see Problem 16.17). The higher the value of δ, the more dense the cluster is expected to be. Consider for example Figure 16.8. The clusters depicted there have a circular shape and their representative circles are of equal radius. Also, the one on the right is denser around its representative than the one on the left. The value of δ for the right cluster is greater than that for the left cluster.

For compact clusters, indices such as the fuzzy hypervolume or the fuzzy density of a cluster can be employed.

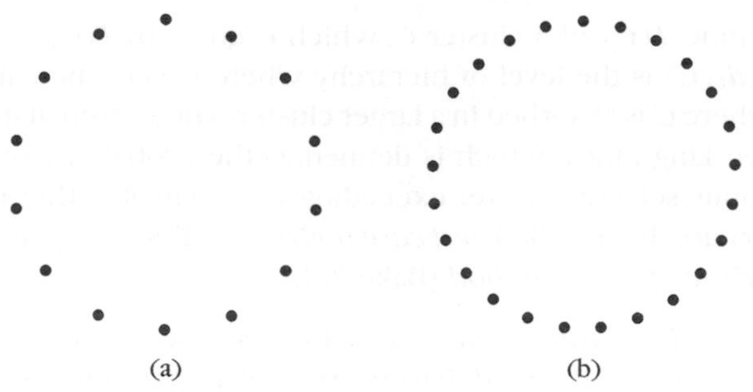

FIGURE 16.8

A sparse and a dense circular cluster.

16.6 CLUSTERING TENDENCY

As discussed in the introduction of the chapter, almost all the clustering algorithms introduced in the previous sections share an annoying feature. That is, they impose a clustering structure on a data set X even though the vectors of X do not exhibit such a structure. Thus, in order to prevent a misleading interpretation of the structure of the data set X, it would be more sensible to check first whether X possesses a clustering structure. If this is the case, then one may proceed by applying a clustering algorithm to X. Otherwise, cluster analysis is likely to lead to misleading results. The problem of determining the presence or the absence of a clustering structure in X is called *clustering tendency*. Usually, this task relies on statistical tests.

Clustering tendency methods have been applied in various application areas (e.g., [Digg 83, Ripl 81]). However, most of these methods are suitable only for $l = 2$. In the sequel, we discuss the problem in the general $l \geq 2$ case. Furthermore, we focus on methods that are suitable for detecting compact clusters (if any).

In this framework, we test the randomness (null) hypothesis (H_0) against the clustering hypothesis and the regularity hypothesis. Let us define these terms more precisely.

- "The vectors of X are randomly distributed, according to the uniform distribution in the sampling window[8] of X" (H_0).
- "The vectors of X are regularly spaced in the sampling window."
  This implies that, they are not too close to each other.
- "The vectors of X form clusters."

If the randomness or the regularity hypothesis is accepted, methods alternative to clustering analysis should be used for the interpretation of the data set X.

[8] In [Smit 84] the sampling window is mathematically defined as the compact convex support set for the underlying distribution of the vectors of the data set X.

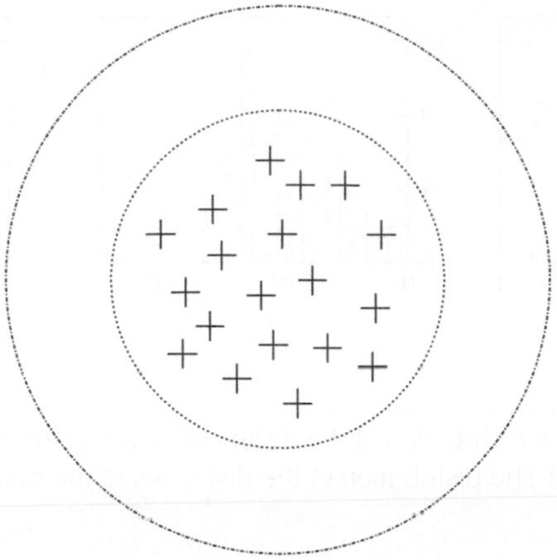

FIGURE 16.9

See text for explanation.

There are two key points that have an important influence on the performance of many statistical tests used in clustering tendency. The first is the dimensionality of the data, l, which affects the performance in a nonobvious way. This dependence can be revealed through simulations [Pana 83].

The other key point is the sampling window. Apart from artificial experiments, in practice, we do not know the sampling window. One of the problems that this may cause is demonstrated in Figure 16.9. The vectors in the dashed circle are uniformly distributed in it. Thus, we expect that tests for randomness will identify this situation. However, if we use as sampling window the region surrounded by the dash-dotted line (for the same data set), the vectors are no longer uniformly distributed and the tests for randomness may fail to accept H_0. Moreover, due to the finite extent of the window, the statistical characteristics of the data are different near the edges of the sampling window than they are in its center. For example, the distribution of the distances of a vector $\boldsymbol{x} \in X$ from the rest of the vectors of X is different when $\boldsymbol{x}$ is in the center than when it is near the border of the sampling window. One way to overcome this situation is to use a periodic extension of the sampling window. Another popular technique is to consider data in a smaller area inside the sampling window, known as *sampling frame*. With this method, we overcome the boundary effects in the sampling frame by considering points outside it and inside the sampling frame, for the estimation of statistical properties.

Example 16.9

Consider a data set X that consists of 100 vectors uniformly distributed in the H_2 hypercube (see Figure 16.10a). Figure 16.10b shows the distribution of the distances between the point $\boldsymbol{x} = [0.5045, 0.4764]^T$ and each of the points of $X - \{\boldsymbol{x}\}$. Also, Figure 16.10c shows the

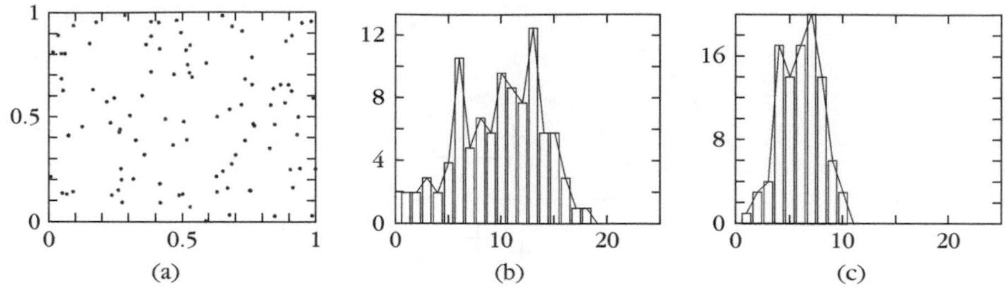

FIGURE 16.10

(a) The data set X. (b) The distribution of the distances of the point $[0.5045, 0.4764]^T$ from the remaining points in X. (c) The distribution of the distances of the point $[0.0159, 0.8089]^T$ from the remaining points in X.

distribution of the distances between the point $y = [0.0159,\ 0.8089]^T$ and each of the points of $X - \{y\}$. Note that x lies close to the center of H_2 and y lies close to its border.

A method for estimating the sampling window is to use the convex hull of the vectors in X. However, the distributions for the tests, derived using this sampling window, depend on the specific data at hand. A second drawback associated with this approach is the high computational cost for computing the convex hull of X. An alternative [Zeng 85, Dube 87b] that seems to work well in practice is to define the sampling window as the hypersphere centered at the mean point of X and including half of its vectors. The fact that half of the vectors are discarded is not so crucial, because in the current framework we want to test only whether the vectors of X possess a clustering structure. If this is the case, then the clusters will be identified by applying a clustering algorithm to all the data of X. Notice the similarity to the sampling frame technique discussed earlier.

In the sequel, we define various test statistics, q, suitable for the detection of clustering tendency. Recall that a crucial quantity we have to determine is $p(q|H_0)$. Moreover, in order to measure the power of q against the regularity and the clustering tendency hypotheses, we also need to determine the respective pdf's under these hypotheses. In the sequel, we provide general guidelines on how to generate clustered and regularly spaced data sets. This is required in order to estimate the pdf's of q under regularity and clustering tendency hypotheses, via Monte Carlo simulations. Randomly spaced data sets may be generated by inserting vectors in the sampling window, according to the uniform distribution.

- *Generation of clustered data.* A well-known procedure for generating (compact) clustered data is the Neyman–Scott procedure [Neym 72]. This procedure assumes that the sampling window is known. It produces a random number of compact clusters, formed at random positions in the sampling window and each consisting of a random number of points. The number of points in each cluster follows the Poisson distribution (Appendix A). The technique requires as inputs the total number of points N of the set, the intensity of the Poisson process λ, and the spread parameter σ that controls the spread of each cluster around its center. According to this procedure, we randomly insert a point y_i in the sampling window, following the uniform distribution. This point serves as the center of the ith cluster, and we determine its number of vectors, n_i, using the Poisson distribution. Then the n_i points around y_i are generated according to the normal distribution with mean y_i and covariance matrix $\sigma^2 I$. If a point turns out to be outside the sampling window, we ignore it and another one is generated. This procedure is repeated until N points have been inserted in the sampling window (see Figures 16.11a and b). In some cases, y_i's are also included as vectors in the set.

- *Generation of regularly spaced data.* Perhaps the simplest way to produce regularly spaced points is to define a lattice in the convex hull of X and to place the vectors at its vertices. An alternative procedure, known as *simple sequential inhibition (SSI)* (see, e.g., [Jain 88, Zeng 85]), is the following. The points y_i are inserted in the sampling window one at a time. For each point we define a hypersphere of radius r centered at y_i. The next point can be placed anywhere in the sampling window in such a way that its hypersphere does not intersect with any of the hyperspheres defined by the previously inserted points. The procedure stops when a predetermined number of points have been inserted in the sampling window, or when no more points can be inserted in the sampling window, after say a few thousand trials (see Figure 16.11c). A variation of this model allows intersection of these hyperspheres up to a certain degree. A measure of the degree of fulfillment of the sampling window is the so-called ***packing density***, which is defined as

$$\rho = \frac{L}{V} V_r \tag{16.55}$$

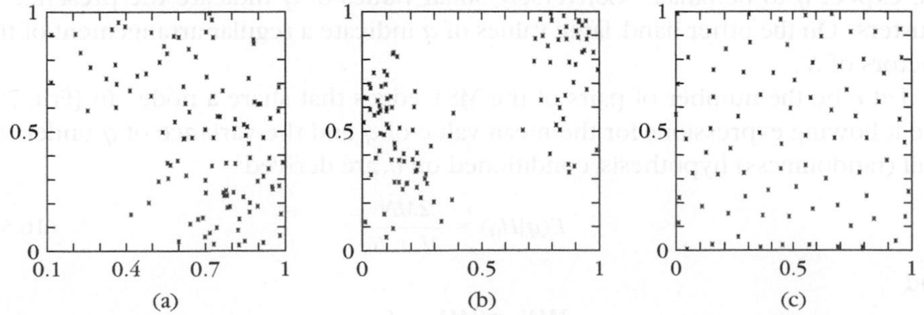

FIGURE 16.11

(a) and (b) Clustered data sets produced by the Neyman–Scott process. (c) Regularly spaced data produced by the SSI model.

where L/V is the average number of points per unit volume and V_r is the volume of a hypersphere of radius r. V_r can be written as

$$V_r = Ar^l \tag{16.56}$$

where A is the volume of the l-dimensional hypersphere with unit radius, which is given by

$$A = \frac{\pi^{l/2}}{\Gamma(l/2 + 1)} \tag{16.57}$$

and $\Gamma(\cdot)$ is the gamma function (Appendix A).

16.6.1 Tests for Spatial Randomness

Several tests for spatial randomness have been proposed in the literature. All of them assume knowledge of the sampling window. The *scan test* ([Naus 82, Cono 79]), the *quadrat analysis* [Grei 64, Piel 69, Mead 74], the *second moment structure* [Ripl 77], and the *interpoint distances* [Ripl 78, Silv 78, Stra 75] provide us with tests for clustering tendency that have been extensively used when $l = 2$. In the sequel, we discuss three methods for determining clustering tendency that are well suited for the general $l \geq 2$ case. All these methods require knowledge of the sampling window.

Tests Based on Structural Graphs

In this section, we discuss a test for testing randomness, that is, based on the idea of the minimum spanning tree (MST) ([Smit 84]). First, we determine the convex region where the vectors of X lie. Then, we generate M vectors that are uniformly distributed over a region that approximates the convex region found before (usually $M = N$). These vectors constitute the set X'. Next we find the MST of $X \cup X'$ and we determine the number of edges, q, that connect vectors of X with vectors of X'. This number is used as the statistic index. If X contains clusters, then we expect q to be small. Conversely, small values of q indicate the presence of clusters. On the other hand, large values of q indicate a regular arrangement of the vectors of X.

Let e be the number of pairs of the MST edges that share a node. In [Frie 79], the following expressions for the mean value of q and the variance of q under the null (randomness) hypothesis, conditioned on e, are derived:

$$E(q|H_0) = \frac{2MN}{M+N} \tag{16.58}$$

and

$$\text{var}(q|e, H_0) = \frac{2MN}{L(L-1)}\left[\frac{2MN-L}{L} + \frac{e-L+2}{(L-2)(L-3)}[L(L-1) - 4MN+2]\right] \tag{16.59}$$

where $L = M + N$. Moreover, it can be shown [Frie 79] that if $M, N \to \infty$ and M/N is away from 0 and ∞, the pdf of the statistic

$$q' = \frac{q - E(q|H_0)}{\sqrt{\text{var}(q|e, H_0)}} \tag{16.60}$$

is approximately given by the standard normal distribution. Thus, we reject H_0 at significance level ρ if q' is less than the ρ-percentile of the standard normal distribution. This test exhibits high power against clustering tendency and little power against regularity [Jain 88].

Tests Based on Nearest Neighbor Distances

Two tests of this kind are the Hopkins test [Hopk 54] and the Cox-Lewis test [Cox 76, Pana 83]. The tests rely on the distances between the vectors of X and a number of vectors which are randomly placed in the sampling window.

The Hopkins Test

Let $X' = \{\boldsymbol{y}_i, i = 1, \ldots, M\}$, $M \ll N$,[9] be a set of vectors that are randomly distributed in the sampling window, following the uniform distribution. Also let $X_1 \subset X$ be a set of M randomly chosen vectors of X. Let d_j be the distance from $\boldsymbol{y}_j \in X'$ to its closest vector in X_1, denoted by $\boldsymbol{x}_j$, and δ_j be the distance from $\boldsymbol{x}_j$ to its closest vector in $X_1 - \{\boldsymbol{x}_j\}$. Then the Hopkins statistic involves the lth powers of d_j and δ_j and it is defined as [Jain 88]

$$h = \frac{\sum_{j=1}^{M} d_j^l}{\sum_{j=1}^{M} d_j^l + \sum_{j=1}^{M} \delta_j^l} \tag{16.61}$$

This statistic compares the nearest neighbor distribution of the points in X_1 with that from the points in X'. When X contains clusters, the distances between nearest neighbor points in X_1 are expected to be small, on the average, and, thus, large values of h are expected. Furthermore, large values of h indicate the presence of a clustering structure in X. When the points in X are regularly distributed in the sampling window, it is expected that, on the average, the term $\sum_{j=1}^{M} d_j^l$ is smaller than $\sum_{j=1}^{M} \delta_j^l$, thus leading to small values of h. Also, small values of h indicate the presence of regularly spaced points. Finally, a value around 1/2 is an indication that the vectors of X are randomly distributed over the sampling window. It can be shown (e.g., [Jain 88]) that if the generated vectors are distributed according to a Poisson random process (hypothesis of randomness) and all nearest neighbor distances are statistically independent, h (under H_0) follows a beta distribution, with (M, M) parameters (Appendix A).

[9] Typically $M = 0.1N$.

Simulation results [Zeng 85] show that this test exhibits high power against regularity for a hypercubic sampling window and periodic boundaries, for $l = 2, \ldots, 5$. However, its power is limited against clustering tendency.

The Cox–Lewis Test

This test is less intuitive than the previous one. It was first proposed in [Cox 76] for the two-dimensional case and it has been extended to the general $l \geq 2$ dimensional case in [Pana 83]. It follows the setup of the previous test with the exception that X_1 need not be defined. For each $\boldsymbol{y}_j \in X'$, we determine its closest vector in X, say $\boldsymbol{x}_j$, and then we determine the vector closest to $\boldsymbol{x}_j$ in $X - \{\boldsymbol{x}_j\}$, say $\boldsymbol{x}_i$. Let d_j be the distance between $\boldsymbol{y}_j$ and $\boldsymbol{x}_j$ and δ_j the distance between $\boldsymbol{x}_j$ and $\boldsymbol{x}_i$. We consider all $\boldsymbol{y}_j$'s for which $2d_j/\delta_j$ is greater than or equal to one. Let M' be the number of such $\boldsymbol{y}_j$'s. Then, an appropriate function R_j of $2d_j/\delta_j$ (see [Pana 83]) is defined for these $\boldsymbol{y}_j$'s. Finally, we define the statistic

$$R = \frac{1}{M'} \sum_{j=1}^{M'} R_j \tag{16.62}$$

It can be shown [Pana 83] that R, under H_0, has an approximately normal distribution with mean 1/2 and variance $12M'$. Small values of R indicate the presence of a clustering structure in X, and large values indicate a regular structure in X. Finally, values around the mean of R indicate that the vectors of X are randomly arranged in the sampling window. Simulation results [Zeng 85] show that the Cox-Lewis test exhibits inferior performance compared with the Hopkins test against the clustering alternative. However, this is not the case against the regularity hypothesis.

Two additional tests are the so called T-squared sampling tests, introduced in [Besa 73]. However, simulation results [Zeng 85] show that the these two tests exhibit rather poor performance compared with the Hopkins and Cox-Lewis tests.

A Sparse Decomposition Technique

This technique begins with the data set X and sequentially removes vectors from it until no vectors are left [Hoff 87]. Before we proceed further, some definitions are needed. A *sequential decomposition* D of X is a partition of X into $L_1, \ldots, L_k$ sets, such that the order of their formation matters. L_i's are also called *decomposition layers*.

We denote by $MST(X)$ the MST corresponding to X. Let $S(X)$ be the set derived from X according to the following procedure. Initially, $S(X) = \emptyset$. We move an end point $\boldsymbol{x}$ of the longest edge, e, of the $MST(X)$ to $S(X)$. Also, we mark this point and all points that lie at a distance less than or equal to b from $\boldsymbol{x}$, where b is the length of e. Then, we determine the unmarked point, $\boldsymbol{y} \in X$, that lies closer to $S(X)$ and we move it to $S(X)$. Also, we mark all the unmarked vectors that lie at a distance no greater than b from $\boldsymbol{y}$. We apply the same procedure for all the unmarked vectors of X. The procedure terminates when all vectors are marked.

Let us define $R(X) \equiv X - S(X)$. Setting $X = R^0(X)$, we define

$$L_i = S(R^{i-1}(X)), \qquad i = 1, \ldots, k \tag{16.63}$$

where k is the smallest integer such that $R^k(X) = \emptyset$. The index i denotes the so-called *decomposition layer*. Intuitively speaking, the procedure sequentially "peels" X until all of its vectors have been removed.

The information that becomes available to us after the application of the decomposition procedure is (a) the number of decomposition layers k, (b) the decomposition layers L_i, (c) the cardinality, l_i, of the L_i decomposition layer, $i = 1, \ldots, k$, and (d) the sequence of the longest MST edges used in deriving the decomposition layers. The decomposition procedure gives different results when the vectors of X are clustered and when they are regularly spaced or randomly distributed in the sampling window. Based on this observation we may define statistical indices utilizing the information associated with this decomposition procedure. For example, it is expected that the number of decomposition layers, k, is smaller for random data than it is for clustered data. Also, it is smaller for regularly spaced data than for random data (see Problem 16.20). This situation is illustrated in the following example.

Example 16.10

(a) We consider a data set X_1 of 60 two-dimensional points in the unit square. The first 15 points stem from a normal distribution, with mean $[0.2, 0.2]^T$ and covariance matrix $0.15I$. The second, the third, and the fourth group of 15 points also stem from normal distributions with means $[0.2, 0.8]^T$, $[0.8, 0.2]^T$, and $[0.8, 0.8]^T$, respectively. Their covariance matrices are also equal to $0.15I$. Applying the sparse decomposition technique on X_1, we obtain 15 decomposition layers.

(b) We consider another data set X_2 of 60 two-dimensional points, which are now randomly distributed in the unit square. The sparse decomposition technique in this case gives 10 decomposition layers.

(c) Finally, we generate a data set X_3 of 60 two-dimensional points regularly distributed in the unit square, using the simple sequential inhibition (SSI) procedure. The sparse decomposition technique gives 7 decomposition layers in this case.

Figures 16.12, 16.13, and 16.14 show the first four decomposition layers for clustered, random, and regularly spaced data. It is clear that the rate of point removal is much slower for the clustered data and much faster for the regular data.

Several tests that rely on the preceding information are discussed in [Hoff 87]. One such statistic that exhibits good performance is the so-called *P statistic*, which is defined as follows:

$$P = \prod_{i=1}^{k} \frac{l_i}{n_i - l_i} \tag{16.64}$$

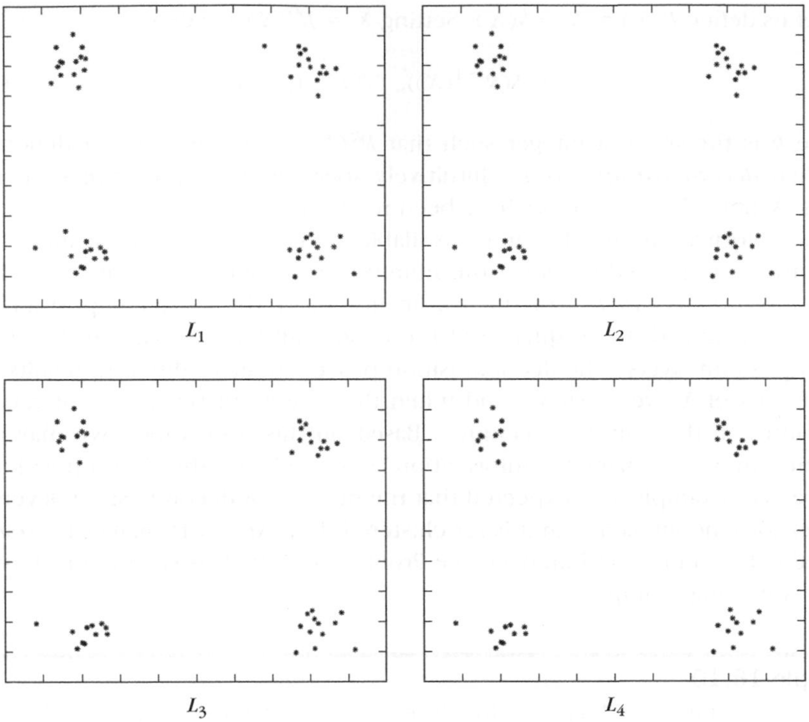

FIGURE 16.12

The first four decomposition layers for clustered data in the unit square (Example 16.10(a)).

where n_i is the number of points in $R^{i-1}(X)$. In words, each factor of P is the ratio of the removed to the remaining points at each decomposition stage.

Preliminary simulation results show high power of P against the clustering alternative. The required pdf's of P under H_0, H_1, and H_2 are estimated using Monte Carlo techniques, since it is difficult to derive theoretical results [Hoff 87].

Finally, tests for clustering tendency for the cases in which ordinal proximity matrices are in use have also been proposed (e.g., [Fill 71, Dube 79]). Most of them are based on graph theory concepts. Let $G_N(v)$ be a threshold graph with N vertices, one for each vector of X (Chapter 13). Then, graph properties, such as the node degree and the number of edges needed for $G_N(v)$ to be connected, are used in order to investigate the clustering tendency of X. Specifically, suppose that we use the number of edges n needed to make $G_N(v)$ connected. Obviously, n depends directly on v. That is, increasing v, we also increase n. Let v^* be the smallest value of v for which $G_N(v^*)$ becomes connected, for the given proximity matrix. Let V be the random variable that models v. Also, let $P(V \leq v|N)$ be the probability that a graph with N nodes and v randomly inserted edges is connected (this is provided from tables in [Ling 76]). Then, for the specific v^*, we determine $P(V \leq v^*|N)$. Very high values of $P(V \leq v^*|N)$ indicate that the proximity matrix

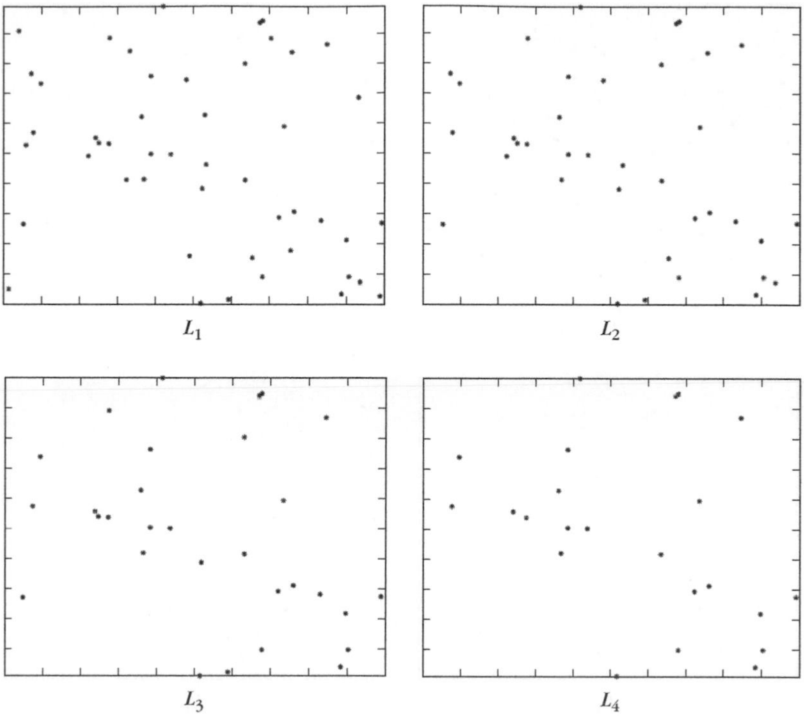

FIGURE 16.13

The first four decomposition layers for randomly distributed data in the unit square (Example 16.10(b)).

was not chosen at random. This is because the within-cluster edges will tend to occur before the between-cluster edges when the data are clustered, thus, delaying the formation of a connected graph.

16.7 PROBLEMS

16.1 Let X be a set of vectors. Show that if the number of clusters in a clustering $\mathcal{C}$ of X is m and the number of groups in a partition $\mathcal{P}$ of X is $q \neq m$, then the maximum values of the Rand, the Jaccard, and the Fowlkes and Mallows statistics are less than 1.

16.2 Prove Eq. (16.10).

16.3 **a.** Repeat Example 16.2 with two-dimensional vectors steming from the normal distributions with means $[0.2, 0.2]^T$, $[0.2, 0.8]^T$, $[0.8, 0.2]^T$, $[0.8, 0.8]^T$, and covariance matrices $0.2^2 I$.

b. Repeat the experiment when all covariance matrices are equal to $0.5^2 I$.

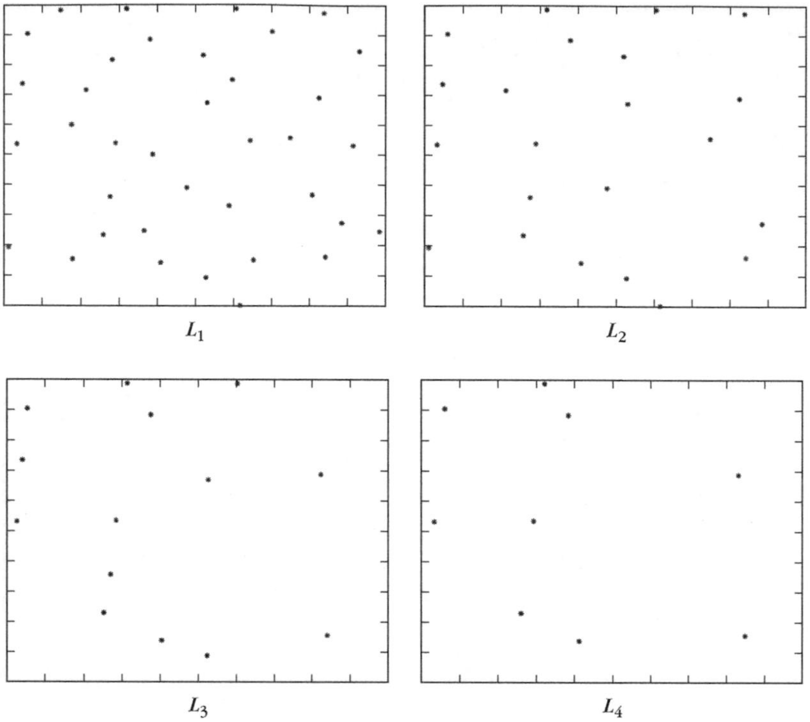

FIGURE 16.14

The first four decomposition layers for regularly spaced data in the unit square (Example 16.10(c)).

16.4 Prove that the values of the *CPCC* in Section 16.3.2 lie in the interval $[-1, 1]$.

16.5 Consider a data set X of six vectors, whose (ordinal) proximity matrix is

$$P = \begin{bmatrix} 0 & 1 & 5 & 7 & 8 & 9 \\ 1 & 0 & 3 & 6 & 10 & 11 \\ 5 & 3 & 0 & 12 & 13 & 14 \\ 7 & 6 & 12 & 0 & 6 & 4 \\ 8 & 10 & 13 & 6 & 0 & 2 \\ 9 & 11 & 14 & 4 & 2 & 0 \end{bmatrix}$$

Run the single and the complete link algorithms on X and compare the resulting dendrograms, using the γ statistic. Comment on the results.

16.6 Let $X = \{\boldsymbol{x}_i, i = 1, \ldots, 12\}$, with $\boldsymbol{x}_1 = [-4, 0]^T$, $\boldsymbol{x}_2 = [-3, 1]^T$, $\boldsymbol{x}_3 = [-3, -1]^T$, $\boldsymbol{x}_4 = [-2, 0]^T$, $\boldsymbol{x}_5 = [2, 0]^T$, $\boldsymbol{x}_6 = [3, 1]^T$, $\boldsymbol{x}_7 = [4, 0]^T$, $\boldsymbol{x}_8 = [3, -1]^T$, $\boldsymbol{x}_9 = [-1, 7]^T$, $\boldsymbol{x}_{10} = [0, 8]^T$, $\boldsymbol{x}_{11} = [1, 7]^T$, $\boldsymbol{x}_{12} = [0, 6]^T$.

a. Let $m = 2$. Consider the vectors $\boldsymbol{w}_1 = [0,0]^T$ and $\boldsymbol{w}_2 = [0,7]^T$, such that the first one represents the points $\boldsymbol{x}_1$ through $\boldsymbol{x}_8$ and the second one represents the rest of the points in X. Compute the values of Γ and $\hat{\Gamma}$ (Section 16.4.1).

b. Let $m = 3$. Consider the vectors $\boldsymbol{w}_1 = [-3,0]^T$ $\boldsymbol{w}_2 = [3,0]^T$, and $\boldsymbol{w}_3 = [0,7]^T$, so that the first one represents the points $\boldsymbol{x}_1$ through $\boldsymbol{x}_4$, the second represents the points $\boldsymbol{x}_5$ through $\boldsymbol{x}_8$, and the third represents the rest of the points of X. Compute the values of Γ and $\hat{\Gamma}$.

c. Let $m = 4$. Define $\boldsymbol{w}_1$ and $\boldsymbol{w}_2$ as in the previous case. Also, define $\boldsymbol{w}_3 = [-0.5, 7.5]^T$ and $\boldsymbol{w}_4 = [0.5, 6.5]^T$, so that the first represents $\boldsymbol{x}_9$ and $\boldsymbol{x}_{10}$, while the second represents $\boldsymbol{x}_{11}$ and $\boldsymbol{x}_{12}$. Compute the values of Γ and $\hat{\Gamma}$.

d. What conclusions can you draw from the comparison of the values of Γ and $\hat{\Gamma}$ obtained from the preceding three cases?

16.7 Estimate the number of operations required for the computation of Dunn's index, D_m, given by Eq. (16.19). What is the total number of computations required for the computation of D_m, for $m = 1, \ldots, N$?

16.8 Define explicitly $diam_i^{GG}$ and $diam_i^{RNG}$ that are involved in the definitions of the GG and the RNG Dunn-like indices. Using these definitions derive explicitly the GG and the RNG Dunn-like indices.

16.9 Show that $D_m^{GG} \leq D_m^{RNG} \leq D_m^{MST}$.

Hint: Use the fact that for a cluster $C_i, E_i^{MST} \subseteq E_i^{RNG} \subseteq E_i^{GG}$.

16.10 **a.** Show that the R_{ij}^{MST} given by Eq. (16.26) satisfies the conditions (C1)–(C5).

b. Taking into account the definition of R_{ij}^{MST}, define R_{ij}^{GG} and R_{ij}^{RNG} and show that they also satisfy the conditions (C1)–(C5).

16.11 Show that $DB_m^{GG} \geq DB_m^{RNG} \geq DB_m^{MST}$.

Hint: Use the fact that for a cluster $C_i, E_i^{MST} \subseteq E_i^{RNG} \subseteq E_i^{GG}$.

16.12 Explain intuitively why the MST DB is more robust to the presence of noisy vectors than the original DB.

16.13 **a.** Prove that, as $q \to 1^+$, PC and PE tend to 1 and 0, respectively.

b. Prove that, as $q \to \infty$, PC and PE tend to $1/m$ and $\log_a m$, respectively.

c. Show that in the latter case, in the plots of PC and PE, the most significant knee is exhibited at $m = 2$.

16.14 Prove that, as $q \to \infty$, the XB index tends to ∞, while XB_q becomes indeterminate.

Hint: Use the following facts: (a) $\lim_{q\to\infty} \boldsymbol{w}_i = \boldsymbol{w}$, where $\boldsymbol{w}$ is the mean vector of all vectors in X, $i = 1, \ldots, m$, and (b) for $q \to \infty$, $u_{ij} = 1/m$.

16.15 **a.** Prove that $\lim_{q\to 1^+} FS_q = 2N\text{trace}(S_w) - N\text{trace}(S_m)$, where S_w and S_m are the within and the mixture scatter matrices defined in Chapter 5.

b. Prove that $\lim_{q\to\infty} FS_q = 0$.

Hint: Use the hints given in the previous problem.

16.16 Prove that the distance of a point $\boldsymbol{x}_i$ from a sphere with center $\boldsymbol{c}_j$ and radius r_j is given by Eq. (16.48).

16.17 Consider a cluster C that consists of the points of a circular arc of radius r, subtending an angle ϕ (of course, this case is of theoretical interest, since the number of vectors in C would be infinite). The covariance matrix of this arc is

$$\Sigma = \frac{1}{L_\phi} \int_{-\phi/2}^{\phi/2} \boldsymbol{x}\boldsymbol{x}^T \, dl - \boldsymbol{m}\boldsymbol{m}^T \tag{16.65}$$

where $\boldsymbol{x} = [r\cos\theta, r\sin\theta]^T$ is a point on the arc, $dl = rd\theta$, and L_ϕ is the arc length. (a) Prove that

$$\delta = \frac{\phi}{2\pi\sqrt{1 - \frac{4sin^2(\phi/2)}{\phi^2}}} \tag{16.66}$$

What is the value of δ when $\phi = 2\pi$?

(b) Repeat for the case where the length of the cluster is a line segment of length L.

16.18 Consider a square of side a. Consider a grid of horizontal and vertical lines in the square so that the distance between two adjacent parallel lines is r. Place in it $(a/2r)^2$ vectors (of course, $a/2r$ is assumed to be an integer) such that each of them lies at an intersection point of a grid and the circles of radius r centered at these points do not intersect at more than one point. (a) Compute the packing density of the square. Repeat the above for the case where r is replaced by $r/2$.

(b) Assuming that no circle is allowed to have a part of it outside the square, is it possible to determine an arrangement of points in the square that results in a higher packing density?

16.19 Sometimes we say that the Hopkins test includes "first-order" information on the data set X and the Cox–Lewis test "second-order" information. Can you justify this proposition?

16.20 Repeat Example 16.10 and explain why (a) the number of decomposition layers is greater in clustered data than in random data and (b) the number of decomposition layers is greater in random data than in regular data.

REFERENCES

[Akai 85] Akaike H.,"Prediction and Entropy," in *A Celebration of Statistics* (Atkinson A.C., Fienberg S.E., eds.), Sprieger-Verlag, New York, pp. 1-24, 1985.

[Back 81] Backer E., Jain A.K. "A clustering performance measure based on fuzzy set decomposition," *IEEE Transactions on Pattern Analysis and Machine Intelligence*, Vol. 3(1), pp. 66-75, 1981.

[Bail 82] Bailey T.A., Dubes R.C. "Cluster validity profiles," *Pattern Recognition*, Vol. 15, pp. 61-83, 1982.

[Bake 74] Baker F.B. "Stability of two hierarchical grouping techniques—Case 1. Sensitivity to data errors," *Journal of the American Statistical Association*, Vol. 69, 440-445, 1974.

[Bake 76] Baker F.B., Hubert L.J. "A graph-theoretic approach to goodness of fitting complete-link hierarchical clustering," *Journal of the American Statistical Association*, Vol. 71, pp. 870-878, 1976.

[Bart 62] Barton D.E., David F.N. "Randomization basis for multivariate tests in the bivariate case—randomness of points in the plane," *Bulletin of the International Statistical Institute*, Vol. 39, pp. 455-467, 1962.

[Beni 94] Beni G., Liu X. "A least biased fuzzy clustering method," *IEEE Transactions on Pattern Analysis and Machine Intelligence*, Vol. 16(9), pp. 954-960, 1994.

[Besa 73] Besag J.E., Gleaves J.T. "On the detection of spatial pattern in plant communities," *Bulletin of International Statistics Institute*, Vol. 45, p. 153, 1973.

[Bezd 74] Bezdek J.C. "Cluster validity with fuzzy sets," *Journal of Cybernetics*, Vol. 3(3), pp. 58-72, 1974.

[Bezd 75] Bezdek J.C. "Mathematical models for systematics and taxonomy," in *Proc. 8th Int. Conf. in Numerical Taxonomy* (Estarook G., ed.), Freeman, San Francisco pp. 143-166, 1975.

[Boug 04] Bouguessa M., Wang S-R. "A new efficient validity index for fuzzy clustering," *Proceedings of the 3rd International Conference on Machine Learning and Cybernetics*, pp. 1914-1919, 2004.

[Bout 04] Boutin F., Hascoët M. "Cluster validity indices for graph partitioning," *Proceedings of the 8th Conference on Information Visualization*, pp. 376-381, 2004.

[Cono 79] Conover W.J., Benent T.R., Iman R.L. "On a method for detecting clusters of possible uranium deposits," *Technometrics*, Vol. 21, pp. 277-282, 1979.

[Cox 76] Cox T.F., Lewis T. "A conditioned distance ratio method for analyzing spatial patterns," *Biometrika*, Vol. 63, p. 483, 1976.

[Cunn 72] Cunningham K.M., Ogilvie J.C. "Evaluation of hierarchical grouping techniques: A preliminary study," *Computer Journal*, Vol. 15, pp. 209-213, 1972.

[Dave 90] Dave R.N., Patel K.J. "Progressive fuzzy clustering algorithms for characteristic shape recognition," *Proc. North American Fuzzy Inf. Process. Soc. Workshop*, Toronto, pp. 121-124, 1990.

[Davi 79] Davies D.L., Bouldin D.W. "A cluster separation measure," *IEEE Transactions on Pattern Analysis and Machine Intelligence*, Vol. 1(2), pp. 224-227, 1979.

[Diac 83] Diaconis P., Efron B. "Computer-intensive methods in statistics," *Scientific American*, May, pp. 116-130, 1983.

[Digg 83] Diggle P.J. *Statistical Analysis of Spatial Point Patterns*, Academic Press, 1983.

[Dube 79] Dubes R.C., Jain A.K. "Validity studies in clustering methodologies," *Pattern Recognition*, Vol. 11, pp. 235–254, 1979.

[Dube 87a] Dubes R.C. "How many clusters are best? An experiment," *Pattern Recognition*, Vol. 20(6), pp. 645–663, 1987.

[Dube 87b] Dubes R.C., Zeng G. "A test for spatial homogeneity in cluster analysis," *Journal of Classification*, Vol. 4, pp. 33–56, 1987.

[Dunn 74] Dunn J.C. "Well separated clusters and optimal fuzzy partitions," *Journal of Cybernetics*, Vol. 4, pp. 95–104, 1974.

[Dunn 76] Dunn J.C. "Indices of partition fuzziness and the detection of clusters in large data sets," *in Fuzzy Automata and Decision Processes* (Gupta M.M., ed.), Elsevier, 1976.

[Efro 79] Efron B. "Bootstrap methods: Another look at jackknife," *Applied Statistics*, Vol. 7, pp. 1–26, 1979.

[Farr 69] Farris J.S. "On the cophenetic correlation coefficient," *Systematic Zoology*, Vol. 18, pp. 279–285, 1969.

[Fill 71] Fillenbaum S., Rapoport A. *Structures in the Subjective Lexicon*, Academic Press, 1971.

[Fowl 83] Fowlkes E.B., Mallows C.L. "A method for comparing two hierarchical clusterings," *Journal of the American Statistical Association*, Vol. 78, pp. 553–569, 1983.

[Fral 98] Fraley C., Raftery A.E., "How many clusters? Which clustering method? Answers via model-based cluster analysis," *The Computer Journal*, Vol. 41, No. 8, pp. 578–588, 1998.

[Frie 79] Friedman J.H., Rafsky L.C. "Multivariate generalization of the Wald–Wolfowitz and Smirnov two-sample tests," *Annual Statistics*, Vol. 7, pp. 697–717, 1979.

[Gath 89] Gath I., Geva A.B. "Unsupervised optimal fuzzy clustering," *IEEE Transactions on Pattern Analysis and Machine Intelligence*, Vol. 11(7), pp. 773–781, 1989.

[Grei 64] Greig-Smith P. *Quantitative Plant Ecology*, 2nd ed., Butterworth, 1964.

[Gord 99] Gordon A. *Classification, 2nd edition*, Chapman and Hall/CRC press, London, 1999.

[Halk 00] Halkidi M., Vazirgiannis M., Batistakis Y. "Quality scheme assessment in the clustering process," *Proceedings of the 4th European Conference on Principles of Data Mining and Knowledge Discovery*, pp. 265–276, 2000.

[Halk 01] Halkidi M., Vazirgiannis M. "Clustering validity assessment: finding the optimal partitioning of a data set," *Proceedings of the International Conference of Data Mining 2001*, pp. 187–194, 2001.

[Halk 02a] Halkidi M., Batistakis Y., Vazirgiannis M. "Cluster validity methods: part 1," *SIGMOD Record*, Vol. 31(2), pp. 40–45, 2002.

[Halk 02b] Halkidi M., Batistakis Y., Vazirgiannis M. "Cluster validity methods: part 2," *SIGMOD Record*, Vol. 31(3), pp. 19–27, 2002.

[Hart 75] Hartigan J.A. *Clustering Algorithms*, John Wiley & Sons, 1975.

[Hast 01] Hastie T., Tibshirani R., Friedman J. *The Elements of Statistical Learning*, Springer, 2001.

[Hoff 87] Hoffman R.L., Jain A.K. "Sparse decompositions for exploratory pattern analysis," *IEEE Transactions on Pattern Analysis and Machine Intelligence*, Vol. 9(4), pp. 551–560, 1987.

[Hopk 54] Hopkins B. "A new method for determining the type of distribution of plant-individuals," *Annals of Botany*, Vol. 18, pp. 213–226, 1954.

[Hube 74] Hubert L.J. "Approximate evaluation techniques for the single-link and complete-link hierarchical clustering procedures," *Journal of the American Statistical Association*, Vol. 69, pp. 698–704, 1974.

[Hube 76] Hubert L.J., Schultz J. "Quadratic assignment as a general data-analysis strategy," *British Journal of Mathemetical and Statistical Psychology*, Vol. 29, pp. 190–241, 1976.

[Hube 85] Hubert L.J., Arabie P. "Comparing partitions," *Journal of Classification*, Vol. 2, pp. 193–218, 1985.

[Hurv 89] Hurvich C.M., Tsai C-L, "Regression and time series model selection in small samples," *Biometrika* Vol. 76, pp. 297–307, 1989.

[Ivch 1991] Ivchenko G., Medvedev Y., Chistyakov A. *Problems in Mathematical Statistics*, Mir Publishers, Moscow, 1991.

[Jain 87a] Jain A.K., Dubes R., Chen C.C. "Bootstrapping techniques for error estimation," *IEEE Transactions on Pattern Analysis and Machine Intelligence*, Vol. 9, pp. 628–633, 1987.

[Jain 87b] Jain A.K., Moreau J.V. "Bootstrap technique in cluster analysis," *Pattern Recognition*, Vol. 20(5), pp. 547–568, 1987.

[Jain 88] Jain A.K., Dubes R.C. *Algorithms for Clustering Data*, Prentice Hall, 1988.

[Kauf 90] Kaufman L., Rousseeuw P. *Finding Groups in Data: an Introduction to Cluster Analysis*, Wiley New York, 1990.

[Kirl 00] Kirlin R.L., Dizaji R.M., "Cluster order using clustering performance index rate, CPIR," NORSIG 2000, Kolmarden, Sweden, June 2000.

[Kris 93] Krishnapuram R., Frigui H., Nasraoui O. "Quadratic shell clustering algorithms and the detection of second-degree curves," *Pattern Recognition Letters*, Vol. 14(7), pp. 545–552, July 1993.

[Kris 95a] Krishnapuram R., Frigui H., Nasraoui O. "Fuzzy and possibilistic shell clustering algorithms and their application to boundary detection and surface approximation—Part I," *IEEE Transactions on Fuzzy Systems*, Vol. 3(1), pp. 29–43, 1995.

[Kris 95b] Krishnapuram R., Frigui H., Nasraoui O. "Fuzzy and possibilistic shell clustering algorithms and their application to boundary detection and surface approximation—Part II," *IEEE Transactions on Fuzzy Systems*, Vol. 3(1), pp. 44–60, 1995.

[Lang 98] Langan D.A., Modestino J.W., Zhang J. "Cluster validation for unsupervised stochastic model-based image segmentation," *IEEE Transactions on Image Processing*, Vol. 7(2), pp. 180–195, 1998.

[Ling 72] Ling R.F. "On the theory and construction of *k*-clusters," *Computer Journal*, Vol. 15, pp. 326–332, 1972.

[Ling 73] Ling R.F. "Probability theory of cluster analysis," *Journal of the American Statistical Association*, Vol. 68, pp. 159–164, 1973.

[Ling 76] Ling R.F., Killough G.S. "Probability tables for cluster analysis based on a theory of random graphs," *Journal of the American Statistical Association*, Vol. 71, pp. 293–300, 1976.

[Lu 00] Lu X. "Comparisons among information-based criteria, a novel modification thereof, and the Monte Carlo Markov chain method," MSc Thesis, University of Victoria, British Columbia, Canada, July 2000.

[Mant 67] Mantel N. "The detection of disease clustering and a generalized regression approach," *Cancer Research*, Vol. 27, pp. 209–220, 1967.

[Mead 74] Mead R. "A test for spatial pattern at several scales using data from a grid of contiguous quadrats," *Biometrics*, Vol. 30, pp. 295–308, 1974.

[Mill 80] Milligan G.W. "An examination of the effect of six types of error perturbation on fifteen clustering algorithms," *Psychometrica*, Vol. 45, pp. 325–342, 1980.

[Mill 83] Milligan G.W., Soon S.C., Sokol L.M. "The effect of cluster size, dimensionality, and the number of clusters on recovery of true cluster structure," *IEEE Transactions on Pattern Analysis and Machine Intelligence*, Vol. 5, pp. 40–47, 1983.

[Mill 85] Milligan G.W., Cooper M.C. "An examination of procedures for determining the number of clusters in a data set," *Psychometrika*, Vol. 50, pp. 159–179, 1985.

[Naus 82] Naus J.J. "Approximations for distributions of scan statistics," *Journal of the American Statistical Association*, Vol. 77, pp. 177–183, 1982.

[Neym 72] Neyman J., Scott E.L. "Processes of clustering and applications," in *Stochastic Point Processes: Statistical Analysis, Theory and Applications* (Lewis P.A.W., ed.), John Wiley & Sons, 1972.

[Pal 95] Pal N.R., Bezdek J.C. "On cluster validity for the fuzzy c-means model," *IEEE Transactions on Fuzzy Systems*, Vol. 3(3), pp. 370–379, 1995.

[Pal 97] Pal N.R., Biswas J. "Cluster validation using graph theoretic concepts," *Pattern Recognition*, Vol. 30(6), pp. 847–857, 1997.

[Pana 83] Panayirci E., Dubes R.C. "A test for multidimensional clustering tendency," *Pattern Recognition*, Vol. 16(4), pp. 433–444, 1983.

[Papo 91] Papoulis A. *Probability, Random Variables and Stochastic Processes*, 3rd ed., McGraw-Hill, 1991.

[Piel 69] Pielou E.C. *An Introduction to Mathematical Ecology*, John Wiley & Sons, 1969.

[Post 93] Postaire J.G., Zhang R.D., Lecocq-Botte C. "Cluster analysis by binary morphology," *IEEE Transactions on Pattern Analysis and Machine Intelligence*, Vol. 15(2), pp. 170–180, 1993.

[Ripl 77] Ripley B.D. "Modelling spatial patterns," *Journal of the Royal Statistical Society*, Vol. B39, pp. 172–212, 1977.

[Ripl 78] Ripley B.D., Silverman B.W. "Quick tests for spatial interaction," *Biometrika*, Vol. 65, pp. 641–642, 1978.

[Ripl 81] Ripley B.D. *Spatial Statistics*, John Wiley & Sons, 1981.

[Riss 78] Rissanen J. "Modeling by shortest data description," *Automatica* 14, pp. 465–471, 1978.

[Riss 89] Rissanen J. "Stochastic complexity in statistical enquiry," *Series in computer science*, 15, World Scientific, Singapore, 1989.

[Rolp 68] Rolph F.J., Fisher D.R. "Tests for hierarchical structure in random data sets," *Systematic Zoology*, Vol. 17, pp. 407–412, 1968.

[Rolp 70] Rolph F.J. "Adaptive hierarchical clustering schemes," *Systematic Zoology*, Vol. 19, pp. 58–82, 1970.

[Schw 76] Schwarz G. "Estimating the dimension of a model," *Annals of Statistics* Vol. 6, pp. 461–464, 1976.

[Sclo 87] Sclove S.L. "Application of model-selection criteria to some problems in multivariate analysis," *Psychometrika*, Vol. 52, pp. 333–343, 1987.

[Sent 07] Sentelle C., Hong S.L., Georgiopoulos M., Anagnostopoulos G.C. "A fuzzy gap statistic for fuzzy c-means," *Proceedings of the 11th IASTED International Conference on Artificial Intelligence and Soft Computing*, pp. 68–73, 2007.

[Shar 96] Sharma S. *Applied Multivariate Techniques*, John Wiley & Sons Inc., 1996.

[Shre 64] Shreider Y.A. *Method of Statistical Testing: Monte Carlo Method*, Elsevier North-Holland, 1964.

[Silv 78] Silverman B., Brown T. "Short distances, flat triangles and Poisson limits," *Journal of Applied Probability*, Vol. 15, pp. 815–825, 1978.

[Smit 84] Smith S.P., Jain A.K. "Testing for uniformity in multidimensional data," *IEEE Transactions on Pattern Analysis and Machine Intelligence*, Vol. 6, pp. 73–81, 1984.

[Snea 77] Sneath P.H.A. "A significance test for clusters in UPGMA phenograms obtained from squared Euclidean distance," *Classification Soc. Bulletin*, Vol. 4, pp. 2–14, 1977.

[Sobo 84] Sobol I.M. *The Monte Carlo Method*, Mir Publishers, Moscow, 1984.

[Stra 75] Strauss D.J. "Model for clustering," *Biometrika*, Vol. 62, pp. 467–475, 1975.

[Tibs 01] Tibshirani R., Walther G., Hastie T. "Estimating the number of clusters in a data set via the gap statistic," *Journal of Royal Statistics Society B*, Vol. 63, pp. 411–423, 2001.

[Wind 81] Windham A.P. "Cluster validity for fuzzy clustering algorithms," *Fuzzy Sets and Systems*, Vol. 5, pp. 177–185, 1981.

[Wind 82] Windham M.P. "Cluster validity for the fuzzy c-means clustering algorithm," *IEEE Transactions on Pattern Analysis and Machine Intelligence*, Vol. 4(4), pp. 357–363, July 1982.

[Xie 91] Xie X.L., Beni G. "A validity measure for fuzzy clustering," *IEEE Transactions on Pattern Analysis and Machine Intelligence*, Vol. 13(8), pp. 841–846, 1991.

[Yarm 87] Yarman-Vural F., Ataman E. "Noise, histogram and cluster validity for Gaussian mixtured data," *Pattern Recognition*, Vol. 20(4), pp. 385–401, 1987.

[Zeng 85] Zeng G., Dubes R.C. "A comparison of tests for randomness," *Pattern Recognition*, Vol. 18(2), pp. 191–198, 1985.

APPENDIX

Hints from Probability and Statistics

A

A.1 TOTAL PROBABILITY AND THE BAYES RULE

Let $\mathcal{A}_i, i = 1, 2, \ldots, M$, be M events so that $\sum_{i=1}^{M} P(\mathcal{A}_i) = 1$. Then the probability of an arbitrary event $\mathcal{B}$ is given by

$$P(\mathcal{B}) = \sum_{i=1}^{M} P(\mathcal{B}|\mathcal{A}_i)P(\mathcal{A}_i) \tag{A.1}$$

where $P(\mathcal{B}|\mathcal{A})$ denotes the conditional probability of $\mathcal{B}$ assuming $\mathcal{A}$, which is defined as

$$P(\mathcal{B}|\mathcal{A}) = \frac{P(\mathcal{B}, \mathcal{A})}{P(\mathcal{A})} \tag{A.2}$$

and $P(\mathcal{B}, \mathcal{A})$ is the joint probability of the two events. Equation (A.1) is known as the *total probability theorem*. From the definition in (A.2) the Bayes rule is readily available

$$P(\mathcal{B}|\mathcal{A})P(\mathcal{A}) = P(\mathcal{A}|\mathcal{B})P(\mathcal{B}) \tag{A.3}$$

These are easily extended to random variables or vectors described by probability density functions and we have

$$p(\boldsymbol{x}|\mathcal{A})P(\mathcal{A}) = P(\mathcal{A}|\boldsymbol{x})p(\boldsymbol{x}) \tag{A.4}$$

and

$$p(\boldsymbol{x}|\boldsymbol{y})p(\boldsymbol{y}) = p(\boldsymbol{y}|\boldsymbol{x})p(\boldsymbol{x}) \tag{A.5}$$

and finally

$$p(\boldsymbol{x}) = \sum_{i=1}^{M} p(\boldsymbol{x}|\mathcal{A}_i)P(\mathcal{A}_i) \tag{A.6}$$

A.2 MEAN AND VARIANCE

Let $p(x)$ be the probability density function (pdf) describing the random variable x. Its mean and variance are defined as

$$E[x] = \int_{-\infty}^{+\infty} xp(x)\,dx, \quad \sigma_x^2 = \int_{-\infty}^{+\infty} (x - E[x])^2 p(x)\,dx \tag{A.7}$$

A.3 STATISTICAL INDEPENDENCE

Two (or more) random variables x and y are statistically independent if and only if

$$p(x, y) = p_x(x)p_y(y) \tag{A.8}$$

It turns out that in this case $E[xy] = E[x]E[y]$. These are generalized to more than two variables.

A.4 MARGINALIZATION

Let x_i, $i = 1, 2, \ldots, l$, be a set of random variables with a joint probability density function $p(x_1, x_2, \ldots, x_l)$. It can be shown that by integrating the joint pdf with respect to some of the variables, over all possible values, the result is the joint pdf of the remaining variables. For example,

$$\int_{-\infty}^{+\infty} \int_{-\infty}^{+\infty} \cdots \int_{-\infty}^{+\infty} p(x_1, \ldots, x_l) dx_{k+1} dx_{k+2} \cdots dx_l = p(x_1, x_2, \ldots, x_k)$$

This calculation is known as *marginalization*. For discrete random variables, marginalization involves probabilities and summations, i.e.,

$$\sum_{x_{k+1}} \sum_{x_{k+2}} \cdots \sum_{x_l} P(x_1, \ldots, x_l) = P(x_1, \ldots, x_k)$$

where summations are over all possible values of the respective variables.

A.5 CHARACTERISTIC FUNCTIONS

Let $p(x)$ be the probability density function of a random variable x. The associated *characteristic function* is by definition the integral

$$\Phi(\Omega) = \int_{-\infty}^{+\infty} p(x) \exp(j\Omega x)\,dx \equiv E[\exp(j\Omega x)] \tag{A.9}$$

If $j\Omega$ is changed into s, the resulting integral becomes

$$\Phi(s) = \int_{-\infty}^{+\infty} p(x) \exp(sx)\,dx \equiv E[\exp(sx)] \tag{A.10}$$

and it is known as the *moment generating function*.

The function

$$\Psi(\Omega) = \ln \Phi(\Omega) \tag{A.11}$$

is known as the *second characteristic function* of x.

The joint characteristic function of l random variables is defined by

$$\Phi(\Omega_1, \Omega_2, \ldots, \Omega_l) = \int_{-\infty}^{+\infty} p(x_1, x_2, \ldots, x_l) \exp\left(j \sum_{i=1}^{l} \Omega_i x_i\right) dx \tag{A.12}$$

The logarithm of the above is the second joint characteristic function of the l random variables.

A.6 MOMENTS AND CUMULANTS

Taking the derivative of $\Phi(s)$ in Eq. (A.10) we obtain

$$\frac{d^n \Phi(s)}{ds^n} \equiv \Phi^{(n)}(s) = E[x^n \exp(sx)] \tag{A.13}$$

and hence for $s = 0$

$$\Phi^{(n)}(0) = E[x^n] \equiv m_n \tag{A.14}$$

where m_n is known as the nth-order moment of x. If the moments of all orders are finite, the Taylor series expansion of $\Phi(s)$ near the origin exists and is given by

$$\Phi(s) = \sum_{n=0}^{+\infty} \frac{m_n}{n!} s^n \tag{A.15}$$

Similarly, the Taylor expansion of the second generating function results in

$$\Psi(s) = \sum_{n=1}^{+\infty} \frac{\kappa_n}{n!} s^n \tag{A.16}$$

where

$$\kappa_n \equiv \frac{d^n \Psi(0)}{ds^n} \tag{A.17}$$

and are known as the *cumulants* of the random variable x. It is not difficult to show that $\kappa_0 = 0$. For a zero mean random variable, it turns out that

$$\kappa_1(x) = E[x] = 0 \tag{A.18}$$

$$\kappa_2(x) = E[x^2] = \sigma^2 \tag{A.19}$$

$$\kappa_3(x) = E[x^3] \tag{A.20}$$

$$\kappa_4(x) = E[x^4] - 3\sigma^4 \tag{A.21}$$

That is, the first three cumulants are equal to the corresponding moments. The fourth-order cumulant is also known as *kurtosis*. *For a Gaussian process all cumulants of order higher than two are zero*. The kurtosis is commonly used as a measure of the non-Gaussianity of a random variable. For random variables described by (unimodal) pdfs with spiky shape and heavy tails, known as leptokurtic or super-Gaussian, κ_4 is positive, whereas for random variables associated with pdfs with a flatter shape, known as platykurtic or sub-Gaussian, κ_4 is negative. Gaussian variables have zero kurtosis. The opposite is not always true, in the sense that there exist non-Gaussian random variables with zero kurtosis; however, this can be considered rare.

Similar arguments hold for the expansion of the joint characteristic functions for multivariate pdfs. For zero mean random variables, x_i, $i = 1, 2, \ldots, l$, the cumulants of order up to four are given by

$$\kappa_1(x_i) = E[x_i] = 0 \tag{A.22}$$

$$\kappa_2(x_i, x_j) = E[x_i x_j] \tag{A.23}$$

$$\kappa_3(x_i, x_j, x_k) = E[x_i x_j x_k] \tag{A.24}$$

$$\kappa_4(x_i, x_j, x_k, x_r) = E[x_i x_j x_k x_r] - E[x_i x_j]E[x_k x_r] \tag{A.25}$$

$$- E[x_i x_k]E[x_j x_r] - E[x_i x_r]E[x_j x_k] \tag{A.26}$$

Thus, once more, the cumulants of the first three orders are equal to the corresponding moments. If all variables coincide, we talk about *auto-cumulants*, and otherwise about *cross-cumulants*, that is,

$$\kappa_4(x_i, x_i, x_i, x_i) = \kappa_4(x_i)$$

that is, the auto-cumulant of x_i is identical to its kurtosis. It is not difficult to see that if the zero mean random variables are mutually independent, their cross-cumulants are zero. *This is also true for the cross-cumulants of all orders.*

A.7 EDGEWORTH EXPANSION OF A PDF

Taking into account the expansion in Eq. (A.16), the definition given in Eq. (A.11), and taking the inverse Fourier of $\Phi(\Omega)$ in Eq. (A.9) we can obtain the following expansion of $p(x)$ for a zero mean unit variance random variable x:

$$p(x) = g(x)\left(1 + \frac{1}{3!}\kappa_3(x)H_3(x) + \frac{1}{4!}\kappa_4(x)H_4(x) + \frac{10}{6!}\kappa_3^2(x)H_6(x) + \frac{1}{5!}\kappa_5(x)H_5(x) + \frac{35}{7!}\kappa_3(x)\kappa_4(x)H_7(x) + \ldots\right) \tag{A.27}$$

where $g(x)$ is the unit variance and zero mean normal pdf, and $H_\kappa(x)$ is the Hermite polynomial of degree k. The rather strange ordering of terms is the outcome of a specific reordering in the resulting expansion, so that the successive coefficients in the series decrease uniformly. This is very important when truncation of the series is required. The Hermite polynomials are defined as

$$H_k(x) = (-1)^k \exp(x^2/2)\frac{d^k}{dx^k}\exp(-x^2/2) \tag{A.28}$$

and they form a complete orthogonal basis set in the real axis, that is,

$$\int_{-\infty}^{+\infty} \exp(-x^2/2)H_n(x)H_m(x)\,dx = \begin{cases} n!\sqrt{2\pi} & \text{if } n = m \\ 0 & \text{if } n \neq m \end{cases} \tag{A.29}$$

The expansion of $p(x)$ in Eq. (A.27) is known as the *Edgeworth expansion*, and it is actually an expansion of a pdf around the normal pdf [Papo 91].

A.8 KULLBACK–LEIBLER DISTANCE

The Kullback-Leibler (KL) distance is a measure of the distance between two probability density functions $p(\boldsymbol{x})$ and $\hat{p}(\boldsymbol{x})$ and is defined as

$$L = -\int p(\boldsymbol{x}) \ln \frac{\hat{p}(\boldsymbol{x})}{p(\boldsymbol{x})} d\boldsymbol{x} \tag{A.30}$$

Sometimes it is referred to as *cross or relative entropy*. The KL distance can be shown to be always nonnegative but it is not a true distance measure, from a mathematical point of view, since it is not symmetric. Sometimes it is referred as the KL divergence.

The KL distance is closely related to the *mutual information* measure, I, between l scalar random variables, $x_i, i = 1, 2, \ldots, l$. Indeed, let us compute the KL distance between the joint pdf $p(\boldsymbol{x})$ and the pdf resulting from the product of the corresponding marginal probability densities, that is,

$$\begin{aligned} L &= -\int p(\boldsymbol{x}) \ln \frac{\prod_{i=1}^{l} p_i(x_i)}{p(\boldsymbol{x})} d\boldsymbol{x} \\ &= \int p(\boldsymbol{x}) \ln p(\boldsymbol{x}) d\boldsymbol{x} - \sum_{i=1}^{l} \int p(\boldsymbol{x}) \ln p_i(x_i) d\boldsymbol{x} \\ &= -H(\boldsymbol{x}) - \sum_{i=1}^{l} \int p(\boldsymbol{x}) \ln p_i(x_i) d\boldsymbol{x} \end{aligned} \tag{A.31}$$

Carrying out the integrations on the right-hand side it is straightforward to see the KL distance is equal to the mutual information, I, defined as

$$I(x_1, x_2, \ldots, x_l) = -H(\boldsymbol{x}) + \sum_{i=1}^{l} H(x_i) \tag{A.32}$$

where $H(x_i)$ is the associated entropy of x_i, defined as ([Papo 91])

$$H(x_i) = -\int p_i(x_i) \ln p_i(x_i)\, dx_i \tag{A.33}$$

It is now easy to see that if the variables $x_i, i = 1, 2, \ldots, l$, are statistically independent their mutual information I is zero. Indeed, in this case $\Pi_{i=1}^{l} p_i(x_i) = p(\boldsymbol{x})$, hence $L = I(x_1, x_2, \ldots, x_l) = 0$.

A.9 MULTIVARIATE GAUSSIAN OR NORMAL PROBABILITY DENSITY FUNCTION

This is defined as a generalization of the univariate normal pdf

$$p(\boldsymbol{x}) = \frac{1}{(2\pi)^{l/2}|\Sigma|^{1/2}} \exp\left(-\frac{1}{2}(\boldsymbol{x}-\boldsymbol{\mu})^T \Sigma^{-1}(\boldsymbol{x}-\boldsymbol{\mu})\right), \tag{A.34}$$

where $\boldsymbol{\mu}$ is the mean vector, that is, $E\left[[x_1, x_2, \ldots, x_l]^T\right] = [\mu_1, \mu_2, \ldots, \mu_l]^T$ and Σ the covariance matrix

$$\Sigma = E[(\boldsymbol{x}-\boldsymbol{\mu})(\boldsymbol{x}-\boldsymbol{\mu})^T] \tag{A.35}$$

and we say that $\boldsymbol{x}$ is normally distributed as $\mathcal{N}(\boldsymbol{\mu}, \Sigma)$. For the one dimensional, $l = 1$, case the covariance matrix becomes the variance σ^2 and the Gaussian density function takes the form

$$p(x) = \frac{1}{\sqrt{2\pi}\sigma} \exp\left(-\frac{(x-\mu)^2}{2\sigma^2}\right)$$

Figure A.1 shows the plots of two Gaussians for the same mean and different variances. For the general l-dimensional case the covariance matrix has the form

$$\Sigma = \begin{bmatrix} \sigma_1^2 & \sigma_{12} & \cdots & \sigma_{1l} \\ \sigma_{21} & \sigma_2^2 & \cdots & \sigma_{2l} \\ \vdots & \vdots & \vdots & \vdots \\ \sigma_{l1} & \sigma_{l2} & \cdots & \sigma_l^2 \end{bmatrix} \tag{A.36}$$

where $\sigma_i^2 = E[(x_i - \mu_i)^2], \sigma_{ij} = \sigma_{ji} = E[(x_i - \mu_i)(x_j - \mu_j)]$. Thus, the main matrix diagonal consists of the respective variances of the elements of the random vector and the off-diagonal elements are the respective covariances between the elements of the random vector. Note that if the random variables x_i are statistically independent, then the mean of the product equals the product of the means, that is, $E[(x_i - \mu_i)(x_j - \mu_j)] = E[(x_i - \mu_i)]E[(x_j - \mu_j)] = 0$, and the covariance matrix is diagonal. However, a diagonal covariance matrix does not, in general, mean that the variables are statistically independent. In the case, though, of multivariate Gaussian

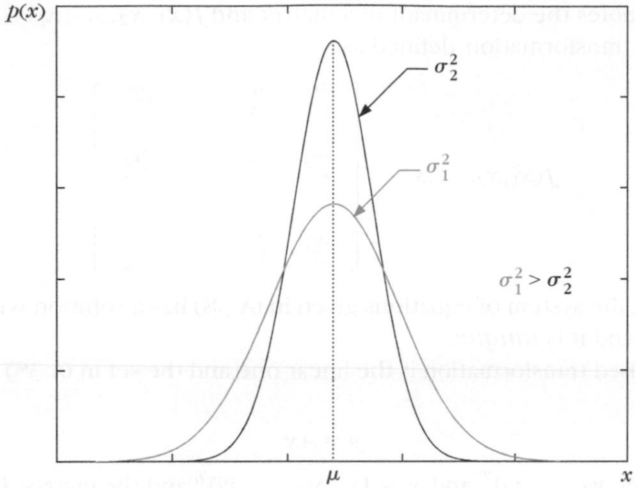

FIGURE A.1

Two Gaussians with the same mean μ and different variances.

densities the opposite is also valid. Indeed, if the covariance matrix is diagonal then it is straightforward to see that

$$p(\boldsymbol{x}) = \prod_{i=1}^{l} p_i(x_i) \tag{A.37}$$

where

$$p_i(x_i) = \frac{1}{\sqrt{2\pi}\sigma_i} \exp\left(-\frac{(x_i - \mu_i)^2}{2\sigma_i^2}\right)$$

which is the univariate Gaussian describing the ith variable (why?). Thus, the joint probability density is the product of the individual (marginal) ones, which is the definition of statistical independence.

A.10 TRANSFORMATION OF RANDOM VARIABLES

Let $X = \{x_1, x_2, \ldots, x_l\}$ be a set of random variables, which are jointly distributed according to the joint pdf $p_X(x_1, x_2, \ldots, x_l)$. We form a new set of random variables by the following transformations

$$y_1 = g_1(x_1),\ y_2 = g_2(x_2), \ldots,\ y_l = g_l(x_l) \tag{A.38}$$

It can be shown [Papo 91] that the joint pdf describing the set Y is given by

$$p_Y(y_1, y_2, \ldots, y_l) = \frac{p_X(x_1, x_2, \ldots, x_l)}{|J(x_1, x_2, \ldots, x_l)|} \tag{A.39}$$

where $|\cdot|$ denotes the determinant of a matrix and $J(x_1, x_2, \ldots, x_l)$ is the Jacobian matrix of the transformation, defined as

$$J(x_1, x_2, \ldots, x_l) = \begin{bmatrix} \frac{\partial y_1}{\partial x_1} & \frac{\partial y_1}{\partial x_2} & \cdots & \frac{\partial y_1}{\partial x_l} \\ \frac{\partial y_2}{\partial x_1} & \frac{\partial y_2}{\partial x_2} & \cdots & \frac{\partial y_2}{\partial x_l} \\ \vdots & \vdots & \vdots & \vdots \\ \frac{\partial y_l}{\partial x_1} & \frac{\partial y_l}{\partial x_2} & \cdots & \frac{\partial y_l}{\partial x_l} \end{bmatrix}$$

provided that the system of equations given in (A.38) has a solution with respect to $x_1, x_2, \ldots, x_l$ and it is *unique*.

A widely used transformation is the linear one and the set in (A.38) is compactly written as

$$\boldsymbol{y} = A\boldsymbol{x} \tag{A.40}$$

where $\boldsymbol{y} = [y_1, y_2, \ldots, y_l]^T$ and $\boldsymbol{x} = [x_1, x_2, \ldots, x_l]^T$ and the matrix A is invertible. Then the system of equations has a unique solution, $\boldsymbol{x} = A^{-1}\boldsymbol{y}$, and the Jacobian is easily shown to be

$$J(x_1, x_2, \ldots, x_l) = A \tag{A.41}$$

The above can be used, for example, in the MATLAB framework to generate jointly Gaussian variables with mean value $\boldsymbol{\mu}$ and covariance matrix Σ, utilizing the normalized Gaussian generator $\mathcal{N}(\mathbf{0}, I)$. The latter is described by

$$p(\boldsymbol{x}) = \frac{1}{(2\pi)^{l/2}} \exp\left(-\frac{1}{2}\boldsymbol{x}^T\boldsymbol{x}\right) \tag{A.42}$$

Let Σ be the covariance matrix of the multivariate Gaussian, which describes the variables to be generated. We know that Σ is a symmetric matrix, $\Sigma = \Sigma^T$, and therefore it can be diagonalized (Appendix B) as

$$\Sigma = P\Lambda P^T$$

where Λ is a diagonal matrix having the eigenvalues of Σ as its elements and P is a unitary matrix ($P^{-1} = P^T$) having as columns the corresponding eigenvectors of Σ. Define, now, the linear transformation

$$\boldsymbol{y} = P\Lambda^{1/2}\boldsymbol{x} \tag{A.43}$$

where $\Lambda^{1/2}$ is the square root of Λ. Assuming the transformation to be invertible (i.e., Σ is invertible) then, recalling (A.39), (A.41), (A.42) and (A.43) we obtain

$$p(\boldsymbol{y}) = \frac{1}{(2\pi)^{l/2}|\Sigma|^{1/2}} \exp\left(-\frac{1}{2}\boldsymbol{y}^T P\Lambda^{-1/2}\Lambda^{-1/2}P^T\boldsymbol{y}\right) \tag{A.44}$$

$$= \frac{1}{(2\pi)^{l/2}|\Sigma|^{1/2}} \exp\left(-\frac{1}{2}\boldsymbol{y}^T\Sigma^{-1}\boldsymbol{y}\right) \tag{A.45}$$

where we have used the fact that the determinant of the Jacobian matrix is $|P\Lambda^{1/2}| = |P\Lambda^{1/2}\Lambda^{1/2}P^T|^{1/2} = |\Sigma|^{1/2}$. Thus, in order to generate a set of random variables described by the multivariate Gaussian $\mathcal{N}(\mathbf{0}, \Sigma)$, it suffices to generate a vector, $\boldsymbol{x}$, using the normalized Gaussian, $\mathcal{N}(\mathbf{0}, I)$, and then transform it, that is, $\boldsymbol{y} = P\Lambda^{1/2}\boldsymbol{x}$. Finally, a further shift by $\boldsymbol{\mu}$, that is, $\hat{\boldsymbol{y}} = \boldsymbol{y} + \boldsymbol{\mu}$, suffices to produce multivariate Gaussian variables described by the Gaussian $\mathcal{N}(\boldsymbol{\mu}, \Sigma)$.

A.11 THE CRAMER–RAO LOWER BOUND

Let $p(\boldsymbol{x}; \boldsymbol{\theta})$ be the pdf of a random vector, parameterized in terms of an r-dimensional vector parameter $\boldsymbol{\theta}$. If X is the set of N observations $\boldsymbol{x}_i$, $i = 1, 2, \ldots, N$, the log likelihood function is the logarithm of the joint pdf of the observations $\ln p(X; \boldsymbol{\theta}) = L(\boldsymbol{\theta})$. The Fisher matrix is defined so that its (i, j) element equals

$$J_{ij} = -E\left[\frac{\partial^2 L(\boldsymbol{\theta})}{\partial\theta_i \partial\theta_j}\right], \quad i, j = 1, 2, \ldots, r \tag{A.46}$$

It can be shown that the ith element of *any* unbiased estimate $\hat{\boldsymbol{\theta}}$, of the parameter $\boldsymbol{\theta}$, based on the observations set X satisfies

$$E[(\hat{\theta}_i - \theta_i)^2] \geq J_{ii}^{-1} \tag{A.47}$$

In other words, its variance is *lower bounded by the (i, i) element of the inverse Fisher matrix*. This is known as the Cramer–Rao bound. If the relation is valid with equality, the corresponding estimator is called *efficient*.

A.12 CENTRAL LIMIT THEOREM

Let $x_1, x_2, \ldots, x_N$ be N independent random variables, with mean and variances μ_i, σ_i^2, respectively. We form the new random variable

$$z = \sum_{i=1}^{N} x_i \tag{A.48}$$

Its mean and variance are given by $\mu = \sum_{i=1}^{N} \mu_i$ and $\sigma^2 = \sum_{i=1}^{N} \sigma_i^2$. The central limit theorem states that as $N \to \infty$, and under certain general conditions, the pdf of the variable

$$q = \frac{z - \mu}{\sigma} \tag{A.49}$$

approaches $\mathcal{N}(0, 1)$, irrespective of the pdfs of the summands [Papo 91]. Thus in practice, for large enough N we can consider z as approximately Gaussian with mean μ and variance σ^2.

A.13 CHI-SQUARE DISTRIBUTION

Let x_i, $i = 1, 2, \ldots, N$, be samples of a Gaussian $\mathcal{N}(0, 1)$ random variable x. The sum of squares variable

$$\chi^2 = x_1^2 + x_2^2 + \cdots + x_N^2 \equiv y \tag{A.50}$$

is a chi-square distributed variable with N degrees of freedom. Its probability density function is given by [Papo 91]

$$p_y(y) = \frac{1}{2^{N/2}\Gamma(N/2)} y^{N/2-1} \exp(-y/2) u(y) \tag{A.51}$$

where

$$\Gamma(b+1) = \int_0^{\infty} y^b \exp(-y)\, dy \quad b > -1 \tag{A.52}$$

where $u(y)$ is the step function (1 for $y > 0$ and 0 for $y < 0$). Recalling the respective definitions, it is easy to show that $E[y] = N, \sigma_y^2 = 2N$.

The chi-square distribution possesses the *additive property*. Let χ_1^2 and χ_2^2 be independent random variables of chi-square distribution with N_1, N_2 degrees of freedom, respectively. Then the random variable

$$\chi^2 = \chi_1^2 + \chi_2^2 \tag{A.53}$$

is a chi-square variable with $N_1 + N_2$ degrees of freedom. Based on these properties, we can show that the variance estimate of Eq. (5.13) is described by a chi-square distribution with $N-1$ degrees of freedom, provided x is Gaussian and the samples x_i are independent. The proof is simple and interesting [Fras 58]. Define the following transformation:

$$y_1 = \sqrt{N}\bar{x} = \frac{x_1 + \cdots + x_n}{\sqrt{N}}$$

$$y_2 = \frac{1}{\sqrt{2}}(x_2 - x_1)$$

$$\vdots$$

$$y_n = \frac{1}{\sqrt{n(n-1)}}[(n-1)x_n - (x_1 + \cdots + x_{n-1})], \quad n = 2, 3, \ldots, N$$

It is easy to show that this transformation is an orthogonal one (Problem 5.5). Thus, the random variables y_i are also Gaussian, statistically independent, and with the same variance σ^2 as x (Problem 5.6). This transformation easily results in

$$\sum_{i=1}^{N} y_i^2 = \sum_{i=1}^{N} x_i^2 \tag{A.54}$$

and of course

$$y_1^2 = N\bar{x}^2 \tag{A.55}$$

Subtracting the two, we obtain

$$\sum_{i=2}^{N} y_i^2 = \sum_{i=1}^{N} (x_i - \bar{x})^2 \equiv (N-1)\hat{\sigma}^2 \tag{A.56}$$

Furthermore, $E[y_i] = 0, i = 2, \ldots, N$. Thus the variable

$$z = \frac{N-1}{\sigma^2}\hat{\sigma}^2 = \sum_{i=2}^{N} \frac{y_i^2}{\sigma^2} \tag{A.57}$$

is a chi-square with $N-1$ degrees of freedom.

A.14 *t*-DISTRIBUTION

Let x and z be two independent random variables with x being $\mathcal{N}(0,1)$ and z a chi-square with N degrees of freedom. Then it can be shown [Papo 91] that the variable

$$q = \frac{x}{\sqrt{z/N}} \tag{A.58}$$

is a so-called t-distributed variable with probability density function given by

$$p_q(q) = \frac{\gamma_1}{\sqrt{(1+q^2/N)^{N+1}}}, \quad \gamma_1 = \frac{\Gamma((N+1)/2)}{\sqrt{\pi N}\Gamma(N/2)}$$

where $\Gamma(\cdot)$ was defined in Eq. (A.44). Thus, from the test statistic in Eqs. (5.14) and (A.57) we have

$$q = \frac{\bar{x}-\mu}{\hat{\sigma}/\sqrt{N}} = \frac{\frac{\bar{x}-\mu}{\sigma/\sqrt{N}}}{\sqrt{z/N-1}} \tag{A.59}$$

Since z is a chi-square with $N-1$ degrees of freedom, q is a t-distributed variable with $N-1$ degrees of freedom. In a similar way we can show that the test statistic in Eq. (5.18) is t-distributed with $2N-2$ degrees of freedom.

A.15 BETA DISTRIBUTION

A random variable follows the Beta distribution with parameters a and $b\,(a, b > 0)$, if its probability density function is defined as

$$p(x) = \begin{cases} \frac{x^{a-1}(1-x)^{b-1}}{B(a,b)}, & 0 < x < 1 \\ 0, & \text{Otherwise} \end{cases} \tag{A.60}$$

where

$$B(a,b) = \int_0^1 u^{a-1}(1-u)^{b-1}\,du \tag{A.61}$$

Its mean and variance are equal to $a/(a+b)$ and $ab/((a+b)^2(a+b+1))$, respectively.

A.16 POISSON DISTRIBUTION

A Poisson distributed random variable X, with parameter a, takes the values $k = 0, 1, 2, \ldots$, with probabilities

$$P(X=k) = e^{-a}\frac{a^k}{k!} \tag{A.62}$$

A *Poisson process* scatters vectors in a Euclidean space in such a way that the random variable X, denoting the number of vectors in a region of volume V, has a Poisson distribution with parameter λV, that is,

$$P(X=k) = e^{-\lambda V}\frac{(\lambda V)^k}{k!}, \quad k = 0, 1, 2, \ldots \tag{A.63}$$

The parameter λ is called the *intensity* of the process and equals the expected number of vectors per unit volume.

A.17 GAMMA FUNCTION

The Gamma function is defined as

$$\Gamma(\alpha) = \int_0^\infty x^{\alpha-1}e^{-x}\,dx$$

If α is an integer, integrating by parts we get

$$\Gamma(n) = (n-1)\Gamma(n-1) = (n-1)!$$

REFERENCES

[Digg 83] Diggle P.J. *Statistical Analysis of Spatial Point Processes*, Academic Press, 1983.

[Fras 58] Fraser D.A.S. *Statistics: An Introduction*, John Wiley, 1958.

[Papo 91] Papoulis A. *Probability Random Variables and Stochastic Processes*, 3rd ed. McGraw Hill, 1991.

[Spie 75] Spiegel M.R. *Schaum's Outline of Theory and Problems of Probability and Statistics*, McGraw Hill, 1975.

Linear Algebra Basics

B

B.1 POSITIVE DEFINITE AND SYMMETRIC MATRICES

- An $l \times l$ real matrix A is called *positive definite* if for *every* nonzero vector $\boldsymbol{x}$ the following is true:

$$\boldsymbol{x}^T A \boldsymbol{x} > 0 \tag{B.1}$$

If equality with zero is allowed, A is called *nonnegative or positive semidefinite*.

- It is easy to show that all eigenvalues of such a matrix are positive. Indeed, let λ_i be one eigenvalue and $\boldsymbol{v}_i$ the corresponding unit norm eigenvector ($\boldsymbol{v}_i^T \boldsymbol{v}_i = 1$). Then by the respective definitions

$$A\boldsymbol{v}_i = \lambda_i \boldsymbol{v}_i \quad \text{or} \tag{B.2}$$

$$0 < \boldsymbol{v}_i^T A \boldsymbol{v}_i = \lambda_i \tag{B.3}$$

Since the determinant of a matrix is equal to the product of its eigenvalues, we conclude that the determinant of a positive definite matrix is also positive.

- Let A be an $l \times l$ symmetric matrix, $A^T = A$. Then the eigenvectors corresponding to distinct eigenvalues are orthogonal. Indeed, let $\lambda_i \neq \lambda_j$ be two such eigenvalues. From the definitions we have

$$A\boldsymbol{v}_i = \lambda_i \boldsymbol{v}_i \tag{B.4}$$

$$A\boldsymbol{v}_j = \lambda_j \boldsymbol{v}_j \tag{B.5}$$

Multiplying (B.4) on the left by $\boldsymbol{v}_j^T$ and the transpose of (B.5) on the right by $\boldsymbol{v}_i$, we obtain

$$\boldsymbol{v}_j^T A \boldsymbol{v}_i - \boldsymbol{v}_j^T A \boldsymbol{v}_i = 0 = (\lambda_i - \lambda_j)\boldsymbol{v}_j^T \boldsymbol{v}_i \tag{B.6}$$

Thus, $\boldsymbol{v}_j^T \boldsymbol{v}_i = 0$. Furthermore, it can be shown that even if the eigenvalues are not distinct, we can still find a set of orthogonal eigenvectors. The same is true for Hermitian matrices, in case we deal with more general complex-valued matrices.

- Based on this, it is now straightforward to show that a symmetric matrix A can be diagonalized by the similarity transformation

$$\Phi^T A \Phi = \Lambda \tag{B.7}$$

where matrix Φ has as its columns the unit eigenvectors ($\boldsymbol{v}_i^T \boldsymbol{v}_i = 1$) of A, that is,

$$\Phi = [\boldsymbol{v}_1, \boldsymbol{v}_2, \ldots, \boldsymbol{v}_l] \tag{B.8}$$

and Λ is the diagonal matrix with elements the corresponding eigenvalues of A. From the orthonormality of the eigenvectors it is obvious that $\Phi^T\Phi = I$, that is, Φ is a unitary matrix, $\Phi^T = \Phi^{-1}$. The proof is similar for Hermitian complex matrices as well.

B.2 CORRELATION MATRIX DIAGONALIZATION

Let $\boldsymbol{x}$ be a random vector in the l-dimensional space. Its correlation matrix is defined as $R = E[\boldsymbol{x}\boldsymbol{x}^T]$. Matrix R is readily seen to be positive semidefinite. For our purposes we will assume that it is positive definite, thus invertible. Moreover, it is symmetric, and hence it can always be diagonalized

$$\Phi^T R \Phi = \Lambda \tag{B.9}$$

where Φ is the matrix consisting of the (orthogonal) eigenvectors and Λ the diagonal matrix with the corresponding eigenvalues on its diagonal. Thus, we can always transform $\boldsymbol{x}$ into another random vector whose elements are uncorrelated. Indeed

$$\boldsymbol{x}_1 \equiv \Phi^T \boldsymbol{x} \tag{B.10}$$

Then the new correlation matrix is $R_1 = \Phi^T R \Phi = \Lambda$. Furthermore, if $\Lambda^{1/2}$ is the diagonal matrix whose elements are the square roots of the eigenvalues of R ($\Lambda^{1/2}\Lambda^{1/2} = \Lambda$), then it is readily shown that the transformed random vector

$$\boldsymbol{x}_1 \equiv \Lambda^{-1/2}\Phi^T \boldsymbol{x} \tag{B.11}$$

has uncorrelated elements with unit variance. $\Lambda^{-1/2}$ denotes the inverse of $\Lambda^{1/2}$. It is now easy to see that if the correlation matrix of a random vector is the identity matrix I, then this is invariant under any unitary transformation $A^T\boldsymbol{x}, A^TA = I$. That is, the transformed variables are also uncorrelated with unit variance. A useful by-product of this is the following lemma.

Lemma *Let $\boldsymbol{x}, \boldsymbol{y}$ be two random vectors with correlation matrices R_x, R_y, respectively. Then there is a linear transformation that diagonalizes both matrices simultaneously.*

Proof. Let Φ be the eigenvector matrix diagonalizing R_x. Then the transformation

$$\boldsymbol{x}_1 \equiv \Lambda^{-1/2}\Phi^T \boldsymbol{x} \tag{B.12}$$

$$\boldsymbol{y}_1 \equiv \Lambda^{-1/2}\Phi^T \boldsymbol{y} \tag{B.13}$$

generates two new random vectors with correlation matrices $R_x^1 = I, R_y^1$, respectively. Now let Ψ be the eigenvector matrix diagonalizing R_y^1. Then the random vectors generated by the unitary transformation ($\Psi^T\Psi = I$)

$$\boldsymbol{x}_2 \equiv \Psi^T \boldsymbol{x}_1 \tag{B.14}$$

$$\boldsymbol{y}_2 \equiv \Psi^T \boldsymbol{y}_1 \tag{B.15}$$

have correlation matrices $R_x^2 = I, R_y^2 = D$, where D is the diagonal matrix with elements the eigenvalues of R_y^1. Thus, the linear transformation of the original vectors by the matrix

$$A^T = \Psi^T\Lambda^{-1/2}\Phi^T \tag{B.16}$$

diagonalizes both correlation matrices simultaneously (one to an identity matrix). All these are obviously valid for covariance matrices as well. □

APPENDIX

Cost Function Optimization

C

In this appendix, we review a number of optimization schemes that have been encountered throughout the book.

Let $\boldsymbol{\theta}$ be an unknown parameter vector and $J(\boldsymbol{\theta})$ the corresponding cost function to be minimized. Function $J(\boldsymbol{\theta})$ is assumed to be differentiable

C.1 GRADIENT DESCENT ALGORITHM

The algorithm starts with an initial estimate $\boldsymbol{\theta}(0)$ of the minimum point and the subsequent algorithmic iterations are of the form

$$\boldsymbol{\theta}(new) = \boldsymbol{\theta}(old) + \Delta\boldsymbol{\theta} \tag{C.1}$$

$$\Delta\boldsymbol{\theta} = -\mu \frac{\partial J(\boldsymbol{\theta})}{\partial \boldsymbol{\theta}}\bigg|_{\boldsymbol{\theta}=\boldsymbol{\theta}(old)} \tag{C.2}$$

where $\mu > 0$. If a maximum is sought, the method is known as *gradient ascent* and the minus sign in (C.2) is neglected.

Figure C.1 shows the geometric interpretation of the scheme. The new estimate $\boldsymbol{\theta}(new)$ is chosen in the direction that decreases $J(\boldsymbol{\theta})$. The parameter μ is very important and it plays a crucial role in the convergence of the algorithm. If it is too small, the corrections $\Delta\boldsymbol{\theta}$ are small and the convergence to the optimum point is very slow. On the other hand, if it is too large, the algorithm may oscillate around the optimum value and convergence is not possible. However, if the parameter is properly chosen, the algorithm converges to a stationary point of $J(\boldsymbol{\theta})$, which can be either, a local minimum ($\boldsymbol{\theta}_1^0$) or a global minimum ($\boldsymbol{\theta}^0$) or a saddle point ($\boldsymbol{\theta}_2^0$). In other words, it converges to a point where the gradient becomes zero (see Figure C.2). To which of the stationary points the algorithm will converge depends on the position of the initial point, relative to the stationary points. Furthermore, the convergence speed depends on the form of the cost $J(\boldsymbol{\theta})$. Figure C.3 shows the constant $J(\boldsymbol{\theta}) = c$ curves, for two cases and for different values of c, in the two-dimensional space, that is, $\boldsymbol{\theta} = [\theta_1, \theta_2]^T$. The optimum $\boldsymbol{\theta}^0$ is located at the center of the curves. Recall that the gradient $\frac{\partial J(\boldsymbol{\theta})}{\partial \boldsymbol{\theta}}$ is always vertical to the tangent to the

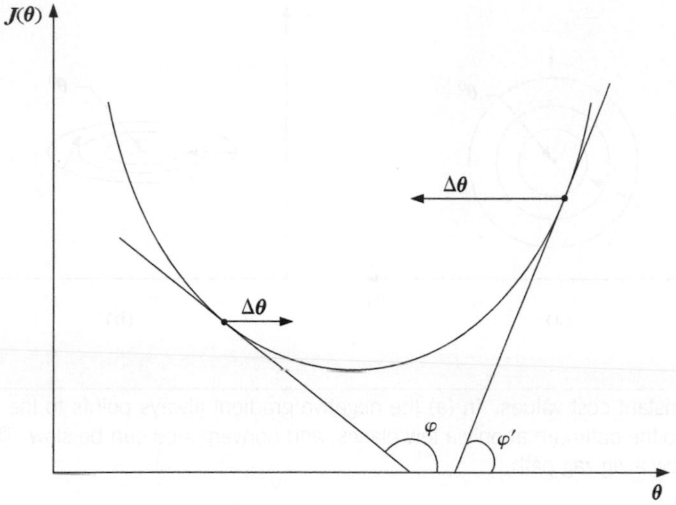

FIGURE C.1

In the gradient descent scheme, the correction of the parameters takes place in the direction that decreases the value of the cost function.

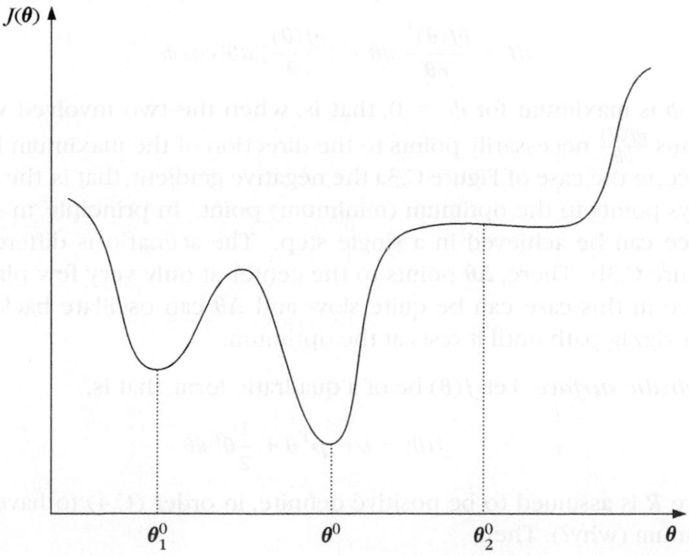

FIGURE C.2

A local minimum, a global minimum, and a saddle point of $J(\theta)$.

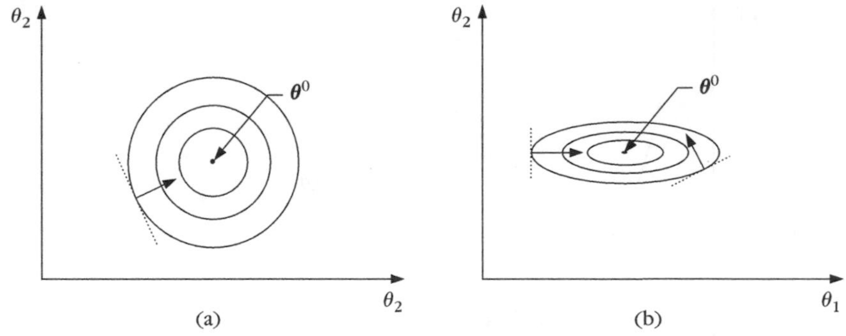

FIGURE C.3

Curves of constant cost values. In (a) the negative gradient always points to the optimum. In (b) it points to the optimum at only a few places, and convergence can be slow. The correction term can follow a zig zag path.

constant J curves. Indeed, if $J(\boldsymbol{\theta}) = c$, then

$$dc = 0 = \frac{\partial J(\boldsymbol{\theta})^T}{\partial \boldsymbol{\theta}} d\boldsymbol{\theta} \Longrightarrow \frac{J(\boldsymbol{\theta})}{\partial \boldsymbol{\theta}} \perp d\boldsymbol{\theta} \tag{C.3}$$

Furthermore, at each point $\boldsymbol{\theta}$ on a curve $J(\boldsymbol{\theta}) = c$, the gradient $\frac{\partial J(\boldsymbol{\theta})}{\partial \boldsymbol{\theta}}$ points to the direction of the maximum increase of $J(\boldsymbol{\theta})$. This is easily seen by writing

$$dJ = \frac{\partial J(\boldsymbol{\theta})^T}{\partial \boldsymbol{\theta}} d\boldsymbol{\theta} = |\frac{\partial J(\boldsymbol{\theta})}{\partial \boldsymbol{\theta}}||d\boldsymbol{\theta}| \cos\phi$$

where $\cos\phi$ is maximum for $\phi = 0$, that is, when the two involved vectors are parallel. Thus $\frac{\partial J(\boldsymbol{\theta})}{\partial \boldsymbol{\theta}}$ necessarily points to the direction of the maximum increase of $J(\boldsymbol{\theta})$. Hence, in the case of Figure C.3a the negative gradient, that is, the correction term, always points to the optimum (minimum) point. In principle, in such cases, convergence can be achieved in a single step. The scenario is different for the case of Figure C.3b. There, $\Delta\boldsymbol{\theta}$ points to the center at only very few places. Thus, convergence in this case can be quite slow and $\Delta\boldsymbol{\theta}$ can oscillate back and forth following a zigzag path until it rests at the optimum.

- *Quadratic surface*: Let $J(\boldsymbol{\theta})$ be of a quadratic form, that is,

$$J(\boldsymbol{\theta}) = b - \boldsymbol{p}^T\boldsymbol{\theta} + \frac{1}{2}\boldsymbol{\theta}^T R\boldsymbol{\theta} \tag{C.4}$$

where R is assumed to be positive definite, in order (C.4) to have a (single) minimum (why?). Then,

$$\frac{\partial J(\boldsymbol{\theta})}{\partial \boldsymbol{\theta}} = R\boldsymbol{\theta} - \boldsymbol{p} \tag{C.5}$$

Thus, the optimum value is given by

$$R\boldsymbol{\theta}^0 = \boldsymbol{p} \tag{C.6}$$

The tth iteration step in (C.1) then becomes

$$\boldsymbol{\theta}(t) = \boldsymbol{\theta}(t-1) - \mu\left(R\boldsymbol{\theta}(t-1) - \boldsymbol{p}\right) \tag{C.7}$$

Subtracting $\boldsymbol{\theta}^0$ from both sides and taking into account (C.6), (C.7) becomes

$$\tilde{\boldsymbol{\theta}}(t) = \tilde{\boldsymbol{\theta}}(t-1) - \mu R\tilde{\boldsymbol{\theta}}(t-1) = (I - \mu R)\tilde{\boldsymbol{\theta}}(t-1) \tag{C.8}$$

where $\tilde{\boldsymbol{\theta}}(t) \equiv \boldsymbol{\theta}(t) - \boldsymbol{\theta}^0$. Now let R be a symmetric matrix. Then, as we know from Appendix B, it can be diagonalized, that is,

$$R = \Phi^T \Lambda \Phi \tag{C.9}$$

where Φ is the orthogonal matrix with columns the orthonormal eigenvectors of R and Λ the diagonal matrix having the corresponding eigenvalues on its diagonal. Incorporating (C.9) into (C.8) we obtain

$$\hat{\boldsymbol{\theta}}(t) = (I - \mu\Lambda)\hat{\boldsymbol{\theta}}(t-1) \tag{C.10}$$

where $\hat{\boldsymbol{\theta}}(t) \equiv \Phi\tilde{\boldsymbol{\theta}}(t)$. Matrix $I - \mu\Lambda$ is now diagonal, and (C.10) is equivalent to

$$\hat{\theta}_i(t) = (1 - \mu\lambda_i)\hat{\theta}_i(t-1) \tag{C.11}$$

where $\hat{\boldsymbol{\theta}} \equiv [\hat{\theta}_1, \hat{\theta}_2, \ldots, \hat{\theta}_l]^T$. Considering (C.11) for successive iteration steps we obtain

$$\hat{\theta}_i(t) = (1 - \mu\lambda_i)^t \hat{\theta}_i(0) \tag{C.12}$$

which converges to

$$\lim_{t\to\infty} \hat{\theta}_i(t) = 0, \Rightarrow \lim_{t\to\infty} \theta_i(t) = \theta_i^0, \quad i = 1, 2, \ldots, l \tag{C.13}$$

provided that $|1 - \mu\lambda_i| < 1, i = 1, 2, \ldots, l$. Thus, we can conclude that

$$\boldsymbol{\theta} \rightarrow \boldsymbol{\theta}^0, \quad \text{if } \mu < \frac{2}{\lambda_{\max}} \tag{C.14}$$

where $\lambda_{\max}$ is the maximum eigenvalue of R (which is positive since R is positive definite). Thus, the convergence speed of the gradient descent algorithm is controlled by the ratio $\lambda_{\min}/\lambda_{\max}$ as (C.12) and (C.14) suggest.

- *Nonquadratic cost functions*: If $J(\boldsymbol{\theta})$ is not quadratic, we can mobilize Taylor's theorem and assume that at some step near a stationary point, $\boldsymbol{\theta}^0$, $J(\boldsymbol{\theta})$ can be written approximately as

$$J(\boldsymbol{\theta}) = J(\boldsymbol{\theta}^0) + (\boldsymbol{\theta} - \boldsymbol{\theta}^0)^T \boldsymbol{g} + \frac{1}{2}(\boldsymbol{\theta} - \boldsymbol{\theta}^0)^T H(\boldsymbol{\theta} - \boldsymbol{\theta}^0) \tag{C.15}$$

where $\boldsymbol{g}$ is the gradient at $\boldsymbol{\theta}^0$ and H is the corresponding Hessian matrix, that is,

$$\boldsymbol{g} = \left.\frac{\partial J(\boldsymbol{\theta})}{\partial \boldsymbol{\theta}}\right|_{\boldsymbol{\theta}=\boldsymbol{\theta}^0}, \quad H(i,j) = \left.\frac{\partial^2 J(\boldsymbol{\theta})}{\partial\theta_i \partial\theta_j}\right|_{\boldsymbol{\theta}=\boldsymbol{\theta}^0} \tag{C.16}$$

Thus, in the neighborhood of $\boldsymbol{\theta}^0$, $J(\boldsymbol{\theta})$ is given approximately by a quadratic form and the convergence of the algorithm is controlled by the eigenvalues of the Hessian matrix.

C.2 NEWTON'S ALGORITHM

The problems associated with the dependence of the convergence speed on the eigenvalue spread can be overcome by using Newton's iterative scheme, where the correction in (C.2) is defined by

$$\Delta\boldsymbol{\theta} = -H^{-1}(old)\frac{\partial J(\boldsymbol{\theta})}{\partial\boldsymbol{\theta}}\bigg|_{\boldsymbol{\theta}=\boldsymbol{\theta}(old)} \tag{C.17}$$

where $H(old)$ is the Hessian matrix computed at $\boldsymbol{\theta}(old)$. Newton's algorithm converges much faster than the gradient descent method and, practically, its speed is independent of the eigenvalue spread. Faster convergence can be demonstrated by looking at the approximation in (C.15). Taking the gradient results in

$$\frac{\partial J(\boldsymbol{\theta})}{\partial(\boldsymbol{\theta})} = \frac{\partial J(\boldsymbol{\theta})}{\partial(\boldsymbol{\theta})}\bigg|_{\boldsymbol{\theta}=\boldsymbol{\theta}^0} + H(\boldsymbol{\theta}-\boldsymbol{\theta}^0) \tag{C.18}$$

Thus, the gradient is a linear function of $\boldsymbol{\theta}$ and hence the Hessian is constant, that is H. Having assumed that $\boldsymbol{\theta}^0$ is a stationary point, the first term on the right-hand side becomes zero. Now let $\boldsymbol{\theta} = \boldsymbol{\theta}(old)$. Then, according to Newton's iteration

$$\boldsymbol{\theta}(new) = \boldsymbol{\theta}(old) - H^{-1}(H(\boldsymbol{\theta}(old) - \boldsymbol{\theta}^0)) = \boldsymbol{\theta}^0 \tag{C.19}$$

Thus, the minimum is found in a single iteration. Of course, in practice, this is not true, as the approximations are not exactly valid. It is true, however, for quadratic costs.

Following a more formal proof (e.g., [Luen 84]), it can be shown that the convergence of Newton's algorithm is quadratic (i.e., the error at one step is proportional to the square of the previous step) while that of the gradient descent is linear. This speedup in convergence is achieved at increased computational cost, since Newton's algorithm requires the computation and then inversion of the Hessian matrix. Furthermore, numerical issues concerning the invertibility of H arise.

C.3 CONJUGATE-GRADIENT METHOD

Discussing the gradient descent method, we saw that, in general, a zigzag path is followed from the initial estimate to the optimum. This drawback is overcome by the following scheme, which results in improved convergence speed with respect to the gradient descent method. Compute the correction term according to the following rule:

$$\Delta\boldsymbol{\theta}(t) = \boldsymbol{g}(t) - \beta(t)\Delta\boldsymbol{\theta}(t-1) \tag{C.20}$$

where

$$\boldsymbol{g}(t) = \frac{\partial J(\boldsymbol{\theta})}{\partial\boldsymbol{\theta}}\bigg|_{\boldsymbol{\theta}=\boldsymbol{\theta}(t)} \tag{C.21}$$

and

$$\beta(t) = \frac{g^T(t)g(t)}{g^T(t-1)g(t-1)} \tag{C.22}$$

or

$$\beta(t) = \frac{g^T(t)\big(g(t) - g(t-1)\big)}{g^T(t-1)g(t-1)} \tag{C.23}$$

The former is known as the Fletcher–Reeves and the latter as the Polak–Ribiere formula.

For a more rigorous treatment of the topic the reader is referred to [Luen 84]. Finally, it must be stated that a number of variants of these schemes have appeared in the literature.

C.4 OPTIMIZATION FOR CONSTRAINED PROBLEMS

C.4.1 Equality Constraints

We will first focus on linear equality constraints and then generalize to the nonlinear case. Although the philosophy for both cases is the same, it is easier to grasp the basics when linear constraints are involved. Thus the problem is cast as

$$\begin{aligned} &\text{minimize} \quad J(\boldsymbol{\theta}) \\ &\text{subject to} \quad A\boldsymbol{\theta} = \boldsymbol{b} \end{aligned}$$

where A is an $m \times l$ matrix and $\boldsymbol{b}, \boldsymbol{\theta}$ are $m \times 1$ and $l \times 1$ vectors, respectively. It is assumed that the cost function $J(\boldsymbol{\theta})$ is twice continuously differentiable and it is, in general, a nonlinear function. Furthermore, we assume that the rows of A are linearly independent, hence A has full row rank. This assumption is known as the *regularity assumption*.

Let $\boldsymbol{\theta}_*$ be a local minimizer of $J(\boldsymbol{\theta})$ over the set $\{\boldsymbol{\theta}\colon A\boldsymbol{\theta} = \boldsymbol{b}\}$. Then it is not difficult to show (e.g., [Nash 96]) that, at this point, the gradient of $J(\boldsymbol{\theta})$ is given by

$$\frac{\partial}{\partial \boldsymbol{\theta}}(J(\boldsymbol{\theta}))|_{\boldsymbol{\theta}=\boldsymbol{\theta}_*} = A^T \boldsymbol{\lambda} \tag{C.24}$$

where $\boldsymbol{\lambda} \equiv [\lambda_1, \ldots, \lambda_m]^T$. Taking into account that

$$\frac{\partial}{\partial \boldsymbol{\theta}}(A\boldsymbol{\theta}) = A^T \tag{C.25}$$

Eq. (C.24) states that, at a constrained minimum, the gradient of the cost function is a linear combination of the gradients of the constraints. This is quite natural. Let us take a simple example involving a single linear constraint, that is,

$$\boldsymbol{a}^T \boldsymbol{\theta} = b$$

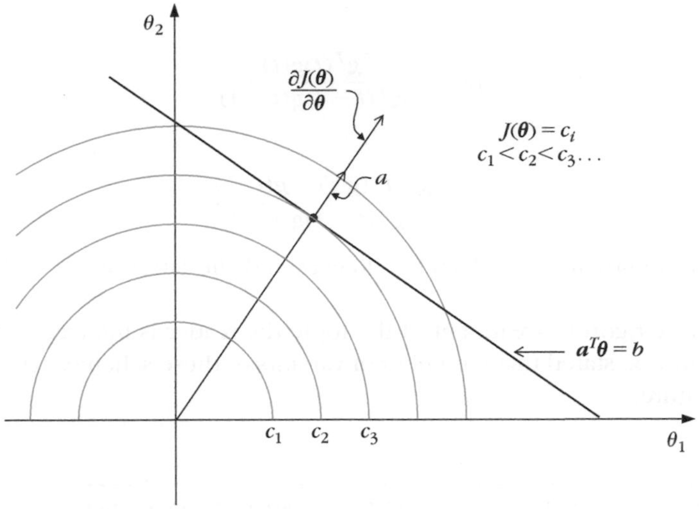

FIGURE C.4

At the minimizer, the gradient of the cost function is in the direction of the gradient of the constraint function.

Equation (C.24) then becomes

$$\frac{\partial}{\partial \boldsymbol{\theta}}(J(\boldsymbol{\theta}_*)) = \lambda \boldsymbol{a}$$

where the parameter λ is now a scalar. Figure C.4 shows an example of isovalue contours of $J(\boldsymbol{\theta}) = c$ in the two-dimensional space ($l = 2$). The constrained minimum coincides with the point where the straight line "meets" the isovalue contours for the first time, as one moves from small to large values of c. This is the point where the line is tangent to an isovalue contour; hence at this point the gradient of the cost function is in the direction of $\boldsymbol{a}$ (see Chapter 3).

Let us now define the function

$$\mathcal{L}(\boldsymbol{\theta}, \boldsymbol{\lambda}) = J(\boldsymbol{\theta}) - \boldsymbol{\lambda}^T(A\boldsymbol{\theta} - \boldsymbol{b}) \tag{C.26}$$

$$= J(\boldsymbol{\theta}) - \sum_{i=1}^{m} \lambda_i(\boldsymbol{a}_i^T\boldsymbol{\theta} - b_i) \tag{C.27}$$

where $\boldsymbol{a}_i^T$, $i = 1, 2, \ldots, m$, are the rows of A. $\mathcal{L}(\boldsymbol{\theta}, \boldsymbol{\lambda})$ is known as the *Lagrangian function* and the coefficients, λ_i, $i = 1, 2, \ldots, m$, as the *Lagrange multipliers*. The optimality condition (C.24), together with the constraints, which the minimizer has to satisfy, can now be written in a compact form as

$$\nabla \mathcal{L}(\boldsymbol{\theta}, \boldsymbol{\lambda}) = \boldsymbol{0} \tag{C.28}$$

where ∇ denotes the gradient operation with respect to both $\boldsymbol{\theta}$ and $\boldsymbol{\lambda}$. Indeed, equating with zero the derivatives of the Lagrangian with respect to $\boldsymbol{\theta}$ and $\boldsymbol{\lambda}$ gives, respectively,

$$\frac{\partial}{\partial \boldsymbol{\theta}} J(\boldsymbol{\theta}) = A^T \boldsymbol{\lambda}$$

$$A\boldsymbol{\theta} = \boldsymbol{b}$$

The above is a set of $m + l$ unknowns, that is, $(\theta_1, \ldots, \theta_l, \lambda_1, \ldots, \lambda_m)$, with $m + l$ equations, whose solution provides the minimizer $\boldsymbol{\theta}_*$ and the corresponding Lagrange multipliers. Similar arguments hold for nonlinear equation constraints. Let us consider the problem

$$\begin{aligned} &\text{minimize} && J(\boldsymbol{\theta}) \\ &\text{subject to} && f_i(\boldsymbol{\theta}) = 0, \quad i = 1, 2, \ldots, m \end{aligned}$$

The minimizer is again a *stationary point* of the corresponding Lagrangian

$$\mathcal{L}(\boldsymbol{\theta}, \boldsymbol{\lambda}) = J(\boldsymbol{\theta}) - \sum_{i=1}^{m} \lambda_i f_i(\boldsymbol{\theta})$$

and it results from the solution of the set of $m + l$ equations

$$\nabla \mathcal{L}(\boldsymbol{\theta}, \boldsymbol{\lambda}) = \mathbf{0}$$

The regularity condition for nonlinear constraints requires the gradients of the constraints $\frac{\partial}{\partial \boldsymbol{\theta}}(f_i(\boldsymbol{\theta}))$ to be linearly independent.

C.4.2 Inequality Constraints

The general problem can be cast as follows:

$$\begin{aligned} &\text{minimize} && J(\boldsymbol{\theta}) \\ &\text{subject to} && f_i(\boldsymbol{\theta}) \geq 0, \quad i = 1, 2, \ldots, m \end{aligned} \tag{C.29}$$

Each one of the constraints defines a region in $\mathcal{R}^l$. The intersection of all these regions defines the area in which the constrained minimum, $\boldsymbol{\theta}_*$, must lie. This is known as the *feasible region* and the points in it (candidate solutions) as *feasible points*. The type of the constraints control the type of the feasible region, that is, whether it is convex or concave. At this point, it will not harm us to recall a few definitions.

Convex functions. Let S be a convex set. A function $f(\boldsymbol{\theta})$

$$f: S \subseteq \mathcal{R}^l \rightarrow \mathcal{R}$$

is called convex in S, if for every $\boldsymbol{\theta}$ and $\boldsymbol{\theta}' \in S$

$$f(\lambda\boldsymbol{\theta} + (1 - \lambda)\boldsymbol{\theta}') \leq \lambda f(\boldsymbol{\theta}) + (1 - \lambda) f(\boldsymbol{\theta}')$$

for every $\lambda \in [0, 1]$. If strict inequality holds, we say that the function is strict convex.

Concave functions. A function $f(\boldsymbol{\theta})$ is called concave, if for every $\boldsymbol{\theta}, \boldsymbol{\theta}' \in S$

$$f(\lambda\boldsymbol{\theta} + (1 - \lambda)\boldsymbol{\theta}') \geq \lambda f(\boldsymbol{\theta}) + (1 - \lambda) f(\boldsymbol{\theta}')$$

for every $\lambda \in [0, 1]$. For strict inequality, the function is known as strict concave.

Figure C.5 shows three functions, one convex, one concave, and one which is neither convex nor concave.

Convex sets. A set $S \subseteq \mathcal{R}^l$ is called convex, if for every pair of points $\boldsymbol{\theta}, \boldsymbol{\theta}' \in S$, the line segment joining these points also belongs to the set. In other words, all points $\lambda\boldsymbol{\theta} + (1-\lambda)\boldsymbol{\theta}'$, $\lambda \in [0, 1]$ belong to the set. Figure C.6 shows two sets, one convex and one nonconvex.

Remarks

- If $f(\boldsymbol{\theta})$ is convex then $-f(\boldsymbol{\theta})$ is concave and vice versa. Furthermore, if $f_i(\boldsymbol{\theta})$, $i = 1, 2, \ldots, m$, are convex, so is the sum $\sum_{i=1}^{m} \lambda_i f_i(\boldsymbol{\theta}), \lambda_i \geq 0$. Similarly, if $f_i(\boldsymbol{\theta})$ are concave, so is their summation.

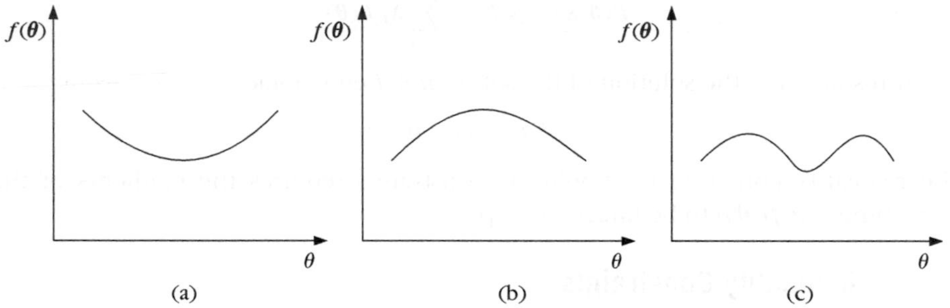

FIGURE C.5

(a) A convex function, (b) a concave function, and (c) a function that is neither convex nor concave.

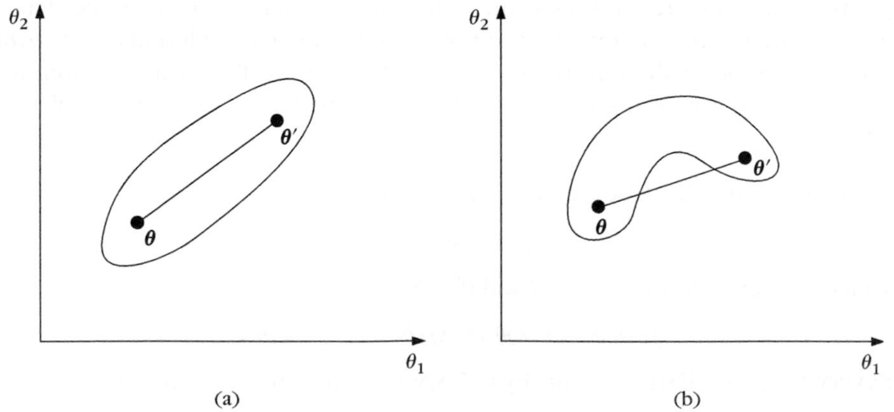

FIGURE C.6

(a) A convex set and (b) a concave set of points.

- If a function $f(\boldsymbol{\theta})$ is convex, it can be shown that *a local minimum is also a global one. This can be easily checked from the graph of Figure C.5. Furthermore, if the function is strict convex then this minimum is unique*. For concave functions, the above also hold true but for points where a maximum occurs.
- A direct consequence of the respective definitions is that if $f(\boldsymbol{\theta})$ is convex then the set

$$X = \{\boldsymbol{\theta} | f(\boldsymbol{\theta}) \leq b,\ b \in \mathcal{R}\}$$

  is convex. Also, if $f(\boldsymbol{\theta})$ is concave then the set

$$X = \{\boldsymbol{\theta} | f(\boldsymbol{\theta}) \geq b,\ b \in \mathcal{R}\}$$

  is also convex.
- The intersection of convex sets is also a convex set.

From the above remarks, one can easily conclude that, if each one of the functions in the constraints in (C.29) is concave, then the feasible region is a convex one. This is also valid if the constraints are linear, since a linear function can be considered either as convex or concave. For more on these issues, the interested reader may refer, for example, to [Baza 79].

The Karush–Kuhn–Tucker (KKT) Conditions

This is a set of *necessary* conditions, which a local minimizer $\boldsymbol{\theta}_*$ of the problem given in (C.29) has to satisfy. If $\boldsymbol{\theta}_*$ is a point that satisfies the regularity condition, then there exists a vector $\boldsymbol{\lambda}$ of Lagrange multipliers so that the following are valid:

$$(1)\quad \frac{\partial}{\partial \boldsymbol{\theta}} \mathcal{L}(\boldsymbol{\theta}_*, \boldsymbol{\lambda}) = \mathbf{0}$$

$$(2)\quad \lambda_i \geq 0, \quad i = 1, 2, \ldots, m$$

$$(3)\quad \lambda_i f_i(\boldsymbol{\theta}_*) = 0, \quad i = 1, 2, \ldots, m \qquad \text{(C.30)}$$

Actually, there is a fourth condition concerning the Hessian of the Lagrangian function, which is not of interest to us. The above set of equations is also part of the sufficiency conditions; however, in this case, there are a few subtle points and the interested reader is referred to more specialized textbooks, for example, [Baza 79, Flet 87, Bert 95, Nash 96].

Conditions (3) in (C.30) are known as *complementary slackness conditions.* They state that at least one of the terms in the products is zero. In the case where, in each one of the equations, only one of the two terms is zero, that is, either λ_i or $f_i(\boldsymbol{\theta}_*)$, we talk about *strict complementarity*.

Remarks

- The first condition is most natural. It states that the minimum must be a stationary point of the Lagrangian, with respect to $\boldsymbol{\theta}$.

- A constraint, $(f_i(\boldsymbol{\theta}_*))$, is called *inactive* if the corresponding Lagrange multiplier is zero. This is because this constraint does not affect the problem. A constrained minimizer $\boldsymbol{\theta}_*$ can lie either in the interior of the feasible region or on its boundary. In the former case, the problem is equivalent to an unconstrained one. Indeed, if it happens that a minimum is located within the feasible region, then the value of the cost function in a region around this point will increase (or remain the same) as one moves away from this point. Hence, this point will be a stationary point of the cost function $J(\boldsymbol{\theta})$. Thus in this case, the constraints are redundant and do not affect the problem. In words, the constraints are inactive and this is equivalent to setting the Lagrange multipliers equal to zero. The nontrivial constrained optimization task is when the (unconstrained) minimum of the cost function is located outside the feasible region. In this case, the constrained minimum will be located on the boundary of the feasible region. In other words, in this nontrivial case, there will be one or more of the constraints for which $f_i(\boldsymbol{\theta}_*) = 0$. These constitute the *active constraints*. The rest of the constraints will be inactive with the corresponding Lagrange multipliers being zero.

  Figure C.7 illustrates a simple case with the following constraints:

$$f_1(\boldsymbol{\theta}) = \theta_1 + 2\theta_2 - 2 \geq 0$$

$$f_2(\boldsymbol{\theta}) = \theta_1 - \theta_2 + 2 \geq 0$$

$$f_3(\boldsymbol{\theta}) = -\theta_1 + 2 \geq 0$$

  The (unconstrained) minimum of the cost function is located outside the feasible region. The dotted lines are the isovalue curves $J(\boldsymbol{\theta}) = c$, with $c_1 < c_2 < c_3$. The constrained minimum coincides with the point where an isovalue curve "touches" the boundary of the feasible region for the first time (smallest value of c). This point may belong to more than one of the constraints, for example, it may be a corner point of the boundary.

- The Lagrange multipliers of the active constraints are *nonnegative*. To understand why this is so, let us consider for simplicity the case of linear constraints $A\boldsymbol{\theta} \geq \boldsymbol{b}$, where A includes the active constraints only. If $\boldsymbol{\theta}_*$ is a minimizer lying on the active constraints, then any other feasible point can be written as

$$\hat{\boldsymbol{\theta}} = \boldsymbol{\theta}_* + \boldsymbol{p}$$

$$A\boldsymbol{p} \geq \boldsymbol{0}$$

  since this guarantees that $A\hat{\boldsymbol{\theta}} \geq \boldsymbol{b}$. If the direction $\boldsymbol{p}$ points into the feasible region (Figure C.7) then $A\boldsymbol{p} \neq \boldsymbol{0}$, that is, some of its components are strictly positive. Since $\boldsymbol{\theta}_*$ is a minimizer, from condition (1) in (C.30) we have that

$$\frac{\partial}{\partial \boldsymbol{\theta}} J(\boldsymbol{\theta}_*) = A^T \boldsymbol{\lambda}$$

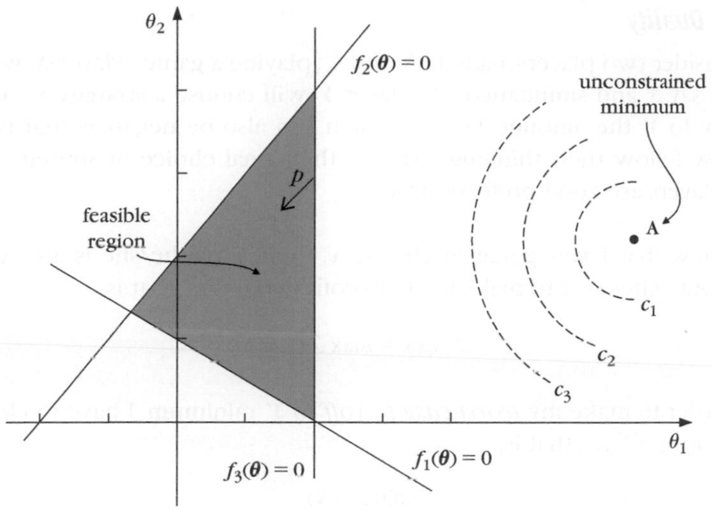

FIGURE C.7

An example of the nontrivial case, where the unconstrained minimum lies outside the feasible region.

The change of the cost function along the direction of $\boldsymbol{p}$ is proportional to

$$\boldsymbol{p}^T \frac{\partial}{\partial \boldsymbol{\theta}}(J(\boldsymbol{\theta})) = \boldsymbol{p}^T A^T \boldsymbol{\lambda}$$

and since $\boldsymbol{\theta}_*$ is a minimizer, this must be a direction of ascent at $\boldsymbol{\theta}_*$. Thus $\boldsymbol{\lambda}$ must be nonnegative to guarantee that $\boldsymbol{p}^T A^T \boldsymbol{\lambda} \geq 0$ for any $\boldsymbol{p}$ pointing into the feasible region. An active constraint whose corresponding Lagrange multiplier is zero is known as *degenerate*.

- It can be shown that, if the cost function is convex and the feasible region is also convex, then a local minimum is also a global one. A little thought (and a look at Figure C.7) suffices to see why this is so.

Having now discussed all these nice properties, the major question arises: how can one compute a constrained (local) minimum? Unfortunately, this is not always an easy task. A straightforward approach would be to assume that some of the constraints are active and some inactive, and check if the resulting Lagrange multipliers of the active constraints are nonnegative. If not, then choose another combination of constraints and repeat the procedure until one ends up with nonnegative multipliers. However, in practice, this may require a prohibitive amount of computation. Instead, a number of alternative approaches have been proposed. In the sequel, we will review some basics from Game Theory and use these to reformulate the KKT conditions. This new setup can be useful in a number of cases in practice.

Min-Max Duality

Let us consider two players, namely X and Y, playing a game. Player X will choose a strategy, say, x and simultaneously player Y will choose a strategy y. As a result, X will pay to Y the amount $\mathcal{F}(x, y)$, which can also be negative, that is, X wins. Let us now follow their thinking, prior to their final choice of strategy, assuming that the players are good professionals.

X: If Y knew that I was going to choose x, then, since he/she is a clever player, he/she would choose y to make his/her profit maximum, that is,

$$\mathcal{F}^*(x) = \max_y \mathcal{F}(x, y)$$

Thus, in order to make my *worst-case payoff* to Y minimum, I have to choose x so as to minimize $\mathcal{F}^*(x)$, that is,

$$\min_x \mathcal{F}^*(x)$$

This problem is known as the *min-max* problem since it seeks the value

$$\min_x \max_y \mathcal{F}(x, y)$$

Y: X is a good player, so if he/she knew that I am going to play y, he/she would choose x so that to make his/her payoff minimum, that is,

$$\mathcal{F}_*(y) = \min_x \mathcal{F}(x, y)$$

Thus, in order to make my *worst-case profit* maximum I must choose y that maximizes $\mathcal{F}_*(y)$, that is,

$$\max_y \mathcal{F}_*(y)$$

This is known as the max-min problem, since it seeks the value

$$\max_y \min_x \mathcal{F}(x, y)$$

The two problems are said to be *dual to each other*. The first is known to be the *primal*, whose objective is to minimize $\mathcal{F}^*(x)$ and the second is the *dual* problem with the objective to maximize $\mathcal{F}_*(y)$.

For any x and y, the following is valid:

$$\mathcal{F}_*(y) \equiv \min_x \mathcal{F}(x, y) \le \mathcal{F}(x, y) \le \max_y \mathcal{F}(x, y) \equiv \mathcal{F}^*(x) \tag{C.31}$$

which easily leads to

$$\max_y \min_x \mathcal{F}(x, y) \le \min_x \max_y \mathcal{F}(x, y) \tag{C.32}$$

Saddle Point Condition

Let $\mathcal{F}(\boldsymbol{x}, \boldsymbol{y})$ be a function of two vector variables with $\boldsymbol{x} \in X \subseteq \mathcal{R}^l$ and $\boldsymbol{y} \in Y \subseteq \mathcal{R}^l$. If a pair of points $(\boldsymbol{x}_*, \boldsymbol{y}_*)$, with $\boldsymbol{x}_* \in X, \boldsymbol{y}_* \in Y$ satisfies the condition

$$\mathcal{F}(\boldsymbol{x}_*, \boldsymbol{y}) \leq \mathcal{F}(\boldsymbol{x}_*, \boldsymbol{y}_*) \leq \mathcal{F}(\boldsymbol{x}, \boldsymbol{y}_*) \tag{C.33}$$

for every $\boldsymbol{x} \in X$ and $\boldsymbol{y} \in Y$, we say that it satisfies the *saddle point condition*. It is not difficult to show (e.g., [Nash 96]) that a pair $(\boldsymbol{x}_*, \boldsymbol{y}_*)$ satisfies the saddle point conditions *if and only if*

$$\max_{\boldsymbol{y}} \min_{\boldsymbol{x}} \mathcal{F}(\boldsymbol{x}, \boldsymbol{y}) = \min_{\boldsymbol{x}} \max_{\boldsymbol{y}} \mathcal{F}(\boldsymbol{x}, \boldsymbol{y}) = \mathcal{F}(\boldsymbol{x}_*, \boldsymbol{y}_*) \tag{C.34}$$

Lagrangian Duality

We will now use all the above in order to formulate our original cost function minimization problem as a min-max task of the corresponding Lagrangian function. Under certain conditions, this formulation can lead to computational savings when computing the constrained minimum. The optimization task of our interest is

$$\begin{aligned} &\text{minimize} \quad J(\boldsymbol{\theta}) \\ &\text{subject to} \quad f_i(\boldsymbol{\theta}) \geq 0, \quad i = 1, 2, \ldots, m \end{aligned}$$

The Lagrangian function is

$$\mathcal{L}(\boldsymbol{\theta}, \boldsymbol{\lambda}) = J(\boldsymbol{\theta}) - \sum_{i=1}^{m} \lambda_i f_i(\boldsymbol{\theta}) \tag{C.35}$$

Let

$$\mathcal{L}^*(\boldsymbol{\theta}) = \max_{\boldsymbol{\lambda}} \mathcal{L}(\boldsymbol{\theta}, \boldsymbol{\lambda}) \tag{C.36}$$

However, since $\boldsymbol{\lambda} \geq \mathbf{0}$ and $f_i(\boldsymbol{\theta}) \geq 0$, the maximum value of the Lagrangian occurs if the summation in (C.35) is zero (either $\lambda_i = 0$ or $f_i(\boldsymbol{\theta}) = 0$ or both) and

$$\mathcal{L}^*(\boldsymbol{\theta}) = J(\boldsymbol{\theta}) \tag{C.37}$$

Therefore our original problem is equivalent with

$$\min_{\boldsymbol{\theta}} J(\boldsymbol{\theta}) = \min_{\boldsymbol{\theta}} \max_{\boldsymbol{\lambda} \geq \mathbf{0}} \mathcal{L}(\boldsymbol{\theta}, \boldsymbol{\lambda}) \tag{C.38}$$

As we already know, the dual problem of the above is

$$\max_{\boldsymbol{\lambda} \geq \mathbf{0}} \min_{\boldsymbol{\theta}} \mathcal{L}(\boldsymbol{\theta}, \boldsymbol{\lambda}) \tag{C.39}$$

Convex Programming

A large class of practical problems obeys the following two conditions:

$$(1) \quad J(\boldsymbol{\theta}) \text{ is convex} \tag{C.40}$$

$$(2) \quad f_i(\boldsymbol{\theta}) \text{ are concave} \tag{C.41}$$

This class of problems turns out to have a very useful and mathematically tractable property.

Theorem *Let $\boldsymbol{\theta}_*$ be a minimizer of such a problem, which is also assumed to satisfy the regularity condition. Let $\boldsymbol{\lambda}_*$ be the corresponding vector of Lagrange multipliers. Then $(\boldsymbol{\theta}_*, \boldsymbol{\lambda}_*)$ is a saddle point of the Lagrangian function, and as we know this is equivalent to*

$$\mathcal{L}(\boldsymbol{\theta}_*, \boldsymbol{\lambda}_*) = \max_{\boldsymbol{\lambda} \geq \mathbf{0}} \min_{\boldsymbol{\theta}} \mathcal{L}(\boldsymbol{\theta}, \boldsymbol{\lambda}) = \min_{\boldsymbol{\theta}} \max_{\boldsymbol{\lambda} \geq \mathbf{0}} \mathcal{L}(\boldsymbol{\theta}, \boldsymbol{\lambda}) \tag{C.42}$$

Proof. Since $f_i(\boldsymbol{\theta})$ are concave, $-f_i(\boldsymbol{\theta})$ are convex, so the Lagrangian function

$$\mathcal{L}(\boldsymbol{\theta}, \boldsymbol{\lambda}) = J(\boldsymbol{\theta}) - \sum_{i=1}^{m} \lambda_i f_i(\boldsymbol{\theta})$$

for $\lambda_i \geq 0$, is also convex. Note, now, that for concave function constraints of the form $f_i(\boldsymbol{\theta}) \geq 0$, the feasible region is convex (see remarks above). The function $J(\boldsymbol{\theta})$ is also convex. Hence, as already stated in the remarks, every local minimum is also a global one; thus for any $\boldsymbol{\theta}$

$$\mathcal{L}(\boldsymbol{\theta}_*, \boldsymbol{\lambda}_*) \leq \mathcal{L}(\boldsymbol{\theta}, \boldsymbol{\lambda}_*) \tag{C.43}$$

Furthermore, the complementary slackness conditions suggest that

$$\mathcal{L}(\boldsymbol{\theta}_*, \boldsymbol{\lambda}_*) = J(\boldsymbol{\theta}_*) \tag{C.44}$$

and for any $\boldsymbol{\lambda} \geq \mathbf{0}$

$$\mathcal{L}(\boldsymbol{\theta}_*, \boldsymbol{\lambda}) \equiv J(\boldsymbol{\theta}_*) - \sum_{i=1}^{m} \lambda_i f_i(\boldsymbol{\theta}_*) \leq J(\boldsymbol{\theta}_*) = \mathcal{L}(\boldsymbol{\theta}_*, \boldsymbol{\lambda}_*) \tag{C.45}$$

Combining (C.43) and (C.45) we obtain

$$\mathcal{L}(\boldsymbol{\theta}_*, \boldsymbol{\lambda}) \leq \mathcal{L}(\boldsymbol{\theta}_*, \boldsymbol{\lambda}_*) \leq \mathcal{L}(\boldsymbol{\theta}, \boldsymbol{\lambda}_*) \tag{C.46}$$

In other words, *the solution $(\boldsymbol{\theta}_*, \boldsymbol{\lambda}_*)$ is a saddle point.* □

This is a very important theorem and it states that the constrained minimum of a convex programming problem can also be obtained as a maximization task applied on the Lagrangian. This leads us to the following very useful formulation of the optimization task.

Wolfe Dual Representation

A convex programming problem is equivalent to

$$\max_{\boldsymbol{\lambda} \geq \mathbf{0}} \mathcal{L}(\boldsymbol{\theta}, \boldsymbol{\lambda}) \tag{C.47}$$

$$\text{subject to } \frac{\partial}{\partial \boldsymbol{\theta}} \mathcal{L}(\boldsymbol{\theta}, \boldsymbol{\lambda}) = \mathbf{0} \tag{C.48}$$

The last equation guarantees that $\boldsymbol{\theta}$ is a minimum of the Lagrangian.

Example C.1
Consider the quadratic problem

$$\begin{aligned}\text{minimize} \quad & \frac{1}{2}\boldsymbol{\theta}^T\boldsymbol{\theta} \\ \text{subject to} \quad & A\boldsymbol{\theta} \geq \boldsymbol{b}\end{aligned}$$

This is a convex programming problem; hence the Wolfe dual representation is valid:

$$\begin{aligned}\text{maximize} \quad & \frac{1}{2}\boldsymbol{\theta}^T\boldsymbol{\theta} - \boldsymbol{\lambda}^T(A\boldsymbol{\theta} - \boldsymbol{b}) \\ \text{subject to} \quad & \boldsymbol{\theta} - A^T\boldsymbol{\lambda} = \mathbf{0}\end{aligned}$$

For this example, the equality constraint has an analytic solution (this is not, however, always possible). Solving with respect to $\boldsymbol{\theta}$, we can eliminate it from the maximizing function and the resulting dual problem involves only the Lagrange multipliers,

$$\begin{aligned}\max_{\boldsymbol{\lambda}} \quad & \left\{-\frac{1}{2}\boldsymbol{\lambda}^T A A^T \boldsymbol{\lambda} + \boldsymbol{\lambda}^T \boldsymbol{b}\right\} \\ \text{subject to} \quad & \boldsymbol{\lambda} \geq \mathbf{0}\end{aligned}$$

This is also a quadratic problem but the set of constraints is now simpler.

REFERENCES

[Baza 79] Bazaraa M.S., Shetty C.M. *Nonlinear Programming: Theory and Algorithms*, John Wiley, 1979.

[Bert 95] Bertsekas, D.P., Belmont, M.A. *Nonlinear Programming*, Athenas Scientific, 1995.

[Flet 87] Fletcher, R. *Practical Methods of Optimization*, 2nd ed., John Wiley, 1987.

[Luen 84] Luenberger D.G. *Linear and Nonlinear Programming*, Addison Wesley, 1984.

[Nash 96] Nash S.G., Sofer A. *Linear and Nonlinear Programming*, McGraw-Hill, 1996.

APPENDIX

D Basic Definitions from Linear Systems Theory

D.1 LINEAR TIME INVARIANT (LTI) SYSTEMS

A discrete linear time-invariant system is characterized uniquely by its *impulse response sequence*, $h(n)$. This is the output of the system when its input is excited by the impulse sequence, $\delta(n)$, that is,

$$\delta(n) = \begin{cases} 1 & \text{for } n = 0 \\ 0 & \text{for } n \neq 1 \end{cases} \tag{D.1}$$

When its input is excited by a sequence $x(n)$, its output sequence is given by the *convolution* of $x(n)$ with $h(n)$, defined as

$$y(n) = \sum_{k=-\infty}^{+\infty} h(k)x(n-k) = \sum_{k=-\infty}^{+\infty} x(k)h(n-k) \equiv h(n) * x(n) \tag{D.2}$$

For continuous time systems the convolution becomes an integral, that is,

$$y(t) = \int_{-\infty}^{+\infty} h(\tau)x(t-\tau)\,d\tau$$

$$= \int_{-\infty}^{+\infty} x(\tau)h(t-\tau)\,d\tau \equiv x(t) * y(t) \tag{D.3}$$

where $h(t)$ is the impulse response of the system, that is, the output when its input is excited by the Dirac delta function $\delta(t)$, defined by

$$\delta(t) = 0, \quad \text{for } t \neq 0, \quad \text{and} \quad \int_{-\infty}^{+\infty} \delta(t)dt = 1 \tag{D.4}$$

Linear time-invariant systems can be:

- *Causal*: Their impulse response is zero for $n < 0$. Otherwise, they are known as *noncausal*. Observe that only causal systems can be realized in real time. This is because for noncausal systems, the output at time n would require knowledge of future samples $x(n+1), x(n+2), \ldots$, which in practice is not possible.

- *Finite impulse response (FIR)*: The corresponding impulse response is of finite extent. If this is not the case, the systems are known as *infinite impulse response* (IIR) systems. For a causal FIR system the input–output relation becomes

$$y(n) = \sum_{k=0}^{L-1} h(k)x(n-k) \tag{D.5}$$

where L is the length of the impulse response. When a system is FIR but noncausal, it can become causal by delaying its output. Take for example the system with impulse response $0, \ldots, 0, h(-2), h(-1), h(0), h(1), h(2), 0, \ldots$. Then,

$$\begin{aligned} y(n-2) &= h(-2)x(n) + h(-1)x(n-1) + h(0)x(n-2) \\ &\quad + h(1)x(n-3) + h(2)x(n-4) \end{aligned} \tag{D.6}$$

That is, at time "n" the output corresponds to the delayed time "$n-2$." The delay is equal to the maximum negative index of nonzero impulse coefficient.

D.2 TRANSFER FUNCTION

The z-transform of the impulse response, defined as

$$H(z) = \sum_{n=-\infty}^{+\infty} h(n)z^{-n} \tag{D.7}$$

is known as the *transfer function* of the system. The free parameter z is a complex variable. The definition in (D.7) is meaningful, provided that the series converges. For most of the sequences of our interest this is true for some region in the complex plane. It can easily be shown that for causal and FIR systems the region of convergence is of the form

$$|z| > |R|, \quad \text{for some } |R| < 1 \tag{D.8}$$

that is, it is the exterior of a circle in the complex plane, centered at the origin, and it contains the unit circle ($|z| = 1$). Let $X(z)$ and $Y(z)$ be the z-transforms of the input and output sequences of a linear time-invariant system. Then (D.2) is shown to be equivalent to

$$Y(z) = H(z)X(z) \tag{D.9}$$

If the unit circle is in the region of convergence of the respective z-transforms (for example, for causal FIR systems), then for $z = \exp(-j\omega)$ we obtain the equivalent Fourier transform and

$$Y(\omega) = H(\omega)X(\omega) \tag{D.10}$$

If the impulse response of a linear time-invariant system is delayed by r samples, for example, to make it causal in case it is noncausal, the transfer function of the delayed system is given by $z^{-r}H(z)$.

D.3 SERIAL AND PARALLEL CONNECTION

Consider two LTI systems with responses $h_1(n)$ and $h_2(n)$, respectively. Figure D.1a shows the two systems connected in serial and Figure D.1b in parallel. The overall impulse responses are easily shown to be

$$\text{Serial} \quad h(n) = h_1(n) * h_2(n) \tag{D.11}$$

$$\text{Parallel} \quad h(n) = h_1(n) + h_2(n) \tag{D.12}$$

D.4 TWO-DIMENSIONAL GENERALIZATIONS

A two-dimensional linear time-invariant system is also characterized by its two-dimensional impulse response sequence $H(m, n)$, which in the case of images is known as a *point spread function*. On filtering an input image array $X(m, n)$ by $H(m, n)$ the resulting image array is given by the two-dimensional convolution

$$\begin{aligned} Y(m,n) &= \sum_k \sum_l H(m-k, n-l)X(k,l) \equiv H(m,n) * * X(m,n) \\ &= \sum_k \sum_l H(k,l)X(m-k, n-l) \end{aligned} \tag{D.13}$$

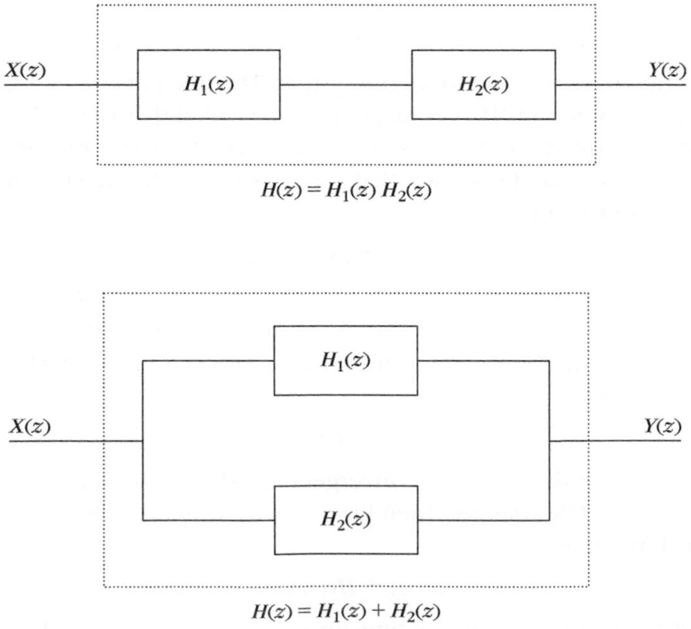

FIGURE D.1

Serial and parallel connections of LTI systems.

Index

D

E

F

G

I

J

K

L

Q

R

S

T

Printed and bound by CPI Group (UK) Ltd, Croydon, CR0 4YY
03/10/2024
01040332-0002